Low-Power CMOS Design

Book of Related Interests from IEEE Press . . .

CMOS: Circuit Design, Layout and Simulation
R. Jacob Baker
1997   Cloth   928 pp   IEEE Order No. PC5689   ISBN 0-7803-3416-7

CIRCUITS AND FILTERS HANDBOOK
Edited by C. H. Chen
1995   Cloth   2896 pp   IEEE Order No. PC5631   ISBN 0-8493-8341-2

NONLINEAR MICROWAVE CIRCUITS
Stephen A. Maas
1997   Paper   496 pp   IEEE Order No. PP5385   ISBN 0-7803-3403-5

MONOLITHIC PHASE-LOCKED LOOPS AND CLOCK RECOVERY CIRCUITS: Theory and Design
Edited by Behzad Razavi
1996   Cloth   512 pp   IEEE Order No. PC5620   ISBN 0-7803-1149-3

Low-Power CMOS Design

Edited by

Anantha Chandrakasan
Massachusetts Institute of Technology, Cambridge

Robert Brodersen
University of California at Berkeley

IEEE
PRESS

A Selected Reprint Volume

The Institute of Electrical and Electronics Engineers, Inc., New York

This book may be purchased at a discount from the publisher
when ordered in bulk quantities. Contact:

IEEE Press Marketing
Attn: Special Sales
445 Hoes Lane, P.O. Box 1331
Piscataway, NJ 08855–1331
Fax: (732) 981–9334

For more information on the IEEE Press,
visit the IEEE home page: http://www.ieee.org/

Printed in the United States of America

10   9   8   7   6   5   4   3   2   1

ISBN 0-7803-3429-9
IEEE Order Number: PC5703

Library of Congress Cataloging-in-Publication Data
Low-power CMOS design / edited by Anantha Chandrakasan, Robert
Brodersen.
    p.   cm.
  Includes bibliographical references and index.
  ISBN 0-7803-3429-9 (alk. paper)
  1. Metal oxide semiconductors, Complementary—Design and
construction—Data processing.   2. Low voltage integrated circuits—
Design and construction—Data processing.   3. Digital integrated
circuits—Design and construction—Data processing.   4. Computer-
aided design.   I. Chandrakasan, Anantha P.   II. Brodersen, Robert
W., (date).
TK7871.99.M44L67   1998
621.381′044—dc21   97–38651
                                            CIP

Contents

Preface

Preface

As one looks over the history of electronics, it becomes clear that one of the primary reasons for major shifts in technology has been the requirement to move toward a more energy efficient operation as the level of complexity increases. This resulted in the movement from vacuum tubes to solid state-logic, from bipolar to MOS and most recently from NMOS to CMOS. Again, however, history is repeating itself and as the complexity of CMOS circuits increases, the issues of power and energy consumption are once again becoming critical. Fortunately, the energy expended per operation for CMOS is continually improving as the technology is advanced. Particularly helpful is the dramatic improvement in energy efficiency that is obtained by reducing the supply voltage, and many of the techniques that are presented in this book exploit that characteristic. However, voltage scaling alone is not enough as the complexity increases and as applications which require the use of a portable energy source require further energy reductions. This book describes a number of these strategies which range over all levels of the design process from optimization of the underlying CMOS technology, logic styles, circuit topologies and architectures, up to the applications and algorithms which are being executed.

The most significant development at the CMOS technology level for low-voltage systems is the reduction of the device threshold voltage. For continuously operated circuits, the increased leakage due to reduced threshold voltages is typically not significant. However, for "event-driven" computation in which intermittent computation activity triggered by external events is separated by long periods of inactivity (e.g., X-server), leakage power is critical. To satisfy the contradicting requirements of high performance during active periods and low-standby leakage, several device technologies and circuit styles have been developed. This includes the dynamic control of threshold voltages in triple-well CMOS using backgate effect, Multiple Threshold CMOS, variable threshold CMOS, and dual-gated SOI technology. All these technologies (Part II) provide a knob to reduce leakage power during idle periods while enabling low-threshold, high-speed operation during active periods.

As supply levels scale into the sub-1V regime exploiting emerging technologies and efficient architectures, there is a need for high-efficiency, low-power DC-DC conversion. In many cases, embedding the power supply control in the processor can save significant power. Rather than designing a system with a static supply to meet a specific timing constraint under worst case conditions (i.e., establishing the feedback around the power converter to fix the output voltage), it is better to allow the voltage to vary such that the timing constraints are just met at any given temperature and operating conditions; this is accomplished by establishing the feedback around a fixed processing rate or delay. Such embedded power supply systems can minimize energy consumption under varying temperature, process parameters, and computational workload. (Part III).

There is a wide range in power dissipation characteristics among the various circuit and logic styles (Part IV). Low-power circuit operation is achieved by reducing critical path delays and avoiding unnecessary transitions beyond what is required to implement the logic function. The ability to power down logic is important, and in some logic styles (e.g., self-timed circuits) this property is implicit. The design of energy recovery logic has gained significant interest, particulary in academia. Energy recovery logic reduces the energy dissipated per switching event by slowing down the transitions to such a level that there are negligible losses in the resistive components.

The scaling of technology into the submicron regime has forced interconnect to be an important factor of overall system power. As a result, it is necessary to develop architectures that minimize wirelengths by exploiting locality and decrease activity through coding techniques and reduced voltage swing (Part V). The voltage swing is decreased by exploiting on-chip regulators, charge sharing, and charge recycling techniques.

Embedded memory is an integral component of digital processors (Part VI). It is important to minimize unnecessary transitions (e.g., using hierarchical decoding schemes and self-timing) and voltage swing on heavily loaded bit-lines. Some techniques that minimize delay also minimize energy per access (e.g., low-swing). The scaling of supply voltages in memory circuits may require exploiting emerging technologies and circuit styles (like MTCMOS or self-reverse biasing techniques). Charge recycling techniques have also been shown to be effective in memory circuits.

By exploiting low-power techniques at all levels of the design, it is possible to reduce power dissipation by orders of magnitude. Several such systems are presented in Part VII, which includes examples of general purpose processors, programmable signal processors, and dedicated signal processors. Most of the gains come from aggressive scaling of supply voltages and avoiding unnecessary transitions (e.g., gated clock).

There has been significant activity in Computer-Aided Design tools for low-power systems. Two fundamentally different approaches for power estimation have been proposed which include probabilistic techniques (in which the node activities are computed in one pass through the logic) and statistical techniques (simulation based approaches). Effective solutions exist at all levels of the design abstraction ranging from logic and circuit to RTL, behavior software levels. Power optimization techniques, however, are not as mature at this point in time.

The second section in Part VIII give an indication of some of the optimization techniques being proposed and the types of power reductions that are presently possible using automated techniques.

It is believed that if the reader applies the appropriate techniques which have been presented here, they will also be able to achieve the orders of magnitude reduction in power consumption that has been shown to be possible. These dramatic results have been obtained by designers for all classes of circuits which range from conventional microprocessors to highly-dedicated signal processing components. It is these kinds of results that indicate—through the era in which CMOS circuits can be designed without consideration for power consumption is gone—there is much that can be done so that the time for CMOS to be supplanted by yet another more energy efficient technology remains far in the future.

Anantha Chandrakasan,
Massachusetts Institute of Technology, Cambridge

Robert Brodersen
University of California at Berkeley

Part I

Overview

Low Power Microelectronics: Retrospect and Prospect

JAMES D. MEINDL, FELLOW, IEEE

The era of low power microelectronics began with the invention of the transistor in the late 1940's and came of age with the invention of the integrated circuit in the late 1950's. Historically, the most demanding applications of low power microelectronics have been battery operated products such as wrist watches, hearing aids, implantable cardiac pacemakers, pocket calculators, pagers, cellular telephones and prospectively the hand-held multi-media terminal. However, in the early 1990's low power microelectronics rapidly evolved from a substantial tributary to the mainstream of microelectronics. The principal reasons for this transformation were the increasing packing density of transistors and increasing clock frequencies of CMOS microchips pushing heat removal and power distribution to the forefront of the problems confronting the advance of microelectronics.

The distinctive thesis of this discussion is that future opportunities for low power gigascale integration (GSI) will be governed by a hierarchy of theoretical and practical limits whose levels can be codified as: 1) fundamental, 2) material, 3) device, 4) circuit, and 5) system. The three most important fundamental limits on low power GSI are derived from the basic physical principles of thermodynamics, quantum mechanics, and electromagnetics. The key semiconductor material limits are determined by carrier mobility, carrier saturation velocity, breakdown field strength and thermal conductivity, and the prime material limit of an interconnect is imposed by the relative dielectric constant of its insulator. The most important device limit is the minimum channel length of a MOSFET, which in turn determines its minimum switching energy and intrinsic switching time. Channel lengths below 60 nm for bulk MOSFET's and below 30 nm for dual gate SOI MOSFET's are projected. Response time of a canonical distributed resistance-capacitance network is the principal device limit on interconnect performance. To insure logic level restoration in static CMOS circuits, a minimum allowable supply voltage is about $4kT/q$. For a conservative 0.1 μm CMOS technology and 1.0 V supply voltage, the minimum switching energy of a ring oscillator stage is about 0.1 fJ and the corresponding delay time is less than 5.0 ps. Five generic system limits are set by: 1) the architecture of a chip, 2) the power-delay product of the CMOS and interconnect technology used to implement the chip, 3) the heat removal or cooling capacity of the packaging technology, 4) the cycle time requirements imposed on the chip and 5) its physical size.

To date, all microchips have been designed to dissipate the entire amount of electrical energy transferred during a binary switching transition. However, new approaches based on the second law of thermodynamics point the way to recycle switching energy by avoiding the erasure of information and switching under quasi-equilibrium conditions. Adiabatic computing technology offers promise of significant new advances in low power microelectronics.

Practical limits are elegantly summarized by Moore's Law which defines the exponential rate of increase with time of the number of transistors per chip. One billion transistor chips are projected for the year 2000 and 100 billion transistor chips are projected before 2020 by joining the results of the analyses of theoretical and practical limits through definition of the chip performance index as the quotient of the number of transistors per chip and the power delay product of the corresponding technology.

I. INTRODUCTION

Low power microelectronics was conceived through the invention of the transistor in 1947 and enabled by the invention of the integrated circuit in 1958. Throughout the following 37 years, microelectronics has advanced in productivity and performance at a pace unmatched in technological history. Minimum feature size F has declined by about a factor of 1/50; die area D^2 has increased by approximately 170 times; packing efficiency PE, defined as the number of transistors per minimum feature area has multiplied by more than a factor of 100 so that the composite number of transistors per chip $N = F^{-2} \cdot D^2 \cdot PE$ has skyrocketed by a factor of about 50×10^6, while the price range of a chip has remained virtually unchanged and its reliability has increased [1]. An inextricable concomitant advance of low power microelectronics has been a reduction in the switching energy dissipation E or power-delay product $Pt_d = E$ of a binary transition by approximately $1/10^5$ times. Consequently, as the principal driver of the modern information revolution, the ubiquitous microchip has had a profound and pervasive impact on our daily lives. Therefore, it is imperative that we gain as deep an understanding as possible of where we have been and especially of where we may be headed with the world's most important technology.

Almost two decades ago Gordon Moore of Intel Corporation observed that the number of transistors per chip had been doubling annually for a period of 15 years [2]. This astute observation has become known as "Moore's Law." With a reduction of the rate of increase to about 1.5 times per year, or a quadrupling every three years, Moore's Law has remained through 1994 an accurate description

Manuscript received November 14, 1994; revised January 12, 1995. This work was supported in part by SRC Contract 93-SJ-374.

The author is with the School of Electrical and Computer Engineering and Microelectronics Research Center, Georgia Institute of Technology, Atlanta, GA 30332-0269 USA.

IEEE Log Number 9409346.

Reprinted from *Proceedings of the IEEE*, pp. 619-635, April 1995.

Hierarchical Matrix of Limits

	Theoretical	Practical
5. System		
4. Circuit		
3. Device		
2. Material		
1. Fundamental		

Fig. 1. Hierarchical matrix of limits on GSI.

of the course of microelectronics. This discussion defines a corollary of Moore's Law which asserts that "future opportunities to achieve multi-billion transistor chips or gigascale integration (GSI) in the 21st century will be governed by a hierarchy of limits." The levels of this hierarchy can be codified as 1) fundamental, 2) material, 3) device, 4) circuit, and 5) system [3]. At each level there are two different kinds of limits to consider, theoretical and practical. Theoretical limits are informed by the laws of physics and by technological invention. Practical limits, of course, must comply with these constraints but must also take account of manufacturing costs and markets. Consequently, the path to GSI will be governed by a hierarchical matrix of limits as illustrated in Fig. 1, which emphasizes the structure of the hierarchy.

Following this introduction, Section II provides a brief retrospective view of low power microelectronics in which the antecedents of many current innovations are cited. Then, Section III treats the most important theoretical limits associated with each level of the hierarchy introduced in the preceding paragraph. In order to elucidate opportunities for low power microelectronics, many of these limits are represented by graphing the average power transfer P during a binary switching transition versus the transition time t_d. For logarithmic scales, diagonal lines in the P versus t_d plane represent loci of constant switching energy. Limits imposed by interconnections are represented by graphing the square of the reciprocal length of an interconnect $(1/L)^2$ versus the response time τ of the corresponding circuit. For logarithmic scales, diagonal lines in the $(1/L)^2$ versus τ plane represent loci of constant distributed resistance-capacitance product for an interconnect. The twin goals of low power microelectronics are to drive both the P versus t_d and the $(1/L)^2$ versus τ loci toward the lower left corners of their allowable zones of operation reflecting switching functions consuming minimal power and time, and communication functions covering maximal distance in minimal time.

Virtually all previous and contemporary microchips dissipate the entire amount of electrical energy transferred during a binary switching transition. This assumption is made in deriving the hierarchy of limits represented in the P versus t_d plane in Section III. However, in Section IV this stipulation is removed in a brief discussion of a new hierarchy of limits on quasi-adiabatic switching operations that recycle, rather than dissipate, a fraction of the energy transferred during a binary switching transition [4], [5].

In Section V practical limits are compactly summarized in a sequence of plots of minimum feature size, die edge, packing efficiency, and number of transistors per chip versus calendar year. Then the results of the discussions of theoretical and practical limits are joined by defining the most important single metric that indicates the promise of a technology for low power microelectronics, and that is the chip performance index or CPI which equals the quotient of the number of transistors per chip and the associated switching energy or CPI $= N/Pt_d$. Section VI concludes with a speculative comment on a paramount economic issue.

II. BACKGROUND

The genesis of low power microelectronics can be traced to the invention of the transistor in 1947. The elimination of the crushing needs for several watts of heater power and several hundred volts of anode voltage in vacuum tubes in exchange for transistor operation in the tens of milliwatts range was a breakthrough of virtually unparalleled importance in electronics. The capability to fully utilize the low power assets of the transistor was provided by the invention of the integrated circuit in 1958. Historically, the motivation for low power electronics has stemmed from three reasonably distinct classes of need [6]–[12]: 1) the earliest and most demanding of these is for portable battery operated equipment that is sufficiently small in size and weight and long in operating life to satisfy the user; 2) the most recent need is for ever increasing packing density in order to further enhance the speed of high performance systems which imposes severe restrictions on power dissipation density; and 3) the broadest need is for conservation of power in desktop and deskside systems where a competitive life cycle cost-to-performance ratio demands low power operation to reduce power supply and cooling costs. Viewed in toto, these three classes of need appear to encompass a substantial majority of current applications of electronic equipment. Low power electronics has become the mainstream of the effort to achieve GSI.

The earliest and still the most urgent demands for low power electronics originate from the stringent requirements for small size and weight, long operating life, utility, and reliability of battery operated equipment such as wrist watches, pocket calculators and cellular phones, hearing aids, implantable cardiac pacemakers, and a myriad of portable military equipments used by individual foot soldiers [6], [7]. Perhaps no segment of the electronics industry has a growth potential as explosive as that of the personal digital assistant (PDA) which has been characterized as a combined pocket cellular phone, pager, e-mail terminal, fax, computer, calendar, address directory, notebook, etc. [8]–[12]. To satisfy the needs of the PDA for low power electronics, comprehensive approaches are proposed that include use of the lowest possible supply voltage coupled with architectural, logic style, circuit, and CMOS technology optimizations [8]–[12]. The antecedents of these concepts are strikingly evident in publications from the 1960's [6], [7], in which several critical principles of low power design

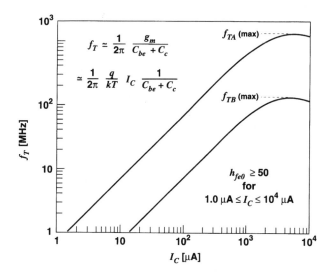

$$f_T \simeq \frac{1}{2\pi} \frac{g_m}{C_{be} + C_c}$$

$$\simeq \frac{1}{2\pi} \frac{q}{kT} I_C \frac{1}{C_{be} + C_c}$$

f_{TA} (max)

f_{TB} (max)

$h_{fe0} \geq 50$
for
$1.0\ \mu A \leq I_C \leq 10^4\ \mu A$

Fig. 2. Transistor gain-bandwidth product versus quiescent collector current. $V_{CE} = 3.0$ V for transistors A and B [7].

were formulated and codified [7]. The first of these was simply to use the lowest possible supply voltage, preferably a single cell battery. The second guideline was to use analog techniques wherever possible particularly in order to avoid the large standby power drain of then available bipolar digital circuits, although the micropower potential of CMOS was clearly articulated by G. Moore *et al.* in 1963 [13].

A third key principle of micropower design that was convincingly demonstrated quite early is the advantage of selecting the smallest geometry, highest frequency transistors available to implement a required circuit function, e.g., a wideband amplifier, and then scaling down the quiescent current until the transistor gain-bandwidth product f_T just satisfies the relevant system performance requirements. (The manifestation of this concept in current CMOS technology is to seek the available technology with the smallest minimum feature size in order to reduce the capacitance that must be charged/discharged in a switching transition.) Bipolar transistor gain bandwidth product is given by [7]

$$f_T = g_m / 2\pi (C_{be} + C_{jc}) \qquad (1)$$

where $g_m = qI_c/kT$ is the transconductance, I_c is the quiescent collector current, $C_{be} = C_{de} + C_{je}$ is the base-emitter capacitance including both junction capacitance C_{je} and minority carrier diffusion capacitance C_{de} (which is proportional to I_c), and C_{jc} is the collector junction capacitance. As illustrated in Fig. 2, suppose that required circuit performance demands a transistor gain bandwidth product $f_T = 120$ MHz which can be satisfied by device A at a collector current $I_{CA} = 0.20$ mA or by device B at a collector current $I_{CB} = 6.0$ mA. The choice of device A for low power design is clear. Moreover, for low current operation of both devices

$$f_T \cong (1/2\pi)(qI_c/kT)/(C_{je} + C_{jc}) \qquad (2)$$

is directly proportional to I_c thus indicating the clear advantage of maximizing gain-bandwidth product per unit of quiescent current drain in all transistors used in ana-

log information processing functions. For example, this concept applies in the design of RF receiver circuits for pocket telephones. It clearly suggests a receiver architecture that minimizes use of high frequency front end analog electronics.

A fourth generic principle of low power design that was clearly articulated in antiquity is the advantage of using "extra" electronics to reduce total power drain [7]. This tradeoff of silicon hardware for battery hardware was demonstrated, e.g., for a multi-stage wideband amplifier in which total current drain was reduced by more than an order of magnitude by doubling the number of stages from two to four while maintaining a constant overall gain-bandwidth product [7]. This concept is rather analogous to the approach of scaling down the supply voltage of a CMOS subsystem to reduce its power drain and speed and then adding duplicate parallel processing hardware to restore the throughput capability of the subsystem at an overall savings in power drain [11], [12].

A final overarching principle of low power design that was rigorously illustrated for a wide variety of circuit functions including dc, audio, video, tuned, and low noise amplifiers, nonlinear mixers and detectors and harmonic oscillators as well as bipolar and field effect transistor digital circuits is that micropower design begins with a judicious specification of the required system performance and proceeds to the optimal implementation that fulfills the required performance at minimum power drain [7].

The advent of CMOS digital technology removed quiescent power drain as an unacceptable penalty for broadscale utilization of digital techniques in portable battery operated equipment. Since the average energy dissipation per switching cycle of a CMOS circuit is given by $E = CV^2$ where C is the load capacitance and V is the voltage swing, the obvious path to minimum power dissipation is to reduce C by scaling down minimum feature size and especially to reduce V. The minimum allowable value of supply voltage V for a static CMOS inverter circuit was derived by R. Swanson and the author in 1972 [14] as

$$V_{\text{smin}} \geq \beta kT/q \qquad (3)$$

where β typically is between 2 and 4.

Early experimental evidence [14] supporting this rigorous derivation is illustrated in Fig. 3, which is a graph of the static transfer characteristic of a CMOS inverter for supply voltages as small as $V_s = 0.10$ V and matched MOSFET's whose threshold voltages of $V_t \cong \pm 0.16$ V were controlled by ion implantation. Further discussion of this topic is presented in Section III of this article.

During the 1970's a variety of new micropower techniques were introduced [15], [16] and by far the most widely used product exploiting these techniques was and is the electronic wristwatch [15]. A striking early application of power management occurred in implantable telemetry systems for biomedical research. It was the use of a 15 μW 500 kHz monolithic micropower command receiver as an RF controlled switch to connect/disconnect a single 1.35 V mercury cell power source to/from an

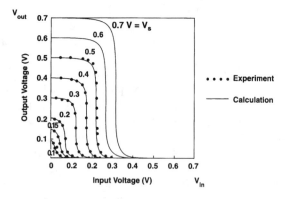

CMOS Inverter Transfer Curves

Fig. 3. Static CMOS inverter transfer characteristic [14].

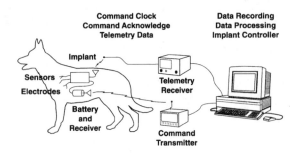

Fig. 4. Implantable telemetry system [16].

A Fundamental Limit from Thermodynamics

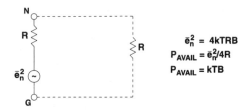

Fig. 5. Model for derivation of fundamental limit from thermodynamics.

implantable biomedical telemetry system. The fabrication processes used to produce the receiver chip were optimized to yield high value diffused silicon resistors [17]. Entire implantable units including active and passive sensors for biopotential, dimension, blood pressure and flow, chemical ion concentrations, temperature and strain were designed and implemented with power conservation as the primary criterion for optimization [16]. In many respects, the overall system operation of an implantable telemetry unit and its deskside external electronics subsystem for data processing, display and storage as illustrated in Fig. 4 is similar to but much smaller in scale than the operation of a modern PDA and its backbone network [18].

In the 1980's, the increasing level of power dissipation in mainstream microprocessor, memory and a host of application specific integrated circuit chips prompted an industry wide shift from NMOS and NPN bipolar technologies to CMOS in order to alleviate heat removal problems. The greatly reduced average power drain of CMOS chips provided a relatively effortless interim solution to the problems of low power design. However, the relentless march of microelectronics to higher packing densities and larger clock frequencies has, during the early 1990's, brought low power design to the forefront as a primary requirement for mainstream microelectronics which is addressed in the remainder of this paper.

III. THEORETICAL LIMITS

A. Fundamental Limits

The three most important fundamental limits on low power GSI are derived from the basic physical principles of thermodynamics, quantum mechanics and electromagnetics [19]. Consider first the limit from thermodynamics. Suppose that the node N illustrated in Fig. 5 is imbedded in a complex microprocessor chip and that between N and ground G, there is an equivalent resistance of value R. Immediately from statistical thermodynamics, it can be shown that the mean square open circuit noise voltage across R is given by [20]

$$\bar{e}_n^2 = 4kTRB \qquad (4)$$

and consequently the available noise power is

$$P_{\text{avail}} = kTB \qquad (5)$$

where k is Boltzmann's constant, T is absolute temperature and B is the bandwidth of the node. Now, it is reasonable to assert that if the information represented at the node is to be changed from a zero to a one, or vice versa, then the average signal power P_s transfer during the switching transition should be greater than (or at least equal to) the available noise power P_{avail} by a factor $\gamma \geq 1$ or

$$P_s \geq \gamma P_{\text{avail}}. \qquad (6)$$

One can then derive an expression for the switching energy E_s transfer in the transition,

$$E_s \geq \gamma kT. \qquad (7)$$

Clearly, Boltzmann's constant k and absolute temperature T are independent of any materials, devices or circuits associated with the node. Consequently, E_s represents a fundamental limit on binary switching energy. For reasons to be cited in the discussion of circuit limits, $\gamma = 4$ will be assumed at this point so that at $T = 300$ K, $E_s \geq 1.66 \times 10^{-20}$ J $= 0.104$ eV. One can interpret this limit as the energy required to move a single electron through a potential difference of 0.104 V, which is a Lilliputian energy expenditure compared with current practice which involves energies greater by a factor of about 10^7. One advantage of larger switching energies is that the probability of error due to internal thermal noise energy E_n, $Pr(E_n > E_s)$, described by a Boltzmann probability distribution function,

$$Pr(E_n > E_s) = \exp(-E_s/kT) \qquad (8)$$

decreases exponentially as E_s/kT increases.

The second fundamental limit on low power GSI is derived from quantum mechanics and more specifically from the Heisenberg uncertainty principle [21], which can be interpreted as follows. A physical measurement associated with a switching transition that is performed in a time Δt must invoke an energy

$$\Delta E \geq h/\Delta t \tag{9}$$

where h is Planck's constant [19]. Consequently, one can show that

$$P \geq h/(\Delta t)^2 \tag{10}$$

is the required average power transfer during a switching transition of a single electron wave packet. Fig. 6 illustrates the fundamental limits from thermodynamics and quantum mechanics in the power-delay plane. Switching transitions to the left of their loci are forbidden, regardless of the materials, devices or circuits used for implementation. The zone of opportunity for low power GSI lies to the right of these limits. As discussed in Section IV, the power treated in Fig. 6 is a rate of energy transfer and not necessarily a rate of energy dissipation, although the later has virtually always been the case in past practice.

The fundamental limit based on electromagnetics simply dictates that the velocity of propagation v of a high speed pulse on a global interconnect must be less than the speed of light in free space c_0 or

$$v = L/\tau \leq c_0 \tag{11}$$

where L is interconnect length and τ is interconnect transit time. As illustrated in Fig. 7, the speed of light limit prohibits operation to the left of the $L = c_0\tau$ locus for any interconnect regardless of the materials or structure used for its implementation.

B. Material Limits

At the second level of the hierarchy, material limits are independent of the macroscopic geometrical configuration and dimensions of particular device structures. The principal material of interest is Si, which is compared here with GaAs. The primary properties of a semiconductor which determine its key material limits are 1) carrier mobility μ, 2) carrier saturation velocity v_s, 3) self-ionizing electric field strength $\mathcal{E}_c$, and 4) thermal conductivity K. In order to calculate a semiconductor material limit that is independent of the macroscopic properties of a specific device, consider a cube of undoped Si of dimension Δx that is imbedded in a three-dimensional matrix of similar cubes. The material limit on switching energy ($E = Pt_d$) can be calculated as the amount of electrostatic energy stored in this cube of Si of dimension $\Delta x = V_0/\mathcal{E}_c$ with a voltage difference V_0 across two of its parallel faces, created by an electric field nearly equal to the self-ionizing value $\mathcal{E}_c$. Thus

$$Pt_d = E = \varepsilon_{Si}V_0^3/2\mathcal{E}_c \tag{12}$$

where ε_{Si} is the permittivity of Si.

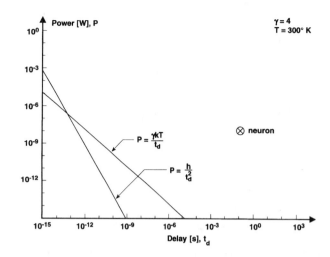

Fig. 6. Average power transfer P during a switching transition versus transition interval t_d for fundamental limits derived from thermodynamics and quantum mechanics.

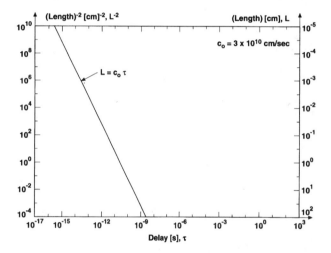

Fig. 7. Square of reciprocal length $(1/L)^2$ of an interconnect versus interconnect circuit response time τ for the fundamental limit from electromagnetics.

The minimum switching time t_d for this stored energy is taken as the transit time of a carrier through the cube, that is

$$t_d \geq V_0/v_s\mathcal{E}_c \tag{13}$$

describes the material transit time limit. For $V_0 = 1.0$ V and $v_s = 10^7$ cm/s, $t_d = 0.33$ ps for Si and 0.25 ps for GaAs. Thus Si bears only a 33% larger electron transit time per unit of potential drop than GaAs for large values of electric field strength typical for GSI. This small disadvantage is a consequence of two factors: The nearly equal saturation velocities of electrons in Si and GaAs as shown in Fig. 8 as well as the 33% larger breakdown field strength $\mathcal{E}_c$ of GaAs. Fig. 8 also illustrates the nearly six-fold advantage in electron velocity and therefore mobility that GaAs enjoys at small values of electric field strength (e.g., < 500 V/cm). In previous generations of technology operating at small values of $\mathcal{E}$, it was carrier mobility rather than saturation velocity at large values of $\mathcal{E}$ (e.g., > 50 000 V/cm), that was

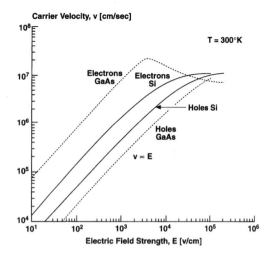

Fig. 8. Carrier velocity versus electric field strength $\mathcal{E}$ for electrons and holes in Si and GaAs.

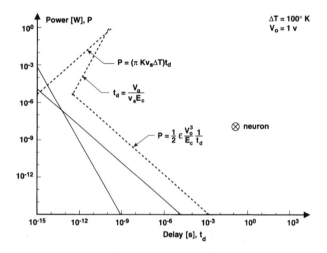

Fig. 9. P versus t_d for fundamental limits and Si material limits based on energy storage, transit time, and heat removal.

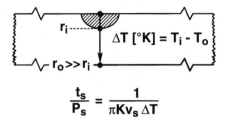

$$\frac{t_s}{P_s} = \frac{1}{\pi K v_s \Delta T}$$

Fig. 10. Model for derivation of material limit based on heat conduction.

the principal determinate of high speed capability, which is no longer the case.

The switching energy limit for Si given by (12) is illustrated in Fig. 9 for a potential swing $V_0 = 1.0$ V, which is presumed to be a minimum acceptable value. Solving (13) for V_0 and substituting into (12) gives the locus of minimum switching times

$$P = \varepsilon_{\text{Si}} \mathcal{E}_c^2 v_s^3 t_d^2 / 2 \qquad (14)$$

designated by (13) i.e., $t_d \geq V_0 / v_s \mathcal{E}_c$ in Fig. 9. As supply voltage is scaled below approximately 1.0 V, the self-ionization effects that determine $\mathcal{E}_c$ no longer persist and the limits described by (12)–(14) no longer apply.

In order to derive the heat removal limit at the material level, Fig. 10 illustrates an isolated generic device which is hemispherical in shape with a radius r_i and located in a chip that is mounted on an ideal heat sink at a temperature T_0. Based on Fourier's law of heat conduction, $Q = -K A dT/dx$ where Q is the heat flow in J/s through an area A in the presence of a thermal gradient dT/dx, the power conducted away from the device of diameter

$2r_i = v_s t_d$ to the heat sink is given by

$$P = \pi K v_s \Delta T t_d \qquad (15)$$

where K is the semiconductor thermal conductivity and T is the temperature difference between the device and the heat sink. Substituting representative values indicates that $t_d / P = 0.21$ ns/W for Si and 0.69 ns/W for GaAs for $\Delta T = 100$ K. This sample calculation indicates that GaAs suffers a switching time per unit of heat removal that is over 300% greater than the corresponding value for Si, when switching time is limited by substrate thermal conductivity, which is about three times larger for Si than GaAs.

If the generic device illustrated in Fig. 10 is surrounded by a hemispherical shell of SiO$_2$ of radius r_s representing an SOI structure, the equivalent thermal conductivity K_{EQ} of the composite structure is given by [22]

$$K_{EQ} = (K_{ox} K_{\text{Si}} r_s / r_i) \{ K_{si} [(r_s / r_i) - 1] + K_{ox} \}^{-1}. \qquad (16)$$

Note that as $r_i \to r_s$, $K_{EQ} \to K_{\text{Si}}$ and as $K_{ox} \to K_{\text{Si}}$, $K_{EQ} \to K_{\text{Si}}$. For $K_{ox} \cong 0.01 K_{\text{Si}}$ and $r_s = 1.5, 2, 4r_i$, $K_{EQ} \cong 0.029, 0.02, 0.013 K_{si}$ which indicates a severe reduction in equivalent thermal conductivity of the SOI structure relative to bulk Si.

In summary, Fig. 9 illustrates a second forbidden zone of operation imposed by the characteristics of Si as a material. Operation to the left of the loci of the three material limits defined by (12), (13), and (15) is proscribed for any Si device whose carriers undergo several scattering events within the active region of the device. That is, the three Si material limits illustrated in Fig. 9 assume that the distance over which bulk carrier transport occurs is greater than several mean free path lengths. For shorter distances, the possibility of quasi-ballistic or velocity overshoot effects is best treated in a particular device context [23], [24].

The primary interconnect material limit is defined by substituting a polymer, with a relative dielectric constant $\varepsilon_r \cong 2$, as the insulator replacing free space in the fundamental speed of light limit, as illustrated in Fig. 11 for the $L = v\tau$ locus. In essence, both the fundamental and material limits assume a uniform lossless transmission line in a homogeneous dielectric.

C. Device Limits

Device limits are independent of the particular circuit configuration in which a transistor or an interconnect is applied. The most important device in modern microelectronics is the MOSFET [25] and the most critical limit

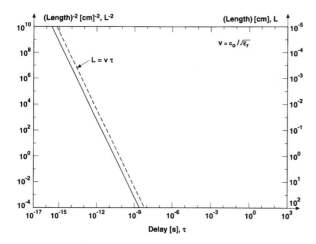

Fig. 11. $(1/L)^2$ versus τ for fundamental limit and material limit with polymer dielectric replacing free space. Both limits are based on the velocity of electromagnetic waves.

on it is its allowable minimum effective channel length $L_{\min}$ [26]. Consider a family of MOSFET's in which all parameters are held constant except effective channel length L, which is allowed to take on a wide range of values, e.g., 3.0 μm $\geq L \geq 0.03$ μm. As L is reduced within this range, eventually so-called short channel effects, are manifest [27]. The source and drain depletion regions begin to capture ion charge in the central region of the channel that is strictly under gate control for longer channels. The salient result of such short channel effects which are aggravated as drain voltage increases [28], is that the threshold voltage V_t is reduced, subthreshold leakage current increases and the MOSFET no longer operates effectively as a switch.

Let us now consider the principal factors which determine the minimum effective channel length $L_{\min}$ of a MOSFET. In order to achieve $L_{\min}$, both gate oxide thickness (T_{ox}) and source/drain junction depth (X_j) should be as small as possible [29], [30]. Gate leakage currents due to tunneling limit T_{ox} [31] and parasitic source/drain resistance limits X_j [32]. In addition, low impurity channels with abrupt retrograde doping profiles are highly desirable for control of short channel effects and high transconductance in bulk MOSFET's [33]. The use of dual gates on opposite sides of a channel ostensibly provides the ultimate structure to contain short channel effects [34]–[36]. Fig. 12 illustrates six different MOSFET structures that have been analyzed consistently to determine their short channel behavior [29] for a very aggressive set of parameters including $T_{ox} = 3.0$ nm for all devices; $X_j = 5.0$ nm for shallow junctions; $X_j = 50$ nm, 100 nm for deep junctions; silicon layer thickness, $d = 5.5$ nm for the SOI single gate device; and silicon channel thickness, $d = 10.9$ nm for the dual gate device. Based on both analytical solutions and numerical solutions, using PISCES and DAVINCI, of the two and three dimensional Poisson equation, the threshold voltage roll-off due to scaling of effective channel length is illustrated in Fig. 13 for each of the MOSFET structures sketched in Fig. 12. Families of curves such as these support the prospect of achieving shallow junction

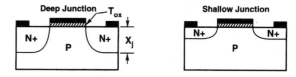

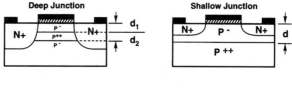

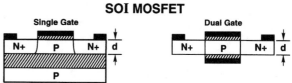

Fig. 12. MOSFET structures.

retrograde channel profile bulk MOSFET's with channel lengths as short as 60 nm [29], [30], [32] and dual gate or DELTA MOSFET's with channel lengths as short as 30 nm [29], [30], [35]. An interesting feature of the analytical formulation of threshold voltage change [29]

$$\Delta V_T \sim \exp\{-(1/\pi)(\varepsilon_{ox}/\varepsilon_{\text{Si}})(L/T_{ox})\} \quad (17)$$

specifically for the case of deep junctions and uniform doping, but also suggestive of other cases [29], [30] indicates the importance of thin gate insulators with high permittivity in the reduction of short channel effects. In addition to threshold voltage shift, other typically more manageable factors such as bulk punchthrough, gate induced drain leakage and impact ionization which contribute to leakage current and the totality of effects that impact reliability must be observed as potential MOSFET limits [31].

Assuming that analysis of the internal physics of the MOSFET serves to define a minimum effective channel length $L_{\min}$, the next stage of effort to define a switching energy limit is to recognize that the relevant energy E is stored on the gate of the MOSFET at the outset of a switching transition. Therefore, given an allowable minimum effective channel length $L_{\min}$, the switching energy limit for a MOSFET is simply

$$E = Pt_d = \frac{1}{2}C_0 L_{\min}^2 V_0^2 \quad (18)$$

where C_0 is the gate oxide capacitance per unit area corresponding to $L_{\min}$. The smallest possible value of the transition time is the channel transit time

$$t_{d\min} = L_{\min}/v_{sc} \quad (19)$$

where v_{sc} is the saturation velocity of carriers in the channel taken as 8×10^6 cm/s for electrons [37]. Assuming 1) minimum feature size $F \cong L_{\min}$, 2) $C_0 = \varepsilon_{ox}/T_{ox} =$

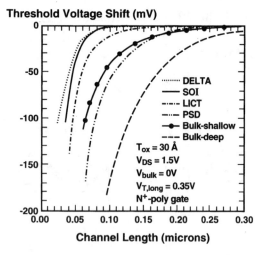

Threshold Voltage Shift (mV)

Channel Length (microns)

- ······ DELTA
- —— SOI
- —·—· LICT
- —··— PSD
- ●—● Bulk-shallow
- —— Bulk-deep

$T_{ox} = 30$ Å
$V_{DS} = 1.5$V
$V_{bulk} = 0$V
$V_{T,long} = 0.35$V
N^+-poly gate

Fig. 13. Short channel threshold voltage versus channel length for the Si devices shown in Fig. 12. Device parameters at 300 K are: 1) Deep-junction bulk MOSFET: $N_A = 8.6 \times 10^{17} \text{cm}^3$, Junction depth = 1000 Å [29]. 2) Shallow-junction bulk MOSFET: $N_A = 8.6 \times 10^{17} \text{cm}^3$ Junction depth = 50 Å. 3) PSD or low impurity channel deep junction MOSFET: $N_A = 10^{16} \text{cm}^3$, $N_A^{++} = 5 \times 10^{18} \text{cm}^3$, $N_A^+ = 5 \times 10^{17} \text{cm}^3$, $d = 218$ Å, Junction depth = 500Å [29]. 4) LICT or low impurity channel shallow junction MOSFET: $N_A = 5 \times 10^{16} \text{cm}^{-3}$, $N_A = 5 \times 10^{18} \text{cm}^3$, $d = 318$ Å, Junction depth = 50 Å. 5) SOI single gate MOSFET: $N_A = 5 \times 10^{18} \text{cm}^{-3}$, $d = 55$ Å. 6) Delta or SOI dual gate MOSFET: $N_A = 5 \times 10^{18} \text{cm}^{-3}$ $d = 109$ Å.

$\varepsilon_{ox}/S_{ox}F$ where S_{ox} is taken as a constant factor relating gate oxide thickness T_{ox} and minimum feature size F, and 3) $V_0 = S_v F$ where S_v is taken as a constant electric field strength relating supply voltage V_0 and F, (18) and (19) are solved simultaneously in order to derive the locus of minimum transition times

$$P = \frac{1}{2}(\varepsilon_{ox}/S_{ox})S_v^2 v_{sc}^3 t_d^2 \quad \text{or}$$
$$P = \frac{1}{2}(C_0 L_{min})(V_0/L_{min})^2 v_{sc}^3 t_d^2 \quad (20)$$

which is designated $t_d = L_{min}/v_{sc}$ in Fig. 14. The MOSFET switching energy limit (18) and the locus of transition time limits are illustrated in Fig. 14, which includes a third forbidden zone of operation to the left of these loci, for all MOSFET's with channel lengths larger than the conservative value $L_{min} = 0.1 \ \mu$m and a minimum gate oxide thickness $T_{ox} = 3.0$ nm. The proximity of the material, (13) and (14), and device, (19) and (20), loci for minimum transition times in Fig. 14 reflects the condition that the electric field strength $\mathcal{E}$ and carrier velocity v_{sc} assumed for the MOSFET's are pressing the material limits of Si. A channel saturation velocity of 8×10^6 cm/s at a tangential field strength of 200 000 V/cm in a 60 nm channel is quite likely to be somewhat underestimated and more refined values that consider velocity overshoot are needed [24], [35].

The key device limit on interconnects is represented by the response time of a canonical distributed resistance-capacitance network driven by an ideal voltage source. The response of such a network to a unit step function is given in the complex frequency(s) domain as [38]

$$v_0(s) = 1/s \cosh[sRC]^{1/2} \quad (21)$$

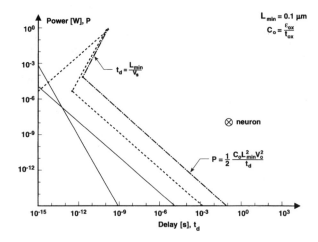

Fig. 14. P versus t_d for fundamental limits, Si material limits and MOSFET device limits derived from gate energy storage and channel transit time.

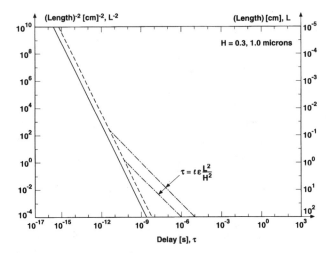

Fig. 15. $(1/L)^2$ versus τ for fundamental limit, material limit and device limits on interconnects for a polymer-copper technology. Device limits represent the response time of a distributed resistance-capacitance network.

where R and C are the total resistance and capacitance respectively and it can be shown that the 0–90% response time is $\tau = 1.0$ RC [38]. In comparison, for a simple RC lumped element model of the distributed network, the 0–90% response time is $\tau = 2.3$ RC. Neglecting fringing effects, for an interconnect of length L

$$\tau \cong (\rho/H_\rho)(\varepsilon/H_\varepsilon)L^2 \quad (22)$$

where (ρ/H_ρ) is the conductor sheet resistance in Ω/square and $(\varepsilon/H_\varepsilon)$ is the sheet capacitance in F/cm². Fig. 15 illustrates (22) for equal metal and insulator thicknesses $H_\rho = H_\varepsilon = H = 0.3 \ \mu$m and 1.0 μm. A third forbidden zone is evident. For example, no polymer-copper interconnect with a thickness H smaller than 0.3 μm can operate to the left of the 0.3 μm locus which represents a contour of constant distributed resistance-capacitance product [39].

Further exploration is needed of MOSFET limits to take account of velocity overshoot and random dopant ion placement as well as other effects and of interconnect limits including, for example, inductance and electromigration.

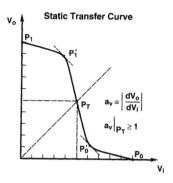

Fig. 16. Static transfer characteristic of a nonideal CMOS inverter.

D. Circuit Limits

The proliferation of limits as one ascends the hierarchy necessitates an increasing degree of selectivity in choosing those to be investigated. At the fourth level, circuit limits are independent of the architecture of a particular system. Four key generic circuit limits that cannot be avoided are discussed hereafter. The initial issue to consider is which logic circuit configurations are the most promising for low power microelectronics. The candidates include, e.g., GaAs direct coupled field effect transistor logic (DCFL), Si bipolar transistor emitter coupled logic (ECL), mainstream CMOS and BiCMOS. Of all the logic families now in use, it appears that common static CMOS has the most promise for low power GSI because 1) it has the lowest standby power drain of any logic family, 2) it has the largest operating margins, 3) it is the most scalable, and 4) it is the most flexible in terms of the circuit functions it can implement. For these reasons, this discussion hereafter focuses exclusively on CMOS logic.

The first and foremost generic circuit requirement that must be met by a logic gate is commonly taken for granted. In pursuing limits this practice cannot be followed. It is important to recognize that signal quantization or the capability to distinguish "zeros" from "ones" virtually without error throughout a large digital system is the quintessential requirement of a logic gate. For static CMOS logic this quantization requirement translates into the necessity for an incremental voltage amplification (a_v) which is greater than unity in absolute value at the transition point P_T of the static transfer characteristic of the gate where input and output signals are equal, as illustrated in Fig. 16. This is a heuristic requirement that can be "seen" by considering the need for $|a_v| > 1$ in order to restore "zero" and "one" levels in an iterative chain of inverters with an arbitrary input level for the initial stage.

An interesting derivative of the requirement for $|a_v| > 1$ is that the minimum supply voltage V_{ddmin} for which a CMOS inverter can fulfill the requirement is [14]

$$V_{dd} \geq (2kT/q)[1 + C_{fs}/(C_0 + C_d)]\ln(2 + C_0/C_d)$$
$$\geq \beta kT/q \approx 0.1\,\text{V} @ T = 300°\text{K} \quad (23)$$

where β typically is between 2 and 4 and C_{fs} is channel fast surface state capacitance, C_d is channel depletion region

capacitance and C_0 is gate oxide capacitance per unit area in each case. This result which was derived prior to 1972 provides a rationale for selecting a switching energy limit $E_s = \gamma kT$ at the fundamental level of the hierarchy (7) by postulating that a signal, carried by a single electron charge q through a potential difference ΔV, requires an energy $q\Delta V = \gamma kT$ and therefore a minimum potential swing $\Delta V = \gamma kT/q$ defined by (23). Given that the presumed minimum acceptable signal swing for defining both the material (12) and device (18) switching energy limits is $V_o = 1.0$ V, the question is, "why not set $V_o = V_{ddmin} = 0.1$ V?" The simple answer to this question is that to do so would require a threshold voltage V_t so small that drain leakage current in the off-state of the MOSFET would be entirely too large for most applications. In considering logic and memory circuit behavior at low supply voltages [40]–[48], a value of supply voltage $V_{dd} = V_o = 1.0$ V appears to be a broadly acceptable compromise between small dynamic power and small static power dissipation, although confirmation of (23) in a low power system with $V_{dd} = 200$ mV is a prominent recent development [48].

Assuming negligible static power drain due to MOSFET leakage currents, a second generic circuit limit on CMOS technology is the familiar energy dissipation per switching transition

$$E = Pt_d = \tfrac{1}{2}C_cV_o^2 \quad (24)$$

where C_c is taken as the total load capacitance of a ring oscillator stage, including output diffusion capacitance, wiring capacitance and input gate capacitance for an inverter which occupies a substrate area of $100\ F^2$ where the minimum feature size $F = 0.1\ \mu$m.

Assuming carrier velocity saturation in the MOSFET's, an approximate value of the drain saturation current is

$$I_{ds} \approx ZC_ov_{sc}(V_g - V_t) \quad (25)$$

where Z is the channel width, V_t is threshold voltage and the gate voltage, $V_g = V_o$. A third generic circuit limit, the intrinsic gate delay can be calculated as $t_d = 1/2[C_cV_o/I_{ds}]$ or using (25)

$$t_d = \tfrac{1}{2}[C_c/Zv_{sc}C_o][V_o/(V_o - V_t)] \quad (26)$$

assuming that the product $Zv_{sc}C_o$ is equal for the N and P-channel MOSFET's and that their threshold voltages are matched. For 1) $C_c \cong S_cF$ where S_c is taken as a constant factor relating load capacitance C_c and minimum feature size F, 2) $V_o = S_vF$ as in the derivation of (20), 3) $Z = S_ZF$ where S_Z is a constant relating channel width Z and F, 4) $C_o = \varepsilon_{ox}/S_{ox}F$ as in the derivation of (20), and 5) $V_o/(V_o - V_t) \cong 1$, (24) and (26) are solved simultaneously for the locus of intrinsic gate delay times,

$$P = 4(V_o/C_c)^2(ZC_ov_{sc})^3t_d^2 \quad (27)$$

which is designated $t_d \cong \tfrac{1}{2}C_c/ZC_ov_{sc}$ in Fig. 17. The CMOS circuit switching energy limit (24) and the locus of intrinsic gate delay times (27) are illustrated in Fig. 17 which includes a fourth forbidden zone, to the left of

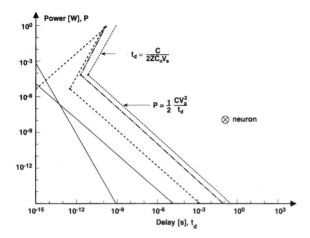

Fig. 17. P versus t_d for fundamental, Si material, MOSFET and CMOS circuit limits. Circuit limits are derived from switching energy and intrinsic gate delay analyses.

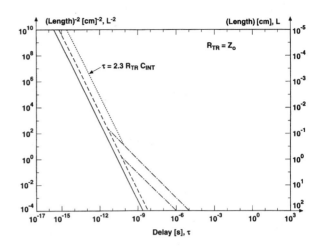

Fig. 18. $(1/L)^2$ versus τ for fundamental, material, device, and circuit limits on interconnects. Circuit limits represent the response time of a circuit consisting of a MOSFET driving a lumped interconnect capacitance.

these loci, for all CMOS circuits using bulk technology with feature sizes larger than $F = 0.1 \ \mu$m. A gate oxide thickness $T_{ox} = 3.0$ nm and a load capacitance $C_c = 0.5$ fF are used in plotting the loci of Fig. 17.

The fourth generic circuit limit applies to a transistor driving a global interconnect presented as a distributed resistance-capacitance network extending, e.g., between opposite corners of a chip. The response time of this global interconnect circuit is [49]

$$\tau \cong (2.3R_{tr} + R_{\text{int}})C_{\text{int}} \qquad (28)$$

where R_{tr} is the output resistance of the transistor driver and R_{int} and C_{int} are the total resistance and capacitance, respectively of the global interconnect. To prevent excessive delay due to wiring resistance, the circuit should be designed so that $R_{\text{int}} < 2.3R_{tr}$ giving

$$\tau \cong 2.3R_{tr}C_{\text{int}} = 2.3R_{tr}c_{\text{int}}L \qquad (29)$$

where c_{int} is the capacitance per unit length of the interconnect. The distributed capacitance of a nearly lossless or TEM-mode transmission line can be expressed as $c_{\text{int}} = 1/vZ_o$ where $v = c_o[\varepsilon_r]^{-1/2}$ is the wave propagation velocity of the line, ε_r is the relative permittivity of its dielectric, $Z_o \approx [\mu_o/\varepsilon_o\varepsilon_r]^{1/2}$ is its characteristic impedance and $c_o = 1/[\mu_o\varepsilon_o]^{1/2}$. This global interconnect response time limit is illustrated in Fig. 18. The region to the left of the

$$\tau \cong 2.3R_{tr}C_{\text{int}} = 2.3(R_{tr}/Z_o)(L/v) \qquad (30)$$

locus is a forbidden zone for driver resistances larger than the designated value for the locus. Although the interconnect models engaged in the preceding discussion are rather elementary, they provide a clear picture of circuit limits which govern global interconnect performance.

Further exploration is needed of limits imposed by other important circuit configurations and by new device structures.

E. System Limits

System limits are the most numerous and nebulous ones in the hierarchy. They depend on all other limits and above all they are the most restrictive ones in the hierarchy. Consequently, it is imperative that these predominant limits be carefully considered. Among the innumerable constraints arising from the fact that each different chip design has its own unique set of limits, there are five inescapable generic system limits that are elucidated in this discussion. These limits are set by 1) the architecture of the chip, 2) the power-delay product of the CMOS technology used to implement the chip, 3) the heat removal or cooling capacity of the chip package, 4) the cycle time requirements imposed on the chip, and 5) its physical size. To illustrate these generic limits it is necessary to select a particular example for a case study, which is intended to be broadly applicable. In keeping with the intent to explore opportunities for low power GSI, salient boundary conditions that are assumed for the case study are: 1) a generic architecture equivalent in complexity to one billion logic gates, i.e., $N_{gT} = 10^9$, 2) CMOS technology with a conservative minimum feature size, $F = 0.1 \ \mu$m, 3) a package cooling coefficient, $Q = 50$ W/cm^2, 4) a clock frequency $f_c = 1.0$ GHz, and 5) a single chip implementation.

A block diagram of the system architecture is illustrated in Fig. 19. It is conceived as a two-dimensional systolic array [50]–[52] of 1024 identical macrocells, each consisting of a number of gates $N_g = 10^9/1024$. Communication between macrocells is assumed to occur only at the physical boundaries of adjacent macrocells. Each macrocell is assumed to receive an unskewed clock signal distributed by a balanced five-level H-tree network [38], [53], to the geometric center of the macrocell. Both logic and timing signals are communicated over arbitrary paths within a square macrocell of dimension L so the maximum path length for clock skew is L and the maximum logic signal path length is $2L$. The macrocell is represented as a random

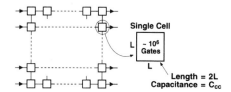

Fig. 19. Systolic array system block diagram.

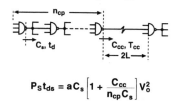

Fig. 20. Critical path used to define the system switching energy limit.

logic network described by Rent's rule [54]

$$N_p = K_p N_g^p \qquad (31)$$

where N_p is the number of signal lines entering or exiting the macrocell, Rent's coefficient $K_p = 0.82$ and Rent's exponent $p = 0.45$ which are empirically determined values for microprocessors [38]. Using Rent's rule as the basis of a stochastic analysis, the average length of an interconnect in gate pitches $\overline{R}$ can be calculated as [55]

$$\overline{R}_{rl} = 2/9\{7(N_g^{p-0.5} - 1)(4^{p-0.5} - 1)^{-1} \\ - (1 - N_g^{p-1.5})(1 - 4^{p-1.5})^{-1}\} \\ \cdot (1 - 4^{p-1})(1 - N_g^{p-1})^{-1} \qquad (32)$$

for $p \neq 0.5$. For the microprocessor-like macrocell with $p = 0.45$, (32) gives $\overline{R}_{rl} \cong 6$. Thus the total wire length loading a gate in the random logic network is

$$1_{rl} = \overline{R}_{rl} \cdot FO \cdot [A_{rl}]^{1/2} \qquad (33)$$

where $FO = 3$ is the fan-out and gate area $A_{rl} = 200$ F^2 is assumed to be limited by transistor packing density. This places a stringent demand on local wiring area which requires a logic gate dimension [38]

$$[A_{rl}]^{1/2} = \overline{R}_{rl} \cdot FO \cdot p_w/e_w n_w \qquad (34)$$

where n_w is the number of wiring levels, p_w is the wiring pitch and e_w is the wiring efficiency factor. For $n_w = 4$, $p_w = 0.2$ μm, and $e_w = 0.75$, as well as for $n_w = 6$, $p_w = 0.2$ μm, and $e_w = 0.5$ transistor packing density limits logic gate area.

As illustrated in Fig. 20 the system switching energy limit is defined by a composite gate which characterizes the critical path of a macrocell. For a logic signal this path is assumed to consist of 1) a chain of n_{cp} random logic gates and 2) one macrocell corner-to-corner global interconnect of length $2L$ [56], [57]. Therefore, the prorata switching energy of the composite gate is given by

$$E = Pt_d = \tfrac{1}{2}C_{rl}[1 + C_{cc}/n_{cp}C_{rl}]V_o^2 \qquad (35)$$

where C_{rl} is the total capacitance loading a random logic gate including MOSFET diffusion capacitance, wiring capacitance for a total interconnect length l_{rl} (33), and MOSFET gate capacitance, and C_{cc} is the total capacitance of the corner-to-corner interconnect circuit. The effective propagation delay time of the composite gate is defined as

$$t_d = t_{drl}(1 + T_{cc}/n_{cp}t_{drl}) \qquad (36)$$

where t_{drl} is the delay time of a random logic gate and T_{cc} is the response time of the corner-to-corner interconnect circuit. In (35), P is the average power dissipation of a composite gate during the propagation delay time t_d.

As illustrated in Fig. 21, the system heat removal limit is defined by the requirement that the average power dissipation of a composite gate $\overline{P}$ must be less than the cooling capacity of the packaging or

$$\overline{P} = aE/T_c \leq QA \qquad (37)$$

where E as in (35) is the switching energy of a composite gate, $a \leq 1$ is the probability that the gate switches during a clock cycle, $T_c = 1/f_c$ is the clock period, Q [W/cm²] is the package cooling coefficient and A is the substrate area occupied by the critical path composite gate. It is assumed that the composite gate area A consists of the area occupied by a random logic gate A_{rl} plus a *pro rata* share of the area of the corner-to-corner driver circuit A_{cc} and that $A_{rl}/A_{cc} = C_{rl}/C_{cc}$ which gives

$$A = A_{rl}(1 + C_{cc}/n_{cp}C_{rl}). \qquad (38)$$

The cycle time can be expressed as

$$T_c = s_{cp}n_{cp}t_d \qquad (39)$$

where it is to be shown that $s_{cp} \geq 1$ accounts for a small clock skew [58]. Combining (35), (37)–(39)

$$P \leq (s_{cp}n_{cp}/a)QA_{rl}(1 + C_{cc}/n_{cp}C_{rl}) \qquad (40)$$

gives the maximum allowable value of P, that is permitted by the cooling capability of the package and therefore the minimum composite gate delay t_d as defined by (35). Assuming 1) $C_{rl} \cong S_{rl}F$ and $C_{cc} = S_{cc}F$ where S_{rl} and S_{cc} are constants relating, respectively, random logic gate load capacitance C_{rl} and corner-to-corner interconnect capacitance C_{cc} to minimum feature size F, 2) $V_o = S_vF$ as in the derivation of (20), and 3) $A_{rl} = S_{rl}F^2$ where S_{rl} is a constant relating random logic gate area A_{rl} to F^2, and solving (35) and (40) simultaneously gives the locus of minimum composite gate delays

$$P = (1 + C_{cc}/n_{cp}C_{rl})(\tfrac{1}{2}C_{rl}V_o^2)^{-2}[(s_{cp}n_{cp}/a)QA_{rl}]^3t_d^2 \qquad (41)$$

which is designated as $\overline{P} \leq QA$ in Fig. 22. In plotting (41), in addition to the parameter values listed in the figure,

13

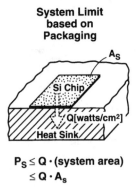

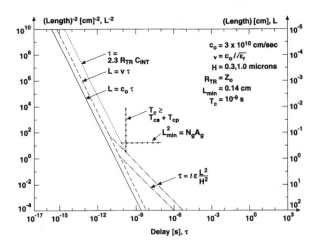

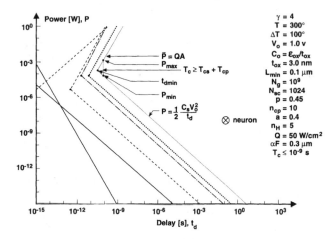

Fig. 21. System heat removal limit based on packaging.

Fig. 22. P versus t_d for fundamental, material, device, circuit, and system limits. System limits are imposed by switching energy $(Pt_d = \frac{1}{2}C_sV_o^2)$, heat removal $(\overline{P} \leq QA)$, and cycle time $(T_c \geq T_{cs} + T_{cp})$ requirements.

$C_{cc} = 100$ fF, $C_{rl} = 3.28$ fF, $A_{rl} = 2 \times 10^{-8}$ cm^2 and $s_{cp} = 1.11$.

The system cycle time limit is given by

$$T_c \geq T_{cs} + T_{cp} \qquad (42)$$

where T_{cs} is the maximum clock skew within a macrocell, $T_{cp} = n_{cp}t_d$ is the critical path delay and t_d is given by (36). From (42)

$$t_{d\max} \leq (T_c - T_{cs})/n_{cp} \qquad (43)$$

is the maximum allowable value of critical path composite gate delay that enables the required cycle time. If the allowed clock skew is $T_{cs} = \eta T_c$ (e.g., $\eta \approx 0.1$) then referring to (39), $s_{cp} = 1/(1 - \eta) \approx 1.11$. To calculate t_d as given by (36) at the system level, appropriate values are used in (26) to compute the random logic gate delay t_{drl} and in (28) for both logic and clock global interconnects assuming $2.3R_{tr} = R_{\text{int}}$ and, referring to (22), $H_\rho = H_\varepsilon = 0.3$ μm. In calculating both T_{cs} and T_{cc} the macrocell size is taken as $L^2 \cong N_gA_{rl}$.

The system critical path composite gate switching energy, heat removal and timing limits are illustrated in Fig. 22. Operation to the left of the switching energy

Fig. 23. $(1/L)^2$ versus τ for all levels of the hierarchy. System limits are imposed by global interconnect response time designated by $T_c \geq T_{cs} + T_{cp}$ and length designated by $L_{\min}^2 = N_gA_g$.

locus designated $P_st_d = 1/2C_sV_o^2$ and the heat removal locus designated $\overline{P} \leq QA$ is forbidden and operation to the right of the timing locus $T_c \geq T_{cs} + T_{cp}$ is also forbidden. The allowable design space for this particular macrocell is the small triangle with vertices 1) $t_{d\min}$ corresponding to minimum achievable propagation delay for the composite gate and therefore to the maximum performance design, 2) $P_{\min}$ corresponding to the lowest composite gate switching power that provides the required clock frequency $f_c = 1.0$ GHz and 3) $P_{\max}$ corresponding to the most mature technology or largest minimum feature size and chip size that provides $f_c = 1.0$ GHz. The three sides of the triangle correspond to contours of 1) constant switching energy E reflecting the performance level of the MOSFET and interconnect technologies, 2) constant heat removal capacity Q reflecting the performance level of the packaging technology, and 3) constant clock period T_c reflecting the performance level required by the design.

At the system level, $(1/L)^2$ versus τ plane limits focus on the longest interconnects since typically they impose the most stringent demands on performance. As illustrated in Fig. 23, the response time of the longest global interconnect, i.e., a logic signal path, of length $2L$ is designated by $T_c \geq T_{cs} + T_{cp}$ since $T_c \geq T_{cs} + T_{cp} = T_{cs} + T_{cc} + n_{cp}t_{drl}$, and $T_{cc} = T_c - T_{cs} - n_{cp}t_{drl}$ is the response time of the longest global interconnect. The actual length of its path is designated $L_{\min}^2 = N_gA_g$, since a smaller area than $L_{\min}^2$ could not accommodate the required number of logic gates N_g using the prescribed technology which requires a gate area A_g. The longest global interconnect cannot have a slower response time nor a smaller length than designated by these two limits. The forbidden zone of operation for the longest interconnects lies external to the small triangle, two of whose sides are defined by the preceding limits. The size of this triangle appears to be almost vanishingly small, particularly as a result of the stringent demands of a 1.0 GHz clock frequency. For smaller values of f_c, the size of the triangle can be enlarged at the expense of reduced performance.

The distinctive feature of the preceding treatment of system limits is that it seeks to describe the unbounded range of options of system architecture (not to mention algorithms) in terms of the absolute minimum number of parameters that enable a concise definition of the generic physical limits on system performance and hence a revealing juxtaposition of these system limits with the full hierarchy of limits which governs opportunities for GSI. A salient feature of the system representation is the definition of a critical path and from that the derivation of a composite logic gate which performs canonical computational operations. Only the first rudimentary results of this approach to low power system simulation, as illustrated in Figs. 22 and 23 are available at this juncture.

IV. QUASI-ADIABATIC MICROELECTRONICS

During an adiabatic process no loss or gain of heat occurs. A quasi-adiabatic process is designed to resemble this ideal behavior. The fundamental opportunity of quasi-adiabatic microelectronics is based on the second law of thermodynamics, which can be stated as follows: In any thermodynamic process that proceeds from one equilibrium state to another, the entropy of a closed system either remains unchanged or increases [59]. Entropy change dS can be expressed as

$$dQ/T = dS \geq 0 \tag{44}$$

where dQ is the heat added to the system and T is its absolute temperature. In a computational process, it is only those steps that discard information or increase disorder and therefore increase entropy ($dS > 0$) which have a lower limit on energy dissipation or heat generation ($dQ > 0$) imposed by the second law of thermodynamics [4], [5]. Consequently, the intriguing prospect of inventing quasi-adiabatic computational technology offers the possibility of reducing power dissipation to levels below those imposed by limits on the nonadiabatic processes discussed in Section III.

To elucidate this principle, consider the circuit operation illustrated in Fig. 24 in which the capacitor C is charged through the resistor R from a voltage source V_{in}. If V_{in} changes as a step function, the energy dissipated in R while charging C to a voltage V_o is given by

$$E_{d1} = \tfrac{1}{2}CV_o^2. \tag{45}$$

However, if V_{in} changes as a very slowly varying ramp function of rise time $T_d \gg 2RC$, then the energy dissipated in R is given by

$$E_{d2} \cong \tfrac{1}{2}CV_o^2[2t_d/T_d] \tag{46}$$

where $t_d = RC$. Since $2t_d/T_d \ll 1, E_{d2} \ll E_{d1}$. In fact, one might say that E_{d2} describes an asymptotically vanishing amount of energy as $T_d \to \infty$ [60]. The reduction in energy dissipation is a consequence of maintaining at all times a quasi-equilibrium condition for which $V_{\text{in}} \to V_{\text{out}}$ to keep the current flow nearly zero so that

$$E_d = \int_0^\infty i^2 R \, dt \to 0. \tag{47}$$

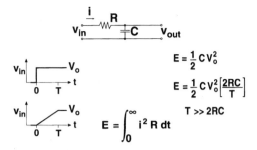

Fig. 24. Quasi-adiabatic switching.

Moreover, the discharge of C through R must be achieved through a very slowly varying ramp function whose fall time is $T_d \gg 2RC$. And, the source voltage generator providing V_{in} must include highly efficient resonant circuits to enable recycling a major fraction of the transferred energy.

The two key requirements for quasi-adiabatic or asymptotically zero dissipation digital microelectronics are summarized by (44) and (47): information cannot be destroyed and quasi-equilibrium operation must prevail [59], [60]. These requirements can be reflected in a hierarchy of limits on quasi-adiabatic microelectronics as illustrated in Fig. 25, which graphs the energy dissipation E_d during a switching transition versus the ratio of external transition time T_d to twice internal transition time $2t_d = 2RC$ or E_d versus $T_d/2t_d$. The salient message that Fig. 25 conveys is that, in principle, external control of the switching transition time (e.g., via a slow ramp of supply voltage as illustrated in Fig. 24) causing $T_d/2t_d \gg 1$ can reduce switching energy dissipation E_d to arbitrarily small amounts. For logarithmic scales, diagonal lines in the E_d versus $T_d/2t_d$ plane represent loci of constant switching energy transfer E, which is precisely the case for the P versus t_d plane. Consequently, it becomes clear that P versus t_d plane limits serve as helpful benchmarks in assessing performance of quasi-adiabatic microelectronics. At the fundamental level of the hierarchy $E_d = \gamma kT$ at $T_d/2t_d = 1$ corresponds to the fundamental limit from thermodynamics (7) as previously shown in the P versus t_d plane by Fig. 6. At the material level, $E_d = 1/2\varepsilon_{\text{Si}}V_o^3/\mathcal{E}_c$ at $T_d/2t_d = 1$ as given by (12) and Fig. 9 and for the device level $E_d = 1/2C_0L_{\min}^2V_0^2$ at $T_d/2t_d = 1$ as given by (18) and Fig. 14. The capability to illustrate fundamental, material and device limits in the E_d versus $T_d/2t_d$ plane is predicated on the assumption that the associated switching behavior can be enabled by means that are unspecified. While this assumption serves to add insight at the first three levels of the hierarchy, it should not be casually engaged at the fourth level because unlike the unchanging materials and device structures of the second and third levels, the circuit configurations used for nonadiabatic operation, such as static CMOS, must change very significantly for quasi-adiabatic operation [60]–[62]. Consequently, without identifying specific quasi-adiabatic circuit topologies and system architectures, the circuit and

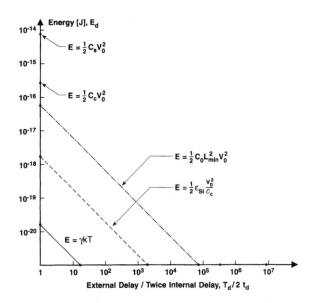

Fig. 25. Energy dissipation E_d during a switching transition versus the ratio of twice internal transition time $2t_d$ to external transition time T_d.

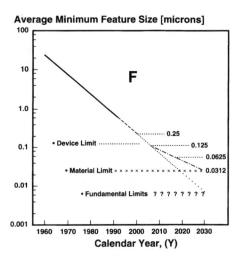

Fig. 26. Average minimum feature size F versus calendar year Y.

system limits that would be displayed in Fig. 25 must be held in abeyance. Invention is expected to abbreviate the delay in completing the E_d versus $T_d/2t_d$ plane hierarchy in which quasi-adiabatic operation will improve on the conventional nonadiabatic circuit and system level benchmarks defined by (24) and (35) respectively and illustrated as single points on the $T_d/2t_d = 1$ axis in Fig. 25 [62].

The current surge of interest in quasi-adiabatic circuit and system techniques [61] underscores the importance of low power microelectronics and thus the establishment of wholistic approaches to minimization of energy expenditure as exemplified by the hierarchy of limits explored in this discussion.

V. PRACTICAL LIMITS

In dealing with practical limits, the key question is "How many transistors can we expect to fabricate in a single silicon chip that will prove to be useful at some specific time in the future?" To gain insight into this issue, the number of transistors per chip N can be elegantly expressed in terms of three macrovariables: $N = F^{-2} \cdot D^2 \cdot PE$ [1], [3].

The evolution of average minimum feature size F for state-of-the-art microchips is described in simplified form in Fig. 26, which is a graph of F versus calendar year Y. In 1960, F was about 25 μm. By 1980, it had scaled down to 2.5 μm. If the historical rate of evolution continues throughout the 1990's, F will be about 0.25 μm in the year 2000. Following that, Fig. 26 illustrates three possible scenarios: 1) the 0.25 μm or pessimistic scenario, 2) the 0.125 μm or realistic scenario, and 3) the 0.0625 μm or optimistic scenario. The pessimistic scenario simply projects no further reduction of F beyond 0.25 μm based on the adverse expectation that the cost per function or the cost per logic circuit of a microchip will reach a minimum for the design, manufacturing, testing and packaging technologies

required by the 0.25 μm generation. This scenario seems unlikely at this time except to pessimists. The realistic scenario projects a further reduction of F at the historic rate until the later years of the first decade of the next century. Then saturation occurs at 0.125 μm again based on the economic expectation that the cost per function of a microchip will reach a minimum for the 0.125 μm generation especially because it could be the last generation for which deep ultraviolet microlithography will suffice. The optimistic scenario projects further reduction of F at a slower rate resulting in about 0.0625–0.0500 μm average minimum feature size during the second decade of the millennium, and then saturation. The slower rate of reduction and saturation of F at 0.0625 μm could be caused by a combination of factors, including astronomical capital costs particularly due to introduction of a radically different microlithography technology (e.g., using X-rays) and a soft collision with the physical limits on dimensions of MOSFET's finally imposing a minimum cost per function on the 0.0625 μm generation. The author's estimate is that CMOS microchips with minimum feature sizes in the 0.0625 μm range will be widely used.

Historically, the advantages of larger chip area have been reduced cost per function, improved performance, enhanced reliability and smaller size and weight at the module, board or box level for microelectronic equipment. The evolution of the square root of microchip area or chip size is illustrated in simplified from in Fig. 27. In 1960, $D = \sqrt{\text{chip area}}$ was about 1.2 mm; in 1980, about 6.5–7.0 mm; and, if the recent historic rate of increase continues throughout the 1990's, D will reach a range around 25 mm in the year 2000. Thereafter, three possible scenarios are again illustrated by segments F, G and H. Scenario F pessimistically projects saturation of D at about 25 mm based on a maximum silicon wafer diameter of 200 mm. A realistic scenario for the 2000–2010 period is that 300 mm wafers will be commonly used and that chip sizes up to 40 mm will be economic. Beyond this a long range optimistic scenario projects 400 mm wafers and over 50 mm chips.

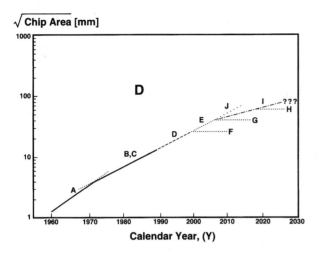

Fig. 27. Square root of die area D versus calendar year Y. $\sqrt{\text{die area}} = D$.

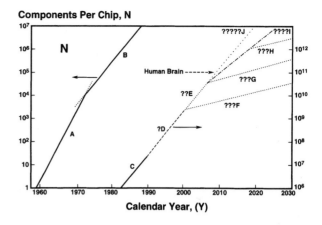

Fig. 29. Number of transistors per chip N versus calendar year Y.

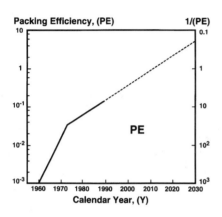

Fig. 28. Packing efficiency PE versus calendar year Y. Note that packing efficiency is defined as the number of transistors per minimum feature area.

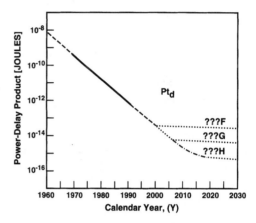

Fig. 30. System level power-delay product Pt_d versus calendar year Y.

The third macrovariable that contributes to the growing number of transistors per chip is their packing efficiency PE, the number of transistors per minimum feature area. The most prominent feature of the evolution of PE, presented in simplified form in Fig. 28, is the abrupt change in the slope of the locus which occurred in the early 1970's. Its cause was the unavailability of silicon real estate on the chip. Prior to about 1972, PE was increased simply by moving transistors and metal interconnects closer together. Since 1972, improvements in PE have been achieved by extending into the third dimension through increasing the number of mask levels in a chip manufacturing sequence. This trend toward clever use of the third dimension is not expected to change. It is interesting that about 2010, PE approaches unity; that is the areal packing efficiency is projected as one transistor per minimum feature area, which is truly a three-dimensional microchip suggesting multiple levels of transistors.

A simplified composite curve illustrating the number of transistors per chip N versus calendar year is shown in Fig. 29. This graph more than any other chronicles the progress of the microchip from its inception in 1959, until

1995 and beyond. The pessimistic scenario denoted by segment F projects a one-billion transistor chip or GSI by the year 2000, a forecast first proposed by the author in 1983 [3]. The realistic scenario projects over 100 billion transistors per chip before the year 2020.

One can also graph switching energy or power-delay product (Pt_d) versus calendar year as illustrated for CMOS technology in Fig. 30, again for three possible future scenarios [62]. Finally, the chip performance index CPI can be calculated as the quotient, of N and Pt_d or $CPI = N/Pt_d$. As illustrated in Fig. 31, the CPI has grown by about twelve decades since 1960 and is realistically projected to grow by about another six decades before 2020. This enormous rate of both productivity and performance enhancements is unprecedented in technological history.

VI. CONCLUSION

Historically there can be no doubt that the predominant pair of forces influencing the explosive growth in the number of transistors per chip has been the technological push of a continuous reduction in the cost per transistor or electronic function performed by a microchip coupled with the pull of ever-expanding markets and revenues.

17

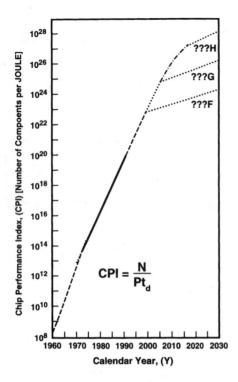

Fig. 31. Chip performance index $CPI = N/Pt_d$ versus calendar year Y.

The paramount issue confronting these positive trends has been, is, and will be the concomitant exponential growth in the capital cost of a new high volume manufacturing line needed for each successive generation of microchips [63]. While this economic issue is well beyond the scope of the current discussion, one relevant hypothesis is explored.

The hierarchy of theoretical limits on microelectronics established over the past decade and more and summarized in this discussion does not indicate that the pessimistic or the realistic or even the optimistic projections of minimum feature size F, die area D^2, packing efficiency PE, number of transistors per chip N, and chip performance index N/Pt_d cannot be achieved. In other words, physical limits per se do not appear to be "show stoppers" over the next two decades. Moreover, assuming that the cost per electronic function performed by a microchip continues to decline, it does appear that market demand will continue to escalate over the next two decades simply because the capacity of the microchip to provide cost effective solutions to the myriad problems of the information revolution is virtually unlimited within this timeframe. Consequently, the paramount issue is unchanged: Will there continue to be sufficient economic incentives to risk the ever growing capital investments required for further reduction of the cost per function of microchips? It is feasible that the response will also be unchanged, especially if the manufacturing cost goals of Sematech are fulfilled [64]. Within the time interval addressed in this discussion, fundamental, material, device, circuit, and system physical limitations may well permit and virtually unbounded market opportunities may well stimulate development of the highly expensive manufacturing technology that will enable continuous reduction, although perhaps at a smaller than historic rate, in the cost per function of microchips. Consequently, it is imperative that we continue to pursue as deep an understanding as possible of the hierarchy of physical limits that govern future opportunities for GSI. The National Technology Roadmap for Semiconductors prepared under the leadership of Sematech, the Semiconductor Research Corporation, and the Semiconductor Industries Association is a laudable contribution toward this effort.

ACKNOWLEDGMENT

The author gratefully acknowledges the encouragement and support of Dr. Robert Burger and Dr. William Lynch in connection with SRC Contract 93–SJ-374. In addition, stimulating discussions with Dr. Vivek De over a period of years are sincerely appreciated. Finally, the author wishes to thank two anonymous reviewers recruited by the guest editor of this issue, Dr. Lewis Terman, for their constructive critiques of the original manuscript.

REFERENCES

[1] J. D. Meindl, "The evolution of solid state circuits: 1958-1992-20??," *1993 IEEE ISSCC Commemorative Suppl.*, pp. 23–26, Feb. 1993.
[2] G. E. Moore, "Progress in digital integrated electronics," *IEEE IEDM Tech. Dig.*, pp. 11–13, 1975.
[3] J. D. Meindl, "Theoretical, practical and analogical limits in ULSI," *IEEE IEDM Tech. Dig.*, pp. 8–13, 1983.
[4] R. Landauer, "Irreversibility and heat generation in the computing process," *IBM J. Res. and Develop.*, vol. 5, no. 3, pp. 183–191, July 1961.
[5] ——, "Dissipation and noise immunity in computation and communication, *Nature*, vol. 335, pp. 779–784, Oct. 27, 1988.
[6] E. Keonjian, Ed., *Micropower Electronics*. London and New York: Pergamon, 1964.
[7] J. D. Meindl, *Micropower Circuits*. New York: Wiley, 1969.
[8] M. Degrauwe *et al.*, "Low power/low voltage: Future needs and envisioned solutions," *1994 IEEE ISSCC Dig. Papers*, pp. 98–99.
[9] S. Kohyama, "Semiconductor technology crises and challenges toward the year 2000," *1994 Symp. VLSI Tech. Dig. of Papers*, pp. 5–8.
[10] D. Singh, "Prospects for low power microprocessor design," *Proc. 1994 Int. Workshop on Low Power Design*, Napa, CA, Apr. 1994, p. 1.
[11] S. Molhi and P. Chatterjee, "I-V microsystems-scaling on schedule for personal communications," *IEEE Circ. and Devices*, Mar. 1994, pp. 13–17.
[12] A. P. Chandrakasan, S. Sheng, and R. Brodersen, "Low-power CMOS digital design, *IEEE J. Solid-State Circ.*, vol. 27, pp. 473–484, Apr. 1992.
[13] G. Moore *et al.*, "Metal-oxide-semiconductor field effect devices for micropower logic circuitry," in *Micropower Electronics*, E. Keonjian, Ed. London/New York: Pergamon, 1964.
[14] R. M. Swanson and J. D. Meindl, "Ion-implanted complementary MOS transistors in low-voltage circuits," *IEEE J. Solid-State Circ.*, vol. SC-7, pp. 146–152, Apr. 1972.
[15] E. A. Vittoz, "Low-power design: Ways to approach the limits," *1994 IEEE ISSCC Dig. Papers*, pp. 14–18.
[16] J. D. Meindl *et al.*, "Implantable telemetry," in *Methods of Animal Experimentation*, vol. 7, W. I. Gay and J. E. Heavner, Eds. New York: Academic, 1986, pp. 37–112.
[17] P. H. Hudson and J. D. Meindl, "A monolithic micropower command receiver," *IEEE J. Solid-State Circ.*, vol. SC-7, pp. 125–134, Apr. 1972.
[18] A. Chandrakasan, A. Burstein, and R. Brodersen, "A low power chipset for portable multimedia applications," *1994 IEEE ISSCC Dig. Papers*, pp. 82–83.

18

[19] R. W. Keyes, "Physical limits in digital electronics," *Proc. IEEE*, vol. 63, pp. 740–766, May 1975.

[20] F. W. Sears, *Thermodynamics*. Reading, MA: Addison-Wesley, 1953.

[21] H. Haken and H. C. Wolf, *Atomic and Quantum Physics*. Berlin: Springer-Verlag, 1984, pp. 83–85.

[22] A. Bhavnagarwala, private communication.

[23] M. V. Fischetti and S. E. Laux, "Monte carlo simulation of transport in technologically significant semiconductors-Part II: Submicrometer MOSFET's," *IEEE Trans. Electron Devices*, vol. 38, pp. 650–660, Mar. 1991.

[24] F. Assaderaghi, "Observation of velocity overshoot in silicon inversion layers," *IEEE Electron Device Lett.*, vol. 14, pp. 484–486, Oct. 1993.

[25] G. Baccarani *et al.*, "Generalized scaling theory and its application to a 0.25 micron MOSFET," *IEEE Trans. Electron Devices*, vol. ED-31, pp. 452–470, Apr. 1984.

[26] K. N. Ratnakumer and J. Meindl, "Short channel MOSFET threshold voltage model," *IEEE J. Solid-State Circ.*, vol. SC-17, pp. 937–947, Oct. 1982.

[27] L. D. Yau, "A simple theory to predict the threshold voltage of short channel IGFET's," *Solid-State Electron.*, vol. 17, pp. 1059–1063, 1974.

[28] R. R. Troutman, "VLSI limitations from drain induced barrier lowering," *IEEE Trans. Electron Devices*, vol. ED-26, pp. 461–469, 1979.

[29] B. Agrawal, V. K. De, and J. D. Meindl, "Opportunities for scaling MOSFET's for GSI," *Proc. ESSDERC 1993*, pp. 919–926.

[30] C. Fiegna *et al.*, "A new scaling method for 0.1–0.25 micron MOSFET," *May 1993 Symp. VLSI Tech. Dig.*, pp. 33–34.

[31] C. Hu, "MOSFET scaling in the next decade and beyond," *Semicon Int.*, June 1994, pp. 105–114.

[32] M. Ono *et al.*, "Sub-59nm gate length N-MOSFET's with 10nm phosphorous S/D junctions," *1993 IEEE IEDM Tech. Dig.*, pp. 119–121.

[33] D. A. Antoniades and J. E. Chung, "Physics and technology of ultra short channel MOSFET's," *1991 IEEE IEDM Tech. Dig.*, pp. 21–24.

[34] D. Hisamoto *et al*, "A fully depleted lean channel transistor (DELTA)-a novel vertical ultrathin SOI MOSFET," *IEEE Electron Device Lett.*, vol. 11, pp. 36–38, Jan. 1990.

[35] D. J. Frank *et al.*, "Monte Carlo simulation of a 30nm dual gate mosfet: How short can Si go?" *1992 IEEE IEDM Dig. Papers*, pp. 553–556.

[36] T. Tanaka *et al*, "Ultrafast low power operation of P+N+ double-gate SOI MOSFETs," *1994 Symp. VLSI Tech. Dig. of Papers*, pp. 11–12.

[37] K. Y. Toh *et al*, "An engineering model for short-channel MOS device," *IEEE J. Solid-State Circ.*, pp. 950–958, Aug. 1988.

[38] B. Bakoglu, *Circuits, Interconnections and Packaging for VLSI*. Reading, MA: Addison-Wesley, 1990, pp. 198–200.

[39] J. Davis, private communication.

[40] M. Nagata, "Limitations, innovations and challenges of circuits and devices into a half micrometer and beyond," *IEEE J. Solid-State Circ.*, vol. 27, pp. 465–472, Apr. 1992.

[41] Y. Nakogame *et al.*, "An experimental 1.5v 64mb DRAM," *IEEE J. Solid-State Circ.*, vol. 26, pp. 465–472, Apr. 1991.

[42] K. Ishibaski *et al.*, "A 1-V TFT-load SRAM using a two-step work-voltage method," *IEEE J. Solid-State Circ.*, vol. 27, Nov. 1992.

[43] T. Sakata *et al.*, "Subthreshold-current reduction circuits for multi-gigabit DRAM's," *IEEE J. Solid-State Circ.*, vol. 29, pp. 761–769, July 1994.

[44] K. Shimohigaski, "Low-voltage ULSI design," *IEEE J. Solid-State Circ.*, vol. 28, pp. 408–413, Apr. 1993.

[45] Y. Nakagome, "Sub-1-V swing internal bus architecture for future low-power ULSI's," *IEEE J. Solid-State Circ.*, vol. 28, pp. 414–419, Apr. 1993.

[46] Y. Taur *et al.*, "High performance 0.1 mm, CMOS devices with 1.5 V power supply," *IEEE IEDM Tech. Dig.*, pp. 127–130, Dec. 1993.

[47] Y. Mu *et al.*, "An ultra low-power 0.1 um CMOS," *Symp. VLSI Tech. Dig.*, pp. 9–10, June 1994.

[48] J. Burr and J. Shott, "A 200 m V self-testing encoder/decoder using standard ultra-low power CMOS," *IEEE ISSCC Dig.*, pp. 84–85, Feb. 1994.

[49] H. B. Bakoglu and J. D. Meindl, "Optimal interconnection circuits for VLSI," *IEEE Trans. Electron Devices*, vol. ED-37, pp. 903–909, May 1985.

[50] H. T. Kung, "Why systolic architectures," *IEEE Compu.*, pp. 37–46, Jan. 1982.

[51] J. A. B. Fortes and B. W. Wah, "Systolic arrays—from concept to implementation," *IEEE Compu.*, pp. 12–17, July 1987.

[52] H. S. Stone and J. Cocke, "Computer architecture in the 1990's," *IEEE Compu.*, pp. 30–38, Sept. 1991.

[53] K. Chin *et al.*, "IBM enterprise system/9000 clock system: A technology and system perspective," *IBM J. Res. Develop.*, vol. 37, no. 5, pp. 867–874, Sept. 1992.

[54] B. S. Landrum and R. L. Russo, "On a pin versus block relationship for partitioning of logic graphs," *IEEE Trans. Compu.*, vol. C-20, pp. 1469–1479, Dec. 1971.

[55] W. E. Donath, "Placement and average interconnection lengths of computer logic," *IEEE Trans. Circ. and Syst.*, vol. CAS-26, pp. 272–277, Apr. 1979.

[56] A. Masaki, "Possibilities of deep-submicrometer CMOS for very high speed computer logic," *Proc. IEEE*, vol. 81, pp. 1311–1324, Sept. 1993.

[57] G. A. Sai-Halasz, "High end processor trends and limits," *Proc. Interconnect Conf. on Advanced Microelectron. Device Proc.*, Sendai, Japan, pp. 753–760, Mar. 3–5, 1994.

[58] J. C. Eble, private communication.

[59] D. Halliday, R. Resnick, and J. Walker, *Fundamentals of Physics*, 4th ed. New York: Wiley, 1993.

[60] S. G. Younis, "Asymptotically zero energy computing using split-level charge recovery logic," Ph.D. dissertation, Dept. of EECS, MIT, 1994.

[61] *Proc. 1994 Int. Workshop on Low Power Design*, Napa, CA, Apr. 1994.

[62] V. De, private communication.

[63] C. R. Barrett, "Silicon valley, what next," *MRS Bull.*, pp. 3–10, July 1993.

[64] W. J. Spencer, "National interests in a global semiconductor industry," *Distinguished Lecturer Series*, Georgia Inst. of Technol., Atlanta, GA, Oct. 3, 1994.

[65] Semiconductor Ind. Assoc., *The National Technology Roadmap for Semiconductors*, 1994.

Micropower IC

ERIC A. VITTOZ

CENTRE ELECTRONIQUE HORLOGER S.A., 2000 NEUCHATEL, SWITZERLAND

ABSTRACT

Various aspects of the realization of micropower LSI circuits are discussed, from the requirements on CMOS technologies to constraints on systems. Available passive and active devices are reviewed, with emphasis on DC, AC and noise characteristics of transistors at very low currents. Problems and solutions encountered in digital and analog circuits are illustrated with a few examples.

1. INTRODUCTION

The power consumed for a given function in an integrated circuit must be reduced for either one of two different reasons.

One reason is to reduce heat dissipation, in order to allow a larger density of functions on the chip. Any amount of power reduction is worthwhile, as long as it does not degrade the overall performance of the circuit.

The other reason is to save energy in battery-operated instruments. This is mandatory in electronic watches, where the average power available is only a few microwatts. Electronic watches have therefore been the driving force in the development of micropower IC's. Other existing or emerging applications are various kinds of pocket instruments, paging systems, implanted bio-medical devices, and devices for environment and security control.

Yet, some reduction in performance must be accepted to achieve micropower operation. The toll is usually a reduced speed for digital circuits and a reduced dynamic range for analog circuits.

For practical reasons, voltage cannot be decreased below that of a single electrochemical cell, which is about 1.5 V. Most of the power reduction must therefore be obtained by reducing the current drain from the usual milliampere level down to the microampere level or below. This brings about two important consequences:

1. Impedance levels are increased by more than three orders of magnitude, which increases the importance of leakage currents and parasitic conductances. Parasitic series resistances are generally negligible, and functional resistances cannot usually be implemented in a simple way.

2. Since the size of the components cannot be reduced, current densities are decreased by more than three orders of magnitude, which results in decreased speed of operation. Moreover, active devices may operate in an unusual range of their characteristics. This allows some new circuit schemes.

This paper will discuss technologies, devices, circuits and systems with respect to the realization of micropower LSI chips.

Reprinted from *Proceedings of the IEEE European Solid State Circuits Conference*, pp. 174-189, September 1980.

2. TECHNOLOGY

Although conventional bipolar [1], [2] or I^2L [3] technologies may achieve interesting results in some specific applications, CMOS is unquestionably the best technology for micropower. The combination of dynamic and static power P consumed by a CMOS gate is roughly given by the well known simple formula

$$P = f \, C \, V_B^2 + V_B \, I_o \qquad \qquad 1)$$

where f is the output frequency of the gate, V_B the supply voltage, C the output capacitance and I_o the average leakage current.

Capacitance C is kept low by keeping substrate and p-well doping low ($1-5 \times 10^{15}$ cm^{-3}). Shallow diffusions(~ 1 μm) help reduce the effective size of sources and drains. Coevaporation of Al and Si may be needed to avoid short circuits during the alloying phase. Gate oxide must not be too thin (70-100 nm) in order to limit gate capacitance.

It can be shown that the smallest possible supply voltage V_B for digital CMOS circuits is about 200 mV [4]. Meanwhile, if the threshold voltage V_T of p and n channel transistors is too low, the residual channel current at zero gate voltage becomes a dominant part of leakage current I_o, and static power consumption is increased. Optimum values of V_T are in the range 0.3 to 0.5 V. An accuracy of $\pm$ 0.1 V can be obtained by ion implantation of both substrate and wells, and by carefully optimized annealing.

Strict control of the whole process is needed to minimize junction reverse currents. They can be kept below 10 nA/mm^2 at 1.5 V and ambient temperature.

Metal gate CMOS was the first MOS technology applied to micropower circuits [5]. It is still used as a standard technology for large scale and low-cost production.

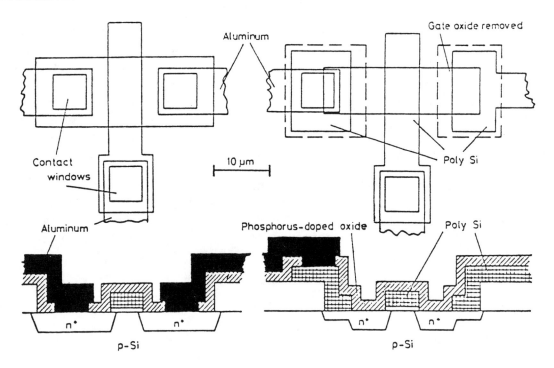

Fig. 1   Standard an polysilicon contact n-channel MOST

Si-gate is preferred [6] in critical micropower applications mainly because of its self-alignment properties. In addition, chip area may be reduced by extensive use of the polysilicon interconnecting layer whose sheet resistance is usually negligible.

Si-gate technology may be further improved by using polysilicon contacts [7], as illustrated in Fig. 1. This technique requires one additional mask but allows a substantial reduction of the size and depth of source and drain diffused areas. This may result in a 40-percent reduction of dynamic power consumption. Futhermore, the possibility of linking gates and drains directly with strips of polysilicon helps reduce the chip area.

Capacitance C can be further reduced by applying local oxidation [8] or silicon-on-sapphire [9], [10] technologies. However, these technologies tend to have higher leakage currents, and can thus only be considered for high frequency applications.

Combinations of CMOS and bipolar transistors have been proposed for micropower applications [11], [12]. Their advantage relies in a higher current handling capability which may be required in the output circuitry.

3. DEVICES

3.1 Transistors

When the drain current of a MOS transistor is decreased by reducing the gate voltage, the device eventually enters the weak inversion region of operation where the usual parabolic transfer characteristics are no more valid. For long channel transistors with a negligible density of fast surface states and negligible leakage current to substrate, the drain current I_D in weak inversion may be expressed as [13]

$$I_D = S\, I_{DO}\, e^{V_G/nU_T} (e^{-V_S/U_T} - e^{-V_D/U_T}) \qquad 2)$$

where S is the effective channel width to channel length ratio; V_G, V_S and V_D are the gate, source, and drain potentials with respect to the substrate, and $U_T = kT/q$. The slope factor n ($n > 1$) is fairly controllable, whereas the characteristic current I_{DO} is very sensitive to process parameters and temperature. Both n and I_{DO} may be considered constant for all transistors of a chip biased by values of V_S that do not differ too much from each other.

This model expresses the fact that a MOS transistor in weak inversion is a barrier-controlled device very similar to a bipolar transistor. Drain current I_{DO} becomes saturated as soon as V_D-V_S exceeds a few U_T. By using this behaviour, one obtains an excellent current source. The saturation value of I_D increases exponentially with $-V_S/U_T$ and V_G/nU_T.

The maximum saturation current in weak inversion is roughly given by [13]

$$I_D = \beta\, U_T^{\,2} \qquad 3)$$

where $\beta = \mu\, C_{ox} S$ is the usual transfer parameter in strong inversion.

The maximum value of β achievable with a transistor of reasonable size (0.02 mm^2) is about 10 mA/V^2 which corresponds to a maximum possible operating current in weak inversion of a few microamperes. For a typical, minimum-sized transistor, this limit is between 10 and 100 nA.

The minimum operating current is limited by carrier generation in the drain and channel depletion layers. It is therefore roughly proportional to the overall area of the transistor. Minimum-sized transistors may be used with a drain current as low as 100 pA at 50°C.

Small-signal gate to drain transconductance is easily derived from the model as

$$g_m = \frac{\partial I_D}{\partial V_G} = \frac{I_D}{nU_T} \qquad 4)$$

Transconductance is proportional to drain current I_D in weak inversion, whereas it is proportional to $I_D^{1/2}$ in strong inversion.

For $V_D - V_S \gg U_T$, a source to drain transconductance may be defined as

$$g_{mS} = -\frac{\partial I_D}{\partial V_S} = \frac{I_D}{U_T} \qquad 5)$$

This transconductance is n times larger than that of the gate and is equal to that of a bipolar transistor operated at the same current.

The maximum small signal amplification achievable with a transistor is limited by the non-zero output conductance g_o of the device, which is due to channel shortening. As shown in figure 2, this conductance is approximately proportionnal to drain current and characterized by an extrapolated voltage V_E :

$$g_o = I_D/V_E \qquad 6)$$

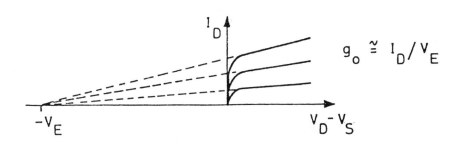

Fig. 2   Output conductance

As shown in Fig. 3, the amplification factor $A_o = g_m/g_o$ is constant and maximum in weak inversion [14] but decreases like $I_D^{-1/2}$ in strong inversion. This maximum possible gain in weak inversion is a consequence of a maximum value of the transconductance-to-current ratio. It may be further increased by increasing the channel length of the transistor.

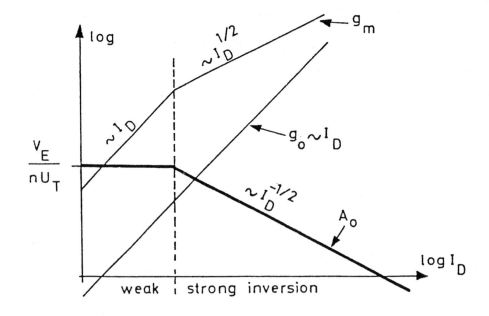

Fig. 3   Amplification factor

Since in weak inversion there is no inverted channel, the major part of the gate capacitance appears between the gate and the substrate and has a value equal to the minimum of the C_G-V_G curve [15]. This value is fairly constant over many orders of magnitude of drain current, and may be 2 to 3 times smaller than the oxide capacitance.

Noise is a limiting factor for transistors used in analog circuits. It may be conveniently characterized by a frequency dependent noise resistance R_n which is a measure of the input noise voltage spectrum.

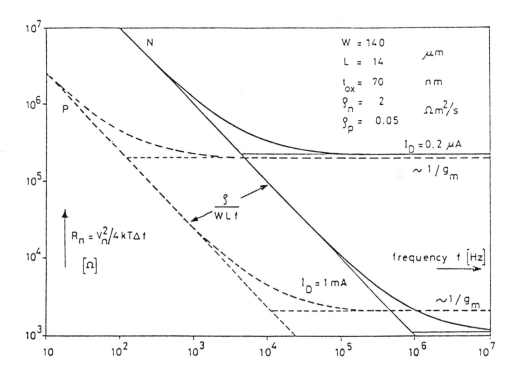

Fig. 4   Noise resistance of MOS transistors

Fig. 4 shows typical noise resistances of large transistors integrated in micropower Si-gate technology [15], [16]. At high frequency, shot noise dominates; the noise resistance is independent of frequency and approximately equal to the inverse of transconductance. It increases if drain current is decreased, but it is minimum at a given current if the transistor is in weak inversion.

At low frequencies, flicker noise dominates. The noise resistance is inversely proportional to frequency and to gate area W L. It also depends on the technology through the parameter ρ .

It may be pointed out that, in the audio frequency range, the noise of the n-channel transistor is independent of drain current down to about 0.2 μA. If gate area is reduced, flicker noise increases and noise stays independent of current up to higher frequencies or lower currents.

Equivalent input noise voltage of this large-size n-channel transistor integrated over the range 100 Hz to 3 kHz is about 8 μV.

The flicker noise resistance of the p-channel transistor of the same size is about 40 smaller.

3.2 Passive devices

Capacitors are very important in the design of modern analog circuits such as switched capacitor (SC) filters.

In metal gate technology, the best capacitor is the aluminum-gate oxide-diffusion structure which yields a specific value of 300 to 500 pF/mm^2. This structure is not available in standard Si-gate technologies due to the self-alignment of gate and diffusions. Nevertheless, a very good capacitor is obtained between the polysilicon and aluminum layers. The vapor-deposited oxide used as the dielectric is thicker than the gate oxide, which reduces the capacitance per unit area to about 100 pF/mm^2. Matching to 1°/$_{oo}$ may be obtained with capacitors of a few picofarads [17].

There is no high resistivity layer available to realize high value resistances. The best layer is the p-well diffusion which reaches 5 to 10 kΩ/$\square$ and allows the realization of a few 100 kΩ on the chip with an accuracy of $\pm$ 20%. This value is much too low for most micropower applications. Higher values may be obtained by lightly doped poly Si-layers, but tolerances become very large for sheet resistivities above 1 MΩ/$\square$.

An interesting feature of most Si-gate processes is the availability of both p and n type polysilicon layers [6]. It allows the realization of the lateral polysilicon diode shown in figure 5 [18], [19]. This diode is perfectly isolated from the substrate by the field oxide layer.

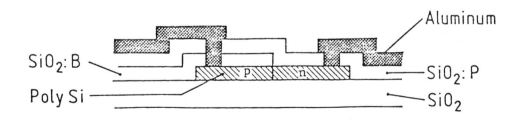

Fig. 5   Lateral polysilicon p-n diode

An important application of this device is the realization of high impedance on-chip voltage multipliers for driving non-volatile memories [20], [21]. It can also be used near the origin of its I-V characteristics as an equivalent resistance of very high value for biasing gates. A 10 μm-wide device has an equivalent resistance ranging from 1 to 100 GΩ . It can also be applied to maintain a logic level against leakage currents [18].

4. DIGITAL CIRCUITS

As expressed by relation 1), the power consumption of logic CMOS circuits may be divided into static and dynamic components.

In order to limit the dynamic power consumption, the average number of nodes which transit during any period of time must be kept as low as possible. High frequency synchronous circuits must therefore be avoided and replaced by asynchronous cells. Each cell should have a minimum number of nodes changing state to achieve a given function. It should furthermore contain a minimum number of minimum sized transistors in order to obtain a low total parasitic capacitance. Logic structures with critical races must therefore be discarded in favour of race-free circuits which work independently of their delays [22].

As an example, Fig. 6 shows a race-free divide-by-2 cell [6], [23] made up of 5 gates, and which has a single input I. The behaviour of this structure is described by the set of equations which defines the 5 internal

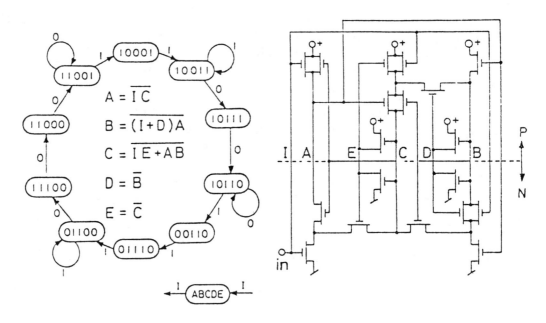

$$A = \overline{I\,C}$$
$$B = \overline{(I+D)A}$$
$$C = \overline{IE+AB}$$
$$D = \overline{B}$$
$$E = \overline{C}$$

Fig. 6   Race-free divide-by-2 cell

variables A to E (output nodes of the gates). The graph of transitions represented on the same figure shows that no more than one variable tends to transit in any given state, hence there can be no race between variables. Each internal variable transits at half the frequency of the input and may thus be used as output of the divider. Since some transistors can be shared by two gates, the total number of transistors per cell is 19.

Careful analysis of this circuit shows that only some of the transistors are needed to change the state of the internal variables. The remaining transistors are only needed to maintain the established states against leakage currents, and may thus be dropped for high frequency circuits. This results in the 9-transistor circuit drawn in solid lines in Fig. 7 [6].

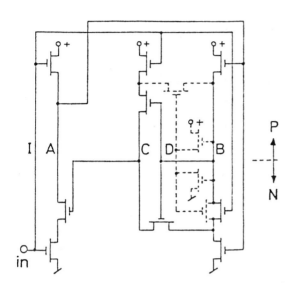

Fig. 7 Dynamic and semi-dynamic divide-by-2 cell

This dynamic version [24], [25], consumes less than half the dynamic power of the static version. Typical values are 0.4 μA/MHz at 1.4 V.

The same principle has been efficiently applied to all kinds of logic cells. It is not really limited to high frequencies, but can be applied as well to low frequency circuits driven by short logic pulses. For example, by keeping the 4 transistors represented by dashed lines in Fig. 7, the stable states for I = 0 are maintained and the 13-transistor circuit operates with short 0 input pulses I of arbitrarily low frequency [18]. This semidynamic scheme helps reduce both power consumption and chip area.

Power may be further reduced by feeding the circuit through a voltage reducer at a voltage close to the sum of p and n thresholds [26]. If the voltage of the energy source is sufficient, various parts of the circuit may even be connected in series [10].

Design complexity and chip area of LSI circuits are greatly reduced by replacing random logic by array logic. The latter can be applied in a quite conventional way to low frequency micropower circuits by taking advantage of the two types of transistors available in CMOS technology [27], [28]. Fig. 8 shows the example of a dynamic PLA combining two NAND arrays, that is controlled by a single clock according to the clocked-CMOS principle [29]. Such configurations may be used directly or as a full ROM with the addition of an output multiplexer. A 1 K bit dynamic ROM occupies about 1 mm^2 in 6 μm technology and consumes approximately 1 nJ at 1.5 V for reading 1 word. This limits the read frequency to 1 kHz for a current drain of 1 μA. ROM's of this size or more are needed in microprocessor circuits for watches [30], [31].

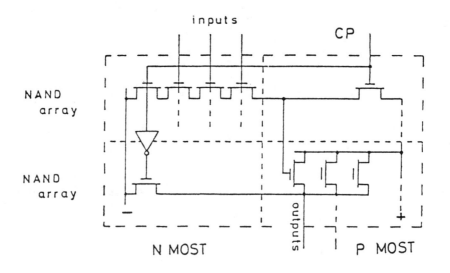

Fig. 8   CMOS dynamic PLA

Micropower static RAM's use standard 6-transistor CMOS cells. Typical power consumption is of the order of 1 μw at 1.5 V for a read/write rate of 5 words/ms. It could be reduced by segmentation of the memory, which reduces the capacitance that must be charged during each read/write cycle. Total area of a 1 K bit static RAM in standard 6 μm technology is about 10 mm^2.

The standard way of reducing the area of a RAM is to use a dynamic single-transistor cell, which must be periodically refreshed. The calculated current required to charge the capacitance of the selection lines during each refresh cycle is shown in Fig. 9, assuming a refresh frequency of 5 kHz and a 1.5 V supply voltage [27]. Typically, a 1 K bit dynamic RAM would consume 40 μA, which is inacceptable for most micropower applications. Paradoxically, this dynamic power consumption can be reduced by increasing the storage capacitance. The refresh frequency may then be decreased whereas capacitance of the selection lines stays fairly constant. The price to pay is an increase in size.

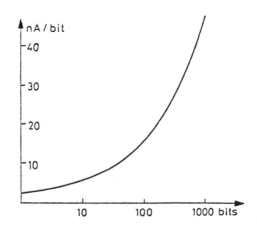

Fig. 9   Power consumption of a dynamic RAM

5. ANALOG CIRCUITS

Although analog circuits do not usually occupy large areas on micropower LSI chips, they may need a large part of the available power and must therefore be carefully optimized in this respect. This is usually done by using simple configurations to implement the power consuming parts of the circuits [32].

One of the first analog micropower circuits was the quartz oscillator included in every electronic watch. This circuit was consuming most of the power in some early superficial designs [23]. Fig. 10 shows the full diagram of a circuit carefully designed for a 32 kHz quartz resonator [33]. The quartz resonator, the transistor T_1 and the two capacitors C_1 and C_2 constitute a simple Pierce oscillator. T_1 operates in weak inversion, where the

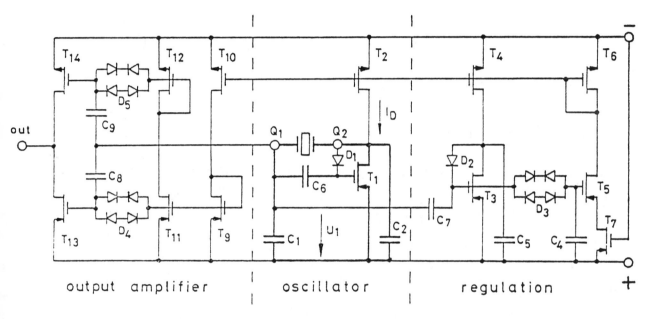

Fig. 10   Quartz oscillator with amplitude
           regulation

drain current required to reach the critical transconductance is minimum. Its gate is biased by a single floating polycrystalline diode D_1, which allows DC isolation from the quartz pin Q_1 by a small capacitance C_6. The drain current I_D is supplied by an amplitude regulator [13], [32] which also uses polycrystalline diodes to replace high-value resistors. The resistor of the non-critical low-pass filter D_3-C_4 is implemented as a symmetrical quad of diodes, to avoid any rectifying effect. This regulator is based on the exponential transfer characteristics of T_3 and T_5 operating in weak inversion. Its output current I_D drops sharply when the amplitude of oscillations coupled through C_7 reaches a critical value. The amplitude of oscillations is thus stabilized at this critical value, which depends solely on nU_T and on the shape factors S of transistors T_3 to T_6. An amplitude of about 150 mV is chosen to limit current drain and distorsions in T_1. Amplification is thus needed to drive logic frequency dividers. It is achieved with a minimum amount of power by the complementary transistors T_{13} and T_{14}, which are separately biased close to their thresholds by means of transistors T_9 to T_{12}. The oscillating signal is capacitively coupled to this amplifier through C_8 and C_9. Equivalent resistances D_4 and D_5 avoid that the oscillator be loaded by T_9 and T_{12}. All coupling capacitors have values below 2 pF.

As shown in Fig. 11, the total current consumed by this oscillator is fairly independent of supply voltage V_B. It is of the order of 50 nA and increases by one to two orders of magnitude if the quartz resonator is removed. Total currents lower than 20 nA have been measured with lower values of capacitors C_1 and C_2.

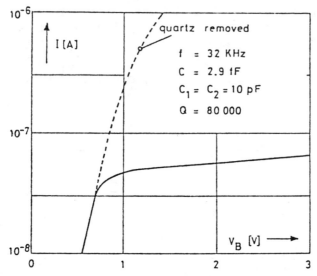

Fig. 11 Current drain of 32 kHz quartz oscillator

Another example of an analog circuit developed for electronic watches is the voltage reference needed to check the state of the battery.

The standard bandgap principle can be applied by using the base-emitter junction of the substrate transistor available in CMOS technologies to realize the diode. A compensation voltage proportional to absolute temperature is obtained by means of the bipolar-like transfer characteristics of MOS transistors in weak inversion [34], [35], [36]. Voltage stability better than $\pm$ 10 mV is easily achieved in the range -20 to +80°C, but initial adjustment at room temperature is needed to compensate mismatches of transistors.

Fig. 12

Gate voltage difference ΔV_G of n-type transistors with n^+ (normal) and p^+ (inverted) doped gates

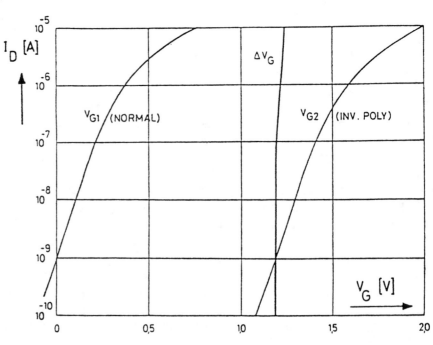

30

Another very interesting solution is based on a transistor pair of the same type, except for the opposite doping type of their polysilicon gates [37]. As shown by Fig. 12, the gate voltage difference ΔV_G of such a pair is nearly independent of current in weak inversion and is close to the bandgap voltage of silicon. The temperature dependence of the bandgap voltage can be easily compensated by choosing different current densities for the transistors of the pair. Voltage stability of $\pm$ 1 mV has been obtained in the range −20 to +80°C, with a total current drain below 100 nA.

These excellent characteristics may find applications much beyond that of battery checking.

Switched capacitor filters find applications in micropower circuits for medical electronics, and they will be needed in future portable devices to process analog signals delivered by sensors.

Most of the power consumed by SC filters goes into the operational amplifiers. These must therefore be specially designed for very low-power filters [17], [38]. A commonly used circuit is shown in figure 13 [14]. It combines a n-channel differential pair T_1-T_3 and a p-channel current mirror T_2-T_4 to achieve voltage amplification at high impedance node (a). A n-channel source follower T_5 is provided to insure low output resistance.

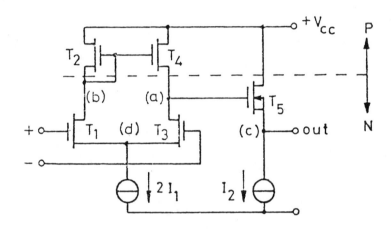

Fig. 13   Simple operational
           amplifier

Figure 14 shows the measured frequency dependence of the gain A of such an amplifier driven at the negative input, with $2I_1 = 0.1\ \mu A$, $I_2 = 1\ \mu A$ and a total capacitive load $C_c = 5$ pF at the output node (c). This circuit behaves essentially as an integrator with a time constant

$$\mathcal{Z}_1 = C_a / g_{m1} \qquad\qquad 7)$$

where C_a is the capacitance loading node (a) and g_{m1} the transconductance of transistors T_1 and T_3. It has a significant pole due to the time constant

$$\mathcal{Z}_2 = C_c / g_{m5} \qquad\qquad 8)$$

at output node (c).

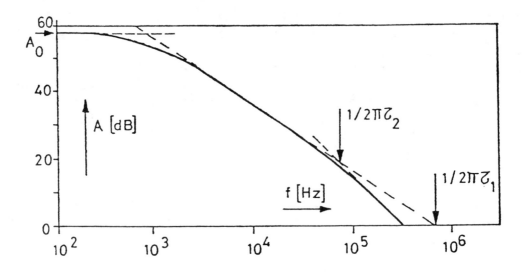

Fig. 14   Gain of operational amplifier

Since the amplifier operates in weak inversion, a fairly high voltage gain A_O, close to 60 dB, is obtained with a single stage, and therefore no compensation is needed for any amount of voltage feedback smaller than 1.

This amount of feedback is always close to 1 in most SC filters. The settling time T_S necessary to reach equilibrium with a residual error $\mathcal{E}$ may then be approximated by [17]:

$$T_s \cong (\mathcal{T}_1 + 2\mathcal{T}_2) \ln \mathcal{E}^{-1} \qquad 9)$$

In a SC filter, all the high frequency noise is transposed to low frequencies by undersampling. For this reason, white shot noise predominates at low currents, even though 1/f flicker noise is dominant in the audio part of the spectrum in the absence of sampling. The noise bandwidth of the amplifier can be shown to be independent of $\mathcal{T}_2$ and equal to

$$\Delta f = 1/4 \mathcal{T}_1 \qquad 10)$$

The equivalent shot noise resistance R_n may be expressed as

$$R_n = \frac{V_n^2}{4 \ kT\Delta f} = \rho / g_{m1} \qquad 11)$$

where $1 < \rho < 4$ depends on the relative contribution to the noise of transistors T_1 to T_4.

Combination of 7), 10) and 11) yields:

$$V_n^2 = \rho \frac{k \ T}{C_a} \qquad 12)$$

Equivalent noise voltage at the input of the amplifier depends thus almost solely on the value of capacitance C_a at amplifying node (a). The maximum possible value of C_a is limited by the settling time T_S given by 9), which must be shorter than the half sampling period $\frac{1}{2\ f_c}$. Application of this condition yields

$$V_n^2 > 2\ kT\ \frac{f_c}{g_{m1}}\, \mu\ln\,\mathcal{E}^{-1} \qquad\qquad 13)$$

Due to the requirement on settling time, the minimum possible shot noise component is thus inversely proportional to transconductance g_{m1}; it is therefore increased at low current.

Fig. 15 shows as an example the circuit diagram of a second order, stray-insensitive filter which has been realized for micropower applications [17].

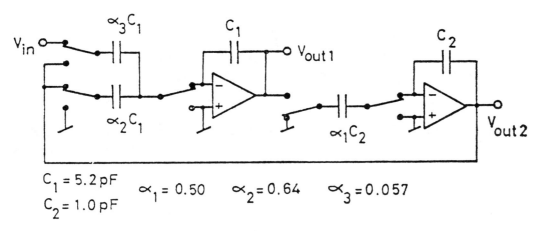

$C_1 = 5.2\,pF$
$C_2 = 1.0\,pF$ $\qquad \propto_1 = 0.50 \qquad \propto_2 = 0.64 \qquad \propto_3 = 0.057$

Fig. 15   Second order stray-insensitive filter

This filter consumes less than 10 µw at 3 V with a resonant frequency at 1.4 kHz. Measured dynamic range is 51 dB, but calculations show that it can be extended to above 60 dB by slightly modifying the amplifiers. Power could be further reduced by using dynamic amplifiers [39], [40], [41].

6. SYSTEMS

Constraints in the realization of complete LSI micropower systems result from the limitations of the subcircuits.

Minimum dynamic power consumption is always achieved by asynchronous circuits, which must therefore be used for high frequency logic functions. The average clock frequency of synchronous processing circuits must be as low as possible. This requires special architectures that allow, for example, to suppress the clock after completion of each operation [30], [31], [42], [43]. Maximum clock frequency is limited by low supply voltage. Fast functions must thus be committed to special peripheral circuits.

Integration of most of the parts of a system on a single chip is desirable, in order to reduce the number of power consuming interfaces, and to save space which is often limited in battery operated instruments.

Fast signal processing must be achieved by analog circuits which must be carefully designed with respect to power consumption.

Latch-up effects which plague many CMOS circuits may be easily avoided by the addition of a single integrated series resistance which limits the current below the holding value (of the order of 1 mA), but which has a negligible effect at nominal current. Battery operation isolates the chip from various noise and impulse signals usually delivered by power networks.

Good examples of micropower capability are given by watch circuits operating at 1.5 V with a standard 32 kHz quartz resonator. Microprogrammed circuits intended for LCD digital watches consume 1 to 6 μA [30]. The best circuits for analog watches need about 0.1 μA. Total current of 2 μA has been reported for a watch circuit designed for 4.2 MHz [26].

7. CONCLUSION

A variety of digital and analog functions can be implemented within a single LSI chip that consumes a few microwatts or less, provided fundamental constraints on speed, bandwidth and dynamic range are accepted. The degradation of performances is less than proportional to the reduction in power consumption. This results from special circuit schemes, some of which take advantage of MOS transistors operating in weak inversion.

Future improvements will be made possible by further refinement of circuit and system designs, and by the availability of submicron technologies boosted by worldwide efforts towards VLSI.

REFERENCES

[1]   M.P. Forrer, "Survey of circuitry for wristwatches", Proc. IEEE, Vol. 60, pp. 1047-1054, Sept. 1972.

[2]   H.W. Rüegg and W. Thommen, "Bipolar micropower circuits for crystal-controlled timepieces", IEEE J. Solid-State Circuits, Vol. SC-7, pp. 105-111, April 1972.

[3]   P.A. Tucci and L.K. Russel, "An I^{2}L watch chip with direct LED drive", IEEE J. Solid-State Circuits, Vol. SC-11, pp. 847-851, Dec. 1976.

[4]   R.M. Swanson and J.D. Meindl, "Ion-implanted complementary MOS transistors in low-voltage circuits", IEEE J. Solid-State Circuits, Vol. SC-7, pp. 146-153, April 1972.

[5]   F. Leuenberger and E. Vittoz, "Complementary MOS low-power low-voltage integrated binary counter", Proc. IEEE, Vol. 57, pp. 1528-1532, Sept. 1969.

[6]   E. Vittoz, B. Gerber and F. Leuenberger, "Silicon-gate CMOS frequency divider for the electronic wrist watch", IEEE J. Solid-State Circuits, Vol. SC-7, pp. 100-104, April 1972.

[7]   B. Gerber, "Performance improvements of complementary dynamic MOS (CODYMOS) frequency dividers with polysilicon source and drain contacts", IEEE Trans. on Electron Devices, Vol. ED-24, pp.1119-1121, Aug. 1977.

[8]   B.B.M. Brandt, W. Steinmaier and A.J. Strachan, "LOCMOS, a new technology", Philips Tech. Rev. No 1, pp. 19-23, 1974.

[9]   A.C. Ipri and J.C. Sarace, "Low-threshold low-power MOS/SOS for high frequency counter applications", IEEE J. Solid-State Circuits, Vol. SC-11, pp. 329-336, April 1976.

[10]   P. Schwob et al, "Silicon-on-sapphire technology opens new possibilities for watch circuits", Proc. Int. Congress of Chronometry, Geneva 1979, pp. 259-264.

[11]   M. Darwish and R. Taubenest, "C-MOS and complementary isolated bipolar transistor monolithic integration process", J. Electrochem. Soc., Vol. 121, pp. 1119-1121, August 1974.

[12]   O.H. Schade, Jr, "BIMOS micropower IC's", IEEE J. Solid-State Circuits, Vol. SC-13, pp. 791-798, Dec. 1978.

[13]   E. Vittoz and J. Fellrath, "CMOS analog integrated circuits based on weak inversion operation", IEEE J. Solid-State Circuits, Vol. SC-12, pp. 224-231, June 1977.

[14]   W. Steinhagen and W.L.Engl, "Design of integrated analog CMOS circuits - a multichannel telemetry transmitter", IEEE J. Solid-State Circuits, Vol. SC-13, pp. 799-805, Dec. 1978.

[15]   J. Fellrath and E. Vittoz, "Small signal model of MOS transistors in weak inversion", Proc. Journées d'Electronique 1977, EPF-Lausanne, pp. 315-324.

[16]   J. Fellrath, "Shot noise behaviour of subthreshold MOS transistors", Rev. de Physique Appliquée, Vol. 13, pp. 719-723, Déc. 1978.

[17]   E. Vittoz and F. Krummenacher, "Micropower SC filters in Si-gate technology", 1980 European Conference on Circuit Theory and Design, Warsaw, Sept. 2-5 1980.

[18]   H. Oguey and E. Vittoz, "Resistance-CMOS circuits", IEEE J. Solid-State Circuits, Vol. SC-12, pp. 283-285, June 1977.

[19]   M. Dutoit and F. Soilberger, "Lateral polysilicon p-n diodes", J. Electrochem. Soc. Vol. 125, pp. 1648-1651, Oct. 1978.

[20]   B. Gerber and J. Fellrath, "Low-power CMOS EEROM for the wrist-watch", Proc. Int. Congress of Chronometry, Geneva 1979, pp. 291-297.

[21]   B. Gerber, J.C. Martin and J. Fellrath, "A 1.5 V single supply, one-transistor CMOS EEPROM", ESSCIRC 80.

[22]   E. Vittoz, C. Piguet and W. Hammer, "Model of the logic gate", Proc. Journées d'Electronique 1977, EPF-Lausanne, pp. 455-467.

[23]   E. Vittoz, "LSI in watches", Solid-State Circuits 1976, (Invited Papers ESSCIRC 80), Ed. Journal de Physique, Paris, pp. 7-27.

[24]   E. Vittoz and H. Oguey, "Complementary dynamic MOS logic circuits", Electronics Letters, Vol. 9, Febr. 22, 1973.

[25]   H. Oguey and E. Vittoz, "CODYMOS frequency dividers achieve low power consumption and high frequency", Electronics Letters, Vol. 9, Aug. 23, 1973.

[26]   H. Oguey and B. Gerber, "Progress in high frequency watch circuits", Proc. Int. Congress of Chronometry, Geneva 1979, pp. 251-257.

[27]   J.C. Martin, "The use of ROM's and RAM's in a wrist-watch", Proc. Int. Congress of Chronometry, Geneva 1979, pp. 285-290.

[28]   J.C. Martin, "Random and programmable logic in watches", From Electronics to Microelectronics, EUROCON 80, North-Holland Publ. 1980, pp. 708-713.

[29]   Y. Suzuki, K. Odagawa and T. Abe, "Clocked CMOS calculator circuitry", IEEE J. Solid-State Circuits, Vol. SC-8, pp. 462-469, Dec. 1973.

[30]   J.C. Martin, C. Piguet and J.F. Perotto, "A review of micro-programmed wrist watches", From Electronics to Microelectronics, EUROCON 80, North-Holland Publ. 1980, pp. 626-631.

[31]   A.E. Waldvogel, "A one chip microcomputer for watch applications", ESSCIRC 80, paper 8.2.

[32]   E. Vittoz, "Micropower switched-capacitor oscillator", IEEE J. Solid-State Circuits, Vol. SC-14, pp. 622-624, June 1979.

[33]   E. Vittoz, "Quartz oscillators for watches", Proc. Int. Congress of Chronometry, Geneva 1979, pp. 131-140.

[34]   Y.P. Tsividis and R.W. Ulmer, "A CMOS voltage reference", IEEE J. Solid-State Circuits, Vol. SC-13, pp. 774-778, Dec. 1978.

[35]   E. Vittoz and O. Neyroud, "A low-voltage CMOS bandgap reference", IEEE J. Solid-State Circuits, Vol. SC-14, pp. 573-577, June 1979.

[36]   G. Tzanateas, C.A.T. Salama and Y.P. Tsividis, "A CMOS bandgap voltage reference", IEEE J. Solid-State Circuits, Vol. SC-14, pp. 655-657, June 1979.

[37]   H.J. Oguey and B. Gerber, "MOS voltage reference based on polysilicon gate work function difference", IEEE J. Solid-State Circuits, Vol. SC-15, pp. 264-269, June 1980.

[38]   R.Dessoulavy et al, "A Synchronous switched capacitor filter", IEEE J. Solid-State Circuits, Vol. SC-15, pp. 301-305, June 1980.

[39]   M.A. Copeland and J.M. Rabaey, "Dynamic amplifier for MOS technology", Electronics Letters, Vol. 15, pp. 301-302, May 1979.

[40]   B.J. Hosticka, "Novel dynamic CMOS amplifier for SC integrators", Electronics Letters, Vol. 15, pp. 532-533, August 1979.

[41]   B.J. Hosticka, "Dynamic amplifiers in CMOS technology", Electronic Letters, Vol. 15, pp. 819-820, Dec. 1979.

[42]   C. Piguet and J.F. Perotto, "Microprocessor for the wrist-watch", Proc. Int. Congress of Chronometry, Geneva 1979, pp. 271-278, (in French).

[43]   G. Wotruba, "1.5 V CMOS Mikrocomputertechnik", LSI Seminar, University of Dortmund, May 1980.

Low-Power CMOS Digital Design

Anantha P. Chandrakasan, Samuel Sheng, and Robert W. Brodersen, *Fellow, IEEE*

Abstract—Motivated by emerging battery-operated applications that demand intensive computation in portable environments, techniques are investigated which reduce power consumption in CMOS digital circuits while maintaining computational throughput. Techniques for low-power operation are shown which use the lowest possible supply voltage coupled with architectural, logic style, circuit, and technology optimizations. An architectural-based scaling strategy is presented which indicates that the optimum voltage is much lower than that determined by other scaling considerations. This optimum is achieved by trading increased silicon area for reduced power consumption.

I. Introduction

WITH much of the research efforts of the past ten years directed toward increasing the speed of digital systems, present-day technologies possess computing capabilities that make possible powerful personal workstations, sophisticated computer graphics, and multimedia capabilities such as real-time speech recognition and real-time video. High-speed computation has thus become the expected norm from the average user, instead of being the province of the few with access to a powerful mainframe. Likewise, another significant change in the attitude of users is the desire to have access to this computation at any location, without the need to be physically tethered to a wired network. The requirement of portability thus places severe restrictions on size, weight, and power. Power is particularly important since conventional nickel-cadmium battery technology only provides 20 W · h of energy for each pound of weight [1]. Improvements in battery technology are being made, but it is unlikely that a dramatic solution to the power problem is forthcoming; it is projected that only a 30% improvement in battery performance will be obtained over the next five years [2].

Although the traditional mainstay of portable digital applications has been in low-power, low-throughput uses such as wristwatches and pocket calculators, there are an ever-increasing number of portable applications requiring low power and high throughput. For example, notebook and laptop computers, representing the fastest growing segment of the computer industry, are demanding the same computation capabilities as found in desktop machines. Equally demanding are developments in personal communications services (PCS's), such as the current generation of digital cellular telephony networks which employ complex speech compression algorithms and sophisticated radio modems in a pocket-sized device. Even more dramatic are the proposed future PCS applications, with universal portable multimedia access supporting full-motion digital video and control via speech recognition [3]. In these applications, not only will voice be transmitted via wireless inks, but data as well. This will facilitate new services such as multimedia database access (video and audio in addition to text) and supercomputing for simulation and design, through an intelligent network which allows communication with these services or other people at any place and time. Power for video compression and decompression and for speech recognition must be added to the portable unit to support these services—on top of the already lean power budget for the analog transceiver and speech encoding. Indeed, it is apparent that portability can no longer be associated with low throughput; instead, vastly increased capabilities, actually in excess of that demanded of fixed workstations, must be placed in a low-power portable environment.

Even when power is available in nonportable applications, the issue of low-power design is becoming critical. Up until now, this power consumption has not been of great concern, since large packages, cooling fins, and fans have been capable of dissipating the generated heat. However, as the density and size of the chips and systems continue to increase, the difficulty in providing adequate cooling might either add significant cost to the system or provide a limit on the amount of functionality that can be provided.

Thus, it is evident that methodologies for the design of high-throughput, low-power digital systems are needed. Fortunately, there are clear technological trends that give us a new degree of freedom, so that it may be possible to satisfy these seemingly contradictory requirements. Scaling of device feature sizes, along with the development of high-density, low-parasitic packaging, such as multichip modules [4]–[6], will alleviate the overriding concern with the numbers of transistors being used. When MOS technology has scaled to 0.2-μm minimum feature size it will be possible to place from 1 to 10 × 10^9 transistors in an area of 8 in × 10 in if a high-density packaging technology is used. The question then becomes how can this increased capability be used to meet a goal of low-power operation. Previous analyses on the question of how to best utilize increased transistor density at the chip level concluded that for high-performance microprocessors the best use is to provide increasing amounts of

Manuscript received September 4, 1991; revised November 18, 1991. This work was supported by DARPA.

The authors are with the Department of Electrical Engineering and Computer Science, University of California, Berkeley, CA 94720.

IEEE Log Number 9105976.

on-chip memory [7]. It will be shown here that for computationally intensive functions that the best use is to provide additional circuitry to parallelize the computation.

Another important consideration, particularly in portable applications, is that many computation tasks are likely to be real-time; the radio modem, speech and video compression, and speech recognition all require computation that is always at near-peak rates. Conventional schemes for conserving power in laptops, which are generally based on power-down schemes, are not appropriate for these continually active computations. On the other hand, there is a degree of freedom in design that is available in implementing these functions, in that once the real-time requirements of these applications are met, there is no advantage in increasing the computational throughput. This fact, along with the availability of almost ''limitless'' numbers of transistors, allows a strategy to be developed for architecture design, which if it can be followed, will be shown to provide significant power savings.

II. SOURCES OF POWER DISSIPATION

There are three major sources of power dissipation in digital CMOS circuits, which are summarized in the following equation:

$$P_{\text{total}} = p_t \left(C_L \cdot V \cdot V_{dd} \cdot f_{\text{clk}} \right)$$
$$+ I_{sc} \cdot V_{dd} + I_{\text{leakage}} \cdot V_{dd}. \quad (1)$$

The first term represents the switching component of power, where C_L is the loading capacitance, f_{clk} is the clock frequency, and p_t is the probability that a power-consuming transition occurs (the activity factor). In most cases, the voltage swing V is the same as the supply voltage V_{dd}; however, in some logic circuits, such as in single-gate pass-transistor implementations, the voltage swing on some internal nodes may be slightly less [8]. The second term is due to the direct-path short circuit current I_{sc}, which arises when both the NMOS and PMOS transistors are simultaneously active, conducting current directly from supply to ground [9], [10]. Finally, leakage current I_{leakage}, which can arise from substrate injection and subthreshold effects, is primarily determined by fabrication technology considerations [11] (see Section III-C). The dominant term in a ''well-designed'' circuit is the switching component, and low-power design thus becomes the task of minimizing p_t, C_L, V_{dd}, and f_{clk}, while retaining the required functionality.

The power–delay product can be interpreted as the amount of energy expended in each switching event (or transition) and is thus particularly useful in comparing the power dissipation of various circuit styles. If it is assumed that only the switching component of the power dissipation is important, then it is given by

$$\text{energy per transition} = P_{\text{total}}/f_{\text{clk}} = C_{\text{effective}} V_{dd}^2 \quad (2)$$

where $C_{\text{effective}}$ is the effective capacitance being switched to perform a computation and is given by $C_{\text{effective}} = p_t \cdot C_L$.

III. CIRCUIT DESIGN AND TECHNOLOGY CONSIDERATIONS

There are a number of options available in choosing the basic circuit approach and topology for implementing various logic and arithmetic functions. Choices between static versus dynamic implementations, pass-gate versus conventional CMOS logic styles, and synchronous versus asynchronous timing are just some of the options open to the system designer. At another level, there are also various architectural/structural choices for implementing a given logic function; for example, to implement an adder module one can utilize a ripple-carry, carry-select, or carry-lookahead topology. In this section, the trade-offs with respect to low-power design between a selected set of circuit approaches will be discussed, followed by a discussion of some general issues and factors affecting the choice of logic family.

A. Dynamic Versus Static Logic

The choice of using static or dynamic logic is dependent on many criteria than just its low-power performance, e.g., testability and ease of design. However, if only the low-power performance is analyzed it would appear that dynamic logic has some inherent advantages in a number of areas including reduced switching activity due to hazards, elimination of short-circuit dissipation, and reduced parasitic node capacitances. Static logic has advantages since there is no precharge operation and charge sharing does not exist. Below, each of these considerations will be discussed in more detail.

1) Spurious Transitions: Static designs can exhibit spurious transitions due to finite propagation delays from one logic block to the next (also called critical races and dynamic hazards [12]), i.e., a node can have multiple transitions in a single clock cycle before settling to the correct logic level. For example, consider a static *N*-bit adder, with all bits of the summands going from ZERO to ONE, with the carry input set to ZERO. For all bits, the resultant sum should be ZERO; however, the propagation of the carry signal causes a ONE to appear briefly at most of the outputs. These spurious transitions dissipate extra power over that strictly required to perform the computation. The number of these extra transitions is a function of input patterns, internal state assignment in the logic design, delay skew, and logic depth. To be specific about the magnitude of this problem, an 8-b ripple-carry adder with a uniformly distributed set of random input patterns will typically consume an extra 30% in energy. Though it is possible with careful logic design to eliminate these transitions, dynamic logic intrinsically does not have this problem, since any node can undergo at most one power-consuming transition per clock cycle.

2) Short-Circuit Currents: Short-circuit (direct-path) currents, I_{sc} in (1), are found in static CMOS circuits.

However, by sizing transistors for equal rise and fall times, the short-circuit component of the total power dissipated can be kept to less than 20% [9] (typically < 5–10%) of the dynamic switching component. Dynamic logic does not exhibit this problem, except for those cases in which static pull-up devices are used to control charge sharing [13] or when clock skew is significant.

3) Parasitic Capacitance: Dynamic logic typically uses fewer transistors to implement a given logic function, which directly reduces the amount of capacitance being switched and thus has a direct impact on the power–delay product [14], [15]. However, extra transistors may be required to insure that charge sharing does not result in incorrect evaluation.

4) Switching Activity: The one area in which dynamic logic is at a distinct disadvantage is in its necessity for a precharge operation. Since in dynamic logic every node must be precharged every clock cycle, this means that some nodes are precharged only to be immediately discharged again as the node is evaluated, leading to a higher activity factor. If a two-input N-tree (precharged high) dynamic NOR gate has a uniform input distribution of high and low levels, then the four possible input combinations (00,01,10,11) will be equally likely. There is then a 75% probability that the output node will discharge immediately after the precharge phase, implying that the activity for such a gate is 0.75 (i.e., $P_{\text{NOR}} = 0.75\, C_L V_{dd}^2 f_{\text{clk}}$). On the other hand, the activity factor for the static NOR counterpart will be only 3/16, excluding the component due to the spurious transitions mentioned in Section III-A-1 (power is only drawn on a ZERO-TO-ONE transition, so $p_{0 \to 1} = p(0)p(1) = p(0)\,(1 - p(0))$). In general, gate activities will be different for static and dynamic logic and will depend on the type of operation being performed and the input signal probabilities. In addition, the clock buffers to drive the precharge transistors will also require power that it not needed in a static implementation.

5) Power-Down Modes: Lastly, power-down techniques achieved by disabling the clock signal have been used effectively in static circuits, but are not as well-suited for dynamic techniques. If the logic state is to be preserved during shutdown, a relatively small amount of extra circuitry must be added to the dynamic circuits to preserve the state, resulting in a slight increase in parasitic capacitance and slower speeds.

B. Conventional Static Versus Pass-Gate Logic

A more clear situation exists in the use of transfer gates to implement logic functions, as is used in the complementary pass-gate logic (CPL) family [8], [10]. In Fig. 1, the schematic of a typical static CMOS logic circuit for a full adder is shown along with a static CPL version [8]. The pass-gate design uses only a single transmission NMOS gate, instead of a full complementary pass gate to reduce node capacitance. Pass-gate logic is attractive as fewer transistors are required to implement important logic functions, such as XOR's which only require two pass tran-

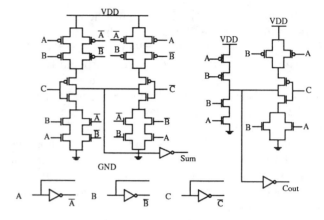

Transistor count (conventional CMOS) : 40

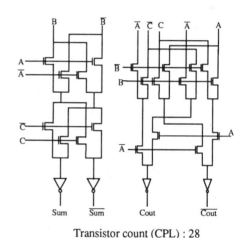

Transistor count (CPL) : 28

Fig. 1. Comparison of a conventional CMOS and CPL adders [8].

sistors in a CPL implementation. This particularly efficient implementation of an XOR is important since it is key to most arithmetic functions, permitting adders and multipliers to be created using a minimal number of devices. Likewise, multiplexers, registers, and other key building blocks are simplified using pass-gate designs.

However, a CPL implementation as shown in Fig. 1 has two basic problems. First, the threshold drop across the single-channel pass transistors results in reduced current drive and hence slower operation at reduced supply voltages; this is important for low-power design since it is desirable to operate at the lowest possible voltages levels. Second, since the "high" input voltage level at the regenerative inverters is not V_{dd}, the PMOS device in the inverter is not fully turned off, and hence direct-path static power dissipation could be significant. To solve these problems, reduction of the threshold voltage has proven effective, although if taken too far will incur a cost in dissipation due to subthreshold leakage (see Section III-C) and reduced noise margins. The power dissipation for a pass-gate family adder with zero-threshold pass transistors at a supply voltage of 4 V was reported to be 30% lower than a conventional static design, with the difference being even more significant at lower supply voltages [8].

C. Threshold Voltage Scaling

Since a significant power improvement can be gained through the use of low-threshold MOS devices, the question of how low the thresholds can be reduced must be addressed. The limit is set by the requirement to retain adequate noise margins and the increase in subthreshold currents. Noise margins will be relaxed in low-power designs because of the reduced currents being switched, however, the subthreshold currents can result in significant static power dissipation. Essentially, subthreshold leakage occurs due to carrier diffusion between the source and the drain when the gate–source voltage V_{gs} has exceeded the weak inversion point, but is still below the threshold voltage V_t, where carrier drift is dominant. In this regime, the MOSFET behaves similarly to a bipolar transistor, and the subthreshold current is exponentially dependent on the gate–source voltage V_{gs}, and approximately independent of the drain–source voltage V_{ds}, for V_{ds} approximately larger than 0.1 V. Associated with this is the subthreshold slope S_{th}, which is the amount of voltage required to drop the subthreshold current by one decade. At room temperature, typical values for S_{th} lie between 60 and 90 mV/decade current, with 60 mV/decade being the lower limit. Clearly, the lower S_{th} is, the better, since it is desirable to have the device "turn off" as close to V_t as possible. As a reference, for an $L = 1.5$-μm, $W = 70$-μm NMOS device, at the point where V_{gs} equals V_t, with V_t defined as where the surface inversion charge density is equal to the bulk doping, approximately 1 μA of leakage current is exhibited, or 0.014 μA/μm of gate width [16]. The issue is whether this extra current is negligible in comparison to the time-average current during switching. For a CMOS inverter (PMOS: $W = 8$ μm, NMOS: $W = 4$ μm), the current was measured to be 64 μA over 3.7 ns at a supply voltage of 2 V. This implies that there would be a 100% power penalty for subthreshold leakage if the device were operating at a clock speed of 25 MHz with an activity factor of $p_t = 1/6$th, i.e., the devices were left idle and leaking current 83% of the time. It is not advisable, therefore, to use a true zero threshold device, but instead to use thresholds of at least 0.2 V, which provides for at least two orders of magnitude of reduction of subthreshold current. This provides a good compromise between improvement of current drive at low supply voltage operation and keeping subthreshold power dissipation to a negligible level. This value may have to be higher in dynamic circuits to prevent accidental discharge during the evaluation phase [11]. Fortunately, device technologists are addressing the problem of subthreshold currents in future scaled technologies, and reducing the supply voltages also serves to reduce the current by reducing the maximum allowable drain–source voltage [17], [18]. The design of future circuits for lowest power operation should therefore explicitly take into account the effect of subthreshold currents.

D. Power-Down Strategies

In synchronous designs, the logic between registers is continuously computing every clock cycle based on its new inputs. To reduce the power in synchronous designs, it is important to minimize switching activity by powering down execution units when they are not performing "useful" operations. This is an important concern since logic modules can be switching and consuming power even when they are not being actively utilized [19].

While the design of synchronous circuits requires special design effort and power-down circuitry to detect and shut down unused units, self-timed logic has inherent power-down of unused modules, since transitions occur only when requested. However, since self-timed implementations require the generation of a completion signal indicating the outputs of the logic module are valid, there is additional overhead circuitry. There are several circuit approaches to generate the requisite completion signal. One method is to use dual-rail coding, which is implicit in certain logic families such as the DCVSL [13], [20]. The completion signal in a combinational macrocell made up of cascading DCVSL gates consists of simply ORing the outputs of only the last gate in the chain, leading to small overhead requirements. However, for each computation, dual-rail coding guarantees a switching event will occur since at least one of the outputs must evaluate to zero. We found that the dual-rail DCVSL family consumes at least two times more in energy per input transition than a conventional static family. Hence, self-timed implementations can prove to be expensive in terms of energy for data paths that are continuously computing.

IV. Voltage Scaling

Thus far we have been primarily concerned with the contributions of capacitance to the power expression CV^2f. Clearly, though, the reduction of V should yield even greater benefits; indeed, reducing the supply voltage is the key to low-power operation, even after taking into account the modifications to the system architecture, which is required to maintain the computational throughput. First, a review of circuit behavior (delay and energy characteristics) as a function of scaling supply voltage and feature sizes will be presented. By comparison with experimental data, it is found that simple first-order theory yields an amazingly accurate representation of the various dependencies over a wide variety of circuit styles and architectures. A survey of two previous approaches to supply-voltage scaling is then presented, which were focused on maintaining reliability and performance. This is followed by our architecture-driven approach, from which an "optimal" supply voltage based on technology, architecture, and noise margin constraints is derived.

A. Impact on Delay and Power-Delay Product

As noted in (2), the energy per transition or equivalently the power–delay product in "properly designed" CMOS circuits (as discussed in Section II) is proportional

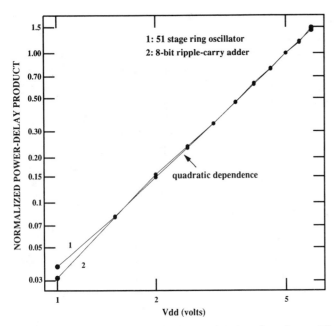

Fig. 2. Power–delay product exhibiting square-law dependence for two different circuits.

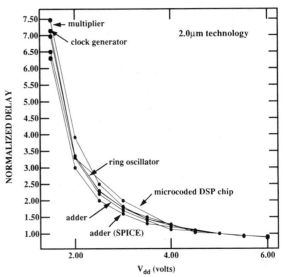

Fig. 3. Data demonstrating delay characteristics follow simple first-order theory.

TABLE I
DETAILS OF COMPONENTS USED FOR THE STUDY IN FIG. 3

Component (all in 2 μm)	# of Transistors	Area	Comments
Microcoded DSP Chip [21]	44 802	94 mm^2	20-b data path
Multiplier	20 432	12.2 mm^2	24 × 24 b
Adder	256	0.083 mm^2	conventional static
Ring Oscillator	102	0.055 mm^2	51 stages
Clock Generator	56	0.04 mm^2	cross-coupled NOR

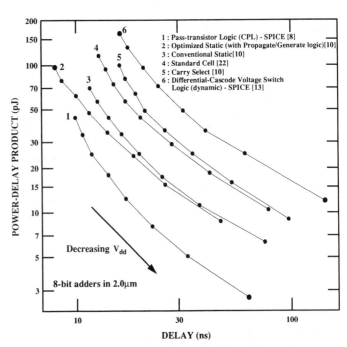

Fig. 4. Data showing improvement in power–delay product at the cost of speed for various circuit approaches.

to V^2. This is seen from Fig. 2, which is a plot of two experimental circuits that exhibit the expected V^2 dependence. Therefore, it is only necessary to reduce the supply voltage for a *quadratic* improvement in the power–delay product of a logic family.

Unfortunately, this simple solution to low-power design comes at a cost. As shown in Fig. 3, the effect of reducing V_{dd} on the delay is shown for a variety of different logic circuits that range in size from 56 to 44 000 transistors spanning a variety of functions; all exhibit essentially the same dependence (see Table I). Clearly, we pay a speed penalty for a V_{dd} reduction, with the delays drastically increasing as V_{dd} approaches the sum of the threshold voltages of the devices. Even though the exact analysis of the delay is quite complex if the nonlinear characteristic of a CMOS gate are taken into account, it is found that a simple first-order derivation adequately predicts the experimentally determined dependence and is given by

$$T_d = \frac{C_L \times V_{dd}}{I} = \frac{C_L \times V_{dd}}{\mu C_{ox} (W/L) (V_{dd} - V_t)^2}. \quad (3)$$

We also evaluated (through experimental measurements and SPICE simulations) the energy and delay performance for several different logic styles and topologies using an 8-b adder as a reference; the results are shown on a log–log plot in Fig. 4. We see that the power–delay product improves as delays increase (through reduction of the supply voltage), and therefore it is desirable to operate at the *slowest* possible speed. Since the objective is to reduce power consumption while maintaining the overall system throughput, compensation for these increased delays at low voltages is required. Of particular interest in this figure is the range of energies required for a transition at a given amount of delay. The best logic family we analyzed (over 10 times better than the worst that we investigated) was the pass-gate family, CPL, (see Section III-B) if a reduced value for the threshold is assumed [8].

Figs. 2, 3, and 4 suggest that the delay and energy behavior as a function of V_{dd} scaling for a given technology is "well-behaved" and relatively independent of logic style and circuit complexity. We will use this result during our optimization of architecture for low-power by treating V_{dd} as a free variable and by allowing the architectures to vary to retain constant throughput. By exploiting the monotonic dependencies of delay and energy versus supply voltage that hold over wide circuit variations, it is possible to make relatively strong predictions about the types of architectures that are best for low-power design. Of course, as mentioned previously, there are some logic styles such as NMOS pass-transistor logic without reduced thresholds whose delay and energy characteristics would deviate from the ones presented above, but even for these cases, though the quantitative results will be different, the basic conclusions will still hold.

B. Optimal Transistor Sizing with Voltage Scaling

Independent of the choice of logic family or topology, optimized transistor sizing will play an important role in reducing power consumption. For low power, as is true for high-speed design, it is important to equalize all delay paths so that a single critical path does not unnecessarily limit the performance of the entire circuit. However, beyond this constraint, there is the issue of what extent the W/L ratios should be uniformly raised for all the devices, yielding a uniform decrease in the gate delay and hence allowing for a corresponding reduction in voltage and power. It is shown in this section that if voltage is allowed to vary, that the optimal sizing for low-power operation is quite different from that required for high speed.

In Fig. 5, a simple two-gate circuit is shown, with the first stage driving the gate capacitance of the second, in addition to the parasitic capacitance C_p due to substrate coupling and interconnect. Assuming that the input gate capacitance of both stages is given by NC_{ref}, where C_{ref} represents the gate capacitance of a MOS device with the smallest allowable W/L, then the delay through the first gate at a supply voltage V_{ref} is given by

$$T_N = K \frac{(C_p + NC_{ref})}{(NC_{ref})} \frac{V_{ref}}{(V_{ref} - V_t)^2}$$

$$= K (1 + \alpha/N) \frac{V_{ref}}{(V_{ref} - V_t)^2} \qquad (4)$$

where α is defined as the ratio of C_p to C_{ref}, and K represents terms independent of device width and voltage. For a given supply voltage V_{ref}, the speedup of a circuit whose W/L ratios are sized up by a factor of N over a reference circuit using minimum-size transistors ($N = 1$) is given by $(1 + \alpha/N)/(1 + \alpha)$. In order to evaluate the energy performance of the two designs at the same speed, the voltage of the scaled solution is allowed to vary as to

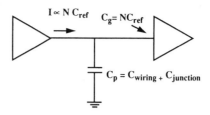

Fig. 5. Circuit model for analyzing the effect of transistor sizing.

keep delay constant. Assuming that the delay scales as $1/V_{dd}$ (ignoring threshold voltage reductions in signal swings), the supply voltage V_N, where the delays of the scaled design and the reference design are equal, is given by

$$V_N = \frac{(1 + \alpha/N)}{(1 + \alpha)} V_{ref}. \qquad (5)$$

Under these conditions, the energy consumed by the first stage as a function of N is given by

$$\text{Energy } (N) = (C_p + NC_{ref}) V_N^2$$

$$= \frac{NC_{ref} (1 + \alpha/N)^3 V_{ref}^2}{(1 + \alpha)^2}. \qquad (6)$$

After normalizing against E_{ref} (the energy for the minimum size case), Fig. 6 shows a plot of Energy (N)/Energy (1) versus N for various values of α. When there is no parasitic capacitance contribution (i.e., $\alpha = 0$), the energy increases linearly with respect to N, and the solution utilizing devices with the smallest W/L ratios results in the lowest power. At high values of α, when parasitic capacitances begin to dominate over the gate capacitances, the power decreases temporarily with increasing device sizes and then starts to increase, resulting in an optimal value for N. The initial decrease in supply voltage achieved from the reduction in delays more than compensates the increase in capacitance due to increasing N. However, after some point the increase in capacitance dominates the achievable reduction in voltage, since the incremental speed increase with transistor sizing is very small (this can be seen in (4), with the delay becoming independent of α as N goes to infinity). Throughout the analysis we have assumed that the parasitic capacitance is independent of device sizing. However, the drain and source diffusion and perimeter capacitances actually increase with increasing area, favoring smaller size devices and making the above a worst-case analysis.

Also plotted in Fig. 6 are simulation results from extracted layouts of an 8-b adder carry chain for three different device W/L ratios ($N = 1$, $N = 2$, and $N = 4$). The curve follows the simple first-order model derived very well, and suggests that this example is dominated more by the effect of gate capacitance rather than parasitics. In this case, increasing devices W/L's does not help, and the solution using the smallest possible W/L ratios results in the best sizing.

From this section, it is clear that the determination of

41

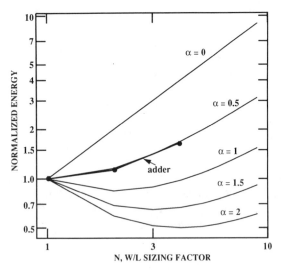

Fig. 6. Plot of energy versus transistor sizing factor for various parasitic contributions.

an "optimal" supply voltage is the key to minimizing power consumption; hence we focus on this issue in the following sections. First, we will review the previous work dealing with choice of supply voltage which were based on reliability and speed considerations [23], [24], followed by an architecturally driven approach to supply voltage scaling.

C. Reliability-Driven Voltage Scaling

One approach to the selection of an optimal power supply voltage for deep-submicrometer technologies is based on optimizing the trade-off between speed and reliability [23]. Constant-voltage scaling, the most commonly used technique, results in higher electric fields that create hot carriers. As a result of this, the devices degrade with time (including changes in threshold voltages, degradation of transconductance, and increase in subthreshold currents), leading to eventual breakdown [11]. One solution to reducing the number of hot carriers is to change the physical device structure, such as the use of lightly doped drain (LDD), usually at the cost of decreased performance. Assuming the use of an LDD structure and a constant hot-carrier margin, an optimal voltage of 2.5 V was found for a 0.25-μm technology by choosing the minimum point on the delay versus V_{dd} curve [23]. For voltages above this minimum point, the delay was found to increase with increasing V_{dd}, since the LDD structure used for the purposes of reliability resulted in increased parasitic resistances.

D. Technology-Driven Voltage Scaling

The simple first-order delay analysis presented in Section IV-A is reasonably accurate for long-channel devices. However, as feature sizes shrink below 1.0 μm, the delay characteristics as a function of lowering the supply voltage deviate from the first-order theory presented since it does not consider carrier velocity saturation under high electric fields [11]. As a result of velocity saturation, the

current is no longer a quadratic function of the voltage but linear; hence, the current drive is significantly reduced and is approximately given by $I = WC_{ox}(V_{dd} - V_t)$ $v_{\max}$ [4]. Given this and the equation for delay in (3), we see that the delay for submicrometer circuits is relatively independent of supply voltages at high electric fields.

A "technology"-based approach proposes choosing the power supply voltage based on maintaining the speed performance for a given submicrometer technology [24]. By exploiting the relative independence of delay on supply voltage at high electric fields, the voltage can be dropped to some extent for a velocity-saturated device with very little penalty in speed performance. This implies that there is little advantage to operating above a certain voltage. This idea has been formalized by Kakumu and Kinugawa, yielding the concept of a "critical voltage" which provides a lower limit on the supply voltage [24]. The critical voltage is defined as $V_c = 1.1E_cL_{\text{eff}}$, where E_c is the critical electric field causing velocity saturation; this is the voltage at which the delay versus V_{dd} curve approaches a $\sqrt{V_{dd}}$ dependence. For 0.3-μm technology, the proposed lower limit on supply voltage (or the critical voltage) was found to be 2.43 V.

Because of this effect, there is some movement to a 3.3-V industrial voltage standard since at this level of voltage reduction there is not a significant loss of circuit speed [1], [25]. This was found to achieve a 60% reduction in power when compared to a 5-V operation [25].

E. Architecture-Driven Voltage Scaling

The above-mentioned "technology"-based approaches are focusing on reducing the voltage while maintaining device speed, and are not attempting to achieve the minimum possible power. As shown in Figs. 2 and 4, CMOS logic gates achieve lower power–delay products (energy per computation) as the supply voltages are reduced. In fact, once a device is in velocity saturation there is a further degradation in the energy per computation, so in minimizing the energy required for computation, Kakumu and Kinugawa's critical voltage provides an *upper* bound on the supply voltage (whereas for their analysis it provided a *lower* bound!). It now will be the task of the architecture to compensate for the reduced circuit speed that comes with operating below the critical voltage.

To illustrate how architectural techniques can be used to compensate for reduced speeds, a simple 8-b data path consisting of an adder and a comparator is analyzed assuming a 2.0-μm technology. As shown in Fig. 7, inputs A and B are added, and the result compared to input C. Assuming the worst-case delay through the adder, comparator, and latch is approximately 25 ns at a supply voltage of 5 V, the system in the best case can be clocked with a clock period of $T = 25$ ns. When required to run at this maximum possible throughput, it is clear that the operating voltage cannot be reduced any further since no extra delay can be tolerated, hence yielding no reduction in power. We will use this as the reference data path for

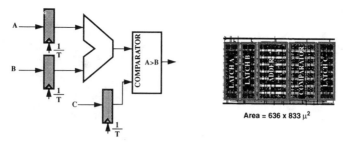

Fig. 7. A simple data path with corresponding layout.

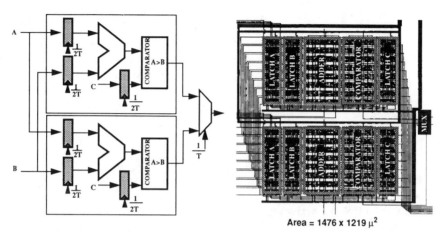

Area = 1476 x 1219 μ²

Fig. 8. Parallel implementation of the simple data path.

our architectural study and present power improvement numbers with respect to this reference. The power for the reference data path is given by

$$P_{\text{ref}} = C_{\text{ref}} V_{\text{ref}}^2 f_{\text{ref}} \qquad (7)$$

where C_{ref} is the total effective capacitance being switched per clock cycle. The effective capacitance was determined by averaging the energy over a sequence of input patterns with a uniform distribution.

One way to maintain throughput while reducing the supply voltage is to utilize a parallel architecture. As shown in Fig. 8, two identical adder–comparator data paths are used, allowing each unit to work at half the original rate while maintaining the original throughput. Since the speed requirements for the adder, comparator, and latch have decreased from 25 to 50 ns, the voltage can be dropped from 5 to 2.9 V (the voltage at which the delay doubled, from Fig. 3). While the data-path capacitance has increased by a factor of 2, the operating frequency has correspondingly decreased by a factor of 2. Unfortunately, there is also a slight increase in the total "effective" capacitance introduced due to the extra routing, resulting in an increased capacitance by a factor of 2.15. Thus the power for the parallel data path is given by

$$P_{\text{par}} = C_{\text{par}} V_{\text{par}}^2 f_{\text{par}}$$
$$= (2.15C_{\text{ref}}) (0.58V_{\text{ref}})^2 \left(\frac{f_{\text{ref}}}{2}\right) \approx 0.36 P_{\text{ref}}. \quad (8)$$

This method of reducing power by using parallelism has the overhead of increased area, and would not be suit-able for area-constrained designs. In general, the parallelism will have the overhead of extra routing (and hence extra power), and careful optimization must be performed to minimize this overhead (for example, partitioning techniques for minimal overhead). Interconnect capacitance will especially play a very important role in deep-submicrometer implementations, since the fringing capacitance of the interconnect capacitance ($C_{\text{wiring}} = C_{\text{area}} + C_{\text{fringing}} + C_{\text{wiring}}$) can become a dominant part of the total capacitance (equal to $C_{\text{gate}} + C_{\text{junction}} + C_{\text{wiring}}$) and cease to scale [4].

Another possible approach is to apply pipelining to the architecture, as shown in Fig. 9. With the additional pipeline latch, the critical path becomes $\max[T_{\text{adder}}, T_{\text{comparator}}]$, allowing the adder and the comparator to operate at a slower rate. For this example, the two delays are equal, allowing the supply voltage to again be reduced from 5 V used in the reference data path to 2.9 V (the voltage at which the delay doubles) with no loss in throughput. However, there is a much lower area overhead incurred by this technique, as we only need to add pipeline registers. Note that there is again a slight increase in hardware due to the extra latches, increasing the "effective" capacitance by approximately a factor of 1.15. The power consumed by the pipelined data path is

$$P_{\text{pipe}} = C_{\text{pipe}} V_{\text{pipe}}^2 f_{\text{pipe}}$$
$$= (1.15C_{\text{ref}}) (0.58V_{\text{ref}})^2 f_{\text{ref}} \approx 0.39 P_{\text{ref}}. \quad (9)$$

With this architecture, the power reduces by a factor of approximately 2.5, providing approximately the same

43

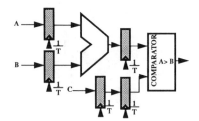 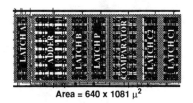

Area = 640 x 1081 μ^2

Fig. 9. Pipelined implementation of the simple data path.

TABLE II
ARCHITECTURE SUMMARY

Architecture Type	Voltage	Area	Power
Simple data path (no pipelining or parallelism)	5 V	1	1
Pipelined data path	2.9 V	1.3	0.39
Parallel data path	2.9 V	3.4	0.36
Pipeline parallel	2.0 V	3.7	0.2

power reduction as the parallel case with the advantage of lower area overhead. As an added bonus, increasing the level of pipelining also has the effect of reducing logic depth and hence power contributed due to hazards and critical races (see Section III-A-1).

Furthermore, an obvious extension is to utilize a combination of pipelining and parallelism. Since this architecture reduces the critical path and hence speed requirement by a factor of 4, the voltage can be dropped until the delay increases by a factor of 4. The power consumption in this case is

$$P_{\text{parpipe}} = C_{\text{parpipe}} V_{\text{parpipe}}^2 f_{\text{parpipe}}$$
$$= (2.5 C_{\text{ref}}) (0.4 V_{\text{ref}})^2 \left(\frac{f_{\text{ref}}}{2} \right) \approx 0.2 P_{\text{ref}}. \quad (10)$$

The parallel-pipeline implementation results in a 5 times reduction in power. Table II shows a comparative summary of the various architectures described for the simple adder–comparator data path.

From the above examples, it is clear that the traditional time-multiplexed architectures, as used in general-purpose microprocessors and DSP chips, are the least desirable for low-power applications. This follows since time multiplexing actually increases the speed requirements on the logic circuitry, thus not allowing reduction in the supply voltage.

V. OPTIMAL SUPPLY VOLTAGE

In the previous section, we saw that the delay increase due to reduced supply voltages below the critical voltage can be compensated by exploiting parallel architectures. However, as seen in Fig. 3 and (3), as supply voltages approach the device thresholds, the gate delays increase rapidly. Correspondingly, the amount of parallelism and overhead circuitry increases to a point where the added overhead dominates any gains in power reduction from

further voltage reduction, leading to the existence of an "optimal" voltage from an architectural point of view. To determine the value of this voltage, the following model is used for the power for a fixed system throughput as a function of voltage (and hence degree of parallelism):

$$\text{Power } (N) = N C_{\text{ref}} V^2 \frac{f_{\text{ref}}}{N} + C_{ip} V^2 \frac{f_{\text{ref}}}{N} + C_{\text{interface}} V^2 f_{\text{ref}}$$

(11)

where N is the number of parallel processors, C_{ref} is the capacitance of a single processor, C_{ip} is the interprocessor communication overhead introduced due to the parallelism (due to control and routing), and $C_{\text{interface}}$ is the overhead introduced at the interface which is not decreased in speed as the architecture is made more parallel. In general, C_{ip} and $C_{\text{interface}}$ are functions of N, and the power improvement over the reference case (i.e., without parallelism) can be expressed as

$$P_{\text{normalized}} = \left(1 + \frac{C_{ip}(N)}{N C_{\text{ref}}} + \frac{C_{\text{interface}}(N)}{C_{\text{ref}}} \right) \left(\frac{V}{V_{\text{ref}}} \right)^2.$$

(12)

At very low supply voltages (near the device thresholds), the number of processors (and hence the corresponding overhead in the above equation) typically increases at a faster rate than the V^2 term decreases, resulting in a power increase with further reduction in voltage.

Reduced threshold devices tend to lower the optimal voltage; however, as seen in Section III-C, at thresholds below 0.2 V, power dissipation due to the subthreshold current will soon start to dominate and limit further power improvement. An even lower bound on the power supply voltage for a CMOS inverter with "correct" functionality was found to be 0.2 V $(8kT/q)$ [26]. This gives a limit on the power–delay product that can be achieved with CMOS digital circuits; however, the amount of parallelism to retain throughput at this voltage level would no doubt be prohibitive for any practical situation.

So far, we have seen that parallel and pipelined architectures can allow for a reduction in supply voltages to the "optimal" level; this will indeed be the case if the algorithm being implemented does not display any recursion (feedback). However, there are a wide class of applications inherently recursive in nature, ranging from the simple ones, such as infinite impulse response and adaptive filters, to more complex cases such as systems solving nonlinear equations and adaptive compression algorithms. There is, therefore, also an algorithmic bound on the level to which pipelining and parallelism can be exploited for voltage reduction. Although the application of data control flowgraph transformations can alleviate this bottleneck to some extent, both the constraints on latency and the structure of computation of some algorithms can prevent voltage reduction to the optimal voltage levels discussed above [27], [28].

Another constraint on the lowest allowable supply voltages is set by system noise margin constraints ($V_{\text{noise margin}}$). Thus, we must lower-bound the "optimal" voltage by

$$V_{\text{noise margin}} \leq V_{\text{optimal}} \leq V_{\text{critical}} \qquad (13)$$

with V_{critical} defined in Section IV-D. Hence, the "optimal" supply voltage (for a fixed technology) will lie somewhere between the voltage set by noise margin constraints and the critical voltage.

Fig. 10 shows power (normalized to 1 at $V_{dd} = 5$ V) as a function of V_{dd} for a variety of cases for a 2.0-μm technology. As will be shown, there is a wide variety of assumptions in these various cases and it is important to note that they all have roughly the same optimum value of supply voltage, approximately 1.5 V. Curve 1 in this figure represents the power dissipation which would be achieved if there were no overhead associated with increased parallelism. For this case, the power is a strictly decreasing function of V_{dd} and the optimum voltage would be set by the minimum value allowed from noise margin constraints (assuming that no recursive bottleneck was reached). Curve 5 assumes that the interprocessor capacitance has an N^2 dependence while curve 6 assumes an N^3 dependence. It is expected that in most practical cases the dependence is actually less than N^2, but even with the extremely strong N^3 dependence an optimal value around 2 V is found.

Curves 2 and 3 are obtained from data from actual layouts [22], and exhibit a dependence of the interface capacitance which lies between linear and quadratic on the degree of parallelism, N. For these cases, there was no interprocessor communication. Curves 2 and 3 are extensions of the example described in Section IV-E in which the parallel and parallel-pipeline implementations of the simple data path were duplicated N times. Curve 4 is for a much more complex example, a seventh-order IIR filter, also obtained from actual layout data. The overhead in this case arose primarily from interprocessor communication. This curve terminates around 1.4 V, because at that point the algorithm has been made maximally parallel, reaching a recursive bottleneck. For this case, at a supply voltage of 5 V, the architecture is basically a single hardware unit that is being time multiplexed and requires about 7 times more power than the optimal parallel case which is achieved with a supply of around 1.5 V. Table III summarizes the power reduction and normalized areas that were obtained from layouts. The increase in

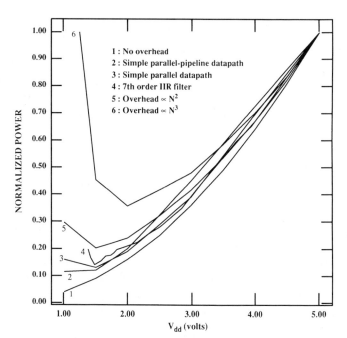

Fig. 10. Optimum voltage of operation.

TABLE III
NORMALIZED AREA/POWER FOR VARIOUS SUPPLY VOLTAGE FOR PLOTS 2, 3, AND 4 IN FIG. 10

Voltage	Parallel Area/Power	Parallel-Pipeline Area/Power	IIR Area/Power
5	1/1	1/1	1/1
2	6/0.19	3.7/0.2	2.6/0.23
1.5	11/0.13	7/0.12	7/0.14
1.4	15/0.14	10/0.11	Recursive bottleneck reached

areas gives an indication of the amount of parallelism being exploited. The key point is that the optimal voltage was found to be relatively independent over all the cases considered, and occurred around 1.5 V for the 2.0-μm technology; a similar analysis using a 0.5-V threshold, 0.8-μm process (with an L_{eff} of 0.5 μm) resulted in optimal voltages around 1 V, with power reductions in excess of a factor of 10. Further scaling of the threshold would allow even lower voltage operation, and hence even greater power savings.

VI. CONCLUSIONS

There are a variety of considerations that must be taken into account in low-power design which include the style of logic, the technology used, and the logic implemented. Factors that were shown to contribute to power dissipation included spurious transitions due to hazards and critical race conditions, leakage and direct path currents, precharge transitions, and power-consuming transitions in unused circuitry. A pass-gate logic family with modified threshold voltages was found to be the best performer for low-power designs, due to the minimal number of transistors required to implement the important logic func-

45

tions. An analysis of transistor sizing has shown that minimum-sized transistors should be used if the parasitic capacitances are less than the active gate capacitances in a cascade of logic gates.

With the continuing trend of denser technology through scaling and the development of advanced packaging techniques, a new degree of freedom in architectural design has been made possible in which silicon area can be traded off against power consumption. Parallel architectures, utilizing pipelining or hardware replication, provide the mechanism for this trade-off by maintaining throughput while using slower device speeds and thus allowing reduced voltage operation. The well-behaved nature of the dependencies of power dissipation and delay as a function of supply voltage over a wide variety of situations allows optimizations of the architecture. In this way, for a wide variety of situations, the optimum voltage was found to be less than 1.5 V, below which the overhead associated with the increased parallelism becomes prohibitive.

There are other limitations which may not allow the optimum supply voltage to be achieved. The algorithm that is being implemented may be sequential in nature and/or have feedback which will limit the degree of parallelism that can be exploited. Another possibility is that the optimum degree of parallelism may be so large that the number of transistors may be inordinately large, thus making the optimum solution unreasonable. However, in any case, the goal in minimizing power consumption is clear: operate the circuit as slowly as possible, with the lowest possible supply voltage.

ACKNOWLEDGMENT

The authors wish to thank A. Burstein, M. Potkonjak, M. Srivastava, A. Stoelzle, and L. Thon for providing us with examples. We also would like to thank Prof. P. Ko, Prof. T. Meng, Prof. J. Rabaey, and Prof. C. Sodini for their invaluable feedback. Lastly, we would also wish to acknowledge the support of S. Sheng by the Fannie and John Hertz Foundation.

REFERENCES

[1] T. Bell, "Incredible shrinking computers," *IEEE Spectrum*, pp. 37–43, May 1991.
[2] Eager, "Advances in rechargeable batteries pace portable computer growth," in *Proc. Silicon Valley Personal Comput. Conf.*, 1991, pp. 693–697.
[3] A. Chandrakasan, S. Sheng, and R. W. Brodersen, "Design considerations for a future multimedia terminal," presented at the 1990 WINLAB Workshop (Rutgers Univ., New Brunswick, NJ), Oct. 1990.
[4] H. B. Bakoglu, *Circuits, Interconnections, and Packaging for VLSI*. Menlo Park, CA: Addison-Wesley, 1990.
[5] D. Benson et al., "Silicon multichip modules," presented at the Hot Chips Symp. III, Santa Clara, CA, Aug. 1990.
[6] G. Geschwind and R. M. Clary, "Multichip modules—An overview," in *Expo SMT/HiDEP 1990 Tech. Proc.*, (San Jose, CA), 1990, pp. 319–329.
[7] D. A. Patterson and C. H. Sequin, "Design considerations for single-chip computers of the future," *IEEE J. Solid-State Circuits*, vol. SC-15, no. 1, Feb. 1980; also *IEEE Trans. Comput.*, vol. C-29, no. 2, pp. 108–116, Feb. 1980.
[8] K. Yano et al., "A 3.8-ns CMOS 16 × 16 multiplier using complementary pass transistor logic," *IEEE J. Solid-State Circuits*, vol. 25, pp. 388–395, Apr. 1990.
[9] H. J. M. Veendrick, "Short-circuit dissipation of static CMOS circuitry and its impact on the design of buffer circuits," *IEEE J. Solid-State Circuits*, vol. SC-19, pp. 468–473, Aug. 1984.
[10] N. Weste and K. Eshragian, *Principles of CMOS VLSI Design: A Systems Perspective*. Reading, MA: Addison-Wesley, 1988.
[11] R. K. Watts, Ed., *Submicron Integrated Circuits*. New York: Wiley, 1989.
[12] D. Green, *Modern Logic Design*. Reading, MA: Addison-Wesley, 1986, pp. 15–17.
[13] G. Jacobs and R. W. Brodersen, "A fully asynchronous digital signal processor using self-timed circuits," *IEEE J. Solid-State Circuits*, vol. 25, pp. 1526–1537, Dec. 1990.
[14] D. Hodges and H. Jackson, *Analysis and Design of Digital Integrated Circuits*. New York: McGraw-Hill, 1988.
[15] M. Shoji, *CMOS Digital Circuit Technology*. Englewood Cliffs, NJ: Prentice-Hall, 1988.
[16] S. Sze, *Physics of Semiconductor Devices*. New York: Wiley, 1981.
[17] M. Aoki et al., "0.1μm CMOS devices using low-impurity-channel transistors (LICT)," in *IEDM Tech. Dig.*, 1990, pp. 939–941.
[18] M. Nagata, "Limitations, innovations, and challenges of circuits & devices into half-micron and beyond," in *Proc. Symp. VLSI Circuits*, 1991, pp. 39–42.
[19] S. C. Ellis, "Power management in notebook PC's," in *Proc. Silicon Valley Personal Comput. Conf.*, 1991, pp. 749–754.
[20] K. Chu and D. Pulfrey, "A comparison of CMOS circuit techniques: Differential cascode voltage switch logic versus conventional logic," *IEEE J. Solid-State Circuits*, vol. SC-22, pp. 528–532, Aug. 1987.
[21] R. W. Brodersen, *Lager: A Silicon Compiler*. Norwell, MA: Kluwer, to be published.
[22] R. W. Brodersen et al., "LagerIV cell library documentation," Electron. Res. Lab., Univ. Calif., Berkeley, June 23, 1988.
[23] B. Davari et al., "A high performance 0.25μm CMOS technology," in *IEDM Tech. Dig.*, 1988, pp. 56–59.
[24] M. Kakumu and M. Kinugawa, "Power-supply voltage impact on circuit performance for half and lower submicrometer CMOS LSI," *IEEE Trans. Electron Devices*, vol. 37, no. 8, pp. 1902–1908, Aug. 1990.
[25] D. Dahle, "Designing high performance systems to run from 3.3 V or lower sources," in *Proc. Silicon Valley Personal Comput. Conf.*, 1991, pp. 685–691.
[26] R. Swansan and J. Meindl, "Ion-implanted complementary MOS transistors in low-voltage circuits," *IEEE J. Solid-State Circuits*, vol. SC-7, pp. 146–153, Apr. 1972.
[27] D. Messerschmitt, "Breaking the recursive bottleneck," in *Performance Limits in Communication Theory and Practice*, 1988, J. K. Skwirzynski, Ed., pp. 3–19.
[28] M. Potkonjak and J. Rabaey, "Optimizing resource utilization using transformations," in *Proc. ICCAD*, Nov. 1991, pp. 88–91.

CMOS Scaling for High Performance and Low Power—The Next Ten Years

BIJAN DAVARI, SENIOR MEMBER, IEEE, ROBERT H. DENNARD, FELLOW, IEEE, AND GHAVAM G. SHAHIDI

Invited Paper

A guideline for scaling of CMOS technology for logic applications such as microprocessors is presented covering the next ten years, assuming that the lithography and base process development driven by DRAM continues on the same three-year cycle as in the past. This paper emphasizes the importance of optimizing the choice of power-supply voltage. Two CMOS device and voltage scaling scenarios are described, one optimized for highest speed and the other trading off speed improvement for much lower power. It is shown that the low power scenario is quite close to the original constant electric-field scaling theory. CMOS technologies ranging from 0.25 μm channel length at 2.5 V down to sub-0.1 μm at 1 V are presented and power density is compared for the two scenarios. Scaling of the threshold voltage along with the power-supply voltage will lead to a substantial rise in standby power compared to active power, and some tradeoffs of performance and/or changes in design methods must be made. Key technology elements and their impact on scaling are discussed.

It is shown that a speed improvement of about 7× and over two orders of magnitude improvement in power-delay product (mW/MIPS) are expected by scaling of bulk CMOS down to the sub-0.1 μm regime as compared with today's high performance 0.6 μm devices at 5 V. However, the power density rises by a factor of 4× for the high-speed scenario. The status of the silicon-on-insulator (SOI) approach to scaled CMOS is also reviewed, showing the potential for about 3× savings in power compared to the bulk case at the same speed.

I. INTRODUCTION

The growth of microelectronics during the last 25 years, from the first successful large-scale integration (LSI) of microprocessor and memory chips to the present ultra large-scale integration (ULSI), has been truly spectacular. The key to this growth has been the drive to much smaller dimensions using the principles of scaling introduced in the early 1970's [1] which we briefly review here. The basic idea of scaling, shown in Fig. 1, is to reduce the dimensions of the MOS transistors and the wires connecting them in integrated circuits. Thus the arrangement on the

Manuscript received October 13, 1994; revised December 8, 1994.

The authors are with the IBM Semiconductor Research and Development Center (SRDC), T. J. Watson Research Center, Yorktown Heights, NY 10598 USA.

IEEE Log Number 9408754.

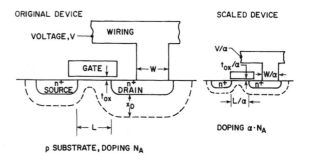

Fig. 1. Principles of constant-electric-field scaling for MOS transistors and integrated circuits.

right is scaled down in size from that on the left by reducing all dimensions by a factor α. The MOS transistor works on the principle of modifying the electric field in the silicon substrate underneath the gate in such a way as to control the flow of current between the source and drain electrodes. Scaling achieves the same electric-field patterns in the smaller transistor by reducing the applied voltage along with all the key dimensions, including the thickness, t_{ox} of the insulating oxide layer between the gate and the silicon substrate. Within the silicon substrate, the electric field patterns are preserved by increasing the impurity doping concentration of the smaller device. Taken along with the reduced applied voltage, this reduces the size of the depletion regions, identified by x_d in Fig. 1, underneath all three transistor electrodes. In general, these depletion regions must be kept separated so that the transistor can be turned off properly by the control gate [2]. The scaled-down depletion regions in the transistor on the right of Fig. 1 allows the separation L between source and drain to be reduced along with the other physical dimensions. In this simple constant-electric-field transformation, the dimension, voltage, and doping are all modified by a common factor α, as noted in the figure.

This constant-electric-field scaling gives three important results. First, the density improves by a factor α^2 due to

Reprinted from *Proceedings of the IEEE*, pp. 595–606, April 1995.

Table 1 Generalized Scaling Relationships

Physical Parameters	Constant-Electric Field Scaling Factor	Generalized Scaling Factor
Linear Dimensions	$1/\alpha$	$1/\alpha$
Electric-Field Intensity	1	ϵ
Voltage (Potential)	$1/\alpha$	ϵ/α
Impurity Concentration	α	$\epsilon\,\alpha$

the smaller wiring and device dimensions. Next, the speed, which is related to g_m/C, improves by a factor α because the capacitance (C) of the shorter wires and smaller devices is reduced by d while the transconductance (g_m) of the devices (scaled in both length and width) remains about the same. Finally, the power dissipation per circuit is reduced by a factor α^2 because of the reduced voltage and current in each device, with the important result that the power density is constant. Thus the increased number of circuits in a given chip area can be accommodated with no increase in the total power dissipation.

Although the original concept of constant-electric-field (CE) scaling is useful and valid, the idea of reducing voltage in proportion to reduced dimensions has not been popular because of reluctance to depart from standardized voltage levels. Also, scaling the threshold voltage of the devices down along with the applied voltage increases the standby leakage current, which limits how far it is practical to scale the power-supply voltage [1]. Therefore, it was useful to broaden the concept of scaling to the more generalized form shown in Table 1, where the electric-field patterns within a scaled device are still preserved, but the intensity of the electric field can be changed everywhere within the device by a multiplicative factor ϵ [3]. Thus the applied voltage, which is given by ϵ/α, can be scaled less rapidly by allowing ϵ to increase. The electric-field patterns within the device are maintained by increasing the doping impurity concentration by a factor ϵ, which preserves the size of the depletion regions, x_d, defined in Fig. 1.

Obviously, there are practical limits to generalized scaling. Increased electric field $(\epsilon > 1)$ is limited by reliability effects during long-term use such as device degradation resulting from hot-carrier mechanisms or gate-insulator failure. Ideally, the current in a scaled device or circuit increases by a factor ϵ^2 and the speed by ϵ, up to the point where it is limited by carrier velocity saturation effects and by the increased series resistance of graded junctions needed for hot-carrier reliability at the higher voltage levels. Because of such reliability considerations, the speed can actually decrease as the electric-field parameter ϵ is increased beyond a practical value, as will be discussed in the next section [4]. Another very significant limit to ϵ is the power dissipation, which increases by ϵ^2 when the speed is constant (as given by the familiar power calculation $CV^2 f$).

The application of scaling to CMOS technology up to now is illustrated in Fig. 2, which shows the typical loaded NAND delay versus L for three logic products. The gate oxide thickness is scaled nearly linearly with channel length as shown, while the voltage levels (including the threshold voltage) are reduced by approximately the square root of L. It is seen that even with the much higher electric

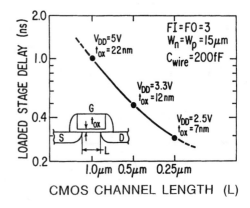

Fig. 2. Scaling of high-performance CMOS technology.

field in the shorter devices, the delay improves only about linearly with channel length due to the velocity saturation and resistance effects discussed above. While the measured performance in Fig. 2 used test circuits with a constant wiring capacitance and device widths, the quoted values are only typical of the $L = 1$ μm generation and will both be scaled in proportion to the lithography dimensions in the real chips in the scaled technologies. Therefore, since both the wiring capacitance and device widths shrink simultaneously in scaled technologies, the delay numbers shown in Fig. 2 remain an accurate measure of the technology's performance.

Although the scaling rules in Fig. 1 show the device and wiring dimensions being scaled by the same factor, they can actually be scaled by different factors. We call this "selective scaling," where the device channel length and t_{ox} are scaled by a factor α_d while the channel width and wiring width are scaled by another factor α_w. The speed is increased in correspondence with the device scaling factor, α_d, while the density improves by α_w^2 and the power per circuit decreases by a factor $\alpha_d \alpha_w / \epsilon^2$ assuming the voltage levels are scaled by ϵ/α_d. For the CMOS logic generations of Fig. 2 the typical minimum lithography dimensions used for interconnections are 1.25, 0.8, and 0.5 μm, respectively.

In the future evolution of CMOS the scaling of voltage levels will become a crucial issue. A new paradigm in the information industry is taking shape which will allow and demand much more frequent voltage reductions down to as low as 1 V or less. The main forces behind this shift are the ability to produce complex, high-performance systems on a single chip and the projected explosion in demand for portable and wireless systems with very low power budgets. Both of these will allow an unprecedented degree of freedom in choosing the supply voltages for the IC's due to self-containment of massive information processing capability. In addition, various memory and ASIC's will also embrace lower supply voltages to maintain manageable power densities. However, the increase in standby leakage current that will result from further scaling of threshold voltage poses a severe challenge, particularly for battery-powered applications.

In the following sections two different scaling scenarios for CMOS devices and logic circuits are discussed as different priorities in the "speed/power/reliability/density/cost" design space. The average longitudinal electric field along the device channel is used as a parameter in projecting the evolution of device miniaturization for a high-performance scenario and for a low-power scenario and in their comparison with the constant electric field (CE) scaling. Next, we discuss the key technology elements which enable the fabrication of the scaled technologies, followed by a discussion of scaled CMOS on silicon on insulator (SOI) which offers significant performance and power improvements over CMOS on bulk.

II. FUTURE CMOS DEVICE SCALING

As miniaturization of CMOS devices progresses without reducing the power-supply voltage as much as proposed by the CE scaling, the electric field increases substantially. This section begins with a discussion of how high the electric field can be without impacting the long term device reliability, while at the same time achieving high performance (speed) and reasonable power. The tradeoff between speed and power dissipation determines how aggressively the supply voltage should be scaled down, while the reliability constraints set an upper limit on the useful power-supply voltage. Scenarios for design point optimization are given for both high-performance and very low-power applications. Then the speed/standby-current tradeoff is addressed, dealing with the issue of nonscalability of the threshold voltage. The impact of the two scaling scenarios on power density is evaluated. Finally, a summary set of CMOS logic scaling guidelines extending for the next ten years is given and the important projected results in terms of performance, power, and density are highlighted.

A. Performance/Voltage/Reliability Tradeoff

The effect of the channel hot-carrier (CHC) limits on the choice of the optimum power-supply voltage for 0.25 μm CMOS is shown in Fig. 3. For this figure, an empirical relationship between CHC limited voltage and device series resistance, added by drain engineering, was developed from measurements on a variety of junction technologies [4]. Using this relationship with a model of the intrinsic device behavior, the circuit performance as a function of the power-supply voltage and CHC margin can be determined. A certain amount of CHC margin, about 0.5 V, depending on the conditions, is needed to allow defect screening with accelerated applied voltage and temperature (burn-in) without damaging the devices. For a constant CHC reliability margin, an optimum supply voltage exists, above which the CMOS performance actually degrades due to the excessive added device series resistance which is needed to support the high voltage operation. At power-supply voltages higher than 2.5 V, in order to limit the long-term reliability impact, drain engineering, e.g., lightly doped drain (LDD), is needed for this 0.25 μm channel length. The LDD structures allow reliable operation at higher electric fields [5], but can result in the reduction of the device performance due to increased device series

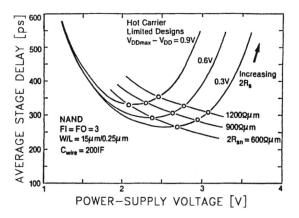

Fig. 3. Performance versus reliability tradeoff for $L = 0.25$ μm CMOS with reliability margin (V_{DD} max $-V_{DD}$) as a parameter.

resistance. Therefore the choice of the supply voltage and the appropriate device profiles represent optimization of a tradeoff between performance and reliability. We should note that the points on the constant reliability margin contours in Fig. 3 are not from the same technology, rather they represent devices with varying series resistance with similar channel lengths. Also, an oxide thickness of 7 nm is used, which gives adequate defect density and reliability at the 2.5 V design point [31]. The reduction of the oxide thickness relative to the previous 3.3 V technologies (Fig. 2) is possible due to the lower power-supply voltage.

As we scale the CMOS technology beyond 0.25 μm, in order to obtain a significant performance improvement of about 1.5× per generation, a reduced source/drain (S/D) spreading resistance is needed which requires S/D junctions with more abrupt profiles [7]. Therefore the drain electric field is increased which results in a lower optimum power-supply voltage for a given reliability margin.

In summary, the CHC degradation dictates an optimum power-supply voltage for high-speed logic technologies. Above this voltage the performance actually degrades due to excessive S/D resistance which is needed to maintain the reliability margin. This optimum voltage is 2.5 V for 0.25 μm devices and it decreases for smaller channel lengths. However, Fig. 3 also clearly shows that even lower voltage can be used with a modest decrease in speed, providing the possibility of significantly lower power consumption.

B. High-Performance and Low-Power Voltage Scaling Scenarios

Taking into account the considerations discussed above and other factors, two scenarios for scaling down power-supply voltage in future scaled CMOS logic technologies are proposed. It is instructive to describe these scenarios in the context of generalized scaling as shown in Fig. 4, which plots the electric field as a function of channel length where the measure of electric field is taken to be the applied power-supply voltage divided by the channel length of the CMOS devices. The three curves in Fig. 4 correspond to the following:

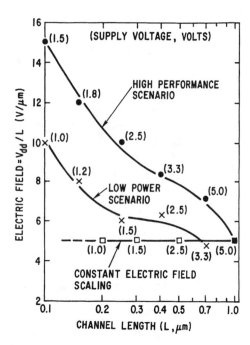

Fig. 4. A measure of the electric field, V_{DD}/L, as a function of channel length for various scaling scenarios.

The first curve gives the high performance scenario, where the power-supply voltage at each channel length is optimized for maximum speed while maintaining adequate long-term device reliability. Other factors in this optimization are the lithography tolerance at the gate level, which is assumed to be 20% of the nominal mask gate length (L_{mask}), and the threshold voltage, which must be high enough for acceptable leakage current (I_{off}) at minimum channel length. Also, the gate insulator thickness and device doping profiles are optimized for both low I_{off} and high drive current (I_{on}) without approaching significant gate oxide tunneling and GIDL current (gate-induced drain leakage caused by band-to-band tunneling in the drain region due to high electric field) problems [8], [9].

The second curve is the low-power scenario, where the power-supply voltage is reduced as compared to the high-performance case at the same channel length. The goal is to lower the power dissipation per device and therefore maintain a power density in the scaled technologies which is similar to that of the starting 1 μm CMOS technology by adhering as closely as practical to CE scaling. At the same time, the speed relative to the high-performance case should not degrade more than an arbitrary 1.5×. The device design modifications and the threshold voltage choice in this case should also provide acceptable leakage current. This will be discussed in more detail in the next section.

The third curve which is given for comparison is the constant electric field (CE) scaling as proposed in [1]. The constant electric field of 5 V/μm is chosen as the starting point, since the CMOS circuits operating at 5 V using 1 μm channel lengths are fairly well optimized with respect to performance versus power dissipation. The electric field strength across the channels are sufficient to bring the operation into the saturation-velocity limited regime (for

example, the average field of 5 V/μm across a single p-channel device with a mobility of 200 cm^2/Vs projects to a velocity of 10^7 cm/s). Also, any increase in vertical electric field is not very productive because parasitic series source and drain resistances would become very significant compared to the inversion layer resistances.

In the high-performance scenario of Fig. 4, the average electric field is twice as high for 0.25 μm CMOS with a 2.5 V power supply compared with 1 μm CMOS at 5 V. However, this higher electric field can be tolerated without undue reliability impact as shown in the previous section. It appears that reducing the applied voltage below the barrier height between the silicon channel and the silicon-dioxide gate insulator is helpful, even though some carriers still attain high enough energy to pass through that barrier [10]. The average electric field increases further as the channel lengths are scaled down to 0.15 μm and 0.1 μm with the power-supply voltages of 1.8 V and 1.5 V, respectively. For the 0.15 μm CMOS technology, it has been demonstrated that the CHC degradation of the nFET is maintained within an acceptable limit for 10 years operation for the minimum channel length of 0.1 μm at 1.8 V [11]. At the same time a performance improvement of 1.5× over the 0.25 μm CMOS technology has been measured. The studies of CHC effects in the 0.1 μm regime indicate that, despite the low power-supply voltage, the degradation caused by CHC cannot be ignored [12], [13]. For the high-performance 0.1 μm CMOS a power-supply voltage of 1.5 V is chosen, which achieves 2× performance improvement over the 0.25 μm CMOS [14]. This performance is limited by the threshold voltage, V_t, which is not fully scaled in order to maintain acceptable I_{off}. The gate oxide thickness for the 0.15 μm and 0.1 μm CMOS technologies are 5 nm and 3.5 nm, respectively. It has been shown that the tunneling current density in the 3.5 nm gate oxide for 1.5 V operation is 10^{13} A/μm^2, which is insignificant for most ULSI circuits and also that the gate-induced drain leakage (GIDL) current is negligible relative to the off-current at minimum channel length, and therefore does not pose any significant limitation on the device design [15].

C. Performance/Power Tradeoff and Nonscalability of the Threshold Voltage

The CMOS circuit power dissipation can be expressed as $P = KCV^2f + I_{off}V$, where K is the switching factor, C is the total load capacitance, and f is the clock frequency. The first term is the active and the second is the standby power dissipation. As CMOS is scaled to small dimensions and the power-supply voltage is reduced to maintain reliability and reasonable active power dissipation, the threshold voltage (V_t) needs to be scaled down at the same rate as the power-supply voltage in order to achieve the desired circuit switching speed. However, lowering the threshold voltage will cause substantial increases in I_{off} and the standby power [16].

The limitation which drives the nonscalability of the V_t is the turnoff behavior of the MOSFET, characterized by the inverse subthreshold slope (S) which is invariant with

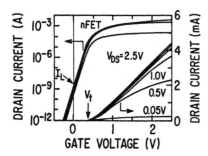

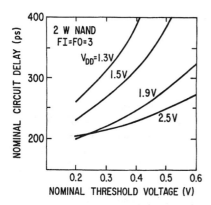

Fig. 5. 0.25 μm nFET device characteristics ($W = 10$ μm). Subthreshold slope = 78 mV/dec.

Fig. 6. 0.25 μm CMOS delay versus threshold voltage and V_{DD}.

scaling [1]. In typical scaled devices S is about 80 mV/*dec* at room temperature, as shown in Fig. 5 for a 0.25 μm nFET. At 85°C the S is about 100 mV/*dec*. Therefore, in general, for every 100 mV reduction in V_t the standby current will be increased by one order of magnitude. This exponential growth of the standby current tends to limit the threshold voltage reduction to about 0.3 V for room temperature operation of conventional CMOS circuits. It should be noted that in this paper the threshold voltage generally refers to the nominal value, extrapolated from the linear drain current versus gate voltage curve at low drain bias as shown in Fig. 5.

The effect of the threshold voltage on performance for various power-supply voltages in 0.25 μm CMOS is shown in the simulation results of Fig. 6 [17], [18]. It can be seen that as the power-supply voltage is reduced, the performance degrades significantly at higher threshold voltages and also becomes more sensitive to tolerances in V_t. Therefore, from the performance point of view, we need to reduce the threshold voltage and also improve the threshold voltage control (reduce V_t variation due to various process tolerances) as the power-supply is scaled down. It should be noted that if we scale the technology below 0.25 μm, the curves in Fig. 6 still apply if the voltages are scaled down with dimensions and the delay axis is multiplied by a constant improvement factor. The optimum V_t can be chosen for different applications, based on the tradeoff requirements between the speed and standby power.

Another critical factor in this tradeoff is the short channel effect (SCE) of the devices. SCE is the lowering of

the threshold voltage at short channel lengths of a given technology relative to the threshold voltage at nominal channel length. Since the standby power is dominated by the shortest channel devices, a high SCE will force a higher nominal device V_t which will result in lower performance (Fig. 6). To some degree SCE reduces naturally in scaled devices due to reduced gate insulator thickness and increased doping levels. Other factors such as nonscalability of the band-bending and built-in junction potentials make it worse [3]. Reduction of the SCE constitutes a major part of the technology optimization and development for scaled CMOS generations. Some of the techniques to reduce SCE are summarized in the next section.

To deal with the problem of nonscalability of the threshold voltage, one possibility is to implement multiple threshold voltages on a chip, at a slightly higher processing cost. Today, dual thresholds are commonly used in DRAM chips where it is possible to raise the V_t of the array devices with a fixed body bias. For logic, low V_t devices can be used where needed for circuit functionality or higher speed, with high V_t devices for the rest of the chip. This can allow the threshold voltage of the low V_t devices to be scaled below 0.3 V. It has also been shown that by active adjustment of the substrate or well bias, including a feedback loop, one can reduce the V_t variation significantly [19]. Although these techniques do not address the fundamental problem of nonscalability of the inverse subthreshold slope, nevertheless they offer possibilities to reduce the threshold voltage without suffering from overwhelming leakage current and standby power. Unfortunately most of the system and architectural power management techniques do not address the issue of the standby power (they address the active power), unless the power management system completely shuts down all power to the chip. This is not usually practical due to the relatively long response time associated with the power management system compared to the system cycle time.

The inverse subthreshold slope of the FET's can be significantly improved (about a factor of 4), by operating the devices at 77 K, liquid nitrogen temperature [20], [21]. In this case the threshold voltages can be scaled with the power-supply voltage without leakage problems, e.g., the V_t can be 0.2 V for a 1.0 V power-supply voltage. The lower V_t along with improved carrier mobility and reduced interconnect resistance, results in significant improvement in performance and power-delay product. However, the application of 77 K CMOS will probably be limited to the highest performance systems. CMOS on silicon on insulator (SOI) offers somewhat improved subthreshold behavior as well as better performance and power-delay product [22]. This will be discussed further in the following sections.

As shown by the generalized scaling theory, the active power density will be considerably higher for the scaled devices in the high-performance scenario (Fig. 4) because of the increased electric field. The relative active power densities for several scaled CMOS technologies are shown in Fig. 7. The relative density for each technology generation is shown in parentheses. This relative density is the

Table 2 CMOS Scaling Guidelines for the Next 10 Years

	Late 1980's	1992	1995	1998	2001	2004
Supply Voltage (V)						
High Performance	5	5/3.3	3.3/2.5	2.5/1.8	1.5	1.2
Low Power	—	3.3/2.5	2.5/1.5	1.5/1.2	1.0	1.0
Lithography Resolution (μm)						
General	1.25	0.8	0.5	0.35	0.25	0.18
Gate Level for Short L	—	0.6	0.35	0.25	0.18	0.13
Channel Length (μm)	0.9	0.6/0.45	0.35/0.25	0.2/0.15	0.1	0.07
Gate Insulator Thickness (nm)	23	15/12	9/7	6/5	3.5	2.5
Relative Density	1.0	2.5	6.3	12.8	25	48
Relative Speed						
High Performance	1.0	1.4/2.0	2.7/3.4	4.2/5.1	7.2	9.6
Low Power	—	1.0/1.6	2.0/2.4	3.2/3.5	4.5	7.2
Relative Power/Function						
High Performance	1.0	0.9/0.55	0.47/0.34	0.29/0.18	0.12	0.077
Low Power	—	0.27/0.25	0.20/0.09	0.08/0.056	0.036	0.041
Relative Power/Unit Area						
High Performance	1.0	2.25/1.38	3.0/2.1	3.7/2.34	3.12	3.70
Low Power	—	0.7/0.63	1.25/0.6	1.02/0.72	0.90	1.97

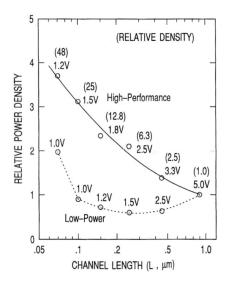

Fig. 7. Relative active power density in CMOS scaled as in Table 2. Relative density is in parentheses.

reciprocal of the square of the relative lithography ground rules for each generation (see Table 2). The upper curve corresponds to the performance driven scenario. In this case, we can see that even with reduced power-supply voltage, the power density increases significantly due to a large increase in the number of devices per unit area. However, one can choose the low power scenario, where the supply voltage is reduced more aggressively, at the expense of the performance (lower curve in Fig. 7). This scenario will be quite attractive for applications where low power and improved power-delay product is the highest priority. As an example, if we compare the 1.5 V operation versus the 2.5 V in 0.25 μm CMOS, there will be over 3.5× reduction in power (Fig. 7) with only about 30% performance degradation (Fig. 6, $V_t \simeq 0.3$ V).

In the low-power scenario, the power density of the scaled technologies goes down relative to the 1 μm CMOS technology until we reach 1.5 V (Fig. 7). This is due to the fact that in that regime the electric field does not go up substantially and is quite close to the constant electric field scaling as shown in Fig. 4, while the interconnection dimensions (which determine the density) are being scaled down less than the channel lengths. However, beyond the $L = 0.25\ \mu$m at 1.5 V point, the relative power density rises (Fig. 7) as we depart from the CE scaling (Fig. 4) because the lower limit on V_t imposes a lower limit of about 1 V for the power-supply voltage without significantly impacting speed.

A key concern in the low-power scenario is the availability of the complete chip set to make up the systems at reduced supply voltage. However, most of the problems can be overcome by various techniques to mix and match different supply voltages on the board or on the chip. Also, the need for a total low power system solution at low cost should drive the semiconductor industry to even a faster pace for offering various memory and ASIC products at reduced supply voltages.

Another key concern is the susceptibility to soft errors due to alpha particles, which is expected to increase due to reduced voltage and capacitance. This could require improved structures such as SOI to reduce the volume of junction area exposed to alpha particle hits.

D. CMOS Scaling Guideline for the Next Ten Years

The above scaling optimizations and limitations are summarized in Table 2 as a guideline for CMOS scaling over the next ten years for logic applications such as microprocessors. The key design parameters are the power-supply voltage, the lithography resolution, the channel length, and the gate insulator thickness. The more aggressively scaled device and power-supply voltages represent the high-performance and low-power scenarios discussed above. In the columns through the 90's less aggressively scaled device design points are included, representing lower cost options with correspondingly lower speed and higher power dissipation. It is assumed that the wiring pitch scales with the general lithography resolution and the relative density is the reciprocal of the square of the relative lithography resolution. The gate level lithography requires higher resolution than the general lithography in these applications, which is in-line with the "selective scaling" that was discussed earlier. However, the critical dimension (CD) requirements of the gate level will be satisfied with the lithography tools that are developed for the density driven products and applications, i.e., DRAM, at any given generation. It is assumed that DRAM products will continue the $4\times$ bits/chip improvement every three years. If that trend is slowed down due to productivity (cell size) and/or economics reasons, it is likely that the pace of the logic scaling proposed in this guideline will also be slowed down.

The effective channel length and the gate oxide thickness in the high-performance scenario are scaled in concert with the power-supply voltage in order to achieve optimum speed, while maintaining acceptable leakage current and reliability margins. In the low-power case, the power-supply voltage is chosen to be lower in order to reduce the power at the expense of some loss of performance. The device dimensions in the low-power scenario are assumed to be similar to the high-performance case. The reason for not using relaxed channel lengths for the low-power case is to minimize the performance degradation. This allows the speed of the low-power case in one generation to be about the same as the speed of the high-performance scenario of the previous generation, with greatly reduced power consumption.

The leakage current is mainly dominated by the devices with the shortest effective channel within the channel length distribution. Therefore, a critical parameter is the channel length tolerance which is assumed to be about 30% of the nominal channel length, leading to a gate lithography tolerance requirement of about 20% of the gate lithography resolution. The gate lithography tolerance is a very key parameter in high-performance CMOS applications since it directly affects the performance at a given leakage current requirement. This parameter is usually the most difficult and expensive feature to control in semiconductor manufacturing. The relative performance of the technology generations are arrived at through various experimental data points [4], [11], [14], [15] and projections based on the scaling principles. Many of the assumptions used in the scaling theory become less accurate at very small dimensions [3]. However, there are compensating effects which tend to make the scaling projections still reasonably accurate. One positive effect is the onset of velocity overshoot behavior which tends to increase the current in very small nFETs [23].

In general it is safe to consider that the devices will meet the long-term CHC and gate oxide reliability requirements for 10 years use even down to sub-0.1 μm channel lengths, because of the much lower operating voltage. Also, the reduction of the device series resistance by sharpening the S/D profiles will not keep pace with the power-supply voltage reduction. This results in lower electric field at the drain region and lower CHC. However, meeting the increasingly demanding yield and defect density requirements of the ultrathin gate oxides and ultrashort devices are some of the key future manufacturing challenges. It

Table 3 Power/Performance/Density Improvement for a RISC Processor with Scaling

	66 MHz	100 MHz	150 MHz
0.65 μm Lith. (120 mm^2)			
$L = 0.45$ μm, 3.6V	7.5 W		
0.5 μm Lith. (74 mm^2)			
$L = 0.25$ μm, 2.5 V		3.4 W	
$L = 0.25$ μm, 1.5 V	0.8 W		
0.35 μm Lith. (36 mm^2)			
$L = 0.15$ μm, 1.8 V			1.8 W
$L = 0.15$ μm, 1.2 V		0.5 W	

appears that fundamental limitations such as statistical dopant fluctuations [24], tunneling through the gate oxide [15], GIDL current [8], and interconnect RC delays [25] do not pose serious barriers for the scaled devices down to 0.05 μm channel length (minimum channel length, year 2004) with 2.5 nm gate oxide, operating at 1 V power-supply voltage.

For the scaled CMOS in the year 2004, a performance gain of about $7\times$, a density improvement of $20\times$, and a power/function reduction of $12\times$ are projected (Table 2, high-performance case) compared with the present 0.6 μm technology at 5 V. However, even in the low-power scenario in the year 2004, the active power density will be higher than today's 5 V technology. The reason is the lower limit of 1 V that we imposed on the power-supply voltage due to nonscalability of the threshold voltage. It has been shown that by operating the 0.1 μm CMOS devices at power-supply voltages lower than 1 V the active power dissipation can be much reduced, but at the cost of increasing the standby power and significantly reducing the performance [26]. In the year 2004, the high-performance and the low-power scenarios exhibit about $80\times$ and $110\times$ improvement in the power-delay product compared with today's 5 V technology, respectively.

An example of the effectiveness of the CMOS scaling in simultaneously improving performance, power and density is shown in Table 3. In this table the operating frequency, chip size, power-supply voltage, and power dissipation of an existing high performance RISC processor and projections into more advanced scaled CMOS are presented. The 66 MHz chip with 120 mm^2 die size at 0.65 μm lithography, consumes about 7 W. The 100 MHz chip, consuming only 3.4 W, is fabricated in selectively scaled 0.25 μm CMOS technology [27] with 74 mm^2 die size, operating at 2.5 V with 3.3 V or 5 V I/O. It is estimated that by operating this chip at 1.5 V the power dissipation will be reduced to 0.8 W at 66 MHz. These numbers are in good agreement with the scaling guidelines in Table 2. It is further estimated that if this processor is fabricated in 0.35 μm lithography with 0.15 μm effective channel length, operating at 1.8 V, the performance will be improved by 50%, the die area reduced by more than $2\times$, and the power dissipation reduced by about $1.8\times$, simultaneously. Dropping the internal supply voltage to 1.2 V will further reduce the power dissipation by another $3\times$, while the performance will degrade by 50%.

The small size of the RISC processor if fabricated in the 0.35 μm lithography generation projected for 1998

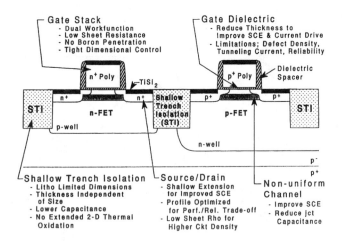

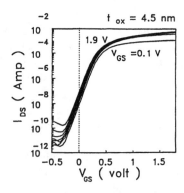

Fig. 9. Turn-on characteristics, drain current versus gate voltage with drain voltage as a parameter (0.3 V steps), for an $L = 0.1$ μm nFET in a 0.15 μm CMOS process ($W = 10$ μm).

Fig. 8. Key technology elements for deep-submicron CMOS.

raises the issue of how the density improvement, projected to increase an additional 4× by 2004, will actually be utilized. Up to now logic chips have grown in size with each successive generation. If this trend is to continue, the boundaries of integration will have to be widened to take in more (or even all) of a system function onto a chip. Some of that function may require other devices and structures to provide analog, dense memory, or nonvolatile memory for example. This is a very important issue, but not one which can be further addressed here. Rather we hope the guideline for CMOS scaling given here will help determine the proper course for the next integration phase of the microelectronic era.

E. Key Technology Elements and Extensions

In order to make the above CMOS scaling a reality, several challenging technology issues must be resolved. It is assumed that the lithography and etching capability for very small dimensions will continue to be driven by the requirements for DRAM as discussed earlier. Therefore, we briefly review here the critical requirements for device and interconnection process developments.

1) Device Technology: The key technology elements for the deep submicron CMOS, excluding the interconnect, are depicted in Fig. 8. They are divided into four modules: Gate stack and gate dielectric, source/drain, isolation, and channel profile. The main requirements for each group are summarized in Fig. 8 [31], [28] and will not be repeated here. Some of the key features are performance driven, e.g., more abrupt source drain (S/D) profiles to lower device series resistance, thinner gate dielectric to increase drive current and improve SCE, and silicided gate to reduce the RC delay for wide devices. It is noteworthy that the reduction of the gate dielectric thickness mainly depends on defect density requirements rather than tunneling current. It has been shown that the defect density is a function of the electric field and that it goes up at higher operating fields [9], [29]. It can be projected that in order to maintain defect densities below 1/cm^2, the oxide field should be below 5 MV/cm in the oxide thickness range below 5 nm [9]. This condition is satisfied in the scaled devices in Table 2.

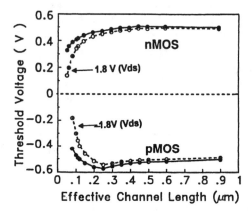

Fig. 10. Threshold voltage roll-off in high-performance 0.15 μm CMOS versus channel length.

Some other technology modules are mainly density driven, e.g., shallow trench isolation (STI), and silicided S/D junctions. STI results in denser circuits, offering lithography limited isolation widths by eliminating bird's beak and offering improved planarity for the critical gate definition step [30]. It should be mentioned that STI also improves circuit performance due to minimizing the perimeter junction capacitance by offsetting the field doping away from the junction edge. The silicided S/D junctions improve the density by reducing the number of contacts required to the S/D area, and therefore freeing up wiring tracks. The main silicidation issues for the reduced dimensions of scaled CMOS are discussed elsewhere [31], [32].

Reducing SCE is of particular importance in scaled CMOS technologies with reduced power-supply voltage, as discussed previously. It has been demonstrated that by using nonuniform channel doping profiles and source/drain extensions with reverse doping preamorphization [33], excellent SCE can be achieved for both nFET and pFET devices as shown in Fig. 9 and Fig. 10 [34]. Less than 200 mV threshold voltage roll-off is achieved down to 0.1 μm at high drain voltage (Fig. 10). Using these devices, high performance CMOS circuits, exhibiting delay per stage of 35 ps at 1.8 V at $L = 0.15$ μm for unloaded, and 200 ps loaded (FI = FO = 3, C_{wire} = 240 fF) have been demonstrated as shown in Fig. 11 [34].

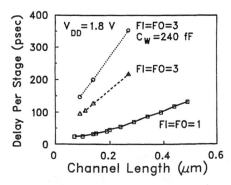

Fig. 11. Measured delay versus channel length for the high-performance $L = 0.15 \ \mu$m CMOS technology ($W_n = W_p/2 = 15 \ \mu$m).

2) Interconnect Issues for High-Performance Applications: With the scaled devices discussed above, CMOS microprocessor speeds should reach in excess of 500 MHz (2 ns cycle time). As the high-end microprocessors get implemented in CMOS technology [35], one needs to pay special attention to the interconnect RC delays, so that the chip performance is not gated by the wiring delay [1], [25]. It has been discussed that due to the nonscalability of the interconnect delays caused by the finite wire resistance, a hierarchy of interconnect levels should be used. First, there are the "short" wires (less than a few hundred microns long) which serve the vast majority of the chip interconnects, and are responsible for the chip wirability by providing a sufficient number of wiring channels. These wires should scale with the lithography as discussed earlier in the scaling guideline. Second, there is a need for "long" wires, where density is secondary to delay considerations [25]. These interconnections were a part of the package in earlier days, but with increased level of integration they are now on the chip as we get closer to the "high-performance systems on a chip" paradigm. These wires run between distant parts of the chip, connecting various functional blocks. The signal propagation delay on the "long" wires should be maintained at a small fraction of the processor cycle time. Therefore, these wires cannot shrink with the rest of the chip dimensions.

This wiring hierarchy is depicted in Fig. 12 [25]. In the top two levels which will be used as long wires, the capacitance per unit length stays constant, while resistance decreases proportionally with the increase in wire cross section. These wires are referred to as "fat" wires by Sai-Halasz [25]. One consequence of having low RC wires is that we will observe transmission line behavior not only on the package, but on the chips as well. For illustration, on a 15 mm long wire the signal flight time cannot be less than 105 ps, which is longer than the switching time of drivers in the scaled technologies below 0.25 μm proposed in the guideline. When the input of a wire is driven with a signal faster than the travel time down that line, we reach the regime where the nonideal transmission line characteristics of the present-day interconnect schemes must be taken into account.

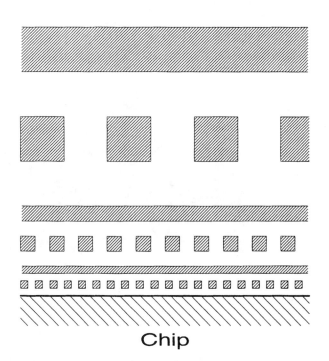

Fig. 12. Example of wiring scheme needed by future high-performance processors to minimize delays due to wire resistance. Three x-y wiring planes are shown in cross section.

Due to the stringent RC requirements for the "fat" wires, it is unlikely that the large width and height of these wires can be significantly reduced by using new conductor and insulator materials with lower resistivity and dielectric constant, respectively. This does not diminish the importance of exploring new interconnect materials with superior properties relative to today's Al/SiO$_2$ system. It should be mentioned that in order to minimize the negative impact of the "fat" wires on the chip wiring density, they should not replace the "short" wiring levels; rather, they should be added to the already existing interconnect scheme. Due to this and also the complicated processing associated with the etch and fill of high aspect ratio via holes, the "fat" wires can potentially increase the processing cost significantly (performance/cost tradeoff). The key elements of the interconnect technology are not discussed here, and can be found elsewhere [36], [37].

Another important function of the interconnect technology is distribution of power to the processing elements. As the chip's complexity grows and the speed increases, supplying peak currents from outside the chip (or even across the chip) becomes very difficult because of inductance effects. Therefore, it may be necessary to provide large amounts of decoupling capacitance distributed throughout the chip to minimize power-supply noise. This is another example of integrating a function which was previously a part of the package.

3) Silicon On Insulator (SOI): Additional significant improvements in power and/or performance can be achieved by implementing scaled CMOS on silicon on insulator (SOI). A schematic cross section of CMOS on SOI is shown in Fig. 13 [38]. The performance improvement of SOI compared to bulk CMOS is mainly due to the reduction

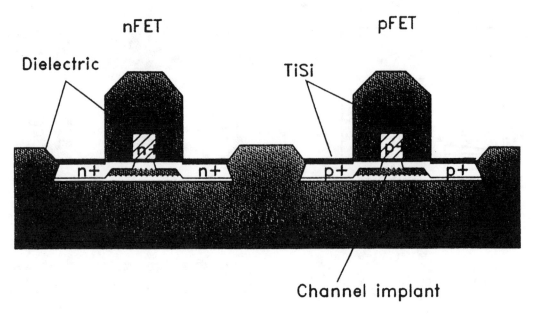

Fig. 13. 0.1 μm CMOS on SOI schematic cross section.

of parasitic capacitances and body effect. Also, in partially depleted device designs, the floating body effect can give rise to a sharper subthreshold slope ($S < 60$ mV/dec) at high drain bias, which effectively reduces the threshold voltage and can actually improve the performance at a given standby leakage current [39]. A major issue here is reduction of the breakdown voltage due to the bipolar action, particularly in the nFET. To reduce the bipolar gain in order to maintain acceptably high breakdown voltage both for normal operation and for burn-in, S/D extensions with halo can be employed [38]. CMOS on SOI can extend the bulk CMOS performance limits as well as improving the performance at a given fab generation. Performance improvements in the range of $1.5\times$–$2.5\times$ have been reported for various CMOS on SOI circuits, relative to CMOS on bulk devices with the same lithography [22], [40]. In addition, CMOS on SOI offers significant reduction in the soft error rate, latch-up elimination, and simpler isolation which results in reduced wafer fabrication steps.

The main challenges and shortcomings of SOI are the availability of low-cost wafers with low defect density, floating body effects on the device and circuit operation, and heat dissipation through the buried oxide. Progress in starting SOI wafers has been made rapidly in the last few years [41]. The process which uses implanted oxygen to form an SiO_2 layer beneath a thin silicon layer (SIMOX) is emerging as a viable process after years of development. The other major approach, which uses bonding of two oxidized wafers and etchback of all but a thin layer of one wafer (BESOI), has also shown capability to make thin uniform layers in the required range of 0.1–0.2 μm thick. Defect density has also been greatly reduced over the last few years and is approaching the same level as that of bulk substrates.

The impact of the SOI shortcomings becomes significantly less as the supply voltage is reduced below 2.5 V, while the impact of the SOI benefits becomes stronger [42]. Therefore, SOI is an excellent candidate for low power/low voltage applications. The I-V characteristics of the nondepleted 0.1 μm devices shown in Fig. 14 are reasonably free of kink effect in the range where they will be operated [39]. By using CMOS on SOI, more than $3\times$ reduction in power-per-stage can be achieved at the same performance and channel length, compared with CMOS on bulk as shown in Fig. 15 for L = 0.15 μm unloaded ring oscillators [42]. Also an SRAM access time of 3.5 ns has been demonstrated at 1.0 V in an experimental 512 Kb circuit [39].

III. SUMMARY AND CONCLUSIONS

Scaled CMOS technology is ideally suited as the engine for both tomorrow's high performance systems and for the coming low power revolution. Within the next decade, there will be an explosive growth of capability in silicon chips, made possible by CMOS scaling. This growth will affect all aspects of human life as the integration of high performance systems on a single chip (as powerful as today's supercomputers) becomes a reality. In this article, a guideline for CMOS scaling over the next 10 years has been presented, with emphasis on the optimization of high-performance and low-power scenarios aimed at logic applications such as microprocessors. After considering key device and technology bottlenecks and various nonscalable elements, it is projected that the performance, density, and active power dissipation will all improve dramatically as scaling proceeds along the path presented in the guideline.

In the year 2004, using sub-0.1 μm devices, speed improvement of about $7\times$, density improvement of about $20\times$, and power/function reduction of more than $10\times$ are expected relative to today's 5 V technology at 0.6 μm. The power-delay product will improve by $80\times$ and $110\times$ for high-performance and low-power scenarios, respectively.

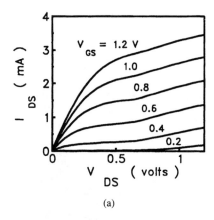

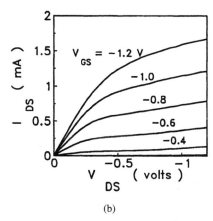

(a)

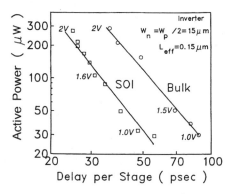

(b)

Fig. 14. Device characteristics of 0.1 μm CMOS on SOI ($W = 10 \mu$m). (a) nFET, (b) pFET.

Fig. 15. Measured power versus delay for $L = 0.15 \mu$m SOI and bulk CMOS with varying voltage.

Reduction of the power-supply voltage is a key element in future CMOS scaling. The increasing demands for reduced power dissipation, as well as the eroding reliability margins, will result in much more frequent power-supply voltage changes in the industry than in the past "constant voltage scaling" era (5 V). Even with the reduction of the voltage, the active power density will grow due to the high rate of density and performance improvements. Standby power will also grow significantly as the threshold voltage is reduced. New design techniques and power management aimed at reduction of active power and possibly standby power are needed [43]. The development of standards and

practices which allow the coexistence of chips with multiple supply voltages at the system level, without compromising speed or cost (e.g., low voltage swing I/O standards) are of high priority. However, it is likely that the great increase in density will cause more and more of the system to be swept up and integrated into a single chip, certainly for the most pervasive, low-cost, and portable applications. Improvements in the interconnection technology for long cross chip wires are necessary to keep up with the advances in device speed.

Most of the projected progress in device technology has already been demonstrated in the laboratory. The work that has been done on SOI is also very promising although it has not yet been readied for products. There are, of course, many concerns about manufacturability for the path shown in the roadmap, including the vital issue of lithography tools. In summary, many practical challenges remain to be faced in the upcoming decade, but the benefits are sure to be very rewarding.

ACKNOWLEDGMENT

The authors wish to acknowledge M. Hakey, T. H. Ning, and M. R. Polcari for their contributions to this work and many very helpful discussions.

REFERENCES

[1] R. H. Dennard, F. H. Gaensslen, H. N. Yu, V. L. Rideout, E. Bassous, and A. R. LeBlanc, "Design of ion-implanted MOSFET's with very small physical dimensions," *IEEE J. Solid-State Circ.*, vol. SC-9, pp. 256–268, May 1974.
[2] B. Hoeneisen and C. A. Mead, "Fundamental limitations in microelctronics -I. MOS technology," *Solid State Electron.*, vol. 15, pp. 819–829, 1972.
[3] G. Baccarani, M. R. Wordeman, and R. H. Dennard, "Generalized scaling theory and its application to a 1/4 micron MOSFET design," *IEEE Trans. Electron Devices*, vol. ED-31, pp. 452–462, Apr. 1984.
[4] B. Davari et al., "A high performance 0.25 μm CMOS technology," *IEDM Tech. Dig.*, pp. 56–59, 1988.
[5] S. Ogura, P. J. Tsang, W. W. Walker, D. L. Critchlow, and J. F. Shepard, "Design and characterization of the lightly doped drain (LDD) insulated gate field effect transistor," *IEEE Trans. Electron Devices*, vol. ED-27, pp. 1359–1367, Aug. 1980.
[6] B. Davari et al., "A high performance 0.25 μm CMOS technology: Part II-Technology," *IEEE Trans. Electron Devices*, vol. 39, pp. 967–975, Apr. 1992.
[7] K. K. Ng and W. T. Lynch, "The impact of intrinsic series resistance on MOSFET scaling," *IEEE Trans. Electron Devices*, vol. ED-34, pp. 503–511, Mar. 1987.
[8] T. Y. Chan, J. Chen, P. K. Ko, and C. Hu, "The impact of gate-induced drain leakage current on MOSFET scaling," *IEDM Tech. Dig.*, pp. 718–721, 1987.
[9] R. Moazzami and C. Hu, "Projecting gate oxide reliability and optimizing reliability screens," *IEEE Trans. Electron Devices*, vol. 37, pp. 1643–1650, July 1990.
[10] J. E. Chung, M. C. Jeng, J. E. Moon, P. K. Ko, and C. Hu, "Low-voltage hot-electron currents and degradation in deep-submicron MOSFET's," *IEEE Trans. Electron Devices*, vol. 37, pp. 1651–1657, July 1990.
[11] G. G. Shahidi et al., "A high performance 0.15 μm CMOS," *1993 Symp. on VLSI Technology*, Kyoto, Japan, pp. 93–94.
[12] M. Dutoit et al., "Experimental study of electron heating in 0.1 μm nMOSFETs," *1993 Symp. on VLSI Tech.*, Kyoto, Japan, pp. 35–36.
[13] T. Mizuno et al., "Hot-carrier effects in 0.1 μm gate length CMOS devices," *IEDM Tech. Dig.*, pp. 695–698, 1992.
[14] Y. Taur et al., "High performance 0.1 μm CMOS devices with 1.5 V power supply," *IEDM Tech. Dig.*, pp. 127–130, 1993.

[15] Y. Mii *et al.*, "High performance 0.1 μm nMOSFET's with 10 ps/stage delay (85 K) at 1.5 V power supply," *1993 Symp. on VLSI Tech.*, Kyoto, Japan, pp. 91–92.

[16] E. J. Nowak, "Ultimate CMOS ULSI performance," *IEDM Tech. Dig.*, pp. 115–118, 1993.

[17] R. H. Dennard, "Power-supply considerations for future scaled CMOS systems," *1989 Symp. on VLSI Tech., Syst. and Applications*, Taipei, Taiwan, pp. 188–192.

[18] B. Davari, R. H. Dennard, and G. G. Shahidi, "CMOS technology for low voltage/low power applications," *Proc. IEEE 1994 CICC*, pp. 3–10.

[19] T. Kobayashi and T. Sakurai, "Self-adjusting threshold-voltage scheme (SATS) for low-voltage high-speed operation," *Proc. IEEE 1994 CICC*, pp. 271–274.

[20] F. H. Gaensslen, V. L. Rideout, E. J. Walker, and J. J. Walker, "Very small MOSFET's for low-temperature operation," *IEEE Trans. Electron Devices*, vol. ED-24, pp. 218–229, Mar. 1977.

[21] J. Y.-C. Sun, Y. Taur, R. H. Dennard, and S. P. Klepner, "Submicrometer-channel CMOS for low-temperature operation," *IEEE Trans. Electron Devices*, vol. ED-34, pp. 19–27, Jan. 1987.

[22] G. G. Shahidi *et al.*, "Fabrication of CMOS on ultrathin SOI otained by epitaxial lateral overgrowth and chemical-mechanical polishing," *IEDM Tech. Dig.*, pp. 587–590, 1990.

[23] G. A. Sai-Halasz, M. R. Wordeman, D. P. Kern, S. Rishton, and E. Ganin, "High transconductance and velocity overshoot in nMOS devices at the 0.1 μm gate-length level," *IEEE Trans. Electron Device Lett.*, vol. 9, pp. 464–466, Sept. 1988.

[24] H. S. Wong and Y. Taur, "Three dimensional atomistic simulation of discrete random dopant distribution effects in sub-0.1 μm MOSFET's," *IEDM Tech. Dig.*, pp. 705–708, 1993.

[25] G. A. Sai-Halasz, "Performance trends in high-end processors," *IEEE Proc.*, vol. 83, pp. 20–36, Jan. 1995.

[26] Y. Mii *et al.*, "An ultra-low power 0.1 μm CMOS," *1994 Symp. on VLSI Tech.*, Hawaii, pp. 9–10.

[27] C. Koburger *et al.*, "Simple, fast, 2.5 V CMOS logic with 0.25 μm channel lengths and damascene interconnect," *1994 Symp. on VLSI Tech.*, Hawaii, pp. 85–86.

[28] B. Davari, "Low voltage/low power device technologies," *IEDM Short Course*, 1993.

[29] C. Hu, "Future CMOS scaling and reliability," *Proc. IEEE*, vol. 81, p. 682, May 1993.

[30] B. Davari *et al.*, "A new planarization technique, using a combination of RIE and chemical mechanical polish (CMP)," *IEDM Tech. Dig.*, pp. 61–64, 1989.

[31] ——, "Shallow junctions, silicide requirements and process technologies for sub-0.5 μm CMOS," *Proc. 22 ESSDERC*, p. 649, 1992.

[32] J. B. Lasky, J. S. Nakos, O. J. Cain, and P. J. Geiss, "Comparison of transformation to low-resistivity phase and agglomeration of $TiSi_2$ and $CoSi_2$," *IEEE Trans. Electron Devices*, vol. ED-38, pp. 262–269, Feb. 1991.

[33] B. Davari, E. Ganin, D. Harame, and G. A. Sai-Halasz, "A new pre-amorphization technique for very shallow p^+/n junctions," *1989 Symp. on VLSI Tech.*, Kyoto, Japan, pp. 27–28.

[34] G. G. Shahidi *et al.*, "Indium channel implant for improved short-channel behavior of submicrometer nMOSFET's," *IEEE Electron Device Lett.*, vol. 14, pp. 409–411, Aug. 1993.

[35] A. Masaki, "Possibilities of CMOS mainframe and its impact on technology R&D," *1991 Symp. on VLSI Technology*, Oiso, Japan, pp. 1–4.

[36] C. Kaanta *et al.*, "Submicron wiring technology with tungsten and planarization," *IEDM Tech. Dig.*, pp. 209–212, 1987.

[37] F. White *et al.*, "Damascene stud local interconnect in CMOS technology," *IEDM Tech. Dig.*, pp. 301–304, 1992.

[38] G. G. Shahidi *et al.*, "A room temperature 0.1 μm CMOS on SOI," *1993 Symp. on VLSI Tech.*, Kyoto, Japan, pp. 27–28.

[39] ——, "SOI For a 1-volt CMOS technology and application to a 512 Kb SRAM with 3.5 ns access time," *IEDM Tech. Dig.*, pp. 813–816, 1993.

[40] A. Kamgar *et al.*, "Ultra-high speed CMOS circuits in thin SIMOX films," *IEDM Tech. Dig.*, pp. 829–832, 1989.

[41] G. W. Cullen, M. T. Duffy, and A. C. Ipri, "Thirty years of silicon on insulators: do trends emerge?," in *Silicon on Insulator Tech. and Devices, Proc. ECS*, vol. 94-11, pp. 5–15, 1994.

[42] G. G. Shahidi, T. H. Ning, R. H. Dennard, and B. Davari, "SOI for low-voltage and high-speed CMOS," *Extended abstracts, 1994 Int. Conf. on Solid-State Devices and Materials*, Yokahama, Japan, pp. 265–267.

[43] A. Chandrakasan, S. Sheng, and R. W. Brodersen, "Low-power CMOS digital design," *IEEE J. Solid State Circ.*, vol. 27, pp. 473–484, Apr. 1992.

Part II

Low Voltage Technologies and Circuits

Low-Voltage Technologies
and Circuits (Invited)

TADAHIRO KURODA, MEMBER, IEEE, AND TAKAYASU SAKURAI,* MEMBER, IEEE

SYSTEM ULSI ENGINEERING LABORATORY, TOSHIBA CORPORATION
(*) INSTITUTE OF INDUSTRIAL SCIENCE, UNIVERSITY OF TOKYO

Abstract—This paper reviews low-voltage device technologies and circuit design techniques for low-power, high- speed CMOS VLSIs. Some of the recent developments, such as employing multiple threshold voltage and controlling threshold voltage through substrate bias in bulk CMOS and Silicon on Insulator (SOI) based technologies, are discussed. Future directions of low-power VLSIs are also described.

1. INTRODUCTION

Lowering both of the supply voltage V_{DD} and threshold voltage V_{th} enables high-speed, low-power operation [1–3]. Figure 1 depicts equi-speed (broken lines) and equi-power (solid lines) curves on a V_{DD}–V_{th} plane [4] calculated from their theoretical models [5]. The contour lines are normalized at a typical design window (box in the figure) where $V_{DD} = 3.3V \pm 10\%$ and $V_{th} = 0.55V \pm 0.1V$. Circuit speed becomes slower at lower V_{DD} and higher V_{th} while power dissipation becomes larger at higher V_{DD} and lower V_{th}. Tradeoffs between speed and power can be explored by sliding the design window on the V_{DD}–V_{th} plane. The upper-left corner shows the worst case speed condition, whereas the lower-right corner shows the worst case power scenario. Results are summarized in Fig. 2 [4]. It suggests that optimum V_{DD} and V_{th} can save the waste (shadows in the figure) caused by the constant V_{DD} of 3.3V and constant V_{th} of 0.55V.

This approach, however, raises three problems, and therefore has not been achieved: 1) degradation of worst-case speed due to V_{th} fluctuation in low V_{DD} [6,7], 2) increase in standby power dissipation in low V_{th} [8–10], and 3) inability to sort out defective chips by monitoring the quiescent power supply current (I_{DDQ}) [11]. Delay variation due to V_{th} fluctuation is increased in low-voltage operation [6,7]. It can be understood in Fig. 1 that the design window should be reduced in size to keep the delay variation percentage constant in low V_{DD}. Its theoretical mode is derived in Kuroda and coworkers [12]. For example, under 50% speed requirement V_{DD} can be reduced to 1V and V_{th} fluctuation should be reduced from $\pm 0.1V$ to $\pm 0.02V$. The second and the third problems come from the increased subthreshold leakage current in low V_{th}. A dotted line in Fig. 1 depicts a equi-power-ratio curve

where power dissipation due to the subthreshold leakage current makes up 10% of the total power dissipation. The design window should be set at high V_{th} regions in a standby mode and testing.

To solve these problems, several circuit schemes are proposed: a multi threshold-voltage CMOS (MTCMOS) scheme [8], a variable threshold-voltage CMOS (VTCMOS) scheme [12], and an elastic-Vt CMOS (EVTCMOS) scheme [13]. The three circuit schemes are discussed and several circuit implementations are presented in Section 2. Device technologies including bulk CMOS and SOI are then reviewed in Section 3. Section 4 is dedicated to discussions on possible future directions and conclusions.

2. LOW-VOLTAGE CIRCUITS

A. Low-Voltage Circuit Schemes

The three circuit schemes which have been proposed to solve the low-V_{DD}/low-V_{th} problems are sketched and compared in Fig. 3. Currently in a system such as a PC, the power supply is turned off by a power management controller when a chip is inactive, but this idea can even be applied at the chip level. The MTCMOS [8] employs two types of V_{th}: low V_{th} for fast circuit operation in logic circuits and high V_{th} for cutting off the subthreshold leakage current in the standby mode. Since parasitic capacitance is much smaller on a chip than on a board, the on-off control of the power supply can be performed much faster on a chip. This takes less than $0.5 \mu s$, which in turn enables frequent power management [14]. This scheme is straightforward and easy to be employed in a current design method. However, it requires very large transistors for the power supply control, and hence imposes area and yield penalties; otherwise circuit speed degrades. These penalties become extreme below 0.9V V_{DD} because (V_{GS}-V_{th}) of the high-V_{th} transistors becomes too small in the active mode. One potential problem is that the MTCMOS cannot store data in the standby mode. Special latches have been developed to solve this problem [14].

While the MTCMOS can solve only the standby leakage problem, the VTCMOS [12] can solve all three problems. It

61

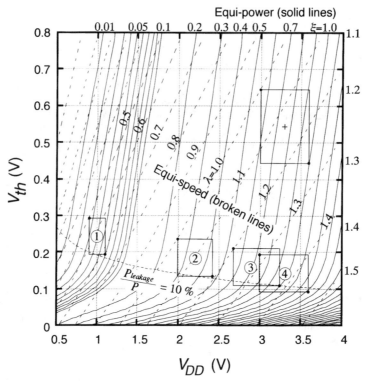

Figure 1 Exploring V_{DD}-V_{th} design space.

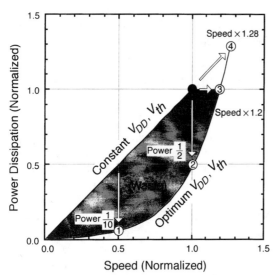

Figure 2 Speed and power saving by optimum V_{DD} and V_{th}.

dynamically varies V_{th} through substrate bias, V_{BB}. Typically, V_{BB} is controlled to compensate for V_{th} fluctuations in the active mode, while in the standby mode and in the I_{DDQ} testing, deep V_{BB} is applied to increase V_{th} and cut off the subthreshold leakage current.

The EVTCMOS [13] controls both V_{DD} and V_{BB} such that when V_{DD} is lowered, V_{BB} becomes that much deeper to raise V_{th} and further reduce power dissipation. Note that internal V_{DD} and V_{SS} are provided by source-follower n- and p-transistors, re-

spectively, whose gate voltages are controlled. In order to control the internal power supply voltage independent from the power current, the source-follower transistors should operate near the threshold. This requires very large transistors.

The essential difference among the three schemes is that the VTCMOS controls the substrate-bias while the others control the power lines. Since much smaller current (almost none) flows in the substrate than in the power lines, a much smaller circuit can control the substrate-bias. This leads to negligible penalties in area and speed in the VTCMOS. Global routing of substrate-contacts, however, may impose an area penalty, which may in turn make the application of the VTCMOS to existing macro designs impractical. It has been experimentally evaluated that the number of substrate (well) contacts can be greatly reduced in low-voltage environments [10–12]. Using a phase-locked loop and an SRAM in a VTCMOS gate-array [11], the substrate noise influence has been shown to be negligible even with 1/400 of the contact frequency compared with the conventional gate-array. A DCT (Discrete Cosine Transform) macro made with the VTCMOS [12] has also been manufactured with substrate- and well-contacts only at the periphery of the macro; it worked without problems realizing power dissipations more than one order of magnitude smaller than a DCT macro in the conventional CMOS design.

B. VTCMOS Circuits

Several variants have been reported [7,9–12] for the VTCMOS scheme, whose salient features are summarized in Fig. 4.

	MTCMOS	VTCMOS	EVTCMOS
Scheme	St'by—[*High-V_{th}] V_{DD}, V_{DD}' ... Low-V_{th} ... V_{SS} Ref.[8]	n-well, St'by—[control (SSB)]—V_{DD} ... V_{SS}, p-well Ref.[12]	[* V_{DD}, St'by—[control (SSB)] ... V_{SS}', V_{SS}, * V_{th}] Ref.[13]
	V_{DD} on-off	V_{BB} control	V_{DD} & V_{BB} control
Effect	+ $I_{st'by}$ reduction	+ ΔV_{th} compensation + $I_{st'by}$ reduction + I_{DDQ} test	+ ΔV_{th} compensation + $I_{st'by}$ reduction + I_{DDQ} test
Penalty	- large serial MOSFET($*$) slower, larger, lower yield - special latch	- triple well (desirable)	- large serial MOSFET($*$)

Figure 3 MTCMOS, VTCMOS, and EVTCMOS.

A self-adjusting threshold-voltage scheme (SATS) [7] reduces the V_{th} fluctuation. When V_{th} is lower than a target value, larger leakage current flows through a leakage current monitor (LCM) and turns on a self-substrate bias (SSB) circuit. As a result, V_{BB} goes deeper and causes V_{th} to increase. Thus the substrate bias is controlled such that the transistor leakage current is adjusted to be a constant value. This means that the V_{th} process fluctuation can be canceled by the SATS. The measured overall V_{th} controllability including static and dynamic effects is ±0.05V while using a bare process V_{th} fluctuation of ±0.15V. The same idea is also presented in Kaenel and coworkers [15].

Two standby power reduction (SPR) circuits [9,10] have been reported to lengthen battery life in mobile applications. One circuit in Seta [9] switches V_{BB} between the power supply and an additional supply for substrate bias. It requires three external power supplies but takes less than $0.1\mu s$ for the substrate bias switching. Triple well technology is a must. The other circuit in Kuroda [10] employs the SSB for the substrate bias. No additional external power supply or additional step in process is required. An active-to-standby mode transition is performed by the SSB, and hence takes about $100\mu s$. On the other hand, a standby-to-active mode transition is carried out by an MOS

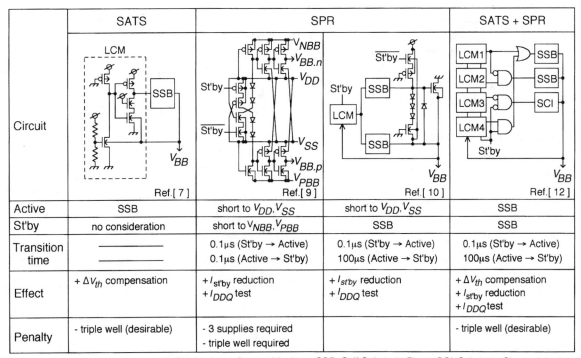

	SATS	SPR		SATS + SPR
Circuit	LCM ... SSB, V_{BB} Ref.[7]	V_{NBB}, $V_{BB.n}$, V_{DD}, St'by, $\overline{\text{St'by}}$, V_{SS}, $V_{BB.p}$, V_{PBB} Ref.[9]	$\overline{\text{St'by}}$, St'by—[SSB], LCM, SSB, V_{BB} Ref.[10]	LCM1—SSB, LCM2—SSB, LCM3—SCI, LCM4, St'by, V_{BB} Ref.[12]
Active	SSB	short to V_{DD}, V_{SS}	short to V_{DD}, V_{SS}	SSB
St'by	no consideration	short to V_{NBB}, V_{PBB}	SSB	SSB
Transition time	———	$0.1\mu s$ (St'by → Active) $0.1\mu s$ (Active → St'by)	$0.1\mu s$ (St'by → Active) $100\mu s$ (Active → St'by)	$0.1\mu s$ (St'by → Active) $100\mu s$ (Active → St'by)
Effect	+ ΔV_{th} compensation	+ $I_{st'by}$ reduction + I_{DDQ} test	+ $I_{st'by}$ reduction + I_{DDQ} test	+ ΔV_{th} compensation + $I_{st'by}$ reduction + I_{DDQ} test
Penalty	- triple well (desirable)	- 3 supplies required - triple well required		- triple well (desirable)

LCM: Leakage Current Monitor,   SSB: Self Substrate Bias,   SCI: Substrate Charge Injector

Figure 4 VTCMOS variants.

switch, and hence is completed in $0.1\mu s$. This "slow falling asleep but quick awakening" feature is acceptable for most applications.

The latest circuit in Kuroda [12] achieves both the SATS and the SPR capability at the same time. Operation principles are the same as the circuits in Kobayashi and Sakuri, and Kuroda [7,10], and hence the same circuit performance.

3. LOW-VOLTAGE TECHNOLOGIES

At room temperature, CMOS circuits operate at supply voltages as low as 0.2V, which is theoretically derived in [16] and experimentally examined in [17]. Optimization of device parameters [18] and tradeoffs between speed and power [3] are reported. The impact of V_{th} variation on circuit speed is experimentally investigated and its model is derived in [6].

Silicon on Insulator (SOI) CMOS processes have several potential advantages over the traditional bulk CMOS technologies. These include a lower parasitic substrate capacitance and a deeper subthreshold cut-off slope. In addition, one circuit family in SOI technology has recently attracted much attention for its unique way to control V_{th}; it involved tying the body to the gate [19]. This can be considered as a kind of VTCMOS in SOI technology. This circuit is called DTMOS (dynamic threshold voltage MOSFET). Consider using an n-channel MOSFET. When the gate of the NMOS increases, the bulk tied to the gate also becomes high, lowering the threshold voltage. Consequently, larger switching current flows to increase the operation speed. On the other hand, when the gate voltage becomes low, the threshold voltage becomes higher and suppresses the leakage current; this assures a low standby current. The issue with this scheme is that the supply voltage should be below the forward voltage of the junctions ($\sim 0.7V$). Otherwise, when V_{DD} is applied to the NMOS gate, bulk-source junctions turn on to increase power dissipation drastically. Another interesting approach is to realize SATS in SOI technology [20]. The tunable range of the threshold voltage with the back-gate of a dual gate SOI technology is wider than the bulk CMOS. An MTCMOS circuit in SOI technology is also reported [21].

4. DISCUSSIONS AND CONCLUSIONS

A possible future low-power VLSI is illustrated in Fig. 5. Some kind of threshold engineering both in circuit and device technology will be essential in the future to enable stable operation in low-voltage environments. Optimum voltages for logic gates, memory circuits, and analog circuits may vary widely. They may change as workload varies with time in data processing. "V_{DD} on demand" in terms of both static and dynamic adjustment of the supply voltage will be carried out on a chip by DC–DC converters [22]. Low-voltage circuit design is another future challenge. Noise issues, especially those induced by inductance and resistance along power lines, should be studied. SOI based technologies are other candidates for achieving ultra low voltage LSIs armed with the DTMOS where gate and bulk are tied together to optimally control the threshold voltage.

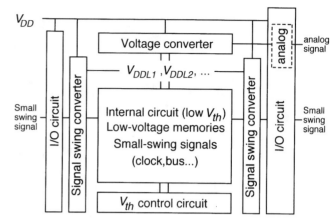

Figure 5 Future low-power VLSI.

5. ACKNOWLEDGMENT

Valuable discussions and constant encouragement by T. Fujita, K. Suzuki, T. Furuyama, and Y. Unno are appreciated.

References

[1] A. P. Chandrakasan, S. Sheng, and R. W. Brodersen, "Low-power CMOS digital design," *IEEE J. Solid-State Circuits,* vol. 27, no. 4, Apr. 1992, pp. 473–484.

[2] T. Kuroda and T. Sakurai, "Overview of low-power ULSI circuit techniques," invited, *IEICE Trans. on Electronics,* vol. E78-C, no. 4, Apr. 1995, pp. 334–344.

[3] D. Liu and C. Svensson, "Trading speed for low power by choice of supply and threshold voltages," *IEEE J. Solid-State Circuits,* vol. 28, no. 1, Jan. 1993, pp. 10–17.

[4] T. Kuroda, "Low power and high performance—Can supply voltage be scaled below 1V?" Rump session in *Symposium on VLSI Circuits,* June 1996.

[5] T. Sakurai and A. R. Newton, "Alpha-power law MOSFET model and its applications to CMOS inverter delay and other formulas," *IEEE J. Solid-State Circuits,* vol. 25, no. 2, Apr. 1990, pp. 584–594.

[6] S. Sun and P. Tsui, "Limitation of CMOS supply-voltage scaling by MOSFET threshold-voltage variation," *Proc. CICC'94,* May 1994, pp. 267–270.

[7] T. Kobayashi and T. Sakurai, "Self-adjusting threshold-voltage scheme (SATS) for low-voltage high-speed operation," *Proc. CICC'94,* May 1994, pp. 271–274.

[8] S. Mutoh et al., "1-V power supply high-speed digital circuit technology with multithreshold-voltage CMOS," *IEEE J. Solid-State Circuits,* vol. 30, no. 8, Aug. 1995, pp. 847–854.

[9] K. Seta et al., "50% active-power saving without speed degradation using standby power reduction (SPR) circuit," *ISSCC Dig. Tech. Papers,* Feb. 1995, pp. 318–319.

[10] T. Kuroda et al., "A high-speed low-power $0.3\mu m$ CMOS gate array with variable threshold voltage (VT) scheme," *Proc. CICC'96,* May 1996, pp. 53–56.

[11] T. Kuroda et al., "Substrate noise influence on circuit performance in variable threshold-voltage scheme," *Proc. ISLPED'96,* August 1996, pp. 309–312.

[12] T. Kuroda et al., "A 0.9V 150MHz 10mW 4mm² 2-D discrete cosine transform core processor with variable-threshold-voltage scheme," *IEEE J. Solid-State Circuits,* vol. 31, no. 11, Nov. 1996, pp. 1770–1779.

[13] M. Mizuno et al., "Elasti-V_t CMOS circuits for multiple on-chip power control," *ISSCC Dig. Tech. Papers,* Feb. 1996, pp. 300–301.

[14] S. Mutoh et al., "A 1V multi-threshold voltage CMOS DSP with an efficient power management technique for mobile phone application," *ISSCC Dig. Tech. Papers,* Feb. 1996, pp. 168–169.

[15] V. Kaenel et al., "Automatic adjustment of threshold & supply voltages for minimum power consumption in CMOS digital circuits," *Proc. Symp. on Low Power Electr.,* 1994, pp. 78–79.

[16] R. M. Swanson and J. D. Meindl, "Ion-implanted complementary MOS transistors in low-voltage circuits," *IEEE J. Solid-State Circuits,* vol. sc-7, no. 2, pp. 146–152, April 1972.

[17] J. B. Burr and T. Shott, "A 200mV self-testing encoder/decoder using Stanford ultra-low-power CMOS," in *ISSCC Dig. Tech. Papers,* Feb. 1994, pp. 84–85.

[18] Z. Chen, J. Burr, J. Shott, and J. Plummer, "Optimization of quarter micron MOSFETs for low voltage/low power application," *Proc. IEDM'95,* Dec. 1995, pp. 63–66.

[19] F. Assaderaghi et al., "A dynamic threshold voltage MOSFET (DTMOS) for very low voltage operation," *IEEE Electron Device Letter,* vol. 15, no. 12, Dec. 1994, pp. 510–512.

[20] I. Yang, C. Vieri, A. P. Chandrakasan, and D. Antoniadis, "Back gated CMOS on SOIAS for dynamic threshold control," *Proc. IEDM'95,* Dec. 1995, pp. 877–880.

[21] T. Douseki et al., "A 0.5V SIMOX-MTCMOS circuit with 200ps logic gate," *ISSCC Dig. Tech. Papers,* Feb. 1996, pp. 84–85.

[22] V. Gutnik and A. Chandrakasan, "An efficient controller for variable supply-voltage low power processing," *Proc. Symposium on VLSI Circuits,* June 1996, pp. 158–159.

Ion-Implanted Complementary MOS Transistors in Low-Voltage Circuits

RICHARD M. SWANSON, MEMBER, IEEE, AND JAMES D. MEINDL, FELLOW, IEEE

Abstract—Simple but reasonably accurate equations are derived, which describe MOS transistor operation in the weak inversion region near turn-on. These equations are used to find the transfer characteristics of complementary MOS (CMOS) inverters. The smallest supply voltage at which these circuits will function is approximately $8kT/q$. A boron ion implantation is used for adjusting MOST turn-on voltage for low-voltage circuits.

INTRODUCTION

RECENTLY, techniques have been developed for fabricating complementary MOS transistors with low turn-on voltages, enabling them to be used in circuits with supply voltages less than 1.35 V [1]–[3]. MOS transistors in low-voltage digital circuits are, by necessity, operating near their turn-on or threshold voltage. Unfortunately, in the vicinity of turn-on, the assumptions commonly used in deriving device characteristic equations are inaccurate [4]. In Section I of this paper, new MOS transistor characteristic equations are derived, which are simple but reasonably accurate in the weak inversion region near turn on, as well as elsewhere.

In Section II these equations are used to find the transfer characteristic of a low-voltage complementary MOS inverter and the minimum supply voltage at which the inverter will function is determined. This limit is important since it serves to define the minimum power-speed product that can be achieved with CMOS digital circuits.

In Section III a technique of adjusting MOST turn-on voltage by ion implantation is described. A theory of the turn-on voltage change as a function of the implantation parameters is developed and a brief description of the MOST fabrication procedure is included. By using ion implantation in conjunction with standard aluminum-gate MOS processing, low-voltage complementary integrated circuits with excellent performance characteristics can be achieved.

I. DEVICE EQUATIONS INCLUDING WEAK INVERSION EFFECTS

An n-channel MOS transistor [Fig. 1(a)] is analyzed. As illustrated in Fig. 1(b), ψ_s is the total band bending and ϕ_f is the potential difference between the intrinsic

Manuscript received October 14, 1971; revised December 1, 1971. This work was partly supported by the Department of Health, Education and Welfare under PHS research Grant 5 PO1 GM17940-02 and by the U. S. Army Electronics Command under Contract DAAB-7-69-C-0192.

The authors are with the Department of Electrical Engineering, Stanford University, Stanford, Calif. 94305.

level (midband gap) and the Fermi level. The potential difference between the electron quasi-Fermi level and the bulk Fermi level is ϕ_c, which is nonzero because there is a transverse electron current flowing (i.e., a drain-source voltage is applied).

Integrating Poisson's equation, the total charge in the semiconductor is given approximately [4] by

$$Q_s = -\sqrt{2q\epsilon_s N_A(\psi_s + kT/q)} \, \exp\,[q(\psi_s - \phi_c - 2\,|\phi_f|)/kT]).$$

(1)

This gives the total semiconductor charge per unit area $Q_s = Q_n + Q_B$ as a function of ψ_s and ϕ_c. Two distinct cases of this equation can be identified. These are 1) a weak inversion case where the inversion layer charge per unit area Q_n is much less than the depletion region charge per unit area Q_B and 2) a strong inversion case where Q_n is much greater than Q_B. Making approximations appropriate in each of these cases, the following formulas, giving Q_n as a function of the gate voltage V_G and the electron quasi-Fermi level ϕ_c are derived in the Appendix.

1) Weak Inversion:

$$-Q_n = C_0\left(n\,\frac{kT}{q}\right) \exp\left(\frac{q}{nkT}\left[V_G - V_T(\phi_c) - n\,\frac{kT}{q}\right]\right),$$

(2a)

valid when $V_G \leq V_T(\phi_c) + n(kT/q)$.

2) Strong Inversion:

$$-Q_n = C_0[V_G - V_T(\phi_c)],$$

(2b)

valid when $V_G \geq V_T(\phi_c) + n(kT/q)$. C_0 is the oxide capacitance per unit area.

$$V_T(\phi_c) = V_{FB} + 2\,|\phi_f| + \phi_c$$
$$+ \frac{1}{C_0}\sqrt{2q\epsilon_s N_A(2\,|\phi_f| + \phi_c)},$$

(3)

is the threshold voltage referenced to the substrate. V_{FB} is the flat-band voltage [5].

N_A is the substrate doping density and ϵ_s is the semiconductor dielectric constant.

$$n = \frac{C_d + C_{fs} + C_0}{C_0}$$

(4)

$$C_d \triangleq \frac{\partial}{\partial \psi_s}\,(Q_B)|_{\psi_s = 2|\phi_f| + \phi_c},$$

Reprinted from *IEEE Journal of Solid-State Circuits*, pp. 146-153, April 1972.

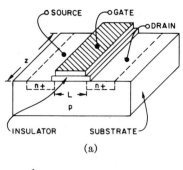

(a)

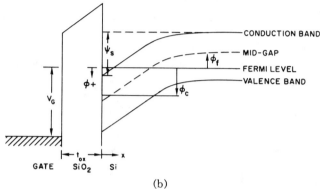

(b)

Fig. 1. (a) Physical structure of n-channel MOST. (b) Band diagram of n-channel MOST.

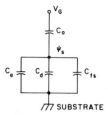

Fig. 2. Model circuit of MOS structure.

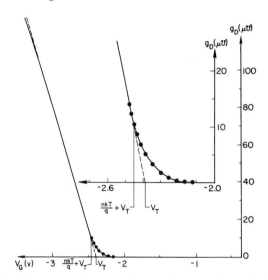

Fig. 3. Channel conductance g_D versus gate voltage V_G for device 1. ——experiment; $\cdots$ theory.

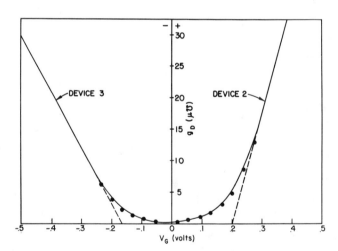

Fig. 4. Channel conductance g_D versus gate voltage V_G for devices 2 and 3. ——experiment; $\cdots$ theory.

and

$$Q_B = \sqrt{2q\epsilon_s N_A(\psi_s)},$$

so that

$$C_d = \frac{\sqrt{2q\epsilon_s N_A}}{2\sqrt{2\,|\phi_f| + \phi_c}}. \tag{5}$$

and $C_{fs} = qN_{fs}$, where N_{fs} is the fast surface state density per electronvolt evaluated at $\psi_s = \phi_c + 2|\phi_f|$.

A simple physical explanation of (2a) is presented here in lieu of the complete derivation given in the Appendix. If the gate voltage is varied by a small amount ΔV_G there will be a small change in surface potential, $\Delta\psi_s$. $\Delta\psi_s$ may be found with the aid of the equivalent circuit shown in Fig. 2. C_e is the capacitance of the mobile electrons, or $(\partial Q_n/\partial\phi_s)$. When the surface is strongly inverted C_e is much larger than C_o, C_d, and C_{fs} so that ψ_s is essentially constant. However, when the surface is weakly inverted C_e is small compared to C_d and C_{fs}. The equivalent circuit then yields $\Delta\psi_s = (1/n)\Delta V_G$. In weak inversion Q_n varies as a constant $\times \exp[q(\psi_s - \phi_c)kT]$. From this, it is apparent that $Q_n \sim \exp[qV_G/nkT - q\phi_c/kT]$. The remaining terms in (2a) are found by considerations presented in the Appendix.

It can be easily shown [5] that when a MOST is operated in the linear region (small drain voltage) the drain-to-source conductance g_D is given by [5],

$$g_D = \frac{Z}{L}\,\mu_n\,|Q_n(\phi_c = 0)|. \tag{6}$$

Plots of the measured versus theoretical linear region g_D

for three devices are shown in Figs. 3 and 4. Table I contains pertinent data about the devices.

The drain current-voltage characteristics of a MOST are now found using the following relation for the drain current I_D [4]

$$I_D = (Z/L)\mu_n \int_0^{V_D} |Q_n(\phi_c)|\,d\phi_c. \tag{7}$$

Here Z is the channel width, L, the channel length, μ_n, the effective electron surface mobility, and V_D, the drain-to-source voltage. The two relations for Q_n, (2a), and (2b), when inserted in (7) generate three distinct modes

67

TABLE I

Devices	Device Constants	Calculated from Plot
Device 1	$N_D = 1 \times 10^{15}$ cm^{-3} $t_{0x} = 1000$ Å $m = 1 + \dfrac{C_d}{C_0} = 1.46$	$V_T = 2.39$ V $n = 3.08$ $N_{fs} = 3.8 \times 10^{11}$ cm^{-2} eV^{-1} $K = 153$ μA/V^2 $\quad = \dfrac{Z}{L} \mu C_0$
Device 2	$N_A = 1.6 \times 10^{16}$ cm^{-3} $t_{0x} = 1000$ Å $m = 2.05$	$V_T = 0.20$ V $n = 2.80$ $N_{fs} = 1.6 \times 10^{11}$ cm^{-2} eV^{-1} $K = 180$ μA/V^2
Device 5	$N_D = 1 \times 10^{15}$ $t_{0x} = 10000$ Å $m = 1.46$	$V_T = -.165$ V $n = 2.70$ $N_{fs} = 2.7 \times 10^{11}$ cm^{-2} eV^{-1} $K = 90$ μA/V^2

of operation for the MOST. The characteristic equations for each mode are presented below.

A. Strong Inversion Only

In this case the gate voltage is large enough to strongly invert the entire channel. Equation (2b) applies to the integration (7). The defining criteria for this region are

$$V_G \geq V_T(0) + n(kT/q)$$

and

$$V_G \geq V_T(V_D) + n(kT/q),$$

where the argument of V_T gives the value of ϕ_c at which (3) is evaluated. Solving the preceding inequality for $V_D{}^*$ gives $V_D \leq V_D{}^*$ where

$$V_D{}^* = V_G - \left(V_T + n\frac{kT}{q}\right) + V_B\left[1 + \frac{V_B}{4|\phi_f|}\right.$$
$$\left. - \sqrt{\left(1 + \frac{V_B}{4|\phi_f|}\right)^2 + \frac{V_G - [V_T + n(kT/q)]}{2|\phi_f|}}\right]. \quad (8)$$

Here V_B is defined as $|Q_b(\psi_s = 2|\phi_f|)|/C_0$ and $V_T = V_T(0)$. Using (2a), (7) becomes

$$I_D = (Z/L)\mu_n C_0\left\{(V_G - V_{FB} - 2|\phi_f|)V_D - \frac{V_D{}^2}{2}\right.$$
$$\left. - V_B\frac{4|\phi_f|}{3}\left[\left(1 + \frac{V_D}{2|\phi_f|}\right)^{3/2} - 1\right]\right\} \quad (9a)$$

valid when $V_D \leq V_D{}^*$ and $V_G \geq V_T + n(kT/q)$. Equations (9a), (9b) can be simplified by assuming that $V_D \ll 2|\phi_f|$ yielding

$$I_D \cong \frac{Z}{L}\mu_n C_0\left[(V_G - V_T)V_D - \frac{V_D{}^2}{2}\right], \quad (9b)$$

valid when $V_D \leq (V_G - (V_T + n(kT/q)))/m$, where $m = (C_0 + C_d)/C_0$.

B. Mixed Strong and Weak Inversion

In this case, the gate voltage is large enough to strongly invert the channel near the source but the drain voltage is also large enough to cause weak inversion near the drain. In other words, the device is operating near saturation. The criteria are $V_G \geq V_T(0) + n(kT/q)$ and $V_G \leq V_T(V_D) + n(kT/q)$. This yields $V_D \geq V_D{}^*$. In this case (2b) applies to that part of the integration where $0 \leq \phi_c \leq V_D{}^*$ and (2a) applies when $V_D{}^* \leq \phi_c \leq V_D$. The major contribution of (1) to the integration is for $V_D{}^* \leq \phi_c \leq V_D{}^* + n(kT/q)$ because the integrand decreases exponentially in ϕ_c. The bulk charge term in $V_T(\phi_c)$, (3), will vary only slightly in this range and thus it is reasonable to expand about $\phi_c = V_D{}^*$. This gives $V_T(\phi_c) = V_T(V_D{}^*) + m(\phi_c - V_D{}^*)$ where $m = (C_0 + C_d)/C_0$. Equation (7) becomes upon integrating and noting that $V_G = V_T(V_D{}^*) + n(kT/q)$,

$$I_D = \frac{Z}{L}\mu_n C_0\left\{(V_G - V_{FB} - 2|\phi_f|)V_D{}^* - \frac{V_D{}^{*2}}{2}\right.$$
$$- V_B\frac{4|\phi_f|}{3}\left[\left(1 + \frac{V_D{}^*}{2|\phi_f|}\right)^{3/2} - 1\right]$$
$$\left. + \frac{1}{m}\left(n\frac{kT}{q}\right)^2\left(1 - \exp\left[\frac{-mq}{nkt}(V_D - V_D{}^*)\right]\right)\right\}, \quad (10)$$

valid when $V_D \geq V_D{}^*$ and $V_G \geq V_T + n(kT/q)$.

The C_d term in n is evaluated at $\rho_c = V_D{}^*$.

C. Weak Inversion Only

In this case the gate voltage is not large enough to strongly invert the channel at any point. The criterion is $V_G \leq V_T + n(kT/q)$.

Integrating as before

$$I_D = \frac{Z}{L}\mu_n C_0 \frac{1}{m}\left(n\frac{kT}{q}\right)^2 \exp\left[\frac{q}{nkT}\left(V_G - V_T - n\frac{kT}{q}\right)\right]$$
$$\cdot \left\{1 - \exp\left[\frac{-mq}{nkT}V_D\right]\right\}, \quad (11)$$

valid when $V_G \leq V_T + n(kT/q)$. The C_d term in n is evaluated at $\phi_c = V_D{}^*$.

Equations (9a), (9b), (10), and (11) characterize the dc behavior of the MOST. The noticeable effects of the weak inversion region are 1) exponential dependence of I_D on V_G and V_D when $V_G \leq V_T + n(kT/q)$ and 2) exponential transition into saturation instead of the classical parabolic form.

The drain characteristics of the device whose Q_n versus V_G relation is shown in Fig. 3 were measured and plotted in Fig. 5. On the same graph, both the classical theory (dashed line) and the theory including weak inversion effects (solid line) are shown. Good agreement is obtained between the experimental data and the predictions of the weak inversion theory.

Figs. 6 and 7 show the experimental and theoretical drain characteristics of a low-threshold voltage complementary pair whose Q_n versus V_G relations are shown in Fig. 4. Only the weak inversion region is plotted. The

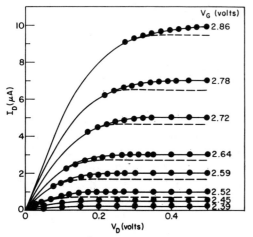

Fig. 5. Drain characteristics for device 1. ——experiment; ----classical theory (9a); ···· theory including weak inversion effects (10a) and (11a).

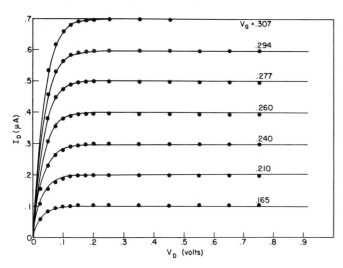

Fig. 6. Drain characteristics for device 2 in weak inversion region. ——experiment; ···· theory (11a).

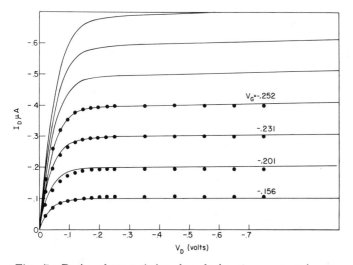

Fig. 7. Drain characteristics for device 3. ——experiment; ···· theory (11a).

values of V_T, n, and $K = (Z/L)\mu C_o$ used in the calculations for Figs. 5, 6, and 7 were taken from Table I. Again, good agreement is obtained for both the n-channel and p-channel devices.

II. LOW-VOLTAGE COMPLEMENTARY INVERTERS

Complementary MOS logic circuits offer distinct advantages in power consumption compared to single-ended circuits. However, to obtain the greatest power savings and lowest power-speed product they should be operated at the lowest practical supply voltage [6]. Realizing that MOS transistors do not turn off as abruptly as the classical equations (9a) and (9b) indicate, but are weakly inverted at gate voltages below V_T, it is desirable to find the minimum supply voltage V_s at which complementary circuits will operate. To this end, the transfer characteristic of a complementary inverter is now determined.

The inverter circuit being analyzed is shown in Fig. 8. The following notations are used. V_{Tp} and V_{Tn} are the magnitude of the thresholds of the n- and p-channel devices, respectively. $K_n = (Z/L)\mu_n C_o$ is the gain constant of the n-channel device and K_p similarly for the p-channel device m_n and n_n, as defined in Section I, refer to the values of m and n for the n-channel device. For the p-channel device m_p replaces m_n and n_p replaces n_n. It is assumed that both transistors are in the weak-inversion-only region. This will be the case if $V_s - V_{Tp} - n_p(kT/q) \leq V_{in} \leq V_{Tn} + n_n(kT/q)$.

Equating the drain current, as found by (11) for both devices, yields

$$
\begin{aligned}
I_D &= K_n\left(n_n\frac{kT}{q}\right)^2\frac{1}{m_n} \\
&\quad \cdot \exp\left[q\left(V_{in} - V_{Tn} - n_n\frac{kT}{q}\right)\Big/ n_n kT\right] \\
&\quad \cdot (1 - \exp[-m_n q V_0/n_n kT]) \\
&= K_p\left(n_p\frac{kT}{q}\right)^2\frac{1}{m_p} \\
&\quad \cdot \exp\left[q\left(V_s - V_{in} - V_{Tp} - n_p\frac{kT}{q}\right)\Big/ n_p kT\right] \\
&\quad \cdot (1 - \exp[-m_p q(V_s - V_0)/n_p kT]).
\end{aligned}
$$

Solving for V_{in} gives

$$
\begin{aligned}
V_{in} &= \frac{kT}{q}\left(\frac{n_n n_p}{n_n + n_p}\right)\ln\left(\frac{K_p}{K_n}\frac{n_p^2 m_n}{n_n^2 m_p}\right) \\
&\quad + \frac{n_n}{n_n + n_p}V_s + \frac{n_n n_p}{n_n + n_p}\left(\frac{V_{Tn}}{n_n} - \frac{V_{Tp}}{n_p}\right) \\
&\quad + \frac{kT}{q}\left(\frac{n_n n_p}{n_n + n_p}\right)\ln\left(\frac{1 - \exp[-m_p q(V_s - V_0)/n_p kT]}{1 - \exp[-m_n q V_0/n_n kT]}\right).
\end{aligned}
$$

$$(12)$$

The measured transfer characteristic for the complementary pair (devices 2 and 3) for various power supply voltages is shown in Fig. 9. Equation (12), using the

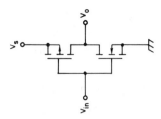

Fig. 8. CMOS inverter circuit.

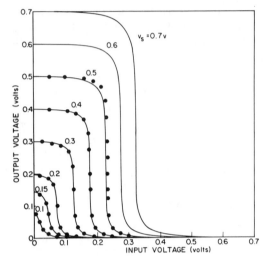

Fig. 9. CMOS inverter transfer characteristics. ——experiment;
···· theory (12a).

device constants from Table I, is also plotted in Fig. 9. It is interesting to note that there is no hysteresis in the transfer characteristic, even when $V_s < |V_{Tn}| + |V_{Tp}|$.

A simple expression for the inverter gain may be found by assuming $m_n = m_p = m$ and $n_n = n_p = n$. Differentiating (12) with respect to V_o gives the reciprocal of the inverter gain. The maximum gain is found to occur at $V_o = V_s/2$ and is given by

$$A_v = \frac{1}{m}\left(\exp\left[\frac{mqV_s}{2nkT}\right] - 1\right). \qquad (13)$$

It is seen that V_s must be at least 3–4 nkT/mq for the inverter to have sufficient gain for use in a digital circuit. Thus

$$V_{s,\min} \approx 4\,\frac{nkT}{mq}. \qquad (14)$$

If $N_{fs} = 0$ then $n/m = 1$ and the inverter displays the maximum nonlinearity obtainable in semiconductor material [7]. Transistors fabricated using conventional clean oxide techniques including a low temperature anneal in forming gas will have n/m in the neighborhood of 2 (see Table I), giving a minimum usable supply of about $8kT/q$ or 0.2 V at 27°C. Standard fabrication techniques yield a spread in turn-on voltages on the order of 0.2 V so a practical supply voltage could never be quite as low as 0.2 V. However, regardless of improvements in control of threshold during fabrication, the supply voltage can never be less than that given by (14).

It is interesting to note that reduced operating temperatures permit lower supply voltages [7].

III. Adjusting MOST Turn-on Voltage by Ion Implantation

In the preceding discussion of low-voltage complementary MOS circuits, it has been assumed that the transistor turn-on voltage could always be adjusted to the desired value, approximately half the supply voltage [8]. Using conventional aluminum gate processing this is possible with n-channel devices. Unfortunately, aluminum gate p-channel devices will always have turn-on voltages of −2 V or less. This section discusses a technique for ion-implanting boron [9] through the gate oxide to form a shallow p-layer in the channel region of a p-channel device. This technique decreases the magnitude of the turn-on voltage from its initial value. As a result, devices whose original thresholds were in the neighborhood of −2–3 V can have their thresholds shifted as close to zero as desired. Device 3, whose transfer and drain characteristics are shown in Figs. 4 and 6, was fabricated by this method, resulting in $V_{TP} = -0.17$ V.

It will be assumed that the implanted boron concentration N_A (at the semiconductor surface), is greater than the n-type substrate doping density N_D creating a p-type surface layer. A depletion region will exist at the p-n junction extending a distance 1_p from the junction toward the surface where from [4]

$$1_p = \frac{N_D}{N_A}\left[\frac{2\epsilon(|\phi_{fp}| + |\phi_{fn}|)}{q}\,\frac{N_A}{(N_A - N_D)N_D}\right]^{1/2}, \qquad (15)$$

assuming that the boron density is uniformly distributed from the surface to a depth W. The total implanted dose per unit area in the silicon is $N_I = N_A W$.

For boron doses of interest N_A will be much greater than N_D. In this case (15) becomes

$$1_p = \frac{W}{qN_1}\sqrt{2qN_D\epsilon_s(|\phi_{fn}| + |\phi_{fp}|)}. \qquad (16)$$

Since $|\phi_{fn}| \approx |\phi_{fp}|$, (16) can be approximated by

$$1_p \simeq W\,\frac{Q_B(\psi_s = 2\,|\phi_f|)}{qN_I}. \qquad (17)$$

Equation (17) reevals that if N_I is greater than $Q_B(\psi_s = 2|\phi_f|)/q$, the junction depletion region will not extend to the semiconductor surface. Q_B/q is about 10^{11} cm^{-2} for the typical n-type substrate resistivity of 5 Ω-cm. From considerations below, it is seen that if N_I is less than 10^{11} cm^{-2} the shift in turn-on voltage is less than 0.5 V. In practice a larger shift in turn-on voltage is usually desired causing 1_p to be less than W. The possibility then exists for an undepleted region of mobile holes to exist near the semiconductor surface. This situation is illustrated in Fig. 10(a). Increasing the gate voltage from the value at flat band creates a surface depletion region in the p-layer. At some voltage the surface depletion region will extend to the junction depletion region removing all mobile holes from the channel region

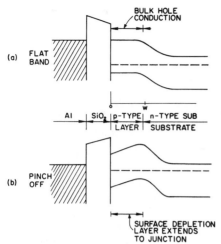

Fig. 10. Band structure for a boron-implanted p-channel MOS transistor.

as illustrated in Fig. 10(b). This is the turn-on voltage for the device.

At this point, calculating the turn-on after implantation V_{TI} is straightforward. The depth of the surface depletion layer 1_s is

$$1_s = \sqrt{\frac{2\epsilon_s \psi_t}{qN_A}}, \quad (18)$$

where ψ_t is the surface depletion region band bending in the p-layer. At turn-on $1_s + 1_p = W$ so that

$$\sqrt{\frac{2\epsilon_s \psi_t}{qN_A}} + \frac{Q_B W}{qN_I} = W. \quad (19)$$

Solving (19) for ψ_t gives

$$\psi_t = \frac{qN_I W(1 - Q_B/qN_I)^2}{2\epsilon_s}. \quad (20)$$

The gate voltage at turn-on, including charge from both the surface and junction depletion regions, is thus

$$V_{TI} = \Phi_{MS} - \frac{Q_{SS}}{C_0} - (|\phi_{fn)}| + |\phi_{fp}|) + \psi_t + \frac{qN_I}{C_0} - \frac{Q_B}{C_0}. \quad (21)$$

Φ_{MS} is the metal-semiconductor work function for the n-type substrate. The turn-on voltage prior to implantation is

$$V_T = \Phi_{MS} - \frac{Q_{SS}}{C_0} - 2|\phi_{fn}| - \frac{Q_B}{C_0}. \quad (22)$$

The shift in turn on ΔV_T is thus

$$\Delta V_T = V_{TI} - V_T = (|\phi_{fp}| - |\phi_{fn}|)$$
$$+ \frac{qN_I}{C_0} + \frac{qN_I W(1 - Q_B/qN_I)^2}{2\epsilon_s}. \quad (23)$$

$(|\phi_{fp}| - |\phi_{fn}|)$ can usually be neglected.

The last term in (23) is the surface band bending and must be less than $2|\phi_{fp}|$ or the surface would become

n-type, prohibiting further band bending. If the surface becomes n-type before the surface depletion region extends deep enough to entirely deplete the p-layer the device cannot be turned off. Since ψ_t must be less than $2|\phi_{fp}|$ for complete turn-off to be possible, (20) gives

$$2|\phi_{fp}| \leq \psi_t = \frac{qN_I W}{2\epsilon_s} \quad (24)$$

assuming $Q_B/qN_I \ll 1$. The maximum possible shift in threshold $\Delta V_{T\,max}$, while having a device that can be turned off, is, using (23) and (24),

$$\Delta V_{T\,max} = 2|\phi_{fp}| + \frac{4|\phi_{fp}|\epsilon_s}{WC_0} = 2|\phi_{fp}|\left[\frac{2\epsilon_s/W}{C_0} + 1\right] \quad (25)$$

Thinner p-layers allow larger shifts in turn-on voltage. When both the gate oxide and W are 1000 Å, $\Delta V_{T\,max}$ is about 5.2 V.

The threshold-shifting boron implantation is done after gate oxidation. Only those ions that pass through the oxide will contribute to N_I. Computer calculations of the Lindhard, Scharff, and Schiott depth-penetration theory [10] predict that the boron will have a Gaussian concentration profile centered about a depth which depends on the implantation voltage and a standard deviation about one third this depth. At a 30-kV implantation voltage, 50 percent of the boron will penetrate a 1000-Å gate oxide, at 40 kV 80 percent will penetrate a 1000-Å gate 93 percent will penetrate. Energies in the range of 30–50 kV have been found suitable for device fabrication provided the doses are adjusted to allow for the partial penetration.

The implantation must be followed by an anneal to restore crystal damage and activate the boron. Annealing for ten minutes at 1050°C appears to give complete activation and gives good SiO_2–Si interface characteristics. Following this anneal the junction depth is about 1500 Å. If the anneal is performed at less than 900°C, the boron will not be totally active and the device properties are degraded. Device 3 received a dose of 1.1×10^{12} cm^{-2} at 30 kV and was annealed at 1050°C. The resulting shift in turn-on voltage was +2 V.

The turn-on voltages of n-channel devices can also be shifted in a positive direction by boron implantation. Of course, in this case there is no junction formed by the implantation. If the depth of the boron is much less than the surface depletion-layer depth, Q_B will not be greatly affected by the implant and the turn-on shift will be $\Delta V_T = qN_I/C_o$.

CMOS integrated circuits with a wide variety of threshold voltages can be fabricated using ion implantation to adjust threshold voltages and to provide a predeposition for doping p-type regions where the n-channel devices are located.

CONCLUSION

The fast surface state density N_{fs} is the most important factor in determining the performance of MOS transistors in the weak inversion region near turn-on. At room temperature CMOS circuits can operate at supply voltages as low as 0.2 V, provided the fast surface state density is low enough. Ion implantation of boron is a convenient method of adjusting the turn-on voltage of MOS transistors, both n- and p-channel, to permit operation at low supply voltages.

APPENDIX

The total charge per unit area Q_s in the semiconductor is given by Sze [4]. For substrate doping levels of interest $q\psi_s/kT > 20$ when ψ_s is near threshold. Q_s is then approximated by

$$Q_s = -\sqrt{2q\epsilon_s N_A(\psi_s + (kT/q) \exp [q(\psi_s - \phi_c - 2|\phi_f|)/kT])}, \quad (26)$$

where N_A is the R-type substrate-doping density and ϵ_s is the dielectric constant of the semiconductor. Assuming that the voltage developed across the surface inversion layer is much less than $2|\phi_f|$, the immobile charge in the surface depletion region Q_B is

$$Q_B = -\sqrt{2q\epsilon_s N_A \psi_s}. \quad (27)$$

This leaves $Q_n = Q_s - Q_B$ mobile electrons per unit area in the inversion layer.

$$Q_n = -[\sqrt{2q\epsilon_s N_A(\psi_s + (kT/q) \exp [q(\psi_s - \phi_c - 2|\phi_f|)/kT])} - \sqrt{2q\epsilon_s N_A \psi_s}] \quad (28)$$

The gate voltage V_G is given by

$$V_G = \Phi_{MS} + \psi_s - \frac{Q_{SS}}{C_0} - \frac{Q_{fs0}}{C_0}$$
$$+ \frac{qN_{fs}}{C_0}(\psi_s - \phi_c - 2|\phi_f|) - \frac{Q_n}{C_0} - \frac{Q_B}{C_0}. \quad (29)$$

Φ_{MS} is the metal semiconductor work function, Q_{SS} the fixed interface charge per unit area, Q_{fs0} the charge in filled fast surface states per unit area when $\psi_s = \phi_c + 2|\phi_f|$, N_{fs} the fast surface state density per electronvolt at $\psi_s = \phi_c + 2|\phi_f|$, and C_0 the oxide capacitance per unit area.

Q_n is found as a function of V_G by eliminating ψ_s from (28) and (29). This cannot be done in closed form. However, from (28) two distinct regions can be identified. A weak inversion region is defined by

$$(kT/q) \exp [q(\psi_s - \phi_c - 2|\phi_f|)/kT] \ll \psi_s,$$

(i.e., the bulk charge dominates the inversion layer charge) and a strong inversion region by $(kT/q) \exp[q(\psi_s - \phi_c - 2|\phi_f|)/kT] \gg \psi_s$. The weak inversion criterion is valid when $\psi_s \leq \phi_c + 2|\phi_f|$. For ψ_s slightly greater than $\phi_c + 2|\phi_f|$ the exponential rapidly increases and the strong inversion criterion is valid.

For the weak inversion region, Q_n is found by a first-order Taylor-series expansion of (28) in powers of $\exp[q(\psi_s - \phi_c - 2|\phi_f|)/kT]$. This gives

$$-Q_n = \frac{\sqrt{2q\epsilon_s N_A}}{2\sqrt{\psi_s}} \frac{kT}{q} \exp [q(\psi_s - \phi_c - 2|\phi_f|)/kT]. \quad (30)$$

Since there will be significant mobile charge only when ψ_s is in the vicinity of $\phi_c + 2|\phi_f|$ it is reasonable to expand the Q_B term in (29) about $\psi_s = \phi_c + 2|\phi_f|$. By the weak inversion criterion Q_n (30) is negligible compared with Q_B and (29) becomes

$$V_G = V_{FB} + \phi_c + 2|\phi_f| + \frac{\sqrt{2q\epsilon_s N_A(\phi_c + 2|\phi_f|)}}{C_0}$$
$$+ \left(1 + \frac{C_d}{C_0} + \frac{C_{fs}}{C_0}\right)(\psi_s - \phi_c - 2|\phi_f|)$$

or

$$\psi_s - \phi_c - 2|\phi_f| = \frac{1}{n}[V_G - V_T(\phi_c)]. \quad (31)$$

where

$$V_{FB} = \Phi_{MS} - \frac{Q_{SS}}{C_0} - \frac{Q_{fs0}}{C_0}$$
$$n = \frac{C_d + C_{fs} + C_0}{C_0}$$

and

$$V_T(\phi_c) = V_{FB} + \phi_c + 2|\phi_f| + \frac{\sqrt{2q\epsilon_s N_A(\phi_c + 2|\phi_f|)}}{C_0} \quad (32)$$

is the gate voltage when $\psi_s = \phi_c + 2|\phi_f|$. C_d and C_{fs} are given by (5).

Inserting (31) into (30) and evaluating the denominator at $\psi_s = \phi_c + 2|\phi_f|$ gives Q_n as a function of V_G in the weak inversion region.

$$-Q_n = C_d(kT/q) \exp q[V_G - V_T(\phi_c)]/nkT. \quad (33)$$

When ψ_s is some small amount (several kT/q) greater than $\phi_c + 2|\phi_s|$ the exponential term in (28) dominates and the strong inversion criterion is satisfied. From (29) it is apparent that ψ_s varies logarithmically with V_G in the strong inversion region. As a first approximation it is assumed that ψ_s is pinned a small potential $\Delta\phi$ above $\phi_c + 2|\phi_f|$. In other words $\psi_s = 2|\phi_f| + \Delta\phi$ when the surface is strongly inverted. Equation (29) then yields, upon solving for Q_n,

$$-Q_n = C_0\left[V_G - V_T(\phi_c) - \left(1 + \frac{C_d}{C_0}\right)\Delta\phi\right]. \quad (34)$$

$\Delta\phi$ is found by assuming that the weak inversion equation (33) must be just tangent to the strong inversion equation (34). The point of tangency V_α is the dividing point above which (34) is valid and below which (33) is valid. Equating (33) and (34) and their derivatives at V_α yields

$$\Delta\phi = \frac{1}{1 + \dfrac{C_d}{nC_0}} \frac{kT}{q} \left[\ln\left(\frac{nC_0}{C_d}\right) - 1 \right] \qquad (35)$$

$$V_\alpha = V_T(\phi_c) + \left(1 + \frac{C_d}{C_0}\right)\Delta\phi + n(kT/q). \qquad (36)$$

Equations (33)–(36) thus furnish an approximate description of the inversion layer charge in both the strong and weak inversion regions.

Usually a turn-on voltage $V'_T(\phi_c)$ is experimentally defined by extension of the linear portion (strong inversion) of the Q_n versus V_G relation to zero charge. From (34) this means that $V'_T(\phi_c) = V_T(\phi_c) + (1 + (C_d/C_0))$ $\Delta\phi$. Substituting this relation for $V'_T(\phi_c)$ into (33)–(36) yields the following convenient equations

$$-Q_n = C_0\left(n\frac{kT}{q}\right)\exp\left[\frac{q}{nkT}\left(V_G - V'_T(\phi_c) - n\frac{kT}{q}\right)\right],$$
$$(37)$$

valid when $V_G \leq V'_T(\phi_c) + n(kT/q)$ and

$$-Q_n = C_0[V_G - V'_T(\phi_c)], \qquad (38)$$

valid when $V_G \geq V'_T(\phi_c) + n(kT/q)$. Since $V'_T(\phi_c)$ differs only by a small constant from $V_T(\phi_c)$, no distinction is made between them in the text.

REFERENCES

[1] K. Nagane and T. Fanak, "Al₂O₃ complementary MOS transistors," in *Proc. 1st Conf. Solid-State Devices*, pp. 132–136, Tokyo, 1969.
[2] T. Leuenberger and E. Vittoz, "Complementary-MOS low-power low-voltage integrated binary counter," *Proc. IEEE* (Special Issue on Materials and Processes in Integrated Electronics), vol. 59, pp. 1528–1532, Sept. 1969.
[3] R. G. Daniels and R. R. Burgess, "The electronic wristwatch; an application for Si Gate CMOS-IC's," *ISSCC Dig. Tech. Papers*, 1971.
[4] S. M. Sze, *Physics of Semiconductor Devices*. New York: Wiley, 1969, p. 517.
[5] A. S. Grove, *Physics and Technology of Semiconductor Devices*. New York: Wiley, 1967.
[6] J. D. Meindl and R. M. Swanson, "Potential improvements in power speed performance of digital circuits," *Proc. IEEE*, vol. 59, pp. 815–816, May 1971.
[7] R. W. Keyes, "Physical problems and limits in computer logic," *IEEE Spectrum*, vol. 6, pp. 36–45, May 1969.
[8] J. R. Burns, "Switching response of complementary symmetry MOS transistor logic circuits," *RCA Rev.*, vol. 24, pp. 627–661, Dec. 1964.
[9] J. F. Gibbons, "Ion implantation in semiconductors,—Part 1: Range distribution theory and experiments," *Proc. IEEE*, vol. 56, pp. 295–319, March 1968.
[10] W. S. Johnson and J. F. Gibbons, "Projected range statistics in semiconductors," Stanford Univ. Press, Stanford, Calif., 1970.

Trading Speed for Low Power by Choice of Supply and Threshold Voltages

Dake Liu and Christer Svensson

Abstract— The trading of speed for low power consumption in CMOS VLSI by using the supply voltage and the threshold voltage as variables has been investigated. It is shown that it is desirable to minimize the supply voltage for minimizing the power consumption. We have investigated both the lower bound of the supply voltage and the possible decrease in power consumption without speed loss under different circuit constraints and calculated the consequences on circuit performance. Results show, for example, that power reductions of about 40 times can be obtained without speed loss by using supply voltages down to about 0.48 V.

I. INTRODUCTION

TODAY CMOS technology has taken over most high-performance complex chips, mainly because of its favorably low power consumption. However, as VLSI technology continues to develop, we are again facing power consumption problems in todays high-end VLSI products. In fact, it seems that power consumption is the most important limitation to the computing capacity of VLSI chips.[1]

An interesting point is that the most important parameter controlling power consumption is the supply voltage. Still, supply voltage is seldom used as a parameter in power optimization. Instead we try to keep the standard 5-V supply voltage as long as we can, and when this becomes impossible due to small dimensions in scaled processes we still try to keep the supply voltage as large as possible [1]. It is the aim of this paper to introduce the supply voltage as the most important parameter controlling CMOS power consumption and to analyze the consequences of this in connection to further technology scaling.

The power consumption in CMOS is quite accurately described by

$$P_g = f_d C_L V_{dd}^2 \qquad (1)$$

where P_g is the power consumption of one gate, f_d is the average operating frequency of that gate, C_L is the total switching capacitance of the gate, and V_{dd} is the supply voltage. For a complete chip system, containing N gates, we may estimate the chip power consumption P from

$$P = N f_d C_L V_{dd}^2. \qquad (2)$$

Manuscript received March 4, 1992; revised July 16, 1992.

The authors are with the LSI Design Center, IFM, Linköping University, S-58183 Linköping, Sweden.

IEEE Log Number 9204134.

[1] Most recent high-performance CMOS chips dissipate very much power. Some examples (7-W signal processor, 13-W vector processor) can be found in the *IEEE International Solid-State Circuits Conference Digest*, 1991, pp. 254–257.

In this expression $N f_d$ is a measure of the total number of bitoperations per second performed by the system and therefore a measure of the system computing capacity. Using this interpretation of (2), we have a very simple expression for the power consumption versus computing capacity of a VLSI system. Thus, if we want to reduce power for a certain computing task, there are only two variables to consider, C_L and V_{dd}. C_L depends on the choice of circuit style and the choice of process technology, which is not discussed in this paper. Reduction of V_{dd} should be a very effective way to reduce the power consumption needed for a certain computing task, according to (2). This is also the general conclusion in [19]. Note, that this conclusion is independent of any assumptions about speed or architecture of the VLSI system. Our conclusion is then quite clear: reduce the voltage as much as possible. In the following we will analyze the lower bound for the supply voltage in CMOS VLSI systems.

The above conclusion may seem too simple. The simplification comes from the fact that we have hidden speed and architecture considerations of the system. The reason why we so far have been reluctant to reduce the supply voltage more than necessary is that it will lead to a reduced speed. In order to keep the same computing capacity under reduced voltage, we thus need to increase parallelism to compensate for the reduced speed. This will require more devices, complicate our algorithms, and give rise to more control overhead [19]. The need for more devices is, however, not necessarily a problem, as continuing VLSI development gives us more and more devices and sometimes we ask ourselves what to use all these devices for [2]. The more complicated algorithms may, of course, be a problem. In conclusion, however, it is our judgement that system power reduction is so important that in many cases the strategy proposed here will prove useful.

Finally, there is a comment on VLSI system clock frequency and interconnections. It is well known that the speed of future VLSI systems using submicrometer technologies is limited by interconnection delays [3]. This is true also for optimistic predictions using several tricks to increase speed [4]. From this point of view, it will be less important to keep up the device speed in future scaled-down processes. Instead, it seems realistic to utilize the benefits from the scaled-down technology not through its increased speed but through its increased device density and to trade increased parallelism for reduced power.

Section II of this paper illustrates the modeling of speed and power consumption used for micrometer and submicrometer technologies. The effects of process-induced statistical errors

Reprinted from *IEEE Journal of Solid-State Circuits*, pp. 10-17, January 1993.

in transistor parameters are discussed in Section III. In Section IV we discuss our method of calculating speed and power consumption and of choosing supply and threshold voltages for optimum performance. In Section V we apply our results to different classes of circuits and give some examples. Section VI, finally, gives some conclusions.

II. Modeling

A. General

The aim of the present work is to investigate the effect of supply voltage on speed and power consumption. There is, however, no need for a detailed analysis of different circuit techniques, different logic styles, or different architectures. We will therefore consider only simple CMOS inverters, as a model for general CMOS logic [3]. We will, in fact, use the inverter to model static logic and use NMOST and PMOST as well as the inverter to model dynamic logic. The speed of the inverter is defined as the average of the delays for rising and falling edges. The signal used should have the same form (rising and falling slopes, etc.) as that which occurs inside normal logic with minimum-size MOST's (assumed to be obtained inside a long inverter chain). The speed of a real system using a certain circuit and logic style is then expected to be proportional to this inverter speed. The power consumption of the inverter is defined for the inverter in drift with a "natural" signal form as above. We believe this power measure can be used for static as well as for dynamic logic. Again, we expect the power of a real system using a certain circuit and logic style to be proportional to this inverter power. Finally, for dynamic logic we must be able to store information on a logic node at least for one clock cycle. This will be modeled as charge storage at a node in an "inverter" chain, assuming that both transistors in the inverter driving the actual node are in their off state. The storage should be correct for a full cycle of a certain minimum clock frequency.

B. MOST Modeling

Submicrometer transistor modeling is important in calculating speed and power consumption in present and future CMOS technologies. The commonly used one-dimensional first-order model may deviate more than 20 times from reality when gate lengths are below 0.5 μm ([5, p. 13]). This is mainly because of mobility degradation, velocity saturation, and drain-induced barrier lowering (DIBL). Two-dimensional modeling consumes much computer time so it cannot be used for circuit simulation directly. Therefore one-dimensional second-order submicrometer MOS models are used in this paper, and the dc model used in this paper is taken from [5] to [7]. Also submicrometer subthreshold current models are used [8], [9] in order to accurately describe also the subthreshold currents. This is important, as we will see, as the leakage currents sets a lower bound to the threshold voltage of dynamic logic and as the MOSFET's sometimes operate in the subthreshold region in this paper.

Threshold voltage modeling is based on the long-channel MOS threshold model in [10, p. 98, eq. (2)]. Three kinds

of short-channel effects are added for short-channel threshold modeling. The first is $\Delta V_{th}(L_{\text{eff}})$ [7], which is called the short-channel effect. The threshold voltage is thus decreased by the value of $\Delta V_{th}(L_{\text{eff}})$, where L_{eff} is the effective gate length. Another correction factor is $\Delta V_{th}(W)$ [5], which makes V_{th} increase by the value of $\Delta V_{th}(W)$, where W is the transistor width. The last part decreases V_{th} by $\Delta V_{th}(\text{DIBL})$ [5] due to the DIBL effect mentioned above.

$$V_{th} = V_{thL} - \Delta V_{th}(L_{\text{eff}}) + \Delta V_{th}(W) - \Delta V_{th}(\text{DIBL}) \quad (3)$$

The V_{thL} in (3) is the long-channel threshold voltage. The drain current calculations are also based on the long-channel model. The long-channel drain current I_{dsL} is defined in the subthreshold region, the linear region, and the saturation region. In the subthreshold and weak inverse regions we have [9]

$$I_{\exp} = \mu_0 C_{ox} \frac{W_{\text{eff}} V_t^2}{L_{\text{eff}}} e^{1.8} e^{[(V_{gs} - V_{th})/(nV_t)]} [1 - e^{-V_{ds}/V_t}]$$
$$(4)$$

$$I_{\text{limit}} = \frac{\mu_0 C_{ox} W_{\text{eff}}}{2 L_{\text{eff}}} (3V_t)^2 \quad (5)$$

$$I_{ds\_weak} = \frac{I_{\exp} I_{\text{limit}}}{I_{\exp} + I_{\text{limit}}}. \quad (6)$$

V_{ds} and V_{gs} are the voltages from the drain and the gate to the source. V_{th} is the threshold voltage and V_t is the thermal voltage. W_{eff} and L_{eff} are effective gate width and length, respectively. The μ_0 is the mobility which depends on the temperature and doping profile (see [18, pp. 85–89, eqs. (4.1.11) and (4.2.9)]). In the linear region $I_{dsL}(V_{gs} > V_{th}, V_{ds} < V_{dsat})$ given in [10], the V_{dsat} is

$$V_{dsat} = \frac{E_{\text{sat}} L_{\text{eff}} (V_{gs} - V_{th})}{E_{\text{sat}} L_{\text{eff}} + (V_{gs} - V_{th})} \quad (7)$$

E_{sat} is defined in [5, pp. 14–16] and the corresponding short-channel currents for the subthreshold and linear regions are given by

$$I_{ds\_strong} = F_\mu F_v I_{dsL} \qquad (V_{gs} > V_{th}) \quad (8)$$

where $F_\mu = \mu_{\text{eff}}/\mu_0$ is the effective μ factor, $F_V = [1 + (V_{dd}/E_{\text{sat}}/L_{\text{eff}})]^{-1}$ is the effective velocity saturation factor, and I_{dsL} is the long-channel drain-to-source current in the linear region [5, p. 6]. In the subthreshold region I_{dsL} is I_{ds_weak}. The short-channel saturation current in strong inversion region is given by

$$I_{ds\_strong} = F_\mu F_v I_{dsat} \left[\frac{L_{\text{eff}}}{L_{\text{eff}} - \Delta L_s} \right]$$
$$\cdot \left[\frac{V_{dsat} + E_{\text{sat}} L_{\text{eff}}}{V_{dsat} + E_{\text{sat}}(L_{\text{eff}} - \Delta L_s)} \right] \qquad (V_{gs} > V_{th}) \quad (9)$$

and ΔL_s is the parameter on [5, p. 29]. Finally, the I_{ds} model of MOS4 has been used [9], the I_{ds_strong} is limited when $V_{gs} > V_{th}$, and $F_\mu F_v I_{ds\_weak}$ is defined for all V_{gs}:

$$I_{ds} = I_{ds\_strong} + F_\mu F_v I_{ds\_weak}. \quad (10)$$

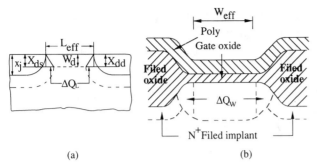

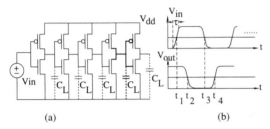

Fig. 1. Short channel length and narrow width effects.

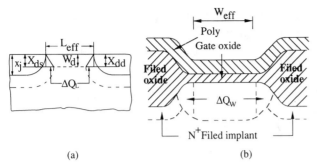

Fig. 2. The inverter chain and the delay time of the inverter.

The load capacitance of a CMOS inverter loaded by another CMOS inverter is composed of the drain-to-bulk capacitances of one n-transistor and one p-transistor (where each diffusion area is large enough to accommodate a metal-to-diffusion contact) and gate capacitances from each n- and p-transistor loading inverter [11], [12]. The gate capacitance is defined as

$$C_g = C_{gs} + C_{gd} + C_{ch}. \qquad (11)$$

C_{gs} is gate-to-source capacitance, C_{gd} is gate-to-drain capacitance, and C_{ch} is the channel capacitance between gate and channel. The drain-to-bulk capacitance C_{db} of a MOST is modeled as the sum of the junction capacitance and sidewall junction capacitance, which is described in detail in [10, p. 106, eq. (5)].

C. Delay and Power Consumption Modeling

The simulation is based on an element inverter coming from an inverter chain, which is shown in Fig. 2(a). The load capacitance of each inverter is defined as

$$C_L = C_{dbni} + C_{gsni+1} + C_{gdni+1} + C_{chni+1} + C_{dbpi} \\ + C_{gspi+1} + C_{gdpi+1} + C_{chpi+1}. \qquad (12)$$

The delay time definition comes from [13], which is shown in Fig. 2(b), and formula (13):

$$t_d = t_2 - t_1 + t_4 - t_3. \qquad (13)$$

The first inverter in the inverter chain is driven by a voltage source with waveform shown in Fig. 2(b). The V_{dd} is the supply voltage, and τ is an initial value for the first inverter in Fig. 2(a). As the power consumption and delay time are from the fourth inverter in the inverter chain, τ does not affect the results if it is defined less than t_d. The output of the first inverter is the input of the second inverter and so on. Every

inverter's output are integrated by formula (14):

$$V_{out} = \int_{t(V_{out}=0)}^{t(V_{out}=V_{dd})} \frac{ABS[I_{ds}(NMOS) - I_{ds}(PMOS)]}{C_L} dt. \qquad (14)$$

The delay time, which is defined by t_1 to t_4, is measured at the fourth inverter in order avoid any influence of τ. The power consumption of an inverter includes three parts, dynamic power consumption, short current power consumption, and static power consumption [14]. The switched energy per bit is defined as E_s in (15) and the dynamic power consumption is accounted for by (16):

$$E_s = V_{dd} \int_0^{V_{dd}} C_L dV \qquad (15)$$

$$P_d = f_{clock} E_s. \qquad (16)$$

The short current power consumption is

$$P_s = f_{clock} I_{mean} V_{dd}. \qquad (17)$$

f_{clock} is the clock frequency of the inverter and I_{mean} is the mean current passing through the drain–source body resistances of the two transistors from V_{dd} to ground in one clock cycle. The static power consumption is

$$P_o = V_{dd} I_{ds}(V_{ds} = V_{dd}, V_{gs} = 0). \qquad (18)$$

The total power consumption for an inverter is

$$P_{tinv} = P_d + P_s + P_o. \qquad (19)$$

D. Modeling of a CPU as a System Example

There are two sets of criteria used in this paper to determine how much the power consumption has been improved. One is the set of elementary inverter parameters (mainly delay time and power consumption, etc.), which are demonstrated above. Another is the set of parameters coming from an example of a CPU, which is taken from the SUSPENS model (Stanford University System Performance Simulator) [3]. This model is based on Rent's rule and we use the model for the chip-level simulation. The example chosen in this paper is that 10 000 gates ($N_g = 10000$) are built in a chip with a dimension of D_c centimeters, where $D_c = \sqrt{N_g} d_g$ is scaled down by pitches (p_w) of the metal interconnection and gate sizes (d_g). The average logic depth (f_{ld}) is 10. The activity ratio A_r (the fraction of gates switching on a given cycle) is 0.15. The dynamic logic simulation has been done on a NORA two-phase clocked domino logic. The average fan-in and fan-out are 3. The static logic simulation is based on the buffered static CMOS with fan-in and fan-out of 3. The Rent's rule constant is 0.4 and the logic gate pitch is 4.2. The utilization efficiency of chip interconnections is defined 0.4. Two layers of metal (Al) interconnections are considered. Parameters of transistors and interconnections are listed in Table I. A possible maximum CPU clock frequency (20) and the total chip driving capacitance (21) are obtained from this model.

$$f_{cmax} = \left(f_{ld} T_g + r_{int} c_{int} \frac{D_c^2}{2} + \frac{D_c}{v_c} \right)^{-1} \qquad (20)$$

TABLE I
CMOS TECHNOLOGY PARAMETERS

Feature size (μm)	2μm	1μm	0.5μm	0.25μm
Gate Length L(μm)	2	1	0.5	0.25
Effective L, Leff(μm)	1.5	0.7	0.3	0.15
Gate width W(μm)	5	3	2	1.5
Gate overlap ΔL(μm)	0.25	0.15	0.1	0.05
Active area depth Xj(μm)	0.25	0.15	0.1	0.05
Gate oxide thick tox(nm)	40	20	10	5
Width of metal Wint(μm)	3	2	1.5	1.25
Metal thickness Hint(μm)	0.5	0.4	0.3	0.2
Metal pitch pw(μm)	6	4	2	1
D & S active area(pm^2)	25	9	4	2.25
Chip size Dchip(cm)	0.945	0.63	0.315	0.157

$$C_{\text{chip}} = \frac{D_c^2 n_w e_w c_{int}}{P_w} + C_{tr} N_g f_g \qquad (21)$$

$$P_{td} = \frac{1}{2} f_{\text{cmax}} A_r C_{\text{chip}} V_{DD}^2 \qquad (22)$$

$$T_g = a_1 f_g R_{\text{gout}} l_{av} c_{int} + a_2 f_g R_{\text{gout}} C_{gin}$$
$$+ r_{int} c_{int} \frac{l_{av}^2}{2} + r_{int} l_{av} C_{gin}. \qquad (23)$$

In formula (20) and in Table I, r_{int} and c_{int} are the interconnection resistance and capacitance per unit length, respectively. v_c is the propagation speed of the electromagnetic wave. T_g is the average gate delay, which includes local interconnections with an average interconnection length of l_{av}. The factors a_1 and a_2 are 1 for dynamic logic. In static logic simulation a_1 is 1.5 and a_2 is 2. f_g is the mean fan-out, R_{gout} is the output resistance of a gate, C_{tr} is the input capacitance of a minimum-sized transistor, and C_{gin} is the average input capacitance of a gate considering all fan-in of this stage and fan-out of the pre-stage. C_{chip} is the total chip capacitance including all MOST capacitances and all interconnection capacitances. n_w is the number of interconnection levels on the chip and p_w is the wiring pitch of chip interconnections. P_{td} is the total dynamic power consumption of the CPU example at the clock frequency of f_{cmax} and, finally, P_t gives the total power consumption of a CPU in the later tables.

E. Modeling of the Lower Bond of the Clock Frequency for Dynamic Logic

It is necessary to decrease the threshold voltage as far as possible in order to keep the speed as high as possible when reducing V_{dd}. When the threshold voltage is reduced, the leakage

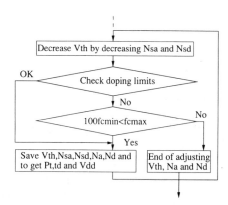

Fig. 3. Method to calculate minimum acceptable V_{th} for dynamic logic.

current of the MOST's is increased. This current discharges the load capacitance during a clock interval and increases the static power consumption in a dynamic synchronous VLSI system. A minimum clock frequency f_{cmin} is taken into account in this paper, which limits the leakage current and illustrates how much the threshold voltage can be reduced. It is defined as if the system clock frequency is lower than this f_{cmin}. The leakage current of the inverter could decrease the charged output voltage on C_L from V_{dd} to an uncertain level (about $1/2\ V_{dd}$) in one clock interval. As the leakage current could have a large deviation, the minimum clock frequency should be defined as (24) for safety:

$$f_{\text{cmin}} = 1/t_{\text{fault}}$$

and

$$t_{\text{fault}} = \text{MIN}\{t_{\text{fault}n}[V_{th}(\text{NMOST})$$
$$= V_{thn} - 3\sigma_{Vth}]t_{\text{fault}p}[V_{th}(\text{PMOST})$$
$$= V_{thp} + 3\sigma_{Vth}]\}. \qquad (24)$$

The t_{fault} is the time interval in which the V_{out} is discharged from V_{dd} or 0 V to a half V_{dd}. Fig. 3 demonstrates the method to calculate this modeling of clock frequency and related minimum threshold voltages for dynamic logic.

III. EFFECTS OF PARAMETER DEVIATIONS

Integrated circuit processes normally give quite large variations of process parameters. It is therefore necessary to take these deviations from nominal values into account. The deviations of the delay time and the power consumption of the inverter are mainly from the deviation of the threshold voltage. All deviations that are considered in integrated circuits relate to local or global variations. Global variations account

$$3\sigma_{Vth} = \sqrt{\sum_i \left(\Delta X_i \frac{\partial V_{th}}{\partial X_i}\right)^2}$$
$$= \sqrt{\left(\Delta W \frac{\partial V_{th}}{\partial W}\right)^2 + \left(\Delta Q_{ss} \frac{\partial V_{th}}{\partial Q_{ss}}\right)^2 + \left(\Delta N_a \frac{\partial V_{th}}{\partial N_a}\right)^2 + \left(\Delta L \frac{\partial V_{th}}{\partial L}\right)^2 + \left(\Delta t_{ox} \frac{\partial V_{th}}{\partial t_{ox}}\right)^2 + \left(\Delta X_j \frac{\partial V_{th}}{\partial X_j}\right)^2} \qquad (25)$$

for the total variation of parameters between chips or wafers fabricated at any occasion. Local variations reflects parameter variations between transistors in the same chip. In our case the goal is to design systems that should work for any fabrication instant, so we need to consider the global variations. Global variations of the threshold voltage in CMOS processes can be expressed as in (25) at the bottom of the previous page, where L and W are the gate length and width, respectively, t_{ox} is the gate oxide thickness, X_j is the drain or source diffusion depth, N_A is the channel surface doping concentration, and Q_{ss} is the oxide–silicon interface charge. As the deviations coming from X_i's represent the contemporary limitation in VLSI fabrication, we will in the following assume that the same relative values ($\Delta X_i/X_i$) can be used to characterize also other processes, including scaled processes. A general motivation for this assumption could be as follows. Any process, including the one used above, is designed with the goal to produce circuits with a reasonable yield. As a result of this, the process will exhibit parameter deviation characteristics for a process quality corresponding to this yield.

It is not possible to discuss in detail all effects of deviations in this paper. Instead we will use the deviation analysis to give limits to the acceptable threshold voltage, taking its effect on the lower frequency range of dynamic circuits and on the static power dissipation of static circuits into account. These are the two effects we have found most critical for the functioning of our circuits. We will thus judge a too large V_{th} deviation as a catastrophic failure, limiting the yield. We will assume that our maximum threshold voltage deviations corresponds to 3σ from the average value. This means that we expect about 0.003 of the fabricated batches to fail because of parameter deviations, which is a very acceptable number. All other parameters calculated in the following will be based on typical parameter values and therefore correspond to typical performance. We will not consider the extreme performance values.

For the values of the deviations in (25) we have chosen to use relative deviations of $\Delta X_i/X_i = 10\%$ for each parameter [15]. These values are assumed to be 3σ values and all parameters in (25) are assumed to be uncorrelated [16]. If we use the 10% deviation value on each parameter in the expression above, we calculate a 3σ threshold voltage deviation of 140 mV for a 1-μm process with a threshold voltage of 0.75 V. This agrees quite well with the maximum deviation normally given by the vendor for such a process.

IV. SIMULATION PROCEDURE

When trading speed for power by reducing the supply voltage, there is no unique goal. Instead the goal depends on the application. We will therefore demonstrate a couple of examples below, rather than proposing some ultimate supply voltage. We will only consider synchronous logic in the present paper, but two different circuit techniques, dynamic logic and static logic. Different processes, optimized in different ways, were simulated in five cases. Case 1 is a fixed technology with fixed threshold voltages and other fixed parameters (fixed for each feature size). This is used as a reference. Case 2 is

optimized for minimum power consumption under constant speed (inverter delay time t_d) for dynamic logic. Case 3 is an optimization of the power consumption under any speed in dynamic logic. Case 4 is aimed at the static logic for minimum power consumption under the same inverter delay time as case 1. Finally, in case 5, we get the lower bound of V_{dd} for the lowest total power consumption under any speed in static logic.

Our reference processes for different feature sizes are presented in Table I. These data represent a compromise between constant voltage scaling and practical considerations. The reference processes (case 1) will use 5-V supply voltage for the 2- and 1-μm processes and 3 V for the 0.5- and 0.25-μm processes. The gate width is chosen as $W_{int} + 2(L/2)$ (as the natural transistor width is given by the size of the drain connection) and L_{eff} is given by $L - 2\Delta L$ with $\Delta L = x_j$. Gate oxide thickness is scaled linearly. As $V_{dd} = 3$ V for $L = 0.25\mu$m is just a reference example, the reliability issue is not discussed and we only take the gate oxide breakdown limitation into account in this paper. Other parameters are taken from [3]. Surface doping is used to control the threshold voltage (and is therefore given by V_{th}) and the substrate and well dopings are assumed 10 times smaller than surface dopings [1].

Delay times and power consumptions below are typical (average) values. Critical parameters that are limiting the gate function are worst case (based on 3σ deviations). In Cases 2 and 3 the leakage current is the limiting factor of the threshold voltage and it is calculated from $V_{th}-3\sigma_{V_{th}}$. In Case 4 the leakage current affects the static power consumption and this limiting factor is again considered for $V_{th}-3\sigma_{V_{th}}$. Furthermore, the inverter output swing and the input/output behavior are limiting factors and it is calculated from $V_{th} \pm 3\sigma_{V_{th}}$. All parameters are calculated at 350 K, assumed to be the worst case from temperature point of view.

Before going into our specific optimized cases, let us study the effect of supply voltage, feature sizes, and threshold voltages on speed. In Figs. 4 and 5 we show our simulated results of speed (t_d) and switch energy (E_s) for the reference processes (case 1 with circle marks) and for processes with optimized threshold voltages (cases 2 and 3 with square marks). For the optimized processes we have chosen individual threshold voltages making $100 \cdot f_{cmin} < f_{cmax}$ at 350 K. The procedure used for these optimizations is demonstrated in Fig. 3. We can see from Fig. 4 the low slope of the speed loss at relatively high supply voltage. This is because of the short-channel effects and it gives more motivation to reduce the supply voltage.

From these curves we see that the power consumption (switch energy) is strongly reduced with reduced supply voltage (as expected) and that a considerable reduction in supply voltage at constant speed can be obtained by threshold voltage optimization. This is further illustrated in Table II, where we compare our reference process with processes optimized for power at constant speed (case 2).

Furthermore, if speed is no constraint, considerably lower supply voltages (and power) can be obtained. In this case we have chosen to present a rather extreme example with no "external" speed constraint. Instead, under the assumption of

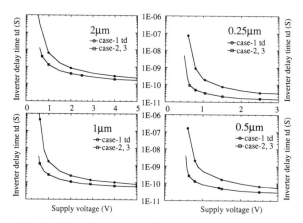

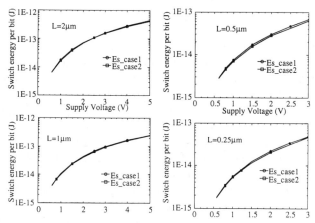

Fig. 4. Delay time versus supply voltage in cases 1, 2, and 3.

Fig. 5. Switch energy per bit of an inverter in cases 1, 2, and 3.

TABLE II
OPTIMIZING POWER CONSUMPTION WITH SAME INVERTER
DELAY TIME (t_d) FOR DYNAMIC LOGIC CIRCUIT

L μm	Delay time(ns)	Before optimization			After optimization				Power reduction
		Vdd	Vth	Es(J)	Vdd	Vthn	Vthp	Es(J)	
1	0.078	5	0.7	2.7E-13	2.80	0.25	-0.31	7.9E-14	3.2
0.5	0.053	3	0.7	6.7E-14	1.40	0.33	-0.22	1.4E-14	4.6
0.25	0.031	3	0.7	4.9E-14	1.05	0.32	-0.18	5.6E-15	8.2

TABLE III
THE LOWER BOUND OF V_{dd} FOR DYNAMIC LOGIC
WHEN $f_{cmax} = 1$ MHz AND $f_{cmin} = 10$ kHz

	GROUP	Dynamic logic Vth =0.7V			Dyn. logic limitation		
		1μm	0.5μm	0.25μm	1μm	0.5μm	0.25μm
INVERTER	Vdd(V)	5.0	3.0	3.0	0.56	0.56	0.535
	Vth when Vds=0 (V)	0.7	0.7	0.7	N:0.48 P:-0.32	N:0.53 P:-0.36	N:0.56 P:-0.37
	Vth when Vds=Vdd	N:0.67 P:-0.47	N:0.66 P:-0.45	N:0.61 P:-0.46	N:0.47 P:-0.28	N:0.52 P:-0.31	N:0.54 P:-0.33
	td(ns)	0.078	0.053	0.031	3.12	3.70	4.70
	Es (J)	2.7E-13	6.7E-14	4.9E-14	3.3E-15	2.3E-15	1.5E-15
	Es reduction [Es(case1)/Es(case3)]				89	29	32
CPU	fcmax(MHz)	94.5	96.8	178	1.0	1.0	1.0
	Pt(W)	4.5E-1	7.2E-2	7.6E-2	5.3E-5	2.8E-5	1.4E-5
	Pt reduction [Pt(case1)/Pt(case3)]				8430	2600	5430

TABLE IV
OPTIMIZING POWER CONSUMPTION WITH SAME INVERTER
DELAY TIME (t_d) FOR STATIC LOGIC CIRCUIT

L μm	Delay time (ns)	Before optimization			After optimization				Power reduction
		Vdd (V)	Vth (V)	Es (J)	Vdd (V)	Vthn (V)	Vthp (V)	Es (J)	
1	0.078	5	0.7	2.7E-13	1.70	0.02	-0.06	2.8E-14	9.64
0.5	0.053	3	0.7	6.7E-14	0.63	0.08	-0.01	2.6E-15	25.8
0.25	0.027	3	0.7	4.9E-14	0.48	0.1	-0.01	1.1E-15	44.5

TABLE V
THE LOWER BOUND OF V_{dd} FOR STATIC LOGIC CHECKED BY 3σ DEVIATION

	GROUP	Dynamic logic Vth =0.7V			Static logic limitation		
		1μm	0.5μm	0.25μm	1μm	0.5μm	0.25μm
INVERTER	Vdd(V)	5.0	3.0	3.0	0.14	0.13	0.12
	Vth when Vds=0 (V)	0.7	0.7	0.7	N:0.10 P:-0.051	N:0.10 P:-0.051	N:0.10 P:-0.051
	Vth when Vds=Vdd	N:0.67 P:-0.47	N:0.66 P:-0.45	N:0.61 P:-0.46	N:0.094 P:-0.043	N:0.092 P:-0.046	N:0.089 P:-0.048
	td(ns)	0.078	0.053	0.031	3.12	1.07	0.59
	Es (J)	2.7E-13	6.7E-14	4.9E-14	2.0E-16	1.2E-16	7.2E-17
	Es reduction [Es(case1)/Es(Case5)]				1275	530	625
CPU	fcmax(MHz)	94.5	96.8	178	0.73	2.93	3.41
	Pt(W)	4.5E-1	7.2E-2	7.6E-2	3.8E-5	8.1E-5	1.6E-4
	Pt reduction [Pt(case1)/Pt(case5)]				1.18E4	880	475

These limitations are checked for an inverter where the n-transistor is 4 times wider than the p-transistor, corresponding to the worst unsymmetry among static two-input gates with minimum transistors. In this extreme case, the static power could be larger than the dynamic power. Because of the very small supply voltage, the total power is very small. Results of these simulations of case 5 are presented in Table V.

V. RESULTS AND DISCUSSIONS

We have investigated the possibilities of reducing the power consumption of CMOS systems by reducing the supply voltage. In order not to lose too much speed it is very important to reduce the threshold voltages at the same time. The investigation has been done by designing optimized processes and carefully simulating these to verify their proper function and get their performances. We will here discuss mainly power consumption in the form of switch energy E_s and SUSPENS CPU total dynamic power P_{td} or total power P_t, and speed in the form of inverter delay t_d and SUSPENS CPU clock rate f_{cmax}.

Case 2 (Table II) demonstrates the effect of optimizing power under the constraint of fixed speed (t_d). By optimizing

dynamic logic, we must be sure that a system can be operated clearly above f_{cmin}. We have therefore chosen a relatively conservative constraint of $f_{cmax} > 100 f_{cmin}$, that is $f_{cmax} = 1$ MHz and $f_{cmin} \leq 10$ kHz, where f_{cmax} is the maximum clock frequency of the SUSPENS CPU model. The results of these simulations (case 3) are presented in Table III.

In case 4 we further notice that the threshold voltage can be reduced even further by removing the constraint of f_{cmin}. This means that we are limited to static CMOS logic. The limiting factors here will be the inverter full swing and gain requirements. All of these limitations are controlled by leakage currents. We will in this case control V_{th} so that all requirements above are fulfilled at worst-case deviation ($V_{th} - 3\sigma_{V_{th}}$). A good improvement is seen in Table IV.

Finally, we go further from step 4 to get the lower limit of the supply voltage in a static logic system. For this case the lower bound of the supply voltage will be given by the demand for a full-swing output voltage from a gate, the voltage gain of gates must be larger than 4, and the noise margin $> 0.15V_{dd}$.

the threshold voltages it is possible to reduce power 3.2–8.2 times. There is not much speed loss at the beginning of the V_{dd} reduction as the mobility degradation and velocity saturation are caused by high supply voltage. Results show that the shorter the gate length is, the better the power consumption improvement will be. The simulation shows that a future 0.25-μm process can have more than 8 times improvement from case 1 and an inverter delay of 31 ps at a switch energy of 5.6 fJ at a supply voltage of 1.05 V. An inverter delay of 31 ps indicates a maximum clocking speed of about 4 GHz ($1/8t_d$, obtained from the TSPC technique [17] in a small circuit).

Case 3 (Table III) demonstrates the lower limit of power consumption for dynamic logic. We have here assumed no speed constraint so that we do have a considerable speed reduction. We find a possible CPU power reduction of $89 \times 94.5 = 8.43 \times 10^3$ times at a cost of a CPU speed decrease of 94.5 times for the 1-μm feature size. There is about a 89 times power reduction from decreasing the switch energy and a 94.5 times power reduction from decreasing the clock frequency. This power reduction can certainly be exploited by compensating the speed loss by an increased parallelism as proposed in the introduction and [19]. The optimum supply voltage proposed in this case is 0.56 V for 1-μm CMOS. Using this strategy, a feature reduction from 1 to 0.25 μm, at constant speed, will improve the power consumption only 3.79 times. Still, a small feature size will certainly increase the packaging density and can be successfully exploited as proposed above. For case 3, dynamic logic at minimum power, we have presented only one relatively extreme case. Of course, other combinations of power and speed, corresponding to cases between Case 2 and Case 3, are possible (see Figs. 4 and 5). In this case, the MOSFET with 0.25-μm gate length operates in the subthreshold region. This conclusion is in agreement with the concepts of electric watch circuits discussed in [20].

Case 4 (Table IV) gives the optimization of the power consumption with the same inverter delay time t_d as of case 1. We get the result that the power consumption could be decreased from 9.64 to 44.5 times when t_d is the same as t_d of case 1 in Table IV. This also indicates that the smaller the feature size is, the larger the power consumption improvement will be. Reducing the supply voltage from 3 to 0.48 V in the 0.25-μm case without speed loss may seem surprising. One reason for this is the decrease of the short-channel effect (mainly F_v). The second reason is that the supply voltage is better utilized when V_{th} is close to zero, as the full swing is used as an effective gate voltage. The third reason is that the relatively large subthreshold current gives extra contribution on charge and discharge of the load capacitances.

In case 5 (Table V), finally, we examine the lower bound for the supply voltage of static logic. Here we get a possible switch energy (E_s) reduction of 1275–625 times with a t_d reduction of 40–19 times in Table V. Voltage gains and noise margins are checked in Fig. 6 by assuming a worst-case supply voltage of $0.9V_{dd}$. The system speed reductions are 33–129 times and the improvements for the system power consumption are from 1.18E4 to 475 times. For 1-μm feature size, the dynamic power is about 10% of the total power consumption. The system speed ($f_{c\max}$) reductions are much larger than the

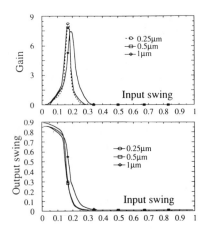

Fig. 6. The inverter gain and input–output characteristics. Note that the voltage scale is relative (V/V_{dd}) and V_{dd}'s are 0.9 of the values in Table V.

reduction of t_d because: 1) the on transistor resistances (R_{tr}) are increased as V_{dd} is scaled down, which increases the delay for global interconnections; and 2) the logic gate capacitances are doubled and the capacitances from local interconnections are larger than those of dynamic logic. Using static logic at 0.25 μm, our results indicate that it is possible to obtain an inverter delay of 0.59 ns at a switch energy of 0.072 fJ at a supply voltage of 0.12 V. An inverter delay of 0.59 ns indicates a maximum clocking speed of about 110 MHz ($1/16t_d$ as above). The SUSPENS CPU model indicates that a processor with 0.5 μm and 10000 gates will clock at 2.93 MHz at a total power consumption of 81 μW (5μW from dynamic power consumption and about 76 μW from static power consumption).

VI. CONCLUSIONS

We have shown that there are good opportunities for reducing the power consumption in present and future digital CMOS systems by the reduction of supply voltage. Without any speed reduction, power savings of the order of 8 times can be obtained by optimizing threshold voltage. By reducing speed (and possibly compensating for this by increased parallelism), power reductions of 2600–8430 times (about 29–89 times from switch energy, others are from speed loss) can be obtained in systems using dynamic logic. In systems using 0.25-μm gate length and static logic only, it is possible to decrease the power consumption about 44 times without inverter speed loss. Finally, using 0.5-μm gate length and static logic, the supply voltage can be decreased even to as low as 0.13 V to get more than 800 times system power reduction at 19 times system speed loss.

REFERENCES

[1] M. Kakumu and M. Kinugawa, "Power supply voltage impact on circuit performance for half and lower submicrometer CMOS LSI," *IEEE Trans. Electron Devices*, vol. 37, no. 8, pp. 1902–1908, Aug. 1990.
[2] J. L. Hennessy and N. P. Jouppi, "Computer technology and architecture: An evolving interaction," *IEEE Trans. Comput*, vol. 24, no. 9, pp. 18–29, 1991.

[3] H. B. Bakoglu, *Circuits, Interconnections, and Packaging for VLSI*, (VLSI Systems Series). Reading, MA: Addison Wesley, 1990.

[4] M. Afghahi and C. Svensson, "Performance of synchronous and asynchronous schemes for VLSI systems," *IEEE Trans. Comput*, vol. 41, no. 7, pp. 858–872, July 1992.

[5] N. G. Einspruch, *VLSI Electronics Microstructure Science*, vol. 18, *Advanced MOS Device Physics*. New York: Academic, 1989, pp. 237–273.

[6] H. Masuda, J.-I. Mano, R. Ikematsu, H. Sugihara and Y. Aoki, "A submicrometer MOS transistor *I–V* model for circuit simulation," *IEEE Trans. Computer-Aided Design*, vol. 10, no. 2, pp. 161–169, Feb. 1991.

[7] A. L. Silburt, R. C. Foss and W. F. Petrie, "An efficient MOS transistor model for computer-aided design," *IEEE Trans. Computer-Aided Design*, vol. CAD-3, no. 1, pp. 104–111, Jan. 1984.

[8] P. C. Chan, P. Liu, S. K. Lau, and M. Pinto-guedes, "A subthreshold conduction model for circuit simulation of submicron MOSFET," *IEEE Trans. Computer-Aided Design*, vol. CAD-6, no. 4, pp. 574–581, July 1987.

[9] Z. Yang, *Introduction for Design Methodology of ULSI*. Bejing: Tsinghua University Publishing, 1990, pp. 98–99 (in Chinese).

[10] P. E. Allen and D. R. Holberg, *CMOS Analog Circuit Design*. New York: Holt, Rinehart and Winston, 1987, p. 55.

[11] S. Shao-Shiun Chung, "A charge-based capacitance model of short-channel MOSFET's," *IEEE Trans. Computer-Aided Design*, vol. 8, no. 1, pp. 1–7, Jan. 1989.

[12] G. W. Taylor, W. Fichtner, and J. G. Simmons, "A description of MOS internal capacitances for transient simulations," *IEEE Trans. Computer-Aided Design*, vol. CAD-1, no. 4, pp. 150–156, Oct. 1982.

[13] D. Deschacht, M. Robert, and D. Auvergen, "Explicit formulation of delay in CMOS data paths," *IEEE J. Solid-State Circuits*, vol. 23, no. 5, pp. 1257–1264, Oct. 1988.

[14] N. Weste and K. Eshraghian, *Principles of CMOS VLSI Design*. Reading, MA: Addison-Wesley, 1985, pp. 144–148.

[15] E. H. Li and H. C. Ng, "Parameter sensitivity of narrow-channel MOSFET's," *IEEE Electron Device Lett.*, vol. 12, no. 11, pp. 608–610, Nov. 1991.

[16] M. Conti, C. Turchetti, and G. Masetti, "A new analytical and statistical-oriented approach for the two dimensional threshold analysis of short-channel MOSFETs," *Solid-State Electron.*, vol. 32, no. 9, pp. 739–747, 1989.

[17] J. Yuan and C. Svensson, "High-speed CMOS circuit technique," *IEEE J. Solid-State Circuits*, vol. 24, no. 1, pp. 62–70, Feb. 1989.

[18] M. Zambuto, *Semiconductor Devices*. New York: McGraw-Hill, 1989.

[19] A. P. Chandrakasan, S. Sheng, and R. W. Brodersen, "Low-power CMOS digital design," *IEEE J. Solid-State Circuits*, vol. 27, no. 4, pp. 473–484, Apr. 1992.

[20] E. Vittoz and J. Fellrath, "CMOS analog integrated circuits based on weak inversion operation," *IEEE J. Solid-State Circuits*, vol. SC-12, no. 3, June 1977.

Limitation of CMOS Supply-Voltage Scaling by MOSFET Threshold-Voltage Variation

Shih Wei Sun and Paul G. Y. Tsui

Motorola, APRDL, 3501 Ed Bluestein Blvd., Austin, Texas 78721, (512) 928-5028

Abstract

A fundamental limit of CMOS supply-voltage (V_{CC}) scaling has been investigated and quantified as a function of the statistical variation of MOSFET threshold-voltage (V_T). Based on the data extracted from a sub-0.5 μm logic technology, the variation of ring-oscillator propagation-delay (T_{PD}) significantly increases as V_{CC} is scaled down towards the MOSFET V_T (Fig. 1). An empirical power-law relationship was then derived to describe the scattering of circuit speed (ΔT_{PD}) as a function of MOSFET V_T variation (ΔV_T) and (V_{CC} - V_T). Agreement between the model and the experimental data was established for V_{CC} values from 4.0 V to 0.9 V. This fundamental limit of CMOS V_{CC} scaling poses an additional challenge for the design and manufacturing of high-performance, low-power portable equipment and battery based systems.

Introduction

For CMOS circuits, the switching power (P) can be described by $P = fC_L V_{CC}^2$, where f is the operating frequency and C_L is the loading capacitance. It is obvious that V_{CC} scaling is the most effective approach in reducing the power consumption [1,2,3]. However, power versus speed trade-off has been one of the major concerns for continued scaling of CMOS supply-voltage. In this work, the emerging issue of CMOS circuit speed scattering at low V_{CC} values has been analyzed from MOSFET V_T variation's point of view. All the experimental results were collected from a modular sub-0.5 μm logic CMOS/BiCMOS technology [4], with n+/p+ dual-poly gate, full titanium self-aligned silicide (SALICIDE), and triple level metallization (Fig. 2). The key process and device parameters of this sub-0.5 μm logic technology are summarized in Table 1.

Model

It is shown in Fig. 1 that the experimentally measured CMOS ring-oscillator (F.I.=F.O.=1) stage delay is tightly distributed at high V_{CC} values (> 2 V). However, T_{PD} variation increases at 2 V V_{CC}, and becomes uncontrollable for V_{CC} less than 1 V. In order to quantify this T_{PD} variation (ΔT_{PD}) at low V_{CC} values, the MOSFET $I_{D,SAT}$ equations were examined. For long-channel MOSFET under the constant-mobility assumption,

$$I_{D,SAT} = (W/2L)\mu C_{OX}(V_G - V_T)^2.$$

For deep sub-micron MOSFET under velocity saturation conditions,

$$I_{D,SAT} = WC_{OX}V_S(V_G - V_T)^1.$$

V_S is the saturation velocity. Based on these MOSFET current-drive ($I_{D,SAT}$) relationships with (V_G - V_T), the CMOS ring-oscillator performance can be empirically modeled with the following power-law relationships. The exponent of 1.5 in the circuit delay versus (V_{CC} - V_T) relationship,

$$T_{PD}^{-1} \propto (V_{CC} - V_T)^{1.5} \quad ...... \quad \{1\},$$

is assumed as an average of the aforementioned (V_G - V_T) exponents of 2 and 1 in the MOSFET $I_{D,SAT}$ equations. By differentiating this T_{PD} expression with respect to the MOSFET V_T, the relationship between

Reprinted from *IEEE Custom Integrated Circuits Conference*, pp. 267-270, 1994.

the circuit delay scattering (ΔT_{PD}) and the MOSFET V_T, ΔV_T can be established:

$$\Delta T_{PD} \propto \Delta V_T / (V_{CC} - V_T)^{2.5} \quad \dots \quad \{2\}.$$

The other process and device parameters affecting the circuit delay variation are included in the pre-proportional constant.

Experimental Results and Discussion

Fig. 3 illustrates the excellent correlation between the measured and calculated ring-oscillator stage delays (using eq. 1, $(V_{CC} - V_T)^{-1.5}$). As shown in Fig. 4, MOSFET V_T plays a key role in circuit performance as V_{CC} is scaled down towards MOSFET V_T. Fig. 5 plots the experimentally measured 1σ R.O. stage delay variation and the calculated ΔT_{PD} using eq. 2, $\Delta V_T / (V_{CC} - V_T)^{2.5}$. Again the data points track the empirical power-law relationship, $\Delta T_{PD} \propto \Delta V_T / (V_{CC} - V_T)^{2.5}$, extremely well for V_{CC} values from 4 V to 0.9 V. Because of the exponent of 2.5 in the denominator, reduction of MOSFET V_T is very effective in reducing ΔT_{PD}. However, the MOSFET off-state leakage and static power consumption become serious concerns at reduced V_T. Fig. 6 plots the 3σ worst-case off-state leakage as a function of MOSFET V_T for three different subthreshold slopes (SS). A 15 mV 1σ V_T variation was assumed. For a MOSFET off-state leakage requirement of 1 pA/μm at room temperature, the V_T has to be 0.56 V for an SS of 85 mV/decade, 0.5 V V_T for an SS of 75 mV/decade, and a V_T of > 0.4 V even for an ideal SS of 60 mV/decade (cf. fully depleted thin-film SOI). Fig. 7 compares the effects of both V_T and ΔV_T on the ring-oscillator stage-delay variation. In order to reduce ΔT_{PD} at low V_{CC} values without any significant increase of the off-state leakage, it is crucial to reduce the MOSFET V_T variation (ΔV_T). As the MOSFET ΔV_T is reduced, not only ΔT_{PD} reduces, the worst-case off-state leakage also becomes lower. Fig. 8 illustrates the percentage variation of the stage-delay ($\Delta T_{PD}/T_{PD}$) as a function of V_{CC}. $\Delta T_{PD}/T_{PD}$ ratio is approximately constant for V_{CC} > 2 V. However, the

ratio increases significantly as V_{CC} is scaled down below 2 V. For current 0.5 μm logic CMOS technologies, the 3σ ΔV_T spec. limit is approximately 10% or less for 3.3 V V_{CC}. In order to keep the same $\Delta T_{PD}/T_{PD}$ ratio at lower V_{CC}'s, the ΔV_T control has to be improved significantly. For example, the 3σ ΔV_T has to be < 5% at 1.8 V V_{CC}, and < 1.5% at 0.9 V V_{CC}. For a 0.5 V MOSFET V_T target, the 1σ ΔV_T has to be controlled less than 2.5 mV at 0.9 V V_{CC}, which represents a definite challenge to the processing capability of ULSI technologies.

Summary

In summary, limitation of CMOS supply-voltage scaling, caused by MOSFET V_T scattering, has been studied. This V_{CC} dependency of ΔT_{pd} was successfully modeled by an empirical power-law relationship: $\Delta T_{PD} \propto \Delta V_T / (V_{CC} - V_T)^{2.5}$. Trade-off has to be considered in reducing either V_T or ΔV_T in order to meet the ΔT_{pd} circuit requirements. For high-performance, low-power product applications with a V_{CC} of 2 V or less, both the on-going improvements in process control and innovations in circuit /system design have to be pursued concurrently.

Acknowledgments

The authors would like to acknowledge the wafer processing, testing, SEM, and management supports of Motorola's Advanced Products Research and Development Laboratory (APRDL).

References

[1] R. Brodersen, A. Chandrakasan, and S. Sheng, IEEE International Solid-State Circuits Conference, p. 168, 1993.
[2] D. Liu and C. Svensson, IEEE J. Solid State Circuits, Vol. 28, No. 1, p. 10, 1993.
[3] K. Shimohigashi and K. Seki, Symp. on VLSI Circuits, p. 54, 1992.
[4] S. W. Sun et al., IEDM Tech. Digest, p. 85, 1991.

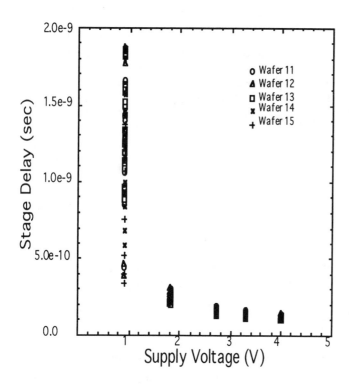

Fig. 1. Experimentally measured CMOS inverter ring-oscillator data ($|V_T| = 0.5$ V, Tox = 105 Å, Leff = 0.4 μm).

CMOS:	NMOS	PMOS
Gate Oxide Tox (Å)	105	105
Threshold Voltage, V_T (V)	0.50	- 0.50
Channel Length, Leff (μm)	0.4	0.4
S/D, Gate Rs (Ohms/sq.)	< 5	< 5
BIPOLAR:	NPN	PNP
Emitter Size (μm)	0.45/1.35	0.6/1.8
Current Gain	140	70
Peak f_T (GHz)	18, 26*	4
BVceo (V)	5.5, 4.0*	- 6.5

* Collector Pedestal Implant

Table 1. Sub-0.5 μm CMOS/BiCMOS parameters.

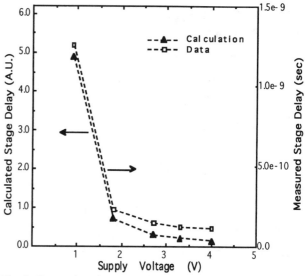

Fig. 3. Comparison between the measured and calculated stage-delay (T_{PD}) using $T_{PD}^{-1} \propto (V_{CC} - V_T)^{1.5}$.

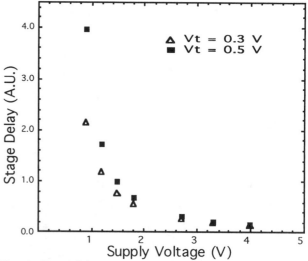

Fig. 4. Stage-delay versus supply voltage for two different threshold voltages (0.3 V and 0.5 V).

Fig. 2. SEM photomicrograph of the full titanium-SALICIDE, and triple-level metallization with stacked and plugged contacts/vias.

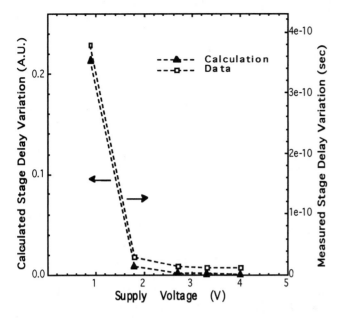

Fig. 5. Comparison of directly measured stage-delay variation and the calculated ΔT_{PD} using $\Delta T_{PD} \propto \Delta V_T/(V_{CC} - V_T)^{2.5}$.

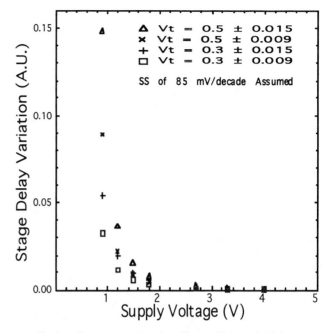

Fig. 7. ΔT_{PD} versus V_{CC} for different V_T's and ΔV_T's.

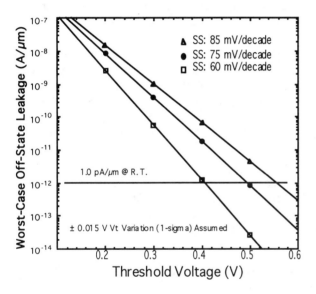

Fig. 6. Limitation of threshold-voltage scaling by MOSFET off-state leakage for subthreshold slopes (SS) of 85, 75, and 60 mV/decade.

$$\Delta T_{PD} \propto \Delta V_T/(V_{CC} - V_T)^{2.5}$$

(1) V_T Reduction: MOSFET off-stage leakage concerns, Higher static power consumption.

(2) ΔV_T Reduction: Need strenuous efforts in controlling all the V_T related process parameters, such as Tox., Nsub, Xj, Lpoly, Qss, W, ..., etc. for very low V_{CC} circuit implementations.

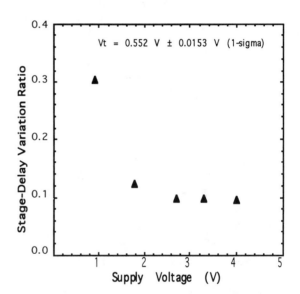

For constant $\Delta T_{PD}/T_{PD}$ Ratio,

V_{CC}	$\Delta V_T/V_T$ (3σ)
3.3 V	< 10 %
1.8 V	< 5 %
0.9 V	< 1.5 %

Fig. 8. Experimentally obtained $\Delta T_{PD}/T_{PD}$ ratio as a function of supply voltage.

1-V Power Supply High-Speed Digital Circuit Technology with Multithreshold-Voltage CMOS

Shin'ichiro Mutoh, *Member, IEEE*, Takakuni Douseki, *Member, IEEE*, Yasuyuki Matsuya, *Member, IEEE*, Takahiro Aoki, *Member, IEEE*, Satoshi Shigematsu, *Member, IEEE*, and Junzo Yamada, *Member, IEEE*

Abstract—1-V power supply high-speed low-power digital circuit technology with 0.5-μm multithreshold-voltage CMOS (MTCMOS) is proposed. This technology features both low-threshold voltage and high-threshold voltage MOSFET's in a single LSI. The low-threshold voltage MOSFET's enhance speed performance at a low supply voltage of 1 V or less, while the high-threshold voltage MOSFET's suppress the stand-by leakage current during the sleep period. This technology has brought about logic gate characteristics of a 1.7-ns propagation delay time and 0.3-μW/MHz/gate power dissipation with a standard load. In addition, an MTCMOS standard cell library has been developed so that conventional CAD tools can be used to lay out low-voltage LSI's. To demonstrate MTCMOS's effectiveness, a PLL LSI based on standard cells was designed as a carrying vehicle. 18-MHz operation at 1 V was achieved using a 0.5-μm CMOS process.

I. INTRODUCTION

A low-power design is essential to achieve miniaturization and long battery life in battery-operated portable equipment. Recently, there has been rapid progress in personal communications service (PCS) based on battery drives, including digital cellular phones, personal digital assistants, notebook, and palm-top computers. Future PCS will be more dedicated to multimedia systems, and thus the LSI's, the key component of the equipment, are desired not only for low-power consumption but also for higher signal or data processing capability [1], [2]. In order to promote this development, the demand for LSI designs achieving both low-power and high-speed performance should become stronger.

Lowering the supply voltage is the most effective way to achieve low-power performance because power dissipation in digital CMOS circuits is approximately proportional to the square of the supply voltage. From the point of view of applications to battery-powered mobile equipment, the supply voltage should be set at 1 V [1]. 1-V operation enables direct battery drive by a single Ni-Cd or Ni-H battery cell even taking the cell's discharge characteristic into account. This provides the smallest size and lightest weight equipment and eliminates the need for a power wasting dc-to-dc voltage converter.

Manuscript received April 25, 1994; revised December 21, 1994.
S. Mutoh, T. Douseki, Y. Matsuya, S. Shigematsu, and J. Yamada are with High-Speed Integrated Circuits Laboratory, NTT LSI Laboratories, Kanagawa 243-01, Japan.
T. Aoki is with Project Team-4, NTT LSI Laboratories, Kanagawa 243-01, Japan.
IEEE Log Number 9412024.

However, it is generally rather difficult to reduce the supply voltage to 1 V. The drastic degradation in speed is the largest problem. Although several studies of high-speed 1-V operating DRAM's have been reported [3], [4], they seem difficult to apply to general logic circuits because they assume stand-by node voltages throughout the entire circuit are predictable in memory LSI's, and utilization of the conventional layout CAD tool is thought to be difficult. Therefore, the development of novel circuit technology that achieves high-speed operation at a low voltage of 1 V with only a single battery drive and can be easily applied to random logic circuits is the key to developing the LSI designs for mobile equipment in the multimedia era.

This paper proposes just such a new 1-V high-speed circuit technology that is applicable to all digital CMOS circuits [5]. We call it multithreshold-voltage CMOS (MTCMOS). Its unique feature is that it uses both high- and low-threshold voltage MOSFET's in a single chip. In the next section, key issues in low-voltage operation are discussed. The MTCMOS technology and its main characteristics are described in Section III. In Section IV, layout schemes based on a standard cell and chip configurations are discussed. Finally, the performance of a PLL LSI designed and fabricated using a 0.5-μm CMOS process as a carrying vehicle for MTCMOS technology is shown in Section V.

II. DESIGN ISSUES FOR LOW VOLTAGE CMOS CIRCUITS

A. Low-Voltage Operation

Power dissipation in digital CMOS circuits is approximately expressed as

$$P \simeq C_L V_{dd}^2 f_{OP} \qquad (1)$$

where C_L is the load capacitance, V_{dd} is the supply voltage, and f_{OP} is the operating frequency. According to this formula, lowering V_{dd} is the most effective way to reduce power dissipation because it is proportional to the square of V_{dd}. Fig. 1 shows the relation between power consumption and supply voltage. It is apparent that lowering V_{dd} contributes significantly to power reduction. Reducing supply voltage from the 3.3 V, widely used at present, to 1 V realizes about 1/10 the power dissipation. Certainly, scaling down C_L or f_{OP} in (1) also contributes to low-power operation. Decreasing capac-

Reprinted from *IEEE Journal of Solid-State Circuits*, pp. 847-854, August 1995.

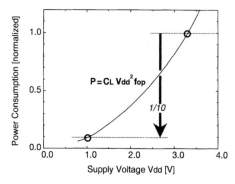

Fig. 1. Relation between power consumption and supply voltage.

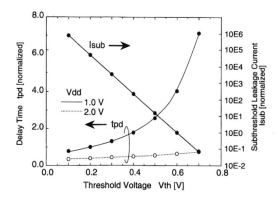

Fig. 2. Gate delay time and subthreshold leakage current dependence on threshold voltage.

itance C_L, however, would be difficult without scaling down the device and wiring, and higher throughput performance usually requires an increase in frequency f_{OP}. Although there have been attempts to lower f_{OP} by introducing parallel processing, this approach generally increases hardware overhead and requires extensive reworking at an architecture or algorithm design level [2].

B. Key Issue for Low-Voltage Operation

Although lowering V_{dd} to 1 V is effective in lowering power dissipation, as previously described, it is generally difficult because the speed performance is dramatically reduced at lower voltages. In CMOS digital circuits, the gate delay time (tpd) is approximately given by

$$tpd \propto \frac{C_L V_{dd}}{I_{DS}} \simeq \frac{C_L V_{dd}}{A(V_{dd} - V_{th})^2} \qquad (2)$$

where C_L is the load capacitance, I_{DS} is the drain current in the saturation region, V_{dd} is the supply voltage, V_{th} is the MOSFET's threshold voltage, and A is a constant. In the above expression, lowering the supply voltage decreases I_{DS} proportional to the square of the voltage difference $V_{dd} - V_{th}$, which results in a drastic increase in gate delay time as V_{dd} approaches V_{th}.

Until now, supply voltage has generally been lowered by scaling down the device feature size to ensure the reliability of thin gate oxides [6], [7]. Speed performance is maintained even at low voltage due to the improvement in transconductance g_m brought about by shrinking feature size to a half or deep submicron size. Considering the increasing demand for extremely low-power operation, however, much lower voltage should be applied to devices in the same generation. In this case, a decrease in delay time at lower voltage must be achieved without relying on device feature size scaling.

One way to overcome the speed degradation problem is to reduce the V_{th} of a MOSFET, as seen clearly in (2). Fig. 2 shows the circuit characteristics dependence on V_{th}. As V_{dd} gets lower from 2 to 1 V, gate delay time tpd becomes more sensitive to V_{th}. Therefore, reducing V_{th} is effective to achieve high-speed operation at a V_{dd} of 1 V. As V_{th} is reduced, however, another significant problem emerges—a rapid increase in stand-by current due to changes in the subthreshold leakage current. Subthreshold leakage current of

a MOSFET I_{sub} at $V_{GS} = 0$ is expressed as

$$I_{sub} \propto \exp\left(\frac{-V_{th}}{S/\ln 10}\right) \qquad (3)$$

where V_{th} is the threshold voltage of a MOSFET, and S is subthreshold swing. Leakage current characteristics at a V_{dd} of 1 V are also shown in Fig. 2. These values are calculated assuming S is 85 mV/decade. As V_{th} is reduced by 0.1 V, I_{sub} becomes about ten times larger. This becomes the source of the large stand-by current. With respect to portable equipment, in particular, the stand-by period is generally much longer than the operating period. Therefore, an increased stand-by current wastes battery power seriously. That is why it has been difficult to satisfy the requirements for both high-speed and low stand-by power at a low supply voltage of 1 V.

III. MTCMOS CIRCUIT TECHNOLOGY

A. Basic Circuit Scheme

The new MTCMOS circuit technology is proposed to satisfy both requirements of lowering the threshold voltage of a MOSFET and reducing stand-by current, both of which are necessary to obtain high-speed low-power performance at a V_{dd} of 1 V.

This technology has two main features. One is that N-channel and P-channel MOSFET's with two different threshold voltages are employed in a single chip. The other one is two operational modes, "active" and "sleep," for efficient power management.

Fig. 3 shows the basic MTCMOS circuit scheme with the NAND gates. The logic gate is composed of MOSFET's with a low threshold voltage of about 0.2–0.3 V. Its power terminals are not connected directly to the power supply lines VDD and GND, but rather to the "virtual" power supply lines VDDV and GNDV. The real and virtual power lines are linked by MOSFET's Q1 and Q2. These have a high threshold voltage of about 0.5–0.6 V and serve as sleep control transistors. Signals SL and $\overline{SL}$, which are connected to the gates of Q1 and Q2, respectively, are used for active/sleep mode control. Circuit operation in each mode at a supply voltage of 1 V is described below.

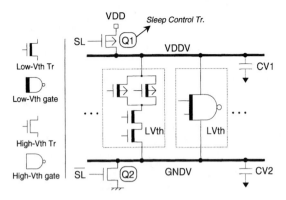

Fig. 3. MTCMOS circuit scheme.

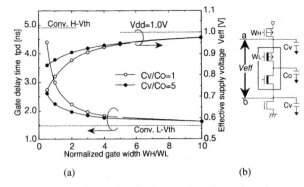

(a)   (b)

Fig. 4. Gate delay time and effective supply voltage dependence on the normalized gate width of the sleep control transistor. (a) Simulation results. (b) Simulation circuit model.

In the active mode, when SL is set low, Q1 and Q2 are turned on and their on-resistance is so small that VDDV and GNDV function as real power lines. Therefore, the NAND gate operates normally and at a high speed because the V_{th} of 0.3 V is low enough relative to the supply voltage of 1 V.

In the sleep mode, when SL is set high, Q1 and Q2 are turned off so that the virtual lines VDDV and GNDV are assumed to be floating. The relatively large leakage current, determined by the subthreshold characteristics of low-V_{th} MOSFET's, is almost completely suppressed by Q1 and Q2 since they have a high V_{th} and thus a much lower leakage current. Therefore, power consumption during the stand-by period can be dramatically reduced by the sleep control.

It should be pointed out that two other factors affect the speed performance of an MTCMOS circuit. One is the size of the sleep control transistors Q1 and Q2, and the other is the capacitances C_{V1} and C_{V2} of the virtual power lines. Q1 and Q2 supply current to the virtual lines. The larger their gate widths are designed, the smaller the on-resistance becomes. C_{V1} and C_{V2} also act as temporary supply sources to internal logic gates. Thus, the voltage rise in GNDV and drop in VDDV caused by the switching of the internal logic gate are suppressed by setting them large enough to maintain high-speed performance.

To confirm the effects, simulations were carried out. Fig. 4 shows the gate delay time tpd and effective supply voltage V_{eff} dependence on the normalized gate width of sleep control transistors W_H/W_L along with the simple single MTCMOS circuit model used for simulations, where V_{eff} is defined as the minimum value of spontaneous voltage difference between VDDV and GNDV (between node a and b in this simulation). It is clear that larger C_V, virtual line capacitance, and W_H, sleep control transistor width, maintain the effective supply voltage V_{eff} for the internal logic gates and enhance the speed performance. For instance, a W_H/W_L of 5 and C_V/C_O of 5 keep the decrease in V_{eff} within 10% of V_{dd} and the degradation in gate delay time within 15% compared to a pure low-V_{th} CMOS. The area penalty for the wider gate transistors is relatively small because they are shared by all the logic gates on a chip. As for C_V, the above condition is generally met in an actual LSI because C_V includes the source capacitances of all the logic gates connected to virtual power lines and wiring capacitances. Therefore, nothing extra need be added.

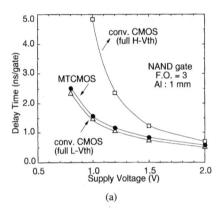

(a)

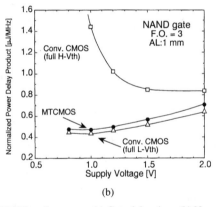

(b)

Fig. 5. MTCMOS performances. (a) Gate delay time. (b) Normalized power delay product dependence on supply voltage.

B. Electrical Performance

The measured MTCMOS logic gate delay time is shown in Fig. 5(a) as a function of supply voltage. Data for the conventional full high-V_{th} and full low-V_{th} CMOS logic gates are also plotted for comparison. It is obvious that the voltage dependence of an MTCMOS gate delay is much smaller than that of a conventional CMOS gate with high-V_{th} and that the MTCMOS gate operates almost as fast as the full low-V_{th} gate. At a 1-V power supply, the MTCMOS gate delay time is reduced by 70% as compared with the conventional CMOS gate with high-V_{th}. The dependence of normalized power-delay product (NPDP) on supply voltage is shown in Fig. 5(b),

TABLE I
CHARACTERISTICS OF MTCMOS CIRCUIT TECHNOLOGY

Power supply voltage	1.0 V
Propagation delay time	1.7 ns/gate
Power dissipation	0.3 μW/MHz/gate
	(2-input NAND with F.O.=3, line = 1 mm)

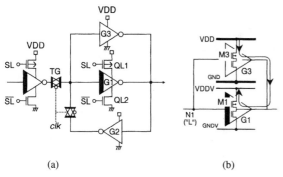

(a)   (b)

Fig. 6. MTCMOS latch circuit. (a) The proposed circuit. (b) The problem of the leakage current path.

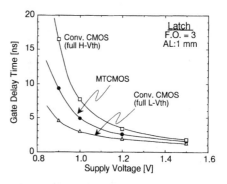

Fig. 7. Latch circuit delay time dependence on supply voltage.

where power consumption is normalized by frequency. At low voltages, especially below 1.5 V, the NPDP of the MTCMOS is much less than that of the conventional high-V_{th} gates, reflecting the improved speed performance at lower voltage. The smallest NPDP is achieved around 1 V in the MTCMOS gates. This shows that power reduction effect proportional to the square of supply voltage overcomes speed degradation in low-voltage operation. In addition, it was confirmed that the stand-by current was reduced three or four orders of magnitude due to the sleep control.

From these results, it is clear that MTCMOS circuit technology achieves both high-speed and low-power operations at a low supply voltage of 1 V or less. The measured characteristics related to this circuit technology are summarized in Table I. At 1 V, NAND gate delay time is typically 1.7 ns per gate, and power consumption is 0.3 μW/MHz per gate with a standard output load of three fanouts and 1 mm of wiring. MTCMOS gate operates about three times faster than conventional 0.5-μm CMOS gate. Power dissipation of 0.3 μW/MHz is 1/10 of the power needed for 3-V operation.

C. Design of Flip-Flop Circuit

Special attention must be paid to the MTCMOS design of latch or flip-flop circuits that have memory functions. This is because memorized data in latch or flip-flop circuits must be retained even in the sleep mode when virtual power lines are floating to cut leakage current completely. The proposed MTCMOS latch circuit is shown in Fig. 6(a), which is used for flip-flop circuits. The features are described below.

1) A conventional inverter G2 and a newly added one G3 are composed of high-V_{th} MOSFET's. They are connected directly to the true power supply lines VDD and GND. The latch path consists of G2 and G3, which are always provided with power. Therefore, data can be retained even in the sleep

mode, when the clock signal CLK is fixed by using the sleep control signal SL. G3 is designed to be smaller to suppress both the increases in the gate delay time and the area.

2) As for the forward path, the inverter G1 and the CMOS-type transmission gate TG are composed of low-V_{th} MOSFET's. This makes high-speed operation possible at 1-V power supply. This circuit also includes local sleep control transistors QL1 and QL2 with high-V_{th}. The reason for including them can be understood with Fig. 6(b), where a node N1 is assumed to maintain a "low" state in the sleep mode. If G1 were connected directly to the virtual power line VDDV, as shown in this figure, VDD and VDDV would be short through M1 and M3, so that stand-by current would be increased in the sleep mode. Therefore, QL1 and QL2 are indispensable for completely cutting the leakage current path. Fig. 7 shows the simulation results for the delay time of the MTCMOS latch circuit. They confirm that the delay time is reduced by 50% at 1 V compared with that of the conventional circuit with high-V_{th}. Furthermore, the stand-by current in the sleep mode was also confirmed to be almost as low as that of the high-V_{th} circuit.

IV. CHIP LAYOUT SCHEME

In order to make this low-voltage technology practical, conventional CAD tools must be applicable to lay out an MTCMOS LSI easily without any special consideration of the particular circuit scheme. To meet this requirement, the MTCMOS standard cell library was developed.

Fig. 8 shows the MTCMOS layout scheme based on a standard cell. The main feature is that the extra components of the MTCMOS circuit are buried in the cells. More specifically, the virtual power supply lines (VDDV and GNDV) and the sleep control signal line (SL) are buried in each cell, while the sleep control transistors Q1 and Q2 with high-V_{th} are buried in the power supply cell that provides the area needed to connect the true and virtual power supply lines to each other in the x and y directions. Power supply cells are placed on both sides of the logic cell based core. The true power supply lines VDD and GND, which are also placed in each cell, fix the voltage of either the substrate or the well and supply current to flip-flop circuits. This layout scheme allows the extra MTCMOS components to be connected automatically throughout the chip by abutting cells with a minimum increase in chip area.

89

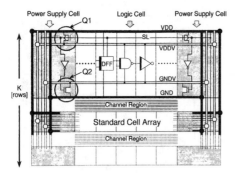

Fig. 8. Chip layout scheme based on a standard cell.

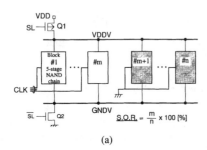

(a)

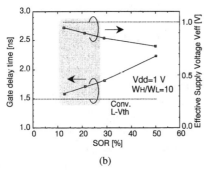

(b)

Fig. 9. Gate delay time and effective supply voltage dependence on Switch-On-Rate (SOR). (a) Simulation circuit model. (b) Gate delay time and effective supply voltage versus SOR.

Because Q1 and Q2 can be placed just under the power supply lines in the power supply cell, their insertion would incur no area penalty. Furthermore, the virtual power lines VDDV and GNDV in each row are connected together so that one cell can be supplied current through all the sleep control transistors within the chip, which contributes to suppress speed degradation.

In this experiment, W_H, the poly-gate width of sleep control transistors Q1 and Q2 in a power supply cell, was design to be ten times larger than that in the logic cells (W_L). Q1 and Q2 are shared by all the logic gates connected to the virtual power lines in this scheme. Therefore, the simultaneous switch-on rate (SOR), which indicates how many logic gates are switched on at almost the same time, seems to affect speed performance, especially in MTCMOS circuits. The amount of current supplied through Q1 and Q2 depends on the switching probability of the internal logic gates. Thus, a high SOR enhances the voltage drops at Q1 and Q2, which consequently reduces the effective supply voltage between VDDV and GNDV. Fig. 9 shows the simulation results for the SOR along

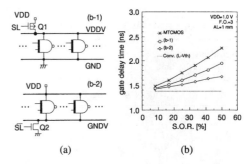

Fig. 10. Influences of virtual power supply lines. (a) Simulation circuit model. (b) Gate delay time versus SOR.

TABLE II
DEVICE TECHNOLOGY

	High-Vth Tr	Low-Vth Tr
Gate length	0.55 μm	0.65 μm
Gate oxide thichness	110 Å	110 Å
N-channel : Vth	0.55 V	0.25 V
P-channel : Vth	-0.65 V	-0.35 V

with the circuit model. Here, the SOR is defined as m/n, where m and n indicate the number of operating logic blocks and the total number of blocks, respectively, assuming a 2-mm block width. As the SOR increases, the voltage on VDDV drops because Q1 has to supply more current to VDDV at one time. For similar reasons, the voltage also rises on GNDV. These voltage changes cause the effective supply voltage V_{eff} between VDDV and GNDV to decrease, extending gate delay time. Generally, however, the SOR is expected to be at most 20 or 30%. In this region, the reduction of V_{eff} is less than 15% of the supply voltage, and the speed performance of 1.7 ns/gate is still quite high.

One way to further decrease the dependence of speed performance on the SOR is to use a sleep control transistor with a wider gate. This is, however, a trade-off between speed and stand-by current because total stand-by current in a chip is approximately given by $2KW_H I_H$. Here, K is the number of rows in the block shown in Fig. 8, while W_H is the gate width of the sleep control transistor, and I_H is its leakage current when the gate's width and length are equal. Another effective way to decrease this dependence is to remove one set of the two sleep control transistors and virtual power lines. The expected improvements in speed are shown in Fig. 10 along with simulation circuits. By removing of GNDV (b-1) and VDDV (b-2), gate delay time can be reduced by 15–25% compared with the basic scheme. The removal of VDDV is clearly more effective in increasing speed. The reason for this is that the threshold voltage of the P-channel MOSFET (Q1) was designed to be higher than that of the N-channel one (Q2) in this study (see Table II).

V. TEST CHIP RESULTS

A. Process and Device Technology

To confirm the effectiveness of MTCMOS circuit technology, a PLL LSI using new MTCMOS standard cells was

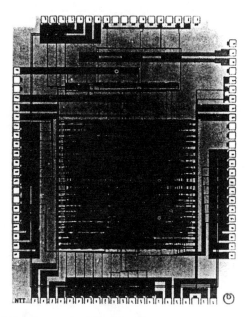

Fig. 11.   Microphotograph of the PLL chip.

TABLE III
AREA PENALTY FACTOR

Combinational Circuit	1.1
Sequential Circuit	2.0
PLL LSI Digital Core	1.3

designed and fabricated. Conventional 0.5-μm CMOS process technology for 3.3-V operation with single-polysilicon and double-metal layers was used. MOSFET's with different V_{th}'s in the same well were formed by optimizing the impurity concentration in the well and controlling the channel doses with two additional masks, which minimizes the increase in the number of process steps. The key device parameters and characteristics are summarized in Table II. The gate length of the low-V_{th} MOSFET is 0.65 μm, which is 0.1 μm longer than that of the high-V_{th} ones. This is preferable to suppress variations in the threshold voltage due to short-channel effects. The gate oxide thickness is 110 Å for both types of MOSFET's. The low-V_{th}'s are 0.25 V for N-channel and -0.35 V for P-channel MOSFET's.

A microphotograph of the PLL chip is shown in Fig. 11. This chip consists of about 5 K gates, including the automatic frequency control circuit and the intermittent operation controller [8]. The whole chip is 4×5 mm^2, and the digital core is about 2×2 mm^2. Table III lists the area penalty factors in this study. An MTCMOS combinational circuit cell has an area about 10% larger than a conventional cell does owing to the insertion of virtual supply lines and the sleep control line. A sequential circuit cell, such as an MTCMOS DFF with clear, needs an area about twice that of a conventional cell in order to store data even in the sleep period. The area increase for the whole digital core, however, is only 30% in spite of the fact that the DFF's occupy a relatively large part (about 50%) of the total gate counts. This is because the channel area is almost unchanged. Moreover, because all DFF's in an actual LSI aren't expected to hold the date during sleep

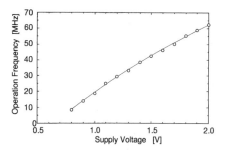

Fig. 12.   Operation frequency of the PLL chip.

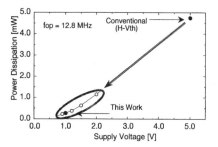

Fig. 13.   Power dissipation of the PLL chip versus supply voltage.

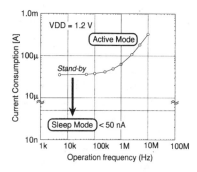

Fig. 14.   Power dissipation of the PLL chip versus operation frequency.

period, the area penalty can be further reduced by appropriately combining the use of a conventional DFF and the DFF with a special memory function.

B. Chip Performance

Fig. 12 shows the measured operation frequency as a function of supply voltage. At 1 V, the chip operates at 18 MHz which is sufficient for many applications.

Fig. 13 shows the power dissipation in the digital core as a function of supply voltage at an operation frequency of 12.8 MHz. The power dissipation of the conventional 5-V operation PLL is also plotted for comparison. At 1 V, power dissipation is drastically reduced to below 1/20 compared with that of the conventional LSI operated at 5 V.

Fig. 14 shows another aspect of the power performance—the operating current versus the operating frequency for the worst case at a supply voltage of 1.2 V. Although the operating current is proportional to the frequency in the region over 1 MHz, it becomes almost constant in the low-frequency region. This is due to the leakage current caused by using low-V_{th} MOSFET's. In the active mode, the leakage current of about 30 μA in this chip is negligible because it is less

TABLE IV
CHARACTERISTICS OF THE PLL CHIP

Power supply voltage	1.0 V
Core size	2.4 mm x 2.3 mm
Cycle frequency	18 MHz
Power dissipation	200 μW (at 10 MHz)
Stand-by current	< 50 nA
Turn-on time	< 500 ns
Gate counts	5 K gates

than 1/10 of the dynamic current consumption at a desired operating frequency of over 10 MHz. In the sleep mode, on the other hand, the current is dramatically reduced to below 50 nA, so that low stand-by characteristics can be obtained.

Typical PLL LSI features are summarized in Table IV. The turn-on time, which is the time needed to switch from sleep to active mode, is less than 500 ns even in the worst case.

VI. CONCLUSION

Multithreshold-voltage CMOS (MTCMOS) circuit technology has been proposed as a way to achieve a 1-V supply voltage high-speed and low-power LSI operation. This technology uses MOSFET's with two different threshold voltages on a single chip and introduces a sleep control scheme for efficient power management. Low-threshold voltage MOSFET's improve the speed performance at a low supply voltage of 1 V, while high-threshold MOSFET's suppress the standby power dissipation. In addition, a standard cell library has been developed to simplify low-voltage LSI designs. To demonstrate the effectiveness of this technology, a PLL LSI based on standard cells was designed as a carrying vehicle using a 0.5-μm CMOS process. High-speed operation of 18 MHz at 1 V confirmed the validity of this new technology.

ACKNOWLEDGMENT

The authors would like to thank S. Horiguchi, E. Arai, N. Ieda, and K. Imai for their suggestions and encouragement.

REFERENCES

[1] R. W. Brodersen, A. Chandrakasan, and S. Sheng, "Design techniques for portable systems," in *ISSCC Dig. Tech. Papers,* pp. 168–169, Feb. 1993.

[2] A. P. Chandrakasan, S. Sheng, and R. W. Brodersen, "Low-power CMOS digital design," *IEEE J. Solid-State Circuits,* vol. 27, pp. 473–484, Apr. 1992.

[3] M. Horiguchi, T. Sakata, and K. Itoh, "Switched-source-impedance CMOS circuit for low standby subthreshold current giga-scale LSI's," *IEEE J. Solid-State Circuits,* vol. 28, pp. 1131–1135, Nov. 1993.

[4] T. Kawahara, M. Horiguchi, Y. Kawajiri, G Kitsukawa, T Kure, and M. Aoki, "Subthreshold current reduction for decoded-driver by self-reverse biasing," *IEEE J. Solid-State Circuits,* vol. 28, pp. 1136–1144, Nov. 1993.

[5] S. Mutoh, T. Douseki, Y. Matsuya, T. Aoki, and J. Yamada, "1-V high-speed digital circuit technology with 0.5-μm multi threshold CMOS," in *Proc. IEEE Int. ASIC Conf.,* Sept. 1993, pp. 186–189.

[6] K. Shimohigashi and K. Seki, "Low-voltage ULSI design," *IEEE J. Solid-State Circuits,* vol. 28, pp. 408–413, Apr. 1993.

[7] *SIA Semiconductor Technology: Workshop Working Group Reports,* Nov. 1992.

[8] M. Ishikawa, N. Ishihara, A. Yamagishi, and I. Shimizu, "A miniaturized low-power synthesizer module with automatic frequency stabilization," in *Proc. IEEE VTC,* 1992, pp. 752–755.

A 1V Multi-Threshold Voltage CMOS DSP with an Efficient Power Management Technique for Mobile Phone Application

SHIN'ICHIRO MUTOH, SATOSHI SHIGEMATSU, YASUYUKI MATSUYA, HIDEKI FUKUDA,
AND JUNZO YAMADA

HIGH-SPEED INTEGRATED CIRCUITS LAB., NTT LSI LABS, ATSUGI, KANAGAWA, JAPAN

A low-power digital signal processor (DSP) is the key component for battery-driven mobile phone equipment since a vast amount of data needs to be processed for multimedia use. Reduced supply voltage is a direct approach to power reduction [1, 2]. This 1V DSPLSI with 26MOPS and 1.1mW/MOPS performance adopts a multi-threshold-voltage CMOS (MTCMOS) technique. A small embedded power-management processor decreases power during waiting periods.

A block diagram of the DSP is shown in Figure 1. To extend battery life, the DSP for a mobile phone requires (a) low-power and high-throughput multiply-accumulation (MAC) during the talking period, and (b) low-power intermittent operation during the waiting period that occupies most of the running time. To meet these requirements, the following are adopted: (1) A 1V MTCMOS circuit with a simple parallel architecture and (2) a power management technique using an embedded processor and a modified DFF suitable for power supply control.

The MTCMOS circuit is shown in Figure 2 [2]. Logic gates are composed of low-threshold voltage (Vth) MOSFETs to decrease delay time at low voltage. However, lower Vth increases the standby power because of subthreshold leakage current. Thus, a high-Vth sleep control MOSFET (SLT) is connected to switch the power supply to the logic core. During the sleep period when the SLT is off, leakage current is drastically reduced. The MTCMOS technique aims to maintain processing throughput at 1V through improved gate speed. Thus, low-power performance is attained without special architectural techniques, shortening the time-to-market. In this study, to further improve MAC performance, MTCMOS and a simple high-speed architecture are combined. Throughput is doubled by the use of the dual data processing units as illustrated in Figure 1.

To save power during the waiting period, a power management scheme combines a small embedded processor (PMP) and MTCMOS sleep control. The PMP uses high-Vth MOSFETs and operates at a lower clock frequency, resulting in low standby and dynamic power. The connection between the PMP and the DSP is shown in Figure 2. Besides working as a general processor, the PMP generates the SL signal and allows the DSP to be supplied with power only when high-throughput processing is required, reducing the DSP core leakage current. The typical running sequence of a mobile phone terminal is shown in Figure 3a. During the talking period, the PMP sets SL low to supply power to the DSP to enable high-speed processing. In the waiting period, the terminal works intermittently to check for calls. As shown in sequence 1 in Figure 3b, the PMP takes charge of all the waiting processing so that the DSP can sleep throughout the waiting period and the leakage current due to the low-Vth MOSFET is eliminated. If waiting processing required is beyond the capability of the PMP, the DSP takes over, as shown in sequence 2. Also in this case, the DSP is intermittently powered only during period "t2" by SL from PMP, and as a result there is much less leakage current than there would be without the PMP. Figure 4 compares energy consumption assuming 1-hour talking time per day and an intermittent ratio γ of 100 in sequence 2. With CMOS, energy is only reduced to 1/3 even though voltage is lowered from 3.3V to 1.0V. This is because standby leakage current increases due to lowered Vth. The design reduces energy to 1/10 by reducing both active and standby power. Particularly noteworthy is that the energy consumed during the waiting period is less than 1/10 of that without MTCMOS and the PMP.

In an LSI with controlled power supply, data must be preserved even during the unpowered period. The modified DFF shown in Figure 5a is adopted [3]. A memory circuit (balloon) has the smallest high-Vth MOSFET connected to a master or slave latch and always supplied with power. This preserves data during a sleep period instead of in the unpowered main latch loop. FF-L, for instance, is used to preserve data in the slave latch (CK is low). The circuit determines which latch holds data by fixing the clock around the sleep period. Figures 5b and 5c show the circuit diagram and timing. When signal SLCK from the PMP is set low, each state of the four-phase clock (∅ A ∅ B and ∅ AN ∅ BN) is fixed at low and high. The FFs clocked by ∅ A and ∅ AN should be FF-L and FF-H (with the balloon connected to the master side), respectively. This circuit does not generate glitches so that the same data before and after a sleep period can be guaranteed.

The DSP chip uses a 0.5μm MTCMOS process. The chip micrograph is shown in Figure 6. The PMP core area is 6.3mm², less than 2.8% of the chip area. Figure 7 shows the speed dependence on supply voltage. 13MHz clock frequency and 26MOPS MAC performance at 1.0V are 4 times faster than standard 0.5μm CMOS. The energy-delay product (EDP) is shown in Figure 8. The energy is only 2.2mW/MHz (1.1 mW/MOPS) at 1V for dual MAC operation. The DSP reduces EDP at 1.0V to 1/6 compared with the EDP at 3.3V. This means that MTCMOS technique achieves a lower EDP DSP independent of architecture. The standby power in the active period is about 350μW, a few % of the total dynamic power. Power can be reduced to 600nW during sleep. This guarantees low-power intermittent wait. Experimental results are summarized in Table 1.

Acknowledgments:

The authors thank S. Horiguchi, S. Date, N. Shibata, K. Ogura, and T. Ishihara for encouragement.

References:

[1] Burr, J. B., et al., "A 200mV Self-Testing Encoder/Decoder using Stanford Ultra-Low-Power CMOS," ISSCC Digest Technical Papers, pp. 84-85, Feb., 1994.

[2] Mutoh, S., et al., "1V Power Supply High-Speed Digital Circuit Technology withMultithreshold-Voltage CMOS," IEEE J. Solid-State Circuits, Vol. 30, pp. 847-854, Aug., 1995.

[3] Shigematsu, S., et al., "A 1-V high-speed MTCMOS circuit scheme for power-down applications," Symp. VLSI Circuits Digest of Technical Papers, pp. 115-116, June, 1995.

Reprinted from *IEEE International Solid-State Circuits Conference*, pp. 168-169, February 1996.

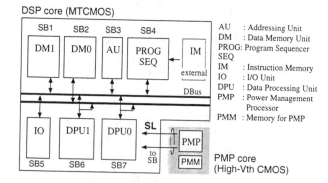

DSP core (MTCMOS)

SB1	SB2	SB3	SB4	
DM1	DM0	AU	PROG SEQ	IM

AU : Addressing Unit
DM : Data Memory Unit
PROG: Program Sequencer
SEQ
IM : Instruction Memory
IO : I/O Unit
DPU : Data Processing Unit
PMP : Power Management Processor
PMM : Memory for PMP

Figure 1: MTCMOS DSP block diagram.

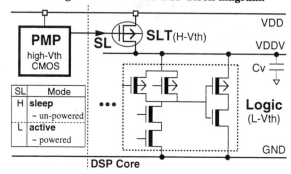

Figure 2: MTCMOS circuit.

SL	Mode	
H	sleep	~ un-powered
L	active	~ powered

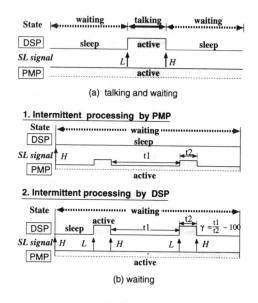

(a) talking and waiting

1. Intermittent processing by PMP

2. Intermittent processing by DSP

(b) waiting

Figure 3: Running sequence.

Process technology	0.5μm MTCMOS
Die size	15x15mm²
Transistor count	logic 267k Tr, RAM 123kb
High Vth	0.55V (n-ch), -0.55V (p-ch)
Low Vth	0.25 V(n-ch), -0.25V (p-ch)
Operating frequency	13.2MHz at 1.0V
MAC performance	26.4MOPS at 1.0V
Power consumption	2.2mW/MHz (1.1mW/MOPS) at 1.0V
Standby power	350μW (active) , 600nW (sleep) at 1.0V

Table 1: Experimental results.

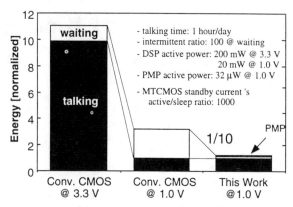

- talking time: 1 hour/day
- intermittent ratio: 100 @ waiting
- DSP active power: 200 mW @ 3.3 V
  20 mW @ 1.0 V
- PMP active power: 32 μW @ 1.0 V
- MTCMOS standby current 's active/sleep ratio: 1000

Figure 4: Low energy effect.

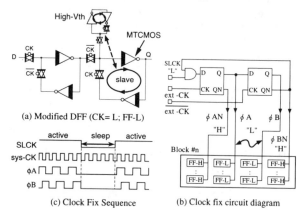

(a) Modified DFF (CK= L; FF-L)

(c) Clock Fix Sequence

(b) Clock fix circuit diagram

Figure 5: Modified FF and clock fix circuit.
Figure 6: See page 438.

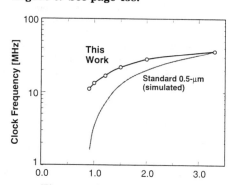

Figure 7: Speed performance.

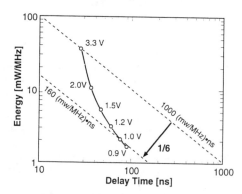

Figure 8: Energy delay product.

94

50% Active-Power Saving without Speed Degradation using Standby Power Reduction (SPR) Circuit

KATSUHIRO SETA, HIROYUKI HARA, TADAHIRO KURODA, MASAKAZU KAKUMU,
AND TAKAYASU SAKURAI

TOSHIBA CORPORATION, KANAGAWA, JAPAN

High-speed and low-power are required for multimedia LSIs, since portability with battery operation is sometimes the key factor for multimedia equipment, while delivering giga operations per second (GOPS) processing power for digital video use [1]. To understand circuit delay and power dissipation dependence on power supply voltage (V_{DD}) and threshold voltage of MOSFETs (V_{TH}), a typical logic circuit shown in Figure 1 is investigated. Fanout is chosen to be 5 which corresponds to the statistical average of gate load in ASICs. Figure 2 shows a simulated delay dependence on V_{DD} and V_{TH}. The same V_{TH} is chosen for nMOS and pMOS. As seen from the figure, if V_{TH} is reduced to 0.3V, V_{DD} can be decreased down to 2V while maintaining the speed at V_{TH} = 0.7V and V_{DD} = 3V which is the typical operation condition for high-speed LSIs. The active power dissipation, in this case, is reduced by more than 50%.

The energy-delay (ED) product is plotted as a function of V_{DD} and V_{TH} in Figure 3. Minimizing the ED product is a good approach for optimizing LSIs for portable use, since the ED product reflects the battery consumption (E) for completing a job in a certain time [2]. The ED product is also minimized at about 0.3V V_{TH} when V_{DD} is 2V.

The only drawback of choosing 0.3V V_{TH} is the increase in standby power dissipation. If this standby power problem can be solved, high-speed and low power operation is achieved just by lowering V_{DD} and V_{TH} at the same time. This paper presents a solution to this problem of standby power increase in the low-V_{TH} region. The main idea of this standby power reduction (SPR) scheme is that a substrate bias is applied in a standby mode to increase the threshold voltage and to lower the subthreshold leak current. While in an active mode, the substrate bias is not applied, assuring high-speed operation. According to measured I_{DS}-V_{GS} characteristics for a 0.3μm nMOS transistor, the threshold voltage can be increased by 0.4V by applying a substrate voltage of -2V. This means that if the substrate bias of -2V is applied in a standby mode, the threshold voltage is increased from 0.3V to 0.7V and thus realizes the same standby current as the 0.7V V_{TH} LSI.

Figures 4 and 5 show a circuit diagram of the proposed SPR circuit and simulated waveforms of the circuit. The circuit consists of a level-shifting part and a voltage-switch part. When CE (Chip Enable) is asserted in an active mode, the n-well bias, V_{NWELL}, becomes equal to V_{DD} which is set at 2V in the test chip design. The p-well bias, V_{pWELL}, becomes V_{SS}. When CE is disabled in standby mode, V_{nWELL} equals V_{NBB} which is set at 4V and V_{pWELL} becomes -2V. Standby-to-active mode transition and an active-to-standby mode transition take about 50ns. The power dissipation of this SPR circuit in the standby mode is 0.1μA, the dominating factor of which is the current through M4 and M5. This current can be reduced by one order of magnitude further if the transition time can be slower. V_{NBB}, V_{PBB}, V_{DD} and V_{SS} are applied from the external source but the power supplies connected to V_{NBB} and V_{PBB} only need to supply 0.1μA or less. Although the SPR scheme can be realized

together with a self sub-bias circuit, the response becomes more than a μs order [3]. The diodes in the circuit are built using a junction-well structure through which current flows only in active mode.

In designing the circuit, care is taken so that no transistor sees high-voltage stress of gate oxide and junctions. V_{GS}-V_{GD} trajectories of MOSFETs used in the SPR circuit do not go beyond $\pm(V_{DD}+\alpha)$, which assures sufficient reliability of gate oxide. On the other hand, V_{SB} (source-bulk voltage) - V_{DB} (drain-bulk voltage) trajectories of MOSFETs in the SPR circuit do not go beyond $\pm(V_{DD}+V_{BIAS})$, where V_{BIAS} signifies the larger voltage of $|V_{NBB} - V_{DD}|$ and $|V_{SS} - V_{PBB}|$. This voltage is applied to junctions but the breakdown voltage of junctions of 0.3μm MOSFETs is more than 9V and hence junction breakdown does not occur for any MOSFETs.

Figure 6 shows a micrograph of the test chip. A ring oscillator constructed with 49 stages of 2-input NAND gates and the SPR circuit are implemented using 0.3μm process technology. The SPR circuit occupies 2500μm² for either n-well or p-well bias circuit. In cases where nMOS circuit determines the speed as in nMOS pass transistor logic environments, only V_{TH} for nMOS should be lowered and hence only p-well bias circuit is needed. If both of the n-well and p-well bias circuits are required as in Figure 4, 5000μm² Si area is occupied and a triple-well technology is to be used. The standby current of less than 0.1μA is measured on the test chip when the chip enable (CE) is disabled. If the CE is asserted in the active mode, the standby current is measured larger by three orders. The speed of the 2-input NAND gate of 300ps is achieved at V_{DD} = 2.0V. Setting time is less than 100ns. The proposed SPR scheme is fully compatible with the existing CAD tools including automatic placement and routers. As for the standard cell library, the cells should be modified to separate substrate bias lines and power supply lines. The area overhead to the total chip, however, is estimated to be less than 5%. The substrate bias lines can be as narrow as possible and can be scaled.

Acknowledgments

Valuable discussions and constant encouragement by H. Shibata, K. Maeguchi, and Y. Unno are appreciated.

References

[1] Matsui, M., et al., "200MHz Video Compression / Decompression Macrocells Using Low-Swing Differential Logic," ISSCC Digest of Technical Papers, pp. 76-77, Feb., 1994.

[2] Burr, J.B., et al., "A 200mV Self-Testing Encoder / Decoder Using Stanford Ultra Low Power CMOS," ISSCC Digest of Technical Papers, pp. 84-85, Feb., 1994.

[3] Kobayashi, T., et al., "Self-Adjusting Threshold-Voltage Scheme (SATS) for Low-Voltage High-Speed Operation," Proc. IEEE CICC'94, pp. 271-271, May, 1994.

Reprinted from *IEEE International Solid-State Circuits Conference*, pp. 318-319, February 1995.

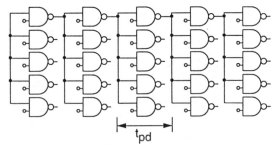

Figure 1: Typical logic circuit used to calculate delay and power vs. supply voltage (VDD) and threshold voltage (VTH).

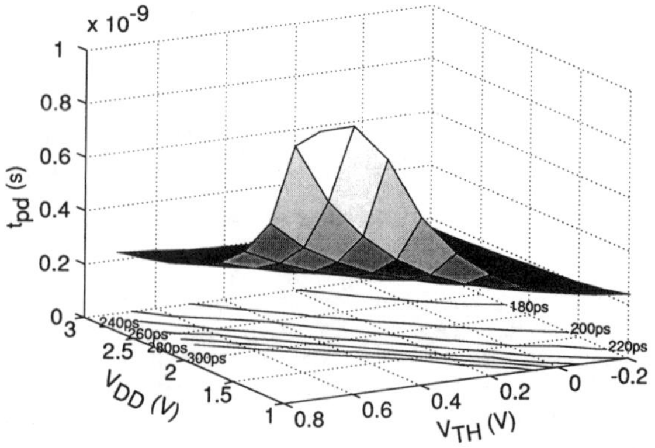

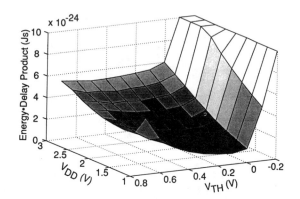

Figure 3: Simulated energy-delay product vs. VDD and VTH.

t_{pd}

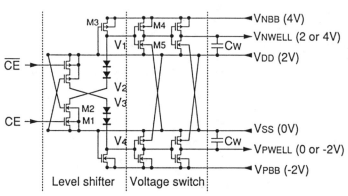

Figure 4: Circuit diagram of proposed standby power reduction (SPR) circuit. Well capacitance (Cw) is supposed to be 1000pF.

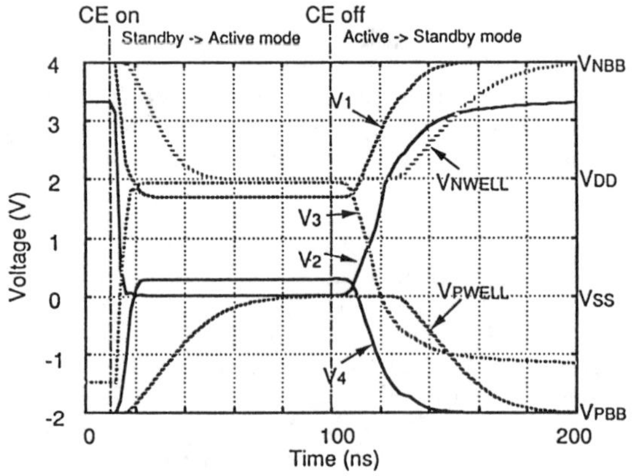

Figure 2: Simulated delay vs. VDD and VTH by SPICE.

Figure 5 : Simulated waveforms of the SPR circuit.

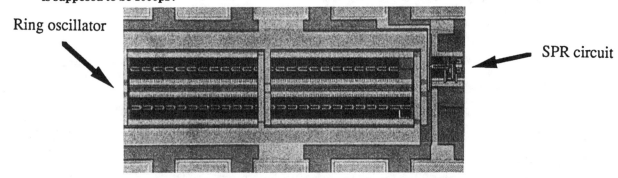

Ring oscillator

SPR circuit

Figure 6: Micrograph of test chip.

A 0.9-V, 150-MHz, 10-mW, 4 mm², 2-D Discrete Cosine Transform Core Processor with Variable Threshold-Voltage (VT) Scheme

Tadahiro Kuroda, *Member, IEEE*, Tetsuya Fujita, Shinji Mita, Tetsu Nagamatsu, Shinichi Yoshioka, Kojiro Suzuki, Fumihiko Sano, Masayuki Norishima, Masayuki Murota, Makoto Kako, Masaaki Kinugawa, *Member, IEEE*, Masakazu Kakumu, *Member, IEEE*, and Takayasu Sakurai, *Member, IEEE*

Abstract— A 4 mm², two-dimensional (2-D) 8 × 8 discrete cosine transform (DCT) core processor for HDTV-resolution video compression/decompression in a 0.3-μm CMOS triple-well, double-metal technology operates at 150 MHz from a 0.9-V power supply and consumes 10 mW, only 2% power dissipation of a previous 3.3-V design. Circuit techniques for dynamically varying threshold voltage (VT scheme) are introduced to reduce active power dissipation with negligible overhead in speed, standby power dissipation, and chip area. A way to explore $V_{DD} - V_{th}$ design space is also studied.

I. INTRODUCTION

LOWERING both the supply voltage V_{DD} and threshold voltage V_{th} enables high-speed, low-power operation [1], [2]. This approach, however, raises two problems [3], [4], 1) degradation of worst-case speed due to V_{th} fluctuation in low V_{DD}, and 2) increase in standby power dissipation in low V_{th}. To solve these problems, several schemes are proposed. A self-adjusting threshold voltage (SAT) scheme [5] reduces V_{th} fluctuation in an active mode by adjusting substrate bias with a feedback control circuit. A standby power reduction (SPR) scheme [6] raises V_{th} in a standby mode by switching substrate bias between the power supply and an external additional supply higher than V_{DD} or lower than GND. A multi threshold voltage CMOS (MT-CMOS) scheme [7] employ low V_{th} for fast circuit operation and high V_{th} for providing and cutting internal supply voltage. The SAT and the SPR are both based upon the same idea that V_{th} is controlled dynamically through substrate bias. However, the two schemes cannot be combined because the SPR requires the external supply for the substrate bias while the SAT generates the substrate bias internally. The MT-CMOS does not solve the first problem. It requires very large transistors for the internal power supply control to impose area and yield penalties, otherwise degrading circuit speed. Furthermore, it cannot be applied to memory

elements without circuit tricks which add another area and speed penalties.

This paper presents a variable threshold voltage scheme (VT scheme) which can solve these two problems uniformly in a unified way by controlling substrate bias with substrate bias feedback control circuits. Unlike the conventional approaches, it requires no external power supply for the substrate bias, leaves no restriction in use, imposes practically no penalty in speed and chip area, and can be applied to both logic gates and memory elements. The VT scheme is employed in a two-dimensional (2-D) 8 × 8 discrete cosine transform (DCT) core processor for portable HDTV-resolution video compression/decompression. This DCT in a 0.3-μm CMOS technology operates at 150 MHz from a 0.9-V power supply and consumes 10 mW, only 2% power dissipation of a previous 3.3-V design [8].

In Section II, low V_{DD}, low V_{th} design space is explored to investigate V_{th} target. In Section III, the VT scheme is presented, followed by descriptions of circuit implementations in Section IV. Section V details the design of the DCT. Experimental results appear in Section VI. Section VII is dedicated for conclusions.

II. EXPLORING LOW-V_{DD} LOW-V_{th} DESIGN SPACE

CMOS power dissipation is given by

$$P = \tfrac{1}{2} \cdot p_t \cdot f_{CLK} \cdot C_L \cdot V_{DD}^2 + I_0 \cdot 10^{-(V_{th}/S)} \cdot V_{DD} \quad (1)$$

where p_t is the switching probability, f_{CLK} is the clock frequency, C_L is the load capacitance, S is the subthreshold swing, and I_0 is a constant which is proportional to total transistor width in a chip. The first term represents dynamic power dissipation due to charging and discharging of the load capacitance, and the second term is leakage current dissipation due to subthreshold conduction. Since the dominant term in a typical CMOS design is the dynamic power dissipation, lowering V_{DD} is effective to low-power design.

Gate propagation delay, on the other hand, is approximately given in [9] by

$$t_{pd} = \frac{k \cdot C_L \cdot V_{DD}}{(V_{DD} - V_{th})^\alpha} \quad (2)$$

Manuscript received April 11, 1996; revised July 23, 1996.
T. Kuroda, T. Fujita, S. Mita, T. Nagamatsu, S. Yoshioka, K. Suzuki, F. Sano, and T. Sakurai are with System ULSI Engineering Laboratory, Toshiba Corp., Kawasaki, Japan.
M. Norishima and M. Kakumu are with LSI Div. II, Toshiba Corp., Kawasaki, Japan.
M. Murota, M. Kako, and M. Kinugawa are with ULSI Device Engineering Laboratory, Toshiba Corp., Kawasaki, Japan.
Publisher Item Identifier S 0018-9200(96)07943-7.

Reprinted from *IEEE Journal of Solid-State Circuits*, pp. 1770–1779, November 1996.

where α is typically 1.3 and k is a constant. Lowering only V_{DD} leads to slower circuit speed, and therefore, both of V_{DD} and V_{th} should be lowered for high-speed, low-power design. When V_{DD} and V_{th} are lowered to V'_{DD} and V'_{th}, and the circuit speed becomes λ times, their relation is given from (2) by

$$\frac{V'_{DD} - V'_{th}}{V_{DD} - V_{th}} = \left(\lambda \cdot \frac{V'_{DD}}{V_{DD}} \right)^{1/\alpha}. \qquad (3)$$

For example, suppose $V_{DD} = 3.3$ V and $V_{th} = 0.6$ V. Under a constant speed condition ($\lambda = 1$), one solution is $V'_{DD} = 2.1$ V and $V'_{th} = 0.2$ V. In this case, the dynamic power dissipation is reduced to 41%. If circuit speed can be reduced to 60% ($\lambda = 0.6$), the dynamic power dissipation can be reduced to 7% at $V'_{DD} = 0.9$ V and $V'_{th} = 0.2$ V.

For more precise estimation, process fluctuation should be taken into account. V_{th} fluctuates typically by ± 0.1 V, which causes t_{pd} variation. From (2), the variation in t_{pd}, K_{VT}, is given by

$$K_{VT} = \frac{\Delta t_{pd}}{t_{pd}}$$
$$= \frac{\alpha \cdot \Delta V_{th}}{V_{DD} - V_{th}}. \qquad (4)$$

In order to assure high yield in production, margin should be incorporated into design so as to satisfy speed specification even with fluctuations in process. Smaller K_{VT} leads to smaller design margin, and therefore, is preferable from area-saving and low-power design point of view. In lowering both V_{DD} and V_{th}, K_{VT} should be kept at least from increasing. From (3) and (4), calculating the condition to keep K_{VT} constant yields

$$\frac{\Delta V'_{th}}{\Delta V_{th}} = \left(\lambda \cdot \frac{V'_{DD}}{V_{DD}} \right)^{1/\alpha}. \qquad (5)$$

In the former examples, under the constant speed condition ($\lambda = 1$) V_{th} fluctuation should be reduced to $\Delta V'_{th}/\Delta V_{th} = 0.71$, and in the 60%-speed condition ($\lambda = 0.6$) it should be reduced to $\Delta V'_{th}/\Delta V_{th} = 0.25$.

But in reality, it is not expected that as V_{th} is lowered, ΔV_{th} is reduced as much. If impurity density in the channel region is simply reduced to lower V_{th} of a surface channel device such as nMOS with n$^+$ polysilicon gates, the short-channel effect degrades to increase ΔV_{th}, reflecting variation of polysilicon gates in size. In a buried-channel device such as pMOS with n$^+$ polysilicon gates, on the other hand, counter doping should be added to lower V_{th}, resulting in higher impurity density and larger ΔV_{th}. It is not that simple to discuss ΔV_{th}, but generally speaking, device researchers expect ΔV_{th} could be increased in low V_{th} and would not be decreased very easily. This is one issue in low-V_{DD}, low-V_{th} CMOS circuit design.

Another issue is the rapid increase in subthreshold leakage in low V_{th} as seen from (1). In portable applications it is clear that large standby leakage becomes a problem. Not only in portable applications but also in desktop applications, the rapid increase in subthreshold current determines the lower limit of V_{th}, and therefore, it is also important.

In order to study these two issues and explore low-V_{DD}, low-V_{th} design space, (1) and (2) are numerically solved with the parameters for this DCT design in a 0.3-μm CMOS technology at junction temperature of 90°C. Contour lines in terms of speed (i.e., maximum operating frequency) and power are drawn on the $V_{DD} - V_{th}$ plane in Fig. 1. In a typical 3.3-V design, V_{DD} is at 3.3 ± 10% V and V_{th} is set to 0.6 ± 0.1 V. The design space is represented by a box in Fig. 1. The maximum operating frequency, f, becomes the slowest, 250 MHz, at $V_{DD} = 3.0$ V and $V_{th} = 0.7$ V. The circuit speed is therefore normalized ($\lambda = 1$) at the upper-left corner of the design-space box. The power dissipation, on the other hand, becomes the largest, 160 mW, at $V_{DD} = 3.6$ V and $V_{th} = 0.5$ V. The power dissipation is therefore normalized ($\xi = 1$) at the lower-right corner of the design-space box. For designing a 150 MHz DCT, which is 60% speed of 250 MHz, the upper-left corner of the design-space box should be placed on the speed contour line with $\lambda = 0.6$. It is found from the lower-right corner of the design-space box that the power dissipation can be reduced to 25% ($\xi = 0.25$), that is 40 mW, by lowering V_{DD} to 1.9 ± 10% V. It can further be reduced to 6% ($\xi = 0.06$), that is 10 mW, by lowering both V_{DD} to 1.0 ± 10% V and V_{th} to 0.27 ± 0.02 V. This supply voltage can be supplied from a single battery source. Reducing ΔV_{th} from ±0.1 V to ±0.02 V also meets the requirement for keeping K_{VT} constant in (5). As shown in Fig. 1, power dissipation due to subthreshold leakage becomes about 1% of the total power dissipation.

To summarize, V_{DD} should be at 1.0 ± 10% V, and V_{th} should be controlled at 0.27 ± 0.02 V in the active mode and higher than 0.5 V in the standby mode.

III. VARIABLE THRESHOLD-VOLTAGE (VT) SCHEME

The VT scheme is conceptually illustrated in Fig. 2. Threshold voltage of a transistor is variable through substrate bias control with a Variable Threshold-voltage circuit (VT circuit). In the active mode, the VT circuit controls the substrate bias, V_{BB}, so as to compensate the V_{th} fluctuation. Even though device V_{th} has 0.1-V fluctuation around 0.15 V, V_{th} is compensated and set at 0.27 ± 0.02 V in the active mode. In the standby mode, the VT circuit applies deeper substrate bias to increase V_{th} to higher than 0.5 V and cut off leakage. Typically, V_{BB} of -0.5 V is applied in the active mode and -3.3 V in the standby mode.

Fig. 3 depicts the VT scheme block diagram. The VT scheme consists of four leakage current monitors (LCM's), the self-substrate bias circuit (SSB), and a substrate charge injector (SCI). The SSB draws current from the substrate to lower V_{BB}. The SCI, on the other hand, injects current into the substrate to raise V_{BB}. The SSB and the SCI are controlled by monitoring where V_{BB} sits in four ranges. Their criteria are specified in the four LCM's; $V_{\text{active}(+)} = -0.3$ V, $V_{\text{active}} = -0.5$ V, $V_{\text{active}(-)} = -0.7$ V, and $V_{\text{standby}} = -3.3$ V. The substrate bias is monitored by transistor leakage current, because the leakage current reflects V_{BB} very sensitively.

Fig. 4 illustrates the substrate bias control. After a power-on, V_{BB} is higher than $V_{\text{active}(+)}$, and the SSB begins to draw

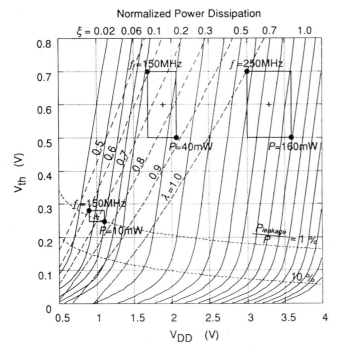

Fig. 1. Exploring low-V_{DD}, low-V_{th} design space. Contour lines in terms of speed (broken lines) and power (solid lines) are drawn.

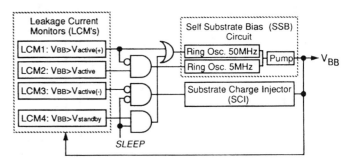

Fig. 3. VT block diagram.

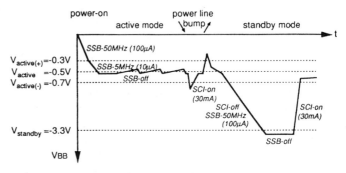

Fig. 4. Substrate-bias control in VT.

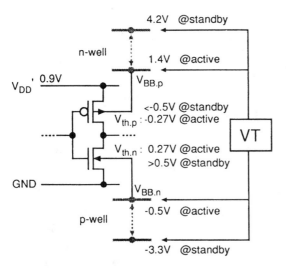

Fig. 2. Variable threshold-voltage (VT) scheme.

100 μA from the substrate to lower V_{BB} using a 50 MHz ring oscillator. This current is large enough for V_{BB} to settle down within 10 μs after a power-on. When V_{BB} goes lower than $V_{active(+)}$, the pump driving frequency drops to 5 MHz and the SSB draws 10 μA to control V_{BB} more precisely. The SSB stops when V_{BB} drops below V_{active}. V_{BB}, however, rises gradually due to device leakage current through MOS transistors and junctions, and reaches V_{active} to activate the SSB again. In this way, V_{BB} is controlled at V_{active} by the on–off control of the SSB. When V_{BB} goes deeper than $V_{active(-)}$, the SCI turns on to inject 30 mA into the substrate. Therefore, even if V_{BB} jumps beyond $V_{active(+)}$ or $V_{active(-)}$ due to a power line bump for example, V_{BB} is quickly recovered to V_{active} by the SSB and the SCI. When "SLEEP" signal is asserted ("1") to go to the standby mode, the SCI is disabled and the SSB is activated again and 100 μA current is

drawn from the substrate until V_{BB} reaches $V_{standby}$. V_{BB} is controlled at $V_{standby}$ in the same way by the on–off control of the SSB. When "SLEEP" signal becomes "0" to go back to the active mode, the SSB is disabled and the SCI is activated. The SCI injects 30 mA current into the substrate until V_{BB} reaches $V_{active(-)}$. V_{BB} is finally set at V_{active}. In this way, the SSB is mainly used for a transition from the active mode to the standby mode, while the SCI is used for a transition from the standby to the active mode. An active to standby mode transition takes about 100 μs, while a standby to active mode transition is completed in 0.1 μs. This "slow falling asleep but fast awakening" feature is acceptable for most of the applications.

The SSB operates intermittently to compensate for the voltage fluctuation in the substrate due to the substrate current in the active and the standby modes. It therefore consumes several microamperes in the active mode and less than one nanoampere in the standby mode, both much lower than the chip power dissipation. Energy required to charge and discharge the substrate for switching between the active and the standby modes is less than 10 nJ. Even when the mode is switched 1000 times in a second, the power dissipation becomes only 10 μW. The leakage current monitor should be designed to dissipate less than 1 nA because it always works even in the standby mode. The low-power circuit design technique is described in the next section.

IV. CIRCUIT IMPLEMENTATIONS

A. Leakage Current Monitor (LCM)

The substrate bias is generated by the SSB which is controlled by the leakage current monitor (LCM). The LCM is therefore a key to the accurate control in the VT scheme. Fig. 5

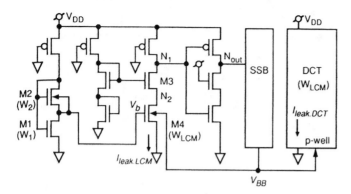

Fig. 5. Leakage current monitor (LCM).

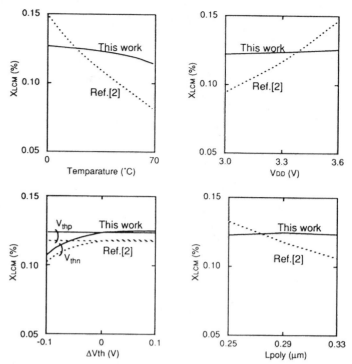

Fig. 6. Current magnification factor of the LCM, X_{LCM}, dependence on circuit condition changes and process deviations simulated by SPICE.

depicts a circuit schematic of the proposed LCM. The circuit works with 3.3-V V_{DD} which is usually available on a chip for standard interfaces with other chips. The LCM monitors leakage current of the DCT, $I_{leak.DCT}$, with a transistor M4 that shares the same substrate with the DCT. The gate of M4 is biased to V_b to amplify the monitored leakage current, $I_{leak.LCM}$. If $I_{leak.LCM}$ is larger than a target reflecting shallower V_{BB} and lower V_{th}, the node N_1 goes "*Low*" and the output node N_{out} goes "*High*" to activate the SSB. As a result, V_{BB} goes deeper and V_{th} becomes higher, and consequently, $I_{leak.LCM}$ and $I_{leak.DCT}$ become smaller. When $I_{leak.LCM}$ becomes smaller than the target, the SSB stops. Then $I_{leak.LCM}$ and $I_{leak.DCT}$ increase as V_{BB} gradually rises due to device leakage current through MOS transistors and junctions, and finally reaches the target to activate the SSB again. In this way, $I_{leak.DCT}$ is set to a target by the on–off control of the SSB with the LCM.

In order to make this feedback control accurately, the current ratio of $I_{leak.LCM}$ to $I_{leak.DCT}$, or the current magnification factor of the LCM, X_{LCM}, should be constant. When an MOS transistor is in subthreshold, its drain current is expressed as

$$I_{DS} = \frac{I_0}{W_0} \cdot W \cdot 10^{(V_{GS} - V_T)/S} \quad (6)$$

where S is the subthreshold swing, V_T is the threshold voltage, I_0/W_0 is the current density to define V_T, and W is the channel width. By applying (6), X_{LCM} is given by

$$X_{LCM} = \frac{I_{leak.LCM}}{I_{leak.DCT}}$$
$$= \frac{W_{LCM}}{W_{DCT}} \cdot 10^{V_b/S} \quad (7)$$

where W_{DCT} is the total channel width of the DCT and W_{LCM} is the channel width of M4. Since two transistors M1 and M2 in a bias generator are designed to operate in subthreshold region, the output voltage of the bias generator V_b is also given from (6) by

$$V_b = S \cdot \log \frac{W_2}{W_1} \quad (8)$$

where W_1 and W_2 is the channel width of M1 and M2, respectively. X_{LCM} is therefore expressed as

$$X_{LCM} = \frac{W_2}{W_1} \cdot \frac{W_{LCM}}{W_{DCT}}. \quad (9)$$

This implies that X_{LCM} is determined only by the transistor size ratio and independent of the power supply voltage, temperature, and process fluctuation. In the conventional circuit [5], on the other hand, where V_b is generated by dividing the V_{DD}-GND voltage with high impedance resistors, V_b becomes a function of V_{DD}, and therefore, X_{LCM} becomes a function of V_{DD} and S, where S is a function of temperature. Fig. 6 shows SPICE simulation results of X_{LCM} dependence on circuit condition changes and process fluctuation. X_{LCM} exhibits small dependence on ΔV_{thn} and temperature. This is because M4 is not in deep subthreshold region. The variation of X_{LCM}, however, is within 15%, which results in less than 1% error in V_{th} controllability. This is negligible compared to 20% error in the conventional implementation.

The four criteria used in the substrate-bias control, corresponding to $V_{active(+)}$, V_{active}, $V_{active(-)}$, and $V_{standby}$ can be set in the four LCM's by adjusting the transistor size W_1, W_2, and W_{LCM} in the bias circuit. For the active mode, with $W_1 = 10\,\mu m$, $W_2 = 100\,\mu m$, and $W_{LCM} = 100\,\mu m$, the magnification factor X_{LCM} of 0.001 is obtained when $W_{DCT} = 1$ m. $I_{leak.DCT}$ of 0.1 mA can be monitored as $I_{leak.LCM}$ of 0.1 μA in the active mode. For the standby mode, with $W_1 = 10\,\mu m$, $W_2 = 1000\,\mu m$, and $W_{LCM} = 1000\,\mu m$, X_{LCM} becomes 0.1. Therefore, $I_{leak.DCT}$ of 10 nA can be monitored as $I_{leak.LCM}$ of 1 nA in the standby mode. The overhead in power by the monitor circuit is about 0.1 and 10% of the total power dissipation in the active and the standby mode, respectively.

The parasitic capacitance at the node N_2 is large because M4 is large. This may degrade response speed of the circuit. The transistor M3, however, isolates the N_1 node from the N_2 node and keeps the signal swing on N_2 very small. This reduces the response delay and improves dynamic V_{th} controllability.

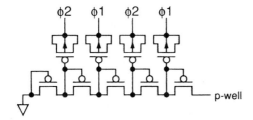

Fig. 7. Pump circuit in SSB.

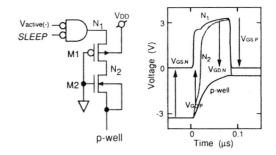

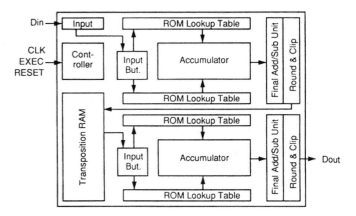

Fig. 8. SCI and its waveforms simulated by SPICE.

Compared with the conventional LCM where V_b is generated by dividing the V_{DD}-GND voltage with high impedance resistors, the V_{th} controllability including the static and dynamic effects is improved from ± 0.05 V to less than ± 0.01 V, response delay is shortened from 0.6 to 0.1 μs, and Si area is reduced from 33 250 to 670 μm^2. This layout area reduction is brought by the elimination of the high impedance resistors by polysilicon.

B. Self-Substrate Bias Circuit (SSB)

Fig. 7 depicts a schematic diagram of a pump circuit in the SSB. PMOS transistors of the diode configuration are connected in series whose intermediate nodes are driven by two signals, $\Phi 1$ and $\Phi 2$, in 180° phase shift. Every other transistor, therefore, sends current alternately from p-well to GND, resulting in lower p-well bias than GND. The SSB can pump as low as -4.5 V. SSB circuits are widely used in DRAM's and E^{2}PROM's, but two orders of magnitude smaller circuit can be used in the VT scheme. The driving current of the SSB is 100 μA, while it is usually several milliamperes in DRAM's. This is because substrate current generation due to the impact ionization is a strong function of the supply voltage. Substrate current in a 0.9-V DCT is considerably smaller than that in a 3.3-V design. Substrate current introduced from I/O pads does not affect the DCT macro because it is separated from peripheral circuits by a triple-well structure. Eventually, no substrate current is generated in the standby mode. From these reasons, the pumping current in the SSB can be as small as several percent of that in DRAM's. Silicon area is also reduced considerably. Another concern about the SSB is an initialization time after a power-on. Even in a 10 mm square chip, V_{BB} settles down within 200 μs after a power-on, which is acceptable in real use.

C. Substrate Charge Injector (SCI)

In the VT scheme, care should be taken so that no transistor sees high-voltage stress of gate oxide and junctions. Transistors are optimized for use at 3.3 V. The gate oxide thickness is 8 nm. The maximum voltage that assures sufficient reliability of the gate oxide is $V_{DD} + 20\%$, or 4 V. The SCI in Fig. 8 receives a control signal that swings between V_{DD} and GND at node N_1 to drive substrate from $V_{standby}$ to V_{active}. In the standby-to-active transition, $V_{DD} + |V_{standby}|$ that is about 6.6 V at maximum can be applied between N_1 and N_2. However, as shown in SPICE simulated waveforms in Fig. 8, $|V_{GS}|$ and $|V_{GD}|$ of M1 and M2 never exceeds the larger of V_{DD} and $|V_{standby}|$. All other transistors in the VT circuit and

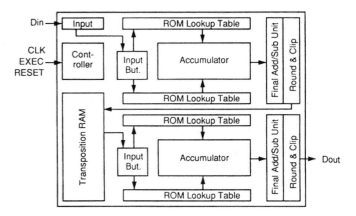

Fig. 9. DCT block diagram.

the DCT macro receive $(V_{DD} - V_{th})$ on their gate oxide when the channel is formed in the depletion and the inversion mode, and less than $|V_{standby}|$ in the accumulation mode. These considerations lead to a general guideline that $V_{standby}$ should be limited to $-(V_{DD} + 20\%)$. $V_{standby}$ of $-(V_{DD} + 20\%)$, however, can shift V_{th} big enough to reduce the leakage current in the standby mode. The body effect coefficient, γ, can be adjusted independently to V_{th} by controlling the doping concentration density in the channel-substrate depletion layer.

V. DCT Design

A. Circuit Design

This DCT core processor executes 2-D 8 $\times$ 8 DCT and inverse DCT. A block diagram is illustrated in Fig. 9. The DCT is composed of two one-dimensional (1-D) DCT and inverse DCT processing units and a transposition RAM. Rounding circuits and clipping circuits which prevent overflow and underflow are also implemented in the cell. The DCT has a concurrent architecture based on distributed arithmetic and a fast DCT algorithm, which enables high throughput DCT processing of one pixel per clock. It also has fully pipelined structure. The 64 input data sampled in every clock cycles are outputted after 112 clock cycle latency.

Various memories which use the same low V_{th} transistors as logic gates are employed in the DCT. Table lookup ROM's (16 b $\times$ 32 words $\times$ 16 banks) employ contact programming and an inverter-type sense-amplifier. Single-port SRAM's (16 b $\times$ 64 words $\times$ 2 banks) and dual-port SRAM's (16 b $\times$ 8 words $\times$ 2 banks) employ a six-transistor cell and a latch sense-amplifier. They all exhibit wide operational margin in

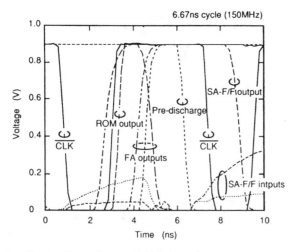

Fig. 10. Simulated waveforms of MAC datapath.

low V_{DD} and low V_{th} and almost behave like logic gates in terms of circuit speed dependence on V_{DD} and V_{th}. No special care is necessary such as word-line boosted-up or a special sense-amplifier.

Small-swing differential pass-transistor logic (SAPL) with sense-amplifying pipeline flip-flop (SA–F/F) [8] is employed for high-speed operation in a 20-b carry skip adder in an accumulator. The SAPL operates and maintains its speed advantage at 0.9 V because the SA–F/F uses a current-mode latch sense-amplifier. As shown in SPICE simulation in Fig. 10, a multiplication and accumulation (MAC) datapath runs at 150 MHz under 0.9 V with no modifications from the 3.3-V design [8].

B. Layout Design

In the conventional CMOS, substrate-contacts are connected to power lines locally, while in the VT scheme they should be interconnected globally for biasing the substrate. This may impose area penalty for separating many substrate contacts, or performance degradation due to substrate noise with few substrate contacts. It is considered, however, that not many substrate contacts are needed in 0.9-V design compared to 3.3-V design because the substrate current generated by the impact ionization becomes several orders of magnitude smaller in 0.9 V. As for substrate noise induced by drain-substrate capacitive coupling, lowering supply voltage is favorable because signal swing as a noise source becomes smaller. It should be effective to add source diffusions because it helps to stabilize V_{BB} by junction capacitance between source diffusions and substrate.

Layout design of the DCT is made in the conventional CMOS fashion, and then it is automatically modified for the VT scheme as illustrated in Fig. 11 by a script of a layout editor. First, have the DCT macro wrapped by deep n-well. Second, generate p-well by inverting n-well data in the deep n-well. Third, replace all the substrate-contacts by source diffusions as long as design rules accept, otherwise remove them. Lastly, place substrate-contacts at the periphery of the deep n-well and the p-well. The p-well becomes one big island and can be connected at periphery. The n-well, on the other hand, becomes many pieces of separated islands.

However, they sit in one deep n-well and can be connected at the periphery, too. Since substrate contacts are only placed at periphery of the 2 mm-square macro, large parasitic substrate resistance is included. Performance degradation or latch up effect due to substrate noise should be examined. Experimental results are presented in the next section. The area penalty, on the other hand, becomes less than 0.1%. This can be done in a p-well or an n-well technology, too, but triple-well structure prevents I/O noise from affecting the DCT macro. The increase in cost and turnaround time by introducing triple-well process is less than 5%. The necessity of the triple-well structure should be examined in the future.

VI. EXPERIMENTAL RESULTS

The DCT core processor is fabricated in a 0.3-μm CMOS, triple-well, double-metal technology. Parameters of the technology and the features of the DCT macro are summarized in Table I. It operates with 0.9-V power supply which can be supplied from a single battery source. Power dissipation at 150 MHz operation is 10 mW. The leakage current in the active mode is 0.1 mA, about 1% of the total power current. The standby leakage current is less than 10 nA, four orders of magnitude smaller than the active leakage current. A chip micrograph appears in Fig. 12(a). The core size is 2 mm square. A magnified picture of the VT control circuit appears in Fig. 12(b). It occupies 0.37 mm × 0.52 mm, less than 5% of the macro size. If additional circuits for testability are removed and the layout is optimized, the layout size is estimated to be 0.3 mm × 0.3 mm. The VT circuit is symmetric for p-well and n-well control. LCM(N), however, occupies more area than LCM(P) because nMOS transistor loads in LCM(N) need longer gate length than pMOS transistor loads in LCM(P) for monitoring the same $I_{\text{leak.LCM}}$.

Fig. 13(a)–(c) shows measured p-well voltage waveforms. Due to large parasitic capacitance in a probe card, the transition takes longer time than SPICE simulation. Just after the power-on, the VT circuits are not activated yet because the power supply is not high enough. As shown in Fig. 13(a), p-well is biased forward by 0.2 V due to capacitance coupling between p-well and power lines. Then the VT circuits are activated and p-well is to be biased at −0.5 V. It tales about 8 μs to be ready for the active mode after the power-on. The active-to-standby mode transition takes about 120 μs as shown in Fig. 13(b), while the standby-to-active mode transition is completed within 0.2 μs as presented in Fig. 13(c).

Compared to the DCT in [8], power dissipation at 150 MHz operation is reduced from 500 mW to 10 mW, that is only 2%. Most of the power reduction, however, is brought by capacitance reduction and voltage reduction by technology scaling. Technology scaling from 0.8 to 0.3 μm reduces power dissipation from 500 to 100 mW at 3.3 V and 150 MHz operation. Without the VT scheme, V_{DD} and V_{th} cannot be lowered under 1.7 and 0.5 V, respectively, and the active power dissipation is to be 40 mW. It is therefore fair to claim that the VT scheme contributes to reduce the active power dissipation from 40 to 10 mW.

The DCT operates at supply voltages from 0.9 to above 3 V. No performance degradation nor latchup effect is observed

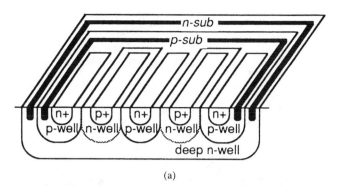

(a)

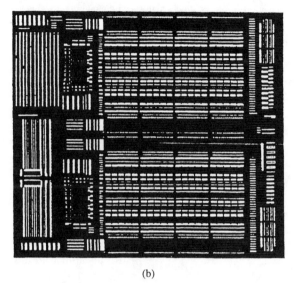

(b)

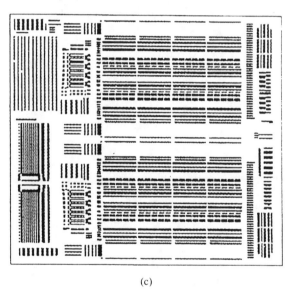

(c)

Fig. 11. DCT layout modification for the VT scheme: (a) device cross-section, (b) p-well (one island), and (c) n-well (pieces of islands) in deep n-well.

TABLE I
FEATURES

Technology	0.3 μm CMOS, triple-well, double-metal, $T_{ox} = 8$ nm, $V_{th} = 0.15$ V $\pm$ 0.1 V
Power supply voltage	1.0 V $\pm$ 0.1 V
Power dissipation	10 mW @ 150 MHz
Standby current	<10 nA @ 70°C
Transistor count	120K Tr
Area	2.0×2.0 mm^2
Function	8×8 DCT and inverse DCT
Data format	9-b signed (pixel), 12-b signed (DCT)
Latency	112 clocks
Throughput	64 clocks/block
Accuracy	CCITT H.261 compatible

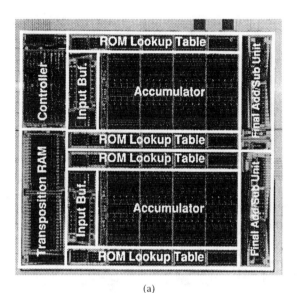

(a)

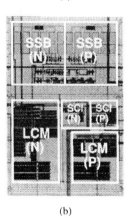

(b)

Fig. 12. Chip micrograph: (a) DCT macro and (b) VT circuits.

even when 100 kΩ resistance is added between the substrate and the output of the SSB.

VII. CONCLUSIONS

A 4 mm^2 2-D DCT core processor for portable multimedia equipment with HDTV-resolution video compression and decompression has been developed in a 0.3-μm CMOS, triple-well, double-metal technology. It operates at 150 MHz from a 0.9 V power supply and dissipates 10 mW, which is only 2% of the previous 3.3 V design. Circuit design techniques for dynamically varying threshold voltage (VT scheme) are introduced to reduce active power dissipation with negligible overhead in speed, standby power dissipation, and chip area. The active-to-standby mode transition takes 120 μs, while the standby-to-active mode transition is completed within 0.2 μs. The VT scheme can be applied to both logic gates and memory elements. Generation of the low-voltage V'_{DD} on chip is a future research work.

103

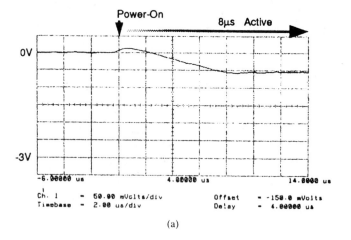

(a)

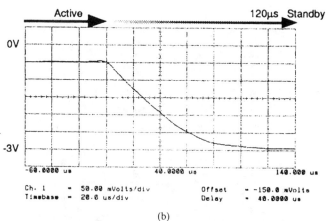

(b)

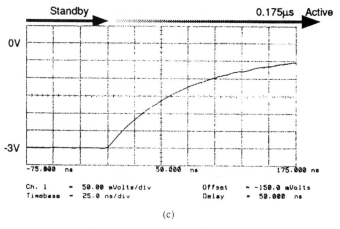

(c)

Fig. 13. Measured p-well V_{BB}: (a) after power-on, (b) active-to-standby, and (c) standby-to-active.

ACKNOWLEDGMENT

The authors would like to acknowledge the encouragement of A. Kanuma, J. Iwamura, K. Maeguchi, O. Ozawa, and Y. Unno throughout the work.

REFERENCES

[1] T. Kuroda and T. Sakurai, "Overview of low-power ULSI circuit techniques," *IEICE Trans. Electron.*, vol. E78-C, no. 4, pp. 334–344, Apr. 1995.

[2] J. B. Burr and T. Shott, "A 200 mV self-testing encoder/decoder using Stanford ultra-low-power CMOS," in *ISSCC Dig. Tech. Papers*, Feb. 1994, pp. 84–85.

[3] T. Kuroda and T. Sakurai, "Threshold-voltage control schemes through substrate-bias for low-power high-speed CMOS LSI design," in *J. VLSI Signal Processing*. Norwell, MA: Kluwer, Special Issue on *Technologies for Wireless Computing*, to be published.

[4] S.-W. Sun and P. G. Y. Tsui, "Limitation of CMOS supply-voltage scaling by MOSFET threshold-voltage variation," *IEEE J. Solid-State Circuits*, vol. 30, no. 8, pp. 947–949, Aug. 1995.

[5] T. Kobayashi and T. Sakurai, "Self-adjusting threshold-voltage scheme (SATS) for low-voltage high-speed operation," in *Proc. CICC'94*, May 1994, pp. 271–274.

[6] K. Seta, H. Hara, T. Kuroda, M. Kakumu, and T. Sakurai, "50% active-power saving without speed degradation using standby power reduction (SPR) circuit," in *ISSCC Dig. Tech. Papers*, Feb. 1995, pp. 318–319.

[7] S. Mutoh, T. Douseki, Y. Matsuya, T. Aoki, S. Shigematsu, and J. Yamada, "1-V power supply high-speed digital circuit technology with multithreshold-voltage CMOS," *IEEE J. Solid-State Circuits*, vol. 30, no. 8, pp. 847–854, Aug. 1995.

[8] M. Matsui, H. Hara, K. Seta, Y. Uetani, L.-S. Kim, T. Nagamatsu, T. Shimazawa, S. Mita, G. Otomo, T. Ohto, Y. Watanabe, F. Sano, A. Chiba, K. Matsuda, and T. Sakurai, "200 MHz video compression macrocells using low-swing differential logic," in *ISSCC Dig. Tech. Papers*, Feb. 1994, pp. 76–77.

[9] T. Sakurai and A. R. Newton, "Alpha-power law MOSFET model and its application to CMOS inverter delay and other formulas," *IEEE J. Solid-State Circuits*, vol. 25, no. 2, pp. 584–594, Apr. 1990.

SOI CMOS for Low-Power Systems

DIMITRI A. ANTONIADIS

MICROSYSTEMS TECHNOLOGY LABORATORIES
MASSACHUSETS INSTITUTE OF TECHNOLOGY
CAMBRIDGE, MA 02139, USA

Abstract—**This paper provides an overview of SOI MOSFET theory and practice with emphasis on circuit applications issues. Fully and partially depleted channel devices are considered, and particular attention is given to describing the so-called floating-body effects that are unique to SOI. Two advanced SOI MOSFET configurations, dual-gate SOI and active-body SOI, specifically developed for low-voltage circuit applications are also introduced. While all of the major SOI issues are covered here, the reader is encouraged to probe deeper in the cited literature.**

INTRODUCTION

Silicon on insulator (SOI) is considered advantageous with respect to bulk silicon as a substrate for CMOS technology, particularly for low-power applications. Figure 1 shows an idealized cross-section through the two transistors of an SOI CMOS inverter. The transistors are entirely isolated from each other with their active areas completely surrounded by a silicon dioxide insulator. Being dielectrically isolated from each other CMOS devices are not subject to CMOS latch-up. In addition, their so-called *body,* or channel area, is inherently floating electrically because it is junction-isolated from the source and drain. This floating-body is the key feature that distinguishes SOI from bulk MOSFET operation.

SOI MOSFETs are broadly classified, depending on the depletion condition of their channel body during operation, into fully depleted (FD) and partially depleted (PD) MOSFETs. While all other SOI technology features, e.g., source/drain capacitance and layout density, are the same, FD and PD devices can exhibit significant operational differences. Also, they have different manufacturability issues. These are all discussed in the following paper.

There are three advantages of SOI MOSFETs relative to bulk MOSFETs for low-power circuit applications: 1) They exhibit significantly reduced source- and drain-to-substrate capacitance because of the elimination of the standard junction of bulk MOSFETs. Depending on the reference bulk technology, as much as tenfold capacitance reduction is achieved. 2) Because of their dielectric isolation SOI MOSFETs do not exhibit the conventional *body-effect* that decreases current drive in stacked devices (e.g., the NMOS devices in a NAND gate), particularly

with a scaled-down power supply voltage, V_{dd}. 3) Because their body floats, it couples to the gate potential resulting in higher ON/OFF current ratio than in their bulk counterparts; therefore, they are more suitable for reduced V_{dd} operation at a given performance. However, as we will see, this floating-body feature makes SOI CMOS device and circuit engineering more complex than for bulk MOSFETS.

Figure 2 illustrates these advantages by comparing delay per stage between SOI and bulk CMOS with approximately the same channel length and threshold voltage. Simple inverter, 2-input NOR, 2-input NAND, and 4-input NAND stages are compared. At $V_{dd} = 1.5$V the relative delay reduction in SOI is 35%, 59%, 63%, and 85%, respectively. The delay reduction in the inverter case partitions approximately 25% due to capacitance reduction and 10% due to gate-body coupling. The increased delay reduction in the other stages is due to the body-effect improvement of stacked devices. Since power dissipation is approximately CV_{dd}^2 the reduced nodal capacitance C leads to a 25% power reduction *at constant* V_{dd}. However, if delay were to be traded-off for power (e.g., the bulk would have to run at $V_{dd} = 2.15$V to achieve same delay as SOI at 1.5V) then the relative power in SOI is $0.75 \times (1.5/2.15)^2 = 0.36$; i.e., power is reduced by a factor of 2.7 at *constant delay*.

In addition to its application in straightforward CMOS, SOI also lends itself to the realization of novel device structures that are not possible in bulk silicon wafers. Such device structures of interest to low power applications are the dual-gate SOI MOSFET, which incorporates a gate under the silicon film, and the active-body MOSFET, in which the channel body voltage is externally controlled. As will be seen later both of these device structures allow dynamic control of the MOSFET threshold voltage and hence even more aggressive power supply voltage scaling than simple SOI with concomitant power savings.

PARTIALLY DEPLETED SOI MOSFETs

CMOS in PD-SOI is easier to manufacture than in FD-SOI. This is because the SOI silicon thickness (t_{si}) can be significantly larger than the depletion depth under the channel, which makes the engineering of SOI devices quite similar to that of their bulk

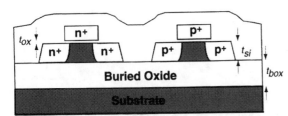

Figure 1 Idealized cross section through two SOI CMOS transistors. Three important film thicknesses are defined: gate oxide thickness, t_{ox}; silicon film thickness, t_{si}; and buried oxide thickness, t_{box}.

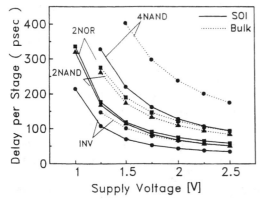

Figure 2 Delay per stage vs. power supply voltage for inverter, 2-input NOR, 2-input NAND, and 4-input NAND stages, implemented in bulk and SOI CMOS with approximately equal channel lengths and threshold voltages. [After Shahidi in "A tutorial on SOI Materials Devices and Technologies," 1995 Int. SOI Conf. Short Course.]

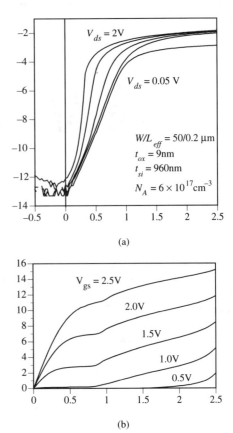

Figure 3 (a) Static transfer characteristics ($\log I_d$ vs. V_{gs}) of PD-SOI nMOSFET. (b) Static output characteristics (I_d vs. V_{ds}).

counterparts. On the other hand, unless the channel body is intentionally tied electrically to the source [1], they exhibit marked *floating-body effects*. Figure 3 shows typical transfer and output *static I-V* characteristics of a PD nMOSFET. Figure 4 (a) depicts a cross section of the device and illustrates the processes taking place at some bias condition $V_{gs} > V_{th}$ and $V_{ds} > 0$. Figure 4 (b) shows a simplified equivalent network for the device [2]. The reason for the two unusual I-V features, the increasing "shoulder" in the transfer characteristics, and the "kink" in the output characteristics is the varying body voltage, V_{bs}, which affects the device V_{th}. V_{bs} under static conditions is established by the balance between impact ionization current and diode leakage to the source. However, it has been shown that these *static floating-body bipolar effects* (bipolar because carriers of both polarities are involved) are absent under normal, rapid switching operation because the impact ionization current is too weak to rapidly change the body charge and hence affect V_{bs} [3]. Therefore, the apparent super-steep subthreshold slope (Fig. 4*a*) cannot be taken advantage of for low-voltage operation.

Of more significance for circuit operation are the *dynamic floating-body effects* which can be distinguished in two categories: *capacitive* and *bipolar turn-on*. These affect the instantaneous value of V_{bs} and hence of the drain current.

Capacitive effects arise from the capacitive coupling of the floating-body to the gate, source, and drain. Contrary to bulk, where charge is free to move in the form of the majority carrier current to or from the *fixed-voltage* body in response to gate,

source, and drain voltages, charges in PD-SOI MOSFETs are near-constant in the time-scale of typical logic clock frequencies; therefore V_{bs} has to *vary* in that time-scale [4,5]. Referring now to Fig. 4*b* and assuming that there is no impact ionization current, V_{bs} for either logic state (i.e., either gate/drain HI/LO or gate/drain LO/HI in *static equilibrium*) is about 0 V, enforced by the source and drain diodes. Since the capacitances are all highly non-linear functions of terminal voltages, the equilibrium body charges in the two states are not equal. It is clear then that under rapid transitions between the two states, the body charge Q_b will assume a *switching-steady-state* value between the two equilibrium values. Because the only current path is through the typically reverse-biased drain diode, and reverse- or slightly-forward-biased source diode (typically $V_{bs} < 0.5V$), the required charge-up or charge-down of the floating body is longer than tens of microseconds; this is very slow compared to clock frequencies. Now allowing impact ionization, there is an additional current source in parallel with the drain diode that pumps a pulse of charge into the body during every transition. This charge is also very small, typically of the order of a fraction of a hole per micron-width per transient, and though it affects the *switching-steady-state* value of Q_b, it does not help reach it much faster [4].

While the slow change of Q_b causes a slow change in the *average value* of V_{bs}, direct capacitive coupling to the gate and drain voltages during transients causes rapid V_{bs} fluctuations. The slow Q_b change can be a concern when a circuit starts switching from an idling state. Effectively, V_{th} would vary in the course of many

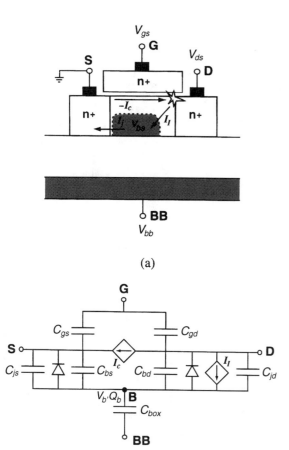

(a)

(b)

Figure 4 (a) Illustration of typical current flows in a PD-SOI nMOSFET at $V_{gs} > V_{th}$, $V_{ds} > 0$. Channel current, I_c (electrons), impact ionization current, I_I (holes), and body-source diode current, I_j (electron-hole recombination current). (b) Simplified (incomplete) equivalent network representation of a SOI nMOSFET. C_{gs}, C_{bs}, C_{gd}, and C_{bd} are the standard intrinsic device capacitances. C_{js} and C_{jd} are the source- and drain-body extrinsic capacitances.

clock cycles rising or decreasing toward its *switching-steady-state* value, depending on the initial state of the device [6,7].

On the other hand, the rapid V_{bs} fluctuation will lead to a more or less repetitive modulation of the device current relative to a fixed V_{bs}. Figure 5 depicts Q_b and V_{bs} as a function of time for a hypothetical inverter that has started switching after idling at the gate/drain HI/LO state. Note that because C_{gb} is largest in the subthreshold region, the gate-voltage ramp couples to V_{bs} producing a positive spike that increases the dynamic subthreshold slope (not shown), which is a beneficial effect. Thus, the net effect of *capacitive dynamic floating-body effects* is to improve the rate of device turn-on, and introduce a slowly hysteretic V_{th}. Proper device engineering can minimize the latter while maintaining the first.

Dynamic bipolar turn-on floating-body effects can arise under conditions that result in V_{bs} sufficiently high to turn-on the source diode. Then the parasitic npn bipolar transistor formed by the source-body-drain turns on and bipolar current flows from drain to source [8,9]. While bipolar turn-on can result under static conditions from exceedingly large impact ionization current (e.g., at a very high V_d), causing what is called *single-device*

latch-up, this is not typically of concern in logic circuit operation. Of more concern is the brief turn-on and current flow that can occur under some pass-transistor conditions. Figure 6a shows such a condition. The source and drain are both at high voltages with the gate at 0V (e.g., right after the pass transistor charged the node capacitance C_n). If V_s and V_d were both high for a sufficiently long time, V_b would also be a high voltage approximately equal to the lesser of the two, brought there by diode leakage. If now V_s transits low, V_{bs} could momentarily be high enough for the diode and hence the parasitic bipolar to turn on. Figure 6b shows one such typical current [8]. Fortunately, the total injected charge is quite small, and the V_s and V_d set-up time required for this effect is long enough so that the effect is manageable.

Summarizing, a PD-SOI CMOS can be built on a variety of different silicon film thicknesses. The key disadvantage for this device type comes from the floating-body effects which need to be carefully managed. Once understood, these devices can be optimized to reduce the severity of these effects; what remains of them is manageable at the circuit design level. Also, floating-body effects can be eliminated by tying the quasi-neutral body of the PD-SOI MOSFET to the source. However, this results in increased area per transistor, as well as electrical asymmetry with respect to the source and drain. Therefore this solution is reserved for special circumstances in circuit design.

FULLY-DEPLETED SOI MOSFETs

When the SOI silicon film thickness (t_{si}) is less than the depletion depth (t_d) under the inversion channel, SOI MOSFETs are said to be fully-depleted. In a FD-SOI MOSFET the body charge cannot be modulated by the terminal voltages. Thus, the charge at either HI/LO or LO/HI gate/drain states is the same. Also, because the whole body is depleted the diode potential barrier is decreased such that the required forward bias for significant conduction can be much smaller than the usual 0.7V. Thus, any injected impact ionization current from the drain is shunted to the source, and thus V_b remains near 0V. Therefore PD-SOI MOSFETs generally exhibit much reduced floating-body effects. Figure 7 shows an example of FD-SOI nMOSFET transfer and output static I-V characteristics. Note the absence of kinks and shoulders, in contrast to Fig. 3.

From the low power application standpoint, another advantage of FD SOI is that the subthreshold slope, $2.3nkT/q$ volts per decade of current, can approach ideal where $n = 1$. This leads to maximal on/off current ratio which allows the minimization of V_{dd}, and hence power, for the given I_{off} and performance. In Fig. 3, this is because the surface potential in weak inversion at the channel-side of C_{gs} is determined by the capacitive divider $C_{gs} - C_{bs} - C_{box}$ and therefore the subthreshold slope factor, n, is given by [10]:

$$n = 1 + (1/C_{gs})/(1/C_{bs} + 1/C_{box}) = 1 + 3t_{ox}/(t_{si} + 3t_{box}) \qquad [1]$$

Since the body is fully depleted it contributes a fixed charge which determines the device V_{th}. Therefore, depending on the channel doping configuration, the body charge, and hence V_{th},

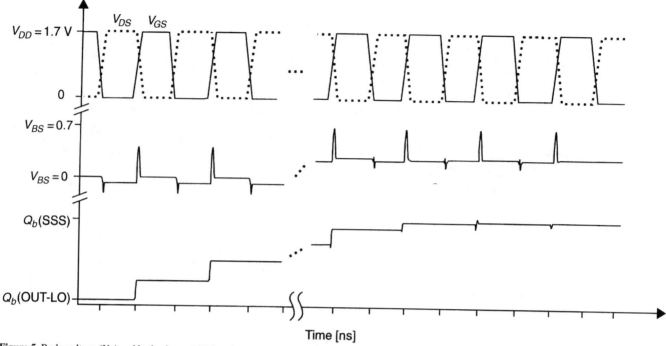

Figure 5 Body voltage (V_{bs}) and body charge (Q_b) for a hypothetical PD-SOI nMOSFET switching with the shown V_{gs} and V_{ds} waveforms after a long idle period at the drain-low state. Because $V_{ds} = 0V$ during the idle state there is no impact ionization current, and Q_b and V_{bs} are both at their minimum steady state values. Transient impact ionization creates the positive steps in Q_b, because of injection of holes, until it reaches its *switching steady-state* value, Q_b (SSS). Accordingly, the average value of V_{bs} rises with the gate coupling spikes superimposed.

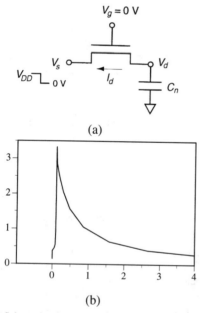

Figure 6 (a) Schematic of a pass-gate configuration. Transient current (I_D) can flow as explained in the text if a PD-SOI nMOSFET had been idling in the initial state ($V_s = V_{dd}$) for a long time. (b) An example of the transient current.

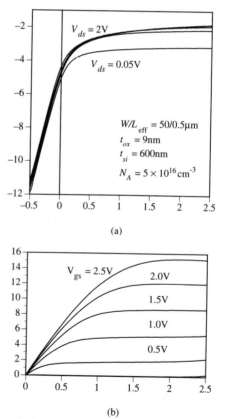

Figure 7 (a) Static transfer characteristics ($\log I_d$ vs. V_{gs}) of FD-SOI nMOSFET. (b) Static output characteristics (I_d vs. V_{ds}).

can depend on the silicon thickness which together with its required thinness places an increased manufacturability burden on FD versus PD CMOS. While a doping scheme that reduces the FD-CMOS V_{th} sensitivity to t_{si} variation has been demonstrated [11], the required t_{si} for FD devices less than 50 nm poses a significant challenge to the formation of low-resistance contacts to the source and drain [12].

As one can imagine, the transition from PD ($t_{si} > t_d$) to FD ($t_{si} < t_d$) behavior is not abrupt. First, it should be noted that since the depletion depth normally increases with distance from the source to the drain, the relevant depletion depth here is the one near the source, i.e., the minimum depletion depth in the channel. Second, since a given doping and t_{si} can produce a marginally PD *long channel* MOSFET, it is quite possible that *short-channel* devices are FD because of increased body depletion provided by the source and drain due to the well-known *charge-sharing effect*. Finally, the diode barrier can be expected to decrease slowly from its normal value ($V_{diode} \sim 0.7V$) as t_{si} decreases below t_d. It is clear then that the degree of floating-body effect amelioration would increase as t_{si} decreases below t_d. This effect is illustrated in Fig. 8, which plots three instantaneous values of V_{th}, each corresponding to nMOSFETs with constant static $I_{off} = 1nA/\mu$m at $V_d = V_{dd}$, versus t_{si} [13]. These values of V_{th} depend on the previous state of the transistor; they switch after a long idle with drain at V_{dd} or 0V, or when they are constantly switching. As can be seen a film thickness of <25nm is required for all these V_{th}s to merge together, indicating complete absence of floating-body effects. This thickness is less than that required for full depletion and is channel-length specific.

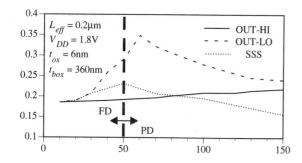

Figure 8 Transient threshold voltage ($V_{T,trans}$) for SOI nMOSFETs with characteristic values shown in the inset, vs. silicon film thickness (t_{si}). Three different values of $V_{T,trans}$ are shown. OUT-HI and OUT-LO are the effective threshold voltages *immediately after* idling at drain high and low states, respectively. After switching steady state (SSS) has been reached (e.g., as in Fig. 4), the effective threshold voltage settles to the SSS values.

Summarizing, while FD-SOI MOSFETs are attractive because of decreased floating-body effects, they are difficult to manufacture because they require very thin t_{si} in order to be truly free of these effects. For this reason a compromise between severity of floating-body effects and manufacturability dictates t_{si} values thicker than optimum (e.g., on order of 50nm for 0.25μm CMOS technology).

SELF-HEATING EFFECTS

Because SOI MOSFETs are separated from the underlying silicon wafer by a thin layer of SiO_2 they are less coupled thermally to the wafer than their bulk MOSFET counterparts. This leads to the well known *self-heating effect*. This effect is evident in the FD and less so in the PD device characteristics (see Fig. 6 at V_{gs} of 2V and 2.5V). Because increasing temperature reduces carrier mobility, self-heating reduces the drain current, and in the more severe cases of high V_{gs} and V_{ds} gives rise to negative differential output conductance. It turns out that in logic applications this effect is not as bad as it looks in the static characteristics; the typical power dissipated per switching cycle in the most heavily loaded logic device is approximately a factor of 10 less than the nominal device $I_{dmax}V_{dd}$, and therefore the device temperature increase due to self-heating is typically less than 8°C [14]. On the other hand, the self-heating effect is also dynamic with time constant of order 1μs, and therefore cannot respond to the typical logic clock rates. For both of these reasons, self-heating is of no concern for logic applications. However, it should be considered carefully in analog applications [15].

DUAL-GATE SOI MOSFETs

Dual-gate SOI MOSFETs are fully-depleted SOI MOSFETs with a gate underneath the channel in addition to the normal top gate. This bottom gate can be either electrically tied directly to the top gate [16,17], or it can be independently controlled [18]. The objective for the top to bottom tied gate concept is to increase the current drive of the MOSFET by providing two parallel inversion channels from source to drain—the top one induced by the top gate and the bottom one induced by the bottom gate. For this to work effectively, the two oxide thicknesses must be equal ($t_{ox} = t_{box}$). Precise alignment of top to bottom gates is necessary to avoid unwanted parasitic overlap capacitances to the source and drain. While this is very difficult to realize, this kind of dual-gate MOSFET can be shown to be the ultimately scaleable MOSFET configuration [19]. From the low-power circuit application standpoint, this dual-gate MOSFET can be considered as a PD or FD-SOI MOSFET, depending on t_{si}, with twice the current drive per unit area, and hence half the source/drain parasitic capacitance per unit current. However, the total gate capacitance per unit current is the same or slightly increased compared to the single-gate SOI MOSFET.

On the other hand, the independently controlled dual-gate MOSFET is specifically designed such that the bottom gate can control the V_{th} of the top-gated MOSFET [18]. This allows the device threshold to be dynamically adjusted during circuit operation. The main low-power application is in logic circuits with variable activity (e.g., event-driven computation) where V_{th} can be adjusted to be low for high performance during computation, and then adjusted to be high for low leakage during idle periods. Other possible applications of dynamic V_{th} control include compensation of current drive variation due to temperature and manufacturing variations. For all these dynamic V_{th} applications, the t_{ox}, t_{si}, and t_{box} must be optimized depending on the target channel length. Depending on particular circuit activity profiles, significant system power reduction at constant performance can be achieved. This subject is covered in detail in [18], which is included later in this volume.

ACTIVE-BODY SOI MOSFETs

Active-body SOI MOSFETs are partially-depleted SOI MOSFETs designed such that their channel bodies can be externally electrically controlled as a fourth terminal [20]. Hence their V_{th} can be adjusted by controlling V_{bs}, and this can be taken advantage of in low-power CMOS circuit optimization [21,22].

A special case of active-body SOI, where the body is directly tied to the gate, has also been proposed as an advantageous configuration for low-power circuit applications [23]. A so-called dynamic-threshold, DT-CMOS is only applicable for very low V_{dd} so that the body-to-source leakage can be negligible. By tying the gate to the body, the DT-CMOS device achieves an ideal subthreshold slope with $n = 1$, which is the same as that of an ideal FD-SOI CMOS, but without the manufacturing difficulties of the very thin t_{si}, discussed under FD-SOI. The ideal subthreshold slope gives rise to maximization of on/off current ratio which allows minimum power for given performance, as outlined in the FD-SOI discussion. Because the body in these devices is tied to the gate, and hence is not floating, they should not exhibit any of the floating-body effects expected in PD-SOI. Details on this technology are covered in paper [23], included later in this volume.

SUMMARY AND CONCLUSIONS

Standard SOI-CMOS, whether with a partially-depleted or fully-depleted channel body, allows power reduction at a given performance relative to bulk CMOS. This power reduction is of order 35%, but the exact amount will depend on the low-power optimization level of the reference bulk technology. PD-SOI is more readily manufacturable than FD-SOI, but it exhibits stronger floating-body effects which must be taken into consideration in circuit design. The SOI configuration opens up the possibility of novel MOSFET structures, the dual-gate SOI and active-body SOI, which can lead to even further reductions of power for given performances by means of new circuit design techniques.

ACKNOWLEDGMENTS

The work of my past and current students Lisa Su, Isabel Yang, Melanie Sherony, Andy Wei and Anthony Lochtefeld, as well as continuing support from the SRC and DARPA, are gratefully acknowledged.

References

[1] T. Iwamatsu, Y. Yamaguchi, Y. Inoue, T. Nishimura, and N. Tsubouchi, "CAD-compatible high-speed CMOS/SIMOX gate array using field-shield isolation," *IEEE Trans. on Elec. Dev.,* vol. 42, no. 11, 1995, pp. 1934–1939.

[2] D. Suh and J. G. Fossum, "A physical charge-based model for non-fully depleted SOI MOSFETs and its use in assessing floating-body-effects in SOI CMOS circuits," *IEEE Trans. on Elec. Dev.,* vol. 42, no. 4, 1995, pp. 728–737.

[3] A. Wei, M. J. Sherony, and D. A. Antoniadis, "Transient behavior of the kink effect in partially-depleted SOI MOSFET's," *IEEE Electron Dev. Letters,* vol. 16, no. 11, 1995, pp. 494–496.

[4] A. Wei, D. A. Antoniadis, and L. A. Bair, "Minimizing floating-body-induced threshold voltage variation in partially-depleted SOI CMOS," *IEEE Electron Dev. Letters,* vol. 17, no. 8, 1996, pp. 391–394.

[5] J. Gautier, K. A. Jenkins, and J. Y. C. Sun, "Body charge related transient effects in floating body SOI NMOSFET's," *Int. Electron Devices Meeting (IEDM) Tech. Dig.,* 1995, pp. 623–626.

[6] D. Suh and J. G. Fossum, "Dynamic floating-body instabilities in partially-depleted SOI CMOS circuits," *Int. Electron Devices Meeting (IEDM) Tech. Dig.,* 1994, pp. 661–664.

[7] A. Wei and D. A. Antoniadis, "Bounding the severity of hysteretic transient effects in partially-depleted SOI CMOS," *Proceedings 1996 IEEE International SOI Conference,* 1996, pp. 74–75.

[8] A. Wei, and D. A. Antoniadis, "Measurement of transient effects in SOI DRAM/SRAM access transistors," *IEEE Electron Dev. Letters,* vol. 17, no. 5, 1996, pp. 193–195.

[9] M. M. Pellela, J. G. Fossum, D. Suh, S. Krishnan, K. A. Jenkins, and M. J. Hargrove, "Low-voltage transient bipolar effect induced by dynamic floating-body charging in scaled PD/SOI MOSFETs," *IEEE Electron Dev. Letters,* vol. 17, no. 5, 1996, pp. 196–198.

[10] H. Lim and J. G. Fossum, "Current-voltage characteristics of thin-film SOI MOSFETs in strong inversion," *IEEE Trans. on Elec. Dev.,* vol. 31, 1985, pp. 401–408.

[11] M. J. Sherony, L. T. Su, J. E. Chung, and D. A. Antoniadis, "Reduction of threshold voltage sensitivity in SOI MOSFETs," *IEEE Electron Dev. Letters,* vol. 16, no. 3, 1995, pp. 100–102.

[12] L. T. Su, M. J. Sherony, H. Hu, J. E. Chung, and D. A. Antoniadis, "Optimization of series resistance in sub-0.2 micron SOI MOSFETs," *IEEE Electron Dev. Letters,* vol. 15, no. 9, 1994, pp. 363–365.

[13] M. J. Sherony, A. Wei, and D. A. Antoniadis, "Effect of body-charge on fully- and partially-depleted SOI MOSFET design," *Int. Electron Devices Meeting (IEDM) Tech. Dig.,* 1996, pp. 125–128.

[14] L. T. Su, J. E. Chung, D. A. Antoniadis, K. E. Goodson, and M. I. Flik, "Measurement and modeling of self-heating in SOI MOSFETs," *IEEE Trans. on Elec. Dev.,* vol. 41, 1994, pp. 69–75.

[15] B. M. Tenbroek, W. Redman-White, M. S. L. Lee, R. J. T. Bunyan, M. J. Uren, and K. M. Brunson, "Characterization of layout dependent thermal coupling in SOI CMOS current mirrors," *IEEE Trans. on Elec. Dev.,* vol. 43, 1985, pp. 2227–2232.

[16] J. P. Colinge, M. Gao, A. Romano-Rodriguez, H. Maes, and C. Clayes, "Silicon-on-insulator 'gate-all-around-device'," *Int. Electron Devices Meeting (IEDM) Tech. Dig.,* 1990, pp. 595–598.

[17] T. Tanaka, H. Horie, A. Ando, and S. Hijiya, "Ananlysis of P+ PolySi double-gate thin-film SOI MOSFETs," *Int. Electron Devices Meeting (IEDM) Tech. Dig.,* 1991, pp. 683–686.

[18] I. Y. Yang, C. Vieri, A. Chandrakasan, and D. A. Antoniadis, "Back-gated CMOS on SOIAS for dynamic threshold voltage control," *Int. Electron Devices Meeting (IEDM) Tech. Dig.,* 1995, pp. 877–880.

[19] H. S. Wong, D. Frank, Y. Taur, and J. Stork, "Design and performance considerations for sub-0.1 micron double-gate SOI MOSFETs," *Int. Electron Devices Meeting (IEDM) Tech. Dig.,* 1994, pp. 747–750.

[20] J. P. Colinge, *Silicon-on-Insulator Technology: Materials to VLSI,* Boston, MA: Kluwer Academic Publishers, 1991.

[21] T. Douseki, S. Shigematsu, Y. Tanabe, M. Harada, H. Inokawa, and T. Tsuchiya, "A 0.5 V SIMOX-MTCMOS circuit with 200 ps logic gate," *Int. Solid-State Circuits Conf. Tech. Dig.,* 1996, pp. 84–85.

[22] T. Fuse, Y. Oowaki, M. Terauchi, S. Watanabe, M. Yoshimi, K. Ohuchi, and J. Matsunaga, "0.5 V SOI CMOS pass-gate logic," *Int. Solid-State Circuits Conf. Tech. Dig.,* 1996, pp. 88–89.

[23] F. Assaderaghi, D. Sinistsky, S. Parke, J. Bokor, P. Ko, and C. Hu, "A dynamic threshold voltage MOSFET (DTMOS) for ultra-low voltage operation," *Int. Electron Devices Meeting (IEDM) Tech. Dig.,* 1994, pp. 809–812.

Back gated CMOS on SOIAS For Dynamic Threshold Voltage Control

Isabel Y. Yang, Carlin Vieri, Anantha Chandrakasan, Dimitri A. Antoniadis

Department Electrical Engineering and Computer Science
Massachusetts Institute of Technology, Cambridge, MA 02139

Abstract

Simultaneous reduction of supply and threshold voltages for low power design without suffering performance losses will eventually reach the limit of diminishing returns as static power dissipation becomes a significant portion of the total power equation. In order to meet the opposing requirements of high performance and low power, a dynamic threshold voltage control scheme is needed. A novel SOI technology was developed whereby a back-gate was used to control the threshold voltage of the front-gate; this concept was demonstrated on a selectively scaled CMOS process.

Introduction

There have been numerous studies on the merits of fully depleted (FD) SOI CMOS and its implications for low power electronics. Various researchers have exploited the use of FD SOI in dual-gated devices in which the top and bottom gates are tied together, resulting in enhanced transconductance [1, 2, 3, 4]. More recently, another group had demonstrated the concept of dynamic threshold voltage control by tying the body to the gate [5]. We have developed a technology, silicon-on-insulator-with-active-substrate (SOIAS), to fabricate back gated FD CMOS devices by capitalizing on existing SIMOX, wafer bonding, and thinning technologies. The back-gate controls the threshold voltage (V_t) of the front-gate device since the surface potentials at the front and back interfaces are coupled in FD SOI devices. The NMOS and PMOS back-gates are switched independently from each other and the front-gate. This paper describes the development of the SOIAS technology, and an evaluation for low-power logic applications.

SOIAS Preparation and Device Fabrication

The SOIAS substrate is a multilayered blanket film stack consisting of the silicon wafer, oxide, intrinsic polysilicon, back-gate oxide, and silicon film. These substrates were prepared using either bonded SIMOX or etched-back bulk wafers. For the bonded SIMOX process, the back gate oxide to be was formed by dry oxidation and intrinsic amorphous silicon was deposited as the

back-gate to be material. The handle wafer had an oxide/nitride stack. The SIMOX wafer was then bonded to the handle wafer, and annealed in N_2 at 1000 ^oC for one hour. The bonded wafers were then etched in 25% TMAH to remove the bulk of the SIMOX wafer, stopping on the buried oxide. Nitride was used on the handle wafer because it has a higher selectivity against silicon in TMAH than oxide and the entire thickness of the wafer had to be etched away. The bulk bonding and etch back process is similar to the SIMOX process except the device wafer was thinned down by chemical/mechanical polishing to a thickness of approximately 0.5 μm. Localized plasma thinning [6] was then used to reduce the thickness to approximately 0.2 μm. Final thinning of the silicon film was accomplished with thermal oxidation and wet oxide strip. Fig. 1 depicts the SOIAS preparation for both processes.

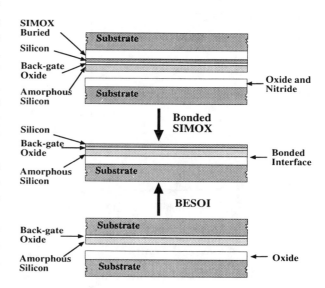

Figure 1: SOIAS preparation using bonded SIMOX and BESOI processes.

The back-gates were formed by ion implantation through the silicon film in two masking steps, resulting in islands of p+ and n+ polysilicon insulated by intrinsic poly. The silicon film was also doped to set the zero back-gate bias value for the front-gate V_t. Using the same type of doping in the back-gate poly and silicon film re-

Reprinted from *IEEE 1995 International Electron Devices Meeting (IEDM)*, pp. 877-880, December 1995.

sulted in near-zero flatband voltage at the back-gate. The front-gate device is then built as in a conventional SOI CMOS process using LOCOS isolation. Fig.2 illustrates the final device schematic. Fig.3 shows the effective electron mobility versus effective transverse electric field of the front-gate device for conventional SIMOX and SOIAS. The universality of the curves indicates no apparent difference between the SOIAS and SIMOX substrates from a device operation point of view. The coupling between the front and back-gates depends on the ratio of the critical film thicknesses: front-gate oxide thickness (t_{fox}), silicon film thickness (t_{si}), and back-gate oxide thickness (t_{box}). We have demonstrated SOIAS with 9 nm t_{fox}, 40 nm t_{si}, and 100 nm t_{box} nominal design parameters in a selectively scaled 1 μm baseline CMOS technology.

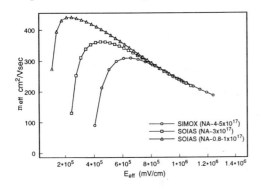

Figure 2: SOIAS back-gated CMOS device schematic.

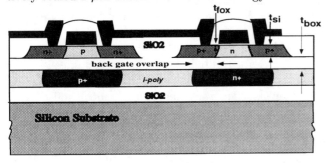

Figure 3: Effective electron mobility for SIMOX and SOIAS.

Device Results

Figures 4-7 show the I-V and subthreshold device characteristics for NMOS and PMOS at two different threshold voltages tuned by biasing the back-gate. A 250 mV change in threshold voltage results in a 3.5-4 decade reduction in off current and a 50-80% current increase at 1 V operation for PMOS and NMOS respectively.

Fig. 8 shows the maximum and minimum tunable V_t limits for the above nominal design parameters. The x-axis is nominal designed V_t which is the threshold voltage at zero back-gate bias. The y-axis, tunable V_t, is obtained by applying various back-gate biases. The tunable V_t

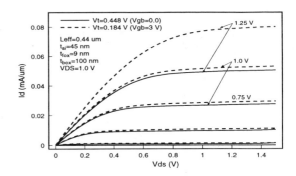

Figure 4: NMOS I-V tuned at different V_t's.

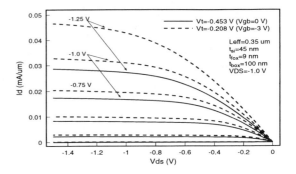

Figure 5: PMOS I-V tuned at different V_t's.

range is quite large for fully depleted back interface as can be seen for the lowest V_t case at zero back-gate bias. Even for the partially depleted highest V_t case, there is still a reasonable tuning range. This has implications for making FD SOI a viable technology since the threshold voltage and the device operating mode can be controlled precisely by the back-gate. As may be seen in Fig. 8, V_t may be fine tuned over a wide range despite variations in t_{si} (average thickness=48.4 nm, maximum thickness= 69.9 nm and minimum thickness=37.6 nm) and L_{eff}. For example, a nominal V_t of 500 mV may be reached even for a $\pm$ 400 mV deviation by using a $\pm$5-6 V back-gate bias. Fig. 9 shows a 101-stage ring oscillator frequency as a function of varying the back-gate-controlled V_t for either the NMOS or PMOS. In Fig.10, both the NMOS and PMOS V_ts were switched low to achieve a 36% increase in speed.

Application to Low Power Systems

We have chosen to assume a model of operation in which "functional units," or modules, share a common V_t. Under this model, an active system's idle components are left in a low-leakage state. In the modeling of a hypothetical microprocessor's energy dissipation, the modules under consideration are the ALU adder unit, the shifter, and the integer multiplier. We have developed total energy equations including switching and static energies for

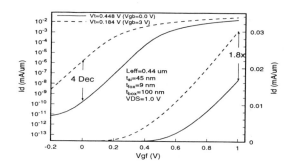

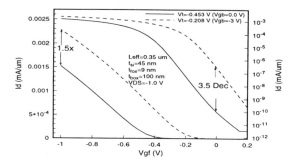

Figure 6: NMOS subthreshold characteristics.

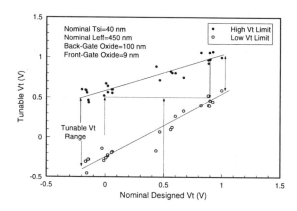

Figure 7: PMOS subthreshold characteristics.

a SOIAS and a benchmark optimized SOI technology in order to analyze the applicability of the SOIAS technology to low power static CMOS logic.

$$E_{SOIAS} = ma(\gamma C_1 V_{dd}^2 + I_{off(low)} V_{dd} t_{cyc})$$
$$+ (1 - ma) I_{off(high)} V_{dd} t_{cyc} + bga C_{box} V_b^2$$
$$E_{SOI} = ma(\gamma C_2 V_{dd}^2) + I_{off(low)} V_{dd} t_{cyc}$$

These equations include:
(a) *Algorithm and architecture parameters:*
ma = module activity factor
bga = back-gate activity factor

Figure 8: Tunable V_t ranges by back-gate biasing.

γ = node switching probability during active period
(b) *Technology and architecture parameter:*
$t_{cycle} = 1/f_{clock}$
(c) *Technology parameters:*
$C_{1,2}$=total switching capacitance (gate capacitance + front-gate overlap capacitance + fringing capacitance + back-gate overlap capacitance)
$I_{off(low)}$=low V_t off current
$I_{off(high)}$=high V_t off current
C_{box}=back-gate oxide capacitance
V_b=back-gate bias

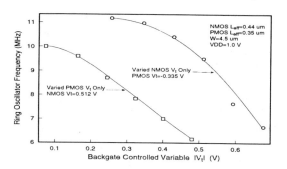

Figure 9: 101 stage ring oscillator output frequency as varied by changing V_t.

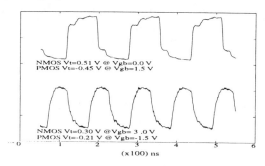

Figure 10: Ring oscillator output at different V_t's tuned by back-gate bias.

The simulation parameters are: t_{si}=40 nm for the practical limit of thinning the silicon film, t_{fox}=7 nm for a 0.25 μm L_{eff} technology; γ, the probability of gate switching in one active cycle is 40% (γ in general is a strong function of bit transition probabilities), 100 MHz clock frequency, and V_{dd} = 1.0 V. The circuit operates by switching the back-gate and lowering V_t when a module is active, and raising the V_t when the module is inactive, satisfying the opposing requirements of high performance and low standby leakage at low power supply voltages. The applicability of SOIAS technology is a function of transistor and functional block usage. To determine functional block usage patterns (ma and bga), we performed a series of program profiling experiments using the ATOM code [7] instrumentation interface for a particular microprocessor implementation, compiler technology, and various

113

algorithms.

The ratio of the total energy dissipation for SOIAS and SOI was analyzed as a function of algorithm and architecture dependent parameters(ma and bga, see Fig.11), as well as technology dependent parameters(t_{box} and source/drain overlap with back gate, Fig.12. The SOIAS technology is most suitable for systems which operate in burst mode, as shown in Fig.11. Near continuous functional block usage which does not exhibit strong temporal locality (e.g adder and shifter in a continuously computing system) does not favor the SOIAS technology. In a system which is frequently idle while awaiting I/O, such an X server which is active 2% of the time, the SOIAS technology dissipates less energy than conventional SOI: 43% for the adder (ma=69.7%, bga=21.3%), 80% for the shifter(ma=10.9%, bga=8.7%) , and 97% for the multiplier(ma=0.83%, bga=0.83%). For low activity modules (the multiplier and shifter), the design space in favor of the SOIAS technology spans the entire parameter range under study for both t_{box} and overlap/L_g (Fig.12a). For high activity modules (the adder), the design space in favor of SOIAS technology strongly depends on the ma/bga ratio. Note for the adder (Fig.12b), there is an optimal range of t_{box} which allows the most overlap where the energy is minimized. Therefore, when the ma and bga values are high, their ratio determines the range of optimal back-gate oxide thicknesses with the maximum design latitude in the back-gate overlap in the technology design space.

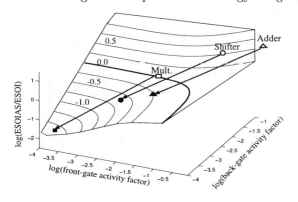

Figure 11: Energy ratio of SOI and SOIAS technologies for systems that are frequently in use (open symbols) and those that are mostly idle (2% usage for X server)(filled symbols).

Conclusions

Dynamic control of threshold voltage has been demonstrated in the novel SOIAS technology developed through bonded SIMOX and BESOI processes. Furthermore, the flexibility in threshold voltage control through back-gate biasing from partially depleted to fully depleted devices provides a viable option for FD SOI. Finally, the applicability of the SOIAS technology for low power design was found to be most suitable for systems which operate in

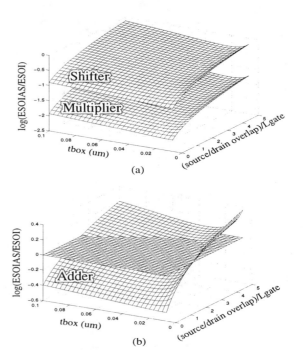

Figure 12: Energy ratio of SOI and SOIAS in the technology design space for functional modules with various activity factors.

burst mode.

Acknowledgments: The authors gratefully acknowledges MIT Lincoln Laboratory for sponsoring this work. In addition, we also appreciate the help of Charles Goodwin of AT&T(Reading, PA) and D.P. Mathur of Hughes(Danbury, Conn.) in wafer thinning. The authors also would like to thank IBIS Corp. for providing all the SIMOX material. I. Yang acknowledges the support of AT&T through a graduate fellowship.

References

(1)T. Tanaka, K. Suzuki, H. Horie, and T. Sugii, "Ultrafast Low-Power Operation of p+-n+ Double-Gate SOI MOSFETs", *Symposium on VLSI Technology*,p11, 1994
(2)T. Tanaka, H. Horie, S. Ando, and S. Hijiya, "Analysis of p+ Poly Si Double-Gate Thin-film SOI MOSFETs", *IEDM Technical Digest'*,p683, 1991
(3)F. Balestra, S. Cristoloveanu, M. Benachir, J. Brini, and T. Elewa, "Double-Gate Silicon-on-Insulator Transistor with Volume Inversion: A New Device with Greatly Enhanced Performance", *Elec. Dev. Lett.*,Vol EDL-8,no. 9, p410, 1987
(4)J.P. Colinge, M.H. Gao, A. Romano-Rodriguez, H. Maes, and C. Claeys, "Silicon-on-Insulator Gate-All-Around Device", *IEDM Technical Digest*, p595, 1990
(5)F. Assaderaghi, D. Sinitsky, S. Parke, J. Bokor, P.K. Ko, and C. Hu, "A Dynamic Threshold Voltage MOSFET (DTMOS) for Ultra-Low Voltage Operation", *IEDM Technical Digest*, p809, 1994
(6) Accu-Thin (Hughes, Danbury Con) was the localized plasma thinning technique used.
(7)A. Srivastava, A. Eustace, "ATOM: A System for Building Customized Program Analysis Tools", *WRL Research Report 94/2*, Digital Equipment Corp. Western Research Laboratory, 1994

Design of Low Power CMOS/SOI Devices and Circuits for Memory and Signal Processing Applications

LARS E. THON AND GHAVAM G. SHAHIDI(*)

IBM ALMADEN RESEARCH CENTER, SAN JOSE, CA 95120 (*) IBM THOMAS J. WATSON RESEARCH CENTER, YORKTOWN HEIGHTS, NY 10598

Abstract—SOI technology has been developed for three decades and in spite of considerable promise has not yet become a mainstream basis for silicon circuits. Bulk CMOS in the deep submicron region faces a number of challenges, e.g., high power consumption in a small chip area. SOI now offers a solution to this problem and also has other applications such as monolithic mixed-signal systems, where the SOI isolation properties can reduce noise transfer between digital and analog components.

This paper starts with a brief treatment of bulk CMOS technology followed by an overview of new directions in SOI device design, including a scalable design point for partially-depleted devices. Unique SOI concerns such as pass-gate leakage and history-dependent device delays are covered, and it is shown that by careful device design these effects can be controlled. The higher performance and reduced power consumption of SOI technology is demonstrated by the comparision of large-scale memory and signal processing functions that have been fabricated in both bulk and SOI technologies. Operating speeds in excess of 600MHz have been observed.

INTRODUCTION

Technology used in microprocessor (μP) fabrication has become the driver of CMOS development. As the effective channel length of mainstream CMOS technologies has been pushed below 0.35μm, the supply voltage has fallen from 5V to 1.8V [1,2]. With channel lengths in the deep submicron region (below 0.5μm), the supply voltage must be reduced for reliability reasons. This trend is marked on Fig. 1. Attempts are generally made to operate each technology at the highest voltage possible in order to get maximum performance. Nevertheless, the supply voltage must be reduced to ensure acceptable device lifetimes. Device performance has been increasing with shorter channel lengths despite the drop in the supply voltage, and there have been marked improvements in switching energy. But as speed and density improve, the power of a chip that fully utilizes the speed-density potential of a technology has increased drastically.

Maximum reported chip power for multiple μP generations is also shown in Fig. 1. For a 0.35μm CMOS the power is 10–100W

and is predicted to be about 100W for a 0.25μm CMOS. A sub-0.25μm CMOS technology is capable of providing a chip with 10–100M transistors, operating at 100–1000MHz if the chip power can be safely dissipated. Thus, it is clear that in addition to hot electron degradation, which from the device design viewpoint is the primary reason to reduce supply voltage, power consumption must be considered in the design point of sub-0.25μm CMOS technology for high-performance logic circuits. Power is of paramount importance in μPs. For most desktop and portable applications, the chip power must be 15W or less. Many of the highest performing μPs in the market today exhibit chip power in the 10–100W region. By reducing the power consumption, markets can be greatly expanded. Clearly there is a big opportunity for technology that preserves μP performance while reducing power consumption.

One effective way of reducing the chip power is to drop the supply voltage faster than otherwise required by lifetime considerations. At lower voltages, the performance of the bulk CMOS is severely affected by the junction capacitances (which are significant in custom designed circuits) and the body-effect (which raises the threshold and causes severe degradation in performance of circuits with stacked- and pass-gate devices). SOI offers means to circumvent obstacles faced by bulk CMOS at low voltages. There is no junction capacitance in SOI. Furthermore, with the proper device design, the subthreshold slope of MOS devices can be lower than 60mV/dec, so that the device's apparent threshold and supply voltage can be further reduced below the trend shown in Fig. 1, while preserving the same leakage as bulk counterparts.

This paper first describes some key features of 0.15μm bulk CMOS technology. The technology has a number of novel device design elements when compared to its predecessors. Next, a design point for high performance CMOS on SOI is described. The major obstacle of SOI is no longer material quality but floating body-effects. Two key floating body-effects are discussed and quantified. Finally, experimental designs of memory and signal processing functions in bulk and SOI CMOS are shown. A 512kB SRAM is described that is fully

115

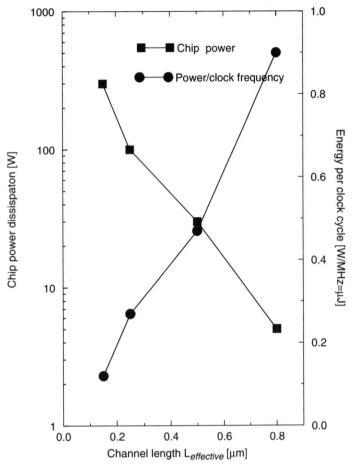

Figure 1 Trends in chip power and energy for bulk CMOS microprocessors.

functional even in the presence of the floating body effects. An 8-tap 12-bit FIR filter (equalizer) design for disk-drive signal processing has been manufactured and evaluated in multiple generations of bulk and SOI CMOS technologies. Speed and power results are presented to quantify how SOI offers a power-delay advantage over bulk CMOS in the deep-submicron region.

SUB-0.25μM DEVICE DESIGN

The state-of-the-art CMOS is a $0.25\mu m$ device operating at 2.5V [1]. Within 1997, parts using $0.15\mu m$ CMOS technology will be offered by semiconductor manufacturers. These deep submicron CMOS technologies utilize shallow trench isolation, a number of novel techniques to overcome the Short Channel Effects (SCE), sub-5nm gate dielectric, salicide technology, sub-$1\mu m$ M1-M2 pitches, and 5 to 6 levels of metal. Among techniques used to improve SCE over the previous generations are highly non-uniform channel doping and ultra-shallow source-drain extension-HALO, obtained by the use of heavy ions (indium and antimony) and low-energy implants. It is desirable that such advances in bulk CMOS processing can be applied also to SOI technology for the deep-submicron regime.

SOI Technology

The "floating-body effects," i.e., those caused by the floating substrate in a MOS device, have appeared as the main impediments to the widespread use of SOI technology. The most prominent of these are the kink effect (increase in output conductance at V_{ds} 1.1V), anomalous subthreshold currents, and early device breakdown. The basic technique applied to eliminate kink effect was the use of fully-depleted SOI films, where the film thickness was less than the depletion width of the device [3]. It has been argued that full depletion not only eliminates many of the floating-body effects, but it also leads to improved SCE. Fully-depleted device designs complicated the SOI technology and resulted in lower device breakdown voltages. The simultaneous rapid improvements in bulk CMOS excluded SOI from the mainstream.

Partially-Depleted SOI Device Design

Probably the largest barrier to the use of SOI has been the lower nFET breakdown voltage, which made operation and the screening of parts very difficult tasks. As the bulk supply voltage dropped, this issue is no longer critical. There are however several present device design issues. The two basic approaches to SOI device design are fully- and partially-depleted devices. Table 1 lists some of the design considerations. The main problem with fully-depleted designs is that there is basically no usable design point, i.e., it is not possible to have a high V_{th} device which is fully-depleted. Fully-depleted designs also have worse SCE and hot electron effects, and they are non-manufacturable. Partially-depleted device designs provide great flexibility for optimizing the device design and manufacture. A partially-depleted deep sub-micron CMOS on SOI was developed [4] for which the devices can turn off completely; their SCE is excellent.

FLOATING-BODY EFFECTS

The body of an SOI FET is floating and may therefore assume any potential within the supply range. The bias within the body is set by the source/drain potentials, generation–recombination rates, and impact ionization within the device. The "time constant" associated with floating-body effects can be 10ms and more. The body bias determines the current of the device. This current can be a complicated function of the device switching history. The key question is whether the effects can be controlled, minimized, and modelled properly. From the circuit designer's point of view, two key effects are *pass-gate leakage* and *history-dependent device delay*.

Table 1. Comparison of Fully- and Partially-depleted SOI Devices.

Fully-depleted	Partially-depleted
threshold design point	any threshold possible
worse SCE, HEF	better SCE, HEF
non-manufacturable	manufacturable
body contact not possible	body contact option
passgate leakage	passgate leakage
	good device turnoff region

116

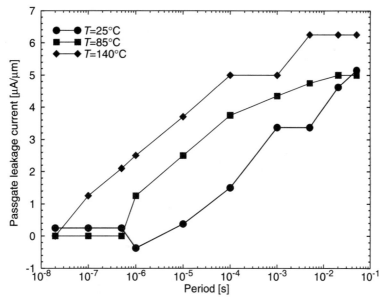

Figure 2 Pass gate leakage current of SOI nFET as a function of biasing period.

Pass-Gate Leakage

When both the source and drain of an SOI device are at the same bias, the body will in time reach that same bias. In particular, if both the source and drain are at high bias then the body will reach that bias. In the case of an nFET with the gate low, the channel is then in accumulation, i.e., there are excess holes in the channel. A change of the source to low bias will produce a small pulse of current that may last a few nanoseconds, associated with the discharge of holes. Thus the device conducts a current, even though the gate is off. This effect can discharge capacitive nodes. Figure 2 shows the peak current as a function of switching frequency [6]. Notice that the peak is about 2 orders of magnitude below the device on-current, and significantly lower than previously reported. Hence proper device design can lower the leakage current to insignificant values for most applications.

History-Dependent Device Delay

This effect is a direct consequence of the variations in drive current of the device due to fluctuating body bias. Figure 3 shows the delay through an SOI inverter as the input frequency is varied. The maximum variation in delay is about 4% [7]. The circuit designer must be aware of the effect so that enough margin is allowed in the critical delay paths.

SOI AND BULK 512KB SRAM

A 512kB SRAM was fabricated to demonstrate the performance of the partially-depleted SOI design point, even in the presence of floating-body effects. The SRAM utilizes dynamic circuits and is self-timed. Because of the pulsed nature of the signaling used in the chip, timing is of great importance. Previously this SRAM had been fabricated in a 0.5μm 3.6V CMOS technology, and an access time of 3.8ns at 3.6V had been obtained [9]. This

time the SRAM was fabricated on SOI with L_{eff} = 0.30μm. Figure 4 shows the access time versus the supply voltage. Neither bulk nor SOI versions worked at supply voltages above 1.6V. Successful demonstration of SRAM shows the real potential of the present SOI device design and provides experimental evidence that partially-depleted SOI is capable of delivering high-performance circuits at low-supply voltages.

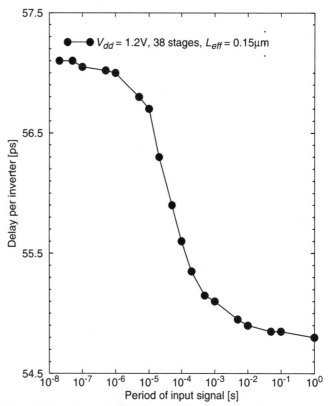

Figure 3 Dependency of delay through SOI inverter on input frequency.

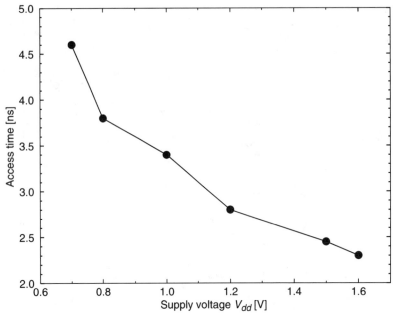

Figure 4 512kB SRAM access time vs. supply voltage.

SOI AND BULK 64kB SRAM

As bulk CMOS chip power has reached 10–50W, reduction of the supply voltage beyond the scaling requirements has been attempted in a number of microprocessors. Scaling of the power supply voltage can significantly reduce power consumption but is usually accompanied by a large drop in performance. The performance loss is a consequence of the junction capacitance (which can be large even in custom-designed microprocessors) and reduced current drive due to MOS body-effect. Both of these effects become more significant at lower voltages. SOI is the ideal substrate for low-voltage CMOS technology. Both the body effect and junction capacitance are nearly absent on SOI. Therefore SOI can operate at low voltages and reduced power without much sacrifice in performance. The power vs. delay of a 64kB SRAM illustrates this point (Fig. 5). The SRAM on SOI can operate at about one third the power of a bulk SRAM with the same performance.

SOI AND BULK FIR FILTERS FOR DISK-DRIVE SIGNAL PROCESSING

A technology-scalable family of FIR filters for mixed-signal disk-drive read channels had previously been designed for bulk CMOS4S [10] 0.8μm and CMOS5X [11] 0.5μm processes [12]. These filters have also been fabricated in experimental 0.25μm

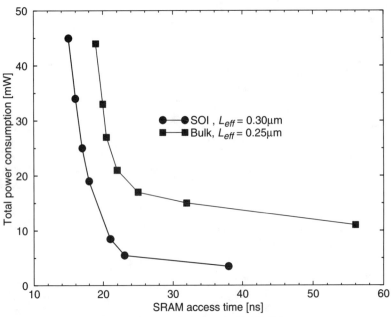

Figure 5 Power vs. delay of 64kB SRAM in bulk and a SOI CMOS.

118

Table 2. Technology Parameters.

Parameter	Technology		
	CMOS4S	CMOS5X	Experiment
L_{drawn}	0.80 μm	0.50 μm	0.50 μm
$L_{effective}$	0.45 μm	0.25 μm	∼0.25 μm
T_{ox}	12 nm	7 nm	∼5 nm
Metal layers	3	3	3

bulk and SOI processes [13]. Key technology parameters are shown in Table 2. The experimental wafers were made with the same mask set as the CMOS5X wafers; hence, the same drawn gate length of 0.5μm is listed for both.

Because of the thinner gate oxide of the experimental process, $t_{ox} = $ ∼5nm, the gate capacitance for same-size transistors is higher than in the CMOS5X technology. With the masks being the same, the total capacitance switched during operation of the chip is higher by a factor of roughly 7/5 in the case of the experimental bulk wafers, or 40%. On the other hand, the drive of each transistor is inversely proportional to t_{ox}, leading us to expect about the same speed from an experimental bulk wafer as from a CMOS5X wafer. In the case of the SOI wafers, additional speed gain is expected due to the large reduction in the source and drain capacitances. Measurements were made on a number of CMOS4S, CMOS5X, and experimental bulk and SOI wafers. Figure 6 shows the maximal speed f_{max} of some filters as a function of the supply voltage V_{dd}. Correct operation up to 620MHz was observed. The energy efficiency of the filter can be defined as the energy per operation (computed sample). Figure 7 shows the energy consumption of the filters as a function

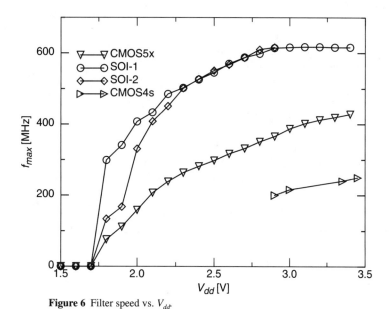

Figure 6 Filter speed vs. V_{dd}.

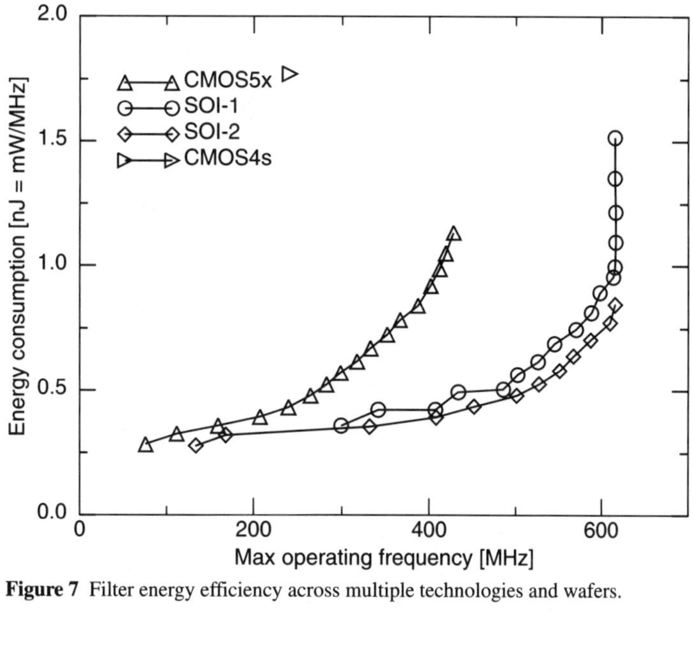

Figure 7 Filter energy efficiency across multiple technologies and wafers.

119

of the maximal operating frequency, with V_{dd} as the underlying parameter. The SOI wafers are considerably more energy efficient than the CMOS5X wafers, differing by as much as a factor of 3 at the 400MHz operating point. Much of this gain is due to the lower V_{dd} required to obtain a given speed for the SOI wafers. At the *same* V_{dd}, the SOI wafers use slightly more energy than the bulk CMOS5X wafer. This is due to the following factors, alluded to earlier: (1) a 7/5 factor increase in gate capacitance per unit area, and (2) the use of the same mask as the CMOS5X wafers (no shrink). With the proper shrink in place, the power consumption for the SOI will be lower for the same V_{dd} and f_{clk}, but again even more significant gains can be obtained via the lower V_{dd} required to obtain a given speed for the SOI circuits.

CONCLUSION

As bulk CMOS moves into deep submicron L_{eff} there is serious concern over the power consumption of chips utilizing such technologies. CMOS on SOI offers the same or better performance at lower power. A new design point for SOI was defined which is manufacturable at 0.25μm and also is extendable below 0.1μm. Floating-body effects can be reduced by careful device design. Implementation examples of memory and signal processing functions verify that the effects are managable for both logic and memory applications. The speed and energy advantages of SOI have been quantified, with power consumption being as low as one third of bulk CMOS at the same speed. Concern about CMOS power consumption now allows SOI to mount a serious challenge to bulk silicon when coupled with lower operating voltages. SOI may ultimately be the substrate of choice in the sub-0.2μm category of CMOS technologies.

ACKNOWLEDGEMENTS

The authors would like to thank their colleagues in IBM for their contributions to this paper.

References

[1] B. Davari et al., "A high performance 0.25um CMOS technology," *IEDM Tech. Dig., IEDM-88,* 1988, p. 56.

[2] G. Shahidi et al., "A high performance 0.15um CMOS," *Tech. Dig. of Symp. on VLSI Tech.,* 1993, p. 93.

[3] J. P. Colinge, *SOI Technology,* Kluwer Press, 1991.

[4] G. Shahidi et al., "A 0.1μm CMOS on SOI," *IEEE Trans. on Electron Devi.,* 1994, p. 2405.

[5] F. Assederaghi et al., "Transient pass-transistor leakage current in SOI MOSFET's," submitted to *IEEE Elec. Dev. Let.*

[6] F. Assederaghi et al., "History dependence of non-fully depleted (NFD) digital SOI circuits," *Tech. Dig. of Symp. on VLSI Tech.,* 1996, p. 122.

[7] T. I.Chappell et al., "A 2ns cycle, 3.8 nS access 512Kb CMOS ECL with a fully pipelined architecture," *IEEE J. of Solid State Circuits,* 1991, p. 1577.

[8] IBM Corporation: CMOS4S technology data. See http://www.chips.ibm.com/products/ams.

[9] C. W. Koburger et al., "A half-micron CMOS logic generation," *IBM J. R&D,* Jan./Mar. 1995, pp. 215–227, vol. 39 no. 1/2. See http://www.chips.ibm.com/products/asics/tech.

[10] L. Thon et al., "A 240 MHz 8-tap programmable FIR filter for disk drive read channels," *Proc. IEEE International Solid State Circuits Conference,* Feb. 1995, pp. 82–83.

[11] L. Thon et al., "250-600MHz 12b digital filters in 0.8-0.25μm bulk and SOI CMOS technologies," *Proc. IEEE ISLPED,* June 1996.

A Dynamic Threshold Voltage MOSFET (DTMOS) for Very Low Voltage Operation

Fariborz Assaderaghi, Stephen Parke, *Member, IEEE,* Dennis Sinitsky, Jeffrey Bokor, *Member, IEEE,* Ping K. Ko, *Senior Member, IEEE,* and Chenming Hu, *Fellow, IEEE*

Abstract— A new mode of operation for Silicon-On-Insulator (SOI) MOSFET is experimentally investigated. This mode gives rise to a Dynamic Threshold voltage MOSFET (DTMOS). DTMOS threshold voltage drops as gate voltage is raised, resulting in a much higher current drive than regular MOSFET at low V_{dd}. On the other hand, V_t is high at $V_{gs} = 0$, thus the leakage current is low. Suitability of this device for ultra low voltage operation is demonstrated by ring oscillator performance down to $V_{dd} = 0.5$ V.

I. INTRODUCTION

DURING the past few years demand for low power and high performance digital systems has grown rapidly. The main approach for reducing power has relied on power supply scaling. Since power supply reduction below $3V_t$ will degrade circuit speed significantly, scaling of power supply should be accompanied by threshold voltage reduction. However, the lower limit for threshold voltage is set by the amount of off-state leakage current that can be tolerated (due to standby power consideration in static circuits, and avoidance of failure in dynamic circuits and memory arrays). To extend the lower bound of power supply, we propose a **D**ynamic **T**hreshold voltage **MOS**FET (DTMOS) with the highest V_t at zero bias and the lowest value at $V_{gs} = V_{dd}$. In the remainder of this paper we will describe the operation of the device, and show its superiority over a regular MOSFET. We will also show some circuit performances using DTMOS.

II. EXPERIMENT AND RESULTS

The SOI devices used in the study are built on SIMOX wafers. Mesa active islands (MESA) were created by plasma-etching a nitride/oxide/silicon stack stopping at buried oxide. P+ polysilicon gate was used for PMOSFET's and N+ for NMOSFET's. The buried oxide thickness was 370 nm–400 nm, and the silicon film thickness was 130 nm–160 nm. A four-terminal layout was used to provide separate source, drain, gate, and body contacts. Besides the four-terminal layout, devices with local gate-to-body connections were also

Manuscript received April 26, 1994; revised September 20, 1994. This project was supported by SRC under Contract 93-DC-324, ISTO/SDIO through ONR under Contract N00014-92-J-1757, and AFOSR/JSEP under Contract F49620-93-C0041.

F. Assaderaghi, D. Sinitsky, J. Bokor, P. K. Ko, and C. Hu are with the Department of Electrical Engineering and Computer Science, University of California at Berkeley, Berkeley, CA 94720 USA.

S. Parke was with the Department of Electrical Engineering and Computer Science, University of California at Berkeley. He is currently with IBM Corporation, East Fishkill, NY USA.

IEEE Log Number 9406930.

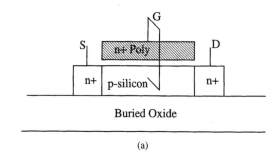

(a)

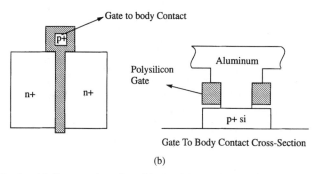

(b)

Fig. 1. (a) Cross section of an SOI NMOSFET with body and gate tied together. (b) Gate to body connection by using aluminum to short the gate and P+ region.

fabricated as illustrated in Fig. 1. This connection uses an oversized metal to P+ contact window aligned over a "hole" in the poly gate [1]. The metal shorts the gate and P+ region. Thus, there is no significant penalty in area.

To operate the DTMOS, floating body and gate of a Silicon-On-Insulator (SOI) MOSFET are tied together. This is not a new configuration, as [1]–[3] have already suggested it. However, [1]–[3] all tried to exploit the extra current produced by the lateral bipolar transistor. This normally requires the body voltage to be larger than 0.6 V. Since current gain of the bipolar device is small, extra drain (collector) current comes at cost of excessive input (base) current, which contributes to the standby current. We will show that most of the improvement can be achieved when gate and body voltages are kept below 0.6 V. This also ensures that base current will stay negligible. Although the same idea can be used in bulk devices, better advantage is reached in SOI, where because of very small junction areas base current and capacitances are appreciably reduced.

Fig. 2 illustrates the NMOS behavior, with a separate terminal used to control the body voltage. The threshold voltage at zero body bias is denoted by V_{to}. Body bias effect is

Reprinted from *IEEE Electron Device Letters,* pp. 510-512, December 1994.

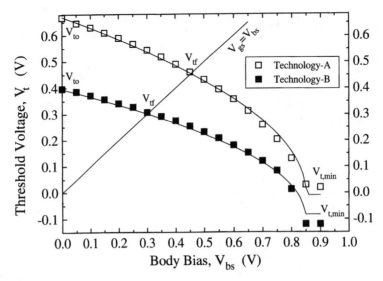

Fig. 2. Threshold Voltage of SOI NMOSFET as a function of body-source forward bias. For Technology-A $T_{ox} = 10$ nm, $N_a = 2.5 \times 10^{17}$ cm^{-3}. For Technology-B $T_{ox} = 6.4$ nm $N_a = 3 \times 10^{17}$ cm^{-3}.

normally studied in the reverse bias regime, where threshold voltage increases as body to source reverse bias is made larger. We propose to use the exact opposite regime. Namely, we "forward bias" the body-source junction (at less than 0.6 V), forcing the threshold voltage to drop.

Specifically, this forward bias effect is achieved by connecting the gate to the body. This is shown as $V_{gs} = V_{bs}$ line in Fig. 2. The intersect of V_t curve and $V_{gs} = V_{bs}$ line determines the point where gate and threshold voltages become identical. This point, which is marked as V_{tf}, is the DTMOS threshold voltage. This lower threshold voltage does not come at expense of higher off-state leakage current, as at $V_{bs} = V_{gs} = 0$ DTMOS and regular device have the same V_t. In fact, they are identical in all respects and consequently have the same leakage. This is clearly seen in Fig. 3. Reduced V_{tf} compared to V_{to} is attained through a theoretically ideal subthreshold swing of 60 mV/dec. Fig. 3 demonstrates this for PMOS and NMOS devices operated in DTMOS mode and in regular mode. Subthreshold swing is 80 mV/dec in the regular devices.

This is not the only improvement. As the gate of DTMOS is raised above V_{tf}, threshold voltage drops further. The threshold voltage reduction continues until $V_{gs} = V_{bs}$ reaches $2\Phi_b$, and threshold voltage reaches its minimum value of $V_{t,min} = 2\Phi_b + V_{fb}$. For example, for technology-B in Fig. 2, at $V_{gs} = V_{bs} = 0.6$ V, $V_t = 0.18$ V compared to $V_{to} = 0.4$ V. In DTMOS operation the upper bound for applied $V_{gs} = V_{bs}$ is set by the amount of base current that can be tolerated. This is illustrated in Fig. 3, where PMOS and NMOS device body (base) currents are shown. At $V_{gs} = 0.6$ V base currents for both PMOS and NMOS devices are less than 2 nA/μm. For larger values of V_{gs} the body current can become excessive. Current drives of DTMOS and regular MOSFET are compared in Fig. 4, for technology-B of Fig. 2. DTMOS drain current is 2.5 times of regular device at $V_{gs} = 0.6$ V, and 5.5 times of regular device at $V_{gs} = 0.3$ V. The improved DTMOS current drive is due to

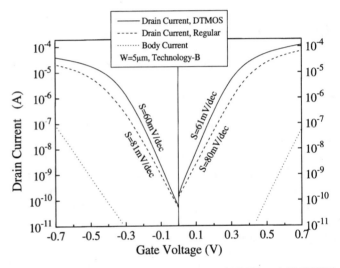

Fig. 3. Subthreshold characteristics of SOI NMOSFET and PMOSFET operated with body grounded and body tied to the gate. Body to source currents are also shown for the case of DTMOS (body tied to the gate).

the following: 1) Inversion charge is increased, as $dQ_n = C_{ox}\left(dV_g - \frac{\partial V_t}{\partial V_g}dV_g\right) \rightarrow C_{eff} = C_{ox}\left(1 + \left|\frac{\partial V_t}{\partial V_g}\right|\right)$. Thus, DTMOS leads to an effectively thinner oxide. 2) DTMOS carrier mobility is higher because the depletion charge is reduced and the effective normal field in the channel is lowered [4]. We realize that DTMOS gate capacitance is larger than regular MOS gate capacitance. However, gate capacitance is only a portion of total capacitance, and current drive of DTMOS is much higher than that of regular MOS. Thus, DTMOS gates are expected to switch faster than regular MOS gates.

AC performance of DTMOS is evaluated by an unloaded 101-stage CMOS ring oscillator. Fig. 5 plots the delay of each stage versus power supply. We emphasize that since the threshold voltages of devices used in the ring oscillator were high (technology-A), the optimum performance was not achieved. For technology-B, ring oscillators are not available.

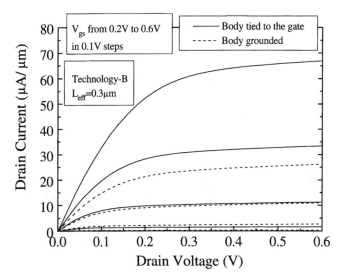

Fig. 4. Drain current of an SOI NMOSFET operated as a DTMOS and as a regular device.

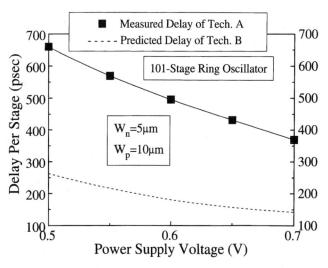

Fig. 5. Delay of a 101-stage ring oscillator. The PMOS and NMOS devices in the ring are DTMOS with $T_{ox} = 10$ nm, and $L_{eff} = 0.3$ μm, $V_{to} = 0.6$ V. The dashed line is prediction of delay for a ring oscillator based on Technology-B ($T_{ox} = 6.4$ nm, $N_a = 3 \times 10^{17}$ cm^{-3}, $N_d = 3 \times 10^{17}$ cm^{-3}) with $L_{eff} = 0.3$ μm.

If the devices based on technology-B are used, the expected delay for unloaded ring oscillator can be calculated by the following equation [5]: $\tau_{pd} = \frac{C}{4} V_{dd} \left(\frac{1}{I_{dsatn}} + \frac{1}{I_{dsatp}} \right)$. This is shown as the dashed line in Fig. 5, where C = 200 fF is used for $W_n = 5$ μm and $W_p = 10$ μm. This value for C was obtained by fitting the equation to the measured τ_{pd} of technology-A.

Finally, in the present form of body-gate connection (Fig. 1) a wide DTMOS may have non-uniform threshold voltage along its width. This warrants investigating other connection schemes.

III. CONCLUSION

For low power operation at very low voltages, a MOSFET should ideally have a high V_t at $V_{gs} = 0$ to achieve low leakage and low V_t at $V_{gs} = V_{dd}$ to achieve high speed. By tying body and gate of an SOI MOSFET together, a dynamic threshold voltage MOSFET (DTMOS) is obtained. This device has ideal 60 mV/dec subthreshold swing. DTMOS threshold

voltage drops as gate voltage is raised, resulting in much higher current drive than regular MOSFET. DTMOS is ideal for very low voltage (< 0.6 V) operation, as demonstrated by ring oscillator data. DTMOS also solves the floating body problems of SOI MOSFET such as kinks and V_t stability. Furthermore, carrier mobility is enhanced.

REFERENCES

[1] S. A. Parke, C. Hu, and P. K. Ko, "Bipolar-FET hybrid-mode operation of quarter-micrometer SOI MOSFET's," *IEEE Electron Device Lett.*, vol. 14, pp. 236–238, May 1993.
[2] J. P. Colinge, "An SOI voltage-controlled bipolar-MOS device," *IEEE Trans. Electron Devices*, vol. ED-34, pp. 845–849, Apr. 1987.
[3] S. Verdonckt-Vandebroek, S. Wong, J. Woo, and P. Ko, "High-gain lateral bipolar action in a MOSFET structure," *IEEE Trans. Electron Devices*, vol. 38, pp. 2487–2496, Nov. 1991.
[4] A. G. Sabnis and J. T. Clemens, "Characterization of the electron mobility in the inverted ⟨100⟩ Si surface," in *Int. Electron Devices Meet. Tech. Dig.*, pp. 18–21, Dec. 1979.
[5] C. Hu, "Low-voltage CMOS device scaling," in *IEEE Int. Solid-State Circuit Conf. (ISSCC) Digest of Technical Papers*, pp. 86–87, Feb. 1994.

A 0.5V SIMOX-MTCMOS Circuit with 200ps Logic Gate

TAKAKUNI DOUSEKI, SATOSHI SHIGEMATSU,
YASUYUKI TANABE, MITSURU HARADA,
HIROSHI INOKAWA, AND TOSHIAKI TSUCHIYA
NTT LSI LABORATORIES, ATSUGI, KANAGAWA, JAPAN

Multi-threshold CMOS (MTCMOS) circuit technology combining low-Vth CMOS logic gates and high-Vth MOSFET is suitable for 1V LSIs for battery-operated portable equipment [1]. A 0.9V 100MHz DSP core uses MTCMOS technology [2]. Improvements in MTCMOS device technology promise to lead to higher operating frequencies. However, higher frequencies will increase power consumption even if the supply voltage is 1V. To reduce the power consumption, it is necessary to lower the supply voltage below 1V, without sacrificing speed. A circuit consisting of depletion-mode MOSFETs operates with 200mV supply [3]. However, it can not be applied to an LSI with more than 1k gates because active-mode leakage current is too large. In addition, the circuit needs back-gate bias, which is much larger than the supply voltage, to increase the threshold voltage and to reduce the leakage current in the sleep mode. To generate the large back-gate bias, multiple supply voltages or a boost circuit are required. This low supply-voltage MTCMOS circuit with SIMOX technology uses enhancement-mode MOSFETs and contains no boost circuit.

The 0.5V SIMOX-MTCMOS circuit is shown in Figure 1. High-speed operation at 0.5V supply is obtained by use of low-Vth CMOS logic gates consisting of fully-depleted body-floating MOSFETs. These MOSFETs improve speed in two ways. First, the delay of the fully-depleted CMOS circuit is not affected by the junction capacitance, which increases significantly at the supply voltage below 1V in bulk MOSFETs. Secondly, the current drive of fully-depleted MOSFETs is better than bulk MOSFETs because the subthreshold swing is small and the threshold voltage can be reduced without any increase in subthreshold current. For high-Vth MOSFETs in the SIMOX-MTCMOS circuit, the MOSFET body is connected to the gate through a reverse-biased MOS diode composed of a low-Vth MOSFET. The body-gate connected MOSFET reduces threshold voltage and increases drain conductance in active mode without increasing the leakage current in sleep mode, making low-supply-voltage operation possible in the MTCMOS circuit. Connecting the body directly to the gate and varying the threshold voltage at the operating mode is used in the well-known DTMOS [4]. However, it can not be applied to a circuit with the supply voltage over 0.8V. The static current characteristics of the DTMOS are shown in Figure 2a. When supply voltage is applied, the pn-diode between the source and body is forward biased. If the supply voltage is over 0.8 V, the diode built-in potential, a large leakage current flows from the body to the gate. To apply the technique over a wide supply-voltage range, a variable high-Vth MOSFET in which a reverse-biased MOS-diode composed of a small low-Vth MOSFET is inserted between the body and the gate. Static current characteristics of the variable high-Vth MOSFET are shown in Figure 2b. The low-Vth reverse-biased MOS-diode is one-tenth the size of the high-Vth MOSFET. For supply voltages over 0.8V, the reverse-biased diode clamps the forward bias of the pn-diode between the body and the source and gate leakage current is suppressed. The drain-conductance increase effect of the variable high-Vth MOSFET is shown in Figure 3. In sleep mode ($V_{GS} = 0V$), the drain-current of the variable high-Vth MOSFET equals that of the conventional MOSFET because the body voltage is equal to the source voltage of 0 V. In active mode ($V_{GS}<0V$), the drain-current of the variable high-Vth MOSFET increases because the threshold voltage is reduced due to the body-voltage reduction. At the supply voltage of 0.5V, the variable high-Vth MOSFET increases the drain conductance to about three times that of the conventional MOSFET.

The sleep control circuit that controls the gate voltage of the variable high-Vth MOSFET in the MTCMOS circuit is shown in Figure 4. To drive the large variable high-Vth MOSFET, the sleep control circuit uses of multistage CMOS inverters combining the low-Vth MOSFET and the variable high-Vth MOSFET with the reverse-biased MOS diode. Since the variable high-Vth MOSFETs are alternatively used in each stage CMOS inverter so they are off in the sleep mode, the nA-order leakage current can be maintained in sleep mode.

To demonstrate the effectiveness of the SIMOX-MTCMOS circuit, a gate-chain TEG comprised of standard cells and including a fundamental logic gate, a 2-input NAND gate with a fanout of 3 and wiring length of 1mm, uses 0.25μm SIMOX-MTCMOS technology. The gate count is 1k gates. The thicknesses of the gate oxide, the active silicon layer, and the buried oxide in the SIMOX-wafer are 5nm, 50nm, and 100nm, respectively. The threshold voltage of the low-Vth MOSFET is 0.13V for the nMOSFET and -0.18V for the pMOSFET. The threshold voltages of the high-Vth MOSFETs, whose body is applied to the supply voltage, are 0.38V and -0.44V, respectively. The channel width of a standard cell is 3μm for the nMOSFET and 6μm for the pMOSFET. The delay times of the 2-NAND logic gate were 80ps at the supply voltage of 1V and 197ps at 0. 5V unloaded. The dependence of delay in the fundamental logic gate on supply voltage is shown in Figure 5. The delay-time characteristic of the conventional bulk-MTCMOS circuit is also shown for comparison. The figure shows that the SIMOX- MTCMOS circuit is most suitable for the supply voltage below 1V. At the supply voltage of 1V, the delay time of the fundamental logic gate is 273ps. The SIMOX-MTCMOS circuit reduces the delay time to less than 50% that of the conventional 0.25μm bulk MTCMOS circuit. At 0.5V supply, delay is 710ps, which is less than 50% that of the conventional 1V 0.5μm bulk MTCMOS circuit [1]. In addition, the dynamic power consumption at the supply voltage of 0.5V is 0.05μW/MHz/gate, less than one-fourth that at the supply voltage of 1V. At the supply voltage of 0.4V, only the SIMOX- MTCMOS circuit can operate and the delay is 1125ps. The maximum delay-time in one wafer is 1273ps and the delay-time variation is <15%. The SIMOX-MTCMOS circuit with variable body-bias scheme can reduce the maximum delay by 15% that of the circuit with fixed body-bias scheme.

An 8k-gate 16b ALU verifies effectiveness of the SIMOX-MTCMOS circuit. A micrograph of the ALU is shown in Figure 6. The chip is 4x4mm² and the core is 1.7x1.7mm². The Shmoo plot for evaluating the maximum operating frequency are shown in Figure 7. At 0.5V supply, maximum operating frequency is 40MHz. Power consumption is 0.35mW. Power consumption in the sleep mode is less than 5nW. Turn-on time is 0.4μs and turn-off time is 8.0μs.

References:

[1] Mutoh, S., et al., "1-V High-Speed Digital Circuit Technology with 0.5μm Multi-Threshold CMOS," Proc. IEEE 1993 International ASIC Conf., pp. 186-189, 1993.

[2] Izumikawa, M., et al., "A 0.9V 100MHz, 4mW, 2mm², 16b DSP Core," ISSCC Digest of Technical Papers, pp. 84-85, Feb., 1995.

[3] Burr, J., et al., "A 200mV Self-Testing Encoder/Decoder using Stanford Ultra-Low-Power CMOS," ISSCC Digest of Technical Papers, pp. 84-85, Feb., 1994.

[4] Assaderaghi, F., et al., "A Dynamic Threshold Voltage MOSFET (DTMOS) for Ultra-Low Voltage Operation," IEDM Tech. Dig., pp. 809-812, 1994.

Reprinted from *IEEE International Solid-State Circuits Conference*, pp. 84-85, February 1996.

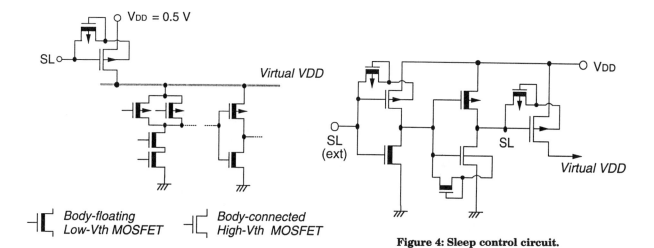

Body-floating
Low-Vth MOSFET

Body-connected
High-Vth MOSFET

Figure 1: 0.5V SIMOX-MTCMOS circuit.

Figure 4: Sleep control circuit.

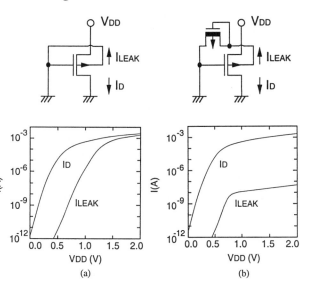

Figure 2: Static current characteristics of high-Vth MOSFET: (a) DTMOS. (b) variable high-Vth MOSFET with reverse-biased MOS diode.

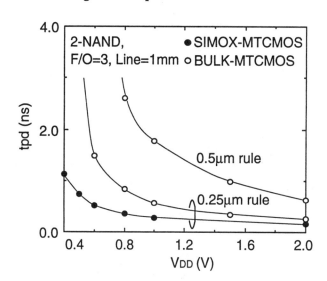

Figure 5: Delay-time of fundamental logic gate.
Figure 6: See page 423.

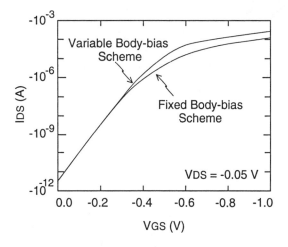

Figure 3: Drain-conductance increase of variable high-Vth MOSFET.

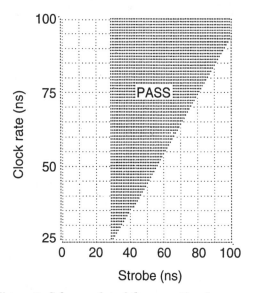

Figure 7: Schmoo plot of the operating frequency.

Part III

Efficient DC-DC Conversion and Adaptive Power Supply Systems

A Low-Voltage CMOS DC-DC Converter for a Portable Battery-Operated System

Anthony J. Stratakos, Seth R. Sanders, and Robert W. Brodersen

University of California, Berkeley
Department of Electrical Engineering and Computer Sciences
Berkeley, CA 94720 USA

Abstract— **Motivated by emerging battery-operated applications that demand compact, light-weight, and highly efficient DC-DC converters, a buck circuit is presented in which all active devices are integrated on a single chip using a standard 1.2μ CMOS process. The circuit delivers 750mW at 1.5V from a 6V battery. To effectively eliminate switching loss at high operating frequencies, the power transistors achieve nearly ideal zero-voltage switching (ZVS) through an adjustable dead-time control scheme. The silicon area and power consumption of the gate-drive buffers are reduced with a tapering factor that minimizes short-circuit current and dynamic dissipation for a given technology and application. Measured results on a prototype IC indicate that on-chip losses at full load can be kept below 8% at 1MHz.**

I. INTRODUCTION

Current trends in consumer electronics demand progressively lower-voltage supplies [1]. Portable electronic devices, such as laptop computers and personal communicators, require ultra low-power circuitry for battery operation. The key to reducing power consumption while maintaining computational throughput in such systems is to use the lowest possible supply voltage and compensate for the resulting decrease in speed with architectural, logic-style, circuit, and other technological optimizations [2].

The multimedia InfoPad terminal is one example of a portable, battery-operated, high-speed personal communication system [3, 4, 5]. The baseband circuitry in the current InfoPad terminal, including the encoder and decoder for data compression, A/D and D/A converters, and spreader and despreader for spread spectrum RF communication, are being designed to operate at 1.5V to minimize power consumption. However, because the system also requires supplies of ±5V to power the RF transceiver circuitry and -17V, and $+12$V for the flat panel display, a number of DC-DC converter outputs are needed to generate these voltages from a single 6V battery. In order to facilitate portability and conserve battery life in In-

foPad, power conversion must be done in minimal space and mass, implying high operating frequencies, while retaining the high efficiency more typical of lower frequency converters.

A completely monolithic supply (active and passive elements) would meet the severe size and weight restrictions of a hand-held device. Because such applications call for low-voltage power transistors, their integration in a standard logic process is tractable. However, existing monolithic magnetics technology cannot provide inductors of suitable size [6]. Fortunately, transformers and inductors fabricated with micron-scale magnetic-alloy and copper thin films are currently being developed [7], indicating that low voltage supplies able to efficiently deliver several watts of power will be realized in a single multi-chip module (MCM) within the next several years.

Motivated by the rigorous requirements of the InfoPad terminal, a 6V to 1.5V, 750mW buck converter has been designed and is presented in this paper. Due to the low voltage and current levels, all of the active devices are integrated on a single chip and fabricated in a standard 1.2μ process through MOSIS[1]. The circuit exhibits nearly ideal zero-voltage switching (ZVS) of the power transistors over a wide range of operating conditions using an adaptive dead-time control scheme, substantially reducing switching losses associated with high-frequency operation. The CMOS gate-drive buffers are designed with a process and application dependent tapering factor which minimizes both their dynamic and short-circuit current losses. The design is scalable, and all of the techniques presented here can be used to improve the efficiency of boost or buck-boost type converters, each of which is typically required in the power distribution scheme of a battery-operated system.

The paper is organized as follows. First, in section II, a brief circuit description is given, including a discussion

---

[1]MOS Implementation System: users submit VLSI designs expressed in digital form, and receive packaged parts in 8 to 10 weeks. MOSIS is under sponsorship of ARPA, and operated by the Information Sciences Institute of USC.

Reprinted from *IEEE Power Electronics Specialists Conference*, pp. 619-626, June 1994.

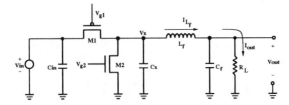

Fig. 1. Low-output-voltage buck circuit.

of relevant timing issues. Section III outlines an adaptive dead-time control scheme. Then, section IV describes the design of the power transistors. An overview of the low-power CMOS gate-drive is presented in section V, including a qualitative loss analysis. In section VI, the key layout and packaging issues are considered. Finally, section VII reports on results measured on a prototype IC.

II. CIRCUIT DESCRIPTION

The power train of the low-voltage buck circuit is shown in Figure 1. All active devices, including the power transistors ($M1$ and $M2$) with their gate-drive, and all regulation and control circuitry, are integrated on a single $4.2mm \times 4.2mm$ die in a 1.2μ CMOS process, and housed in a 64-pin PGA package. Currently, the design requires four passive components at the board level: a decoupling input capacitor (C_{in}), a snubber capacitor (C_x), and an output filter inductor and capacitor (L_f and C_f).

The operation of the circuit is described with reference to the waveforms of Figure 2, which represent the steady-state behavior of an ideal circuit with lossless filter components. The inverter node voltage, V_x, is quasi-square with an operating frequency of f_s =1 MHz, and a nominal duty cycle, D, of 25%. The soft-switching behavior is similar to that described by [8] and other authors. The converter delivers 500mA at 1.5V from a 6V battery.

Assume that at a given time (the origin in Figure 2), the rectifier ($M2$) is on, shorting the inverter node to ground. Since by design, the output is assumed DC and greater than zero, a constant negative potential is applied across L_f, and I_{L_f} is linearly decreasing. By choosing the filter inductor small enough such that the current ripple exceeds the average load current, I_{L_f} ripples below zero.

If the rectifier is turned off after the current reverses (and the pass device, $M1$, remains off), L_f acts as a current source, charging the inverter node. To achieve a lossless low to high transition at the inverter node, the pass transistor is turned on when $V_x = V_{in}$. In this scheme, a pass device gate transition occurs exactly when $V_{ds_1} = 0$.

With the pass device on, the inverter node is shorted to V_{in}. Thus, a constant positive voltage is applied across L_f, and I_{L_f} linearly increases. The high to low transition at V_x is initiated by turning $M1$ off. As indicated by Figure 2, at this time, the sign of current I_{L_f} is positive.

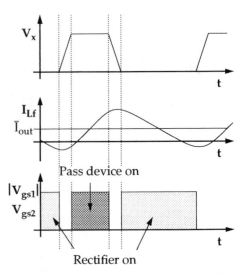

Fig. 2. Nominal waveforms of the buck circuit in Figure 1.

Again, L_f acts as a current source, discharging C_x. If the rectifier is turned on with $V_x = 0$, a lossless high to low transition of the inverter node is achieved, and the rectifier is switched at $V_{ds_2} = 0$.

In this scheme, the filter inductor is used to charge and discharge the parasitic and snubber capacitance at the inverter node in a lossless manner, allowing the power transistors to be switched at zero drain-source potential (ZVS). This essentially eliminates switching loss, a factor which may otherwise preclude the use of high-frequency converters in low-power systems.

A. Timing Issues

Because the inverter node transition intervals are designed to be small compared to the switching period, I_{L_f} is assumed triangular with peak values $\bar{I}_{out} - \frac{\Delta I_{L_f}}{2}$ and $\bar{I}_{out} + \frac{\Delta I_{L_f}}{2}$ which are constant over the entire dead-time. The ratio of inverter node transition times is given by the ratio of currents available for each commutation:

$$\frac{\tau_{xLH}}{\tau_{xHL}} = \frac{\frac{\Delta I_{L_f}}{2} + \bar{I}_{out}}{\frac{\Delta I_{L_f}}{2} - \bar{I}_{out}}, \qquad (1)$$

and approaches unity for large inductor current ripple. Here, τ_x indicates an inverter node transition time, with subscripts LH and HL denoting low to high and high to low transitions respectively, $\bar{I}_{out}$ is the average load current, and $\frac{\Delta I_{L_f}}{2}$ is the zero-to-peak inductor current ripple. Choosing a maximum asymmetry in the transition intervals of $\frac{\tau_{xLH}}{\tau_{xHL}} = 4$ at full load results in a minimum zero-to-peak ripple of

$$\frac{\Delta I_{L_f}}{2} = \frac{5}{3}\bar{I}_{out} = 833.3mA, \qquad (2)$$

130

and requires a filter inductor of

$$L_f = \frac{V_{in}D(1-D)}{f_s\Delta I_{L_f}} = 675\text{nH}. \quad (3)$$

Allowing for a 2% peak to peak ac ripple in the output voltage,

$$C_f \approx \frac{25\Delta I_{L_f}}{3f_s} = 13.9\mu\text{F} \quad (4)$$

is selected.

To slow the inverter node transitions, a snubber capacitor is added at V_x. The total capacitance required to achieve $\tau_{x_{LH}} = 0.1T_s = 100\text{ns}$ is

$$C_x = \frac{\bar{I}_{out}}{15V_{in}f_s} = 5.56\text{nF} \quad (5)$$

where C_x includes the snubber and all parasitic capacitance at the inverter node, V_x.

III. ADAPTIVE DEAD-TIME CONTROL

To ensure ideal ZVS of the power transistors, each dead-time when neither device conducts, τ_D, must equal the corresponding inverter node transition time:

$$\tau_{D_{LH}} = \tau_{x_{LH}} \quad (6)$$
$$\tau_{D_{HL}} = \tau_{x_{HL}}.$$

In practice, the relationships of (6) may not always hold true. As indicated by Figure 2, the inductor current ripple is symmetric about the average load current. As the average load varies, the dc component of the I_{L_f} waveform is shifted, and the current available for commutating the inverter node, and ultimately, the inverter node transition time, is modified.

In one approach to soft-switching, a value of average load may be assumed, yielding estimates of the inverter node transition times. Gate delays are then used to determine fixed dead-times based on these estimates. In this way, switching loss is reduced, yet perhaps not to negligible levels. To keep switching losses at a reasonable level in these circuits, they must be operated at lower frequencies, thereby increasing the size of the passive filter components, and decreasing the power density.

In portable, battery-operated applications where area and battery-life are at a premium, this approach to soft-switching may not be adequate. To illustrate the potential hazards of fixed dead-time operation, Figure 3 shows the impact of non-ideal ZVS on conversion efficiency through reference to a high to low transition at the inverter node. In Figure 3a, $\tau_{x_{HL}} > \tau_{D_{HL}}$, causing the rectifier turn-on to occur at $V_{ds_2} > 0$, partially discharging C_x through a resistive path and introducing losses. In Figure 3b, $\tau_{x_{HL}} < \tau_{D_{HL}}$, such that the inverter node rings below zero

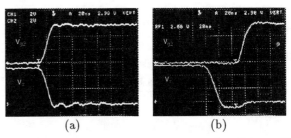

(a)   (b)

Fig. 3. Non-ideal ZVS and its impact on conversion efficiency. The upper trace is V_{g2}, the lower trace is V_x, the vertical scale is 2 V/div, and the horizontal scale is 20 ns/div.

until the drain-body junction of $M2$ becomes forward biased. In low-voltage applications, the forward-bias diode voltage is a significant fraction of the output voltage; thus, body diode conduction must be avoided for efficient operation. When the rectifier ($M2$) turns on, it must remove the excess minority carrier charge from the body diode and charge the inverter node back to ground, dissipating additional energy.

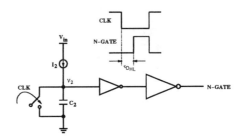

Fig. 4. Variable turn-on delay generator for the rectifier.

To provide effective ZVS over a wide range of loads, an adaptive dead-time control scheme is outlined here. Figures 4 and 5 show the $\tau_{D_{HL}}$ generator and adaptation scheme. Assuming the dead-time is large compared to an inverter delay:

$$\tau_{D_{HL}} \approx \frac{C_2 V_M}{I_2}, \quad (7)$$

where V_M is the switching threshold of an inverter.

To make the dead-time control scheme adaptive, current source I_2 is adjusted based on the relative timing of V_x and V_{gs_2}. In Figure 5, an error voltage proportional to the difference between $\tau_{x_{HL}}$ and $\tau_{D_{HL}}$ is generated on integrating capacitor C_I. The error voltage is sampled and held, and used to update I_2 in the variable turn-on delay generator of Figure 4. Using this technique, effective ZVS is ensured over a wide range of operating conditions and process variations. A nearly ideal ZVS high to low inverter node transition is shown in Figure 6.

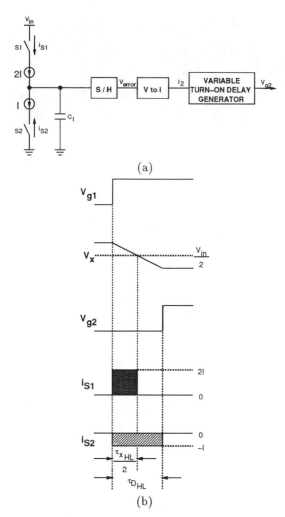

(a)

(b)

Fig. 5. Rectifier turn-on delay adjustment loop: (a) possible circuit implementation, (b) associated waveforms.

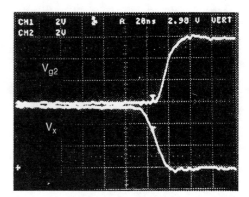

Fig. 6. A zero-voltage-switched high to low inverter node transition. The vertical scale is 2 V/div; the horizontal scale is 20 ns/div.

IV. FET Design

During their conduction intervals, the power transistors operate exclusively in the ohmic region, where $r_{ds} = R_0 \cdot \frac{1}{W}$ (the channel resistance is inversely proportional to gate-width, W, with constant of proportionality R_0). Thus, conduction loss in the FET, Mj, is given by

$$P_{q_j} = \frac{I_{D_j - rms}^2 R_0}{W_j}. \tag{8}$$

It is further assumed that the power transistors exhibit ideal lossless zero-voltage switching. Therefore, their gate-drain (Miller) and drain-body capacitance are switched with negligible loss. Since the gate capacitance increases linearly with increasing gate width ($C_g = C_{g0}W$), and the power consumed by the gate-drive buffers varies linearly with the capacitive load, the gate-drive dissipation due to

Mj is of the form:

$$P_{g_j} = E_{g0} f_s W_j, \tag{9}$$

where E_{g0} is the total gate-drive energy consumed in a single low-high-low gate transition cycle. Using an algebraic minimization [9], the optimal gate width of Mj,

$$W_{j-opt} = \sqrt{\frac{I_{D_j - rms}^2 R_0}{E_{g0} f_s}} \tag{10}$$

is found to give equal conduction loss and gate-drive loss, where:

$$P_{q_j - opt} = P_{g_j - opt} = \sqrt{I_{D_j - rms}^2 R_0 E_{g0} f_s} \tag{11}$$

and $P_{q_j} + P_{g_j}$ is at its minimum.

The layout of the power transistors is discussed in section VI.

V. Gate-Drive

In CMOS circuits, a power transistor is conventionally driven by a chain of N inverters which are scaled with a constant tapering factor, u, such that

$$u^N = \frac{C_g}{C_i}. \tag{12}$$

Here, C_g is the gate capacitance of the power transistor and C_i is the input capacitance of the first buffering stage. This scheme, depicted in Figure 7, is designed such that the ratio of average current to load capacitance is equal for each inverter in the chain. Thus, the delay of each stage and rise/fall time at each node are identical. It is a well known result that under some simplifying assumptions, the tapering factor u that produces the minimum propagation delay is the constant e [10]. However, in power

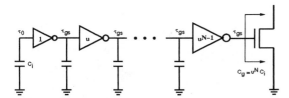

Fig. 7. CMOS gate-drive scheme.

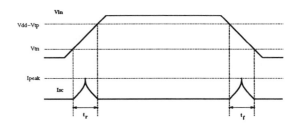

Fig. 8. Short-circuit current in an inverter.

circuits, the chief concern lies not in the propagation delay of the gate-drive buffers, but the energy dissipated during a gate transition.

In a ZVS power circuit, the following timing constraints are desired:

$$\tau_x \gg \tau_{gs} \approx u\tau_0 \tag{13}$$

where τ_x is the inverter node transition time, τ_{gs} is the maximum gate transition time which ensures effective ZVS of the power transistor, τ_0 is the output transition time (rise/fall time) of a minimal inverter driving an identical gate, and u is the tapering factor between successive inverters in the chain. In general, it is desirable to make τ_{gs} as large as possible (yet still a factor of five to ten less than τ_x), minimizing the gate-drive dissipation. Given τ_{gs} and τ_0, if there exists some $u > e$ such that the criteria given by (13) are met, the buffering scheme of Figure 7 will provide a more energy efficient CMOS gate-drive than that obtained through minimization of delay.

A. Determination of the Inverter Chain

In this analysis, a minimal CMOS inverter has an NMOS device with minimum dimensions (W_o/L) and a PMOS device whose gate width is $\mu_n/\mu_p \approx 3$ times that of the NMOS device. It has lumped capacitances C_i at its input and C_o at its output. Given that the pull-down device operates exclusively in the triode region during the interval of interest, it can be shown that the output fall time of a minimal inverter driving an identical gate from $V_{out} = V_{dd} - |V_{tp}|$ to $V_{out} = V_{tn}$ is:

$$t_f \equiv \tau_0 = \frac{C_o + C_i}{W_o}K, \tag{14}$$

which is linearly proportional to the capacitive load, inversely proportional to the gate-width of the n-channel device, and directly related to the application and technology dependent constant:

$$K \equiv \frac{2L}{\mu_n C_{ox}(V_{dd} - V_{tn})} \bullet$$
$$\ln\left[\frac{(2V_{dd} - 3V_{tn})}{(V_{tn})}\frac{(V_{dd} - V_{tp})}{(V_{dd} - 2V_{tn} + V_{tp})}\right]. \tag{15}$$

A similar expression can be found for the rise time.

The factor u which results in an output transition time τ_{gs} is found by solving:

$$\tau_{gs} = \frac{K(C_o + uC_i)}{W_0} \approx u\tau_0, \tag{16}$$

yielding a corresponding tapering factor of

$$u = \frac{\tau_{gs}W_0 - KC_o}{KC_i} \tag{17}$$

between successive buffers. Given u, the number of inverters in the chain is:

$$N = \frac{\ln(C_g/C_i)}{\ln(u)}. \tag{18}$$

The inverter chain guarantees a gate transition time of τ_{gs} with minimum dissipation, and a propagation delay of

$$t_p \approx Nut_{p0} \tag{19}$$

where t_{p0} is the propagation delay of a minimal inverter loaded by an identical gate.

B. Loss Analysis

There are two components of power dissipation in the inverter chain:

$$P_{dyn} = C_T V_{dd}^2 f_s \tag{20}$$

$$P_{sc} = \sum_{i=1}^{N} \bar{I}_{sc_i} V_{dd} \tag{21}$$

where $\bar{I}_{sc_i}$ is the mean short-circuit current in the ith inverter in the chain, and the total switching capacitance, including the load, is

$$C_T = (1 + u + u^2 + \cdots + u^{N-1})(C_o + C_i) + C_g$$
$$= \left[\frac{u^N - 1}{u - 1}\right](C_o + C_i) + C_g. \tag{22}$$

Since u^N is the constant given by (12), C_T and thus, the dynamic dissipation, is minimized for large u.

Though the dynamic component is readily calculated from (20) and (22), the short-circuit dissipation is more

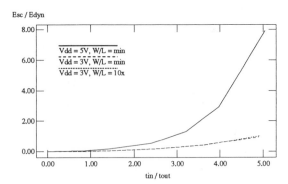

Fig. 9. Short-circuit energy versus input transition time.

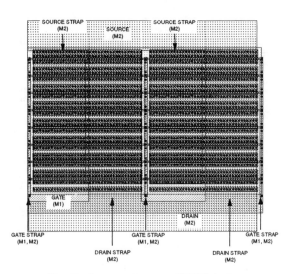

Fig. 10. Layout of a power PMOS device.

difficult to quantify. From Figure 8, it can be seen that short-circuit current exists in a CMOS inverter while the n- and p-channel devices conduct simultaneously ($V_{tn} < V_{in} < V_{dd} - |V_{tp}|$), and the total energy consumed during an input transient is proportional to both the input transition time and the peak short-circuit current (which in turn, is related to the output transition time [11]). Figure 9 plots simulation results of the ratio of short-circuit to dynamic dissipation per cycle versus the ratio of input to output transition times for a minimal inverter operated at $V_{dd} =$5V and $V_{dd} =$3V, and a ten times minimal inverter operated at $V_{dd} =$3V. These results illustrate three key points regarding short-circuit dissipation in a CMOS inverter:

- The normalized E_{sc} is seen to increase exponentially with normalized input edge rate, but is negligible for equal input and output transition times.

- While the magnitude of short-circuit current is dependent on device dimensions (I_{peak} increases linearly with device size), the ratio of E_{sc} to E_{dyn} appears to be independent of size.

- For $V_{dd} \gg V_t$, there is an approximately linear increase in E_{sc} with increasing supply voltage. While the duration of short-circuit current flow decreases linearly, the magnitude increases quadratically.

Therefore, because the tapering factor u is constant throughout the inverter chain, providing equal transition times τ_{gs} at each node, the short-circuit dissipation is made negligible. Furthermore, for $u > e$, less silicon area will be devoted to the buffering; thus parasitics, and ultimately, dynamic energy loss, are reduced as compared to the conventional CMOS gate-drive.

To make a first-order estimate of the total gate-drive loss as a function of power transistor gate width, (20), (21), and (22) are used in conjunction with the values of u and N derived in subsection A, giving:

$$P_g \approx WC_{g0}V_{in}^2 f_s \bullet \qquad (23)$$
$$\left[\frac{K(C_o + C_i)}{\tau_{gs}W_o - KC_o - KC_i} + 1 \right]$$

for a power transistor of gate-width W. To obtain (23), it is assumed that all capacitances increase linearly with gate-width and $u^N \gg 1$. In this way, gate-drive losses are expressed as a linear function of gate-width, identical in form to (9).

VI. LAYOUT AND PACKAGING ISSUES

The power transistors are arrayed as a number of parallel fingers whose gates are strapped in metal1 and metal2 (Figure 10). The length of each finger is determined by the maximum tolerable distributed RC delay of the gate structure. In order to reduce series source and drain resistance, the diffusion is heavily contacted, with source-body and drain running horizontally in metal1, and vertically in metal2. Because the ZVS of the power devices effectively eliminates their drain-body junction capacitance switching losses, increased drain diffusion area is traded for more reliable and efficient body diode conduction. Rather than sharing diffusion between adjacent fingers in the layout, well contacts are placed at a minimum distance from drain contacts, minimizing both the series resistance and transit time [12] of the body diode. Due to the potentially large magnitude of substrate current injection, healthy well contacts are placed throughout the structure. A guard ring surrounds each power device, eliminating latch-up and decoupling the ground and V_{dd} bounce from the analog regulation circuitry that can result from substrate current injection.

134

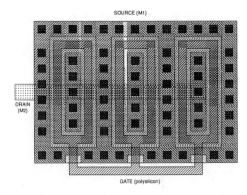

SOURCE (M1)

DRAIN (M2)

GATE (polysilicon)

Fig. 11. Ring structure layout in the gate-drive buffers.

In contrast to the power transistors, the ideal layout of the gate-drive buffers minimizes their output diffusion capacitance. Thus, drain diffusion is shared wherever possible. For the larger buffers, a ring structure is employed [13]. In this scheme, illustrated in Figure 11, the gate polysilicon of two adjacent fingers is connected inside the diffusion, forming a ring of polysilicon which maximizes gate-width for a given drain diffusion area. The diffusion surrounding the ring is chosen to be the source of the device, leaving the small diffusion area within the ring as the drain. Not only is the drain area capacitance at a minimum for a given gate-width in this structure, but the sidewall capacitance, typically the dominant factor in processes with a sub-micron feature size, is completely eliminated.

In this design, the critical problems associated with packaging effects are the stray inductances in series with loop formed by the input decoupling capacitor C_{in} and the power FETs ($L_s I^2 f_s$ loss), and series resistance in V_{in}, ground, and the inverter node ($I^2 R$ loss). Therefore, a large number of pins are dedicated to reduce loss in the package: 14 V_{in} pins, 13 ground pins, and 16 inverter node pins. Because the power transistors are operated in a ZVS scheme, the pin capacitance at the inverter node is used as snubber capacitance and does not contribute additional loss.

VII. MEASURED RESULTS

A prototype IC (Figure 12) was fabricated in a standard 1.2μ CMOS process through the MOSIS program. The circuit delivers 750mW at 1.5V from a 6V battery. Figure 13 shows the steady-state V_{g1}, V_{g2}, I_{L_f}, and V_x waveforms at full load. Tables 1 and 2 summarize the features of the prototype design and report the sources of dissipation.

While power transistor gate-drive and conduction loss are balanced and predicted well by theory and simulation, the measured efficiency of 79% at full load is substantially lower than anticipated. This can be attributed

to several factors. First, due to an undetected layout error in the turn-on delay generators, dead-time adjustment was achieved at the board level with a biasing resistor. Because of the associated increase in capacitive parasitics over the monolithic implementation, comparatively large static currents were required to obtain the desired dead-times. Thus, the power consumption of the ADTC circuitry was greater than an order of magnitude larger than anticipated, comprising nearly 30% of the overall loss. Secondly, throughout the design, efficiency was traded for testability: a number of intermediate signals were brought off-chip at the expense of additional switching capacitance, resulting in a severe penalty in dynamic power consumption. For example, the dissipation of the VCO increased by a factor of three in order to enhance its testability. Finally, a major component of loss is accredited to the package. In particular, the series resistance in the V_{in}, ground, and V_x bonding wires contributed a total of 47.3mW of loss (28% of the total loss), and the stray inductance in the loop formed by the input decoupling capacitor and the power transistors contributed an additional 20mW of loss (10% of total loss). These components of dissipation will virtually be eliminated using a custom lead frame in an MCM or chip-on-board (COB) technology.

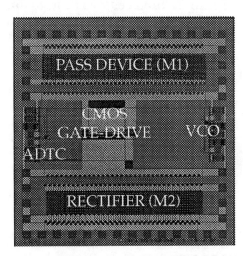

PASS DEVICE (M1)

CMOS GATE-DRIVE VCO

ADTC

RECTIFIER (M2)

Fig. 12. Chip photograph: 6V to 1.5V 750mW DC-DC converter. Die size = 4.2mm × 4.2 mm.

The results measured on the prototype indicate that in this circuit, on-chip losses (including losses in the power transistors, and regulation and control circuitry) can be kept below 8% at full load. With a custom package, an efficiency greater than 90% may be achieved. The approach presented is evidently viable for realizing a high efficiency and compact power converter for portable battery-operated applications. When completed, this circuit would require one custom IC, three small ceramic chip capacitors, and a small inductor. The inductor might be realized with a multi-turn air-core spiral in a custom substrate.

Table 1: Prototype Design Summary

Technology	1.2μ CMOS
	1 poly, 2 metal
Gate Oxide Thickness	21.4 nm
Minimum Gate Length	$L_{eff} = 0.9\mu$
Filter Inductor	$L_f = 675$ nH
Filter Capacitor	$C_f = 20\ \mu$F
Off-chip Snubber Capacitor	$C_{snubber} = 4$ nF
$M1$ Gate Width	$W_1 = 10.2$ cm
$M1$ Buffering	$u = 5.2$
	$N = 4$
$M2$ Gate Width	$W_2 = 10.5$ cm
$M2$ Buffering	$u = 8.7$
	$N = 4$
Operating Frequency	$f_s = 1$ MHz
Battery Input Voltage	$V_{in} = 6$ V
Output Voltage	$V_{out} = 1.5$ V
Output Power at Full Load	$P_{out} = 750$ mW
Predicted Efficiency at Full Load (Unpackaged)	92%
Measured Efficiency at Full Load (Packaged)	79%

Table 2: Sources of Dissipation

	$M1$	$M2$
Gate-Drive Loss	11.2 mW	13.9 mW
Channel r_{ds} Conduction Loss	10.1 mW	14.0 mW
Bonding r_{ds} Conduction Loss	5.1 mW	42.2 mW
Total Loss	3.5 %	9.3 %
L_s Stray Inductance		≈ 20 mW
L_f Series Resistance		16.9 mW
C_f esr		2.3 mW
C_{in} esr		< 1 mW
VCO (On-chip and Off-chip Pins)		6.2 mW
ADTC (Off-chip)		48.4 mW

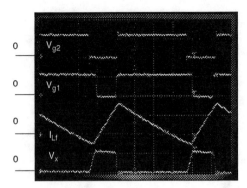

Fig. 13. Measured steady-state waveforms: V_{g2}, V_{g1}, I_{L_f}, V_x (top to bottom). The horizontal scale is 20 ns/div. The vertical scale is 2 V/div for the voltage waveforms, and 1 A/div for the inductor current waveform.

REFERENCES

[1] M. Kakumu and M. Kingugawa, "Trading Speed for Low Power by Choice of Supply and Threshold Voltages", *IEEE Journal of Solid-State Circuits*, vol. 28, pp. 10–17, Jan. 1993.

[2] A. Chandrakasan, S. Sheng, and R. Brodersen, "Low-Power CMOS Digital Design", *IEEE Journal of Solid-State Circuits*, vol. 27, Apr. 1992.

[3] A. Chandrakasan, T. Burd, A. Burstein, S. Narayanaswamy, and R. Brodersen, "System Design of a Multimedia I/O Terminal", *in Proc. IEEE Workshop on VLSI Signal Processing*, 1993.

[4] R. Brodersen, A. Chandrakasan, and S. Sheng, "Low-Power Signal Processing Systems", *in Proc. VLSI Signal Processing Workshop*, pp. 3–13, 1992.

[5] A. Burstein, A. Stoelze, and R. Brodersen, "Using Speech Recognition in a Personal Communications System", *in Proc. International Communications Conf.*, 1992.

[6] W. Baringer and R. Brodersen, "MCMs for Portable Applications", *in Proc. IEEE Multi-Chip Module Conf.*, 1993.

[7] C. Sullivan and S. Sanders, "Microfabrication of Transformers and Inductors for High Frequency Power Conversion", *in Proc. IEEE Power Electronics Specialists Conf.*, 1993.

[8] D. Maksimovic, "Design of the Zero-Voltage Switching Quasi-Square-Wave Resonant Switch", *in Proc. IEEE Power Electronics Specialists Conf.*, 1993.

[9] L. Casey, J. Ofori-Tenkorang, and M. Schlect, "CMOS Drive and Control Circuitry for 1-10 MHz Power Conversion", *IEEE Transctions on Power Electronics*, vol. 6, pp. 749–758, Oct. 1991.

[10] C. Mead and L. Conway, *Introduction to VLSI Systems*, Addison-Wesley, Reading, MA, 1980.

[11] H. Veendrick, "Short-Circuit Dissipation of Static CMOS Circuitry and Its Impact on the Design of Buffer Circuits", *IEEE Journal of Solid-State Circuits*, vol. sc-19, pp. 468–473, Aug. 1984.

[12] G. Rittenhouse and M. Schlecht, "A Low-Voltage Power MOSFET with a Fast-Recovery Body Diode for Synchronous Rectification", *in Proc. IEEE Power Electronics Specialists Conf.*, pp. 96–105, 1990.

[13] J. Rabaey, *Digital Integrated Circuits: A Design Perspective*, Prentice-Hall, Englewood Cliffs, NJ, to be published in 1994.

Ultra Low-Power Control Circuits for PWM Converters

ABRAM P. DANCY AND ANANTHA CHANDRAKASAN

BOTH AUTHORS ARE AFFILIATED WITH THE DEPARTMENT OF ELECTRICAL ENGINEERING AND COMPUTER SCIENCE AT THE
MASSACHUSETTS INSTITUTE OF TECHNOLOGY, CAMBRIDGE, MA 02139 USA

Abstract—Low-voltage high-efficiency DC–DC conversion circuits are integral components of battery-operated portable systems. This paper presents the design of a pulse width modulated (PWM) controller for a milliwatt level DC–DC converter. Low-power control circuitry is constructed by utilizing a delay line based PWM generator and low-resolution feedback. A DC–DC converter has been fabricated which delivers 5mW at 1V, while achieving an efficiency of 88%. The PWM consumes only 10μW.

1. INTRODUCTION

Electronic circuits for mobile systems can be designed to operate over the range of the voltages supplied by the battery over its discharge cycle. However, adding some form of power regulation can significantly increase battery life, since it allows circuitry to operate at the "optimal" supply voltage from a power perspective. As the power dissipation of electronic circuits drop (exploiting low-voltage process technology and other power management techniques), there is a need for high-efficiency DC–DC conversion circuits to deliver low power levels. A high-efficiency low-voltage down converter that delivered power levels of 750mW has been previously reported [1], and there are numerous commercial controllers available to create logic level outputs in the 100mW to 1W range [2,3].

This work addresses high-efficiency conversion techniques at power levels on the order of hundreds of microwatts to tens of milliwatts. While this power level seems very low, it is actually possible to implement complex DSP, such as video compression, with only a few milliwatts [4,5]. Also, since digital circuits can have a wide range of switching activity (from power-down mode to full-active mode), it is necessary to design the converter to operate efficiently over a wide range of output power levels. Figure 1 shows a block diagram of our low-power buck regulator with integrated controller and power switches. Through a combination of digital PWM circuitry, efficient feedback techniques (i.e., low resolution feedback), and integrated switches with synchronous rectification, we are able

to achieve efficiencies of 88% while delivering a load of 5mW at 1V output.

2. LOW-RESOLUTION DIGITAL FEEDBACK

Part of the motivation for using a digital PWM circuit is the hypothesis that a very simple A/D conversion with some digital processing could accomplish the voltage feedback function with far less power dissipation than that required by even the simplest analog feedback circuits, due to the static current draw of the analog circuits. Minimally, the A/D conversion can be accomplished by a comparator, which indicates whether the sensed output voltage is above or below a reference voltage. The comparator is implemented as a dynamic comparator, which evaluates only upon command, and does not dissipate any power outside of the brief evaluation period. In this single-bit feedback case, the duty cycle digital word is created by a counter which counts up when the output is lower than the reference and counts down when the output is higher than the reference. This digital counter accomplishes the functionality of an integrator.

There are certain disadvantages to using a single-bit feedback signal. First, the output will never reach an equilibrium voltage; rather it will approach a limit cycle, since the output is always measured to be in error (either too high or too low) and the error signal is not proportional to the magnitude of the error. Second, the response of the control loop must be slow in order to limit the magnitude of the steady-state limit cycle. After the duty cycle is changed, it should not be modified again until the dynamic response of the output filter has subsided. If the duty cycle is adjusted faster than the filter-response time, the magnitude of the steady-state limit cycle will be large (several duty-cycle increments) or unbounded.

The existence of a steady-state limit cycle is not in itself a severe problem. If the duty-cycle increments are small enough, the magnitude of the limit cycle can be on the order of the switch-

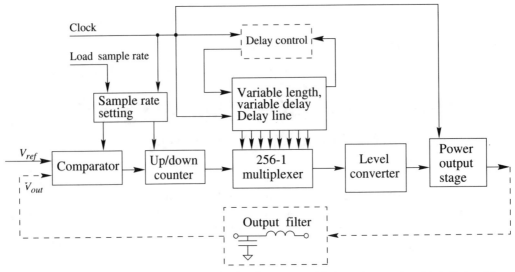

Figure 1 Block diagram of implemented converter. Dashed lines represent blocks not integrated on chip in first prototype.

ing frequency ripple. The limit cycle could also be eliminated by generating a three-state feedback representing too high, too low, and acceptable. In some cases, the slow response time of the control loop may be an important consideration. This can be addressed by performing comparisons between the output voltage and other reference values, thus creating an error signal with more than just a single bit of resolution. Additional comparisons could eliminate the limit-cycle behavior and allow faster response time.

3. DELAY LINE BASED PWM

When using analog circuits, a PWM signal is typically created by comparing a ramp signal to a reference value with a static comparator. Digital PWM circuits can avoid the problem of static power dissipation. A previously reported method of creating a PWM signal from a digital command is to use fast-clocked counters [6], but the power of the reported controller alone is on

the order of milliwatts. A clock frequency f_{clk} is chosen to be 2^N times the switching frequency (f_{SW}) where N is the number of bits in the digital command word. The clock is used to divide the switching period into 2^N increments. This method requires the use of very fast clocks for large values of N, and the need to run an N-bit counter at f_{clk} precludes the use of low supply voltages, which would otherwise provide low power operation. The fast-clocked counter approach for PWM generation is shown in Fig. 2. Another way to create a pulse-width modulated signal from an N-bit digital value is to use a tapped delay line. Since this approach operates from the switching frequency clock, the power is significantly reduced relative to the fast-clocked counter approach.

Figure 3 shows a schematic for the proposed digital word-to-PWM circuit. The essential components of a tapped delay line PWM circuit are the delay line and a multiplexer. A pulse from a reference clock starts a cycle and sets the PWM output to go high (after a delay designed to match the propagation delay ex-

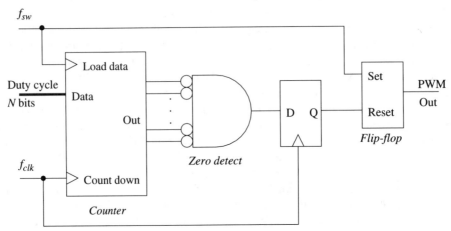

Figure 2 Fast-clocked counter approach for creating PWM signals. f_{clk} is $2^N f_{SW}$, and there are 2^N duty-cycle increments.

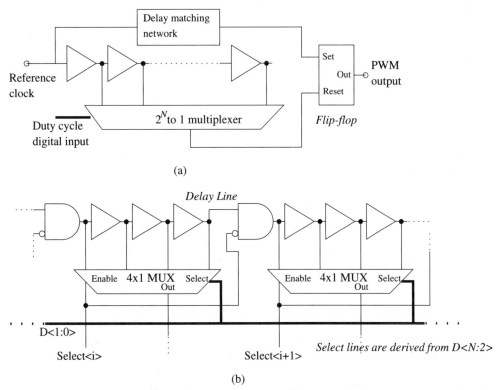

(a)

(b)

Figure 3 A PWM circuit created from a tapped delay line and a multiplexer. (*a*) High-level view of tapped delay line. (*b*) Selection logic used to prevent propagation of input pulse. Outputs of 4 × 1 muxes are ORed together.

perienced through the multiplexer). The reference pulse propagates down the delay line, and when it reaches the output selected by the multiplexer, it is used to set the PWM output low. The total delay of the delay line is adjusted so that the total delay is equal to the reference clock period. That is, feedback is used to turn the delay line into a delay-locked-loop (DLL), which locks to the period of the input clock.

In order to avoid superfluous switching activity (and the resulting power dissipation), the multiplexer is flat (as opposed to a tree structure), so that only the selected input propagates (i.e., it causes switching transitions) on the outputs of the first level of gates. The flat multiplexer increases the energy dissipated when the selection signals change (relative to a more hierarchical, tree structured multiplexer); however, due to the slow rate at which the duty cycle will change, this increase will not increase the power of the PWM circuit significantly. It is also possible to avoid switching transitions on the delay line buffers past the selected tap. As a consequence of the design and layout of the delay line and multiplexer, selection signals are available to indicate which block of four buffers contains the tap which the multiplexer is configured to select. These selection signals take the form of 64 (for a 256-stage delay line) "one-hot" signals, each available in the proximity of their respective set of buffers. These selection signals can be used to gate the pulse propagating down the delay line to save power. The pulse is prevented from propagating past the set of four buffers which is selected by the multiplexer (see Fig. 3b), thus eliminating unnecessary switching events in the nodes downstream from the selected taps. Another signal is available to

override this feature so that all pulses propagate to the end of the delay line. This is necessary to measure and control the total delay of the delay line.

4. DELAY LINE LENGTH CONTROL

The total delay of the delay line can be controlled if the buffers in the delay line are "starved," i.e., the current available to switch their outputs is limited by a series MOS device in the subthreshold or linear region. The schematic of a starved buffer, as well as measured data concerning the aggregate delay of a chain of such elements are shown in Figs. 4 and 5. The delay decreases exponentially with the starvation gate input voltage for

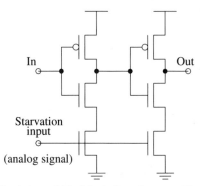

Figure 4 Circuit for variable delay buffer using starved inverters. The delay is controlled by the voltage level on the gates of the lowest two NMOS devices.

139

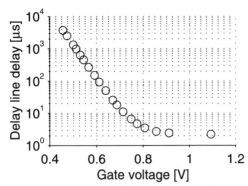

Figure 5 The measured delay of 256-stage chain of starved buffers at a supply voltage of 1.1V. The NMOS threshold is about 0.78V.

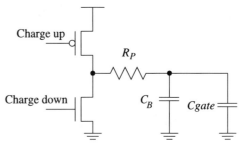

Figure 7 Circuitry to charge rate control gate of delay line. C_{gate} is the distributed gate capacitance of the delay line; R_p and C_B are added for stabilization.

below the threshold (for the data shown, $V_{T_n} \approx 0.78$V), indicating that in this region, the starvation NMOS devices, are operating in subthreshold conduction.

The voltage level on the starvation gates is controlled so that the total delay of the chain of buffers matches the period of the input reference clock (at the switching frequency). In order to match this delay to the clock period, the time of arrival of a pulse to the last tap of the delay line is compared to the time of the next clock pulse. The control circuitry determines which pulse arrived first, then charges or discharges the starvation gate voltage until the other pulse arrives (see Fig. 6).

The control signals can modify the gate-control voltage with the circuit of Fig. 7. The gain of this control loop is determined by the charging resistance (R_p), the total capacitance on the signal node ($C_{gate} + C_B$), the supply voltage, and the operating point DC voltage on the output. C_{gate} is the distributed gate capacitance of the delay line, and C_B is a capacitor added for loop stabilization. Because the charging or discharging current depends on the voltage drop across R_p, the gain differs for positive and negative errors. The power dissipation of this feedback loop is small because only a few gates switch at each evaluation, and operation is at a fraction of the switching frequency.

For a fully-integrated controller, it is prudent to have the "charge up/down" control signals gate a current source to increase or decrease the voltage on the output node. This configuration avoids the dependence of the gain on the supply voltage and the bias voltage, and thus avoids the problem of building ei-

ther a large capacitor or a large resistor on chip, this ultimately saves the die area. For a practically-sized integrated capacitor, the current source current will be very small (on the order of 0.1μA to 1μA) so the additional power dissipation (from the current source) will be minimal.

Gardner [7] discusses the stability of phase-locked loops controlled by a charge pump. Controlling a delay-locked loop as proposed here is fundamentally less difficult, since a DLL does not integrate phase errors as a PLL. (A DLL has a lower order than a PLL.) In order to generate a model for the DLL, we consider it to be a discrete-time system and linearize about the operating point of the sub-threshold starvation devices:

$$D[n] = T_0 - Kv[n] \qquad [1]$$

$$v[n+1] = v[n] - \frac{I_c}{C_B + C_{gate}}(T_0 - D[n]) \qquad [2]$$

Here, $D[n]$ is the delay of the delay line in a given cycle, K is the incremental gain of the delay line in the units S/V, $v[n]$ is the small-signal value of the incremental rate control voltage, and I_c is the charging current. For the case shown in Fig. 7, I_c will depend on the power supply, the bias point on the rate control gate, the resistance R_p, and whether the delay error is positive or negative. The Z-transform of this first order system yields the following characteristic equation:

$$z = 1 - \frac{I_c}{C_B + C_{gate}}K, \qquad [3]$$

and the system is stable if $-1 < z < 1$.

It is assumed that the length comparison is not done every cycle. This helps to avoid the problem that the delay of the delay line may be a multiple of the clock period. If the delay of the delay line is twice the clock period, pulses at the last tap may align perfectly with input clock pulses, and the control circuitry will not converge correctly. When the length comparison is not done every cycle, pulses are only allowed to propagate to the last tap of the delay line when a comparison is desired, and the threat of converging to the wrong delay length is reduced. The sample rate for the length comparison is determined by the leakage on the control gate, the tolerable ripple of the output

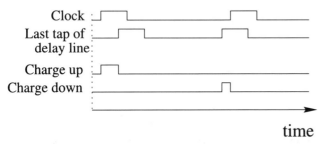

Figure 6 Sample timing diagram showing generation of control signals.

duty cycle (caused by delay length jitter), and the necessary transient response for the delay length.

5. RESULTS

To test and evaluate the digital PWM circuit discussed above, the circuit was fabricated on an integrated circuit, along with output switches to create a down converter. The circuit was built on a 0.6μm CMOS process. The output switches are a 2mm-wide NMOS device and a 3.8mm-wide PMOS device, constructed with many parallel fingers (see [1] for a discussion of the layout of CMOS power devices). The intended application is a 5mW, 3V in, 1V out DC–DC converter, and the output switches were sized for this load. The delay line has 256 buffers, giving 8-bit resolution for the value of the duty cycle. Output feedback was accomplished by comparing the output voltage with an off-chip reference using a dynamic comparator. The dynamic comparator consists of NMOS input transistors with cross-coupled pull-up devices. The comparator has a precharge phase, and then evaluates, after which time there is no current draw from the comparator. Figs. 8, 9, and 10, and Table 1 summarize the performance of the fabricated system and show some typical output waveforms.

As shown in Table 1, the power dissipation of the control circuitry is on the order of a mere 10μW at low supply voltages. This power varies depending on the switching frequency and the control-circuit supply voltage. The power dissipation also depends on the duty cycle selected, since nodes downstream from the selected tap of the delay line are not activated, except during

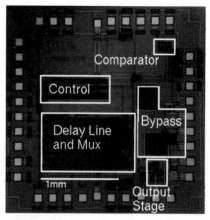

Figure 8 Photograph of fabricated chip and summary of results. Chip area is 2.3mm $\times$ 2.4mm.

cycles where the delay line total delay is adjusted. For the runs shown, the total delay is adjusted every other cycle. Therefore, a variation in the duty cycle from 100% to 0% should halve the power dissipation. With a 2V supply, the controller is suitable for switching frequencies as high as 1MHz, and the lower limit on the switching frequency is less than 1kHz.

The low power of this controller enabled the construction of an 88%-efficient power converter. No special techniques (such as ZVS) were applied to the output stage to increase efficiency. The magnitude of the limit cycle ripple at steady state is 50mV, and the switching frequency ripple is 13mV. The response to a change in the reference voltage (not shown) is essentially slew limited because the error signal is not proportional to the magnitude of the error.

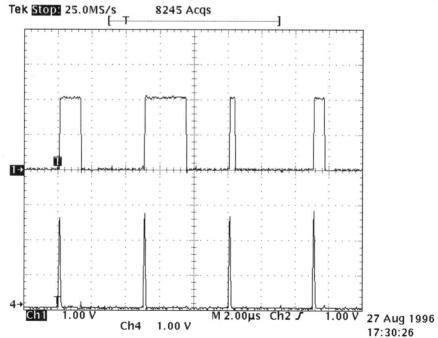

Figure 9 Pulse-width modulated signal, with clock. Duty cycle is varied between duty cycles 25%, 50%, 6.75%, and 12.5%.

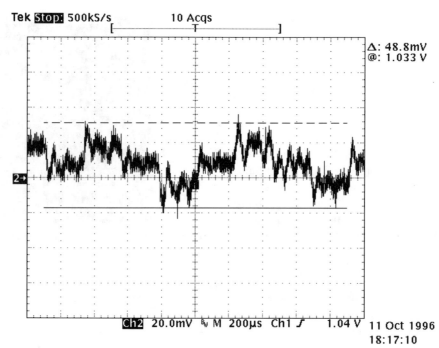

Figure 10 Single-bit feedback limit cycle ripple. The system is under closed-loop feedback control, at 5mW output power. The arrow on the left indicates 1.05V level.

Table 1. Summary of PWM and Power-Converter Performance

Minimum PWM V_{DD}: 1.1V

PWM Current with (V_{DD} = 1.2V, f_{SW} = 330kHz)
   D = 12%: I = 7.1μA
   D = 50%: I = 8.8μA

Delay Range: <1kHz to 1MHz
         (V_{DD} = 2V)

Efficiency with 5mW load: 88%
(V_{IN} = 3V), V_{OUT} = 1V, f_{SW} = 330kHz, L = 220μH, C = 0.22μF, Filter area is 0.024in²)

6. CONCLUSIONS

We have shown that it is possible to achieve stable operation with only a single bit of feedback regarding the output voltage, albeit with degradation of output response time. A circuit was presented which creates a pulse-width modulated signal at frequencies appropriate for very small low-power voltage converters, with a power dissipation on the order of 10μW.

7. ACKNOWLEDGMENTS

This work is funded by ARPA contract #DAAL-01-95-K3526. Abram Dancy also receives funding from an NSF Graduate Research Fellowship.

References

[1] A. Stratakos, S. Sanders, and R. Brodersen, "A low-voltage CMOS DC-DC converter for a portable battery-operated system," *IEEE Power Electronics Specialists Conference,* 1994, pp. 619–626.

[2] Linear Technology, *LT1073 Datasheet, Micropower DC-DC Converter.* See www.linear.com.

[3] Maxim Integrated Products, *MAX 756 Datasheet, Adjustable-Output Step-Up DC-DC Converter,* January 1995. See www.maxim-ic.com.

[4] A. P. Chandrakasan and R. W. Brodersen, *Low Power Digital CMOS Design,* Norwell, MA: Kluwer Academic Publishers, 1995.

[5] B. Gordon and T. Meng, "A 1.2mW video-rate 2-D color subband decoder," *IEEE Journal of Solid-state Circuits,* Dec. 1995, pp. 1510–1516.

[6] G. Wei and M. Horowitz, "A low power switching power supply for self-clocked systems," *1996 International Symposium on Low Power Electronics and Design,* 1996, pp. 313–318.

[7] F. M. Gardner, "Charge-pump phase-lock loops," *IEEE Transactions on Communications,* vol. COM-28, no. 11, Nov. 1980, pp. 1849–1858.

A Voltage Reduction Technique for Battery-Operated Systems

VINCENT VON KAENEL, PETER MACKEN, AND MARC G. R. DEGRAUWE, MEMBER, IEEE

Abstract —A self-regulating voltage reduction circuit is presented which adjusts the internal supply voltage at the lowest value compatible with the speed requirements of the chip, taking temperature and technology parameters into account. Besides enhancing reliability, this new technique also saves power, which is important for battery-operated systems such as laptop computers and portable instrumentation for radio-communication systems.

I. INTRODUCTION

SUPPLY voltage reduction of VLSI chips is important to save power and to avoid undesirable effects, such as punchthrough and substrate currents, when small-geometry technologies are used. In order to remain compatible with the external supply voltage standards and the technology requirements, on-chip voltage reduction techniques can be used.

In the past such techniques have been applied to memories [1]–[4], watches, etc., where the on-chip voltage is reduced to a fixed value. Such static systems of reduction can neither take into account speed requirements nor all technology parameters.

In this paper, a new voltage reduction technique based on functional comparison is presented. With this new technique, a regulated system adjusts the supply voltage of a digital circuit at the functional boundary for the speed requirements, the temperature, and the technology parameters.

In the next section, the principle of the self-adjusted technique is discussed. In Section III the fixed-voltage technique with temperature compensation is presented. The results of measurements are given in Section IV. Finally, in Section V possible applications are discussed.

Manuscript received April 18, 1990; revised May 28, 1990. This work was supported by the Centre Electronique Horloger S.A.
V. Von Kaenel and M. G. R. Degrauwe are with Recherche et Développement, Centre Suisse d'Electronique et de Microtechnique S.A., CH-2000 Neuchâtel 7, Switzerland.
P. Macken was with Recherche et Développement, Centre Suisse d'Electronique et de Microtechnique S.A., CH-2000 Neuchâtel 7, Switzerland. He is now with the University of Chicago, Chicago, IL 60637.
IEEE Log Number 9037773.

Reprinted from *IEEE Journal of Solid-State Circuits*, pp. 1136-1140, October 1990.

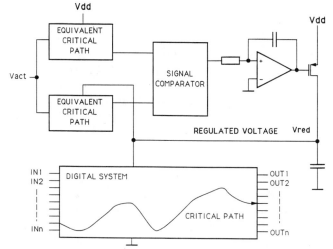

Fig. 1. Principle of a self-adjusted voltage reducer.

II. PRINCIPLE OF SELF-ADJUSTED VOLTAGE REDUCTION

The power consumption of a digital circuit is given by the well-known formula

$$\text{power} = f \cdot C \cdot V_{DD}^2 + V_{DD} \cdot I_{DC} \qquad (1)$$

where f is the operating frequency, C is the equivalent capacitance of the circuit, and I_{DC} is the static current. Reducing the supply voltage will decrease power consumption, however, it will also slow down the critical path of the circuit. If the supply voltage is further reduced, the critical path finally becomes too slow to assure the correct functioning of the chip. The general voltage reduction technique presented in Fig. 1 is based on this last observation. The following functional blocks need to be added to the circuit: two "equivalent critical paths," a signal comparator, and an integrator.

The "equivalent critical path" is a small circuit with electrical properties which are comparable to those of the actual critical path of the original circuit.

The signal comparator will detect if there is a difference between the signals coming out of the two equivalent critical paths, one of which is biased at the full voltage supply while the other is biased at a reduced voltage. Due to the feedback loop the system will adjust the reduced voltage so that one "equivalent critical path" is biased at the minimum required value to produce a correct output signal.

Since the equivalent critical path has the same behavior as the actual critical path of the original circuit, one can use the reduced voltage as a supply for the original circuit. Eventually a small voltage shift can be implemented to obtain a security margin.

In order to reduce the overall power consumption and to have a reliable circuit, it is mandatory to have a small circuit which can model the actual critical path in an accurate way.

It can be shown that, in a first-order approximation, the ratio of the delay of a critical path to the period of a ring oscillator is a constant which depends only on the number of stages, the dimensions of the transistors, and the load capacitances. That means that a ring oscillator can be used as an equivalent critical path for all digital circuits. Since by changing the supply voltage of a ring oscillator the frequency changes (VCO), one can implement the system shown in Fig. 1 with a phase-locked loop (PLL) as shown in Fig. 2. The fixed-frequency divider (N) is used to implement the ratio between the delay of the VCO and that of the actual critical path. The input signal f_{in}, which affects the reduced voltage value, represents the speed requirement. The chip's general clock frequency can be used as f_{in}.

By adjusting the VCO supply voltage, the PLL causes the VCO to oscillate at $N \cdot f_{in}$. Supposing that the dimensions of the VCO transistors are such that the critical path functions correctly at the regulated voltage, the digital circuit will always function correctly, since changing technology parameters, temperature, or the frequency f_{in} affects the VCO in the same way as it affects the circuitry. So it is sufficient to measure or simulate the delay of the critical path at a certain supply voltage and calculate the dimensions of the VCO and the division factor N accordingly. One is then assured that the digital circuit will always work properly.

III. Fixed-Voltage Reductor

This technique consists of imposing a fixed reduced supply voltage to the circuit [5]–[7]. In this case there is no activation signal, so the work frequency is fixed. Only static technology parameters can be taken into account by this technique. There is no feedback from the digital circuit to the voltage generator, so there is no functional verification of the circuit. In this way, the reliability is not guaranteed.

In the simple circuit like that shown in Fig. 3, the reduced voltage is approximately equal to the maximum

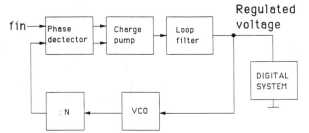

Fig. 2.   Implementation of the supply voltage reducer by a PLL.

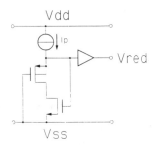

Fig. 3.   Generator of fixed reduced voltage.

of both threshold voltages. The main advantages of this approach are the small surface and the low consumption of this circuit. As shown hereunder, some security margins must be employed to ensure that the digital circuit still works in the worst case of technology parameters and temperature.

The reduced voltage takes into account technology parameters such as the V_T's and β's, however, capacitor variation must be considered separately. It is interesting to note that the oxide thickness T_{ox} does not influence the speed of a digital circuit in first approximation when the gate capacitors are the dominant load capacitance.

For temperature effect, the current reference versus temperature can be modeled as follows:

$$I_p = I_{p0} \cdot \left(\frac{T}{T_{ref}} \right)^a \qquad (2)$$

where a is the temperature coefficient of the reference, T_{ref} is some reference temperature, T is the temperature value, and I_{p0} is the current reference at T_{ref}.

The β's are modeled as follows:

$$\beta = \beta_0 * \left(\frac{T}{T_{ref}} \right)^{-\alpha} \qquad (3)$$

where α is the temperature coefficient of the mobility ($\alpha = 1.5$) and β_0 is the beta at T_{ref}.

It is assumed that the V_T's vary linearly with a temperature coefficient of -2 mV/°C. These laws applied to the circuit shown in Fig. 3, and to a digital circuit they give an important characteristic of the fixed reductor circuit, namely the slope of V_{red} versus temperature as a function of V_{red} (Fig. 4). By adjusting the ratio of well resistance to polysilicon resistance used for R in the current reference (Fig. 5), it is possible to obtain a similar temperature coefficient for reduced voltage and the digital circuit.

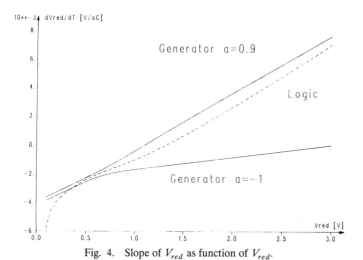

Fig. 4. Slope of V_{red} as function of V_{red}.

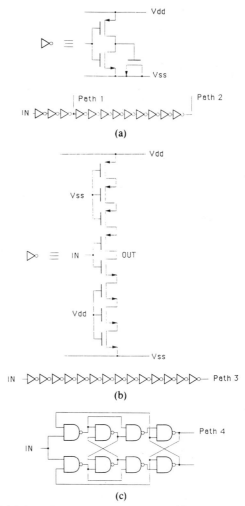

Fig. 6. (a) Schematic of critical paths 1 and 2. (b) Schematic of critical path 3. (c) Schematic of critical path 4.

Fig. 5. Current reference.

Since variations of resistor values affect the current and thus also the value of the reduced voltage, this technique demands the use of large security margins during the design.

Because of the low consumption and small area used for the fixed-voltage reduction, this technique is suitable for small digital circuits with low power consumption.

IV. MEASUREMENTS

In order to show the validity of the proposed self-adjusted principle, the system in Fig. 2 has been integrated in a 2-μm n-well CMOS process. To model the critical path, four test circuits have been implemented:

1) a short series of inverters with the same dimensions as those of the VCO, but charged with an extra gate capacitance at the inverter output (Fig. 6(a));
2) a long series of these inverters, to see the effect of a long path (Fig. 6(a));
3) a series of inverters with four NMOS and four PMOS transistors in series, to test the influence of putting transistors in series (Fig. 6(b)); and
4) a frequency divider, which is a small digital circuit (Fig. 6(c)).

Furthermore, in order to be able to show the influence of technology variation, a "split lot" integration was done

which changed the oxide thickness over 10% and the sum of the threshold voltages of the n- and p-transistor MOS from 1.07 to 1.60 V.

As shown in Fig. 7(a) and (b), the ratio of delay of the critical path integrated to the period of the VCO depends weakly on the supply voltage (V_{red}) and the temperature. This confirms that a VCO can be used as an equivalent critical path. The variations are caused by foreseeable changes in the gate capacitance near the threshold voltage. If one wants the digital circuit to operate correctly in a range of frequencies (f_{in}), the operating points should be chosen where the critical path is the slowest relative to the VCO. As a consequence, the regulated voltage will be too high for all other frequencies.

Fig. 8 shows the ratio of the regulated supply voltage to the minimum supply voltage, for frequencies of f_{in} ranging from 400 Hz to 4 MHz. For the four cases of critical paths integrated, the regulated voltage is at worst 10% too high, which means that the circuit will always work near the point of lowest power consumption.

Technology parameters are accurately tracked, as shown in Fig. 9, which represents the regulated voltage (V_{red}) for two different wafers. The ratio of the delay of a critical

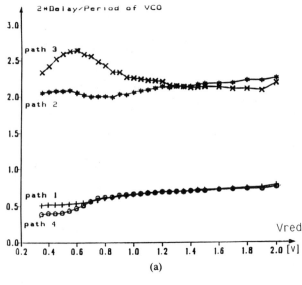

(a)

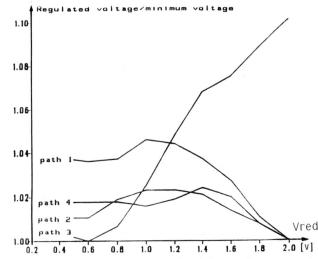

Fig. 8. Ratio of regulated voltage to the minimum voltage necessary to fulfill speed requirements.

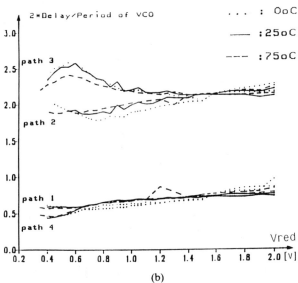

(b)

Fig. 7. (a) Ratio of delay of a critical path to the period of the VCO. (b) Ratio of delay of a critical path to the period of the VCO, for three different temperatures.

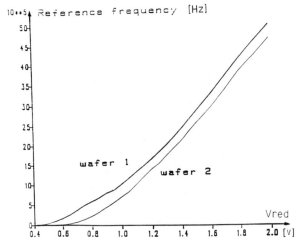

Fig. 9. Reference frequency as function of regulated voltage V_{red}, for two different wafers.

path to the period of the VCO varies less than 20% while the sum of the threshold voltage changes 250 mV.

Power consumption for the whole system is only a few microamperes. Tests have shown that it can be reduced to the nanoampere level by operating the voltage regulation system intermittently. Indeed, for a stable f_{in} signal, the rate of variation of the reduced voltage is very slow because of the slow variation of the temperature and technology parameters. The system can operate intermittently by opening the regulation loop for a given time and closing it only in order to compensate for the leakage currents that empty the loop filter capacitance (Fig. 10).

V. POSSIBLE OTHER APPLICATIONS

In battery-operating systems it is important to detect when the battery should be changed in order to avoid reliability problems.

A common way to detect the end of life (EOL) of a battery is to compare the battery voltage to a fixed boundary value. This fixed value of comparison must be calculated in the worst case of technology and temperature. However, the actual speed requirements, temperature, and technology parameters of the circuit are not taken into account in the detection. That means a fixed value cannot be the optimum value for all cases of technology and temperature because the security margin alters the EOL detection.

The best way to detect the EOL of the battery is to use a threshold value that depends on the minimum supply voltage of the digital circuit. When using the proposed voltage reduction technique, the EOL detection is very simple and reliable. It suffices to compare the battery voltage to the reduced voltage supply with a small margin for detection just before the limit as shown in Fig. 11. In this way, the EOL detection is dependent on the technology parameters and the temperature. This allows the

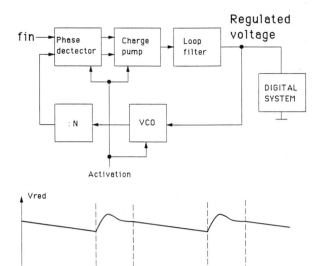

Fig. 10. Supply voltage reducer for intermittent operation.

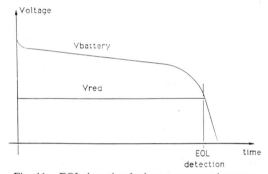

Fig. 11. EOL detection for battery-operated systems.

complete use of the battery with respect to the actual needs of the digital circuit.

VI. Conclusion

In this paper, two techniques of voltage reduction have been presented. Both techniques can significantly reduce the power consumption of digital CMOS circuits.

The fixed reduction of voltage is usable for small systems with a low initial consumption. In this case, the optimum voltage is not reached and the correct operation of the circuit is not guaranteed.

The self-adjusted reduction of voltage is adapted to bigger digital systems. The correct operation of the digital circuit is guaranteed. The supply voltage is near the optimum for the speed requirements. This new technique is more versatile, more accurate, and more reliable than the fixed one.

References

[1] H. Fukuda *et al.*, "A BICMOS channelless masterslice with on-chip voltage convertor," in *ISSCC Dig. Tech. Papers*, Feb. 1989, pp. 176–177.
[2] T. Furuyama *et al.*, "A new on-chip voltage convertor for submicron high density DRAM's," in *Proc. ESSCIRC*, Sept. 1989, pp. 10–12.
[3] K. Itoh *et al.*, "An experimental 1Mb DRAM with on-chip voltage limiter," in *ISSCC Dig. Tech. Papers*, Feb. 1984, pp. 282–283.
[4] T. Mano *et al.*, "Circuit techniques for a VLSI memory," *IEEE J. Solid-State Circuits*, vol. SC-18, pp. 463–470, Oct. 1983.
[5] E. Vittoz, "The design of high-performance analog circuit on digital CMOS chips," *IEEE J. Solid-State Circuits*, vol. SC-20, no. 3, pp. 522–528, June 1985.
[6] M. Degrauwe *et al.*, "Adaptive biasing CMOS amplifier," *IEEE J. Solid-State Circuits*, vol. SC-17, no. 3, pp. 657–665, June 1982.
[7] E. Vittoz, M. G. R. Degrauwe, and S. Bitz, "High-performance crystal oscillator circuits: Theory and application," *IEEE J. Solid-State Circuits*, vol. 23, no. 3, pp. 774–783, June 1988.
[8] S. Chou *et al.*, "Chip voltage: Why less is better," *IEEE Spectrum*, vol. 24, no. 4, Apr. 1987.

Automatic Adjustment of Threshold & Supply Voltages for Minimum Power Consumption in CMOS Digital Circuits

Vincent R.von Kaenel, Matthijs D. Pardoen, Evert Dijkstra, Eric A. Vittoz

CSEM Centre Suisse d'Electronique et de Microtechnique SA
Maladière 71, 2007 Neuchâtel, Switzerland, Fax : +41 38 205 630

Abstract

Decreasing the power consumption of CMOS digital circuits by both supply voltage (V_B) and threshold voltage (V_T) reduction (by technology) is limited by the V_T spread (technology and temperature dependent). This paper presents a control circuit that minimizes the sum of the dynamic and static power consumption by reducing and controlling both the supply and threshold voltages. To compensate for temperature and technology variations, the latter is electrically adjusted by bulk biasing. The control loop has been designed such that the speed is maintained.

Introduction

The power consumption (P) of a CMOS digital gate equals the sum of dynamic and static power (P_{dyn} and P_{stat}). Assuming a mean output signal duty cycle of 50%, a total load capacitance C, a maximum frequency of operation f_{max}, a standby current of the digital gate I_0, the specific current of the MOS $I_{sn,p}$ ($I_s = 2n\mu C_{ox} \frac{W}{L_{eff}} U_T^2$), the slope factor n and U_T the thermal voltage, the power consumption for one gate equals:

$$P = P_{dyn} + P_{stat} = V_B^2 \alpha_t f_{max} C + V_B I_0$$

$$I_0 = \frac{1}{2} I_{sn} e^{\frac{-V_{Tn}}{n_n U_T}} + \frac{1}{2} I_{sp} e^{\frac{-|V_{Tp}|}{n_p U_T}}$$

(1,2)

The activity ratio α_t is defined as the ratio between the repetition frequency of operation f and the maximum frequency of operation f_{max}:

$$\alpha_t = f/f_{max} \leq 1 \quad \text{with} \quad f_{max} = 1/\left(t_{dn} + t_{dp}\right) \quad (3,4)$$

Assuming the MOS always operates in saturation, the gate delay $t_{dn,p}$ can be approximated by:

$$t_{dn,p} = \left. \frac{C V_B}{I_{dn,p}} \right|_{Vgs=V_B} \quad (5)$$

To avoid the classical discontinuity in the MOS drain current model $I_{dn,p}$ near $Vgs=V_T$, we use the following model that gives an acceptable precision in a very wide range of currents [1] (NMOS):

$$I_d = I_s L n^2 \left[1 + e^{\frac{Vgs - V_T}{2n U_T}} \right] \quad (6)$$

The formulas (4,5,6) show that by decreasing both V_B and V_T, the speed (f_{max}) can be maintained. For a large number of gates (N_g), the total power consumption P_t is :

$$P_t = \sum_{i=1}^{N_g} \left(P_{dyni} + P_{stati} \right) \approx N_g \left(V_B^2 \alpha f_{max} \overline{C} + V_B \overline{I_0} \right) \quad (7)$$

The approximation introduced by (7) uses the averaged values of gate capacitance $\overline{C}$ and standby current of the gate $\overline{I_0}$. With m the number of gate transition cycles to carry out an operation, the average gate activity is defined as :

$$\alpha = \frac{m}{N_g} \frac{f}{f_{max}} \quad (8)$$

Minimizing Power consumption

The goal of our circuit is to minimize the power consumption without speed loss. Figure 1 shows the power consumption of an existing microprocessor as function of the supply voltage V_B. The maximum frequency is maintained for both curves by adjusting V_T. The upper curve has a larger average gate activity α than the lower one. For V_B lower than $\approx 4U_T$ [1], no signal regeneration is possible and so digital circuits can not work. For a high V_B value, the power consumption is dominated by the dynamic power. For a low V_B value the static power is dominant. An minimum power consumption exists between those two regions.

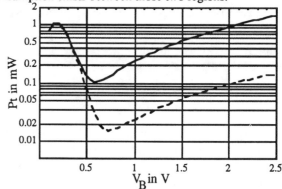

Fig.1 Calculated power as function of V_B for the same f_{max} (100MHz) with $\alpha_1 = 1\%(f=1MHz)$ for the upper curve and $\alpha_2 = 0.1\%(100kHz)$ for the dashed line.

Corresponding value of V_T are plotted as a function of V_B in fig.2. From those two figures it can be seen that the actual power consumption may increase significantly above the minimum power consumption because of a typical uncontrolled V_T spread of ± 200mV (due to technology, temperature). Therefore, when lowering both supply and

Reprinted from *IEEE Symposium on Low Power Electronics*, pp. 78-79, October 1994.

threshold voltage, it is absolutely necessary to control the V_T spread [1,2]. One way to electrically control the value of V_T is to bias the bulk. As a first approximation, the technological threshold voltage V_{T0} can be modified by the bulk-source voltage V_{bs} as: $V_T \approx V_{T0} - V_{bs}(n-1)$ (9)
V_{bs} can not be higher than $\approx 0.4V$ (NMOS) because of the forward base current in the parasitic bipolar associated with each MOS. However in reverse biasing, V_{bs} can be as low as -5V. So, it is necessary to have very low V_{T0} and to bias the bulk in reverse to get the optimum V_T value.

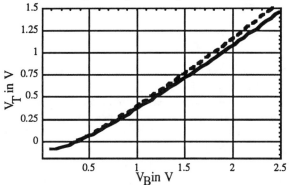

Fig.2 V_T as function of V_B for the same conditions as in Fig.1, the dashed line shows the V_{Tp} and the solid one the V_{Tn}.

Control Circuit

The control circuit automatically finds the optimum V_T for both p-and n-channel MOS and optimum V_B [5]. It can be shown that the minimum power is related to the ratio P_{dyn}/P_{stat} consumed by the digital circuit (fig.1). By fixing the ratio I_{dyn}/I_{stat} to a well-chosen value, the resulting total power consumption approaches the minimum. We propose a system that adapts the V_T (thereby the static current) as a function of the dynamic power consumption by keeping the ratio of I_{dyn}/I_{stat} constant.

The reference current I_{ref}, in the static current loop (fig.3), for a known ratio I_{dyn}/I_{stat} is derived from a measurement of the total current consumed by the digital circuit under control. A separate static current control loop is needed for each type of MOS (n- and p-channel).

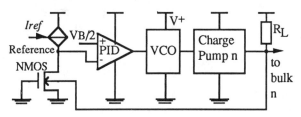

Fig.3   Static current control loop for the NMOS.

The static current control loop generates a bulk bias voltage such that the static current in the current reference is equal to that in the reference MOS. We use a charge pump, with variable output impedance[4], loaded by a fixed resistor (or current source) to generate a continuously variable bias voltage from one battery cell. The output impedance of the

charge pump is a function of the input frequency [4]. This frequency is generated by a VCO that is controlled by the error signal resulting from the comparison between the reference and the actual static current obtained by a reference MOS.

To ensure the speed of the circuit, we need to adapt V_B to the required value. This can be done by a simple circuit as shown in fig.4, or by a more accurate yet more complex circuit as described in [3] which takes into account dynamic parameters as parasitic capacitances and frequency of operation.

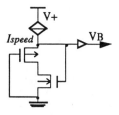

Fig.4   Simple schematic for V_B generation

The circuit that provides V_B to the digital circuit can be a DC-DC converter or a linear voltage regulator. In the first case, the power is reduced with the square of supply voltage, but at least an external component is needed. In the second case, the power is only reduced proportionally to the supply voltage.

Conclusions

A significant power reduction can be achieved by reducing both supply and threshold voltages without speed performance loss. This requires a precise adjustment of threshold voltages, which can only be obtained by an electrical control. To approach the minimum power consumption, the proposed control loop balances the static and dynamic power consumption, by adjusting the bulk voltage with an on-chip multiplier. To fully use the system's potential, a very low threshold voltages and triple well technology is necessary.

Acknowledgements

We thank EM Marin for supporting this program.

References

[1] Eric A.Vittoz, "Low-power design : ways to approach the limits", Plenary address *ISSCC'94*, February 16-18 1994, San Francisco.

[2] James Burr, "Stanford ultra low power CMOS", *Hot Chips Symposium V*, August 10, 1993

[3] V.von Kaenel et al., "A voltage reduction technique for battery-operated systems", *IEEE JSSC-25*, No5, October 1990, pp.1136-1140.

[4] John F.Dickson, "On-chip high-voltage generation in MNOS integrated circuit using an improved voltage multiplier technique", *IEEE JSSC-11*, No3, June 1976.

[5] V.R.von Kaenel, M.D.Pardoen, Patent pending

Low-Power Operation Using Self-Timed Circuits and Adaptive Scaling of the Supply Voltage

Lars S. Nielsen, Cees Niessen, Jens Sparsø, and Kees van Berkel

Abstract— Recent research has demonstrated that for certain types of applications like sampled audio systems, self-timed circuits can achieve very low power consumption, because unused circuit parts automatically turn into a stand-by mode. Additional savings may be obtained by combining the self-timed circuits with a mechanism that adaptively adjusts the supply voltage to the smallest possible, while maintaining the performance requirements. This paper describes such a mechanism, analyzes the possible power savings, and presents a demonstrator chip that has been fabricated and tested. The idea of voltage scaling has been used previously in synchronous circuits, and the contributions of the present paper are: 1) the combination of supply scaling and self-timed circuitry which has some unique advantages, and 2) the thorough analysis of the power savings that are possible using this technique.

I. INTRODUCTION

THE DOMINANT source of power dissipation in digital CMOS circuits is the dynamic power dissipation:

$$P_{\text{dynamic}} = a \cdot f_{\text{clk}} \cdot C_L \cdot V_{DD}^2 \tag{1}$$

where f_{clk} is the switching frequency, C_L the total node capacitance in the circuit, a the average fraction of the total node capacitance being switched (also referred to as the activity factor), and finally V_{DD} the supply voltage.

For a given technology and application, the power consumption can be minimized by reducing V_{DD} and/or a.

Reducing V_{DD} leads to an increase in circuit delays. With good accuracy, the circuit delay can be estimated using the following equation [1], where μ is the mobility, C_{ox} the oxide capacitance, V_t the threshold voltage, and W/L the width to length ratio of transistors.

$$T_d = \frac{C_L \cdot V_{DD}}{\mu C_{\text{ox}}(W/L)(V_{DD} - V_t)^2}. \tag{2}$$

The activity factor a can for example be reduced by avoiding glitches or by gating the clock. While synchronous circuits

Manuscript received June 15, 1994; revised August 23, 1994.
L. S. Nielsen and J. Sparsø is with the Department of Computer Science, Technical University of Denmark, DK-2800 Lyngby, Denmark.
C. Niessen and C. H. van Berkel are with Philips Research Laboratories, 5656 AA Eindhoven, The Netherlands.
IEEE Log Number 9406368.

require special design effort and clock gating circuitry, self-timed circuits inherently avoid redundant transitions. For this reason, self-timed circuits have attracted more attention in recent years, particularly in areas where the computational complexity is strongly data dependent. An example of work in this area is the error corrector for the Digital Compact Cassette (DCC) player developed at Philips Research Laboratories [2]. This circuit dissipates about 80% less power than its synchronous counterpart. For an introduction to self-timed circuit design we refer to [3]–[5].

Another advantage of self-timed circuits is the ability to exploit variations in fabrication process and operating conditions in the best possible way. The performance of the chip depends on actual circuit delays, rather than on worst-case delays.

This paper describes a technique that combines self-timed circuitry with a mechanism that adaptively adjusts the supply voltage to the minimum possible, taking into account: process variations, operating conditions, and data dependent computation times. Adaptive supply scaling or "just-in-time processing" has been studied both at Philips Research Laboratories [6] and at the Technical University of Denmark [7] and this paper combines the experiences of the two parties.

The idea of voltage scaling has been used previously in synchronous circuits, and the contributions of the present paper are: 1) the combination of supply scaling and self-timed circuitry which has several unique advantages, and 2) the thorough analysis of the power savings that are possible using this technique.

The paper is organized as follows. Section II presents the concept of adaptive supply scaling. Section III provides an analysis of the power savings that can be obtained, and Section IV extends the analysis to include the effects of velocity saturation and short-circuit currents. Section V presents a small demonstrator chip that has been fabricated and tested. Section VI concludes the paper and discusses some open questions.

II. ADAPTIVE SUPPLY SCALING

In this section a system architecture for adaptive supply scaling is proposed and suitable applications are discussed. In the end of the section the approach is related to voltage scaling in synchronous circuits.

Reprinted from *IEEE Transactions on VLSI Systems*, pp. 391–397, December 1994.

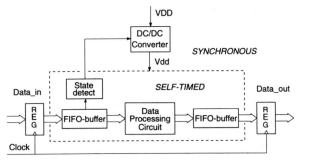

Fig. 1. Self-timed circuit in synchronous environment using adaptive supply scaling.

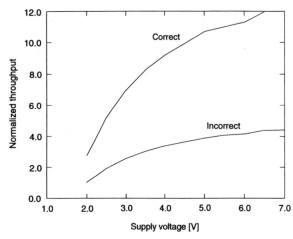

Fig. 2. Normalized throughput versus supply voltage for the DCC error corrector.

A. System Architecture

The proposed system using adaptive supply scaling in combination with self-timed circuitry is shown in Fig. 1. The system consists of the data processing circuit itself, two FIFO-buffers, a state detecting circuit, and a DC–DC converter for scaling down the supply voltage. The converter can be anything from a resistive device (a transistor on the chip) to a more sophisticated lossless device.

The actual design of the DC–DC converter is beyond the scope of this paper. Current research in low-power portable electronics includes design of low-voltage on-chip DC-DC converters, and efficiencies above 90% have been reported. A few pointers to this literature are: [8],[9].

The self-timed circuit is operating in a synchronous environment and the requirements are therefore, that the input buffer never runs full, and that the output buffer never becomes empty. With this requirement, synchronization problems will not occur at the synchronous/asynchronous interface.

The state detecting circuit monitors the state of one of the buffers, for example the input buffer as shown in Fig. 1. In this case, if the buffer is running empty, the circuit is operating too fast, and the supply voltage can be reduced. If, on the other hand, the buffer is running full, the supply voltage must be increased. The alternative is to monitor the output FIFO, and the state of the buffer must then be interpreted in a complementary way: a buffer running full should lead to a lower supply voltage, and vice versa. In this way the supply voltage will be adjusted to the actual workload, at all times maintaining the throughput requirements at lowest possible supply voltage.

The synchronous embedding shown in Fig. 1 was used for illustration purposes. Adaptive supply scaling may be used in a wider range of applications. Furthermore the architecture in Fig. 1 uses two FIFO's. In many cases one of the FIFO's may be omitted, because of particular characteristics of the algorithm/application to be implemented, or because buffering is provided by the environment [6]. This leaves the other FIFO to be part of the feed-back loop.

B. Suitable Applications

Adaptive supply scaling is particularly useful in systems with highly sequential algorithms that perform a large number of computation steps per data item, and where the computation time is data dependent. In addition, many systems are designed for worst-case conditions in order to guarantee response time, and therefore they possess a great unused speed potential. A safety margin of 2.5 is common in synchronous circuits, to accommodate variations in process and operating conditions. The idea is to convert this speed potential into a corresponding power saving, by reducing the power supply until the delay of the computation just fits the available time slot. The FIFO-buffers allow for averaging, which enables the system to exploit data dependencies.

Two factors limit the usefulness of the approach, 1) the FIFO-buffers add to the latency as seen by the environment, and 2) V_{dd} should only vary at a slow rate relative to the internal operational speed of the circuit, otherwise it may interfere with the operation of the circuit (signal levels, noise margin, etc.). In many applications latency is not a critical issue, and even in real time audio systems a latency in the order of a few milliseconds is acceptable. The limitations on the dynamics of V_{dd} makes the technique most suitable for applications with moderate throughput requirements where the external and internal frequency of operation differ by one or more orders of magnitude.

Examples of algorithms/applications that are particularly suited for adaptive supply scaling are, sampled audio systems, floating-point units, and error correction. For instance, the DCC error corrector described in [2] processes code words of 32 bytes and the processing time of a code word depends critically on its correctness. The measured throughput for correct code words is three times that for incorrect code words (cf. Fig. 2). Given that over 95% of the code words are correct, the DCC error corrector can operate below 2 V most of the time. Only a sequence of incorrect codewords would scale up V_{dd}.

C. Relation to Existing Techniques

Having presented the key ideas, it is relevant to relate the approach to voltage scaling in synchronous circuits.

One approach is to derive the supply voltage from the clock frequency as described in [10]. Here, a self-regulating voltage reduction circuit adjusts the internal supply voltage to the lowest value compatible with chip speed requirements, taking temperature and technology parameters into account. This is done using a phase locked loop where the clock signal is compared with the output of an on chip ring oscillator, whose delay-V_{dd} properties are assumed to be proportional to those of the actual circuit.

This mechanism obviously has some resemblance to the method described in this paper. However, there are some important differences and advantages that originate from the very nature of self-timed circuits — the handshaking that signals when computations have finished:

- In the self-timed approach the feedback is based on the actual delays in the circuit. This makes it more robust. The ring oscillator may provide a good match for static CMOS circuitry, but for circuits including pass-logic, memories and other irregular parts the matching may be less accurate. This means that a safety margin has to be introduced, reducing the power savings.
- The self-timed approach takes advantage of data dependencies, and that can contribute significantly to the power savings (cf. Section III-D).
- The feed-back signal controlling the DC-DC converter is easily derived from the FIFO's, and the FIFO's further smoothens fluctuations in workload, which again tends to filter out fluctuations in V_{dd}.

III. ANALYSIS OF POWER SAVINGS

In this section, the power savings made possible by the use of adaptive supply scaling will be estimated based on first order approximations of circuit delays. In Section IV the analysis is extended to include the effects of short-circuit currents and velocity saturation. First, the fabrication process, operating conditions, and data dependencies are considered with a lossless DC-DC converter, and second, the power loss related to the converter is taken into account. In order for the results to be independent of the fabrication process, the supply voltage V_{DD} will be normalized with respect to the threshold voltage V_t.

It should be noted that all estimations in the analysis will be based on a self-timed circuit with a constant throughput requirement, and no comparison between self-timed circuits and synchronous circuits is made.

A. Power Versus Delay

A circuit designed for worst-case conditions, allows for supply scaling, when worst-case conditions are not present. Operating the circuit at a fixed supply voltage V_{DD}, leads to the power consumption $P(V_{DD})$ and scaling the supply voltage to V_{dd}, leads to $P(V_{dd})$. The power reduction γ can thus be expressed:

$$\gamma = \frac{P(V_{dd})}{P(V_{DD})} = \left(\frac{V_{dd}}{V_{DD}}\right)^2. \tag{3}$$

In the typical case (typical process and operating conditions), the supply voltage can be reduced until circuit delays $T_{d,\text{typ}}$ match those determined by the worst-case conditions

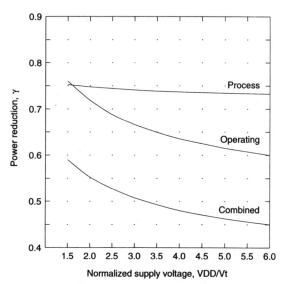

Fig. 3. Power savings due to variations in process, in operating conditions, and both.

$T_{d,\text{worst}}$. Using (2) the reduced supply voltage is found by solving the following equation for V_{dd}:

$$T_{d,\text{typ}}(V_{dd}) = T_{d,\text{worst}}(V_{DD})$$
$$\Rightarrow \frac{V_{dd}}{\mu_{\text{typ}}C_{\text{ox}}(W/L)(V_{dd} - V_{t,\text{typ}})^2}$$
$$= \frac{\alpha_{VDD}V_{DD}}{\alpha_T\alpha_p\mu_{\text{typ}}C_{\text{ox}}(W/L)(\alpha_{VDD}V_{DD} - \alpha_{\text{th}}V_{t,\text{typ}})^2}. \tag{4}$$

A number of coefficients are introduced in this equation to accommodate process variations and operating conditions:
Operating:

$$\alpha_{VDD} = V_{DD,\text{worst}}/V_{DD}$$
$$\alpha_T = \mu_{\text{typ}}(T)/\mu_{\text{typ}}(T_0).$$

Process:

$$\alpha_{\text{th}} = V_{t,\text{worst}}/V_{t,\text{typ}}$$
$$\alpha_p = \mu_{\text{worst}}(T_0)/\mu_{\text{typ}}(T_0).$$

where T is the temperature. As it can be seen, only variations in μ, V_{DD} and V_t are included in the analysis, and a distinction between the influence of operating conditions and process variations is made.

B. Process Variations

To estimate the amount of power that can be saved due to a typical process outcome, (4) is solved with a 15% variation on both μ and V_t, leading to $\alpha_p = 0.85$ and $\alpha_{\text{th}} = 1.15$. These values stem from the technology used for the demonstrator circuit described in Section V and are representative for typical 1 μm CMOS processes. The result is shown in Fig. 3 labeled "Process".

The figure shows, that the power reduction is approximately constant over the supply voltage range, and that the dissipated power, for the typical case, is 3/4 of that dissipated in the worst-case. In case of the best fabrication process, the dissipation will be approximately half of that in the worst case (not shown in Fig. 3).

C. Operating Conditions

Operating conditions influence circuit delays through α_T (temperature) and α_{VDD} (supply voltage) in (4). When temperature rises, mobility decreases [11]:

$$\mu(T) = \left(\frac{T}{T_0}\right)^{-\frac{3}{2}} \mu(T_0) = \alpha_T \cdot \mu(T_0). \qquad (5)$$

The exponent is an empirical value, and values ranging from 1.5 to 2 are reported in [11]. In order not to overestimate the possible power savings, the value 1.5 is used in this analysis. Using $T_0 = 300$ K as typical operating condition and $T = 350$ K as worst case gives $\alpha_T = 0.80$, and with a 10% tolerance on the supply voltage, $\alpha_{VDD} = 0.9$. With these numbers, an estimation of the power dissipation in the typical case, compared to that in the worst case, can be made. The result is shown in Fig. 3 labeled "Operating".

The combined effects of process variations and operating conditions, can also be found in Fig. 3 with the label "Combined". At $V_{DD} = 3V_t$, the power dissipation can be approximately halved.

D. Data Dependencies

A simple model is introduced to quantify the possible power savings due to data dependencies, i.e. variations in workload: For each input data, the system makes a sequence of computations, which is data dependent. Using this model, the workload can be expressed as a "cycle utilization" or "duty cycle" factor d, corresponding to the average number of computation steps divided by the worst-case number.

With duty cycle d, the circuit delay can be scaled by $1/d$, yielding a cycle utilization equal to 1. Including this dependency in (4), gives:

$$T_{d,\text{typ}}(V_{dd}) = \frac{1}{d} \cdot T_{d,\text{worst}}(V_{DD})$$

$$\Rightarrow \frac{V_{dd}}{\mu_{\text{typ}} C_{\text{ox}}(W/L)(V_{dd} - V_{t,\text{typ}})^2}$$

$$= \frac{1}{d} \cdot \frac{\alpha_{VDD} V_{DD}}{\alpha_T \alpha_p \mu_{\text{typ}} C_{\text{ox}}(W/L)(\alpha_{VDD} V_{DD} - \alpha_{\text{th}} V_{t,\text{typ}})^2} \qquad (6)$$

from which V_{dd} can be derived. The reduction of V_{dd}, caused by the reduced workload, is not the only effect that will influence the power reduction γ. When $d < 1$, less work is being done, leading to a linear reduction of the power dissipation based on this effect alone. The power reduction can thus be expressed as a combination of the two effects:

$$\gamma = d \cdot \frac{P(V_{dd})}{P(V_{DD})} = d \cdot \left(\frac{V_{dd}}{V_{DD}}\right)^2 \qquad (7)$$

To estimate the influence of data dependencies on power reduction, the α-coefficients in (6) are all set to one, and V_{dd} is found and inserted into (7). In Fig. 4 the power reduction γ is plotted as a function of d for two examples: $V_{DD} = 3V_t$ and

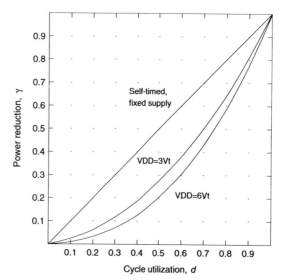

Fig. 4. Power savings due to data dependencies.

$V_{DD} = 6V_t$. For comparison the figure also shows the power reduction in a self-timed circuit with a fixed supply voltage.

It is notable that for large values of V_{DD} the execution frequency is proportional to the supply voltage (refer to (2)), and since the execution frequency scales with d, $V_{dd}(d)$ can be expressed:

$$V_{dd}(d) = d \cdot \frac{C_L}{\mu C_{\text{ox}}(W/L)} \quad \text{for } V_{DD} \gg V_t \qquad (8)$$

which is linear in d. Combining this result with (7):

$$\gamma \approx d^3 \quad \text{for } V_{DD} \gg V_t. \qquad (9)$$

For $V_{DD} = 3V_t$ in Fig. 4, $(V_{dd}/V_{DD})^2 \approx d$ and therefore:

$$\gamma \approx d^2 \quad \text{for } V_{DD} = 3V_t.$$

In summary, the power reduction in a self-timed circuit with fixed V_{DD} is proportional to d, and the power reduction in a self-timed circuit with adaptive supply scaling can range from d^2 to d^3 when $V_{DD} > 3V_t$.

E. Circuitry for Supply Scaling

Adaptive supply scaling involves two power losses: one corresponding to the circuit overhead, and another to the efficiency of the DC-DC converter. The power loss in the circuit overhead (the FIFO-buffers and state detecting circuit) can be relatively small and is ignored in the analysis. The power loss in the DC-DC converter, on the other hand, can be quite significant, depending on the type of converter being used. In the analysis a resistive approach is used as the worst case and a lossless converter as the best case. Using a resistive approach the power saving γ is reduced to (cf. (3)):

$$\gamma = \frac{V_{dd}}{V_{DD}}. \qquad (10)$$

153

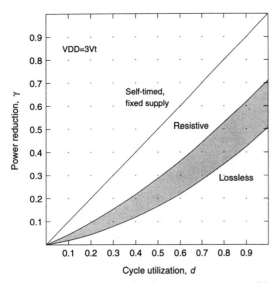

Fig. 5. The combined effects of process variations, operating conditions, and data dependencies for the typical case.

F. Summary — Combining the Effects

Fig. 5 shows the possible power savings, using both a lossless and a resistive converter, when all the effects described above are combined: 1) typical process, 2) typical operating conditions, and 3) data dependencies. A supply voltage of $V_{DD} = 3V_t$ is assumed (corresponding to $V_{DD} = 2.4$ V in a typical CMOS process). For larger values of V_{DD} the possible power saving increases, and for smaller values it decreases. As V_{DD} approaches $2V_t$ the circuit delays increase drastically, enabling only marginal power savings.

At $d = 1$ the supply voltage is reduced to $V_{dd} = 2.1V_t$ due to typical process and operating conditions. At $d = 0.35$ the supply voltage is reduced to $V_{dd} = 1.6V_t$. As a reference, Fig. 5 also shows the power savings in a self-timed circuit with fixed supply voltage.

Comparing a self-timed circuit using adaptive supply scaling with a self-timed circuit using a fixed supply, two interesting cases are:

- For a worst-case computation ($d = 1$) and $V_{DD} = 3V_t$, the power saving, using a resistive supply scaling, is a factor of 1.4. This is a lower bound on the saving.
- For a computation with data dependency ($d = 0.35$) and $V_{DD} = 3V_t$, the power saving using a lossless supply scaling is a factor of 3.6, and for $V_{DD} = 6V_t$ ($= 5$ V for $V_t = 0.83$ V) the power saving is a factor of 6.4.

This latter example corresponds to the cycle utilization factor of the DCC error corrector described in Section II-B. As this chip has a rather low cycle utilization, the factor of 6.4 may be considered as an upper bound on the possible power savings in general.

IV. REFINING THE ANALYSIS

In this section the analysis is extended to include the effects of short-circuit currents and velocity saturation. Both effects lead to additional power savings, but it should be noted that the impact on power reduction is strongly technology and design dependent.

A. Short-Circuit Dissipation

In (1) short-circuit dissipation was ignored as a contribution to dynamic power dissipation. This form of dissipation occurs during a gate-output transition, when both the n-path and the p-path conduct. The short-circuit current may be substantial (both transistor paths are in saturation), but this lasts only for the duration of the corresponding input transition. For carefully designed circuits the short-circuit dissipation is typically about 20% of the dynamic dissipation for a channel length of 1 μm and $V_{DD} = 5$ V. The short-circuit dissipation for $V_{dd} \geq 2V_t$ is given by [12]:

$$P_{\text{short}} = \frac{\beta}{12} \cdot (V_{dd} - 2V_t)^3 \cdot \frac{\tau}{T_p} \tag{11}$$

where β is the gain factor of a MOS transistor, τ the rise or fall time, and T_p the clock period. Both τ and T_p scale with the supply voltage and therefore:

$$P_{\text{short}} \sim V_{dd}^3 \quad \text{for } V_{dd} \gg 2V_t. \tag{12}$$

Hence, the power reduction due to down scaling of the supply voltage is even more attractive than implied by (3). For $V_{dd} < 2V_t$ short-circuit dissipation is negligible.

B. Velocity Saturation

Velocity saturation [13] is a phenomenon that is becoming more and more significant as technology is being scaled down. Due to velocity saturation the performance of CMOS circuits grows less than linearly with $V_{dd} - V_t$, as suggested by (2). For some technologies the velocities of electrons and holes in MOS channels tend to saturate beyond an electric field $\mathcal{E}$ of 2–6 V/μm. At $V_{DD} = 5$ V this effect may reduce saturation currents (and therefore the performance) by more than a factor of two! The good side of velocity saturation is that, when V_{dd} is scaled down the corresponding performance loss will be modest. This implies that the power savings can be substantially better than estimated in the previous section.

The significant impact velocity saturation can have on circuit performance, is well illustrated by the throughput of the DCC error corrector shown in Fig. 2. The technology used for this chip has a critical electric field $\mathcal{E}_c = 1.7$ V/μm which leads to substantial performance degradation at high supply voltages. In the figure this is seen by the rapid decline of the slope of the normalized throughput versus V_{DD} graph.

Modifying (2) for velocity saturation with L being the length of transistor channels we get [13, Eqn. 5.3.10]:

$$\begin{aligned} T_d' &= T_d \cdot \left(\frac{V_{dd}}{L\mathcal{E}_c} + 1 \right) \\ &= \frac{C_L \cdot V_{DD}}{\mu C_{ox}(W/L)(V_{DD} - V_t)^2} \cdot \left(\frac{V_{dd}}{L\mathcal{E}_c} + 1 \right). \end{aligned} \tag{13}$$

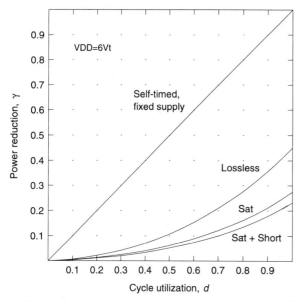

Fig. 6. Power reduction due to typical process and operating conditions when considering the effect of velocity saturation, labeled "Sat", and the combined effect of velocity saturation and short-circuit dissipation, labeled "Sat + Short". In all cases a lossless supply scaling is assumed.

C. An Example

As mentioned before the effects of short-circuit currents and velocity saturation depends very much on the technology and the particular design. To provide a quantitative analysis of the power savings we assume a technology with $V_{DD} = 6V_t$, $L\mathcal{E}_c = 2V_t$, and a design where short-circuit dissipation equals to 20% of the dynamic power dissipation at $V_{DD} = 6V_t$.

Using these figures and (13), the power savings due to the presence of velocity saturation may be estimated. The result is shown by the graph labeled "Sat" in Fig. 6. For comparison the figure also shows the power reduction in a self-timed circuit using a fixed supply voltage, and the estimated power reduction using adaptive supply scaling with a lossless DC-DC converter, when typical process, typical operating conditions, and data dependencies are considered (similar graphs are found in Fig. 5 for $V_{DD} = 3V_t$). The figure shows a substantial power reduction. At worst case computation ($d = 1$) the velocity saturation leads to additional power savings of 40%.

The effect of short-circuit dissipation can be estimated using (11). Combining this with the effect of velocity saturation the graph labeled "Sat + Short" in Fig. 6 is obtained. As V_{dd} is scaled down the short-circuit current is reduced, but even for worst case computations ($d = 1$) the scaling of the supply voltage is significant enough to make short-circuit dissipation negligible ($P_{\text{short}}(V_{dd})/P_{\text{short}}(V_{DD}) = 0.02$). It is now possible to reduce the power dissipation by a factor of 13 at $d = 0.35$.

V. THE DEMONSTRATOR CHIP

A test chip has been fabricated via Eurochip in a 1.0 μm CMOS process. The chip contains a system for adaptive supply scaling, identical to the self-timed circuit in Fig. 1. It is noted that the supply scaling is performed off-chip, thereby allowing for experimentation with different circuit configurations.

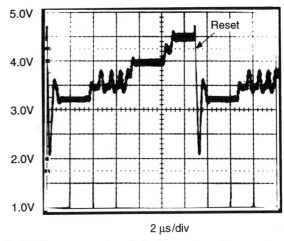

Fig. 7. Oscilloscope snap shot of adaptive supply scaling.

The input FIFO is a 10 word deep buffer implemented using the latch type in [4]. From this, 9 state bits are generated in the state detecting circuit and fed to the external supply scaling circuit. The off-chip scaling circuit, used in the test setup, is a simple D/A converter, which scales the supply voltage linearly depending on the number of data words in the input FIFO.

The main circuit is a delay-insensitive circuit that implements a 16-bit dual-rail ripple-carry adder. In the adder circuit two Cascode Voltage Switch Logic (CVSL) function blocks [14] are used, one for the *sum* function and one for the *carry*. The scheme of indication in the ripple adder is identical to the one used by Martin in [15], which utilizes the carry-kill and carry-generate properties of the full-adder. Using this approach, the delay of the addition is data dependent, ranging from 1 to 16 times the delay of one full-adder. The circuit is therefore suitable for validation of the power estimations, based on data dependencies, found in Section III.

Fig. 7 shows an oscilloscope picture obtained by cyclically applying input data composed of sequences of operands that cause the carry to ripple 4, 8, 12, and 16 positions, respectively. Each cycle is initiated by a reset and the operands are input at a 16 MHz rate.

The figure shows that the first, third, and fourth sequence lead to stable supply voltages, whereas the second sequence (where the carry ripples 8 positions) requires a supply voltage different from the discrete voltages available by the supply scaling in this configuration. Therefore the supply voltage fluctuates between two adjacent supply voltage levels.

VI. CONCLUSION

In this paper we have described a technique that can increase the power savings for self-timed circuits as much as an order of magnitude or more. The technique is called "adaptive supply scaling" or "just-in-time processing" and is particularly useful in systems that implement sequential algorithms with a data dependent computation time.

The fabricated test chip nicely demonstrates the feasibility of adaptive supply scaling, but is clearly not a practical application (the relative cost of the overhead circuitry, is prohibitive in this case). For "adaptive supply scaling" or "just-in-time processing" to pay off, larger subsystems must be considered. An example is the DCC error corrector discussed above. On the other hand the subsystem can also be too big. For example, it is sometimes the case that substantial but *local* variations in workload only lead to minor variations in workload at the subsystem level. The granularity is therefore a topic that requires further research. Other open questions are: 1) Are circuits that operate on multiple adaptively scaled supply voltages stable? If not in general, under which conditions? 2) What are minimal and optimal buffer sizes, given the time constants involved? 3) Is latch-up a problem after a sudden drop in V_{dd}?

ACKNOWLEDGMENT

The ESPRIT Basic Research Working Group 7225 (ACiD-WG) has provided a forum for exchange of ideas and has helped foster this joint paper.

REFERENCES

[1] A. P. Chandrakasan, S. Cheng, and R. W. Brodersen, "Low-power CMOS digital design," *IEEE J. Solid-State Circ.*, vol. 27, no. 4, pp. 473–483, April 1992.
[2] C. H. van Berkel, R. Burgess, J. Kessels, A. Peeters, M. Roncken, and F. Schalij, "Asynchronous circuits for low power: a DCC error corrector," *IEEE Design & Test*, vol. 11, no. 2, pp. 22–32, 1994.
[3] I. E. Sutherland, "Micropipelines," *Communications ACM*, vol. 32, no. 6, pp. 720–738, June 1989.
[4] J. Sparsø and J. Staunstrup, "Delay-insensitive multi-ring structures," *Integration, the VLSI J.*, vol. 15, no. 3, pp. 313–340, Oct. 1993.
[5] G. Birtwistle and A. Davis, Ed., *Proc. Banff VIII Workshop: Asynchronous Digital Circ. Design, Banff, Alberta, Canada, Aug. 28–Sept. 3, 1993.*, Springer Verlag, Workshops in Computing Science, 1995. Contributions from: S. Furber, "Computing without clocks: Micropipelining the ARM processor." A. Davis, "Practical asynchronous circuit design: Methods and tools." K. van Berkel, "VLSI programming of asynchronous circuits for low power." J. Ebergen, "Parallel program and asynchronous circuit design."
[6] C. H. van Berkel and C. Niessen, "An apparatus featuring a feedback signal for controlling a powering voltage for asynchronous electronic circuitry therein," European Patent Application 92203949.0, Published June 1993.
[7] L. S. Nielsen and J. Sparsø, "Low-power operation using self-timed circuits and adaptive scaling of the supply voltage," in *1994 Int. Workshop Low Power Design*, April 22–27, 1994. Unpublished proceedings.
[8] J. G. Kassakian and M. F. Schlecht, "High-frequency high-density converters for distributed power supply systems," *Proc. IEEE*, vol. 76, no. 4, pp. 362–376, 1988.
[9] F. Goodenough, "Off-Line and one-cell IC converters up efficiency," *Electronic Design*, pp. 55–64, June 27, 1994.
[10] P, Macken, M. Degrauwe, M. van Paemel, and H. Oguey, "A voltage reduction technique for digital systems," in *ISSCC 1990 Dig. Tech. Papers*, pp. 238–239, 1990.
[11] L. A. Glasser and D. W. Dobberpuhl, *The Design and Analysis of VLSI Circuits.* Reading, MA: Addison-Wesley, 1985.
[12] H. J. M. Veendrick, "Short-circuit dissipation of static CMOS Circuitry and Its Impact on the Design of Buffer Circuits," *IEEE J. Solid-State Circ.*, vol. SC-19, no. 4, pp. 468–473, Aug. 1984.
[13] Y. P. Tsividis, *Operation and Modeling of the MOS Transistor.* New York: McGraw-Hill Int., 1988.
[14] L. G. Heller and W. R. Griffin, "Cascode voltage switch logic: A differential CMOS logic family," *IEEE ISSCC Dig. Tech. Papers*, Feb. 1984.
[15] A. J. Martin, "Asynchronous datapaths and the design of an asynchronous adder," *Formal Methods in System Design*, vol. 1, no. 1, pp. 119–137, July 1992.

A Low Power Switching Power Supply for Self-Clocked Systems

GU-YEON WEI AND MARK HOROWITZ

COMPUTER SYSTEMS LABORATORY
STANFORD UNIVERSITY
STANFORD, CA 94305

Abstract—This paper presents a digital power supply controller for variable frequency and voltage circuits. By using a ring-oscillator as a method of dynamically predicting circuit performance, the regulated voltage is set to the minimum required to operate at a reference frequency which maximizes energy efficiency. Our initial test silicon, implemented with a fixed-frequency controller, is analyzed and reveals that the controller's power consumption is a major limitation for such a design. We introduce a new architecture with variable frequency control, which allows the controller's supply and frequency to scale along with the load device and to achieve a conversion efficiency of greater than 90% over a dynamic range of frequency and voltage.

1. INTRODUCTION

Power consumption in clocked digital CMOS circuits is dominated by dynamic power dissipation, and is given by:

$$P_{dynamic} = \alpha\, C_{Tot} V_{dd}^2 f_{clk} \qquad [1]$$

where α is the activity factor, C_{tot} is the total switched capacitance, V_{dd} is the supply voltage, and f_{clk} is the clock frequency. While α and C_{Tot} are generally fixed for a specific application, reducing supply voltage and clock frequency can result in dramatic power savings. A self-clocked system [1–3] achieves maximum savings by dynamically adjusting the clock frequency depending on the performance requirements of the system, which varies depending on workload. By also dynamically adjusting the power supply voltage to the minimum required to meet these changing performance requirements (i.e., desired clock frequency), considerable power savings are possible. Figure 1 provides a graphical representation of this idea, where normalized dynamic power is plotted versus normalized frequency. Given dynamic power's quadratic dependence on voltage, reducing the voltage along with the frequency allows for much more efficient power consumption in comparison to a fixed supply voltage. Building such a system requires two components different from a conventional fixed-voltage synchronous system: a method of dynamically predicting the chip's

performance at different operating voltages, and an adaptive power supply regulator.

A number of researchers have proposed using a delay-chain/ring-oscillator to model the critical path of a circuit and also using the frequency of the ring as an indicator of the chip's performance [2,3]. The chip and the oscillator are built on the same die so that they closely track over process, temperature, and voltage. This paper explores the design of the other component that is needed in these self-clocked systems—an adaptive supply regulator.

While many papers have been written on building efficient power converters [4,5] to reduce power consumption, power supply regulators for self-clocked systems provide interesting new opportunities and constraints. The largest difference from conventional regulator design is that the inputs to this controller arrive as two frequencies in digital form. This makes a digital control loop appealing, since no A/D conversion is required. Furthermore, if the digital controller can use the same frequency and voltage as the load circuit, its power will scale with the load's, making the converter efficient over a wide operating range. This range is needed since the digital load's power is likely to vary with the supply voltage to the third or higher power. Having the controller's power consumption scale is critical for maintaining high conversion efficiency, especially under low-power conditions. This paper describes the issues and trade-offs with this type of adaptive supply regulator and shows how an efficient converter can be built.

The following section starts by reviewing the issues in designing a switching power supply, and then describes the overall architecture, including the conflicting constraints that the designer must try to satisfy. The section discusses both the difficulties with using an adaptive supply and the limitations of the fixed frequency architecture which we initially built. Analysis of the measured data has led us to develop an improved architecture, which is described in a following section. In the last section, we reevaluate the use of a digital controller and conclude that this design approach holds promise for self-clocked systems.

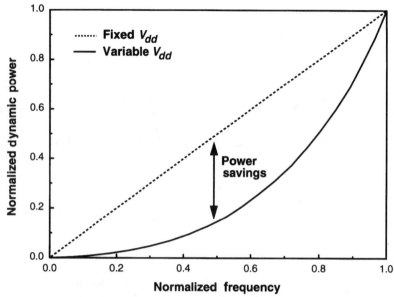

Figure 1 Normalized dynamic power vs. frequency.

2. ARCHITECTURE

An adaptive power supply regulator consists of three main building blocks: a power converter, a mechanism for predicting circuit performance, and a controller. A synchronous buck converter circuit [4], shown in Fig. 2, is employed for efficient power conversion. The output of this type of converter is simply the average voltage of a pulse-width modulated (PWM) square wave input supplying power to the load. Since the averaging is done by ideally lossless components (the L and C) acting as a low-pass filter, the power delivered to the load is close to the power that is pulled from the external supply (BV_{dd}). Losses in this type of converter come from many sources, including resistive losses in the transistor switches, capacitive losses due to the charging of the transistor gates and parasitic capacitances, short circuit current through the transistor switches, and parasitic losses of the filter's inductor and capacitor. Additional power is consumed by the buffers required to drive the CMOS power transistor switches. For low power conversion, the power needed to operate the controller is also significant.

In order to regulate the output voltage, the duty cycle of the square wave to the buck converter is set by a feedback control loop (frequency-locked loop), which tracks CV_{dd} with respect to a reference. A difficulty associated with designing this controller arises from the LC resonance of the buck converter. From an open loop frequency-response analysis, a resonant peak at the cut-off frequency of the LC filter is observed. A sharp peak, quantified by the quality factor (Q), is desired for efficient power conversion. With a simple integral control, this resonant peak must be kept below unity gain in the open loop frequency response to ensure stability. As a result, loop bandwidth must be set very low. Thus, a more complex proportional, integral, and derivative (PID) control (Fig. 3) is required for stability and improved transient response to load changes. The compensation zeros introduced by the proportional and derivative control allow the unity gain point to be set beyond the resonant peak, improves the transient response, and eliminates the limitations due to the Q of the resonant circuit.

Compared to an analog controller, a digital implementation imposes additional constraints on the design. Since the digital

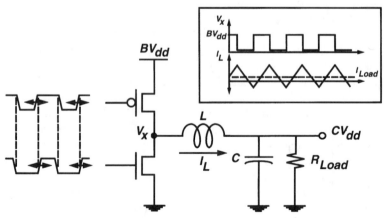

Figure 2 Buck converter.

158

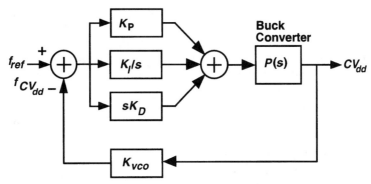

Figure 3 Control loop frequency domain model.

control is a sampled system, negative phase shift due to the delay through the loop adversely affects stability. This phase shift is linearly proportional to the bandwidth and inversely proportional to the sampling rate (which is equivalent to the switching frequency). While a higher switching frequency is desirable for smaller external filter components, it is limited by voltage resolution requirements. As a consequence, bandwidth must be traded for stability. For low-power systems, however, a slower switching frequency is advantageous for reducing the converter's power consumption.

Another design constraint is power efficiency at low loads. A simple buck converter consumes a minimum energy each cycle. This is set by the CV^2 energy needed to switch the transistors, and the I^2R losses to support the current supplied (Fig. 2). Current through the inductor is the sum of the load current and a ripple current (the magnitude of the ripple depends on the L and C parameters and the switching frequency). At low load currents, the I^2R losses from the ripple current through the inductor and transistors can limit efficiency. Thus, for high efficiency under low-load conditions, the converter must both reduce its switching frequency (to reduce the CV^2 energy) and change from a continuous to a discontinuous mode of operation; this eliminates losses due to a circulating current in the LC tank of the buck converter. Our initial approach for the control loop was a digital implementation with variable frequency operation, support for low-power operation, and one that used the adaptive regulated supply as the supply voltage for the controller itself. The original architecture is shown in Fig. 4. The ring-oscillator frequency is used to 'run' the converter. It sets the switching fre-

quency of the buck converter, and it is also used as a time-base for the rest of the system. This ring-oscillator, which also tracks the chip's performance, is counted over a switching period and compared with the reference frequency, also counted over the same period. A binary equivalent of the error between the two frequencies then feeds into the PID control block to make the appropriate corrections at the output. Thus, the output of the buck converter is regulated to the appropriate voltage for the load, modeled by the ring oscillator, to operate at the reference frequency. The digital implementation of the PID blocks uses shifters, adders, and registers to mimic equivalent analog blocks. Since this digital control loop was planned to operate at a variable frequency, the power consumption of the controller could be maintained at a constant fraction of the power delivered if it is also run off the regulated supply.

Taking advantage of the relative ease with which non-linear methods can be implemented in digital controllers, a pulse-squashing technique is used to further reduce power consumption of the converter. Given very low load conditions, ripple current through the inductor can cause the overall current to switch polarity so that I^2R power is unnecessarily wasted through the power transistors. In order to minimize this power dissipation, when the controller detects current circulating back from the capacitor, subsequent pulses to the power transistors are squashed until the regulated voltage falls below a specified threshold. Current polarity through the power transistors is detected at the end of each switching period by momentarily turning off both NMOS and PMOS transistors and sensing the drain voltage. In addition to eliminating power dissipated by the cir-

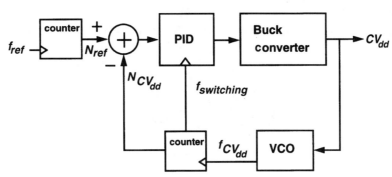

Figure 4 Original architecture block diagram.

culating current, the buck converter's switching frequency effectively decreases, reducing the power consumed by switching the power transistors.

Although a variable operating frequency for the digital control loop is ideal for minimizing its power consumption, it is not ideal for loop stability. The PID time constants are proportional to the sampling frequency while the complex poles of the LC filter are fixed, which makes the loop parameters vary with operating frequency. Given the additional constraint of the maximum ripple tolerable at the regulated supply, which is a function of the filter components' sizes and operating frequency, a stable loop over the dynamic range of the variable frequency could not be achieved. In order to verify the feasibility of a digital control loop and evaluate the performance of the rest of the system, we have opted for a fixed-frequency control loop for our first test silicon. Instead of using the variable-frequency output of the ring oscillator, a fixed external frequency is used to generate the operating frequency. The trade-off is that the fixed-frequency digital control consumes a fixed overhead power and no longer tracks the power consumption of the load.

Our fabricated version of the switching power supply controller architecture consists of a PID control, operates at a fixed frequency, and utilizes a pulse-width modulation technique with a buck converter to regulate the supply voltage with respect to a desired reference frequency. A non-linear pulse-squashing method is also implemented to reduce the controller's power consumption at low loads.

3. Fixed-Frequency Controller Test Chip

A block diagram of the test chip is shown in Fig. 5. The output and reference voltages are converted to equivalent 9-bit binary values by counting the pulses out of the ring oscillators over a fixed sampling period. A fixed external clock (f_{sys}) feeds into the synchronous counter and the toggling highest order bit sets the sampling frequency. The difference between the binary representation of the output and reference voltages feeds into the PID block. Shifters are used to set the appropriate gain coefficients for each of the control blocks. To avoid quantization error, the adders are 16-bit values and accommodate for additional bits created by the shifters. A subsequent output D/A conversion is effectively implemented by comparing the binary representation of the duty cycle, calculated by the PID block, to the output of the free-running synchronous counter. The comparator generates a fixed-frequency square wave whose pulse width is modulated proportional to the varying output of the PID block.

4. Measured Results

The digital PID controller was fabricated in a MOSIS 1.2μm CMOS process and tested with off-chip power transistors and filter components. Having verified loop stability, where CV_{dd} tracks the input reference voltage, the performance of the supply was evaluated by measuring the power conversion efficiency and observing its transient response to step changes in the input reference voltage.

Figure 6 presents the measured results of the conversion efficiency with respect to different input reference voltages. Although an actual application would have a capacitive load with a variable-frequency clock, these measurements were performed with a resistive load to obtain a first-order approximation of performance. With this configuration, the fabricated architecture achieves conversion efficiency greater than 90% while delivering 850mW at 3.5V. However, the efficiency rolls off at lower levels of regulated voltage and power delivered to the load since the digital controller and buck converter consume some fixed overhead power. Figure 7 shows a decomposition of the different power consumption components as a percentage of the total power supplied to the power supply. As shown by the graph, the percentage of overhead power consumed increases as the regulated voltage (CV_{dd}) decreases. The largest component of overhead power is due to the controller. While parts of the controller operating at the slow switching frequency could be powered off the regulated voltage, components operating at the high fixed frequency (f_{sys}) and a fixed voltage (BV_{dd}) consume a fixed amount of power. As was intended in the original

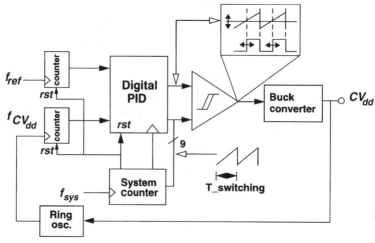

Figure 5 Digital controller block diagram.

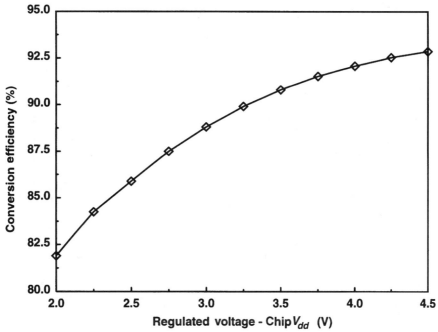

Figure 6 Measured conversion efficiency.

design, this power can be reduced with a variable frequency and voltage operation. An improved architecture, described later in this paper, has been developed to achieve this goal. The next largest component of overhead power is due to the losses of the buck converter. This unavoidable component of overhead power can be further optimized with a more careful design of the converter, given specific power goals. Lastly, power consumed by the buffers driving the power transistors is indicated. Although the power savings achieved in the discontinuous mode is not reflected in the measured data presented, power can be reduced for low loads by squashing pulses, effectively reducing the switching frequency.

The transient response of the stable control loop was obtained by using a square wave as the reference voltage and observing the regulated output voltage waveform. The non-ideal response (Fig. 8) exhibits a rapid initial response to 60% of the final value and a slow settling time. This phenomenon, known as the doublet problem, is caused by the derivative control's zero close to, but not exactly cancelling out, the dominant pole. A discrepancy between the initial and final values of the pole-zero pair allows a fast initial response corresponding approximately to the loop bandwidth, but it exhibits slow settling at the time constant of the dominant pole. This transient response limitation has been eliminated in the new design.

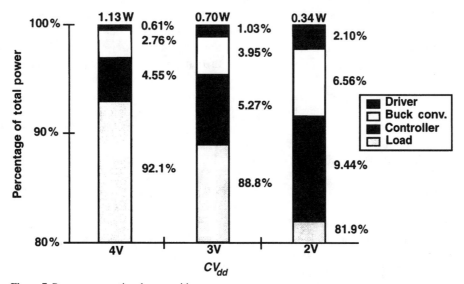

Figure 7 Power consumption decomposition.

161

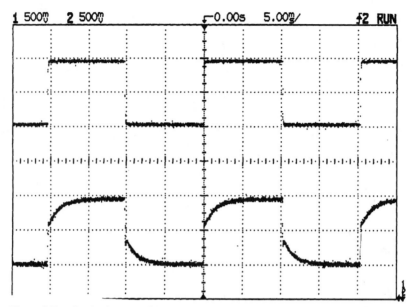

Figure 8 Regulated voltage transient response.

5. NEW ARCHITECTURE

The key to improving efficiency while maintaining loop stability is to allow the internal frequency and voltage of the controller to change while keeping the controller's loop parameters relatively constant. Since these parameters are proportional to the update rate (or switching frequency) of the controller, it must be kept relatively constant (within a factor of two). To accomplish this, a frequency-detection circuit is required. In the previous design, the switching frequency was set by the binary ramp wave, generated by a 9-bit synchronous system counter, which was clocked with an external fixed frequency, shown in Fig. 5 as f_{sys}. In our new design, the system counter is clocked by a variable-frequency clock, and the number of bits that the counter counts up to is also variable. Thus, by detecting the frequency of the clock to the counter, the bits of the counter can be shifted to maintain a switching frequency which only varies by a factor of two. For example, when the frequency is half the maximum rate, the counter will only use 8 bits, restoring the switching frequency back to the maximum rate. To further fix the loop parameters, the PID blocks' gain coefficients, implemented with shifters and adders, can be dynamically adjusted within the factor of two change in frequency, thus limiting the loop parameters' excursions with frequency. By incorporating these two methods into the controller, a stable configuration for the loop can be achieved. This is verified by the simulated open loop frequency response shown in Fig. 9. A simulated transient response of the loop to step changes in the reference is also provided in Fig. 10.

Given a variable-frequency controller powered off the regulated supply, it is possible to keep the controller's power consumption overhead at a fixed percentage of the total power consumed. Components of the controller which operate at the switching frequency can also be powered off the regulated supply as long as the timing requirements of this much slower frequency can be met at the lowest regulated voltage level. Since this controller's power is nominally proportional to V^2f, we project that the power consumed should be on the order of 1mW at 1.5V and no longer be the dominant limitation of efficiency.

The buck converter and buffers must operate at a relatively constant frequency and fixed supply voltage, which can limit efficiency. However, since the optimal sizing for the power transistors of the buck converter depends on the power being delivered to the load, on-chip segmented power transistors have been implemented. By turning off sections of the power transistor at varying levels of regulated voltage, the switched capacitance of the buck converter's power transistors and buffers adaptively varies.

Thus, power consumed by the buck converter is no longer fixed, but decreases for lower levels of regulated voltage and further improves efficiency. Furthermore, for low voltage operation, the CV^2f power of the load should be sufficiently low so that the controller enters the discontinuous pulse-squashing mode of operation and the effective switching frequency is also reduced. Therefore, even at low operating voltages, high-conversion efficiency can be achieved.

The layout of the new design has been completed and simulation results present the improvement in conversion efficiency (Fig. 11). Using IRSIM, a switch level simulator, the projected CV^2f power of the variable-frequency architecture is compared to the measured results of the fixed-frequency design. Given a variable-frequency implementation, the controller's power consumption exhibits a cubed dependence on the regulated voltage. Thus, the conversion efficiency improves dramatically (Fig. 12), with greater than 90% efficiency over the dynamic range of regulated voltage levels.

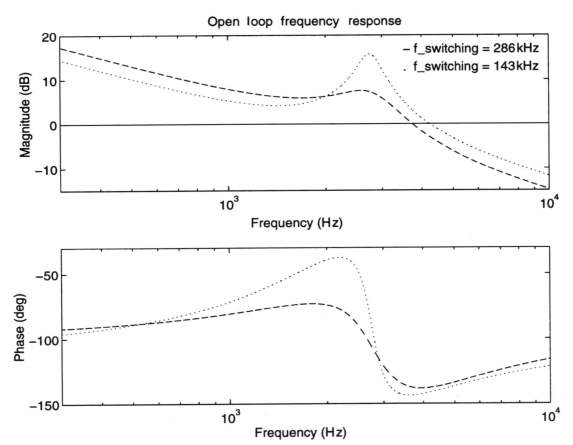

Figure 9 New architecture—simulated open-loop frequency response.

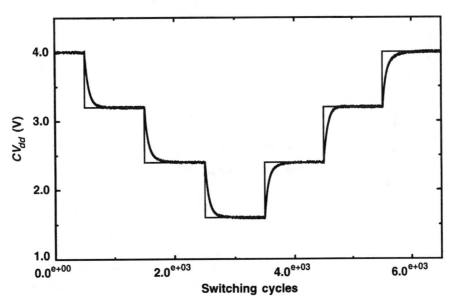

Figure 10 New Architecture—simulated closed-loop transient response.

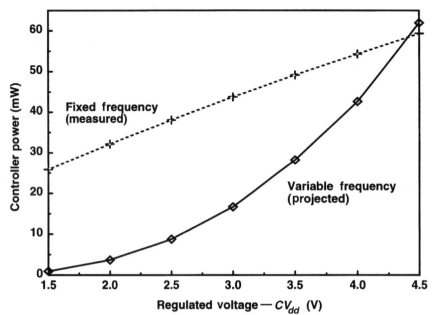

Figure 11 Controller power comparison.

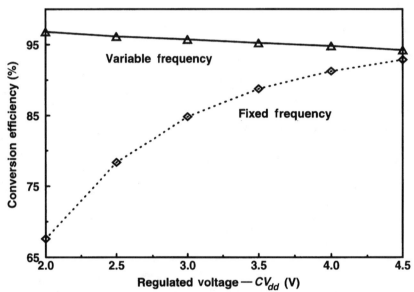

Figure 12 Projected conversion efficiency.

6. CONCLUSION

In this paper we have explored the potential of designing a digital power supply controller for self-clocked systems to improve energy efficiency. A digital controller which can be embedded within the digital system to which power is supplied has the advantage of tracking the performance and power consumption of the rest of the system. The test silicon demonstrated the feasibility of a digital implementation but showed that fixed-frequency operation limited its effectiveness due to its fixed overhead power consumption. In order to maintain high conversion efficiency over a wide range of voltage and power, a variable-frequency controller is required. By dynamically adjusting the switching frequency and gain coefficients to maintain relatively constant loop parameters, a variable frequency controller design which takes advantage of the power-saving potential of self-clocked systems can be achieved. Additional power-saving techniques implemented as segmented on-chip power transistors and pulse-squashing further improve efficiency. Simulation results of the new variable frequency design show that the controller's power dynamically tracks the power consumed by the load and conversion efficiency greater than 90% can be achieved over a wide range of voltages.

7. Acknowledgments

We would like to thank Bharadwaj Amrutur, Ken Yang, Stefanos Sidiropoulos, Dan Weinlader, and Don Ramsey for invaluable discussions and insight. This research was supported by ARPA under contract J-F81-92-194.

References

[1] L. Nielsen, C. Niessen, J. Sparso, and K. van Berkel, "Low-power operation using self-timed circuits and adaptive scaling of supply-voltage," *IEEE Trans. on VLSI Systems,* vol. 2, no. 4, Dec. 1994, pp. 391–397.

[2] P. Macken, M. Degrauwe, M. van Paemel, and H. Oguey, "A voltage reduction technique for digital systems," *IEEE ISSCC 1990 Dig. Tech. Papers,* Feb. 1990, pp. 238–239.

[3] M. E. Dean, "STRiP: a self-timed RISC processor," Ph.D. Thesis, Stanford University, 1992.

[4] A. J. Stratakos, R. W. Broderson, and S. R. Sanders, "High-efficiency low-voltage DC-DC conversion for portable application," *IEEE Power Electronics Specialists Conference,* vol. 1, Jun. 1994, pp. 619–626.

[5] A. P. Chandrakasan and R. W. Brodersen, "Low power digital CMOS design," Boston, MA: Kluwer Academic Publishers, 1995.

Variable-Voltage Digital Signal Processing

VADIM GUTNIK, ANANTHA CHANDRAKASAN

THE MASSACHUSETTS INSTITUTE OF TECHNOLOGY, CAMBRIDGE, MA

Abstract—**The use of dynamically adjustable power supplies as a method to lower power dissipation in DSP is analyzed. Power can be reduced substantially without sacrificing performance in applications where the computational workload changes with time. If latency can be tolerated, buffering data and averaging processing rates can yield further power reduction. Continuous variation of the supply voltage can be approximated by very crude quantization and dithering. A chip has been fabricated and tested to verify the closed-loop functionality of a variable-voltage system. The control framework developed is applicable to generic DSP applications.**

1. INTRODUCTION

A. Motivation

Most techniques to lower power consumption of ICs assume static behavior; i.e., circuit and system parameters are chosen at design time to minimize power dissipation. In fact, in some applications adjusting the circuit *during operation* could save more power.

The number of operations performed per sample in many DSP systems can be minimized dynamically by exploiting time-varying signal statistics. Digital video is one such example—the amount of information per frame in an MPEG encoder varies dramatically. A more generic application is described in [1]: the number of taps of an FIR filter is varied based on the power of the out-of-band noise. The idea is to keep just enough taps in the FIR such that the stopband energy in the output is below a specified limit. In such applications where the computational workload per sample (henceforth simply called "workload") varies with time, power-down techniques reduce power in a linear fashion.

The power can be reduced further if a variable power supply is used in conjunction with a variable switching-speed processor. The aim is to lower supply voltage and slow the clock during reduced workload periods instead of working at a fixed speed and idling.

B. Proposed System

Self-timed systems have been suggested to take advantage of data dependencies [2]. Although such systems track actual delays well, the overhead to generate completion signals for each logic operation can be substantial, in terms of both delay and switching activity.

We propose using a synchronous design instead. A synchronous system cannot take advantage of bit-level dependencies in the computation; however, algorithmic variations in workload can still be exploited to lower power. Like the self-timed variable-voltage system, some buffering would be needed on the input and output for synchronization, but a synchronous system could be designed *without* the completion generation signals, using static CMOS or any other logic families, and thus with very low overhead. A block diagram of the system developed in this work is shown in Fig. 1. Input data is buffered in the FIFO; the buffered data can be used to estimate the minimum sufficient processing rate. A loop around the voltage regulator and the ring oscillator establishes a supply voltage at which the critical path just meets timing requirements.

2. VARIABLE SUPPLY FUNDAMENTALS

Since the goal of a variable supply system is to minimize power, it is best to start by considering what sets the power dissipation of a digital system. If both the technology and architecture are fixed, the determining factor in power dissipation is timing; at higher voltages, the circuits operate faster but use more power.

A. Expected Power Savings

A convenient formalism is that digital signal processing deals with data as discrete *samples*. Hence, the average power and computation for the processor can be related to the energy per sample and the computation performed therein.

First-order CMOS delay models [3] can be manipulated to yield the energy dissipation per sample [4]:

$$E(r) = CV_0^2 T_s f_r r \left[\frac{V_t}{V_0} + r/2 + \sqrt{r \frac{V_t}{V_0} + (r/2)^2} \right]^2, \qquad [1]$$

where C is the average switched capacitance, $V_0 = [(V_{ref} - V_t)^2]/V_{ref}$, T_s is the sample period, f_r is the clock speed at $V_{DD} = V_{ref}$, and r is the *normalized sample processing rate,* or the clock speed normalized to f_r. For comparison, conventional digital

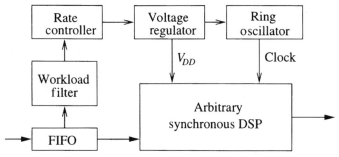

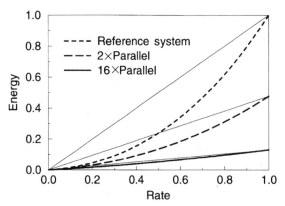

Figure 1 Synchronous, workload-dependent variable voltage system.

Figure 3 Effect of parallelization on power.

logic works at a fixed voltage and idles if the computation finishes early, so $E_{fixed}(r) = E(1)r$.

Figure 2 shows a plot of Eq. 1, for $V_{ref} = 2V$ and $V_t = 0.4V$ along with the fixed-supply line. The ratio of the two curves $[E(r)/E(1)r]$ gives the energy savings ratio per block for any given r, but the ratio changes with r. At high rates the power is the same because the voltage is the same; at very low rates the variable voltage approaches V_t so the ratio is $(V_t/V_{DD})^2$. The area between the curves is a convenient measure of the energy savings achievable by varying supply voltage.

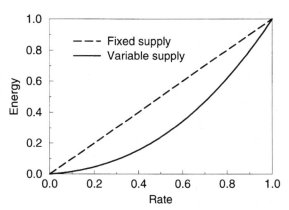

Figure 2 Energy versus Rate for variable and fixed supply systems.

B. Parallelization

The plot of energy versus rate is a convenient tool to analyze the influence of different architectures on power. In particular, the effect of parallelization is easy to show. The idea is to copy a circuit block N times and divide the system clock by N; since the clock slows down, the voltage can be lowered, and the system runs at lower power [3]. Algebraically, the parallel system has an energy–rate relation given by

$$E_{par}(r; N) = NE(r/N) + \text{Overhead}(N) \qquad [2]$$

The "overhead" consists of the extra routing and multiplexing necessary for a parallelized system. Figure 3 shows an E–r plot of Eq. 2 for $N = 1$ (the reference system) and for $N = 2$ and 16, again with the fixed-supply lines for comparison, ignoring overhead.

There are two important points to be made from the plot. First, note that at relatively low rates the curves are coincident;

if typical workload is a small fraction of the maximum workload, the variable-voltage system performs as well as the parallelized, but it has a much smaller area. The second is that the E–r curve has a smaller curvature as N increases. Equivalently, the area between the constant supply and variable supply curves is relatively smaller, so the power savings achievable by varying voltage diminish as N increases. This is exactly the same effect that limits the energy savings from further parallelization; in both cases the energy savings arise because energy per clock cycle increases with speed. As V_{DD} approaches V_t, the circuit delay increases rapidly, so slowing the clock allows only small changes in V_{DD}; if voltage doesn't change, the energy per clock cycle doesn't change with rate. In fact, due to subthreshold conduction, Eq. 1 does not model device performance well if V_{DD} approaches V_t. However, gate delays increase so rapidly near the threshold voltage that the result is nearly correct.

3. RATE CONTROL

The previous equations define the static relationship between power and processing rate, i.e., the instantaneous power as a function of rate. The dynamic behavior, or the average power as a function of sequence of rates, is equally important.

A. Averaging Rate

There is a subtle but significant distinction between the processing rate r and what will be called the *computational workload,* denoted w. The rate r is the processing speed, or more simply the system clock frequency, while w is a measure of how much processing needs to be done on the incoming blocks of data. For example, a digital video application may have a maximum clock rate of 50MHz and computation dependent on what fraction of the image changed. Half of the image changing on a certain frame corresponds to $w = 1/2$. If the clock frequency happens to be 30MHz during the computation of that frame, $r = 0.6$; as long as both r and w are normalized, the computation on that sample will finish for $r \geq w$. Even in the cases where it is not necessary, it is often advantageous to buffer the workload so that r need not follow w exactly. Why incur the overhead of a buffer when r can be set equal to w?

Motivation. Consider three sequences of processing rates, $r_1[n]$, $r_2[n]$, and $r_3[n]$, with the following probability mass functions, or *pmf*s. Note that the average rate is the same for all three distributions ($\bar{r}_1 = \bar{r}_2 = \bar{r}_3 = 0.5$), so in the long run the same number of operations is performed in all cases; yet the expected energies turn out quite different. Substitution into Eq. 1 shows that $\overline{E(r_1)} = 0.40$ while $\overline{E(r_2)} = 0.31$ and $\overline{E(r_3)} = 0.21$.

Table 1. Three Sequences of Processing Rates.

R	$Pr(r_1 = R)$	$Pr(r_2 = R)$	$Pr(r_3 = R)$
0.1	0.5	0.25	0
0.5	0	0.5	1
0.9	0.5	0.25	0

Convexity and Jensen's Inequality. The dependence of power on *pmf* arises from the fact that energy is a *convex* (or "concave up") function of rate. Since both the energy per block and the number of operations done per block add linearly, the point describing the average rate and dissipated power for any two samples lie along the chord connecting their respective operating points; e.g., if the rate goes from 0.1 to 0.9, the average power lies on the line connecting the points $[0.1, E(0.1)]$ and $[0.9, E(0.9)]$ as shown in Fig. 4. But since the E–r curve is convex, any such chord always lies above the curve! So, if we average the workload over two sample periods and work at a fixed rate, the power will be lower. In the example above, the sequence r_2 corresponds to r_1 averaged two blocks at a time, and indeed the power dissipation is lower.

Of course, this process can be continued by averaging workload over longer periods and lowering the energy further. In the limit, if we could average workload over all the samples first and then process them at a fixed rate, the average power would be at the minimum; in fact, $E(r_3)$ was much lower than $E(r_1)$. This result is summarized in *Jensen's inequality* [5]:

$$\text{for convex } E(\cdot): \qquad \overline{E(r)} \geq E(\bar{r}) \qquad [3]$$

In short, averaging the rate lowers power. It is this result that makes filtering the rate important for variable-supply systems; for comparison, note that the E–r curve for fixed-supply systems is straight, so averaging the rate does not save power.

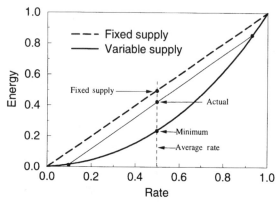

Figure 4 Averaging example.

B. Constraints

As argued above, the lowest possible power is achieved by averaging the rate over all samples. In practice, latency and buffer-size constraints limit the number of samples over which the computation can be averaged. The buffer over- and under-flow constraints on the rate $r[n]$ given a workload $w[n]$ are

$$\forall n_0: \qquad \sum_{n=-\infty}^{n_0} w[n] \geq \sum_{n=-\infty}^{n_0} r[n] \geq \sum_{n=-\infty}^{n_0-B} w[n] \qquad [4]$$

As described by the constraints in Eq. 4, the optimum $r[n]$ depends on the *pmf* of the workload. Even for simple distributions the optimization rapidly becomes unmanageable. Furthermore, even when a functional form exists, the calculation of $r[n]$ from a $w[n]$ sequence could be prohibitive in terms of area and power. Fortunately, approximations exist.

FIR Filter of the Workload. When $w[n]$ is known *a priori*, a moving average of the workload is a simple, good approximation to $r[n]$. It can be shown that any weighted average satisfies the required constraints for any $w[n]$:

$$r[n] = \sum_{k=0}^{B-1} a_k w[n-k] \qquad [5]$$

This average can be written as an FIR filter with constraints as shown below.

$$0 \leq a_k \leq 1 \qquad \sum a_k = 1 \qquad [6]$$

The filter is causal by construction. It is also conveniently time-invariant and linear. Non-negativity of $r[n]$ is guaranteed by choosing the a_k as prescribed in Eq. 6. That this satisfies the over- and under-flow constraints is proven in [4]. Intuitively, the buffers do not overflow because the system is linear and any single data sample is processed correctly. That is, if only one data sample required computation and all the others were zero, by the Bth sample time, exactly enough operations would have been completed to finish processing that sample. Since the $r[n]$ superpose, the system satisfies the processing constraints for all the data samples.

C. Update Rate

It is clear that data buffers allow r to be somewhat decoupled from w; as shown above, the rate can be averaged over several samples to lower power. It is a small step to push the buffering idea further: if the power (and hence the rate) are averaged over B samples, shouldn't we only have to update r at $1/B$ of the sample frequency? As will be shown later, slowing the update rate can save power in the power supply. Unfortunately, this can have unwelcome repercussions in the power dissipated in the DSP. Two examples are analyzed below: a subsampled FIR controller and quasi-Poisson queue.

Subsampled FIR. Figure 5 compares the processing rates on a sample $w[n]$ sequence achieved by two $B = 2$ controllers: system A updates on every sample, while system B updates on every other sample. Label the workload of the data sitting in the ith position of the input buffer w_i. By setting the rate equal to the average

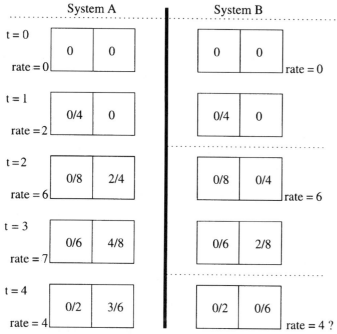

Figure 5 Rate averaging example.

workload in the buffer $r = (w_1 + w_2)/2$ system A never idles, and at the same time guarantees that there is always room in the buffer for the next data sample when it arrives. This is a specialization of the "weighted average FIR" controller analyzed above.

The added constraint on system B appears when $w_1 > w_2$, which happens in the example at $t = 4$. If the system works at the average rate, the first element will not be out of the queue before the next data sample arrives. In fact, to guarantee that the queue does not overflow the rate controller has to work at $r = \max(w_1, (w_1 + w_2)/2)$. In order to work at the average rate, system B would need a buffer of length 3. This extra buffering is needed by any system that slows down the update rate; in order to update at $1/C$ of the sample arrival time, the buffer length must be increased to $B_s \geq B + C - 1$.

Quasi-Poisson Queues. Poisson queues, or queues with exponentially-distributed service and arrival times, may be used to model a variety of physical processes, including some signal processing applications. For example, a packet-switched network may be modeled as a Poisson queue; the packets arrive independently, and are decoded and routed faster if they have fewer errors. We may be interested in finding the processing rate at which the probability P_L of losing packets due to buffer overflow is less that same constant.

Constant Processing Rate. The probability of overflow can be derived easily by treating this as a Markov process [6]. If λ is the arrival rate and μ the service rate, P_L is

$$P_L = \frac{\left(\frac{\lambda}{\mu}\right)^{n-1}}{\sum_{k=0}^{n-1} \left(\frac{\lambda}{\mu}\right)^k} \qquad [7]$$

The higher the processing rate, the lower the probability of losing a packet; for a buffer of length 4 and $P_L = 0.1\%$, $\mu/\lambda \geq 9.6$.

Variable Processing Rate. The high μ/λ ratio means that the packets must be serviced at a rate much higher than the average arrival rate, so the processor is idle nearly 90% of the time. This seems like an excellent application for a variable supply system: process quickly only when the queue starts to fill up.

Again solving the state equation we find that the steady-state probabilities are

$$p \propto [\mu_2\mu_3\mu_4 \quad \lambda\mu_3\mu_4 \quad \lambda^2\mu_4 \quad \lambda^3]^T \qquad [8]$$

The expected power[1] is given by $[0, \mu_2^2, \mu_3^2, \mu_4^2] \cdot p$. Optimizing the μ_i for minimum power (with the same 0.1% overflow constraint) gives

$$\mu_2 \approx 3 \quad \mu_3 \approx 7.5 \quad \mu_4 \approx 32 \quad \text{and} \quad \overline{E} \approx 5 \qquad [9]$$

Sampled Queue. The system presented above requires the rate to change at arbitrary times—any time a sample arrives or is serviced, the rate changes. A better model for a realizable queue would have the rate changing at discrete times.

Evolution of the state of the queue from one sample to the next is derived in [4]; unfortunately, the optimization is non-algebraic, so an approximate solution was obtained numerically. The minimum power solution for the length 4, 0.1% overflow case is plotted in Fig. 6 for a range of sample times.

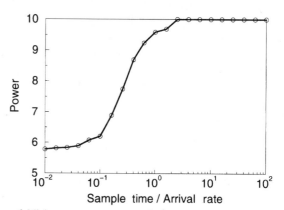

Figure 6 Minimum power versus sample time.

For update times much faster than the arrival rate, the minimum power solution approaches the unconstrained minimum of Eq. 9. For times much slower than the arrival rate, the system cannot vary the rate as fast as the queue fills up, so the minimum is simply the fixed-rate minimum. The key observation is that for supply voltage variation to be effective, the voltage update time must be comparable to the workload variation time. In other words, the bandwidth of the power supply must exceed the rate at which data arrive.

4. VARIABLE SUPPLY AND CLOCK GENERATION

In variable supply systems, power savings are achieved by lowering the supply voltage as the system clock slows down; indeed, this is the *only* reason such a system saves power. In the

---

[1]The energy terms should be $E[\mu_i]$, where $E(\cdot)$ is defined by Eq. 1 rather than by μ_i^2, but the optimization is much simpler with the simpler expression, and the results are essentially the same.

context of this paper, the term "power supply" is used to mean the power converter that draws current from some battery or rectified source, filters and regulates the voltage level, and outputs a supply voltage. The words "static" and "dynamic" will be used to describe the voltage level that is generated, *not* the internal operation of the converter; thus a switching converter that produces 5V will be considered static.

A. Specifications

In the simplest static supply systems, the power supply is trivial—the battery voltage is used to power the chip directly. However, in most low power systems some regulation is required, and especially in the case of switching converters, low-pass filtering is needed as well. The output voltage of dynamic converters also needs to be filtered and regulated, but the performance criteria are somewhat altered from those for a conventional, static power supply.

First, the desired transient response is markedly different. Since the ideal static supply has no variations on the output, the low-pass filter cut-off frequency is designed to be as low as volume and cost constraints allow. A dynamic supply still needs a low-pass filter to attenuate ripple, but also needs fast-step response to allow rate changes as previously described.

The second difference relates to the DC voltage level. A system with a static supply is typically designed to meet timing constraints at a specific voltage. At a lower level of abstraction, this means that feedback is established around the power converter to fix the output voltage as shown in the left half of Fig. 7.

A more efficient approach for fixed-rate systems was presented in [7,8], where the feedback around the entire systems establishes a fixed circuit delay rather than a fixed voltage, as shown in the right half of Fig. 7.

Several systems have been designed to compensate for process and temperature variations [7–9]. In each case the circuit delays are measured indirectly by means of a ring oscillator with a period matched to circuit delays. The ring oscillator is used as the VCO of a phase- or frequency-locked loop; when the loop settles the supply voltage for the oscillator it is at the exact point where circuit delays equal the clock period, so it can be used as a reference for the supply voltage for the chip.

The power and area overhead is small; a single loop may be sufficient for the entire chip. Stability concerns limit the speed

of the loop, but since temperature changes slowly, the bandwidth is adequate.

The main advantage of the fixed-throughput approach is that the safety margin delay, the extra time allocated to make sure the chip meets timing constraints, needs only to account for small intra-die process variations since the inter-die delay variations are measured and compensated.

Active Damping. To prevent ringing on the output of the switching converter, the *LC* filter should be damped. In most cases, the series resistance of the switches or the parallel resistance of the load presents enough damping. If that is not the case, extra damping needs to be added to limit oscillations.

A simple approach that works well for fixed supplies is to add a resistor R_p in parallel with the load; to avoid the DC dissipation, a large capacitor can be added in series with R_p, as shown in Fig. 8.

This turns out not to be an efficient method for variable supplies because the parallel resistor dissipates energy every time the voltage changes. It is possible to avoid this by *actively damping* the loop; i.e., introducing feedback into the system to emulate a resistor. The transfer function from the input to the output with a forward gain a and feedback gain f is $A = a/(1 + af)$. The forward path is a second-order filter, so a has the form $1/(s^2 + 2\alpha s + \omega_0^2)$ where α is a damping term and $\omega_0 = \sqrt{1/(LC)}$ is the resonance frequency. Hence, the total response is

$$a = \frac{1/(s^2 + 2\alpha s + \omega_0^2)}{1 + f/(s^2 + 2\alpha s + \omega_0^2)}$$

$$= \frac{1}{s^2 + 2\alpha s + \omega_0^2 + f} \tag{10}$$

To increase the damping, the feedback needs to be of the form βs for some constant β; the current through a capacitor is of this form. A capacitor cannot be used directly because the feedback signal must interface to the digital part of the signal, so an A/D converter is needed. Figure 9 shows the block diagram of an actively-damped supply. The current through the capacitor is digitized and passed back to the pulse-width modulator. The power cost of this scheme is in the A/D, but very low resolution is needed so the A/D can be small and low-power.

Simulated waveforms are shown in Fig. 10. Without any damping, the voltage would oscillate at each step; with active damping, the circuit behaves just as it would if it were passively damped, except that the power dissipation can be lower.

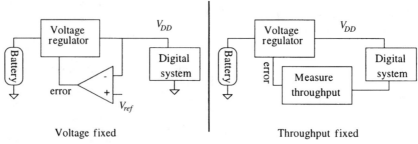

Figure 7 Feedback loops around the power supply versus the system.

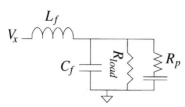

Figure 8 Passive damping.

B. Loop Stability and Step Response

Open Loop. Two methods to establish a voltage appropriate for a given rate have been mentioned previously. The first is an extension of the static supply approach; for every desired rate an adequate duty cycle level would be read out of a ROM and passed to the voltage regulator, as shown in Fig. 11.

This data is passed to pulse-width modulator. When the desired rate increases, the voltage goes up first to ensure that timing constraints will be met, and then the clock speeds up; when rate falls, the clock slows down and then the voltage drops. This has the advantage that changes in rate yield the fastest possible transitions on the output voltage, but as mentioned above, process and temperature variations would force the voltage to be high enough that delays meet worst-case, rather than actual, timing constraints.

Closed Loop: PLL. The phase-locked loop approach avoids this problem. Rather than having a separate clock and power supply, both the clock and power converter are part of a feedback loop, and the system clock is based on chip delays. A system inspired directly by [7] is shown in Fig. 12.

Just as in the case of the fixed supply system, the PLL adjusts supply voltage to the lowest possible level compatible with the required number of gate delays between registers. When the desired rate changes, the reference clock that the PLL sees changes and the loop re-locks. The drawback is in the time constant of the changes.

The time constant is determined by the bandwidth of the power supply, the loop filter, and the extra pole introduced by the integration of frequency to phase. With no loop filter, the feedback pole at DC combines with the second-order pole from the power supply to give a peaked response loop gain. A loop filter with a bandwidth significantly less than that of the power-

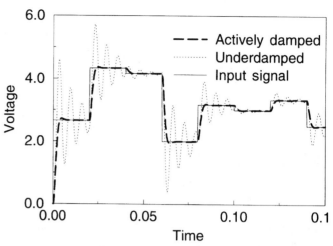

Figure 10 Active damping waveforms.

supply output filters can be added, but then this bandwidth limits how fast the loop locks.

A similar approach is presented in [8], showing that by incorporating a PID controller it is possible to circumvent some of the problems with the straightforward PLL. However, the resulting controller is much more complicated and dissipates significantly more power than the one proposed here.

Proposed Hybrid Approach. The two previous schemes, the PLL and the lookup table, adjust the output voltage at different rates because they were originally intended for different functions. The PLL does best in tracking process variations where its low bandwidth is sufficient, and the lookup table is well suited to making fast voltage steps to a predefined level. In fact, the characteristics are not mutually exclusive; it is possible to merge the lookup-table approach with the phase-locked loop to get fast voltage steps and process tracking at the same time.

The lookup table should still be used to get the fastest possible voltage changes; however, if the voltage levels are stored in RAM instead of ROM, they can be updated to track process and temperature. A schematic is shown in Fig. 13.

The rate comparison and updates can be done very slowly compared to the bandwidth of the power supply. For example, if a buck converter switches at 1MHz, the output filters can have

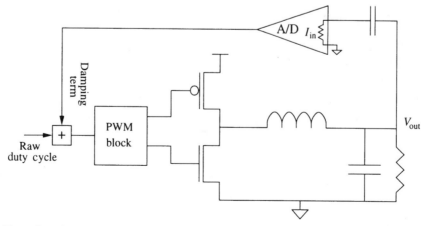

Figure 9 Active damping.

171

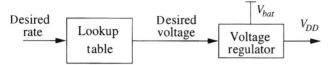

Figure 11 Open-loop variable supply system.

a bandwidth of ≈100kHz. For comparison, temperature compensation can be done at frequencies below 1kHz. Thus, the dynamics of the power supply are insignificant in the feedback loop so there are no instabilities.

Quantization and Dithering. Both the open-loop and the hybrid implementation have a look-up table to translate from rate to voltage. Since the overhead of the controller scales with the number of entries in the lookup table, a smaller table is preferable to a larger one. Figure 14 shows the $E–r$ curve for four-level voltage quantization. The lowest curve is the theoretical minimum $E–r$ as predicted by Eq. 1; the area between it and the fixed-supply line is a measure of the power savings achievable by varying power.

If each sample must be processed at a fixed voltage and in one sample period (i.e., without buffering), the rate must be the next highest available rate. So, if the available rates are 0.25, 0.5, 0.75 and 1 and the sample workload is 0.6, the controller would have to choose the 0.75 rate and idle for part of the cycle. This gives the "stair-step" curve in Fig. 14.

If the voltage can be changed during the processing of one sample, the voltage can be dithered. In the example above, by processing for 40% a sample time at rate of 0.75 and 60% at 0.5, the average rate can be adjusted to the 0.6 that is needed. This

dithering leads to an $E–r$ curve that connects the quantized (E,r) points. As the figure shows, even a four-level lookup table is sufficient if dithering is used.

If the sample period is too short to allow dithering within the sample, the same effect can be achieved by allowing processing of one sample to extend beyond one sample period; in fact, this was the assumption made in the earlier FIR filter analysis. If one sample is processed at a rate higher than its workload because of quantization, the next will be processed at a lower rate.

5. Implementation and Testing

A. Chip Design

A chip was designed to test the stability of the feedback loop and to verify that timing constraints are met as the modified phase-locked loop changes the clock and supply voltage. Since the focus is on the variable supply voltage control, only a token amount of processing is done by the DSP subsection, but it can be reconfigured to emulate applications with long sample periods (i.e., computationally-intensive cases like video processing) as well as applications with shorter sample periods. Similarly, the rest of the circuitry is designed for flexibility rather than efficiency.

Block Diagram. The block diagram of the chip is shown in Fig. 15. At the highest level of abstraction, the test chip consists of four blocks. The FIFO and the DSP make up the datapath, while the supply controller and ring oscillator are part of the control loop that generates both the supply voltage and the clock for the circuit.

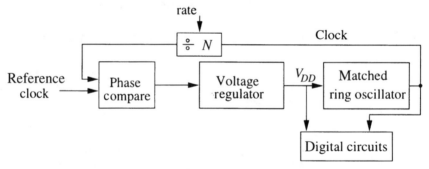

Figure 12 Block diagram of a phase-locked loop.

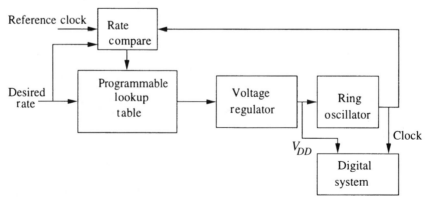

Figure 13 Hybrid system.

172

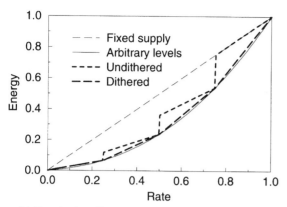

Figure 14 Quantization effect.

During operation, input data is buffered in the FIFO until it is needed by the DSP. The control loop controls the processing rate to avoid queue overflow and underflow; as the queue fills up, the clock speed increases to cope with the higher workload, and as the queue empties the clock slows. Thus, the FIFO acts as both a buffer for the data and as the workload-averaging mechanism. This control loop, which keeps the FIFO from overflowing by varying the processing rate, forms the "outer" control loop.

A second, or "inner," control loop is formed between the power supply controller and the ring oscillator. This is the quasi-PLL that adjusts the voltage steps in the controller to match processing rates. It is in this loop that temperature and processing variations are measured and corrected.

Supply Control. The supply controller has several functions: it assigns an operating rate based on workload; it translates that desired rate into a digital word proportional to the voltage level; and finally, it generates a pulse-width modulated signal that is filtered and buffered off chip and fed back as the supply voltage for the DSP and ring oscillator.

Setting the Desired Rate and Voltage. The desired rate is periodically updated based on the number of words in the queue. This is a direct implementation of the sampled queue controller described earlier. Workload is averaged implicitly by considering the queue length in the FIFO as the processing rate.

The desired processing rate is translated to a supply voltage by means of a lookup table. The desired rate forms the address,

and the 4-bit output gives the fraction of the battery voltage needed to achieve the desired rate. The current table entry is updated at the end of each sample period by comparing the number of actual clock edges to the desired number, so the voltage levels track temperature and battery voltage (as well as process variations).

The 4-bit word output from the register file is converted into a pulse width by means of a counter. A minimum DC level is established by using a 5-bit counter so that on every cycle the output pulse would be high for several clock periods plus the number of periods specified in the lookup table.

DSP. The only function of the DSP on the test chip is to emulate the throughput and timing constraints of a general variable-workload signal processor. This functionality was implemented as two separate parts.

Any algorithm that terminates on a data-dependent condition generates a variable workload. For example, an algebraic approximation iteration that terminates when the desired precision is reached, or a video decompression procedure that processes data until it finds an end-of-frame marker, both appear as variable workload algorithms. This chip uses a counter to count from zero until it matches the input data. When the count matches, the data word is considered processed, and another word is fetched from the FIFO. To verify that the right words are being fetched and executed, the counter state is an output of the chip. The block diagram of the DSP is shown in Fig. 16.

The critical path consists of strings of inverters placed between the counter and output latches. The four delay lines actually have different numbers of inverters. The difference allows calibration of the safety margin available in the clock cycle; if all delay lines shorter than a cut-off length consistently latch correctly and all longer lines do not, the clock is correctly tuned to model delays of the cut-off length. If errors are intermittent, the delay is not well matched.

B. Test Results

The chip was fabricated in a 0.8μm CMOS process. The final layout consumed 2.2.mm $\times$ 3mm; the area was determined by the 40-pad pad ring (Fig. 17).

The basic functionality of both the DSP and the control loop has been verified. Figure 18 shows measured waveforms for

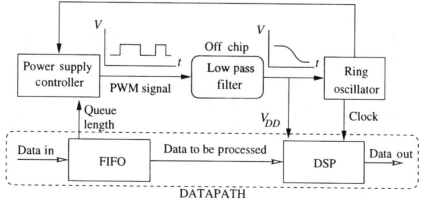

Figure 15 Dynamic supply voltage chip block diagram.

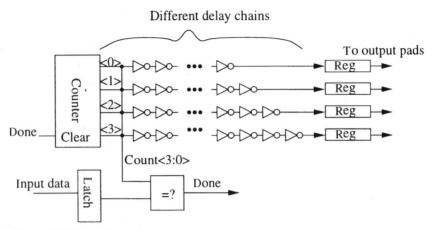

Figure 16 DSP block diagram.

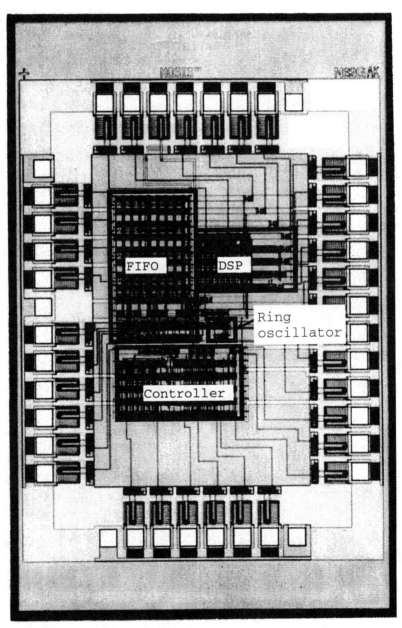

Figure 17 Variable supply controller chip plot.

174

two bits of the queue length and the supply voltage. The top two bits of the queue length are used to determine at what rate to process. As the figure shows, when the queue is full, the controller raises the voltage to a level determined by the delays; and as the queue empties, the voltage drops in discrete levels. Timing analysis (presented and analyzed in [4]) confirms that the critical path is well matched as voltage varies.

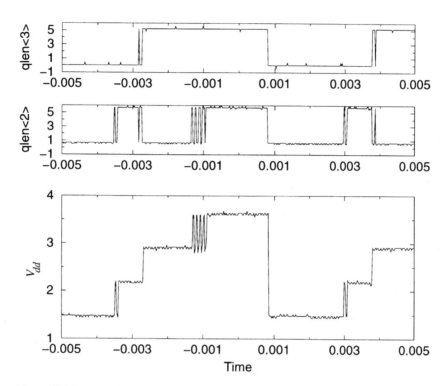

Figure 18 Measured voltage levels.

6. CONCLUSIONS

For applications where workload changes with time, the power consumed by the DSP can be lowered by varying supply voltage. Video and communications *IC*s are particularly likely to benefit from variable voltage supplies. The possibility of varying voltage dynamically may allow some fixed-workload algorithms to be replaced by variable-workload algorithms with lower mean power requirements.

Where latency and extra buffering can be tolerated, averaging workload can lower power even further. When the workload can be predicted *a priori,* explicit low-pass filtering can be done; otherwise, the data can be queued and the processing adjusted on the basis of queue length. Optimum filtering algorithms are difficult to find and dependent on the statistics of the input data, but easily computed approximations exist.

Continuous variation of the supply voltage can be approximated by very crude quantization and dithering. As long as the power supply is damped correctly, the voltage transitions do not cause extra dissipation. Since the power and area of the supply control circuitry scales with the number of voltage levels, quantization to only a few bits minimizes overhead.

Finally, the control circuitry developed is applicable to general DPS circuitry. No assumptions have been made about the DPS in the development of the control; the DSP is synchronous. All that is required for the control is that the critical timing path is known.

7. ACKNOWLEDGMENTS

V. Gutnik is supported by an NDSEG Fellowship. This research is funded by ARPA Contract #DAAL01-95-K-3526.

References

[1] J. Ludwig, H. Nawab, and A. Chandrakasan, "Low power digital filtering using approximate processing," *IEEE Journal of Solid-State Circuits,* Mar. 1996.

[2] L. S. Nielsen et al., "Low-power operation using self-timed circuits and adaptive scaling of the supply voltage," *IEEE Transactions on VLSI,* vol. 2, no. 4, Dec. 1994, pp. 391–397.

[3] A. P. Chandrakasan, S. Sheng, and R. W. Brodersen, "Low-power CMOS digital design," *IEEE J. Solid-State Circuits,* vol. 27, no. 4, Apr. 1992, pp. 473–483.

[4] V. Gutnik, "Variable supply voltage for low power DSP," Master's Thesis, Massachusetts Institute of Technology, 1996.

[5] T. M. Cover and J. A. Thomas, *Elements of Information Theory.* Wiley Series in Telecommunications. John Wiley & Sons, 1991.

[6] A. W. Drake, *Fundamentals of Applied Probability Theory.* New York, NY: McGraw-Hill Publishing Company, 1988.

[7] P. Macken et al., "A voltage reduction technique for digital systems," *Digest of Technical Papers," IEEE International Solid-State Circuits Conference,* February 1990, pp. 238–239.

[8] G. Wei and M. Horowitz, "A low power switching power supply for self-clocked systems," *International Symposium on Low Power Electronics and Design,* 1996, pp. 313–318.

[9] V. R. von Kaenel et al., "Automated adjustment of threshold and supply voltages for minimum power consumption in CMOS digital circuits," *IEEE Symposium on Low Power Electronics,* 1994, pp. 78–79.

Scheduling for Reduced CPU Energy

Mark Weiser, Brent Welch, Alan Demers, Scott Shenker

Xerox PARC
3333 Coyote Hill Road
Palo Alto, CA 94304
{weiser,welch,demers,shenker}@parc.xerox.com

Abstract

The energy usage of computer systems is becoming more important, especially for battery operated systems. Displays, disks, and cpus, in that order, use the most energy. Reducing the energy used by displays and disks has been studied elsewhere; this paper considers a new method for reducing the energy used by the cpu. We introduce a new metric for cpu energy performance, millions-of-instructions-per-joule (MIPJ). We examine a class of methods to reduce MIPJ that are characterized by dynamic control of system clock speed by the operating system scheduler. Reducing clock speed alone does not reduce MIPJ, since to do the same work the system must run longer. However, a number of methods are available for reducing energy with reduced clock-speed, such as reducing the voltage [Chandrakasan *et al* 1992][Horowitz 1993] or using reversible [Younis and Knight 1993] or adiabatic logic [Athas *et al* 1994].

What are the right scheduling algorithms for taking advantage of reduced clock-speed, especially in the presence of applications demanding ever more instructions-per-second? We consider several methods for varying the clock speed dynamically under control of the operating system, and examine the performance of these methods against workstation traces. The primary result is that by adjusting the clock speed at a fine grain, substantial CPU energy can be saved with a limited impact on performance.

1 Introduction

The energy use of a typical laptop computer is dominated by the backlight and display, and secondarily by the disk. Laptops use a number of techniques to reduce the energy consumed by disk and display, primarily by turning them off after a period of no use [Li 1994][Douglis 1994]. We expect slow but steady progress in the energy consumption of these devices. Smaller computing devices often have no disk at all, and eliminate the display backlight that consumes much of the display-related power. Power consumed by the CPU is significant; the Apple Newton designers sought to maximize MIPS per WATT [Culbert 1994]. This paper considers some methods of reducing the energy used for executing instructions. Our results go beyond the simple power-down-when-idle techniques used in today's laptops.

We consider the opportunities for dynamically varying chip speed and so energy consumption. One would like to give users the appearance of a 100MIPS cpu at peak moments, while drawing much less than 100MIPS energy when users are active but would not notice a reduction in clock rate. Knowing when to use full power and when not requires the cooperation of the operating system scheduler. We consider a number of algorithms by which the operating system scheduler could attempt to optimize system power by monitoring idle time and reducing clock speed to reduce idle time to a minimum. We simulate their performance on some traces of process scheduling and compare these results to the theoretical optimum schedules.

2 An Energy Metric for CPUS

In this paper we use as our measure of the energy performance of a computer system the MIPJ, or millions of instructions per joule. MIPS/WATTS = MIPJ. (Of course MIPS have been superseded by better metrics, such as Specmark: we are using MIPS to stand for any such workload-per-time benchmark). MIPJ is not improving that much for high-end processors. For example, a 1984 2-MIPS 68020 consumed 2.0 watts (at 12.5Mhz), for a MIPJ of 1, and a 1994 200-MIPS Alpha chip consumes 40 watts, so has a MIPJ of 5. However, more recently lower speed processors used in laptops have been optimized to run at low power. For example, the Motorola 68349 is rated at 6 MIPS and consumes 300 mW for 20 MIPJ.

Other things being equal, MIPJ is unchanged by changes in clock speed. Reducing the clock speed causes a linear reduction in energy consumption, but a similar reduction in MIPS. The two effects cancel. Similarly, turning the computer off, or reducing the clock to zero in the "idle-loop", does not effect MIPJ, since no instructions are being executed. However, a reduced clock speed creates the opportunity for quadratic energy savings; as the clock speed is reduced by n, energy per cycle can be reduced by n^2. Three methods that achieve this are voltage reduction, reversible logic, and adiabatic switching. Our simulations assume n^2 savings, although it is really only important that the energy savings be greater than the amount by which the clock rate is reduced in order to achieve an increase in MIPJ.

Voltage reduction is currently the most promising way to save energy. Already chips are being manufactured to run at 3.3 or 2.2 volts instead of the 5.0 voltage levels commonly used. The intuition behind the power savings comes from the basic energy equation that is proportional to the square of the voltage.

$$E/clock \propto V^2$$

The settling time for a gate is proportional to the voltage; the lower the voltage drop across the gate, the longer the gate takes to stabilize. To lower the voltage and still operate correctly, the cycle time must be lowered first. When raising the clock rate, the voltage must be increased first. Given that the voltage and the cycle time of a chip could be adjusted together, it should be clear now that the lower-voltage, slower-clock chip will dissipate less energy per cycle. If the voltage level can be reduced linearly as the clock rate is reduced, then the energy savings per instruction will be proportional to the square of the voltage reduction. Of course, for a real chip it may not be possible to reduce the voltage linear with the clock reduction. However, if it is possible to reduce the voltage at all by running slower, then there will be a net energy savings per cycle.

Currently manufacturers do not test and rate their chips across a smooth range of voltages. However, some data is available for chips at a set of voltage levels. For example, a Motorola CMOS 6805 microcontroller (cloned by SGS-Thomson) is rated at 6 Mhz at 5.0 Volts, 4.5 Mhz at 3.3 Volts, and 3 Mhz at 2.2 Volts. This is a close to linear relationship between voltage and clock rate.

The other important factor is the time it takes to change the voltage. The frequency for voltage regulators is on the order of 200 KHz, so we speculate that it will take a few tens of microseconds to boost the voltage on the chip.

Finally, why run slower? Suppose a task has a deadline in 100 milliseconds, but it will only take 50 milliseconds of CPU time when running at full speed

to complete. A normal system would run at full speed for 50 milliseconds, and then idle for 50 milliseconds (assuming there were no other ready tasks). During the idle time the CPU can be stopped altogether by putting it into a mode that wakes up upon an interrupt, such as from a periodic clock or from an I/O completion. Now, compare this to a system that runs the task at half speed so that it completes just before its deadline. If it can also reduce the voltage by half, then the task will consume 1/4 the energy of the normal system, even taking into account stopping the CPU during the idle time. This is because the same number of cycles are executed in both systems, but the modified system reduces energy use by reducing the operating voltage. Another way to view this is that idle time represents wasted energy, even if the CPU is stopped!

3 Approach of This Paper

This paper evaluates the fine grain control of CPU clock speed and its effect on energy use by means of trace-driven simulation. The trace data shows the context switching activity of the scheduler and the time spent in the idle loop. The goals of the simulation are to evaluate the energy savings possible by running slower (and at reduced voltage), and to measure the adverse affects of running too slow to meet the supplied demand. No simulation is perfect, however, and a true evaluation will require experiments with real hardware.

Trace data was taken from UNIX workstations over many hours of use by a variety of users. The trace data is described in Section 4 of the paper. The assumptions made by the simulations are described in Section 5. The speed adjustment algorithms are presented in Section 6. Section 7 evaluates the different algorithms on the basis of energy savings and a delay penalty function. Section 8 discusses future work, including some things we traced but did not fully utilize in our simulations. Finally, Section 9 provides our conclusions.

4 Trace Data

Trace data from the UNIX scheduler was taken from a number of workstations over periods of up to several hours during the working day. During these times the workloads included software development, documentation, e-mail, simulation, and other typical activities of engineering workstations. In addition, a few short traces were taken during specific workloads such as typing and scrolling through documents. Appendix I has a summary of the different traces we used.

Table 1: Trace Points

SCHED	Context switch away from a process
IDLE_ON	Enter the idle loop
IDLE_OFF	Leave idle loop to run a process
FORK	Create a new process
EXEC	Overlay a (new) process with another program
EXIT	Process termination
SLEEP	Wait on an event
WAKEUP	Notify a sleeping process

The trace points we took are summarized in Table 1. The idle loop events provide a view on how busy the machine is. The process information is used to classify different programs into foreground and background types. The sleep and wakeup events are used to deduce job ordering constraints.

In addition, the program counter of the call to sleep was recorded and kernel sources were examined to determine the nature of the sleep. The sleep events were classified into waits on "hard" and "soft" events. A hard event is something like a disk wait, in which a sleep is done in the kernel's biowait() routine. A soft event is something like a select that is done awaiting user input or a network request packet. The goal of this classification is to distinguish between idle time that can be eliminated by rescheduling (soft idle) and idle that is mandated by a wait on a device (hard idle).

Each trace record has a microsecond resolution time stamp. The trace buffer is periodically copied out of the kernel, compressed, and sent over the network to a central collection site. We used the trace data to measure the tracing overhead, and found it to range from 1.5% to 7% of the traced machine.

5 Assumptions of the Simulations

The basic approach of the simulations was to lengthen the runtime of individually scheduled segments of the trace in order to eliminate idle time. The trace period was divided into intervals of various lengths, and the runtime and idletime during that interval were used to make a speed adjustment decision. If there were excess cycles left over at the end of an interval because the speed was too slow, they were carried over into the next interval. This carry-over is used as a measure of the penalty from using the speed adjustment.

The ability to stretch runtime into idle periods was refined by classifying sleep events into "hard" and "soft" events. The point of the classification is to be fair about what idle time can be squeezed out of the simulated schedule by slowing down the processor. Obviously, running slower should not allow a disk request to be postponed until just before the request

completes in the trace. However, it is reasonable to slow down the response to a keystroke in an editor such that the processing of one keystroke finishes just before the next.

Our simulations did not reorder trace data events. We justify this by noting that only if the offered load is far beyond the capacity of the CPU will speed changes affect job ordering significantly. Furthermore, the CPU speed is ramped up to full speed as the offered load increases, so in times of high demand the CPU is running at the speed that matches the trace data.

In addition, we made the following assumptions:

The machine was considered to use no energy when idle, and to use energy/instruction in proportion to n^2 when running at a speed n, where n varies between 1.0 and a minimum relative speed. This is a bit optimistic because a chip will draw a small amount of power while in standby mode, and we might not get a one-to-one reduction in voltage to clock speed. However, the baseline power usages from running at full speed (reported as 1.0 in the graphs) also assume that the CPU is off during idle times.

It takes no time to switch speeds. This is also optimistic. In practice, raising the speed will require a delay to wait for the voltage to rise first, although we speculate that the delay is on the order of 10s of instructions (not 1000s).

After any 30 second period of greater than 90% idle we assumed that any laptop would have been turned off, and skipped simulating until the next 30 second period with less than 90% idle. This models the typical power saving features already present in portables. The energy savings reported below does not count these off periods.

There was assumed to be a lower bound to practical speed, either 0.2, 0.44 or 0.66, where 1.0 represents full speed. In 5V logic using voltage reduction for power savings, these correspond to 1.0 V, 2.2 V and 3.3V minimum voltage levels, respectively. The 1.0 V level is optimistic, while the 2.2 V and 3.3V levels are based on several existing low power chips. In the graphs presented in section 7, the minimum voltage of the system is indicated, meaning that the voltage can vary between 5.0 V and the minimum, and the speed will be adjusted linearly with voltage.

6 Scheduling Algorithms

We simulated three types of scheduling algorithms: unbounded-delay perfect-future (OPT), bounded-delay limited-future (FUTURE), and bounded-delay limited-past (PAST). Each of these algorithms adjust the CPU clock speed at the same time that scheduling decisions are made, with the goal of decreasing time wasted in the idle loop while retaining interactive response.

OPT takes the entire trace, and stretches all the run times to fill in the idle times. Periods when the machine was "off" (more that 90% idle over 30 seconds) were not considered available for stretching runtimes into. This is a kind of batch approach to the work seen in the trace period: as long as all that work is done in that period, any piece can take arbitrarily long. OPT power savings were almost always limited by the minimum speed, achieving the maximum possible savings over the period. This algorithm is both impractical and undesirable. It is impractical because it requires perfect future knowledge of the work to be done over the interval. It also assumes that all idle time can be filled by stretching runlengths and reordering jobs. It is undesirable because it produces large delays in runtimes of individual jobs without regard to the need for effective response to real-time events like user keystrokes or network packets.

FUTURE is like OPT, except it peers into the future only a small window, and optimizes energy over that window, while never delaying work past the window. Again, it is assumed that all idle time in the next interval can be eliminated, unless the minimum speed of the CPU is reached. We simulated windows as small as 1 millisecond, where savings are usually small, and as large as 400 seconds, where FUTURE generally approaches OPT in energy savings. FUTURE is impractical, because it uses future knowledge, but desirable, because no realtime response is ever delayed longer than the window.

By setting a window of 10 to 50 milliseconds, user interactive response will remain high. In addition, a window this size will not substantially reduce a very long idle time, one that would trigger the spin down of a disk or the blanking of a display. Those decisions are based on idle times of many seconds or a few minutes, so stretching a computation out by a few tens of milliseconds will not affect them.

PAST is a practical version of FUTURE. Instead of looking a fixed window into the future it looks a fixed window into the past, and assumes the next window will be like the previous one. The PAST speed setting algorithm is shown at the top of the next column.

There are four parts to the code. The first part computes the percent of time during the interval when the

Speed Setting Algorithm (PAST)

`run_cycles` is the number of non-idle CPU cycles in the last interval.

`idle_cycles` is the idle CPU cycles, split between hard and soft idle time.

`excess_cycles` is the cycles left over from the previous interval because we ran too slow. All these cycles are measured in time units.

```
idle_cycles = hard_idle + soft_idle;
run_cycles += excess_cycles;
run_percent = run_cycles /
    (idle_cycles + run_cycles);

next_excess = run_cycles -
    speed * (run_cycles + soft_idle)
IF excess_cycles < 0. THEN
    excess_cycles = 0.

energy = (run_cycles - excess_cycles) *
    speed * speed;

IF excess_cycles > idle_cycles THEN
    newspeed = 1.0;
ELSEIF run_percent > 0.7 THEN
    newspeed = speed + 0.2;
ELSEIF run_percent < 0.5 THEN
    newspeed = speed -
        (0.6 - run_percent);

IF newspeed > 1.0 THEN
    newspeed = 1.0;
IF newspeed < min_speed THEN
    newspeed = min_speed;

speed = newspeed;
excess_cycles = next_excess;
```

CPU was running. The `run_cycles` come from two sources, the runtime in the trace data for the interval, and the `excess_cycles` from the simulation of the previous interval.

The `excess_cycles` represents a carry over from the previous interval because the CPU speed was set too slow to accommodate all the load that was supplied during the interval. Consider:

```
next_excess = run_cycles -
    speed * (run_cycles + soft_idle)
```

The `run_cycles` is the sum of the cycles presented by the trace data and the previous value of `excess_cycles`. This initial value is reduced by the soft idle time and the number of cycles actually performed at the current speed. This calculation represents the ability to squeeze out idle time by lengthening the runtimes in the interval. Only "soft" idle, such as waiting for keyboard events, is available for elimination of idle. As the soft idle time during an interval approaches zero, the excess cycles approach:

```
run_cycles * (1 - oldspeed)
```

The energy used during the interval is computed based on an n^2 relationship between speed and power consumption per cycle. The cycles that could not be serviced during the interval have to be subtracted out first. They will be accounted for in the next interval, probably at a higher CPU speed.

The last section represents the speed setting policy. The adjustment of the clock rate is a simple heuristic that attempts to smooth the transitions from fast to slow processing. If the system was more busy than idle, then the speed is ramped up. If it was mostly idle, then it is slowed down. We simulated several variations on the code shown here to come up with the constants shown here.

7 Evaluating the Algorithms

Figure 1 compares the results of these three algorithms on a single trace (Kestrel March 1) as the adjustment interval is varied. The OPT energy is unaffected by the interval, but is shown for comparison. The vertical access shows relative power used by the scheduling algorithms, with 1.0 being full power. Three sets of three lines are shown, corresponding to three voltage levels which determine the minimum speed, and the three algorithms, OPT, FUTURE, and PAST. The PAST and FUTURE algorithms approach OPT as the interval is lengthened. (Note that the log scale for the X axis.) For the same interval PAST actually does better than FUTURE because it is allowed to defer excess cycles into the next interval, effectively lengthening the interval. The intervals from 10 msec to 50 msec are considered in more detail in other figures.

Figure 2 shows the excess cycles that result from setting the speed too slow in PAST when using a 20 msec adjustment interval and the same trace data as Figure 1. Note that the graph uses log-log scales. Cycles are measured in the time it would take to execute them at full speed. The data was taken as a histogram, so a given point counts all the excess cycles that were less than or equal that point on the X axis, but greater than the previous bucket value in the histogram. Lines are used to connect the points so that the spike at zero is evident. The large spike at zero indicates that most intervals have no excess cycles at all. There is a smaller peak near the interval length, and then the values drop off.

As the minimum speed is lowered, there are more cases where excess cycles build up, and they can accumulate in longer intervals. This is evident Figure 2

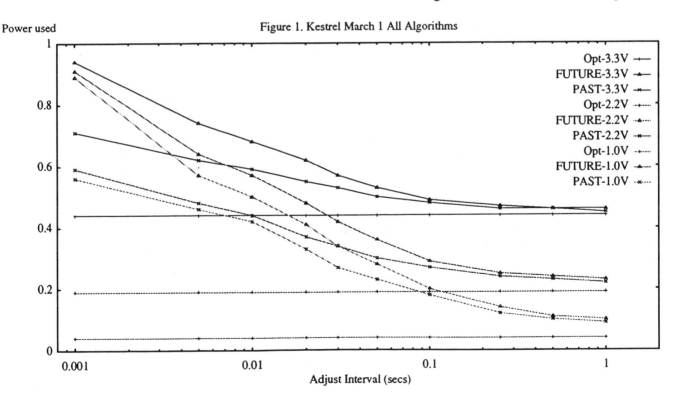

Figure 1. Kestrel March 1 All Algorithms

Power used

Opt-3.3V
FUTURE-3.3V
PAST-3.3V
Opt-2.2V
FUTURE-2.2V
PAST-2.2V
Opt-1.0V
FUTURE-1.0V
PAST-1.0V

Adjust Interval (secs)

where the points for 1.0 V are above the others, which indicates more frequent intervals with excess cycles, and the peak extends to the right, which indicates longer excess cycle intervals.

Figure 3 shows the relationship between the interval length and the peak in excess cycle length. It compares the excess cycles with the same minimum voltage (2.2 V) while the interval length varies. This is from the same trace data as Figures 1 and 2. The main result here is that the peak in excess cycle lengths shifts right as the interval length increases. All this means is that as a longer scheduling interval is chosen, there can be more excess cycles built up.

Figure 4 compares the energy savings for the bounded delay limited past (PAST) algorithm with a 20 msec adjustment interval and with three different min-

imum voltage limits. In this plot each position on the X access represents a different set of trace data. The position corresponding to the trace data used in Figures 1 to 3 is indicated with the arrow.

While there is a lot of information in the graph, there are two overall points to get from the figure: the relative savings for picking different minimum voltages, and the overall possible savings across all traces.

The first thing to look for in Figure 4 is that for any given trace the three points show the relative possible energy savings for picking the three different minimum voltages. Interestingly, the 1.0 V minimum does not always result in the minimum energy. This is because it. has more of a tendency to fall behind (more excess cycles), so its speed varies more and the power consumption is less efficient. Even when 1.0 V does

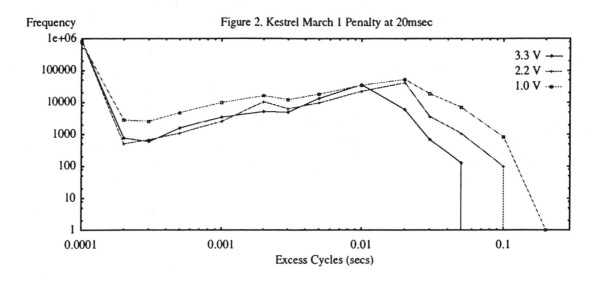

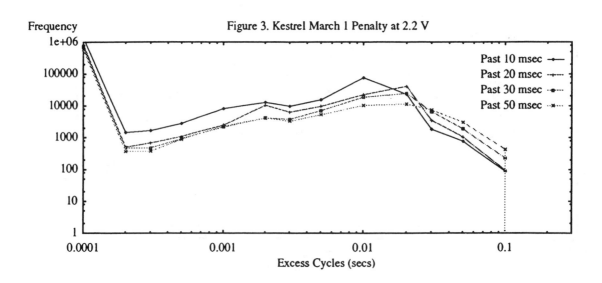

182

provide the minimum energy, the 2.2 V minimum is almost as good.

The other main point conveyed by Figure 4 is that in most of the traces the potential for energy savings is good. The savings range from about 5% to about 75%, with most data points falling between 25% to 65% savings.

Figure 5 fixes the minimum voltage at 2.2 V and shows the effect of changing the interval length. The OPT energy savings for 2.2 V is plotted for comparision.Again, each position on the X axis represents a different trace. The position corresponding to the trace data used in Figures 1 to 3 is indicated with the arrow.

In this figure the main message to get is the difference in relative savings for a given trace as the interval is varied. This is represented by the spread in the points plotted for each trace. A longer adjustment period results in more savings, which is consistent with Figure 1.

Figures 6 and 7 show the average excess cycles for all trace runs. These averages do not count intervals with zero excess cycles. Figure 6 shows the excess cycles at a given adjustment interval (20 msec) and different minimum voltages. Figure 7 shows the excess cycles at a given minimum voltage (2.2 V) and different intervals. Again, the lower minimum voltage

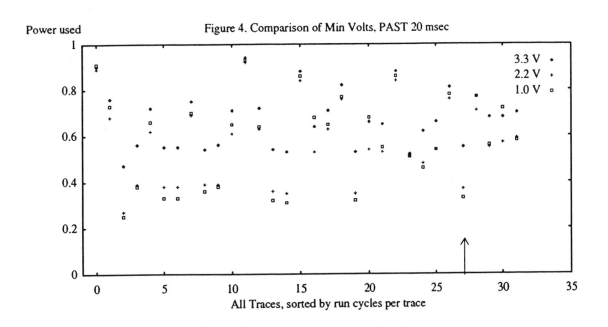

Figure 4. Comparison of Min Volts, PAST 20 msec

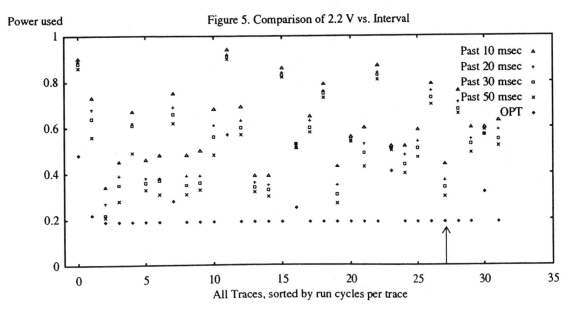

Figure 5. Comparison of 2.2 V vs. Interval

183

results show more excess cycles, and the longer intervals accumulate more excess cycles.

There is a trade off between the excess cycles penalty and the energy savings that is a function of the interval size. As the interval decreases, the CPU speed is adjusted at a finer grain and so it matches the offered load better. This results in fewer excess cycles, but it also does not save as much energy. This is consistent with the motivating observation that it is better to execute at an average speed than to alternate between full speed and full idle.

8 Discussion and Future Work

The primary source of feedback we used for the speed adjustment was the percent idle time of the system. Another approach is to classify jobs into background, periodic, and foreground classes. This is similar to what Wilkes proposes in his schemes to utilize idle time [Wilkes 95]. With this sort of classification the speed need not be ramped up when executing background tasks. Periodic tasks impose a constant, measurable load. They typically run for a short burst and then sleep for a relatively long time. With these tasks there is a well defined notion of "fast enough",

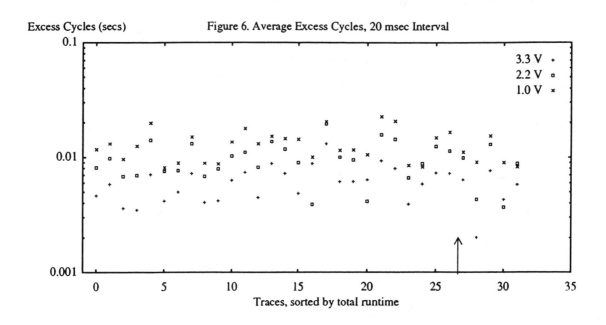

Figure 6. Average Excess Cycles, 20 msec Interval

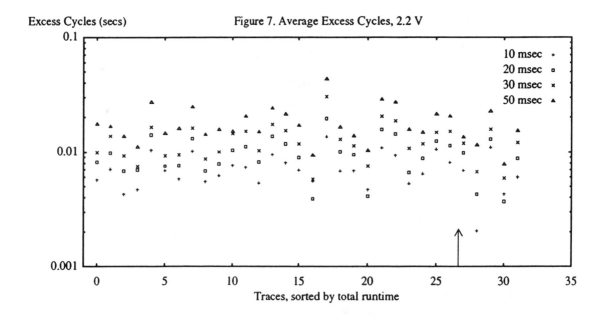

Figure 7. Average Excess Cycles, 2.2 V

and the CPU speed can be adjusted to finish these tasks just in time. When there is a combination of background, periodic, and foreground tasks, then the standard approach is to schedule the periodic tasks first, then fit in the foreground tasks, and lastly fit in the background tasks. In this case there would be a minimum speed that would always execute the periodic tasks on time, and the system would increase the speed in response to the presence of foreground and background tasks.

The simulations we performed are simplified by not reordering scheduling events. In a real rate-adjusting scheduler, the change in processing rates will have an effect on when jobs are preempted due to time slicing and the order that ready jobs are scheduled. We argue that unless there is a large job mix, then the reordering will not be that significant. Our speed adjustment algorithm will ramp up to full speed during heavy loads, and during light loads the reordering should not have a significant effect on energy.

In order to evaluate more realistic scheduling algorithms, it would be interesting to generate an abstract load for the simulation. This load includes CPU runs with preemption points eliminated, pause times due to I/O delays preserved, and causal ordering among jobs preserved. Given an abstract load, it would be possible to simulate a scheduler in more detail, giving us the ability to reorder preemption events while still preserving the semantics of I/O delays and IPC dependencies.

We have attempted to model the I/O waits by classifying idle time into "hard" and "soft" idle. We think this approximation is valid, but it would be good to verify it with a much more detailed simulation.

9 Conclusions

This paper presents preliminary results on CPU scheduling to reduce CPU energy usage, beyond the simple approaches taken by today's laptops. The metric of interest is how many instructions are executed for a given amount of energy, or MIPJ. The observation that motivates the work is that reducing the cycle time of the CPU allows for power savings, primarily by allowing the CPU to use a lower voltage. We examine the potential for saving power by scheduling jobs at different clock rates.

Trace driven simulation is used to compare three classes of schedules: OPT that spreads computation over the whole trace period to eliminate all idle time (regardless of deadlines), FUTURE that uses a limited future look ahead to determine the minimum clock rate, and PAST that uses the recent past as a predictor of the future. A PAST scheduler with a 50 msec window shows power savings of up to 50% for conservative circuit design assumptions (e.g., 3.3 V), and up to 70% for more aggressive assumptions (2.2 V). These savings are in addition to the obvious savings that come from stopping the processor in the idle loop, and powering off the machine all together after extended idle periods.

The energy savings depends on the interval between speed adjustments. If it is adjusted at too fine a grain, then less power is saved because CPU usage is bursty. If it is adjusted at too coarse a grain, then the excess cycles built up during a slow interval will adversely affect interactive response. An adjustment interval of 20 or 30 milliseconds seems to represent a good compromise between power savings and interactive response.

Interestingly, having too low a minimum speed results in less efficient schedules because there is more of a tendency to have excess cycles and therefore the need to speed up to catch up. In particular, a minimum voltage of 2.2 V seems to provide most of the savings of a minimum voltage of 1.0 V. The 1.0 V system, however, tends to have a larger delay penalty as measured by excess cycles.

In general, scheduling algorithms have the potential to provide significant power savings while respecting deadlines that arise from human factors considerations. If an effective way of predicting workload can be found, then significant power can be saved by adjusting the processor speed at a fine grain so it is just fast enough to accommodate the workload. Put simply, the tortoise is more efficient than the hare: it is better to spread work out by reducing cycle time (and voltage) than to run the CPU at full speed for short bursts and then idle. This stems from the non-linear relationship between CPU speed and power consumption.

Acknowledgments

This work was supported in part by Xerox, and by ARPA under contract DABT63-91-C-0027; funding does not imply endorsement. David Wood of the University of Wisconsin helped us get started in this research, and provided substantial assistance in understanding CPU architecture. The authors benefited from the stimulating and open environment of the Computer Science Lab at Xerox PARC.

Appendix I. Description of Trace Data

The table on the next page lists the characteristics of the 32 traces runs that are reported in the figures. The table is sorted from shortest to longest runtime to match the ordering in Figures 4 through 7. The elapsed time of each trace is broken down into time spent running the CPU on behalf of a process (Runtime), time spent in the idle loop (IdleTime), and time when the machine is considered so idle that it would be turned off by a typical laptop power manager (Offtime). The short traces labeled mx, emacs, and fm are of typing (runs 1 and 2) and scrolling (run 3) in various editors. The remaining runs are taken over several hours of everyday use

References

William C. Athas, Jeffrey G. Koller, and Lars "J." Svensson. "An Energy-Efficient CMOS Line Driver Using Adiabatic Switching", 1994 IEEE Fourth Great Lakes Symposium on VLSI, pp. 196-199, March 1994.

A. P. Chandrakasan and S. Sheng and R. W. Brodersen. "Low-Power CMOS Digital Design". JSSC, V27, N4, April 1992, pp 473--484.

Michael Culbert, "Low Power Hardware for a High Performance PDA", *to appear* Proc. of the 1994 Computer Conference, San Francisco.

Fred Douglis, P. Krishnan, Brian Marsh, "Thwarting the Power-Hungry Disk", Proc. of Winter 1994 USENIX Conference, January 1994, pp 293-306

Mark A. Horowitz. "Self-Clocked Structures for Low Power Systems". ARPA semi-annual report, December 1993. Computer Systems Laboratory, Stanford University.

Kester Li, Roger Kumpf, Paul Horton, Thomas Anderson, "A Quantitative Analysis of Disk Drive Power Management in Portable Computers", Proc. of Winter 1994 USENIX Conference, January 1994, pp 279-292.

S. Younis and T. Knight. "Practical Implementation of Charge Recovering Asymptotically Zero Power CMOS." 1993 Symposium on Integrated Systems (C. Ebeling and G. Borriello, eds.), Univ. of Washington, 1993.

Wilkes, John "Idleness is not Sloth", *to appear*, proc. of the 1995 Winter USENIX Conf

Table 2: Summary of Trace Data

I	Trace	Runtime	Idle	Elapsed	Offtime
0	feb28klono	0.906	29.094	9H 24M 20S	33828.9
1	idle1	1.509	28.653	39S	9.05
2	heur1	7.043	3.103	10S	0
3	emacs2	7.585	31.719	40S	0
4	emacs1	8.060	32.273	40S	0
5	mx2	8.362	30.916	39S	0
6	mx1	9.508	30.871	41S	0
7	fm1	9.544	10.594	20S	0
8	em3	11.669	27.580	40S	0
9	fm2	16.679	23.770	41S	0
10	mx3	20.738	18.642	39S	0
11	feb28dekanore	30.548	541.045	9H 24M 40S	33307.8
12	fm3	30.626	9.942	41S	0
13	mar1klono	41.822	1011.251	9H 55M 46S	34690.6
14	feb28mezzo	61.940	449.717	9H 24M 20S	33346.1
15	mar1cleonie	214.656	1321.591	9H 50S	30913.0
16	feb28kestrel	510.259	3362.222	1H 4M 33S	0
17	feb28corvina	524.248	768.857	9H 24M 41S	32588.0
18	mar1mezzo	686.340	673.204	9H 55M 36S	34375.7
19	mar1egeus	695.409	4774.911	9H 55M 35S	30263.6
20	feb28ptarmigan	1497.908	2207.005	1H 1M 41S	0
21	feb28fandango	1703.037	3489.760	9H 24M 17S	28665.0
22	feb28zwilnik	4414.429	29448.058	9H 24M 21S	0
23	mar1zwilnik	4914.787	30823.917	9H 55M 38S	0
24	mar1kestrel	5135.297	30599.364	9H 55M 34S	0
25	feb28siria	6714.109	27146.678	9H 24M 20S	0
26	mar1siria	8873.114	26868.738	9H 55M 37S	0
27	feb28egeus	9065.477	13500.028	6H 16M 6S	0
28	mar1corvina	10898.545	24648.883	9H 55M 57S	210.202
29	mar1ptarmigan	12416.924	23319.178	9H 55M 34S	0
30	mar1fandango	20101.182	15638.594	9H 55M 38S	0
31	mar1dekanore	25614.651	14168.562	9H 55M 58S	7191.81

Part IV

Circuit and Logic Styles

Silicon-Gate CMOS Frequency Divider for the Electronic Wrist Watch

ERIC VITTOZ, ASSOCIATE MEMBER, IEEE, BERNARD GERBER, AND FRITZ LEUENBERGER, MEMBER, IEEE

Abstract—An integrated complementary MOS transistor scale-of-two counter for applications in electronic wrist watches has been realized. Silicon-gate technology applied to a very simple but safe dividing circuit has resulted in a substantial reduction of the total area of the integrated structure with the following performance. At a supply voltage of 1.35 V the maximum frequency is 2 MHz and the dynamic power consumption per stage is 1.6 nW/kHz. The complementary substrate is obtained by a sealed-capsule low-surface concentration diffusion and doped oxides as impurity sources are used to allow simultaneous diffusion of both types of MOS transistors. Finally, a simple dynamic circuit derived from the basic structure is described.

I. INTRODUCTION

THE MAIN part of the electronics of a quartz-crystal wristwatch is the frequency divider. It is well known that the power consumption of these circuits must be kept below a few microwatts to ensure a sufficient life time of the available energy source. Bipolar transistors of a single type [1] or complementary [2] have been successfully used for this purpose. Meanwhile, complementary MOS logic circuits are particularly suitable for this application. Their power consumption P is essentially proportional to the frequency f according to the relation $P = fCV^2$, where V is the power supply voltage and C the sum of the different capacities modified by suitable weighting factors. The length of the dividing chain can be increased while the total power consumption stays below twice that of the first stage.

It is desirable to work at a high input frequency to be able to use a small low-cost quartz crystal with a small temperature coefficient. It is therefore important to reduce as much as possible the dynamic power consumption of the first dividing stages. The dynamic consumption of the circuits described in the literature up to now has ranged from 10 to 30 nW/kHz per stage [3], [4] at 1.35 V.

To ensure enhancement mode operation, the supply voltage V cannot be reduced to a value much below 1 V. The only way to reduce dynamic consumption, therefore, is to reduce the equivalent capacity C. Silicon-gate technology applied to a very simple but safe dividing circuit, involving no logical hazards, allows a substantial reduction of the total area of the integrated structure, and hence a decrease in capacity C, without making the technological processes critical.

Manuscript received July 1, 1971; revised October 28, 1971.
The authors are with the Centre Electronique Horloger S.A., Neufchâtel, Switzerland.

II. DESCRIPTION OF THE CIRCUIT

The logical structure chosen for the scale-of-two cell may be defined by the following set of Boolean equations:

$$A = \overline{IC}$$
$$B = \overline{(I + D)A}$$
$$C = \overline{IE + AB}$$
$$D = \bar{B}$$
$$E = \bar{C}$$

where I is the input variable. This structure is the first original result of an attempt to synthesize all possible counters realizable with a given number of two-level gates. It has been obtained manually by means of a special algorithm based on the standard Huffman [5] method. This algorithm is presently applied systematically on a computer.

Each of these equations can be realized in CMOS technology as an inverter or as a single- or two-level gate as shown in Fig. 1. The complete switching cycle of each variable is given in Table I. In the state corresponding to the first line of this table, each equation of the set is satisfied. It is thus a stable state which will be preserved as long as the input variable I is equal to zero. The moment it switches to 1 (line 2), the equation for B is no longer satisfied. This new state is therefore unstable and B tends to change state leading to another unstable state where the equation for D is not satisfied (line 3). Transition of D leads to a second stable state (line 4) which will be maintained as long as the input variable I is equal to 1.

The cycle then goes on through two more stable states before returning to the first one. It can be seen that this structure does not involve any essential logical hazards: for every unstable state, a single equation of the set is not satisfied, hence a single variable tends to change state. This dividing structure works independently of the delay associated with each gate. Moreover, only the uncomplemented form of the input variable I is required. This avoids the use of an inverter to drive the first stage and thus contributes to the reduction of power drain. Each of the variables A, B, C, D, and E switches at a frequency half that of the input I and may thus be used as the output variable. The $\frac{3}{4}$ duty cycle of A may be useful for some applications.

Examination of Fig. 1 shows that transistors 2 and 6 have their source and drain, respectively, in common.

Reprinted from *IEEE Journal of Solid-State Circuits*, pp. 100-104, April 1972.

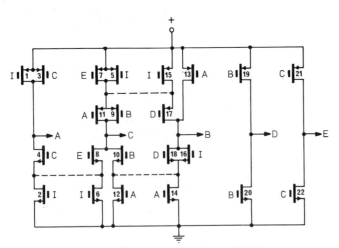

Fig. 1. Basic circuit of scale-of-two CMOS stage.

The condition necessary to allow a common drain as well, as represented by the dotted line, is that simultaneous conduction of transistors 4 and 8 should not affect the state of variables A and C. This condition is realized, since in the only state of the cycle for which transistors 4 and 8 are simultaneously conducting ($C = E = 1$), the variables A and C are both equal to 1. Transistors 2 and 6 have then their three corresponding electrodes in common and transistor 6 may therefore be eliminated. The same argument is applicable to pairs 5–15 and 12–14, respectively. This leads to the elimination of transistors 14 and 15. One obtains then the circuit of Fig. 2, consisting of 10 p-channel and 9 n-channel MOS transistors.

III. TECHNOLOGY

A. Substrate Preparation

A number of different approaches are known to prepare complementary substrates with the necessary low-impurity surface concentration.

1) Preferential expitaxial layer deposition [3], [4].

2) Ion implantation followed by a suitable drive-in diffusion [7].

3) Outdiffusion from a buried p-type layer into epitaxially grown n-type layer [6].

4) Solid–solid vacuum diffusion [8], [9].

The circuits described in this paper have been realized with the last-mentioned technique. One possible approach is to perform the predeposition in a vacuum furnace [9]. We have chosen to use a sealed system. The dopant is provided by boron-doped ($N_A = 2.2 \times 10^{16}$ cm^{-3}) single-crystal wafers. The ratio of source wafers to wafers to be diffused is four. Fig. 3 shows the wafer arrangement within the quartz capsule. The capsule is evacuated to 10^{-6} torr and subsequently sealed. Predeposition time and temperature are 10 h and 1200°C, respectively. The predeposition is followed by a drive-in diffusion at 1200°C in a nonoxidizing ambient. This step serves to adjust the surface concentration to the desired value, e.g., 0.8×10^{16} cm^{-3}. The well is 10–12 μm deep. Wells prepared in such a way had been successfully used for com-

TABLE I
SWITCHING CYCLE OF VARIABLES

I	A	B	C	D	E	Type of state stable	unstable	Path of the switching current
O	I	I	O	O	I	●		
I	I	I	O	O	I		●	12,16
I	I	O	O	O	I		●	19
I	I	O	O	I	I	●		
O	I	O	O	I	I		●	5,9
O	I	O	I	I	I		●	22
O	I	O	I	I	O	●		
I	I	O	I	I	O		●	2,4
I	O	O	I	I	O		●	13
I	O	I	I	I	O		●	20
I	O	I	I	O	O	●		
O	O	I	I	O	O		●	I
O	I	I	I	O	O		●	10,12
O	I	I	O	O	O		●	21
O	I	I	O	O	I	●		

The last column indicates the transistors through which the current charging or discharging the switching node is flowing.

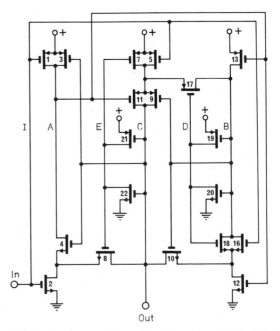

Fig. 2. Final circuit of divider stage involving 19 CMOS transistors.

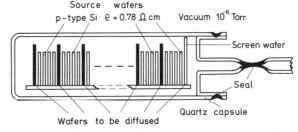

Fig. 3. Quartz capsule and wafer arrangement used for diffusion of p-type wells.

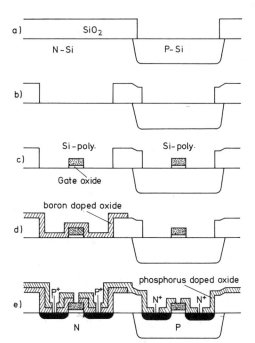

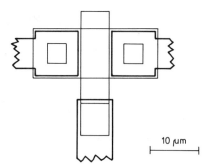

Fig. 5. Top view (to scale) of p-channel and n-channel transistors used in circuit.

Fig. 4. Schematic drawing illustrating major steps of fabrication sequence.

TABLE II

	p Channel	n Channel
Substrate doping level	10^{15} cm^{-3}	
Surface concentration of p-type well		8.6×10^{15} cm^{-3}
Depth of well		10–12 μm
Thickness of gate oxide	0.1 μm	0.1 μm
Source and drain junctions		
junction depth	1.1 μm	1.1 μm
sheet resistivity	150 $\Omega/\square$	100 $\Omega/\square$
Extrapolated threshold voltage	-0.4 ± 0.05 V	$+0.95 \pm 0.15$ V
Drain leakage current at $V_D = 1.5$ V	10 pA	10 pA
Voltage $V_G = V_D$ at $I_D = 100$ pA		
field MOS aluminum gate	7 ± 0.4 V	3.3 ± 0.2 V
field MOS silicon gate	5.2 ± 0.1 V	3.2 ± 0.1 V
Drain junction breakdown voltage	-75 V	$+45$ V
Mobilities (at $V_D = 10$ mV)	152 cm^2/V·s	545 cm^2/V·s

plementary metal gate circuits where source and drain were diffused to a depth of approximately 4 μm. In our silicon-gate circuits, source and drain are only about 1 μm deep and the depth of the p-well diffusion could therefore be reduced accordingly. Surface concentration varies not more than ±5 percent from wafer to wafer.

B. Fabrication Sequence for Complementary Silicon-Gate Circuits

Starting with a 5-Ω·cm n-type substrate of (100) orientation, p-type wells are diffused according to the procedure outlined in Section III-A. The essential wafer fabrication steps are illustrated in Fig. 4.

First, the masking oxide for the p wells is reduced to a thickness of about 0.2 μm [Fig. 4(a)]. A thermal oxide is then regrown to a thickness of about 0.8 μm within the well region and slightly more on the n-type substrate. The second photolithographic step serves to open the windows delineating source drain and gate regions for both the p-channel and the n-channel devices [Fig. 4(b)]. The 0.1-μm-thick gate oxide is now grown in dry oxygen at 1200°C followed by 15-min *in situ* annealing in helium at the oxidation temperature.

A 0.6-μm-thick layer of polycrystalline silicon is now deposited from the vapor phase. Deposition rate is 0.06 μm/min at a substrate temperature of 850°C. This is followed by selective etching of the polycrystalline silicon and removal of the 0.1-μm-thick oxide in the source and drain regions [Fig. 4(c)]. Next, a boron-doped oxide is deposited over the entire surface and then selectively removed over the p-type substrate [Fig. 4(d)]. A phosphorous-doped oxide is deposited, again over the entire wafer surface, and the source and drain regions of both the p-channel and n-channel devices are diffused simul-

taneously for 10 min at 1200°C. This step serves at the same time to densify the deposited doped oxides.

Contact windows are now opened [Fig. 4(e)] and 0.6 μm of aluminum are evaporated from an electron gun operated in a conventional vacuum system. The sixth and final masking step defines the interconnection pattern. Annealing in forming gas (20 percent H$_2$, 80 percent N$_2$) for one hour at 400°C completes the wafer processing sequence.

C. Device Dimensions and Electrical Characteristics

Low dynamic power dissipation requires using the smallest possible device dimension compatible with acceptable fabrication yield. Simultaneous diffusion of the p-channel and n-channel devices using doped oxides as impurity sources is a significant step in this direction [10]. The lateral diffusions extend less than 1 μm underneath the gate oxide. This results in very small overlap- and drain-substrate capacitances. Fig. 5 shows the dimensions of the devices used in our circuit. Channel width is 10 μm and effective channel length is about 5 μm. The aluminum interconnections are 6 μm wide. Table II lists some of the important device parameters.

The extrapolated threshold voltage for the n-channel field MOS transistor is about +4.5 V. While this value may seem rather low, it is nevertheless acceptable when we consider that these circuits are used with supply

193

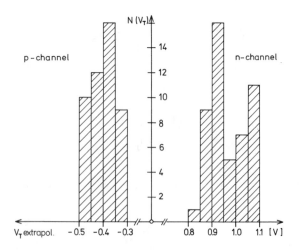

Fig. 6. Distribution of extrapolated threshold voltages of test transistors.

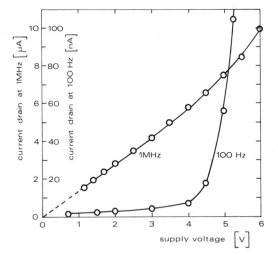

Fig. 7. Photomicrograph showing stage two of three-stage frequency divider.

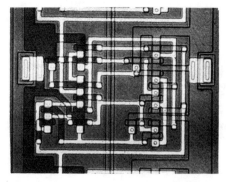

Fig. 8. Current drain of three-stage divider measured as a function of supply voltage at low and high input frequency.

voltages in the range from 1.3 to 1.5 V. Fig. 6 shows the distributions, over one wafer, of extrapolated threshold voltages for both p-channel and n-channel test devices. This threshold voltage is defined by the intercept of the straight line portion of the $I_D^{\frac{1}{2}} = f \; (V_G \equiv V_D)$ curve with the voltage axis. Considering the ±5-percent spread in surface concentration for the p wells, the spread of ±15 percent in threshold voltage for the n-channel de-

vices seems rather high. Local variations of boron redistribution and oxide charge might be responsible for this discrepancy. The results reported in Section IV have been obtained on circuits fabricated on the same wafer.

IV. CIRCUIT EVALUATION

An experimental circuit comprising a cascade of 3 divider stages according to Fig. 2 and an output inverter separately powered has been realized on the basis of the typical MOS transistor shown in Fig. 5. The layout rules have been chosen in order to minimize the total capacitance and thus the power consumption, with no effort to maximize the possible frequency of operation. The capacitance per unit area of the drain–substrate junction being larger than those of the gate–thick oxide–silicon and metal–thick oxide–silicon regions, the areas of all p⁺ and n⁺ islands have been minimized. Interconnections and crossovers are provided by the metallic and polycrystalline gate layers.

Fig. 7 is a photomicrograph of the medium stage occupying an area of 0.17 by 0.26 mm².

The measured current drain of the three stages is shown in Fig. 8 as a function of the supply voltage. The current drain at 1-MHz input frequency is essentially dynamic and therefore proportional to the voltage. After adding the easily calculated contribution of the input capacitance of the first stage, it corresponds to an equivalent capacity C of 0.9 pF per stage or a total dynamic consumption of 1.6 nW/kHz per stage at 1.35 V. The current drain at 100-Hz input frequency is essentially due to leakage currents. It corresponds to the average static consumption of all logic states. This current increases rapidly above 4 V due to the appearance of parasitic channels below the thick oxide. Fig. 9 shows the maximum input frequency measured as a function of the supply voltage. The circuit is working up to 2 MHz at 1.35 V and up to 10 MHz at 3 V.

V. DISCUSSION AND POSSIBLE IMPROVEMENTS

Compared to previously published results [3], [4], the reported performance corresponds to a reduction by a factor of about 10 in the dynamic power consumption and allows the realization of a cascade of frequency dividers consuming just a few microwatts for an input frequency of 1 MHz. It is possible to decrease this power further by reducing the line width of the interconnections.

On the other hand, it may become desirable to increase the maximum frequency at low voltage. An immediate means is a reduction of the thresholds by increasing the resistivity of both the n substrate and p wells, while making sure that the static current drain does not increase beyond reasonable limits.

Considering the relatively high drain breakdown voltages, it seems feasible to reduce further the channel lengths of the transistors. Furthermore, one may optimize the width of each channel to increase the frequency

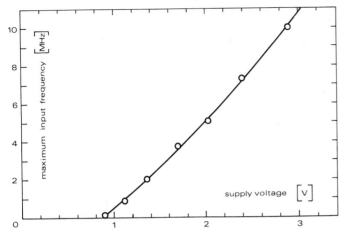

Fig. 9. Maximum input frequency as a function of supply voltage.

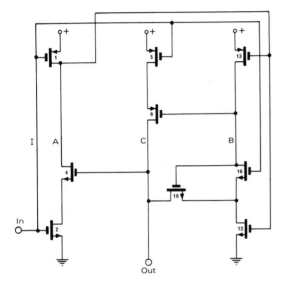

Fig. 10. Dynamic divider stage.

nodes. Moreover, variables D and E are used only to drive some of these 6 transistors and are therefore no longer needed. Thus, the inverters 19–20 and 21–22 may be canceled as well. This leads to the very simple dynamic circuit consisting of 10 transistors only, illustrated in Fig. 10.

This structure has not yet been integrated, but a layout sketch, following the same rules as those observed in laying out the static circuit has been drawn, and capacitances at the various nodes have been computed.

A breadboard simulation has been carried out according to the following procedure. The transistors have identical dc characteristics as those to be used in the integrated structure, but the capacitances calculated at the various nodes are multiplied by a factor F (20 000 in our example) in order to render negligible all stray capacitances associated with the breadboard. At the same time the frequency is scaled down by the same factor F.

This simulation test has shown that dynamic power consumption is reduced roughly by a factor of 2, corresponding to a reduction of the total capacitance C in the same ratio.

ACKNOWLEDGMENT

D. Chauvy and his co-workers have developed the closed-capsule diffusion process while R. Taubenest and his staff have developed the techniques for depositing polycrystalline silicon layers as well as doped oxides. We also wish to thank Dr. H. Oguey for suggestions regarding the dynamic structure.

REFERENCES

[1] R. Guye, D. Chauvy, and K. Hübner, "Circuits intégrés d'une montre-bracelet à quartz," in *Proc. Coll. Int. Chronométrie*, ser. B2, B245, Paris 1969.
[2] H. Rüegg, W. Thommen, and P. Sauthier, "A saturation-controlled flip-flop for low voltage micropower systems," in *ISSCC Dig. Tech. Papers*, pp. 60–61, 1971.
[3] F. Leuenberger and E. Vittoz, "Complementary-MOS low-power low-voltage integrated binary counter," *Proc. IEEE*, vol. 57, pp. 1528–1532, Sept. 1969.
[4] R. G. Daniels and R. R. Burgess, "The electronic wristwatch: an application for Si-gate CMOS IC's" in *ISSCC Dig. Tech. Papers*, pp. 62–63, 1971.
[5] D. A. Huffman, "The synthesis of sequential switching circuits," *J. Franklin Inst.*, vol. 257, pp. 169–190, Mar. 1954, and pp. 275–303, Apr. 1954.
[6] F. Leuenberger, "Verfahren zur Herstellung einer integrierten Schaltung," Swiss Patent 482 305, filed July 19, 1968; issued Nov. 30, 1969.
[7] T. N. Toombs, R. M. Finnila, H. G. Dill, and L. O. Bauer, "High-density ion-implanted C-MOS technology," presented at the Int. Electron Devices Meeting, Washington, D. C., Oct. 11–13, 1971, paper 21.3.
[8] W. J. Armstrong and M. C. Duffy, "A closed-tube technique for diffusing impurities into silicon," *J. Electrochem. Technol.*, vol. 4, p. 475, 1969.
[9] R. Dahlberg *et al.*, "Technique and kinetics of solid-solid vacuum diffusion processes in silicon," in *Semiconductor-Silicon*, R. R. Haberecht and E. L. Kern, Eds. New York: Electrochemical Soc., 1969, pp. 458–468.
[10] R. W. Ahrons and P. D. Gardner, "Interaction of technology and performance in complementary symmetry MOS integrated circuits," *IEEE J. Solid-State Circuits*, vol. SC-5, pp. 24–29, Feb. 1970.

limit at the price of a very small increase in power. Modified annealing treatments might lead to increased carrier mobilities and thus to higher ratios of transconductance to input capacity.

Besides these suggested technological improvements, it will be shown now that the circuit itself can be improved with a view to high-frequency performance.

The last column of Table I indicates the transistors through which the current charging or discharging a switching node flows. It can be seen that only 13 among the 19 transistors do participate in the process of switching from each stable state to the next one; the remaining 6 transistors, numbered 3, 7, 8, 11, 17, and 18, are only used to maintain the established stable states. For sufficiently high frequencies, one obtains a dynamic structure by eliminating these 6 transistors. The stable states are now maintained by the capacitances of the various

CODYMOS Frequency Dividers Achieve Low Power Consumption and High Frequency

H. OGUEY AND E. VITTOZ

CENTRE ELECTRONIQUE HORLOGER S.A., 2000 NEUCHATEL, SWITZERLAND

Indexing terms: Frequency dividers, Integrated circuits

Frequency dividers made with complementary dynamic m.o.s. (CODYMOS) circuits require only a small number of transistors and interconnections, a single input signal and operate with a minimum number of successive transitions. This leads to a drastic reduction in stray capacitance and current consumption, and to an increase in speed. A simplified analysis of these quantities is given for binary and ternary dividers. An experimental circuit, integrated with silicon-gate technology, operates from a 1·35 V battery, divides by 3 up to an input frequency in excess of 20 MHz and draws a current of 0·4 μA/MHz.

CODYMOS (complementary dynamic metal–oxide semiconductor) frequency dividers are obtained by interconnecting p- and n-channel field-effect transistors according to their logical equations.[1] Fig. 1a shows a by-3 divider. It consists of 3 stages A, B and C, each of two p-channel and two n-channel transistors. The input signal drives two complementary transistors of each stage, whereas each internal variable drives two complementary transistors of the next stage. This structure is ripple- and hazardfree, as every change of the input variable I induces a change in one output variable only. Accordingly, it is very fast. The highest frequency f_{max} of the input variable I is roughly given by

$$f_{max} \simeq \frac{\beta_p(V_B - V_{Tp})^2}{8C_N V_B}$$

where β_p denotes the saturation-current factor and V_{Tp} the threshold voltage of the p transistors, V_B is the battery voltage and C_N is the highest node capacitance. The highest frequency corresponds to the equality between one halfperiod of the input signal and the transition of a node capacitance charged by a constant current of $\frac{1}{4}\beta_p(V_B - V_{Tp})^2$ (the factor $\frac{1}{4}$ results from the series connection of two transistors).

The current consumption of a simple divider stage is given by

$$I = f(C_N + C_{IN})V_B$$

The geometry of the individual transistors has the following effect: if we assume that $\beta \simeq W/L$ and $C_N \simeq WL$, with W and L the channel width and length, respectively, it is easy to see that, at a given frequency, the current I is proportional to WL, whereas the frequency limit is proportional to $1/L^2$ and roughly independent of W. Both I and f_{max} are improved by a reduction in the dimensions.

This circuit has been integrated with the aid of known silicon-gate technology.[2] The n and p channel transistors have identical geometries, with $L = 5 \mu$m and $W = 8 \mu$m on the masks. Measurements of the p-channel transistors yield $\beta_p = 35 \mu$A/V^2 and $V_{Tp} = 0.3$ V. At $V_B = 1.2$ V, the series connection of two transistors is able to draw a current of 7·1 μA. A node capacitance $C_N = 0.12$ pF is charged in

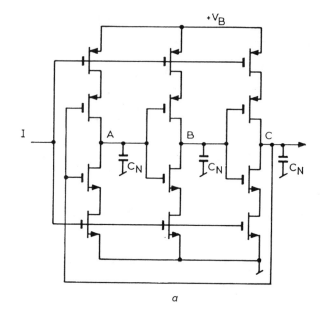

a

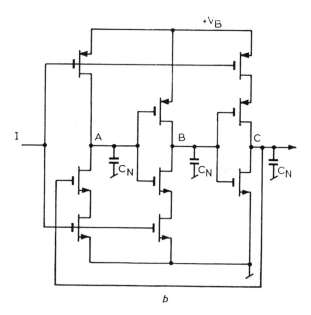

b

Fig. 1 CODYMOS *frequency divider*
a By 3
b By 2

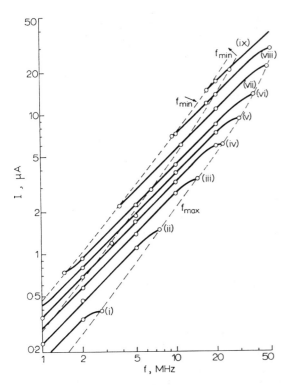

Fig. 2 *Direct current and limits against frequency for an integrated* codymos *frequency divider*

(i) $V_B = 0.6$ V
(ii) $V_B = 0.8$ V
(iii) $V_B = 1.0$ V
(iv) $V_B = 1.2$ V
(v) $V_B = 1.4$ V
(vi) $V_B = 1.6$ V
(vii) $V_B = 2.0$ V
(viii) $V_B = 2.5$ V
(ix) $V_B = 3.0$ V

Fig. 2 shows, for a high-performance unit, the current consumption of one stage, including an input driver, as a function of input-signal frequency. This circuit reaches frequencies in excess of 20 MHz at a battery voltage of 1·2 V, and draws a current of less than 4 μA at 10 MHz and 1·4 V. A dynamic power consumption of less than 0·6 μW/MHz per stage is obtained.

The dynamic nature of this circuit prevents its use at low frequencies; at an intermediate frequency, its behaviour may be dependent on the initial conditions. However, for a wide range of frequencies and battery voltages, the circuit divides correctly for every initial condition, making it very attractive as an input stage of a divider chain. Among a series of circuits made from four different wafers, none has been found with an upper frequency limit of less than 8 MHz.

A by-2 frequency divider is shown in Fig. 1*b*. It was originally derived from its static counterpart,[2] but it can also be regarded as a contraction of the circuit of Fig. 1*a*. The elimination of three transistors transforms two of the six stable states into transitory states, so that the division ratio becomes 1 : 2. This circuit shows a performance only slightly inferior to that of the circuit of Fig. 1*a*. Moreover, it is completely insensitive to initial conditions, and divides correctly down to frequencies below 1 kHz.

codymos circuits, even in this early stage of development, look very promising for such applications as quartz wrist watches incorporating a small high-frequency AT quartz.

H. OGUEY
E. VITTOZ

Centre Électronique Horloger S.A.
Rue A-L Breguet 2
Neuchâtel, Switzerland

23rd July 1973

20·3 ns, allowing for an input frequency f_{max} of 24·6 MHz. The input capacitance C_{IN} is 0·18 pF, including the gate capacitance of the driver stage. The current consumption should be $I/f = 0.42$ μA/MHz at 1·4 V.

References

1 VITTOZ, E., and OGUEY, H.: 'Complementary dynamic m.o.s. logic circuits', *Electron. Lett.*, 1973, **9**, pp. 77–78
2 VITTOZ, E., GERBER, B., and LEUENBERGER, F.: 'Silicon-gate c.m.o.s. frequency divider for the electronic wrist watch', *IEEE J. Solid-State Circuits*, 1972, **SC-7**, pp. 100–104

Short-Circuit Dissipation of Static CMOS Circuitry and Its Impact on the Design of Buffer Circuits

HARRY J. M. VEENDRICK

Abstract —This paper gives a detailed discussion of the short-circuit component in the total power dissipation in CMOS circuits, on the basis of an elementary CMOS inverter. Design considerations are given for CMOS buffer circuits, based upon the results of the dissipation discussion, to increase circuit performance.

LIST OF PARAMETERS USED

A	process-determined constant defined in (11)
a	ratio between L_p and L_n [(A9)]
b	ratio between parasitic nodal capacitance and load capacitance
β	gain factor (μA/V^2) of an MOS transistor
β_n	β of NMOS transistor
β_p	β of PMOS transistor
β_N	β of nMOST and pMOST of the Nth (symmetrical) inverter of a string
$\beta_\square$	β of a transistor with equal channel length and channel width
C_{gN}	input gate capacitance of the Nth inverter of a string
C_L	load capacitance
C_N	total capacitance on node N
C_o	input capacitance of the first inverter of a string
C_{ox}	gate oxide capacitance
$C_{par\,N}$	parasitic capacitance on node N
f	frequency ($=1/T$)
I	short-circuit current
I_{mean}	mean value of the short-circuit current
I_{max}	maximum value of the short-circuit current
L_n	gate length of the nMOST
L_p	gate length of the pMOST
ΔL_n	gate length minus effective channel length of the nMOST
ΔL_p	gate length minus effective channel length of the pMOST
N	number of inverters
P	total power dissipation
P_1	dynamic power dissipation
P_2	short-circuit power dissipation
T	period-time of a signal ($=1/f$)

t	time
τ	rise or fall time of a signal[1]
τ_f	fall time of a signal[1]
τ_i	rise or fall time of an input signal[1]
τ_o	rise or fall time of an output signal[1]
τ_r	rise time of a signal
V_{dd}	supply voltage
V_{in}	input voltage
V_{out}	output voltage
V_T	threshold voltage
V_{T_n}	threshold voltage of nMOST
V_{T_p}	threshold voltage of pMOST
W_{nN}	channel width of the nMOST of the Nth inverter
W_{pN}	channel width of the pMOST of the Nth inverter

INTRODUCTION

DURING the last five years CMOS technology has become one of the most dominant technologies for VLSI circuits.

The most important reason for this is its low static power dissipation, due to the absence of dc currents during periods when no signal transients occur. However, during an edge of an input signal there will always be a short-circuit current flowing from supply to ground in static CMOS circuits. So far only limited analyses and discussions have appeared in the literature on this power component of static CMOS circuits [1].

In integrated circuits it is always necessary to drive large capacitances (bus lines, "off-chip" circuitry, etc.), often at high clock frequencies. Such driving circuits (buffers) will take a relatively large part of the total power consumption of the chip. It is clear that optimization of such circuits requires a different approach as compared to optimization of CMOS logic [2]. These buffer circuits need extra attention to obtain minimum power dissipation. Therefore, a detailed discussion on power dissipation of a basic CMOS inverter will be given first.

Manuscript received October 19, 1983; revised December 28, 1983.
The author is with Philips Research Laboratories, 5600 JA Eindhoven, The Netherlands.

---

[1]Although the rise and fall times are commonly defined to be the time between the 10 and 90 percent level of the signal extremes, in this paper these parameters are defined as the total duration of a linearized edge.

Reprinted from *IEEE Journal of Solid-State Circuits*, pp. 468–473, August 1984.

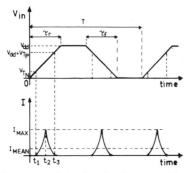

Fig. 1. Basic CMOS inverter.

Fig. 2. Current behavior of an inverter without load.

DISSIPATION OF A BASIC CMOS INVERTER

A static CMOS inverter does not dissipate power during the absence of transients on the input: when the input (Fig. 1) is at high level (V_{dd}), only the NMOS transistor conducts, and when the input is at low level, only the pMOST will conduct. However, during a transient on the input, there will be a time period in which both the nMOST and pMOST will conduct, causing a short-circuit (I) to flow from supply to ground, as shown in Fig. 2 for an inverter without load. This current flows as long as the input voltage (V_{in}) is higher than a threshold voltage (V_{T_n}) above V_{ss} and lower than a threshold ($|V_{T_p}|$) below V_{dd}.

If we load the inverter output of Fig. 1 with a capacitance C_L, then the dissipation of the circuit consists of two components:

$$\text{dynamic dissipation: } P_1 = C_L \cdot V^2 \cdot f \text{ and} \quad (1)$$

$$\text{short-circuit dissipation: } P_2 = I_{mean} \cdot V. \quad (2)$$

Clearly, the dynamic component P_1 does not depend on the inverter design (apart from contributions due to parasitic output capacitances, such as junction capacitances). The second component P_2, however, strongly depends on the inverter design.

Since there is a difference in the short-circuit dissipation of an inverter without load and that of an inverter with load, we start our discussions on the basis of an inverter with zero load capacitance. For simplicity we assume that the inverter is symmetrical (an asymmetrical inverter is not fundamentally different), which means that

$$\beta_n = \beta_p = \beta \text{ and } V_{T_n} = -V_{T_p} = V_T. \quad (3)$$

During the period ($t_1 - t_2$; Fig. 2) in which the short-circuit current I increases from 0 to I_{max}, the output voltage (V_{out}) will be larger than the input voltage (V_{in}) minus the threshold voltage (V_T) of the nMOST. As a consequence,

the NMOS transistor will be in saturation during this period of time.

Using the simple MOS formula, this leads to

$$I = \frac{\beta}{2}(V_{in} - V_T)^2 \text{ for } 0 \leqslant I \leqslant I_{max}. \quad (4)$$

This current will reach its maximum value when V_{in} equals half the supply voltage ($V_{in} = V_{dd}/2$), due to the assumption that the inverter was symmetrical. Another result of this assumption is that the current behavior during the time period $t_1 - t_3$ will be symmetrical with respect to the time t_2.

The mean current during a time T (equal to one period of the input signal) can thus be written as

$$I_{mean} = 2 * \frac{2}{T} \int_{t_1}^{t_2} I(t)\, dt = \frac{4}{T} \int_{t_1}^{t_2} \frac{\beta}{2}(V_{in}(t) - V_T)^2\, dt. \quad (5)$$

Assuming equal rise and fall times[2] ($\tau_r = \tau_f = \tau$) of the input signal (symmetrical) and a linear relation between the input voltage (V_{in}) and time (t) during its transients

$$V_{in}(t) = \frac{V_{dd}}{\tau} \cdot t, \quad (6)$$

it can be derived from Fig. 2 that

$$t_1 = \frac{V_T}{V_{dd}} \cdot \tau \quad \text{and} \quad t_2 = \frac{\tau}{2}. \quad (7)$$

Equations (5), (6), and (7) lead to

$$I_{mean} = \frac{2\beta}{T} \int_{\tau/2}^{V_T \cdot \tau/V_{dd}} \left(\frac{V_{dd}}{\tau} \cdot t - V_T\right) d\left(\frac{V_{dd}}{\tau} \cdot t - V_T\right), \quad (8)$$

which has the solution

$$I_{mean} = \frac{1}{12} \cdot \frac{\beta}{V_{dd}} \cdot (V_{dd} - 2V_T)^3 \cdot \frac{\tau}{T}. \quad (9)$$

From (2) and (9) the following expression can be derived for the short-circuit dissipation of a CMOS inverter without load:

$$P_2 = \frac{\beta}{12} \cdot (V_{dd} - 2V_T)^3 \cdot \frac{\tau}{T}. \quad (10)$$

As $1/T = f$, (10) shows that this dissipation component is also proportional to the frequency of switching. Because V_{dd} and V_T are process-determined, the only design parameters that affect P_2 are β and the input rise and fall times (τ) of the inverter.

For an inverter with capacitive load, the β's of the transistors are determined by requirements on output rise and fall times. In this case the short-circuit dissipation depends only on the duration of the input signal edges. As will be shown further on, these edges should not be too long, especially in the case of driver circuits that have a large β value. In the derivation of (10), we started with an

---

[2] The definitions of rise and fall times used here are different from those in common use (see the list of parameters used).

199

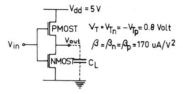

Fig. 3. The inverter used in the example.

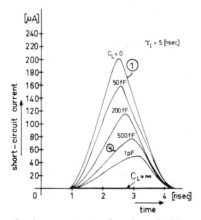

Fig. 4. Short-circuit current as a function of different inverter load capacitances.

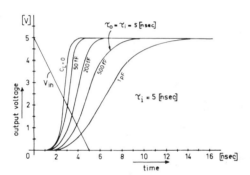

Fig. 5. Inverter output voltage behavior for different inverter load capacitances.

inverter without load. The following example examines what happens when we load the inverter with different capacitances.

Example

The discussions that follow are based upon the inverter whose parameter values are shown in Fig. 3. Its operation was simulated with a circuit analysis program. Some results are presented in Fig. 4. The figure shows the short-circuit current behavior, during a time interval $t_1 - t_3$ (see Fig. 2), as a function of the load capacitance C_L, for input rise and fall times of 5 ns. Curve ① shows the behavior of the inverter without load. At any time this current is the maximum short-circuit current that can occur. This means that all other current characteristics for different load capacitances must be within this curve. Curve ④ shows the short-circuit current behavior of the inverter when it is loaded with a characteristic capacitance C_L of 500 fF. In this case the rise and fall times on the output node are equal to the rise and fall times on the input. Fig. 5 shows

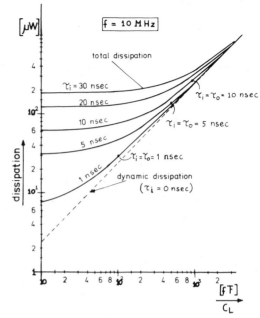

Fig. 6. Inverter dissipation as a function of the inverter load capacitance.

the output transient curves for different values of the load capacitances when the input is switched from a high level to ground with a fall time of 5 ns.

As expressed in (10) for a CMOS inverter with zero load capacitance, it is obvious that the dissipation versus load capacitance characteristic will depend on the rise and fall times of the input signal. Fig. 6 shows this characteristic for different values of the input rise and fall times (τ_i). The dashed line shows the dynamic dissipation ($f = 10$ MHz), while the solid lines show the actual inverter dissipation (dynamic plus short-circuit dissipation). The points where the load capacitance corresponds to equal input and output rise and fall times for the different characteristics are indicated on the figure.

From these characteristics we can conclude that if the operation of the inverter is such that the output signal and input signal have equal rise and fall times, the short-circuit dissipation will be only a fraction (< 20 percent) of the total dissipation. However, if the inverter is more lightly loaded, causing output rise and fall times that are relatively short as compared to the input rise and fall times, then the short-circuit dissipation will increase to the same order of magnitude as the dynamic dissipation. Therefore, to minimize dissipation, an inverter used as part of a buffer should be designed in such a way that the input rise and fall times are less than or about equal to the output rise and fall times in order to guarantee a relatively small short-circuit dissipation.

Fig. 7 shows the linear relationship [according to (10)] between the short-circuit dissipation and the input rise and fall times (τ_i) derived by means of circuit simulations. In this case the inverter of Fig. 3 was loaded with a capacitance of 500 fF. From Fig. 5 it was known that a load capacitance of 500 fF causes 5 ns rise and fall times of the output signal, when the inverter input rise and fall times

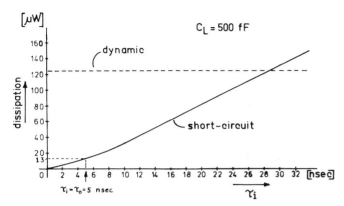

Fig. 7. Inverter dissipation as a function of the input rise and fall times.

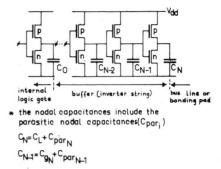

Fig. 8. Inverter string acting as a buffer circuit.

are equal to 5 ns. This point, $\tau_i = \tau_0 = 5$ ns, is indicated on Fig. 7 and the corresponding short-circuit dissipation is only about 10 percent of the dynamic dissipation.

This result of designing a string of inverters in such a way that the input and output rise and fall times of each inverter are equal to obtain minimum dissipation can very well be applied to the design of static CMOS buffers.

Although (10) was derived for an inverter with zero load capacitance and therefore is for the maximum short-circuit dissipation, it can also be applied to inverters designed with $\tau_i = \tau_0$. It has empirically been found that for such designs the short-circuit dissipation is half of the maximum, calculated with (10).

DESIGN CONSIDERATIONS

In integrated circuits it is always necessary to drive large capacitances, like bus lines or "off-chip" circuitry. Moreover, this must often occur at high speed, which will take a relatively large part of the total power dissipation of the chip. Particularly in the case of bus lines, which control a large number of inputs of the different subcircuits on a chip, it is necessary to have short signal rise and fall times [(10)] to minimize dissipation.

Suppose we want to drive such a bus interconnection line or "off-chip" circuitry with a signal coming from an internal node A. Let us assume that the logic gate, having node A as output, is capable of charging a capacitance C_0 in a time t to 95 percent of the supply voltage. From the foregoing results we know that if we use an inverter string as a buffer circuit between node A and the bus line, the rise and fall times on each node of the string should be equal to the required rise and fall times on the bus line (or bonding pad) to be driven.

The problem now is how to design an inverter string (Fig. 8) loaded with a capacitance C_L, with τ ns rise and fall times on each node, driven from an internal logic gate capable of charging and discharging the input capacitance C_0 in the same time. In [3] it was derived that a factor of e between the β's of the successive inverters (tapering factor) was needed to guarantee a minimum propagation delay time for such an inverter string. However, it is well understood that in terms of dissipation and silicon area this will

not lead to an optimum design. Design optimization for minimum dissipation and silicon area requires a different approach, as will be shown in the following.

It has been derived in the Appendix that a "minimum dissipation design" of the inverter string is completely determined by the following three equations:

$$\beta_N = \frac{C_N}{\tau_N} * A,$$

where

$$A = \frac{1}{V_{dd} - V_{T_n}} \cdot \left\{ \frac{2V_{T_n}}{V_{dd} - V_{T_n}} + \ln\left[\frac{2(V_{dd} - V_{T_n}) - V}{V} \right] \right\} \tag{11}$$

is constant for a given technology.

$$\frac{\beta_{N-1}}{\beta_N} = \frac{A}{\tau \cdot \beta_\square} \cdot (1+b) \cdot C_{ox} \cdot L_{n_N}$$
$$\cdot \left\{ L_{n_N}(1+3a^2) - \Delta L_{n_N} - 3a \cdot \Delta L_{p_N} \right\} \tag{12}$$

and

$$\left(\frac{\beta_N}{\beta_{N-1}} \right)^N = \frac{C_N}{C_0} \tag{13}$$

where β_N represents the β of the last inverter stage, β_{N-1}/β_N is the tapering factor at equal input and output rise and fall times and N is the number of inverters of the string. For a practical application the following assumptions are made for the parameters of (11), (12), and (13):

$$\left. \begin{array}{l} V_{T_n} = -V_{T_p} = 1 \text{ V} \\ V = 0.05 * V_{dd} \text{ [in (11)]} \\ V_{dd} = 5 \text{ V} \end{array} \right\} A \approx 1$$

$$\beta_{n_\square} = 42 \ \mu\text{A/V}^2 \qquad \beta_{p_\square} = 14 \ \mu\text{A/V}^2$$

$$C_{ox} = 700 \ \mu\text{F/m}^2$$

$$L_n = 2.5 \ \mu\text{m} \qquad \Delta L_n = 0.5 \ \mu\text{m} \qquad \Delta L_p = 1 \ \mu\text{m}.$$

Required rise and fall times: $\tau = 5$ ns. Practical values for the constants a and b are $a = 3/2.5$ and $b = 0.1$.

201

With (12) this leads to

$$\frac{\beta_N}{\beta_{N-1}} = 11.5.$$

This tapering factor, as it is often called, is strongly process dependent; nearly all parameters in (12) are determined by the process. Different CMOS processes may therefore lead to different tapering factors.

If, in a practical situation, $C_o = 100$ fF and C_L (Fig. 8) equals 10 pF and we want 5 ns rise and fall times (τ) on the output of the driver (inverter string), then the design procedure is as follows (with the above assumptions):

according to Fig. 8:

$$C_N = C_L + C_{par_N} = (1+b)\cdot C_L = 11 \text{ pF}$$

according to (11):

$$\beta_N = \frac{C_N}{\tau_N}\cdot A = \frac{C_N}{\tau}\cdot A = 4.4 * 10^{-3} \text{ A/V}^2.$$

Thus, the last inverter stage (N) is determined by

$$\beta_{n_N} = \beta_{p_N} = \beta_N = 4.4 * 10^{-3} \text{ A/V}^2,$$

or (see Appendix)

$$\left(\frac{W_n}{L_n - \Delta L_n}\right)_N = \frac{\beta_N}{\beta_{n_\square}} = 105, \quad \text{so} \left(\frac{W_n}{L_n}\right)_N = \frac{210 \ \mu\text{m}}{2.5 \ \mu\text{m}}.$$

and

$$\left(\frac{W_p}{L_p - \Delta L_p}\right)_N = \frac{\beta_N}{\beta_{p_\square}} = 315, \quad \text{so} \left(\frac{W_p}{L_p}\right)_N = \frac{630 \ \mu\text{m}}{3 \ \mu\text{m}}.$$

The $(N-i)$ inverters are now determined by the tapering factor

$$\frac{\beta_N}{\beta_{N-1}} = 11.5.$$

With the given $C_0 = 100$ fF and $C_N = 11$ pF we find from (13) that the number of inverters needed for this example will be equal to $N = 1.83$. (This, of course, has to be rounded off to $N = 2$.)

For this example, therefore, the inverter string should be designed as shown in Fig. 9 to guarantee a very small short-circuit dissipation and a minimum area consumption. The parasitic nodal capacitances are also depicted in Fig. 9.

By means of circuit simulations the mean power dissipation has been calculated at a clock frequency of $f = 1/T = 10$ MHz. Table I shows the results of a comparison of two tapering factors: a factor of 11.5, which is derived in this paper (process-determined) from optimization of power dissipation and area, and a factor e, which is derived [3] from optimization of the propagation delay.

In this example the most important improvement due to choosing a tapering factor equal to 11.5 instead of a factor e is a much smaller area ($<1/4$) and a reduced parasitic

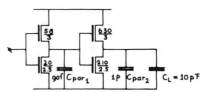

Fig. 9.   The designed inverter string for a practical example.

TABLE I
COMPARISON OF THE PERFORMANCES OF TWO INVERTER STRINGS
WITH DIFFERENT TAPERING FACTORS

	factor 11.5	factor e
number of inverters	2	5
size of the PMOST in the last inverter	630/3	2660/3
dyn. power dissipation	2.5 mW	2.5 mW
par. ,, ,,	1.4 mW	5.4 mW
propagation delay	6.5 nsec	5.5 nsec

power consumption ($\approx 1/4$). This parasitic power consumption is the total power consumption minus the power dissipation which was actually meant ($C_L V^2 f$).

In our case, the propagation delay has only been increased by 1 ns as compared to a tapering factor equal to e.

In summarizing, we can state that optimization of the power dissipation of CMOS driving circuits, like buffers, will lead to a better overall circuit performance (power, delay, and area) then can be achieved by optimization of the propagation delay.

SUMMARY AND CONCLUSIONS

In this paper a simple formula is derived for a quick calculation of the maximum short-circuit dissipation of static CMOS circuits. A detailed discussion of this short-circuit dissipation is given based upon the behavior of the inverter when loaded with different capacitances. It was found that if each inverter of a string is designed in such a way that the input and output rise and fall times are equal, the short-circuit dissipation will be much less than the dynamic dissipation (< 20 percent). This result has been applied to a practical design of a CMOS driving circuit (buffer), which is commonly built up of a string of inverters. An expression has also been derived for a tapering factor between two successive inverters of such a string to minimize parasitic power dissipation. Finally, it is concluded that optimization in terms of power dissipation leads to a better overall performance (in terms of speed, power, and area) than is possible by minimization of the propagation delay.

APPENDIX

It can easily be derived [1] what transistor is needed to discharge a capacitive load C_N from the supply voltage V_{dd}

202

to a voltage V in the time τ_N:

$$\beta_N = \frac{1}{\tau_N} \cdot \frac{C_N}{V_{dd} - V_{T_n}} \cdot \left\{ \frac{2V_{T_n}}{V_{dd} - V_{T_n}} + \ln\left[\frac{2(V_{dd} - V_{T_n}) - V}{V} \right] \right\}$$

(A1)

where

$$\frac{1}{V_{dd} - V_{T_n}} \cdot \left\{ \frac{2V_{T_n}}{V_{dd} - V_{T_n}} + \ln\left[\frac{2(V_{dd} - V_{T_n}) - V}{V} \right] \right\} = A$$

(A2)

is a constant for a given technology. Thus,

$$\beta_N = \frac{C_N}{\tau_N} \cdot A$$

(A3)

and for the $(N-1)$th inverter:

$$\beta_{N-1} = \frac{C_{N-1}}{\tau_{N-1}} \cdot A.$$

(A4)

Again, assuming the inverter to be symmetrical:

$$\beta_n = \beta_p = \beta; \qquad V_{T_n} = -V_{T_p} \quad \text{and} \quad \tau_r = \tau_f = \tau \quad \text{(A5)}$$

and, because of the difference in mobility of holes and electrons:

$$\frac{W_p - \Delta W_p}{L_p - \Delta L_p} = 3 * \frac{W_n - \Delta W_n}{L_n - \Delta L_n}.$$

(A6)

As $W_n \gg \Delta W_n$ and $W_p \gg \Delta W_p$, (A6) reduces to

$$\frac{W_p}{L_p - \Delta L_p} = 3 * \frac{W_n}{L_n - \Delta L_n}.$$

(A7)

With

$$W_n = (L_n - \Delta L_n) \frac{\beta_N}{\beta_{n_\square}}$$

(A8)

and given a linear relation between L_n and L_p

$$L_p = a \cdot L_n$$

(A9)

we find

$$W_p \cdot L_p = W_n \cdot L_n \cdot 3a \cdot \frac{a \cdot L_n - \Delta L_p}{L_n - \Delta L_n}.$$

(A10)

From Fig. 8 it is known that

$$C_{N-1} = C_{g_N} + C_{\text{par}_{N-1}}.$$

(A11)

In a practical design the parasitic capacitance $C_{\text{par}N-1}$ of node $N-1$ will be proportional to its load capacitance C_{gN}, so that

$$C_{\text{par}_{N-1}} = b \cdot C_{g_N} \quad \text{and} \quad C_{\text{par}_N} = b \cdot C_L.$$

(A12)

From (21) and (22) we derive

$$C_{N-1} = (1 + b) \cdot (W_{p_N} \cdot L_{p_N} + W_{n_N} \cdot L_{n_N}) \cdot C_{\text{ox}}.$$

(A13)

This, combined with (A10) yields

$$C_{N-1} = (W_{n_N} \cdot L_{n_N})$$
$$* \left\{ 1 + a \cdot \frac{(3a \cdot L_{n_N} - 3\Delta L_{p_N})}{L_{n_N} - \Delta L_{n_N}} \right\} \cdot (1 + b) \cdot C_{\text{ox}}.$$

(A14)

Equations (A4), (A5), and (A14) result in

$$\beta_{N-1} = \frac{A}{\tau} \cdot (1 + b) \cdot C_{\text{ox}} \cdot \frac{W_{n_N} \cdot L_{n_N}}{L_{n_N} - \Delta L_{n_N}}$$
$$\cdot \left(L_{n_N} - \Delta L_{n_N} + 3a^2 \cdot L_{n_N} - 3a \cdot \Delta L_{p_N} \right).$$

(A15)

Finally, from (18) and (25) we derive

$$\frac{\beta_{N-1}}{\beta_N} = \frac{A}{\tau \cdot \beta_{n_\square}} * (1 + b) \cdot C_{\text{ox}} \cdot L_{n_N}$$
$$\cdot \left\{ L_{n_N}(1 + 3a^2) - \Delta L_{n_N} - 3a \cdot \Delta L_{p_N} \right\}.$$

(A16)

With equations (A1) and (A16) the inverter string is completely determined.

The number of inverters (N), which depends on the ratio in (26), is now determined by

$$\left(\frac{\beta_N}{\beta_{N-1}} \right)^N = \frac{C_N}{C_0}$$

(A17)

where C_0 and C_N represent the input capacitance and the output load capacitance of the inverter string, respectively (Fig. 8).

ACKNOWLEDGMENT

The author wishes to acknowledge the helpful suggestions and contributions made by A. T. van Zanten and C. Hartgring.

REFERENCES

[1] M. I. Elmasry, "Digital MOS integrated circuits: A tutorial," *Digital MOS Integrated Circuits*. New York: IEEE Press pp. 4–27.
[2] A. Kanuma, "CMOS circuit optimization," *Solid-State Electron.*, vol. 26, no. 1, pp. 47–58, 1983.
[3] C. Mead and L. Conway, *Introduction to VLSI Systems*. New York: pp. 12–15.

A 3.8ns CMOS 16 x 16 Multiplier Using Complementary Pass Transistor Logic

KAZUO YANO, TOSHIAKI YAMANAKA, TAKASHI NISHIDA, MASAYOSHI SAITOH,
KATSUHIRO SHIMOHIGASHI, AND AKIHORO SHIMIZU*

CENTRAL RESEARCH LABORATORY, HITACHI LTD., KOKUBUNJI, TOKYO, JAPAN
*HITACHI VLSI ENGINEERING CORP., TOKYO, JAPAN

ABSTRACT

A 3.8ns 257mW CMOS 16x16 multiplier with a supply
voltage of 4V is described. A complementary pass
transistor logic(CPL) is proposed and applied to almost
the entire critical path. The CPL consists of
complementary input/output, nMOS-pass-transistor logic
network and CMOS output inverters. The CPL is twice as
fast as the conventional CMOS due to lower input
capacitance and higher logic construction ability. Its
multiplication time is the fastest ever reported,
including times of bipolar and GaAs ICs, and it is also
shown to be further enhanced to 2.6ns 60mW at 77K.

1. INTRODUCTION

The speed of CMOS devices, which were used mainly in
low-power high-densisty chips, has increased
drastically with the rapid progress in miniaturization.
CMOS speed is getting close to that of Si bipolar
technology; for example, with submicron CMOS technology,
a 15ns 1Mb SRAM[1] and a 7.4ns 16×16b multiplier[2] were
actually realized. However, the recent progress in large
-scale numerical simulations and the real-time digital
-signal processing requires further improvement in CMOS
speed.

A multiplier is an essential element in any digital
signal processing circuit, and constitutes the critical
path in DSP and FPU LSIs. Therefore, the demand for
multiplier-performance improvement is increasing.
Consequently, many fast multipliers based on various
device technologies, e.g. Si bipolar, GaAs, JJ, have
been reported. This paper describes a 0.5μm-CMOS
16x16b multiplier with a multiplication time of 3.8ns
at a supply voltage of 4V, which is the fastest ever
reported including bipolar[3] and HEMT[4] ICs. The circuit
techniques and the technology used in this multiplier
have the potential for 100MHz operation of 32-64bit
floating point multiplication, thus realizing very fast
DSP, FPU and ASICs.

A new family of advanced differential CMOS logic,
complementary pass transistor logic(CPL) is proposed
and fully utilized on almost the entire critical path
to achieve this very fast speed. First, the CPL is
introduced and then CPL implementation of the
multiplier is described.

2. CPL: THE CONCEPT AND EXAMPLES

There have been several differential CMOS logic
families proposed, such as CVSL[5] and DSL[6], for CMOS
-circuit-speed improvement. These have the common
features of complementary-data input/output, nMOS logic
tree, and pMOS cross-coupled load, which can reduce
input capacitance and increase logic-construction
ability; thus, they were claimed to enhance speed.
However, the actual advantage of these circuits is less
than that anticipated in the original paper, as
clarified in Ref.7. This is because the pMOS cross
-coupled latch cannot easily be inverted due to the
regenerative property of the latch.

The main concept behind CPL is the utilization of
nMOS pass-transistor network for logic organization,
instead of source-grounded nMOS trees in the
conventional differential logic, as shown in Fig.1. CPL
consists of complementary input/output, nMOS-pass
-transistor logic network and CMOS output inverters.
The pass transistors perform both pull-down and pull-up,
thus pMOS latch can be eliminated, allowing the
advantages of the differential circuits to be fully
enjoyed in CPL. The CMOS output inverters amplify the
output signal, increase the logic threshold voltage,
and drive capacitive load. The logic-threshold shift is

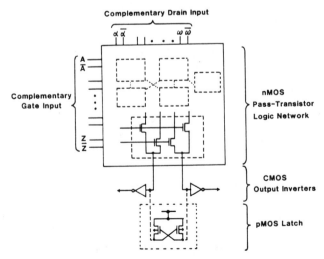

Fig.1 Basic circuit configuration in CPL.

Reprinted from *IEEE Custom Integrated Circuits Conference*, pp. 10.4.1-10.4.4, 1989

necessary, because the high level of the pass
-transistor output is lowered from the supply voltage
by the threshold voltage of the nMOS. These output
inverters are "overhead", in the sense that they are
needed whether the circuit has 1, 2, or many inputs.
Therefore the overall device count will be minimized by
designing with complex logic functions. The pMOS latch
can also be added to CPL, as shown in Fig.1, in order
to only decrease static power consumption, as opposed
to the conventional pull-up function.

Arbitrary Boolean functions can be constructed from
the pass-tranistor network by combining four basic
circuit modules: an AND/NAND module, an OR/NOR module,
an XOR/XNOR module and a wired-AND/NAND module, as
shown in Fig.2, in which the XOR/XNOR module is used
once[8]. Note that these various functions are produced
by an identical circuit configuration with the change
of input configuration. This property of CPL is
apparently suitable for masterslice design. The 3- and
the more-way input logic functions can also be easily
constructed similarly to the 2-way logic in Fig.2.

The CPL full adder, which is used in the present
multiplier design, is shown in Fig.3(a). Since pMOS can
be eliminated in logic construction, the input
capacitance is about half that of the conventional CMOS
configuration[9] shown in Fig.3(b), thus achieving higher
speed and lower power dissipation. Moreover, the
powerful logic-construction ability of CPL due to the
multi-level pass-transistor network realizes complex
Boolean functions efficiently in a small number of MOS
transistors, thus reducing area and delay time. In fact,
the transistor count in CPL full adder is 28 in the
active area of $3.7 \times 11.4mm^2$, whereas in CMOS it is 40
in the area of $5.5 \times 8.6mm^2$. Further, the complementary
-input/output function eliminates the internal inverter
to provide XOR input in the full adder, thus reducing
critical-path gate count. The measured worst delay time
of the CPL full adder in a 10-stage full-adder chain is
as small as 0.31ns/b, as shown in Fig.4. On the other
hand, the conventional CMOS full adder, which is
designed and fabricated on the same wafer, requires 0.
63ns/b. Thus the CPL is twice as fast as the
conventional CMOS.

Another example of CPL in the multiplier is as a 4
-bit carry-look-ahead(CLA) unit used in a final 32b
adder. The circuit of this unit is shown in Fig.5. The
32b adder includes 6 CLA units. The unit employs CPL to
quickly transfer the carry-output to the upper unit(C4
and C4) after receiving the carry-input from the lower
unit(C0 and C0). Complementary carry data can be
inverted simply by twisting the carry lines. Thus the
circuit requires only one pass transistor and a simple
inverter to generate next-unit input. The measured
delay time is only 0.15ns/4b.

3. MULTIPLIER ARCHITECHTURE

The 16×16-bit multiplier was designed using a
parallel multiplication architecture, as shown in Fig.6.
A Wallace-tree adder array and a CLA adder were used to
minimize the critical-path gate stages. The critical
path consists of a partial product generator, 7 full
adders, a look-ahead carry generator, and 2 carry
-propagate circuits (Fig.5).

4. DEVICE FABRICATION

The multiplier and the full-adder chains are
fabricated with the double-level-metal $0.5 \mu m$-CMOS
technology. The minimum feature size is $0.5 \mu m$, and
the gate oxide thickness is 12.5nm. In this fabricated 0.
$5 \mu m$-CMOS device optimized for CPL, the pass-transistor
-nMOSs are designed to have a threshold voltage of 0V
as shown in Figs.3 and 5. This reduces the static power
dissipation and delay time. The interconnection metal
consists of 1st-level W and 2nd-level Al. The W was
adopted for its high immunity to electromigration.

5. PERFORMANCE RESULTS

A microphotograph of the multiplier is shown in Fig.7.
The multiplier has 8500 transistors in an active area
of $1.3 \times 3.1mm^2$, whereas the area including bonding pads
is $1.6 \times 4.5mm^2$. The performance was measured using the
worst case pattern, $FFFF \times 8001-7FFF \times 8001$. Circuit
simulation was also performed to confirm that this
pattern consists of a series of worst case operations
of the full adder and the CLA. The multiplier chip was
mounted on the 68-pin pin-grid-array ceramic package,
followed by waveform observation of the clock inputs
and the product outputs through the source follower
circuit on the chip. The multiplication time was
measured at both room and liquid nitrogen temperatures.
The latter was considered for high-end applications,
such as super computer. The maximum multiplication time
was 3.8ns and 2.6ns at room and liquid-nitrogen
temperatures, respectively, as shown in Figs.8, 9, and
10. Power dissipation was 257mW and 60mW at 300K and 77K,
respectively, for 10MHz operation in a pattern of $FFFF \times$
$FFFF-0000 \times FFFF$.

Multiplication times at both room and liquid-nitrogen
temperatures are faster than those of any other devices,
including bipolar transistor, GaAs-MESFETs, and HEMTs,
as is shown in Fig.11. In addition, the power
dissipation of the multiplier is much lower than that
of the other devices. The features of this multiplier
are summarized in Table 1.

6. CONCLUSIONS

This paper described a fast 16×16-bit multiplier
using a new differential CMOS logic family, CPL. In CPL,
differential logics are constructed without pMOS
latching load, thus the speed is twice as fast as

conventional CMOS. The multiplier is the fastest ever reported at both 300K and 77K, proving that the half micron CMOS technology fully utilizing CPL at least has a speed which is competitive with those of other fast devices with a much smaller power dissipation at room temperature, and is faster at liquid-nitrogen temperature.

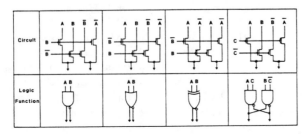

Fig.2 Circuit modules in CPL.

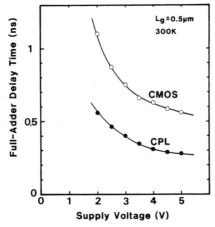

Fig.4 Full-adder delay time vs. supply voltage.

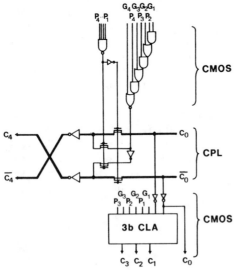

Fig.5 4-bit carry-look-ahead circuit using both CMOS and CPL. Cj: carry, Gj: generator, Pj: propagator.

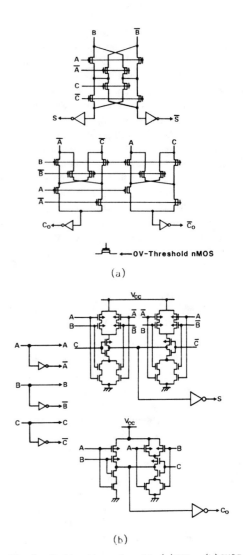

(a)

(b)

Fig.3 Full-adder circuit (a)CPL, (b)CMOS.

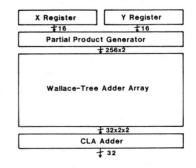

Fig.6 Block diagram of 16×16b multiplier.

Fig.7 Microphotograph of the 16×16b multiplier chip(left) and 10-stage full-adder chains(right)

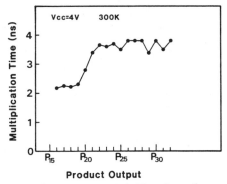

Fig.8 Measured multiplication time.

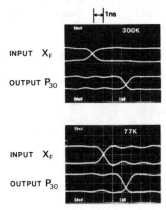

Fig.9 Measured waveforms

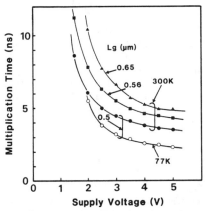

Fig.10 Measured multiplication time vs. supply voltage.

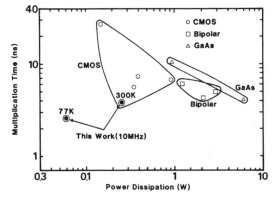

Fig.11 Comparison of delay and power dissipation of high speed 16×16b multipliers.

Table 1. Features of 16×16 bit parallel multiplier

Architecture	Wallace Tree + Carry Look Ahead		
Technology	0.5μm CMOS		
Gate Length	0.5μm		
Gate Oxide Thickness	12.5nm		
Metal Line/Space W	0.8/0.8		
Al	1.0/1.0		
Active Area	1.3x3.1mm²		
Transistor Count	8500		
Multiplication Time	3.8ns (@4V, 300K),	2.6ns	(@4V, 77K)
Power Dissipation(10MHz)	257mW (@4V, 300K),	60mW	(@4V, 77K)

ACKNOWLEDGEMENT

The authers wish to thank Yoshio Sakai from Hitachi Takasaki Works, Osamu Minato from Hitachi Musashi Works, and Toshiaki Masuhara, Tetsuya Nakagawa, Tohru Baji, Kenji Kaneko, Terumi Sawase, Koichiro Ishibashi from Central Research Laboratory for their useful discussions. The authers are also greatly indebted to Kunihiro Yagi and the device processing staff members for their support throughout the sample fabrication, and to Atsushi Kawamata from Hitachi VLSI Engineering Corp. and Kiyotsugu Ueda for their layout design.

REFERENCES

1) K.Sasaki, S.Hanamura, K.Ueda, T.Ohno, O.Minato, K. Nishimura, Y.Sakai, S.Meguro, M.Tsunematsu, T.Masuhara, M.Kubotera, and H.Toyoshima, ISSCC, pp174(1988)
2) Y.Oowaki, K.Numata, K.Tsuchiya, K.Tsuda, A.Nikayama, and S.Watanabe, ISSCC, pp52(1987)
3) M.Suzuki, M.Hirata and S.Konaka, ISSCC, pp70(1988)
4) K.Kajii, Y.Watanabe, M.Suzuki, I.Hanyu, M.Kosugi, K. Odani, T.Mimura and M.Abe, 1987 CICC, pp199(1987)
5) L.G.Heller, W.R.Griffin, J.W.Davis, and N.G.Thoma, ISSCC, pp16(1984)
6) L.C.M.G.Pfennings, W.G.J.Mol, J.J.J.Bastiaens, and J. M.F.van Dijk, ISSCC, pp212(1985)
7) K.M.Chu and D.L.Pulfrey, IEEE J.Solid State Circuits, Vol.SC-22, pp528(1987)
8) T.Kengaku, Y.Shimazu, T.Tokuda, and O.Tomisawa, IECE of Japan, 2-83(1987)
9) M.Uya, K.Kaneko, J.Yasui, ISSCC, pp90(1984)

A HIGH SPEED, LOW POWER, SWING RESTORED PASS-TRANSISTOR LOGIC BASED MULTIPLY AND ACCUMULATE CIRCUIT FOR MULTIMEDIA APPLICATIONS

Akilesh Parameswar, Hiroyuki Hara, Takayasu Sakurai

Toshiba Corporation
1, Komukai Toshiba-cho, Saiwai-ku, Kawasaki, Japan 210

Abstract

Swing Restored Pass-transistor Logic (SRPL), a high speed, low power logic circuit technique for VLSI applications is described. By the use of a pass-transistor network to perform logic evaluation, and a latch type swing restoring circuit to drive gate outputs, this technique renders highly competitive circuit performance. An SRPL based Multiply and Accumulate Circuit for multimedia applications is implemented in double metal $0.4\mu m$ CMOS technology.

Introduction

To date, the most widely used VLSI circuit design technique has been full CMOS. It has been attractive because it makes it easy to implement reliable circuits that have excellent noise margins. However, the continuing push for higher performance systems has, in recent years, brought the disadvantages of full CMOS to the fore, and a number of researchers have proposed alternative logic techniques (1-3). The majority of these have been static techniques because dynamic logic styles still suffer from charge sharing and noise margin problems, and difficulties in design and design for testability.

Complimentary Pass-transistor Logic (CPL) (1), uses a complimentary output pass-transistor logic network to perform logic evaluation, and CMOS inverters for driving of the outputs. This arrangement however suffers from leakage current through the inverter. Double Pass-transistor Logic (DPL) (2), uses both pMOS and nMOS devices in the pass-transistor network to avoid non-full swing problems, but it has high area and high power drawbacks. As the name suggests, Differential Cascode Voltage Switch with Pass Gate (DCVSPG) (3) is the same as the cascode voltage switch logic proposed in (4), but uses a pass-transistor network for logic evaluation. This logic style suffers from degraded pull down performance when used in a long chain.

In this paper, we propose a high speed, low power logic circuit technique that attempts to overcome these problems.

Swing Restored Pass-transistor Logic

Basic Circuit

The generic Swing Restored Pass-transistor Logic (SRPL) gate consists of two main parts as shown in Fig. 1. A complimentary output pass-transistor logic network that is constructed of n channel devices, and a latch type swing restoring circuit consisting of two cross coupled CMOS inverters. The gate inputs are of two types, pass variables that are connected to the drains of the logic network transistors, and control variables that are connected to the gates of the transistors. The logic network has the ability to implement any random Boolean logic function. Fig.2, for instance, shows the implementation of an SRPL full adder. The complimentary outputs of the pass-transistor logic network are restored to full swing by the swing restoration circuit.

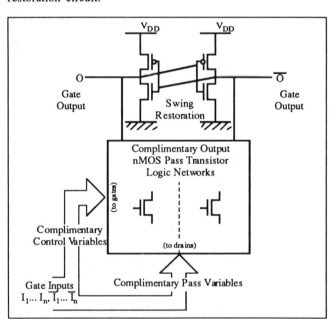

Figure 1 : Generic SRPL Gate

Gate Optimization

We have found that in the interests of speed, the nMOS transistors of the logic network farther away from the output should have larger drivability (i.e. size) than those closer to the output. This is because the transistors closer to the output pass smaller swing high signals due to the voltage drop across the transistors farther away from the output. The precise values for a given circuit depend on layout and other circuit considerations, so they must be determined by case by case simulation. Typical values are indicated in Fig. 2.

Reprinted from *Proceedings of Custom Integrated Circuits Conference*, pp. 278-281, 1994.

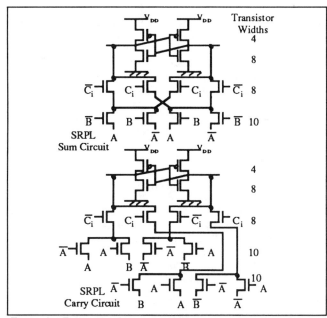

Figure 2 : Full Adder Circuit in SRPL

The optimization of the swing restoring latch is an important determinant of overall gate speed. If high speed latch inversion is required, the pMOS transistors should not be made too large. However, a large pMOS transistor size means that faster driving of the load is possible. Hence a trade-off exists, which is qualitatively demonstrated by the graph in Fig. 3. Simulations were performed on identical cascaded SRPL gates, each with a fan-out of 2, and assuming typical pass network transistor sizes shown in Fig 2. Simulations were done with SPICE, using parameters of Toshiba's 0.4μm CMOS process. The x-axis of Fig. 3 plots the ratio of the size of the pMOS transistor to the size of the topmost pass network nMOS transistor, while the y-axis plots the delay from the $0.5V_{DD}$ mark of a pass input of the gate to the $0.5V_{DD}$ mark of the output of the *subsequent* gate in the cascade.

For very small values of the $p_{latch}/n_{network}$ ratio, the gate output load becomes too large for the pMOS to be able to drive efficiently, and for very large values, the latch requires an inordinate amount of time to flip, reaching infinity (i.e. doesn't invert) over a certain limit. There exists a further dimension to the optimization, in that the curve of Fig. 3 moves up for very small or very large values of the $n_{latch}/n_{network}$ ratio. If the latch nMOS device is too small, discharging is penalized, whereas if it is too large, it introduces undue capacitive loading.

On the whole, the graph shows that though there is a trade off in determining the size of the pMOS transistor in the latch, there exists substantial design margin, making it easy to design circuits in SRPL. This design margin also means that SRPL circuits are quite robust against process variations, which might cause the threshold voltages of the transistors to fluctuate.

Performance Comparison with Competing Techniques

Full adders in CMOS, CPL, DPL, DCVSPG and SRPL were constructed, and simulated in the cascaded conditions shown in Fig. 4. Again, 0.4μm CMOS process parameters were used to perform SPICE simulations. The worst case waveforms for each

of the full adders are shown in Fig. 5. Other performance values are recorded in Tab. 1.

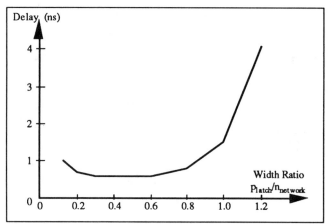

Figure 3 : Dependence of Delay on Transistor Widths

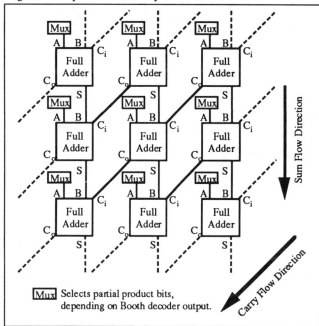

Figure 4 : Carry Save Addition of Partial Products

As Fig. 5 shows, CMOS has the slowest speed. Moreover, power consumption is quite high. The main reason for these poor performance figures is that the inefficient pMOS network of CMOS leads to a higher transistor count, larger gate area, and larger input capacitances due to the poor drivability of the pMOS transistor. DPL proves to be about 30% faster than CMOS, but this is at the expense of a higher transistor count, and more power consumed. DCVSPG is much faster than CMOS, but suffers from the problem that it cannot be used in the array structure of Fig. 4. The reason for this is that there is no pull down mechanism other than that through the pass-transistor networks. Thus for long chains of cascaded gates, the pull down becomes severely degraded as shown by the dotted line of Fig. 5.

CPL, as Fig. 5 clearly shows is the fastest of the five techniques. However, this is achieved at the expense of high power consumption. Furthermore, CPL suffers from the major drawback that it is a non-full swing technique. The non-full swing signals at the inputs of the inverters mean poor noise margins, particularly as the inverter threshold is susceptible to process variations. Moreover, CPL circuits consume static power because of the leakage current that is always flowing through one of the inverters of a gate. The inverter output never quite reaches V_{SS} as the curve of Fig 5. shows. Because a V_{DD} of 3.3V is high relative to a channel width of 0.4μm, the speed degrading effects of the leakage are not prominent. However, when V_{th} is a significant fraction of V_{DD}, as it will certainly be in the future, the fall time of the output lengthens, and CPL becomes slower than SRPL.

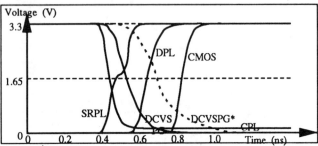

Figure 5 : Full Adder Worst Case Waveforms

SRPL has good speed performance. In the simulated conditions of Fig. 4, each SRPL circuit within the full adder fans out to only two other similarly sized circuits (carry and sum). This implies relatively light loading conditions, much less than the usual CMOS stage ratio of 3.5 or 4. It is important to note that this condition is not restricted to the simulated case. Low fanout is a very common occurrence in the design of VLSI circuits, particularly in data paths. In such conditions, it makes sense to connect the pass-transistor network output to the gate output, and to restore the swing with the cross coupled pair of inverters. The initial rise in voltage caused by the pass network output takes the gate output voltage a good margin above the $V_{th,n}$ of the transistors of the following gate, speedily setting up the correct logical path. Also because of the relatively light loading conditions, the inversion of the latch is faster, and so the $P_{latch}/n_{network}$ ratio can be made slightly larger. Thus a good pull up time through the *a priori* set up logical path of the following gate is achieved.

Table 1 : Comparison of Full Adder Circuits

	CMOS	CPL	DPL	DCVS PG	SRPL
Speed (ns)	0.82	0.44	0.63	0.53	0.48
Power at 100Mz (mW)	0.52	0.42	0.58	0.3	0.19
Power-Delay Product (normalized)	1.0	0.43	0.86	0.37	0.21
Transistor Count	40	28	48	24	28

As Tab. 1 shows, SRPL has the lowest power consumption and the lowest power-delay product of the different techniques. The main reasons for the low power are the low transistor count and the low input capacitance. Also, the fast inversion action of the latch quickly cuts off any d.c. path through the pass network.

In summary, SRPL shows itself to be a very competitive low power, high speed circuit technology. SRPL circuits will also occupy less area because of the lower transistor count. Particularly because the number of p channel devices is small, less area will be wasted on well boundary separation. The pMOS transistors are also smaller than, for instance, the CPL case, leading to slightly better area performance. This promising logic technique was used to construct a Multiply and Accumulate Circuit (MAC) for multimedia applications.

Multiply and Accumulate Circuit

The multiply and accumulate operation is crucial to a wide range of signal processing applications. With the increasing level of integration of processors dedicated to multimedia, it has become essential that high speed MAC macrocells be provided on chip. However, high speed is not the sole imperative. System portability is also a key issue, and hence low power is also very important. The MAC presented in this paper was designed with these requirements foremost in mind.

MAC Architecture

The overall circuit is shown in Fig. 6. The multiplier and multiplicand are 16-bit wide, whereas the accumulated result has a bit width of 32. A pipelined scheme was not implemented because the frequency of operation was expected to be more than sufficient to cover even the most advanced multimedia applications. Furthermore, pipelining introduces problems of complicated control and timing, and extra area and power required by the pipeline registers.

A Booth decoding scheme was used to obtain 8 partial products, which are added in a carry save manner as shown in Fig. 4. Each full adder row receives a running sum and carry from the row above. The very top adder of each column of the summation receives one of its inputs from the accumulated total of the previous cycle, which is fed back as shown in Fig. 6. A Wallace tree architecture for partial product addition was not used because the such an architecture would lead to larger power consumption due to the larger area and wiring requirements. Each of the full adders in the partial product summation array is constructed using the SRPL technique described above.

The final CLA adder cum register to which the partial product summation array outputs its carry saved result uses the same design as that of (5), where a dynamic sense amplifying scheme is used to perform both carry propagation, and latching of the final result. This design is ideally suited to the MAC design because of the high speed addition followed by the instantaneous latching. The complimentary outputs of the SRPL based summation array perfectly match the complimentary input requirements of the sense amplifying technique used by the final adder. It should be noted though that the dynamic sense amplifying technique used in (5) is completely different from the static swing restoring technique proposed in this paper.

Performance

The MAC was fabricated using a double metal 0.4μm process as summarized in Tab. 2. The chip photomicrograph is shown in Fig. 7. As Tab. 2. shows, the MAC operates at a maximum frequency of 150MHz, which more than sufficient for multimedia applications. Moreover, the power consumed is only 34mW at this frequency, satisfying the other important multimedia requirement. The 150MHz operating frequency translates to a one cycle delay time of 6.7ns. For comparison, the MAC was simulated with a CPL partial product addition array. The simulated delay time was 6.3ns.

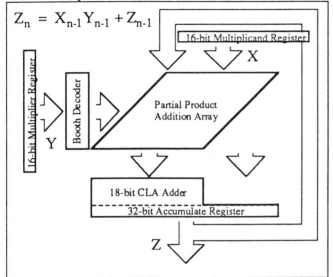

$$Z_n = X_{n-1} Y_{n-1} + Z_{n-1}$$

Figure 6 : Multiply and Accumulate Circuit

Though the SRPL MAC is 0.4ns slower than the CPL version, it should be remembered that CPL is the fastest technique ever reported, being nearly twice as fast as CMOS. Moreover, the power consumed by the CPL version was estimated to be more than twice that consumed by the SRPL MAC. In addition, as has been mentioned, CPL suffers from margin problems that will be exacerbated by the future reduction of the supply voltage, and this reduction in V_{DD} will also lead to speed degradation.

Conclusion

A new high speed, low power logic circuit technology was proposed, and used to implement a multiply and accumulate circuit in double metal 0.4μm CMOS. The MAC achieves a frequency of 150MHz, and 34mW and shows much promise for multimedia applications.

Table 2 : MAC Characteristics

Technology	CMOS process
n Channel Length	0.4μm (Eff. 0.39μm)
p Channel Length	0.5μm (Eff. 0.47μm)
Gate Oxide Thickness	9 nm
No. of Metal Layers	2
Power Supply Voltage	3.3 Volts
Operating Frequency	150MHz
Latency	0 cycles
Power Consumed at 150MHz	34 mW
Active Area	0.98 mm²

Figure 7 : MAC Photomicrograph

Acknowledgements

The authors would like to gratefully thank the assistance and encouragement of Fumihiko Sano, Yoshinori Watanabe, Masataka Matsui, Hidetoshi Koike, Fumitomo Matsuoka, Masakazu Kakumu and Kenji Maeguchi

References

(1) K. Yano et al, "A 3.8ns CMOS 16x16 Multiplier Using Complimentary Pass-transistor Logic," vol. 25, no. 2, pp.388-395, April 1990.

(2) M. Suzuki et al, "A 1.5ns 32bit CMOS ALU in Double Pass-transistor Logic," 1993 IEEE International Solid-State Circuits Conference, pp. 90-91.

(3) F.S. Lai and W. Hwang, "Differential Cascode Voltage Switch with Pass Gate Logic Tree for High Performance CMOS Digital Systems," 1993 International Symposium on VLSI Technology, Systems and Applications, pp. 358-362.

(4) L.G. Heller, W.R. Griffin, J.W. Davis and N.G. Thoma, "Cascode Voltage Switch Logic : A Differential CMOS Logic Family," 1984 IEEE International Solid-State Circuits Conference, pp. 16-17

(5) M. Matsui et al, "Sense-amplifying pipeline flip-flop scheme for 200MHz video de/compression macrocells," in press (1994 IEEE International Solid-State Circuits Conference).

Static power driven voltage scaling and delay driven buffer sizing in Mixed Swing QuadRail for sub-1V I/O swings

Ram K. Krishnamurthy, Ihor Lys, L. Richard Carley

Department of Electrical and Computer Engineering
Carnegie Mellon University, Pittsburgh, PA 15213

Abstract

This paper describes and explores the design space of a four power-supply rail methodology (called Mixed Swing QuadRail) for performing low voltage logic in a high threshold voltage CMOS fabrication process. Power and delay trade-offs are studied to suggest approaches for efficient selection of voltage levels and buffer transistor sizes. Posynomial models for QuadRail power and delay are derived to show that at reduced I/O swings (sub-1V), both under- and over-sizing of transistors can lead to steeply increased delays. Transistor sizing techniques are proposed for optimizing delay and energy per logic operation as a function of load capacitance and voltage levels. Experimental results from detailed HSPICE simulations and an And-Or-Invert (AOI222) QuadRail test chip fabricated in the Hewlett-Packard 0.5μm process are presented to support the models and demonstrate significant power reduction compared to static CMOS.

1 Introduction

The fast-growing portable communications industry, driven by the need for performing high throughput DSP and multimedia tasks at low power, has spawned great interest in innovative low power circuit design techniques [1]. Most of these techniques have focused on using standard CMOS logic circuits and lowering the power supply voltage (voltage scaling) [2], because of its quadratic influence on dynamic power consumption. For instance, a recently published low power microprocessor [3] and DSP embedded processor [4] targeted at portable applications operate at power supply voltages significantly below the process-permitted maximum voltages. However, when power supply voltages are scaled below the sum of the threshold voltages of a NMOS and PMOS device ($V_{tn} + |V_{tp}|$), gate delays increase steeply making them a substantial critical path delay contributor. Furthermore, variations in device threshold voltages due to inevitable variations in the IC fabrication process have limited the lowest possible operating voltage to slightly above the larger of V_{tn} or V_{tp} [5]. In a CMOS process with nominal V_t's of 0.75V, this lower bound is about 1V. Random variations in transistor V_t are inversely proportional to $\sqrt{L \cdot W}$, and the constant of proportionality can be as high as 30mV-μm [6]. With rapidly reducing feature sizes, it is

obvious these variations are becoming even higher. Reducing nominal V_t's and electronically controlling their variations [7], [2] are possible solutions, but they are both difficult and expensive.

In this paper, we describe the Mixed Swing QuadRail methodology, which addresses maximum possible voltage scaling with little or no reduction in operating speed, in a high threshold voltage CMOS fabrication process. The described architecture requires four power supply rails to be distributed to all circuits sharing this signalling methodology [8], and logic is performed by intermixing high and low swinging voltage signals. The design space of Mixed Swing QuadRail is explored to study the power and delay trade-offs. A static power driven voltage scaling approach is described for selection of high and low swing voltage levels, by evaluating the ratio of off- to on-drive currents for a worst-case on-drive scenario. Posynomial power and delay models for QuadRail are derived to study device sizing tradeoffs and techniques are proposed for efficient transistor sizing to optimize delay and energy per logic operation. Comparisons of our power and delay models to HSPICE simulations using Level 13 BSIM1 models in the Hewlett-Packard CMOS14TB 0.5μm process are performed. Experimental results from detailed HSPICE simulations and an And-Or-Invert (AOI222) QuadRail test chip fabricated in the same process are presented to demonstrate significant power reduction compared to static CMOS.

2 Mixed Swing QuadRail gate architecture

The essence of the Mixed Swing QuadRail methodology, is that it allows the designer to exploit the best aspects of both voltage scaling, and full-swing CMOS. As the name suggests, this requires an additional pair of power supply rails and special low swing drivers. Unlike other low swing transceiver techniques, we propose efficient off-chip low voltage supplies. Dynamic power reduction is obtained by driving loads at reduced voltage swings while performing logic at high swings. For typical digital ICs, the power used to drive interconnect loads can be 50-80% of total power. As feature sizes shrink, this percentage will grow because interconnect capacitance (due to fringing) falls more slowly than gate capacitance (due to active area). Thus the technique will be even more applicable at reduced feature sizes.

Fig. 1 shows a two stage Mixed Swing QuadRail gate, consisting of a logic stage and driver/buffer stage. The buffer stage is a CMOS inverter, with high swinging inputs (Vdd1-Vss1 = V_l) and low swinging outputs (Vdd2-Vss2 = V_b), both centered to maximize noise margins. No DC path exists between the supplies. PMOS devices in both stages are 2.5X wider than the NMOS to equalize rise/fall times. The buffer transistors are ra-

Reprinted from *International Symposium on Low Power Electronics and Design,* pp. 381-386, August 1996.

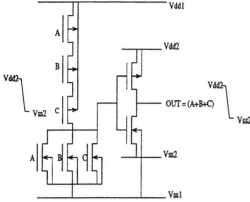

Figure 1: Mixed Swing (two-stage) QuadRail 3-input OR gate.

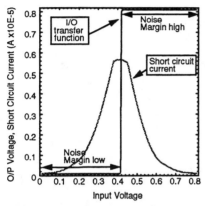

Figure 2: Mixed Swing QuadRail DC transfer chanracteristics.

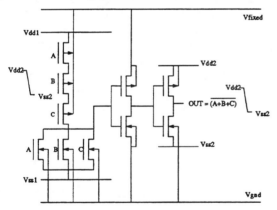

Figure 3: Mixed Swing (three-stage) QuadRail 3-input NOR gate.

tioed by a factor k ($>=1$) relative to logic stage transistors to improve current drive. Each stage has its own N-well in order to minimize body effect. Since the devices are driven by $(V_l + V_b)/2$, switching performance remains good. The ratio of load to driver input capacitance (which in typical ICs is large), sets an upper bound on power savings compared to static CMOS.

The logic stage is identical to a CMOS inverting gate topology, except its inputs have reduced swings. This is tolerable because the transition region in a CMOS gate is smaller than the input range. Noise margins are smaller in absolute terms, but still large compared to some logic families, e.g., ECL. However, care must be taken to control both power supply noise and crosstalk, especially from high swing lines to low swing lines.

Both these problems are best addressed by developing CAD tools that can assess these problems and can design to meet noise margin specifications. Noise margins are set by maximum allowable totempole currents in the logic stage, and are approximately $V_t/2$ as shown in Fig 2. The multi-stage gate has high gain, fully restored outputs, and essentially Class B power/switching characteristics.

This technique can be extended to three (or more) stages as shown in Fig. 3 to allow larger voltage differences between the highest and lowest swing stages by using intermediate stages. These are also CMOS inverters (or gates), and can perform logic (not shown here). Because the buffer's input swing is increased, the gate's output drive is greater for a given buffer size. Although the extra stage adds delay, the increase in drive can result in a 4X reduction in the driver's size and input load. Alternately, the steeper slope at the output can cause lower short circuit power in the fanout gates, and lower delay even with the added stage. Any number of high or reduced voltage logic stages can be cascaded to form more complex functions, and followed by a buffer to deliver the output to the next gate. This is desirable because grouping of several clustered gates into a single more complex gate reduces delay, power and area. These added intermediate gates consume negligible static power, because of their full swing inputs. The avoided buffer stages would have added input load (driven full swing) like the extra intermediate gate itself, as well as parasitic output load.

The analysis presented in the ensuing sections is for the two stage architecture in Fig 1, but can be extended to more number of stages.

3 Mixed Swing QuadRail power and delay models

The dynamic power dissipated by a QuadRail gate driving a load capacitance C_{load} can be expressed as the sum of the energies drawn by each stage from their respective supply rails over one clock cycle [2], i.e.,

$$P_{dyn} = \alpha \cdot k \cdot C_{in} \cdot \left(V_l\right)^2 \cdot f_{clk} + \alpha \cdot C_{load} \cdot \left(V_b\right)^2 \cdot f_{clk} \tag{1}$$

where, C_{in} is the gate capacitance per input, α is the switching activity, f_{clk} is the input signal frequency, and k is as defined in section 2. Parasitic source/drain capacitances for each stage are accounted for in C_{load} and kC_{in}. The static and short circuit power components in the logic stage are as given in [9], and are negligible for the buffer stage. As the buffer transistor size increases, logic stage loading increases, increasing the dynamic power. This decreases the buffer's switching time, reducing short circuit power in all receivers (this reduction is more significant for large fanouts). Thus *QuadRail circuit power dissipation is a posynomial [10] function of buffer transistor size.*

The quadratic relationship between dynamic power and V_b suggests that smaller V_b is desirable for minimal power. This is limited by the smallest I/O swings possible under noise margin constraints [11]. The maximum separation between logic and buffer swings is limited by totempole off-currents in the logic stage. Thus, selection of high and low voltage levels involves careful consideration of the off-drive currents. Section 4 explains our static power driven approach for optimal volt-

213

age level selection.

Transistor level optimization problems have typically adopted RC-tree delay models [9], which deviate from SPICE simulations by 10-20%, yielding suboptimal solutions [12]. This is primarily due to not considering input waveform slope and short channel effects, both of which become significant at submicron feature sizes. Further, at reduced power supply voltages, transistors are predominantly operating in the saturation region, and a resistance approximation of a transistor during switching is inadequate. We have derived an analytical gate delay model for QuadRail gates taking into consideration both input waveform slope (approximated as a ramp signal [13]) and channel length modulation, a dominant short channel effect [9]. The expression for the 50% rising/falling delay of each stage is as follows (mathematical derivations are omitted due to space constraints):

$$Delay_{logic} = \frac{2 \cdot k \cdot C_{in}}{\beta_1 \cdot \lambda} \cdot \frac{1}{\left(\Delta + V_b - V_{t1}\right)^2} \cdot \ln\left(\frac{V_l + \frac{1}{\lambda}}{\frac{V_l}{2} + \frac{1}{\lambda}}\right) +$$

$$\frac{t_T}{} -$$

$$\frac{t_T}{3 \cdot V_b} \cdot \frac{1}{\left(\Delta + V_b - V_{t1}\right)^2} \cdot \left(\left(\Delta + V_b - V_{t1}\right)^3 - \left(\Delta - V_{t1}\right)^3\right) \quad (2)$$

$$Delay_{buffer} = \frac{C_{load}}{k \cdot \beta \cdot \left(\Delta + V_b - V_{t2}\right)} \cdot \ln\left(\frac{4 \cdot \left(\Delta + V_b - V_{t2}\right) - V_b}{2 \cdot \left(\Delta + V_b - V_{t2}\right) - V_b}\right) +$$

$$m \cdot t_{1 (r/f)} \quad (3)$$

where, $t_{1(r/f)}$ is the first stage output's rise/fall time, given by:

$$t_{1 (r/f)} = \frac{2 \cdot k \cdot C_{in}}{\beta_1 \cdot \lambda} \cdot \frac{1}{\left(\Delta + V_b - V_{t1}\right)^2} \cdot \ln\left(\frac{0.9V_l + \frac{1}{\lambda}}{\Delta + V_b - V_{t1} + \frac{1}{\lambda}}\right) +$$

$$\frac{t_T}{} -$$

$$\frac{t_T}{3 \cdot V_b} \cdot \frac{1}{\left(\Delta + V_b - V_{t1}\right)^2} \cdot \left(\left(\Delta + V_b - V_{t1}\right)^3 - \left(\Delta - V_{t1}\right)^3\right) +$$

$$\frac{k \cdot C_{in}}{\beta_1 \cdot \left(\Delta + V_b - V_{t2}\right)} \cdot \ln\left(\frac{2 \cdot \left(\Delta + V_b - V_{t2}\right) - 0.1V_l}{0.1V_l}\right) \quad (4)$$

Δ is the seperation between rails, i.e., Vdd1-Vdd2 = Vss2-Vss1, t_T is the input rise/fall time, λ is the channel length modulation factor [9], β_1 and β are the transconductance gain factors of the logic stage and a unit-sized (1X NMOS, 2.5X PMOS) transistor respectively, V_{t1} and V_{t2} are the logic and buffer stage threshold voltages[1], and m is an empirically fitted constant for a given set of voltage levels[2].

Increasing the buffer transistor size (k) leads to increased load on the logic stage while improving the buffer current drive, i.e., *QuadRail delay is also a posynomial function of buffer transistor size*. This suggests that there exists an optimal buffer

---
1. logic and buffer stage threshold voltages are different because opposite type devices are in conduction for any input combination that causes a transition at the output.

2. since only a portion of the logic stage output's slope affects the buffer stage delay, the input waveform slope's contribution is empirically fitted through HSPICE Level13 BSIM1 models in our analysis.

transistor size for a gate (for minimal delay) as a function of voltage levels and load capacitance. Section 5 explores the selection of buffer transistor size for delay and energy per logic operation optimization.

4 Static power driven voltage level selection

As mentioned in section 3, selection of high and low voltage levels in QuadRail is critical for minimizing static power as well as noise margin degradation. In order to ensure strongly turned-off devices in the logic stage, we must restrict the off-currents (static power) to an extremely small fraction of the average on-drive currents (dynamic power). Fig. 4 shows the ratio of logic stage totempole off-current to the worst-case on-drive current vs. high voltage swing for buffer swings of 0.4-1.0V for the 3-input OR gate (Fig. 1). It is observed that all graphs have two distinct regions - a steeply falling region, where I_{off} (I_{on}) falls quadratically (linearly) with V_{logic}, and a flat region where I_{off} falls exponentially with V_{logic}, due to sub-threshold conduction. Selecting a I_{off}/I_{on} ratio defines unique logic voltage swings at these buffer voltage swings; the smaller this ratio, the tighter the turn-off. As an example, selecting the "knee" of these graphs as operating points, the static currents are less than 2.5% of the on-drive currents. Fig. 5 shows these "knee" points on a buffer swing vs logic swing plot. It is observed that the graph is approximately linear and corresponds to roughly $V_{logic} = V_{buffer} + 2V_t$. Any operating point above this line implies a larger static dissipation ($I_{off} > 2.5\%$ of I_{on}) and any operating point below this line implies even tighter turn-off at the cost of increased logic and buffer stage delays. Thus, scaling down operating buffer and logic voltage levels along this line offers an efficient technique for simultaneous reduction of static and dynamic power dissipation, without degrading the switching characteristics and noise margins. As an example, at $V_{logic} = 2.0V$ and $V_{buffer} = 0.6V$, a dynamic power reduction of 10X compared to static CMOS operating at 2.0V is obtained for the same load capacitance, buffer size, and clock frequency, while ensuring sufficient turn-off characteristics.

5 Delay driven buffer size selection

While optimal logic and buffer swings are set by static power driven techniques, selection of buffer transistor sizes is determined by delay constraints. From (3) it is seen that for large loads, unit-sized buffers have inadequate current drives and high delays. Since QuadRail delay is posynomial, there exists an optimal buffer transistor size (guaranteed to be a global minima [10]), for which delay is minimized. This *delay optima* is determined by differentiating (2)+(3) with respect to k. The optimal buffer transistor size depends on $\sqrt{C_{load}}$, $\sqrt{\beta_1}$, and approximately on the square root of the ratio of V_{buffer} to V_{logic}. Since QuadRail power is also a posynomial function of buffer size, there exists a value of k, for which power is also minimized. This *power optima* can be determined if the fanout and interconnect loading on a gate, and the buffer transistor sizes of those fanout gates are known. In general, larger the fanout, larger the power reduction obtained due to sizing the driving buffers at this *power optima*. Thus, increasing the buffer transistor size towards the *delay optima* simultaneously of-

214

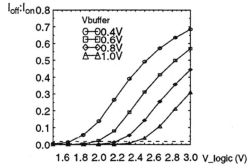

Figure 4: Off- to on-drive current ratios vs. logic stage voltage.

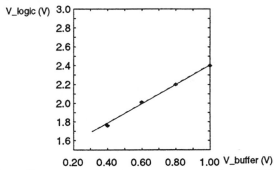

Figure 5: Logic vs. buffer stage voltage swing with $I_{off}/I_{on} < 0.025$.

fers a power reduction, until power starts to increase beyond the *power optima*.

6 Analysis of QuadRail power and delay models

To analyze the QuadRail power and delay models, we have considered a 6-input And-Or (AO222) gate cascade circuit. Fig. 6 and 7 show the gate and experimental circuit setup. The driving gate drives all the fanout gates' inputs in addition to a capacitive load of 300fF (corresponding to approximately 3000μm of metal interconnect in the HP 0.5um process). The fanout gates have unit sized buffer transistors. Fig. 8 shows the power and delay for this setup obtained from our model with $V_{buffer} = 0.8V$, k varying from 1-10 and V_{logic} varying from 1.5-3.0V. Some important conclusions can be drawn from these graphs:

- As V_{logic} is scaled towards $V_{buffer} + 2V_t$, i.e., more tighter logic stage turn-off, non-optimal sizing can cause steep delay penalties, both for over- and under-sized buffers. As $V_{logic} \rightarrow 3.0V$, buffer overdrive increases and optimal sizing does not significantly impact delay. We conclude that optimal buffer transistor sizing is critical at reduced power supply voltage swings.

- As $V_{logic} \rightarrow 3.0V$, non-optimal sizing of buffer transistors incurs a power penalty because of the high short circuit power component with minimum sized buffers. This penalty depends on the driving gate's fanout and interconnect loading. As V_{logic} is scaled, the short circuit power diminishes cubically, and power penalty due to minimum sized buffer transistors also diminishes. We conclude that minimum sized buffer transistors are best for optimal power at reduced power supply

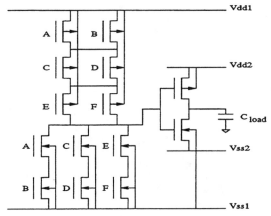

Figure 6: QuadRail 6-input AND-OR (AO222) gate.

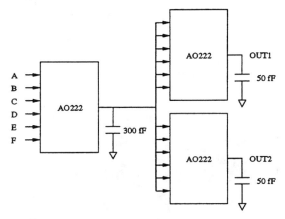

Figure 7: AO222 experimental circuit setup.

voltage swings, although there still exists a *power optima* very close to unit size.

Fig. 9 shows the power and delay for the same circuit setup obtained at one operating point: $V_{logic} = 2.2V$ and $V_{buffer} = 0.8V$ (i.e., a snapshot of Fig. 8 with $V_{logic} = 2.2V$). Our models show good agreement to HSPICE simulation results; the optimal buffer transistor size predicted by our models is within 2% of HSPICE results over many operating voltage levels and capacitive loads. Notice that both Fig. 8 (our models) and Fig. 9 (HSPICE simulation) correctly show a less steeper delay penalty for over-sizing than under-sizing as expected. This is due to the relative dominance of the logic and buffer stage delays in the total delay (eqn. (2) and (3) respectively).

7 CMOS vs. QuadRail comparison results

In this section we present power and delay comparisons between QuadRail and static CMOS for the AO222 gate. QuadRail operates at the same operating point as in section 6 ($V_{logic} = 2.2V$ and $V_{buffer} = 0.8V$) and CMOS operates at $V_{dd} = 3.0V$ and at a V_{dd} for which QuadRail and CMOS delays are approximately equalized, i.e., difference in delays is less than 1.5ns ($V_{dd} = 1.6V$). The comparison results are obtained through HSPICE simulations in the 0.5μm process. Load capacitances in the range 0-1pF and buffer sizes of 1X, 2X, and 4X are considered for both cases. Fig. 10 shows the worst-case delay of the CMOS and QuadRail AO222. Fig. 11 shows the power

AO222 circuit falling delay (s)

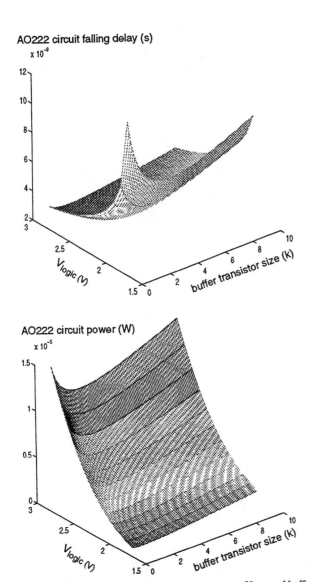

Figure 8: AO222 circuit falling delay and power vs. V_{logic} and buffer transistor size (k) [f_{clk} = 50 MHz; α = 1].

AO222 circuit power (W)

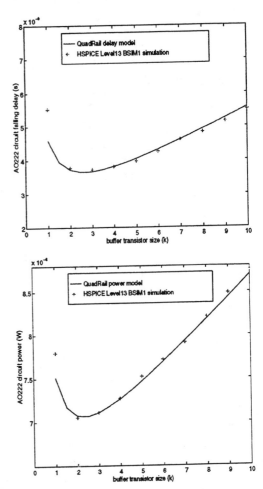

Figure 9: QuadRail delay and power models compared to HSPICE results [V_{logic} = 2.2V, V_{buffer} = 0.8V, f_{clk} = 50 MHz, α = 1]..

consumption of both CMOS and QuadRail gates at 50MHz and α = 1. With increasing loads, both QuadRail and CMOS delays increase with the same steepness, but QuadRail's power increases less steeply than CMOS due to reduced load voltage swing. Thus, at a load capacitance of 1pF, with equal delays a 3X power reduction is obtained compared to CMOS, and a 10X power reduction is obtained compared to CMOS at V_{dd} = 3.0V (corresponding delay penalty = 3X), when both are sized optimally for that load. The power savings is even higher as load capacitance increases beyond our range of analysis. At small loads (< 150fF), CMOS power dissipation is better compared to QuadRail when their delays are equal: this is due to QuadRail's *not-fully-turned-off-logic-stage* subthreshold power dissipation. Since static CMOS inputs swing rail to rail, the only off currents (and hence static power) is due to leakage currents of parasitic p-n junctions formed by the transistor diffusion regions and well/substrate. Fig. 10 re-emphasizes the importance of selecting the optimal size for buffer transistors in QuadRail - a 2.2X delay penalty if the buffers are 1X for

AO222 falling delay (ns)

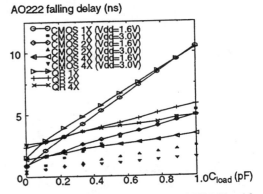

Figure 10: CMOS vs. QuadRail comparison: AO222 falling delay vs. C_{load} for 1X, 2X, and 4X buffers.

C_{load} = 1pF as opposed to their optimal size of 4X.

8 AOI222 Test-chip measurement results

A 6-input And-Or-Invert (AOI222) QuadRail test chip was fabricated in the HP 0.5µm process, to compare power and delay of QuadRail vs. static CMOS. 17 AOI222 gates were cascaded together with each AOI222 driving the next AOI222's 6 inputs and an additional load of 0.25mm, 0.50mm, 1.0mm, and 2.0mm of metal interconnect capacitance. The buffer

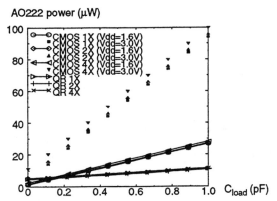

AO222 power (μW)

Figure 11: CMOS vs. QuadRail comparison: AO222 power vs. C_{load} for 1X, 2X, and 4X buffers.

stage:logic stage transistor size ratio is 2.5X. The AOI222 gate is constructed in a NAND-NAND-INVERT configuration as in Fig. 3. For the QuadRail AOI222s, the first (preamplifier) stage 2-input NAND gates and second (logic) stage 3-input NAND gate operate at supply voltage swings of 2.0V and 3.0V respectively. The buffer stage supply voltage and I/O swings are 1.0V. For the CMOS AOI222 all three stages operate at $V_{dd} = 3.0$V. Table 1 summarizes the measured power (with α = 1) and input-pin to output-pin delay for the QuadRail and CMOS AOI222 blocks. The input signal frequency is 10MHz. A 3.1X power savings is achieved compared to CMOS for a 2mm interconnect loading. At this load, QuadRail AOI222 delay is 1.06X higher than CMOS, offering an overall power-delay product reduction of 2.92X, at the same operating frequency. HSPICE full-chip simulation results show good agreement (within 10%) to these experimental measurements. The AOI222 test chip microphotograph is shown in Fig. 12.

Table 1. QuadRail AOI222 test chip measurement results.

interconnect length (mm)	QuadRail power (μW)	CMOS power (μW)	QuadRail delay (ns)	CMOS delay (ns)
0.25	206	383	32.82	18.24
0.50	214	418	33.80	19.24
1.00	275	450	34.99	20.81
2.00	289	896	39.81	37.62

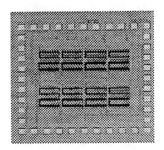

Figure 12: Microphotograph of the QuadRail AOI222 test chip.

9 Conclusion

Mixed Swing QuadRail approach presents an effective methodology for low voltage (sub-1V) logic in a high threshold voltage CMOS fabrication process, while maintaining high performance. Static power driven selection of the high and low swing voltage levels in QuadRail offers simultaneous reduction of static and dynamic power without degradation of switching characteristics and noise margins. QuadRail delay and power models reveal the importance of optimal selection of buffer transistor sizes at sub-1V I/O swings: both under- and over-sized buffer transistors can lead to steeply increased delay penalties. Comparison results between static CMOS and QuadRail show that significant power savings can be achieved through optimal selection of voltage levels and buffer sizes. Detailed HSPICE simulations using Level 13 BSIM1 models and test results from a 6-input AOI222 chip fabricated in the HP 0.5μm process substantiate our models and demonstrate significant power reduction compared to static CMOS.

10 Acknowledgments

We would like to acknowledge many insightful discussions with faculty at CMU, particularly Rob Rutenbar and Donald Thomas. This work was supported in part by the Defense Advanced Research Projects Agency and the National Science Foundation. The government has certain rights to this material.

References

[1] P.R. Gray, H.S. Lee, J.M. Rabaey, C.G. Sodini, and B.A. Wooley, "Challenges and Opportunities in Low Power Integrated Circuit Design", *SRC Report S94019*, November 1994.

[2] A.P. Chandrakasan and R.W. Broderson, *Low Power Digital CMOS Design*, Kluwer Academic Publishers, 1995.

[3] J. Montenaro et al, "A 160MHz 32b 0.5W CMOS RISC Microprocessor". *Digest of technical papers, ISSCC*, February 1996, pp.214-215.

[4] C. Crippa et al., "A 2.7V CMOS Single Chip Baseband Processor for CT2/CT2+ Cordless Telephones", *Proc. CICC*, May 1996, pp. 123-126.

[5] S.W. Sun and P.G.Y. Tsui, "Limitation of CMOS Supply Voltage Scaling by MOSFET Threshold Voltage Variation", *Proc. CICC*, May 1994, pp. 267-270.

[6] M.J.M. Pelgrom, A.C.J. Duinmaijer, and A.P.G. Welbers, "Matching Properties of MOS Transistors", *IEEE J. Solid-State Circuits*, Vol. 24, October 1989, pp. 1433-1440.

[7] J.B. Burr and J. Shott, "A 200mV Self-Testing Encoder/ Decoder using Stanford Ultra Low Power CMOS", *Digest of technical papers, ISSCC*, February 1994, pp. 84-85.

[8] L.R. Carley and I. Lys, "QuadRail: A Design Methodology for Ultra Low Power Integrated Circuits", *Proc. NAPA Valley Workshop on Low Power IC Design*, April 1994.

[9] H.B. Bakoglu, *Circuits, Interconnections, and Packaging for VLSI*, Addison Wesley Publishers, 1990.

[10] J. Ecker, "Geometric Programming: methods, computations, and applications", *SIAM Review*, July 1980, pp. 338-362.

[11] M. Kakumu and M. Kinugawa, "Power Supply Voltage Impact on Circuit Performance for Half and Lower Submicrometer CMOS LSI", *IEEE Trans. Electron Devices*, Vol. 37, August 1990, pp. 1902-1908.

[12] B. Hoppe, G. Neuendorf, D.S. Landsiedel, and W. Specks, "Optimization of High-Speed CMOS Logic Circuits", *IEEE Trans. CAD*, Vol. 9, March 1990, pp. 236-247.

[13] N. Hedenstierna and K.O. Jeppsen, "CMOS Circuit Speed and Buffer Optimization", *IEEE Trans. CAD*, Vol. 6, March 1987, pp. 270-281.

The Power Consumption of CMOS Adders and Multipliers

THOMAS K. CALLAWAY

SILICON GRAPHICS, INC.

MOUNTAIN VIEW, CA 94306

EARL E. SWARTZLANDER, JR.

DEPARTMENT OF ELECTRICAL AND COMPUTER ENGINEERING

UNIVERSITY OF TEXAS AT AUSTIN

AUSTIN, TEXAS 78712

Abstract—Minimizing the power consumption of circuits is important for a wide variety of applications, both because of increasing levels of integration and the desire for portability. Since adders and multipliers are so widely used in computers, it is also important to minimize the delay. This research reports on the dynamic power dissipation and delay of CMOS implementations of six different adders and four multipliers. Simulation and direct measurement of the performance of a test chip were used to evaluate their switching characteristics. The results are used to rank the adders and multipliers on area, delay, and dynamic power dissipation.

1. INTRODUCTION

An important attribute of arithmetic circuits for most applications is minimizing the delay or latency. For a growing number of applications, minimizing the power consumption is of equal or greater importance. The most direct way to reduce power is to use CMOS circuits, which generally dissipate less power than their bipolar counterparts. Even for CMOS, the use of arithmetic circuits with minimum power consumption is attractive.

In static CMOS the dynamic power dissipation of a circuit depends primarily on the number of logic transitions per unit time [1]. As a result, the average number of logic transitions per operation can serve as the basis for comparing the efficiency of a variety of arithmetic circuit designs. Callaway and Swartzlander have investigated the average power dissipation of adders [2] and multipliers [3]. Their work on adders has been extended by Nagendra, Owens, and Irwin [4].

The results presented in this research were obtained from three different sources. A gate-level simulator was used to generate a first estimate of each circuit's dynamic power consumption. For that gate-level simulator, the circuits are constructed using only AND, OR, and INVERT gates. The circuits are then subjected to 10,000 pseudo-random inputs. For each input, the number of gates that switched during the addition is counted, and an average number of gate-output transitions is computed for that circuit.

The second estimate is derived from detailed circuit simulation using a program called CAzM, which is similar to SPICE. A 16-bit version of each circuit was designed in 2-micron, 2-level metal static CMOS using MAGIC, and the resulting layout was the basis for the circuit simulation. Transistor parameters are those of a 2-micron MOSIS fabrication run. Each adder and multiplier is presented with 1,000 pseudo-random inputs, and the average power dissipation waveform is plotted. Each addition is allowed a time of 50 ns, and each multiplication is allotted 100 ns, so that the voltages will stabilize completely before the next operation is performed.

The third method for obtaining results relies on direct physical measurement. A chip containing 16-bit implementations of the six different adders has been designed, fabricated, and tested. Each of the six adders has its own power pin, which enables direct measurement of each adder's dynamic power dissipation without having to estimate other power sinks.

2. ADDERS

The following types of adders were simulated: ripple carry [5], constant block-width single-level carry skip [5], variable block-width multi-level carry skip [6], carry lookahead [7], carry select [8], and conditional sum [9].

In choosing an adder for a particular application, several things must be considered, including delay, size, and dynamic power consumption. Table 1 presents the worse-case number of gate delays for the six adders. This is a first-level estimate of worst-case delay, which neglects gate fanin and fanout, gate

Table 1. Worst-Case Delay.

Adder type	Adder size (bits)		
	16	*32*	*64*
Ripple carry	36	68	132
Constant carry skip	23	33	39
Variable carry skip	17	19	23
Carry lookahead	10	14	14
Carry select	14	14	14
Conditional sum	12	15	18

sizing, and interconnections; however, it does serve to estimate the relative delays of the 6 adders.

The size of each adder is approximated to a first order by the number of gates, which is given in Table 2. The actual area required for a given adder will depend on the types of gates (e.g., three or four input gates require more area than two input gates), the amount of wiring area, and the available form factor. However, the number of simple logic gates is a good model for the relative area of arithmetic circuits [10]. Adders which have a larger gate count generally occupy more area than circuits with a smaller gate count.

2.1 Gate-Level Simulation

Gate-level simulation is performed using a 3-value (0,1,X) event-driven simulator. For the gate-level simulations, each gate is assumed to have unit delay and unit-power dissipation. Each

circuit is subjected to 10,000 pseudo-random input patterns. For each pattern, the number of gate-output transitions is counted.

Before the first pseudo-random pattern is applied, the adder is initialized by applying zero inputs and allowing it to stabilize. Any gate transitions during this initialization are not counted. After that, pseudo-random inputs are presented one after the other, with no zero inputs in between. Table 3 gives the average number of gate-output transitions per addition observed for each adder in 10,000 randomly distributed input patterns with a random distribution of carry inputs.

Figure 1 shows the average number of gate output transitions versus time for the 16-bit adders, based on the unit-delay, unit-power gate model with 10,000 pseudo-random inputs. The number of transitions for each adder except the conditional sum adder increases linearly with the word size. For the conditional sum adder, the increase is more than linear. Figure 2 shows the probability distribution of the number of gate-output transitions per addition, and indicates that the power dissipation is normally distributed.

2.2 Circuit Simulation Results

The simulator used to obtain the previous results uses a rather simple model, and provides little time information. In order to verify those results, 16-bit versions of the six different adders were designed in 2-micron CMOS. Netlists were extracted from

Table 2. Number of Gates

Adder type	Adder size (bits)		
	16	*32*	*64*
Ripple carry	144	288	576
Constant carry skip	156	304	608
Variable carry skip	170	350	695
Carry lookahead	200	401	808
Carry select	284	597	1228
Conditional sum	368	929	2139

Table 3. Number of Gate Output Transitions.

Adder type	Adder size (bits)		
	16	*32*	*64*
Ripple carry	90	182	365
Constant carry skip	99	197	389
Variable carry skip	108	220	438
Carry lookahead	100	202	405
Carry select	150	317	652
Conditional sum	231	590	1464

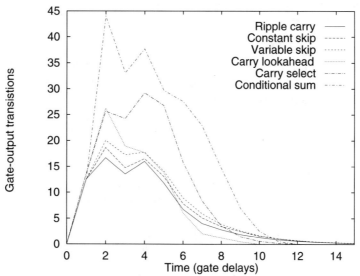

Figure 1 16-bit adder gate-output transitions.

219

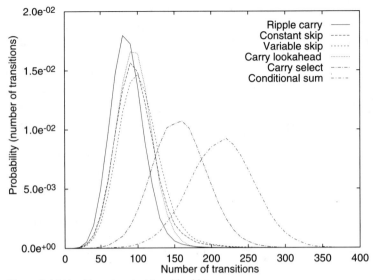

Figure 2 16-bit adder gate-output transition histogram.

the layout, and CAzM was then used to estimate the worst-case delay and the average power dissipation.

Table 4 shows the worst-case delay of the six 16-bit adders as estimated with CAzM. The pattern which results in the worst-case delay is not the same for all of the adders, but the basic pattern causes a carry to be generated in the first bit position, which must then be propagated through the adder to the most-significant output.

There are two entries of interest in Table 4. The first is the carry lookahead adder. Based on the number of gate delays, it should have the lowest delay. However, the gates in the worst-case path include large fanin and fanout gates, so the unit-delay assumption is inaccurate. The second is for the conditional sum adder, which has no large fanin gates. However, there are several large fanout gates, so again, the unit-delay assumption is inaccurate.

Table 5 shows the area occupied by each adder and the number of gates. The actual areas exhibit a wider range (approximately 4:1) than the gate counts from Table 2 (approximately 2.5:1), but the relative sizes are consistent with the gate counts.

The average power dissipation per addition shown in Table 6 is obtained by simulating the addition of 1,000 pseudo-random inputs with a 50-ns clock period, and averaging the results. This table shows that there is a significant difference in the power dissipation for the various adders.

Table 4. Worst-Case Delay Estimated with CAzM.

Adder type	Delay (ns)	Gate delays
Ripple carry	51.4	36
Constant carry skip	28.6	23
Variable carry skip	22.8	17
Carry lookahead	22.5	10
Carry select	18.6	14
Conditional sum	21.2	12

Table 5. 16-Bit Adder Area.

	Area (mm^2)	Gate count
Ripple carry	0.26	144
Constant carry skip	0.33	156
Variable carry skip	0.49	170
Carry lookahead	0.53	200
Carry select	0.88	284
Conditional sum	1.14	368

Table 6. Average 16-Bit Adder Power Dissipation from CAzM.

Adder type	Power (mW)	Logic transitions
Ripple carry	1.7	90
Constant carry skip	1.8	99
Variable carry skip	2.2	108
Carry lookahead	2.7	100
Carry select	3.8	150
Conditional sum	5.4	231

It is useful to look at the distribution in time of the average power dissipation, as shown in Fig. 3 for each of the six adders. The shapes of these curves are quite similar to those from the simple unit-power model in Fig. 1.

2.3 Physical Measurement

The third method of obtaining results is to directly measure the power dissipation. The test chip shown in Fig. 4 was constructed such that each adder has a separate power pin. Because of this, measuring the power dissipation is relatively easy. A simple test board, which operates at 2MHz, was constructed to provide inputs and functionality testing for the chips. Linear-feedback shift registers on the test board are used to generate the pseudo-random operands.

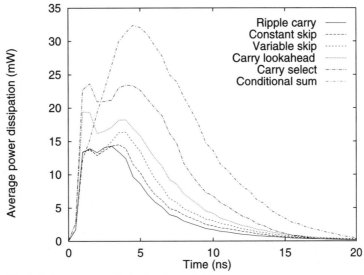

Figure 3 Average power dissipation from CAzM.

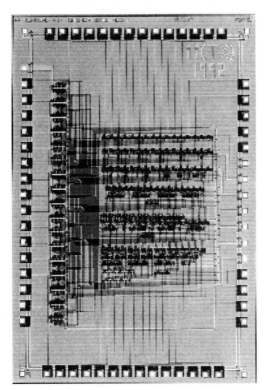

Figure 4 Die photo of test chip.

A set of measurements was made by connecting a multimeter in series with the power supply and measuring the current while the adders are accepting pseudo-random inputs at 2MHz. The average power dissipated by each adder is presented in Table 7. The problem with this measurement is that it is difficult to tell over exactly what period of time the multimeter is measuring the current. These measurements are in agreement with the results from the simple gate-level model and the results from CAzM, although the ratio between the measured and simulated power dissipation values decreases with the adder area. This is most likely due to a simulated leakage current which is too high.

Table 7. Measured Power Dissipations at 2MHz.

Adder type	Simulated (mW)	Measured (mW)
Ripple carry	0.21	0.43
Constant carry skip	0.23	0.49
Variable carry skip	0.27	0.50
Carry lookahead	0.32	0.58
Carry select	0.43	0.69
Conditional sum	0.60	0.84

3. MULTIPLIERS

Four different multipliers are analyzed: Array [11], Split Array [12], Wallace Tree [13], and Radix-4 Modified Booth Recoded Wallace Tree [7]. The same simulations were performed for both the multiplier and the adder circuits. However, no multiplier circuits were fabricated, so there are no measurements for the multipliers.

The worst-case delay for the multipliers is approximated by the number of gate delays from the inputs to the slowest output. Table 8 shows the worst-case delay (in gate delays) for 8-, 16-, and 32-bit versions of each multiplier. Here, the multipliers are described using only AND, OR, and INVERT gates.

The area of each multiplier circuit is approximated to a first order by the number of gates, as given in Table 9. The actual area of the multiplier circuits depends on several things, including the available form factor and the number of levels of metal available for wiring.

Table 8. Worst-Case Delay.

| Multiplier type | Multiplier size (bits) | | |
	8	16	32
Array	50	98	198
Split array	38	64	114
Wallace	35	51	63
Modified booth	32	42	54

Table 9. Number of Gates.

	Multiplier size (bits)		
Multiplier type	*8*	*16*	*32*
Array	546	2,378	9,897
Split array	709	2,754	10,691
Wallace	559	2,544	10,430
Modified booth	975	3,375	12,353

Table 10. Average Number of Gate-Output Transitions.

	Multiplier size (bits)		
Multiplier type	*8*	*16*	*32*
Array	570	7,224	99,906
Split array	569	4,874	52,221
Wallace	549	3,793	20,055
Modified booth	964	3,993	19,542

3.1 Gate-Level Simulation

Gate-level simulation for the multipliers is performed in the same manner as for the adders. The circuit is subjected to a stream of 10,000 pseudo-random inputs. During each input, the number of gates that switch output states is recorded, and an average number of gate-output changes per multiplication is computed. Table 10 shows the average number of gate output transitions per multiplication.

Figures 5 and 6 show the average number of gate output transitions as a function of time for 16- and 32-bit versions of each of the 4 multipliers, based on a unit-gate delay, and a unit-power dissipation. As can be seen from the figures, there is quite a difference in the curves for the 16-bit and 32-bit implementations. Figure 7 shows that the average number of logic transitions per multiplication is normally distributed.

3.2 Circuit-Simulation Results

In order to provide actual numbers for the worst-case delay and the average power dissipation of the multiplier circuits, a 16-bit version of each multiplier was laid out in a 2-micron,

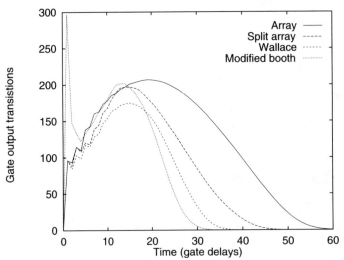

Figure 5 16-bit multiplier gate-output transitions.

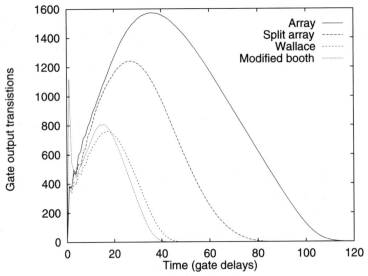

Figure 6 32-bit multiplier gate-output transitions.

222

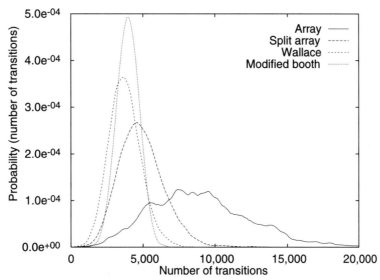

Figure 7 16-bit multiplier gate-output transition histogram.

2-level metal CMOS process. Transistor netlists were extracted from the layouts, and CAzM was used to simulate both the worst-case delay and the average power dissipation.

Table 11 shows the worst-case delay for each of the 16-bit multipliers, computed using CAzM. For the array multiplier, the worst-case pattern causes a carry to be generated in the upper left corner of the array, which will then ripple down the diagonal and across the fast adder to the most significant output. Slight variations are used for the other 3 multiplier circuits. In this case, the number of gate delays corresponds well to the simulated worst-case delay.

Table 12 gives the actual silicon area of each 16-bit multiplier, computed by measuring the area of the layout for each multiplier. Although the array and the Wallace multipliers have a similar gate count, the Wallace multiplier requires twice the area.

The average power dissipation per multiplication is shown in Table 13. The values are obtained by simulating the multiplication of 1,000 pseudo-random inputs with a clock period of

Table 11. Worst-Case Display Estimated with CAzM.

Multiplier type	Delay (ns)	Gate delays
Array	92.6	50
Split array	62.9	38
Wallace	54.1	35
Modified booth	45.4	32

Table 12. 16-Bit Multiplier Area.

Multiplier type	Area (mm^2)	Gate count
Array	4.2	2,378
Split array	6.0	2,754
Wallace	8.1	2,544
Modified booth	8.5	3,375

Table 13. Average Power Dissipation from CAzM.

Multiplier type	Power (mW)	Logic transitions
Array	43.5	7,224
Split array	38.0	4,874
Wallace	32.0	3,793
Modified booth	41.3	3,993

100ns. This shows that there is a significant difference in the power dissipation of the various multipliers. Also, the multiplier with the lowest average power dissipation (Wallace) is neither the smallest nor the slowest.

Figure 8 shows the average power dissipation as a function of time. The curves in Figs. 5 and 8 are quite similar, even to the initial peak for the modified Booth multiplier.

4. CONCLUSIONS

In this research we have examined 6 types of adders and 4 types of multipliers in an attempt to model their power dissipation. We have shown that the use of a relatively simple model provides results that are qualitatively accurate when compared to more sophisticated models and to physical implementations.

The main discrepancy between the simple model and the physical measurements seems to be the assumption that all gates will consume the same amount of power when they switch, regardless of their fanin or fanout. Because the carry lookahead adder has several gates with a fanout and fanin higher than 2, the simple model underestimates its power dissipation.

The array multiplier can be laid out efficiently, but it is slow and consumes quite a bit of power. Its power consumption grows in proportion to the cube of the word size. The Wallace multiplier is less regular and requires more area as a result, but it dissipates less power (and its power dissipation grows as the square of the word size).

223

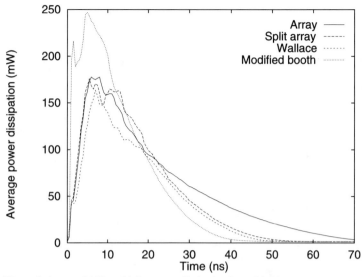

Figure 8 Average 16-bit multiplier power dissipation from CAzM.

ACKNOWLEDGMENTS

The authors would like to thank Crystal Semiconductor, and in particular David Medlock, for help in testing the chips.

References

[1] N. Weste and K. Eshraghian, *Principles of CMOS VLSI Design: A Systems Perspective.* Reading, MA: Addison-Wesley Publishing Company, 1988.

[2] T. K. Callaway and J. E. E. Swartzlander, "Estimating the power consumption of CMOS adders," *Proceedings of the 11th Symposium on Computer Arithmetic,* 1993, pp. 210–216.

[3] T. K. Callaway and E. E. Swartzlander, Jr., "Optimizing multipliers for WSI," *IEEE International Conference on Wafer Scale Integration,* 1993, pp. 85–94.

[4] C. Nagendra, R. M. Owens, and M. J. Irwin, "Power-delay characteristics of CMOS adders," *IEEE Transactions on VLSI Systems,* vol. 2, 1994, pp. 377–381.

[5] D. Goldberg, "Computer arithmetic," *Computer Architecture: A Quantitative Approach,* Morgan Kaufmann Publishers, 1990, pp. A2, A3, A31–A39.

[6] S. Turrini, "Optimal group distribution in carry-skip adders," *Proceedings of the 9th Symposium on Computer Arithmetic,* Sep. 1989, pp. 96–103.

[7] O. L. MacSorley, "High-speed arithmetic in binary computers," *IRE Proceedings,* vol. 49, 1961, pp. 67–91.

[8] O. J. Bedrij, "Carry-select adder," *IRE Transactions on Electronic Computers,* vol. EC-11, Jun. 1962, pp. 340–346.

[9] J. Sklansky, "Conditional-sum addition logic," *IRE Transactions on Electronic Computers,* vol. EC-9, 1960, pp. 226–231.

[10] T. K. Callaway, *Area, Delay, and Power Modeling of CMOS Adders and Multipliers.* Ph.D. thesis, The University of Texas at Austin, 1996.

[11] A. D. Pezaris, "A 40ns 17-bit by 17-bit array multiplier," *IEEE Transactions on Computers,* vol. C-20, 1971, pp. 442–447.

[12] J. Iwamura, K. Suganuma, M. Kimura, and S. Taguchi, "A CMOS/SOS multiplier," *Proceedings of the ISSCC,* 1984, pp. 92–93.

[13] C. S. Wallace, "A suggestion for a fast multiplier," *IEEE Transactions on Electronic Computers,* vol. EC-13, 1964, pp. 14–17.

Delay Balanced Multipliers for Low Power/Low Voltage DSP Core

Toshiyuki Sakuta*, Wai Lee and Poras T. Balsara**

Integrated Systems Laboratory, Texas Instruments Inc., P.O.Box 655474, MS446, Dallas, TX 75265
* Permanent address: Research & Development Division, Hitachi America, Ltd., 50 Prospect Avenue, Tarrytown, NY 10591
** Permanent address: Department of Electrical Engineering, University of Texas at Dallas, P.O.Box 830688, EC33, Richardson, TX 75083
e-mail addresses : sakuta@hc.ti.com, lee@hc.ti.com, poras@utdallas.edu

1. Introduction

The choice between an array and a Wallace-tree multiplier architecture depends on the tradeoffs between area and speed. Wallace-tree multipliers offer higher speed but consume larger areas than array multipliers. As for the power dissipation, recent research on signal transition activity, which directly influences the dynamic power dissipation, indicated that the array multiplier had a architectural disadvantage [1]. Delay imbalance of the signals in a full adder array causes many spurious transitions and consequently unnecessary power dissipation. In this work, the potentials for power reduction of both Wallace-tree multiplier and array multiplier were explored. By implementing a few delay buffer circuits appropriately, the delays in the full adder array can be balanced and the majority of the spurious transitions are suppressed. Thus, significant reduction in power dissipation of the array multiplier can be achieved.

2. Delay balanced array multiplier

As shown in the lower half of Fig. 1, multiplicand bits are simultaneously input to all the partial product generators at every stage in the conventional array multiplier. All full adders start computing at the same time without waiting for the propagation of sum and carry signals from the previous stage. This results in spurious transitions at the output and wastes power [2]. Furthermore, as shown in Fig. 2, since these spurious transitions are propagated to the next stage continuously, their numbers grow stage by stage like a snow ball. This causes a significant increase in power dissipation.

The following two countermeasures, shown in Fig. 1, were taken to solve this problem. One is that the delay circuit which has the same delay property as that of a full adder circuit was inserted in the multiplicand signal path at each full adder stage. Another is that a register was placed at the input of booth encoder so that the timing of multiplier encoding could be controlled by the clock. The same delay circuit as the one used in the first measure was inserted in the clock signal path at each full adder stage. By these measures, the output transition of a partial product generator at each full adder stage and the transitions of the sum and the carry output of a full adder at the previous stage are synchronized. Therefore, the significant reduction of spurious transitions was achieved with virtually no penalty in performance. The area increase for these additional circuits in a 24 bit x 24 bit Booth multiplier is about 8.5%.

3. Delay balanced Wallace-tree multiplier

A Wallace-tree multiplier occupies about 40% more area than a corresponding array multiplier. It also has a problem of high wiring density of internal signals. On the other hand, it has low probability of occurrence of spurious transitions since most inputs to full adders at each stage are naturally synchronized due to its inherent parallel structure. In addition, fast multiplication is achieved because of the small number of necessary full adder stage [3, 4]. Figure 3 shows a Wallace-tree structure for a 13 bit column compressor required for a 24 bit x 24 bit Booth multiplier.

Signals A and B, which jump over one full adder stage and are the inputs to the full adder of the next stage, cause spurious transitions inside this multiplier. A delay balanced Wallace-tree multiplier is shown in Fig. 4. As shown in this figure, delay circuits which have the same delay as a full adder circuit were inserted in the signal paths of A and B so that A and B are synchronized with the other inputs of the corresponding adders. The penalty in performance is also negligible as in the array multiplier's case.

4. SPICE simulation results

SPICE simulations of both multipliers were performed utilizing random input patterns as multiplicands and multipliers. Table 1 shows the device technology used for this study. The number of spurious transitions (S) and valid transitions (V) generated at the output of each full adder and the carry lookahead adder (CLA) in 64 clock cycles were extracted from simulation results. S/V represents the ratio of wasted power to necessary power. 2D contour map of the values of S/V was plotted at the corresponding position on the floor plan as shown in Fig. 5 (a) through (d). Figure 5 (a) shows that a large number of spurious transitions occurs all over the conventional array multiplier. S/V increases rapidly as the number of stages increases. The average value of S/V over the entire multiplier is 2. The remarkable reduction in spurious transitions by the techniques stated in section 2 can be seen in Fig. 5 (b). The average value of S/V in this case is only 0.3. In Fig. 5 (c), many spurious transitions can also be observed at the stage 2.3 and the stage 5 of Wallace-tree multiplier where the inputs to full adders are not synchronized. The effect of the delay circuit for suppression of spurious transitions in this multiplier is shown in Fig. 5 (d). Since a Wallace-tree multiplier consumes relatively less power by nature, the effect of matching delay in power saving is only 6.5% while 36% of power saving was attained in the delay balanced array multiplier. The SPICE simulation results at different operating voltages are summarized in Fig. 6 (a) and (b). Finally, a comparison of layout area, delay and power-delay product among three multiplier architectures is shown in Table 2.

5. Conclusion

A simple but effective technique, which synchronizes the propagation of signals at each full adder stage, has cut the power dissipation of an array multiplier down to equal to or less than that of a Wallace-tree multiplier with a minimal penalty in performance and layout area. This delay balanced array multiplier is a strong candidate for low power and small area DSP core for portable equipment.

Reference

[1] J.Leijten, et al., "Analysis and Reduction of Glitches in Synchronous Networks", European Design & Test Conf., Dig. Tech. papers, pp.398-403, Mar. 1995.
[2] C. Lemonds, et al., "A Low Power 16 by 16 Multiplier Using Transition Reduction Circuitry", Intl. Workshop on L/P Design, Dig. Tech. papers, pp.13-142, Apr. 1994.
[3] C. S. Wallace, "A Suggestion for a Fast Multiplier", IEEE Trans. Electron. Computer, vol.EC-13, pp.14-17, 1964.
[4] L. Dadda, "Some Schemes For Parallel Multipliers", Alta Freq., vol.34, p.349-356, 1965.

Reprinted from *1995 IEEE Symposium on Low Power Electronics: Digest of Technical Papers,* pp. 36-37, October 1995.

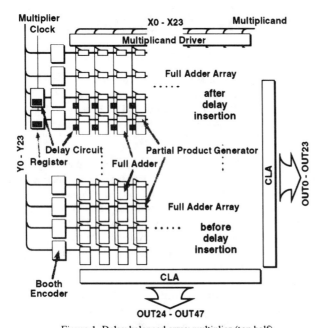

Figure 1 Delay balanced array multiplier (top half) and conventional array multiplier (bottom half)

Labels in Figure 1: Multiplier Clock; X0 - X23 Multiplicand; Multiplicand Driver; Full Adder Array; after delay insertion; Delay Circuit; Register; Partial Product Generator; Full Adder; Y0 - Y23; CLA; OUT0 - OUT23; Full Adder Array before delay insertion; Booth Encoder; CLA; OUT24 - OUT47

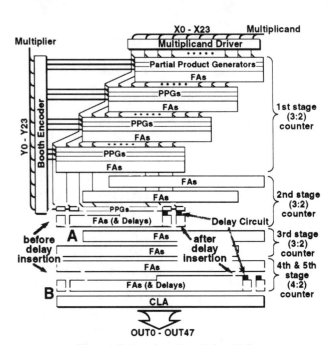

Figure 4 Delay balanced parallel multiplier

Labels in Figure 4: Multiplier; X0 - X23 Multiplicand; Multiplicand Driver; Partial Product Generators; FAs; PPGs; FAs; 1st stage (3:2) counter; Y0 - Y23; Booth Encoder; FAs; 2nd stage (3:2) counter; PPGs; FAs (& Delays); Delay Circuit; before delay insertion; A; FAs; after delay insertion; 3rd stage (3:2) counter; FAs; 4th & 5th stage (4:2) counter; B; FAs (& Delays); CLA; OUT0 - OUT47

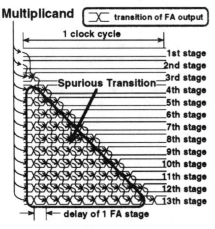

Figure 2 Spurious transition in a conventional array multiplier

Labels in Figure 2: Multiplicand; transition of FA output; 1 clock cycle; 1st stage; 2nd stage; 3rd stage; Spurious Transition; 4th stage; 5th stage; 6th stage; 7th stage; 8th stage; 9th stage; 10th stage; 11th stage; 12th stage; 13th stage; delay of 1 FA stage

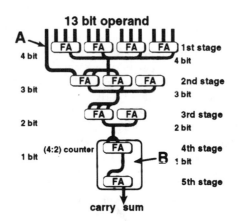

Figure 3 Wallace tree structure for 13 bit column compressor

Labels in Figure 3: 13 bit operand; A; FA FA FA FA 1st stage; 4 bit; FA FA FA 2nd stage; 3 bit; FA FA 3rd stage; 2 bit; (4:2) counter FA 4th stage; 1 bit; B; FA 5th stage; carry sum

Table 1 Device technology

Lg	0.35 μm
Vthn/Vthp	+0.5 V / -0.5 V (+0.3 V / -0.3 V @Vdd = 0.9 V)
tox	6 nm

Table 2 Comparison of three different architectures (normalized with conventional array multiplier's values)

Type \ Item	Conventional Array Multiplier	Delay Balanced Array Multiplier	Delay Balanced Parallel Multiplier
Layout Area	1.00	1.09	1.40
Delay	1.00	1.02	0.75
PxD product	1.00	0.64	0.66

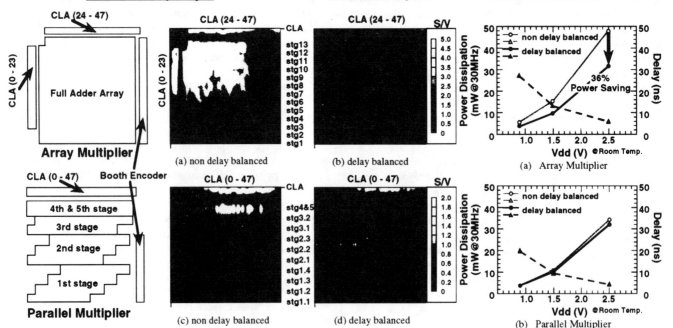

Labels in Figure 5/6: CLA (24 - 47); CLA (0 - 23); Full Adder Array; Array Multiplier; CLA (0 - 47); Booth Encoder; 4th & 5th stage; 3rd stage; 2nd stage; 1st stage; Parallel Multiplier

(a) non delay balanced; (b) delay balanced; stg13...stg1; S/V 5.0 ... 0

(c) non delay balanced; (d) delay balanced; stg4&5, stg3.2, stg3.1, stg2.3, stg2.2, stg2.1, stg1.4, stg1.3, stg1.2, stg1.1; S/V 2.0 ... 0

Power Dissipation (mW @30MHz); non delay balanced; delay balanced; 36% Power Saving; Vdd (V) @Room Temp.; Delay (ns)
(a) Array Multiplier

Power Dissipation (mW @30MHz); non delay balanced; delay balanced; Vdd (V) @Room Temp.; Delay (ns)
(b) Parallel Multiplier

Figure 5 Ratio of spurious transition to valid transition (S/V)

Figure 6 SPICE simulation result

226

Asynchronous Does Not *Imply* Low Power, But . . .

KEES VAN BERKEL*   HANS VAN GAGELDONK*   JOEP KESSELS   CEES NIESSEN
AD PEETERS*   MARLY RONCKEN   RIK VAN DE WIEL*

PHILIPS RESEARCH LABORATORIES

PROF HOLSTLAAN 4, 5656 AA EINDHOVEN, THE NETHERLANDS (BERKEL@NATLAB.RESEARCH.PHILIPS.COM)

Abstract—**Asynchronous circuits are often praised for their low-power properties. Although they definitely have a low-power potential it is generally non-trivial to achieve this. We identify five low-power opportunities of asynchronous VLSI circuits and illustrate how these can be realized, by a DCC error corrector and standby circuits for pagers.**

1. INTRODUCTION

In the recent revival of asynchronous VLSI circuits, a number of research groups have investigated and demonstrated the potential of asynchronous circuit techniques for low-power applications. Low-power asynchronous microprocessors have been reported by CalTech [1] and Manchester University [2]. Other demonstrators include a DCC error corrector [3–5], DSP applications for hearing aids [6], and pager standby circuits [7]. Overviews on the low-power potential of asynchronous circuits can be found in [8,9].

Ironically, the low-power potential of asynchronous circuits has been known for a long time in the digital wristwatch industry. In most electronic watches the 32-kHz signal from a quartz crystal is divided to a frequency of a few Hertz. This divider is an asynchronous circuit and is essential to the ultra low-power dissipation ($<1\mu W$) of a digital wristwatch. Asynchronous dividers are also applied in the low-power PLL for the StrongARM [10].

Before we review a number of asynchronous low-power opportunities, let us first examine the clock power in synchronous circuits. A popular expression for the power dissipation of a clocked circuit is $f_c \alpha C V^2$, where f_c is the clock frequency, C the total circuit capacitance, V the supply voltage, and α the circuit activity (i.e. the average fraction of nodes that make a transition during a clock period). An important limitation of this expression is that the contribution of the clock power is implicit. The capacitance of the clock wire, including the direct and indirect loads contributed by latches and flip-flops, often amounts to

*Partly at Eindhoven University of Technology.

several nanofarads, and its activity equals 2 by definition. This makes the clock a major power sink, in particular in applications with a modest α.

In Fig. 1 the clock power is plotted as function of α for various values of k: the ratio of the number of combinational gate equivalents over the number of flip-flops. Typical values of α lie in the range of 10–20%, implying that 30–80% of the power dissipation is clock related. Circuit activity is often that low for a variety of reasons. Sub-blocks, for instance, may have to deal with exceptions only, or operate on lower sample rates than other blocks. In audio functions, the ratio of clock and sample frequency may easily exceed a factor 100 to allow the reuse of resources in sequential algorithms. A target in the design of low-power asynchronous circuits is to exploit this low activity.

2. ASYNCHRONOUS LOW-POWER OPPORTUNITIES

Redesigning a clocked VLSI circuit as an asynchronous circuit does not automatically reduce power dissipation. Especially when the architecture is copied at the register-transfer level, the resulting circuit is probably somewhat bigger at virtually the same dissipation. However, by exploiting certain characteristics of an application, asynchronous circuits can create low-power opportunities that are beyond the synchronous reach. "Asynchronous" does not *imply* "low power," but it does do the following:

- often allows the elimination of most clock power;
- encourages distributed, localized control;
- offers more architectural options/freedom;
- offers automatic, instantaneous stand-by mode, at arbitrary granularity in time and function; and
- offers more freedom to adapt supply voltages.

2.1 Reduced Clock Power

In asynchronous design, latches and flip-flops are enabled only when new values arrive, thereby avoiding a waste of clock power. Structured design of such asynchronous circuits requires

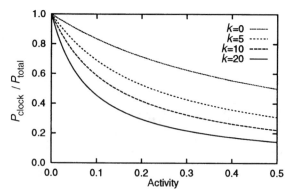

Figure 1 Clock power as a fraction of the total power in a synchronous circuit as a function of the circuit activity α for various k.

a timing discipline to replace that of a global clock. Handshake signaling is such a discipline.

Handshake signaling involves two parties, viz., an active and a passive one. The handshake channel that connects both parties consists of a request wire (from active to passive) and an acknowledge wire (vice versa). A four-phase handshake is a signaling sequence along these wires as depicted in Fig. 2. The request can be interpreted as a local clock signal. The acknowledge is then an explicit indication of the completion of the action triggered by the request. In synchronous circuits this completion is left implicit.

Asynchronous VLSI systems can be composed of elementary building blocks ("handshake components") that interact by means of handshake signaling exclusively [3]. An example of a handshake component is the so-called sequencer of Fig. 3. It has passive port a and active ports b and c. A request through a triggers first a complete handshake through b, followed by one through c, followed by the completion of the handshake through a. The active ports of the sequencer can be used to control other parts of the circuit. Note that when the first side is being served, the second side is standing by (dissipating virtually zero in CMOS), and vice versa. The other handshake component in Fig. 3 is a duplicator, which for each handshake through a_0 encloses two handshakes through a_1.

A cascade of K duplicators (as in Fig. 4) yields a 2^K-fold repeater, with 2^i handshakes on channel a_i enclosed by a single handshake on a_0. This repeater suggests an interesting low-

Figure 3 Symbol for a handshake sequencer (the asterisk labels the active port that is served first) and a duplicator.

power argument. Assume that duplicator K runs at a fixed frequency f, dissipating P. Then duplicator $K - 1$ runs at half that frequency and hence dissipates only $\frac{1}{2}P$. Duplicator $K - 2$ runs at only $\frac{1}{4}f$, etc. Since the series $1 + \frac{1}{2} + \frac{1}{4} + \ldots$ is bounded by 2, the power dissipation of the repeater is independent of the value of K. Compare this to a synchronous solution, with all state-holding elements clocked, where the dissipation is proportional to K.

For modulo-N counters several designs with both bounded power dissipation and bounded response time (independent of N) have been presented, e.g. in [8].

2.2 Distributed Control

Asynchronous architectures tend to have distributed control, in which redundancy in the state encoding is used to increase locality, resulting in shorter wires and lower gate fan outs. Whereas the required additional state-holding elements tend to increase power in clocked architectures, they may actually reduce power in asynchronous ones. This is illustrated with the N-way sequencer of Fig. 5.

The N-way sequencer consists of $N - 1$ binary sequencers, where the active handshakes are issued sequentially, starting with b_1. A full cycle involves each binary sequencer only once, making its energy dissipation proportional to $N - 1$. A centralized controller would use fewer flip-flops, say $[\log N]$, which makes the energy dissipation for a full cycle proportional to $N[\log N]$.

In a more complex system, each subsystem may be controlled by its own multi-way sequencer, possibly with widely varying values of N. In an asynchronous implementation synchronization between parallel modules occurs only when required, allowing each module to make progress corresponding to its local needs. In a clocked implementation, synchronization essentially takes place at each clock tick, resulting in an excess of clock ticks for less-demanding modules.

2.3 Architectural Freedom

The universe of asynchronous circuits is vastly larger than that of synchronous circuits. The associated additional design freedom can often be exploited to reduce power consumption. One example of this is the aforementioned distribution of control, another example is based on the latch-based character of many asynchronous circuits.

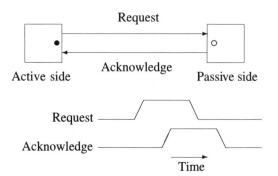

Figure 2 A four-phase handshake-signaling sequence: request up; acknowledge up; request down; and acknowledge down. The symbols ● and ○ denote the active and passive sides respectively.

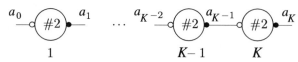

Figure 4 2^K-fold repeater composed of K duplicators.

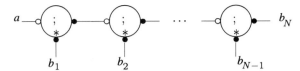

Figure 5 *N*-way sequencer composed of $N - 1$ binary sequencers.

In synchronous circuits, most state-holding elements are implemented as master-slave flip-flops ("D-types"). During one clock phase, data is read from the slaves and (after possible modification) written into the master latches. In the other clock phase, data is transferred from the master latches to the slave latches.

In asynchronous circuits the much finer resolution in time can be used to sequence data transfers among latches, thereby avoiding many of the slave latches. This is illustrated by a 4-place shift register:

forever do

$b!x3; x3 := x2; x2 := x1; x1 := x0; a?x0$

od

where the *W*-bit wide variables *xi* are implemented by *W* latches. Here $b!x3$ denotes the output of the value stored in *x3* through handshake port *b*. Similarly, $a?x0$ stands for an input through handshake port *a*, where the incoming value is stored in *x0*. The elimination of slave latches saves both area and power. For large *W* the overhead in the control part (a multi-way sequencer) can be neglected, and the power reduction may approach 50%. Note that, due to the sequential propagation of the vacancy from the output variable *x3* toward the input, the cycle time grows proportional with the number of stages. Implementations for higher throughput and still lower power can be found in [8].

2.4 Elimination of Standby Power

Asynchronous circuits provide an automatic and instantaneous stand-by mode at arbitrary granularity in time and function. In this mode the circuit dissipates leakage power only, typically a few μW. In a clocked circuit a similar state can be reached by gating the clock. However, the clock crystal, oscillator circuitry, and possibly PLLs continue to dissipate power, typically 0.1mW/MHz or more. Shutting these circuits down is often not acceptable because start-up would require milliseconds before the oscillator is operating in a stable regime.

In applications where duty cycles are low and where fast response times are required, this may reduce overall power consumption considerably. For example, in hand-held telecom applications, a microprocessor is often used for the higher layers of the protocol stack, typically for a few milliseconds each second. An asynchronous microprocessor such as the AMULET2e [2] would then display a remarkable power efficiency.

During standby, interrupt detection generally is an essential function. In asynchronous circuits this can be performed using a simple (leakage-only) edge-detector, as opposed to the constant sampling/polling of the interrupt lines that is required in a synchronous circuit.

2.5 Adaptive Scaling of the Supply Voltage

Asynchronous circuits can be implemented such that they automatically adjust their operating speed when the supply voltage varies. This property can be exploited to choose the lowest possible supply voltage for a given function in a very simple way. Two variants can be distinguished, namely, *feedback* and *feed forward*.

An example of a feedback variant is shown in Fig. 6. In this design, *f* is a continuous varying signal that is a measure for the phase difference between the input and the output of the FIFO. When the FIFO comprises two stages, for instance, *f* represents a value between 0 and 4π.

If the processor is operating too fast, the buffer becomes more than half full, and the DC/DC converter is requested to reduce V_{dd}. Similarly, if the processor is running too slow, V_{dd} is increased.

When the embedded circuit is faced with a higher workload, the supply voltage will be increased, and hence the power consumption also temporarily increases. If such a stepwise increase (or decrease) can be predicted, it may pay off to prevent such an increase (or decrease) in power consumption. In such a situation, feed-forward adaptation of the supply voltage is an interesting option.

Suppose that a sequential computation consisting of two parts, say *A* and *B*, and that a time *T* is available for execution. The dependence on the supply voltage of the execution time for *A* is a simple model given by $T_a = t_d/V_a$, whereas for energy we have $E_a = C_a V_a^2$, in which C_a is the total effective switched capacitance during *A*.

We can minimize the energy required for the computation by choosing possibly different supply voltages V_a and V_b for these two parts. The energy required for the computation is given by $E = C_a V_a^2 + C_b V_b^2$. Substitution of the time-dependence relation, and choosing $T_a = \alpha T$ and $T_b = (1 - \alpha)T$ results in

$$E = C_a \left(\frac{t_a}{T}\right)^2 \left(\frac{1}{\alpha}\right)^2 + C_b \left(\frac{t_b}{T}\right)^2 \left(\frac{1}{1 - \alpha}\right)^2$$

The minimum of the above function is found by determining α such that $dE/d\alpha = 0$, which gives

$$\left(\frac{\alpha}{1 - \alpha}\right)^3 = \frac{C_a}{C_b} \left(\frac{t_a}{t_b}\right)^2$$

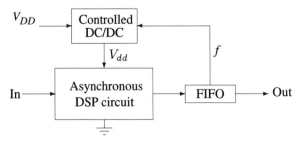

Figure 6 Just-in-time DSP [10]. Supply voltage V_{dd} is regulated such that FIFO is kept about half full.

From this, α can be determined and, subsequently, the supply voltage for A (and similar for B):

$$V_a = \frac{t_a}{T}\left[1 + \sqrt[3]{\frac{C_b}{C_a}\left(\frac{t_b}{t_a}\right)^2}\right]$$

This feed-forward choice of the supply voltage minimizes the expected energy consumption. It turns out that the supply voltages are such that the power consumption is constant, i.e., $E_a/T_a = E_b/T_b$. This implies that the higher the workload, the lower the supply voltage (in contrast to feedback). If, for instance, $t_a = t_b$ and $C_a > C_b$, then the supply voltages are chosen such that $V_a < V_b$ and $T_a > T_b$.

An implementation of feed-forward adaption based on two separate power supplies is shown in Fig. 7. This power-supply switch is used in the DCC Error Detector [4,5]. The Error Detector normally operates on the low supply voltage V_{LL} and in case of exceptions (which are predicted by the Detector and require more processing) switches to the high supply voltage V_{HH} by making signal *high* high.

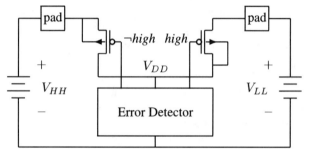

Figure 7 Switching between power supplies V_{LL} and V_{HH}, assuming $V_{LL} \leq V_{HH}$.

The feed-forward solution shown in Fig. 7 does not eliminate derating margins for variations in processing, temperature, and supply voltage, in contrast to feedback.

In a synchronous circuit, additional circuitry is needed to adjust the speed of the circuit to the supply voltage. This can be done in discrete steps, by dividing or multiplying a crystal oscillator output, or in a continuous way, for instance, by varying the duty cycle of a digital PLL [11].

3. DCC Error Corrector

Two synchronous ICs for an experimental DCC player have been designed in Tangram [3]. These circuits are used in play mode and their main function is de-interleaving and error correction of the bit-stream from tape before it is offered to the audio source decoder, which then produces a stereo audio signal. Here, we focus on one of the chips, namely the error detector.

The DCC error detector (DDD) is a Reed-Solomon decoder that processes codewords consisting of 32 or 24 8-bit symbols and containing 4 or 6 parity symbols. Its function and implementation are described in detail in [4,5].

The DDD algorithm is highly sequential. The backbone of the program actually consists of a sequence of 48 steps, which is implemented as a distributed 48-way sequencer and guarantees very local activity.

Two asynchronous variants of the DDD have been fabricated, one double-rail (2 wires per data bit) and one single-rail (1 wire per data bit). A synchronous realization of the same function, which was optimized for low-power operation, was also available. The important performance characteristics of these circuits are listed in Table 1.

Table 1. Comparison of Three DCC Chips (All in 0.8μ CMOS; Measurements at 5V for Typical DCC Input Data).

Quantity	Unit	Sync	Double	Single
Core area	mm²	3.3	7.0	3.9
Cycle time	μs		83	40
Power	mW	2.7	2.3	0.37

The single-rail version turns out to be superior to the double-rail circuit in all dimensions; it is faster, consumes less energy and requires less circuit area. An additional advantage of the single-rail circuit is that it is realized in a generic standard-cell library, whereas the double-rail circuit is based on dedicated asynchronous cells, such as C-elements and double-rail data operators. Single-rail circuits thus remove a serious roadblock toward exploitation of asynchronous circuits [12].

At a supply voltage of 5V, under typical operating conditions, the DDD needs only 282ms for the processing of worst-case input data (when full correction is required), where it may actually take 1s. For a typical set of input data, it will then consume 0.74mW. To remove this excess performance, the supply voltage may be lowered to 2.8V (assuming a factor 2 for derating). The power consumption then drops to 0.24mW.

Adaptive supply-voltage scaling can now be applied to reduce the power consumption further. For the handling of correct words for one second of music, the DDD requires 69ms and 0.58mJ (at 5V). If all codewords are incorrect, an additional 214ms and 3.2mJ are required. For typical DCC input data, however, only 5% of the codewords are incorrect.

In the feed-forward variant this implies that 42% of the time should be spent on processing correct code words at a supply voltage of 1.7V. The remaining time can then be filled with the additional processing time required for incorrect words at 3.7V.

The feedback variant can be used to eliminate the derating margin and thus further lower the supply voltage. The DDD algorithm allows for the adjustment of the supply voltage at the codeword rate. For typical input data (5% incorrect) the DDD will then operate 95% of the time at 1.3V and 5% at 2.2V.

Power consumption in the various modes of operation is listed in Table 2.

Table 2. DDD Power Consumption for Typical Double-speed DCC Input Data in Various Environments.

Quantity Unit	V_{low} V	V_{high} V	Power mW
Standard	5.0	5.0	0.74
Min. volt.	2.8	2.8	0.24
Feed forward	1.7	3.7	0.15
Feedback	1.3	2.2	0.08

4. PAGER STANDBY CIRCUITS

The use of radio paging, which typically allows one-way communication of numeric or text information, continues to grow rapidly as a low-cost alternative to cellular or mobile telephony. Figure 8 shows a common pager architecture that comprises a discrete or integrated receiver (Rx), a digital IC to decode the protocol (Dec), a microcontroller (μC), and a user interface (IO) offering key input, display output, and an alert facility. Since pager units must be small and battery operated, low-power consumption is a prime consideration in pager design. Pagers are, however, high-volume consumer products for which chip area is still an important issue. For these products, asynchronous designs have a drawback when compared to synchronous ones as they tend to be larger.

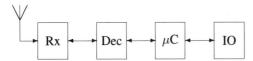

Figure 8 Pager architecture.

Recently-proposed pager codes, such as APOC and Flex, allow the pager to enter a so-called standby mode for long periods of time. In standby mode the receiver section is disabled and only a small part of the system (the standby circuitry) is active. In taking the cost factors (i.e., area and power) into account to come to an optimal design, the standby circuitry is crucial since its relative contribution is *marginal* with respect to *area,* but *substantial* with respect to *power dissipation.* Therefore by designing the standby circuitry asynchronously, a noticeable reduction of the power dissipation can be obtained for a negligible increase of the area.

Table 3 shows an estimation of the average power dissipation for an APOC pager in which we distinguish the four main functions as well as three modes of operation. In this estimation we assume a synchronous design with extensive clock gating. The table shows that the standby circuitry in the decoder consumes 50% of the total power. Therefore, when looking for further reductions of the overall power consumption, it makes sense to begin with the standby circuitry.

Table 3. Distribution Average Power Dissipation (%).

Mode	Rx	Dec	μC	IO	Total
Standby	1	50	2	1	54
Rx active	28	0	0	0	28
IO active	0	0	6	12	18
Total	29	50	8	13	100

We have designed asynchronous standby circuits for an APOC pager decoder [7] which, when compared to synchronous designs, dissipate four times less power and are 40% larger in size. Therefore by designing only the standby circuitry asynchronously, we have extended the battery life of a product for

nearly no cost in area. The decoder chip, which apart from the standby circuits is completely synchronous, has been fabricated and is currently being tested.

5. CONCLUSION

Circuit activity in many systems-on-silicon is low, and varies over time and place. Asynchronous design styles exploit this by enabling local memory elements only when required by the underlying algorithm. In a clocked circuit, a similar result can be obtained by locally gating the clock signal. Asynchronous circuits, however, avoid distribution of the "hot wire," viz., the wire with the highest capacitance and an activity of two.

In addition to exploiting low circuit activity, asynchronous circuits can also be used to *reduce* activity. Distribution of control is in essence a form of sparse-state encoding in which fewer flip-flops change state during each algorithmic step. In clocked circuits this would increase power (more flip-flops), whereas in asynchronous circuits this reduces power (fewer changes).

A further reduction in activity can be realized by removing slave latches. This fits well with asynchronous control, because of the availability of very fine time granularity. Rather than a high-frequency global clock, local control signals can be created by, for example, an N-way sequencer.

The demand-driven nature of asynchronous circuits becomes most clear in applications were the standby (idle) state dominates and an instantaneous response to interrupts is required. In an asynchronous circuit a transition on an interrupt wire can trigger full (processor) activity immediately. In a clocked circuit, in contrast, a crystal, oscillator, and PLLs must be continuously active to sample interrupt wires and enable an instantaneous response.

All of the aforementioned ideas concentrate on reducing the number of gate-input transitions. These measures are orthogonal to reducing the energy *per* such a transition. The latter can most effectively be realized by adaptively scaling the supply voltage using either a feed-forward or feedback approach. Asynchronous implementations tend to be simpler since scaling of the clock frequency is not an issue.

Asynchronous is more than low power. Since in future CMOS technologies wire delays will tend to dominate over switching delays, IC-wide synchrony cannot be maintained. Moreover, in more and more applications, the environment provides more than one independent time reference. Such specifications naturally lead to systems-on-silicon with asynchronous, data-driven communication.

ACKNOWLEDGEMENTS

Part of this work was supported by the European Commission through ESPRIT project 6143, EXACT (EXploitation of Asynchronous Circuit Technologies) and basic research working group 7225, ACiD.

References

[1] A. J. Martin et al., "The design of an asynchronous microprocessor." In C. L. Seitz, editor, *Advanced Research in VLSI: Proc of the Decennial Caltech Conf on VLSI*, MIT Press, 1989, pp. 351–373.

[2] S. B. Furber, J. D. Garside, S. Temple, and J. Liu. "AMULET2e: An asynchronous embedded controller," *Advanced Research in Asynchronous Circuits and Systems*, IEEE Comp Soc Press, 1997.

[3] K. van Berkel et al., "A fully-asynchronous low-power error corrector for the DCC player." *IEEE Journal of Solid-State Circuits*, vol. 29, no. 12, 1994, pp. 1429–1439.

[4] K. van Berkel et al., "A single-rail re-implementation of a DCC error detector using a generic standard-cell library." *Asynchronous Design Methodologies*, IEEE Comp Soc Press, 1995, pp. 72–79.

[5] J. Kessels. "VLSI programming of a low-power asynchronous Reed-Solomon decoder for the DCC player." *Asynchronous Design Methodologies*, IEEE Comp Soc Press, 1995, p. 44–52.

[6] L. S. Nielsen and J. Sparsø, "A low-power asynchronous data-path for a FIR filter bank," *Advanced Research in Asynchronous Circuits and Systems*, IEEE Comp Soc Press, 1996.

[7] J. Kessels and P. Marston. "Designing asynchronous standby circuits for a low-power pager." *Advanced Research in Asynchronous Circuits and Systems*, IEEE Comp Soc Press, 1997.

[8] K. van Berkel and M. Rem, "VLSI programming of asynchronous circuits for low power." In G. Birtwistle and A. Davis, editors, *Asynchronous Digital Circuit Design*, Workshops in Computing, Springer-Verlag, 1995, p. 152–210.

[9] J. Tierno. R. Manohar, and A. Martin, "Energy and entropy measures for low power design," *Advanced Research in Asynchronous Circuits and Systems*, IEEE Comp Soc Press, 1996.

[10] V. van Kaenel, D. Aebischer, C. Piguet, and E. Dijkstra, "A 320-MHz, 1.5mW at 1.35V CMOS PLL for microprocessor clock generation," *IEEE Journal of Solid-State Circuits*, vol. 31, no. 11, 1996, pp. 1715–1722.

[11] A. Chandrakasan, V. Gutnik, and T. Xanthopoulos, "Data driven signal processing: An approach for energy efficient computing." *Low Power Workshop*, 1996, p. 347–352.

[12] A. Peeters and K. van Berkel, "Single-rail handshake circuits," *Asynchronous Design Methodologies*, IEEE Comp Soc Press, 1995, pp. 53–62.

[13] L. S. Nielsen et al., "Low-power operation using self-timed and adaptive scaling of the supply voltage," *IEEE Trans on VLSI Systems*, vol. 2, no. 4, 1994, pp. 391–397.

Latches and Flip-flops for Low Power Systems

CHRISTER SVENSSON AND JIREN YUAN

DEPT. OF PHYSICS, LINKÖPING UNIVERSITY, 58183 LINKÖPING, SWEDEN.

EMAIL: CHS/JRY@IFM.LIU.SE

Abstract—The power consumption of latches and flip-flops is described, including its effect on the clock distribution power consumption. Several circuit techniques for latches and flip-flops are compared for power and speed using analog simulation, and the results are discussed.

INTRODUCTION

Power consumption in digital systems is normally dominated by the energy dissipated in connection to signal transitions. As the clock is the signal with the most transitions, it, and the circuits connected to it, becomes very important from the power consumption point of view. Quite often, the clocking system dominates the total power consumption [1]. In this short paper, we will analyze the power consumption in connection to latches and flip-flops and discuss means to reduce power consumption in this context.

POWER CONSUMPTION OF LATCHES AND FLIP-FLOPS

The power consumption in CMOS circuits is dominated by the dynamic power, which can be expressed as [2,3]:

$$P = \Sigma \, \alpha f_c C V_{dd}^2$$

where the power consumption is a sum over the electrical nodes in the system, and for each node α is the activity (the number of complete transitions per clock cycle), f_c is the clock frequency, C is the capacitance of the node, and V_{dd} is the supply voltage.

Let us discuss this expression in some detail [3]. The activity, α, is thus the number of complete transitions (up and down) during one clock cycle. This means that the clock itself has $\alpha = 1$. This is a high value, so the clock is always expensive in power. We may therefore conclude that the clocked capacitance should be minimized in order to save power. An ordinary data signal may have various values of α. The maximum value is 0.5, as data can change once per clock cycle at maximum. The minimum value is of course zero. A random binary signal has $\alpha = 0.25$. Practical signals have values between 0 and 0.25, depending on the situation. A "natural" signal (e.g., speech) nor-

mally have an activity of 0.25 of its least significant bits (random-like), whereas its most significant bits have very low values; most of the time the speech signal is nearly zero or relatively weak. The large variation in activity of data signals must be taken into account when judging power consumption. In dynamic circuits, we sometimes utilize precharging. This means that a data node is charged, e.g., high for each clock cycle. This leads to an activity of such a node of 0.5, independent of the data activity (assuming that the probabilities for data high and low are equal). Precharged nodes therefore cannot utilize low data activity for low power. We may also have unintentional activity in our circuit, which we call glitches. Glitches may occur if a node is expected to stay constant through a clock transition, but instead it changes twice, thus returning to its original value. Such a glitch may not be important for the functionality of our system, but it will increase the power consumption, as it calls for charging and discharging a node capacitance. Circuits where glitches can occur are precharged stages, e.g., the PTSPC stage in Fig. 1b.

The key signal in synchronous systems is obviously the clock. The clock signal is normally heavily loaded by all latches and flip-flops in the chip and by the clock distribution network, covering the whole chip. Clock driving thus becomes very important from the power consumption point of view, as the clock power may be 20–40% of the total chip power [1]. The standard method to drive large loads is to use a tapered inverter chain [2]. Such a driver will have a total (switching) capacitance, C_{driver}, of [3]:

$$C_{driver}/C_L = (1 + C_o/C_i)(1 - 1/Y)/(f - 1)$$

where C_L is the total clock network load capacitance, C_o/C_i is the ratio of the output and input capacitance of the driver, $Y = C_L/C_i$ and f are the tapering factor. Here f is given by $f^N = Y$ where N is the number of stages in the driver. Note the sensitivity for f-values close to 1. For such a chain, it can be shown that there exists an optimum value of f for which the total delay of the chain is minimum. For $C_o = C_i$, this optimum will occur at $f = 3.5$. The delay is however a relatively flat function of f, so also larger values of f may be acceptable. On the other hand, if we need

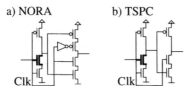

Figure 1 Precharged logic.

very short rise- and fall-times, or very high frequency, we may need $f < 3.5$. Using the above optimum value of $f = 3.5$ leads to a power consumption of the driver which is about 80% of the power consumption in the load; in other words, the driver causes an excess power consumption of 80% over the load. This is quite a large overhead. It can be reduced by using larger tapering factors (i.e., fewer stages), but for very fast systems we may need to have smaller f, instead increasing the overhead further. From the latch and flip-flop perspective we may conclude the following: 1) as clock network must be minimized in capacitance to save power, we should avoid multiple clocks; and 2) the clock load in latches and flip-flops should be minimized, as any such load is multiplied by the driver overhead, considering power consumption.

DYNAMIC LATCHES

Dynamic latches are the simplest and most efficient timing circuits. Normally we use two kinds: n-type, transparent for clock high, and p-type, transparent for clock low. Below we show the n-types. Let us start with a discussion of single-phase clocked, single-rail latches ("single rail" means that data occur on a single wire). We will then discuss 4 different circuits (Fig. 2) [2]. The first example, the classic latch, is the oldest one, and also the fastest [4]. It can also be seen as an inverter followed by a transmission gate. It is very fast because it behaves as a true single stage. The delay introduced by the clock inverter to the p-transistor is unimportant, as the n-transistor will start to charge the output before the p-transistor turns on. The second

example, C²MOS latch, is very similar to the classic one, the only difference is one wire connecting the upper and the lower part. This is slower but more robust than the Classic. Both of these have four clocked transistors (including the inverter), three of which are loading the clock input. The third circuit is a TSPC half latch ("TSPC" is True Single Phase Clocked; the clocks do not need to be inverted) [5]. This latch will only isolate high inputs at clock low. To obtain a full latch we need to have two of these, the nonprecharged TSPC latch (Fig. 2d). The last latch is slower than the classic one (because of two stages), more robust and has only two clocked transistors.

In order to better understand the role of the half latch, we may observe the latches in the context of precharged logic (Fig. 1). Here the bold transistors are to be replaced by a network performing the desired logical function. For low clock the logic stage is precharged, i.e., the intermediate nodes in Fig. 1 are precharged high. Both latches are nontransparent; the TSPC half latch is closed because the input is high. For clock high, the logic stage evaluates and the intermediate node has valid data. This valid data is inverted to the output, as both latches now are transparent. We may note that the precharged stage (with a single input transistor) plus the TSPC half latch in Fig. 1b behaves similar to a latch. We call it a precharged TSPC latch (PTSPC).

Another class of circuits use dual rail data (i.e., data that is represented by its value and its inverted value). In Fig. 3 we show some dual rail latches. Figure 3a shows a latch derived from CVSL logic [2]. Normally clock and data transistors are interchanged compared to Fig. 3a, but our choice here is somewhat better for latches. This latch depends on transistor ratios, i.e., the n-transistors must be stronger than the p-transistors in order to flip the latch. This is not a big problem for the n-latch, but considerably worse for the p-latch. In Fig. 3b we show a precharged form of CVSL latch [6]. It needs an extra stage to keep output data during precharge. As it is precharged and relatively complex, it will use more power so we will not discuss it further. In Fig. 3c we show a new variation of the CVSL latch, the dynamic single transistor clocked (DSTC) latch, using a common clocked transistor to save power [7]. This latch is both fast and power-efficient, but sensitive to input glitches when in hold state.

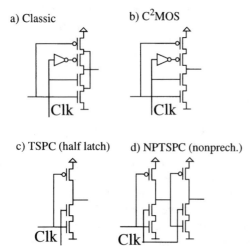

Figure 2 Single rail dynamic latches.

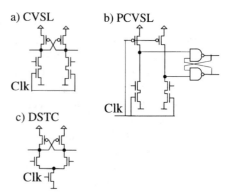

Figure 3 Double rail dynamic latches.

234

Static latches always need some form of positive feedback (like a crosscoupled inverter pair), automatically making them have complementary outputs [2]. Some are normally designed with dual-rail inputs, some with single-rail inputs.

In Fig. 4a we show a single input latch based on the classical set-reset flip-flop, 3GATE latch. Note that the AND-NOR gate pair is implemented as a simple static CMOS gate (using six transistors). This latch looks complex, but has relatively low power consumption due to only four clocked transistors. In Fig. 4b we demonstrate the classical transmission-gate based static latch, TGATE latch, using six clocked transistors. Both of these are used as basis for flip-flops in standard CMOS cell libraries, the latter being more common than the former. Both are thus very well established in industry. Another static latch is the RAM-type, based on the 6-transistor SRAM memory cell (Fig. 4c) [2]. It can also be derived from the dynamic CVSL latch in Fig. 3a. Like the CVSL latch, this latch depends on transistor ratios, with the same problem mentioned above. In Fig. 4d, finally, we show a static version of the DSTC latch (Fig. 3c) [7].

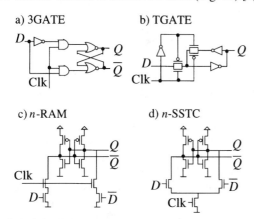

a) 3GATE
b) TGATE
c) *n*-RAM
d) *n*-SSTC

Figure 4 Static latches.

FLIP-FLOPS

Flip-flops are normally derived from two latches, e.g., a *p*-latch followed by an *n*-latch to obtain a positive edge-triggered flip-flop. We may also add an inverter at the output to achieve a complementary output. This scheme is useful for Classic, C2MOS, and NPTSPC latches forming corresponding flip-flops. Concerning PTSPC, it turns out that a very nice flip-flop can be formed from a *p*-half-latch and a PTSPC latch, here called 9T (nine transistors in the basic unit) (Fig. 5a) [5]. When using CVSL-latches, all signals are double-rail, so we do not need the output inverter. As mentioned above, the *p*-CVSL latch is not very good, so we could also use two *n*-latches and a clock inverter, thus clocking the first *n*-latch with a complementary clock. Concerning the DSTC latch, it is also not very nice in its *p*-latch form, and it has some problems with input glitches. It can however be combined with a new circuit, replacing the *p*-DSTC latch, forming a very nice and efficient flip-flop (Fig. 5b) [7] using only 2 clocked transistors.

In the same way as we formed dynamic flip-flops from latch pairs, we may use static latch pairs to form static flip-flops,

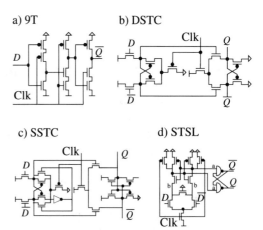

a) 9T
b) DSTC
c) SSTC
d) STSL

Figure 5 Flip-flops.

e.g., for the static latch types 3GATE and TGATE. There is also a static version of the DSTC flip-flop, called SSTC flip-flop (Fig. 5c) [7]. Like the dynamic version, it uses only 2 clocked transistors, making it a good candidate for very-low power consumption. Finally, we will also discuss a static flip-flop based on a self-timed slave latch, STSL (Fig. 5d) [8]. When clock is low, the intermediate nodes a are precharged high, making the output set-reset stage keep its previous value. When the clock goes high, the input *n*-transistors are activated and one of the nodes b becomes low, making the corresponding a-node low. This can however only happen once as the crosscoupled *n*-pair prevents new input data from propagating to nodes a, as the *n*-transistor which was not active initially, now is off and therefore not active until the next precharge. The valid data on nodes a is finally propagating to the output, as the low value will reset the output stage. This flip-flop was used in a recent attempt to create a high-performance, low-power microprocessor, called StrongARM [8].

All flip-flops demonstrated so far have utilized one of the clock edges to move data from input to output. We may also design flip-flops which utilize both clock edges, double-edge triggered flip-flops [9,10]. Such flip-flops may save power in the clock distribution network as the clock frequency is halved for fixed throughput, compared to single-edge triggered flip-flops. The principle is to have one positive-edge and one negative-edge triggered flip-flop in parallel, with some means for multiplexing their outputs. The power consumption will be approximately double that of a single-edge triggered flip-flop (for fixed frequency). In Fig. 6 we show an example of an efficient double-edge triggered flip-flop [10].

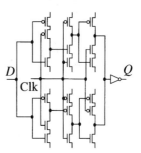

Figure 6 A double-edge triggered flip-flop.

Comparison of different circuit techniques is always very difficult [11]. There are many reasons for this. The result depends on many parameters, including the technology used, transistor sizes, data statistics, supply voltage, capacitive loads, and the context in which the circuit is used. It is seldom possible to use literature data in comparisons because there isn't enough control of these parameters. We therefore have chosen to compare different latches and flip-flops by making our own simulations of all circuits, trying to keep critical parameters as equal as possible. Still, comparisons are risky, as it is not obvious how to compare circuits with quite different properties or what parameters to choose. We have however tried our best using the following principles.

All circuits are simulated using the same fabrication process (i.e., 0.8μm CMOS process from AMS [12]), and including transistor parasitics but not wiring parasitics. All transistor sizes are the same ($w_p = 6\mu$m, $w_n = 3\mu$m), except special cases where only minimum transistors are needed (e.g., for holding data only, here we use $w_{n,p} = 2\mu$m) or where the circuit calls for different sizes for its reliable function (in n-CVSL, n-RAM, and n-STC latches, $w_p = 4\mu$m, $w_n = 6\mu$m, and $w_{min} = 2\mu$m, respectively, whereas in p-CVSL, p-RAM, and p-STC latches $w_p = 12\mu$m, $w_n = 3\mu$m and $w_{min} = 2\mu$m, respectively).

All outputs are loaded by 2 standard inverters (with $w_p = 6\mu$m and $w_n = 3\mu$m). In our power figures, the power consumption of these loads is included. All circuits use one external clock. If inverted clocks are needed they are generated by inverters in the circuits. For MS flip-flops, each latch has its own inverter (if needed). All inputs (including clock) are driven by standard inverters with zero output capacitance (drain- and source-capacitance parameters are set to zero). By including the power consumption of these inverters, we automatically include the input capacitance power consumption of our circuits. The input slopes to these inverters are sharp enough to make their short circuit current negligible.

Some circuits have complementary outputs in their basic versions and some have not. We have chosen to give most circuits a complementary output (except in the dynamic latch case). If not there from the beginning, we have added an inverter at the output. Each output is loaded as above. This point is further discussed below.

Each circuit was simulated in HSPICE, and the delay (worst delay from clock edge to output) and the power consumption were recorded. Delays are defined from 50% to 50% of a full swing. For power consumption we used a clock frequency of 100MHz and recorded the value for input data constant ($\alpha = 0$, sometimes for both high- and low-input values) and for input data changing each clock cycle ($\alpha = 0.5$). The results of our comparisons are demonstrated in Figs. 7 through 10. We have here shown the power consumption for $\alpha = 0$ and $\alpha = 0.5$, as the endpoints of vertical bars, versus the worst-case delay. We have also given the calculated power consumption for $\alpha = 0.25$ (random data) as a circle on each bar. In some cases there are two values of power for $\alpha = 0$, for data high and low, respectively, shown as an extra horizontal bar in the diagram. Each vertical bar corresponds to one circuit. Note that data normally has activities between 0 and 0.25, i.e., in the lower part of the vertical bars.

In Fig. 7 we show simulation results, from left to right, of n-Classic, n-NPTSPC, n-C2MOS, n-PTSPC, n-DSTC, n-CVSL, p-CVSL, and p-DSTC dynamic latches. Note, that here n-Classic, n-NPTSPC, n-C2MOS, and n-PTSPC have single-ended outputs, whereas n-CVSL, p-CVSL, n-DSTC, and p-DSTC have complementary outputs. If adding an inverter to the output of the single-rail types, their worst-delay increases about 0.4ns and their power consumption increases about 60μW. NPTSPC has the lowest power consumption of these, which is explained by its very few clocked transistors (only two)

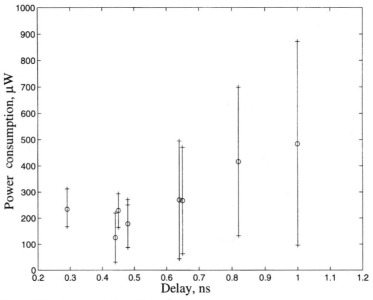

Figure 7 Power consumption of dynamic latches.

236

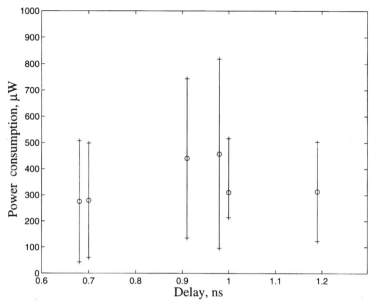

Figure 8 Power consumption of static latches.

and the lack of precharging. The precharged version of TSPC and PTSPC is next. Each has a reasonable good speed. All others have comparable power consumption, depending on activity. The clearly highest speed is demonstrated by the Classic latch, as also pointed out above.

Figure 8 shows simulation results from n-SSTC, n-RAM, p-RAM, p-SSTC, TGATE, and 3GATE static latches. The lowest power consumption (and highest speed) are demonstrated by n-SSTC and n-RAM. However p-RAM and p-SSTC are considerably worse, both from the point of view of power and speed. The reason for this is that they depend on transistor ratios, which means that when the p-transistors must be stronger than the n-transistors, their sizes have to increase substantially. Of the other two, the 3GATE latch has somewhat less power consumption than the TGATE latch, but worse speed [3].

In Fig. 9 we show simulation results from CVSL, DSTC, Classic, NPTSPC, 9T, and C2MOS dynamic flip-flops, where all have complementary outputs. Here the groups DSTC, NPTSPC, and 9T have the lowest power consumption and reasonable speeds. CVSL has the highest speed. Note, however, that NPTSPC and 9T may be designed without complementary outputs. If so, they are considerably improved to less than a 0.5ns delay. The other three flip-flops have comparable power consumption and quite different speeds. Note again that Classic and C2MOS have noncomplementary versions with better performance (Classic, 0.33ns delay). As mentioned above, an alternative to the CVSL design with a p-latch followed by an n-latch (as used here), we may consider using two n-latches, of which the first one has an inverter on the clock input. It turns out that the two versions are comparable in performance because

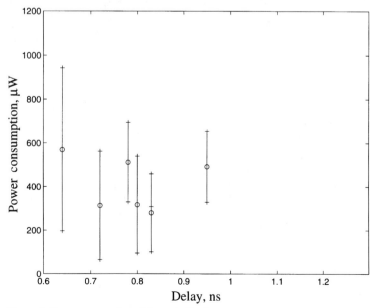

Figure 9 Power consumption of dynamic flip-flops.

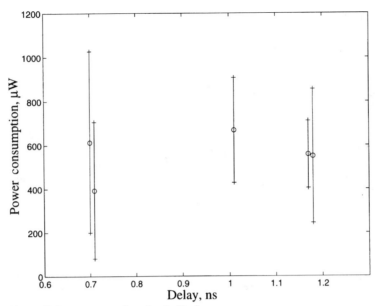

Figure 10 Power consumption of static flip-flops.

the extra clocked inverter causes some delay increase and a relatively large increase in power (especially at low activity).

In Fig. 10 we show simulation results from RAM, SSTC, TGATE, STSL, and 3GATE type static flip-flops. From a power point of view, the SSTC flip-flop is obviously superior to the others. It also has a nearly minimum delay. The low-power consumption of this flip-flop is due to simplicity and very few clocked transistors. This is also important from a clock driver point of view. Among the others, RAM has both low power and low delay. Again, a RAM version with two n-latches has nearly the same performance. Also the 3GATE flip-flop has a reasonable power but a larger delay. In spite of the goals of the StrongARM project [8], the STSL flip-flop seems inferior to the others in our comparison (power may not be the largest for random data, but the minimum power is quite large and the delay is bad).

CONCLUSIONS

The power consumption of latches and flip-flops is strongly dependent on the clock load and on the data activity. The effect of the clock load on system power consumption is further enhanced by the clock-buffer power consumption. Low power circuits should therefore have as few clocked transistors as possible. They should also avoid multiple clocks and precharging. Low supply voltage is an efficient method to save power, but may lead to speed loss. A low-power circuit technique should therefore be suitable for low supply voltages and have good speed. When comparing different circuit topologies for latches and flip-flops we find no correlation between power and speed. It is therefore quite possible to combine low power with high speed. The best static flip-flop seems to be the recently-proposed static-single transistor clocked type, which has about 40% lower power than

commonly used flip-flops (at α about 0.15), combined with high speed. Among the different dynamic latches investigated, the non-precharged true single-phase clocked type has about 60% less power consumption than the standard C2MOS type (at α about 0.15), at a very reasonable speed.

References

[1] D. Liu and C. Svensson, "Power consumption estimation in CMOS VLSI chips," *IEEE Journal of Solid State Circuits,* vol. 29, 1994, pp. 663–670.

[2] N. Weste and K. Eschraghian, *Principles of CMOS VLSI Design,* 2nd ed., Addison-Wesley, 1993.

[3] C. Svensson and D. Liu, "Low power circuit techniques," in *Low Power Design Methodologies,* M. Pedram and J. Rabaey, editors, Kluwer, 1996.

[4] W. Bowhill et al., "A 300 MHz 64b quad-issue CMOS microprocessor," *ISSCC Digest of Technical Papers,* 1995, pp. 182–183.

[5] J. Yuan and C. Svensson, "High speed CMOS circuit technique," *IEEE Journal of Solid State Circuits,* vol. 24, 1989, pp. 62–70.

[6] D. R. Renshaw and C. H. Lau, "Race-free clocking of CMOS pipelines using a simple global clock," *IEEE Journal of Solid State Circuits,* vol. 25, 1990, pp. 766–769.

[7] J. Yuan and C. Svensson, "New single-clock CMOS latches and flipflops with improved speed and power savings," *IEEE Journal of Solid State Circuits,* vol. 32, 1997, pp. 62–69.

[8] J. Montanaro et al., "A 160MHz 32b 0.5W CMOS RISC microprocessor," *ISSCC Digest of Technical Papers,* 1996, pp. 214–215.

[9] R. Hossain, L. D. Wronski, and A. Albicki, "Low power design using double edge triggered flip-flops," *IEEE Tr. on VLSI Systems,* vol. 2, 1994, pp. 261–265.

[10] M. Afghahi and J. Yuan, "A novel implementation of double-edge triggered flip-flop for high speed CMOS circuits," *IEEE Journal of Solid State Circuits,* vol. 26, 1991, pp. 1168–1170.

[11] R. Zimmermann and W. Fichtner, "Low-power logic styles: CMOS versus pass-transistor logic," *IEEE Journal of Solid State Circuits,* vol. 32, 1997, pp. 1079–1090.

[12] "0.8 μm CMOS process parameters," Austria Microsystems International, Dec. 1994.

ZIG-ZAG PATH TO UNDERSTANDING

ROLF LANDAUER

IBM Thomas J. Watson Research Center

Yorktown Heights, NY 10598

ABSTRACT

Our understanding of the fundamental physical limits of information handling has developed along a very convoluted path. Most of the initially plausible physical conjectures have turned out to be wrong. A participant's personal view of these events is not a disciplined contribution to the history of science. I do, however, list my own mistakes along with those of others.

1: Computation

The attempt to understand the ultimate physical limits of the computational process is almost as old as the modern electronic digital computer. This search was stimulated by the earlier examples of thermodynamics which arose from the attempt to understand the limits of steam engines, and by Shannon's channel capacity theory. These examples suggested that the computer could be characterized in a similar way. The earlier prototypes had derived optimum performance limits in a way which transcended particular technologies and design choices. Scientists and engineers take pride in the ability to do back–of–the–envelope calculations, to quickly reach to the critical aspects without encumbering details. Yet, in this field of ultimate physical limits of information handling, the quick and dirty approaches have turned out to be wrong in a remarkably consistent way. All of the first answers have been misleading.

In the 1950's it was natural to associate a binary degree of freedom with kT and to assume that a minimal energy of that order had to be associated with an elementary logic step, to provide noise immunity. The identification of "associate with" with required energy dissipation tended to be made in casual blackboard discussions. It was not until 1961 [1] that it became clear that the process which really required a minimal and unavoidable energy penalty was the discarding of information.

The assumption that an energy dissipation of the order of kT was required by every logic step seemed, in the 1950's, to be a natural consequence of "known" results. It was "known," after all, that it took $kTln2$ to send a bit along a communications line, and computation required the frequent transmission of signals. It was also "known" that measurement required energy dissipation. Unfortunately, as we now know, the communication results were really more limited than generally presumed. Furthermore the sophisticated literature on the measurement process often left it unclear what a measurement really is and how to tell a measurement from an elephant's trunk. Additionally, as is now known, classical measurement theory as then accepted suffered from its own blemishes which were not widely understood until the 1980's.

This subject, during the 1950's was mostly one for casual unpublished speculations. Ref. [2] tells us, for example, that " ...von Neumann [3] and Brillouin [4] conjectured that $kTln2$ minimum energy must be spent for each step of information processing". Many other authors have related statements, e.g. Igeta's reference to Brillouin [5]. Brillouin's famous book [4], despite a chapter *The Problem of Computing,* does not allude to the actual logic processes in a computer, e.g. to a logical *and* or a logical *or*, and contains no references to a total working computer, such as a Turing machine or a cellular automaton. Arthur Burks [3], nine years after von Neumann's death, and five years after Ref. [1], credits von Neumann with the notion that a computer must dissipate at least $kTln2$ at room temperature "per elementary act of information, that is per elementary decision of a two–way alternative and per elementary transmittal of one unit of information." This comes closer to a sensible discussion than can be found in Ref. [4], but still involves the ambiguous *per elementary decision of a two-way alternative.* Indeed, we now know [6], the mere transmission of information does not require any minimal energy dissipation [7].

von Neumann's insight into the computational process and associated physics was impressive. One of his patents [8], assigned to my company, has exerted a tremendous influence on my own work. I do not

Reprinted from *IEEE PhysComp '94*, pp. 54-59, 1994

question that von Neumann *could* have answered the energy dissipation question properly, and perhaps did. But the written record demonstrating that has not been published.

In 1961, when Ref. [1] appeared, it was natural to assume that discarding information was an inevitable part of computation. After all, computers did lots of that. Ref. [1] already appreciated that information discarding operations could be imbedded in larger operations which were 1:1. But that did not constitute an understanding of reversible computation; that had to await Charles Bennett's paper [9] which Wheeler and Zurek have labelled *epoch–making* [10]. In particular, I had assumed (quite incorrectly), that a computation which runs along a chain of 1:1 transformations is a table look–up device, where the designer has to anticipate every possible computational trajectory. Bennett's insight that computation can utilize a series of steps, each logically reversible, and that this in turn allows physical reversibility, was counter–intuitive when it first appeared. (Reversible computation, without understanding of its physical significance, was first described by Lecerf in 1963 [11].) That computation could be done with arbitrarily little dissipation, per step, was not actually in contradiction to the principal thesis of Ref. [1], but seemed to go remarkably beyond that. It also distinguished the computational process in a surprising way from our perceptions, at that time, about communications and measurement.

The notions expressed by Charles Bennett and myself have been elaborated in many ways by others. Despite that, some grumbling persists as evidenced by two papers in the 1992 version of this workshop [5, 12]. This is best characterized by the excerpts from Ref. [13], based in turn on my own unpublished comments in the electronic forum resulting from the 1992 version of this workshop.

Nothing in science is ever settled totally, and beyond all question. In 1994 we will again see some challenges to the second law, and some proposals for superluminal signal propagation. A few of these may be genuine open–minded and scholarly attempts to probe the limits of our certainty. A larger number will reflect honest attempts to respond to a poor exposition with the challenger unaware of the diversity of paths that has led to the established conclusions. But the large majority...

... I did not believe or understand Charles [Bennett] when he first explained his emerging notions to me, in 1971. It took me some months to come to understanding and agreement.

It is no longer 1961 or 1973. Our concepts have *been explored, expanded, and reformulated by many colleagues with different backgrounds and differing motivations. I, therefore, believe that our results are established as well as most scientific results can be, even though there will continue to be a modest flow of objections....*

For a number of years all reversible computer embodiments were either impractical mechanical machinery, or else were based on nucleic–acid inspired chemical reaction sequences. I found myself searching for deep reasons for that; why were there no electrical versions? That was solved eventually, when Likharev [14] came out with a Josephson junction version. There was no deep mystery! On the other hand, Likharev's proposal was very demanding on the allowed variation of components. Furthermore it was tied to a technology which was about to disappear from the list of serious candidates.

The 1992 version of this meeting showed through realistic CMOS circuit proposals, that reversible computation was not just a tool for answering conceptual questions about limits, but a tool for reducing power consumption in reality [15]. Once again an unexpected turn, even though partially anticipated by the earlier work of Seitz [16]. Normally, in CMOS circuits, the energy stored in capacitances is discharged and dissipated in each switching event. In the proposals of Ref. [15] the charging process is reversed, and the energy taken back into (the suitably designed) power supply. The forward propagation of the signal, between latches, is simply reversed. Whether the added delays and complexity that come with these proposals leave them advantageous is a question to be settled by further exploration. These proposals, in turn, have inspired a version [17] which avoids the need for detailed reversal, but still saves most of the power.

Why was the possibility of reversible CMOS logic appreciated at such a late stage? In part perhaps, because the people interested in reversible computation were largely computer scientists or basically oriented physicists, rather than circuit or device technologists. But there were additional errors in perception, at least in my own case, and I suspect that others shared my mistake, which will be explained. Reversible computation as invented and described by Bennett [9], requires every single step to be reversible. It is not enough to save a copy of the initial state of the computer. Now an *and* or *or*, for example, are not 1:1 operations; the input cannot be deduced from the output, and therefore one might (and I did) assume that such operations cannot be utilized in reversible computing. Reversible computation is as shown in Fig. 1a, in contrast to

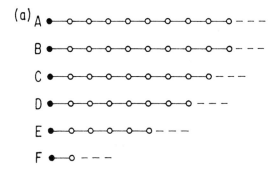

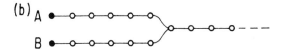

Figure 1: a. One–to–one computation. The left–hand end of a horizontal chain represents the initial state, and forward computation represents motion to the right, through a sequence of states represented by successive open circles. Different capital letters correspond to different initial states, that is, different programs. b. Information–discarding junction. Two computational paths, moving to the right, merge into one.

Fig. 1b where trajectories merge.

Now, however if, as in *Hansel and Gretel* you leave markers behind to identify your trail, Fig. 1b *is* reversible. You only need to spot and pick up the markers on the return trip. In the recent reversible CMOS proposals the markers are replaced by the input signal, which stays there. It remains there until the output reversal is completed. Thus, the usual simple logic circuitry with *and* and *or* gates can be used. Of course, if we want to save the capacitive energy in the latches it gets more complex; that will not be discussed here.

2: Measurement and communication

The historical zig–zag pattern also applies to two auxiliary fields: the energy required for measurement and the energy required for the communications channel. The history of Maxwell's demon which caused us to focus on the energy needed in the classical measurement, has been documented in detail in Refs. [13, 18]. Maxwell's demon utilizes the fact that if we have information about the motion of individual molecules, we can extract their kinetic energy to do useful work. Thus, we must make measurements on individual

molecules. For a good many years it was assumed that the transfer of information from the object to be measured, to the meter, required a minimal and unavoidable energy dissipation, and that this dissipation saved the second law. Instead it is the resetting of the meter, to a standardized state, after each use (and before the first use), that requires the energy dissipation needed to save the second law. An adapted quotation from Ref. [19] describes this setting: *Well known discussions, by Brillouin [4], followed by a refined version due to Gabor [20], invoke the fact that to "see" a molecule a photon must be used. In order to distinguish it from the surrounding black body radiation, the photon must have an energy $h\nu > kT$. This energy is assumed to be lost in the process. A great deal of later literature of which we cite only a few examples, echoes these notions [21]. In retrospect, this acceptance of the Brillouin–Gabor view appears as one of the great puzzles in the sociology of science. If someone proposes a method for executing the "measurement,"... which consumes a certain amount of energy, why should we believe that the suggested method represents an optimum?*

The communications channel represents another similar episode. There is a widespread presumption that it takes $kTln2$ to send a bit from one place to another. It is implied that this follows directly from Shannon's results for the linear transmission line

$$C = Wlog_2 \frac{P+N}{N} \qquad (1)$$

where W is the line's bandwidth, P the signal power, and $N = kTW$ the thermal equilibrium noise power. Shannon never claimed such a universal applicability for his result. I have supplied a number of counter-examples to show that a bit can be transmitted with arbitrarily little dissipation, if we are willing to do it slowly [19, 22]. Despite the fact that my analysis ran contrary to the prevailing wisdom, it has been ignored. Only Porod [23] has paid me the compliment of a debate, and Ref. [24] correctly characterizes the limits of the widely accepted viewpoint.

3: Quantum computers

Finally let me allude to the totally quantum mechanical computational process. I had attempted, in unpublished work, to describe such a process, e.g. quantum versions of the Bennett–Fredkin–Turing machine [25], but got hopelessly bogged down in the complexity of that. Eventually Benioff saw the way to do that [26]. You invoke a Hamiltonian (or a unitary time evolution) which causes the information bearing degrees of freedom to interact, and to evolve with time, as they do in a computer. You introduce no other parts or degrees of freedom. I was too engineering oriented to see that possibility; I assumed that you had to describe the apparatus and not just the Hamiltonian. Benioff's idealization was of course just that. A penalty paid to permit a theory. Benioff's work led to a wide literature which I have assessed elsewhere [6, 7, 27]. Feynman was a particularly significant contributor to the stream [28] that followed Benioff. Feynman was present at the 1981 workshop at M.I.T. [29] where many of us discussed Benioff's notions and the paper Benioff presented there [30]. Did we understand and believe Benioff? Feynman did not need much of a clue, and as a result generated his own very appealing and effective view [28] of quantum mechanical computation. Unfortunately, Feynman failed to cite Benioff. This meeting also provided Feynman's first exposure to Bennett's notion of reversible computation. Feynman understood immediately, an impressive feat in 1981. For an alternative and complementary view of Feynman's role in this field see Ref. [31].

Recently there has been a good deal of excitement about quantum parallelism, undoubtedly reflected in some companion papers in this volume, as well as in the earlier 1992 proceedings [32]. I cannot but help contrast that field to the discussions of the energy requirements of the communications channel. The proponents of quantum parallelism have provided us with a description of the computational Hamiltonian, not of apparatus. They have not, up to now, inquired about the consequences of Hamiltonians which deviate somewhat from the exact desired value. This generosity of spirit contrasts strikingly with that found in the communications channel literature. The latter concentrates on the linear boson channel, despite the fact that written communication, as well as the shipment of floppy disks, are well established ways of sending messages. Levitin's contribution [33] to the earlier 1992 workshop is typical of this literature.

4: Overview

The path to understanding in science is often difficult. If it were otherwise, we would not be needed. This field, however, seems to have suffered from an unusually convoluted path.

Conference records [29, 34] demonstrate that a good many perceptive investigators enter a field without any attempt to read the existing literature. That is a contributing factor to the difficulty we have noted.

The lack of experimental data also permits us to stick to erroneous concepts. But these are auxiliary sources; the zig-zag path, ultimately, arises from the difficulty of the problem when compared to our ability.

References

[1] R. Landauer, "Irreversibility and Heat Generation in the Computing Process," *IBM J. Res. Dev.* vol. 5, 1961, pp. 183-191. [Reprinted in: *Maxwell's Demon.* H.S. Leff and A.F. Rex, eds. Princeton: Princeton Univ. Press, 1990, pp. 188-196].

[2] E. Goto, W. Hioe and M. Hosoya, "Physical Limits to Quantum Flux Parametron Operation," *Physica C* vol. 185–189, 1991, pp. 385–390.

[3] J. von Neumann, *Theory of Self-Reproducing Automata*, A.W. Burks, ed. Urbana: Univ. Illinois Press, 1966.

[4] L. Brillouin, *Science and Information Theory*. New York: Academic Press, 1956.

[5] K. Igeta, "Physical Meaning of Computation," In: *Proc. Workshop on Physics and Computation.* Los Alamitos: IEEE Computer Society Press, 1993, pp. 184–191.

[6] R. Landauer, "Information is Physical," *Phys. Today* vol. 44, May 1991, pp. 23–29.

[7] R. Landauer, "Information is Physical," In: *Proc. Workshop on Physics and Computation.* Los Alamitos: IEEE Computer Society Press, 1993, pp. 1–4.

[8] J. von Neumann, "Non-linear Capacitance or Inductance Switching, Amplifying and Memory Organs," *U.S. Patent 2,815,488*, filed 4/28/54, issued 12/3/57, assigned to IBM.

[9] C.H. Bennett, "Logical Reversibility of Computation," *IBM J. Res. Dev.* vol. 17, 1973, pp. 525–532.

[10] J.A. Wheeler and W.H. Zurek, eds. *Quantum Theory and Measurement.* Princeton: Princeton Univ. Press, 1983. p. 782.

[11] Y. Lecerf, "Machines de Turing Réversibles. Récursive insolubilité en $n \epsilon N$ de l' équation $u = \theta^n u$, o ù θ est un 'isomorphisme de codes.' " *Compt. Rend.* vol. 257, 1963, pp. 2597-2600; "Récursive insolubilité de léquation générale de diagonalisation de deux monomorphismes de

monoïdes libres $\phi\chi = \psi\chi$," *Compt. Rend.* vol. 257, 1963, pp. 2940–2943.

[12] T. Schneider, "Use of Information Theory in Molecular Biology," In: *Proc. Workshop on Physics and Computation.* Los Alamitos: IEEE Computer Society Press, 1993, pp. 102–110.

[13] H.S. Leff and A.F. Rex, "Maxwell's Demon: Entropy Historian," In: *Proc. Entropy Conference, April, 1994. Cambridge: MIT Press, to be published.*

[14] K.K. Likharev, "Classical and Quantum Limitations on Energy Consumption in Computation," *Int. J. Theor. Phys.* vol. 21, 1982, pp. 311–326.

[15] R.C. Merkle, "Towards Practical Reversible Logic," In: *Proc. Workshop on Physics and Computation.* Los Alamitos: IEEE Computer Society Press, 1993, pp. 227–228; J.S. Hall, "An Electroid Switching Model for Reversible Computer Architectures," *ibid.*, pp. 237–247; J.G. Koller and W.C. Athas, "Adiabatic Switching, Low Energy Computing, and the Physics of Storing and Erasing Information," *ibid.*, pp. 267–270.

[16] C.L. Seitz, A.H. Frey, S. Mattisson, S.D. Rabin, D.A. Speck and J.L.A. van de Sneepscheut, "Hot-Clock nMOS," *1985 Chapel Hill Conference on VLSI.* Chapel Hill: Computer Science Press, pp. 1–17.

[17] A. Kramer, J.S. Denker, S.C. Avery, A.G. Dickinson and T.R. Wik, "Adiabatic Computing with the 2N-2N2D Logic Family," preprint.

[18] H.S. Leff and A.F. Rex, eds. *Maxwell's Demon.* Princeton: Princeton Univ. Press, 1990.

[19] R. Landauer, "Computation, Measurement, Communication and Energy Dissipation," In: *Selected Topics in Signal Processing.* S. Haykin, ed. Englewood Cliffs: Prentice–Hall, 1989, pp. 188–196. (Note: Some diagrams printed with wrong orientation.)

[20] D. Gabor, "Light and Information," In: *Progress in Optics.* E. Wolf, ed. Amsterdam: North–Holland Publishing Co. vol. I, 1961, pp. 109–153.

[21] F.T.S. Yu, *Optics and Information Theory.* New York: John Wiley, 1976, Ch. 5, pp. 93-111; A.F. Rex, "The Operation of Maxwell's Demon in a Low Entropy System," *Am. J. Phys.* vol. 55, 1987, pp. 359–362; J.D. Barrow and

F.J. Tipler, *The Antropic Cosmological Principle.* Oxford: Clarendon Press, 1986, See pp. 179, 662; K.G. Denbigh and J.S. Denbigh, *Entropy in Relation to Incomplete Knowledge.* Cambridge: Cambridge Univ. Press, 1985, pp. 1–118; E.E. Daub, "Maxwell's Demon," *Stud. Hist. Phil. Sci.* vol. 1, 1970, pp. 213–227.

[22] R. Landauer, "Energy Requirements in Communication," *Appl. Phys. Lett.* vol. 51, 1987, pp. 2056–2058;

[23] W. Porod, "Comment on 'Energy Requirements in Communication ' " *Appl. Phys. Lett.* vol. 52, 1988, p. 2191; R. Landauer, "Response to 'Comment on Energy Requirements in Communication ' " *Appl. Phys. Lett.* vol. 52, 1988, pp. 2191–2192.

[24] C.M. Caves and P.D. Drummond, "Quantum Limits on Bosonic Communication rates," *Rev. Mod. Phys.* vol. 66, 1994, pp. 481–537

[25] R. Landauer, "Fundamental Limitations in the Computational Process," *Ber. Bunsenges. Phys. Chem.* vol. 80, 1976, pp. 1048–1059.

[26] P. Benioff, "The Computer as a Physical System: A Microscopic Quantum Mechanical Hamiltonian Model of Computers as Represented by Turing Machines," *J. Stat. Phys.* vol. 22, 1980, pp. 563–591; "Quantum Mechanical Hamiltonian Models of Turing Machines," *J. Stat. Phys.* vol. 29, 1982, pp. 515–546; "Quantum Mechanical Models of Turing Machines that Dissipate No Energy," *Phys. Rev. Lett.* vol. 48, 1982, pp. 1581–1585.

[27] R. Landauer, "Is Quantum Mechanics Useful?" *Proc. R. Soc. London A,* 1994, *in press*; "Is Quantum Mechanically Coherent Computation Useful?" *Proc. 4th Drexel Symposium on Quantum Nonintegrability, Sept. 8-10, 1994.* D.H. Feng and B.-L. Hu, eds. (International Press), *to be published.*

[28] R. Feynman, "Quantum Mechanical Computers," *Opt. News* vol. 11(2), 1985, pp. 11-20; [Reprinted in: *Found. Phys.* vol. 16, 1986, pp. 507–531.]

[29] Papers presented at the 1981 Conference on Physics of Computation, Part I: Physics of Computation, *Int. J. Theor. Phys.* vol. 21, 1982, pp. 165–350; Part II: Computational Models of Physics, *ibid.*, 1982, pp. 433–600; Part III: Physical Models of Computation, *ibid.*, 1982, pp. 905–1015.

[30] P. Benioff, "Quantum Mechanical Hamiltonian Models of Discrete Processes That Erase Their Own Histories: Application to Turing Machines," *Int. J. Theor. Phys.* vol. 21, 1982, pp. 177–201.

[31] J. Mehra, *The Beat of a Different Drum.* Oxford: Oxford Univ. Press, 1994, Ch. 24, pp. 525–542.

[32] R. Jozsa, "Computation and Quantum Superposition," In: *Proc. Workshop on Physics and Computation.* Los Alamitos: IEEE Computer Society Press, 1993, pp. 192–194; A. Berthiaume and G. Brassard. "Oracle Quantum Computing," *ibid.,* pp. 195–199.

[33] L.B. Levitin, "Physical Information Theory Part I. Quasiclassical Systems," In: *Proc. Workshop on Physics and Computation.* Los Alamitos: IEEE Computer Society Press, 1993, pp. 210–214.

[34] *Proc. Workshop on Physics and Computation.* Los Alamitos: IEEE Computer Society Press, 1993.

A Low-Power Multiphase Circuit Technique

BOYD G. WATKINS, MEMBER, IEEE

Abstract—The principle of multiphase MOS digital circuits is briefly discussed and the features of some presently used multiphase schemes are given. The basic theory and implementation of a six-phase scheme, which make use of a basic principle not normally considered in low-power digital circuitry, are then described, and the first-order equations for power dissipation are derived. The theoretical power dissipation of a six-phase shift register is compared to the power dissipation of equivalent shift registers using other low-power circuit techniques, including the complementary MOS transistor technique, and it is shown that the six-phase technique has the lowest power dissipation from very low frequencies up to a limiting high frequency. Finally, the power dissipation of an actual six-phase circuit is compared to the dissipation predicted from the derived equations.

I. INTRODUCTION

THE DEVELOPMENT of large scale integrated arrays has put added emphasis on the development of new circuit techniques, which makes it possible to build circuits of small size, high initial yield, and low power dissipation. The first two features have always been desirable for making low-cost integrated circuits. The low power dissipation has been important in aerospace applications and portable battery-operated data processing and communication equipment; however, in any practical fabrication of these complex arrays, low power dissipation is essential for reasonable chip area and packaging considerations. The added emphasis on low power has stimulated new thinking in the area of digital circuit techniques. This is especially true in MOS field-effect transistor circuit development, since the MOS circuit technology seems to be the most promising for the development of very complex large scale arrays. This paper is primarily concerned with low power dissipating MOS circuit techniques. A general discussion of the existing MOS techniques is included for background and prospective.

II. MOS DIGITAL CIRCUIT TECHNIQUES

There are presently three basic categories of switching circuit techniques in use in MOS circuit development.[1],[2] These are: the resistor pullup scheme which would include dc, and 2Ø and 3Ø techniques; the push-pull switching scheme which would include complementary MOS transistors and 4Ø techniques; and the capacitor-pullup scheme.

In the first category, the basic dc inverter consists simply of a switch tied to ground and a load-resistor tied to the power supply. In the 2Ø and 3Ø techniques, the resistor is clocked on for a convenient period of time.

Manuscript received June 12, 1967.
The author is with the Microelectronic Division, Philco-Ford Corporation, Santa Clara, Calif.

The clock can also be used to sample and hold the inverter information onto a storage capacitor. This capacitor can be the gate of an MOS transistor or a *p-n* junction capacitor. This scheme is shown in Fig. 1. Circuits utilizing the clock resistor for multiphase techniques have certain advantages over circuits employing only the dc technique. These are the following.

1) The delay function, or shift-register function, can be realized with less transistors.
2) Information coordinating or timing problems are simplified.
3) In special applications or situations where fast output-voltage transitions are needed, the necessary driver can be designed to have lower power dissipation.

In the second category a basic push-pull inverter consists of a switch tied to the power supply and a switch tied to ground that would alternately switch a load capacitance between the power supply and ground, in accordance with the information which was present. The exception to this is that in the case of 4Ø techniques, a series of two switches is tied to ground and the load capacitance is always switched to the power supply, and then switched to ground if the proper information is present (Fig. 2). The four-phase circuit technique possesses some advantages and some disadvantages in relation to the complementary transistor technique. The advantages are that push-pull switching can be accomplished with MOS transistors of one type. This means that, in general, the processing of 4Ø circuits is simpler than the processing of complementary MOS transistor circuits. Another advantage is that in simple gates such as a three-input NOR, the number of transistors to accomplish the gate is less with the 4Ø technique than with the complementary transistor technique. These two advantages combine to make a third advantage, that being that in certain types of large scale arrays a higher density of logic can be achieved with the four-phase technique. The difficulties are that in certain control applications the timing problem is much more complicated with the 4Ø technique than with the complementary technique. Another disadvantage would be in the area of memories, where the complementary technique can exhibit very low standby power but the 4Ø technique does not have a standby mode; that is to say, the basic technique requires that the load capacitors be constantly precharged and discharged, regardless of information change. As in the first category, it is difficult to meaningfully compare the complementary and four-phase techniques, since they are not applicable to the same type of logic applications.

Reprinted from *IEEE Journal of Solid-State Circuits*, pp. 213-220, December 1967.

245

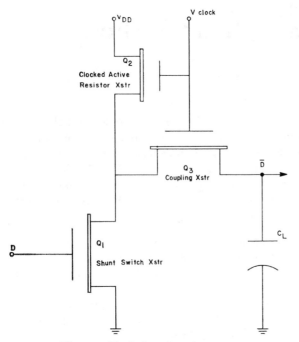

Fig. 1. Clocked resistor-inverter.

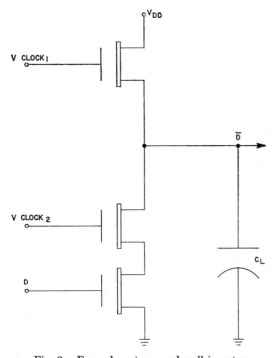

Fig. 2. Four-phase-type push-pull inverter.

In the third category, a simple stage consists of a charging capacitor which is tied to a clock signal, a switch to ground and a load capacitance. Depending on the information present, the charging capacitor is either shunted to ground or connected in series with a load capacitance, at which time some portion of the clock signal is coupled onto the load capacitance. The details of this circuit scheme will be described in the following section. The advantages of this multiphase circuit scheme is that it can offer very low power dissipation. In some

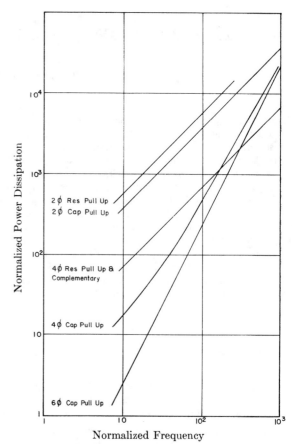

Fig. 3. Power dissipation versus frequency comparison for six types of MOS one-bit delay circuits.

cases it will have attractive layout features for achieving very high density circuits. In order to emphasize the power-dissipation advantage of this last technique in special applications, such as shift registers, a theoretical comparison of shift-register power dissipation is made for shift registers utilizing the previously mentioned MOS techniques. (See Fig. 3.) Certain simplifying assumptions must be made in order to realistically compare these various techniques. The assumptions are:

1) all the shift registers have equal load capacitances;
2) the information to the shift registers is changed every cycle;
3) the amplitude of the signal swings is identical; and
4) the size of the MOS transistor is compatible with the load capacitance chosen, etc.

In general, the comparison shows that the 6Ø technique which will be described here has the lowest power dissipation over a certain range of frequencies. In addition, other variations of the capacitance pullup scheme, namely the 2Ø and 4Ø clock-pullup techniques, have lower power dissipation than the other techniques, with the exception of the complementary circuit techniques.

III. 6Ø CIRCUIT-TECHNIQUE THEORY

The 6Ø circuit technique is based on the following simple theory (Fig. 4). If an uncharged capacitance C is switched to a dc voltage V by a series combination of an ideal switch and resistor R, then when the capacitor is charged to V, the resistor will have dissipated an amount of energy equal to $\frac{1}{2}CV^2$. This is true even in the case when the resistance approaches zero. On the other hand, if an uncharged capacitor is switched to a voltage V through a series combination of an ideal switch and finite resistance, and if that voltage V is zero at the time of switching and then approaches some steady-state value V at a finite rate, the power dissipated in the finite resistance will be less than $\frac{1}{2}CV^2$. In the limit, when R goes to zero, the energy dissipated will go to zero. In order to see how this basic theory can be utilized in a useful circuit configuration, an explanation of Fig. 5 must be given. Here we have a combination of ideal switches and ideal diodes and capacitances. The ideal switch is one which has no resistance when the contact is closed; the ideal diode is one which has neither forward voltage drop when it is conducting forward current, nor reverse leakage. The explanation of the circuit operation is as follows.

If we look at the waveforms shown in Fig. 6, we can see that after $\emptyset_1$ occurs, C_2 will be discharged to zero through CR_1. Next $\emptyset_2$ occurs, which closes $SW2$, then as $\emptyset_3$ occurs if $SW1$ is open, some portion of $\emptyset_3$ is coupled onto C_2 through C_1 and $SW2$. When $\emptyset_3$ reaches its maximum value, $\emptyset_2$ returns to zero. Switch $SW2$ opens when $\emptyset_2$ goes to zero, which leaves C_2 charged to $\emptyset_3(C_1/C_2 + C_1)$. When $\emptyset_3$ returns to zero, C_1 is recharged to its initial state through CR_2. Had SW1 been closed, the $\emptyset_3$ signal would have been shunted to ground and C_2 would have been left uncharged. After the next three phases (4, 5, and 6), which are identical to phases 1, 2, and 3, C_4 will either be charged if C_2 had been previously uncharged, and vice versa. Thus, if the period necessary for all six phases to occur is considered as a bit time, each set of three phases (1, 2, and 3 and 4, 5, and 6) performs the functions of inversion and $\frac{1}{2}$ bit of delay.

The net result is that after the six phases have occurred, information which is on the control of switch 1 is delayed by one bit time and transferred to C_4. The ideal circuit can perform this delay function without dissipating power. The practical realization of this circuit is shown in Fig. 7. Here the switches are realized with MOS FET transistors; the diodes CR_1 and CR_3 are normal $P+$, N diodes; and diodes CR_2 and CR_4 are the drain to body diodes of the MOS FET transistors. In this circuit, the switches do not have zero forward voltage. However, if the transition rates of $\emptyset_1$, $\emptyset_3$, $\emptyset_4$, and $\emptyset_6$ are slow in relation to their respective circuit time constants, then the power dissipation will be low. In particular, the dissipation will be lower than if the load capacitances C_2 and C_4 had been charged with push-pull switches or complementary MOS transistors, since the integral of the product current and voltage over the charging time will be less for the switches and diodes. The conditions for low power dissipation are itemized in Table I and the simplified power-dissipation formula based on these conditions is as follows.

If we operate with the ideal waveforms shown in Fig.

6, the energy per cycle dissipated in the ON Q_1 during the $\emptyset_3$ transition is

$$E_{Q_1} = 2C_1^2 V_x^2 r_{\text{on}Q_1} m_1 f,$$

where m_1 relates the transition time of $\emptyset_3$ to the operating frequency. (See Appendix.) Expressing this in terms of the MOS transistor device parameters we have

$$E_{Q_1} = \frac{C_1^2 V_x^2 m_1 f}{k_1\left(\dfrac{V_x C_1}{C_1 + C_2} - V_T\right)}.$$

When Q_1 is ON there is effectively no dissipation in Q_2; if Q_1 is OFF, the energy per cycle dissipated in Q_2 is roughly

$$E_{Q_2} \sim \left(\frac{C_1 C_2}{C_1 + C_2}\right)^2 V_x^2 r_{\text{on}Q_2} m_1 f.$$

Expressing the energy more accurately in terms of the transistor parameter and taking into account the variation of $r_{\text{on}Q_2}$ with output voltage, we have

$$E_{Q_2} = \frac{V_x C_1 C_2^2 m_1 f}{(C_1 + C_2)2k_2} \ln\left[\frac{V_y - V_T'}{V_y - \dfrac{V_x C_c}{C_c + C_L} - V_T'}\right].$$

After Q_2 has charged the load capacitance C_2 to a one level, there will be dissipation in diode CR_1 during the next $\emptyset_1$ time. The energy dissipated per cycle is

$$E_{CR_1} = \frac{V_x C_1 C_2}{C_1 + C_2} V_{\text{fwd}CR_1}.$$

Expressing this in terms of the diode parameters we have

$$E_{CR_1} = \frac{V_x C_1 C_2 KT}{(C_1 + C_2)q} \ln\left(\frac{V_x C_1 m_2 f}{(C_1 + C_2)I_s} + 1\right),$$

where m_2 relates the transition time of $\emptyset_1$ to the operating frequency. When the pullup capacitor C_1 is recharged to its initial condition, after charging C_2 to a one level, there will be energy dissipated in diode CR_2. This occurs when $\emptyset_3$ returns to zero. The energy per cycle is

$$E_{CR_2} = \frac{V_x C_1 C_2}{C_1 + C_2} V_{\text{fwd}CR_2}.$$

Expressing this in terms of the diode parameters we have

$$E_{CR_2} = \frac{V_x C_1 C_2 KT}{(C_1 + C_2)q} \ln\left(\frac{V_x C_1 m_1 f}{I_s} + 1\right).$$

In considering the energy per cycle dissipated in a whole bit, it can be seen that when Q_1 is ON there will be energy dissipated in Q_1, Q_4, CR_3, and CR_4 during the six phase period, and when Q_1 is OFF there will be energy dissipated in Q_2, CR_1, CR_2, and Q_3. Thus the energy dissipated per cycle for one bit is constant and equal to

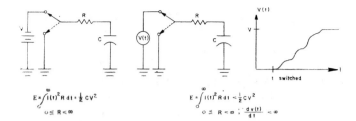

$$E = \int_0^\infty I(t)^2 R\, dt = \frac{1}{2}CV^2$$
$$0 \leq R < \infty$$

$$E = \int_0^\infty I(t)^2 R\, dt < \frac{1}{2}CV^2$$
$$0 \leq R < \infty; \quad \frac{dv(t)}{dt} < \infty$$

Fig. 4. Switching theory.

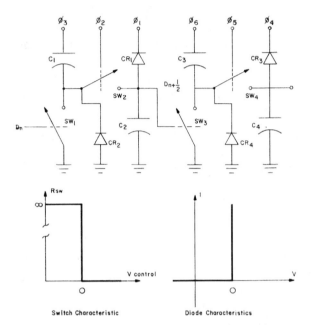

Fig. 5. Ideal six-phase one-bit delay circuit.

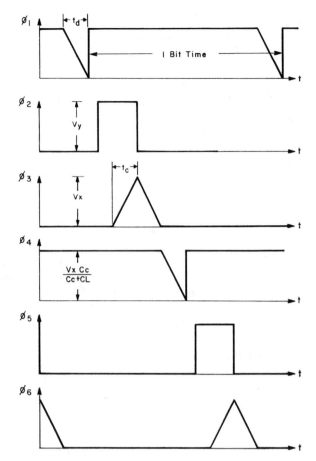

Fig. 6. Ideal six-phase waveforms.

$$E_T = \frac{C_c^2 V_x^2 m_1 f}{k_I \left(\dfrac{V_x C_c}{C_c + C_L} - V_T\right)} + \frac{V_x C_c^2 C_L m_1 f}{2k_c(C_c + C_L)}$$

$$\cdot \ln\left[\frac{V_y - V_T'}{V_y - \dfrac{V_x C_c}{C_c + C_L} - V_T'}\right] + \frac{V_x C_c C_L KT}{(C_c + C_L)q}$$

$$\cdot \left[\ln\left(\frac{V_x C_c m_2 f}{(C_c + C_L)I_s} + 1\right) + \ln\left(\frac{V_x C_c m_1 f}{I_s} + 1\right)\right],$$

where

$$V_x = |\emptyset_1| = |\emptyset_4|, \quad V_y = |\emptyset_2| = |\emptyset_5|, \quad k_I = k_1 = k_3,$$

$$k_c = k_2 = k_4, \quad C_c = C_1 = C_3, \quad \text{and} \quad C_L = C_2 = C_4.$$

The total energy dissipated per cycle is a function of frequency and goes to zero as frequency approaches zero. At very high frequencies the energy approaches a value

$$E_T \sim V_x^2 \left(C_c + \frac{C_c C_L}{C_c + C_L}\right).$$

Now consider an equivalent one-bit delay circuit in which the load capacitors C_L are switched in a push-pull manner

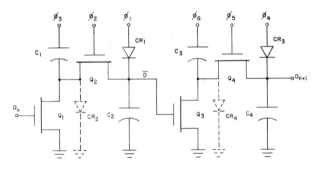

Fig. 7. Six-phase capacitor-pullup one-bit delay circuit.

between the one level $V_{(1)} = (V_x C_L/C_c + C_L)$ and ground. If the information at the input to the circuit changes each cycle, the energy dissipated per cycle will be

$$E_T = V_{(1)}^2 C_L = \frac{V_x^2 C_L^3}{(C_c + C_L)^2}$$

and is constant with operating frequency. Therefore, it can be seen that for operating frequencies between zero and some critical frequency the six-phase circuit will dissipate less energy per cycle than the equivalent push-pull-type circuit. Since the power dissipation is $P = E_T f$ the six-phase circuit will dissipate less power at operating

TABLE I

CONDITION FOR LOW-POWER OPERATION

Device	Conditions	Explanation
Q_1	$\dfrac{d\emptyset_3}{dt} C_1 r_{on Q_1} \ll V(1)_{C_2}$	The voltage which occurs across the ON Q_1 during the $\emptyset_3$ transitions must be much smaller than the output one level to which C_2 is charged.
Q_2	$\dfrac{d\emptyset_3}{dt} \dfrac{C_1 C_2}{C_1 + C_2} r_{on Q_2} \ll V(1)_{C_2}$	The voltage which occurs across the ON Q_2 during $\emptyset_3$ transitions must be much smaller than the output one level to which C_2 is charged.
CR_1	$\dfrac{d\emptyset_1}{dt} C_2 r_{fwd CR_1} \ll V(1)_{C_2}$	The voltage which occurs across the conducting diode CR_1 during the $\emptyset_1$ transitions must be much smaller than the output one level to which C_2 is charged.
CR_2	$\dfrac{d\emptyset_3}{dt} C_1 r_{fwd CR_2} \ll V(1)_{C_2}$	The voltage which occurs across the conducting diode CR_2 during the $\emptyset_3$ transitions must be much smaller than the output one level to which C_2 is charged.
Q_3	$\dfrac{d\emptyset_6}{dt} C_3 r_{on Q_3} \ll V(1)_{C_4}$	The voltage which occurs across the ON Q_3 during $\emptyset_6$ transitions must be much smaller than the output one level to which C_4 is charged.
Q_4	$\dfrac{d\emptyset_6}{dt} \dfrac{C_3 C_4}{C_3 + C_4} r_{on Q_4} \ll V(1)_{C_4}$	The voltage which occurs across the ON Q_4 during $\emptyset_6$ transitions must be much smaller than the output one level to which C_4 is charged.
CR_3	$\dfrac{d\emptyset_4}{dt} C_4 r_{fwd CR_3} \ll V(1)_{C_4}$	The voltage which occurs across the conducting diode CR_3 during the $\emptyset_4$ transitions must be much smaller than the output one level to which C_4 is charged.
CR_4	$\dfrac{d\emptyset_6}{dt} C_3 r_{fwd CR_4} \ll V(1)_{C_4}$	The voltage which occurs across the conducting diode CR_4 during the $\emptyset_6$ transitions must be much smaller than the output one level to which C_4 is charged.

Note:

$$V(1)_{C_2} = |\emptyset_3| \frac{C_1}{C_1 + C_2} \quad \text{and} \quad V(1)_{C_4} = |\emptyset_6| \frac{C_3}{C_3 + C_4}.$$

frequencies up to some critical frequency, as shown in Fig. 3.

This realization of the 6$\emptyset$ technique with MOS transistors and *p-n* diodes involves a MOS double-diffused process which is more complicated than the MOS process. If process simplicity is paramount, the basic technique can still be used with certain sacrifices. For instance, discharge diodes $CR_{1,3}$ can be omitted at the cost of more power dissipation. That is to say, we can transfer information onto the load capacitance in a low dissipative manner, but when we remove information that is a "1" to a "0" signal, we will do this in a dissipative manner as in push-pull switching. This technique, without the diode, would be 4$\emptyset$ capacitor pullup technique (Fig. 8). If it is required to further simplify the technique, so that only two phases are necessary, the gating signals and capacitor-charging signals can be combined. This will result in additional power dissipation, since the coupling transistor will not be on at the beginning of the charge phase and we will be transferring charge to the load capacitor in a manner that is more like the push-pull switching. This would constitute the 2$\emptyset$ capacitor pullup technique (Fig. 9).

The six-phase waveforms presented so far are idealized waveforms. For instance, phase 3 and 6, the triangle-shaped waveforms, could be RC discharge waveforms with

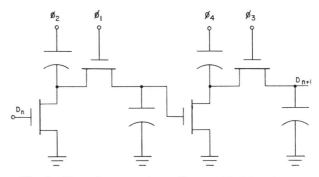

Fig. 8. Four-phase capacitor-pullup one-bit delay circuit.

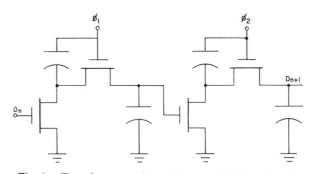

Fig. 9. Two-phase capacitor-pullup one-bit delay circuit.

little sacrifice in power dissipation. The main criterion is that the rate of change of the waveform be finite. In this respect, we could use a sine-wave shape for phase 3 and 6. Phase 2 and 5, the gating signals, should be primarily a square wave but they need not be sharp square waves. Phase 1 and 3 again do not have to be triangular, but must have finite rates of change going from $-V$ to zero. The possibility of using a sine-wave shape for phase 3 and 6 is attractive in terms of keeping the clock-generator power dissipation down. In some applications, especially shift registers, all the charging capacitances taken in parallel would appear as a fixed capacitance to ground. This capacitance could be nulled by an inductor, so that the sum of the charging capacitances and inductor would look like a tuned circuit to the generator. Here we would be accomplishing two goals. The primary goal is keeping chip power very low, since in the very complex large array chip power is the most important thing to reduce. The second feature is that we can reduce the clock-generator power, since the generator is driving a tuned circuit as opposed to load capacitance. This would not be true for phases 1, 2, 4, and 5, however.

IV. The Results of the 6∅ Circuit Techniques

The power-dissipation performance of the 6∅ technique was evaluated with an integrated test vehicle (Fig. 10) which consisted of, among other things, a one-bit delay ring oscillator (Fig. 11) so designed as to enable us to add known amounts of load and charging capacitance and measure the transconductances of the MOS transistors. The waveforms of the ring oscillator are shown in Fig. 12 for the case of the ideal waveforms. A special apparatus was used for the power-dissipation measurements, consisting of a low-quality vacuum system and thermistor temperature sensor. Here the thermal impedance between the circuit and the ambient can be made high, and this gave us the capability of detecting power dissipation in the microwatt range. The results of the power-dissipation measurements with the ideal waveforms are shown in Fig. 13, and these are compared with the theoretical calculations of power dissipation.

Discrepancies between the measured power dissipation and calculated power dissipation ($\sim$25 percent) were partly due to the fact that the theoretical expression for power dissipation was derived from the low-level assumption in Table I. These assumptions were not completely valid for the measurement conditions. Another major cause of the discrepancy was the reduction of "one" levels by the lateral p-n-p which is inherent with the coupling transistor. Here the drain acts as the emitter, the substrate as the base, and the source as the collector. When the load capacitor ($C_{1,3}$) gets recharged through the drain to substrate diode, this causes injection into the lateral p-n-p transistor. That causes collector current to flow, which partly discharges the load capacitor ($C_{2,4}$) and reduces the one level. That effect can be reduced by laying out the circuit to minimize the lateral p-n-p beta of the coupling transistors $Q_{2,4}$.

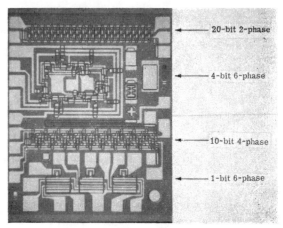

Fig. 10. Integrated test vehicle.

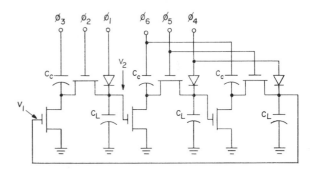

Fig. 11. One-bit delay-ring counter.

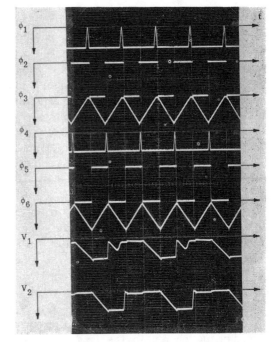

Fig. 12. One-bit delay-ring counter waveforms.

In a typical six-phase circuit (Fig. 14) there would be an additional one level loss, due to clock coupling through the gate overlap capacitance of the coupling transistor.

The integrated test vehicle included a four-bit six-phase capacitor pullup shift register, a ten-bit four-phase

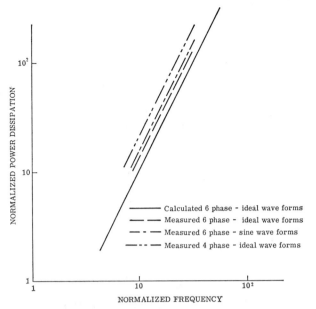

Fig. 13. Comparison of theoretical power dissipation with actual power dissipation for one-bit ring counter.

Calculated 6 phase - ideal wave forms
Measured 6 phase - ideal wave forms
Measured 6 phase - sine wave forms
Measured 4 phase - ideal wave forms

NORMALIZED POWER DISSIPATION

NORMALIZED FREQUENCY

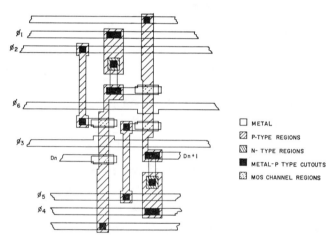

☐ METAL
▨ P-TYPE REGIONS
▧ N- TYPE REGIONS
■ METAL-P TYPE CUTOUTS
▦ MOS CHANNEL REGIONS

Fig. 14. Layout of six-phase one bit of delay.

capacitor pullup shift register, and a 20-bit two-phase capacitor pullup shift register. The two-phase and four-phase registers were designed for high speed. They both operated up to 20 MHz. The six-phase register was designed for low speed to facilitate the study of parasitic effects. All shift registers indicated that parasitic effects could be observed, but did not seriously degrade circuit operation. The two major results of the parasitic effect were an increase in power dissipation and a reduction of the maximum operating speed.

The power dissipation of the one-bit delay-ring counter was then measured, using first a sine wave of equal amplitude in place of $\emptyset_3$ and $\emptyset_6$ and then using the ideal wave forms with $\emptyset_1$ and $\emptyset_4$ missing (four-phase operation).

From the results (Fig. 13), it can be seen that the sine-wave operation caused little increase (~20 percent) in the circuit dissipation, but going to four-phase operation caused a significant increase in the dissipation, as was predicted in Fig. 3.

V. Conclusions

The major type of MOS circuitry in use today has been qualitatively compared and the advantages of using multiphase techniques have been described. It is shown that the six-phase techniques offer a power-dissipation advantage in certain circuit applications which cannot be equaled by any of the other techniques. In particular, results of the theoretical comparison show that the 6Ø techniques can have a power-dissipation advantage over the complementary technique in a shift-register application. The characterization of the 6Ø test vehicle shows that the theoretical equations are substantially valid.

In regard to applicability of 6Ø techniques to present integrated-circuit needs, it appears that the added process complexity needed to fabricate 6Ø circuits may be a significant disadvantage at present; however, the 2Ø and 4Ø clock pullup circuit techniques that are derived from 6Ø techniques can be implemented presently, and have significant power-dissipation and layout advantages in various applications. With the advent of an economical MOS double-diffused process, it appears that 6Ø techniques may offer very attractive possibility in the fabrication of large scale arrays where minimum power dissipation is of major importance.

Appendix

If we consider the energy dissipated in the on switch transistor Q_1 during $\emptyset_3$ we have

$$E_{Q_1} = \int_0^{2t_c} i_{C_1}^2 r_{on\,Q_1}\, dt, \tag{1}$$

where t_c is the time it takes for $\emptyset_3$ to go from zero to its peak value V_X. If we assume that the voltage across r_{on} is much smaller than V_X (as in Table I), then

$$i_{C_1} = C_1 \frac{V_X}{t_c}. \tag{2}$$

From MOS transistor theory,[3] $r_{on\,Q_1}$ would equal

$$r_{on\,Q} = \frac{1}{2k_1(V_G - V_T)} = \frac{1}{2k_1\left(\dfrac{V_X C_1}{C_1 + C_2} - V_T\right)}, \tag{3}$$

where

$$k = \frac{\epsilon_{ox}\mu W}{L t_{0x}}.$$

Solving (1) in terms of (2) and (3)

$$E_{Q_1} = \frac{(C_1 V_X)^2}{k_1\left(\dfrac{V_X C_1}{C_1 + C_2} - V_T\right)t_c}. \tag{4}$$

If we define $t_c = 1/m_1 f$, then the equation can be rewritten for convenience as

$$E_{Q_1} = \frac{C_1^2 V_X^2 m_1 f}{k_1\left(\dfrac{V_X C_1}{C_1 + C_2} - V_T\right)}. \tag{5}$$

251

If we consider the energy dissipated in the coupling transistor Q_2 when Q_1 is off we have

$$E_{Q_2} = \int_0^{t_c} i_{C_2}^2 r_{onQ_2(t)}\, dt, \qquad (6)$$

where

$$i_{c_2} = \frac{C_1 C_2}{C_1 + C_2} \frac{V_X}{t_c}. \qquad (7)$$

If we assume that the voltage drop across r_{on} is much smaller than V_X (as in Table I), then

$$r_{onQ_2(t)} = \frac{1}{2k_2(V_G - V_T')} = \frac{1}{2k_2\left(V_y - \dfrac{V_X C_1 t}{(C_1 + C_2)t_c} - V_T'\right)}, \qquad (8)$$

where V_y is the peak value of $\emptyset_2$; ($V_X C_1 t/(C_1 + C_2)t_c$ is the voltage on C_2 during the charging time t_c and V_T' is the effective threshold voltage when the source of the coupling transistor is not at the same potential as the body;

$$V_T' \sim V_T + \frac{1}{2}\sqrt{\frac{V_X C_1}{C_1 + C_2}}.$$

Solving (6) in terms of (7) and (8) we have

$$E_{Q_2} = \frac{V_X C_2^2 C_1}{2k(C_1 + C_2)t_c} \ln \frac{(V_y - V_T')}{\left(V_y - \dfrac{V_X C_1}{C_1 + C_2} - V_T'\right)}. \qquad (9)$$

We can rearrange terms and express this in frequency as

$$E_{Q_2} = \frac{V_X C_2^2 C_1 m_1 f}{2k(C_1 + C_2)} \ln \frac{(V_y - V_T')}{\left(V_y - \dfrac{V_X C_1}{C_1 + C_2} - V_T'\right)}. \qquad (10)$$

The energy dissipated in the diode CR_1 when C_2 is discharged after having been charged to a one level by Q_2 is

$$E_{CR_1} = \int_0^{t_d} i_{C_2} V_{fwd}\, dt, \qquad (11)$$

where t_d is the time it takes for $\emptyset_1$ to go from its peak value $V_{(1)} = (V_X C_1/C_1 + C_2)$ to zero. If we assume that the fwd diode drop is small in relation to $V_{(1)}$ (as in Table 1)

$$i_{c_2} = C_2 \frac{V_X C_1}{(C_1 + C_2)t_d} \qquad (12)$$

and

$$V_{fwd} = \frac{KT}{q} \ln\left(\frac{i_{c_2}}{I_s} + 1\right). \qquad (13)$$

Solving (11) in terms of (12) and (13).

$$E_{CR_1} = \frac{V_X C_1 C_2 KT}{(C_1 + C_2)q} \ln\left(\frac{V_X C_1 C_2}{(C_1 + C_2)t_d I_s} + 1\right). \qquad (14)$$

Letting $t_d = 1/m_2 f$, we can rewrite (14) as

$$E_{CR_1} = \frac{V_X C_1 C_2 KT}{(C_1 + C_2)q} \ln\left(\frac{V_X C_1 C_2 m_2 f}{(C_1 + C_2)I_s} + 1\right). \qquad (15)$$

The energy dissipated in the diode CR_2 when C_1 is recharged to its initial condition is

$$E_{CR_2} = \int_0^{t_{RC}} i_{c_1} V_{fwd}\, dt, \qquad (16)$$

where t_{RC} is the time it is being recharged. Capacitor C_1 loses a charge Q_X when it charges capacitor C_2 to a one level $(V_X C_1/C_1 + C_2)$ where

$$Q_X = \frac{V_X C_1 C_2}{C_1 + C_2}. \qquad (17)$$

Since C_1 is recharged by i_C, where

$$i_{c_1} = \frac{V_X}{t_c} C_1 = V_X C_1 m_1 f \qquad (18)$$

the time required to recharge C_1 is

$$t_{RC} = Q_X/i_{c_1} = \frac{t_c C_2}{C_1 + C_2}. \qquad (19)$$

The forward diode drop is

$$V_{fwd} = \frac{KT}{q} \ln\left(\frac{V_X C_1 m_1 f}{I_s} + 1\right). \qquad (20)$$

Solving (16) in terms of (18), (19), and (20)

$$E_{CR_2} = \frac{V_X C_1 C_2 KT}{(C_1 + C_2)q} \ln\left(\frac{V_X C_1 m_1 f}{I_s} + 1\right). \qquad (21)$$

ACKNOWLEDGMENT

The author would like to thank A. Thompson and E. Tang for their many contributions to this work.

REFERENCES

[1] J. Karp and E. deAtley, "Use four-phase MOS IC logic," *Electronic Design*, vol. 15, pp. 62–66, April 1967.
[2] M. M. Mitchell and R. W. Ahrons, "MOS micropower complementary transistor logic," *Computer Design*, pp. 28–38, February 1966.
[3] S. R. Hofstein and F. P. Heiman, "The silicone insulated-gate field-effect transistor," *Proc. IEEE*, vol. 51, pp. 1190–1202, September 1963.

252

Asymptotically Zero Energy
Split-Level Charge Recovery Logic

Saed G. Younis   younis@ai.mit.edu
Thomas F. Knight, Jr.   tk@ai.mit.edu

M.I.T. Artificial Intelligence Laboratory
545 Technology Square, Cambridge, MA 02139

Abstract [1]

As the clock and logic speeds increase, the energy requirements of CMOS circuits are rapidly becoming a major concern in the design of personal information systems and large computers. Earlier, we presented a new form of CMOS Charge Recovery Logic (CRL), with a power dissipation that falls with the square of the operating frequency, as opposed to the linear drop of conventional CMOS circuits [1]. The technique relied on constructing an explicitly reversible pipelined logic gate, where the information necessary to recover the energy used to compute a value is provided by computing its functional inverse. Information necessary to uncompute the inverse is available from the subsequent inverse logic stage.

In this paper, we present a new and greatly simplified form of CRL, we call it Split-Level CRL (SCRL). Using split-level voltages, SCRL differs from our original CRL in that it needs only 2, instead of 16, times as many devices as conventional CMOS, requires only one wire for every signal, instead of 4, and actively drives all outputs during sampling. Furthermore, we will show how to construct SCRL circuits using only 2 external inductors for every chip.

Depending on the Q of available inductors, we anticipate energy savings of over 99% per logic operation for logic which is clocked sufficiently slowly.

Introduction

Power dissipation in conventional CMOS primarily occurs during device switching. One component of this dissipation is due to charging and discharging the gate capacitances through conducting, but slightly resistive, devices. We note here that it is not the charging or the discharging of the gate that is necessarily dissipative, but rather that a small time is allocated to perform these operations. In conventional CMOS, the time constant associated with charging the gate through a similar transistor is RC, where R is the ON resistance of the device and C its input capacitance. However, the cycle

[1] This research is supported in part by the Advanced Research Projects Agency under contracts N00014-91-J-1698 and DABT63-92-C-0039.

time can be, and usually is, much longer than RC. An obvious conclusion is that energy consumption can be reduced by spreading the transitions over the whole cycle rather than "squeezing" it all inside one RC.

To successfully spread the transition over periods longer than RC, we insist that two conditions apply throughout the operation of our circuit. Firstly, we forbid any device in our circuit from turning ON while a potential difference exists across it. Secondly, once the device is switched ON, the energy transfer through the device occurs in a controlled and gradual manner to prevent a potential from developing across it. These conditions place some interesting restrictions on the way we usually perform computations. To perform a non-dissipative transition of the output, we must know the state of the output *prior to and during* this output transition. Stated more clearly, to non-dissipatively reset the state of the output we must at all times have a copy of it. The only way out of this circle is to use *reversible logic*. It is this observation that is the core of our low energy charge recovery logic. By low energy, or non-dissipative, we mean that the energy per computational step can be made arbitrarily small by spreading the computation over a longer period.

The paper starts by examining the topology and functionality of a simple SCRL gate, an inverter. We will then show how a general SCRL gate fits into a computational pipeline. We will proceed to show how to generate the swings on the rails using 2 external inductors. Finally, we will discuss how SCRL relaxes some of the device sizing constraints that apply to conventional CMOS circuits as well as illustrate some techniques through which the cost of irreversibility, in energy and complexity, could be minimized.

Split-Level CRL Gate

In this section we describe the topology and operation of an SCRL inverter. Extension to other gates is straight forward. A device-level diagram of the SCRL inverter is shown in Figure 1. It is identical to a conventional inverter except for the addition of a pass gate at the output and that the top and bottom rails are now driven by clocks rather than

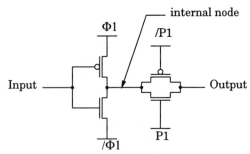

Figure 1: SCRL inverter.

V_{dd} and GND. We call the clock controlling the top rail ϕ_1 and that controlling the bottom rail $/\phi_1$. We refer to the clocks that control the pass gate as P_1 and $/P_1$.

Initially, the input, ϕ_1, $/\phi_1$, the output, and all internal nodes are at $V_{dd}/2$. In addition, P_1 is at GND and $/P_1$ is at V_{dd}, *i.e.*, the pass gate is turned off. After accepting a valid input, V_{dd} or GND, we turn the pass gate on by gradually swinging P_1 and $/P_1$ to V_{dd} and GND respectively. We now gradually swing ϕ_1 to V_{dd} and $/\phi_1$ to GND. The fact that both ϕ_1 and $/\phi_1$ start at $V_{dd}/2$ and *split* towards V_{dd} and GND respectively is the reason we call this family Split-Level Charge Recovery Logic (SCRL). If the input to the gate was at V_{dd} then the output would follow $/\phi_1$ to GND. If the input was at GND then the output would follow ϕ_1 to V_{dd}. We note that at the end of the ϕ_1, $/\phi_1$ swings, the output is the NOT of the input. The output is also actively driven and could now be sampled by another gate later in the pipeline.

After the output is sampled by a later gate, the pass gate of this inverter is turned off thus tri-stating the output. Following that, we return ϕ_1 and $/\phi_1$ to $V_{dd}/2$. This in effect restores all the nodes except the output to $V_{dd}/2$. We are now ready to accept a new input. Please note that allowing the input to change prior to resetting all the nodes to $V_{dd}/2$ could turn some devices on while there is a potential difference across them leading to dissipation.

Remember that the output is still at a valid logic level, not $V_{dd}/2$, and before turning on the pass gate we must restore the level of this output to $V_{dd}/2$ to prevent dissipative charge sharing. The promise is that at the point that the pass gate disconnected the output from the inverter, the output was connected to a different gate that has the job of restoring its level to $V_{dd}/2$. We will show how this is done in the following section.

Reversible Pipeline

The reason for not letting an SCRL restore its own output to $V_{dd}/2$ is to allow pipelining. Note that to non-dissipatively restore the output to $V_{dd}/2$, the input to the gate must be held constant during the splitting and restoration of its rails. The same restriction dictates that this gate does not restore itself before the subsequent gate in the pipeline restores itself and so on. This means that a new input to a pipeline must be held constant until

Figure 2: Abstraction box.

the effect of this input propagates all the way to the end of the pipeline and until the restoration of the pipeline starting from the last stage reaches back to the first gate. This form of "pipelining" is obviously not very useful.

In this section we show how to connect SCRL gates, or stages, in a non-dissipative pipeline. The main purpose of this method of interconnection is to provide a way of restoring the level of gate outputs to $V_{dd}/2$ with the right timing. We build the pipeline out of copies of an abstraction box shown in Figure 2.

We think of this box as containing a parallel set of SCRL gates performing any logical function of an arbitrary number of inputs. Symbolically, the output of the box represents a bundle containing the outputs of the SCRL gates internal to the box. The input to the box represents a bundle containing all the inputs of the gates internal to the box. The function computed by the box is identified by the letter in the center of the box. Finally, indicated at the bottom of the box are the clocks used to control both the split-level rails and the pass gate controls of all the SCRL gates internal to that box. A clock of ϕ_1 in the lower right corner indicates that the top rail is connected to ϕ_1 and the bottom rail to $/\phi_1$, while a clock of P_1 in the lower left corner indicates that the pass gate is on when P_1 is high.

Using this abstraction, Figure 3 illustrates how SCRL gates are connected to produce a non-dissipative pipeline. Note that the box with a function F^{-1} performs the inverse operation of the box with a function F. The computation proceeds from left to right in the top half of the pipeline and the "uncomputation" proceeds from right to left on the bottom half of the pipeline.

Each line linking SCRL gates is connected to the outputs of two different SCRL gates. For example, node (a) is connected to the output of F_1 and to the output of F_2^{-1}. There are two reasons why no logic fights occur between the gates driving the same line. The first is that when one gate is driving the line the other is tri-stated and visa-versa. The second is that during hand off, the voltages at the output of the gates is guaranteed to be equal. In this pipeline, the forward gates are responsible for gradually swinging an output line from $V_{dd}/2$ to V_{dd} or GND depending on the computation. The reverse pipeline is responsible for restoring the output line from the active levels to $V_{dd}/2$.

To avoid dissipation, the backward gates have to determine the value of the output that they are about to restore to $V_{dd}/2$ and set their internal node to that level before their pass gate is switched on, *i.e.*, before the line in handed off from the forward gate. To see how this works, we go through the events that occur after a new input, say a_0 is

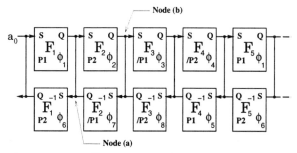

Figure 3: Non-dissipative multi-stage pipeline connection.

presented to the pipeline. First P_1 turns on the pass gate of F_1 and turns off the pass gate of F_2^{-1}. Next ϕ_1 splits setting node (a) to the valve $F_1(a_0)$. F_2 goes through similar transitions and produces $F_2(F_1(a_0))$ at node (b). Similarly F_2^{-1} produces $F_2^{-1}(F_2(F_1(a_0))) = F_1(a_0)$. Note that at this point the voltage levels at the outputs of F_1 and F_2^{-1} are at the same level which means that it is now safe to hand off node (a) to F_2^{-1} from F_1 by swinging P_1 low. After the hand off, we can restore F_1 by restoring ϕ_1. This could occur even without having to wait until F_2 is restored because F_2^{-1} is still holding node (a) at its valid value. After F_2 is restored, F_2^{-1} gradually restores node (a) to $V_{dd}/2$ and hands it over to F_1. The timing diagram for a four phase clocking scheme is shown in Figure 4. For $\phi_1 \ldots \phi_8$ in the figure, a high indicates when they are split and a low when they are restored. For $P_1 \ldots P_2$, a high is V_{dd} and a low is GND. With this pipeline, we are able to accept a new input every ϕ_1 without needing to wait for the restoration of later stages.

There remains one problem however. At the end of the pipeline, the input to F_5^{-1} is not restored and hence driving this line is dissipative. Furthermore, it could not be generated, as this is the place where reversibility is broken. This implies the fundamental limit that links information entropy with thermodynamic entropy. If at any moment a piece of information that is vital to reconstruct the past is lost, energy is dissipated. Fortunately, this dissipation occurring only at the end of a long pipeline and is negligible.

Rail Driver Circuit

As shown before, for a 4-phase SCRL implementation, we need a total of 20 clocks, or swinging rails. The previous implementation required a single inductor for every swinging rail and used the circuit shown in Figure 5.

A rail could be approximated by a capacitor in series with a resistor. The capacitor is the sum of the capacitances that the rail is driving and the resistor is the equivalent resistance of the devices through which the capacitances are driven. Suppose that the initial voltage on the rail was V_{init} and we want to swing the rail to V_{fin}. To start the swing, we connect the rail through an inductor

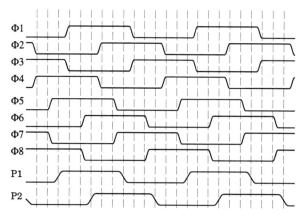

Figure 4: Rail timing for 4 phase SCRL.

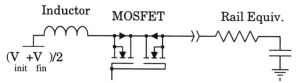

Figure 5: CRL rail driver circuit.

to a DC power supply at $(V_{init} + V_{fin})/2$. Current starts to build up in the inductor and the rail starts the swing towards V_{fin}. At the moment that the current drops back again to zero we disconnect the inductor. The rail should now be at V_{fin}. The action of connecting and disconnecting the rail is performed by the power MOSFET. Please note that the inductor is only necessary during the transition and is otherwise disconnected from the rail. Note further that the current in a disconnected inductor is zero. With this in mind, we should be able to multiplex the inductor among multiple rail circuits so long as these multiplexed rails do not have simultaneous transitions. Examining the timing diagram of Figure 4, we see that no more than two transitions occur simultaneously at any moment. From this we see that the maximum number of required external inductors is 2. Integrating everything but the inductors inside the silicon chip results in a chip with SCRL behavior with very little external components.

There are two difficulties in fabricating circuits with high energy recovery factors. The first is the Q of available inductors. While our derivation assumed an infinite inductor Q, realistic high frequency inductors at room temperature have Q's limited to the range of 100-200, placing an upper limit on the achieved energy recovery.

The second and more serious difficulty is the dissipation associated with controlling the action of the power MOSFET. Koller and Athas [2] have studied this problem and have shown that dissipation in the power switch forces the energy dissipation of the overall system to follow $f^{1/2}$ instead of the theoretical minimum that drops linearly with f. Fortunately, we can partially recover the linear rela-

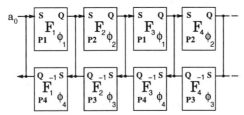

Figure 6: Two-Phase SCRL pipeline.

tionship between energy dissipation and operating frequency by employing a number of techniques described in [3].

SCRL Clocking Variants

In what follows, we will describe a number of alternatives for clocking SCRL circuits. These circuits differ primarily by the number of required phases and/or rails that are needed to control their operation. The pipeline described previously in this paper required four-phase clocking. This used four different clock phases in the forward pipeline and four others for the reverse pipeline. By four-phase we mean that the shortest feedback path in the pipeline has to span a minimum of four pipeline stages. Here, we will show how to construct SCRL circuits using two-phase clocking. Generalizing to three-phase, five-phase and six-phase clocking is a little tricky but not overly difficult and will not be included here for lack of space. One might simplistically think that less phases lead to less required rails. This unfortunately is not true because for some implementations the required phases are not symmetric and therefore the complement of a phase cannot be used to drive an oppositely timed rail. For this reason, the primary reason for reducing the number of phases is to minimize the number of stages for the shortest feedback path and to increase the throughput of the pipeline. Additionally, the lower the number of phases that an SCRL circuit uses, the easier it is to understand and apply.

Two-Phase SCRL

For the implementation that we will describe in the this sections the basic gate is the same as the one described in the "SCRL Gate" section of this paper. Figure 6 shows a pipeline of a two-phase SCRL implementation. The timing relationships among the rails are shown in Figure 7. Two-phase SCRL forfeits the benefit of always actively driving the nodes whenever they are sampled in exchange for achieving two-phase pipelining.

For $\phi_1 \ldots \phi_4$ in the timing diagram, a high indicates the time when ϕ and $/\phi$ are split and a low indicates when they are at $V_{dd}/2$. For $P_1 \ldots P_4$, a high indicates that P is at V_{dd} and $/P$ at GND while a low indicates that P is at GND and $/P$ at V_{dd}. The bottom two timing lines indicate the states of outputs driven by ϕ_1 and ϕ_2 gates. A high indicates when the output is at an valid level of V_{dd} or GND, while a low indicates that the output is at $V_{dd}/2$. The shaded regions in the timing diagrams indicate the times at which the signals are not being actively driven, *i.e.*, floating at an active level.

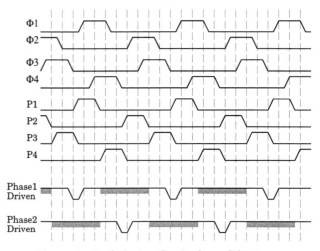

Figure 7: Rail timing for 2 phase SCRL.

We note here that clocking schemes with more than two phases always have actively driven nodes.

Non-Inverting Stage

Since the basic SCRL gate mimics that of conventional CMOS we find out that it is not possible to pass a signal through an SCRL stage without inverting it. For some circuits it is necessary to receive both the true and complement of a logical signal simultaneously at the inputs of a logic gate. Starting with a single signal, it is not possible to have its true and complement arrive at a later stage simultaneously, given the circuits we have described so far. In order to pass a signal without inversion we substitute the basic SCRL gate with the one shown in Figure 8. Please note that this buffer requires an additional set of controlling clocks we call "Fast ϕ_1" and "Fast $/\phi_1$" for a ϕ_1 gate. The restriction on Fast ϕ_1 is that is splits *immediately after* ϕ_1 splits and that is restores *just before* ϕ_1 restores. In other words, the transitions of ϕ_1 contain within them the transitions of Fast ϕ_1. For stages where we want to pass a signal without inversion, we use a gate similar to the one in Figure 8 and we clock its fast clocks according to the relations described.

In place of the inverters in Figure 1 and Figure 8 one can put any CMOS gate such as NAND, NOR etc. We can see that an additional benefit of a non-inverting SCRL gate, is that it allows each functional block to have a 2-level logic implementation. This generally aids in reducing the storage buffers that are sometimes needed for reversibility.

Another benefit of having a 2-level SCRL has to do with optimal step-up ratio of logic gates. It is well known that a CMOS inverter made out of the minimum size devices can optimally drive between 3-5 other inverters of the same size. A minimum sized inverter driving more than this optimal step-up number of loads similar to its size would have a larger delay. Since the energy saving of SCRL is referenced to the maximum operating frequency of a similar circuit in conventional CMOS, this longer delay leads to less energy savings. For an inverter

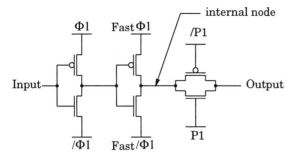

Figure 8: Non-Inverting SCRL Gate.

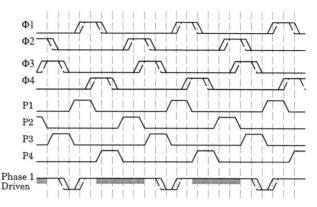

Figure 9: Timing diagram of Two-phase SCRL with fast rails for non-inverting stages.

to drive more than 3-5 loads and maintain the same speed, it must be made out of larger sized devices. Unfortunately, larger devices have larger input gate capacitances and hence present a larger load to the gates that are driving them. To see how this could be a problem, let us consider building a multiplier out of an array of identical 1-Bit SCRL adder gates. The multiplier would consists of an array of gates in which each gate takes its inputs from a previous *identical* gate and provides on its output the data for the inputs of another *identical* gate. Typically in these arrangements, each output would fan-out to drive more than 3-5 loads because each input to a gate feeds a number of devices internal to that gate. For SCRL, just as for CMOS, having an output drive more than 3-5 loads its size is not optimal. As mentioned earlier, increasing the driving capability of a gate so as to be able to drive the loads, i.e. by doubling the width of the devices used in it, also increases the input capacitance, and hence the load, that this gate presents to the *identical gate* driving it. By attempting to increase the driving capability, we also increased the loads, and thus lost the benefit that we where attempting to gain.

Having 2-level SCRL allows for increasing the driving capability of a gate without increasing the load it presents to the other gates. This is done by performing most of the computations in the first level and then using the second level to provide the buffering. For this reason the first stage can consist primarily of minimum sized devices, and thus present the minimum load for the previous gate, while the second stage is made of devices 3-5 times the minimum size to give optimal driving capability.

Finally, the timing diagram of a two-phase SCRL with Fast clocks is shown in Figure 9. The figure indicates the position of the transitions of the fast rails using the dashed lines.

Device Sizing in SCRL Gates

In conventional CMOS, the devices within a gate are sized to accommodate the worst case delay in the system. This is because the system clock is directly related to the worst case propagation delay in the system. If the propagation delay of a CMOS gate located between 2 registers is longer than the clock cycle, the circuit will not work. Note however that in SCRL the clock period is always considerably longer than the worst case propagation delay

and therefore the circuit will always function correctly. The reason why we are concerned with optimal sizing for SCRL gates is because the energy savings is directly related to the ratio of the maximum possible operating frequency of the circuit to the intentionally much lower frequency at which we will operate the circuit. Therefore, if we wanted to operate at 1 MHz, then building circuits that can potentially run at 100 MHz would save us 10 times more energy than if we did a poor job and build a circuit that could operate up to 10 MHz only.

There is however one subtle, but *very important* difference. Since it is the average, not the instantaneous, power than we are concerned with, sizing of devices in SCRL gates should be optimized for the *average, not the worst,* case. For example, it is considered poor design practice to use carry-ripple adders in conventional CMOS circuits. An 8-bit carry-ripple adder consists of 8 1-bit adders. There, it is possible that the carry out of the 1-bit adder in the least significant position affect the result out of the 1-bit adder in the most significant position. To guarantee correct operation in conventional CMOS, the designer has to size his devices, or slow down the system clock, to accommodate the case in which the carry propagates from the least significant bit to the most significant bit, *i.e.,*the worst case. For an 8-bit adder this could correspond to 8 gate delays. Statistically however, a carry in an adder propagates 0.6 bit positions on the average and only makes it all the way across about 1/1600th of the time in an 8-bit carry-ripple adder. Therefore a properly sized SCRL would be optimized for 0.6 carry propagate and would only consume more energy whenever the very infrequent worst case occurs. The importance of this cannot be over emphasized. The same could be applied to sizing paralleled NMOS transistors in a NOR gate. Unlike conventional CMOS where the sizes of paralleled devices are not decreased because of the worst case of only one device being ON, paralleled devices in SCRL could be made smaller because *on the average* more than one of the paralleled devices would be on at any given time. This would improve the energy savings factor as well as require less device area to implement.

Lowering Irreversibility Cost

In CRL and SCRL the reverse pipeline is required to accurately provide a delayed copy of the inputs that were used in the forward pipeline. Without the reverse pipeline we do not have enough information to always correctly compute the delayed copy of the inputs that are required to non-dissipatively reset the stages in the forward pipeline. The penalty of erroneously computing a delayed copy of these inputs is to dissipate energy similar to conventional CMOS for every erroneous bit. Unfortunately, there are situations in which providing the inverse of a function in the forward pipeline is not feasible or at least cumbersome. Luckily all is not lost since in most of these cases we could apply a number of techniques that would make the dissipation associated with this irreversibility minimal.

The first relies on the observation that we are mainly concerned in reducing the *average, not the instantaneous*, power consumption. Gates that computed the inverse function of the gates in the forward direction produced a correct copy of the inputs all the time. Without these inverse gates, we cannot guarantee to be correct all the time. In certain applications however, we can guarantee to be correct *most* of the time. Since dissipative events only occur whenever we guess wrong, being correct most of the time results in substantial energy savings when compared to conventional CMOS *without* the need for reversibility. To illustrate this we consider an example of an 8-Input NAND gate. This gate outputs a FALSE if and only if all the inputs where TRUE. Otherwise, this gate outputs a TRUE. Assuming that the input bits are random, the probability of the output of this gate being at TRUE is $255/256 = 0.996$. If we always assume the output to be at TRUE, then we will have a dissipative event, caused by a wrong prediction, only 0.3% of the time. In the pipeline in Figure 3 let the F_1 be this 8 input NAND gate. Then we can omit F_2^{-1}, assume that this omitted inverse gate output a FALSE all the time, and be right 99.6% of the time. This could be important in situations in which the computation of F_2^{-1} is not feasible or otherwise cumbersome.

The second technique concerns the way multistage buffering is done in CMOS and in SCRL. To drive a large load in CMOS, one must go through a number of progressively larger devices with each device driving another that is slightly larger than itself until the last one in the chain is large enough to drive the load. In SCRL, each larger stage would be paralleled by another stage of comparable size in the reverse direction. If reversibility was broken immediately after the largest stage then dissipation would be large owing to the fact that the input capacitance of the large reverse gate is significant. To alleviate the problem, we must proceed with the pipeline beyond the last stage with inverters in the forward and reverse direction scaling down the size at each successive stage until we reach the minimum size possible. If reversibility is broken immediately after this minimum size stage, dissipation would be minimized due to the much smaller input

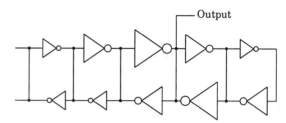

Figure 10: Scaling down stage sizes before breaking reversibility in SCRL.

capacitances of the reverse stage. This is shown in Figure 10.

Demonstration Chip

To verify the concepts in this paper, we have completed a design for an 8x8 bit multiplier chip using SCRL techniques outlined above. Among the concepts included in this chip is the use of carry-ripple adders in the forward direction and carry-ripple subtractors in the reverse direction, as well as the sizing of devices within the chip to optimize for the average, instead of the worst, case. The chip also uses predictor buffers where reversibility is broken, and employs the scale-up scale-down technique in sizing the pad driver chains. The chip is currently being fabricated by MOSIS.

Conclusion

We have demonstrated a new techniques for constructing charge recovery logic. The new form, SCRL, enjoys the same asymptotically near-zero dissipation property as its CRL predecessor. Key ideas include the reduction in the number of devices per gate and the number of wires per signal by using split-level clock rails and by distributing the tasks of splitting and restoration among different gates. We believe that the simplicity of SCRL gate design, nearly identical to conventional CMOS, makes its adoption more acceptable when compared to our previous implementation. A demonstration chip employing SCRL techniques is currently in fabrication.

References

[1] Younis, S., Knight, T., F., "Practical Implementation of Charge Recovering Asymptotically Zero Power CMOS", *Proceedings of the 1993 Symposium on Integrated Systems*, MIT Press, 1993, pp. 234-250.

[2] Athas, W. C., Koller, J. G., and Svensson, L., "An Energy-Efficient CMOS Line Driver Using Adiabatic Switching," *Proc. 1994 IEEE Great Lakes Symposium on VLSI*, March, 1994.

[3] Younis, S., "Asymptotically Zero Energy Computing with Split-Level Charge Recovery Logic", PhD Thesis, June 1994 (in preparation).

Low-Power Digital Systems Based on Adiabatic-Switching Principles

William C. Athas, Lars "J." Svensson, *Member, IEEE*, Jeffrey G. Koller,
Nestoras Tzartzanis, and Eric Ying-Chin Chou, *Student Member, IEEE*

Abstract— Adiabatic switching is an approach to low-power digital circuits that differs fundamentally from other practical low-power techniques. When adiabatic switching is used, the signal energies stored on circuit capacitances may be recycled instead of dissipated as heat. We describe the fundamental adiabatic amplifier circuit and analyze its performance. The dissipation of the adiabatic amplifier is compared to that of conventional switching circuits, both for the case of a fixed voltage swing and the case when the voltage swing can be scaled to reduce power dissipation. We show how combinational and sequential adiabatic-switching logic circuits may be constructed and describe the timing restrictions required for adiabatic operation. Small chip-building experiments have been performed to validate the techniques and to analyse the associated circuit overhead.

Index Terms— Adiabatic amplification, adiabatic charging, adiabatic switching, low-power CMOS, reversible computation, switching energy reduction with preserved signal energies.

I. INTRODUCTION

THE IMPORTANCE of reducing power dissipation in digital systems is increasing as the range and sophistication of applications in portable and embedded computing continues to increase. System-level issues such as battery life, weight, and size are directly affected by power dissipation. Inroads into reducing power dissipation of the digital systems only serves to improve the performance and capabilities of these systems.

CMOS has prevailed as the technology of choice for implementing low-power digital systems. One of the most important reasons is the reduction in switching energy per device caused by the continually shrinking feature sizes. Another important reason is the almost ideal switch characteristics of the MOS transistor, which translates into a negligible static power dissipation compared to the switching (transient) power dissipation.

Basic energy and charge conservation principles can be used to derive the switching energy and power dissipation for static, fully restoring CMOS logic. Throughout this article, we will frequently refer to such CMOS logic as "conventional" logic. Consider the generic CMOS gate shown in Fig. 1. A load capacitance C_L, representing the input capacitance of the next logic stage and any parasitic capacitances, is connected to the dc supply voltage V_{dd} through a pull-up block composed of pFET's and to ground through a pull-down block of nFET's. With the pull-down network tied and the pull-up network cut,

Manuscript received June 15, 1994; August 23, 1994. This work was supported by ARPA under Contract DABT63-92-C0052.

The authors are with the Information Sciences Institute, University of Southern California, Marina del Rey, CA 90292 USA.

IEEE Log Number 9406367.

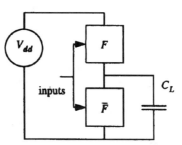

Fig. 1. Generic, conventional CMOS logic gate with pull-up and pull-down networks.

the output is discharged to ground. When the inputs change so that the pull-up network is tied and the pull-down network cut, charge will flow out of V_{dd} and into the load capacitance until the output reaches V_{dd}.

In the process of charging the output, a charge of size $Q = C_L V_{dd}$ is delivered to the load. The power supply must supply this charge at voltage V_{dd}, so the energy supplied is $Q \cdot V_{dd} = C_L \cdot V_{dd}^2$. The energy stored on a capacitance C_L charged to V_{dd} is only half of this, or $(1/2) C_L V_{dd}^2$. Because energy is conserved, the other half must be dissipated by the pFET's in the pull-up network. The same amount of energy is dissipated, regardless of the make-up of the network, the resistance of the pFET's, and the time taken to complete the charging. Similarly, when the inputs change again causing the output to discharge, all of the signal energy stored on the capacitance is inevitably dissipated in the pull-down network, because no energy can enter the ground rail ($Q \cdot V_{gnd} = Q \cdot 0 = 0$).

Thus, from an energy conservation perspective, the conventional case represents a maximum of wastefulness. All charge is input to the circuit at voltage V_{dd} and exists at voltage 0. The energy of the charge upon entry is $C_L V_{dd} \cdot V_{dd}$. The energy of the charge upon exit is 0. Energy dissipation from delivery to removal of the charge is $C_L V_{dd}^2$. All of this energy is dissipated as heat.

Since the energy dissipated when a signal capacitance is cycled is fixed at twice the signal energy, the only way to reduce the energy dissipation in a conventional CMOS gate is to reduce the signal energy. This increases the sensitivity to background noise and thus the probability of malfunction.

Switching energy can be made to be considerably less than signal energy. From the theory of charge control [13], charges can be distinguished as either controlling or controlled charge.

Reprinted from *IEEE Transactions of VLSI Systems*, pp. 398-407, December 1994.

For MOSFET's, the controlling charge is on the gate, while the controlled charge flows through the channel. Dissipation is caused by the resistance encountered by the controlled charge. If charge transport is slowed down, energy that would otherwise be dissipated in the channel can be conserved for later reuse. The energy advantage can be readily understood by assuming a constant current source that delivers the charge $C_L V_{dd}$ over a time period T. The dissipation through the channel resistance R is then:

$$E_{\text{diss}} = P \cdot T = I^2 \cdot R \cdot T = \left(\frac{C_L V_{dd}}{T}\right)^2 \cdot R \cdot T$$
$$= \left(\frac{RC_L}{T}\right) \cdot C_L V_{dd}^2. \tag{1}$$

Equation (1) shows that it is possible to charge and discharge a capacitance through a resistance while dissipating less than $C_L V_{dd}^2$ of energy. It also suggests that it is possible to reduce the dissipation to an arbitrary degree by increasing the switching time to ever-larger values. We refer to this as the principle of *adiabatic charging*. We use the term "adiabatic" to indicate that all charge transfer is to occur without generating heat. As is the typical usage of the term in thermodynamics, fully adiabatic operation is an ideal condition that is asymptotically approached as the process is slowed down. To the best of our knowledge, Seitz and co-workers [1] were the first to formulate the relationship of (1) and to use the effect in practical digital circuits.

Switching circuits that charge and discharge their load capacitance adiabatically are said to use *adiabatic switching*. This article describes and analyzes the power dissipation and performance of adiabatically-switching logic circuits built in CMOS. The circuits rely on special power supplies that provide accurate pulsed-power delivery. It is important to note that adiabatic switching techniques can be an attractive alternative to other low-power design approaches only if the supplies can deliver power efficiently and recycle the power fed back to them. The circuits described here are compatible with energy-efficient, resonant power supplies that we have developed and prototyped [2].

Many factors must be considered when determining the conditions for which adiabatic switching offers superior low-power operation to other approaches. When signal-voltage swing is significantly greater than the threshold voltages of the CMOS devices, the advantages can be readily determined from the analysis of the *adiabatic amplifier*, which is discussed in detail in Section II. On the other hand, when signal-voltage swing can be scaled down to reduce power dissipation, the advantages of adiabatic amplification rapidly diminish for all but the slowest applications, as shown in Section III.

The adiabatic amplifier and the simple gate structure on which it is based can be straightforwardly generalized so that in addition to amplification or buffering, it can implement arbitrary logic functions. Section IV describes the transition from combinational logic to sequential logic. The transition constitutes a turning point, since conventional storage elements cannot be made fully adiabatic. Reversible-logic techniques may be used to avoid storage elements and thus make it possible to build fully adiabatic sequential systems. To this end, we have developed techniques for constructing reversible logic pipelines similar in organization to those developed and demonstrated by Younis and Knight [9]. Section IV also includes a description of a design exercise, where the reversible pipeline structure is used to construct a highly pipelined adder. The results of this exercise indicate that for reversible logic to be a competitive approach to low-power CMOS, either the premium on reducing the energy dissipated per operation must be extremely high, or logic styles and synthesis procedures need to be invented which result in circuit designs with much less logical overhead.

II. ADIABATIC AMPLIFICATION

The adiabatic amplifier, which is the fundamental circuit to our approach, is a simple buffer circuit that uses adiabatic charging to drive a capacitive load. It is useful in a stand-alone configuration, and also as a part of more sophisticated circuits. In this section, we analyse its efficiency and describe its stand-alone use as a line driver for an address bus on a memory board.

Equation (1) assumes that C_L is charged or discharged through a constant resistance. Conventional gates, such as that in Fig. 1, use pull-up and pull-down networks composed of nonlinear FET devices. Fortunately, a switch network can be linearized to a first approximation by replacing each FET with a fully restoring CMOS transmission gate (T-gate). The T-gate is built from an nFET and a pFET connected in parallel. To tie the T-gate with minimal on-resistance, the gate of the pFET is grounded and the gate of the nFET is tied to V_{dd}. For a small voltage drop across the T-gate, which is the intended region for adiabatic circuits, we may model the conductances, G_p and G_n, of the FET's as:

$$G_p = \frac{C_p}{K_p}(V_{\text{ch}} - V_{\text{th}}) \tag{2}$$

$$G_n = \frac{C_n}{K_n}(V_{dd} - V_{\text{ch}} - V_{\text{th}}). \tag{3}$$

C_p and C_n are the gate capacitances of the FET's, V_{th} is the threshold voltage of the nFET, and K_p and K_n represent process constants independent of gate capacitance and voltage. V_{ch} is the average channel voltage. We assume that the channel length is held constant, so that the gate capacitance is directly proportional to the channel width. These equations do not take into account body effects or the difference in threshold voltage between nFET's and pFET's. Accuracy is therefore limited to a factor of two.

By selecting the widths of the two FET's such that $K_n/C_n = K_p/C_p$, we may simplify the sum of the two conductances to:

$$G_p + G_n = \frac{C_n}{K_n}(V_{dd} - V_{\text{ch}} - V_{\text{th}} + V_{\text{ch}} - V_{\text{th}})$$
$$= \frac{C_n}{K_n}(V_{dd} - 2V_{\text{th}}) \tag{4}$$

The on-resistance of the T-gate is the reciprocal of this sum:

$$R_{\text{tg}} = \frac{K_n}{C_n(V_{dd} - 2V_{\text{th}})}. \tag{5}$$

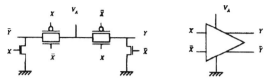

Fig. 2. Adiabatic amplifier: schematic and logic symbol. V_A should be connected to a pulsed-power supply.

This formulation assumes that both devices are conducting, which is not the case when V_{ch} is within one threshold voltage of either supply rail. However, at these extremes the on-resistance will be less than the value given by (5), so the formulation can be used as a worst-case estimate.

The adiabatic amplifier of Fig. 2 is a simple but important circuit element, which may be used to drive a capacitive load by adiabatic charging. The amplifier consists of two T-gates and two nFET clamps. The input is dual-rail encoded, since both signal polarities are needed to control the T-gates. The output is also dual-rail encoded; this choice helps keep the capacitive load seen by the power supply data-independent, which simplifies its design, and is also required when other T-gates (such as other amplifiers) are to be controlled by the output signal. The operation of the amplifier is straightforward. First, the input is set to a valid value. Next, the amplifier is "energized" by applying to V_A a slow voltage ramp from 0 to V_{dd}. If the ramp is slow compared to $R_{tg}C_L$, one of the load capacitances will be adiabatically charged through one of the T-gates. The output signal is now valid and can be used as an input to other circuits. Next, the amplifier is de-energized by ramping the voltage on V_A back to 0. The signal energy that was stored on the load capacitance flows back into the power supply connected to V_A. It falls on the power supply to recycle this returned energy.

We now analyze the energy efficiency of the amplifier, using (1) and (5). The dissipation caused directly by charging and discharging the load capacitance, E_{load}, can be expressed by combining (1) and (5). To account for a nonconstant charge current, the dissipation of (1) must be multiplied by a constant shape factor ξ (which takes the value $\pi^2/8 \approx 1.23$ for a sine-shaped current). The simple formula of (1) becomes:

$$E_{load} = 2\xi \frac{R_{tg}C_L}{T} C_L V_{dd}^2 = \frac{2\xi}{T} \frac{K_n}{C_n(V_{dd} - 2V_{th})} C_L^2 V_{dd}^2. \quad (6)$$

Additionally, parasitic effects such as diffusion capacitance of the T-gates and the clamp nFET's will contribute to the total load capacitance. To account for these contributions would unduly complicate the equations and would not contribute to an understanding of the general result. For the analysis that follows, these capacitances will be treated as negligible by assuming that total input capacitance is small relative to the effective load.

Energy is also dissipated to drive the input capacitance. For the moment, we will analyse the case when the inputs are driven conventionally to V_{dd}, so that the input energy, E_{in}, is dissipated. We furthermore assume that the total input capacitance is proportional to the gate capacitance of the nFET of the T-gate, with a constant of proportionality α;

if no parasitics are considered, $\alpha = (C_n + C_p)/C_n$. Thus, $E_{in} = \alpha C_n V_{dd}^2$. The total energy dissipated per cycle is then:

$$E_{total} = E_{in} + E_{load} = \alpha C_n V_{dd}^2 + \frac{2\xi}{T} \frac{K_n}{C_n(V_{dd} - 2V_{th})} C_L^2 V_{dd}^2. \quad (7)$$

Equation (7) defines an important trade-off in adiabatic CMOS circuits. When the channel width, and thereby the input capacitance, is increased, the energy dissipated in charging the load decreases, but the energy dissipated in charging the inputs increases proportionally. If C_n is a free parameter in the design of the amplifier, the minimal energy dissipation is achieved when the two terms of (7) are equal:

$$C_{n_{opt}} = \sqrt{\frac{2\xi}{\alpha T} \frac{K_n}{(V_{dd} - 2V_{th})} C_L}. \quad (8)$$

Inserting (8) into (7) yields the following expression for the minimum dissipation:

$$E_{total_{min}} = 2 \cdot \alpha C_{n_{opt}} V_{dd}^2 = \sqrt{\frac{8\alpha\xi}{T} \frac{K_n}{(V_{dd} - 2V_{th})}} C_L V_{dd}^2. \quad (9)$$

Equation (9) illustrates a fundamental limitation[1] for adiabatic charging with conventionally driven control signals: the switching energy will scale as $T^{-1/2}$, as opposed to the T^{-1} of (1). An obvious improvement is to use two amplifier stages, where a smaller adiabatic amplifier is used to drive the input signals of the final stage. Let C_{n1} and C_{n1} be the gate capacitances of the nFET's of the T-gates of the first and second amplifier, respectively. The input capacitance of the second amplifier, αC_{n2}, then constitutes the load of the first amplifier. Allotting half of the total charging time to each stage, we get:

$$E_{total} = \alpha C_{n1} V_{dd}^2 + \frac{K_n}{(V_{dd} - 2V_{th})} \frac{4\xi}{T} \left(\frac{\alpha^2 C_{n2}^2}{C_{n1}} + \frac{C_L^2}{C_{n2}} \right) V_{dd}^2. \quad (10)$$

With both C_{n1} and C_{n2} as free variables, the total minimum energy dissipation is:

$$E_{total_{min}} = 8 \left(\frac{\alpha\xi}{T} \frac{K_n}{(V_{dd} - 2V_{th})} \right)^{3/4} C_L V_{dd}^2. \quad (11)$$

From (9) to (11), the energy scaling improves from $T^{-1/2}$ to $T^{-3/4}$. It can be shown that for a cascading of n such amplifiers, the energy dissipation scales as:

$$E_{total_{min}} \sim T^{2^{-n}-1}. \quad (12)$$

This result is not very useful in practice, since the parasitic capacitances can no longer be neglected when the capacitances of the final gate-drive FET's become comparable to that of the load. Also, if the switching is to be performed within a constant time interval, the ramp time for each of the steps must decrease at least linearly in the number of cascaded amplifier stages.

These results present a dilemma. The cost of using conventionally driven inputs with adiabatic logic is a scaling in switching energy which decreases sublinearly with increases

---
[1] A similar relationship has been derived previously for the case of resonant power supplies based on power MOSFET circuits [16].

261

in switching time. T^{-1} scaling may be asymptotically approached, even when the inputs of the amplifier cascade are driven conventionally, but to immediately achieve linear scaling requires the use of all-adiabatic circuits, where the input signals of all amplifiers are also switched adiabatically. The implications for all-adiabatic operation are serious and are discussed in detail in Section IV.

A. The Adiabatic Line Driver

As shown above, adiabatic amplifiers make it possible to reduce switching energy while signal energy is held constant, something which cannot be done with conventional CMOS switching. To investigate how well the theory reflects reality, we designed, implemented, and tested an adiabatic line driver chip (ALDC) for driving an address bus of a memory board [2]. This application involves relatively large driven capacitances and moderate speed requirements, which translates to considerable theoretical power savings. The signal energies are fixed by voltage swing requirements and input capacitances of the memory chips, so no improvements are possible with conventional drivers. The ALDC is also very simple, allowing us to characterize it fully and isolate the non-ideal phenomena.

The ALDC comprises eight adiabatic amplifiers in a 40-pin DIP, sharing the same supply voltage. It was designed to use 5% of the power of a conventional driver (according to (1)) when driving 100 pF per output line at 1 MHz, and is implemented in a MOSIS-supplied 2 μm process. A simple pulsed-power supply, outlined in Fig. 3, was used to test the circuit. The power supply and some parasitics not present in a conventional driver limited the actual dissipation gain from a factor of 20 to a factor of 6.3. It is instructive to consider the distribution of the dissipation over the different parts of the circuit: 51% of the power dissipation is in the ALDC, 29% is in the channel of the power supply FET switch used to time the energy transfer, 17% is in the circuit driving the gate of that switch, and 3% is in the parasitic resistance in the inductor. This breakdown emphasizes the importance of taking the dissipation of the entire system (including the power supply) into the account, not only that of the logic circuit.

III. SUPPLY VOLTAGE SELECTION

For many applications, signal energy levels are determined by the input/output requirements of the digital system. Industry standards for interfacing (such as for the ALDC) or the peculiar requirements of the physical properties of the interface device (such as for liquid crystal displays) mandate certain voltage swings. For these applications, the benefits of adiabatic switching improve with increasing voltage swing, as can be seen by comparing the energy required to cycle a signal conventionally to the adiabatic dissipation of (6):

$$\frac{E_{\text{conv}}}{E_{\text{load}}} = \frac{C_L V_{dd}^2}{E_{\text{load}}} = \frac{C_n T}{2\xi K_n C_L}(V_{dd} - 2V_{\text{th}}). \quad (13)$$

The energy savings ratio increases linearly with V_{dd} for an all-adiabatic system and increases as $V_{dd}^{1/2}$ or $V_{dd}^{3/4}$ for the mixed approaches described by (9) or (11).

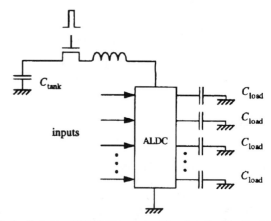

Fig. 3. Test setup of ALDC. The capacitor C_{tank} is held at V_{dd}. The ALDC inputs are set with all output capacitances discharged; the nFET switch is pulsed; the current pulse through the inductor is steered to the intended output capacitances. The nFET is pulsed again for the charge to return to the tank capacitor. Only thereafter may the inputs change.

On the other hand, if we are free to choose the voltage swing for minimum dissipation, we might try to scale down signal energies (by increasing V_{dd}) as far as the application permits and then apply adiabatic charging techniques to further reduce switching energies. This is, however, an unattractive strategy that does not lead to minimal dissipation for T-gate based circuits. In fact, there is an optimal supply voltage swing for T-gate-based adiabatic circuits. This voltage will vary with the approach taken; for simplicity, we will analyse the all-adiabatic approach. Equation (6) can be rewritten:

$$E_{\text{load}} = \frac{2\xi}{T} \frac{K_n}{C_n(V_{dd} - 2V_{\text{th}})} C_L^2 V_{dd}^2$$
$$= \frac{2\xi}{T} \frac{K_n C_L^2}{C_n} \left[\frac{V_{dd}^2}{(V_{dd} - 2V_{\text{th}})} \right]. \quad (14)$$

The energy has a minimum for $V_{dd} = 4V_{\text{th}}$. Again, neither the body effect nor the regions where only one FET device is turned on are accounted for. SPICE simulations and more detailed analysis which takes into account body effects, subthreshold regions, and waveform shapes all show that the minimum is reasonably shallow and that $4V_{\text{th}}$ yields close to the optimal dissipation. When $4V_{\text{th}}$ is substituted for V_{dd} in (6), the energy dissipation of the adiabatic amplifier reduces to:

$$E_{V_{\text{opt}}} = \frac{C_L^2}{C_n} \cdot \frac{16\xi K_n V_{\text{th}}}{T}. \quad (15)$$

We now wish to compare this dissipation to that of a conventional inverter driving the same capacitance, but where the voltage swing can be chosen freely. The conventional circuit will use devices of the same size as the adiabatic one, and its supply voltage will be selected to equalize the speed of the two drivers. To model switching delay as a function of supply voltage for conventional circuits, we use the following well-known approximation [3]:

$$T = \frac{C_L V_{dd}}{I_{\text{sat}}} = \frac{2K_n V_{dd}}{(V_{dd} - V_{\text{th}})^2} \frac{C_L}{C_n}. \quad (16)$$

This expression is based on the drain current of a device in saturation. Velocity saturation is unlikely to influence the delay for the low supply voltages we are interested in and is consequently not taken into account. Mathematical analysis confirmed by SPICE simulations of a ring oscillator indicates that this formula is accurate down to $V_{dd} \approx 1.3 V_{\mathrm{th}}$. From (16), we can now express the required supply voltage and the resulting switching energy as functions of the allowed switching time:

$$V_{dd} = V_{\mathrm{th}} + \frac{rK_n + \sqrt{(rK_n)(rK_n + 2TV_{\mathrm{th}})}}{T} \quad (17)$$

$$E_{\mathrm{conv}} = C_L V_{dd}^2$$
$$= C_L \left(V_{\mathrm{th}} + \frac{rK_n + \sqrt{(rK_n)(rK_n + 2TV_{\mathrm{th}})}}{T} \right)^2.$$
$$(18)$$

The parameter r relates the output and input capacitances ($r = C_L/C_n$). We see from (18) that when switching time is increased in a conventional CMOS circuit by reducing supply voltage, dissipation will initially fall off as T^{-2}. It will, however, tend asymptotically to $C_L V_{\mathrm{th}}^2$ when T grows very large and V_{dd} approaches V_{th}. The adiabatic case has no such asymptote, as is seen in (15). Fig. 4 is a graph of the dissipation of the two approaches for a MOSIS 2 μm CMOS process and $r = 10$. Below 5 ns, (15) is inapplicable because the charging time is no longer sufficiently long compared to $R_{tq}C_L$ at 4 V_{th}. At 25 ns, V_{dd} for (18) equals 1.3 V_{th}, which is the lower limit for using (16).

We assumed in this analysis that the device sizes for both approaches are equal. Thus, the input capacitances for both circuits are approximately equal, if the conventional driver is ratioed according to carrier mobility. However, the adiabatic driver requires that dual-rail signalling be used, while conventional one does not. This requirement for dual-rail signalling tends to proliferate throughout the system, doubling the hardware amount. Assuming that both signal polarities of the input are available, the choice between the two approaches depends on the allowable switching time and on the input/output requirements. Initially, supply voltage reduction is very attractive, since there is a quadratic decrease in energy dissipation as switching time is increased. However, its scalability is limited and depends on having the flexibility to set supply voltages based on circuit latency requirements.

In practical circuits, several additional sources of overhead must be considered for an adiabatic circuit to offer lower energy dissipation than a conventional one with a freely chosen supply voltage. These include the FET clamps on undriven outputs and additional sequencing requirements so that energy can be efficiently delivered and recovered. The modest gains indicated by Fig. 4 may easily be overwhelmed. Furthermore, the situation becomes more severe when sequential functions with adiabatic charging are attempted, as will be seen in the next section.

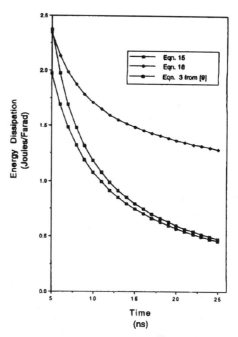

Fig. 4. Trading energy for delay for a conventional (18) and an adiabatic (15) driver. The transition time of the conventional driver is set by adjusting the supply voltage (V_{dd}). The voltage swing of the adiabatic driver is set at 4V_{th}. Equation 3 from Younis [3] is the exact formula for energy dissipation when $T \gg RC$ does not hold.

IV. ADIABATIC LOGIC CIRCUITS

A straightforward extension of the adiabatic amplifier lets us implement arbitrary combinational logic functions that use adiabatic charging. The circuit of Fig. 1 may be transformed into an adiabatically-switched circuit by replacing each of the pFET's and nFET's in the pull-up and pull-down networks with T-gates, and by using the expanded pull-up network to charge the true output and the expanded pull-down network for the complement output. Fig. 5 depicts this circuit transformation. Both networks in the transformed circuit are used to both charge and discharge the output capacitances. The dc V_{dd} source of the original circuit must be replaced by a pulsed-power source to allow adiabatic operation. The optimal sizes of the T-gates of the function networks can be determined from the equations of Section II.

As with conventional CMOS logic, it is desirable for reasons of performance and complexity management to partition a large block of logic into smaller ones and then compose them to implement the original larger function. However, if values are allowed to ripple through a chain of cascaded adiabatic logic gates, non-adiabatic flow of energy will occur. Adiabatic operation is possible only if the inputs of a gate are held stable while the gate is energized. As observed by Hall [5], a cascade of initially de-energized circuits may be energized in succession, but must then be de-energized in reverse order before the input values to the cascade may change. These "retractile cascades" are impractical for several reasons: they require a large and possibly indeterminate number of supply voltage waveforms; these waveforms all have different pulse widths;

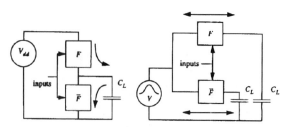

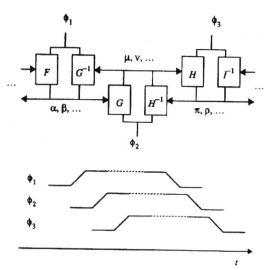

Fig. 5. (a) A conventional CMOS logic gate. (b) A corresponding adiabatic gate may be constructed by reorganizing the switch networks as shown, replacing the pFET's and nFET's of the conventional gate with T-gates. The arrows indicate the charge flow when the load capacitances are charged and discharged.

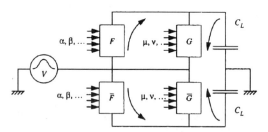

Fig. 6. A pipelinable adiabatic gate. The inputs $\alpha, \beta, \ldots$ control the F paths through which one of the load capacitances is charged. The G paths, used for discharge, are controlled by a different set of signals: $\mu, \nu, \ldots$. As in Fig. 5, arrows indicate charge flow.

Fig. 7. Conceptual adiabatic pipeline using invertible functions. For simplicity, the multiple switch networks needed for multirail signaling are not shown. α, β, μ, etc. are logic variables computed by the pipeline. The corresponding pulse-power/clock signals, denoted by ϕ_i, are also shown. One stage must be completely energized before the charging of the next stage commences.

and since an N stage cascade requires time proportional to N to produce each result, the latency is proportional to N and the throughput is proportional to $1/N$.

Pipelining can be advantageously used to improve the throughput of the system. By using latches to hold the inputs of a retractile stage, we could circumvent the requirement that the preceding stage stays energized until the current stage has been de-energized. Conventional latches, would, however, result in the unfavorable $T^{-1/2}$ or $T^{-3/4}$ scaling of the mixed adiabatic and conventional approaches. If the retractile stage is expanded to include an explicit discharge path for de-energizing, as shown in Fig. 6, the controlling inputs of the energizing paths need not be kept stable to handle de-energizing. A means for pipelining then exists that does not use latches, and T^{-1} scaling can be retained. The control signals for the de-energizing path must be stable throughout the discharge to ensure adiabatic operation. Therefore, they cannot be derived directly from the outputs of the stage being de-energized [6]. Any attempt at schemes based on the same idea is bound to fail for thermodynamic reasons [8]. The de-energizing path must instead be controlled by signals derived from the output of the *following* stage in the pipeline. A pure copy of the signal is sufficient, but this method only defers the problem and is not a general solution when resources are limited. The ability to derive control signals for the de-energizing path is guaranteed if all logic blocks implement functions that are invertible, so that their inputs may be recomputed from their output.

By constructing all of the logic stages according to Fig. 6 and restricting the function blocks to invertible functions only, it is possible to assemble a fully adiabatic pipeline as shown in Fig. 7. (A pipeline structure similar in organization but differing in the details has been presented by Younis and Knight [9].) Once energized, the output values of a stage are held for a certain time before de-energizing commences; an idle period separates subsequent energy cycles. During the hold period, the output values are used to set the next pipeline stage and to reset the previous stage. The pulsed-power supply voltages for the pipeline are shown in the figure. The requirement that the outputs of one stage are held while the surrounding stages are energized and de-energized naturally leads to set of supply voltage waveforms partially overlapping in time. Since they not only power the computations but also pace them, the supply waveforms may equally well be thought of as clock signals. The use of exclusively invertible functions makes the pipeline reversible: if the clocks are reversed in time, the pipeline runs backwards. Reversible logic has been studied in theoretical physics since Bennett showed that computations need not destroy information [7]. Since then, a large body of theory has been developed [10]–[12].

The complexities of two mutually inverse function blocks are usually similar, so the typical circuit overhead for the separate discharge path is a factor of two. This overhead is avoided if the de-energizing paths are replaced by diodes. The diodes would be connected with their anodes to the respective output node and their cathodes to the supply. They will not influence circuit behavior while the stage is energized, being back-biased for the uncharged output and shorted by the switch network for the charged output. During de-energizing, the high output will dump its charge back to the supply through the diode. Attractively simple and fast circuits using a similar approach have recently been published by Kramer *et al.* [14]. The dissipation caused by passing the output charge through a diode drop is a drawback, since it is essentially independent of the ramp time.

To validate the reversible pipeline structure shown in Figs. 7 and 8, we designed a small shift register as a CMOS chip. Shifting is an inherently reversible operation, and the simple structure allowed us to stay in the realm of small, highly controllable experiments. The small size also meant that all signal nodes inside the chip could be brought out to the off-chip pads for observation. Following the adiabatic amplification and switching principles outlined above, the shift register was implemented with dual-rail logic and T-gates for the separate charge and discharge paths. Four symmetric clock/supply signals, ninety degrees out of phase, were used, together with off-chip storage capacitances on all circuit nodes to preserve the dynamic data values during the clock transitions. Each "phase stage" of the register required eight transistors, and each "bit stage" was composed for four phase stages, for a total of thirty-two transistors for each stage. A complete cycle of a clock signal would move a bit from one bit stage to the next.

The principle of reversible pipelines assumes that every phase stage is followed by another stage that controls its de-energizing. In practical circuits, the last stage of the pipeline produces a final result of the computation, which is then output and discarded. This discharge of the output nodes of the last pipeline stage will therefore always be dissipative, but the dissipation will be proportional to the number of output bits rather than to the total number of bits in the pipeline. In order to characterize the efficiency of the shift register without having to consider these secondary effects, the input to the shift register was connected to the output, making it a circular and completely reversible machine with no net flow of information into or out of the system.

The shift register, although very simple, is to the best of our knowledge the first working sequential adiabatic CMOS circuit. Its performance was determined simply by measuring the dc current fed to the pulsed-power supply. For a four bit-stage chip, power dissipation (including that of the pulsed-power unit) was 130 μW at 63 kHz driving 82 pF loads to 4.5 V. The slow clocking speed was due in part to the timing generator used in the test setup. The ramp times were set at approximately 1 μs, which suggests that the design could have run at 250 kHz with little modification. At 63 kHz, the energy dissipation per clock cycle was 2.3 nJ. The electrical work done per cycle was to completely charge and discharge eight 82 pF loads to 4.5 V. The equivalent conventional work for driving these loads would be 13.2 nJ.

When logic functions more general than shifting are implemented in a reversible pipeline, signal encoding brings additional overhead compared to the conventional case. The outputs of a gate carry invalid values when the corresponding supply voltage is in its idle phase and therefore de-energized. To prohibit these values from activating devices in the switch networks of the neighboring stages, quad-rail encoding of the signal values can be used [9]. Active-low signals then control the P-channel devices, while active-high signals control the

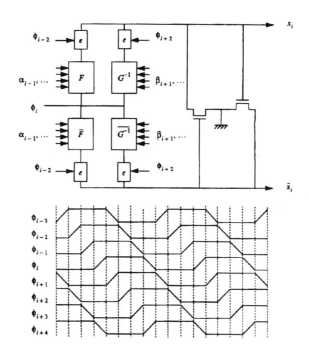

Fig. 8. Detailed dual-rail logic gate for use in reversible pipeline, and its eight-phase adiabatic clock signals. The "enable switches," marked e, are transmission gates, tied when the indicated clock signal is high. For data signals (α and β), the sub-index indicates phase of validity. The clamp devices at the output of the gate ensures that the undriven output stays at ground.

N-channel devices. One pair of signals change from their idle values to represent a zero, the other pair to represent a one. To avoid the extra complexity factor of two brought by quad-rail signalling, enable switches may be placed in series with the switch networks to prevent charge flow during the idle phase. This results in the gate style and clocking scheme depicted in Fig. 8 (two devices have been added to clamp the undriven output to ground).

As a design exercise, we used the gate style of Fig. 8 for a highly pipelined FIR filter. We believe a filter, specified by transfer function and throughput rate, to be a realistic early application for adiabatic logic circuits: the switching time can easily be increased to allow operation at lower energy levels without changing the specification. This property was also noted by Duncan *et al.*, who demonstrated trading switching time for energy dissipation by varying the supply voltage of a filter implemented with conventional logic [15].

The adder is the central combinational building block of a FIR filter. Following the example of Duncan, we designed a bit-level-pipelined adder circuit to allow the largest possible switching time for a given filter throughput. Fig. 9 shows a gate-level diagram of the adder, complete with the delay cells needed to preserve a bit of information until it may be reversed. The three-bit version shown in the figure was fabricated and its functionality verified; the extension to a wider version is straightforward. To indicate the total circuitry overhead caused by the use of reversibility, those elements that would be needed in a conventional adder have been shaded in Fig. 9. Because

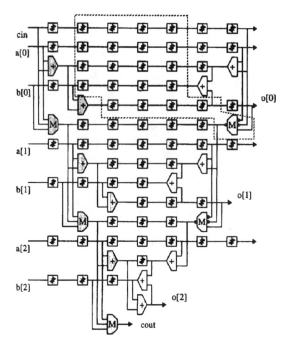

Fig. 9. Pipelined ripple-carry adder, implemented with reversible logic. Each circuit node is driven by two gates: one for charging and one for discharging. Blocks marked "+" are XOR gates; "M" denote majority gates; and bidirectional delay cells are shown with reciprocating arrows. The carry chain mandates a large number of delay cells. Only the shaded elements would be present in a conventional adder.

of dual-rail signalling, separate charge-up and charge-down paths, and the use of T-gates for switches, each gate is also more complex than its conventional counterpart. As a result, the three-bit reversible adder requires 20 times the number of devices and 32 times the area of a conventional adder using the same technology and laid out by the same designer. The overhead clearly gets worse for a wider adder, since the number of delay cells grows as the square of the adder width.

We may decrease the hardware overhead of the adder significantly by using reversibility more judiciously. For example, by allowing at least part of the signal energy of the carry signal from the least significant bit in Fig. 9 to dissipate (such as by using a passive diode charge-down path), we may remove all the elements within the dashed frame. Furthermore, the dissipation caused by operating these elements may actually outweigh the signal energy of the carry signal, unless the adder is operated *very* slowly. Layout extraction and circuit simulation of the reversible adder (to account for parasitics) indicate that a four-bit adder exhibits lower dissipation than two cascaded two-bit adders only if the charging times are thousands of times slower than those possible in a conventional adder. This comparison assumes that the energy of the non-reversed carry signal interconnecting the two two-bit adders is completely dissipated. Charging has to be much slower yet for a 32-bit adder to outperform two cascaded 16-bit adders.

It may be argued that the bit-level-pipelined adder is a particularly bad example for reversible computing because of the long carry-chain dependency. However, a version using carry-lookahead over two bits fares no better because of increased capacitance in the lookahead network itself. Similar assessments apply to other modified versions that retain the basic structure of Fig. 9. We conjecture that new breakthroughs in the composition of reversible logic gates are necessary for a fully-reversible 16- or 32-bit pipelined adder to be competitive with circuits that use reversibility on a lesser scale.

V. CONCLUSION

With the adiabatic switching approach, circuit energies are conserved rather than dissipated as heat. Depending on the application and the system requirements, this approach can sometimes be used to reduce the power dissipation of digital systems. For situations where the voltage swing is given by external constraints, adiabatic switching is the only known scalable approach to trade off energy dissipation for switching speed. Since the reduction in switching energy is linear in supply voltage (13), adiabatic switching is a promising approach for circuits that drive large capacitive loads to voltages above four times the threshold voltage.

The situation changes when supply voltages can be freely chosen to minimize dissipation for a given performance level. The optimum voltage swing for T-gate-based adiabatic circuits is close to $4V_{\text{th}}$ regardless of switching speed, whereas conventional circuits can reasonably operate with V_{dd} approaching V_{th}. This corresponds to an energy dissipation factor of 16 in favor of the conventional approach. However, the dissipation of adiabatic switching is not limited by the $C_L V_{\text{th}}^2$ asymptote of (18), and for sufficiently slow switching, the adiabatic circuit will operate at lower power levels than its conventional counterpart.

Other sources of overhead must also be considered when analyzing the energy efficiency of adiabatic circuits. The shape factor of the input waveform (ξ) accounts for an additional 23% of dissipation when sine-shaped currents (such as those generated by resonant power supplies) are used. The requirement for dual-rail signals and T-gates where single-rail signals and individual FET's could otherwise be used roughly doubles the area required for logic. With dual-rail signals, half of the circuit nodes will switch each cycle, while for single-rail signals, this number would be the worst case. Also, a significant part of the additional area will contribute to load capacitance that would not be present in a conventional circuit. A final source of overhead are the timing constraints placed on the inputs and outputs.

In total, the circuit overhead for combinational logic is at least 250%, not including the difference in duty factor between the dual-rail and single-rail signals or the added load capacitance due to the additional dual-rail circuitry. To compensate for this, switching time must be increased proportionately. If adiabatic logic is driven by conventional logic, the minimal increase in switching time will be greater because of the $T^{-1/2}$ and $T^{-3/4}$ scaling properties.

The desirable T^{-1} scaling, which can only be achieved with all-adiabatic operation, mandates that the logic be fully reversible. Requiring that inputs be held constant across multiple stages of cascaded logic would preclude using adiabatic switching for complex logic functions. Pipelining is possible and has been demonstrated in practice; if each combinational logic block is replaced by two, one for charging the output and one for discharging, the inputs may change while the output is valid. The inputs for the discharge block must then be driven by the outputs of the following pipeline stage. The stages can be sequenced by overlapping clock signals, which also supply energy to the circuits, to implement reversible pipelines. The overhead in logic is approximately a factor of two because of the need for separate charge and discharge blocks. However, since these blocks are activated at different times, each path is used only once for each full compute cycle. Hence, the energy dissipation is comparable to that of the single block which is used for charging in both directions.

Finally, the use of only reversible logic for fully-adiabatic pipelines puts additional constraints on logic implementation. This is especially noticeable when adiabatic switching and reversible logic are applied to logic functions with sequential dependencies across many stages, as in the example of the carry chain for a bit-level pipelined adder. The overhead of storing the intermediate bits so that they may be later reversed grows as the square of the adder width. Maintaining these bits does not contribute to the computation proper but only to the overhead of performing the computation reversibly. The energy needed for temporarily storing the bits in reversible latches may be greater than the energy the arrangement was supposed to recover.

In summary, if adiabatic switching is to be useful for realizing low-power digital systems in practice, it will most likely be in hybrid configurations which include conventional or partially adiabatic [6] latches and logic. The non-adiabatic latches and logic will have to operate with the smallest energy levels possible, since these energies will be dissipated in all or part. As was speculated about the applications for hot-clock nMOS [1], those parts of a chip which involve driving large capacitive loads, i.e., pads, data buses, and globally decoded signals, are the most promising candidates for adiabatic switching techniques. The analyses and experiments presented in this article are a first step towards understanding where these techniques can be practically used.

ACKNOWLEDGMENT

The authors wish to thank Prof. S. Mattisson of Lund University for many helpful discussions.

REFERENCES

[1] C. L. Seitz, A. H. Frey, S. Mattisson, S. D. Rabin, D. A. Speck, and J. L. A. van de Snepscheut, "Hot-clock NMOS," in *Proc. 1985 Chapel Hill Conf. VLSI*, 1985, pp. 1–17.

[2] W. C. Athas, J. G. Koller, and L. "J." Svensson, "An energy-efficient CMOS line driver using adiabatic switching," *Proc. Fourth Great Lakes Symp. VLSI Design*, pp. 196–199, Mar. 1994.

[3] A. P. Chandrakasan, S. Sheng, and R. W. Brodersen, "Low-power CMOS digital design," *IEEE J. Solid-State Circ.*, vol. 27, no. 4, pp. 473–484, April 1992.

[4] C. A. Mead and L. Conway, *Introduction to VLSI systems.* Reading, MA: Addison-Wesley, 1980.

[5] J. S. Hall, "An electroid switching model for reversible computer architectures," in *Proc. ICCI'92, 4th Int. Conf. on Computing and Information*, 1992.

[6] J. G. Koller and W. C. Athas, "Adiabatic switching, low energy computing, and the physics of storing and erasing information," in *Proc. Workshop on Physics and Computation*, PhysCmp '92, Oct. 1992; IEEE Press, 1993.

[7] C. H. Bennett, "Logical reversibility of computation," *IBM J. Res. Dev.*, vol. 17, pp. 525–532, 1973.

[8] R. Landauer, "Irreversibility and heat generation in the computing process," *IBM J. Res. Dev.*, vol. 5, pp. 183–191, 1961.

[9] S. G. Younis and T. F. Knight, "Practical implementation of charge recovery asymptotically zero power CMOS," in *Proc. 1993 Symp. on Integrated Syst.*, MIT Press, 1993, pp. 234–250.

[10] C. H. Bennett and R. Landauer, "The fundamental physical limits of computation," *Scientific American*, pp. 48–56, July 1985.

[11] R. Merkle, "Reversible electronic logic using switches," *Nanotechnology*, vol. 4, pp. 21–40, 1993.

[12] C. H. Bennett, "Time/space trade-offs for reversible computation," *SIAM J. Computing*, vol. 18, pp. 766–776, 1989.

[13] E. M. Cherry and D. E. Hooper, *Amplifying Devices and Low-Pass Amplifier Design.* New York: Wiley, 1968.

[14] A. Kramer, J. S. Denker, S. C. Avery, A. G. Dickinson, and T. R. Wik, "Adiabatic computing with the 2N-2N2D logic family," in *1994 Symp. VLSI Circ.: Digest of Tech. Papers*, IEEE Press, June 1994.

[15] P. J. Duncan, S. Swamy, and R. Jain, "Low-power DSP circuit design using bit-level pipelined maximally parallel architectures," in *Proc. 1993 Symp. on Integrated Syst.*, MIT Press, 1993, pp. 266–275.

[16] D. Maksimovic, "A MOS gate drive with resonant transitions," in *Proc. IEEE Power Electron. Specialists Conf.*, IEEE Press, 1991, pp. 527–532.

Adiabatic Dynamic Logic

Alex G. Dickinson and John S. Denker

Abstract— With *adiabatic* techniques for capacitor charging, theory suggests that it should be possible to build gates with arbitrarily small energy dissipation. In practice, the complexity of adiabatic approaches has made them impractical. We describe a new CMOS logic family—adiabatic dynamic logic (ADL)—that is the result of combining adiabatic theory with conventional CMOS dynamic logic. ADL gates are simple, general, readily cascadable, and may be fabricated in a standard CMOS process. A chain of 1000 ADL inverters has been constructed in 0.9 μm CMOS and successfully tested at 250 MHz. This result, together with comprehensive circuit simulation, suggest that ADL offers an order of magnitude reduction in power consumption over conventional CMOS circuitry.

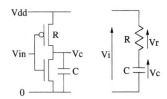

Fig. 1.   A static CMOS inverter and its equivalent circuit for the case where the capacitor C is being charged through a device of on-resistance R.

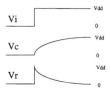

Fig. 2.   Voltage waveforms present in the equivalent circuit when charging the capacitor from 0 V to V_{dd} in a conventional manner.

I. INTRODUCTION

POWER consumption has become a critical concern in both high-performance and portable applications. Methods for power reduction based on the application of *adiabatic* techniques to CMOS circuits have recently come under renewed investigation. In thermodynamics, an adiabatic energy transfer through a dissipative medium is one in which losses are made arbitrarily small by causing the transfer to occur sufficiently slowly.

If circuits can be made to operate in an adiabatic regime with consequently low energy dissipation, then the energy used to charge the capacitive signal nodes in a circuit may be recovered during discharge and stored for reuse. The efficiency of such a circuit is then limited only by the "adiabaticity" of the energy transfers.

Conventional CMOS circuits are pathologically nonadiabatic. Capacitive signal nodes are rapidly charged and discharged (the energy transfer) through MOS devices (the dissipative medium). At times, the full supply potential appears across the channel of the device, resulting in high device current and energy dissipation.

Younis and Knight [1] and Merkle [2] have proposed adiabatic logic families with near zero dissipation. However, these schemes require reversible logic gates, and in the case of Younis and Knight, each gate requires 16 times the number of devices as conventional logic. Koller and Athas [3], [4] have described a scheme based on transmission gates with limited cascadability. Hinman and Schlecht [5] describe a means for performing adiabatic logic with bipolar devices, or the complex emulation of bipolar devices with NMOS devices. Seitz *et al.* [6] recognized the benefits of adiabatic charging, but used a two-phase clocking strategy that precluded adiabatic operation.

We have developed a new logic family, adiabatic dynamic logic (ADL), based on the application of adiabatic techniques

Manuscript received August 11, 1994; revised November 9, 1994.
The authors are with AT&T Bell Laboratories, Holmdel, NJ 07733 USA.
IEEE Log Number 9408736.

to dynamic CMOS logic structures. ADL is not *completely* adiabatic, but offers an order of magnitude reduction in power consumption using simple, readily cascadable gates that may be fabricated in a standard CMOS process. In this paper, we will describe adiabatic charging, ADL gates, a resonant power supply, and the results of our simulations and test fabrications.

II. ADIABATIC SWITCHING

A. Conventional Charging

The dominant factor in the dissipation of a CMOS circuit is the *dynamic* power required to charge capacitive signal nodes within the circuit. Fig. 1 shows a basic CMOS inverter, together with an equivalent circuit of the charging mechanism. Fig. 2 shows the voltage waveforms present when the input of the inverter swings from *high* to *low,* causing the capacitor C to begin charging. At the instant of switching, the full supply potential appears across the on-resistance R of the p-type device—the waveform then decays as the capacitor is charged to V_{dd}. To charge the signal node capacitance C from a supply of potential V_{dd}, a charge $q = CV_{dd}$ is taken from the supply through the p-type device. The total energy $E_T = qV_{dd} = CV_{dd}^2$.

Only half of the energy is applied to storing the signal on the capacitor—the other $\frac{1}{2}CV_{dd}^2$ is dissipated as heat, primarily in the device on-resistance R. Note that the dissipation is independent of this resistance: it is a result of the capacitor charge being obtained from a constant voltage source V_{dd}.

To drive the inverter output low, the n-type device is used to discharge the $\frac{1}{2}CV_{dd}^2$ energy stored in capacitor C by short circuiting the capacitor and dissipating energy as heat.

Reprinted from *IEEE Journal of Solid-State Circuits,* pp. 311-315, March 1995.

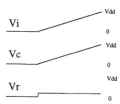

Fig. 3. Voltage waveforms present in the equivalent circuit when charging the capacitor from 0 V to V_{dd} in an adiabatic manner.

Hence, the total charge/discharge cycle has required an energy CV_{dd}^2—half being dissipated in charging, and half being used for information storage before it too is dissipated during discharge.

B. Adiabatic Charging

Adiabatic switching can be achieved by ensuring that the potential across the switching devices is kept arbitrarily small. In Fig. 2, it can be seen that the potential V_r across the switch resistance is high in the conventional case because of the abrupt application of V_{dd} to the RC circuit.

Adiabatic charging may be achieved by charging the capacitor from a time-varying source, as shown in Fig. 3. This source has an initial value of $V_i = 0$ V—the ramp increases towards V_{dd} at a slow rate that ensures $V_r = V_i - V_c$ is kept arbitrarily small. This rate is set by ensuring that the period of the ramp $T \gg RC$.

In fact, the energy dissipated is (From Athas [4])

$$E_{\text{diss}} = I^2 RT = \left(\frac{CV_{dd}}{T}\right)^2 RT = \left(\frac{RC}{T}\right)CV_{dd}^2. \quad (1)$$

A linear increase in T causes a linear decrease in power dissipation. Adiabatic discharge can be arranged in a similar manner with a descending ramp.

Now, if T is sufficiently larger than RC, energy dissipation during charging $E_{diss} \rightarrow 0$, and so the total energy removed from the supply is $\frac{1}{2}CV_{dd}^2$—the minimum required to charge the capacitor and hence hold the logic state. This energy may be removed from the capacitor and returned to the power supply adiabatically by ramping V_i back down from V_{dd} to 0 V. As a result, given a suitable supply, it should be possible then to charge and discharge signal node capacitances with only marginal net losses.

Note that the RC time constant of a typical CMOS process is about 100 ps. If we set T to ten time constants, the resulting delay through an adiabatic gate would be 1 ns.

III. AN ADIABATIC DYNAMIC LOGIC GATE

A. Conventional Dynamic Logic

A conventional dynamic CMOS inverter is shown in Fig. 4. When the clock Φ is *low*, the output is always pulled to V_{dd}; when the clock goes *high*, the output remains *high* if the input is *low*, or goes *low* if the input is *high*. Note that if the output remains *high*, it does so not as a "driven" value, but as charge stored on the output node capacitance C.

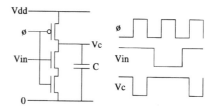

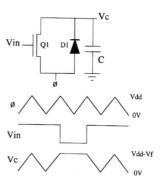

Fig. 4. A conventional dynamic CMOS inverter and associated clock, input, and output waveforms for a 1–0–1 input and resulting 0–1–0 output.

Fig. 5. An adiabatic dynamic logic CMOS inverter and associated clock, input, and output waveforms for a 1–0–1 input and resulting 0–1–0 output.

Clearly, there is abrupt (nonadiabatic) charging and discharging of the output node in this circuit. Also (as with static CMOS circuits), there is an additional loss mechanism resulting from direct current flowing from the supply to ground during the switching operation, when both n-type and p-type devices are conducting simultaneously.

In working to develop an adiabatic logic family, our attention was drawn to dynamic circuits because of their time-sequential nature. All nodes are charged *(precharged)* during one half cycle of the clock. The same nodes then remain charged or are discharged *(evaluated)* during the next half cycle. Conceptually, at least, if one half cycle of the clock were a rising ramp, and the other a falling ramp, one might be able to produce adiabatic operation.

B. The ADL Inverter

The ADL inverter is shown in Fig. 5. It is comprised of a diode $D1$ with forward drop V_f, a transistor $Q1$ of threshold V_t, and an implicit capacitance C that represents the load of subsequent gates. Note the simplicity of the circuit, and the lack of dc paths to ground. All power is supplied by the time-varying "clock supply" Φ.

As with a conventional dynamic CMOS gate, there are two basic stages of operation. In the *precharge* stage, the clock-supply ramps from 0 to V_{dd}, precharging the output through the forward-biased diode $D1$ to $V_{out} = V_{dd} - V_f$. In the *evaluate* stage, the clock supply ramps down from V_{dd} to 0. If the input is *high*, $Q1$ conducts, and so the output is driven *low*, $V_{out}=0$. If the input is *low*, the output remains *high* at $V_{out} = V_{dd} - V_f$.

Clearly, an inversion has occurred between input and output. In addition, all transitions of the output node are driven by a ramped voltage supply, and are hence potentially adiabatic.

Fig. 6. The four-phase clock waveform required for interconnecting ADL gates. The vertical arrows indicate the flow of data from the *hold* time of one gate to the *evaluate* time of a subsequent gate.

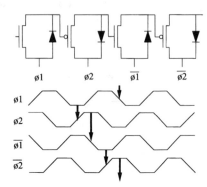

Fig. 7. A "cycle" of four interconnected ADL inverters and their associated four-phase clocks.

IV. SYSTEMS OF ADL GATES

A. Cascaded Gates

Adiabatic circuit operation requires that only small potentials appear across the channel of a device while it is conducting. This must be taken into consideration when interconnecting gates, as it implies that devices should never "turn on" while there is a nonzero voltage present across the device channel.

Between *precharge* and *evaluate,* there are only small potential differences present in the ADL inverter circuit: both clock-supply and output nodes are near V_{dd}. We make use of this observation by adding a stage of constant clock voltage between these two stages. During this *input* stage, the input signal may safely transition without causing a nonadiabatic transition within the circuit. Similarly, we add a *latch* stage between *evaluate* and *precharge* to ensure that the output is latched for a finite time. The resulting clock waveform is shown in Fig. 6.

To connect two gates, it is necessary to synchronize their respective clock supplies such that the *latch* stage of the first gate coincides with the *evaluate* stage of the second. However, as the first gate begins precharging, it will cause the input of the second gate to undergo a *low* to *high* transition, resulting in a nonadiabatic transition in the second gate. The solution to this problem is shown in Fig. 7: the second stage is constructed as an inverted version of the first, precharging *low* and evaluating *high*. The precharge of the first gate can then only cause the input device of the second gate to *stop* conducting: an allowable adiabatic transition.

Because there are four stages in a single clock cycle *(precharge, input, evaluate, latch)*, it is necessary to place four gates in series before the last gate may feed the first. The clock-supply waveforms consists of two trapezoidal waveforms

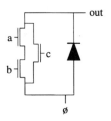

Fig. 8. A complex ADL gate implementing $\overline{a \cdot b + c}$.

separated by a quarter period and their complements. A four-gate "cycle" and its associated clock waveforms are shown in Fig. 7.

B. Clock Voltage Considerations

A requirement of the particular system of gate interconnection that we have described is that when a gate is in its *precharge* state, the voltage present at its output must be such that it not turn on the input device of the subsequent gate, i.e., $V_f < V_{th}$. In the case of a precharge *high* gate, for example, if the precharge voltage $(V_{dd} - V_f)$ is less than the turn-on point of the subsequent p-type device $(V_{dd} - V_{th})$, it will turn on the input p-device of the subsequent gate during precharge. We have found two means of the avoiding this problem. The first is to ensure that the forward drop of the diode V_f is less than the threshold voltage V_{th} of the p-type or n-type device to which it is connected. The second technique is to offset the clock voltages: for example, the first and third gates in Fig. 7 would have a positive offset compared to the clocks for the second and fourth gates to ensure correct operation.

C. Complex ADL Gates

Extension of the ADL inverter to more complex NAND/NOR gates is shown in Fig. 8. The complexity of an ADL gate is low: n FET's $+ 1$ diode for an n input gate compared to $2n$ FET's and $(n + 2)$ FET's for conventional static and dynamic CMOS, respectively. In terms of routing, an ADL system requires four clock lines and two tub bias lines—comparable to some conventional CMOS systems based on two clocks their complements, and two power lines.

In addition to the synchronous interconnection of ADL gates, it is possible under certain conditions to combine gates asynchronously, as shown in Fig. 9, if they have the same precharge type. This arrangement has the advantage of adding an optional inversion function at the output or input of a gate, and potentially doubling the number of gates available for operations in a four-phase clock cycle (eight gates rather than four). The arrangement is, however, sensitive to the rise and fall time of the clock edges. If, for example, the right-hand gate is being left *high* during evaluation, the output of the left-hand gate must be evaluating *low*. If the output of the left-hand gate lags the clock as it falls towards 0, the input device of the right-hand gate may turn on, causing the output of that gate to go *low* erroneously. However, given that ADL in general is based on using relatively slow clock edges to ensure adiabatic operation, this has not been found to be a major limitation.

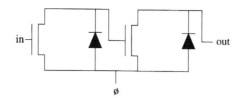

Fig. 9. A pair of gates with identical precharge types (positive) connected to a single clock phase.

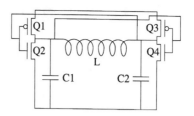

Fig. 10. A resonant oscillator "clock supply" for an ADL system. Note that there are no active devices in the main current path between the two capacitive loads.

V. ENERGETICS

Further insight into an ADL circuit can be gained by thinking of its *evaluate-precharge* sequence as being one of a "write" being followed by an "erase." During *evaluate,* the output node is either left alone or discharged (written). During *precharge,* the diode $D1$ observes the state of the output, and in effect makes a decision as to whether or not the output needs to be erased. It is the use of the diode in this decision-making process that dissipates most of the energy in the gate.

The object of using reversible logic (Younis and Knight [1], for example) is to eliminate the cost of erasing the information held in latches. In ADL, this erasure does not come for free, but it is cheap: the diode makes a purely *local* erasure decision.

As charge q is drawn from the supply V_{dd} to recharge C_{ADL}, the diode drop V_f causes an energy loss:

$$E_{\text{pre}} = qV_f = C_{\text{ADL}}V_{dd}V_f. \quad (2)$$

The energy dissipated during evaluation (discharge) is (1)

$$E_{\text{eval}} = \left(\frac{RC_{\text{ADL}}}{T}\right)C_{\text{ADL}}V_{dd}^2 \quad (3)$$

which will be much smaller than $C_{ADL}V_{dd}V_f$ for $T \gg RC_{ADL}$. The ratio of conventional to ADL energy dissipation is then

$$\frac{E_{\text{conv}}}{E_{\text{ADL}}} = \frac{C_{\text{conv}}V_{dd}^2}{C_{\text{ADL}}V_{dd}V_f} = \left(\frac{V_{dd}}{V_f}\right)\left(\frac{C_{\text{conv}}}{C_{\text{ADL}}}\right). \quad (4)$$

For typical values of $V_{dd} = 5$ V, $V_f = 0.5$ V, and for our measured capacitance ratios of between 1 and 2, we find that the potential power savings cover the range of 10–20 over conventional CMOS logic, before allowing for clock generation losses.

VI. ADIABATIC CLOCK GENERATION

By itself, the use of adiabatic charging will only improve power consumption by a factor of two: the $\frac{1}{2}CV_{dd}^2$ of resistive heat dissipation is removed, but the $\frac{1}{2}CV_{dd}^2$ stored in the capacitor is still lost during discharge. If we are to realize the full potential of the technique, it is necessary to return that energy to the supply for use in a subsequent cycle.

A resonant circuit will serve to drive the largely capacitive load presented by the ADL system if a sinusoid may be substituted for the trapezoidal supply waveform. We have, in fact, verified that ADL gates will operate from sinewave sources in both simulations and circuit testing.

Fig. 10 shows a simple resonant oscillator for driving two complementary clock-supply nodes of an ADL system, represented by the capacitors C_1 and C_2. If, at time t_0, $V_1 = 5$ V and $V_2 = 0$ V, then the inductor L will serve to pump charge from C_1 to C_2 and back again. The waveform presented across each capacitor will be a sinewave with an amplitude that decreases with time due to the resistive losses in the inductor and ADL circuits. To maintain the amplitude of the supply waveform, switches to the dc supply and ground are provided as shown ($Q1Q4$), and are activated at appropriate moments to "top-up" the system. The gates on these devices are driven by the resonant circuit itself.

Note that in this circuit, there is only an inductor present in the high-current supply path between the two capacitors. The switches only carry the relatively low top-up current, minimizing the potential resistive losses in the circuit.

The four phases required for an ADL system may be generated by two of these oscillators running with a quarter-period phase difference. Each oscillator may be built on-chip with the circuit, with only the small inductor (about 50 nH) connected between two package pins off-chip.

A difficulty with this resonant oscillator approach is that the load capacitance of an ADL circuit is data-dependent. As these loads change, the oscillator frequency must be actively stabilized to maintain a constant clock rate. One means of avoiding this problem is to adopt fully differential signaling in which, for every high-going node, there is a low-going node resulting in a constant capacitive load to the oscillator.

VII. SIMULATION AND IMPLEMENTATION

Our early detailed simulations of ADL inverter chains indicated that operation was possible over a frequency range from less than 1 MHz to greater than 1 GHz. An average factor of 15 improvement in power dissipation over conventional CMOS was found in the 1–100 MHz frequency range. For example, at 100 MHz, an ADL inverter consumed 1.7 μW (assuming a 100% efficient resonant power supply) compared to 26 μW for a conventional inverter. These results are for $V_{dd} = 5$ V; however, the inverters operated down to $V_{dd} = 2$ V in simulation (with consequently less power).

We have been fabricating a series of ADL test circuits in 0.9 μm CMOS. Because suitable diodes are not readily available in this process, we have been using FET's connected as diodes (gate tied to one end of the channel), with the penalty that the effective forward drop across these devices is quite high—typically about 1 V.

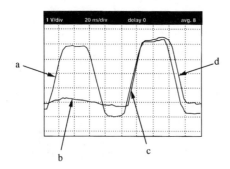

Fig. 11. Electron beam probed output of a 0.9 μm precharge low CMOS ADL inverter with a trapezoidal clock supply **a**. Note that the output of the gate precharges to within V_f of the clock low point **b**. During an evaluate, the output closely tracks the clock **c**, indicating that there is very little potential across the channel of the input device. During precharge **d**, the output again tracks the clock with a V_f drop.

In our first fabrication run, we built a chain of 64 inverters, which was successfully tested to 250 MHz the limit of our test equipment. Our second fabrication run included several chains of 1000 inverters, and established that ADL circuits could maintain good logic levels through many levels of logic. Clock skew tolerance was found to be in excess of 15% between adjacent phases.

We have used electron beam probing to obtain waveforms from the functioning circuits. Fig. 11 shows the operation of an ADL inverter at 10 MHz with alternating input. The output tracks the supply ramp very accurately during evaluation, and trails it by about 1 V during precharge, as expected. Because of the difficulty of accurately measuring the net power (the small difference between large input and output powers) drawn from a high-frequency supply, we have not yet been able to obtain accurate figures for power consumption. We are examining calorimetric approaches to this problem.

We have simulated and built a clock supply circuit (described earlier) from discrete components, and have found it to operate with 85–90% efficiency. Other more sophisticated and easily synchronized circuits have been simulated with similar results, and are presently being fabricated on-chip. We intend to use these to directly drive ADL circuits and accurately measure power consumption by monitoring the dc supply to the oscillators.

VIII. CONCLUSIONS

Our experience with ADL circuits indicates a factor of 15 reduction in dissipation in 0.9 μm CMOS. When combined with our 85–90% efficient resonant clock-supply circuits, the logic family offers an order of magnitude reduction in power consumption over conventional CMOS.

ADL circuit design will benefit from the addition of low-threshold diodes and transistors to the CMOS process. At the same time, the low field strengths present in devices used in an adiabatic regime may allow simplification of FET device design.

ACKNOWLEDGMENT

The authors wish to think B. Ackland, S. Avery, D. Inglis, L. Jackel, A. Kramer, and T. Wik for their many valuable contributions to this work.

REFERENCES

[1] S. G. Younis and T. F. Knight, "Practical implementation of charge recovering asymptotically zero power CMOS," in *Research on Integrated Systems: Proc. 1993 Symp.* Cambridge, MA: M.I.T. Press, 1993.
[2] R. C. Merkle, "Reversible electronic logic using switches," *Nanotechnol.,* vol. 4, pp. 21–40, 1993.
[3] J. G. Koller and W. C. Athas, "Adiabatic switching, low energy computing, and the physics of storing and erasing information," in *Proc. Phys. Computation Workshop,* Dallas, TX, Oct. 1992.
[4] W. C. Athas, J. G. Koller, and L. J. Svensson, "An energy-efficient CMOS line driver using adiabatic switching," Inform. Sci. Inst., CA, Tech. Rep. ACMOS-TR-2, July 1993.
[5] R. T. Hinman and M. F. Schlect, "Recovered energy logic—A highly efficient alternative to today's logic circuits," in *Proc. IEEE Power Electron. Specialists Conf.,* 1993, pp. 17–26..
[6] C. L. Seitz *et al.,* "Hot-clock nMOS," in *Proc. 1985 Chapel Hill Conf. VLSI.* Rockville, MD: Computer Science, 1985.

Part V

Driving Interconnect

Sub-1-V Swing Internal Bus Architecture for Future Low-Power ULSI's

Yoshinobu Nakagome, *Member, IEEE*, Kiyoo Itoh, *Senior Member, IEEE*, Masanori Isoda,
Kan Takeuchi, *Member, IEEE*, and Masakazu Aoki, *Member, IEEE*

Abstract— A new bus architecture is proposed for reducing the operating power of future ULSI's. This architecture will relieve the constraint of the conventional supply voltage scaling, which makes it difficult to achieve both high speed and a low standby current if the supply voltage is scaled to less than 2 V. It employs new types of bus driver circuits and bus receiver circuits to reduce the bus signal swing while maintaining a low standby current. The bus driver circuit has a source offset configuration through the use of low-V_T MOSFET's and an internal supply voltage corresponding to the reduced signal swing. Bus delay is almost halved with this driver when operated at 0.6-V swing and 2-V supply. The bus receiver circuit has a symmetric configuration with two-level conversion circuits, each of which consists of a transmission gate and a cross-coupled latch circuit. Fast level conversion is achieved without increasing the standby current. The combination of new bus driver and bus receiver enables the bus swing to be reduced to one-third that of the conventional architecture while maintaining a high-speed data transmission and a low standby current. A test circuit is designed and fabricated using 0.3-μm processes. The operation of the proposed architecture was verified, and further improvements in the speed performance are expected by device optimization. The proposed architecture is promising for reducing the operating power of future ULSI's.

I. INTRODUCTION

REDUCING operating power is one of the most important design issues that will confront future high-density and high-performance ULSI's such as CPU's and memories [1]. Performance improvement causes a drastic increase in power dissipation due to increased operating frequency and increased chip size. For example, the recent trend in the operating power for reduced-instruction set computer (RISC) CPU chips is a drastic increase when the performance improves, as shown in Fig. 1. The trend for 3.3 and 1.5 V is estimated under the assumption that the dynamic power dissipation is dominant. In order to achieve a 1000-MIPS CPU with a reasonable operating power comparable to that of the state-of-the-art CPU chip, or in order to achieve a 100-MIPS CPU with an operating power of a few watts, the power must be reduced by 1/10 that of the current trend. There are two ways to meet this requirement: 1) reduce the operating voltage to around 1.5 V, and 2) introduce a new circuit technology that reduces power while using a higher supply voltage such as 2 V.

Manuscript received August 24, 1992; revised October 21, 1992.
Y. Nakagome, K. Itoh, K. Takeuchi, and M. Aoki are with the Central Research Laboratory, Hitachi Ltd., Kokubunji, Tokyo 185, Japan.
M. Isoda is with Hitachi VLSI Engineering Corporation, Kodaira, Tokyo 187, Japan.
IEEE Log Number 9207065.

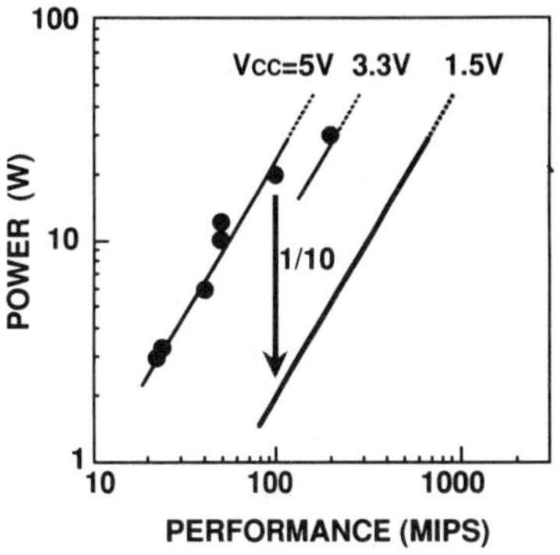

Fig. 1. Trend in operating power of RISC chips.

A supply voltage reduction to 1.5 V, however, may cause operating speed degradation. Fig. 2 shows the trend in normalized delay calculated for two cases, both ignoring the effect of velocity saturation [2]. We assume that the device dimensions such as the gate length, L_G, are scaled proportionally to V_{CC}, and that the total transistor width on a chip, W_{TOTAL}, is quadrupled every generation. The lower line is for the case where the threshold voltage, V_T, is scaled proportionally to the device dimensions and the supply voltage. The speed continues to improve in this case, but the standby current, I_{SB}, reaches as high as 3 A for the worst V_T variation and the highest temperature. The major reason for this increase in I_{SB} is the reduction of V_T rather than the increase in W_{TOTAL}. This high I_{SB} cannot be allowed for low-power applications, especially for battery-based ones. The upper line is for the case where V_T slightly increases as the device dimensions are scaled to maintain I_{SB} as low as 0.2 mA. In this case, the speed will not be improved for operating voltages of less than 2 V. This implies that we cannot have the advantages of scaling if the operating voltage is 2 V or less. The purpose of this work is to propose a new circuit technology that reduces power while operating from the supply voltage of about 2 V.

This paper presents a new internal bus architecture that reduces operating power by suppressing the bus signal swing to less than 1 V while using a higher supply voltage such as 2 V [3]. This architecture can achieve a low power dissipation comparable to the conventional architecture operated from 1.5-V power supply, while maintaining high speed and low standby current under 2-V power supply. The new concept is described in Section II. The reduced-swing circuit techniques are discussed in Section III. Finally, the experimental results using test circuits are presented.

Reprinted from *IEEE Journal of Solid-State Circuits*, pp. 414–419, April 1993.

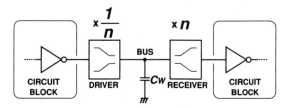

Fig. 2. Trend in normalized delay.

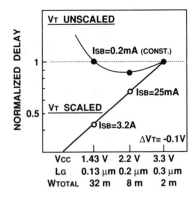

Fig. 3. Proposed architecture for internal bus swing reduction.

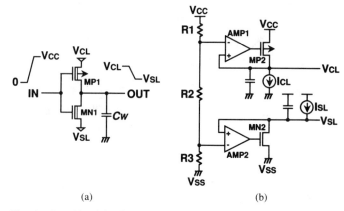

Fig. 4. Bus driver circuit: (a) source offset driver and (b) internal supply generator.

II. CONCEPT FOR INTERNAL BUS SWING REDUCTION

Fig. 3 shows a schematic illustration of the proposed architecture. It consists of circuit blocks and internal buses, i.e., data transmission lines. As the performance of ULSI's increases, the chip size and the bus width will also increase, and the total wiring capacitance of the bus lines will become considerably larger. Moreover, the number of internal bus lines will increase with macrocell-based ULSI designing, which integrates large-scale functional blocks such as a CPU, a floating-point processing unit (FPU), and memories on a chip. One estimate shows that charging and discharging the bus lines will take up to half or more the total chip power for 0.1-μm ULSI's [4]. When it is assumed that the ratio of the bus power to the total chip power is 0.5, and the bus swing is one-third of the supply voltage, the total chip power can be reduced to 66%. Thus, reducing the bus-signal swing is an effective means of reducing chip power. In addition to the increase in chip power, another concern for the design of ULSI's is the reliability problems of interconnections. This new architecture relieves electromigration problems due to the reduced charging and discharging current.

There has been a similar approach to reduce the internal signal swing for high-speed data transmission and power reduction [5], [6]. But it cannot be widely applied to ULSI's because it consumes a considerable amount of dc current [5], or it needs an additional clock [6].

The proposed architecture features a bus driver and bus receiver circuit. The bus driver drives a bus line with reduced output swing, V_{CC}/n. The bus receiver amplifies the reduced-swing signal to a full-swing signal. With this architecture, the gate-level circuit in the circuit blocks operates faster because the supply voltage can be higher than that determined by the power limitation. The key issues for this architecture are to achieve high-speed level conversion both from high to low

and from low to high, and to achieve a low standby current comparable to that in the conventional scheme.

III. SUB-1-V SWING CIRCUIT TECHNIQUES

A. Source Offset Driver

Fig. 4(a) shows the source offset bus driver circuit, and Fig. 4(b) is the internal supply voltage, V_{CL} and V_{SL}, generator. The source offset driver has a conventional CMOS inverter configuration. The differences from the conventional one are: 1) it is operated with the internal supply voltages, V_{CL} and V_{SL}, while the input is driven by the full-swing signal of a logic circuit operated with V_{CC}; and 2) the threshold voltages for both MOSFET's are lowered such that $\Delta V_{TN} = V_{SL}$ and $\Delta V_{TP} = V_{CC} - V_{CL}$. With this source offset scheme, the effective gate voltages, $V_{GS} - V_T$, for both on- and off-states are the same with the conventional CMOS inverter operated on V_{CC}. Consequently, a low standby current can be maintained while reducing the bus rise/fall times. The internal supply voltage generator consists of a voltage divider using a resistor string and buffers. The resistor string generates reference voltages, $V_{CC} \times (R2 + R3)/(R1 + R2 + R3)$ and $V_{CC} \times R3/(R1 + R2 + R3)$, for V_{CL} and V_{SL}, respectively. The buffer for V_{CL} consists of a differential amplifier AMP1, a p-channel MOSFET $MP2$, and a current bias source I_{CL}. This configuration is the same as that used for a 16-Mb DRAM [7]. It features a high charging current while maintaining a low standby current, which is almost equal to I_{CL}, when the load current is not consumed. A compensation capacitor on the V_{CL} supply line enhances the stability of the feedback loop. The buffer for V_{SL} has a symmetric configuration to that of V_{CL}.

The bus delays for the proposed and conventional schemes were simulated. The MOSFET parameters are those determined for a 1.5-V operated 64-Mb DRAM [8]. The gate length is 0.6 μm for both n- and p-channel MOSFET's, the gate-oxide thickness is 6.5 nm, and the threshold voltages are 0.5 and -0.5 V for the standard n- and p-channel MOSFET's and -0.1 and 0.1 V for the low-V_T MOSFET's, when they are tuned for 0.6-V swing operation with a 2-V power supply. The 10 to 90% rise delay and the 90 to 10% fall delay are plotted against bus swing in Fig. 5. The size of the MOSFET's of the

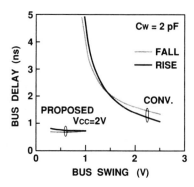

Fig. 5. Simulated bus delay versus bus signal swing.

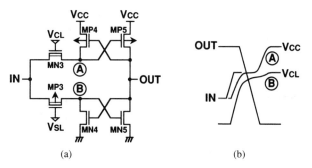

Fig. 6. Symmetric level converter: (a) circuit schematic and (b) schematic illustration of operation.

driver are $W_N = 15\,\mu$m and $W_P = 50\,\mu$m for both cases. In the conventional scheme, the bus signal swing is equal to the supply voltage, while the supply voltage for the reduced swing scheme is fixed to 2 V. Serious speed degradation is observed for the conventional scheme when the supply voltage is less than 2 V. On the other hand, the bus delays are almost constant from 0.3 to 1 V in the reduced-swing scheme. This is because the reduction in signal swing compensates for the reduction in drive current. The delay for the proposed scheme with a 0.6-V swing is 0.7 ns, about half that of the conventional scheme with a supply voltage of 2 V.

B. Symmetric Level Converter

One of the key circuits in this architecture is a bus receiver that converts a reduced-swing signal to a full-swing signal for the next logic stage. Fig. 6(a) shows the symmetric level converter (SLC) circuit used for the bus receiver. It has a symmetric configuration consisting of two level conversion circuits. The upper half converts an intermediate high level (i.e., V_{CL}) at the input to V_{CC} at the output, and the lower half converts an intermediate low level (i.e., V_{SL}) to ground. Each level conversion circuit consists of a transmission gate ($MN3$ or $MP3$) and a cross-coupled latch circuit. The operation of this circuit is shown in Fig. 6(b). We assume here that the input makes the transition from low (V_{SL}) to high (V_{CL}). The voltage at node A rises toward V_{CL}. The voltage at node B also rises toward the value determined by the conductance ratio of $MP3$ to $MN4$, because both MOSFET's turn on at this stage. This voltage becomes very close to V_{CL} because the conductance of $MN4$ is much smaller than that of $MP3$. Consequently, the node-B voltage is high enough to turn on $MN5$, and the output goes to ground. This turns on $MP4$, and the node-A voltage reaches V_{CC}. This turns off $MP5$, and no dc current path is formed from V_{CC} to ground.

Simulated output voltage V_{OUT} and the dc current from V_{CC} to ground, I_{CC}, are plotted against input voltage V_{IN} in Fig. 7. The conventional receiver consists of a CMOS inverter with MOSFET's having the same size as that of $MP5$ and $MN5$ in Fig. 6(a). In the conventional case, a dc current as high as several ten microamperes is obtained for a reduced swing input voltage of 0.7 or 1.3 V, although V_{OUT} makes almost a full transition from V_{CC} to ground. On the other hand, I_{CC} for the proposed scheme is efficiently suppressed to as low as 1 pA for the reduced input signal swing of 0.6 V. Due

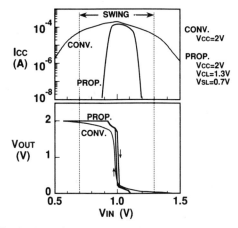

Fig. 7. Simulated transfer characteristics for the symmetric level converter.

to positive feedback from the output to the input, a hysteresis characteristic is observed. But the hysteresis window is as small as 0.05 V because the conductance ratio of transmission gate to feedback transistor is larger than 10.

One of the concerns in the reduced-swing architecture is the operating margin. Sensitivity to the variations in MOSFET characteristics and to the variations in supply voltages is important. Fig. 8 shows the simulated critical input voltages for the symmetric level converter plotted as a function of the threshold voltage variations of MOSFET's. The critical input voltage is defined as the input voltage which gives $V_{OUT} = V_{CC}/2$. The V_T's for both n- and p-channel MOSFET's are shifted in the same direction to give the worst case. The V_T variations affect the hysteresis window as well as the critical input voltages. This is due to the imbalance in conductance between the transmission gate and the feedback transistor. The worst variation of the critical input voltage ranges from 0.88 to 1.11 V within ΔV_T of ±0.1 V. This result is almost identical to that of the conventional CMOS inverter. The critical voltages are hardly affected by V_{CL} or V_{SL} fluctuation, but vary with V_{CC} or ground-level fluctuation. The critical input voltage variation is about half that of the V_{CC} or ground-level fluctuation. This is because the critical voltage is mainly determined by the transfer characteristics of the CMOS inverter, which is affected by the V_{CC} and ground-level fluctuation. These results imply that a sufficient operating margin can be maintained if the threshold voltages are controlled within ±0.1 V, and if the V_{CC} or ground fluctuation is suppressed below 0.2 V, 10% of the supply voltage.

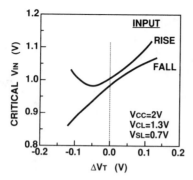

Fig. 8. Simulated critical input voltages versus threshold voltage variations.

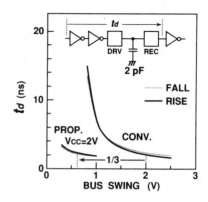

Fig. 10. Simulated transmission delay versus bus signal swing.

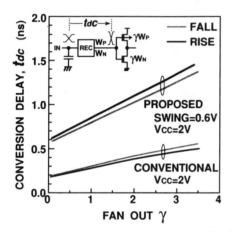

Fig. 9. Simulated level conversion delay versus fan-out.

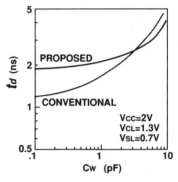

Fig. 11. Simulated transmission delay versus wiring capacitance of bus.

Another issue for the bus receiver design is the increase in conversion delay due to a reduced output drive current of the level converter. This is because the drive current of the level converter is determined by the effective gate voltages of the output drive MOSFET's, $V_{CL} - V_{TN}$ for n-channel and $V_{SL} - V_{CC}$ for p-channel MOSFET's, and the effective gate voltages are small compared with that of the conventional CMOS inverter operated with V_{CC}. Thus, a careful design is needed to reduce the fan-out as much as possible. Fig. 9 shows the simulated level conversion delay plotted against fan-out. Fan-out, γ, is defined here as the transistor width ratio of the next stage inverter to the output inverter of the level converter. The conversion delay is defined as the delay from the input (50%) to the receiver output (50%). The input transition time is 1 ns for both fall and rise cases. The conversion delay is almost three times that of the conventional CMOS inverter. However, the delay time difference can be minimized to about 0.4 ns through output loading reduction. Moreover, the increased level conversion delay can be compensated for by the reduction of the bus delay as will be explained in the next section.

C. Performance

The overall transmission delay is simulated for the proposed and the conventional architecture, as shown in Fig. 10. The transmission delay includes: 1) the delay for the two inverters that drive the gate of the driver, 2) the bus driver delay and the bus delay, and 3) the receiver delay. The wiring capacitance, C_W, is 2 pF. The delay for the conventional scheme increases drastically when V_{CC} is less than 2 V. On the other hand, the delay for the proposed scheme with a 0.6-V swing is 2.3 ns, which is comparable to that for the conventional scheme with $V_{CC} = 2$ V. The simulated consuming current was 489 μA for the conventional scheme and 165 μA for the proposed scheme for the operating frequency of 100 MHz. Thus, the signal swing and the power can be reduced by about 1/3, while maintaining a high speed, without an increase in the standby current.

The effect of the wiring capacitance on the transmission delay is simulated as shown in Fig. 11. With the proposed scheme, the delay time is less sensitive to C_W. This is because the ratio of the bus delay to the total transmission delay is small compared with the conventional scheme. This indicates that the proposed scheme is suitable for data transmission through heavily capacitive wiring.

The effect of wiring resistance on the transmission delay and the bus delay is simulated as shown in Fig. 12. For both the proposed and conventional schemes, the delay time increases when the resistance is larger than a few hundred ohms, and the wiring RC dominates when the resistance is larger than 1 kΩ. The proposed scheme is more sensitive to resistance increases than the conventional scheme for the same reason as the dependence on wiring capacitance. Thus, the advantage of the proposed scheme can be maintained, if the wiring RC is smaller than the bus delay shown in Fig. 5.

IV. EXPERIMENTAL RESULTS

To evaluate the feasibility of the proposed scheme, test circuits with a 30-stage chain of receivers and drivers were

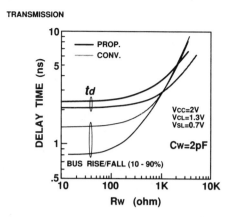

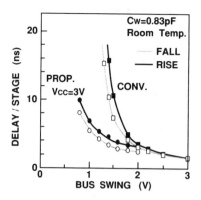

Fig. 12. Simulated transmission and bus delays versus wiring resistance of bus.

Fig. 14. Measured delay versus bus signal swing.

CONVENTIONAL

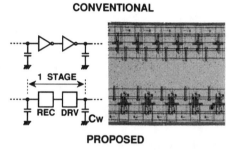

PROPOSED

Fig. 13. Microphotograph of the test circuits.

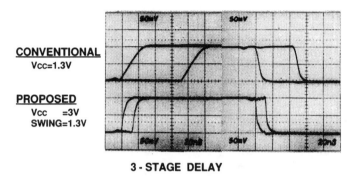

3 - STAGE DELAY

Fig. 15. Source-follower output waveforms for driver output.

designed and fabricated using 0.30-μm CMOS processes. A microphotograph of the test circuits is shown in Fig. 13. A dummy wiring capacitance of 0.83 pF was loaded on each driver output. The area penalty due to this proposed architecture is small, especially in the case of heavy loading, because the area is mainly determined by that of the driver, which is the same for both schemes. The external power supplies are used for V_{CL} and V_{SL}.

Fig. 14 shows the measured delay per stage against the bus swing. The supply voltage was 3 V for the proposed scheme. The minimum swing to achieve a 10-ns delay was 0.8 V for the proposed scheme, but it was 1.5 V for the conventional scheme. This improvement is not as much as was expected from the simulation results. This is because the threshold voltages were 0.5 V larger than the designed value, and low-V_T MOSFET's were not formed. Thus, further improvements are expected by device optimization.

Fig. 15 shows the operating waveforms of the test circuits, corresponding to a three-stage delay. The bus signal was 1.3 V for both cases. Fast bus rise and fall times were obtained for the proposed scheme as expected from the simulation.

These experimental results confirm the feasibility of the proposed bus architecture.

V. CONCLUSION

A new bus architecture is proposed for reducing the operating power of future ULSI's. This architecture will relieve the constraint of the conventional supply voltage scaling. It can efficiently reduce the bus swing by employing a source offset bus driver and symmetric level converter. The source offset driver employs low-V_T MOSFET's and an internal supply voltage corresponding to the reduced signal swing. Bus delay is almost halved with this driver when operated at 0.6-V swing and 2-V supply. The symmetric level converter consists of a transmission gate and a cross-coupled latch circuit on each half, and enables fast level conversion without increasing the standby current. The combination of new bus driver and symmetric level converter can reduce the bus swing by 1/3 that of the conventional architecture while maintaining a high-speed transmission and a low standby current. The operation of the proposed architecture was verified through test circuits fabricated using 0.3-μm processes, and further improvements in the speed performance are expected by device optimization.

ACKNOWLEDGMENT

The authors wish to thank K. Shimohigashi, E. Takeda, and M. Ishihara for their helpful suggestions They would also like to thank T. Kaga, T. Kisu, Y. Kawamoto, T. Nishida, and the fabrication staff in the laboratory for the device fabrication.

REFERENCES

[1] A. P. Chandrakasan *et al.*, "Low-power CMOS digital design," *IEEE J. Solid-State Circuits,* vol. 27, pp. 473–484, Apr. 1992.
[2] J. R. Burns, "Switching response of complementary-symmetry MOS transistor logic circuits," *RCA Rev.,* pp. 627–661, Dec. 1964.
[3] Y. Nakagome *et al.*, "Sub-1-V swing bus architecture for future low power ULSIs," in *Symp. VLSI Circuits Dig. Tech. Papers,* June 1992, pp. 82–83.
[4] M. Fukuma, "Limitations on MOS ULSIs," in *Symp. VLSI Technology Dig. Tech. Papers,* June 1987, pp. 7–8.
[5] H. J. Shin *et al.*, "A 250-Mbit-s CMOS crosspoint switch," *IEEE J. Solid-State Circuits,* vol. 24, pp. 478–486, Apr. 1989.

[6] H. B. Bakoglu *et al.,* "New CMOS driver and receiver circuits reduce interconnection propagation delays," in *Symp. VLSI Technology Dig. Tech. Papers,* May 1985, pp. 54–55.

[7] M. Horiguchi *et al.,* "A tunable CMOS DRAM voltage limiter with stabilized feedback amplifier," *IEEE J. Solid-State Circuits,* vol. 25, pp. 1129–1135, Oct. 1990.

[8] Y. Nakagome *et al.,* "An experimental 1.5-V 64-Mb DRAM," *IEEE J. Solid-State Circuits,* vol. 26, pp. 465–472, Apr. 1991.

Data-Dependent Logic Swing Internal Bus Architecture for Ultralow-Power LSI's

Mitsuru Hiraki, *Member, IEEE,* Hirotsugu Kojima, Hitoshi Misawa, Takashi Akazawa, and Yuji Hatano

Abstract—A reduced-swing internal bus scheme is proposed for achieving ultralow-power LSI's. The proposed data-dependent logic swing bus (DDL bus) uses charge sharing between bus wires and an additional bus wire to reduce its voltage swing. With this technique, the power dissipation of the proposed bus is reduced to 1/16 that of a conventional bus when used for a 16-b-wide bus. In addition, a dual-reference sense-amplifying receiver (DRSA receiver) has been developed to convert a reduced-swing bus signal to a CMOS-level signal without loss of noise margin or speed. Experimental circuits fabricated using 0.5-μm CMOS process verify the low-power operation of the proposed bus at an operating frequency of 40 MHz with a supply voltage of 3.3 V.

I. INTRODUCTION

RECENTLY, the demand for low-power LSI's has been increasing rapidly [1]–[5] with the increasing popularity of many kinds of battery-operated portable systems, such as notebook computers and digital cellular telephones, because increasing the battery lifetime is extremely important for these portable systems. To achieve low-power LSI's, reducing bus power dissipation is one of the most efficient ways because a large portion of chip power, in general, is dissipated to charge and discharge bus wires which usually comprise numerous bits of long wires, and thus are inevitably accompanied by large wiring capacitance. To reduce bus power dissipation, a reduced-swing low-power bus architecture has been proposed [6]. However, the low-power benefit of this scheme is considerably offset by the dc power consumed in the internal supply voltage system essential to its architecture.

This paper describes a new reduced-swing internal bus architecture which achieves extremely low-power operation [7], with no dc power consumption within any of our bus architecture. In our bus scheme, the voltage swing is dramatically reduced by charge sharing between bus wires and an additional bus wire that we call a dummy ground. Deterioration of noise margin has to be avoided while reducing the voltage swing of the bus signals. To meet this requirement, we also developed a sense-amplifying receiver which is practically immune to noise. In Section II, the large amount of power dissipated in a conventional bus is briefly discussed to explain our motive for this research, and the concept of our reduced-swing bus architecture is presented. Some simulation results

Manuscript received August 26, 1994; revised December 12, 1994.

M. Hiraki and Y. Hatano are with the Central Research Laboratory, Hitachi, Ltd., Tokyo 185 Japan.

H. Kojima is with Hitachi America Ltd., R&D, San Jose, CA 95134 USA.

H. Misawa is with Hitachi Microcomputer Systems, Ltd., Yamanashi 400-01 Japan.

T. Akazawa is with the Semiconductor and Integrated Circuits Division, Hitachi, Ltd., Kodaira 187 Japan.

IEEE Log Number 9409856.

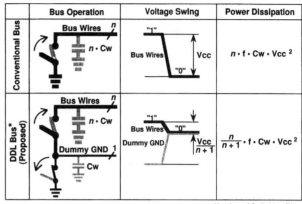

Fig. 1. Basic concept of the conventional bus and the proposed bus (DDL bus).

for this bus scheme are shown in Section III. Section IV presents a receiver circuit technique that we developed for use in our bus architecture to convert a reduced-swing bus signal to a CMOS-level signal. Section V describes some measured results obtained with experimental circuits. Section VI is dedicated to the conclusion.

II. CONCEPT OF DATA-DEPENDENT LOGIC SWING BUS

The concept of the bus scheme that we propose is illustrated in Fig. 1. A conventional bus scheme is also shown for comparison and for use in explaining how a conventional bus dissipates a large amount of power.

In a conventional bus, the voltage swing of bus signals coincides with the supply voltage V_{cc}. Thus, the bus power dissipation required to switch n bits of the bus signals is given by

$$P = n \cdot f \cdot C_w \cdot V_{cc}^2 \tag{1}$$

where f and C_w are the switching frequency and wiring capacitance, respectively. Since increasing the number of switching bits n causes a proportional increase in power dissipation, and buses generally include numerous bits of bus wires (16 b, 32 b, or 64 b for example), the conventional bus dissipates a considerably larger amount of power.

In our bus scheme, the voltage swing is greatly reduced by using an additional bus wire that we call a dummy ground (denoted as Dummy GND in Fig. 1). This dummy ground is initially discharged to the real ground level and

Reprinted from *IEEE Journal of Solid-State Circuits,* pp. 397-401, April 1995.

then immediately isolated from the ground. The charge of bus wiring capacitance is discharged to the dummy ground instead of the real ground. When n bits of the bus signals switch from "1" (V_{cc} level) to "0," the voltage swing is reduced to

$$V_{swing} = V_{cc}/(n+1) \qquad (2)$$

because the charge initially held by the wiring capacitance of the n bus wires $n \cdot C_w \cdot V_{cc}$ is shared by the wiring capacitance $(n+1) \cdot C_w$ (i.e., the wiring capacitance of both the n bus wires and the dummy ground). Thus, the bus power dissipation required to switch n bits of the bus signals is given by

$$\begin{aligned} P &= n \cdot f \cdot C_w \cdot V_{swing} \cdot V_{cc} \\ &= n/(n+1) \cdot f \cdot C_w \cdot V_{cc}^2. \end{aligned} \qquad (3)$$

The voltage swing $V_{cc}/(n+1)$ is not fixed, but is further reduced as the number of switching bits n increases. Therefore, no matter how many bits are switched, the total bus power dissipation $n/(n+1) \cdot f \cdot C_w \cdot V_{cc}^2$ does not exceed $1 \cdot f \cdot C_w \cdot V_{cc}^2$, which is the amount needed for a conventional bus to switch a single bit. No dc power is consumed within any of the proposed bus system. Since the logic swing changes according to the number of switching bits, i.e. depending on the data contents that are transmitted by the bus, the proposed bus scheme is called a data-dependent logic swing bus (DDL bus).

Fig. 2 shows the architecture and timing charts of the DDL bus. D_{in}, BD, and D_{out} denote the CMOS-level input to the bus, reduced-swing bus signal, and CMOS-level output from the bus, respectively. $D_{in}[1:N]$ expresses N-bit data composed of $D_{in}[1], D_{in}[2], \cdots, D_{in}[N]$, where N stands for the data bit width. The bus operation in each cycle is as follows. Before a cycle starts, BD is precharged to the supply voltage, and the dummy ground is predischarged to the ground level, both with the precharge control signal denoted as $\overline{\Phi PRE}$. When the driver enable signal $\overline{\Phi DE}$ falls, the drivers are enabled and the bits of BD are either discharged to the dummy ground or stay at the supply voltage, depending on each driver input. At the falling edge of the receiver enable signal $\overline{\Phi RE}$, receivers are enabled and convert the reduced-swing bus signals BD into CMOS-level outputs D_{out}.

Since the voltage swing of BD is greatly reduced, we have to be very careful about noise. To prevent deterioration of the noise margin, we use a differential amplifying scheme for the receiver. As a reference signal, differential amplifiers usually use an intermediate level between "1" and "0." In the DDL bus, however, it is difficult to generate this intermediate because the "0" level of BD changes over a wide voltage range. To solve this problem, we developed a receiver which internally creates a virtual intermediate using the "1" and "0" level references that are denoted as "1"Ref and "0"Ref. These two references are easily provided by extending the bus width by two bits, one of which simply stays at the supply voltage (thus, no driver is needed) to provide "1"Ref, and the other of which has its driver input fixed at "0" to provide "0"Ref. Since this receiver uses two references, it is called a dual-reference sense-amplifying receiver (DRSA receiver). Section IV covers the DRSA receiver in more detail.

The "1" level of BD is susceptible to crosstalk from its adjacent bits because it is floating. The "1" bit undergoes the

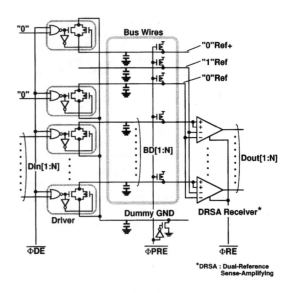

(a)

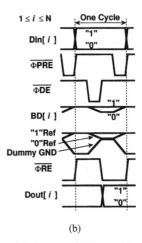

(b)

Fig. 2. Architecture and timing of the DDL bus. (a) Architecture. (b) Timing.

largest crosstalk when both its upper and lower adjacent bits switch from "1" to "0." To compensate for this crosstalk, therefore, the equivalent crosstalk is intentionally given to "1"Ref by laying out the bus wire of "1"Ref between those of the two "0" level reference bits ("0"Ref and "0"Ref+). "0"Ref+ is provided in exactly the same way as "0"Ref.

III. SIMULATED RESULTS FOR DDL BUS

Fig. 3 shows simulated waveforms of 16-b DDL bus signals with a supply voltage of 3.3 V. 0.5-μm device parameters were used in this simulation. The number of switching bits in BD is changed from 0 to 16. The total number of switching bits is increased by 2 b because "0"Ref and "0"Ref+ switch in every cycle. Thus, the total number of switching bits changes from 2 to 18. As the number of switching bits in BD increases from 0 to 16, the voltage swing is reduced from 1.1 V to 0.18 V. This intuitively explains how the power dissipation is kept small irrespective of the number of switching bits. For example, even if all 16 b in BD are switched, the power dissipation of the DDL bus is small because the voltage swing

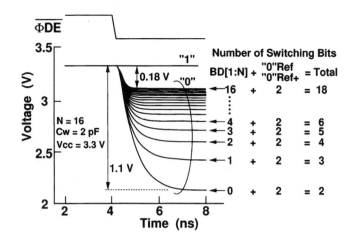

Fig. 3. Simulated waveforms of a DDL bus signal.

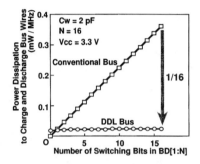

Fig. 4. Simulated power dissipation to charge and discharge bus wires.

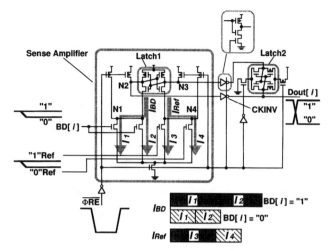

Fig. 5. Circuit diagram of the DRSA receiver.

is reduced to as small as 0.18 V. The power dissipation is small also in the case that the voltage swing increases up to 1.1 V because there are only 2 switching bits ("0"Ref and "0"Ref+).

To verify the low-power characteristics of the DDL bus, the power dissipation needed to charge and discharge the bus wires was simulated for both the DDL bus and a conventional precharge-type bus. The results shown in Fig. 4 assume a data bit width of 16 b. The simulation results show that the maximum power dissipation of the DDL bus is reduced to as small as 1/16 that of the conventional bus. This is because the DDL bus switches all bits of the bus signals while consuming only about the same amount of power as the conventional bus needs to switch a one-bit signal.

IV. RECEIVER CIRCUIT TECHNIQUE

The circuit diagram of the DRSA receiver is shown in Fig. 5. This receiver converts the reduced-swing bus signal BD to the CMOS-level output D_{out} by referring to the "1" and "0" level references ("1"Ref and "0"Ref). The receiver operates as follows. When the receiver enable signal $\overline{\Phi RE}$ is high, nodes $N1$, $N2$, $N3$, and $N4$ are precharged to the supply voltage. When $\overline{\Phi RE}$ falls, the sense amplifier is activated, and the logic value of BD is converted into the current signal difference between I_{BD} and I_{Ref}. Bar charts are provided for clarification. The current I_{Ref} is the sum of I_3 and I_4, one of which is large and the other of which is small, corresponding to "1"Ref and "0"Ref, respectively. On the other hand, the

current I_{BD} is the sum of I_1 and I_2, both of which are either large and small, depending on the logic value of BD. Thus, the current I_{BD} becomes larger or smaller than the current I_{Ref} depending on BD. Latch1 senses this current signal difference and immediately converts it into complementary CMOS-level signals. One of these signals is output through the clocked inverter CKINV which is enabled when $\overline{\Phi RE}$ is low. When $\overline{\Phi RE}$ rises again, Latch2 is activated to hold the output. Static power dissipation is avoided in the DRSA receiver because Latch1 cuts off the current paths (I_{BD} and I_{Ref}) of the sense amplifier.

Simulated performance of the DRSA receiver was compared with that of a conventional receiver. As the conventional receiver, we chose a master-slave flip-flop, a widely used receiver in CMOS logic design.

Fig. 6 shows the simulated fan out dependence of the receiver delays for the DRSA receiver and the conventional receiver in both the rise and fall cases. The receiver delay is defined as the delay from the receiver activation time to the time when the receiver output is determined. The fan out is defined as the ratio of the gate capacitance driven by the receiver to the gate capacitance of the final stage inverter of the receiver. The simulation results show that the DRSA receiver is nearly as fast as the conventional receiver. This may be rather surprising because the speed of the DRSA receiver should be slowed by the level-conversion overhead caused by its sense amplifier. We analyzed the critical paths of the receivers and found that this overhead is considerably relieved for two reasons, which are explained in Fig. 7. The first reason is that the critical paths of the DRSA receiver are shorter by one stage than those of the conventional receiver. In the conventional receiver, the critical paths include 3 or 4 stages, while there are only 2 or 3 stages in the DRSA receiver, including the sense amplifier which is expressed with a symbol here. The second reason is that the critical path of the DRSA receiver does not pass through the sense amplifier when the receiver output falls. This is because the sense amplifier output denoted as $\overline{OUT}$ does not change but simply stays at the precharged level in this case.

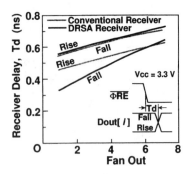

Fig. 6. Simulated receiver delay.

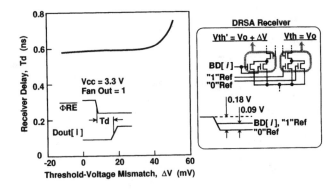

Fig. 9. Simulated dependence of receiver delay on threshold-voltage mismatch.

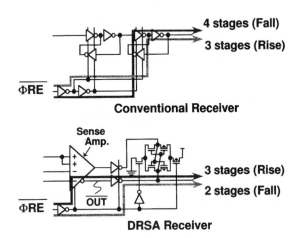

Fig. 7. Critical paths of the DRSA and conventional receivers.

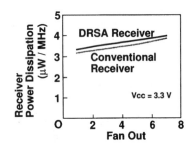

Fig. 8. Simulated receiver power dissipation.

Fig. 8 shows simulated fan out dependence of power dissipation for the DRSA receiver and conventional receiver. The DRSA receiver dissipates almost as little power as the conventional one. This is because dc current paths are completely excluded from the DRSA receiver, as in the conventional receiver.

Since the DRSA receiver amplifies a very small voltage difference, threshold-voltage mismatch of the MOSFET's used in the receiver circuit is one of the most crucial design issues. Fig. 9 shows the simulated dependence of the DRSA receiver delay on the threshold-voltage mismatch. In this simulation, the mismatch ΔV was given to the threshold voltage of the two n-channel MOSFET's that receive the bus signal BD. The voltage swing of the bus signal was assumed to

be 0.18 V, which is the minimum voltage swing for a 16-b DDL bus operating with a supply voltage of 3.3 V. The "1" level of the bus signal was assumed to be pulled down by 0.09 V, as measured with an experimental circuit, in worst case crosstalk. The simulation results show that the threshold-voltage mismatch has almost no effect on the receiver delay as long as the mismatch is smaller than 40 mV. We think this is a sufficiently large operating margin because threshold-voltage mismatch within a small area, such as a basic cell, is typically reduced to about 10 mV [8].

When the threshold-voltage mismatch exceeds 40 mV, however, the receiver delay increases significantly. This is because the receiver output flips to the wrong value when the mismatch exceeds half the input voltage difference (90 mV × 1/2 = 45 mV in this case). The potential for such a fatal error increases if we implement a 32-b or 64-b DDL bus by simply extending the bus width because the minimum voltage swing of the bus signal is reduced even further than that of the 16-b DDL bus. However, we can keep the same operating margin as that of a 16-b DDL bus if we use double or quadruple 16-b DDL buses as a 32-b or 64-b DDL bus, respectively.

V. EXPERIMENTAL RESULTS

Experimental circuits of a 16-b DDL bus and a conventional precharge-type bus were fabricated using 0.5-μm CMOS technology with a double-metal process. The experimental circuits consist of drivers, receivers, bus wires, precharging/predischarging circuits, and small logic circuits to control the other components. The bus wires were laid out with a pitch of 2.4 μm, where the width of the wires was 1.35 μm and the space between the wires was 1.05 μm, for both the first and second metal. A photomicrograph of a test chip is shown in Fig. 10. The bus wires were 21.6-mm long. The first metal was used for the majority of the bus wires.

With these experimental circuits, operation of the 16-b DDL bus was verified at an operating frequency of 40 MHz with a supply voltage of 3.3 V. Fig. 11 shows measured waveforms of the control signals ($\overline{\Phi PRE}$, $\overline{\Phi DE}$, and $\overline{\Phi RE}$), the twelfth-bit CMOS-level input, the twelfth-bit reduced-swing bus signal, the "0" level reference, the dummy ground, and the twelfth-bit CMOS-level output, when the input data in Cycles 1, 2, 3, 4, and 5 are 0000, 0010, AAAA, FFFF, and FFEF (hexadecimal

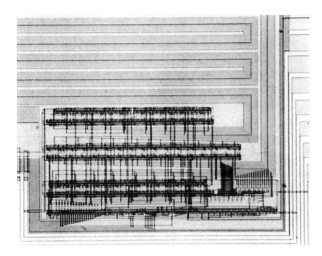

Fig. 10. Photomicrograph of the test chip.

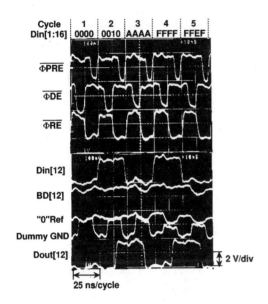

Fig. 11. Measured waveforms for the DDL bus (@V_{cc} = 3.3 V).

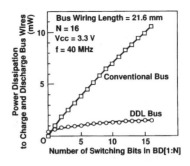

Fig. 12. Measured power dissipation to charge and discharge bus wires.

TABLE I
COMPARISON OF CELL AREA

	DDL bus	Conventional bus	DDL/conventional
Driver	918 μm^2	810 μm^2	1.13
Receiver	1593 μm^2	1269 μm^2	1.26

expression). In Cycles 4 and 5, the "0"Ref and Dummy GND are not completely equalized by the time when the receiver is activated because the voltage swings of the "0"Ref are comparatively large. This, however, does not lead to a fatal error because the "0" level of the bus signals (*BD* [12] in Cycle 5, for example) is pulled down nearly to the same level as that of "0"Ref even in these cases.

Fig. 12 compares the measured power dissipation for the 16-b DDL bus and the conventional bus operating at a frequency of 40 MHz with a supply voltage of 3.3 V. The measured results show that the power dissipation of the DDL bus is reduced to about 1/8 that of the conventional bus. The deviation from the simulated results, which showed that the power dissipation of the DDL bus was reduced to 1/16 that of the conventional bus (Fig. 4), results from the voltage swing increase due to the effect of the wiring capacitance between the Dummy GND and its adjacent bus wires (see Appendix). Therefore, this deviation can be minimized by laying out the Dummy GND such that comparatively large

space is given only between the Dummy GND and its adjacent wires (doubling their layout pitch, for example).

The cell areas of the drivers and receivers used in the experimental circuits are summarized in Table I. The cell area of the DDL bus driver is 13% larger than that of the conventional bus driver and the cell area of the DDL bus receiver (DRSA receiver) is 26% larger than that of the conventional receiver. This is because the driver and receiver of the DDL bus need more devices than those of the conventional bus. However, this area increase is not serious since the layout area increase is very small compared with a typical chip size. In our estimation, the penalty of chip size increase is less than 0.5%.

VI. CONCLUSION

We have proposed an extremely low-power bus architecture called the data-dependent logic swing bus or DDL bus. The key to low-power operation is the reduced-swing bus scheme that uses the dummy ground. With this scheme, power dissipation is reduced to 1/16 that of a conventional bus when used for 16-b-wide buses. In addition, a newly developed dual-reference sense-amplifying receiver or DRSA receiver is used to convert the reduced-swing bus signal of the DDL bus into CMOS-level output. The DRSA receiver uses a differential amplifying scheme that is essential in preventing noise margin deterioration. The DRSA receiver is almost as fast as a conventional CMOS receiver because its level-conversion overhead is relieved by its shorter critical paths. 40-MHz operation of the experimental DDL bus was verified at a supply voltage of 3.3 V. These simulated and measured results together demonstrate that the DDL bus scheme is a very promising technique for realizing ultralow-power LSI's.

APPENDIX

Since the capacitance between the Dummy GND and its adjacent wires is charged and discharged in every cycle, it increases the effective wiring capacitance of the Dummy GND. For the other bus wires, on the other hand, the effect of the capacitance between adjacent wires is relieved as the number of switching bits is increased. For example, when all bits of *BD* switch from "1" to "0," no capacitance between the wires is charged or discharged except in the most and least significant bits. Thus, the increase in the effective wiring capacitance of these bus wires can be neglected. When the effective wiring capacitance of the Dummy GND is increased by α times ($\alpha > 1$), the voltage swing of the bus signals which was given by (2) is increased nearly by α times as follows.

$$V_{\text{swing}} = V_{\text{cc}} \cdot \alpha/(n + \alpha). \tag{4}$$

This results in the increase nearly by α times in the power dissipation compared with the ideal case of $\alpha = 1$. The value of α is estimated to be increased to about 2 in this test chip because the Dummy GND and its adjacent wires were laid out with the minimum pitch of 2.4 μm.

ACKNOWLEDGMENT

The authors wish to thank M. Hotta for his support and encouragement, T. Noguchi, M. Tonomura, M. Itoh, J. Kimura, H. Suzuki, and M. Kokubo for their useful discussions. We are greatly indebted to M. Otsuka, K. Akama, and S. Kato for their layout design of the test chip.

REFERENCES

[1] Y. Nakagome *et al.*, "An experimental 1.5-V 64-Mbit DRAM," *IEEE J. Solid-State Circuits*, vol. 26, pp. 465–472, Apr. 1991.

[2] M. Muller, "The ARM600 processor and FPA," in *HOT Chips IV Symp. Record*, Aug. 1992, pp. 3.3.1–3.3.11.

[3] K. Ueda *et al.*, "A 16b low-power-consumption digital signal processor," in *ISSCC Dig. Tech. Papers*, Feb. 1993, pp. 28–29.

[4] A. Chandrakasan *et al.*, "A low power chipset for portable multimedia applications," in *ISSCC Dig. Tech. Papers*, Feb. 1994, pp. 82–83.

[5] T. Matsuura *et al.*, "1.2-V mixed analog/digital circuits using 0.3-μm CMOS LSI technology," in *ISSCC Dig. Tech. Papers*, Feb. 1994, pp. 250–251.

[6] Y. Nakagome *et al.*, "Sub-1-V swing internal bus architecture for future low-power ULSI's," in *Symp. VLSI Circuits Dig. Tech. Papers*, June 1992, pp. 82–83.

[7] M. Hiraki *et al.*, "Data-dependent logic swing internal bus architecture for ultra-low-power LSI's," in *Symp. VLSI Circuits Dig. Tech. Papers*, June 1994, pp. 29–30.

[8] K. Ishibashi *et al.*, "A 12.5-ns 16-Mb CMOS SRAM with common-centroid-geometry-layout sense amplifiers," *IEEE J. Solid-State Circuits*, vol. 29, pp. 411–418, Apr. 1994.

An Asymptotically Zero Power Charge-Recycling Bus Architecture for Battery-Operated Ultrahigh Data Rate ULSI's

Hiroyuki Yamauchi, Hironori Akamatsu, and Tsutomu Fujita

Abstract—An asymptotically zero power charge recycling bus (CRB) architecture, featuring virtual stacking of the individual bus-capacitance into a series configuration between supply voltage and ground, has been proposed. This CRB architecture makes it possible to reduce not only each bus-swing but also a total equivalent bus-capacitance of the ultramultibit buses running in parallel. The voltage swing of each bus is given by the recycled charge-supplying from the upper adjacent bus capacitance, instead of the power line. The dramatical power reduction was verified by the simulated and measured data. According to these data, the ultrahigh data rate of 25.6 Gb/s can be achieved while maintaining the power dissipation to be less than 100 mW, which corresponds to less than 10% of the previously reported 0.9 V suppressed bus-swing scheme, at $V_{cc} = 3.6$ V for the bus width of 512 b with the bus-capacitance of 14 pF per bit operating at 50 MHz.

I. INTRODUCTION

THE ultrahigh data rate of 25 Gb/s and beyond is one of the most important design requirements in realizing the future ULSI's for super high-definition (HD) moving pictures and three-dimensional (3-D) computer graphics applications in consumer electronics and handheld personal equipments, such as palmtop PC's and portable multimedia access supporting full-motion digital video. The most effective means to achieve such a data rate is to employ a large number of buses, corresponding to the parallelism of N-bit, interconnecting the embedded memory, the graphics controller, etc., on a ULSI system chip, instead of increasing operating frequency as shown in Fig. 1. For example, even at an operating frequency of 50 MHz, parallel buses of more than 512 b are required to achieve the ultrahigh data rate of 25 Gb/s and beyond as shown in Fig. 2. However, a drastic increase of the power dissipation which is direct propotion to the number of bus width N is inevitable due to the increased total bus-capacitance $(N \bullet C_{\text{bus}})$, where C_{bus} is the bus-capacitance of each bus. Even if suppressing the bus-swing voltage of V_{bus} from 3.6 V down to 0.9 V in order to reduce the charging and discharging amount [1], there still remains a bus-power dissipation of a far above 300 mW at V_{cc} of 3.6 V for the bus width of 512 b with bus-capacitance of 14 pF per bit operating at 50 MHz, which is intolerable for battery operation as shown in Fig. 2. This is because of the restrictions on battery-life, battery

Manuscript received September 22, 1994; revised December 12, 1994.
The authors are with the Semiconductor Research Center, Matsushita Electric Industrial Co., Ltd., Osaka, Japan.
IEEE Log Number 9409859.

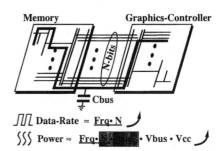

Fig. 1. Background for this low-power bus architecture.

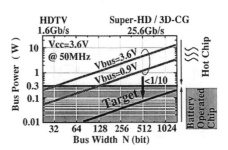

Fig. 2. Target on bus-power consumption for this work.

size and weight for requirements of adequate portability [2]. This paper proposes a complete Charge-Recycling Bus (CRB) architecture that can reduce the bus-power dissipation to less than 10% of the conventional suppressed bus-swing scheme while realizing the data rate of 25 Gb/s [3]. This asymptotically zero power CRB architecture features virtual stacking of the individual bus-capacitance into a series configuration between supply voltage and ground. It enables to reduce not only each bus-swing but also a total equivalent bus-capacitance of the ultramultibit buses running in parallel. The voltage swing of each bus is given by the recycled charge-supplying from the upper adjacent bus capacitance, instead of the power line. According to the simulated and measured data of the test chip fabricated by using 0.5 μm CMOS technology, the ultrahigh data rate of 25.6 Gb/s can be accomplished while maintaining the power dissipation to be less than 100 mW at $V_{cc} = 3.6$ V for the bus width of 512 b with the bus-capacitance of 14 pF per bit operating at 50 MHz.

In the following section, the concept of CRB architecture is explained. In Section III, the principle of charge-recycle operation in CRB is discussed. The circuit configuration of CRB is described in Section IV. The circuit operation and

Reprinted from *IEEE Journal of Solid-State Circuits*, pp. 423–431, April 1995.

performance are demonstrated in Section V. Conclusions are given in Section VI.

II. CONCEPT OF CHARGE RECYCLING BUS (CRB) ARCHITECTURE

As mentioned in Section I, to achieve the ultrahigh data rate of 25 Gb/s and beyond, parallel buses of more than 512 b are required, even at an operating frequency of 50 MHz as shown in Fig. 2. Thus, in order to reduce the charging and discharging amount, suppressing the bus-swing voltage of V_{bus} down to less than 1 V is necessary. However, even if using conventional suppressed bus-swing ($V_{\text{bus}} = 0.9$ V) scheme coupled with a down converter, the bus power dissipation cannot be reduced adequately for battery operation. This is because there still remain a contribution from the total bus-capacitance ($N \bullet C_{\text{bus}}$), where N and C bus are the number of total bus pairs and the capacitance of the individual bus, respectively.

Before the charge recycling bus (CRB) architecture is conceptually described, what causes wasteful-power dissipation in the conventional scheme is discussed by comparing the conventional two types of full-swing and suppressed-swing bus schemes.

A. Coventional Bus Scheme

Schematics of conventional "full-swing" and "suppressed swing" bus scheme are shown in Fig. 3(a) and (b), respectively. In these figures, the operating waveforms of the three-parallel complementary bus pairs in the three cycles are illustrated. Each operating waveform in one cycle is divided into two periods. One half of the cycle is for the equalizing and the precharging at the midpoint of the swing voltage of each complementary data bus pair, and another is for the developing of the complemetary bus-swing voltage and establishing of the transferred data.

In this paper, unless otherwise noted, the conventional "full-swing" bus scheme shown in Fig. 3(a) features the following two bus controls: 1) half-level precharging in the former half of the cycle; and 2) developing from the 1/2 V_{cc} to the "High" (V_{cc}) or "Low" (V_{ss}) level in the latter half of the cycle. This bus control can save the bus power by a half of that compared with a output bus power of static CMOS logic gate undergoing of the "High" to "Low" CMOS level transition.

In Fig. 3(b), the two simbols of resistance between V_{cc} and V_{bus} and between V_{bus} and V_{ss} represent the turn-on-resistance of MOSFET's composing the output-driver and the reference voltage generator, respectively, in the voltage down-converter. This voltage down-converter including reference voltage generator basically functions as a resistance potential divider suppressing the bus swing voltage from V_{cc} down to V_{bus}. However, this scheme induces a large amount of the power-loss within the resistance. For example, that value corresponds to 75% of the total power consumption, when the V_{bus} is 0.9 V for $V_{\text{cc}} = 3.6$ V. This is because of the Joul-Heat power loss between the V_{cc} and the V_{bus}. In addition, dc idling current to V_{ss} (I_{dc}) also induce a wasteful dc power-loss as shown in Fig. 3(b).

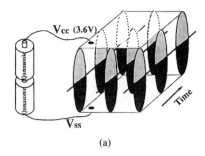

(a)

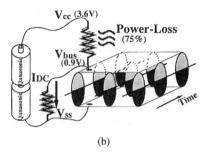

(b)

Fig. 3. Comparison of conventional bus schemes. (a) Full-swing bus scheme. (b) Suppressed-swing bus scheme with on-chip voltage down converter.

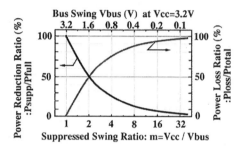

Fig. 4. Ratios of power reduction and power loss as a function of suppressed-swing ratio (m).

As mentioned before, the suppressed bus swing voltage of V_{bus} is a key factor in reducing the bus-power consumption. For example, the power reduction ratio of the suppressed bus scheme to the full-swing bus scheme ($P_{\text{supp}}/P_{\text{full}}$) can be reduced with the suppressed bus swing ratio m ($V_{\text{cc}}/V_{\text{bus}}$) in Fig. 4. However, the power loss ratio of the Joul-Heat power to the total bus power ($P_{\text{loss}}/P_{\text{total}}$) increases with the m ($V_{\text{cc}}/V_{\text{bus}}$). For the $m = 16$, corresponding to the $V_{\text{bus}} = 0.2$ V for the $V_{\text{cc}} = 3.2$ V, the ($P_{\text{loss}}/P_{\text{total}}$) reaches over 90%. This result implies that a new suppressed bus scheme is required for realizing a dramatic bus power reduction to an asymptotically zero compared with the conventional one, without a resistance voltage divider inducing a Joul-Heat power loss.

B. Charge Recycling Bus (CRB) Architecture

The proposed approach to low-power bus design features the CRB architecture, making it possible to meet both of the following two key bus challenges simultaneously.

1) The first key point is the suppressed bus swing scheme using the capacitance-voltage-divider, which no longer requires wasteful dc idling current and Joul-Heat power

288

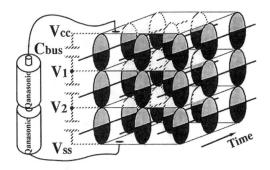

Fig. 5. Concept of charge-recycling bus architecture.

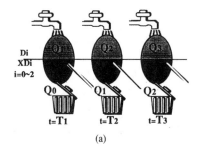

(a)

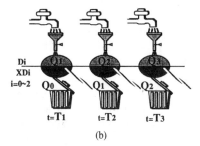

(b)

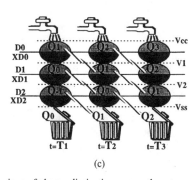

(c)

Fig. 6. Comparison of charge-dissipation among three types of bus schemes. (a) Full-swing bus scheme. (b) Suppressed-swing bus scheme with on-chip down converter. (c) Charge-recycling bus scheme.

loss, unlike conventional resistance voltage divider, such as voltage down converter.

2) Another key point is the charge-recycling [3], [4] among the multibit buses running in parallel in every cycle, that is, the used charge for establishing of high level in each bus is never dumped to ground every time.

The key concept of the CRB architecture is "virtual stacking" of the individual bus-capacitance C_{bus} into a series configuration between V_{cc} and V_{ss}, in order to reduce not only the bus-swing but also the total equivalent bus-capacitance. "Virtual stacking" means the following points.

1) In fact, the individual bus-capacitance is mainly composed of the bus wiring capacitance to substrate and the junction capacitance of the MOSFET's that is connected to the bus. Thus, each bus capacitance C_{bus} cannot be directly connected in series configuration as shown in Fig. 5.

2) However, a simple bus control, featuring the stepwise precharging of each bus pair and interbus charge sharing among the stepwise precharged adjacent bus capacitances in the former and the latter half of the clock cycle, respectively, makes it possible to realize the "virtual stacking" of each bus capacitance. The conceptual description is illustrated in Fig. 8 and the detail discussion about that is made in Section III.

The point in favor of this bus architecture is that the resistance voltage divider is no longer necessary to realize the suppressed small bus swing. This is because that this virtual stacking of the individual bus capacitance C_{bus} can play a role of virtual capacitance voltage divider, just like a three capacitors connected in series between V_{cc} and V_{ss} as shown in Fig. 5. Basically, each stepwise-divided voltage $V1$ and $V2$ is determined only by the ratio of the distributed bus capacitance values of C_{bus}. This is because that an establishing of the high or low level of each bus pair is given by the charge sharing with the adjacent capacitance, which is precharged to one-step higher or lower than own potential, respectively.

C. Dissipated Charge Amount Comparison

To compare the charge dissipation among the conventional full-swing and suppressed bus scheme and the proposed charge-recycling bus architecture, the illustrations of operating waveforms for three parallel complementary bus pairs ($Di, XDi, \ i = 0, 1, 2$) in three cycles of $T1$, $T2$, and $T3$ are shown in Fig. 6, emphasizing on charge supplying from the power line and charge dumping to ground. In these illustrations, let water-faucet represents the power supplying line of V_{cc} and let garbage-can represents ground line of V_{ss}. Funnel with valve represents the voltage down converter making it possible to reduce the charging and discharging amount by suppressing the bus swing voltage as shown in Fig. 6(b).

The important point in the conventional bus scheme shown in Fig. 6(a) and (b) is that every bus undergoing of transition from the precharged level of midpoint to high level or low level (that is to say, charging and discharging) in every cycle, the new charge supplying of $Q1$, $Q2$, and $Q3$ are required at $T1$, $T2$, and $T3$, respectively, and the used charge, after only one playing a role of high-level establishing in the former cycle, is thrown away to the ground line just like garbage. This implies that every time charge supplying from the power source line is required to transmit every one bit in every cycle, just like $Q1$, $Q2$, and $Q3$ shown in Fig. 6. Thus, for example, in order to transmit the 25 Gb/s, an intolerable large amount of charging and discharging is needed in the conventional bus architecture shown in Fig. 6(a) and (b).

On the other hand, in the case of the charge recycling bus (CRB) architecture shown in Fig. 6(c), from the top data bus

TABLE I
POWER COMPARISON AMONG THREE TYPES AS FOLLOWS.
1) CONVENTIONAL FULL-$V_{cc-swing}$. 2) CONVENTIONAL
SUPPRESSED $V_{cc/m}$ SWING. 3) M-STACKED CRB ARCHITECTURE

	Conv. full	Conv. suppr.	CRBm
Total-Dissipated Q_{total}	$m \cdot Ci \cdot Vcc$	$m \cdot Ci \cdot Vcc /m$	$C_1 \cdot Vcc /m$
	all for (1 ~ m-th)	all for (1 ~ m-th)	only for top (1-st)
	full-Vcc-swing	1/m-Vcc-swing	1/m-Vcc-swing
Norm. Power	1	$1 / m$	$1 / m^2$

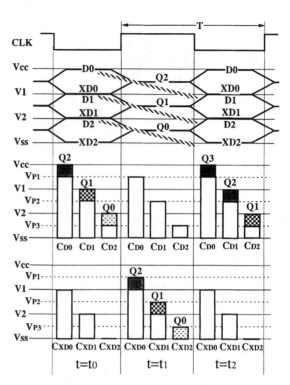

Fig. 7. Concept of charge-recycling operation among the stacking bus capacitance.

pair $(D0, XD0)$ through the bottom data bus pair $(D2, XD2)$, are virtually connected in series between V_{cc} and V_{ss}. Thus, for example, the intermediate bus pair $(D1, XD1)$ and the bottom bus pair $(D2, XD2)$ never connect to the power line directly. Instead of the power line, each bus pair except for the top bus pair recycle the used-charge stored on the adjacent upper bus capaicitance pair, in order to undergo each high-level transition. For example, the charge $Q1$ can be recycled again and again by rolling down to the lower adjacent bus pair in every cycle, $T2$ and $T3$, until this $Q1$ reaches the ground line as shown in Fig. 6(c). As a result, regarding the parallel transmission of the three bits in the cycle of $T2$, only one charge supplying of $Q2$ is needed for the top bus pair and the other two bits can be transmitted by using the recycled charge $Q1$ and $Q0$ for the intermediate bus pair $(D1, XD1)$ and the bottom bus pair $(D2, XD2)$, respectively, where $Q0$ represents the charge supplied from the power line at the one cycle before the cycle of $T1$. This implies that additional charge supplying is no longer necessary for stacking bus pairs except for the top bus pair, and an asymptotically zero power data transmission can be realized with the increasing number m of stacking bus pairs. This is because the bus swing voltage of the top bus pair can be reduced with the increasing number of the stacking bus pairs and the amount of only one charge supplying can be suppressed to the negligible small value (that is asymptotically zero) compared with that of the conventional full swing bus scheme.

Table I shows the power comparison based on the dissipated charge amount among the following three types: 1) conventional full-swing of V_{cc} scheme (Conv. full); 2) conventional suppressed swing of $V_{cc/m}$ scheme (Conv. suppr.); and 3) the CRB architecture with stacking bus number of m (CRB_m).

Regarding the total amount of the dissipated charge for the CRB architecture, except for the charge supplying for the $C1$ of the first top data bus pair, a new charge supplying is no longer necessary for the other bus pairs as mentioned before repeatedly. In addition, the bus swing voltage is suppressed to the $1/m$ of V_{cc}. Thus, the total dissipated charge (Q total) can be expressed as shown in Table I. On the other hand, for the conventional two bus schemes, all parallel bus pairs numbering from the first through the mth require the new charge supplying simultaneously. In addition, the bus swing voltage is full-swing of V_{cc} for the full swing bus scheme, and the $1/m$ of V_{cc} for the suppressed swing bus scheme. Therefore, the Q total for each conventional one can be shown by each expression in Table I. According

to these expressions, the power reduction to $1/m^2$ of that compared with the conventional full swing scheme can be realized by using the CRB architecture. Even comparing with the conventional suppressed swing ($V_{cc/m}$) scheme, the CRB scheme can reduce the bus power dissipation to $1/m$ of the conventional one.

III. PRINCIPLE OF THE CRB OPERATION

In this section, why the CRB control makes it possible to realize the suppressed small bus swing and charge recycling among the parallel bus pairs, simultaneously, is discussed. In addition, how "virtual stacking" of the bus capacitance to substrate in series configuration between V_{cc} and V_{ss} can be provided, is described.

A. Charge-Recycling Operation Among the Stacking BusCapacitance

The charge recycling operation among the stacking of numerous bus capacitances in series configuration between V_{cc} and V_{ss} is discussed by using the illustration of behavior of the stored charge on each bus shown in Fig. 7.

Fig. 7 shows the behavior of the stored charge on each bus in the period of the time t_0, t_1, and t_2. The latter half of the cycle (t_0 and t_2) are for establishing of the complementary high and low levels of each bus pair. The former half of the cycle (t_1) is for equalizing and precharging of each bus pair. The C_{Di} and C_{XDi} of $i = 0, 1, 2$ shown in Fig. 7 represent the bus capacitance-name of the top bus pair $(D0, XD0)$, the intermediate bus pair $(D1, XD1)$, and the bottom bus pair $(D2, XD2)$, respectively.

The height of each bar-graph represents the charged potential level of each bus. For example, the potential values of C_{D0} and C_{XD0} are V_{cc} and $V1$ at t_0, respectively, and when the time becomes t_1, the potential values of those become the same precharged level V_{P1} of the top bus pair. When the time becomes t_2, the potential values of C_{D0} and C_{XD0} become $V1$ and V_{cc}, respectively, where new charge $Q3$ added on the capacitance C_{XD0} is supplied from V_{cc} line, and the used charge $Q2$ added on the bus capacitance C_{D1} is transferred from the the upper adjacent bus capacitance C_{XD0}. In the same way, the recycled charge $Q1$ added on the bus capacitance C_{D2} is given from the upper adjacent bus capacitance C_{XD1}. The used charge $Q0$ stored on the bus capacitance C_{XD2} at t_1 is dumped to the ground line at t_2.

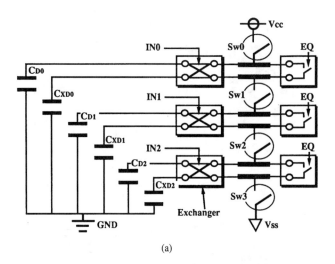

(a)

B. Concept of Bus Control for the CRB Architecture

The concept of bus control for CRB architecture are shown in Fig. 8. Schematic in Fig. 8(a) shows how the bus capacitance to substrate of the three bus pairs running in parallel (C_{Di} and C_{XDi} of $i = 0 \sim 2$) connect in series each other between V_{cc} and V_{ss}, where the input of the bus exchanger (INi of $i = 0 \sim 2$) controls either of each bus pair connects to either of the upper or lower adjacent bus pair. As mentioned before, in the former half of the clock cycle, each bus pair is equalized by the signal of EQ (shown in Fig. 8(a)) and each stepwise equalized potential level (5/6, 3/6, or 1/6 of V_{cc}) is retained as shown in Fig. 7. Fig. 8(b) shows the timing diagram of the CRB operation. In the latter half of the clock cycle, one of each bus pair connects to one of the adjacent bus pair by controlling of bus exchanger, depending on the input signal (INi of $i = 0 \sim 2$), for example, when INi of $i = 0 \sim 2$ becomes (0, 1, 1), one of the top bus pair (C_{XD0}) connects to V_{cc}, and another (C_{D0}) connects to one of the intermediate bus pair (C_{D1}) as shown in Fig. 8(b). The switches (SWi of $i = 0 \sim 2$) for connecting between adjacent bus pairs and between the top/bottom bus and V_{cc}/V_{ss} line, are turned on in the latter half of the clock cycle, and are turned off during the equalizing period.

The block diagram of the CRB architecture is shown in Fig. 9(a). The timing diagram and the truth table of the bus control signal are shown in Fig. 9(b).

Regarding the ith complementary bus pair in the CRB architecture, the input INi and the equalization signal EQ establish not only the output pair, Di and XDi, bus also the bus-level signals at nodes H and L as follows.

1) In the former half of the clock cycle, the EQ signal sychronized with the system clock equalize Di and XDi of the complementary bus-output pair, while the nodes of the bus-level signals H and L become open, that is Hi-Z state.

2) In the latter half of the clock cycle, the input signal INi switches higher (or lower) level of the bus-output pair Di and XDi to the node of the bus-level signal H (or L) according to the the truth table in Fig. 9(b).

The series connection of the numerous bus capacitance (C_{Di} and C_{XDi} of $i = 0 \sim (m - 1)$) is realized by joining the node H of the $(i + 1)$th bus pair to the node L of the ith

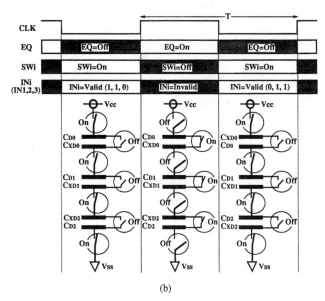

(b)

Fig. 8. Concept of charge-recycling bus (CRB) architecture. (a) Schematic of CRB. (b) Timing diagram of CRB operation.

bus pair. For example, when IN$i = 0$ and IN$i + 1 = 1$, the bus-output Di connects the nodes of L of the #i block, and the bus-output $Di + 1$ connects the nodes of H of the #$i + 1$ block. As a result, the Di and $Di+1$ are connected each other and become the same potential level, that is, the stored charge Qi and $Qi + 1$ on the capacitances (C_{Di} and C_{Di+1}) of the bus Di and $Di + 1$, respectively, are shared between the two. If the values of the both capacitances are equal, the Di and $Di + 1$ undergo the low and high transition synmetrically.

As a result, by using the above mentioned bus control and configuration, the bus capacitance to substrate of the three bus pairs running in parallel (C_{Di} and C_{XDi} of $i = 0 \sim 2$) can be virtually connected in series between V_{cc} and V_{ss} as shown in Fig. 8(b). This bus architecture performs charge-recycling between the upper and lower adjacent bus pairs in every cycle and the equivalent bus capacitance of the parallel operating buses can be reduced to $1/m^2$ of that compared with full swing bus scheme, where m is the number of the buses running in parallel. In addition, this charge-recycling bus architecture play a role of virtual capacitance voltage divider making it

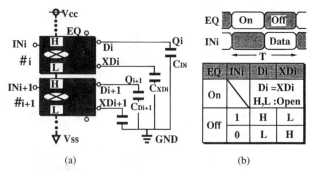

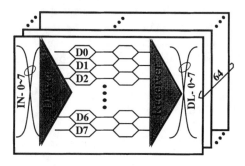

Fig. 9. Bus controle for CRB architecture. (a) Block diagram of CRB. (b) Timing diagram and truth-table.

Fig. 11. Bus configuration of CRB architecture for ultramultibit (512 b) buses.

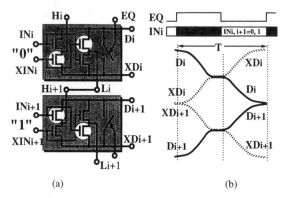

Fig. 10. Transistor-level circuit configuration of CRB architecture. (a) Circuit configuration of CRB. (b) Timing diagram and operating waveforms.

possible to realize the suppressed bus swing without using the conventional equivalent resistance voltage divider that may induce the intolerable dc idling current [5].

IV. CIRCUIT CONFIGURATION OF THE CRB

A. Transistor Level Circuit Configuration of CRB Driver

The transistor level circuit configuration of the CRB driver and their timing diagram and schematic operating waveforms are shown in Fig. 10.

This driver circuit consists of the bus switch and equalizer. The bus switch is composed of the two MOSFET pairs, and each gate electrode is connected to the complementary input pair ($\text{IN}i/\text{XIN}i$), respectively, and each common source of MOSFET pair is connected to the bus-level node (Hi or Li) as shown in Fig. 10(a).

Taking a look at the operating waveforms of the output pairs ($Di/XDi, Di + 1/XDi + 1$), in the former half of the clock cycle, the EQ signal is turned on and the intrabus pairs (Di/XDi and $Di + 1/XDi + 1$) are equalized and in the latter half of the clock cycle, the bus switches are controlled by the complementary input pair ($\text{IN}i/\text{XIN}i$ and $\text{IN}i + 1/\text{XIN} + 1$) as shown in Fig. 10(b). For example, when the input of the upper and lower driver circuits become "0" and "1," respectively, the four circled transistors are turned on, and the bus Di in the upper circuit and the bus $Di+1$ in the lower circuit are connected and equalized based on the charge sharing between each bus capacitance. This charge sharing between the Di and $XDi + 1$ of the upper and lower circuits makes it possible to establish the low level of the Di for the upper circuit and the high level of the $XDi + 1$ for the lower

circuit without a new charge supplying from the power line, simultaneously as shown in Fig. 10(b). This charge transfer and recycling from the upper to the lower bus capacitance can be realized for the all cases in this same way.

B. Bus Configuration for Ultramultibit Buses

The bus configuration of the CRB architecture with ultra-multibits buses, such as 512 b bus pairs, is shown in Fig. 11.

Parallel buses of 512 b are divided into 64 blocks. One block includes the driver and receiver circuit for the bus wirings for 8 b, numbering from $D0$ through $D7$. The bus driver suppress the bus swing voltage to 1/8 of V_{cc} and the receiver amplifies the suppressed bus swing to a full swing of V_{cc}. The most important issue in realizing charge-recycling is prevention of wasteful charge dissipation, such as dc idling current to ground. To solve this, the input gate pairs of the upper and lower drivers are controlled by the signals of fulll-swing of V_{cc} ($\text{IN}i$ of $i = 0 \sim 7$) so that a pair of the input gates never turn on simultaneously [6].

C. CMOS Driver and Receiver Configurations

The relation between the gate types of the driver and receiver in the upper and lower circuits is complementary as shown in Figs. 12(a) and 13(a), that is, PMOS gate drivers and NMOS gate reciver are used for the upper bus pairs whose each operating bus potential is higher than $1/2$ V_{cc}, and NMOS gate drivers and PMOS gate receiver are used for the lower bus pairs whose each operating bus potential is lower than $1/2$ V_{cc}. The circuit configurations of NMOS and PMOS gate drivers and the complementary input and output signal waveforms for the NMOS and PMOS drivers are shown in Fig. 12(b) and 12(c), respectively.

In the case of PMOS driver, to prevent the input gate pair from turning-on simultaneously, the voltage of the input signal pair never undergo a transition of V_{cc} to V_{ss} simultaneously, that is, only one of the complementary signals retains the V_{cc} level constantly. In the equalizing period, both of the two signals set the V_{cc} level, and only one of the complementary input undergoes a high-to-low transition in the rest of the cycle. For the NMOS driver, only one of the input pair maintain the V_{ss} level every time, and after equalizing period, only one input signal goes high level of V_{cc}.

The circuit configurations and the complementary input and output signal waveforms for the NMOS and PMOS receivers are shown in Fig. 13(b) and (c), respectively. The

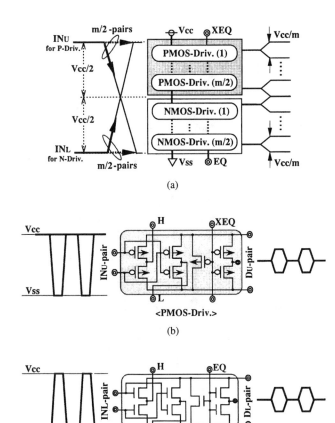

(a)

(b)

<PMOS-Driv.>

(c)

<NMOS-Driv.>

Fig. 12. Transistor-level circuit configuration of CRB architecture. (a) CMOS driver configuration. (b) PMOS driver. (c) NMOS driver.

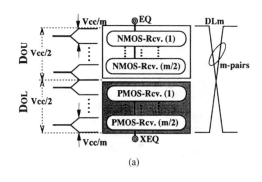

(a)

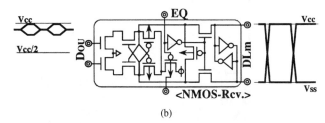

<NMOS-Rcv.>

(b)

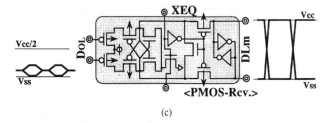

<PMOS-Rcv.>

(c)

Fig. 13. Transistor-level circuit configuration of CRB architecture. (a) CMOS receiver configuration. (b) PMOS receiver. (c) NMOS receiver.

bus receiver is required for stable amplifications and low-power operations over the wide input-voltage range of V_{cc}. To meet this requirement, the receiver circuit is composed of a dynamically latched type current sense amplifier and a gate-receiver that functions as a voltage to current converter. The CMOS configuration in the upper and lower receiver circuits can provide a fast operation by obtaining of a larger V_{GS} more than $V_{cc}/2$ for both types of the circuits. In addition, since the parasitic capacitance value of the output node in the receiver circuit and the input gate capacitance value in the driver circuit are less than 5% of that compared to the bus capacitance of several pF, the total power consumption is hardly increased even if including the receiver's and driver's one.

V. CIRCUIT OPERATION AND PERFORMANCE

A. Simulated Results and Discussions of the CRB Operation

The simulated operating waveforms of the 8 b stacking buses, numbering from $D0$ through $D7$, are shown in Fig. 14(a). When the operating voltage V_{cc} is 3.3 V, the swing voltage of each bus is automatically fixed to about 0.4 V as shown in Fig. 14(a). Stable data transfer can be achieved at the operating frequency of 50 MHz. The important concern in this architecture is how the precharged potential of each bus pair is determined between V_{cc} and V_{ss}. Basically, each precharged potential is determined only by the distributed bus capacitance values. When the capacitance values of all buses are equaled, a

difference of each precharged potential is distributed equally, just like a staircase.

This is because potential transition of high-to-low or low-to-high in each bus is undergone by only the charge sharing between the adjacent bus capacitance precharged at a different potential, except for each one of the top and bottom bus pairs connecting V_{cc} and V_{ss}, respectively. Needless to say, when the "power-on state," that is, when the potential of power supply line is changed from V_{ss} to V_{cc}, several dummy clock cycles are required for achieving the above mentioned stable precharge operation. This is because it takes several clock cycles corresponding to the number of stacking bus pairs, for the stored charge on the top bus capacitance supplied from the power line, to be transferred from the top bus pair down to the bottom bus pair.

When introducing the CRB architecture to the practical chip is considered, several ways can overcome the above mentioned "power-on" problem (for example, the simple resistance voltage divider that is activated only in the power-on state can supply the suitable initial potential for each bus).

Another important concern in this architecture is how the charge sharing and transferring speed between both adjacent bus pairs is distributed. Basically, the required time for charge sharing and tranferring from the upper to the lower bus capacitance is depended on the drive-ability of the bus driver MOSFET. The speed differences, among the top bus pair $D0$ through the intermediate bus pair $D3$ for PMOS type driver and among the intermediate bus pair $D4$ through the

293

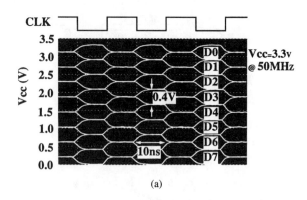

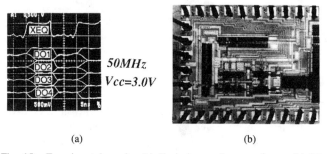

(a)

(b)

Fig. 14. Simulated results. (a) 50 MHz Operating waveforms. (b) Comparison of operating current waveforms.

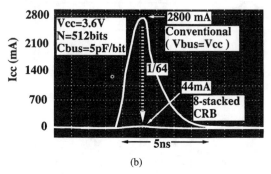

(a)   (b)

Fig. 15. Experimental results. (a) Typical operating waveforms. (b) Microphotograph of CRB test-site.

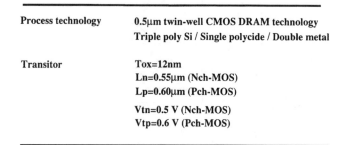

TABLE II
DEVICE CHARACTERISTICS

Process technology	0.5μm twin-well CMOS DRAM technology
	Triple poly Si / Single polycide / Double metal
Transitor	Tox=12nm
	Ln=0.55μm (Nch-MOS)
	Lp=0.60μm (Pch-MOS)
	Vtn=0.5 V (Nch-MOS)
	Vtp=0.6 V (Pch-MOS)

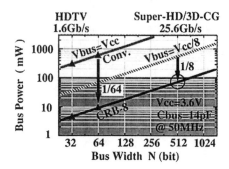

Fig. 16. Comparison of bus-power consumption between conventional and CRB.

bottom bus $D7$ pair for NMOS driver are suppressed to less than 2 ns, which is negligibly small compared to 10 ns of half of the clock cycle corresponding to 50 MHz as shown in Fig. 14(a). This speed difference is caused by the body effect (threshold voltage V_t increasing with voltage of source to substrate potential) of each driver MOSFET. For example, regarding the PMOS driver, source potential of each driver MOSFET is as follows: 1) $V_s = V_{cc}$ for $D0$ bus pair driver; and 2) $V_s = 5/8 \, V_{cc}$ for $D3$ bus pair driver. Thus, the threshold voltage difference between the two becomes about 100 mV for $V_{cc} = 3.3$ V, and results in inducing of the speed difference of less than 2 ns.

Next, operating current amount comparison between the conventional full-swing bus scheme and the CRB architecture with stacked bus structure every 8 b are given in Fig. 14(b). The condition parameter is assumed as follows: 1) $V_{cc} = 3.6$ V; 2) the total operating bus number $N = 512$ b; 3) bus capacitance per bit $C_{bus} = 5$ pF; and 4) stacking bus number for CRB architecture $m = 8$.

The peak current of about 2800 mA (2.8 A) for the conventional full-swing bus scheme can be reduced to 44 mA

by using the CRB architecture, that corresponds to 1/64 of the conventional one. This is because that regarding the 8-stacked parallel bits in each block of 64 blocks, zero power data transmission can be realized based on the charge-recycling bus architecture, except for the only one bit of the top data bus pair. In other words, the CRB architecture can make the most of the "hot" new charge supplied from power line in each block, without waste charge dissipation, such as every time charge dumping to ground.

B. Measured Results and Discussion

The measured typical operating waveforms of the 4-stacked CRB ($m = 4$)is shown in Fig. 15(a). In addition, the microphotograph of the test-site of the CRB circuits is given Fig. 15(b). This test-site was fabricated by using the 0.5 μm CMOS DRAM technology for a fabrication convenience. The features of process technolgy and transistor parameter such as threshold voltage and channel length, etc, are shown in Table II.

A stable precharging and developing of each bus pair can be observed at the operating frequency of 50 MHz and $V_{cc} = 3.0$ V. Each precharged potential and each swing voltage can be achieved constantly, just same as simulated results.

C. Comparison of Bus Power Consumption

In Fig. 16, the comparison of the bus power consumption among the three types bus scheme of the conventional full-swing bus, the suppressed swing bus, and the CRB, are shown as a fuction of number of the bus width.

The most upper line show for the conventional full-swing bus scheme with V_{bus} of full V_{cc} (3.6 V). The intermediate line show the bus power consumption for the conventional suppressed bus scheme with $V_{\text{bus}} = 1/8 \ V_{\text{cc}}$ (0.45 V). The lower line represent the dissipated bus power of the 8-stacked CRB ($m = 8$) architecture. The measured data are denoted by using triangular marks.

This CRB scheme can reduce the bus power to 1/64 and 1/8 of that compared with conventional full swing and suppressed bus scheme, respectively. According to these data, the ultra-high data rate of 25 Gb/s can be achieved, while maintaining the power dissipation to be less than 100 mW at V_{cc} of 3.6 V, where the bus width is 512 b and the bus capacitance is 14 pF/b, and the operating frequency is 50 MHz. The proposed CRB architecture is the most promising candidate for interconnecting the embedded memory and graphics controller, etc., on battery-operated ULSI's system chip for super HD applications.

VI. CONCLUSION

A CRB architecture, featuring virtual stacking of the individual bus-capacitance into a series configuration between supply voltage and ground, has been proposed. This can reduce not only each bus-swing but also a total equivalent bus-capacitance of the ultramultibit buses running in parallel. The charging and discharging of each bus are based on the charge-sharing among the upper and lower adjacent bus capacitance equivalently connected in series, instead of the power line. According to the simulated and measured data, the data rate of 25.6 Gb/s can be realized while maintaining the power dissipation to be less than 100 mW at $V_{\text{cc}} = 3.6$ V for the bus width of 512 b with the bus-capacitance of 14 pF per bit operating at 50 MHz.

After the several problems, such as the noise issue and the CAD issue in introducing to the practical ULSI's, have been overcome, this proposed architecture will become the most promising candidate for interconnecting the embedded memory and graphics controller, etc., on battery-operated ULSI's, making it possible to realize future personal communications services (PCS's) applications with universal portable multimedia access supporting full motion high-definition digital video.

ACKNOWLEDGMENT

The authors would like to thank the other personnel of Semiconductor Research Center for their dedicated contribution to this project. Also, the authors are grateful to Dr. T. Takemoto, Dr. M. Inoue and Dr. S. Hashimoto for their continuous support and encouragement.

REFERENCES

[1] Y. Nakagome et al., "Sub-1 V swing bus architecture for future low-power ULSI's," in Symp. VLSI Circuits Dig. Tech. Papers, pp. 29–30, June 1992.
[2] T. Bell, "Incredible shrinking computers," IEEE Spectrum, vol. 28, no. 5, pp. 37–43, May 1991.
[3] H. Yamauchi et al., "A low power complete charge-recycling bus architecture for ultra-high data rate ULSI's," in Symp. VLSI Circuits Dig. Tech. Papers, pp. 21–22, June 1994.
[4] T. Kawahara et al., "A charge recycle refresh for Gb-scale DRAM's in file applications," in Symp. VLSI Circuits Dig. Tech. Papers, pp. 41–42, May 1993.
[5] D. Takashima et al, "Low power on-chip supply voltage conversion scheme for 1G/4G bit DRAM's," in Symp. VLSI Circuits Dig. Tech. Papers, pp. 114–115, June 1992.
[6] H. Yamauchi et al., "A 20 ns battery-operated 16 Mb CMOS DRAM," in ISSCC Dig. Tech. Papers, pp. 44–45, Feb. 1993.

Bus-Invert Coding for Low-Power I/O

Mircea R. Stan, *Member, IEEE,* and Wayne P. Burleson, *Member, IEEE*

Abstract— Technology trends and especially portable applications drive the quest for low-power VLSI design. Solutions that involve algorithmic, structural or physical transformations are sought. The focus is on developing low-power circuits without affecting too much the performance (area, latency, period). For CMOS circuits most power is dissipated as dynamic power for charging and discharging node capacitances. This is why many promising results in low-power design are obtained by minimizing the number of transitions inside the CMOS circuit. While it is generally accepted that because of the large capacitances involved much of the power dissipated by an IC is at the I/O little has been specifically done for decreasing the I/O power dissipation.

We propose the *Bus-Invert* method of coding the I/O which lowers the bus activity and thus decreases the I/O peak power dissipation by 50% and the I/O average power dissipation by up to 25%. The method is general but applies best for dealing with buses. This is fortunate because buses are indeed most likely to have very large capacitances associated with them and consequently dissipate a lot of power.

Index Terms—low-power dissipation, CMOS VLSI, coding.

I. Introduction

CMOS VLSI is intrinsically a low-power technology [23]. When compared to TTL, ECL or GaAs at similar levels of integration the power dissipated by CMOS is several orders of magnitude lower. This is one of the major reasons for the widespread acceptance of CMOS in all kinds of applications from consumer appliances to some supercomputers. The power dissipated in a CMOS circuit can be classified as static power dissipation (overlap current and DC static) and dynamic power dissipation. The static power dissipated by a CMOS VLSI gate is in the nanowatt range [23], [9] and for the purpose of this paper will be ignored. The dynamic power dissipated by a CMOS circuit is of the form [23], [14]:

$$P_{chip} \propto \sum_{i=1}^{N} C_{load_i} \cdot V_{dd}^2 \cdot f \cdot p_{t_i} \qquad (1)$$

where the sum is done over all the N nodes of the circuit, C_{load_i} is the load capacitance at node i, V_{dd} is the power supply voltage, f is the frequency and p_{t_i} is the activity factor at node i.

The special interest in low-power CMOS design is relatively recent and is due to an increased interest in portable applications. For achieving low-power in CMOS circuits one or more of the terms V_{dd}, C_{load_i}, f and p_{t_i} must be minimized. Decreasing V_{dd} has a quadratic effect and is thus a very efficient way of decreasing the power dissipation [3]. Even then the decrease in power obtained by lower V_{dd} is not of several

Manuscript received April 26, 1993; revised April 15, 1994 and August 11, 1994. This work was supported in part by the NSF under Grant MIP-9 208267.

The authors are with the Department of Electrical and Computer Engineering, University of Massachusetts, Amherst, MA 01003 USA.

IEEE Log Number 9408370.

orders of magnitude as required by portable applications. This is why lowering V_{dd} must be done in conjunction with other methods for decreasing the power dissipation even further. Lowering the activity factor is a very promising way of further decreasing the power dissipation. It can be viewed at different *levels of abstraction* [3]:

- In the *behavioral* domain advances at the *algorithmic* level can dramatically decrease the number of transitions by lowering the number of steps necessary for a given computation [3].

- In the *structural* domain at the *logic* level there are approaches [18], [11] that consider the switching probabilities associated with different gate implementations. For random inputs the output of an AND gate is more likely to be 0 than 1 while the output of an OR gate is more likely to be 1 than 0. An XOR gate has equal probabilities for 0 or 1. By properly choosing the multilevel implementation of a boolean function the circuit can be optimized such that the total number of transitions is minimized.

- At the *logic* and *layout* levels the effect of *glitches* must be considered. Glitches represent unnecessary and unwanted transitions that contribute among other things to increased noise and power dissipation. Among general circuit topologies balanced-trees are less likely to generate glitches than chain topologies [4] when properly balanced also at the layout level.

The *peak* power dissipation is determined by the maximum instantaneous value of the activity factor while the *average* power is determined by the average activity factor. Decreasing the average power dissipation is generally the target of low-power techniques but the maximum activity factor is also important because of its relationship with the simultaneous switching noise and the resulting ground bounce [13]. Figs. 1 and 2 illustrate two cases that have the same average but very different maximum switching activities.

In Section II we propose *coding* as a method of decreasing the activity factor on buses. Section III explains the *Bus-Invert* method suitable for data buses while section IV considers coding on an address bus. Coding the I/O was proposed in [21], [16] as a noise reduction method. Codes were developed for reducing the I/O transmitted noise for both terminated and unterminated lines. Since transitions are a common cause in CMOS for both noise and power dissipation many of the methods used in [21], [16] can be applied for low-power design. This was independently done in [20]. There is also a patent [8] that proposes coding the I/O for reduced state changes. The *Bus-Invert* method can be characterized as an example of what in [21] is called "starvation coding" and in [20] is called "limited-weight coding".

Reprinted from *IEEE Transactions of VLSI Systems*, pp. 49-58, March 1995.

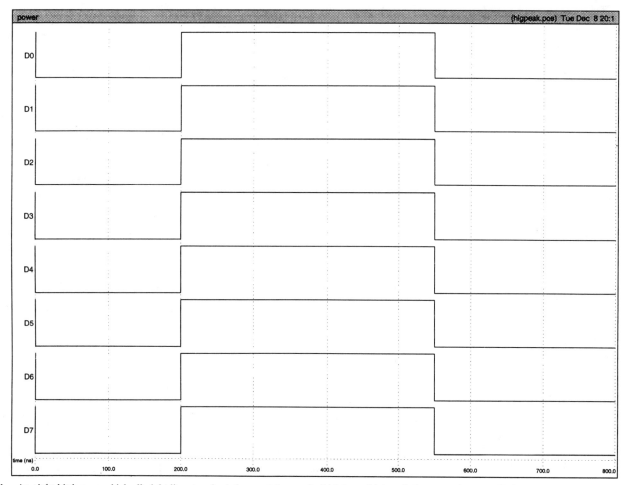

Fig. 1. An eight-bit bus on which all eight lines toggle at the same time and which has a high peak (worst-case) power dissipation. There are 16 transitions over 16 clock cycles (average 1 transition per clock cycle).

II. CODING FOR LOW-POWER I/O

The floorplan of a VLSI IC is being formed of the internal circuit surrounded by the I/O padframe. The power dissipated at the I/O can be as low as 10% [7] and as high as 80% [15] of the total power dissipation. For circuits optimized for low-power the power dissipated at the I/O is typically around 50% of the total [24]. This large I/O power dissipation is a consequence of the huge (compared with the internal circuit) dimensions of the devices in the I/O pads and of the external capacitances due to I/O pins, wires and connected circuits. The devices in the I/O pads need to be large in order to drive the large external capacitances and this further increases their own parasitic capacitances. The effect of the large capacitances is twofold: long delays (which lead to lower performance) and high-power dissipation. The usual way of addressing I/O pads performance and power dissipation is at the layout and circuit level [13]. In this paper we consider I/O power dissipation at a higher level of abstraction.

The research on low-power design has focused on the internal circuit [3]–[6], [11], [18] despite the fact that the I/O power dissipation is at least an important part if not the dominant factor in power dissipation [15]. One reason is that the I/O is considered somehow ''fixed'' or ''given.'' In high-level synthesis the I/O is likely to be a constraint on the design, part of the initial specification. The chip is designed such that it obeys a predefined I/O operation. When this ''black-box'' approach is taken there is little freedom to modify the I/O for low-power. We change that and look at coding in order to decrease the I/O activity.

In order to make (1) tractable a simplification that is generally made is to consider that each node has the same ''average'' load capacitance [18]. If V_{dd} and f are the same for all gates (1) becomes:

$$P_{chip} \propto C_{average} \cdot V_{dd}^2 \cdot f \cdot N(transitions) \qquad (2)$$

where $N(transitions)$ is the maximum (for peak power) or average (for average power) total number of transitions for the entire chip.

Considering the same ''average'' load capacitance for all the nodes can sometimes lead to wrong conclusions. Nodes with very large capacitances can dominate the total power dissipation. For example in [2] the power dissipation of different adders is first estimated with a simplified model and then actually measured on a fabricated chip. The estimation for the carry look-ahead adder was far from the actual measured value exactly because the influence of the large capacitance look-ahead tree was oversimplified. In this paper a simplified model is also used but this time with *two* different types of nodes: a small capacitance node typical for the internal circuit and a large capacitance node typical for the I/O. We also

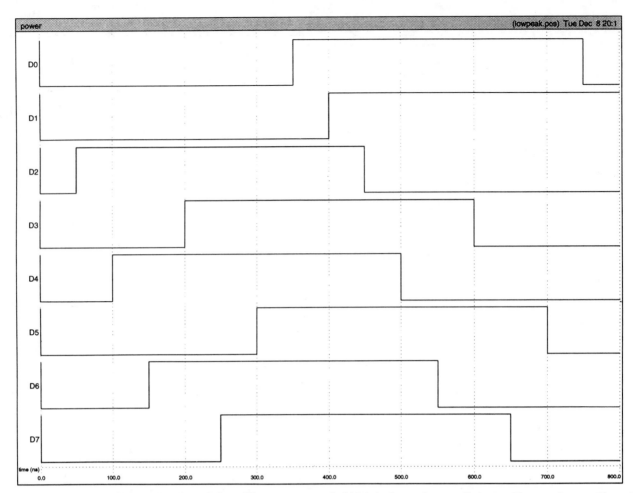

Fig. 2. An eight-bit bus on which the eight lines toggle at different moments and which has a low peak power dissipation. There are the same 16 transitions over 16 clock cycles and thus the same average power dissipation as in Fig. 1.

assume that the V_{dd} and f have already been optimized for low-power and thus the only way of further decreasing the power dissipation is to lower the number of transitions. This assumption is in the spirit of [18]. With these assumptions the total power can be written as a sum of the power dissipated by the internal circuit and the power dissipated at the I/O:

$$P_{chip} \propto C_{int} \cdot N(transitions)_{int} + C_{I/O} \cdot N(transitions)_{I/O}$$

The internal number of transitions $N(transitions)_{int}$ is generally much larger than $N(transitions)_{I/O}$ because the number of internal nodes is much larger than the number of I/O nodes, while C_{int} is generally much smaller than $C_{I/O}$. The total power dissipated by the chip is then the sum of two comparable terms:

- P_{int} (average or peak): the product of a small internal capacitance and a large (average or maximum) number of transitions and,
- $P_{I/O}$ (average or peak): the product of a large I/O capacitance and a small (average or maximum) number of transitions.

We are now ready to state the idea behind our proposed method of decreasing the power dissipation: Code the data in order to decrease the number of transitions on the large capacitance side (I/O) even at the expense of slightly increasing the number of transitions on the low capacitance side (internal circuit).

Intuitively, the total power dissipation will decrease by decreasing the number of transitions on the high capacitance side. Because the number of internal transitions is already large, increasing it by a comparatively small amount is likely to be insignificant. A numerical example will help illustrate this. Let's consider an 8-bit wide data bus on which coding is used in order to decrease power dissipation. By coding (see section III) the maximum bus activity can be reduced by 50% and the average bus activity by at most 25%. The increase in the internal number of transitions is of the order of $n \log n$ where n is the bus width (see section III-D). With the power dissipated at the I/O 50% of the total we'll consider: $V_{dd} = 1V$, $f = 1MHz$, $N(transitions)_{I/O} = 8$, $C_{I/O} = 2500$ and $N(transitions)_{int} = 2500$, $C_{I/O} = 8$ (a conservative estimate that the I/O capacitances are two orders of magnitude larger than the internal capacitances). The power dissipated in the uncoded case will then be: $P_{uncoded} = 2500 \cdot 8 + 8 \cdot 2500 = 40000$.

If the above $N(transitions)_{I/O} = 8$ is the *average* number of transitions by coding it can be reduced to 6 (25%). At the same time $N(transitions)_{int}$ will be increased with $n \log n = 8 \cdot 3 = 24$. The power dissipated becomes:

298

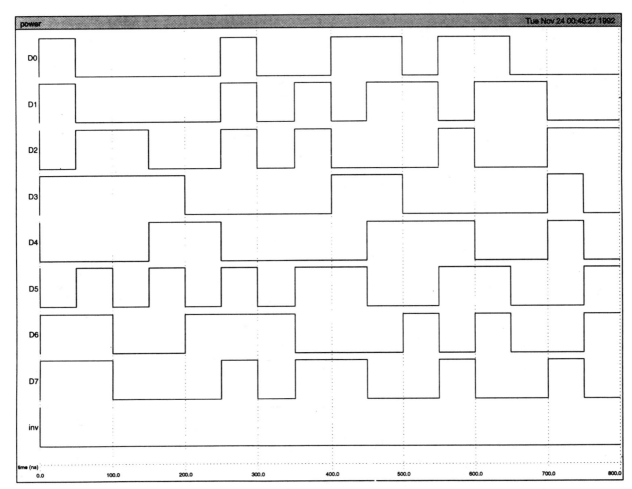

Fig. 3. A typical eight-bit synchronous data bus. The transitions between two consecutive time-slots are "clean." There are 64 transitions for a period of 16 time-slots. This represents an average of 4 transitions per time-slot, or 0.5 transitions per bus line per time-slot.

TABLE I
TYPICAL VALUES OF 33%, 50% OR 67% FOR THE PERCENTAGE OF I/O POWER
DISSIPATION FROM THE TOTAL POWER DISSIPATION ARE CONSIDERED. WITH
CODING THE MAXIMUM I/O POWER DISSIPATION CAN BE REDUCED BY HALF
AND THE AVERAGE I/O POWER DISSIPATION CAN BE REDUCED BY 25%

If the amount of I/O power dissipation from total is:	If the I/O maximum power is lowered to:	the total maximum power is lowered to:	If the I/O average power is lowered to:	the total average power is lowered to:
33%		84%		91%
50%	50%	75%	75%	88%
67%		67%		83%

$$P_{average} = 2500 \cdot 6 + 8 \cdot 2524 = 35192 \approx 88\% \text{ of } P_{uncoded}.$$

If $N(transitions)_{I/O} = 8$ is the *maximum* number of transitions by coding it can be reduced to 4 (50%). The increase in $N(transitions)_{int}$ will be the same $n \log n = 8 \cdot 3 = 24$. The maximum power dissipated after coding is:

$$P_{maximum} = 2500 \cdot 4 + 8 \cdot 2524 = 30192 \approx 75\% \text{ of } P_{uncoded}.$$

As can be seen the savings in I/O power dissipation translate almost entirely into savings at the chip level with very little penalty due to the increase in the internal activity factor.

Table I gives an idea of the approximate decrease in total power dissipation when the I/O number of transitions is cut by 50% (for maximum) or by 25% (for average) by considering some typical values for the percentage of I/O power dissipation from the total power.

III. POWER DISSIPATION ON A DATA BUS

Buses are a typical model for I/O communication. Furthermore buses are generally heavily loaded and consequently dissipate a lot of power. Glitches can be a serious cause of increased power dissipation because they represent unnecessary transitions [3], [5]. This is why we consider a synchronous bus model in which the transitions between two consecutive time-slots are "clean" (Fig. 3). Latches are needed on the bus-drivers side for such "clean" transitions.

We'll consider the activity on a typical data bus to be characterized by a *random uniformly distributed* sequence of values. Of course this is just an approximation, in general there is a degree of redundancy which can be exploited for example by lossless compression techniques [25]. Data compression is an efficient method to decrease power dissipation and by removing the extra redundancy the data can be accurately approximated as a random sequence. Even if compression is not used we believe that a random process is a reasonable approximation for the sequence on a data bus for our purpose. The assumption of random uniformly distributed inputs is also conveniently made by most of the statistical power estimation methods [18]. With this assumption for any given time-slot the data on an n-bit wide bus can be any of 2^n possible values with equal probability. The average number of transitions per time-

299

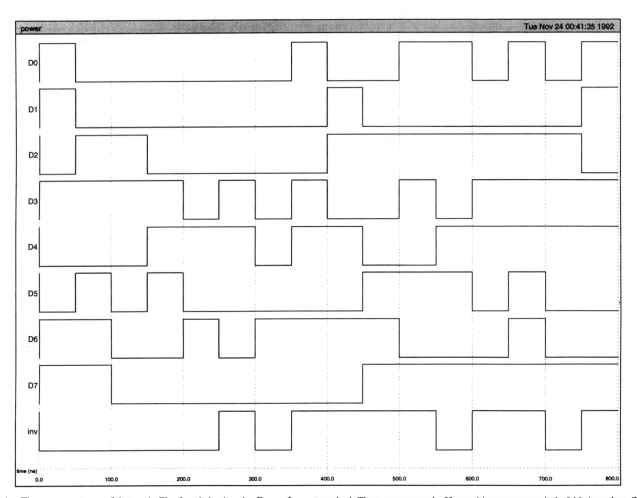

Fig. 4. The same sequence of data as in Fig. 3 coded using the $Bus-Invert$ method. There are now only 53 transitions over a period of 16 time-slots. This represents an average of 3.3 transitions per time-slot, or 0.41 transitions per bus line per time-slot. The maximum number of transitions for any time-slot is now 4.

slot will be $n/2$. For example on an eight-bit bus there will be an average of 4 transitions per time-slot or 0.5 transitions per bus-line per time-slot. Thus the average power dissipation for the I/O will be proportional with $n/2$. When all the bus-lines toggle at the same time (the probability of this happening in any time-slot is $1/2^n$) there will be a maximum of n transitions in a time-slot and thus the peak (worst-case) power dissipation is proportional with n. The *ground-bounce* is also worst in this case [13].

Fig. 3 shows a sequence of 16 time-slots for an 8-bit data bus. The values were pseudo-randomly generated and will be used further for illustration purposes:

$$D0 : 1000010011011000$$
$$D1 : 1000010101101100$$
$$D2 : 0110010100010011$$
$$D3 : 1111000011000010$$
$$D4 : 0001100001110010$$
$$D5 : 0101010110011001$$
$$D6 : 1100111000101001$$
$$D7 : 1100010110010010$$

The total number of transitions for these pseudo-random values for the 16 time-slots is 64 or an average of 4 transitions per time-slot, exactly as in the theoretical analysis. Because of the short time-frame the maximum number of transitions for

this example is only 6. The question is: can we do better? Can we decrease the number of transitions for the I/O while transferring the same amount of information? If we consider the I/O as something "fixed" or as a "constraint" on the design then there is nothing we can do. But if we consider the I/O as another design variable then the problem becomes a typical coding case: Code the I/O in order to transfer the same amount of information in the same amount of time but with a lower number of transitions and thus decrease the total peak and average power dissipation.

A. Bus-Invert—Coding for Low-Power on a Data Bus

Let's denote as the *data value* the piece of information that has to be transmitted over the bus in a given time-slot. Then the *bus value* will denote the coded value (the actual value on the bus). Typically a code needs extra *control bits*. The *Bus-Invert* method proposed here uses one extra control bit called *invert*. By convention then $invert = 0$ the bus value will equal the data value. When $invert = 1$ the bus value will be the *inverted* data value. The peak power dissipation can then be decreased by half by coding the I/O as follows (*Bus-Invert* method):

1) Compute the Hamming distance (the number of bits in which they differ) between the present *bus value* (also counting the present invert line) and the next *data value*.

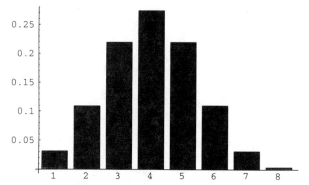

Fig. 5. The binomial distribution for the Hamming distances of the next data value. The maximum is at n/2=4.

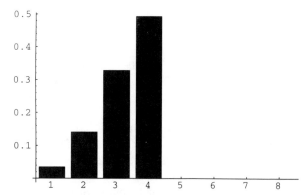

Fig. 6. The probability distribution for the Hamming distances of the next bus value for an 8-bit data bus encoded with the $Bus-Invert$ method. The transitions of the $invert$ signal are also counted.

2) If the Hamming distance is larger than $n/2$, set $invert = 1$ and make the next bus value equal to the inverted next data value.

3) Otherwise let $invert = 0$ and let the next bus value equal to the next data value.

4) At the receiver side the contents of the bus must be conditionally inverted according to the $invert$ line, unless the data is not stored encoded as it is (e.g., in a RAM). In any case the value of $invert$ must be transmitted over the bus (the method increases the number of bus lines from n to $n + 1$).

The *Bus-Invert* method generates a code that has the property that the *maximum* number of transitions per time-slot is reduced from n to $n/2$ and thus the peak power dissipation for the I/O is reduced by half. From the coding theory point of view the *Bus-Invert* code is a time-dependent Markovian code. Fig. 4 shows the same data sequence as Fig. 3 only this time using the *Bus-Invert* coding in order to decrease the number of transitions:

```
D0 : 1000000100110101
D1 : 1000000010000001
D2 : 0110000011111110
D3 : 1111010100101111
D4 : 0001110110011111
D5 : 0101000001110100
D6 : 1100101111000100
D7 : 1100000001111111
inv : 0000010111101101
```

While the maximum number of transitions is reduced by half the decrease in the *average* number of transitions is not as good. For an 8-bit bus for example the average number of transitions per time-slot by using the *Bus-Invert* coding becomes 3.27 (instead of 4), or 0.41 (instead of 0.5) transitions per bus-line per time-slot. This means that the *average* number of transitions is 81.8% as compared with an unencoded bus. There are two reasons why the performance of the *Bus-Invert* coding for decreasing the *average* number of transitions is not as good as for decreasing the *maximum* number of transitions:

- the *invert* line contributes itself with some transitions,
- the distribution of the Hamming distances for the next data values is *not* uniform.

Fig. 5 shows the distribution of the Hamming distance for the next *data value* with the probabilities that the next value

will differ from the present one in 1, 2, 3, 4, 5, 6, 7, or 8 b positions. It is a *binomial* distribution where the maximum probability is for a Hamming distance of 4 (and is equal to $\frac{C_8^4}{2^8} = 0.27$). Unfortunately 4 (or $n/2$ in the general case) is exactly the value for which there is no gain in inverting the contents of the bus and thus the coding is ineffective for that value.

The distribution of the Hamming distances for the next *bus value* when using the *Bus-Invert* method is shown in Fig. 6. The very large probability (0.5) at 4 means that it is likely that the next value on the bus will differ from the present one in 4 positions. This is why the average number of transitions per time-slot is large (3.27), and thus the decrease in average power dissipation is less than for peak power dissipation.

Until now the particular case of an 8-bit bus was considered. How does the method perform on other bus widths? It turns out that as the width of the bus n increases the term corresponding to $n/2$ in the binomial distribution becomes dominant and the decrease in average power dissipation becomes even smaller. For example for a 16-bit data bus the average number of transitions per time-slot becomes 6.8 (compared to 8 for an unencoded bus) for an average of 0.43 transitions per bus-line per time-slot, which is 85% of the unencoded case (compare with 82% for an 8-bit bus). For an even wider bus (e.g., 64) the decrease in average power dissipation gets even smaller. The decrease in peak power dissipation remains the same 50% for any n.

B. The Bus-Invert Method is Optimal

With the assumption of random distribution for the sequence of data the *Bus-Invert* method is optimal in the sense that given the same redundancy (1 extra bus line) no other coding can achieve better reduction in the number of transitions. This can be seen by considering all possible next values on the bus. The next value can be the same as the present one and this will correspond to no transitions on the bus. The next value can differ in only one position (one transition) and there will be $C_{n+1}^1 = n+1$ such possible next values (there will be $C_n^1 = n$ values with $invert = 0$ and another one (all 1s data value) with $invert = 1$). Similarly there will be C_{n+1}^2 values that will generate two transitions, C_{n+1}^3 values that will generate three transitions, up to $C_{n+1}^{n/2}$ values that will generate $n/2$ transitions (assume n even without loss of generality). As can

301

be seen:

$$C_{n+1}^0 + C_{n+1}^1 + C_{n+1}^2 + \cdots + C_{n+1}^{n/2} = 2^n$$

and all 2^n possible next values are taken into account. The average number of transitions per bus cycle will be:

$$\frac{C_{n+1}^0 \cdot 0 + C_{n+1}^1 \cdot 1 + C_{n+1}^2 \cdot 2 + \cdots + C_{n+1}^{n/2} \cdot n/2}{2^n}$$

It can be seen that the *Bus-Invert* method uses *all* the patterns with at most $n/2$ transitions. This is what in [20] is called a *perfect limited-weight code* and in [21] is called a *starvation code*. Any other code with 2^n codewords can either use a permutation of these same patterns and then will exhibit the same average bus activity or use patterns with more than $n/2$ transitions and generate a larger bus activity. From this point of view the *Bus-Invert* method is optimal and any other coding method must use more than one extra line in order to further decrease the bus activity.

C. Partitioned Code for Lower Average Power

In order to decrease the average I/O power dissipation for wide buses the observation that the *Bus-Invert* method performs better for small n can be used to partition the bus into several narrower *subbuses*. Each of these subbuses can then be coded independently with its own *invert* signal. For example a 64-bit bus could be partitioned into eight 8-bit subbuses with a total of eight *invert* signals. Because of the assumption that the data to be transferred over the wide bus is random uniformly distributed, the statistics for the narrower subbuses will be independent and the sequence of data for each subbus will be random uniformly distributed. For example for a 64-bit bus partitioned into eight 8-bit subbuses the average number of transitions per time-slot will be 26.16 (8 times 3.27, the average for one 8-bit subbus) and the average number of transitions per bus-line per time-slot will be 0.41 (as for an 8-bit bus with one *invert* line). The maximum number of transitions is not improved by partitioning the bus and remains the same $n/2$.

Partitioning in order to get a lower average number of transitions has the obvious drawback of needing the extra *invert* lines, but as was shown in section 3.2 *any* code that tries to improve on the *Bus-Invert* has to use extra bits. On the other hand a partitioned code is no longer *optimal* and there are other codes that using the same number of extra bits as a partitioned code can achieve a lower bus activity. These are the codes that systematically use patterns with m or less number of transitions where $m < \frac{n}{2}$ and which are called "*m*-limited weight codes" in [20]. Unfortunately the *algorithmic* generation of such codes is not well understood at this moment and using look-up tables in ROM as suggested in [21] is out of the question because of the extra area and power dissipation.

For a partitioned *Bus-Invert* code the lowest number of transitions is obtained for subbuses of width 2. That means that for a *minimum* average number of transitions a 64 b bus would need to be partitioned into 32 2 b subbuses. Obviously the penalty induced by the large number of extra bus lines ($n/2$) will generally favor a coarser partitioning. The average number of transitions per bus-line per time-slot for a 2 b bus is 0.375 (as compared to 0.5 for an unencoded bus) or 75%

TABLE II
COMPARISON OF UNENCODED I/O AND CODED I/O WITH ONE OR MORE *invert* LINES. THE COMPARISON LOOKS AT THE AVERAGE AND MAXIMUM NUMBER OF TRANSITIONS PER TIME-SLOT, PER BUS-LINE PER TIME-SLOT, AND I/O POWER DISSIPATION FOR DIFFERENT BUS-WIDTHS

nr. of bus lines	mode	average transitions /time-slot	average transitions /bus-line	average I/O power dissipation	maximum transitions /time-slot	maximum transitions /bus-line	peak I/O power dissipation
2	unencoded	1	0.5	100%	2	1	100%
	1 *invert*	0.75	0.375	75%	1	0.5	50%
8	unencoded	4	0.5	100%	8	1	100%
	1 *invert*	3.27	0.409	81.8%	4	0.5	50%
	4 *invert*	3	0.375	75%			
16	unencoded	8	0.5	100%	16	1	100%
	1 *invert*	6.83	0.427	85.4%	8	0.5	50%

of the unencoded case. Fig. 7 considers the previous example (see Figs. 3 and 4) of an 8 b bus, this time partitioned into four 2 b subbuses.

As can be seen from Table II the decrease in average power dissipation for the *Bus-Invert* method with or without partitioning is not very large but at least is consistent. It would be extremely "expensive" in terms of extra code bits (and consequently extra pins) to decrease the bus activity much more. Returning to our 8-bit data bus we saw that by using the *Bus-Invert* method with only one extra pin the maximum bus activity can be cut to 4. If it is needed to further cut it to 3 for example a *3-limited-weight code* is needed. The inequality that needs to be satisfied is [20], [21]:

$$C_{8+k}^0 + C_{8+k}^1 + C_{8+k}^2 + C_{8+k}^3 \geq 2^8 = 256 \qquad (3)$$

where k represents the extra code bits, $C_{8+k}^0 + \cdots + C_{8+k}^3$ is the number of codewords of length $8+k$ with at most 3 transitions and 2^8 is the required number of data patterns. By simple inspection it can be seen that the minimum k that satisfies (3) is $k = 4$. So 3 more bits are needed compared to the *Bus-Invert* method for a small extra decrease. In general the number of necessary extra code bits grows exponentially [21] and the method becomes nonpractical when a very small transition activity is needed. Another big problem for the limited-weight codes is that no general algorithmic way of generating them is known yet.

C. Implementation of the Bus-Invert Method

In order to implement the *Bus-Invert* coding scheme some extra circuitry is needed on the data-path which means extra area and sometimes lower performance. For an 8 b nonpartitioned bus a possible circuit for the driver side is shown in Fig. 8. There are 16 extra XOR gates and a majority voter (5 out of 9) circuit (because the *invert* line contributes with transitions the voter must also take it into account). The bus-drivers, bus-receivers and the latches are needed for the unencoded case also and do not represent an overhead. The majority voter can be implemented digitally with a tree of full adders (Fig. 9). Thus the circuit will generate on the order of $O(n \log n)$ extra internal transitions.

If the delay and extra power consumption of the digital majority voter is considered too large an alternate analog circuit can be used. The main observation here is that the computation of the *invert* line is not critical for data integrity.

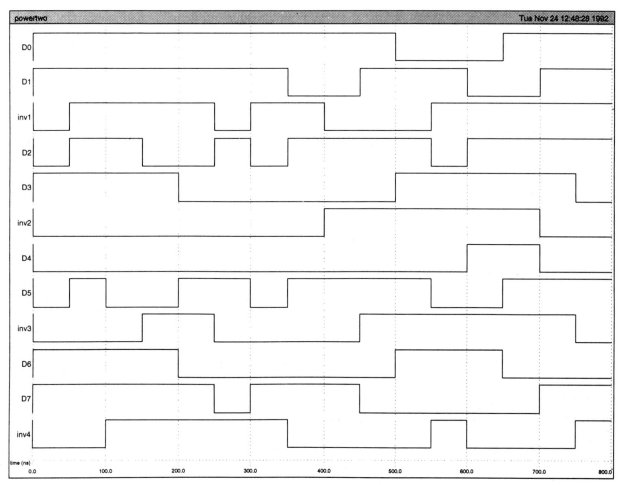

Fig. 7. An 8 b data bus partitioned into four 2 b subbuses. There are 48 transitions for a period of 16 time-slots. This represents an average of 3 transitions per time-slot, or 0.375 transitions per bus line per time-slot. The maximum number of transitions in any time-slot is 4. There is at most one transition per time-slot in each of the four subbuses.

If for example the Hamming distance is larger than $n/2$ but the majority voter decides that *invert* should be a 0, the data on the bus will not be inverted. This will only mean that the potential power savings will be lost but still the data will not be corrupted. With this observation the simpler analog circuit in Fig. 9 can be used [21] without fearing the possible "impreciseness" of such analog circuitry.

The receiver side is much simpler because it only needs to conditionally invert the bus contents in order to get the correct data value. The delay of the data-path is slightly increased but this tradeoff of performance versus low-power is common to almost all methods of decreasing power dissipation. If so desired the encoding and decoding operations can be pipelined for keeping the overall throughput unchanged.

IV. LOW POWER FOR AN ADDRESS BUS

In section III the case of a random uniformly distributed sequence of values was considered and it was claimed that this is close to what happens on a data bus. For an address bus this assumption is generally no longer valid. In the extreme case of a circular FIFO implemented with a RAM and a counter [10] the RAM addresses are sequential. Generally, an address bus will tend to have a sequential behavior. Most of the time the number of transitions will be small and the overall average will be

also small. For example for $n = 8$ and a pure sequential behavior the average number of transitions will be 1.99 per time-slot or only 0.25 per bus line per time-slot, much smaller than 0.5 as was the case with random uniformly distributed values. The *Bus-Invert* method will have a minimal impact on the average power dissipation in this case (1.82 average number of transitions per time-slot). The idea of *coding* can still be used in order to decrease the power dissipation for a sequential behavior.

A. Coding for Low-Power on an Address Bus

The Gray code is perfect for an address bus. With only one transition per time-slot or $1/n$ (0.125 for an 8-bit bus) transitions per bus line per time-slot it represents an improvement of 45% over the unencoded sequential case. The Gray code is easy to generate but the extra circuit will require extra area and will affect the performance. As for the *Bus-Invert* method the Gray coding can be pipelined for keeping the throughput constant. Although convenient for simplifying the theoretical analysis the model of sequential values on an address bus is generally too simplistic for a real system. Only some percentage of bus addresses are typically sequential [12] with the others being essentially random. In that case a mixed coding, Gray and *Bus-Invert* will give the best results for both peak and average power dissipation.

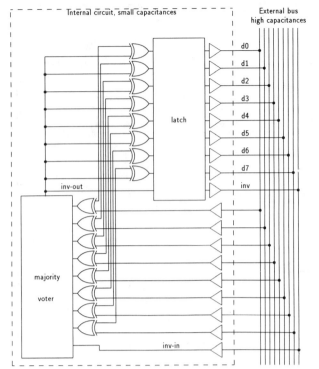

Fig. 8. Possible circuit for the driver side on an 8 b bus. There are 16 XOR gates, 8 for conditionally inverting the bus contents and 8 for comparing the present bus-value with the next data-value. The majority voter circuit decides according to the Hamming distance whether to invert or not the next value. The value of the present *invert* line must be considered by the majority voter.

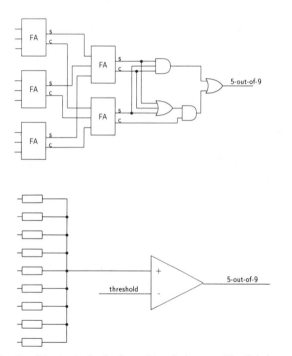

Fig. 9. Possible circuits for the 5 out of 9 majority voter. The digital voter is implemented as a tree of full-adders with the last stage of the tree simplified by taking advantage of "don't care" terms. The analog voter is simpler but less accurate and uses only resistors and a voltage comparator.

As an example of what it means to ignore the impact of design on power dissipation consider the idea of using a LFSR instead of a counter for generating addresses for a circular FIFO [19]. Because of its pseudo-random behavior the average

number of transitions per bus line per time-slot for LFSR addressing will be 0.5 like in the case of a "random" data bus. This is much higher than for a counter and thus such a circuit would dissipate more power.

V. CONCLUSIONS AND FUTURE WORK

Although the *Bus-Invert* method was explained in the particular setting of dynamic I/O power dissipation the same methods can be applied in any case where large capacitances are involved. Examples are: multichip module (MCM) interconnections, wafer-scale integration (WSI) intermodule connections and even on-chip buses. Furthermore it is likely that the method will also reduce the total I/O overlap current (because it reduces the number of switching I/O lines) which was ignored in our analysis but which in general contributes to additional power dissipation because of the large I/O loads. Special connections like the RGB connection to the display of a portable polygon render described in [22] can also use the technique. Depending on the ratio between the small (what was called "internal capacitance" in the paper) and large (what was called "I/O capacitance" in the paper) capacitances the applicability of the method can be considered in many other cases.

The proposed *Bus-Invert* method of decreasing power dissipation like other methods [3] represents a trade-off between performance and power dissipation. The performance decreases because the XOR's and the majority voter circuit increase the area and delay of the data-path. The majority voter has a delay of $O(\log n)$. The good result is that the smallest decrease in performance is obtained for $n = 2$ (e.g., for buses partitioned into 2 b subbuses) for which the lowest number of transitions is also obtained. Another trade-off is represented by the extra number of I/O pins needed. Answers to some of the above problems are as follows.

• The increase in the delay of the data-path. As was mentioned before lower performance is a common problem with all methods of decreasing power dissipation. Anyhow the *Bus-Invert* method represents an "absolute" method of decreasing power dissipation. By looking at the power-delay product which removes the effect of frequency (delay) on power dissipation, a clear improvement is obtained in the form of an absolute lower number of transitions. It is also relatively easy to pipeline the bus activity. The extra pipeline stage and the extra latency must then be considered.

• The increased number of I/O pins. As was mentioned before *ground-bounce* is a big problem for simultaneous switching in high speed designs [13]. That is why modern microprocessors use a large number of V_{dd} and *GND* pins. The *Bus-Invert* method has the side-effect of decreasing the maximum ground-bounce by approximately 50%. Thus circuits using the *Bus-Invert* method can use a lower number of V_{dd} and *GND* pins and by using the method the total number of pins might even decrease. Still a problem with the increased number of I/O pins is their additional *static* power dissipation which was ignored in our analysis.

- Both the driver and the receiver on the bus must use the *Bus-Invert* method in order to code and decode correctly the information on the bus. This means that the method must be applied at the *system* level, on all the circuits connected to the bus. An option here is to store the data already encoded in RAM in order to keep the memory subsystem unchanged, with the coding and decoding done only at the master. The power savings would be entirely obtained at write and since the sequence at read is likely to be the same as at write (e.g., for a block-oriented memory subsytem) the power savings will be obtained also at read.

In the future we plan to apply coding and in particular the *Bus-Invert* method in some practical application. For example the *Rambus* emerges as a promising bus standard [17] for very high-speed applications and is particularly attractive for I/O coding because it is relatively narrow (byte-wide), block-oriented (easy to pipeline) and proposed for portable applications. We also plan to further study the theory of limited-weight codes and in particular their relationship to error-correcting codes.

As a last comment, it is interesting to mention that the *Bus-Invert* method decreases the total power dissipation although both the *total* number of transitions increases (by counting the extra internal transitions) and the *total* capacitance increases (because of the extra circuitry). This is possible because the transitions get redistributed very nonuniformly, more on the low-capacitance side and less on the high-capacitance side. This unintuitive result is powerful and it contradicts a simple analysis where only the total number of transitions is considered without taking into account how large the node capacitances are.

ACKNOWLEDGMENT

The authors would like to thank the anonymous reviewers for their constructive criticism and comments.

REFERENCES

[1] H. B. Bakoglu, *Circuits, Interconnections and Packaging for VLSI.* Reading, MA: Addison-Wesley, 1990.

[2] T. K. Callaway and E. E. Swartzlander, "Estimating the power consumption of CMOS adders," in *11th Symp. Comp. Arithmetic*, Windsor, Ont., 1993, pp. 210-216.

[3] A. P. Chandrakasan, S. Sheng, and R. W. Brodersen, "Low-power CMOS digital design," *IEEE J. Solid-State Circ.*, pp. 473-484, Apr. 1992.

[4] A. P. Chandrakasan, M. Potkonjak, J. Rabaey, and R. W. Brodersen, "HYPER-LP: A system for power minimization using architectural transformations," *ICCAD-92*, Santa Clara, CA, Nov. 1992, pp. 300-303.

[5] ——, "An approach to power minimization using transformations," *IEEE VLSI for Signal Processing Workshop*, 1992, CA.

[6] S. Devadas, K. Keutzer, and J. White, "Estimation of power dissipation in CMOS combinational circuits," in *IEEE Custom Integrated Circ. Conf.*, 1990, pp. 19.7.1-19.7.6.

[7] D. Dobberpuhl *et al.*, "A 200-MHz 64-bit dual-issue CMOS microprocessor," *IEEE J. Solid-State Circ.*, pp. 1555-1567, Nov. 1992.

[8] R. J. Fletcher, "Integrated circuit having outputs configured for reduced state changes," U.S. Patent 4,667,337, May, 1987.

[9] D. Gajski, N. Dutt, A. Wu, and S. Lin, *High-Level Synthesis, Introduction to Chip and System Design.* Norwell, MA: Kluwer Academic Publishers, 1992.

[10] J. S. Gardner, "Designing with the IDT SyncFIFO: The architecture of the future," *1992 Synchronous (Clocked) FIFO Design Guide*, Integrated Device Technology AN-60, Santa Clara, CA, 1992, pp. 7-10.

[11] A. Ghosh, S. Devadas, K. Keutzer, and J. White, "Estimation of average switching activity in combinational and sequential circuits," in *Proc. 29th DAC*, Anaheim, CA, June 1992, pp. 253-259.

[12] J. L. Hennessy and D. A. Patterson, *Computer Architecture - A Quantitative Approach.* Palo Alto, CA: Morgan Kaufmann Publishers, 1990.

[13] S. Kodical, "Simultaneous switching noise," *1993 IDT High-Speed CMOS Logic Design Guide*, Integrated Device Technology AN-47, Santa Clara, CA, 1993, pp. 41-47.

[14] F. Najm, "Transition density, a stochastic measure of activity in digital circuits," in *Proc. 28th DAC*, Anaheim, CA, June 1991, pp. 644-649.

[15] C. A. Neugebauer and R. O. Carlson, "Comparison of wafer scale integration with VLSI packaging approaches," *IEEE Trans. Components, Hybrids, and Manufact. Technol.*, pp. 184-189, June 1987.

[16] A. Park and R. Maeder, "Codes to reduce switching transients across VLSI I/O pins," *Computer Architecture News*, pp. 17-21, Sept. 1992.

[17] *Rambus-Architectural Overview*, Rambus Inc., Mountain View, CA, 1993.

[18] A. Shen, A. Ghosh, S. Devadas, and K. Keutzer, "On average power dissipation and random pattern testability," in *ICCAD-92*, Santa Clara, CA, Nov. 1992, pp. 402-407.

[19] M. R. Stan, "Shift register generators for circular FIFOs," *Electronic Engineering*. London, England: Morgan Grampian House, Feb. 1991, pp. 26-27.

[20] M. R. Stan and W. P. Burleson, "Limited-weight codes for low power I/O," International Workshop on Low Power Design, Napa, CA, Apr. 1994.

[21] J. Tabor, "Noise reduction using low weight and constant weight coding techniques," Master's thesis, EECS Dept., MIT, May 1990.

[22] W.-C. Tan and T. H.-Y. Meng, "Low-power polygon renderer for computer graphics," *Int. Conf. A.S.A.P.*, 1993, pp. 200-213.

[23] N. Weste and K. Eshraghian, *Principles of CMOS VLSI Design, A Systems Perspective.* Reading, MA: Addison-Wesley Publishing Company, 1988.

[24] R. Wilson, "Low power and paradox," *Electronic Engineering Times*, pp. 38, Nov. 1, 1993.

[25] J. Ziv and A. Lempel, "A universal Algorithm for sequential data compression," *IEEE Trans. Inform. Theory*, vol. 23, pp. 337-343, 1977.

A sub-CV^2 pad driver with 10 ns transition time

L. "J." Svensson, W.C. Athas, and R.S-C. Wen

{svensson,athas}@isi.edu, shwen@pcocd2.fm.intel.com

http://www.isi.edu/acmos

University of Southern California/Information Sciences Institute

4676 Admiralty Way, Marina del Rey, California 90292-6695

Abstract

We describe an application of stepwise charging for driving off-chip signals. Our driver, designed for a chip built in a 0.8 μm bulk CMOS process, dissipates significantly less than CV^2 per cycle while driving a 100 pF load to a 5 V swing with transition times as low as 10 ns. The transition time is adjustable; dissipation measured on a test chip ranges from 53% of CV^2 at 16 ns to 70% at 10 ns. The target chip is approximately 6% larger with our driver than with a conventional one.

1. Introduction

Low-power drivers for off-chip loads (pad drivers) present special challenges to the silicon-chip designer. The driving energies for external signals are typically orders of magnitude larger than for internal signals. Consequently, off-chip loads often dominate the total dissipation of a chip, even though there are likely many more internal than external nodes, and despite higher internal switching frequencies.

Pad-driver dissipation is difficult or expensive to reduce significantly by the conventional methods of voltage and capacitance reduction. The voltage swing is typically set by the interface level specifications of the surrounding chips. The capacitance is largely determined by packaging. A reduction of either may incur a prohibitive cost or be infeasible for other reasons. Thus, the design space for a pad driver is severely restricted by system-level constraints. It is possible, when designing a tapered-inverter-chain driver, to trade dissipation for delay to some extent [1], but the total dissipation per charge-discharge cycle never goes below CV^2.

In this paper, we present a pad driver designed to drive a load capacitance of 100 pF with a 5 V swing, with transition time less than 15 ns, and with total dissipation well below CV^2. The target chip uses 5 V external signalling with a 3.3 V core supply voltage. As a result, 80% of the dissipation is due to off-chip signalling. Significant further reduction of the total dissipation required a sub-CV^2 driver.

The driver described here uses *stepwise charging* to accomplish the dissipation goal. The principles of stepwise charging and a proof-of-concept driver have been described previously [2]; related ideas have also been published [3,4]. The driver presented here is capable of edge rates 50 times faster than those of the proof-of-concept driver, while still dissipating well under CV^2 per cycle. The increased level of performance required the implementation to be quite different from the proof-of-concept driver: here, the control pulses for the switches are generated asynchronously.

The remainder of this paper is organized as follows. In Section 2, we give a brief background on the theory of stepwise charging. In Section 3, we describe the design of the pad driver, based on our mathematical modelling and on the application specifications. In Section 4, we describe a test chip, present performance data measured from that chip, and compare theory with simulation and measurements. Section 5 contains our conclusions.

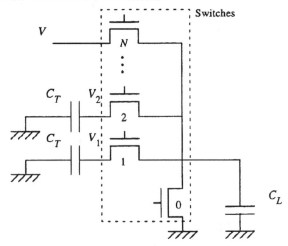

Figure 1. Simplified schematic of a stepwise-charging driver for a capacitive load C_L. To charge the load from 0 to V, the switches 1, 2, ..., N are pulsed in succession. Thus, the load is charged in N steps. To discharge the load, the switches are pulsed in the reverse order: $N-1, N-2, ..., 0$.

Reprinted from *IEEE International Symposium on Low Power Electronics and Design '96*, pp. 105-108, 1996.

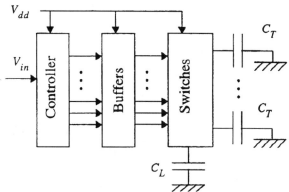

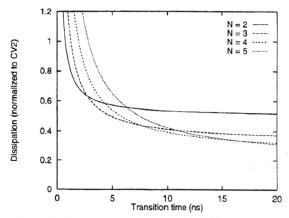

Figure 2. Block diagram of stepwise pad driver. A transition on the input causes the controller to generate control pulses to turn on and off the switches (cf. Figure 1) in succession. The buffers match the drive capabilities of the controller outputs to the rather large gate capacitances of the switches.

2. Background

The idea behind stepwise charging is to charge and discharge a capacitive load successively, in several steps, by moving charge from and to intermediate potentials. A conventional CMOS inverter is a one-step charger: it delivers all charge directly to the load at potential V_{dd} and returns all charge from the load directly to ground at potential 0.

Figure 1 depicts a simplified schematic of a stepwise driver with N steps. Tank capacitors, C_T, provide the intermediate potentials $V_1 ... V_{N-1}$, given by $V_i \approx (iV)/N$. With these potential values, the average voltage drop across each switch when the tanks are successively connected to the load is limited to $V/(2N)$. Dissipation per cycle is therefore reduced by a factor of N compared to the conventional, one-step case. The tanks need not be preset to V_i; under weak conditions, the potentials provably converge to the ideal values when the driver repeatedly cycles the load

A complete pad driver requires a number of blocks in addition to the switches. Figure 2 shows a high-level block diagram of the driver described in this paper. The tank capacitors are shared by several drivers. Each tank must then be larger than the sum of all load capacitances; tanks are therefore placed off-chip, as they would be in most cases.

3. Driver design

The first decision in the design of a stepwise driver is how many charging steps to use. For low-power purposes, we wish to find the number that gives the lowest overall dissipation. The best number of steps (based on a simple linear model of the switch) is given by the following equation [2]:

$$N_{opt} = \sqrt[3]{\frac{t_{rise}}{4m\bar{\rho}}},$$ (1)

Figure 3. Theoretical dissipation of the stepwise pad driver as function of transition time and number of charging steps, normalized to CV^2. Note that dissipation exceeds CV^2 at sufficiently short transition times: the supplementary energy needed to drive several very large switch gate capacitances outweighs the gains in output energy.

Here, $\bar{\rho}$ models the intrinsic speed of the process; for the 0.8 μm process used for this design, $\bar{\rho}$ = 15 ps. The parameter m is selected by the designer. It quantifies the exponential nature of the charging process: it is the number of RC time constants used for each charging step. Suitable values range from 2 to 4. We used m = 3 in this design.

Figure 3 shows the theoretical dissipation as a function of transition time for different numbers of charging steps. With the circuit and process parameters listed above, N_{opt} = 4. The minimum is shallow: only a few percent is gained by going from three to four steps. Moreover, the total gate area of the switches increases as the *square* of the number of steps [2]: thus, a four-step design would require almost twice as much gate area as a three-step design. For these reasons, we chose three steps for this design.

Figure 1 shows each switch represented as a single nFET. The present design is different in two important ways. First, each tank capacitor is connected to the load with two switches in parallel: one for charging and the other for discharging. This arrangement allows each switch to be optimally sized for its unique voltage range. In contrast, with single switches, each switch must be sized for the worse of the two transitions, and thus an unnecessarily large gate capacitance is driven during the other transition. Single switches suffice for the ground and V_{dd} connections, which each take part in only one transition. The second difference from Figure 1 is that two switches (the V_{dd} switch and the discharge switch for the top tank capacitor) are pFETs. These have a lower on-resistance per gate capacitance for the upper third of the voltage swing, where these switches operate.

The switch sizes were chosen to meet a target transition time of 10 ns as a safety margin against a slower-than-usual run. Switch device widths range from 250 μm to 1000 μm.

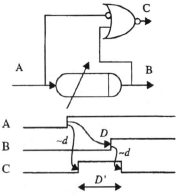

Figure 4. The delay line has a tunable delay D. The intrinsic delay through the gate is constant and similar for both paths (A→C and B→C), so the output pulse width D' is close to D.

The controller of Figure 2 generates the gate pulses for the switches. With a transition time of 15 ns and three charging steps, the gate pulses should have a duration of 5 ns. No 200 MHz clock is available on the target chip, so we designed an asynchronous controller, whose core building block is outlined in Figure 4. The input signal is routed to a voltage-controlled delay line. The original and delayed versions of the signal are combined to yield a pulse of duration close to the delay when a positive-going edge appears on the input. Identical delay blocks are appended for more steps, and mirror-image logic generates pulses on negative-going edges. This arrangement also allows precise adjustment of the transition time during testing, which would not have been possible with a simpler feedback scheme [5].

4. Test chip

We have compared circuit simulations with measurements from a test-chip implementation of the driver. Before fabrication, the dissipation of the complete driver, including the controller, was estimated with HSPICE to be 60% of CV^2 at a transition time of 10 ns. This is worse than the theoretical estimate of 41% of CV^2 (Figure 3). Note, however, that the theory is based on an idealized device model. It also assumes that the output is allowed to fully reach each intermediate voltage before the next charging step commences. Both of these assumptions affect prediction accuracy. Furthermore, controller dissipation is not included in the theoretical figure.

Figure 5 shows lab measurements of the test chip driver dissipation as a function of transition time. The results of HSPICE simulations using device parameters from the actual fabrication run are also shown (at around 65% of CV^2, these values are somewhat higher than for the pre-fabrication simulations). Intriguingly, the measured dissipation is *lower* than the simulated values. This effect appears to be due to parasitic inductance in the test setup. Figure 5 includes simulated figures with a 30 nH inductance added in each bonding wire. The resulting figures are closer to the measured ones, although we have not yet been able to recre-

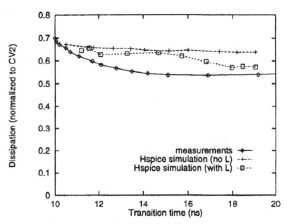

Figure 5. Measured and simulated dissipation for pad driver as a function of transition time. The simulated waveforms with added inductance exhibit ringing not present in the lab setup. Since this ringing introduces a quasi-random element in the measurements, the curve is not as smooth for this case.

ate the detailed behavior of the test setup in simulation. It should be noted that lumped inductances are an overly simplified model of the actual current loops involved.

Figure 6 shows oscilloscope traces of the output signal of the test chip. Two traces with different transition times are shown. The edges appear smooth in the faster case because of test setup inductances and finite transition times of the gate pulses controlling the switches.

A photomicrograph of the test chip pad driver is shown in Figure 7. As can be seen in the figure, the switches occupy a large part of the area. They would be less prominent if a smaller load and/or a longer transition time were specified. The buffer section contains tapered inverter chains matching the drive capabilities of the controller outputs to the large gate capacitances of the switches. The input stage handles level conversion from the core voltage swing (3.3 V) to the 5 V used in the rest of the driver.

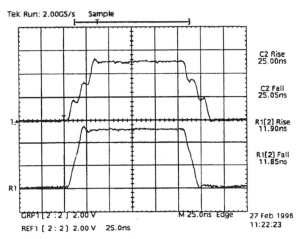

Figure 6. Scope traces of output waveforms for two different transition times. Rise and fall time values are also shown.

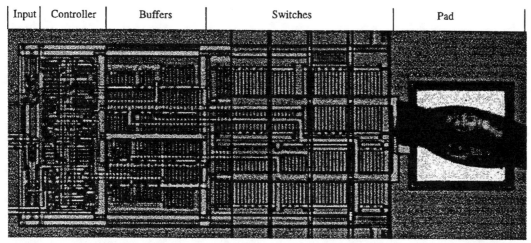

| Input | Controller | Buffers | Switches | Pad |

Figure 7. Pad driver (test chip photo). Four pad rings run across the switches; in addition to the supply voltage and ground, they carry the two intermediate voltages. The tank capacitances are connected to the rings via dedicated pads (not shown). For size reference, the bonding area is 90 μm square.

5. Conclusion

Pad drivers based on stepwise charging offer the unique combination of moderately fast transition times and sub-CV^2 dissipation. The design described here reaches transition times as low as 10 ns while dissipating only 70% of CV^2. This is in the same range as that of typical slewrate-limited pads. We believe that performance could be improved somewhat by fine-tuning the design and by exploiting better packaging.

The RC-based design procedure [2] for stepwise-charging buffers is adequate for determining starting points for circuit simulation iterations, even for drivers with relatively fast transition times such as this one. It should be realized, however, that Equation (1) is based on a highly simplified model of the MOS device, where gate capacitance and on-resistance are linearly related. Neither short-channel effects such as velocity saturation nor packaging inductance are taken into account. Refinements of the theory would be useful, but detailed simulations would probably still be necessary to determine the actual performance of the driver, particularly for short-channel devices.

Stepwise pad drivers incur a penalty in terms of silicon area and I/O pins. Figure 7 indicates that the total area of driver and pad is approximately twice that of a conventional design. The overall area of the target chip, which is approximately 6 mm square, is about 6% larger with stepwise pad drivers than with conventional ones. The target chip is not I/O-limited, and the extra pads needed to connect the drivers to the off-chip tank capacitors do not further increase the pad frame. The number of tank capacitor pins required per intermediate voltage depends largely on package and bond wire inductance. The inductance improves the dissipation figures somewhat, but it also degrades the transition time. We currently rely on circuit simulations to estimate its influence.

In summary, we have applied our theory and practice of stepwise charging to the specific design problem of a low-power pad driver for a target chip specification. The resulting prototype has demonstrated the viability of stepwise charging as a practical means to reduce power dissipation when conventional techniques cannot be leveraged, as is the case for pad drivers, where voltage swing and transition time requirements severely constrain the design space.

6. Acknowledgment

This research was supported by ARPA under contracts number MDA903-92-D-0020, DABT63-19-C0052, and DAAL01-95-K3528. We thank Sanjaya Dharmasena for helping out with simulations, and Erlend Olson for valuable comments.

7. References

[1] Brian S. Cherkauer and Eby G. Friedman. A unified design methodology for CMOS tapered buffers. *IEEE Transactions on VLSI Systems*, vol. 3, no. 1, pages 99–111, March 1995.

[2] L."J." Svensson and J.G. Koller. Driving a capacitive load without dissipating fCV^2. In *Proceedings of IEEE Symposium on Low Power Design*, 1994, pages 100–101.

[3] Mark Hahm. Modest power savings for applications dominated by switching of large capacitive loads. In *Proceedings of IEEE Symposium on Low Power Design*, 1994, pages 60–61.

[4] Kei-Yong Khoo and Alan N. Willson, Jr. Charge recovery on a databus. In *Proceedings of ISLPD '95*, Dana Point, 1995, pages 185–189.

[5] Mircea R. Stan and Wayne P. Burleson. Low-power CMOS clock drivers. In *Proceedings of TAU '95*, pages 149–156.

Part VI

Memory Circuits

Reviews and Prospects of Low-Power Memory Circuits

KIYOO ITOH

1. INTRODUCTION

A major area of low-power LSI research is circuit technology which reduces RAM chip power [1–4]. This forms a basis not only for other LSI memory chips such as flash and ROM, but also for on-chip memory subsystems such as embedded DRAMs (merged DRAM and logic) and caches that have both become increasingly important for modern memory systems. In fact, without detailed knowledge regarding RAMs and their low-power circuits, not even MPU designers are able to design innovative system-on-chip LSIs that miss the Memory Wall [5]. A low-power circuit, for both active and data-retention modes, has been developed to confine the chip-power which increases with increasing memory capacity and/or speed. The circuit technology developed so far has focused on three key issues: reductions in charging capacitance, operating voltage, and DC current. Of these issues, a reduction in the operating voltage has become relatively important not only to reduce power, but also to ensure device reliability in scaled-down devices, and to extend the use of LSIs to battery-based portable systems. Recent exploratory achievements in ultralow-voltage operations of 1V or less suggest the great potential of CMOS circuits even in the 0.1–μm era, although the voltage reduction inevitably needs memory-cell development focusing on a wider voltage-margin design.

Here, low-power CMOS circuit technology for DRAM and SRAM chips is reviewed and investigated, emphasizing low-voltage because this is becoming one of the biggest challenges in LSI design.

2. DRAM TECHNOLOGY AND CIRCUITS

A. Low-Power Circuits

Power reduction in the active mode is essential to realize high-speed, highly-reliable chip operation, and low cost through the use of cheap plastic-package. In addition, this lowers the junction temperature, and thus reduces the cell leakage-current of one-transistor one-capacitor DRAM cells. In the 1970s DRAM chips reduced power with NMOS dynamic-circuit techniques which featured node-precharging and gate-voltage boosting above the power-supply voltage, V_{DD} (or V_{CC}), with a MOSFET

capacitor. They were applied to I/O buffers, drivers, decoders, and other circuits throughout the chip. Dynamic NMOS cross-coupled sense-amplifiers on each data (bit) line also contributed to power reduction. In the 1980s, coupled with the low-power advantage of CMOS circuits, two techniques heavily contributed to power reduction: lower charging capacitance due to the partial activation of multi-divided data-lines by both multi-level metal wiring and longer data-retention time (refresh time); and lower operating voltage resulting from V_{DD} reduction, half-V_{DD} data-line precharging, and an on-chip voltage-down converter. In particular, vertical-capacitor memory cells such as trench- and stacked-cells helped to lower the data-line operating voltage while maintaining a high enough signal-charge. In the 1990s memory subsystem technologies, which are familiar to system designers, such as pipeline-, synchronous-, prefetch-, and multi-bank interleave-operations have even been incorporated into general purpose DRAMs. They enhance throughput while maintaining low power, allowing products to be successively developed such as extended data-out DRAMs (EDOs), synchronous DRAMs (SDRAMs), and Rambus DRAMs. Small swing impedance-matched interfaces and high-density packaging are also crucial to reduce I/O power. Embedded DRAMs drastically reduce I/O power through the inherently small internal-capacitance of the chip.

Power reduction in the data-retention mode is also a major concern in terms of the battery operation of small memory-systems. There have been many proposals to reduce both AC and DC currents from major power sources in the retention mode: reductions in the data-line charge consumed by refresh operations through refresh-time extension and a charge-recycle refresh scheme; DC current reduction of on-chip voltage converters (i.e., voltage-down and -up converters, substrate back-bias generator) through shutting down a higher power converter, which is dedicated only to the active mode, of parallel-connected converters, or through reducing the cycle time for a charge pumping circuit. More details on low-power DRAM circuits are in [1–4].

B. Ultralow-Voltage Circuits

There are many design issues to realize an ultralow-voltage operation of less than 2V: The development of scaled-down FETs with sufficiently low V_T (MOSFET threshold voltage) and shallow junction is essential. A shallow junction reduces the V_T variations and offset voltage of sense amplifiers, and is formed by reductions of ion-implantation energy and process temperature. A high signal-to-noise ratio (S/N) cell to cope with voltage reduction is important. Thus vertical-capacitor memory cells and high-permittivity materials to obtain a higher cell-capacitance are indispensable. Noise suppression is also important because the noise margin decreases with V_{DD} reduction. The following sub-threshold current issue is also the key to a low-voltage operation.

There has been an ever-increasing number of DRAM papers on ultra-low-voltage design focusing on the subthreshold DC current issue, although this design is still in its infancy. At present attention to this issue arising from the scaled-down V_T is only being paid to the data-retention mode, since the V_T is still too high. With a further reduction in V_T, however, even the numerous circuits inactive during the active mode will start to generate subthreshold current, eventually dominating the active current of the chip. Thus the lack of a solution to this issue is a major obstacle to ultralow-voltage operation. The subthreshold-current reduction circuits could be created more effectively and easily than MPUs or ASICs. This results from the following two features of DRAM chips: 1) The use of many kinds of iterative circuit blocks such as memory-cell arrays, decoders and their relevant drivers, sense amplifiers, address buffers, and I/O buffers. The blocks are major sources of the subthreshold current, which is proportional to the channel width of the chip because of their overwhelmingly large total channel width. 2) Incorporation of input-predetermined logics, which are governed by few external clocks, for the peripheral circuit. Thus a designer can predict which FETs in the chip will cut off during the standby mode and cause standby subthreshold-current. The reduction circuits proposed so far are based on the gate-source backbiasing scheme, the static or dynamic multi-V_T scheme, and the over-drive (node-boost) scheme.

Iterative Circuit Blocks. The largest channel-width block is the DRAM array. A small V_T for each cell-transistor, however, causes cell-information destruction rather than an increase in DC current, unlike SRAMs. This is because during active periods, resultant subthreshold-current flows from each non-selected-cell node to a corresponding floating data-line node which has stayed in a completely discharged state of 0V after selected-cell activation. Current flow continues only for a short period until both nodes have been equilibrated. Solutions are gate-source backbiasing schemes applied to the cell transistor such as boosted sense-ground and word-line negative-biasing scheme, as shown in Fig. 1. The decoders and their associated drivers have the second largest channel-width to consume sub-threshold current. The self-biasing scheme [1] and partial activation of the multi-divided power line [1] drastically reduce the current. The area and speed penalty are negligibly small because only a few FETs are added for each iterative-circuit block. Numerous CMOS sense amplifiers [3] enable low-voltage and high-speed operations through a well-driving scheme combined with a triple-well structure. The scheme drives the well so that V_Ts are lowered at the initial stage of sense operation, and then raised to reduce the subthreshold current. An over-drive scheme

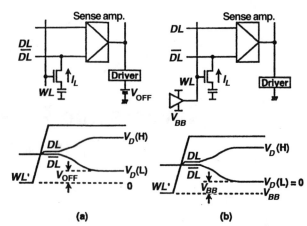

Figure 1 Cell-transistor biasing scheme. [3]. (a) Boosted sense-ground. (b) Word-line negative biasing. *DL, $\overline{DL}$:* data line, *WL:* non-activated word line, *WL':* activated word line, I_L subthreshold-current flowing from each non-selected cell when the data line ($\overline{DL}$) is discharged by amplification of a selected-cell signal.

[3] allows the use of a high V_T for sense amplifiers because boosting their common-source nodes eliminates the V_T drop.

Other Peripheral Circuits. Other peripheral logic-circuits reduce their power with switched-source impedance and multi-V_T schemes [1,6]. In a multi-V_T scheme a high-V_T switching MOS-FET cuts off the leakage current during the inactive period while a low-V_T MOSFET is used for the main signal path during the active period. A well-driving scheme [3] changes the V_T dynamically so that V_T in the standby period is higher than that in the active period. Though it has the disadvantage of a slow response time due to driving a highly capacitive well, it has the advantage of a multi-V_T scheme. An offset driving scheme [3] allows the use of a low V_T by offsetting the source level of the driver by the difference between the normal (high) V_T and the low V_T. This necessitates an on-chip voltage-down converter or two power supplies.

3. SRAM TECHNOLOGY AND CIRCUITS

A. Low-Power Circuits

The power sources of SRAMs are memory-cell arrays, data-line loads, column/sense circuits, and other peripheral circuits. There have been memory-cell innovations to reduce the total current from the many non-selected cells in an array such as a high-resistance polysilicon load NMOS cell, a TFT-load NMOS cell, and a full CMOS cell that enables low-voltage/high-speed operation and makes its fabrication process compatible with that of microprocessors. Partial activation of a multi-divided word line and pulse operation of a word-line circuit with address transition detection (ATD) drastically reduce the ratio current established between each selected cell along a word line and the corresponding data-line static load. This is because the activations reduce the number of selected cells and the pulse width of the ratio current. Cross-point cell selection [7] minimizes the cell number although this necessitates a word-line boost and modification of the cell-β-ratio for successful write-operation of the single-ended mode. A PMOS cross-coupled latch-type amplifier combined with ATD pulse operation reduces the power of column/sense circuits that increase in number according to movement toward a wide-bit I/O configuration. More details on low-power SRAM circuits are in [1]. A typical application of the circuits above is a low-power TLB (translation look-aside buffer) [16]. A synchronous power supply reduces both the currents of a differential static-amplifier on each match-line of the CAM array and each data-line load of the data-SRAM array.

B. Ultralow-Voltage Circuits

For ultralow-voltage operation the SRAM array is a major concern because of the largest channel-width block involving the subthreshold current, unlike that in DRAMs. Note that the current of the remaining circuit blocks could be suppressed with circuits [10] similar to those of the DRAMs previously described.

A high V_T is needed for cell driver-FETs to suppress the huge subthreshold current, as shown in Fig. 2. A high V_T is also nec-

essary for cell transfer-FETs to avoid the leak current flowing to either of the data-lines. Thus the V_T decreases the cell voltage-margin as the cell-supply voltage (V_{DD}) approaches the V_T, limiting the minimum V_{DD} of the chip. Figure 3 shows various cell-driving schemes to overcome this limitation. Two-step word driving [8] enables high-speed write-operation directly from the data lines by boosting the word line to eliminate the V_T drop of the transfer FETs. The poor driving capability of TFT cell loads needs the above word driving despite a write-speed penalty caused by the necessity of a preceding read-operation. The raised DC voltage [9], generated by an on-chip voltage-up converter, and supplied to the cell loads, allows the storage node voltage of the cell to quickly rise in the write-operation because of an increased TFT conductance. Both schemes, however, suffer from a slow read-operation because of high V_T of the transfer FETs. The step-down boosted word line [10] offers high-speed operation as well as low power despite the additional delay for boosting, and variations in the amplitude and duration of the boosted pulse for process variations. The negative source-line scheme [11] also achieves high speed with a reduced V_T for the cell driver-FETs and an effective boost for the cell transfer-FETs. However, heavy capacitance, which almost equals data-line capacitance multiplied by the number of selected cells, established at the source-line prevents single V_{DD} operation. This is because an on-chip negative-voltage generator comprising charge-pumping circuits can never manage such heavy capacitance. The offset source-line scheme [12] solves the heavy capacitance issue, requiring an on-chip voltage-up converter instead of the above negative-voltage generator. In the above cells the minimum V_{DD} may be around 1V for practical designs in which soft error, V_T variations, and V_T mismatch in addition to a high V_T are considered. The boosted storage-node scheme [13] lowers the minimum V_{DD} down to less than 0.5V with the help of an on-chip voltage-up converter. A low V_T for the transfer FETs combined with negative word-line biasing also helps high-speed operation. However, the challenge is to realize sufficiently high voltage generation with low power.

The dynamic data-line load [8,11,14] that is particular to DRAMs is best in terms of power and speed, although the

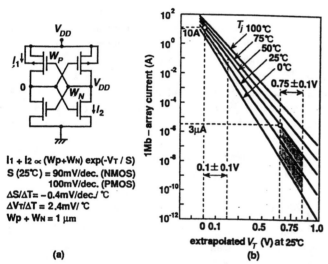

$I_1 + I_2 \propto (W_P + W_N) \exp(-V_T / S)$
$S\ (25°C) = 90mV/dec.\ (NMOS)$
$100mV/dec.\ (PMOS)$
$\Delta S/\Delta T = -0.4mV/dec./\ °C$
$\Delta V_T/\Delta T = 2.4mV/\ °C$
$W_P + W_N = 1\ \mu m$

(a)

(b)

Figure 2 Calculated subthreshold current of a SRAM cell-array. (a) The current sources in a full CMOS cell. (b) Subthreshold current of a 1Mb SRAM array versus the extrapolated V_T.

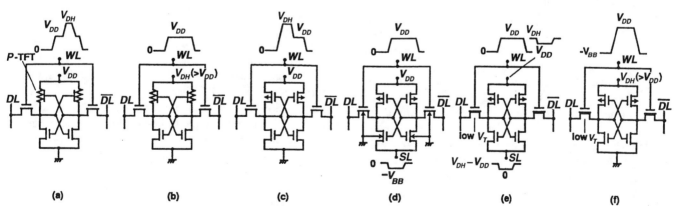

(a)   **(b)**   **(c)**   **(d)**   **(e)**   **(f)**

Figure 3 SRAM cell driving schemes. (a) Two-step word-voltage [8]. (b) Raised V_{DD} ($= V_{DH}$) [9]. (c) Step-down boosted wordline [10]. (d) Negative source-line [11]. (e) Offset source-line [12]. (f) Boosted storage node [13].

static (ratio) load has been common in SRAMs. The voltage swing on the data line, however, must be sufficiently reduced. The write power is greatly reduced by a small-signal differential-write operation combined with the half-V_{DD} data-line precharge [11]. In this operation a small signal written into the cell is amplified by turning on a cell feedback-loop after isolating the datalines. Control of the word-line replica feedback [14] precisely limits the read signal amplitude against various design-parameter variations. Regarding other peripheral circuits a CMOS latch-type amplifier [11], a high-V_T boosted NMOS I/O driver instead of a low-V_T PMOS driver, and a charge recycling I/O [10] are also proposed for ultralow-voltage operation.

4. Perspective

Here a perspective on low-power/low-voltage circuits is given with emphasis on ultralow-voltage operation.

A. Power-Supply Standardization

In the course of lowering the operating voltage of general purpose RAMs the external-power-supply standardization issue [1,3] is encountered, unlike with custom LSIs. The power-supply is still controversial even for 256-Mb and 1-Gb DRAM chips, although we are presently at 3.3V for 64-Mb DRAMs. However a gradual transition toward internal power-supplies of sub-V levels that are generated by an on-chip voltage-down converter [1] seems inevitable in terms of the ever-decreasing sizes of device features. Thus the converter is an important power-supply standardization circuit. Custom RAMs in which first priority is placed on ultralow-power rather than standardization, however, will not use an on-chip converter to avoid power loss at the converter. In any event, a landmark will be passed at around 1V which is suitable for one cell battery-operation. At a level of 0.5V, even one solar-cell operation [13] may become possible.

B. Ultralow-Voltage Operation

The above-described subthreshold-current issue should be solved through improvements in the existing subthreshold-current-reduction circuits. Note that the resulting circuits will be common in almost all DRAMs and SRAMs since the difference in circuit techniques between both types of RAMs will disappear, as expected from the above discussion. For example, in the past, DRAM products have needed the word-line boost while SRAM products have never needed it. In the ultralow-voltage era, however, both RAMs will need bootstrapping circuit techniques in the cell to eliminate the effect of high V_T. Multi-V_T logic circuits are also common. However, more attention should be paid to the following: boosting gives relatively ever-increasing stress voltage to high-V_T FETs as operating voltage decreases; the high V_T eventually restricts ultralow-voltage operation although an operation of at most 1V could be managed. This is because the conductance of high-V_T FETs decreases relatively as the V_{DD} approaches high V_T. This implies that the additional area and clock-power due to FET insertion increase relatively to

compensate for conductance decrease. Thus in the long run low-voltage operations of less than 1V are a real challenge.

C. On-Chip Voltage Generators

On-chip voltage generators support ultralow-voltage operations. The voltage-down converter offers single- and standard-V_{DD} operation by adjusting the internal supply voltage. The substrate-bias generator can control the V_T. The voltage-up converter directly raises voltage for boosted nodes without boosting capacitance. All generators necessitate high conversion efficiency, precise control, and trimming of internal voltages according to design-parameter variations which will be described later. Their power consumption is another concern. For example, a higher boost-ratio for a voltage-up converter consumes higher power because of lower conversion efficiency. Thus low-power/low-voltage analog circuits are expected to be increasingly important.

D. Design-Parameter Variations

Suppression or compensation circuits against variations in design parameters regarding V_T, channel length, temperature, and V_{DD} are important. Device-parameter variations are inevitably introduced during volume production. Even a fixed variation makes chip-to-chip speed-variations wider with lowering V_{DD}. Furthermore, miniaturization of FETs enhances variations, causing unexpectedly wide speed-variations at ultralow V_{DD}. Unregulated power supply provided from batteries makes the design more complicated with further speed-variations. On-chip voltage generators tracking the variations may partly solve the problem.

E. Others

In addition to noise suppression, and the development of a short channel FET and a higher S/N cell, as described earlier, SOI device-development [15] may be essential if the floating substrate problem is to be solved.

These are topics for future research.

5. Conclusion

A review and a perspective on the low-power/low-voltage RAM circuits were presented. The results obtained are as follows. 1) reductions in charging capacitance, operating voltage, and DC current reduced the DRAM and SRAM power. 2) In addition to a higher S/N cell design and miniaturization of devices, power-supply standardization circuits, on-chip low-power voltage generation, and suppression and compensation circuits against expected wide design-parameter variations are keys to ultralow-power/ultralow-voltage design.

References

[1] K. Itoh et al., "Trends in low-power RAM circuit technologies," IEEE Proc., vol. 83, no. 4, Apr. 1995, pp. 524–543.

[2] K. Itoh, "Low power memory design," *Low Power Design Methodologies,* J. M. Rabaey and M. Pedram, editors, Kluwer Academic Publishers, Oct. 1995, pp. 201–251.

[3] K. Itoh et al., "Limitations and challenges of multi-gigabit DRAM circuits," in *Symp. VLSI Circ. Dig. Tech. Papers,* June 1996, pp. 2–7.

[4] K. Itoh, "Trends in megabit DRAM circuit design," *IEEE J. Solid-State Circuits,* vol. 25, no. 3, June 1990, pp. 778–789.

[5] A. Saulsbury et al., "Missing the memory wall: the case for processor/memory integration," *23rd Int. Symp. Computer Architecture,* June 1996.

[6] D. Takashima et al., "Standby/active mode logic for sub-1-V operating ULSI Memory," *IEEE J. Solid-State Circuits,* vol. 29, no. 4, April 1994, pp. 441–447.

[7] M. Ukita et al., "A single-bit line cross-point cell activation (SCPA) architecture for ultra-low-power SRAM's," *IEEE J. Solid-State Circuits,* vol. 28, no. 11, Nov. 1993, pp. 1114–1118.

[8] K. Ishibashi et al., "A 1-V TFT-load SRAM using a two-step word-voltage method," *ISSCC Dig. Tech. Papers,* Feb. 1992, pp. 206–207.

[9] K. Ishibashi et al., "A 6-ns 4–Mb CMOS SRAM with offset-voltage-insensitive current sense amplifiers," *IEEE J. Solid-State Cirsuits,* vol. 30, no. 4, Apr. 1995, pp. 480–486.

[10] H. Morimura et al., "A 1-V 1-Mb SRAM for portable equipment," *Symp. Low Power Electronics and Design,* Aug. 1996, pp. 61–66.

[11] H. Mizuno et al., "Driving source-line (DSL) cell architecture for sub-1-V high-speed low-power applications," *Symp. VLSI Circ. Dig. Tech. Papers,* Jun. 1995, pp. 25–26.

[12] H. Yamauchi et al., "A 0.8 V/100 MHz/sub-5 mW-operated mega-bit SRAM cell architecture with charge-recycle offset-source driving (OSD) scheme," *Symp. VLSI Circ. Dig. Tech. Papers,* Jun. 1996, pp. 126–127.

[13] K. Itoh et al., "A deep sub-V, single power-supply SRAM cell with multi-V_T, boosted storage node and dynamic load," *Symp. VLSI Circ. Dig. Tech. Papers,* Jun. 1996, pp. 132–133.

[14] B. S. Amruturet et al., "Techniques to reduce power in fast wide memories," *Symp. Low Power Electronics,* Oct. 1994, pp. 92–93.

[15] S. Kuge et al., "SOI-DRAM circuit technologies for low power high speed multigiga scale memories," *IEEE J. Solid-State Circuits,* vol. 31, no. 4, Apr. 1995, pp. 586–591.

[16] H. Higuchi et al., "A 2-ns, 5-mW, synchronous-powered static-circuit fully associative TLB," *Symp. VLSI Circ. Dig. Tech. Papers,* Jun. 1995, pp. 21–22.

Trends in Low-Power RAM Circuit Technologies

KIYOO ITOH, SENIOR MEMBER, IEEE, KATSURO SASAKI, MEMBER, IEEE, AND
YOSHINOBU NAKAGOME, MEMBER, IEEE

Invited Paper

Trends in low-power circuit technologies of CMOS RAM chips are reviewed in terms of three key issues: charging capacitance, operating voltage, and dc current. The discussion includes a general description of power sources in a RAM chip, and covers both DRAM's and SRAM's. In DRAM's, successive circuit advancements have produced a power reduction equivalent to two to three orders of magnitude over the last decade for a fixed memory capacity chip. Coupled with the low-power advantage of CMOS circuits, two technologies have been the major contributors to power reduction: lower charging capacitance due to partial activation of multi-divided arrays that use multi-divisions of data and word lines; and lower operating voltage resulting from external power supply reduction, half-V_{DD} precharging, and on-chip voltage down converting scheme. In SRAM's, partial activation of a multi-divided word line drastically reduces the dc current from the data-line load to the selected cell. In addition to advances in the sense amplifier circuit, an auto power down scheme that uses address transition detection for word driver and column circuitry further reduces the dc current. It is also shown that to design ultralow voltage DRAM's and SRAM's, the application of subthreshold current reduction circuits (such as source-gate back biasing) to cell and iterative circuit blocks will be indispensable in the future.

I. INTRODUCTION

Low-power LSI technology is becoming an increasingly important and growing area of electronics. In particular, low-power RAM is a major area of this technology in which there has been rapid and remarkable progress in power reduction, with strong potential for future improvements [1], [2]. In large memory capacity RAM chips, active power reduction is vital to realizing low-cost, high-reliability chips because it allows plastic packaging, low operating current, and low-junction temperature. Hence, various low power circuit technologies concerning reductions in charging capacitance, operating voltage, and static current have been developed. As a result, active power has been reduced at

Manuscript received May 3, 1994; revised September 8, 1994.
K. Itoh and Y. Nakagome are with the Central Research Laboratory, Hitachi, Ltd., Kokubunji, Tokyo 185, Japan.
K. Sasaki is with Hitachi America, Ltd., Semiconductor Research Laboratory, San Jose, CA 95134 USA.
IEEE Log Number 9408303.

every generation despite a fixed supply voltage, increased chip size, and improved access time, as exemplified by the DRAM chips shown in Fig. 1(a).

Recent exploratory achievements in the movement toward low-voltage operation (departing from the standard power supply of 5 V) seemingly give promise of future improvements. Low-voltage operation has already been one of the most important design issues for 64-Mb or larger DRAM chips, since it is essential not only to further reduce power dissipation, but also to ensure reliability for miniaturized devices. This movement implies that the state-of-the-art device technology in miniaturization has at last progressed to the extent of using operating voltages of 3 V or less. In fact, a 64-Mb DRAM chip with an operating voltage of 1.5 V has been reported [3], affording an active current of 23 mA at a 230 ns cycle time. Reducing the data-retention power in DRAM's has been increasingly important for battery backup applications (where SRAM's are normally used), since DRAM's are inherently less expensive to produce than SRAM's. A data-retention current as low as 3 μA at 2.6 V has been reported for 4-Mb chip [4], as shown in Fig. 2. Such low current finally allows the design of DRAM battery operation, even under active operation mode, which is a key in battery-based handheld equipment such as mobile communication systems. The ever-decreasing supply voltage requirements could result in ultralow power and extend battery life. One target is 0.9 V, the minimum voltage of a NiCd cell. Even in this case, higher performance will eventually be required due to the ever-increasing demand for digital signal processing capability in such mobile equipment. To simultaneously achieve low voltage and high-speed operation, however, threshold voltage (V_T) scaling is highlighted as an emerging issue. This is because the MOSFET subthreshold dc current [5], even in CMOS chips, increases with decreasing V_T. The resultant dc current eventually dominates both active and retention currents, losing the low-power advantage of CMOS circuits that we take for granted today. The issue is essential not only for designing multi-gigabit DRAM chips

Reprinted from *Proceedings of the IEEE*, pp. 524-543, April 1995.

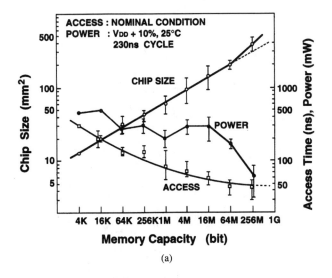

(a)

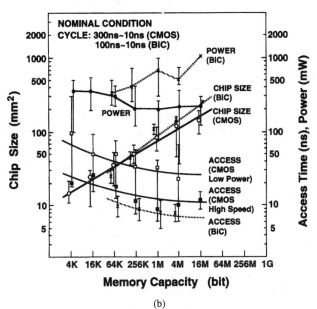

(b)

Fig. 1. Trends in RAM chip performance. Bit-width for I/O pin is mostly 1 b or 4 b for DRAM's and high-speed SRAM's, and mostly 8 b for low-power SRAM's. Data are from *Dig. Tech. Papers of ISSCC and Symp. VLSI Circuits.* For DRAM, data are added to [1]. (a) DRAM. (b) SRAM.

with a feature size of 0.1 μm or less in the future, but also for designing ultralow voltage multi-megabit DRAM chips with an existing fabrication process tailored to scaled V_T.

For SRAM's, there has been also a strong requirement for low power. In early days, SRAM chip development was focused on low-power applications especially with very low standby and data-retention power, while increasing memory capacity with high-density technology. Nowadays, however, more emphasis has been placed on high-speed rather than large memory capacity [6], [7], primarily led by cache applications in high-speed microprocessors. ECL BiCMOS SRAM chips with access times of 5–10 ns are good examples. In this case, power and chip size are of less concern, as shown in Fig. 1(b). For low-power applications, however, the primary concern is for CMOS technology to realize high-speed with minimum power. Tremendous efforts have

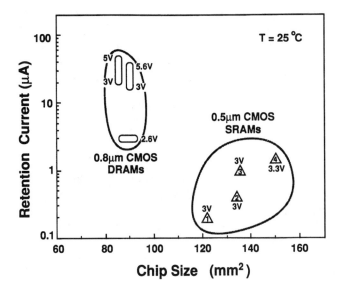

Fig. 2. Data retention current versus chip size for 4-Mb RAM chips [2] (4-Mb SRAM: 1, 2, 3: TFT load cell; 4: Poly load cell).

been made to minimize power with more emphasis on static current reduction and low voltage operation on long signal transmission lines. Consequently, there has been a decrease in CMOS SRAM power dissipation, in spite of a rapid increase in chip size and improved operating speed. However, this reduction trend does not look very significant compared with the trend in DRAM's. This is because high-speed characteristics have been more emphasized in the recent development of 1-Mb or larger SRAM chips. A typical example of a low-power high-density SRAM chip is a 16-Mb CMOS SRAM that achieves an access time of 15 ns and power dissipation of 165 mW at 30 MHz with 3-V power supply [8]. As for data-retention current, recent 4-Mb CMOS SRAM chips achieve sub-μA data-retention currents, as shown in Fig. 2, which are still about one order of magnitude less than those of 4-Mb DRAM chips, because they don't require a refresh operation. Ultralow voltage operation is also essential in SRAM design. This poses serious concerns for a subthreshold current and a V_T mismatch developed in a memory cell.

DRAM's and SRAM's both have reduced power dissipation through the use of common low-power circuit designs. However, there are few papers clarifying which circuits are common or different based on a systematic comparison.

This paper describes trends in low-power circuit technologies of CMOS RAM chips in terms of three key issues: charging capacitance, operating voltage, and dc static current, differentiating between DRAM's and SRAM's. First, power sources for active and data retention modes in a chip are described and essential differences in power sources between DRAM's and SRAM's are clarified. Next, the power reduction circuits for each power source are separately reviewed in the following sections, first for DRAM's, then for SRAM's. In Section III, DRAM circuits are discussed, with emphasis on charging capacitance for the memory array. In Section IV, SRAM circuits are intensively discussed with emphasis on dc current reduction for memory array and column circuitry. Perspectives on ultralow voltage design are also described in each section, mainly in terms of subthreshold current, although the design is still in its infancy.

II. SOURCES OF POWER DISSIPATION IN DRAM AND SRAM

A. Active Power Sources

Fig. 3 shows a simplified memory chip architecture for investigating its power dissipation. The chip comprises three major blocks of power sources: memory cell array, decoders (row and column), and periphery. Note that all the m cells on one word line are simultaneously activated in this logical array model. A unified active power equation for modern CMOS DRAM's and SRAM's is approximately given by

$$P = V_{DD}I_{DD} \tag{1}$$
$$I_{DD} = mi_{act} + m(n-1)i_{hld} + (n+m)C_{DE}V_{INT}f$$
$$+ C_{PT}V_{INT}f + I_{DCP} \tag{2}$$

for normal read cycle, where V_{DD} is an external supply voltage, I_{DD} is a current of V_{DD}, i_{act} is an effective current of active or selected cells, i_{hld} is an effective data retention current of inactive or nonselected cell, C_{DE} is an output node capacitance of each decoder, V_{INT} is an internal supply voltage, C_{PT} is a total capacitance of CMOS logic and driving circuits in periphery, I_{DCP} is a total static (DC) or quasi-static current of periphery, and f is an operating frequency ($f = 1/t_{RC}$: t_{RC} is a cycle time). Major sources of I_{DCP} are column circuitry and differential amplifiers on the I/O lines. Other contributors to I_{DCP} are the refresh-related circuits and on-chip voltage converters essential to DRAM operation; these include a substrate back-bias generator, a voltage-down converter, a voltage-up converter, a voltage reference circuit, and a half-V_{DD} generator. The dc currents of these circuits are virtually independent of the operating frequency. Hence, for high-speed operation I_{DCP} becomes relatively small compared with other ac current components. At high frequencies, the data retention current, $m(n-1)i_{hld}$, is negligible—evidenced by the small cell leakage current and small periphery current necessary for the refresh operation in DRAM's. In SRAM's, i_{hld} is also quite small, as described later. Decoder charging current, $(n+m)C_{DE}V_{INT}f$, is also negligibly small in modern RAM's incorporating a CMOS NAND decoder [9], because only one out of n or m (nodes) is charged at every selection: $n+m=2$. The active current for both DRAM's and SRAM's is shown in Fig. 4. Note that I_{DD} increases with increasing memory capacity, that is, with increased m and n.

1) DRAM: Destructive readout characteristics of a DRAM cell necessitate successive operations of amplification and restoration for a selected cell on every data line. This is performed by a latch-type CMOS sense amplifier on each data line. Consequently, data line is charged and

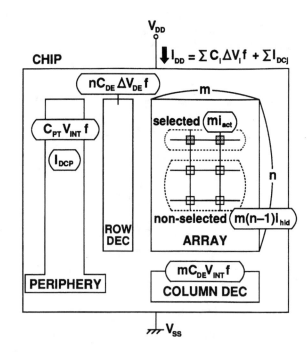

Fig. 3. Sources of power dissipation in a RAM chip.

discharged with a large voltage swing of ΔV_D (usually 1.5 ~ 2.5 V) and with charging current of $C_D\Delta V_D f$ where C_D is the data line capacitance. Hence, the current is expressed as

$$I_{DD} \cong [mC_D\Delta V_D + C_{PT}V_{INT}]f + I_{DCP}. \tag{3}$$

(1) and (3) show that the following issues are the keys to reducing active power for a fixed cycle time: 1) reducing charging capacitance (mC_D, C_{PT}), 2) lowering the external and internal voltages (V_{DD}, V_{INT}, ΔV_D), and 3) reducing static current (I_{DCP}). In particular, emphasis must be placed on reduction of total data-line dissipation charge ($mC_D\Delta V_D$), since it dominates the total active power, as described in Section III. However, the dissipation charge must be reduced while maintaining an acceptable signal-to-noise-ratio (S/N)–a compromise, since they are closely related to each other. S/N is an extremely important issue [1] for stable operation. This stems from the principle of DRAM cell operation that cell signal is not only small, but also read out on the floating data line, which is susceptible to noise. The signal, v_s, is approximately expressed as

$$v_s \cong (C_S/C_D)V_{DD}/2 = (C_S/C_D)\Delta V_D = Q_S/C_D \tag{4}$$

for the half-V_{DD} precharging scheme ($\Delta V_D = V_{DD}/2$) described in Section III, where C_S and Q_S are cell capacitance and cell signal charge, respectively. It is obvious from (3) and (4) that reducing C_D is effective for both reducing I_{DD} and increasing v_s, while reducing ΔV_D degrades v_s, despite the I_{DD} reduction. This implies the importance of increasing C_S and/or decreasing noise instead.

2) SRAM: Nondestructive readout characteristics of SRAM never require restoration of cell data, allowing the elimination of a sense amplifier on each data line. To obtain a fast read, the cell signal on the data line is made as small

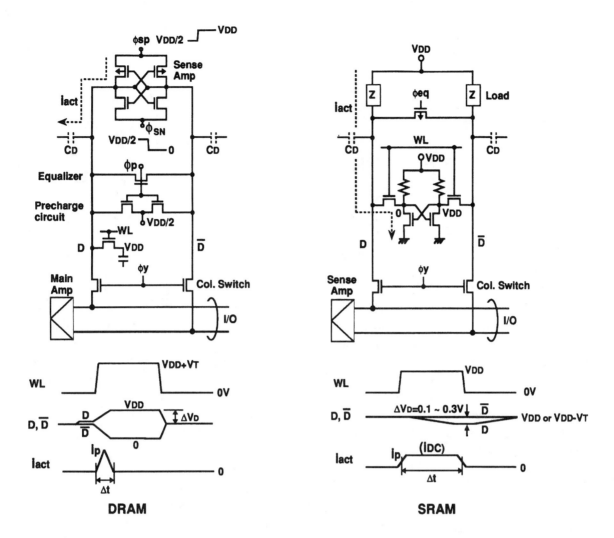

Fig. 4. Comparison between DRAM and SRAM read operation. Half-V_{DD} precharging scheme is assumed in DRAM's.

as possible, transmitted to the common I/O line through the column switch, and amplified by a sense amplifier. Since the cell signal is developed as the ratio voltage of a data-line load impedance to cell transistors, a ratio current i_{DC} flows along the data line during word-line activation, Δt. Here, data-line charging current is negligibly small due to a very small ΔV_D (= 0.1 V ~ 0.3 V), although it is prominent for write operation, as described in Section IV. Thus the current for read operation is expressed as

$$I_{DD} \cong [mi_{DC}\Delta t + C_{PT}V_{INT}]f + I_{DCP}. \quad (5)$$

To reduce active power, the three issues of static current, voltage, and capacitance, as in DRAM's, are vital. However, static current charge, $mi_{DC}\Delta t$, should be reduced more intensively because it dominates the total active current, which differs between SRAM's and DRAM's. Obviously, the S/N issue is not so serious as in DRAM's because of the ratio operation.

Eventually, DRAM's and SRAM's have evolved to use similar circuit techniques, although emphasis on each of the three issues is different between both types of RAM,

as described in the following sections. To clearly show the state-of-the-art RAM designs, the range of cell design parameters for the active current in DRAM's and SRAM's are compared in Table 1. Note the peak current, mi_P, which is a good measure of active power. A partial activation scheme of a multi-divided word line, as described later, reduces the DRAM current down to the SRAM level, although it is still in an experimental stage for DRAM's. In addition to the active current described above, the subthreshold dc current of a MOSFET will be a source of active power dissipation for future DRAM's and SRAM's supplied with ultralow voltages of less than 2 V. As V_T becomes small enough to no longer cut off the transistor, the subthreshold current increases exponentially with decreasing V_T. Temperature and V_T variations enhance the current, causing an active current increase and cell stability degradation. Moreover, the subthreshold current is proportional to total channel width of the FET's. Therefore, more attention should be paid to iterative circuit blocks such as decoder, word driver, and memory cell. Subthreshold current generated from an overwhelmingly large number of inactive (nonselected)

321

Table 1 Comparison of Determinants of Active Cell Current Between DRAM's and SRAM's (Estimated based on data in ISSCC Digests. DRAM's are 1 Mb to 16 Mb and SRAM's are 256 Kb to 4 Mb)

	DRAM	SRAM
Number of cells on a data-line pair	256~512	256~1024
C_D (pF)	0.2~0.3	1.0~2.0
v_s (V)	0.1~0.2	0.1~0.3
ΔV_D (V)	1.5~2.5	0.1~0.3
$i_p/i_{DC}(r)$ (μA)	20~50	100~200
Δt (ns)	10~20	5~50
m	2k~8k (128~512*)	64~128*
$mi_p/mi_{DC}(r)$ (mA)	100~160 (6~10*)	6~25*

*Partial activation of multi-divided word line

circuits would be eventually larger than ac current from the small number of active (selected) circuits, dominating the total chip active current. All the SRAM flip-flop cells could be dc substrate current sources [10].

B. Data Retention Power Sources

1) DRAM: In the data retention mode, a memory chip has no access from outside and the data are retained by the refresh operation. The refresh operation is performed by reading data of the m cells on a word line and restoring them for each of the n word lines in order. Note that n in the logical array in Fig. 3 corresponds to the number of refresh cycle in catalog specification. A current given by (3) flows every time m cells are refreshed at the same time. The frequency f at which the refresh current flows is n/t_{REF}, where t_{REF} is refresh time of cells in the retention mode and increases with reducing junction temperature. Thus from (3) data retention current is given by

$$I_{DD} \cong [mC_D\Delta V_D + C_{PT}V_{INT}](n/t_{REF}) + I_{DCP}. \quad (6)$$

The t_{REF} is much longer than the $t_{REF\,max}$ which is guaranteed in catalog specification, as shown in Fig. 10. This is because $t_{REF\,max}$ is for active mode operating at the maximum frequency of around 10 MHz where cell leakage current is maximized with highest junction temperature. On the other hand, t_{REF} is for extremely slow refresh frequency ($n/t_{REF} \ll 62$ KHz) where the current is minimized with lowest junction temperature. In any event, I_{DCP} becomes relatively large for other ac current components because of small n/t_{REF}. This implies the necessity of reducing both ac and dc components.

2) SRAM: In low power CMOS SRAM's static cell leakage currents, mni_{hld}, is the major source of retention current because of negligibly small I_{DCP}. The leakage current or retention current have been maintained to a small value by the following memory cell innovations. A high-resistance polysilicon load cell has been widely used for its high-density and low i_{hld} characteristics, although a

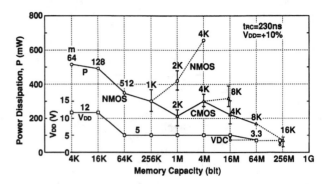

Fig. 5. Trend in power dissipation of DRAM chips. Data for 256-Mb are added to those in [1].

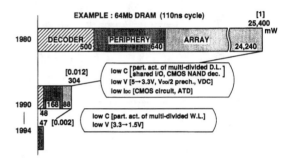

Fig. 6. Low-power circuit advancement of DRAM chip over the last decade.

full CMOS 6-T cell provides the smallest i_{hld}. For 4-Mb or higher density SRAM's, however, the polysilicon load cell starts to be replaced by a TFT-load cell since it faces difficulty in keeping low-power retention characteristics [11], [12].

For ultralow voltage operation involving scaled V_T, subthreshold currents from almost all circuits in DRAM's and SRAM's could drastically increase the retention current, again posing a serious concern.

III. LOW POWER DRAM CIRCUITS

A. Active Power Reduction

The power has been gradually decreased in spite of the increase in memory capacity, as shown in Figs. 1 and 5 [1]. This is due to low-power circuits developed at each generation. For a given memory capacity chip, successive circuit advancements have produced a power reduction equivalent to 2–3 orders of magnitude over the last decade. Fig. 6 shows the power dissipation of a 64-Mb DRAM, hypothetically designed with the NMOS circuit of the 64-Kb generation in 1980, compared with the CMOS circuit presented in 1990 [3]. Almost the same process and device technologies are assumed. The drastic reduction in power by about two orders of magnitude is due to many sophisticated circuits: partial activation of a multi-divided data-line and shared I/O which reduce mC_D in (3); CMOS NAND decoder which reduces $(n + m)C_{DE}$; external

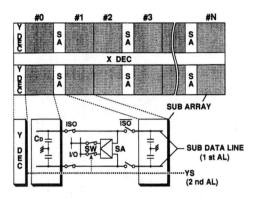

Fig. 7. Multi-divided data-line architecture with shared SA, shared I/O and shared Y decoder scheme [13].

supply voltage (V_{DD}) reduction from 5 to 3.3 V, half-V_{DD} data line precharge and on-chip voltage-down conversion which reduce V_{DD}, ΔV_D, and V_{INT} in (1) and (3); and CMOS drivers and pulse operation of column circuitry and amplifiers which reduce periphery static current, I_{DCP}. An exploratory 64-Mb DRAM chip, also hypothetically designed using the state-of-the-art technology, could further reduce operating power to about 1/10 that of the 1990 chip. Subthreshold current reduction to enable 1.5-V V_{DD} and partial activation of a multi-divided word line are responsible for this reduction. In this section, these key circuit technologies are reviewed. Here, the details of CMOS circuits are omitted because they are well known, although they are crucial to reduce the power of mega-bit DRAM's [9].

1) Charging Capacitance Reduction: The charging capacitance of a data-line is reduced by partial activation of multi-divided data lines, which is now widely accepted in commercial 16-Mb chips. Partial activation of multi-divided word lines further reduces the capacitance, although this is still in the experimental stage. Thus a combination of multi-division of the data-line and word-line minimizes the capacitance.

Partial Activation of Multi-divided Data Line: With increasing memory capacity, the ever-increasing number of memory cells connected to one data line causes an increased C_D and thus increased power as well as a poor signal-to-noise ratio, as expressed by (3) and (4). A practical solution is to divide one data line into several sections and to activate only one section. In the early days, the number of divisions was increased by simply increasing the number of Y decoder by using additional chip area. This Y decoder division was applied to various sense amplifier (SA) arrangements such as one SA at each division (normal arrangement) and a shared SA [1]. However, the division has almost completed at the 16-Mb generation with the combination of shared SA, shared I/O, and shared Y decoder shown in Fig. 7 [13]. Shared I/O further divides a multi-divided data line into two, which are selected by the isolation switches, ISO and $\overline{\text{ISO}}$. Shared SA provides an almost doubled cell signal with halved C_D. A shared Y decoder can be made without any increase in area

by using the second-level metal wiring for the column selection line (YS). The partial activation is performed by activating only one sense amplifier along the data line. Fig. 8(a) shows trends in total data-line charging capacitance, $C_{DT}(= mC_D)$. The C_{DT} has been minimized by sharing SA, I/O, and Y decoder, and by increasing the number of n, as described later. Fig. 8(b) shows trends in total data-line dissipating charge, $Q_{DT}(= mC_D\Delta V_D)$. The Q_{DT} has been suppressed as much as possible with the help of ever-reducing operating voltage, as shown in Fig. 5 and described later.

Partial Activation of Multi-divided Word Line: Fig. 9 shows the hierarchical word-line structure recently proposed in an experimental 256-Mb DRAM [14]. One word line is divided into several by the subword line (SWL) drivers. Any SWL is selected by both the main word line (MWL) and the row select line (RX), allowing partial activation of word lines. Note that the number of cells connected to one SWL in this physical array corresponds to m in the logical array in Fig. 3. The MWL pitch can be relaxed to 1/4 with this configuration. SWL drivers are placed alternately to relax the tight layout pitch of the DRAM, although the architecture is similar to that of SRAM, as described later. The architecture seemingly does not meet requirements for the traditional address multiplexing scheme since it increases the number of row address signals or introduces a speed penalty involved in additional selection. However, as far as power reduction is concerned, it has great potential. For example, the C_{DT} could be reduced down to about 100 pF at 256 Mb, as shown in Fig. 8, with assuming a C_D of 200 fF, a ΔV_D of 1 V and the number of activated data lines reduced by 1/32 as in [14]. Consequently, the architecture almost halves the chip power of the partial activation of the data line.

Refresh Time Increase: Even though reduction of C_D is achieved by the use of partial activations described above, reduction of m is never achieved without the help of increasing the maximum refresh time of the cell, $t_{REF \max}$. This stems from the necessity to preserve the refresh-busy rate [1], [15], γ, expressed by

$$\gamma = t_{RC \min}/(t_{REF \max}/n)$$
$$= (M/m)(t_{RC \min}/t_{REF \max}) \qquad (7)$$

where $t_{RC \min}$ and M are the minimum cycle time and memory capacity, respectively. It expresses what percentage of the time is not accessible from outside of the chip. A smaller γ is preferable since it involves less conflict between refresh operation and normal operation. Hence, for a fixed M, it is necessary to maintain $mt_{REF \max}$ to keep γ constant assuming a fixed $t_{RC \min}$. This implies a reduced m accompanied by an increased $t_{REF \max}$. Moreover, to quadruple M, $mt_{REF \max}$ must be quadrupled. This has been achieved by almost doubling both m and $t_{REF \max}$, which results from compromising the power with cell leakage current [1]. As a result, $n(= M/m)$ has gradually increased with each successive generation with increased $t_{REF \max}$ for commercial DRAM's [16],

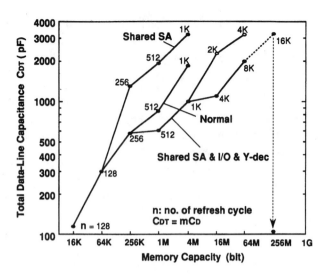

Total data-line capacitance

(a)

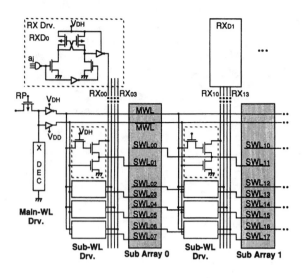

Fig. 9. Hierarchical word line architecture [14].

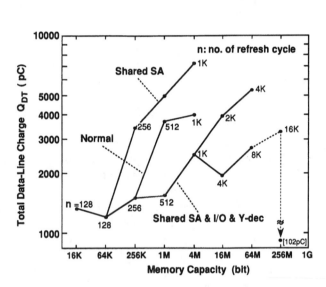

Total data-line dissipating charge

(b)

Fig. 8. (a)(b) Trends in total charging capacitance and dissipating charge of data lines. Data from [1] are revised with data from *Dig. Tech. Papers of ISSCC and Symp. on VLSI Circuits*. C_D of 200 fF and ΔV_D of 1 V are assumed for 256 Mb.

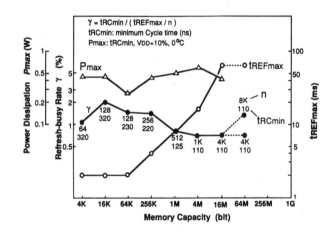

Fig. 10. Trends in power dissipation (P_{max}), refresh-busy rate (γ), and maximum refresh time ($t_{REF\,max}$) [16].

as shown in Fig. 10. Alternative choices for n and m, as shown in Figs. 5 and 8, have been eventually rejected by this compromise. However, it seems difficult to keep the pace of doubling $t_{REF\,max}$ at each generation because it is determined by cell leakage current. One solution is the use of a new refreshing scheme [14], [17] favorable to partial activation of a multi-divided word line. Note that the traditional scheme uses the same n, that is, the same m for both normal and refresh operations. The new scheme, however, uses a reduced m for normal operation which is

a determinant of maximum power, while maintaining the same m as in the traditional scheme for refresh operation to preserve γ. The resulting power reduction allows an increased $t_{REF\,max}$ with a reduced junction temperature.

2) Operating Voltage Reduction: V_{DD} reductions from 12 to 5 V at the 64-Kb generation and then to 3.3 V at the 64-Mb generation, as shown in Fig. 5, and half-V_{DD} data-line precharge which has been widely used since commercial 1-Mb DRAM's are well known as contributors to low power. On-chip voltage down converters [13] which are employed in commercial 16-Mb DRAM's to maintain a standard 5-V supply also contribute to low power. This low-power operation necessitates an improvement in signal-to-noise ratio, as described in Section II.

Half-V_{DD} Data Line Precharge: An excellent circuit for reducing the array operating current is half-V_{DD} data-line precharge [18], [19]. Table 2 shows full-V_{DD} precharge with an NMOS sense amplifier (SA) which was popular until the 256-Kb generation and half-V_{DD} data-line precharge favorable to CMOS SA. In full-V_{DD} precharge,

Table 2 Half-V_{DD} Precharging Sheme [9]

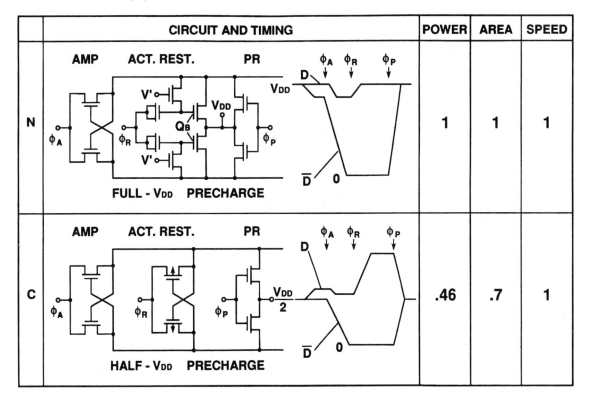

	CIRCUIT AND TIMING	POWER	AREA	SPEED
N	AMP ACT. REST. PR ϕ_A ϕ_R ϕ_P FULL - V_{DD} PRECHARGE	1	1	1
C	AMP ACT. REST. PR ϕ_A ϕ_R ϕ_P HALF - V_{DD} PRECHARGE	.46	.7	1

the voltage difference on the data lines is amplified by applying pulse ϕ_A and then the resultant degraded high level is restored to full-V_{DD} level by applying pulse ϕ_R to the active restore circuit. In contrast, the restore operation in half-V_{DD} precharge is simply achieved by applying pulse ϕ_R to the cross coupled PMOSFET's. In principle, half-V_{DD} precharge halves the data-line power of full-V_{DD} precharge with halved data-line voltage swing. A large spike current caused during restoring or precharging periods is also halved with less noise generation, allowing a quiet array.

On-chip Voltage Down Converter (VDC): The keys to designing VDC are provision of a stable and accurate output voltage under rapidly changing load current and provision of on-chip burn-in capability. Fig. 11 shows a schematic of a typical VDC [20] and the step response for the load current, I_L. The almost fixed output voltage, V_{DL}, is about 3.3 V for $V_{DD} = 5$ V, 16-Mb DRAM's. For accuracy and load current driving capability, it consists of a current-mirror differential amplifier (Q_1–Q_4, Q_S) and common-source drive transistor (Q_6). As shown in Table 1, the array current for a DRAM is fairly large compared with that of an SRAM. The peak height is more than 100 mA with a peak width of around 20 ns. Thus the gate width of Q_6 has to be more than 1000 μm. In order to minimize the output voltage drop ΔV_{DL}, the gate voltage of Q_6, V_G, has to respond quickly when the output goes low. An amplifier current I_S of 2–3 mA enables such a fast response time. Bias current source I_B is needed to clamp the output voltage when the load current

becomes almost zero. To ensure enough loop stability with minimized area and operating current, phase compensation [20] is indispensable. The reference voltage V_{REF} must be accurate over wide variations of V_{DD}, process, and temperature for stable operation, because the voltage level determines the amount of cell signal charge as well as the speed performance. Bandgap V_{REF} generator is well known for this purpose although the formation of bipolar transistors is the key when the CMOS process is used [21], [22]. Instead of this, a CMOS V_{REF} generator utilizing the threshold voltage difference has been proposed [23]. It employs a voltage up converter with trimming capability for precise adjustment of the output voltage. A burn-in operation with the application of a high stress voltage to devices is indispensable in VLSI production both for reliability testing and for chip screening. For this purpose, the V_{REF} generator is designed to output a raised voltage when V_{DD} is higher than the value for normal operation [22], [23]. Otherwise, the fixed voltage fails to apply a higher stress voltage.

Signal-to-Noise Ratio Improvement: There have been many important circuit techniques to maintain a sufficient signal-to-noise-ratio [1] such as a folded data-line arrangement to minimize array noise, word-line bootstrapping to store a full-V_{DD} level, and multi-division of data lines as described earlier. A half-V_{DD} cell plate is also an important concept for enabling doubled storage capacitance for a fixed electric field across the capacitor insulator [15]. It turns out that for a 16-Mb DRAM the significant increase in data-line interference noise [24] poses other constraints on

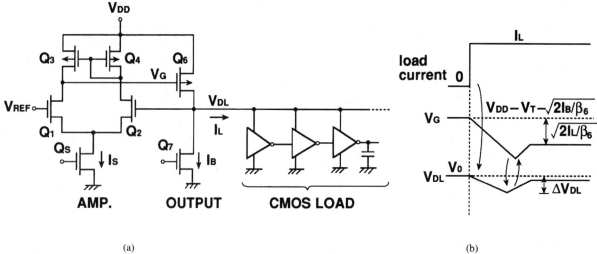

(a)   (b)

Fig. 11.   Voltage down converter [20]. (a) Circuit schematic. (b) Step response for loading.

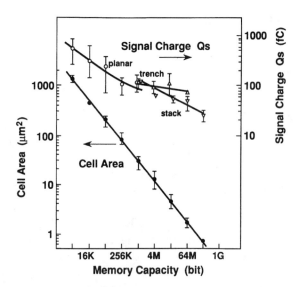

Fig. 12.   Trends in memory cell area and signal charge. Data of 256-Mb DRAM are added to those in [16].

cell and array design. To overcome this issue, a transposed data-line structure [25] or a shielded data-line structure has been employed. Memory cell developments focusing on maintaining a high enough signal charge and high immunity to noise charge induced by alpha particle hits [26] are also essential. From around the 4-Mb generation, to increase the capacitor area, vertical capacitor memory cells such as trench and stacked cells have been used in commercial DRAM's instead of planar cells. Recent 64–256 Mb stacked cells feature a storage node formed on the dataline and an advanced structure increasing the surface area [27]. In addition, the incorporation of new insulating films having a higher dielectric constant has been studied [28]. Due to such innovations in cell structure, signal charge reduction from the 64-K to 64-Mb generations is less than 1/5 while the cell area becomes 1/100, as shown in Fig. 12.

3) DC Current Reduction: Fig. 13 shows column signal path circuitry [29] which is a main source of static current.

It consists of a pair of data lines, a column switch, a pair of I/O lines, a load circuit for the I/O lines, and a differential amplifier ("main amp") for detecting the small signal voltage on the I/O lines. The dc current flows from the I/O line load to data lines while the column switch is on. Also, the main amplifier consumes dc current for amplifying the signal, because it employs the conventional current mirror differential amplifier, as described in Section IV. The dc currents can be shut down with pulse-operation technique. For example, in the static column mode where a DRAM operates as an SRAM for the column address signals, the column switch and the main amplifier is activated only when address-signal transition occurs. The ATD generates such control pulses, as described in Section IV.

B. Data Retention Power Reduction

Reduction of both dc and ac current components in data-retention mode is a prime concern. Minimizing the power of on-chip voltage converters such as VDC, voltage-up (V_{DH}) converter, substrate back-bias (V_{BB}) generator, V_{REF} generator, and half-V_{DD} generator reduces the dc current component. Extending the refresh time and reducing refresh charge reduce the ac current component. They reduce I_{DCP}, $1/t_{REF}$, and $mC_D\Delta V_D$ in (6), respectively.

1) Voltage Conversion Circuits: A precisely controlled V_{DH} level is important for eliminating V_T loss in memory cells and the periphery. However, a dynamic type booster which utilizes a capacitor to overdrive the output to the V_{DH} level suffers from uncontrollable output voltage for variation of V_{DD}, because V_{DH} is determined only by the capacitance ratio. A possible solution is a static word driver which directly operates from a V_{DH} dc power supply [30].

Fig. 14 shows a V_{DH} generator [31]. It features the use of two kinds of charge-pump circuits to provide charges for a pure capacitive output load: a main pump and an active kicker. The main pump compensates for a small charge loss due to the leakage current of the load. It is driven by the ring oscillator which is activated when V_{DH} is lower than

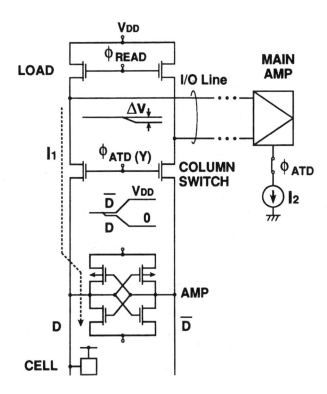

Fig. 13. Pulse operation of column signal path circuitry [29]. Typical parameters for 16-Mb DRAM's are: $I_1/I_2 = 0.5/1$ mA, $\Delta v = 0.5$ V, $C_{I/O} = 2$ pF.

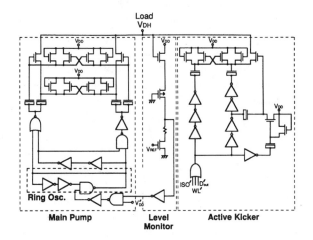

Fig. 14. V_{DH} generator [31].

the level determined by the level monitor. The active kicker operates synchronously with the load-circuit operation such as ISO driving (Fig. 7), word-line driving, and output buffer driving. This circuit compensates for a large charge loss due to load circuit operations. Non operation of the main pump and extremely slow cycle operation of the active kicker in the data-retention mode provide a minimized retention current.

A V_{BB} generator is indispensable for stable operation of a DRAM, especially for the array. This provides a negative dc voltage of around -2 to -3 V to the P-substrate which is almost capacitive. Fig. 15 shows the V_{BB} generator

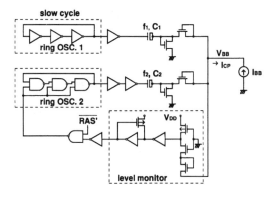

Fig. 15. Power saved V_{BB} generator [29].

featuring two sets of charge pump circuits [29]: a slow cycle ring oscillator 1 for supplying a small current during retention and standby modes and a fast cycle ring oscillator 2 for supplying a sufficiently larger current during the active cycle ($\overline{RAS}$ on) or when the level monitor detects the V_{BB} level is high. Thus it minimizes the retention current by shutting down the fast cycle circuit. A similar approach is useful for a VDC design [21]. Another alternative is to stop the oscillation of the V_{BB} generator while the DRAM is not in an active cycle [32]. This configuration is also useful for a VDC design.

With regard to a V_{REF} generator, a static V_{REF} generator has been proposed. However, a practical generator consumes more than 10 μA, because it needs a gain-trimable amplifier circuit in order to get an accurate voltage level [23]. Gain trimming with polysilicon resistance makes it difficult to get high resistance values of over 1 MΩ without a large area and poor noise immunity against substrate bounce. Thus an exploratory dynamic V_{REF} generator, as shown in Fig. 16, has been proposed. It features dynamic operation like a sample and hold technique [33]. Reference current I_R, which is determined by the threshold-voltage difference ΔV_T and resistance R_R, is mirrored to the output node and converted to the voltage determined by the value of resistance R_L. The output voltage is determined only by the V_T difference and resistance ratio. Thus an accurate voltage can be obtained by trimming R_R, with well known polysilicon fuses, even when ΔV_T fluctuates due to process variations. The current path is enabled while ϕ_1 is applied and the output voltage is sampled on the hold capacitance C_H while ϕ_2 is applied. These control pulses are generated from an on-chip self refresh circuit. For a pulse width ϕ_1 of 200 ns and a sampling interval of more than 100 ms, the total current is reduced to as low as 0.3 μA.

2) Refresh Time Extension: To extend the refresh time t_{REF} according to a reduced junction temperature in data-retention mode, a self-refresh control with an on-chip temperature detection circuit [34] and the use of a cell-leakage monitor circuit on the chip [35], [36] have been proposed.

3) Refresh Charge Reduction: One practical way to reduce the refresh charge is to reduce n (increase m) from

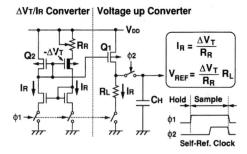

Fig. 16. Dynamic V_{REF} generator [33].

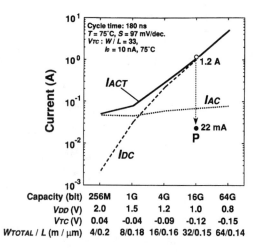

Fig. 17. Trends in active current of future DRAM chips [38]. Threshold voltage (V_{TC}) is defined for a constant drain-source current of 10 nA.

that for active operation [17]. This effectively reduces the operating frequency for the periphery while maintaining the array power constant. Another possible way is to reduce the voltage swing of data lines in the data-retention mode [36]. The resultant reduced signal charge and increase in soft error rate are additional issues with this scheme. A charge recycle refresh scheme [37] has been proposed to reduce the data-line dissipating charge. In this scheme, the charges used in one array, conventionally poured out in every cycle, are transferred to another array and used, enabling the data-line charging current to be halved.

C. Perspective

As is described in Section I, V_T scaling is the major concern for achieving ultra-low-voltage and hence ultra-low-power VLSI's. The dc chip current caused by subthreshold current [5], I_{DC}, increases exponentially with V_T reduction accompanied by lowering V_{DD}, and would increase not only data-retention current, but also active current. As a result, I_{DC} would even exceed the ac chip current, I_{AC}, which is the total charging current for capacitive loading, and eventually dominates the active current of the chip, I_{ACT}, as shown in Fig. 17 [38]. Subthreshold current from iterative circuit blocks is mainly responsible for this, as described in Section II. Note that the definition of V_T in the figure is for a constant drain-source current of 10 nA. This constant current V_T is estimated to be smaller by about 0.2 V than the V_T defined by extrapolation of the saturation current which is familiar to circuit designers. Attempts proposed so far to solve the problem are: source-gate back biasing and cutting off the leakage path with a high-V_T MOSFET while using a low-V_T MOSFET for the main signal path during the active period. These techniques reduce the dc current during the inactive period. This section gives details from a circuit design standpoint.

1) CMOS Basic Circuit: Fig. 18(a) shows a schematic of the switched-source-impedance (SSI) CMOS circuit [39]. It features a switched impedance (Sw-Z) at the source of the N-MOSFET Q_N, consisting of a resistor R_S for limiting the subthreshold current and a switch S_S for bypassing R_S. The switch keeps on during the active period so that the circuit works as a conventional CMOS circuit. For the inactive period when the switch is off, the subthreshold current raises the source voltage to V_{SL}, giving a gate-source back-biasing so that the current is reduced. This is just a negative

feedback effect. Thus the circuit features immunity to V_T fluctuations as a result of negative feedback. Resistor R_S may be implemented as a highly resistive poly-Si wire or as a MOSFET with a small channel width/length ratio. In fact, an intentional resistor is not necessarily needed. When the resistor is eliminated, R_S is regarded as the leakage resistance of switch S_S. This scheme is also applicable to other logic gates as long as the input voltages are predictable. Another circuit for reducing the subthreshold current is to use a switched-power-supply inverter with a level holder [40], as shown in Fig. 18(b). This is useful for some applications in which input voltage is not predictable as in the active mode of a DRAM, although there is a speed degradation due to Q'_N and Q'_P, and high-V_T eventually restricts the lower limit of V_{DD}. The power supply of the CMOS circuit is controlled by FET switches Q'_N and Q'_P. The V_T's of all FET's except Q_N and Q_P are high enough, allowing negligible subthreshold current. As soon as the input level has been evaluated at high speed due to low-V_T of Q_N and Q_P and the resultant output is held in the holder, the switches are turned off. Thus the output level is maintained without subthreshold current. The level holder can be laid out with minimized area since it plays only a role of level holding.

2) Iterative Circuit Blocks: One of the most efficient applications of the SSI scheme is to iterative circuit blocks. To reduce the subthreshold current in standby mode, it has been implemented in the decoded word driver of an experimental 256-Mb DRAM [41], [42], as shown in Fig. 19. A P-ch switching FET, Q_S, was inserted between the power-supply line V_{DH} and the driver-FET's (Q's) common-source terminal. The node PSL decreases in the standby mode by V_{SL} related to the subthreshold current (nI) in the P-MOSFET's in the driver circuits. V_{SL} is proportional to $\ln(nW_D/W_S)$. The channel width of the switching transistor W_S can be narrowed to an extent comparable to that of the driver FET W_D without speed degradation in active mode, since only one of n ($n = 256$,

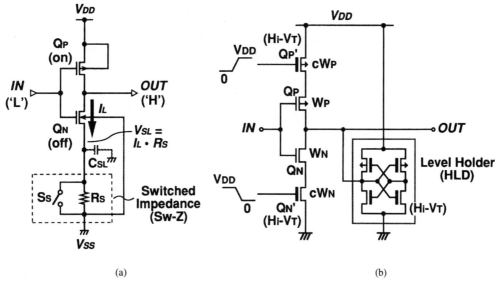

Fig. 18. Subthreshold current reduction for basic CMOS circuit [39], [40]. (a) Switched-source impedance. (b) Switched power supply inverter with level holding.

Table 3 Partial Activation of Multi-Divided Power Line [38]

		Conventional	1-D Selection	2-D Selection
Configuration		$m \cdot n$ Ckt.	Activated Inactivated	k
Active Subthreshold Current	Ideal	$m \cdot n \cdot l$	$n \cdot l$	$(n/k) \cdot l$
	Actual	$m \cdot n \cdot l$	$n \cdot l + (m-1) \cdot a \cdot l$	$(n/k) \cdot l + (m-1) \cdot a \cdot l + (k-1) \cdot b \cdot l$

for example) driver FET's turns on. Then, the subthreshold current decreases exponentially with V_{SL}, and the descent of the node PSL stops within 200–300 mV below V_{DH}. This enables a high-speed recovery (2–3 ns) of the node PSL back to the V_{DH} level in the transition from the standby mode to the active mode. The subthreshold current in active mode is another concern for iterative circuit blocks, although that in standby mode is reduced by the circuit described above. After one selected word line is activated, all the drivers are sources of subthreshold current, eventually dominating the total active current. The issue is overcome by partial activation of multi-divided power-line [40], as shown in Table 3. One-dimensional (1D) power-line selection features a selective power-supply to part of the circuit block by dividing it into m sub-blocks each consisting of n circuits. The operation is performed by

turning on a switch corresponding to a selected (activated) sub-block, while the others remain off. All the nonselected (inactivated) sub-blocks substantially have no subthreshold current since the same voltage relationship as in standby mode in Fig. 19 is established in each sub-block. This reduces the current to $n \cdot I$ with an m-fold reduction. For further reduction of the current, two-dimensional (2D) selection [38] has been reported. In this configuration, a circuit block having $m \cdot n$ circuits is divided into $m \cdot k$ sub-blocks, m in a row and k in a column. The subthreshold current is reduced, in inverse proportion to the number of sub-blocks, to $(n/k) \cdot I$ with an $m \cdot k$-fold reduction. In the conventional scheme, active dc current is as high as 1.2 A for a hypothetically designed 16-Gb DRAM [40]. The dominant factor is the subthreshold current in the iterative circuit blocks such as word drivers, decoders, and

329

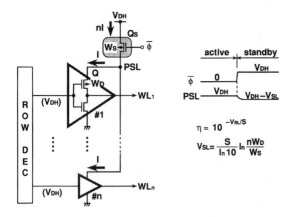

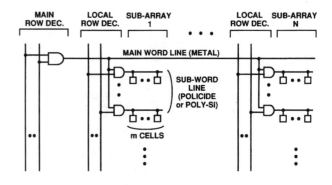

Fig. 21. Divided word line (DWL) structure [45].

Fig. 19. Subthreshold current reduction of DRAM word driver [41], [42]. S denotes the subthreshold swing, and η is the current reduction ratio.

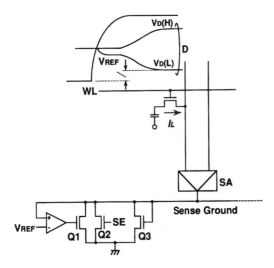

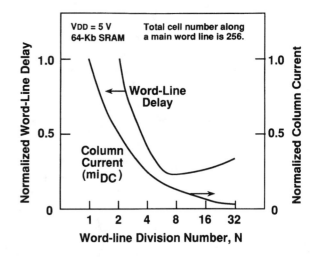

Fig. 22. Word-line delay and column current decrease by divided word line (DWL) structure [45].

Fig. 20. Boosted sense ground scheme [43].

IV. Low-Power SRAM Circuits

A. Active Power Reduction

SRAM chip development has been driven by low-power circuit applications, although the emphasis has recently been on more enhanced speed. Partial activation of a multi-divided word line and pulse operation of word-line circuitry are typical examples of low-power circuits, which drastically reduce the dc current that dominates total active current with decreasing m and Δt in (5). Pulse operation of the column/sense circuitry is another example of low-power techniques. Column/sense circuitry inevitably includes differential circuits, which unfortunately consumes much dc current to achieve high speed. Therefore, the pulse operation is essential to reducing I_{DCP} in (5). Address transition detection (ATD) plays an important role in the pulse operations for word line and column/sense circuitry. For column/sense circuitry, lowering the operating voltage while maintaining high-speed amplification capability for small signals is also critical.

1) DC Current Reduction: Partial Activation of Multidivided Word Lines: The multiple row decoder scheme and double-word line scheme [44], which divides a word line into subword lines, greatly reduces the static current of SRAM's. A more sophisticated word-line division, called

SA driving circuits. These dc currents could be reduced with a 2-D selection scheme down to 22 mA, as shown by point P in Fig. 17. When this scheme is combined with partial activation of multi-divided word line, an additional ac power and layout area due to 2-D control is negligible. This is because sub-array decode signal can be utilized for driving switches.

3) Memory Cell: The subthreshold current of a cell transistor would flow from the cell storage node to the data line while the data line is held at the low level. The current flow can be prevented by the use of a source-gate back-biasing scheme. Fig. 20 is an example of the application to a 256-Mb DRAM [43]. It is called the boosted sense-ground scheme. In an active cycle, the CMOS latch-type amplifier (SA) and Q_1 are activated and the voltage of the sense ground becomes V_{REF}. Q_2 is turned on by using signal SE at the beginning of the sensing operation in order to accelerate data-line discharging. Both Q_1 and Q_2 are turned off during the standby period and Q_3 is used to clamp the voltage level of the sense ground to minimize the standby current.

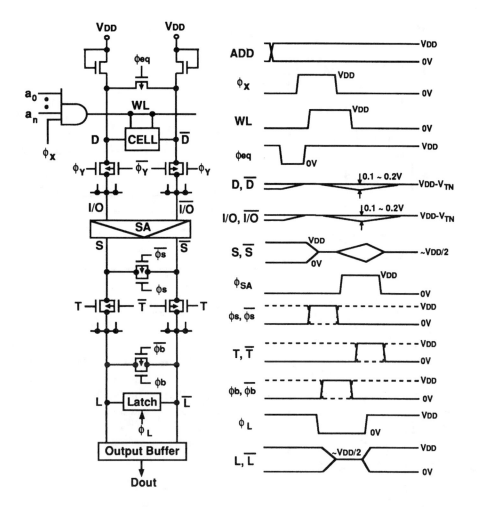

Fig. 23. Simplified view of pulsed operation of word line, sense amplifier SA, and latch circuit [48].

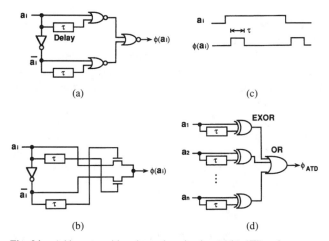

Fig. 24. Address transition detention circuits. (a)(b) ATD pulse generating circuits [49], [50]. (c) ATD pulse waveform. (d) Summation circuit of all ATD pulses generated from all address transitions.

divided word line (DWL), adopts a two-stage hierarchical row decoder structure [45], as shown in Fig. 21. The DWL scheme requires two levels of metal layers; one for a main word line and the other for a data line. The number of sub-word lines connected to one main word line in the data-line direction is generally four (at most eight), compromising an area to a main row decoder with an area to a local row decoder. DWL features two-step decoding to select one word line, greatly reducing the capacitance of the address lines to a row decoder and the word-line RC delay. Fig. 22 [45] shows that the column current (mi_{DC}) and word-line delay decrease as the word-line division number increases. The DWL scheme has been used in most high-density SRAM's of 1 Mb and greater. In a recent 16-Mb SRAM [8], a word line was divided into 32 sub-word lines: $N = 32$ and $m = 256$. The cell current was reduced to one thirty-second of its original level with the DWL scheme. However, this scheme eventually results in a long main word line having a load capacitance that increases due to the increase in the number of local row decoders. This is because the word-line division number N must increase while keeping m (number of cells selected simultaneously) small. To overcome the problem, two approaches have been proposed: a combination of a multiple row decoder and a DWL [46], and a three-stage hierarchical row decoder scheme [47]. An experimental 4-Mb SRAM with the three-stage hierarchical decoder reveals a reduction of total capacitance in the decoding pass by 30% and a reduction in delay by 20% compared with the DWL scheme.

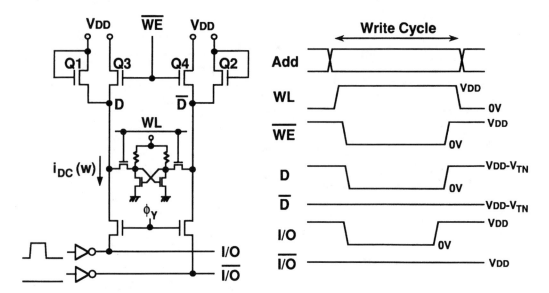

Fig. 25. Simplified diagram of data line load control with write enable signal /WE (Variable impedance data-line load) [51].

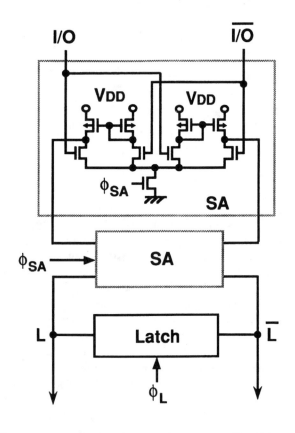

Fig. 26. Power switching of current-mirror sense amplifier [48].

Pulse Operation of Word-Line Circuitry. The duration of the active duty cycle can be shortened by pulsing the word-line for the minimum time required for reading and writing in a cell array, as shown in Fig. 23 [48]. This reduces the power by the duty ratio of the pulse duration to the cycle time. The word-activating pulse, ϕ_x, is obtained by lengthening out the original ATD pulse, described shortly, enough to build up the data-line signal and latch the

amplified signal by ϕ_L. This scheme is usually employed with a pulsed sense amplifier and a latch circuit, as shown in Fig. 23. One of data line signal pairs (D, $\overline{D}$) is selected by ϕ_y and $\overline{\phi}_y$, and transmitted to I/O and $\overline{I/O}$, respectively. The I/O signals are amplified by a sense amplifier SA. The amplified S and $\overline{S}$ signals from a subarray are selected by signals T and $\overline{T}$ and then are transmitted to the latch where the signals are latched to keep the data output valid after the word line and sense amplifier are inactivated. An on-chip pulse-generating scheme using ATD appeared in SRAM's first, then in DRAM's. An ATD circuit comprises delay circuits and an exclusive OR circuit, as shown in Fig. 24(a) and (b) [49], [50]. An ATD pulse $\phi(a_i)$ is generated by detecting "L" to "H" or "H" to "L" transitions of any input address signal a_i as shown in Fig. 24(c). All the ATD pulses generated from all the address input transitions are summed up to one pulse, ϕ_{ATD}, as shown in Fig. 24(d). This summation pulse is usually stretched out with a delay circuit and used to reduce power or speed up signal propagation.

An additional dc current in the write cycle, as shown in Fig. 25 [51], is another concern. An accurate array current expression for the read cycle, exclusive of periphery ac and dc currents in (5), is given by

$$I_{DDA}(r) = [mi_{DC}(r)\Delta t + mC_D\Delta V_r]f \qquad (8)$$

and a corresponding current for the write cycle is given by

$$\begin{aligned} I_{DDA}(w) = [&(m - p)i_{DC}(r)\Delta t \\ &+ pi_{DC}(w)\Delta t + pC_D\Delta V_w]f \end{aligned} \qquad (9)$$

where $i_{DC}(r)$ and $i_{DC}(w)$ are data-line static currents in the read and write cycles; ΔV_r and ΔV_w are data-line voltage swings in the read and write cycles, respectively; p is the number of data which are simultaneously written into cells. In a typical 5 V 4-Mb SRAM, $I_{DDA}(r)$ and $I_{DDA}(w)$ are 4.1 mA and 6.4 mA, respectively, assuming

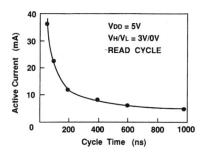

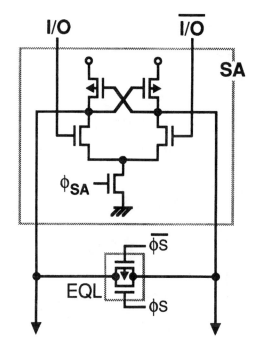

Fig. 27. Measured active current in a 4-Mb CMOS SRAM with pulse word line and pulse sense amplifier control [11]. V_H and V_L are high and low-input voltage levels, respectively.

Fig. 28. PMOS cross coupled sense amplifier [55].

$i_{DC}(r) = 100\ \mu A$, $i_{DC}(w) = 1.0$ mA, $\Delta V_r = 0.2$ V, $\Delta V_w \cong V_{DD}, C_D = 1$ pF, $m = 128$, $p = 8$, $\Delta t = 30$ ns, and $f = 10$ MHz. Inherently large $pi_{DC}(w)$ and $pC_D\Delta V_w$ make $I_{DDA}(w)$ larger than $I_{DDA}(r)$, especially for multibit SRAM's which make p large. To reduce $I_{DDA}(w)$, both $i_{DC}(w)$ and C_D must be decreased. To reduce $i_{DC}(w)$, the variable impedance load [51], [52] makes all the data-line load impedances high in the write cycle by cutting off Q_3 and Q_4 with the write enable signal, $\overline{WE}$, as shown in Fig. 25. In some SRAM's the loads are entirely cut off during the write cycle to stop any dc current [53]. C_D is reduced by partially activating the data lines that are divided into two or more portions, just as in DRAM's.

Pulse Operation of Column/Sense Circuitry: A dc current of 1 mA to 5 mA flows in a sense amplifier on the I/O line. The power dissipation becomes the larger portion of the total chip power as the number of I/O lines increases to obtain higher data throughput for high-speed processors. Fig. 26 shows a switching scheme of well known current-

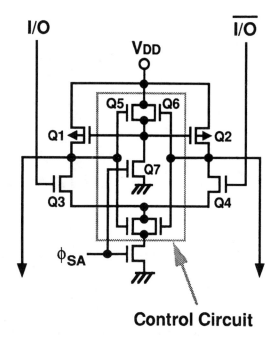

Fig. 29. Differential sense amplifier with current source control [57].

mirror sense amplifiers [48]. Two amplifiers are serially connected to obtain a full supply-voltage swing output since one stage of the amplifier does not provide a gain enough for a full swing. A positive pulse, ϕ_{SA}, activates the sense amplifiers just long enough to amplify the small input signal; then the amplified output is latched. Hence, the switching scheme combined with a pulsed word line scheme reduces the power, especially at relatively low frequencies, as shown in Fig. 27 [11]. Pulse operation of word-line circuitry and sense circuitry was also applied to fast SRAM's [67], achieving 24-mA power consumption at 40 MHz and 13-ns access time. Further current reduction is gained by sense amplifier current control [54], which switches the current to the minimum level required for maintaining data from a high level necessary only during amplification. Moreover, a latch-type sense amplifier such as a PMOS cross coupled amplifier [55], as shown in Fig. 28, greatly reduces the dc current after amplification and latching, since the amplifier provides a nearly full supply-voltage swing with positive feedback of outputs to PMOSFET's. As a result, the current in the PMOS cross coupled sense amplifier is less than one-fifth of that in a current-mirror amplifier, with affording very fast sensing speed. Equalizers EQL equilibrate the paired outputs of the amplifier before amplification, which requires much more accurate timings for stable operation.

2) Operating Voltage Reduction: Low Voltage Sense Amplifier: A 5-V power supply and NMOS data-line loads, as shown in Fig. 23, have been widely used for SRAM's. An NMOS V_T drop provides intermediate input voltages, essential to obtain large gains and fast sensing speeds, to sense amplifiers. For lower V_{DD} operation, however, if a resultant data-line signal voltage close to V_{DD} is

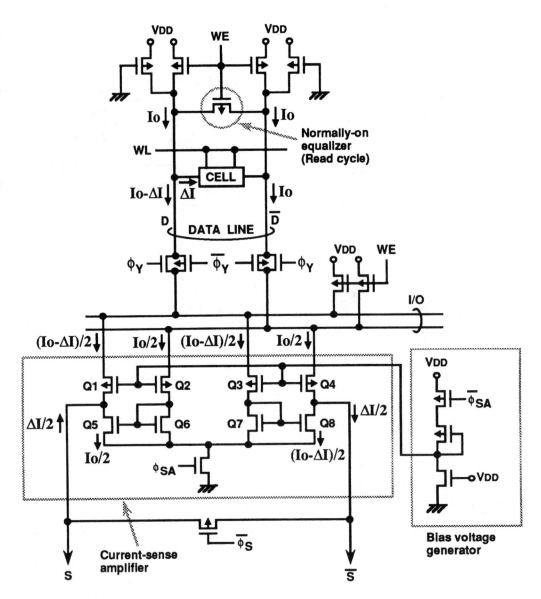

Fig. 30. Current sense amplifier [59].

amplified, PMOS data-line loads without the V_T drop are more suitable. This is made possible by level-shifting the data-line voltages. A resulting intermediate voltage allows it to use conventional voltage amplifiers for the succeeding stage. The use of NMOS source followers [47] and scaled low-V_T NMOS's [56] are good examples of level shifting. The other method which features a level shifting with amplification function is shown in Fig. 29 [57]. Despite an input level close to V_{DD}, an amplified output is developed on an intermediate level since the circuit operates as an equivalent level shifter as well as an amplifier. The currents of Q_1–Q_4 are controlled with the amplifier outputs to provide a stable operation. Additional power due to the control circuit is negligibly small because a small size transistor is used for Q_7 and the control circuit can be used commonly for more than two sense amplifiers.

A current sense amplifier combined with PMOS data-line loads [58], [59], as shown in Fig. 30, facilitates low-voltage, high-speed operation. A normally on equalizer

makes a small current difference between the paired data lines during a read cycle, depending on cell information. The current-mirror configuration of the amplifier circuit makes a current difference of $\Delta I/2$ between Q_1 and Q_5 (Q_4 and Q_8), which eventually discharges and charges the outputs, S and $\overline{S}$. The bias voltage generator provides an intermediate voltage of 1 V to 1.5 V at 3-V V_{DD} to increase gain by operating Q_1–Q_4 close to the saturation region. Current sensing provides the advantage of an extremely small data-line voltage swing of less than 30 mV, and eliminates the need for pulsed data-line equalization, allowing for fast sensing. Note that the required voltage swing in a conventional voltage amplifier is 100 mV to 300 mV, as shown in Table 1. For a fixed delay of 1.2 ns, the amplifier reduces the current consumption by about 2 mA compared with a conventional current-mirror voltage amplifier.

On-chip Voltage Down Conversion. On-chip power supply conversion was first used for a 256-Kb SRAM [60] to internally supply 3.3 V to the 0.7-μm devices with a

5-V V_{DD}. The SRAM has a power down mode to reduce standby chip current to 5 μA. A 4-Mb SRAM, which turns off one of two VDC's (voltage down converters) to provide a 50-μA standby current, automatically shut down the two VDC's when the external supply voltage is reduced to 3.3 V to obtain a 1-μA data retention current [12]. An experimental VDC which achieves sub-μA standby current has been reported [61]. The VDC circuits in SRAM's are basically the same as DRAM VDC's, as shown in Section III.

3) Charging Capacitance Reduction: Charging capacitance reduction techniques, initially targeted at obtaining high speed in SRAM's, also contribute to power reduction, which is more of an issue in DRAM's. These techniques include data-line division, I/O line division, and pre-decoding scheme. Inserting a pre-decoding stage between an address buffer and a final decoder optimizes both speed and power, and has been used in most SRAM's.

B. Data Retention Power Reduction

On-chip voltage converters in commercial SRAM's have not been widely used, although there have been many attempts which differ from the DRAM approach. This is mainly due to an inherently wider voltage margin and a different operating principle of the SRAM cell. Thus data retention current has been sufficiently reduced solely by the memory cell improvement, as described in Section II. Switching the power supply voltage to 2 V–3 V, from 5 V at normal operation, further reduces the data retention current. However, this voltage switching approach is eventually restricted by the ultralow voltage ability of SRAM's, described in the following section.

C. Perspective

Low-voltage operation capability of an SRAM memory cell is a critical issue. A full CMOS memory cell with an inherently wide voltage margin can be operated at the lowest supply voltages, as exemplified by a 1-Mb SRAM with a 200-ns access time at 1 V [62]. For polysilicon-loads or polysilicon PMOS load cells, a two-step word voltage scheme [56] and an array operational scheme with a raised dc voltage [63] have been proposed for widening the voltage margin. The raised dc voltage supplied to the source of PMOSFET's in a cell allows the storage node voltage of the cell to quickly rise in the write operation. However, the low-voltage SRAM's mentioned above are quite slow because of relatively high V_T of 0.5 V and more. Thus the low V_T designs are indispensable for high-speed operation. However, this in turn tremendously increases memory cell leakage currents [10] due to subthreshold current, although subthreshold current increases in the periphery could be overcome by the use of reduction circuits, as described in Section III. In addition to a soft error increase due to a decreased signal charge, the V_T mismatch that continues to increase between paired FET's in a cell [64], and V_T variation [65], are also prime concerns. Circuit techniques to suppress the mismatch in sense amplifiers with a 3

V power supply have been recently reported [63], [66]. However, there are still challenges facing ultralow voltage SRAM's.

V. CONCLUSION

The remarkable progress in power reduction of CMOS RAM chips and supporting circuit technologies were reviewed in terms of three key issues: charging capacitance, operating voltage, and dc current. Using the format of a general description of power sources in a RAM chip, power reduction circuit technologies were separately reviewed for DRAM's and SRAM's. In DRAM's, reducing the charging capacitance by partially activating the multi-divided array is extremely important. Lowering the operating voltage using external power supply reduction, half-V_{DD} precharging, and on-chip voltage down conversion, combined with high signal-to-noise-ratio technologies in cells, also contribute to power reduction. These technologies have produced a power reduction equivalent to two to three orders of magnitude over the last decade for a fixed memory capacity chip. In SRAM's, partially activating a multi-divided word line is essential in reducing dc current. In addition to advances in the sense amplifier circuit, pulsing the word driver and column circuitry through address transition detection further reduces the dc current. It was also shown that to design ultralow voltage DRAM's and SRAM's, the application of subthreshold current reduction circuits, such as source-gate back biasing, to cell and iterative circuit blocks will be indispensable in the future. Continuing challenges facing ultralow power RAM's will open the door to an exciting era in low-power electronics.

ACKNOWLEDGMENT

The authors wish to thank M. Ishihara, A. Endoh, K. Shimohigashi, E. Takeda, and M. Aoki, Hitachi, Ltd. for their helpful suggestions and discussions. The authors are also indebted to J. A. Whitacre, Hitachi America, Ltd. for assisting with the manuscript.

REFERENCES

[1] K. Itoh, "Trends in megabit DRAM circuit design," *IEEE J. Solid-State Circ.*, vol. 25, pp. 778–789, June 1990.
[2] Y. Nakagome and K. Itoh, "Review and prospects of DRAM technology," *IEICE Trans. Electron.*, vol. E74-C, no. 4, pp. 799–811, Apr. 1991.
[3] Y. Nakagome *et al.*, "An experimental 1.5-V 64-Mb DRAM," *IEEE J. Solid-State Circ.*, vol. 26, pp. 465–472, Apr. 1991.
[4] K. Satoh *et al.*, "A 4 Mb pseudo-SRAM operating at 2.6 ± 1 V with 3 μA data-retention current," in *ISSCC Dig. Tech. Papers*, Feb. 1991, pp. 268–269.
[5] S. M. Sze, *Physics of Semiconductor Devices*, 2nd ed. New York: Wiley, 1981.
[6] M. Takada *et al.*, "Reviews and prospects of SRAM technology," *IEICE Trans.*, vol. E74, no. 4, pp. 827–838, Apr. 1991.
[7] K. Sasaki, "High-speed, low-voltage design for high-performance static RAM's," in *Proc. Tech. Papers, VLSI Tech., Syst., and Appl.*, May 1993, pp. 292–296.
[8] M. Matsumiya *et al.*, "A 15 ns 16Mb CMOS SRAM with reduced voltage amplitude data bus," in *ISSCC Dig. Tech. Papers*, Feb. 1992, pp. 214–215.

[9] K. Kimura *et al.,* "Power reduction techniques in megabit DRAM's," *IEEE J. Solid-State Circ.,* vol. SC-21, pp. 381–389, June 1986.

[10] M. Aoki and K. Itoh, "Low-voltage, low-power ULSI circuit techniques," *IEICE Trans. Electron.,* vol. E77-c, no. 8, pp. 1351–1360, Aug. 1994.

[11] K. Sasaki *et al.,* "A 23-ns 4-Mb CMOS SRAM with 0.2-μA standby current," *IEEE J. Solid-State Circ.,* vol. 25, pp. 1075–1081, Oct. 1990.

[12] S. Hayakawa *et al.,* "A 1 μA retention 4 Mb SRAM with a thin-film-transistor load cell," in *ISSCC Dig. Tech. Papers,* Feb. 1990, pp. 128–129.

[13] K. Itoh *et al.,* "An experimental 1 Mb DRAM with on-chip voltage limiter," in *ISSCC Dig. Tech. Papers,* Feb. 1984, pp. 106–107.

[14] T. Sugibayashi *et al.,* "A 30 ns 256 Mb DRAM with multi-divided array structure," in *ISSCC Dig. Tech. Papers,* Feb. 1993, pp. 50–51.

[15] K. Fujishima *et al.,* "A 256K dynamic RAM with page-nibble mode," *IEEE J. Solid-State Circ.,* vol. SC-18, pp. 470–478, Oct. 1983.

[16] K. Itoh, "Reviews and prospects of deep sub-micron DRAM technology," in *Int. Conf. Solid-State Devices and Materials,* Yokohama, Extended Abs., Aug. 1991, pp. 468–471.

[17] K. Kenmizaki *et al.,* "A 36 μA 4 Mb PSRAM with quadruple array operation," in *Symp. VLSI Circ. Dig. Tech. Papers,* May 1989, pp. 79–80.

[18] N. C. C. Lu and H. Chao, "Half-V_{DD} bit-line sensing scheme in CMOS DRAM," *IEEE J. Solid-State Circ.,* vol. SC-19, pp. 451–454, Aug. 1984.

[19] H. Kawamoto *et al.,* "A 288-K CMOS pseudostatic RAM," *IEEE J. Solid-State Circ.,* vol. SC-19, pp. 619–623, Oct. 1984.

[20] H. Tanaka *et al.,* "Stabilization of voltage limiter circuit for high-density DRAM's using pole-zero compensation," *IEICE Trans. Electron.,* vol. E75-C, no. 11, pp. 1333–1343, Nov. 1992.

[21] D. Chin *et al.,* "An experimental 16-Mbit DRAM with reduced peak-current noise," *IEEE J. Solid-State Circ.,* vol. 24, pp. 1191–1197, June 1989.

[22] R. S. Mao *et al.,* "A new on-chip voltage regulator for high density CMOS DRAM's," in *Symp. VLSI Circ. Dig. Tech. Papers,* June 1992, pp. 108–109.

[23] M. Horiguchi *et al.,* "Dual-regulator dual-decoding-trimmer DRAM voltage limiter for burn-in test," *IEEE J. Solid-State Circ.,* vol. 26, no. 11, pp. 1544–1549, Nov. 1991.

[24] Y. Nakagome *et al.,* "The impact of data-line interference noise on DRAM scaling," *IEEE J. Solid-State Circ.,* vol. 23, pp. 1120–1127, Oct. 1988.

[25] T. Yoshihara *et al.,* "A twisted bit line technique for multi-Mb DRAM's," in *ISSCC Dig. Tech. Papers,* Feb. 1988, pp. 238–239.

[26] N. C. C. Lu, "Advanced cell structures for dynamic RAM," *IEEE Circ. and Devices Mag.,* pp. 27–36, Jan. 1989.

[27] T. Ema *et al.,* "3-dimensional stacked capacitor cell for 16M and 64M DRAM's," in *IEDM Tech. Dig.,* Dec. 1988, pp. 592–595.

[28] T. Horikawa *et al.,* "(Ba$_{0.75}$Sr$_{0.25}$)TiO$_3$ films for 256 Mbit DRAM," *IEICE Trans. Electron.,* vol. E77-C, no. 3, pp. 385–391, Mar. 1994.

[29] K. Satoh *et al.,* "A 20 ns static column 1 Mb DRAM in CMOS technology," in *ISSCC Dig. Tech. Papers,* Feb. 1985, pp. 254–255.

[30] G. Kitsukawa *et al.,* "A 23-ns 1-Mb BiCMOS DRAM," *IEEE J. Solid-State Circ.,* vol. 25, pp. 1102–1111, Oct. 1990.

[31] D. Lee *et al.,* "A 35 ns 64 Mb DRAM using on-chip boosted power supply," in *Symp. on VLSI Circ. Dig. Tech. Papers,* June 1992, pp. 64–65.

[32] Y. Konishi *et al.,* "A 38 ns 4 Mb DRAM with a battery back-up mode," in *ISSCC Dig. Tech. Papers,* Feb. 1989, pp. 230–231.

[33] H. Tanaka *et al.,* "Sub-1-μA dynamic reference voltage generator for battery-operated DRAM's," in *Symp. VLSI Circ. Dig. Tech. Papers,* May 1993, pp. 87–88.

[34] D. C. Choi *et al.,* "Battery operated 16M DRAM with post package programmable and variable self refresh," in *Symp. VLSI Circ. Dig. Tech. Papers,* June 1994, pp. 83–84.

[35] K. Sawada *et al.,* "Self-aligned refresh scheme for VLSI intelligent dynamic RAM's," in *Symp. VLSI Tech. Dig. Tech. Papers,* May 1986, pp. 85–86.

[36] M. Tsukude *et al.,* "Automatic voltage-swing reduction (AVR) scheme for ultra low power DRAM's," in *Symp. VLSI Circ. Dig. Tech. Papers,* June 1994, pp. 87–88.

[37] T. Kawahara *et al.,* "A charge recycle refresh for Gb-scale DRAM's in file applications," in *Symp. VLSI Circ. Dig. Tech. Papers,* May 1993, pp. 41–42.

[38] T. Sakata *et al.,* "Two-dimensional power-line selection scheme for low subthreshold-current multi-gigabit DRAM's," in *ESS-CIRC Dig. Tech. Papers,* Sept. 1993, pp. 33–36.

[39] M. Horiguchi *et al.,* "Switched-source-impedance CMOS circuit for low standby subthreshold current giga-scale LSI's," *IEEE J. Solid-State Circ.,* vol. 28, pp. 1131–1135, Nov. 1993.

[40] T. Sakata *et al.,* "Subthreshold-current reduction circuits for multi-gigabit DRAM's," in *Symp. VLSI Circ. Dig. Tech. Papers,* May 1993, pp. 45–46.

[41] G. Kitsukawa *et al.,* "256 Mb DRAM technologies for file applications," in *ISSCC Dig. Tech. Papers,* pp. 48–49, Feb. 1993.

[42] T. Kawahara *et al.,* "Subthreshold current reduction for decoded-driver by self biasing," *IEEE J. Solid-State Circ.,* vol. 28, pp. 1136–1144, Nov. 1993.

[43] M. Asakura *et al.,* "A 34 ns 256 Mb DRAM with boosted sense-ground scheme," in *ISSCC Dig. Tech. Papers,* Feb. 1994, pp. 140–141.

[44] O. Minato *et al.,* "A 42 ns 1 Mb CMOS SRAM," in *ISSCC Dig. Tech. Papers,* Feb. 1987, pp. 260–261.

[45] M. Yoshimoto *et al.,* "A 64 Kb CMOS RAM with divided word line structure," in *ISSCC Dig. Tech. Papers,* pp. 58–59, Feb. 1983.

[46] F. Miyaji *et al.,* "A 25 ns 4 Mb CMOS SRAM with dynamic bit line loads," in *ISSCC Dig. Tech. Papers,* pp. 58–60, Feb. 1983.

[47] T. Hirose *et al.,* "A 20-ns 4-Mb CMOS RAM with hierarchical word decoding architecture," *IEEE J. Solid-State Circ.,* vol. 25, pp. 1068–1074, Oct. 1990.

[48] O. Minato *et al.,* "A 20 ns 64K CMOS RAM," in *ISSCC Dig. Tech. Papers,* pp. 222–223, Feb. 1984.

[49] T. Tsujide *et al.,* "A 25 ns 16 × 1 static RAM," in *ISSCC Dig. Tech. Papers,* pp. 20–21, Feb. 1981.

[50] K. Hardee *et al.,* "A fault-tolerant 30 ns/375 mW 16K × 1 NMOS static RAM," *IEEE J. Solid-State Circ.,* vol. SC-16, pp. 435–443, Oct. 1981.

[51] S. Yamamoto *et al.,* "A 256K CMOS RAM with variable-impedance loads," in *ISSCC Dig. Tech. Papers,* pp. 58–59, Feb. 1985.

[52] Y. Kobayashi *et al.,* "A 10 μW standby-power 256K CMOS SRAM," in *ISSCC Dig. Tech. Papers,* Feb. 1985, pp. 60–61.

[53] S. Murakami *et al.,* "A 21 mW 4 Mb CMOS SRAM for battery operation," in *ISSCC Dig. Tech. Papers,* Feb. 1991, pp. 46–47.

[54] K. Sasaki *et al.,* "A 15 ns 1 Mb CMOS RAM," in *ISSCC Dig. Tech. Papers,* pp. 174–175, Feb. 1988.

[55] ——, "A 9-ns 1-Mbit CMOS RAM," *IEEE J. Solid-State Circ.,* vol. 24, pp. 1219–1225, Oct. 1989.

[56] K. Ishibashi *et al.,* "A 1 V TFT-load SRAM using a two-step word-voltage method," *IEEE J. Solid-State Circ.,* vol. 27, pp. 1519–1524, Nov. 1992.

[57] K. Sasaki *et al.,* "A 16 Mb CMOS SRAM with a 2.3 μm^2 single-bit-line memory cell," in *ISSCC Dig. Tech. Papers,* Feb. 1993, pp. 250–251.

[58] E. Seevinck, "A current sense-amplifier for fast CMOS SRAM," in *Symp. VLSI Circ. Dig. Tech. Papers,* May 1990, pp. 71–72.

[59] K. Sasaki *et al.,* "A 7 ns 140 mW/Mb CMOS SRAM with current sense amplifier," in *ISSCC Dig. Tech. Papers,* Feb. 1992, pp. 208–209.

[60] A. Roberts *et al.,* "A 256K SRAM with on-chip power supply conversion," in *ISSCC Dig. Tech. Papers,* pp. 252–253, Feb. 1987.

[61] K. Ishibashi *et al.,* "A voltage down converter with submicroampare standby current for low-power static RAM's," *IEEE J. Solid-State Circ.,* vol. 27, pp. 920–926, June 1992.

[62] T. Yabe *et al.,* "High-speed and low-standby-power circuit design of 1–5 V operating 1 Mb full CMOS SRAM," in *Symp. VLSI Circ. Dig. Tech. Papers,* May 1993, pp. 107–108.

[63] K. Ishibashi *et al.,* "A 6-ns 4-Mb CMOS SRAM with offset-voltage-insensitive current sense amplifiers," in *Symp. VLSI Circ. Dig. Tech. Papers,* June 1994, pp. 107–108.

[64] D. Burnett *et al.,* "Implication of fundamental threshold voltage variations for high-density SRAM and logic circuits," in *Symp. VLSI Tech. Dig. Tech. Papers,* May 1994, pp. 15–16.

[65] S. Sun *et al.,* "Limitation of CMOS supply-voltage scaling by MOSFET threshold-voltage variation," in *CICC Proc.,* May 1994, pp. 267–270.

[66] K. Seno *et al.,* "A 9 ns 16 Mb CMOS SRAM with offset reduced current sense amplifier," in *ISSCC Dig. Tech. Papers,* pp. 248–249, Feb. 1993.

[67] S. Flannagan *et al.,* "Two 13-ns 64K CMOS SRAM's with very low active power and improved asynchronous circuit techniques," *IEEE J. Solid-State Circ.,* vol. 21, pp. 692–703, Oct. 1986.

Standby/Active Mode Logic for Sub-1-V Operating ULSI Memory

Daisaburo Takashima, Shigeyoshi Watanabe, Hiroaki Nakano,
Yukihito Oowaki, Kazunori Ohuchi, and Hiroyuki Tango

Abstract—New gate logics, standby/active mode logic I and II, for future 1 G/4 Gb DRAM's and battery operated memories are proposed. The circuits realize sub-1-V supply voltage operation with a small 1-μA standby subthreshold leakage current, by allowing 1 mA leakage in the active cycle. Logic I is composed of logic gates using dual threshold voltage (Vt) transistors, and it can achieve low standby leakage by adopting high Vt transistors only to transistors which cause a standby leakage current. Logic II uses dual supply voltage lines, and reduces the standby leakage by controlling the supply voltage of transistors dissipating a standby leakage current. The gate delay of logic I is reduced by 30–37% at the supply voltage of 1.5–1.0 V, and the gate delay of logic II is reduced by 40–85% at the supply voltage of 1.5–0.8 V, as compared to that of the conventional CMOS logic.

I. INTRODUCTION

LOWERING the supply voltage to 1.0 V or below is the most effective way to reduce power consumption and to assure the reliability of miniaturized MOS transistors in the future gigabit DRAM's, and it also meets the requirement of 1.0–1.5 V battery operation for portable equipment [1], [2]. However, the threshold voltage Vt cannot be simply scaled down as the supply voltage does due to the subthreshold leakage current problem. Fig. 1 compares the gate delay in each DRAM generation, assuming a constant standby leakage current of 1 μA per chip. With device integration, the total channel width, "Total W," of transistors in a chip which cause the standby leakage current increases, and it becomes 16 m in a 4 Gb DRAM. So the threshold voltage of transistors for a 4 Gb DRAM must be higher than those for previous generation DRAM's to keep 1 μA standby leakage current. Assuming subthreshold swings for nMOS and pMOS transistors are 110 mV/dec and 120 mV/dec at 85°C, respectively, the required threshold voltages become 0.65 V for nMOS transistor and -0.7 V for pMOS transistor. Therefore, the voltage difference between the supply voltage and Vt becomes smaller. As a result, the gate delay in gigabit DRAM generation will increase in spite of device miniaturization [3] as shown in Fig. 1.

In this paper, two new circuit techniques, standby/active mode logic I and logic II, for sub-1 V supply voltage ULSI memories are proposed. Logic I, II can reduce gate delay at a low supply voltage while maintaining a low standby leakage current by using dual Vt transistors [4], [5] or by introducing dual supply voltage lines and controlling one of

Manuscript received August 18, 1993; revised December 10, 1993.

The authors are with the ULSI Research Center, Toshiba Corporation, 1, Komukai, Toshiba-cho Saiwai-ku, Kawasaki 210, Japan.

IEEE Log Number 9215693.

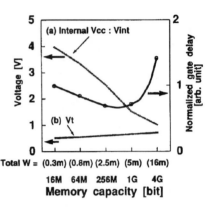

Fig. 1. Gate delay of an inverter when standby leakage current is maintained at 1 μA in a chip.

the supply voltage lines [4], [6]–[8]. In Section II, the concept of standby/active mode logic I, II is described. Section III presents the performance of logic I and its application to DRAM circuits for low voltage operation. Section IV presents the performance of logic II, and a distributed supply voltage line driver scheme to fit gigascale DRAM's. In Section V, performance of two new logic schemes and their dependence on process variations are discussed. Finally, Section VI gives a conclusion.

II. CONCEPT OF STANDBY/ACTIVE MODE LOGIC I, II

ULSI memories, especially DRAM's, are different from other ULSI logics such as MPU's and ASIC logics in the following two features.

1) The levels of nodes are set to a determined level, "high" or "low" in the standby cycle. Therefore, a circuit designer can predict which transistors in the memory circuits are cut off during the standby cycle and cause standby leakage current.

2) The DRAM active current is of an order of 100 mA, which is 10^4–10^5 times larger than the standby current. Therefore, the transistor leakage current at the active cycle is negligible even when it becomes 10^3 times larger than that in the standby cycle.

To make the most of these two features, two new gate logics, standby/active mode logic I and logic II, for ULSI memories are designed. Fig. 2 shows the concept of the new logic I, II. The concept of the two new logics is to reduce gate delay by allowing 1 mA leakage current, I(leak), per chip in the active cycle, while keeping the low standby leakage current of 1 μA per chip. Logic I realizes this concept by introducing dual V_t's,

Reprinted from *IEEE Journal of Solid-State Circuits*, pp. 441-447, April 1994.

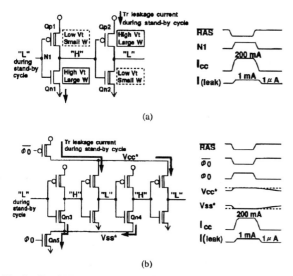

(a)

(b)

Fig. 2. Circuit diagrams and timing charts of standby/active mode logics:
(a) logic I, (b) logic II.

and logic II also realizes this concept by using dual supply
voltage lines and by controlling them during the standby cycle.

A. Logic I Using Dual Vt Transistors

Fig. 2(a) shows inverters using logic I. The logic I is
composed of logic gates using high and low Vt transistors.
High threshold voltage is introduced to all the transistors $Qn1$
and $Qp2$ which are cutoff in the standby cycle, in order to
suppress the subthreshold leakage current below 1 μA per
chip. On the contrary, low Vt transistors are adopted for
transistors $Qn2$ and $Qp1$ which are turned on in the standby
cycle to decrease the gate delay. The lower limit of Vt of
these transistors is designed to allow 1 mA leakage current
per chip at the active cycle.

B. Logic II Using Dual Supply Voltage Lines

Fig. 2(b) shows inverters using logic II. In logic II, the Vt of
every transistor is designed low to reduce gate delay. In order
to reduce standby leakage current, the internal supply voltage
line is divided to that connected to the cutoff transistors and
that connected to the turned-on transistors. The voltage of the
ground line Vss^{**} for the cutoff nMOS transistors is designed
to be higher than the gate voltage of these nMOS transistors
in the standby cycle, and the voltage of the supply voltage line
Vcc^{**} for the cutoff pMOS transistors is designed to be lower
than the gate voltage of these pMOS transistors. For example,
in the standby cycle, the supply voltage line driver $Qn5$ for
Vss^{**}, which is the source line for the cutoff nMOS transistors
$Qn3$ and $Qn4$, is turned off. Therefore, the voltage of Vss^{**}
rises gradually due to the leakage current of these transistors,
and it results in an improvement of the cutoff characteristics
of these transistors, and the suppression of the leakage current.

III. Performance of Logic I and Its Application

In this section, first, the dual Vt design of logic I is
discussed. Next, the dependency of the gate delay of the new
logic on Vcc and channel width is estimated by simulation,

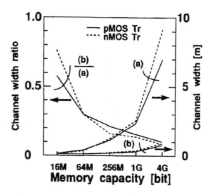

Fig. 3. Trend of each nMOS and pMOS transistor channel width ratio (b)/(a),
where (a) is the total channel width of transistors which are off in the standby
cycle, and (b) is the total channel width of transistors which turn off in the
active cycle.

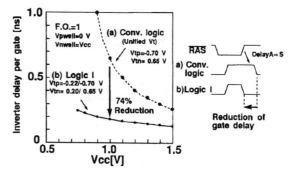

Fig. 4. Simulated gate delay, Delay.$A \to S$, of logic I when changing from
active cycle to standby cycle.

and the optimization of channel width is discussed. Finally,
some applications of logic I to DRAM circuits are presented.

A. Vt Design

In logic I, the Vt of high Vt transistors depends on both
the allowable standby leakage current, $I_s(\text{leak})$, of 1 μA, and
the total channel width W_s of these transistors which are off
during the standby cycle. On the other hand, the Vt of low
Vt transistors depends on the allowable active leakage current
$I_{A(\text{leak})}$ of 1 mA and also on the total channel width W_A of
the transistors which turn off in the active cycle. Fig. 3 shows
the trends of the ratio W_A/W_s. With device integration, this
total W_A shown in line (b) becomes smaller than the total
channel width W_S shown in line (a) because of the partial
activation of the DRAM memory cell array. The activation
ratios of W_A/W_S are estimated 0.08 for nMOS transistor and
0.1 for pMOS transistor in a 4 Gb DRAM. Considering the
subthreshold leakage in the active cycle and the activation
ratio, the threshold voltage difference between low and high
Vt transistors in a 4 Gb DRAM are given as below.

$$VtL - VtH = -S * \log\left[(I_{A(\text{leak})}/I_{s(\text{leak})})/(W_A/W_S)\right]$$
$$= -0.33\,\text{V} - 0.12\,\text{V} = -0.45\,\text{V}$$
for nMOS transistor
$$= -0.36\,\text{V} - 0.12\,\text{V} = -0.48\,\text{V}$$
for pMOS transistor.

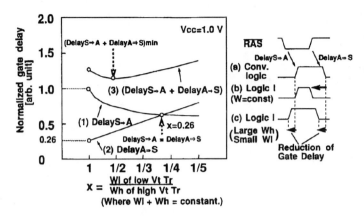

Fig. 5. Channel width dependency of gate dealy of logic I, where $\text{Delay} S \rightarrow A$ is the delay changing from standby cycle to active cycle, and $\text{Delay} A \rightarrow S$ is that from the active cycle to standby cycle.

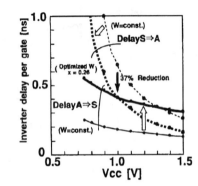

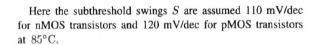

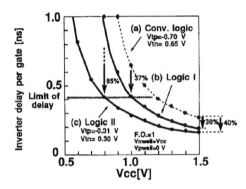

Fig. 6. Simulated gate delay of logic I with channel width W optimized at $V cc = 1.0$ V.

Fig. 7. Simulated gate delay for conventional logic and proposed logic I, II.

Here the subthreshold swings S are assumed 110 mV/dec for nMOS transistors and 120 mV/dec for pMOS transistors at 85°C.

B. Gate Delay Estimation

If the Vt of transistors which are turned on in the standby cycle is simply lowered, the gate delay reduces only when the low Vt transistor turns on. That means the gate delay of gates activated when the chip changes from the active cycle to the standby cycle, $\text{Delay} A \rightarrow S$, is reduced. Our simulation tells the $\text{Delay} A \rightarrow S$ is reduced by 74% as compared to that of the conventional logic at a $V cc = 1.0$ V as shown in Fig. 4. However, $\text{Delay} S \rightarrow A$, the gate delay when the chip changes from the standby cycle to the active cycle, stays unchanged that of the conventional logic, due to the nonlowered Vt of the transistors. For an effective application of logic I to DRAM circuitry, not only $\text{Delay} A \rightarrow S$ but also $\text{Delay} S \rightarrow A$ should be reduced to achieve fast access time. Therefore, to reduce $\text{Delay} S \rightarrow A$, the channel widths Wh of the high Vt transistors are enlarged, whereas the channel widths Wl of the low Vt transistors are reduced, maintaining

total channel width, $Wh + Wl$, constant. Fig. 5 shows the dependence of $\text{Delay} S \rightarrow A$ and $\text{Delay} A \rightarrow S$ on the channel width ratio; $x = Wl/Wh$ at $V cc = 1.0$ V. The delay ratio $(\text{Delay} A \rightarrow S)/(\text{Delay} S \rightarrow A)$ is $A = 0.26$ when x is 1 as shown in Fig. 4. $\text{Delay} S \rightarrow A$ is decreased with a decrease in x while $\text{Delay} A \rightarrow S$ is increased, and when $x = A^{1/2}$, the sum of $\text{Delay} A \rightarrow S$ and $\text{Delay} S \rightarrow A$ has a minimal value. When $x = A$, the delay ratio $(\text{Delay} A \rightarrow S) / (\text{Delay} S \rightarrow A)$ reaches 1. Fig. 6 shows the $V cc$ dependence of the gate delay of logic I when the channel wdith ratio; x is $A = 0.26$ assuming 1 V $V cc$ operation. Both $\text{Delay} S \rightarrow A$ and $\text{Delay} A \rightarrow S$ is reduced by 37% as compared to those of the conventional logic at $V cc = 1.0$ V, using the optimized value of $x = 0.26$. This channel width ratio x may be optimized for the lowest $V cc$ of a designed chip. Curve (a) in Fig. 7 shows the gate delay with the optimized x for each $V cc$, under the condition that $\text{Delay} S \rightarrow A$ is equal to $\text{Delay} A \rightarrow S$. Thirty to 37% reduction of the gate delay is achieved in the operation range of 1.5–1.0 V. $V cc$ min is also reduced to 1.0 V assuming that $V cc$ min is determined as a supply voltage giving a certain limit for the delay, as shown in Fig. 7. By reducing power supply voltage to 1.0 V the power consumed by charging/discharging bit-lines is reduced by 30% as compared to the conventional logic.

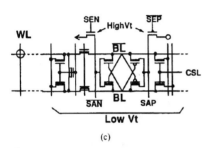

(a)

(b)

(c)

Fig. 8. Applications of logic 1 to a DRAM chip, (a) NAND gate, (b) ring oscillator in pump circuit, and (c) DRAM core circuits.

C. Application to DRAM Circuits

Logic I can be easily applied to almost all the internal circuits in DRAM's. Fig. 8 shows three examples application of logic I to DRAM circuits. Logic I is applicable to NAND and NOR gates as well as inverter, as shown in Fig. 8(a). When two or more cutoff transistors are connected in series, the Vt of only one of them must be high as shown in Fig. 8(a). Circuits operating in asynchronous with a /RAS clock such as a ring oscillator in the pump circuit can also adopt logic I using a /STOP clock generated internally as shown in Fig. 8(b). In a DRAM chip, during the standby cycle, all bit-lines and sense amplifier drive-lines, /SAN and SAP, are precharged to a half Vcc level, as shown in Fig. 8(c). The transistors of the precharge circuit and the column switch do not cause standby leakage current because they are turned on during the standby cycle. The transistors of the sense amplifiers which are cut off during the standby cycle also do not cause standby leakage current because the source and drain voltages are the same $(Vcc/2)$. Therefore, the Vt of these circuits can be reduced, except for the transistors to drive sense amplifiers. The leakage current of these transistors during the active cyce is included in Fig. 3(b), and it can be suppressed small enough by the partial activation technique.

IV. Performance of Logic II and New Driver Scheme

In this section , a new supply voltage line scheme to take full advantage of logic II in a DRAM chip is introduced, and the performance of logic II with the new scheme is also estimated.

A. Distributed Supply Voltage Line Driver Scheme

Logic II is useful for further scaled DRAM's operating at less than 1.0 V, because the Vt value of every transistor can be decreased. However, there are a few concerns when applying logic II to scaled DRAM's. The total channel width of cutoff transistors connected to internal supply voltage lines Vss^{**} and Vcc^{**} becomes 16 m in a 4 Gb DRAM. This results in large parasitic capacitances of about 10 nF for the supply voltage lines. Therefore, if logic II is simply applied to all DRAM circuits, it will cause a larger power dissipation and a long setup time to drive $Vss^*(Vcc^*)$ to the voltage of $Vss(Vcc)$ when changing to the active mode. Moreover, it causes a large subthreshold leakage current of all cutoff transistors at the active cycle.

To overcome this problem, a distributed supply voltage line driver scheme is introduced to the DRAM chip, as shown in Fig. 9. DRAM row decoders and sense amplifier drivers, and column decoders are all divided into multiple segments. The internal supply voltage lines $Vss^{**}(Vcc^{**})$ is also divided into distributed supply voltage lines, $Vssx1 - n(Vccx1 - n)$ for the row decoders, $Vssz1 - n(Vccz1 - n)$ for the sense amplifier drivers, $Vssy1 - m(Vccy1 - m)$ for the column decoders, and a common supply voltage line $Vss^{**}(Vcc^{**})$, whereas the supply voltage line $Vss(Vcc)$ connected to the nMOS (pMOS) transistors, which are turned on in the standby cycle, is not divided. At the standby cycle, the leakage current can be limited only by two common supply voltage drivers $Qn0$ and $Qp0$. In the active cycle, the common supply voltage driver $Qn0(Qp0)$ is activated, and also three distributed supply voltage drivers, $Qnx1(Qpx1), Qny1(Qpy1)$, and $Qnz1(Qpz1)$, for a selected array are activated, as shown in Fig. 9. Therefore, the charges of the core circuits driven in the active cycle can be reduced to $1/n$, and leakage current in the active cycle is also suppressed to $1/n$ in the case of $n = m$.

The bounces of the internal supply voltage lines in the active cycle can also be a serious problem in logic II. They cause an unwanted delay, or even cause a malfunction in the active cycle. Fig. 10 shows the dependence of the voltage of each internal supply voltage line on the channel width, $Wn(Wp)$, of each nMOS (pMOS) driver transistor, in the case that the channel width $Wn(Wp)$ for $Qn0(Qp0)$ is equal to that for $Qny1(Qpy1)$ and the sum of those for $Qnx1(Qpx1)$ and $Qnz1(Qpz1)$. In the 4 Gb DRAM, a channel width of $Wn = 22\,000\,\mu m(Wp = 48\,000\,\mu m)$ is necessary to suppress the voltage bounce within 0.05 V, except supply voltage lines, $Vssz1 - n(Vccz1 - n)$, for charging/discharging bit-lines. The drain current for nMOS transistor of $Wn/Ln = 10\,\mu m/0.12\,\mu m$ and that for pMOS transistor of $Wp/Lp = 10\,\mu m/0.15\,\mu m$ discussed in this section are 1.02 mA and 0.58 mA at $Vgs = Vds = 1.0$ V, respectively [9].

Fig. 11 shows the total nMOS and pMOS channel widths driven by the common supply voltage driver and the increase in chip size versus the number of segmentations $n(= m)$, in a 4 Gb DRAM. As previously described, the channel width $W1$ of the peripheral and core circuits driven by the common

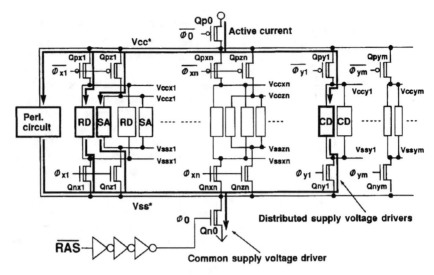

Fig. 9. Distributed supply voltage line driver scheme applied to logic II.

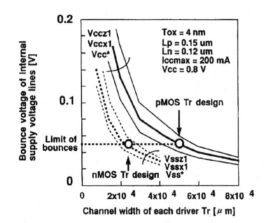
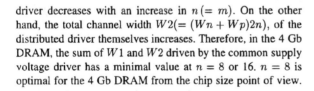

Fig. 10. The voltage bounces of internal supply voltage lines versus channel width of each nMOS and pMOS driver transistor.

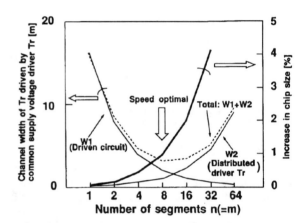

Fig. 11. Channel width of nMOS and pMOS transistors driven by common supply voltage driver and increase in chip size versus the number of segmentations.

driver decreases with an increase in $n (= m)$. On the other hand, the total channel width $W2 (= (Wn + Wp)2n)$, of the distributed driver themselves increases. Therefore, in the 4 Gb DRAM, the sum of $W1$ and $W2$ driven by the common supply voltage driver has a minimal value at $n = 8$ or 16. $n = 8$ is optimal for the 4 Gb DRAM from the chip size point of view.

B. Performance of Logic II

The gate delay of logic II is shown as curve (c) in Fig. 7. The 0.05 V bounces of supply voltage lines are considered here. Forty to 85% reduction in the gate delay of logic II is achieved as compared with the conventional logic in the range of 1.5 V to 0.8 V Vcc. Vcc min is also reduced to 0.8 V. Therefore, the power consumed by charging/discharging bit-lines is reduced by 58%. In logic II, the lower limit of Vt is determined by the sum of the channel width of $1/n$ of line (a) in Fig. 3 and that of line (b) in Fig. 3, considering the tolerant voltage of 0.05 V bounces of the internal supply voltage lines. The setup time

to drive the supply voltage lines to $Vss\,(Vcc)$ when changing to an active cycle is only 4.6 ns at $Vcc = 0.8\,V, T = 25°C$ in the 4 Gb DRAM, by optimizing the number of segmentation of $n = 8$. The setup time of 4.6 ns is relatively small, when considering that the access time will be reduced from 210 to 35 ns in the 0.8 V, 4 Gb DRAM using logic II. Moreover, the power consumed to set up internal lines by using this scheme is 2.9% of the total active power consumption of the 4 Gb DRAM.

V. DISCUSSION

Table I summaries the features of the logic I and II. The logic II achieves faster speed and lower power dissipation due to lower Vcc min as compared with the logic I. However, the logic II has a delay penalty to set up internal lines to $Vss\,(Vcc)$ when changing to the active cycle. Therefore, the logic II can be applied to further integrated memories to take advantage of the better performance at lower Vcc as compared to the

[5] T. Yabe *et al.*, "High-speed low-standby-power circuit design of 1V to 5V operating 1Mb full CMOS SRAM," in *Symp. VLSI Circuits Dig. Tech. Papers,* May 1993, pp. 107–108.
[6] G. Kitsukawa, "256Mb DRAM technologies for file applications," in *ISSCC Dig. Tech. Papers,* Feb. 1993, pp. 48–49.
[7] M. Horiguchi *et al.,* "Switched-source-impedance CMOS circuit for low standby subthreshold current giga-scale LSI's," in *Symp. VLSI Circuits Dig. Tech. Papers,* May 1993, pp. 47–48.
[8] T. Sakata *et al.,* "Subthreshold-current reduction circuits for multi-gigabit DRAM's," in *Symp. VLSI Circuits Dig. Tech. Papers,* May 1993, pp. 45–46.
[9] A. Toriumi *et al.,* "High speed 0.1um CMOS devices operating at room temperature," in *Extended Abstracts Conf. Solid State Dekvices and Materials,* Aug. 1992, pp. 487–489.

TABLE I
SUMMARY OF LOGIC I AND LOGIC II

	Logic I	Logic II
Target Device	1.5–1 V 1 Gb DRAM	1.5–0.8 V 4 Gb DRAM
Speed	30–37% Faster	40–85% Faster
$V cc$ min	1.2 V → 1.0 V	1.2 V → 0.8 V
Power	30% Reduction	56% Reduction
Advantage	Easy Implementation	Good Performance
Drawback	2 Extra Masks	Set up Time (4.6 ns), Chip Size Penalty (1%)

logic I. The logic I can be easily applied to many kinds of memories using two extra masks for nMOS and pMOS low Vt transistors.

In this paper, the performances of the two new logics are evaluated at the constant Vt condition. The Vt is determined from the subthreshold swing at 85°C. In an actual chip, the variations in process and operating condition cause about ±0.1 V Vt variation. The leakage current in the active cycle is less than 1/20 of the active current of 200 mA, even considering the leakage current variation in the range of 0.1 mA to 10 mA caused by ±0.1 V Vt variation. Although the Vcc min of the two new logics varies in proportion to Vt variation as the conventional logic does, the gate delay reduction of the new logics is almost constant.

VI. CONCLUSION

New standby/active mode logic I and II have been developed for future 1 G/4 Gb DRAM's. The proposed logic I, II can achieve sub-1 V supply voltage operation with a small 1 μA subthreshold leakage current during the standby cycle, with an allowance of a 1 mA transistor leakage current during the active cycle. The gate delay of logic I is reduced by 30–37% with the optimized channel widths in the range of 1.5 V to 1.0 V Vcc, as compared to that of the conventional logic. The gate delay of logic II is also reduced by 40–85% as compared to that of the conventional logic in the range of 1.5–0.8 V Vcc. The proposed logic I, II are easily applicable not only to 1 G/4 Gb DRAM's but also to other kinds of memories such as SRAM's, and battery-operated memories.

ACKNOWLEDGMENT

The authors would like to thank F. Masuoka for his encouragement. They would also like to thank Y. Watanabe, T. Hara, and the DRAM group for their helpful suggestions and useful discussions, and would like to thank S. Inaba for the device data.

REFERENCES

[1] K. Ishibashi *et al.,* "A 1V TFT-load SRAM using a two-step word-voltage method," in *ISSCC Dig. Tech. Papers,* Feb. 1992, pp. 206–207.
[2] A. Sekiyama *et al.,* "A 1-V operating 256-kb full-CMOS SRAM, *IEEE J. Solid-State Circuits,* vol. 27, no. 5, pp. 776–782, May 1992.
[3] Y. Nakagome *et al.,* "Sub-1-V swing internal bus architecture for future low-power ULSI's," *IEEE J. Solid-State Circuits,* vol. 28, no. 4, pp. 414–419, Apr. 1993.
[4] D. Takashima *et al.,* "Stand-by/active mode logic for sub-1V 1G/4Gb DRAMs," in *Symp. VLSI Circuits Dig. Tech. Papers,* May, 1993, pp. 83–84.

A Charge Recycle Refresh for Gb-Scale DRAM's in File Applications

Takayuki Kawahara, *Member, IEEE*, Yoshiki Kawajiri, Masashi Horiguchi, *Member, IEEE*, Takesada Akiba, Goro Kitsukawa, Tokuo Kure, *Member, IEEE*, and Masakazu Aoki, *Member, IEEE*

Abstract— A charge recycle refresh for low-power DRAM data-retention, featuring alternative operation of two memory arrays, is proposed, and demonstrated using a 64 kb test chip with 0.25 μm technology. After amplification in one array, the charges in that array are transferred to another array, where they are recycled for half amplification. The data-line current dissipation is only half that of the conventional refresh operation, and the voltage bounce of the power supply line is 60% of the conventional. This scheme is further extended for application to n arrays with $1/n$ data-line current dissipation. Moreover, the multi-array activation with Charge Recycle Refresh is proposed, in which the same peak current as in the conventional scheme is achieved with a small number of refresh cycles for refreshing all the cells.

I. INTRODUCTION

THE introduction of handheld personal electronic equipment such as palmtop PC's has encouraged the development of low-power yet large capacity semiconductor memory devices. The reduction of data-retention current, in which data-line charging current has a major role, is one of the most important issues in enhancing Gb-scale DRAM capabilities for file applications [2]. Since the number of refreshing cells has been quadrupled with each new generation, the total data-line charges contributing to power dissipation has increased sharply with increasing density (Fig. 1). This has occurred even though the data-line voltage has been reduced. One effective way to reduce data-line charging current is to extend the refresh period during the data-retention mode [3], but the effect of this scheme depends strongly on the low-leakage process technology, in which progress has been relatively slow.

In addition, since the refresh period is limited by the junction leakage current I_j of the memory cell while the number of refreshing cells is quadrupled in every generation, the activation of a larger number of memory cell arrays in the refresh operation than that in the normal operation [4] tends to be introduced to maintain the number of refresh cycles and the refresh period. However, that scheme causes large peak current. Thus, the reduction of peak current associated with the data-retention is also important in Gb-scale DRAM's.

Manuscript received September 3, 1993; revised February 21, 1994. Part of this paper was presented at the 1993 Symposium on VLSI Circuits [1].

T. Kawahara, M. Horiguchi, G. Kitsukawa, and T. Kure are with Central Research Laboratory, Hitachi Ltd., Kokubunji, Tokyo 185, Japan.

Y. Kawajiri is with the Semiconductor and Integrated Circuits Division, Hitachi Ltd., Kodaira, Tokyo 187, Japan.

T. Akiba is with Hitachi Device Engineering Company, Ltd., Mobara, Chiba 297, Japan.

IEEE Log Number 9402123.

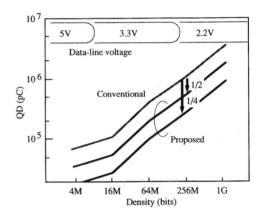

Fig. 1. Projected total data-line charges contributing to power dissipation.

Moreover, the I_j itself tends to increase in every generation as follows. The peripheral of the junction region can only be scaled by a factor of $1/k$ (k: scaling factor) on the contrary to the $1/k^2$ for area of that as shown in Table I. Therefore, the peripheral component ratio of the junction leakage current to the area component increases by a factor of k in every generation. Thus, the $1/k$ scaled peripheral component becomes dominant. And the leakage current per unit length and area also increases gradually in every generation because the impurity concentration is increased by a factor of k even though the voltage at the junction is decreased by $1/\sqrt{k}$ with commonly used quasi-constant voltage scaling [5]. From this point of view, the scheme of the activation of a larger number of memory cell arrays in the refresh operation becomes the key issue.

This paper therefore proposes a charge recycle concept that directly reduces the data-line charging current. It also reduces the peak current associated with charging up a larger number of memory cell arrays in the refresh operation.

The Charge Recycle operation concept and a DRAM circuit based on this concept are described in Section 2. The performance of the proposed circuit is discussed in Section 3, and some experimental results are shown in Section 4. An extension of this concept is also described in Section 5. Multi-array activation with the proposed Charge Recycle Refresh is presented in Section 6.

II. CIRCUIT AND TIMING DIAGRAMS

The Charge Recycle Refresh is a refresh scheme in which the charges used in one array, conventionally poured out in

Reprinted from *IEEE Journal of Solid-State Circuits,* pp. 715-722, June 1994.

TABLE I
JUNCTION LEAKAGE CURRENT OF MEMORY CELLS. JUNCTION REGION IS
APPROXIMATED AS A SQUARE. α: JUNCTION LEAKAGE CURRENT PER
UNIT AREA, β: JUNCTION LEAKAGE CURRENT PER UNIT LENGTH.

Memory Cell Transistor	Reference Generation	Next Generation
	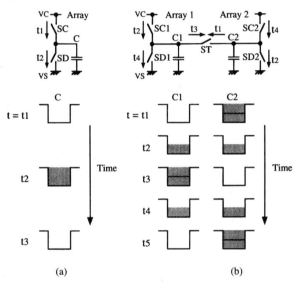	Scaled k (=1.6)
Junction Area (A)	1	$1/k^2$
Junction Peripheral Length (4L)	1	$1/k$
Junction Leakage Current	$\alpha A + 4\beta L$	$\alpha A/k^2 + 4\beta L/k$
Current at peripheral / at bottom (area)	1	k
Impurity Concentration	1	k
Voltage at Junction	1	$1/\sqrt{k}$

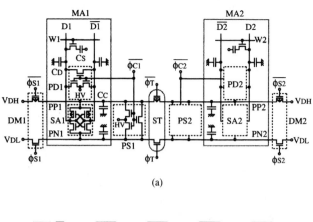

(a)

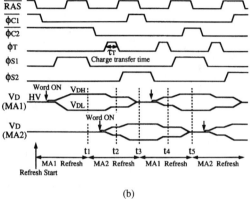

(b)

Fig. 3. Circuit and timing diagram of two-array Charge Recycle Refresh.

Fig. 2. Principle of charge recycle operation.

every cycle, are transferred to another array and used there. The principle of the Charge Recycle Refresh is shown in Fig. 2(b), and the conventional scheme is shown in Fig. 2(a).

In the conventional scheme, at time t_2, the data-line load C in the array charges through the switch SC. At time t_3, the load discharges through the switch SD. These refresh operations are repeated independently in each array.

In the proposed scheme, first at t_1, the data-line load $C2$ in the array 2 is charged up by the switch SC2. At t_2, half the charge in $C2$ (the array 2) is transferred to $C1$ (the array 1) through the transfer switch ST located between the two arrays. The result is that $C1$ is charged up to half the original level and $C2$ is half discharged. After this new operation, the remaining charge in $C2$ is discharged by switch SD2 at t_3, while additional charge is supplied by switch SC1 to supplement the load of $C1$ up to the full level. In comparison with the conventional operation, only half as much current is dissipated for charging $C1$ by SC1 because half charging has already been completed. Next, the charge in $C1$ is reused in $C2$ at t_4, so charging $C2$ at t_5 also requires only half as much current. Repeating these operations thus results in a data-line charging current that is about half the conventional charging

current, and the small peak current reduces the voltage bounce in the power supply line.

A DRAM circuit based on the two-array Charge Recycle Refresh is shown in Fig. 3(a). The parasitic capacitance of the data-line pair $D1$, $\overline{D1}$ and the sense amplifier driving lines PP1, PN1 in MA1 corresponds to that of $C1$ in Fig. 2(b), and the capacitance of the data-line pair $D2$, $\overline{D2}$ and the sense amplifier driving line PP2, PN2 in MA2 corresponds to $C2$. The transfer switch ST is supervised by ϕ_T, $\overline{\phi_T}$. The sense amplifier drivers DM1 and DM2 respectively correspond to SC1 and SC2 in Fig. 2(b). Since the discharging operation in Fig. 2(b) is equivalent to the short operation of the data-line pair, the short circuits PD1, PS1 act the same as SD1, and PD2, PS2 the same as SD2.

The internal timings of $\overline{\phi_{C1}}$ to ϕ_{S2}, generated by the $\overline{RAS}$ clock, and the data-line voltages V_D are shown in Fig. 3(b). The states of data-line pairs at times t_1–t_5 correspond to the states in Fig. 2(b) at these times. When the ϕ_T is changed at t_2 and t_4, the data-line charges in one array are transferred to the other array, and the other reuses these charges to amplify the data-line signal voltage to half the full level. Note that no special voltages are necessary, since these half levels are generated automatically. This operation differs from the conventional scheme in the timing of signals $\overline{\phi_{C1}}$, $\overline{\phi_{C2}}$, in addition to the introduction of ϕ_T, $\overline{\phi_T}$. The level of $\overline{\phi_{C1}}$, $\overline{\phi_{C2}}$ becomes low after a certain $\overline{RAS}$ falls, and the level becomes high again after the $\overline{RAS}$ rises in the second cycle. The phase of these two signals differs by one $\overline{RAS}$ cycle. These signals can be easily generated using the $\overline{RAS}$ clock and a clock

345

with a period twice as long as that of the $\overline{\text{RAS}}$ clock. In Fig. 3(b), the $\overline{\text{RAS}}$ clock is an example of a control clock. The self-refresh operation using an internal timing generator for a control clock can also be used.

The actual operation of the charge recycle refresh in a DRAM array is illustrated in Fig. 4. The operations consist of the initial refresh in MA1, the recycle from MA1 to MA2, the refresh in MA2, the recycle from MA2 to MA1, and returning to the refresh in MA1. These operations are repeated cyclically. First, the word line W in MA1 is selected, and the data-line signal corresponding to the charge storage in the cell appears as shown in Fig. 4(a). This data-line signal is amplified by the sense amplifier SA1 at t_1 when the DM1 switches are ON. The current from the power supply line V_{DH} is supplied to the sense amplifier driving line to amplify the signal. This full signal is, therefore, rewritten to the selected cell. Therefore, the refresh operation in MA1 is completed. During these operations in MA1, the data-line voltage in MA2 remains at the precharged level. Next, the charge recycle operation from MA1 to MA2 begins, as shown in Fig. 4(b). The word line in the array MA2 is selected and a small data-line signal appears. At time t_2, the transfer switches ST turn on. The charges in the array MA1 are transferred to the array MA2, and are recycled in MA2 to amplify the small data-line signal. Therefore, the data-line signal in MA2 is half amplified, while the data-line signal in MA1 is half discharged. Note that the internal data-line voltage shown here is generated automatically.

The refresh operation in MA2 proceeds as shown in Fig. 4(c). After half amplification by the recycle operation, the transfer switches ST turn off. Then, the DM2 switches turn on at t_3. Therefore, the data-line signal is fully amplified in MA2. Since the data-lines were already amplified to the half level, the additional charge supplied to the data-line from the V_{DH} is half of the conventional amount. Therefore, the refresh operation in MA2 is accomplished half by the charge recycle and half by the charge from the power supply line. During or after these operations, the data-line in MA1 is shorted and precharged.

The charge recycle operation from MA2 to MA1 is shown in Fig. 4(d). A different word line W' is selected in MA1. Then, the transfer switches ST turn on again at t_4. The charge in MA2 is transferred to MA1. The data-line signal in MA1 is half amplified, and the data-line signal in MA2 is half shorted or discharged. This cycle repeats as shown in Fig. 4(e). This is the refresh operation in MA1 by supplying the additional charge from the V_{DH}. When the switches DM1 turn on at t_5, the half amplified data-line is amplified to the full level. The charges supplied by the V_{DH} are also half. In addition, the half discharged data-line signal in MA2 is shorted and precharged. In this way, the cyclic refresh charge transfer operation is repeated between two arrays.

The total charge dissipations in one array are estimated as follows for the conventional and the proposed schemes.

$$Q_{\text{conv.}} = \frac{1}{2} \cdot (V_{\text{DH}} - V_{\text{DL}}) \cdot (C_C + n \cdot (C_D + C_S)) \cdot m \tag{2.1}$$

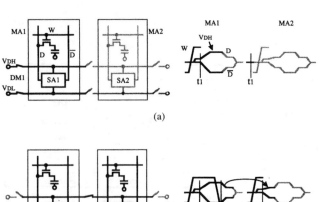

(a)

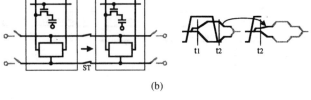

(b)

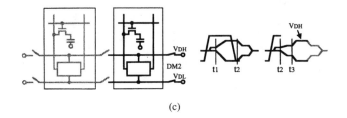

(c)

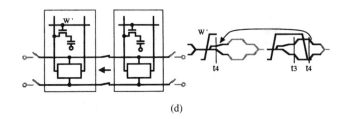

(d)

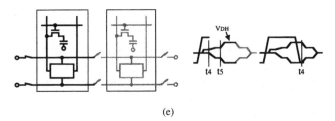

(e)

Fig. 4. Charge Recycle Refresh Operation.

$$
\begin{aligned}
Q_{\text{proposed}} = & \frac{1}{2} \cdot (V_{\text{DH}} - V_{\text{DL}}) \cdot (C_C + n \cdot (C_D + C_S)) \\
& + \frac{1}{2} \cdot \frac{1}{2} \cdot (V_{\text{DH}} - V_{\text{DL}}) \\
& \cdot (C_C + n \cdot (C_D + C_S)) \cdot (m - 1)
\end{aligned} \tag{2.2}
$$

where,

n number of data-line pairs,
m number of cells per sense amplifier,
C_C and C_D parasitic capacitances of the sense amplifier driving line and the data-line,
C_S cell storage capacitance.

In a typical scheme, m is set to as large as 256 or 512. Therefore, the total charge dissipation of the proposed scheme is about half that of the conventional scheme.

346

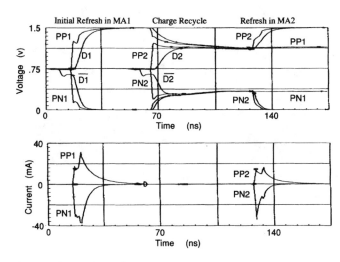

Fig. 5. Simulated waveforms.

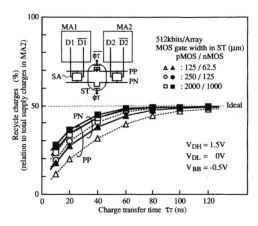

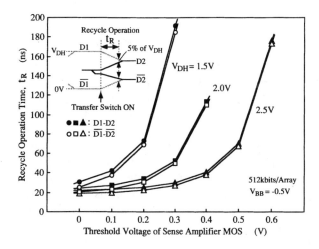

Fig. 6. Ratio of recycle charges.

Fig. 7. Recycle operation time dependence on threshold voltage of sense amplifier MOS.

III. PERFORMANCE OF THE PROPOSED CIRCUIT

Fig. 5 shows the simulated waveforms of the data-lines and sense amplifier driving lines, and the current flow from the external power supply lines (V_{DH} and V_{DL} in Fig. 3(a)) through the sense amplifier drivers (DM1 and DM2 in Fig. 3(a)) for each PP1-PN2. The data-line voltage is 1.5 V, and the MOS's in transfer switch ST are 500 μm in pMOS and 250 μm in nMOS. The charge transfer time, where ST is activated, is 60 ns. In the initial refresh in MA1, which is identical to the conventional operation, large currents flow. During the recycle operation from MA1 to MA2, data-lines in MA2 are amplified to about half level by the recycled charges. No current flows from the external voltage lines. Moreover, in the amplification from these half-level data lines $D2$ and $\overline{D2}$ to full swing by the additional charges from the external power supply line, the current flows are almost half. Therefore, the voltage bounce in the power supply lines is also reduced.

Since the charges are transferred by ST in the Charge Recycle Refresh, the amount of charge that can be transferred and the charge transfer time are important design issues. Fig. 6 shows how, in the simulation, the ratio of recycled charges to the total supply charges in MA2 depends on the charge transfer time, with the gate width of the MOS in ST as a parameter. The data-line voltage is 1.5 V and the threshold voltage of the sense amplifier MOS is 0.1 V. As the transfer time increases, the charge dissipation in MA2 decreases because the amplification in MA2 is accomplished by reusing these charges. The recycle ratio is ideally 50%. This ratio seems to saturate after the transfer time of 60 ns with a gate width of more than 250/125 μm This is because the ON resistance of switch ST reaches sufficiently low with a gate width of this value, and the operation is determined by the half amplifying speed of the sense amplifier in this region.

In the Charge Recycle Refresh scheme, the data-line signal is amplified to half level by the sense amplifier driving line, which finally reaches to the half level. This means that the sense amplifier should operate at lower voltage than in the conventional scheme. Therefore, the threshold voltage of the sense amplifier MOS is another important design issue. The

simulated recycle operation time dependence on the threshold voltage is shown in Fig. 7, with the data-line voltages as parameters. Here, the recycle operation time is defined from the start of sense amplifier operation to the time at which the data-line voltage difference between the original data-line and the transferred data-line reaches 5%. The ratio of data-line parasitic capacitance to cell storage capacitance is constant in these simulations. As the data-line voltage is set higher, a higher threshold voltage can be used. This is because the initial small data-line signal corresponding to charge storage in the cell is larger when the data-line voltage is set higher with the constant ratio of data-line parasitic capacitance to cell storage capacitance. Although the higher operating voltage reduces the recycle operation time, it causes large power dissipation again. In the data-retention mode of the DRAM, however, the refresh operation can be relatively slow because all the memory cells may be refreshed during the given refresh interval. Therefore, the Charge Recycle Refresh under low voltage operation with low threshold voltage of the sense amplifier MOS is practical for low power dissipation.

In the Charge Recycle Refresh, the data-lines and the sense amplifier driving lines are kept in a floating state at the half amplified level by the recycle operation. The subthreshold current with low threshold voltage MOS in

347

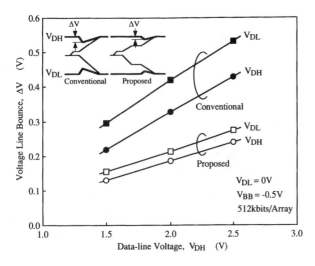

Fig. 8. Voltage line bounce dependence on data-line voltage.

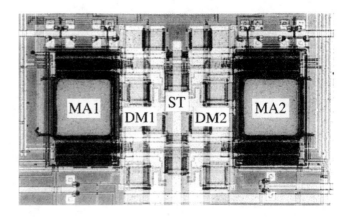

Fig. 9. Microphotograph of 64 kb test chip.

the sense amplifier seems to crush the half-amplified signal. However, in a practical use, this half-level state emerges only during a period of less than 100 ns including, the period of change to the half state. In addition, since the p-well and n-well voltages are kept at the conventional level, the subthreshold current is reduced as the threshold voltage is increased by the body effect.

The voltage transient K of the sense amplifier driving line to amplify the small data-line voltage is $1/2$ compared to the conventional scheme due to charge sharing in the Charge Recycle Refresh operation unlike in the conventional scheme in which the line is directly driven by the sense amplifier driver (DM1, DM2, as shown in Fig. 3(a)). The sense amplifier noise V_{se} is dependent on the voltage transient K of the sense amplifier driving line, expressed by the following [6].

$$V_{se} = A \cdot \sqrt{K} \cdot \left(\frac{\Delta C_D}{C_D} + \frac{\Delta \beta}{\beta} \right) + \Delta V_T \qquad (3.1)$$

where, $\Delta C_D/C_D$ is the imbalance in data-line pair capacitance, $\Delta \beta/\beta$ is the imbalance in β of the flip-flop pair of MOS's in the sense amplifier, and A is a constant. Since the K is small in the Charge Recycle Refresh scheme, the sense amplifier noise is reduced. Therefore, smaller signal can be amplified by the sense amplifier than that in the conventional scheme. This enables to set the refresh interval longer to reduce power, which is important for file applications of DRAM's.

The voltage bounce in the power supply lines is expected to be reduced in the Charge Recycle Refresh operation from the simulated current waveforms shown in Fig. 5. The simulated voltage bounce versus the data-line voltage is shown in Fig. 8. The low voltage of V_{DL} is set at 0, and the high voltage of V_{DH} changes from 1.5 V to 2.5 V. The voltage bounce at V_{DH} is reduced to 60% at 1.5 V operation, and reduced to 56% at 2.5 V operation. In an actual chip, the small voltage bounce is expected to stabilize the circuit operation, especially for an on-chip voltage limiter. This is an important feature of the Charge Recycle Refresh for multi-array activation, to reduce the number of refresh cycles, as described in the Section 6.

The area and speed penalty in the Charge Recycle Refresh scheme is negligible as follows. The transfer switches are located under the wiring area and area overhead of the interconnection is lighten by the appropriate layout. The cycle time in the Charge Recycle Refresh scheme is longer than that in the conventional scheme due to the additional transfer time as mentioned before. In the data-retention mode, however, the increase in refresh overhead is negligible as 2.6% (conventional) to 3.8% (proposed) with 16 k cycles/64 ms. And a few additional control circuit are needed for Charge Recycle Refresh operation.

IV. EXPERIMENTAL RESULT

To evaluate the circuit for the Charge Recycle Refresh, a 64 kb test chip was fabricated using 0.25 μm technology (Fig. 9). Two sets of 32 kb arrays, MA1 and MA2, had sense amplifier drivers DM1 and DM2 and a transfer switch ST between them. Since the small arrays were prepared and the large ST with DM1 and DM2 were located between the two array in the test chip, the area penalty seems large. In an actual chip, 512 kb $\sim$ 1 Mb arrays are used with the same size of ST, DM1 and DM2 in this figure.

The proposed transfer operations of data-lines were observed to directly measure the waveforms of the sense amplifier driving lines by using a small capacitance FET probe, as shown in Fig. 10. The operating voltage was 1.5 V. The sense amplifier driving lines of PP1 and PN1 were first amplified to half level by the recycle operation, and then amplified to the full level. Then the charge in PP1 and PN1 was transferred to PP2 and PN2, and recycled. Thus, the original driving lines were half discharged, and the driving lines of PP2 and PN2 were half amplified.

V. EXTENDED CHARGE RECYCLE OPERATION

The concept of Charge Recycle operation originally proposed for two arrays as described above, can be further extended to multiple arrays as shown in Fig. 11 (four-array Charge Recycle). In this case, six transfer switches ST12 $\sim$ ST41 are required. This number equals the number of combinations for selecting two arrays among four arrays. By transferring charges from one of the four arrays to another, the

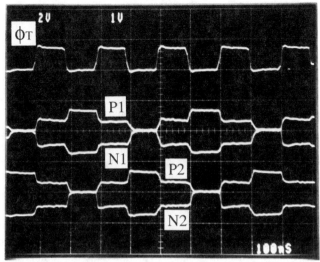

H: 100ns/div.   V: 2V/div., 1V/div.

Fig. 10.   Measured voltage waveforms.

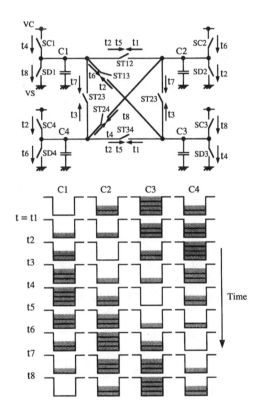

Fig. 11.   Extended charge recycle operation.

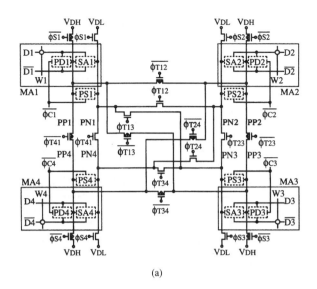

(a)

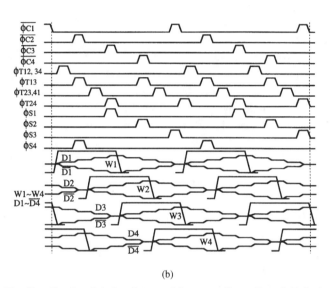

(b)

Fig. 12.   Circuit and timing diagram of four-array Charge Recycle Refresh.

charging current on the data-line is reduced to about $1/4$ of the conventional charging current. The total charge dissipation can be reduced to $1/n$ of the conventional scheme by Charge Recycle operations among the n pieces of memory arrays connected by $_nC_2$ pairs of transfer switches.

A DRAM circuit and timing diagrams based on the four-array Charge Recycle is shown in Fig. 12. Four pairs of sense amplifier driving lines are connected by six pairs of transfer switches. After the small data-line signal appears in $D1$ and

$\overline{D1}$, this signal is amplified by the charge in $D2$ and $\overline{D2}$ to $1/4$ of full amplitude. The data-line signal in $D2$ and $\overline{D2}$ is also shorted to the same amplitude. Then, the $1/4$ amplified signal in $D1$ and $\overline{D1}$ is amplified to $1/2$ of full amplitude by the charge in $D3$ and $\overline{D3}$. The half-amplified signal in $D1$ and $\overline{D1}$ is amplified to $3/4$ by the charge in $D4$ and $\overline{D4}$. Finally, the $3/4$ amplified signal in $D1$ and $\overline{D1}$ is amplified to the full amplitude by the V_{DH}. Therefore, the current dissipation from the external voltage is only $1/4$ that of a conventional scheme. The peak current is also reduced to that ratio.

VI. MULTI-ARRAY ACTIVATION WITH THE CHARGE RECYCLE REFRESH

Multi-array activation in the data retention mode, which is different from the normal cycle, is useful [4] in maintaining the number of refresh cycles and the refresh period. However, since multi-array activation causes a large peak current during the refresh operation, the design of power supply line in the DRAM chip is difficult. Therefore, a multi-array activation with the Charge Recycle Refresh is proposed, in which the

349

TABLE II
REFRESH CURRENT COMPARISON.

	Features	Schematic Refresh Current	I_{self}	Activated Sub-Array
(a)	Conventional Refresh	Time → Array Peripheral #1 #2 #3 #4 #5 .... #n	1	Sub-Array / Chip
(b)	Two-array Charge Recycle Refresh	#1 #2 #3 #4 #5 .... #n	0.63	Charge Recycle
(c)	Two-array Charge Recycle Refresh with double-array activation	#1 #2 #3 .... #n/2	0.5	
(d)	Four-array Charge Recycle Refresh with quadruple-array activation	#1 #2 .... #n/4	0.25	

same peak current as in the conventional scheme is achieved with a small number of refresh cycles for refreshing all of the cells.

The refresh current comparison is shown in Table II. Here, it is assumed that the initial refresh is treated in the same way as the others in the Charge Recycle Refresh operation. In the conventional refresh, the selected array is equal to that in the normal cycle. The self-refresh current I_{self} is expressed as follows [4] , ignoring the DC current.

$$I_{\text{self}} = \left(I_{\text{peri}} + \frac{I_{\text{cell}}}{N_{\text{ref}}} \right) \cdot \frac{T_{\text{rc}}}{T_{\text{ref}}} \cdot N_{\text{ref}} \qquad (6.1)$$

where,

I_{peri} current of peripheral circuits,

I_{cell} total data-line current consumed by refreshing all the cells,

T_{rc} cycle time during normal operation,

T_{ref} refresh interval,

N_{ref} number of refresh cycles required for refreshing all the cells.

To reduce I_{self}, two alternatives exist. They are increasing T_{ref} and reducing N_{ref} on the sacrifice of increasing $I_{\text{cell}}/N_{\text{ref}}$. However, T_{ref} cannot be increased because of memory cell retention time. When introducing the multi-array activation with Charge Recycle Refresh, I_{self} can be reduced by decreasing the N_{ref} without increasing $I_{\text{cell}}/N_{\text{ref}}$ as follows.

Using the two-array Charge Recycle Refresh, and ignoring slight increases in peripheral current, I_{self} is expressed as

$$I_{\text{self}} = \left(I_{\text{peri}} + \frac{\frac{1}{2} \cdot I_{\text{cell}}}{N_{\text{ref}}} \right) \cdot \frac{T_{\text{rc}}}{T_{\text{ref}}} \cdot N_{\text{ref}}. \qquad (6.2)$$

This results in an I_{self} of 0.63, compared to that of a conventional. However, as shown in Table II(b), the peripheral component becomes dominant in this scheme, since the I_{peri} cannot be recycled.

When introducing the double-array activation with two-array Charge Recycle Refresh (Table II(c)), I_{self} can be

expressed as

$$I_{\text{self}} = \left(I_{\text{peri}} + \frac{\frac{1}{2} \cdot I_{\text{cell}}}{\frac{1}{2} \cdot N_{\text{ref}}} \right) \cdot \frac{T_{\text{rc}}}{T_{\text{ref}}} \cdot \frac{1}{2} \cdot N_{\text{ref}}$$

$$= \left(I_{\text{peri}} + \frac{I_{\text{cell}}}{N_{\text{ref}}} \right) \cdot \frac{T_{\text{rc}}}{T_{\text{ref}}} \cdot \frac{1}{2} \cdot N_{\text{ref}}. \qquad (6.3)$$

The result is reducing the N_{ref} by half without increasing the $I_{\text{cell}}/N_{\text{ref}}$. That means the I_{self} is reduced to half by decreasing the N_{ref} with the same peak current as in the conventional scheme. The quadruple-array activation for the four-array Charge Recycle Refresh (Table II(d)) has an I_{self} of 0.25 that of a conventional refresh with the same peak current. Moreover, the small power dissipation implies that a low junction temperature is achieved even though the number of refresh cycles is reduced to $1/2$ or $1/4$, conventionally which caused large power dissipation. This low junction temperature results in a longer refresh period, because of the smaller cell leakage current. That enables the longer refresh interval for low power.

VII. CONCLUSION

A Charge Recycle Refresh for low-power DRAM data-retention, featuring alternative operation of two memory arrays was proposed, and demonstrated using a 64 kb test chip with 0.25 μm technology. The data-line current dissipation is only half that of the conventional refresh operation, and the voltage bounce of the power supply line is also reduced to 60% of the conventional. The original scheme was further extended for application to n arrays with $1/n$ data-line current dissipation. Moreover, the multi-array activation with Charge Recycle Refresh was proposed with a small number of refresh cycles and a small peak current. Thus, the reduction of data-retention current is achieved to enhance the Gb-scale DRAM capabilities for file applications, and the multi-array activation with small peak current is also achieved to relax the requirement for the refresh characteristic which is limited by the junction leakage current of memory cells.

ACKNOWLEDGMENT

The authors wish to thank M. Ishihara, T. Matsumoto, R. Hori, K. Kajigaya, K. Itoh, K. Shimohigashi, T. Nishida, K. Ohyu and J. Etoh for their helpful suggestions. The authors would also like to thank S. Shukuri, T. Kisu and Y. Yamashita for their support in fabricating the test chip.

REFERENCES

[1] T. Kawahara et al., "A charge recycle refresh for Gb-scale DRAM's in file applications," Symp. on VLSI Circuits, Dig. of Tech. Papers, pp. 41–42, May 1993.

[2] G. Kitsukawa et al., "256 Mb DRAM technologies for file applications," ISSCC Dig. of Tech. Papers, pp. 48–49, Feb. 1993.

[3] K. Sato et al., "A 4-Mb pseudo SRAM operating at 2.6 ± 1 V with 3 μa data retention current," IEEE J. Solid-State Circuits, vol. 26, no. 11, pp. 1556–1562, Nov. 1991.

[4] K. Kenmizaki et al., "A 36 μA 4 Mb PSRAM with quadruple array operation," Symp. on VLSI Circuits, Dig. of Tech. Papers, pp. 79–80, May 1989.

[5] P. K. Chatterjee *et al.*, "The impact of scaling laws on the choice of *n*-channel or *p*-channel for MOS LSI," *IEEE Electron Device Lett.*, vol. 1, no. 10, pp. 220–223, Oct. 1980.
[6] H. Masuda *et al.*, "A 5 V-only 64 K dynamic RAM based on high S/N design," *IEEE J. Solid-State Circuits*, vol. 15, no. 5, pp. 846–854, Oct. 1980.

A 1-V 1-Mb SRAM for Portable Equipment

Hiroki MORIMURA, and Nobutaro SHIBATA

NTT LSI Laboratories

3-1, Morinosato Wakamiya, Atsugi-shi, Kanagawa-ken 243-01, JAPAN

Abstract

Low-power and high-speed circuit techniques are described for 1-V battery operated SRAMs. A design concept is shown that uses two kinds of MOSFETs different in threshold voltages to reduce power dissipation due to the subthreshold leakage current in both standby and active modes. We propose a step-down boosted wordline scheme to reduce power dissipation in the memory array to 57% while accelerating the sensing speed. A novel bidirectional differential internal-bus architecture provides data transmission that is 45% faster than in the conventional architecture, yet without area or power penalty. A charge-recycling I/O buffer incorporating a data transition detector reduces the power dissipation of the I/O buffer by 30%. A 1-Mb SRAM designed using these techniques and 0.5-μm CMOS technology demonstrated 4.8-mW power dissipation and a 75-ns address access time with standby power of 1.2-μW at a 1-V power supply.

1 Introduction

Low-voltage digital signal processors (DSPs) have been developed for battery-operated portable equipment [1-2]. To create high-performance systems using these DSPs requires the development of low-power external memories. SRAM macrocells based on MTCMOS technology [3] and a cache memory using low-Vth MOSFETs in the peripheral circuitry [4] have been proposed for 1-V high-speed operation. However, these techniques are not suitable for low-power high-speed large-scale memories [5] because the subthreshold leakage current in the active mode is estimated to exceed

the sub-mA level, amounting to 10 to 20% of the total active power. For long battery life, power consumption including leakage current must be reduced in not only standby but also active mode.

In this paper, we describe a design concept that reduces leakage current in both standby and active modes, and propose a step-down boosted wordline scheme, a novel bidirectional differential internal-bus architecture, and a charge-recycling I/O buffer incorporating a data transition detector. We also present experimental results for a 1-Mb SRAM fabricated using 0.5-μm single-polysilicon triple-metal CMOS technology.

2 Circuit Techniques

2.1 Design concept for low-power and fast memories

In memory LSIs, the critical path for read-out and data-write is clear because almost all nodes are controlled by address and $\overline{\text{WE}}$-signal. It is thus possible to reduce the leakage current and access time by using the design concept shown in Fig. 1. A high-Vth MOSFET is applied to the cascode-connected MOSFET connected to the power or ground line directly. This is an effective way to suppress the threshold-voltage increase in high-Vth MOSFETs due to the bodyeffect. To reduce area, high-Vth MOSFETs Q_A and Q_B are shared among logic gates that have the same input through a local powerline. High-Vth MOSFETs are applied on the noncritical paths, and low-Vth MOSFETs are applied on the critical path. This reduces delay time in the critical path and cuts off leakage current in all logic gates except those on the critical path. Drivers, consisting of a low-Vth Pch MOSFET and a high-Vth Nch MOSFET, are applied to achieve large drivability for heavy capacitance loads. Their output is set high to cut off leakage current during standby.

A main-decoder configuration based on this concept is shown in Fig. 2. The outputs of the address-bus drivers

Reprinted from *International Symposium on Low Power Electronics and Design*, pp. 61-66, August 1996.

are set high during standby. A conventional NAND gate is not applicable because the leakage current of the Pch MOSFETs is not cut off in this situation. In the NAND gate in Fig. 2, Q1's source node is connected to an address bus and Q2 is controlled by the reverse signal. The NAND gate therefore has no leakage path during standby. Since the main-decoder is not on the critical path for memory-cell access, high-Vth Nch MOSFETs can be used in all of the cascode-connected MOSFETs in the NAND circuit to reduce leakage current in the address decoding.

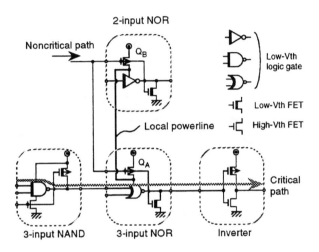

Fig. 1. Design concept for low-power and fast memories.

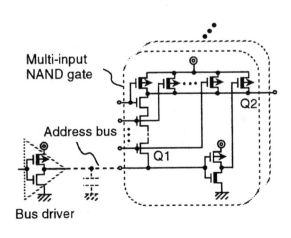

Fig. 2. Main decoder.

2.2 Step-down boosted wordline scheme

Memory cells are composed of high-Vth MOSFETs to reduce leakage current. Although it is possible to use the boosted wordline technique to reduce access time, using it increases the power dissipation in the memory array. We thus propose a step-down boosted wordline scheme to reduce the power dissipation while accelerating sensing speed. This scheme is shown in Fig. 3.

The scheme features a boosted-pulse generator and wordline voltage selectors. The boosted voltage is generated using metal-insulator-metal (MIM) capacitors. These capacitors are embedded along the word driver array to get enough capacity and to reduce area by using a capacitor electrode as a transmission line of the boosted voltage. The selected-wordline voltage is controlled by Q3 and Q4. To activate Q3, the boosted voltage is applied to the gate node to prevent Vth drop. The boost period is determined by the width of the address transition detection (ATD) pulse. Except in the circuit selected by the main-wordline, the leakage current is cut off by the high-Vth MOSFETs.

A simulated waveform of the sub-wordline is shown in Fig. 4. At the beginning of memory-cell access, the

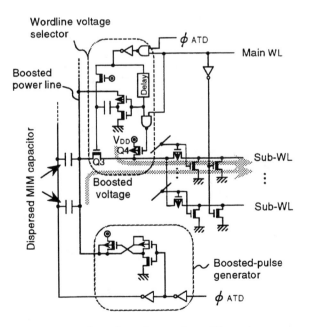

Fig. 3. Step-down boosted wordline scheme.

wordline is boosted to more than 1.4 V at a 1-V power supply. Since the memory-cell drivability is improved by a factor of two, the bitline transition is accelerated. After the sensing operation, Q3 is turned off and Q4 is turned on. The wordline voltage is then set to the supply voltage. This scheme is useful for reducing the power consumption in the memory array because the cell current is reduced by half after the stored data are output to the bitlines. Figure 5 shows the current consumption for various operating frequencies. The simulation results indicate that the power dissipation can be reduced by 43% at 10-MHz compared with a conventional boosted wordline scheme.

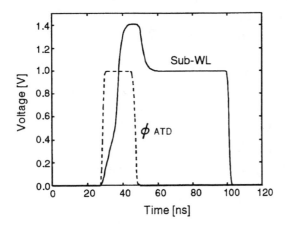

Fig. 4. Simulated waveform of sub-wordline.

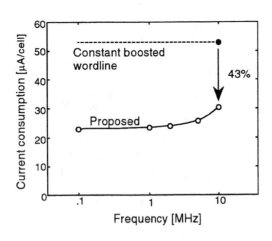

Fig. 5. Current consumption of a selected memory cell.

2.3 Low-Power differential internal-bus architecture

Differential data transmission can be used to decrease the delay time between a memory array and I/O buffers. It is important to reduce both the area and the power dissipation. Our proposed differential internal-bus architecture is shown in Fig. 6.

The bidirectional transmission contributes to area reduction. However, the conventional bidirectional buffer increases stray capacitance because both the driver and receiver are connected to the data bus. Our bidirectional buffer features Q5 and Q6, which work as pulldown transistors during write operations and as transfer gates during read operations. The input capacitance of the receivers is thus separated from the data bus. Furthermore, this architecture consumes no static current. A high-level signal, which is provided through a data-transmission line, controls the gate-input Pch MOSFET and the source-input Nch MOSFET in the receiver to the off-state, cutting the current path from V_{DD} to the ground.

Figure 7 shows the relationship between delay time and current consumption at 10-MHz operation. The proposed architecture reduces delay time to 55-60% in contrast to a conventional single-rail transmission with the same current consumption .

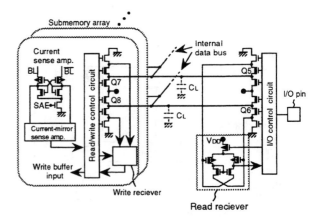

Fig. 6. Differential internal-bus architecture.

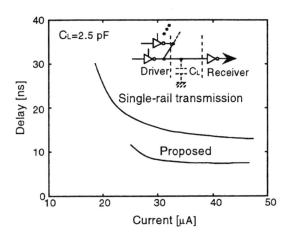

Fig. 7. Relationship between delay time and current consumption.

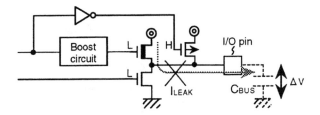

Fig. 8. Output buffer for low supply voltages.

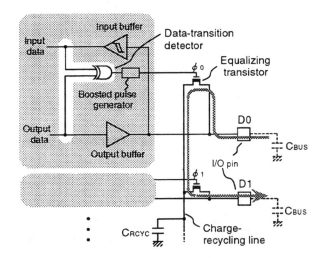

Fig. 9. Charge-recycling I/O buffer.

2.4 charge-recycling I/O buffer

A CMOS interface is used to reduce the power dissipation due to the static current. It is necessary to enhance the drivability of the output buffer and to cut off the leakage current during high-impedance mode because the I/O buses have large stray capacitance of 30-100 pF. It is also important to reduce the dynamic power dissipation in the I/O buffer.

Low-Vth PMOS pull-up are not suitable for use in the output buffers because current leakage into bus stray capacitance is inevitable in high-impedance-mode operation. Figure 8 shows an output buffer configuration for low supply voltages. A low-Vth Nch MOSFET with a boosted gate voltage is employed as a pull-up transistor. When the output buffer is in high-impedance mode, the source-node potential of the Nch pull-up transistor gets higher than the gate potential because of leakage current. This causes reverse bias across the gate-to-source, increasing the threshold voltage due to the bodyeffect. Therefore, leakage current to the bus capacitance is cut off. The high-Vth Pch MOSFET is employed to guarantee high-level output because the boosted voltage falls off gradually due to leakage current.

Our new charge-recycling I/O buffer is shown in Fig. 9. It features a bus-data transition detector. Each I/O pin is connected to a charge-recycling line through an equalizing transistor. The bus data are monitored by Schmidt-type input buffers. The equalizing transistor is activated in the I/O buffer that output data different from the bus data; and it is not activated in the I/O buffer whose output data is the same as the bus data. A charge on the bus transiting from high to low is recycled to a charge on the bus transiting from low to high through the charge-recycling line. Since the charge is recycled only among I/O buffers that transit data, power dissipation is reduced efficiently. To enhance recycling efficiency for timing skew, capacitor C_{RCYC} is added to the charge-recycling line.

Figure 10 shows the simulated waveforms of the I/O buffer; ϕ_0 and ϕ_1 are equalize pulses boosted to more than 1.6 V at a 1-V power supply. The charge on the D0 bus is recycled to the D1 bus. Charge recycling achieves, in theory, a maximum efficiency of 50%. The pulse width for charge recycling is set to 5 ns to suppress access time increase. Therefore, the recycle efficiency of this circuit is about 30%.

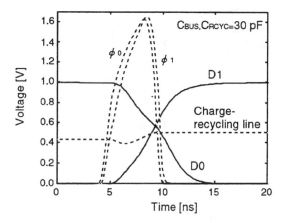

Fig. 10. Simulated waveforms of I/O buffer.

3 Experimental Results

We have designed a 1-V operation 1-Mb SRAM, using 0.5-μm CMOS technology, to confirm the circuit techniques discussed in this paper. Current sense amplifiers were adopted to reduce the sensing delay. Total simulated power dissipation during a read cycle (Fig. 11) was reduced to 77% by using a step-down boosted wordline scheme and charge-recycling I/O buffers.

A microphotograph of the test chip is shown in Fig. 12. The chip size was 16-mm×12.65-mm. The operating waveforms are shown in Fig. 13. The measured address-access time was 75-ns. Power consumption at 10-MHz operation with 30-pF load was 3.6 mW in the write cycle and 4.8 mW in the read cycle. A low standby power of 1.2-μW was achieved by using high-Vth MOSFETs. These power dissipations were low enough to guarantee long battery life. Table 1 summarizes the chip characteristics.

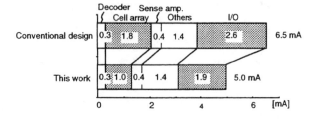

Fig. 11. Power dissipation during read cycle.

Table 1. Chip characteristics.

Process technology	0.5-μm CMOS single-polysilicon triple-metal MIM capacitor
Gate length	
High-Vth FET	0.55 μm
Low-Vth FET	0.65 μm
Gate oxide	11 nm
Threshold voltages	
N-ch FET	0.25 V, 0.55 V
P-ch FET	-0.35 V, -0.65 V
Cell size (pure CMOS)	11μm × 10 μm
Chip size	16 mm × 12.65 mm
Organization	64 Kwords × 16 bits
Supply voltage	1 V
Address access time	75 ns
Power dissipation (@ 10 MHz)	3.6 mW (write) 4.8 mW (read)

4 Summary

We have described a design concept for reducing leakage current in both the active and standby mode for 1-V battery-operated SRAMs. A step-down boosted wordline scheme, a novel bidirectional differential internal-bus architecture, and a charge-recycling I/O buffer with a data-transition detector were proposed. The measured power consumption and address access time were 4.8-mW and 75-ns, and the standby power of 1.2-μW was achieved for a 1-V 1-Mb SRAM fabricated using 0.5-μm CMOS technology.

Acknowledgment

We are indebted to Hideki Fukuda and Shigeru Date for their encouragement and useful advice and to Mayumi Watanabe for her technical support on the layout design.

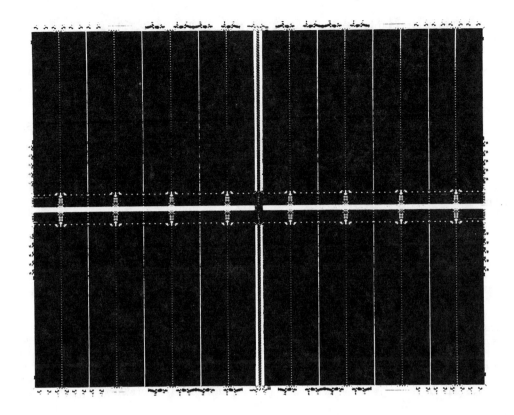

Fig. 12. Microphotograph of 1-V 1-Mb SRAM.

References

[1] Mutoh, *et al.*, "A 1-V Multi-Threshold Voltage CMOS DSP with an Efficient Power Management Technique for Mobile Phone Application," ISSCC Dig. Tech. Papers, pp. 168-169, Feb. 1996.

[2] Izumikawa, *et al.*, "A 0.9-V 100-MHz 4-mW 2-mm² 16-b DSP Core," ISSCC Dig. Tech. Papers, pp. 84-85, Feb. 1995.

[3] Date, *et al.*, "1-V 30-MHz Memory-Macrocell-Circuit Technology with a 0.5-μm Multi-threshold CMOS," IEEE Symp. on Low Power Electronics Dig. Tech. Papers, pp. 90-91, Sept. 1994.

[4] Mizuno, *et al.*, "A 1-V 100-MHz 10-mW Cache using Separated Bit-Line Memory Hierarchy and Domino Tag Comparators," ISSCC Dig. Tech. Papers, pp. 152-153, Feb. 1996.

[5] Takashima, *et al.*, "Standby/Active Mode Logic for Sub-1V Operating ULSI Memory" IEEE J. Solid-State Circuits, vol. 29, no. 4, pp. 441-447, Apr. 1994.

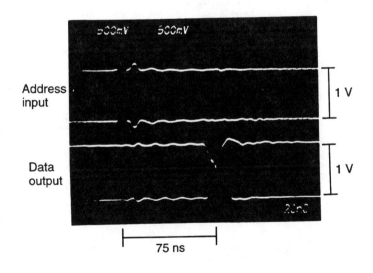

Fig. 13. Measured waveforms of 1-V 1-Mb SRAM.

A Single Bitline Cross-Point Cell Activation (SCPA) Architecture for Ultra Low Power SRAMS

MOTOMU UKITA, SHUJI MURAKAMI, TADATO YAMAGATA, HIROTADA KURIYAMA,
YASUMASA NISHIMURA, AND KENJI ANAMI

MITSUBISHI ELECTRIC CORP., ITAMI, JAPAN

Single bitline cross-point cell activation (SCPA) architecture reduces active power consumption and reduces chip size of high-density SRAMs. The architecture enables the smallest column current possible without increasing the block division of the cell array. Since the decoder area is reduced due to less block division, the memory core can be smaller than with a conventional divided word line (DWL) structure [1].

Figure 1 shows the circuit used in SCPA architecture. The Y-address controls access transistors as well as the X-address. Since only one memory cell at the cross point of X- and Y-addresses is activated, column current is drawn by only this cell, and thus is minimized. Moreover, SCPA allows the number of blocks to be reduced because the column current level does not depend on the number of block divisions in the SCPA.

The load element is a pMOS thin-film transistor (TFT). Since SCPA architecture uses a single-bitline structure, a memory cell needs only one bitline and two serial-access transistors [2]. Therefore, one memory cell consists of 6 elements and occupies an area equivalent to that of a conventional cell. The cell layout of the experimental device is shown in Figure 2. Because the bitline is shared by neighboring memory cells, three metal lines run every two memory cells. The size of the memory cell is reduced by using stacked-gate transistor technology in which two serial access transistors are stacked next to each other. The experimental device incorporating a $2.2 \times 3.9 \mu m^2$ memory cell is fabricated using 5-level-polysilicon (including stacked gate), single-alminum, $0.4\mu m$ CMOS technology.

Figure 3 shows the electrical characteristics of the inverters used in the SCPA cell. Inverter-A, without an access transistor, and inverter-B with series access transistors exhibit characteristics similar to those of a simple CMOS inverter and an nMOS inverter using an enhancement nMOS load resistor, respectively. In a read operation, there must be two stable points. In a conventional cell, the ß-ratio of the driver transistor and the access transistor must be more than 2.5 in order to have two stable points. The SCPA cell does not require such a large ß-ratio because inverter-A is CMOS. The driver transistor size is reduced, reducing the memory cell area.

In a "Low" write cycle, the data is conveyed to storage node N1 through the access transistors after the bitline is forced low. Then, node N2 is charged "High" by the pMOS-TFT load. The time required to write "Low" depends on the on current of the TFT load. The experimental device had a 2.5µs write time due 6nA on-current. However, if a 1µA-on-current TFT is used, the time required to write "Low" can be below 15ns [3].

In a "High" write cycle, the high level is conveyed through the access transistors. However, changing the memory cell data from low to high is impossible because the memory cell has two stable points when the bitline is high. To write "High", one of the two stable points of the memory cell must be eliminated. To accomplish this, the X- and Y-wordlines are boosted in a write cycle. The dashed curve in Figure 3 shows the characteristic of inverter-B with only one stable point when the wordline is boosted higher than Vcc.

Figure 4 shows the circuit diagram of the wordline-boost circuit. This circuit is activated only during the write cycle. The supply voltage is 3V because the SCPA architecture is intended for 16Mb SRAM. In a conventional structure, the boosted node is precharged to Vcc-Vth, not enough to charge the MOS boost capacitor under low-supply voltage. To raise the precharge level to Vcc, a pMOS transistor is used to charge the boost capacitor, as shown in Figure 4. This enables the boost circuit to work with a low-voltage power supply. The n-well is connected to the boosted node and the pMOS gate is regulated so that it turns off during a boost cycle to avoid reverse current from the boosted node flowing to Vcc. The circuit allows the wordline level to be equal to Vcc during a read cycle (non-boost cycle). This circuit in a 16Mb SRAM consumes 1.8mA (Vcc=3V,Tc=100ns).

The circuit shown in Figure 5 is used to sense data on a single bitline. The dummy cell is placed at each bitline and activated when the neighboring bitline is selected. The amplifier senses a differential voltage between the selected bitline and the dummy cell bitline. The dummy cell bitline is adjusted to a middle level by controlling the dummy cell and the bitline load.

Table 1 compares SCPA architecture with conventional architecture for a 16Mb SRAM organized as 2Mwx8b, which is divided into 64 blocks in the conventional case and 8 blocks in the case of SCPA. The column current is reduced from 12.8mA to 0.8mA. The total current is reduced by 10.2mA including 1.8mA consumed at the boost circuit. In the SCPA, the total active current is 15.9mA while in conventional architecture it is 26.1mA. In SCPA, the memory cell size is equal to that in conventional architecture, however the number of local decoders is reduced from 64 to 8. Moreover, the number of GND lines is reduced because of a much smaller column current in SCPA. As a result, the area of the memory core is reduced by 10% in a 16Mb SRAM.

Figure 6 shows a micrograph of the experimental device composed of 32b memory cells and peripheral circuits including X,Y-decoders, a boost circuit, sense amplifiers and other features. Stacked-gate transistors are used as access transistors. Figure 7 shows the waveform for a read cycle. Typical read access time is 15ns. All data are measured with one 3V supply. Figure 8 shows the relation between supply voltage and measured boost level. The boost circuit operates successfully at 1V.

Acknowledgments

The authors thank H. Komiya, T. Yoshihara, H. Miyoshi and Y. Kohno for encouragement.

References

[1] Yoshimoto, M., et al., "A divided word-line structure in static RAM and its application to 64k full CMOS RAM", IEEE J. Solid-State Circuits, vol. SC-18, pp. 479-485, Oct. 1983.

[2] Stewart, R. G., A. G. F. Dingwall, "16K CMOS/SOS Asynchronous Static RAM", ISSCC DIGEST OF TECHNICAL PAPERS, pp. 104-105, p.286, Feb. 1979.

[3] Hayashi, F., M. Kitakata, "A High Performance Polysilicon TFT Using RTA and Plasma Hydrogenation Applicable to Highly Stable SRAMs of 16Mbit and beyond", Symp. VLSI Technology Dig. Tech. Papers, pp. 36-37, June 1992.

Reprinted from *IEEE International Solid State Circuits Conference*, pp. 252-253, February 1994.

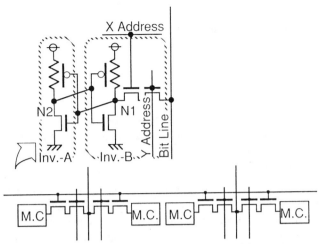

Figure 1: Memory cell used for SCPA architecture.

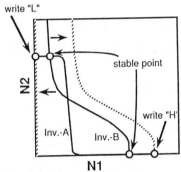

Figure 3: Characteristics of SCPA cell inverters.

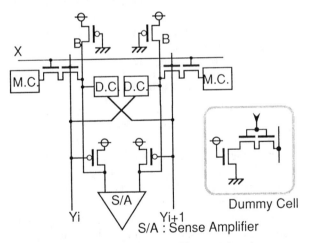

S/A : Sense Amplifier

Figure 5: Dummy cell sense structure.

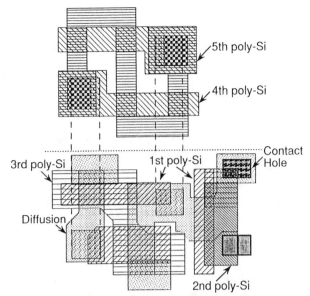

Figure 2: Memory cell layout for SCPA architecture.

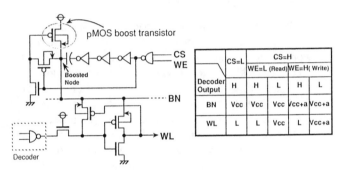

Figure 4: Wordline boost circuit.

	CS=L	CS=H		
		WE=L (Read)	WE=H(Write)	
Decoder Output	H	H	L	H	L
BN	Vcc	Vcc	Vcc	Vcc+a	Vcc+a
WL	L	L	Vcc	L	Vcc+a

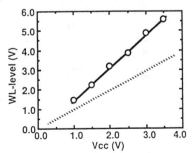

Figure 7: Waveform in the read operation.

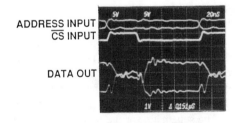

Figure 8: Measured boost level vs. supply voltage.
Figure 6: See page 298.

	Conventional	SCPA
Process	4poly-Si 2Al	5poly-Si 2Al
Block	64 128-col. blocks	8 1024-col. blocks
GND line every	16 columns	256 columns
Column current	12.8mA	0.8mA
Current drawn	26.1mA	15.9mA
Memory core area	100%	89.8%
Memory cell size	$2.2 \times 3.9 \mu m^2$	$2.2 \times 3.9 \mu m^2$

Table 1: Comparison of conventional and SCPA architectures in 16Mb SRAMs.

Techniques To Reduce Power In Fast Wide Memories[1]

Bharadwaj S. Amrutur and Mark Horowitz

Center for Integrated Systems, Stanford University, CA 94305

Abstract

Memories contain large arrays with high capacitance bitlines and IO lines. To reduce the power of memory accesses we limit the swings on these by controlling the time the lines are driven by using a replica feedback. The swings are set to 10% of the supply over a wide range of process and operating conditions.

Introduction

Memory power is an important component of the power budget in today's electronic systems. This paper looks at circuit techniques that can be used to reduce the power requirements of a wide access width memory while having minimum effect on its access time.

In CMOS memories the access path can be broken into two parts: from address to local wordline select, and from local wordline to sensed data. It is usually the latter half of the access (which involves driving the bitlines and local bus lines and sensing them) that consume the most power. These lines are heavily loaded and thus require a large amount of energy each time they are changed. While various techniques [1,2] have been used to partition the array so only the desired bit lines move, for wide access SRAMs this is not enough. This paper shows how power can be further reduced by limiting the energy consumed by each bitline. The key idea is to limit the array swings by controlling the drive signal (for bitlines this is the local wordline) rather than electrical clamping. The desired pulse is created by using a replica feedback, to ensure good tracking.

Limited Bitline Swings

In our memory, local word lines are generated by ANDing the block select and the global word lines. Since the block select signal arrives later than the global wordlines, we control the width of the local wordline pulse using the block select signal. The schematic for the block select circuitry is shown in Figure 1. To set the pulse width a dummy memory cell is also activated, which pulls down a reference bitline. This bitline is approximately one tenth the height of a regular bitline. This memory cell delay sets the block-select pulse width. Since the real bitlines swing about 1/10 of the reference, they swing about 10% of the supply voltage. Figure 2 shows the simulated swings on the bitlines normalized to the supply for different supply voltages and different process corners.

The timing signal from the reference bit line also provides the needed edge to trigger the clocked bitline sense amplifer, and eventually the precharge circuits.The circuitry we use is very similar to that shown in [3].

The area overhead of this approach is very small. It requires two extra columns and rows to implement the reference cell and bitline. Their location in a block is shown in Figure 3. The reference cell is located at the intersection of the inner row and column to create better matching with the rest of the memory cells in the array.

Limited Local Data Bus Swings

Having reduced the bitline power, we also need to limit the energy dissipated in driving the local data bus which connects the blocks to the memory IO ports. Once again we use a limited pulse width signal to create small swing (10% of Vdd) signals. Figure 4 shows the column circuit in a block. It consists of the precharge circuit, read senseamps and write senseamps. The read sense amps are latched crosscoupled structures with a pulsed sense signal. A replica sense amp is used to control the pulse width. To reduce write power, the write data is also transmitted as low swing signals and is amplified and driven onto the bitlines by another crosscoupled senseamp. The IO circuits, shown in Figure 5, consist of global sense amps, to amplify the read data, a pulsed write driver to transmit write data as low swing signals and precharge devices for the local data bus.

Test SRAM

To test these ideas we have designed a 2k x 32bit SRAM in a 1.2μ CMOS technology. The memory is partitioned into 8 blocks each 256 rows and 32 cells. To minimize power used in the rest of the RAM we were careful to reduce unneeded transitions. For example only the precharge control in the selected block is activated and the precharge in all other blocks do not change. As another example, we separate the precharge devices into two groups and activate only one for reads and both for writes, as the bitline swings are bigger for writes (see Figure 4).

The simulated access times and power dissipation is shown in Table 1 for typical conditions. The power estimate is for a cycle time of 22 gatedelays.The test chip is currently being fabricated.

References

[1] Chu S. T et al, "A 25nS low-power full-CMOS 1Mbit (128kx8) Sram," *JSSC,* vol 23, pp. 1078-1084, Oct 88.

---

1. The funding for this research is provided by ARPA under contract # J-FBI-92-194

Reprinted from *Proceedings of the 1994 Symposium on Low Power Electronics*, pp. 92-93, 1994.

360

[2] Wong, D.T. et al, "A llnS 8kx18 CMOS static ram with 0.5um devices," *JSSC*, vol 23, pp. 1095-1103, Oct 88.

[3] Braceras, G. et al, "A 200 Mhz internal/66Mhz external 64kB embedded virtual three-port cache sram," *ISSCC digest of technical papers*,1994, pp. 262-263.

Acknowledgments

We would like to thank Tom Chanak, John Maneatis, Drew Wingard and Stefanos Sidiropoulos for invaluable discussions and help.

TABLE 1. Power and access time versus supply

Supply (V)	T access (Gatedel)	Power (mW)	Gatedel (nS)
1.5	22.3	5.2	2.3
3.0	21.0	75.0	0.62
5.0	19.4	366.0	0.38

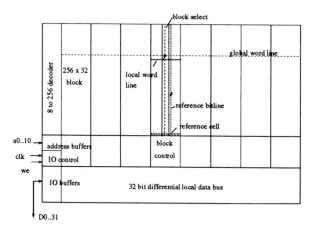

Figure 3 Memory architecture

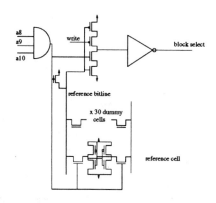

Figure 1 Block Select

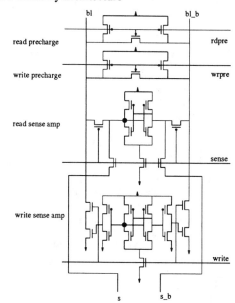

Figure 4 Column circuit

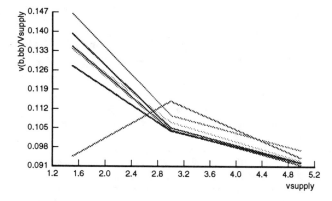

Figure 2 Bitline swing vs supply for different processes

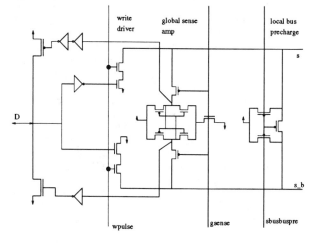

Figure 5 IO circuit

361

A 2-ns, 5-mW, Synchronous-Powered Static-Circuit Fully Associative TLB

H. Higuchi, S. Tachibana, M. Minami, and T. Nagano

Central Research Laboratory, Hitachi, Ltd.

Kokubunji, Tokyo 185, Japan

1 Introduction

Virtual memory is used in most high-performance computer systems to extend the address space. Virtual addresses are translated by the system into physical addresses at run-time. The translation is usually accelerated by special hardware called a translation look-aside buffer (TLB). Thus, TLBs are required for high-speed operation. In conventional high-speed TLBs, set-associative memories are utilized. But they need a large chip area. Fully associative TLBs which use content addressable memories (CAM) realize smaller chip areas. But slow circuit speed and large power dissipation are drawbacks in large entry-TLBs (1).

Recently, a high-speed, 4-ns, 64-entry fully associative TLB was reported (2), using fast differential amplifiers with reference signal generators. The TLB can compete with the set-associative TLBs in circuit speed. However, the circuit requires two additional signal-lines and a reference signal generator on each row of CAM arrays, which enlarge a chip area. This paper describes high-speed, low-power fully associative TLBs which do not need any signal lines added to conventional TLBs by using a newly developed matched signal and reference signal generator circuits.

2. TLB Circuits

The TLB circuits are shown in Fig. 1. CAM arrays are arranged on both sides of the data-RAM array to reduce metal-line capacitances. The most prominent circuit of this TLB is the match signal and reference signal generators. In conventional TLBs and also in the TLB refered to above, matched signals are generated passively, i.e., the matched signal is defined as the signal that remains at a precharged high level after all unmatched lines are pulled down to a low level by the current in the unmatched CAM cells. In contrast, the proposed circuits generate the matched signals actively, i.e., the matched signal is defined as the signal which is pulled up to the highest voltage in all match-lines by the current supplied through MOSFETs MP1 and MP2 in Fig.1. The gate width of MN1 is designed to be half that of the MOSFET of the CAM cells, and the Cd made of the dummy MOSFETs has almost the same capacitance as the match-line.

This generator circuit makes it possible for TLBs to be composed of static circuits which require no timing margins. In other words, they can operate in both asynchronous mode and synchronous mode. First, the asynchronous operation mode is explained using the simulator-generated wave forms shown in Fig. 2(a). Vcsml and Vcsp in Fig. 1 are held low and Vcsn is held high. These values set the match-line on which all input-data match with the CAM data to a high level of Vml. They also set match-lines on which input-data is unmatched with the CAM data to a low level, and set reference-lines to a level between the matched and unmatched levels. When the input-data changes, as shown in Fig. 2(a), match-lines which match with the input-data, are raised to high level and unmatched ones fall to or stay at low level because the supplied currents flow through MOSFETs in unmatched CAM cells. The potential differences between the match-lines and reference-lines are detected with differential amplifiers

and inverters drive the word-lines of the data RAM array. Although this mode eliminates the timing margin, it requires constant currents through MP1, MP2, and MN3.

To reduce the currents, synchronous power supply (synchronous mode) is proposed. In this mode, as shown in Fig. 2(b), Vcsml is held high during steady states, and is set to a low level just after the input-data are fixed. After a period needed for the circuit to generate the matched and reference signals, the Vcsml is raised to a high level. In this operation, currents are supplied to the circuits only in a short period just after the input-data is fixed. The rising edge of Vcsml determines the end of the reading mode of the data RAMs, namely, word-lines are at a high level only when the Vcsml is at low level. This control method is successfully used to reduce power dissipation by shortening the power supply period to only when the current is required. A power-supply period is indicated in Fig. 2(b). This power supply method allows low power dissipation without speed loss.

This synchronous mode additionally increases the speed of the circuit by differential sensing of the pre-equalized potentials between the match-lines and reference-lines.

3. Experiments and Results

These circuits were implemented in a 32-bit input-data, 16-entry, 32-bit data-out TLB TEG, which was fabricated with the 0.25 μm CMOS process. A photograph of the chip is shown in Fig. 3. The wave forms shown in Fig. 4 are of two input-data and a matched signal which appeared on the word line of the data RAM. Wave forms were photo-graphed using pico-probe. The match signal was output at around 1.2 ns after the input-data change. The delay time is 30% larger than that generated by the circuit simulator. The RAM data is expected to appear on Do terminals in less than 2 ns.

The power dissipation of operation with two input-data, two CAM rows, and two columns of data RAM was evaluated by using a circuit simulator. The total power of 5 mW was calculated by multiplying the power dissipation obtained by the simulator by 16. The cycle time used in power dissipation evaluation was 10 ns.

4. Conclusions

A high-speed, low-power fully associative TLB composed of static circuits is proposed and evaluated by using a circuit simulator and a fabricated 16-entry TLB TEG. Newly developed match/reference signal generators made it possible to use static differential amplifiers which eliminate the timing margin required in conventional TLBs. The synchronous power-supply reduce power consumption. The performance -delay time of 2 ns and total power of 5 mw- of a 32-bit input-data, 16-entry, 32-bit data-out TLB, was confirmed by a fabricated TLB TEG.

References

1) M. Milenkovic: IEEE Micro pp.70-85 (April 1990)
2) R. A. Heald, et. al.: IEEE J. Solid State Circuits 28 pp.1078 - 1083 (1993)

Reprinted with permission from *1995 Symposium on VLSI Circuits*, H. Higuchi, et al., "A 2-ns 5-mW Synchronous-Powered Static-Circuit Fully Associative TLB," pp. 21-22, June 1995. © Japan Society of Applied Physics.

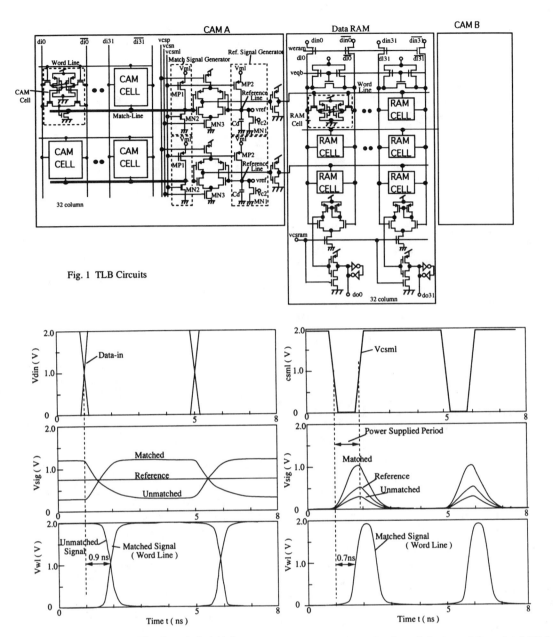

Fig. 1 TLB Circuits

Fig, 2 (a) Simulated Wave Forms in Static Mode

Fig, 2 (b) Simulated Wave Forms in Synchronously Power Supplied Mode

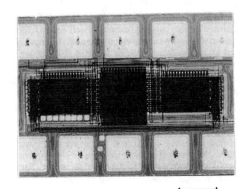

Fig. 3 Chip Photograph

100 μm

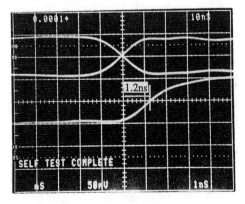

Fig. 4 Wave Forms of TLB TEG

Driving Source-Line (DSL) Cell Architecture for Sub-1-V High-Speed Low-Power Applications

Hiroyuki Mizuno and Takahiro Nagano

Central Research Laboratory, Hitachi, Ltd.

Kokubunji, Tokyo, 185 Japan

Abstract

A novel SRAM cell architecture for sub-1-V high-speed operation is proposed without using either low-V_{th} MOSFETs or modifying the cell layout pattern. A source-line connected to the source terminals of driver MOSFETs is controlled to be negative and floating in the read- and write-cycles, respectively. The cell-access time is reduced to 1/4-1/2 at a supply voltage of 0.5-1.0V. Limiting the bit-line swing reduces the writing power needed to charge the bit-lines to 1/10, and it realizes a faster write-recovery. The feasibility of low-power 100-MHz operation over a wide range of supply voltages is demonstrated.

1 Introduction

The growing demand for battery-operated devices such as personal digital assistant has led to intensive study of low-power high-speed memory/logic LSI chips [1]. However, a practical circuit technology for sub-1-V fast-access memory has not been reported. Additionally, power dissipation in the write-cycle will inevitably become an issue in future highly distributed cache memory systems [2]. We propose a novel driving source-line (DSL) cell architecture ideally suited for sub-1-V applications that maintains high throughput over a wide range of supply voltages and that reduces power dissipation during the write-cycle.

2 Driving Source-Line Cell Architecture

Our proposed DSL cell architecture is shown in Fig. 1. The source-line (SL) is connected to the source terminals of the driver MOSFETs; it is controlled to be negative during the read-cycle and floating during the write-cycle. The layout of the SL is the same as that of a conventional cell connected to a ground (V_{SS}) [3]. The threshold voltages (V_{th}) of the MOSFETs constituting the DSL cell range from 40 to 60% of the supply voltage, V_{DD}. Standby power due to subthreshold leakage current is thus negligible in the DSL cell architecture. Figure 2 illustrates the waveforms of the conventional and the DSL cell.

The boosting word-line (WL) voltage technique has been reported for low-voltage operation [4]. In this scheme, power dissipation is increased by a large WL voltage swing, but only the drain current of the transfer MOSFET is increased. This, in turn, reduces the static noise margin (SNM). In contrast, the SNM for the DSL cell does not change because the driver and transfer MOSFET currents are simultaneously increased.

2.1 Read-Cycle

The SL is driven to be negative when the WL is high. This gives rise to the higher drain currents of the transfer and driver MOSFETs resulting from both higher gate to source voltage and lower threshold voltage caused by for-ward substrate bias [5]. The DSL cell thus has a shorter access time. The delay time (t_{WL-BL}), defined as the time interval for a bit-line (BL) voltage difference of 0 to 100 mV, was simulated by using a spice circuit simulator. The cells are organized as 256 words × 32 bits; 0.25-μm CMOS devices are used.

Figure 3 shows t_{WL-BL} dependence on negative SL bias V_{BB} as a parameter of V_{DD}. A delay time for V_{BB} of 0 V corresponds to that of a conventional cell. A significant decrease in delay time is obtained even at a smaller V_{BB} of -0.5 V, which results in negligible power dissipation when driving an SL of around 50μW at 50 MHz.

Figure 4 represents t_{WL-BL} versus V_{DD} for the conventional and DSL cell as a parameter of V_{th}. The DSL cell architecture maintains fast access time over a wide range of V_{DD}. The cell-access time of the DSL cell are 1/4-1/2 those of the conventional cell at a V_{DD} of 0.5-1.0 V, respectively. Immunity against threshold voltage variation affecting access time is an advantage of the DSL cell architecture.

2.2 Write-Cycle

In the write-cycle, the SL is driven to be floating at the beginning and back to V_{SS} at the end. Since floating the SL makes the driver MOSFET inactive, a smaller bit-line voltage swing is adequate for controlling the node potentials in the cell. At the end of the write-cycle, the cell itself works as a latch-type sense amplifier to raise the node potential to V_{DD}. Because the write-cycle power mainly dissipates in bit-line capacitance, a smaller bit-line voltage swing follows a lower power dissipation. The DSL cell architecture is thus advantageous for wide-bit memories. Figure 5 shows the simulated power advancement in the write-cycle. Power dissipation is reduced to 1/10 for bit-line driving (P_{BL}) and to 1/3 in total at 50 MHz. Small bit-line swing inherently realizes a faster write-recovery.

3 Implementation Results

The circuit used is shown in Fig. 6. A latch-type sense amplifier is utilized. The bit-line is precharged to $V_{DD}/2$. The bit-line voltage swings are controlled by the WE-pulse width generated by using small MOSFETs (M1 and M2). The SL is controlled by a newly developed source driver constructed with high-V_{th} MOSFETs. Figure 7 shows the simulated waveforms at a V_{DD} of 0.5 V. Fast access time and writing with a small bit-line voltage swing, only about 100 mV, demonstrate 100-MHz operation feasibility.

4 Conclusion

The proposed SRAM cell using a driving source-line cell architecture reduces cell-access time to 1/4-1/2 at a supply voltage of 0.5-1.0V. Limiting the bit-line swing reduces the writing power needed for charging bit-lines to

1/10, and it realizes a faster write-recovery. A 256-word × 32-bit SRAM using a driving source-line cell architecture was designed; it verifies the feasibility of 100-MHz operation over a wide range of supply voltages.

Acknowledgement

The authors would like to thank Drs. K. Ishibashi, H. Higuchi and E. Takeda for their valuable discussions.

References

[1] A.P. Chandrakasan, *et al.*, *IEEE Journal of Solid State Circuits*, vol. 27, no. 4, pp. 473-484, Apr., 1992.

[2] D.B. Lidsky, *et al.*, *1994 IEEE Symposium on Low Power Electronics*, pp. 16-17, 1994.

[3] A. Sekiyama, *et al.*, *1990 Symposium on VLSI Circuits*, pp. 53-54, June, 1990.

[4] K. Ishibashi, *et al.*, *1992 ISSCC*, pp. 206-207, Feb., 1992.

[5] F. Assaderaghi, *et al.*, *1994 IEEE Symposium on Low Power Electronics*, pp. 58-59, Oct., 1994

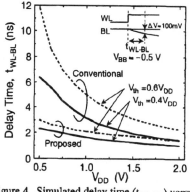

Figure 4. Simulated delay time (t_{WL-BL}) versus V_{DD} when using two different V_{th}.

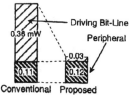

Figure 5. Power advancement at $V_{DD} = 1.0$ V.

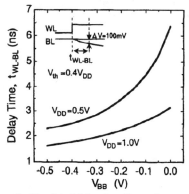

Figure 1. Driving source-line (DSL) cell architecture.

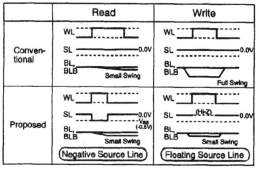

	Read	Write
Conventional	WL, SL 0.0V, BL, BLB Small Swing	WL, SL 0.0V, BL, BLB Full Swing
Proposed	WL, SL 0.0V V_{BB} (-0.5V), BL, BLB Small Swing (Negative Source Line)	WL, SL (Hi-Z) 0.0V, BL, BLB Small Swing (Floating Source Line)

Figure 2. Comparison with conventional cell architecture.

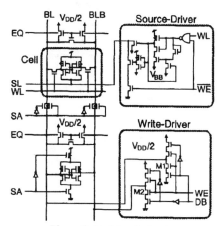

Figure 6. Peripheral circuitry.

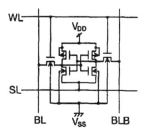

Figure 3. Simulated delay time (t_{WL-BL}) versus V_{BB}.

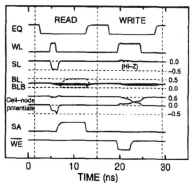

Figure 7. Simulated waveforms at $V_{DD} = 0.5$ V. Dashed lines in BL show the output of the sense amplifier.

Part VII

Portable Terminal Electronics

Energy Dissipation In General Purpose Microprocessors

Ricardo Gonzalez and Mark Horowitz

Abstract—In this paper we investigate possible ways to improve the energy efficiency of a general purpose microprocessor. We show that the energy of a processor depends on its performance, so we chose the energy-delay product to compare different processors. To improve the energy-delay product we explore methods of reducing energy consumption that do not lead to performance loss (i.e., wasted energy), and explore methods to reduce delay by exploiting instruction level parallelism. We found that careful design reduced the energy dissipation by almost 25%. Pipelining can give approximately a 2× improvement in energy-delay product. Superscalar issue, however, does not improve the energy-delay product any further since the overhead required offsets the gains in performance. Further improvements will be hard to come by since a large fraction of the energy (50–80%) is dissipated in the clock network and the on-chip memories. Thus, the efficiency of processors will depend more on the technology being used and the algorithm chosen by the programmer than the micro-architecture.

I. Introduction

THE interest in lowering the power of a processor has grown dramatically over the past few years. This interest in power dissipation is fueled partly by the high power levels (greater than 50 W [4]) of today's state-of-the art processors, as well as the growing market for portable computation devices. The result of this attention to power is an increasing diverse space of processors that a user can choose between, from simple machines with modest performance and power consumption under one Watt to processors with sophisticated architectural features and power dissipation in the tens of Watts. Yet for processors, power and performance seem to be strongly correlated. Low-power invariably means lower performance. This correlation is different from some published results for signal processing applications, where power has been reduced by three orders of magnitude while maintaining the same performance [6], [9]. This paper will show that the reason for the slow progress in processors is fundamental—barring a radical new architecture, the energy-delay product of a processor will be roughly set by the energy-delay product of the underlying technology.

The next section describes how the power and delay of the underlying CMOS technology used for processors are related, and shows how it is easy to trade increased delay for reduced energy. Thus the energy needed to complete a program (in

Manuscript received June 21, 1995; revised March 22, 1996. This research was supported by the Advanced Research Projects Agency under contract J-FBI-92-194.

The authors are with the Computer Systems Laboratory, Stanford University, Stanford, CA 94305 USA.

Publisher Item Identifier S 0018-9200(96)06477-3.

Jules/instruction or its inverse SPEC/W) is not a good metric to compare two designs. Instead, one needs to consider both energy and delay simultaneously, which is the reason we use the energy-delay product (in Jules/SPEC or its inverse $SPEC^2/W$) in the rest of this paper.

Section III then looks at a lower bound on the energy and energy-delay product for a processor by investigating a number of ideal machines, including an unpipelined machine, a simple RISC machine, and a superscalar machine. In these machines only the energy of data storage is accounted for, which includes essential latches, cache accesses, and register file accesses. Since the control and computation are free, these machines form a lower bound on what can be achieved in any real machine.

Then, Section IV looks at two real processors that we have designed, a simple RISC machine and a superscalar processor called TORCH. Starting from an unoptimized design, conventional optimizations can be used to save about 25% of the machine power. After these optimizations, we show that the real machines are within a factor of two from the ideal machine given in the previous section. Since this power is distributed in a number of small units, reducing the overhead will be fairly difficult.

II. Energy-Delay Product

Until recently, performance was the single most important feature of a microprocessor. Today, however, designers have become more concerned with the power dissipation, and in some cases low power is one of the key design goals. This has led to an increasing diversity in the processors available. Comparing processors across this wide spectrum is difficult, and we need to have a suitable metric for energy efficiency. Table I shows several possible metrics for a few of the processors available today [4], [5], [8], [12], [15], and [19]. Power is not a good metric to compare these processors since it is proportional to the clock frequency. By simply reducing the clock speed we can reduce the power dissipated in any of these processors. While the power decreases, the processor does not really become "better."

Another possible metric is energy, measured in Jules/Instruction or its inverse SPEC/W. While better than power, this metric also has problems. It is proportional to cv^2 so one can reduce the energy per instruction by reducing the supply voltage or decreasing the capacitance—with smaller transistors. Both of these changes increase the delay of the circuits, so we would expect the lowest energy processor to also have very low performance, as shown in Table I. Since

Reprinted from *IEEE Journal of Solid-State Circuits*, pp. 1277-1284, September 1996.

369

TABLE I
CURRENT MICROPROCESSORS

	Dec 21164	UltraSPARC	P6	R4600	R4200	Power 603
SPECint92	346.00	250.00	200.00	110.00	55.00	75.00
SPECfp92	506.00	360.00	150.00	84.00	30.00	85.00
Average	426.00	305.00	175.00	97.00	42.50	80.00
Frequency	300.00	167.00	133.00	150.00	80.00	80.00
Power	50.00	30.00	15.00	5.500	1.80	3.00
SPEC/W	8.51	10.17	8.75	17.64	25.00	26.67
SPEC^2/W	3621.01	3100.83	2041.70	1710.33	1003.47	2133.33
L_{min}	0.50	0.45	0.60	0.64	0.64	0.50
SPEC^2/Wλ^2	4470.38	3100.83	3629.69	3459.51	2029.73	2633.74

we usually want minimum power at a given performance level, or more performance for the same power, we need to consider both quantities simultaneously. The simplest way to do so is by taking the product of energy and delay (in Jules/SPEC or its inverse SPEC^2/W). Although there is a very wide distribution in both performance and power among processors—as shown in Table I—in terms of energy-delay product, all processor are remarkably close to one another.

To improve the energy-delay product of a processor we must either increase its performance or reduce its energy dissipation without adversely affecting the other quantity. Many of the common low-power design techniques do not reduce the energy-delay product, they simply allow the designer to trade-off performance for energy [11]. By reducing the supply voltage, for example, one can reduce the energy and the performance by almost one order of magnitude with only small changes in the energy-delay product.

An effective way to reduce the energy-delay product is to shrink the technology. If the scaling factor is λ, under ideal scaling conditions [9] the energy-delay product scales with λ^4. However, most technologies do not scale ideally since the threshold voltage often does not scale [7]. Moreover, the performance of a processor is also limited by the external memory system. So we would expect the energy-delay product to scale somewhere between λ^2 and λ^3. In Table I we also show the energy-delay product scaled by λ^2, under the row labeled "SPEC^2/Wλ^2." The spread between high-performance and low-power processors becomes even narrower. Thus, the energy efficiency of a processor is highly dependent on the efficiency of the underlying technology. As technology scales, the energy delay product of processors will continue to improve. Since the improvement should be approximately constant for all processors, we will ignore technology scaling for the remainder of this paper.

Once technology scaling has been factored in, the energy-delay of the processors shown in Table I are within a factor of two from each other, while the performance and energy vary by more than an order of magnitude. To understand why processors are so similar, the next section investigates what are the intrinsic factors that set the energy and energy-delay product for a processor. This simplified model indicates that architectural changes like superscalar issue have only a modest effect on a processor's energy-delay product.

III. LOWER BOUND ON ENERGY AND ENERGY-DELAY

Although there have been several papers that have presented dramatic reductions in the power dissipation of processors, there has been no work to investigate what is the lower bound on the energy and energy-delay product, and therefore we cannot determine if the reductions accomplished represent a large fraction of the potential improvements. In this section we will investigate what these bounds are by looking at three idealized machines: an unpipelined processor, a simple pipelined RISC processor, and a superscalar processor.

All processors fundamentally perform the same operations. They fetch instructions from a cache, use that information to determine which operands to fetch from a register file, then either operate on this data, or use it to generate an address to fetch from the data cache, and finally store the result. These operations are sequenced using a global clock signal. High performance processors have sophisticated architectural features that improve the instruction fetch and data cache bandwidth and exploit more instruction level parallelism.[1] But they must still perform these basic operations for each instruction. In a real processor, of course, there is much more overhead. For example, often many functional units are run in parallel and only the "correct" result is used, and there is considerable logic that is needed to control the pipeline during stalls and exceptions. However, since this overhead depends on the implementation, and in some sense is not essential to the functionality of the processor, we will neglect it. Most of the operations outlined above require either reading or writing one of the on-chip memories—instruction and data caches, and the register file. Only one of the steps requires computation. So we will aggressively assume that the energy cost of performing computation is zero, and we will also assume that communication costs within the datapath are also zero. We will only consider the energy needed to read and write memories, and, where required, the energy to clock storage elements, such

---

[1] Program semantics require that instruction i be executed before instruction i + 1. However, if the second instruction does not depend upon the first, they can both be executed simultaneously. We call this instruction level parallelism.

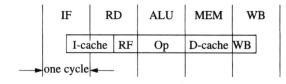

IF   instruction fetch
RD   register fetch and instruction decode
ALU operate on operands
MEM memory access
WB   register write back

Fig. 1. Pipeline diagram.

as pipeline latches. Furthermore, these machines only dissipate energy that is absolutely necessary. Most processors perform some operations speculatively, in the hope that the result is useful. Our ideal machines have perfect knowledge so there is never a need for speculative operations. And finally, there are no unwanted transitions, such as glitches. We use a trace-based simulation to estimate the energy and delay of these idealized machines. For a more detailed description of our simulation methodology please refer to the Appendix.

A. Machine Models

The simple machine is the most basic compute element, similar to DLX [10]. The processor consists of a simple state machine that fetches an instruction, fetches the operands, performs the operation, and stores the result. The processor is not pipelined so we model the execution time by assuming control transfer instructions (CTI) take two cycles, ALU operations take three cycles, and cache operations take four cycles. Since this machine is not pipelined, we assume no energy is dissipated in clocking. Only the energy required to read and write the caches and register file is taken into consideration. This implementation should have the lowest average energy per instruction; however, since it does not take advantage of the available parallelism, it will have relatively poor performance and high energy-delay product.

To improve performance we created a pipelined processor that is similar to the previous one, except that we use pipelining to exploit instruction level parallelism (ILP). That is, more than one instruction is executed concurrently. Thus, the change in energy-delay product of this machine compared to the ideal will be due entirely to pipelining. We use the traditional MIPS or DLX [10] five-stage pipeline shown in Fig. 1. Since it is not possible to build a pipelined machine without some kind of storage element, in addition to the energy required to read and write memories, we also take into account the energy required to clock the latches in the pipeline and the PC chain.

The ideal superscalar processor is similar to the pipelined machine, except that it can execute a maximum of two instructions per cycle. Since we are not concerned with the energy dissipated in the control logic, how the instruction level parallelism is found is only significant in that it sets the amount of parallelism exploited. We use an aggressive static scheduler from TORCH [16] which gives comparable performance to a dynamic scheduled machine. Since most

integer programs do not have enough parallelism to fill both issue slots regularly, we assume that this machine supports some conditional execution of instructions as directed by the compiler. This machine is an idealization of the TORCH machine that is described in a later section. The difference in energy-delay product between this machine and the previous one will be due to the ability to execute two instructions in parallel.

B. Comparison of Energy and Energy-Delay Product

Fig. 2 shows the total energy required to complete a benchmark. All numbers in this section have been normalized to the corresponding value for the superscalar processor. As we would expect, the unpipelined machine has the lowest energy dissipation, and the superscalar machine has the highest. The execution time, shown in Fig. 3, is reversed. The superscalar processor has the highest performance, while the unpipelined machine has the lowest performance. In the figure we have explicitly shown the time spent stalled due to the memory system. Fig. 4 shows the energy-delay product. Pipelining provides a big boost in performance for very little energy cost, so it gives almost a 2× improvement in energy-delay product. Superscalar issue, on the other hand, only gives a small improvement in energy-delay product. The performance gain is small and the energy cost is higher. In this simple model, a processor with a wider parallel issue would be of limited utility. The basic problem is that the amount of instruction level parallelism (in the integer benchmarks that we use) is limited,[2] so the effective parallelism is small. Yet increasing the peak issue rate increases the energy of every operation, even in this ideal machine, because the energy of the memory fetches and clocks will increase. Furthermore, the speedup of the superscalar processor is limited by Amdahl's law. As the performance of the processor increases, the stall time becomes a larger fraction of the total execution time, so that the corresponding decrease in energy-delay product will be smaller. As we will show in the following section, the overhead in a superscalar machine is actually worse than that of a simple machine, and overwhelms the modest gain in energy-delay product gotten from parallelism.

IV. Current Implementations

To see how real machines compare with the ideal machines described in the previous section, this section looks at two processors that we have designed, a simple RISC machine and a superscalar processor we call TORCH. In the three idealized machines presented earlier we focused only on the essential operations and ignored all sources of overhead. For the real processors we will include all sources of overhead. This simulation method is also described in the Appendix.

As has been shown before [2], [3], [15], and [19], some simple optimizations can yield significant gain in energy dissipation. With careful design it is possible to reduce the

---

[2]There have been several reports on the amount of ILP available in typical integer programs [13], [16]–[18]. These studies have found that for realistic machines the ILP is severely limited and is most cases smaller than two. Even for very aggressive machines the average number of instructions issued per cycles is less than two.

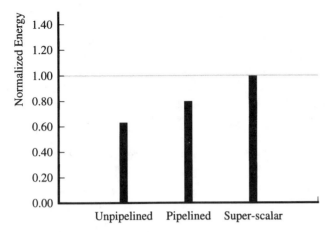

Fig. 2. Normalized energy of ideal machines.

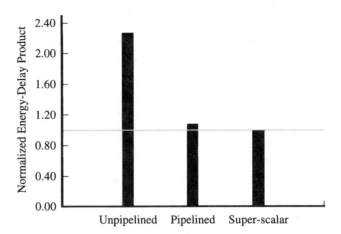

Fig. 4. Normalized energy-delay product of ideal machines.

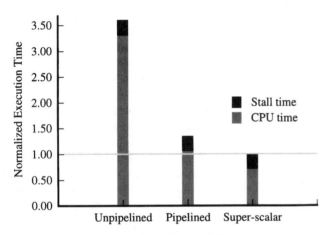

Fig. 3. Normalized execution time of ideal machines.

energy waste considerably without much change to the critical path. We will briefly discuss some of the optimizations used in our designs and what improvements we observed. We then compare the results from the optimized real machines to the ideal machines of the previous section.

A. Machine Models

The simple RISC processor is similar to the original MIPS R3000 described by Kane [14], except that it includes on-chip caches. The processor has the same five-stage pipeline that was shown in Fig. 1. This processor is similar to the ideal pipelined machine except that it accounts for all the energy required to complete the instruction and it includes all the overhead associated with the architecture, such as the TLB and co-processor 0. The difference in energy-delay product between this processor and the pipelined ideal machine represents a limit on possible improvements in efficiency if we only focus on the logic required to execute an instruction.

TORCH [16] is a statically scheduled two-way superscalar processor, which uses the same five-stage pipeline that was shown in Fig. 1. TORCH supports conditional execution of instruction directed by the compiler. The processor includes shadow state that can buffer the results of these conditional instructions until they are committed. To improve the code density a special NOP bit is used to indicate that an instruction

should be held for a cycle before it is executed. This is important in programs that do not have enough parallelism to keep both datapaths busy because it improves the instruction cache performance and reduces the number of cache accesses needed to complete a program. Instructions are encoded in a 40-b word which are packed in main memory using a special format that improves the instruction fetch efficiency. Instructions are unpacked as they are written in the first level instruction cache.

We use TORCH as an example because of its smaller overhead compared to that of an aggressive dynamically scheduled processor. Due to a larger overhead, such a processor would have a higher energy per instruction but equivalent performance [16], and thus have a higher energy-delay product.

B. Energy Optimizations

The additional energy associated with a particular architectural feature can be divided into overhead and waste. We consider overhead that part of the energy that cannot be eliminated with careful design. For example, adding additional functional units will increase the average wire length, thus increasing the overhead of the architecture. Other sources of energy can simply be designed away. For example, in a pipelined design, clock gating can be used to eliminate unwanted transition of the clock during interlock cycles. A good implementation will reduce waste as much as possible. However, the designer must carefully weigh the gains in power dissipation versus the cost in complexity or cycle time cost. If a particular optimization causes a large increase in cycle time, then dissipating slightly more energy but running at a higher frequency may be better. Following are a few techniques that we used to reduce waste in our implementations.

Clock gating can be used to eliminate transitions that should never have happened. In our implementation we qualify latches in the datapath when the instruction does not produce a result. In TORCH, densely coded NOP's improve the code density and reduce the number of instruction cache accesses. However, they introduce extra instructions into the pipeline which can cause spurious transitions. Using clock gating we can eliminate these spurious transitions. We also qualify all latches that are not in the main execution datapath. These latches tend to be

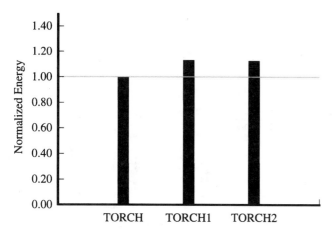

Fig. 5. Reduction in energy from simple optimizations.

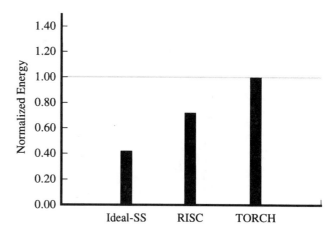

Fig. 6. Normalized energy for ideal superscalar, simple RISC, and TORCH processors.

enabled only infrequently. For example, latches in the address translation datapath only need to be enabled on a load or store instruction, or during an instruction cache miss.

Most of the latches in the control sections are not qualified since this can introduce clock skew. Control sections are automatically synthesized and we do not have very fine control over the implementation of clock gating. However, the power dissipated in the control sections is only a small fraction of the total power, and the clock power is only a fraction of that. We do qualify some control signals to prevent spurious transitions in the datapath buses. If an instruction has been dynamically NOPed, then preventing the datapath latches from opening is not sufficient. If the control signals to the datapath change, this can cause spurious transitions on some of the datapath buses, which can cause significant energy consumption.

Through the use of clock gating we saved approximately one third (33%) of the clock power or close to 15% of the total power. Most of the latches in datapath blocks have an enable signal. To further reduce the energy we would need to either make the enabling signal more restrictive or qualify latches in the control sections. In either case, further improvements would require a larger effort and provide smaller returns.

We use selective activation of the large macrocells to reduce power dissipation. Obviously the most important step was to eliminate accesses to the caches and the register file when the machine is stalled. Also important was to eliminate accesses to the instruction cache when an instruction is dynamically NOPed. In this case, the instruction has already been read on the previous cycle so there is no need to reread the cache. By doing so we saved approximately 8% of the total power.

Speculative operations are commonly used to improve the performance of microprocessors. For example, two source operands are read from the register file before the instruction is decoded. This reduces the critical path but consumes extra energy, since one or both operands may not be needed. One simple way to eliminate the waste is to pre-decode the instruction as they are fetched from the second-level cache and store a few extra bits in the instruction cache. Each bit would indicate whether a particular operand should be fetched from the register file. Since the energy required to perform a cache access is not a strong function of the number of bits accessed, the cost will be small.

The designers of the DEC 21164 chip used a similar idea [4]. The second level cache is 96 kB three-way set-associative. To reduce the power dissipation, the cache tags are accessed first. Once it is known in which bank, if any, the data is resident in, only that bank is accessed. This reduces the number of accesses from four to two, reducing the power dissipation. However, the first-level cache refill penalty increases by one cycle.

The idea of reducing speculative operations should not be taken too far. In TORCH, as in most microprocessors, computing the next PC is usually one of the critical paths. The machine must determine whether a branch is taken, and then perform an addition to either compute the branch target or to increment the current PC. To remove the adder from the critical path, we perform both additions in parallel, while the outcome of the branch is determined, and then use a multiplexer to select one of the two results. The energy we save by eliminating the extra addition is negligible, while the cycle time penalty is large. Thus, designers should carefully consider the trade-off when eliminating speculative operations.

Fig. 5 shows the energy saved using the simple optimizations presented in this section. TORCH refers to the fully optimized processor. In TORCH1 we have disabled selective activation of the instruction cache and always speculatively read the register file. In TORCH2 we have removed all qualification from the clocks. We normalized to the energy of TORCH. Thus, simple optimizations can save almost 20–25% of the total power. However, further gains would be harder to come by because—as we will show later—the energy is dissipated in a variety of units, none of which accounts for a significant fraction of the total energy.

C. Comparison to Ideal Machines

Fig. 6 shows the energy for the ideal superscalar machine, the simple RISC machine, and TORCH. All figures in this sections have been normalized to the corresponding value for TORCH. As expected, TORCH requires the most energy to execute a program, closely followed by the RISC processor. The difference in energy between the ideal machine and TORCH represents potential future improvements. Since we

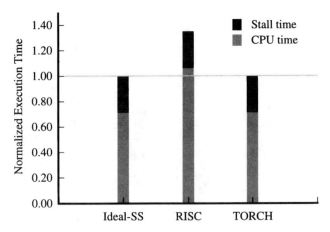

Fig. 7. Normalized execution time for ideal super-scalar, simple RISC, and TORCH processors.

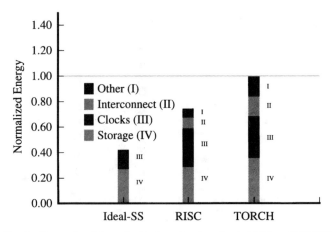

Fig. 9. Energy breakdown for ideal superscalar, simple RISC, and TORCH processors.

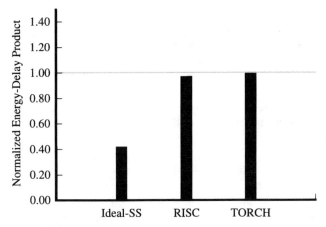

Fig. 8. Normalized energy-delay product for ideal super-scalar, simple RISC, and TORCH processors.

were very aggressive in our simulations, we would expect the gains to be even smaller than this.

Fig. 7 shows the total execution time for the same three machines. Since TORCH and the ideal superscalar machine have identical execution models, they have the same execution time. The superscalar processors have a 1.35× speedup when compared to the single issue machine. With an ideal memory system, the speedup would have been 1.6×. This highlights the importance of the external memory system, even in low-power applications.

Finally, Fig. 8 shows the energy-delay product for all three machines. As expected from the two previous figures, TORCH and the simple RISC processor have nearly the same efficiency. Super-scalar issue provides a small gain in performance. Unfortunately, it also increases the energy cost of all instructions. The net result is no significant improvement in energy-delay product.

The ideal super-scalar processor is roughly twice as efficient as the real machine. Since we want to understand how to make TORCH approach the ideal processor, we look at where the extra energy is dissipated.

D. Energy Breakdown

We chose to divide the energy dissipation into four categories. The first is energy dissipated in reading and writing memories, such as the caches, and the register file. The second category is energy dissipated in the clock network. The third is energy dissipated in global interconnect nodes. The final category comprises everything not in one of the previous categories.

Fig. 9 shows the energy breakdown for three of the five machines we simulated. The figures have been normalized to the total energy dissipation of TORCH. The energy dissipated in the on-chip memories is almost the same in the ideal machines as in our implementations. This means that we were successful in reducing the number of unnecessary cache accesses. The clock power in the real machines is slightly higher since there is a large amount of state that we did not consider in the ideal machines. Although each storage element dissipates very little energy, the sum total represents a significant fraction. Global interconnect and computation each represent about 15% of the total energy. The most important point in this graph is that further improvements in energy-delay product will be hard to come by because the energy is dissipated in many small units. To reduce the energy significantly one would have to reduce the energy of many of these small units.

V. CONCLUSION

We have shown that the performance and energy of a processor are related. Over a wide spectrum of performance and energy, processors seem to have quite similar energy-delay products. We have shown there are two main reasons for this. First, all processors perform similar functions. Two of the most important operations, accessing on-chip memories and clocking, account for a large (50–80%) fraction of the total energy dissipation. Thus, doubling the energy dissipated in the combinational logic will only slightly change the overall energy dissipation. Second, only pipelining provides a large boost in performance for a low cost in energy. Other architectural features do not seem to significantly change the energy-delay product. Given that most processors today are pipelined, we would expect all of them to have a similar energy-delay product.

Also, although it is possible to reduce the energy requirements by careful design, this will only provide a one-time improvement. We have shown that using easy-to-implement optimizations, the energy can be reduced by approximately 25%. Further improvements would require much more sophisticated optimizations since the remaining energy is dissipated in many small units, none of which account for a significant fraction of the total energy dissipation.

APPENDIX
SIMULATION METHODOLOGY

There are two sources of energy dissipation in CMOS circuits, static energy which results from resistive paths from the power supply to ground, and dynamic energy, which results from switching capacitive loads between two voltage levels. Static energy dissipation exists whenever the chip is powered on and usually represents a small fraction of the total energy dissipated, unless the chip is idle for large periods of time. Ignoring static energy dissipation, we can estimate the energy dissipation as

$$E = \sum_i \frac{n_i}{2} c_i V^2 \qquad (1)$$

where n_i is the number of times the ith node toggles, and c_i is the capacitive load of that node. Thus, to compute the energy dissipation, we need to accurately estimate the capacitance and the toggle count for each node.

Estimating the energy dissipation of some macrocell blocks, like SRAMS, is easier because most of the energy in these blocks is dissipated in pre-charge/discharge differential structures. Since one of the two signals in a differential pair is guaranteed to discharge on every access, the energy dissipated is independent of the actual data values read or written. Thus, we can approximate the energy dissipated in the SRAMS by simply counting the number of read accesses and the number of write accesses and then multiplying by the energy dissipated per access [1].

To estimate the energy dissipation of the idealized machines, we use a simple trace-based method. In these machines all the energy is dissipated in the memories or the storage elements, so it is only necessary to compute the toggle count and capacitive load of the clock. Given a trace of all of the instructions executed, we can compute the number of accesses to each memory and the total energy dissipated. We find the clock energy by determining the capacitance of a single latch and multiplying by the minimum number of latches to maintain the processor state.

To estimate the energy dissipation of the simple RISC machine and TORCH, we must accurately determine the toggle count and the capacitance for each of the nodes. We have developed a complete Verilog model of TORCH which includes the main datapath, instruction and data caches, coprocessor 0, TLB, and external interface; although, it does not include a floating point unit. The model can stall, take exceptions, or take external interrupts. The model can also run in MIPS mode in which case it behaves like the simple RISC machine. We use this model to estimate both the toggle counts and the capacitance of each node.

To determine the toggle count we run a simulated program on the Verilog model. We have written several routines which use Verilog's programming language interface (PLI) to gather this information. The PLI routines also traverse the network to determine the total capacitance of each node. The gate and diffusion capacitance is annotated in each of our library cells. We have written several programs which allow us to estimate the wire capacitance for both standard cell and datapath blocks. We estimate the length of global wires by generating a floorplan of the chip and then extracting the layout.

Since the performance and the energy dissipation of the processor depend on the test being run, we decided to use *compress* and *jpeg,* from the SPEC benchmark suite. The first is representative of traditional integer programs normally run on a processor. The second program is more representative of multimedia type applications. The results for both applications are so similar, both in terms of performance and energy, that we show the results for *compress only*. We compile the benchmarks using the TORCH optimizing C compiler or the MIPS C compiler, depending on the target machine.

REFERENCES

[1] B. S. Amrutur and M. Horowitz, "Techniques to reduce power in fast wide memories," in *Symp. Low Power Electr.,* IEEE, Oct. 1994, pp. 92–93.
[2] R. Bechade *et al.,* "A 32 b 66 MHz 1.8 W microprocessor," in *IEEE Int. Solid-State Circuits Conf.,* Feb. 1994, pp. 208–209.
[3] T. Biggset *et al.,* "A 1 Watt 68040-compatible microprocessor," in *Symp. Low Power Electr.,* IEEE Solid-State Circuits Council, Oct. 1994, vol. 1, pp. 12–13.
[4] W. J. Bowhill *et al.,* "A 300 MHz 64 b quad-issue CMOS RISC microprocessor," in *IEEE Int. Solid-State Circuits Conf.,* Feb. 1995, pp. 182–183.
[5] A. Chamas *et al.,* "A 64 b microprocessor with multimedia support," in *IEEE Int. Solid-State Circuits Conf.,* Feb. 1995, pp. 178–179.
[6] A. P. Chandrakasan, S. S. Brodersen, and R. W. Brodersen, "Low-power CMOS digital design," *IEEE J. Solid-State Circuits,* vol. 27, no. 4, pp. 473–483, Apr. 1992.
[7] Z. Chen, J. Shott, J. Burr, and J. D. Plummer, "CMOS technology scaling for low voltage low power applications," in *Symp. Low Power Electr.,* Oct. 1994, pp. 56–57.
[8] R. P. Colwell and R. L. Steck, "A 0.6 μm BiCMOS processor with dynamic execution," in *IEEE Int. Solid-State Circuits Conf.,* Feb. 1995, pp. 176–177.
[9] B. M. Gordon and T. H.-Y. Meng, "A low power subband video decoder architecture," in *Int. Conf. Acoustics, Speech and Signal Processing,* Apr. 1994, pp. 409–412.
[10] J. L. Hennessy and D. A. Patterson, *Computer Architecture A Quantitative Approach,* 1st ed. San Mateo, CA: Morgan Kaufmann, 1990.
[11] M. Horowitz, T. Indermaur, and R. Gonzalez, "Low-power digital design," in *Symp. Low Power Electr.,* Oct. 1994, pp. 8–11.
[12] Integrated Device Technology, Inc., *Orion Product Overview,* Mar. 1994.
[13] N. P. Jouppi and D. Wall, "Available instruction-level parallelism for superscalar and superpipelined machines," in *Int. Conf. Architec. Sup. Programming Language and Operating Systems,* Apr. 1989, pp. 272–282.
[14] G. Kane, *MIPS RISC Architecture.* Englewood Cliffs, NJ: Prentice-Hall, 1988.
[15] D. Pham *et al.,* "A 3.0 W 75SPECint92 85SPECfp92 superscalar RISC microprocessor," in *IEEE Int. Solid-State Circuits Conf.,* Feb. 1994, pp. 212–213.
[16] M. D. Smith, "Support for speculative execution in high-performance processors," Ph.D. thesis, Stanford University, Stanford, CA, Nov. 1992.
[17] M. D. Smith, M. Johnson, and M. Horowitz, "Limits on multiple instruction issue," in *Int. Symp. Computer Architecture,* Boston, MA, Apr. 1989, pp. 290–302.
[18] D. Wall, "Limits of instruction-level parallelism," in *Int. Conf. Architec. Sup. Programming Language and Operating Systems,* Santa Clara, CA, Apr. 1991, pp. 176–188.
[19] N. K. Yeung *et al.,* "The design of a 55SPECint92 RISC processor under 2 W," in *IEEE Int. Solid-State Circuits Conf.,* Feb. 1994, pp. 206–207.

Energy Efficient CMOS Microprocessor Design

Thomas D. Burd and Robert W. Brodersen

University of California, Berkeley

Abstract

Reduction of power dissipation in microprocessor design is becoming a key design constraint. This is motivated not only by portable electronics, in which battery weight and size is critical, but by heat dissipation issues in larger desktop and parallel machines as well. By identifying the major modes of computation of these processors and by proposing figures of merit for each of these modes, a power analysis methodology is developed. It allows the energy efficiency of various architectures to be quantified, and provides techniques for either individually optimizing or trading off throughput and energy consumption. The methodology is then used to qualify three important design principles for energy efficient microprocessor design.

1: Introduction

Throughput and area have been the main forces driving microprocessor design, but recently the explosive growth in portable electronics has forced a shift in these design optimizations toward more power conscious solutions. Even for desktop units and large computing machines, the cost of removing the generated heat and the drive towards "green" computers are making power reduction a priority.

An energy-efficient design methodology has been developed for signal processing applications, resulting in a strategy to provide orders of magnitude of power reduction [1]. These applications have a fixed throughput requirement due to a real-time constraint given by the application (e.g. video compression, speech recognition). Microprocessors targeted for general purpose computing, however, generally operate in one of two other computing modes. Either they are continuously providing useful computation, in which case maximum throughput is desired, or they are in a user interactive mode, in which case bursts of computation are desired.

A framework for an energy-efficient design methodology more suitable for a microprocessor's two operating

modes will be presented. Using simple analytic models for delay and power in CMOS circuits, metrics of energy efficiency for the above modes of operation will be developed and their implications on processor design will be presented. This paper will conclude with the application of these metrics to quantify three important principles of energy-efficient microprocessor design.

2: CMOS Circuit Models

Power dissipation and circuit delays for CMOS circuits can be accurately modelled with simple equations, even for complex microprocessor circuits. These models are dependent upon six variables which an IC designer may control to either individually minimize or trade off power and speed. These models hold only for digital CMOS circuits, and are thus not applicable to bipolar or BiCMOS circuits.

2.1: Power Dissipation

CMOS circuits have both static and dynamic power dissipation. Static power arises from bias and leakage currents. While statically-biased gates are usually found in a few specialized circuits such as PLAs, their use has been dramatically reduced in CMOS design. Furthermore, careful design of these gates generally makes their power contribution negligible in circuits that do use them [2]. Leakage currents from reverse-biased diodes of MOS transistors, and from MOS subthreshold conduction [3] also dissipate static power, but are insignificant in most designs.

The dominant component of power dissipation in CMOS is therefore dynamic, and arises from the charging and discharging of the circuit node capacitances found on the output of every logic gate. This capacitance, C_L, can be expressed as:

$$C_L = C_W + C_{FIX} \qquad \text{(EQ 1)}$$

Reprinted from *Proceedings of the 28th Annual HICSS Conference*, pp. 288-297, January 1995.

C_W is the product of a technology constant and the device width, W, over which the designer has control. C_W is composed of the subsequent gates' input capacitance and part of the diffusion capacitance on the gate output. C_{FIX} is composed of the remaining part of the diffusion capacitance which is purely technology dependent, and the capacitance of the wires interconnecting these gates which may be minimized by efficient layout.

For every low-to-high logic transition in a digital circuit, C_L incurs a voltage change ΔV, drawing an energy $C_L \Delta V \, V_{DD}$ from the supply voltage at potential V_{DD}. For each node $n \in N$, these transitions occur at a fraction α_n of the clock frequency, f_{CLK}, so that the total dynamic switching power may be found by summing over all N nodes in the circuit:

$$Power = V_{DD} \cdot f_{CLK} \cdot \sum_{i=1}^{N} \alpha_i \cdot C_{L_i} \cdot \Delta V_i \qquad (EQ\ 2)$$

Aside from the memory bitlines in CMOS circuits, most nodes swing a ΔV from ground to V_{DD}, so that the power equation can be simplified to:

$$Power \cong V_{DD}^2 \cdot f_{CLK} \cdot C_{EFF} \qquad (EQ\ 3)$$

where the effective switched capacitance, C_{EFF}, is commonly expressed as the product of the physical capacitance C_L, and the activity weighting factor α, each averaged over the N nodes.

During a transition on the input of a CMOS gate both p and n channel devices may conduct simultaneously, briefly establishing a short from V_{DD} to ground. In properly designed circuits, however, this short-circuit current typically dissipates a small fraction (5-10%) of the dynamic power [4] and will be omitted in further analyses.

2.2: Circuit Delay

To fully utilize its hardware, a digital circuit's clock frequency, f_{CLK}, should be operated at the maximum allowable frequency. This maximum frequency is just the inverse of the delay of the processor's critical path. Thus, the circuit's throughput is proportional to 1/delay.

Until recently, the long-channel delay model (in which device current is proportional to the square of the supply voltage) suitably modelled delays in CMOS circuits [5]. However, scaling the minimum device channel length, L_{MIN}, to below 1 micron (which is common in today's process technology), degrades the performance of the device due to velocity saturation of the channel electrons. This phenomenon occurs when the electric field (V_{DD} / L_{MIN}) in the channel exceeds 1V/um [6].

$$Delay \cong \frac{C_L}{I_{AVE}} \cdot \frac{V_{DD}}{2} \cong \frac{C_L \cdot V_{DD}}{k_V \cdot W \cdot (V_{DD} - V_T - V_{DSAT})} \qquad (EQ\ 4)$$

The change in performance can be analytically characterized by what is known as the short-channel or velocity-saturated delay model shown in Equation 4. I_{AVE} is the average current being driven onto C_L, and is proportional to W, the technology constant k_V, and to first-order, V_{DD}. V_T is the threshold voltage (typically 0.5 - 1.0 volts) and is the minimum V_{DD} for which the device can still operate. For large V_{DD}, V_{DSAT} is constant, with typical magnitude on order of V_T. For V_{DD} values less than $2V_T$, V_{DSAT} asymptotically approaches $V_{DD} - V_T$.

2.3: Circuit Design Optimizations

The designer can minimize some of the variables in Equations 3 and 4 to either individually or simultaneously minimize the delay and power dissipation. This can be accomplished by minimizing C_{EFF}, while keeping W constant. The following three methods for minimizing C_{EFF} can be correlated so they may not always be individually optimized.

First, the switching frequency, α, can be minimized to reduce power dissipation without affecting the delay. This optimization shows the best promise for significant power reduction. The α factor can be minimized by a number of techniques such as dynamically gating the clock to unused processor sections and selecting cell/module topologies that minimize switching activity.

Another approach is to minimize the number of nodes, which will minimize the total physical capacitance, and likewise reduce the power. However, this method may come at the expense of computation per cycle, as in the example of reducing a 32-bit datapath to a 16-bit datapath.

The third approach is to reduce C_{FIX} by minimizing the interconnect capacitance, which optimizes both power and delay. Since speed has always been a primary design goal, this is already done as general practice.

Although it beyond the control of the designer, it is worth noting the impact of technology scaling on power and delay. Capacitances scale down linearly with technology parameter L_{MIN} while transistor current stays approximately constant if V_{DD} remains fixed (constant-voltage technology scaling). Thus, delay scales linearly with L_{MIN} for constant power; likewise, power scales linearly with L_{MIN} for constant delay. Essentially, technology scaling is always beneficial.

2.4: Trading off Delay and Power

The remaining three variables under the designer's control can only be used to trade-off delay and power. As the voltage is reduced, the delay increases hyperbolically as the supply voltage approaches V_T. Meanwhile, the power drops due to the product of the squared voltage term and the frequency (inverse of the delay) term. Thus, by operating at various values of supply voltage, a given processor architecture can be made to cover a large range of operating points as demonstrated in Figure 1.

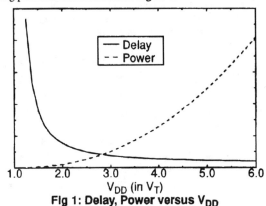

Fig 1: Delay, Power versus V$_{DD}$

Another method for trading off delay and power is to vary the device width W. However, the trade-off is dependent on how large C_{FIX} is with respect to C_W. Empirical data shows these capacitances are typically on the order of the same size for transistors with minimum width. As the width is increased above the minimum value, the power increases but the delay is decreased. However, C_W soon dominates C_{FIX}, and then power scales linearly with width, while the delay remains approximately constant. Thus, there is a delay-power trade-off, but only over a small range of delay values.

Scaling the clock frequency is a third approach which is most beneficial if it is coupled with voltage scaling. If the clock frequency is reduced, the delay may be increased (keeping it equal to $1/f_{CLK}$) by reducing the supply voltage and thus saving power. If the voltage is kept constant, then power and throughput reduce linearly with clock frequency.

3: Energy Efficiency

No single metric quantifies energy efficiency for all digital systems. The metric is dependent on the type of computation the system performs. We will investigate the three main modes of computation: fixed throughput, maximum throughput, and burst throughput. Each of these modes has a clearly defined metric for measuring energy efficiency, as detailed in the following three sections.

Throughput, T, is defined as the number of operations that can be performed in a given time. When clock rate is inversely equal to the critical path delay, throughput is proportional to the amount of concurrency per clock cycle (i.e. number of parallel operations) divided by the delay, or equivalently:

$$Throughput \equiv T = \frac{Operations}{Second} \propto \frac{Concurrency}{Delay} \quad \text{(EQ 5)}$$

Operations are the units of computation at the highest level of the design space. Valid measures of throughput are MIPS (instructions/sec) and SPECint92 (programs/sec) which compare the throughput on implementations of the same instruction set architecture (ISA), and different ISAs, respectively.

3.1: Fixed Throughput Mode

Most real-time systems require a fixed number of operations per second. Any excess throughput cannot be utilized, and therefore needlessly dissipates power. This property defines the fixed throughput mode of computation. Systems operating in this mode are predominantly found in digital signal processing applications in which the throughput is fixed by the rate of an incoming or outgoing real-time signal (e.g.: speech, video).

$$Metric|_{FIX} = \frac{Power}{Throughput} = \frac{Energy}{Operation} \quad \text{(EQ 6)}$$

Previous work has shown that the metric of energy efficiency in Equation 6 is valid for the fixed throughput mode of computation [5]. A lower value implies a more energy efficient solution. If a design can be made twice as energy efficient (i.e. reduce the energy/operation by a factor of two), then its sustainable battery life has been doubled; equivalently, its power dissipation has been halved. Since throughput is fixed, minimizing the power dissipation is equivalent to minimizing the energy/operation.

3.2: Maximum Throughput Mode

In most multi-user systems, primarily networked desktop computers and supercomputers, the processor is continuously running. The faster the processor can perform computation, the better. This is the defining characteristic of the maximum throughput mode of computation. Thus, this mode's metric of energy efficiency must balance the need for low power and high speed.

$$Metric|_{MAX} = ETR = \frac{E_{MAX}}{T_{MAX}} = \frac{Power}{Throughput^2} \quad \text{(EQ 7)}$$

A good metric for measuring energy efficiency for this mode is given in Equation 7, henceforth called the Energy to Throughput Ratio, or ETR (E_{MAX} is the energy/operation, or equivalently power/throughput, and T_{MAX} is the throughput in this mode). A lower ETR indicates lower energy/operation for equal throughput or equivalently indicates greater throughput for a fixed amount of energy/

operation, satisfying the need to equally optimize through-put and power dissipation. Thus, a lower ETR represents a more energy efficient solution. The Energy-Delay Product [7] is a similar metric, but does not include the effects of architectural parallelism when the delay is taken to be the critical path delay.

In most circuits, however, ETR is not constant for different values of throughput. The throughput can be adjusted with the delay-power trade offs shown in Section 2.4; but, unfortunately, none of the methods perform linear trade-offs between energy/operation and throughput, and only V_{DD} allows the throughput to be adjusted across a reasonable dynamic range.

Because energy/operation is independent of clock frequency, the clock should never be scaled down in maximum throughput mode; only the throughput scales with clock frequency, so the ETR actually increases. Increasing all device widths will only marginally decrease delay while linearly increasing energy. Generally, it is optimal to set all devices to minimum size and only size up those lying in the critical paths.

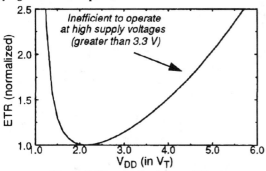

Fig 2: ETR as a function of V_{DD}.

As shown in Figure 2, V_{DD} can be adjusted by a factor of almost three ($1.4V_T$ to $4V_T$) and the ETR only varies within 50% of the minimum at $2V_T$. However, outside this range, the ETR rapidly increases. Clearly, for supply voltages greater than 3.3V, there is a rapid degradation in energy efficiency.

To compare designs over a larger range of operation for the maximum throughput mode, a better metric is a plot of the energy/operation versus throughput. To make this plot, the supply voltage is varied from the minimum operating voltage (near V_T in most digital CMOS designs) to the maximum voltage (3.3V- 5V), while energy/operation and delay are measured. The energy/operation can then be plotted as a function of delay, and the architecture is completely characterized over all possible throughput values.

Using the ETR metric is equivalent to making a linear approximation to the actual energy/operation versus throughput curve. Figure 3 demonstrates the error incurred in using a constant ETR metric. For architectures with

similar throughput, a single ETR value is a reasonable metric for energy efficiency; however, for designs optimized for vastly different values of throughput, a plot may be more useful, as Section 5.1 demonstrates.

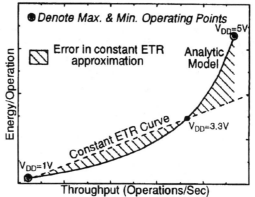

Fig 3: Energy vs. Throughput metric.

3.3: Burst Throughput Mode

Most single-user systems (e.g. stand-alone desktop computers, portable computers, PDAs, etc.) spend a fraction of the time performing useful computation. The rest of the time is spent idling between user requests. However, when bursts of useful computation are demanded, e.g. spread-sheet updates, the faster the throughput (or equivalently, response time), the better. This characterizes the burst throughput mode of computation. The metric of energy efficiency used for this mode must balance the desire to minimize power dissipation, while both idling and computing, and to maximize throughput when computing.

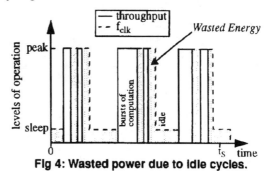

Fig 4: Wasted power due to idle cycles.

Ideally, the processor's clock should track the periods of computation in this mode so that when an idle period is entered, the clock is immediately shut off. Then a good metric of energy efficiency is just ETR, as the power dissipated while idling has been eliminated. However, this is not realistic in practice. Many processors do not having a power saving mode and those that do so generally support only simple clock reduction/deactivation modes. The hypothetical example depicted in Figure 4 contains a clock reduction (sleep) mode in which major sections of the processor are powered down. The shaded area indicates the processor's idle cycles in which power is needlessly dissipated, and whose magnitude is dependent upon whether the processor is operating in the "low-power" mode.

$$E_{MAX} = \frac{Total\ Energy\ Consumed\ Computing}{Total\ Operations} \quad \text{(EQ 8)}$$

$$E_{IDLE} = \frac{Total\ Energy\ Consumed\ Idling}{Total\ Operations} \quad \text{(EQ 9)}$$

Total energy and total operations can be calculated over a large sample time period, t_S. T_{MAX} is the peak throughput during the bursts of computation (similar to that defined in Section 3.2), and T_{AVE} is the time-average throughput (total operations / t_S). If the time period t_S characterizes the computing demands of the user and/or target system environment (T_{AVE}), then a good metric of energy efficiency for the burst throughput mode is:

$$Metric|_{BURST} = METR = \frac{E_{MAX} + E_{IDLE}}{T_{MAX}} \quad \text{(EQ 10)}$$

This metric will be called the Microprocessor ETR (METR); it is similar to ETR, but also accounts for energy consumed while idling. A lower METR represents a more energy efficient solution.

Multiplying through Equation 8 by t_S•(fraction of time computing) shows that E_{MAX} is the ratio of compute power dissipation to peak throughput T_{MAX}, as previously defined in Section 3.2. Thus, E_{MAX} is only a function of the hardware and can be measured by operating the processor at full utilization.

E_{IDLE}, however, is a function of t_S and T_{AVE}. The power consumed idling must be measured while the processor is operating under typical conditions, and T_{AVE} must be known to then calculate E_{IDLE}. However, expressing E_{IDLE} as a function of E_{MAX} better illustrates the conditions when idle power dissipation is significant.

Equation 9 can be rewritten as:

$$E_{IDLE} = \frac{[Idle\ Power\ Dissipation] \cdot [Time\ Idling]}{[Average\ Throughput] \cdot [Sample\ Time]} \quad \text{(EQ 11)}$$

With the Power-Down Efficiency, β, is defined as:

$$\beta = \frac{Power\ dissipation\ while\ idling}{Power\ dissipation\ while\ computing} = \frac{P_{IDLE}}{P_{MAX}} \quad \text{(EQ 12)}$$

E_{IDLE} can now be expressed as a function of E_{MAX}:

$$E_{IDLE} = \frac{[\beta \cdot E_{MAX} \cdot T_{MAX}] \cdot [(1 - T_{AVE}/T_{MAX}) \cdot t_S]}{[T_{AVE}] \cdot [t_S]} \quad \text{(EQ 13)}$$

Equation 14 shows that idle power dissipation dominates total power dissipation when the fractional time spent computing (T_{AVE}/T_{MAX}) is less than the fractional power dissipation while idling (β).

$$METR = ETR\left[1 + \beta\left(\frac{T}{T_{AVE}} - 1\right)\right], \quad T \geq T_{AVE} \quad \text{(EQ 14)}$$

The METR is a good metric of energy efficiency for all values of T_{AVE}, T_{MAX}, and β as illustrated below by analyzing the two limits of the METR metric.

Idle Energy Consumption is Negligible ($\beta << T_{AVE}/T_{MAX}$): The metric should simplify to that found in the maximum throughput mode, since it is only during the bursts of computation that power is dissipated and operations performed. For negligible power dissipation during idle, the METR metric in Equation 14 degenerates to the ETR, as expected. Likewise, for perfect power-down ($\beta = 0$) and minimal throughput ($T_{MAX} = T_{AVE}$), the METR is exactly the ETR.

Idle Energy Consumption Dominates ($\beta >> T_{AVE}/T_{MAX}$): The energy efficiency should increase by either reducing the idle energy/operation while maintaining constant throughput, or by increasing the throughput while keeping idle energy/operation constant. While it might be expected that these are independent optimizations, E_{IDLE} may be related back to E_{MAX} and the throughput by β (T_{AVE} is fixed):

$$\frac{E_{IDLE}}{E_{MAX}} \cong \frac{P_{IDLE}/T_{AVE}}{P_{MAX}/T_{MAX}} = \beta \cdot \frac{T_{MAX}}{T_{AVE}} \quad \text{(EQ 15)}$$

Expressing E_{IDLE} as a function of E_{MAX} yields:

$$METR \cong \frac{\beta \cdot E_{MAX}}{T_{AVE}}, \quad \text{(Idle Energy Dominates)} \quad \text{(EQ 16)}$$

If β remains constant for varying throughput (and E_{MAX} stays constant), then E_{IDLE} scales with throughput as shown in Equation 15. Thus, the METR becomes an energy/operation minimization similar to the fixed throughput mode. However, β may vary with throughput, as will be analyzed further in Section 4.4.

4: Design Optimizations

Many energy efficiency optimization techniques developed for the fixed throughput mode of computation are applicable to the a microprocessor operating in either the maximum or burst throughput modes though not always yielding equal gains. Those techniques that successfully apply to microprocessors are outlined below

If processors are compared without a target system's requirements in mind, then ETR is a reasonable metric of comparison. However, if the targeted system resides in the single-user domain and the required average operations/second T_{AVE} can be characterized, then METR is a better metric of comparison.

4.1: Fixed Throughput Optimization

Orders of magnitude of power reduction have been achieved in fixed throughput designs by optimizing at all levels of the design hierarchy, including circuit implementation, architecture design, and algorithmic decisions [1].

One such example is a video decompression system in which a decompressed NTSC-standard video stream is displayed at 30 frames/sec on a 4" active matrix color LCD. The entire implementation consists of four custom chips that consume less than 2mW [1].

There were three major design optimizations responsible for the power reduction. First, the algorithm was chosen to be vector quantization which requires fewer computations for decompression than other compression schemes, such as MPEG. Second, a parallel architecture was utilized, enabling the voltage to be dropped from 5V to 1.1V while still maintaining the throughput required for the real-time constraint imposed by the 30 ms display rate of the LCD. This reduced the power dissipation by a factor of 20. The reduction in clock rate compensated for the increased capacitance; thus, there was no power penalty due to the increased capacitance to make the architecture more parallel (though it did consume more silicon area). Third, transistor-level optimizations yielded a significant power reduction.

4.2: Fixed vs. Max Throughput Optimization

Since the fixed throughput mode is a degenerate case of the max throughput mode, the low-power design techniques used in the fixed throughput mode are also applicable to the maximum throughput mode. This is best visualized by mapping the procedure for exploiting parallelism onto the Energy/operation vs. Throughput plot. There is a two-step process to exploit parallelism for the fixed throughput mode as shown in Figure 5.

Step 1: Ideally, doubling the hardware (arch1->arch2), doubles the throughput. Although the capacitance is doubled, the energy per operation remains constant, because two operations are completed per cycle

Step 2: Reduce the voltage to trade-off the excess speed and achieve the original required throughout. With this parallel architecture, the clock frequency is halved, but the throughput remains constant with respect to the original design.

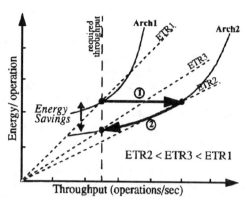

Fig 5: Energy minimization for fixed throughput.

For a processor operating in the maximum throughput mode of computation, it is favorable to reduce the ETR as much as possible, thus, only step 1 is required. In the above example, the final efficiency may even be reduced by decreasing the voltage as was done in step 2.

The energy-throughput trade-off of the maximum throughput operation allows one processor to address separate market segments that have different throughput and power dissipation requirements by simply varying the value of the supply voltage. In essence, a high-speed processor may be the most energy efficient solution for a low throughput application, if the appropriate supply value is chosen. Unfortunately, there are practical bounds that limit the range of operation (e.g. the minimum and maximum supply voltages), preventing one processor from spanning all possible values of throughput.

4.3: Maximum Throughput Optimization

Simply scaling voltage and device sizes are methods of trading off throughput and energy; but they are not energy efficient if used alone since they do not reduce the ETR. However, the architectural modifications that allowed the voltage to be reduced in the fixed throughput mode, do reduce ETR. Conversely, clock frequency reduction, as is common in many portable computers today, does not decrease ETR (it actually increases it) as shown in Section 3.2.

There are three levels in the digital IC design hierarchy. Energy efficient design techniques can drastically increase energy efficiency at all levels as outlined below. Many techniques have corollaries between the fixed and maximum throughput modes, although some differences arise, as noted.

Algorithmic Level: While the algorithm is generally implemented in the software/code domain, which is removed from processor design, it is important to understand the efficiency gains achievable. Similar gains are possible in hardware-implemented algorithms found in signal processing applications.

By using an algorithm implementation that requires fewer operations, both the throughput is increased, and less energy is consumed because the total amount of switched capacitance to execute the program has been reduced. A quadratic improvement in ETR can be achieved [7]. This does not always imply that the program with the smallest dynamic instruction count (path length) is the most energy efficient, since the switching activity per instruction must be evaluated. What needs to be minimized is the number of primitive operations: memory operations, ALU operations, etc. In the case of RISC architectures, the machine code closely resembles the primitive operations, making this optimization possible by minimizing path length. However, in CISC architectures, the primitive operations per each machine instruction need to be evaluated, rather than just comparing path length.

The design of the ISA, however, which does impact the hardware, may be optimized for energy efficient operation. Each instruction must be evaluated to determine if it is more efficient to implement it in hardware, or to emulate it in software.

Architectural Level: The predominant technique to increase energy efficiency is architectural concurrency; with regards to processors, this is generally known as instruction-level parallelism (ILP). Previous work on fixed throughput applications demonstrated an energy efficiency improvement of approximately N on an N-way parallel/ pipelined architecture [5]. This assumes that the algorithm is fully vectorizable, and that N is not excessively large.

Moderate pipelining (4 or 5 stages), while originally implemented purely for speed, also increases energy efficiency, particularly in RISC processors that operate near one cycle-per-instruction. More recent processor designs have implemented superscalar architectures, either with parallel execution units or extended pipelines, in the hope of further increasing the processor concurrency.

However, an N-way superscalar machine will not yield a speedup of N, due to the limited ILP found in typical code [8][9]. Therefore, the achievable speedup, S, will be less than the number of simultaneous issuable instructions, and yields diminishing returns as the peak issue rate is increased. S has been shown to be between two and three for practical hardware implementations in current technology [10].

If the code is dynamically scheduled in employing superscalar operation, as is currently common to enable backwards binary compatibility, the C_{EFF} of the processor will increase due to the implementation of the hardware scheduler. Even in statically scheduled architectures such as VLIW processors, there will be extra capacitive over-

head due to branch prediction, bypassing, etc. There will be additional capacitance increase because the N instructions are fetched simultaneously from the cache, and may not all be issuable if a branch is present. The capacitance switched for unissued instructions is amortized over those instructions that are issued, further increasing C_{EFF}.

The energy efficiency increase can be analytically modelled. Equation 17 gives the ETR ratio of a superscalar architecture versus a simple scalar processor; a value larger than one indicates that the superscalar design is more energy efficient. The S term is the ratio of the throughputs, and the C_{EFF} terms are from the ratio of the energies (architectures are compared at constant supply voltage). The individual terms represent the contribution of the datapaths, C_{EFF}^{Dx}, the memory sub-system, C_{EFF}^{Mx}, and the dynamic scheduler and other control overhead, C_{EFF}^{Cx}. The 0 suffix denotes the scalar implementation, while the 1 suffix denotes the superscalar implementation. The quantity C_{EFF}^{C0} has been omitted, because it has been observed that the control overhead of the scalar processor is minimal: $C_{EFF}^{C0} << C_{EFF}^{D0,M0}$ [11].

$$ETR|_{RATIO} = \frac{S\left(C_{EFF}^{D0} + C_{EFF}^{M0}\right)}{\left(C_{EFF}^{C1} + C_{EFF}^{D1} + C_{EFF}^{M1}\right)} \quad \text{(EQ 17)}$$

Whether ILP architecture techniques can yield significant energy efficiency improvement is not inherently clear. The presence of C_{EFF}^{C1} due to control overhead, and increase of C_{EFF}^{M1} (with respect to C_{EFF}^{M0}) due to unissued instructions, may even negate the increase due to S. Current investigation is attempting to quantify these terms and the resulting efficiency increase for a a variety of superscalar implementations.

Other aspects of architecture design can be optimized for improved efficiency. One example is the reduction of extraneous switching activity by gating the clock to various parts of the processor when possible [12]. Another example is the minimization of the lengths of the most active busses.

Circuit level: The design techniques implemented at this level are similar for the fixed and maximum throughput modes. For example, the topologies for the various subcells (e.g. ALU, register file, etc.) should be selected by their ETR, and not solely for speed. Low-swing bus drivers are currently being investigate for high-speed operation; but, these drivers are also applicable to low power design because the energy per transition drops linearly with the voltage swing (ΔV is reduced, not V_{DD}).

The basic transistor sizing methodology is to reduce every transistor not in the critical path to minimum size to minimize C_{EFF}. There are a number of other techniques to minimize effective capacitance which are also viable for optimizing energy efficiency [1].

4.4: Burst Throughput Optimization

If the energy consumed while idling is negligible compared to that consumed during bursts of computation, then the METR metric simplifies to the ETR metric, and all the design optimizations to increase energy efficiency in the maximum throughput mode are equivalently valid in the burst throughput mode.

However, if the energy consumption during idle dominates the total energy consumption, then different optimizations are required. As was shown in Equation 14, this occurs when the fractional time spent computing (T_{AVE}/T_{MAX}) is less than the fractional power dissipation while idling (β). Then the METR optimization is to minimize β and E_{MAX}, as seen from Equation 16. Furthermore, the exact optimization depends on whether β changes as the throughput T is varied as shown below.

β *is independent of throughput*: This case generally applies to processors with no power-down mode for idle periods; if the clock frequency remains the same (or proportional) during both computation and idle periods, then the idle power dissipation tracks the compute power dissipation. If the throughput is now increased while E_{MAX} is held constant (e.g. using parallelism, as was shown in Figure 5 to decrease the ETR), the METR remains constant because the increase in compute power dissipation causes the idle power dissipation to increase as well.

The METR can be optimized by minimizing E_{MAX}, similar to the fixed throughput mode optimization. By reducing V_{DD} to scale down both throughput and E_{MAX}, the processor's efficiency can be maximized. The efficiency will keep improving as V_{DD} is reduced until $E_{IDLE} < E_{MAX}$, and the energy consumption during idle is not dominant anymore; decreasing V_{DD} any further will have little effect on the energy efficiency.

β *varies with throughput*: This is common for processors that implement idle power down modes; the power dissipation during idle is not proportional to the power dissipation during computation. So, for constant idle power dissipation, or equivalently, constant E_{IDLE}, it is most efficient to deliver as much throughput as possible, since any increase in computational power dissipation is negligible compared to the total power dissipation. In practice, β will be less than inversely proportional to throughput (e.g. due to latency switching between operating modes) so that E_{IDLE} is not independent of throughput. However, energy efficiency will continue to increase with throughput until idle power dissipation is no longer dominant.

By itself, the processor hardware can only provide a moderate value of β, which is the ratio of the power dissipated executing a nop instruction to the power dissipated executing a typical instruction. While executing nop instructions, the internal state of the controller and datapath is not changing ($\alpha = 0$), but the clock line is still transitioning every cycle. The processor's clock dissipates a sizable fraction of the total power, anywhere from 10% to 50%. Even if the clock is gated to those pipeline sections executing nop instructions, the instruction-memory access per cycle will continue to dissipate power. If the processor is used in a laptop computer, and T_{AVE} is on the order of 1 SPECint92 (high estimate for user's average operations/second) and β is reasonably estimated as 0.2, it is not energy efficient to increase the peak throughput of the processor beyond 5 SPECint92. Thus, to deliver a more tolerable response time to the user, energy efficiency will have to be degraded.

It is imperative that the operating system intervene to provide further reductions in β. In doing so, β will typically become a function of throughput because the operating system can decouple the compute and idle regimes' power dissipation. The hardware can enable software power down modes by providing instructions to halt either parts of the processor or the entire thing, as is becoming common in embedded microprocessors.

Independent of β's relation to throughput, the METR metric indicates poor energy efficiency whenever the energy/operation consumed idling dominates total energy consumption. If β is constant, the energy efficiency is maximized by reducing E_{MAX} through V_{DD} reduction, until the throughput is roughly T_{AVE}/β. If β is inversely dependent on throughput, then the energy efficiency is maximized by increasing the throughput, and possibly V_{DD}, until E_{MAX} is roughly equal to E_{IDLE}.

5: Design Principles

A few examples are presented below to demonstrate how energy efficiency can be properly quantified. In doing so, three design principles follow from the optimization of the previously defined metrics: a high-performance processor is usually an energy-efficient processor; reducing the clock frequency does not increase the energy efficiency; and lastly, idle power dissipation limits the efficiency of increasing deliverable throughput.

5.1: High Performance is Energy Efficient

Table 1 lists two hypothetical processors that are similar to ones available today -- B targets the low-power market, and A targets the high-end market; both are fabricated in the same technology, which allows an equal comparison. SPEC is either SPECint92, or if a floating point unit is present, the average of SPECint92 and SPECfp92 [7]. A misused metric for measuring energy efficiency is SPEC/Watt (or Dhrystones/Watt, MIPS/Watt, etc.). Processor B may boast a SPEC/Watt eight times greater than A's, and declare that it is eight times as energy efficient. This metric only compares operations/energy, and does not weight that B has 1/15th the performance.

Table 1: Comparison of two processors

Proc.	SPEC (T_{MAX})	Watts	V_{DD}	SPEC/Watt ($1/E_{MAX}$)	ETR (10^{-3})
A	150	30.0	3.3	5.0	1.33
B	10	0.25	3.3	40.0	2.50

The ETR (Watts/SPEC^2) metric indicates that processor A is actually *more* energy efficient than processor B. To quantify the efficiency increase, the plot of energy/operation versus throughput in Figure 6 is used because it better tracks processor A's energy at the low throughput values. The plot was generated from the delay and power models in Section 2.

According to the plot, processor A would dissipate 0.154 W at 10 SPEC, or 60% of processor B's power, despite the low V_{DD} ($1.31V_T$) for A. Conversely, A can deliver 31 SPEC at 0.25W (V_{DD}=$1.66V_T$), or 310% of B's throughput. This does assume that processor A can be operated at this supply voltage.

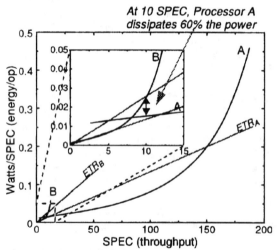

Fig 6: Energy vs Throughput of Processors A & B.

While the ETR correctly predicted the more energy-efficient processor at 10 SPEC, it is important to note that processor A is not more energy efficient for all values of SPEC, as the ETR metric would indicate. Because the nominal throughput of the processors is vastly different, the Energy/Operation versus Throughput metric better tracks the efficiency, and indicates a cross-over throughput of 8 SPEC. Below this value, processor B is more energy efficient.

5.2: Clock Reduction is not Energy Efficient

A common fallacy is that reducing the clock frequency f_{CLK} is energy efficient. In maximum throughput mode, it is quite the opposite. At best, it allows an energy-throughput trade-off when idle energy consumption is dominant in burst throughput mode.

In the maximum throughput mode, energy is independent of f_{CLK}, so as the latter is scaled down, the throughput decreases, and the ETR increases, indicating a less efficient design. If f_{CLK} is halved, the power is also halved. However, it takes twice as long to complete any computation, so the energy/operation consumed is constant. Thus, if the energy source is a battery, halving f_{CLK} is equivalent to doubling the computation time, while maintaining constant computation per battery life.

In the burst throughput mode, clock reduction may trade-off throughput and operations per battery life (i.e. energy/operation), but only when E_{IDLE} dominates total energy consumption (β>>T_{AVE}/T_{MAX}) and β is independent of throughput such that E_{IDLE} scales with throughput. When this is so, halving f_{CLK} will double the computation time, but will also double the amount of computation per battery life. If the user is engaged in an application where throughput degradation is acceptable, then this is a reasonable trade-off. If either E_{MAX} dominates total energy consumption, or β is inversely proportional to throughput, then reducing f_{CLK} does not affect the total energy consumption, and the energy efficiency drops.

If V_{DD} were to track f_{CLK}, however, so that the critical path delay remains inversely equal to the clock frequency, then constant energy efficiency could be maintained as f_{CLK} is varied. This is equivalent to V_{DD} scaling (Section 3.2) except that it is done dynamically during processor operation. If E_{IDLE} is present and dominates the total energy consumption, then simultaneous f_{CLK}, V_{DD} reduction may yield a more energy efficient solution.

5.3: Faster Operation Can Limit Efficiency

If the user demands a fast response time, rather than reducing the voltage, as was done in Section 5.1, the processor can be left at the nominal supply voltage, and shut down when it is not needed.

For example, assume the target application has a T_{AVE} of 10 SPEC, and both processor A and B have a β factor of 0.1. If the processors' V_{DD} is left at 3.3V, B's METR is exactly equal to its ETR value, which is 2.5×10^{-3}. It remains the same because it never idles. Processor A, on the other hand, spends 14/15ths ($1 - T_{AVE}/T_{MAX}$) of the time idling, and its METR is 3.2×10^{-3}. Thus, for this scenario, processor B is more energy efficient.

However, if processor A's β can be reduced down to

0.05, then the METR of processor A becomes 2.26×10^{-3}, and it is once again the more energy efficient solution. For this example, the cross-over value of β is 0.063.

This example demonstrates how important it is to use the METR metric instead of the ETR metric if the target application's idle time is significant (i.e. T_{AVE} can be characterized). For the above example, a β for processor A greater than 0.063 leads the metrics to disagree on which is the more energy efficient solution. One might argue that the supply voltage can always be reduced on processor A so that it is more energy efficient for any required throughput. This is true if the dynamic range of processor A is as indicated in Figure 6. However, if some internal logic limited the value that V_{DD} could be dropped, then the lower bound on A's throughput would be located at a much higher value. Thus, finite β can degrade the energy efficiency of high throughput circuits due to excessive idle power dissipation.

6: Conclusions

Metrics for energy efficiency have been defined for three modes of computation in digital circuits. The appropriate metric for the fixed throughput mode, typical of most digital signal processing circuits, is energy/operation. Two other modes which apply to the operation of a microprocessor are maximum throughput and burst throughput modes.

A good energy efficiency metric for the maximum throughput mode is energy/operation versus throughput, which can be approximated with a constant ETR value. Many of the techniques developed for low power design in the fixed throughput mode can be successfully applied to the energy efficient design of a processor in the maximum throughput mode.

However, a better metric to describe more typical processor usage is the Microprocessor ETR, or METR; it includes the energy consumption of the idle mode, which can dominate total energy consumption in user-interactive applications. Decreasing the energy consumption of the idle mode is critical to the design of a energy efficient processor and complete shut down of the clock while idling is optimal. If this cannot be accomplished, then it is imperative that the operating system implement a power down mode so that the idle power dissipation becomes independent of the computing power dissipation. Then the METR optimization will maximize the throughput delivered to the user in an energy efficient manner. Otherwise, if idle power dissipation is proportional to the compute power dissipation, achieving energy efficient operation requires the throughput to be minimized.

An organized analytical approach to the optimization of power in microprocessor design, based on metrics that include the requirement of both throughput and energy, as well as actual application operation, allow the designer to quantify energy efficiency and provide insights into the design issues of energy efficient processor design.

This research is sponsored by ARPA.

References

[1] A. Chandrakasan, A. Burstein, R.W. Brodersen, "A Low Power Chipset for Portable Multimedia Applications", *Proceedings of the IEEE International Solid-State Circuits Conference*, Feb. 1994, pp. 82-83.

[2] T. Burd, *Low-Power CMOS Cell Library Design Methodology*, M.S. Thesis, University of California, Berkeley, 1994.

[3] S. Sze, *Physics of Semiconductor Devices*, Wiley, New York, 1981.

[4] H. Veendrick, "Short-Circuit Dissipation of Static CMOS Circuitry and Its Impact on the Design of Buffer Circuits", *IEEE Jour. of Solid State Circuits*, Aug 1984, pp. 468-473

[5] A. Chandrakasan, S. Sheng, R.W. Brodersen, "Low-Power CMOS Digital Design", *IEEE Journal of Solid State Circuits*, Apr. 1992, pp. 473-484.

[6] R. Muller, T. Kamins, *Device Electronics for Integrated Circuits*, Wiley, New York, 1986

[7] M. Horowitz, T. Indermaur, R. Gonzalez, "Low-Power Digital Design", *Proceedings of the Symposium on Low Power Electronics*, Oct. 1994.

[8] D. Wall, *Limits of Instruction-Level Parallelism*, DEC WRL Research Report 93/6, Nov. 1993.

[9] M. Johnson, *Superscalar Microprocessor Design*, Prentice Hall, Englewood, NJ, 1990.

[10] M. Smith, M. Johnson, M. Horowitz, "Limits on Multiple Issue Instruction", *Proceedings of the Third International Conference on Architectural Support for Programming Languages and Operating Systems*, Apr. 1989. pp 290-302

[11] T. Burd, B. Peters, *A Power Analysis of a Microprocessor: A Study of the MIPS R3000 Architecture*. ERL Technical Report, University of California, Berkeley, 1994.

[12] S. Gary, et. al.; "The PowerPC 603 Microprocessor: A Low-Power Design for Portable Applications", *Proceedings of the Thirty-Ninth IEEE Computer Society International Conference*, Mar. 1994, pp. 307-315

A 160-MHz, 32-b, 0.5-W CMOS RISC Microprocessor

James Montanaro, Richard T. Witek, Krishna Anne, Andrew J. Black, *Member, IEEE*, Elizabeth M. Cooper,
Daniel W. Dobberpuhl, Paul M. Donahue, Jim Eno, Gregory W. Hoeppner, David Kruckemyer,
Thomas H. Lee, *Member, IEEE*, Peter C. M. Lin, Liam Madden, Daniel Murray, Mark H. Pearce, Sribalan Santhanam,
Kathryn J. Snyder, Ray Stephany, and Stephen C. Thierauf, *Associate Member, IEEE*

Abstract— This paper describes a 160 MHz 500 mW StrongARM®[1] microprocessor designed for low-power, low-cost applications. The chip implements the ARM® V4 instruction set [1] and is bus compatible with earlier implementations. The pin interface runs at 3.3 V but the internal power supplies can vary from 1.5 to 2.2 V, providing various options to balance performance and power dissipation. At 160 MHz internal clock speed with a nominal Vdd of 1.65 V, it delivers 185 Dhrystone 2.1 MIPS while dissipating less than 450 mW. The range of operating points runs from 100 MHz at 1.65 V dissipating less than 300 mW to 200 MHz at 2.0 V for less than 900 mW. An on-chip PLL provides the internal clock based on a 3.68 MHz clock input. The chip contains 2.5 million transistors, 90% of which are in the two 16 kB caches. It is fabricated in a 0.35-μm three-metal CMOS process with 0.35 V thresholds and 0.25 μm effective channel lengths. The chip measures 7.8 mm × 6.4 mm and is packaged in a 144-pin plastic thin quad flat pack (TQFP) package.

I. INTRODUCTION

AS personal digital assistants (PDA's) move into the next generation, there is an obvious need for additional processing power to enable new applications and improve existing ones. While enhanced functionality such as improved handwriting recognition, voice recognition, and speech synthesis are desirable, the size and weight limitations of PDA's require that microprocessors deliver this performance without consuming additional power. The microprocessor described in this paper[2] directly addresses this need by delivering 185 Dhrystone 2.1 MIPS while dissipating less that 450 mW. This

TABLE I
PROCESS FEATURES

Feature Size	0.35 μm
Channel Length	0.25 μm
Gate Oxide	6.0 nm
Vtn/Vtp	0.35V/-0.35V
Power Supply	2.0V (nominal)
Substrate	P-epi with n-well
Salicide	Cobalt-disilicide in diffusions and gates
Metal 1	0.7 μm AlCu, 1.225 μm pitch (contacted)
Metal 2	0.7 μm AlCu, 1.225 μm pitch (contacted)
Metal 3	1.4 μm AlCu, 2.8 μm pitch (contacted)
RAM Cell	6 transistor, 25.5μm^2

represents a significantly higher performance than is currently available at this power level.

II. CMOS PROCESS TECHNOLOGY

The chip is fabricated in a 0.35-μm three-metal CMOS process with 0.35 V thresholds and 0.25 μm effective channel lengths. Process characteristics are shown in Table I. The process is the result of several generations of development efforts directed toward high-performance microprocessors. It is identical to the one used in Digital Equipment Corporation's current generation of Alpha chips [2] except for the removal of the fourth layer of metal and the addition of a final nitride passivation required for plastic packaging.

The factors which drive process development for low-power design are similar to those which drive the process for pure high-performance although the motivation sometimes differs. For example, while both types of designs benefit from maximizing Idsat of the transistors at the lowest acceptable Vdd, the motivation for a pure high-performance design is reducing power distribution and thermal problems rather than extending battery life. Similar arguments apply to minimizing transistor leakage and on-chip variation of transistor parameters. This convergence of goals has been essential to our ability to develop one process to satisfy the requirements of both low-power and high-performance families.

III. POWER DISSIPATION TRADEOFFS

RISC microprocessors operating at 160 MHz are fairly common using current CMOS process technology. The novel

Manuscript received April 1, 1996; revised July 3, 1996.

J. Montanaro, R. T. Witek, K. Anne, J. Eno, G. W. Hoeppner, K. J. Snyder, and R. Stephany are with Digital Equipment Corporation, Austin, TX 78759 USA.

A. J. Black, E. M Cooper, and L. Madden were with Digital Equipment Corporation, Palo Alto, CA 94301 USA. They are now with Silicon Graphics, Inc. Mountain View, CA 94043 USA.

D. W. Dobberpuhl, P. M. Donahue, D. Kruckemyer, D. Murray, M. H. Pearce, and S. Santhanam are with Digital Equipment Corporation, Palo Alto, CA 94301 USA.

T. H. Lee is with Stanford University, Palo Alto, CA 94305 USA.

P. C. M. Lin was with Digital Equipment Corporation, Palo Alto, CA 94301 USA. He is now with C-Cube Microsystems, Milpitas, CA 95035 USA.

S. C. Thierauf is with Digital Equipment Corporation, Hudson, MA 01749 USA.

Publisher Item Identifier S 0018-9200(96)07777-3.

[1]ARM and StrongARM are registered trademarks of Advanced RISC Machines Ltd.

[2]The microprocessor described is the Digital Equipment Corporation SA-110, the first microprocessor in the StrongARM® family.

Reprinted from *IEEE Journal of Solid-State Circuits*, pp. 1703-1714, November 1996.

TABLE II
POWER DISSIPATION TRADEOFFS

Start with Alpha 21064: 200MHz @ 3.45V. Power Dissipation = 26W

Vdd reduction:	Power reduction =	5.3x	⇒ 4.9 W
Reduce functions:	Power reduction =	3x	⇒ 1.6 W
Scale process:	Power reduction =	2x	⇒ 0.8 W
Reduce clock load:	Power reduction =	1.3x	⇒ 0.6 W
Reduce clock rate:	Power reduction =	1.25x	⇒ 0.5 W

aspect of this design is the ability to achieve this operating frequency at power levels which are low enough for handheld applications. Several design tradeoffs were made to achieve the desired power dissipation. In order to illustrate their effect on the design, it is interesting to imagine applying these tradeoffs to an earlier design whose power dissipation occupies the opposite end of the power spectrum, the first reported Alpha microprocessor [3]. This Alpha chip was fabricated in a 0.75-μm CMOS process and operated at 200 MHz dissipating 26 W at 3.45 V. The impact of these tradeoffs is summarized in Table II.

The first decision is to reduce the internal power supply to 1.5 V. This change cuts the power by a factor of 5.3. While this has the desired effect, it has implications for the cycle time which are considered in Section VII.

The next step is to reduce the functionality. As compared to the early Alpha chip, the most obvious sections missing in this design are the floating point unit and the branch history table. Floating point is not required in the target applications and the low branch latency of this design eliminates the need for the branch history table. Less obvious, but very important, is reduced control complexity. This is a simple machine and we have worked hard to keep it so. We estimated that the reduced functionality would cut power by a factor of three.

Process scaling reduces node capacitances and therefore chip power. Note that although the area components of the capacitance will decrease as the square of the scale factor, the total capacitance change with scaling will be less dramatic primarily due to the effect of periphery capacitance. We estimate that scaling from 0.75 μm of the early Alpha chip to our current 0.35 μm process results in a power reduction of about a factor of two, a linear reduction with scale factor. Once again, coupled with this positive effect of process scaling are a host of other issues. Some of those issues are considered in Section VI.

Next, consider the clock power. The clock power of the Alpha chips is fairly large and while that clocking strategy works well for Alpha machines, it is not appropriate for a low-power chip. Our clocking strategy and our latch circuits are described in some detail later. One major change from the Alpha design was to reject the pair of transparent latches per cycle used on the Alpha design. Instead, on this design, we switched to a single edge-triggered latch per cycle to reduce clock load and latch delay. Our estimate is that the changes in the clocking reduced the clock power by a factor of two. Since the clock power was about 65% of the total power on the first Alpha chip, this results in a reduction of about 1.3.

Finally, the reduction in clock frequency from 200 MHz to 160 MHz drops the power by 1.25.

Clearly, this analysis is not rigorous, but it suggests that it is reasonable to build a 160 MHz processor chip that dissipates around half a watt. A similar analysis was performed at the beginning of the project to select the power supply voltage and operating frequency and to determine whether significant changes in design method would be required to meet the performance and power goals. It is interesting to note that with the exception of the clocking changes, the design methods and philosophy used on this design were very similar to that used on the Alpha chips.

IV. INSTRUCTION SET

The microprocessor implements the ARM® V4 [1] instruction set. The architecture defines 30 32-b general purpose registers and a program counter (PC). Registers are specified by a 4-b field where registers 0 to 14 are general purpose registers (GPR) and register 15 is the PC. The current processor status register contains a current mode field which selects either an unprivileged user mode or one of six privileged modes. The current mode selects which set of GPR's is visible. In addition to basic RISC features of fixed length instructions and simple load/store architecture, the architecture implemented includes several features to improve code density. These include conditional execution of all instructions, load and store multiple instructions, auto-increment and auto-decrement for loads and stores, and a shift of one operand in every ALU operation. The architecture supports loads and stores of 8-, 16-, and 32-b data values. In addition to the standard 32-b computations, there is a 32-b × 32-b multiply accumulate with a 64-b product and accumulator.

V. CHIP MICROARCHITECTURE

As shown in Fig. 1, the chip is functionally partitioned into the following major sections: the instruction unit (IBOX), integer execution unit (EBOX), integer multiplier (MUL), memory management unit for data (DMMU), memory management unit for instructions (IMMU), write buffer (WB), bus interface unit (BIU), phase locked loop (PLL), and caches for data (Dcache) and instructions (Icache). To minimize pin power and support the high-speed internal core, one half of the chip area is devoted to the two 16 K caches. The pad ring occupies one-third of the chip area and the processor core fills the remaining one-sixth of the chip area.

The processor is a single issue design with a classic five-stage pipeline—Fetch, Issue, Execute, Buffer, and Register File Write (Fig. 2). All arithmetic logic unit (ALU) results can be forwarded to the ALU input and there is a one-cycle bubble for dependent loads.

For example, the pipeline diagram in Fig. 2 shows a SUB-TRACT followed by a dependent LOAD. Note that at the end of cycle 3, we bypass the result from the SUBTRACT back into the ALU to compute the load address in cycle 4 without stalling the pipe. The third instruction is an ADD which uses the result of the previous LOAD. The ADD is held in the

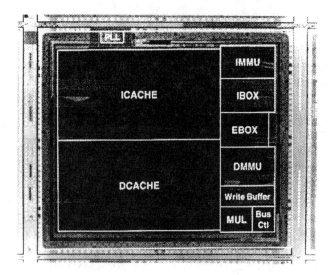

Fig. 1.   Chip photo with overlay.

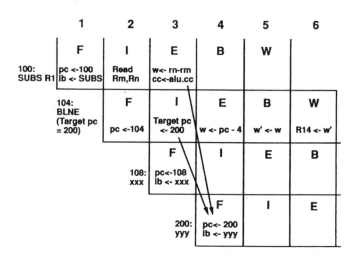

Fig. 3.   Pipeline diagram of a branch.

	1	2	3	4	5	6
100: **SUBS R1**	F pc <-100 ib <- SUBS	I Read Rm,Rn	E w <- rn-rm cc <-alu.cc	B w' <- w	W R1<-w'	
104: **LDR R2, [R1,d]!**		F pc <-104 ib <- LDR	I Read Rm, Rn	E w <- d+R1 la <- d+R1	B L<-mem(la) w' <- w	W R2 <- L R1 <-w'
108: **ADD x,R2,y**			F pc <-108 ib <-ADD	I Read Rm, Rn	I Read Rm, Rn	E w <-R2+y

Fig. 2.   Basic pipeline diagram.

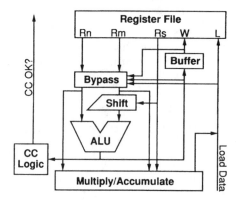

Fig. 4.   EBOX block diagram.

Issue stage for one additional cycle until the LOAD data is available at the end of cycle 5.

The IBOX can resolve conditional branches in the Issue stage even when the condition codes are being updated in the current Execute cycle. By providing this optimized path, the IBOX incurs only a one-cycle penalty for branches taken, so the chip does not require branch prediction hardware. For example, in the pair of instructions shown in Fig. 3, the BRANCH and LINK instruction at the (program counter) PC of 104 depends on the condition codes which are being generated by the SUBTRACT in the previous instruction. The condition codes from the Execute stage of the SUBTRACT are available at the end of cycle 3, in time to swing the PC multiplexer in the IBOX to point at the branch target PC during the next Fetch cycle.

The optimization of the branch path represents a power versus performance tradeoff in which performance won. In our effort to hold the one cycle branch penalty, we included a dedicated adder in the IBOX to calculate the branch target address and consumed additional power in the EBOX adder to meet the critical speed path to control the PC multiplexer. Due to critical path constraints, the adder in the IBOX must run every cycle, even if the instruction is not a branch.

In the early stage of the design, one of our concerns was whether the decision to pursue this optimized branch path

would increase our cycle time. As the design turned out, our best efforts in this ALU path and in the cache access path resulted in nearly identical delays for these two longest critical speed paths.

Data for integer operations comes from a 31-entry register file with three read and two write ports. Sixteen of the registers are visible at any time with 15 additional shadow registers specified by the architecture to minimize the overhead associated with initiating exceptions. The EBOX contains an ALU with a full 32-b bidirectional shifter on one of the input operands. It includes bypassing circuitry to forward the data from the Dcache or the ALU output to any of the read ports. Fig. 4 shows the circuit blocks involved in the branch path. The path starts at a latch in the bypassers and, in a single cycle, includes a 0- to 32-b shift, a 32-b ALU operation, and a condition code computation to swing the PC multiplexer for the next cycle. The registers to hold the condition codes were implemented in the EBOX so that this path could be locally optimized. Analysis of code traces indicated that most ALU operations included a shift of zero, so for this case, the shifter is disabled and bypassed to reduce power.

The EBOX also contains a 32-b multiply/accumulate unit. The multiplier consists of a 12- by 32-b carry-save multiplier array which is used for one to three cycles depending on the value of multiplicand and a 32-b final adder to reduce the carry-save result. For multiply accumulate operations, the accumulate value is inserted into the array so that an additional cycle is not required for the Multiplies with Accumulate. Multiply Long instructions require one additional cycle. This results in a MULTIPLY or MULTIPLY/ACCUMULATE in two to four cycles and MUL LONG or MUL LONG/ACCUMULATE in three to five cycles.

The Wallace tree implementation was chosen to minimize the delay through the array. This implementation required careful floor planning and custom layout to keep the wiring under control. The decision to perform 12 b of multiply per cycle was based on wiring tradeoffs made during the physical planning phase of the design rather than critical path concerns. When the multiplier is not in use, all clocks to the section stop and the input operands do not toggle.

The chip features separate 16 kByte, 32-way set associative virtual caches for instructions and data. Each cache is implemented as 16 fully associative blocks. Each cache is accessed in a single cycle for both reads and writes, providing a two-cycle latency for return data to the register file. One eighth of each cache is enabled for a cache access.

The Dcache is writeback with no write allocation. The block size is 32 bytes with dirty bits provided for each half block to minimize the data which needs to be castout in the event of a dirty victim. The physical address is stored with the data to avoid address translation during castouts.

Given the size of the caches and the low power target for the chip, it was important that we have fine granularity of bank selection. In addition, we required associativity of at least four-way for cache efficiency and it was important to performance that we maintain a single cycle access. We considered several solutions to this problem, including traditional four-way set associative caches, and decided that the simplest approach which satisfied all the requirements was to implement the caches as smaller, bank-addressed, fully associative caches. This resulted in 32-way associativity but this level of associativity was a side effect of the implementation used, not the result of a goal to get associativity significantly above four-way.

The chip includes separate memory management units (MMU) for instructions and data. Each MMU contains a 32-entry fully associative translation lookaside buffer (TLB) with entries which can map either 4 kB, 64 kB, or 1 MB pages. TLB fills are implemented in hardware. In addition to the standard memory management protection mechanisms, the ARM® architecture defines an orthogonal memory protection scheme to allow the operating system easy access to large sections of memory without manipulating the page tables. This functionality requires a set of additional checks which must be performed after the TLB lookup. The resulting critical path was sufficiently long that we self-timed the RAM access in the TLB to allow us to perform the lookup and complex protection checks in a single cycle.

A write buffer with eight 16-byte entries handles stores and castouts from the Dcache. The write buffer includes a single-entry merge latch to pack up sequential stores to the same entry.

During normal operations, an external load request takes priority over stores on the pin bus. However, in the event of a load which hits in the write buffer, the chip executes a series of priority stores which raises the priority of the Write Buffer on the external bus above that of any loads. External stores occur and the write buffer empties until the store which was pending at the load address completes. At this point, top priority reverts back to loads.

VI. POWER DOWN MODES

There are two power down modes supported by the chip—Idle and Sleep.

Idle mode is intended for short periods of inactivity and is appropriate for situations in which rapid resumption of processing is required. In Idle mode, the on-chip PLL continues to run but the internal clock grid and the bus clock stop toggling. This eliminates most activity in the chip and the power dissipation drops from 450 mW to 20 mW. Return from Idle to normal mode is accomplished with essentially no delay by simply restarting the bus clock.

Sleep mode is designed for extended periods of inactivity which require the lowest power consumption. The current in Sleep mode is 50 μA which is achieved by turning off the internal power to the chip. The 3.3 V I/O circuitry remains powered and the chip is well behaved on the bus, maintaining specified levels if required by the drive enable inputs. Return from Sleep to normal operation takes approximately 140 μs.

As was noted earlier, a low voltage process is key to the design of a microprocessor which will run at 160 MHz while dissipating less than 450 mW. However, the same low device thresholds which allow the reduction of Vdd also result in significant device leakage. While this leakage is not large enough to cause a problem for normal operation, it does pose problems for standby current, especially if the process skews toward short channel devices. Our initial analysis indicated that the chip would dissipate over 100 mW in Idle mode with the clocks stopped. To reduce this leakage, we lengthened devices in the cache arrays, the pad drivers, and certain other areas. This brought the leakage power to within the required value of 20 mW in the fastest process corner. As a backup, we relaxed our design rules to allow the remaining gate regions, which are drawn with a standard 0.35 μm gate length, to be biased up algorithmically without violating design rules in case it was necessary to meet the leakage requirements.

The requirement for standby power in Sleep is more than two orders of magnitude lower than the Idle power. To meet the power limit in Sleep, we considered a variety of options including integrated power supply switches and substrate biasing schemes before choosing the simple approach of turning off the internal supply. This approach is reasonable for this generation of parts since they have a dedicated low voltage supply. As more parts of the system shift to the low voltage supply, this may no longer be acceptable. The conflicting requirements of high performance at low voltage and low standby current promise to create interesting challenges in future designs.

The power switch to turn off the internal power supply during Sleep is implemented off-chip as part of the power supply circuit for the low voltage supply. No state is stored internally during Sleep since in typical PDA systems, the Sleep

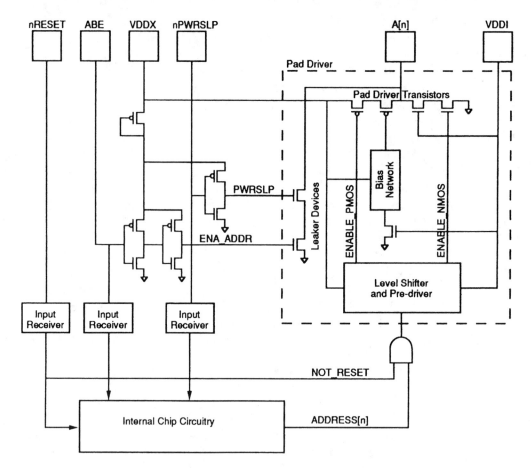

Fig. 5. Pad circuitry.

state corresponds to the user turning the system off. Therefore, the time associated with reloading the cache upon return from Sleep is acceptable.

The requirements in Idle and Sleep complicated the design of the bus interface circuits. This section includes the level-shifting interface between the internal low voltage (1.5 to 2.2 V) signals and the 3.3 V external pin bus. The bus interface circuits must drive and receive signals which are higher voltage than those nominally supported by the 0.35-μm process without using circuits which would cause us to exceed the current limit specified by the Idle spec. In addition, during Sleep the pads must be able to sustain the value on the output pins despite the loss of internal Vdd (Vddi) from the low voltage supply which is powered off by the system. The circuitry used to implement this function is shown in Fig. 5.

Since Vddi will be driven to zero by the system during Sleep, it is used not only as a power supply but also as a logic signal. All circuitry which must be active in Sleep is driven from the external, 3.3 V supply (Vddx) which has been dropped through diode-connected PMOS devices to reduce the stress on the oxide of these devices. Before signaling the chip to enter Sleep, the system asserts the nRESET pin (active low) which drives all enabled outputs to a specified state—disabled for control signals and zero for addresses and data. It then asserts nPWRSLP (active low) which is ANDed with the appropriate output enable control to turn on small leaker devices which will hold the output pin in the appropriate state during Sleep. In the circuit shown in Fig. 5, the output

is an address. Therefore, the address bus enable (ABE) pin is the control pin on the lower NMOS leaker and a buffered version of nPWRSLP controls the top device. Finally, the Vddi pins are actively driven to zero by the system. This action disables the output stage of the pad driver circuit by turning off the transistors closest to the pad—the NMOS directly and the PMOS via the bias network whose output goes to Vddx when its path to Vss is cut off. Note that for any input whose value is required during Sleep (ABE and nPWRSLP in the example described), a separate parallel input receiver must be implemented since the normal input receiver requires Vddi.

VII. CIRCUIT IMPLEMENTATION

The circuit implementation is pseudostatic and allows the internal clock to be stopped indefinitely in either state. Use of circuits which might limit low voltage operation was strictly controlled and the design was simulated to ensure operation significantly below the nominal 1.5 V level of the low voltage supply. The values of the internal supply and operating frequency were optimized to achieve maximum performance for less than half a watt.

The vast majority of the design is purely static, composed of either complementary CMOS gates or static differential logic. In certain situations, wide NOR functions were required and these were implemented in a pseudostatic fashion using either static weak feedback circuits or self-timed circuits to latch the output data and return the dynamic node to its precharged state.

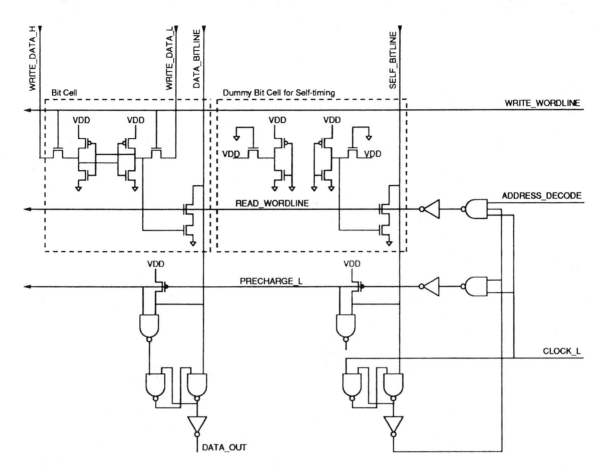

Fig. 6. Self-timed RF precharge.

The register file (RF) uses the self-timed approach to return the bit lines to the precharged state after an access (Fig. 6). In this circuit, an extra self-timing column of bit cells with a dynamic bit line was implemented to mimic the timing of the data bit lines. Fig. 6 shows one cell from a column of register file data bit cells and one cell from the extra self-timing column (only one read port is shown). The bit cells in this extra column are all tied off so that the SELF_BITLINE signal will always discharge when the READ_WORDLINE goes high. When the SELF_BITLINE falls, it will set an RS latch causing the SELF_ENABLE signal to fall. This will disable the READ_WORDLINE and cause the bit lines to be precharged high when the read access is complete. Since the DATA_BITLINE's are received by low sensitive RS latches, the output data will be held when the bit line is precharged high. The self-timing RS latch is cleared when CLOCK_L goes low. This causes the SELF_ENABLE signal to go high, enabling the read port for the access in the next clock cycle. A separate SELF_BITLINE signal is implemented for each of the three register file ports so that the clocks for the three ports can be enabled independently.

The transistor leakage associated with the low threshold voltages is problematic for these pseudostatic circuits. If a weak feedback circuit is used in a NOR structure which is precharged high, excessive leakage in the parallel NMOS pulldowns would require that the feedback be fairly strong, which in turn would reduce the speed of the circuit. In the limit of very wide NOR's, it may not be possible to size a PMOS leaker so that it can supply the leakage of all the off NMOS pulldowns without making the leaker too large to be overpowered by a single active pulldown. In the case of a self-timed approach, a similar problem exists but it usually is manifested as a vanishingly small timing margin for the self-timed circuit to fire before the data on the dynamic node decays away. In either case, we addressed this issue by requiring the length of pulldowns on dynamic nodes to be slightly larger than minimum. Transistor leakage current is a strong function of channel length so a 12% increase in device length results in a leakage reduction in the worst case of about a factor of 20. The resulting leakage makes implementation of either weak feedback or a self-timed approach very reasonable.

The operating frequency at 1.5 V can be roughly derived by starting with the frequency of the Alpha processor in the same process technology [2] and scaling for the use of a longer tick model and then Vdd. Since the long tick design requires the chip to perform a full SHIFT and a full ADD in a single cycle, this approximately doubles the cycle time required. The effect of Vdd scaling is roughly linear for this range of Vdd. Combining these effects results in an operating frequency at 1.5 V given by

$$433 \text{ MHz} * 0.5 * (1.5 \text{ V}/2.0 \text{ V}) = 162 \text{ MHz}.$$

This pair of voltage and frequency values agrees well with the power estimate outlined in Section III. Note that for power supply voltages much lower than 1.5 V, the operating frequency decreases with voltage in a manner which is sig-

391

nificantly stronger than linear. This fact sets a practical lower limit on the power supply voltage in most applications.

Power estimates made early in the design are prone to errors in either direction. In the case of this design, the power dissipated at 1.5 V was lower than the 450 mW target, so we shifted the nominal internal Vdd to 1.65 V to increase the yield in the 160 MHz bin.

A. Clock Generation

An on-chip PLL [4] generates the internal clock at one of 16 frequencies ranging from 88 to 287 MHz based on a fixed 3.68 MHz input clock. Due to internal resource constraints and our limited experience with low-power analog circuits, we contracted with Centre Suisse d'Electronique et de Microtechnique (CSEM) from Neuchâtel, Switzerland, to design the PLL and engaged Professor T. Lee from Stanford as a consultant on the project. Our initial feasibility work resulted in several design tradeoffs.

First, while there was a system requirement that the chip return quickly from the Idle state to normal operation, there was no such constraint on returning from the Sleep state. Based on this determination and our 20 mW power budget in Idle, we concluded that if we could keep the PLL power below 2 mW, we could leave the PLL running in Idle and remove the requirements on the PLL lock time. Thus, the need for a very low power PLL is dictated by the power budget in Idle, not in normal operation.

Next, we had specified a large frequency multiplication factor to allow the use of a common and cheap low frequency crystal clock source for consumer products. Early investigations indicated that this would make tight phase locking difficult. However, when we looked at target systems, we found no pressing need for phase locking. Consequently, we removed phase locking as a design criteria and concentrated our efforts and design tradeoffs on minimizing phase compression.

Finally, while the PLL was designed to handle the noise expected on the chip power supplies, we discovered toward the end of the design that the PLL was under its area budget and there was additional space available in the vicinity. We took advantage of this opportunity to provide cleaner power to the PLL by RC filtering our internal supply and we dedicated 1 nF of on-chip decoupling cap to this purpose.

CSEM performed the circuit and layout design and we placed the completed block into the microprocessor. Since we anticipated that the characterization of the PLL integrated in the chip would present some difficulties, we reserved one of the six die sites on our first pass reticle set for a test chip which contained several variants of the full PLL and interesting subblocks. These circuits allowed access to a variety of nodes in the PLL without compromising the design of the PLL instantiated in the chip. The results of the PLL characterization are reported in [4].

B. Clock Distribution

The chip operates from two clocks as shown in Fig. 7. An internal clock, called DCLK, is usually generated by the PLL.

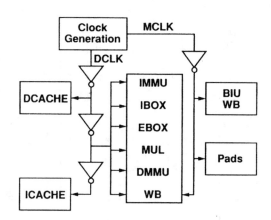

Fig. 7. Clock regimes.

The second clock is a bus clock, known as MCLK which operates up to 66 MHz. MCLK can be supplied by an external asynchronous source or by the chip based on a division of the PLL clock signal.

There are five clock regimes in the chip. The first two regimes are sourced by MCLK and consist of the pad ring which receives MCLK directly and the bus interface unit (BIU) and part of the write buffer which receive MCLK through conditional clock buffers. The last three regimes are sourced by the internal DCLK clock tree and contain the Dcache, the Icache, and the core. In this case, the core includes the IBOX, EBOX, MUL, IMMU, DMMU, and part of the write buffer.

Both MCLK and DCLK are distributed by buffered H-trees to conditional clock buffers in the various sections of the chip. The buffers in the H-tree allow the use of smaller lines for distribution and result in lower clock power. Although the three internal clock regimes are all sourced by the same H-tree, the topology of the chip did not allow corresponding sections of the H-tree to be routed in the same metal. This resulted in an increase in the expected skew between the caches and the core. In addition, we discovered that we could squeeze a bit more performance from the chip if we intentionally offset the clock in the caches relative to the clock in the core. Consequently, we used the clock buffers in the H-tree to tune the clock so that the Dcache receives a clock which is one gate delay earlier than the core and the Icache receives a clock which is one gate delay later than the core.

Fig. 8 shows the physical routing of the internal clock tree. The buffer stages are not shown but they exist in the center of the chip and in four symmetric locations—two in the center of the I and D caches and two in locations at the cache/core interface. The final leg of the H-tree is tied to conditional clock buffers in the caches and the core. The problems associated with clock skew within the caches are reduced by the fact that only a single bank in each cache is enabled. This limits the physical distance over which tightly-controlled clocks need to be delivered in the cache regions.

The clocks in the core present a more interesting problem. The final leg of the clock tree in the core stretches the full height of the chip and tight control of skew along this node is required for speed and functionality. It is implemented as a vertical, metal 2 line driven from four nominally equidistant points. The clock buffers are standard cells of varying drive

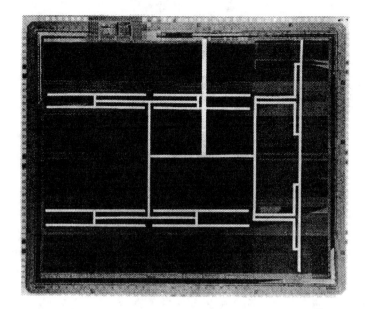

Fig. 8. Physical routing of clock tree.

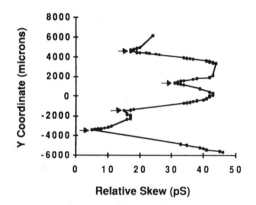

Fig. 9. Clock arrival time in the core.

strength built directly under this M2 line to minimize local variation in delay.

Circuit simulations of the H-tree were performed using SPICE to determine the skew between clock regions and within each of the clock regions. The nodes in the grid were extracted from layout and contained more than 30 000 R and C elements. Fig. 9 shows the relative clock arrival time versus the Y coordinate for each conditional clock buffer on the vertical leg of the clock tree in the core. The four arrows on the graph indicate the points from which the final leg is driven. The data points are the relative arrival times of the clock input to the conditional clock buffers sourced by the clock tree. The total simulated skew is 41 pS assuming maximum metal resistance.

C. Clock Switching

One additional complication related to the internal clock tree is that it is not always driven by the clock from the PLL, known as CCLK. During cache fills, the clock source for the internal sections of the chip switches over to MCLK so that the whole chip is running synchronous to the bus (Fig. 10). This simplifies fills and it reduces power since the bus clock is significantly slower than CCLK. Note that since this machine has a blocking cache, not much happens while waiting for a

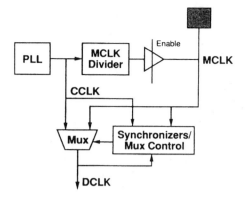

Fig. 10. Clock switching circuit.

cache fill. Therefore, running on the slower bus clock during fills has essentially no performance impact.

Since MCLK and CCLK might be asynchronous, switching the driver of DCLK quickly between the two clock sources is difficult. Careful attention must be paid to the synchronization of the Mux control signals to prevent glitch pulses on the clock during the transition between the clock sources.

Clock switching is only used during fills. Stores which miss in the cache and castouts are written to memory through the write buffer without switching the internal clock over to MCLK. The write buffer receives both DCLK and MCLK and passes the data for external stores across the DCLK/MCLK interface with proper attention to synchronization issues between the two clock regimes. One interesting characteristic of clock switching is that it gives the system designer another option to save power in situations for which the full performance of the chip is not required. By disabling clock switching on the fly, you can configure the chip to run off the bus clock. There is no limit on asymmetry or maximum pulse width of the bus clock, so the chip can be operated at very low frequencies if desired.

D. Conditional Clock Buffers

Conditional clock buffers are simple NAND/invert structures with an integral latch on the condition input. The buffers must be matched to their load to minimize skew. Since adding dummy clock loads is contrary to the low-power design philosophy, we created scaled clock buffers which would produce matched clocks for a wide range of loads and only needed to add dummy clock loads for a small number of very lightly loaded clock nodes. The task of matching the clock buffers to the load was greatly simplified by the fact the clock load presented by our standard latches is largely data-independent.

While the use of conditional clock buffers is central to the design method used on the chip, it should be noted that the critical paths to generate the condition input to these buffers represent some of the most difficult design problems in the chip. In this case, we decided that the power saving associated with the conditional clocking was worth the additional design effort and possible performance reduction.

E. Latch Circuits

The standard latches used in the design are differential edge-triggered latches (Fig. 11). The circuit structure is a precharged

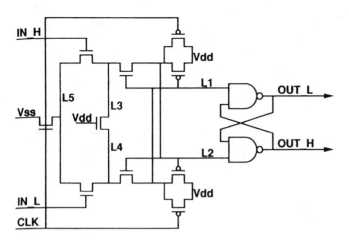

Fig. 11. Latch circuit.

differential sense amp followed by a pair of cross-coupled NAND gates. The sense amp need not be particularly well balanced because the inputs to the latch are full CMOS levels. The NMOS shorting device between nodes L3 and L4 provides a dc path to ground for leakage currents on nodes L1 and L2 in case the inputs to the latch switch after the latch evaluates. At normal operating frequencies, this device is not particularly important but it is required for the latch to be static. Note that since the dc current flowing is due only to device leakage, the magnitude of the current is insignificant to the power in normal operation.

F. Testability

The chip supports IEEE 1149.1 boundary scan for continuity testing. In addition, it has two hardware features to aid in manufacturing testing. The first is a bypass to allow CCLK to be driven from a pin synchronous to MCLK. This allows the tester to control the timing between CCLK and MCLK to make the asynchronous sections appear to be deterministic. The second test feature provides a linear feedback shift register (LFSR) that can be loaded with instruction data from the Icache. Loading the LFSR can be conditioned based on the value of address bit 2 and the Icache hit signal. The LFSR is loaded after the Fetch stage to allow the instruction following a branch to be read from the Icache and loaded into the LFSR. This feature allows any random pattern to be loaded into the Icache and then read out by alternating branch instructions with data patterns words.

VIII. POWER DISSIPATION RESULTS

A. Measured Results

Power dissipation data was collected on an evaluation board running Dhrystone 2.1 with the bus clock running at one-third of the PLL clock frequency. Dhrystone fits entirely in the internal caches so, after the first pass through the loop, pin activity is limited. This is the highest power case because cache misses cause the internal clocks to run at the bus speed and result in a lower total power. For both sets of measurements, external Vdd is fixed at 3.3 V. For an internal Vdd of 1.5 V, the total power is 2.1 mW/MHz. If the internal supply is set to 2.0 V, the total power is 3.3 mW/MHz. Note that the ratio

of the power at 1.5 and 2.0 V does not track Vdd^2 because it contains a component of external power and the external Vdd is fixed.

B. Simulated Power Dissipation by Section

An analysis of node transitions based on simulation was performed to estimate the power dissipation associated with the various major sections of the chip (Table III). Toggle information was collected based on 160 000 cycles of Dhrystone and combined with extracted node capacitances to estimate power dissipation by node and this data was further grouped by section. The clock power listed in Table III is due only to the global clock circuits.

A few points are worth noting.

- First, the power is dominated by the caches as you might expect given their size. This is despite our efforts to reduce their power through bank selection and other means. The Icache burns more power than the Dcache because it runs every cycle.
- Next, the PLL power is insignificant in normal operation. As was noted earlier, its low power characteristics are only important in Idle.
- Finally, since reduction in clock power was one of our explicit goals, it is interesting to consider the total clock power. If you extract the local clock power from the nonclock sections and sum it, you get a total clock power, including the global clock trees, the local clock buffers, and the local clock loads. This power is 25% of the total chip power, significantly less than the 65% consumed by the clocks in the Alpha microprocessor used in our initial feasibility studies.

Conditional clocking was an integral part of the design method, so it is difficult to determine the power saving associated with it. However, the power associated with driving the conditional clocks is 15% of the chip power and if the conditions on all the conditional clock buffers were always true, this power would quadruple. This does not account for the additional power savings that has been achieved by blocking spurious data transitions.

IX. CAD TOOLS

The CAD tools used on this chip were largely the same as those used on our Alpha designs [5]. This is not surprising

TABLE III
SIMULATED POWER DISSIPATION BY SECTION

ICACHE	27%
IBOX	18%
DCACHE	16%
CLOCK	10%
IMMU	9%
EBOX	8%
DMMU	8%
Write Buffer	2%
Bus Interface Unit	2%
PLL	< 1%

since the performance target of the chip roughly parallels that of the Alpha family as noted in Section VII. The most significant departure was in the area of static timing verification and race analysis where the adoption of edge-triggered latching required significant modifications to the tools used in the Alpha designs.

X. PROJECT ORGANIZATION

One of the challenging aspects of this project was geographical. The detailed design was performed at four sites across a nine hour time zone range. The initial feasibility work and architectural definition was done at Digital Semiconductor's design center in Austin with on-site participation by personnel from Advanced RISC Machines Limited (ARM). The implementation was more widely distributed with the caches, MMU's, write buffer, and bus interface unit at Digital Semiconductor's design center in Palo Alto, the instruction unit, execution unit, and clocks in Austin, the pad driver and ESD protection circuits at Digital Semiconductor's main facility in Hudson, MA, and the PLL at the CSEM design center in Neuchâtel, Switzerland. In addition, we consulted with Hudson for CAD and process issues, with ARM in Cambridge, England, for all manner of architectural issues and implementation tradeoffs associated with ARM® designs and with T. Lee from Stanford on the PLL. The implementation phase of the project took less than nine months with about 20 design engineers.

XI. CONCLUSION

The microprocessor described uses traditional high performance custom circuit design, an intentionally simple architectural design, and advanced CMOS process technology to produce a 160 MHz microprocessor which dissipates less than 450 mW. The internal supplies can vary from 1.5 to 2.2 V while the pin interface runs at 3.3 V. The chip implements the ARM® V4 instruction set and delivers 185 Dhrystone 2.1 MIPS at 160 MHz. The chip contains 2.5 million transistors and is fabricated in a 0.35-μm three-metal CMOS process. It measures 7.8 mm $\times$ 6.4 mm and is packaged in a 144-pin plastic thin quad flat pack (TQFP) package.

ACKNOWLEDGMENT

The authors would like to acknowledge the contributions of the following people:

F. Aires, M. Bazor, G. Cheney, K. Chui, M. Culbert, T. Daum, K. Fielding, J. Gee, J. Grodstein, L. Hall, J. Hancock, H. Horovitz, C. Houghton, L. Howarth, D. Jaggar, G. Joe, R. Kaye, J. Kapp, I. Kim, Y. Lou, S. Lum, D. Noorlag, L. O'Donnell, K. Patton, J. Reinschmidt, S. Roberts, A. Silveria, P. Skerry, D. Souyadalay, E. Supnet, L. Tran, D. Zoehrer, and the PLL design team at CSEM.

The support which they received on many aspects of the design from the people at Advanced RISC Machines, Ltd. was very important and keenly appreciated.

REFERENCES

[1] ARM Architecture Reference, Advanced RISC Machines, Ltd., Cambridge, U.K., 1995.
[2] P. Gronowski et al., "A 433 MHz 64 b quad-issue RISC microprocessor," in ISSCC Dig. Tech. Papers, Feb. 1996, pp. 222–223.
[3] D. Dobberpuhl et al., "A 200 MHz 64 b dual-issue CMOS microprocessor," IEEE J. Solid-State Circuits, vol. 27, no. 11, Nov. 1992.
[4] V. von Kaenel et al., "A 320 MHz, 1.5 mW CMOS PLL for microprocessor clock generation," in ISSCC Dig. Tech. Papers, Feb. 1996, pp. 132–133.
[5] T. F. Fox, "The design of high-performance microprocessors at Digital," in 31st ACM/IEEE Design Automation Conf., San Diego, CA, June 1994, pp. 586–591.

A 320 MHz, 1.5 mW @ 1.35 V CMOS PLL for Microprocessor Clock Generation

Vincent von Kaenel, *Member, IEEE*, Daniel Aebischer, Christian Piguet, and Evert Dijkstra

Abstract— This paper describes a low-power microprocessor clock generator based upon a phase-locked loop (PLL). This PLL is fully integrated onto a 2.2-million-transistors microprocessor in a 0.35-μm triple-metal CMOS process without the need for external components. It operates from a supply voltage down to 1 V at a VCO frequency of 320 MHz. The PLL power consumption is lower than 1.2 mW at 1.35 V for the same frequency. The maximum measured cycle-to-cycle jitter is ± 150 ps with a square wave superposed to the supply voltage with a peak-to-peak amplitude of 200 mV and rise/fall time of about 30 ps. The input frequency is 3.68 MHz and the PLL internal frequency ranges from 176 MHz up to 574 MHz, which correspond to a multiplication factor of about 100.

TABLE I
PROCESS PARAMETERS SUMMARY

0.35μm CMOS Parameters summary	
Feature size	0.35μm
Channel length	0.25μm
Threshold voltages	± 0.35V
Gate oxide	6.5nm
Substrate	p-epi with n-well

I. INTRODUCTION

THIS paper reports on the design of a phase-locked loop (PLL) for on-chip clock generation in a high performance microprocessor (μP) [1]. The μP is targeted for portable computing applications where power consumption must be minimized. The primary motivations for the PLL design are as follows.

First, the PLL is used as a frequency multiplier to generate a high-speed internal clock for the microprocessor from a low-frequency, low-cost, 3.68 MHz quartz oscillator. The multiplication factor is in the range of 100 X. Additionally, the microprocessor requires a precisely controlled 50% duty cycle clock. The PLL incorporates a common technique of generating a clock at twice the microprocessor clock frequency that is then divided by two to achieve an accurate 50% duty cycle. The integration of the PLL and divide-by-two circuitry on-chip eliminates the need to drive the high capacitance board level interconnects with high frequency signals and effectively lowers the power consumption of the system board.

Second, the PLL must allow fast recovery from the μP's idle (low power) mode to normal (full speed) mode. Two design approaches were considered. The first is a PLL that is disabled during idle mode. This approach minimizes power during idle mode but requires a low settling time for the PLL during the transition from idle to normal mode. Alternatively, a low-power PLL design that runs continuously during idle mode was considered. This approach allows instantaneous transition between the modes but requires the power consumption of the PLL to be minimized. This paper describes such a PLL.

Manuscript received May 1, 1996; revised August 5, 1996. This work has been carried out under a contract of Digital Equipment Corporation.

The authors are with CSEM, Swiss Center for Electronics and Microtechnology, Inc.. 2007 Neuchâtel, Switzerland.

Publisher Item Identifier S 0018-9200(96)08095-X.

Finally, when looking at target systems for the μP, we found no pressing need for phase locking. Consequently, we removed this as a design criteria and concentrated our efforts on minimizing clock cycle-to-cycle jitter.

The integration of the PLL onto a microprocessor complicates the design as the PLL is exposed to the microprocessor power supply switching noise. This noise induces output clock period variation (jitter). To obtain the highest performance from the microprocessor, the output jitter of the PLL has to be as low as possible. The power consumption of the microprocessor has been further reduced by scaling the supply voltage down to 1.35 V. The microprocessor and the PLL have been implemented in a 0.35-μm CMOS process which features low threshold MOS devices to maintain the speed performance at low supply voltage (Table I).

In summary, the challenge was to design a PLL with a high multiplication factor that combines limited output jitter, low supply voltage, and low power consumption.

II. SYSTEM TRADEOFFS

In our case, the major noise source for the output jitter is the power supply switching noise of the digital circuits (estimated to be 10% of the supply voltage, so the absolute minimum supply voltage is 1.2 V). This noise has a very large bandwidth, even larger than the μP clock frequency. So, the power supply noise rejection (PSNR) of analog circuits inside the PLL must be maximized.

For low-power PLL's, the second jitter source is the intrinsic noise of MOS devices in the VCO. This noise can be reduced by increasing the current flowing into sensitive devices, thus increasing the power consumption. On the other hand, to obtain low-voltage analog circuits, the saturation voltage of MOS devices needs to be reduced by using wider devices. This results in a larger parasitic capacitance between the supply voltage and the internal sensitive analog nodes, which might decrease the PSNR for the same current consumption. This paper describes how these contradictory effects have been

Reprinted from *IEEE Journal of Solid-State Circuits*, pp. 1715-1722, November 1996.

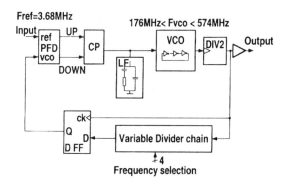

Fig. 1. PLL Architecture.

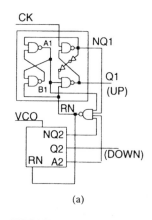

(a)

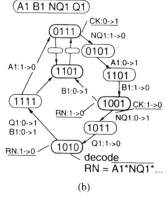

(b)

Fig. 2. Conventional NAND-Based PFD with flow graph.

taken into account to optimize the current consumption with a high PSNR at low-supply voltage [12].

The PLL architecture is depicted on Fig. 1. The reference frequency is delivered externally by a quartz oscillator at 3.68 MHz. This signal is the reference frequency for the phase and frequency detector (PFD). The feedback of internal clock is compared to the reference clock for phase and frequency error. The PFD generates Up or Down pulses to the charge pump circuit (CP) which is followed by the loop filter (LF). This filter stabilizes the PLL. The loop filter voltage controls the voltage controlled oscillator (VCO). The output frequency of the VCO is divided by two in order to deliver a 50% duty cycle clock signal. The output frequency of the VCO can be programmed from a minimum frequency of 176 MHz to a maximum frequency of 574 MHz. The divider ratio can be selected between 16 values ranging from 48 to 156.

III. PLL BLOCKS

A. Phase and Frequency Detector

The PFD uses a dual D-flip-flop (DFF) structure [2], [3]. As the phase difference between the inputs decreases, the pulse width on Up or Down also decreases. To avoid the appearance of a dead zone (range of phase difference where no PFD output is generated), it is usual to have a minimum pulse width on Up and Down even if there are no phase differences between the two inputs. These simultaneous Up and Down signals in the steady state of the PLL create a short circuit in the charge pump which results in a perturbation on the LF voltage, and hence produces jitter. To limit the LF voltage perturbation without having a dead zone, we have reduced the minimum Up and Down pulse widths by reducing the reset delay in the PFD. Moreover, the linearity of the PFD and charge pump is affected by this minimum pulse width on Up and Down when there is no phase difference between the inputs of the PFD. The new PFD has no dead zone and the nonlinearity near the steady state of the PLL is reduced.

Fig. 2(a) shows the conventional NAND-based PFD [10]. It is constructed with two flip-flops which are set when the clock transition goes low. Fig. 2(b) shows the flow graph of one of these flip-flops. The reset signal RN has to be generated as soon as the flip-flop reaches the stable state 1010 ($A1$, $B1$, $NQ1$, $Q1$) from the preceding unstable state 1011. The variable NQ, switching before Q (Up or Down), is used to

generate the reset signal. We consider now the steady state of the PLL (perfect synchronization between VCO and CK). Up ($Q1$) and Down ($Q2$) rise. Then $NQ1$ and $NQ2$ fall after a gate delay. As $A1$ and $A2$ are already active, the reset is decoded. The assertion of reset causes the Up and Down signals to transition low. So, the minimum pulse width of Up and Down is at least the delay three gates (two inputs NAND, four inputs NAND and three inputs NAND). These simultaneous pulses on Up and Down are a contradictory signal for the charge pump resulting in a perturbation on the loop filter voltage and finally in jitter. Moreover, due to the chosen state assignment, one has to use both internal variables NQ and A to generate properly the reset signal RN. If not, the state 0111 will generate a reset signal when it is not wanted. Neither NQ nor A alone can be used to generate the reset. It results in a NAND gate RN with four inputs ($NQ1$, $A1$, $NQ2$, $A2$). Such a gate is slow, as the reset is achieved with $RN: 1 \rightarrow 0$, through four MOS transistors in series. Designers of this PFD have inserted two inverters between the transitions of NQ and Q in such a way as to create a delay between NQ and Q due to this slow reset gate. The proposed design (Fig. 5) solves this problem while using a fast reset gate with only two MOS device in series.

The design of the new PFD has been performed using an asynchronous race-free design method. A detailed description of this design methodology is presented in [5] and [6]. A basic schematic of such a circuit is shown in Fig. 3. This circuit has the same basic function as the NAND-Based PFD, but the state assignment is different. Two D-flip-flops with $D = 1$ are clocked with the two clock signals which are compared. If CK is active before VCO, the Up output is generated while Down

397

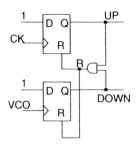

Fig. 3. PFD schematic.

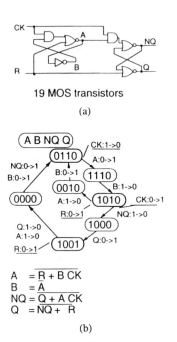

19 MOS transistors

(a)

$$A = \overline{R + B\,CK}$$
$$B = \overline{A}$$
$$NQ = \overline{Q + A\,CK}$$
$$Q = \overline{NQ + R}$$

(b)

Fig. 4. D-flip-flop with $D = 1$ with flow graph.

is produced if VCO is active before CK. As soon as the two clock signals are simultaneously active, the two D-flip-flops are reset. If CK and VCO are active at the same time, ideally neither Up nor Down has to be set. In such a case, the reset signal R has to be fast enough to minimize the Up and Down activation time when the PLL is in the steady state (perfect synchronization between CK and VCO).

Fig. 4 shows an implementation of a D-flip-flop with $D = 1$ having a noncritical race. The idea is to minimize the simultaneous activation time of Up and Down. As both A and Q have to go Down, while B and NQ have to switch on, both A and Q can switch simultaneously when the reset is active, producing the simultaneous switching on of B and NQ. As two variables switch simultaneously (A and Q between states 1001 and 0000, 1000 or 0001 are possible intermediate states), there is a race. However, the final stable state (0110) does not depend on the order of switching of A and Q. Hence, the result of the race is always the same (noncritical race) and is layout independent (it does not depend on the value of parasitic capacitances on internal nodes nor of the size of MOS devices).

Based on this D-flip-flop implementation, Fig. 5 shows the phase frequency detector accordingly to Fig. 3. To speed Up the reset even more, a NOR gate producing the reset signal R is controlled by the NQ signals of the D-flip-flops. These NQ signals switch before the Q (Up or Down) outputs between

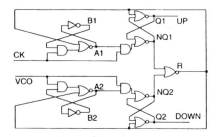

Fig. 5. Phase and frequency detector.

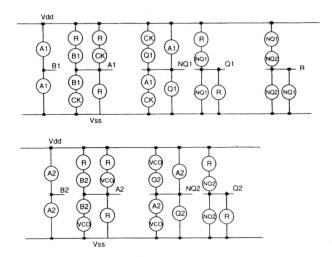

Fig. 6. PFD MOS circuit (MOS devices represented by circles. NMOS devices are between internal nodes (A1, A2, B1, B2, . . .) and VSS, PMOS are between internal nodes and VDD).

states 1010 and 1001 (see Fig. 4). As soon as the NQ signal is low, the NOR gate generates a reset signal for the D-flip-flops. As the Up or Down signal is generated after $NQ1$ or $NQ2$, the minimum pulse width on Up or Down ($Q1, Q2$) is the delay of only one gate (two inputs NOR). By a careful sizing of the Q NOR gate and the reset gate, it is possible to have a very low pulse width on Up and Down in the steady state of the PLL. This results in a lower jitter and better linearity near the steady state without dead zone.

Each D-flip-flop contains 19 transistors in a branch-based implementation [8], [9]. Fig. 6 depicts the PFD MOS circuit with 42 transistors. The delay has been reduced compared to the conventional NAND-based PFD circuit [10] due to the fact that the number of serial transistors in branches is limited to two for all gates. It is possible to invert the MOS schematic of Fig. 6 in order to have a NAND gate based PFD. However, no significant speed advantage has been observed between these two version since the maximum number of PMOS in series in each branch remains the same. A careful analysis has been performed on the complete PFD circuit of Fig. 5. This design requires that both flip-flops are simultaneously reset. If one flop-flop resets early it can result in the reset signal returning to $R = 0$ before the second flip-flop has reset. Such a behavior is inherent to the basic structure of Fig. 3 and can be controlled with careful matching of the two flip-flops

B. Charge Pump

The charge pump circuit (CP) uses a switch structure that cancels the charge injection by using dummy devices (Fig. 7) [6]. Even with low-threshold voltage MOS devices, the static

Jitter contributor without supply noise	Jitter (p-p)	unit
White Noise in VCO	±30	ps
Dead zone of PFD	<±10	ps
Leakage on LF and Charge injection	±15	ps
Total Jitter without supply noise	±55	ps
Jitter due to a 0.2V supply jump in 30ps		
VCO induced jitter	±80	ps
Jitter induced by the change of the LF voltage	±10	ps
Total Jitter due to a 0.2V supply jump	±90	ps
Jitter due to a 10mV substrate jump in 30ps		
VCO induced jitter	<±5	ps
Total jitter due to a 10mV substrate jump	±5	ps
Total Jitter(sum of the above contributors)	**±150**	**ps**

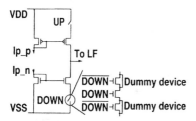

Fig. 7. Charge pump and switch detail.

phase error due to channel leakage current is lower than 4 mrad (2 ps @ 320 MHz). This jitter source is negligible compared to the supply noise contribution (Table II). The specified lock-time (100 μs) made it possible to choose a low PLL natural frequency (100 kHz), thus reducing the ripple voltage on the LF, and hence the output jitter. However, in order that the PLL compensates for the $1/f$ noise in the MOS devices of the VCO, the natural frequency of the loop should not be lower than 100 kHz. The LF realizes a zero and two poles, which results in a third-order system [2]–[4].

C. VCO

As far as the PSNR and low supply voltage operation are concerned, the VCO is the most critical block because its internal noise results directly in jitter. Moreover, low-voltage operation limits the design options. To obtain a fully integrated PLL, a current controlled ring oscillator (CCO) is the basic element of the VCO (Fig. 8). It allows high-frequency operation at low-voltage operation, since no additional capacitances have been used on internal nodes other than the ones created by the inverter devices.

The design procedure for the CCO is described in the following lines. The CCO MOS devices are in the strong inversion region and

$$\beta_n = \beta_p = \beta$$
$$n_n = n_p = n$$
$$V_{Tp} = V_{Tn} = V_T$$
$$I_D = \frac{\beta}{2n}(V_{gs} - V_T)^2 \qquad (1)$$

where β is the gain of the MOS device, n the weak inversion slope, V_T the threshold voltage, and I_D the drain current in

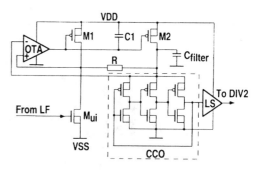

Fig. 8. VCO detailed schematic.

saturation. As this is a low-power, low-voltage CCO, the MOS device are not in velocity saturation. Assuming that only one inverter is active at a time

$$\frac{I_{cco}}{\beta} \propto (V_{cco} - V_T)^2. \qquad (2)$$

The minimum voltage across the CCO is obtained when I_{cco}/β is minimum for a given process. The β is maximum when the length of the MOS device is minimum for a given width. The CCO current can be calculated from the well known formula of consumption in digital circuits $I_{cco} = fC_{eq}V_{cco}$ (3). In our case, the gate and the drain capacitance are much larger than the interconnect capacitance, therefore the equivalent capacitance on each internal node of the CCO C_{eq} is proportional to the width of MOS device, hence proportional to β for a minimum length MOS device. The CCO current over β is also

$$\frac{I_{cco}}{\beta} \propto fV_{cco}. \qquad (3)$$

If the CCO voltage is much larger than the threshold voltage, from (2) and (3) we can say that (for a given β)

$$f \propto \sqrt{I_{cco}} \propto V_{cco}. \qquad (4)$$

Equations (3) and (4) are not sufficient to calculate the CCO MOS devices β value for a given frequency and process. The white noise (thermal noise in MOS devices) affects the cycle-to-cycle jitter of the VCO through two major contributions. First, the voltage V_{cco} across the CCO collects noise coming from the voltage to current converter, assuming the CCO with C_{filter} as an equivalent parallel RC load. This causes a rather low frequency modulation to the VCO period. The second source of jitter results from the CCO internal charging and discharging time uncertainty of the node capacitors by the noisy MOS currents.

To limit the jitter due to the MOS device channel white noise (thermal noise), the current flowing in the CCO must be sufficiently high. The output period variance σ_T decreases when the current flowing in the CCO increases [11]

$$I_{cco} \propto \left(\frac{1}{\sigma_T^2}\right)^a. \qquad (5)$$

Allowing a limited amount of jitter (due to thermal noise) in the CCO and given an operating frequency, the minimum current flowing in the CCO can be calculated. Then, using (2) and (3) the β can be derived. To have a minimum CCO

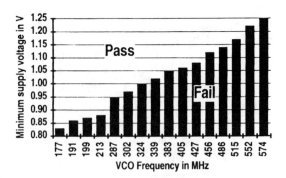

Fig. 9. Measured minimum PLL supply voltage.

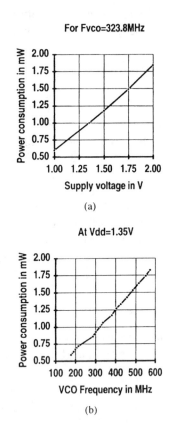

For Fvco=323.8MHz

(a)

At Vdd=1.35V

(b)

Fig. 10. Measured PLL power consumption.

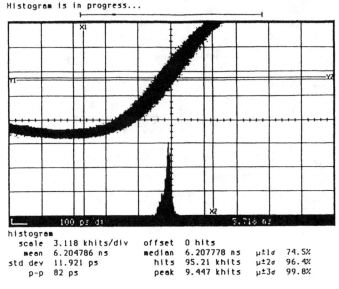

histogram
scale   3.118 khits/div   offset   0 hits
mean   6.204786 ns   median   6.207778 ns   μ±1σ   74.5%
std dev   11.921 ps   hits   95.21 khits   μ±2σ   96.4%
p-p   82 ps   peak   9.447 khits   μ±3σ   99.8%

Fig. 11. Measured jitter without supply noise.

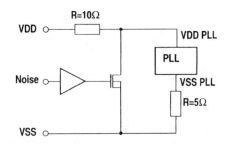

Fig. 12. Integrated supply noise generator.

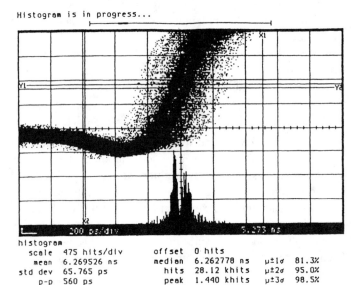

histogram
scale   475 hits/div   offset   0 hits
mean   6.269526 ns   median   6.262778 ns   μ±1σ   81.3%
std dev   65.765 ps   hits   28.12 khits   μ±2σ   95.0%
p-p   560 ps   peak   1.440 khits   μ±3σ   98.5%

Fig. 13. Measured jitter with supply noise (436 mV in 30 ps).

voltage in a given process, a minimum length for MOS devices should be chosen and no additional capacitances should be added on internal nodes of the CCO (C_{eq} minimum).

Then, as the supply voltage is specified, the maximum saturation voltage of the mirror supplying the current to the CCO is determined. It is not possible to use a cascoded current mirror for operation as low as 1.2 V because the saturation voltage of such a mirror is too high for the expected PSNR. Therefore, to ensure a maximum PSNR even at low supply voltages, a new circuit called active cascode is used (Fig. 8). Two characteristics linked to the VCO sensitivity to supply noise are interesting: first, the sensitivity to very low frequency or dc variations on the supply voltage; second, the sensitivity to high frequency variation on the supply voltage. The VCO dc sensitivity to supply voltage is low (1 ps per 25 mV of supply variation for $f_{vco} = 320$ MHz by worst case simulation at a supply voltage of 1.2 V and 1 ps per at least 50 mV in the same condition has been measured) due to the

feedback loop, with an operational transconductance amplifier (OTA) that maintains the voltage equality across the mirror devices ($M1$ and $M2$). The high frequency PSNR is mainly determined by the ratio of the drain parasitic capacitance of the mirror MOS devices and the filtering capacitor (C_{filter}) across the CCO. This last capacitor cannot be increased too much because it introduces a pole in the PLL which can make it unstable. The size of the mirror devices is determined by the maximum allowable saturation voltage. The simulated and

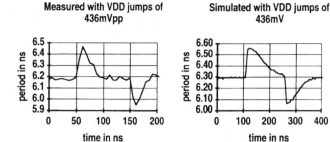

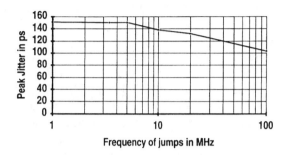

Fig. 14. Cycle-to-cycle period measurements and simulation.

Fig. 15. Measured jitter with supply noise (200 mV).

TABLE IV
PLL PERFORMANCE SUMMARY FOR $F_{\text{vco}} = 320$ MHz

	Measured	Calculated
Supply voltage range	1.0 - 2.2V	1.2 - 2.2V
Power consumption (V_{DD}=1.5V)	1.2mW	<1.5mW
VCO period sensitivity to supply voltage ($V_{DD min}$=1.2V)	1ps/52mV	<1ps/25mV
Jitter without supply noise	±41ps	±55ps
Jitter with V_{DD} jumps of 200mVpp, 30ps edge time ($V_{DD min}$=1.2V)	±150ps	±150ps
Settling time	60µs	<100µs
Area	0.21mm^2	

TABLE III
SIMULATED MAXIMUM CURRENT CONSUMPTION AT 320 MHz AND 1.35 V

Block	I	unit
VCO, current reference	800	µA
Divider chain	240	µA
Divider by two	50	µA
Charge Pump	10	µA
Phase and Frequency Detector	2	µA
Total	1102	µA

measured high frequency PSNR is 1 ps per 2.5 mV of supply variation in closed loop for a VCO output frequency of 320 MHz and with a supply voltage of 1.2 V. When the supply voltage is higher, the PSNR is also higher, the given values are worst case. The stability of this active cascode is ensured by the resistor R between the input and output of the mirror. The relation between current and frequency in this CCO is not linear. To a first approximation, the frequency depends on the square root of the current for a voltage across the CCO much larger than the maximum threshold voltage of P or N-MOS devices (4). To obtain a linear gain for the VCO, the voltage to current converter (V/I) should have a quadratic transfer function. This is achieved by using a M_{ui} in the strong inversion region (Fig. 8).

The level shifter (LS) provides a full signal swing to the first divide-by-two circuit to generate a 50% duty cycle signal for the μP.

D. Dividers

In order to reduce the power consumption, the divider chain has been realized with asynchronous dividers (or ripple counters). The jitter introduced by these dividers is cancelled by a D-flip-flop that resynchronizes the output signal of the dividers with the output frequency of the PLL.

The divider ratio can be selected between 16 values ranging from 48 to 156. In order to have low-power programmable

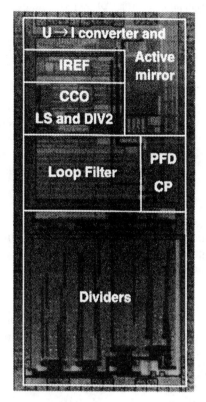

Fig. 16. PLL micrograph.

dividers, we have implemented 16 independent ripple counters that can be selected by a multiplexor and a demultiplexor. The area of these 16 dividers is larger than the area of a single programmable divider, however, the power consumption is lower because of the simplicity of the ripple counter.

IV. MEASUREMENT RESULTS

The minimum operating supply voltage versus the VCO frequency is represented on Fig. 9. For these measurements, the process of the measured PLL was approximately slow for both N and P-MOS devices. So this is a worst case. For the nominal VCO frequency (320 MHz) the minimum supply voltage is 1 V which is well below the minimum specified operating voltage of 1.2 V.

The power consumption versus supply voltage for the nominal frequency (320 MHz) is depicted in Fig. 10(a). At the nominal supply voltage (1.35 V), the power consumption is lower than 1.2 mW. Fig. 10(b) shows the power consumption as a function of the VCO frequency with a supply voltage of 1.35 V. At the maximum frequency, which is 574 MHz, the power consumption is lower than 2 mW.

The jitter without supply noise has been measured and plotted on Fig. 11 for the nominal frequency and supply voltage. This measurements gives ±41 ps of cycle-to-cycle jitter (PLL in closed loop). The worst case calculated value is ±55 ps.

The noise spectrum generated by the digital part is related to the internal speed of the process. Hence, as a model for this noise, we have used a supply voltage square wave with rise and fall times given by the maximum speed of the process and an amplitude given by the simulations of the package response. In our case, we have assumed 200 mV peak-to-peak amplitude of noise and rise and fall times of 30 ps (simulated). This noise model is really more severe than an externally generated sine or square wave on the supply voltage up to the frequency of the VCO because of the wide noise spectrum generated. As it is not possible to generate externally such a noisy supply voltage, we have integrated a noise generator to characterize the PSNR of this PLL. Fig. 12 shows the integrated supply noise generator. A switch controlled by an external signal induces a voltage drop across a serial resistor with the PLL supply voltage. When the switch is off, the serial resistance produce a negligible voltage drop because of the low-power consumption of the PLL. The simulated rise and fall time of the square wave produced by this generator is 30 ps, and in any case related to the maximum speed of the process. These rise and fall times are difficult to measure. However, with an oscilloscope we have observed rise and fall times lower than 1 ns (by on-chip probing) which is lower than one cycle of the CCO. One advantage of this noise model is that it can be used in simulation during the design. The serial resistor and the width of the switch has been designed to produce a voltage jump of at least 200 mV. In our case of process, the voltage drop was around 400 mV. Then, by linear interpolation between the measured jitter without supply noise and the measured jitter with supply noise, we have calculated the jitter for a jump of 200 mV. Improvement in the noise generator made it possible to electrically adjust the amplitude of the square wave to 200 mV. The jitter measurements with these improved test structure show that the linear interpolation gives accurate results. Fig. 13 shows the measured jitter with a repetition frequency of jumps of 1 MHz asynchronous to the reference frequency and to the output frequency, an amplitude of jumps of 436 mV (the supply voltage jumps between 1.2 V and 1.636 V), and for the nominal VCO frequency of 320 MHz. The measured jitter with the PLL in closed loop is ±280 ps which corresponds to ±150 ps for supply voltage jumps of 200 mV. The two peaks on the histogram of Fig. 13 are not due to a high sensitivity to dc supply voltage, but due to the shape of the excitation waveform (noise generator) used on the supply voltage (square wave).

To closely analyze the output jitter due to supply noise, we have used a high sampling rate single-shot oscilloscope (8 GS/s) and a dedicated program that calculates each PLL output period duration by interpolation between the sampled points. The accuracy of this analysis can be as good as ±1 ps RMS [14]. Fig. 14(a) shows the period as function of the time during a period of a square wave on the supply voltage. Fig. 14(b) shows the simulated response of the PLL to the same supply voltage noise (including all parasitic coupling capacitances extracted from layout). The simulations and the measurements results are close. So by this simulation method it is possible to have an accurate prediction of the measured PSNR of the PLL in closed loop.

The measured jitter for jumps of 200 mV as a function of the repetition frequency of the jumps is depicted on Fig. 15. It is important to note that Fig. 15 is not a Bode plot of the noise response of the PLL because the input signal is a square wave superposed on the supply voltage and the output signal is the jitter. The maximum measured jitter for the nominal output frequency of 160 MHz (nominal VCO frequency of 320 MHz divided by two) with a minimum supply voltage of 1.2 V is ±150 ps.

The current consumption of each subblock is presented in Table III. The PLL area is 0.21 mm^2 (Fig. 16). The temperature range is from 0°C to 100°C. Even with a better process (0.18-μm CMOS) and for a VCO operating frequency of 200 MHz, the lowest power consumption reported so far is 2 mW for a comparable jitter without supply noise [13]. At the same VCO operating frequency of 200 MHz, our PLL consumes only 1 mW [15]. The performances summary is presented in Table IV.

V. CONCLUSION

The designed circuit shows that it is possible to overcome the issue of PLL short settling time by using a very low power PLL generating the clock signal even in idle mode. So, the recovery time from idle mode to the normal mode is virtually zero.

The resulting PLL power consumption is lower than 1.2 mW @ 1.35 V. The measured output cycle-to-cycle jitter with a square wave of 200 mVpp with 30 ps rise and fall time as a supply voltage noise is ±150 ps. The jitter measured with the integrated noise generator is really close to the simulation result.

The measured PLL minimum supply voltage for a VCO frequency of 320 MHz is 1.0 V with a slow process corner.

The association of very low operating supply voltage and high PSNR has been achieved by using the active cascode in the VCO. The low power consumption has been achieved by a careful analysis of the jitter created by thermal noise in the VCO and choosing the minimum VCO current that allows to reach our goal.

This PLL is integrated on a microprocessor [1] that is now in full production. The PLL met all the design specifications.

ACKNOWLEDGMENT

The collaboration of DEC's StrongArm design team and T. Lee of Stanford University has been extremely valuable.

REFERENCES

[1] James Montanaro *et al.*, "A 160 MHz 32b 0.5 W CMOS RISC microprocessor," in *ISSCC'96*, Feb. 8–10, 1996, San Francisco CA, pp. 214–215.
[2] F. M. Gardner, *Phaselock Techniques* 2nd ed.   New York: Wiley, 1979.
[3] ——, "Charge-pump phase-lock loops," *IEEE Trans. Commun.*, vol. COM-28, no. 11, pp. 1849–1858, Nov. 1980.
[4] M. van Pamel, "Analysis of a charge pump PLL's," *IEEE Trans. Commun.*, vol. 42, no. 7, pp. 2490–2498, July 1994.

[5] C. Piguet, "Logic synthesis of race-free asynchronous CMOS circuits," *IEEE J. Solid-State Circuits*, vol. 26, no. 3, pp. 271–380, Mar. 1991.

[6] Vittoz, E. Dijkstra, Shiels, Eds., *Low Power Design: A Collection of CSEM Papers*. Cleveland, OH: Electronic Design Books, A Division of Penton, 1995.

[7] T. Ibaraki and S. Muroga, "Synthesis of networks with a minimal number of negative gates," *IEEE Trans. Comput.*, vol. C-20, no. 1, Jan. 1971.

[8] J.-M. Masgonty et al., "Technology- and power-supply-independent cell library," in *IEEE CICC'91*, May 12–15, 1991, San Diego, CA, USA, Conf. 25.5

[9] C. Piguet et al., "Low-power low-voltage digital CMOS cell design," in *Proc. PATMOS'94*, Oct. 17–19, 1994 Barcelona, Spain, pp. 132–139.

[10] I. A Young et al., "A PLL clock generator with 5 to 100 MHz of lock range for microprocessor," *IEEE J. Solid-State Circuits*, vol. 27, no. 11, pp. 1599–1607, Nov. 1992.

[11] J. A. Mcneil, "Jitter in ring oscillators," Ph.D. dissertation, Boston College of Engineering, 1994

[12] E. A. Vittoz, "Low power design: Ways to approach the limits," Plenary address ISSCC 1994, San Francisco, CA.

[13] M. Mizuno et al., "A 0.18 μm CMOS hot-standby PLL using a noise-immune adaptive gain voltage-controlled oscillator," presented at *ISSCC'95*, Feb. 15–17 1995, San Francisco CA.

[14] M. K. Williams, "A discussion of methods for measuring low-amplitude jitter," presented at *Int. Test Conf.*, Washington DC, 1995.

[15] V. von Kaenel et al., "A 320 MHz, 1.5 mW @ 1.35 V CMOS PLL for microprocessor clock generation," in *ISSCC'96*, Feb. 8–10, 1996, San Francisco CA, pp. 132–133.

A Low-Power Chipset for a Portable Multimedia I/O Terminal

Anantha P. Chandrakasan, Andrew Burstein, and Robert W. Brodersen, *Fellow, IEEE*

Abstract— This paper presents the design of a low-power chipset for a portable multimedia terminal that supports pen input, speech I/O, text/graphics output, and one-way full-motion video. Its power consumption was minimized using an approach that involves optimization at all levels of the design, including extended voltage scaling, reduced swing logic, and switched capacitance reduction through operation reduction, choice of number representation, exploitation of signal correlations, self-timing to eliminate glitching, logic design, circuit design, and physical design. The entire chipset, which performs protocol conversion, synchronization, error correction, packetization, buffering, video decompression, and D/A conversion, is implemented in 1.2 μm CMOS and operates from a 1.1 V supply while consuming less than 5 mW.

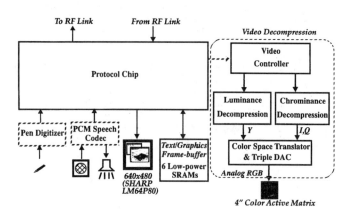

Fig. 1. Overview of the InfoPad Multimedia Terminal.

I. INTRODUCTION

THE NEAR future will bring the fusion of three rapidly emerging technologies: personal communications, portable computing, and high bandwidth communications. Over the past several years, the number of personal communications services and technologies has grown explosively. In portable computing, "notebook" computers more powerful than the desktop systems of a few years ago are commonplace. Wide area high bandwidth networks that are in the planning stages will provide the backbone to interconnect multimedia information servers. Wireless communications through radio and IR links are being used as LAN replacement as well as to provide ubiquitous audio communications. However, there has yet to be an integration of these diverse services, in part because of the difficulty in providing a portable terminal that can process the high speed multimedia data provided by the network servers. The circuitry to support the wireless link for such a terminal is being investigated and a low-power solution to this problem is felt to be feasible [1]. The focus of this paper is the processing between this wireless modem and the terminal I/O devices with emphasis on how this processing can be realized with minimal possible power consumption.

II. SYSTEM OVERVIEW OF THE PORTABLE MULTIMEDIA TERMINAL

The portable multimedia terminal described here provides untethered access to fixed multimedia information servers. The

Manuscript received May 16, 1994; revised August 7, 1994. This project was supported by ARPA. The work of A. Chandrakasan was supported by a Fellowship from IBM.

A. P. Chandrakasan is with the Department of Electrical Engineering, Massachusetts Institute of Technology, Cambridge, MA 02139 USA.

A. Burstein and R. W. Brodersen are with the EECS Department, Cory Hall, University of California, Berkeley, CA 94720 USA.

IEEE Log Number 9406261.

terminal is designed to transmit audio and pen input from the user to the network on a wireless uplink and to receive audio, text/graphics and compressed video from the backbone on the downlink. The portability requirement forces a design focus on power reduction. The availability of communications between the terminal and computational resources on the wired network provides a major degree of freedom for optimizing power; i.e. any computation that does not have to be performed on the terminal, including both general purpose applications and certain I/O tasks such as speech and handwriting recognition, can be moved to the backbone network.

Since the general purpose applications are performed by compute servers on the backbone network, the terminal electronics only has to provide the interface to I/O devices, as shown in Fig. 1. Six chips (a protocol chip, a bank of six 64 kbit SRAM's for text/graphics frame-buffering, and four custom chips for the video decompression) provide the interface between a high speed digital radio modem and a commercial speech codec, pen input circuitry, and LCD panels for text/graphics and full-motion video display. The chips provide protocol conversion, synchronization, error correction, packetization, buffering, video decompression, and digital-to-analog conversion. Through extensive optimization for power reduction, the total power consumption is less than 5 mW.

The protocol chip communicates between the various I/O devices in the system. On the uplink, 4 kbps digitized pen data and 64 kbps, μ-law encoded speech data are buffered using FIFO's, arbitrated and multiplexed, packetized, and transmitted in a serial format to the radio modem. On the down link, serial data from the radio at a 1 Mbs rate is depacketized and demultiplexed, the header information (containing critical information such as data type and length) is

Reprinted from *IEEE Journal of Solid-State Circuits*, pp. 1415–1428, December 1994.

error corrected and transferred through FIFO's to one of the three output processing modules: speech, text/graphics, and video decompression.

The speech module reads parallel data from a FIFO and transmits it in a serial format to a codec at a rate of 64 kbs. The text/graphics module handles the protocol used to transmit the bitmaps for text/graphics. The text/graphics module corrects error sensitive information using a Hamming (7, 4, 1) code, and generates the control, timing and buffering for conversion of the 32 bit wide data from the text/graphics frame-buffer (implemented using six low-power 64 kbit SRAM chips) to the 4 bits at 3 MHz required by a 640×480, passive LCD display. The final output module is the video decompression, which is realized using four chips. The decompression algorithm is vector quantization, which involves memory lookup operations from a codebook of 256 4×4 pixel patterns. Compressed YIQ video is buffered using a ping-pong scheme (one pair of memories for Y and one pair for IQ), providing an asynchronous interface to the radio modem and providing immunity against bursty errors. The amount of RAM required is reduced over conventional frame-buffer schemes by a factor of 24 by storing the video in compressed format. The YIQ decompressed data is sent to another chip that converts this data to digital RGB and then to analog form using a triple DAC that can directly drive a 4 inch active matrix color LCD display. A controller chip performs the video control functions, including the synchronization of the various chips and the color LCD display, control of the ping-pong memories, and loading of the code-books. It uses an addressing scheme that eliminates the need for an output line buffer.

III. LOW-POWER IMPLEMENTATION OF THE I/O PROCESSING MODULES

The power consumption in CMOS circuits is primarily attributed to charging of load capacitors and is given by:

$$P_{\text{switching}} = \alpha_{0 \to 1} C_L V_{dd}^2 f_{\text{clk}} \qquad (1)$$

where f_{clk} is the clock frequency, V_{dd} is the operating supply voltage, C_L is output load capacitance, and $\alpha_{0 \to 1}$ is the node transition activity factor, the fraction of the time the node makes a power consuming transition (i.e., a $0 \to 1$ transition) inside the clock period. For example, Fig. 2 shows the activity factor, $\alpha_{0 \to 1}$, for a 2-input NOR gate as a function of the input signal probabilities, P_a and P_b,

$$\alpha_{0 \to 1} = P_0 P_1 = (1 - (1 - P_a)(1 - P_b))(1 - P_a)(1 - P_b). \qquad (2)$$

From this plot, it is clear that understanding the signal statistics and their impact on switching events can be used to significantly reduce the power dissipation. Gate-level power analysis tools have been developed that use probabilistic approaches to estimate the internal node activities of a network given the distribution of the input signals [2],[3]. This chipset exploited signal statistics to reduce the total number of transitions. The physical capacitance is reduced by utilizing a low-power cell library that features minimum sized transistors and optimized layout. In addition to minimizing the switched capacitance, an architecture driven voltage scaling strategy is

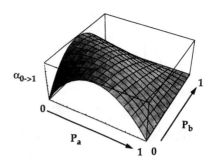

Fig. 2. Influence of signal statistics on switching activity: Transition probability for 2-input NOR gate.

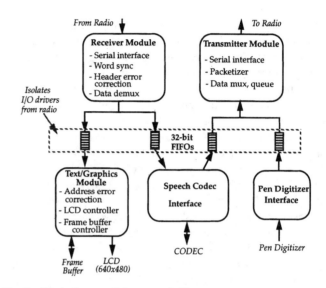

Fig. 3. Block diagram of the protocol chip.

used to scale the supply voltage to the 1–1.5 V range while meeting the computational throughput requirements. Since the circuits operate at a supply voltage that is less than the sum of the NMOS and PMOS threshold voltages ($V_{tn} = 0.7$, $V_{tp} = -0.9$), the devices can never conduct simultaneously for any possible input voltage, *eliminating* short-circuit power. Also, due to the high values of the threshold voltage, the leakage power is negligible.

A. Protocol Module

The protocol chip, shown in Fig. 3, communicates between the various I/O devices in the system. It provides the interface between a commercial radio modem and the pen digitizer, the speech codec, and the 640×480 text/graphics LCD display. It also controls the custom frame-buffer of the text/graphics display and provides the necessary interface between the frame-buffer and the LCD display. The wireless modem is isolated from the I/O processing modules using 32-bit asynchronous FIFO's.

1) Parallelism Enables Operation in the 1–1.5 V Range: Since power is proportional to the square of the supply voltage, it is only necessary to reduce the supply voltage for a *quadratic* improvement in the energy consumption. Unfortunately, this simple solution to low power design comes at the cost of increased gate delays and therefore lower

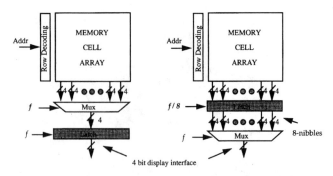

Fig. 4. Parallel memory access enables low-voltage operation.

functional throughput. As presented in [4], an architecture driven voltage scaling strategy using parallelism and pipelining to compensate for the increased gate delays at reduced supply voltages can maintain functional throughput. Parallelism and pipelining were used extensively in this chipset for both arithmetic computation and memory access, enabling supply voltage scaling into the 1–1.5 V range.

To illustrate this strategy, consider the example of reading black and white pixel data from the text/graphics frame-buffer to the 640 × 480 LCD display. The LCD display is a split-panel display in which the top half (640 × 240) and bottom half are addressed in parallel. Each half requires 4 bits (or 4-pixel values) at a rate of 3 MHz. Fig. 4 shows two alternate schemes for reading the required 4 bits of data from memory at the throughput rate, f. On the left side is the serial access scheme in which the 4 bits of data are read out in a serial format and the memory is clocked at the throughput rate f. For this single-ported SRAM frame-buffer implementation (in which the read/write data and address busses are shared) utilizing a serial access scheme requires a supply voltage of 3 V. Another approach (shown on the right side) is to exploit the sequential data access pattern for the frame-buffer and access several 4-bit nibbles in parallel, allowing the memory to be clocked at a lower rate. For example, reading eight nibbles in parallel from the memory allows the memory to be clocked at $f/8$ without loss in throughput. The latched data (8 nibbles) in the protocol chip is then multiplexed out at the throughput rate. This implies that the time available for the memory read operation for the parallel implementation is 8 times longer than the serial scheme and therefore the supply voltage can be dropped for a fixed throughput. The parallel version can run at a supply voltage of 1.1 V (which corresponds to the voltage at which the gate delays increase by a factor of 8 relative to the serial access scheme running at 3 V) while meeting throughput requirements. It is interesting that architectural techniques can be used to drive the supply voltage to such low levels even with a process that has $V_{tn} = 0.7$ V and $V_{tp} = -0.9$ V. If the process could be modified to reduce the threshold voltage, the power supply voltage and therefore the power consumption can be further reduced. An optimum threshold voltage compromises between switching power and leakage power and was found to be around 0.3 V [5].

In general, for signal processing applications that do not have feedback in the computation, arbitrary amount of parallelism can be exploited to reduce the power supply voltage.

However, exploiting parallelism often comes at the expense of increased silicon area (for example, there is an increase in the number of I/O pins for the SRAM chip going from a serial memory architecture to a parallel memory access architecture) and capacitance. The increase in capacitance must be traded off against lower voltage to arrive at an optimum level of parallelism and voltage [4]. For DSP applications that have feedback in the computation, algorithmic transformations are required to exploit parallelism and reduce power consumption [6]. Architecture driven voltage scaling can also be applied to general purpose computation (by using superscalar or super-pipelining architectures) but the amount of parallelism in the computation is often limited [7].

2) Reduced Swing Circuitry: Reducing the power supply voltage is clearly a very effective way to reduce the energy per operation since it has a quadratic impact on the power consumption. At a give supply voltage, the output of CMOS logic gates make rail to rail transitions; an approach to reducing the power consumption further is to reduce the voltage swing on large capacitance nodes. For example, using a NMOS device to pull up a node will limit the swing to $V_{dd} - V_t$, rather than rising all the way to the supply voltage. The power consumed for a $0 \rightarrow V_{dd} - V_t$ transition will be $C_L V_{dd}(V_{dd} - V_t)$, and therefore the power consumption reduction over a rail to rail scheme will be $\propto V_{dd}/(V_{dd} - V_t)$. This scheme of using an NMOS device to reduce the swing has two important negative consequences: first, the noise margin for output high (NM_H) is reduced by the amount V_T, which can reduce the margin to 0 V if the supply voltage is set near the sum of the thresholds. Second, since the output does not rise to the upper rail, a static gate connected to the output can consume static power for a high output voltage (since the PMOS of the next stage will be "on"), increasing the effective energy per transition.

Therefore, to utilize the voltage swing reduction, special gates are needed to restore the noise margin to the signal, and eliminate short-circuit currents. These gates require additional devices that will contribute extra parasitic capacitances. Fig. 5 shows a simplifed schematic of such a gate, used in the FIFO memory cells of the low-power cell library used in this chipset [8]. This circuit uses a precharged scheme that clips the voltage of the bit-line (which has several transistors similar to M5 connected to it) to $V_{dd} - V_t$ where $V_t > V_{t0}$ due to the body effect. The devices M1 and M4 precharge the internal node, $\overline{\text{out}}$, to V_{dd}, and the bit-line to $V_{dd} - V_T$. During evaluation ($\varphi =$"1"), if V_{in} is high, the bit-line will begin to drop, as shown in the SPICE output next to the schematic. Because the capacitance ratio of the bit-line to the internal node is very large, once the bit-line has dropped roughly 200 mV to sufficiently turn on M3, the internal node quickly drops to the potential of the bit-line, providing signal amplification. Thus, this circuit conserves energy greatly by reducing the voltage swing on the high-capacitance bit-line, and reduces delay by providing signal amplification.

3) Level-Conversion Circuitry to Interface with I/O Devices: Although the protocol chip operates at a supply voltage of 1.1 V, it has to communicate with I/O devices running at higher voltages—for example and text/graphics 640 × 480

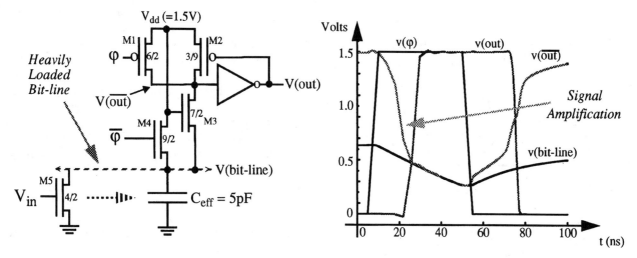

Fig. 5. Signal swing reduction for memory circuits: FIFO example.

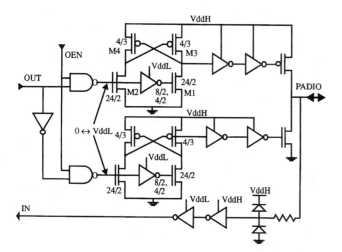

Fig. 6. Level-conversion I/O pad buffer.

LCD display running at 5 V and the speech codec running at 5 V. Level-conversion I/O pads convert the low voltage-swing signals from the core of the low-power chips (e.g., 1.5 V) to the high-voltage signal swings required by I/O devices or vice-versa. Fig. 6 shows the schematic of a level converting I/O pad driver.

This tri-state output buffer uses two supply voltages: the low-voltage supply, V_{ddL}, that is tied to the pad-ring and the high-voltage supply, V_{ddH}, coming the through another unbuffered pad. The low-to-high conversion circuit is a PMOS cross-coupled pair (M3, M4) connected to the high supply voltage, driven differentially (via M1, M2) by the low-voltage signal from the core. The N-device pulldowns, M1 and M2, are DC ratioed against the cross-coupled P-device pullups, M3 and M4, so that a low-swing input ($V_{ddL} = 1$ V) guarantees a correct output transition ($V_{ddH} = 5$ V). That is, the PMOS widths are sized so that the drive capability of the NMOS can overpower the drive of the PMOS, reversing the state of the latch. The ratio is larger than just the ratio of mobilities because the PMOS devices are operating with $V_{GS} = V_{ddH}$, and the NMOS are operating with $V_{GS} = V_{ddL}$. This level-converting pad consumes power only during transitions and

consumes no dc power. The remaining buffer stages and output driver and supplied by V_{ddH}. This level-conversion pad will work only with an NWELL process since it requires isolated wells for the PMOS devices that are connected to the high voltage.

4) Use of Gated Clocks to Reduce the Switching Activity: At the logic level, gated clocks are used extensively to power down unused modules. For example, consider the error correction in the text/graphics module. The basic protocol for the text/graphics module that is sent over the radio link and through the text/graphics FIFO is a base address for the frame-buffer followed by bit-mapped data—the length information is decoded in the depacketizer module and a control field in the FIFO (a thirty-third bit indicating End of Packet—EOP) is used to delimit the packets. The base address is the starting location in the frame-buffer for the bit-mapped data. While address information is sensitive to channel errors (since errors can cause the data to be written in the wrong area of the screen), bit-mapped data is not very sensitive since errors in data appear as dots on the screen. Therefore, to enable a bandwidth efficient wireless protocol, only the address information is error corrected and the bit-mapped data is left unprotected. Fig. 7 shows a block diagram of a power efficient implementation of this function. A register is introduced at the output of the FIFO and a grated clock enables the error correction module when processing the address information; the ECC is shut down during the rest of the time. The gated clock is generated by a controller which uses the state information of the FIFO (which is based on the $\overline{\text{EMPTY}}$ signal and the "End of Packet" signal from the FIFO). In this manner, the inputs to the ECC are not switching when the data portion of the protocol is accessed from the FIFO. Since typically the address is only a small portion of the text/graphics packet, significant power savings is possible (the amount of power savings is variable depending on packet size). Gated clocks were used in many circuits to tune the frequency and hence clock load for each module.

5) Low-Power Cell Library: The entire chipset was designed using a cell library that was optimized for low-power operation. The library contains parameterized datapath cells

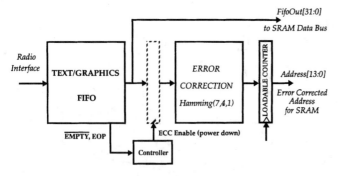

FifoOut[31:0]
to SRAM Data Bus

Radio Interface

TEXT/GRAPHICS FIFO

ERROR CORRECTION
Hamming(7,4,1)

LOADABLE COUNTER

Address[13:0]
Error Corrected
Address
for SRAM

EMPTY, EOP

ECC Enable (power down)

Controller

Fig. 7.   Gated clocks are used to shut down modules when not used.

(adders, registers, shifters, counters, register files, buffers, etc.), multiple strength and multiple input standard cells for random logic (NAND, NOR, registers, buffers, etc.), low-power memories (FIFO, SRAM), and low-voltage pads (clock pads, low-voltage I/O pads, level-converting pads, etc.). Some of the key design principles used in the cell library are outlined below:

- Device sizing: minimum sized devices were used in most datapath and standard cells. Minimum sized devices should be used to minimize the physical capacitance for circuitry inside the datapath blocks where self-load, rather than interconnect capacitance dominates [4]. When driving large interconnect capacitances, for example when communicating between datapaths, large devices are used.
- Reduced swing circuitry: as described in Section III-A-2, reduced swing circuitry is used limit the voltage swing in memory circuits (FIFO and SRAM) and reduce the power consumption in a linear fashion.
- Self-timed circuits: the memory circuits use self-timing to eliminate spurious transitions on high capacitance data busses. This guarantees the minimum number of transitions per memory access.
- Clocking strategy: the true single phase clock (TSPC) methodology was used for pipeline registers [9]. The schematic along with the device sizes are shown in Fig. 8. The rise-time of the clock signal is critical to proper operation and therefore a control slice attached to each datapath register provides local clock buffering. TSPC was chosen over a two-phase scheme since it reduces the power consumed in the interconnect since only one clock has to be distributed. Fig. 8 also shows a small modification to the TSPC register which can be used by controllers to generate gated clocks. As shown in the timing on the right of Fig. 8, with the extra PMOS device M1, the output (which is the gated clock and is used to clock other registers) can be forced to a ZERO during the low phase of the clock. Without this PMOS device, it will not be possible to generate rising edges for the gated clock on two consecutive rising edges of the system clock.

6) Optimizing Placement and Routing for Low Power: At the layout level, the place and route is optimized such that signals that have high switching activity (such as clocks) are assigned short wires and signals that have low switching activity are allowed to have long wire. Fig. 9 shows an example that involves the routing of large data and control busses from the text/graphics module of the chip core to the pads on the south side of the protocol chip, shown in Fig. 10. The text/graphics module on the protocol chip communicates to both the text/graphics frame-buffer and to the text/graphics 640×480 LCD display. The split-panel display requires 8 bits (4 for the top half and 4 for the bottom half) at a rate of 3 MHz, while each SRAM module uses 32 bits clocked at 375 kHz (using the parallel access scheme described in Section III-A-1). An activity factor (for both $0 \rightarrow 1$ and $1 \rightarrow 0$ transitions) of 1/2 is assumed for the data. The address bits are also clocked at 375 kHz but they have a much lower switching activity since the accesses are mostly sequential, coming from the output of a counter. The address bus is time-multiplexed between the read and write addresses for the SRAM, but since the write into the text/graphics frame-buffer is relatively infrequent, the address bus usually carries the read address and is therefore very temporally correlated (the number of transition for a counter output per cycle is $1 + 1/2 + 1/4 + \cdots \approx 2$ since the lsb switches every cycle, the next lsb switches every other cycle, etc.). As seen from the plot of physical capacitance as a function of distance from core to pads, the display data and display clock, which have high switching activity, are assigned the shortest wire lengths, while the SRAM address, which has an activity factor 16 times lower, is allowed to have the longest wires.

B. Text/Graphics Frame-Buffer Module

The frame-buffer for the split panel text/graphics display contains six 64 kbit SRAM chips (Fig. 11), which were synthesized from the low-power cell library's tileable SRAM module. This module meets several constraints to make it a useful component of a wide variety of low power systems.

First, the SRAM module builds memories over a wide variety of sizes. At one extreme, it can create entire memory chips that are only limited by maximum die size, such as the frame-buffer memories used for the text/graphics display. At the other extreme, the same module synthesizes smaller (hundreds or thousands of bit), area and power-efficient memories that are placed on the same chip as datapath and control systems (as in the video decompression chips described in Section III-C). The designer specifies the number of words, which may be any size over two, and the number of bits per word, which may be from two to sixty four. The designer can also control the aspect ratio by specifying the number of blocks at the architecture level, as detailed below.

Next, the SRAM must meet the throughput requirements at reduced supply voltages. The cycle time for an on-chip memory access is 100 ns at 1.5 V in a 1.2 μm process. As described earlier, parallel word access can be used to operate the SRAM at lower voltages without loss in throughput.

The final requirement is to minimize the external timing and control of the SRAM since these signals will appear on the high capacitive external pad and interconnect I/O. The SRAM is synchronized to the rising edge of a single clock; all other timing signals are generated internal to the SRAM, using self-timing circuitry that scales with the size of the SRAM. Also,

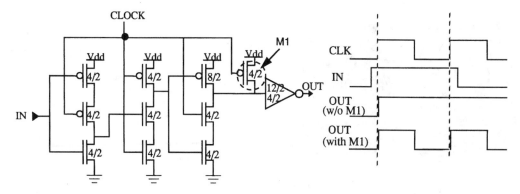

Fig. 8. Schematic of a modified TSPC latch that can be used to generate gated clocks.

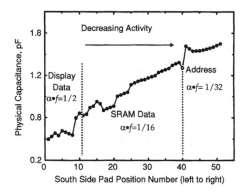

Fig. 9. Optimizing placement for low-power: Example of routing large data/control busses.

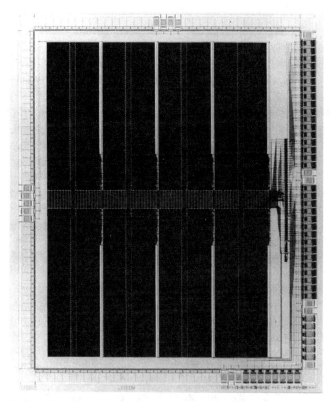

Fig. 11. Die photo of the text/graphics frame-buffer which stores data for the 640 x 480 LCD display.

Fig. 10. Die photo of the protocol chip.

the SRAM can use either a single, bidirectional bus or can use separate input and output buses.

1) Memory Architecture: The architecture of the SRAM is designed both to minimize power consumption and to enhance scalability. At the highest level of the architecture, the SRAM is organized into a parameterizable number of independent, self-contained blocks, each of which reads and writes the full bit width of the entire memory. For example, the 64 kbit (2 k word by 32 bit) frame buffer chips contain eight 8 kbit blocks, each organized as 256 words by 32 bits. Since one of the fundamental power saving techniques is to minimize the effective capacitance switching per clock cycle, the SRAM only activates circuits in one of these blocks per clock cycle (Fig. 12). The power savings groom this architecture are twofold. First, there is less overhead capacitance, since only one set of control signals and decoders switch at one time. Second, and more fundamentally, by having wide data widths into each block, the memory has minimal column decoding, the hence less column capacitance to switch per bit. From an ideal power consumption perspective, it would be best to have no column decoding at all; as a practical concession to pitch matching, this design uses 2-to-1 column decoding.

409

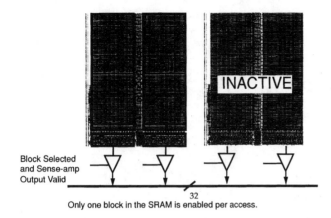

Fig. 12. SRAM block organization.

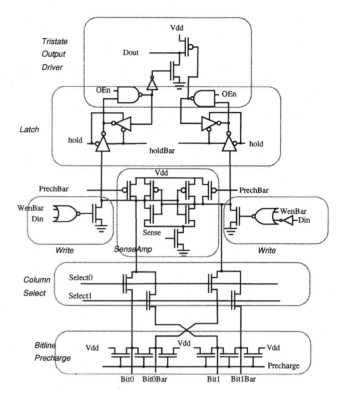

Fig. 13. Precharge, sense-amp, and "glitch"-free tri-state buffer.

Power consumption is greatly reduced at the architectural level by beginning each cycle with the block address decoding and then enabling only one block. Memory access would be faster—but power consumption much higher—if all blocks activate at the start of each cycle and the block level decoder selects the output from among all the blocks after the data has been read. In essence, it is not worthwhile to use speculative execution (in this case, speculating on the address) because the gain in speed does not allow enough of a reduction in supply voltage to offset the increase in effective capacitance switched. Controlling the activation of the SRAM at the block level also allows further optimization at the circuit level within the blocks. For example, because no control or clock signals make transitions within a block unless that block is selected, it is often optimal for both speed and power to use fully dynamic circuits; the architecture ensures that dynamic circuits make the same number of transitions as static circuits, so the lower capacitance of the dynamic circuits enables higher speed and lower power. In essence, judicious use of gated clocks (i.e. block level decoding) allowed greater choice of circuit styles.

Using self-contained blocks also makes the architecture flexible and expandable. Since the number of words per block and the total number of blocks are parameterized, the designer can control the aspect ratio by trading off between the number and size of the blocks. Adding blocks to the memory increases its number of words, with minimal effects on power consumption and circuit design. Since circuits in only one block switch at a time (except for the block level decoders, which consume minimal power) the only increase in power consumption is from the increased wiring capacitance between blocks. The only circuits that need to change are the block level decoders. All of the other control, timing, decoding, and sensing circuitry in the blocks are insensitive to the number of blocks in the memory.

2) Circuitry: At the circuit level, the SRAM's most important power saving technique is to reduce the voltage swing on the bitlines. As shown in Fig. 13, the bitlines are precharged through NMOS devices, so the maximum bitline voltage is the apply voltage minus the NMOS threshold voltage (with body effect). Compared to full swing bitlines, this reduces power consumption by only 20% for $V_{dd} = 5$ V, but as much as 50% for $V_{dd} = 1.5$ V and 75% for $V_{dd} = 1.2$ V. (It was

not practical to limit the minimum voltage on the bitlines by timing the wordline signals because the timing would have to work for many sized memories, over many voltages, and for different fabrication technologies.) This precharge strategy also allows the column select transistors to double as cascode amplifiers, creating voltage gain between the bitlines and the input to senseamp.

Another important power saving technique is to eliminate glitching on the data bus during read operations. The output driver of the senseamp shown in Fig. 13 is tri-stated at the beginning of each clock cycle. Even after the output enable signal is true, the output remains tri-stated until the senseamp's cross-coupled latch has resolved the data value, at which time either the NMOS or the PMOS output device drives the data bus.

One of the major handicaps of using a high threshold process ($V_{tp} = -0.9$ V, $V_{tn} = 0.7$ V) at low supply voltages is that the low gate overdrive voltage for the PMOS ($V_{gs} - V_t$ exacerbates their lower carrier mobility, creating a large imbalance between the current drive of the PMOS compared to the NMOS devices. A partial remedy is to make sure that large capacitances are never driven by more than one series connected PMOS device. The senseamp output driver meets this requirement, in addition to eliminating glitches. Another strategy is to use a non-minimum device ratio ($Wp : Wn$) for circuitry in the critical path.

3) Tiling: To create an SRAM, the designer gives the layout generation tool two parameters—the number of words and the number of bits per word—and may provide and third, optional parameter—the number of blocks. The first step for the layout generator is to tile the individual blocks, each of which has the same layout. Most of the block layout is

410

straightforward: the memory cells are tiled in 2 dimensional arrays (bits, words); the senseamps are tiled in 1 dimensional arrays (words) as are address latches (address lines) and word line buffers (words); a single control slice is added per block. Tiling the address decoding logic is more complex because it employs a tree structure nand decoder. The number of columns in the address decoder is $\log_2$ the number of word lines in the memory cell array. The address decoder contains one row for each word line. However, only the first column contains one transistor for each word line; each subsequent column contains half as many transistors as its predecessor, until the final column has only 2 transistors. Thus, the number of transistors in the decoder tree is approximately twice the number of word lines. The address decoders are placed in the middle of the blocks in order to reduce the RC delay of the poly word lines by a factor of 4.

Between each pair of blocks is an interconnect channel that is also created by tiling. Data, address, control, and power lines are taken from the blocks in first layer metal and connected to second layer metal buses. The interconnect channels are identical except for the block select logic, which decodes the high order address bits to activate a single block. The individual blocks and interconnect channels are tiled together to form the entire SRAM.

C. Video Decompression Module

The final output module is the video decompression module which includes all of the circuitry required to take a compressed video stream from the radio and convert it to the analog data format required by a $4''$ (128×240 pixels) color active matrix display. This section presents the design of a set of 4 chips (as shown on the block diagram in Fig. 1) to perform video decompression for this low-resolution display; one chip performs all of the control functions for the video decompression and interface to the radio, two chips perform the frame-buffering and the video decompression based on the vector quantization table lookup algorithm, and one chip performs the color space conversion and analog interface to the display.

1) Color (Luminance and Chrominance) Video Decompression: The choice of algorithm is the most highly leveraged decision in meeting the power constraints. The ability for an algorithm to be parallelized is critical and the basic complexity of the computation must be highly optimized. Minimizing the number of operations to perform a given function is critical to minimizing the overall switching activity and therefore the power consumption. The task of selecting the algorithm for the portable terminal depends only on the traditional criteria of achievable compression ratio and the quality of reconstructed images, but also on computational complexity (and hence power), and robustness to high bit error rates.

Most compression standards (for example, JPEG and MPEG) are based upon the Discrete Cosine Transform (DCT). The basic idea in intra-frame schemes such as JPEG is to apply a two-dimensional DCT on a blocked image (typically 8×8) followed by quantization to remove correlations within a given frame. In the transform domain, most of the image energy is

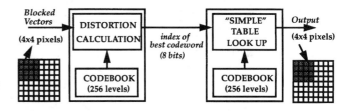

Fig. 14. Video compression/decompression using vector quantization.

packed into only a few of the coefficients, and compression is achieved by transmitting only a carefully chosen subset of the coefficients. One main characteristic of the DCT is the symmetric nature of the computation; i.e., the coder and decoder have equal computational complexity. Although the computational complexity of the DCT can be optimized by restructuring algorithms, it still requires several arithmetic and memory operations per pixel [10].

An alternative compression scheme is vector quantization (VQ) coding, which is asymmetrical in nature and has been unpopular due to its complex coder requirements. Fig. 14 shows the basic idea behind intraframe vector quantization compression. On the encoder, a group of pixels is blocked into a vector and compared (using some metric such as Mean Square Error) against a set of predetermined reproduction vectors (a set of possible pixel patterns) and the index of the best match is output from the encoder. The decoder has a copy of all possible reproduction vectors (codebook) and the index of the best codeword is used to reconstruct the image using a simple lookup table operation.

For this implementation, the image is segmented into 4×4 blocks (i.e., the vector size is 16) and there are 256 entries in the codebook. The original image on the encoder side is represented in the RGB domain using 6 bits for each color plane—using 6 bits to represent video data instead of 8 bits results in very little visual distortion on this low-resolution display. Color information is transformed to the YIQ domain and each plane is individually coded with separate codebooks. The I and Q color components (called the chrominance components) are subsampled in both the horizontal and vertical dimensions. Therefore, the YIQ representation ($Y : 1, I : 1/4, Q : 1/4$) gives $2 : 1$ compression over the RGB representation ($R : 1, G : 1, B : 1$). On each plane, VQ results in $12 : 1$ compression since only 8 bits are transmitted for each 4×4 block (choosing 1 out of 256 codes) instead of 16×6 bits for the true data. A total of $24 : 1$ compression is achieved ($32 : 1$ if the quantization from 8 bit to 6 bit representation is taken into account) and therefore to support full-motion color video at 30 frames per second, this system consumes a bandwidth of 690 kbits/s on the wireless link.

In addition to the simplicity of the decoder, intraframe VQ has two other advantages. First, VQ localizes errors in space; i.e. errors in the VQ codeword data appear as small corrupted 4×4 blocks on the screen and don't corrupt large portions of the screen. Simulations indicate that even with Bit Error Rates as high as 10^{-3}, the image looks reasonable and therefore no error correction was applied on the codebook data, significantly enhancing the bandwidth efficiency of the wireless protocol. Run-length coding, as used in JPEG, is not suitable

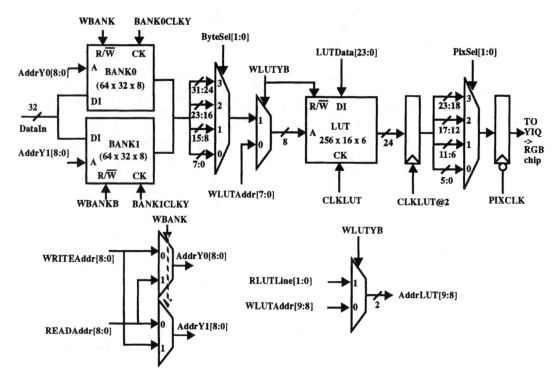

Fig. 15. Block diagram of the luminance decompression chip.

for wireless operation since errors can cause loss of synchronization and can corrupt large portions of the image. Second, intraframe compression localizes errors in time: errors don't accumulate from frame to frame, unlike differential schemes such as MPEG. In differential schemes, errors stay on the screen until a full new image is sent (in an intra frame mode). This chipset does support interframe mode, in which only portions of the screen that are changed are updated. However, this mode can be used only when the bit error rates are low.

Fig. 15 shows a block diagram of the luminance decompression chip. The incoming compressed video data (VQ codewords for the image) is stored using a ping-pong addressing scheme. For example, when the compressed video coming from the RF link is being stored in BANK0, the compressed data stored in BANK1 is read out to be decompressed using the lookup table (LUT). After a full frame of compressed video is assembled in BANK0, the $R/\overline{W}$ signal (signal *WBANK*) of the two frame-buffers toggles, resulting in data being read from BANK0 to the lookup table while new compressed video data is written into BANK1. This ping-pong addressing scheme provides an asynchronous interface between the radio clock and the video system and provides immunity against bursty channel errors; if a frame of compressed video is dropped or if higher priority data is sent over the link (such as text/graphics data or speech which require data to be sent with minimal latency), the complete compressed frame that is already stored in the terminal is decompressed and displayed until a new frame of compressed video is assembled in the other frame-buffer. Also, since the refresh rate of the color display is 60 Hz while the image is updated only at 30 Hz, the ping-pong frame-buffer is actually required unless the bandwidth on the link is doubled.

The video is stored in a compressed rather than uncompressed format to reduce the amount of memory in the system by a factor of 24. Rather than decompressing and storing the data once and displaying the decompressed image twice (since as mentioned above, the refresh rate is 60 Hz while the image is updated at 30 Hz), the image is stored in a compressed format and is read out and decompressed twice; that is, there is no decompressed frame-buffer. The compressed frame-buffer is read out in parallel fashion, where four 8-bit VQ codewords are read in parallel (once again the access pattern of data is know and is exploited) though only one is used at a given time; this once again enabled operation at supply voltage as low as 1.1 V. In this implementation, the video frame-buffer was clocked at 156 kHz while meeting the throughput rate of 2.5 MHz. The output of the frame-buffer is multiplexed at 4 : 1, and is used to index the lookup table which generates the decompressed data.

The chip uses an addressing scheme that eliminates the need for an output line buffer which is typically used to convert a block data format to the raster format required by the display. Fig. 16 shows the numbering of codewords in the frame-buffer and the ordering of pixels inside each block. Each time a codeword is accessed from the frame-buffer, only 4 pixels are read out from the lookup table, creating a raster output which can be sent directly to the display. That is, pixels $P0$–$P3$ corresponding to each codeword between CW0 and CW31 are read from the lookup table and then $P4$–$P7$ are read once again for codewords CW0 through CW31, and then $P8$–$P11$, and finally $P12$–$P15$. Thus, each codeword is accessed four times per image. This approach increases the frequency of codeword access relative to a scheme which reads each codeword only once (and stores $P0$–$P15$ in a line-

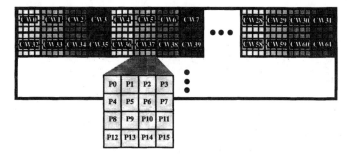

Fig. 16. Addressing of the frame-buffers and lookup table.

buffer), but since the codewords are accessed at a much lower frequently relative to the lookup table data, and since the line-buffer access has been eliminated, the overall power due to memory access is reduced by approximately a factor of 1.5. This is an example of architectural restructuring to reduce the number of operations.

The control of the luminance decompression chip (address generation, clock generation, multiplexor selection, etc.) is performed by the video controller chip. The multiplexors for read/write address lines are integrated in the decompression chip. RLUT Line [1 : 0] controls the row of pixels that are read for a given codeword. The lookup table is programmable over the radio link through the video controller chip. The color video decompression chip is implemented using a very similar architecture to the luminance architecture shown in Fig. 15. Since I and Q data are each sub-sampled by a factor 4, the amount of frame-buffer memory required is reduced by a factor of 4 for each component. Also, since they use the same addressing scheme, the I and Q data are stored in the same physical frame-buffer and the data is interleaved; i.e. the 32-bit frame-buffer word is organized as: CWI0, CWI1 CWQ0 CWQ1, and the 4 : 1 mux to select the codeword in the Y decompression chip is replaced by two 2 : 1 muxes, one for I (to select between CWI0 and CWI1) and one for Q (to select between CWQ0 and CWQ1). The chrominance (I) decompression chip also has two separate (256 × 16 × 6) lookup table memories. Fig. 17 shows the die photograph of the luminance decompression chip. The chip consumes only 115 μW running at 1.5 V to support a video throughput rate of 2.5 MHz for the 4″ color display.

2) Video Controller: The video controller performs all the control functions for the video decompression module. It generates all of the timing for the NTSC display, interfaces to the radio, controls the frame-buffers and lookup tables and performs synchronization for the system. A summary of the functions performed is outlined below:

• NTSC sync generation for the 4″ display: The LQ4RA01 4″ color display is responsive to a standard composite sync signal with negative polarity of the same amplitude level as that of the video composite signal. The standard sync found in NTSC format has extra timing information such as an extra half line in one field (to distinguish between even and odd fields), and pre and post equalization half-line pulses during vertical sync that are not a necessity for proper operation of the LCD. A significant simplification of this protocol which involves block sync

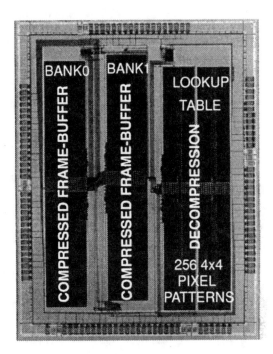

Fig. 17. Die photo of the luminance decompression chip.

(with no equalization pulses) and the elimination of the half-line can be used to obtain a sync that still provides adequate synchronization information. The implemented sync signal is simply a scaled digital combination of VSYNC (vertical sync) and HSYNC (horizontal sync). The original and modified sync signals are shown in Fig. 18.

• Decodes and demultiplexes the radio data: similar to the function performed in the protocol chip (Section III-A), the video controller chip decodes packets from the radio and demultiplexes between a frame-buffer FIFO and LUT FIFO. Contained inside each frame-buffer packet (that is sent over the frame-buffer FIFO) is the encoding of type information (TYPE $= 0 \implies$ data is for the Y frame-buffer, and TYPE $= 1 \implies$ data is for the IQ frame-buffer). Similarly, inside the LUT packet is encoding information about the LUT data type (Y, I or Q).

• Controls reading and wiring of the frame-buffer memories for both Y and I decompression chips: it generates the read and write addressed for the ping-pong frame-buffers. It also generates the multiplexor control signals that selects the output of the frame-buffers (ByteSel in Fig. 15). It also performs the $R/\overline{W}$ control (i.e., controls when the frame-buffers switch between reading to writing) and generates the clocks for the SRAMs.

• Controls loading and reading of the lookup table memories for Y, I and Q data: it generates the write address (WLUTAddr in Fig. 15) and part of the read address (RLUTLine in Fig. 15) for the lookup table memories. It also generates the multiplexor control signals that selects the output of the LUT (PixSel in Fig. 15).

413

Field One (Odd) Composite Timing

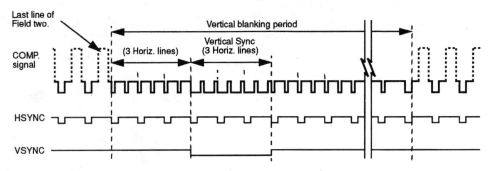

Field Two (Even) Composite Timing

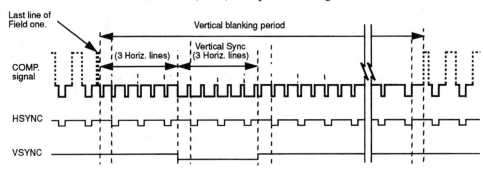

(Both Fields) Modified (Simplified) Timing

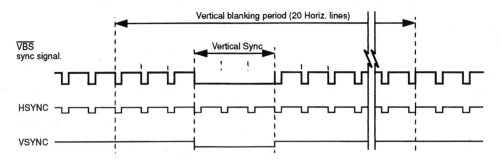

Fig. 18. Simplification of the NTSC system timing.

• Support for variable sized packets: a decompressed frame of video can be broken into multiple packets and the size of the packets is variable, providing a flexible platform to test the effects of packet sizes. Also, this allows a form of interframe coding in which only the differences between the current frame and the frame corresponding to two image ago is sent. This is effective only if the BER is fairly low for the reasons explained earlier.

• Matches pipelined delays in the system: since the system is pipeline, the output sync signals for the display must be delayed to avoid offsets of pixel data on the display. The timing signals needed by the color space converter are also generated in this chip.

3) Color-Space Converter and Digital-to-Analog Converter: The digital YIQ information from the video decompression chips are sent to a color space converter which converts it to analog RGB to drive the 4″ color LCD display

from SHARP. The digital YIQ is first converted to digital RGB and then a triple digital-to-analog converter directly drives the display.

a) Digital YIQ-to-digital RGB conversion: In the YIQ to RGB translation, which involves multiplication with constant coefficients, the switching events are minimized at the algorithmic level by substituting multiplications with hardwired shift–add operations (in which the shift operations degenerated to wiring) and by optimal scaling coefficients. In this way, the 3×3 matrix conversion operation degenerated to 8 addition. The implementation was fully parallel and therefore there was no controller. For I/O communication (between the decompression chips and the color space chip) and in the matrix computation, sign-magnitude representation is chosen over two's complement to reduce the toggle activity in the sign bits [11]. For example, going from -1 to 0 with result in all the bits toggling in two's complement representation.

414

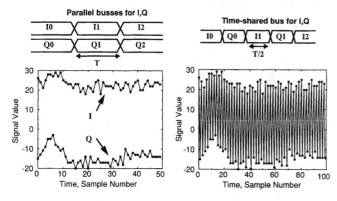

Fig. 19.   Time-multiplexing can increase the switching activity.

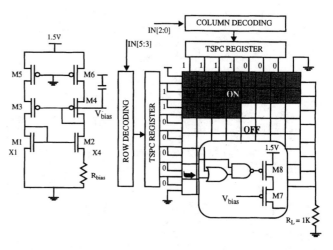

Fig. 20.   Low-voltage 6–bit digital-to-analog converter.

TABLE I
SUMMARY OF CHIPSET FOR THE INFOPAD TERMINAL

Chip Description	Area (mmxmm)	Minimum Supply Voltage	Power at 1.5V
Protocol	9.4 x 9.1	1.1V	1.9mW
Frame-buffer SRAM (with loading)	7.8 x 6.5	1.1V	500μW
Video Controller	6.7 x 6.4	1.1V	150μW
Luminance Decompression	8.5 x 6.7	1.1V	115μW
Chrominance Decompression	8.5 x 9.0	1.1V	100μW
Color Space Conversion and Triple DAC	4.1 x 4.7	1.3V	1.1mW

At the architecture level, time-multiplexing was avoided as it can destroy signal correlations, increasing the activity. Fig. 19 shows two alternate schemes for transmitting the I and Q data from the decompression chips to the color space converter chip. On the left is a fully parallel version in which I and Q have separate data busses. Also shown is the data for I and Q for a short segment in time. As shown, the data is slowly varying and therefore has low switching activity in the higher order bits. On the right is a time-multiplexed version in which there is a single time-shared bus in which the I and Q samples are interleaved. As seen from the signal value on the data bus, there is high switching activity resulting in higher power.

This digital YIQ to digital RGB implementation consumes 100 μW, which is two to three orders of magnitude lower than a design which is not optimized for power. This difference in power is attributed to 1.5 V operation instead of 5 V operation (which provides a $\times 11$ reduction in power), using hardwired optimized shift–add multipliers instead of true-multipliers (which reduces switched capacitance by $\times 5$), using an optimized cell library that uses minimum sized devices and single phase clock methodology (which reduces switched capacitance by $\times 2$–3), using an optimized number representation in I/O and matrix multiplication ($\times 1.2$), integrating the DAC with the signal processing (so that the output interchip power is eliminated and reduces power by $\times 1.3$), using an hardware assignment which keeps uncorrelated data on different units ($\times 1.5$), and using 6 bits/plane instead of 8 bits/plane ($\times 1.3$).

b) Low-voltage digital-to-analog converter: A low-voltage, low-throughput 6-bit DAC has been developed to drive the 4" SHARP LCD. The LCD display takes pixel data in the analog R, G, B format which has voltage levels compatible with the NTSC format (i.e., $V_{pp} = 0$ to 0.7 V). The DAC, shown in Fig. 20, has an architecture based on a conventional nonweighted current switched array [12]. Based on a decoded 6-bit digital word, an appropriate number of current sources are turned ON and summed. The output voltage is obtained by passing current through an external resistor. Since the settling time requirement is quite low and since the LCD display has a high impedance capacitive load, the external resistor was chosen to be approximately 1 kΩ rather than 75 Ω. This reduces the power consumption by more than an order of magnitude since the average current drawn from the supply is reduced by more than order to

magnitude. The row decoding and column decoding logic is identical to the one presented in [12]. The decoding logic was implemented using minimum sized low-power standard cells.

A current source that operates at a reduced supply voltage of 2.7 V has been developed [13]. The current source used in this DAC is similar. However, due to the modest DAC throughput requirements for this 4" display, and to statistically reduce the power consumption by a factor of 2 (on average only half of the current sources will be ON) a single ended architecture is used instead of the differential scheme. The current cell consists of stacked PMOS devices (M7 and M8) with the top transistor M8 being digitally switched, thus operating in the linear region. Therefore, effectively, the output resistance of the current source is the output resistance of a single transistor (M7) degenerated with a source resistor. To increase the output resistance of M7, the length of the device was made nonminimum. In order to operate the DAC down to a supply voltage of 1.5 V and to meet the 0.7 V_{pp} requirement for the output, the W/L of M8 was made large to keep the voltage dropped across M8 to less than 100 mV. Also, the bias voltage was set up so that the $V_{gs} - V_t$ for M7 was approximately 200 mV, allowing supply voltages to be as low as 1.2–1.3 V. The power consumption of the DAC

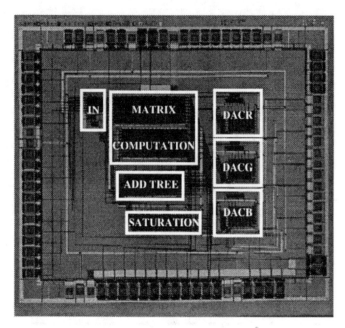

Fig. 21. Die photo of the color-space converter and digital-to-analog converter.

(a)

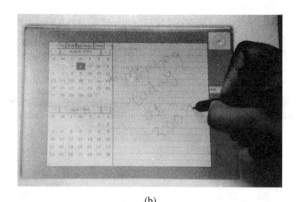

(b)

Fig. 22. (a) The IPGraphics terminal (pen, speech I/O, and text/graphics). (b) A notebook application running on the IPGraphics terminal.

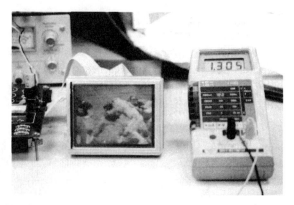

Fig. 23. Output of video decompression chips running at a supply voltage of 1.3 V.

is dominated by the analog power and for a 1.5 V supply: $P_{\mathrm{avg}} = V_{dd} \bullet I_{\mathrm{avg}} = 1.5 \bullet 1/2 \bullet 0.7/1.2 \text{ K} = 440 \ \mu\text{W}$. The power measured in the actual system was lower since the DAC was shut down (the digital input was forced to 0) during the horizontal and vertical blanking periods.

The digital decoding inside each current cell is implemented using a single complex CMOS gate $\overline{(A + B) \bullet C}$ to eliminate glitching on the gate of M8 as compared to path balancing approaches [14]. The device sizes on the complex gate are minimum to minimize the dynamic power consumption.

Fig. 20 also shows the schematic of the bias circuitry which is a bootstrapped current reference [15]. The top current mirror, M3 through M6, forces the bias currents in M1 and M2 to be equal. Transistors M5 and M6 operate in the linear region and emulate the voltage drop across the switch transistor in the current cell array (M8). Simple first-order analysis shows that the reference current is given by:

$$I_{\mathrm{ref}} = \frac{2}{K_n R_{\mathrm{bias}}^2} \left(\frac{1}{\sqrt{\frac{W}{L1}}} - \frac{1}{\sqrt{\frac{W}{L2}}} \right)^2 . \qquad (3)$$

Note that the reference current $(I_{\mathrm{ref}} = I_1 = I_2)$ is to first order independent of supply variations (this is due to the bootstapped techniques used). The bias current is set by sizing M1 and M2 and by choosing R_{bias}. The current source can operate at low supply voltages even with a process that has a standard threshold voltage. The minimum operating voltage for this current reference is given by:

$$V_{dd\mathrm{min}} = V_{ds6} + |V_{tp4}| + V_{d\mathrm{sat4}} + V_{d\mathrm{sat2}} + I_{\mathrm{ref}} R_{\mathrm{bias}}. \qquad (4)$$

In the above equation, V_{ds6} is very small through device sizing (<100 mV since M6 is in the linear region), V_{tp} for the MOSIS 1.2 μm CMOS process is typically 0.9 V, and the last three terms can be made small (100–200 mV) by device sizing and choice of bias current. Therefore, the current source can operate down to the 1.2–1.5 V range.

Fig. 21 shows the die photo of the color space translator and triple DAC. Table I shows the summary of the chipset implemented in 1.2 μm CMOS technology.

IV. System Implementation

A PCB containing the protocol chip, 6 SRAM chips, a speech codec, and pen interface logic has been fabricated and tested. Various power supply voltages needed for the design including −17 V (adjustable using a trimpot) for display drive, 12 V for dc-to-ac inverter for the backlight, 1.5 V for the custom chips, and −5 V for the speech codec have been realized using commercial chips. Custom low-voltage high efficiency switching regulators have been fabricated and will be integrated into the next generation terminal [16]. This board is integrated with a commercial radio modem to realize a complete I/O terminal with a 1 Mbit/s wireless channel. Fig. 22 shows the photograph of this terminal, IPGraphics, which supports pen input, speech I/O and text/graphics output. The next version will also provide support for one-way full motion video and Fig. 23 shows the output of the decompression chips (running at 1.3 V) on a 4" SHARP active matrix display.

V. Conclusion

A low-power chipset for a portable multimedia terminal which supports pen input, speech I/O and one-way full-motion video has been presented. A system level approach which involves optimizing the circuitry, the architectures and the algorithms was used to minimize power consumption. The biggest power savings came from an architecture driven voltage scaling strategy that allowed supply voltages as low as IV. The memory circuits used reduced swing circuits that clamped the voltage swings further, resulting in further power reduction. The capacitance switched was minimized through operation reduction, choice of number representation to statistically reduce the number of transitions, exploitation of signal correlations, self-timing to eliminate glitching, use of gated clocks, optimized transistor sizing, and optimizing place and route. The entire chipset that performs protocol conversion, synchronization, error correction, packetization, buffering, video decompression and D/A conversion operates from a 1.1 V supply and consumes less than 5 mW.

Acknowledgment

The authors thank R. Allmon, T. Burd, S. Narayanaswamy, R. Neff, I. O'Donnell, and B. Richards for their invaluable contributions. The mechanical packing for the terminal was designed by R. Doering, C. Smith, and Prof. P. Wright.

References

[1] S. Sheng, A. P. Chandrakasan, and R. W. Brodersen, "A portable multimedia terminal," *IEEE Commun. Mag.*, vol. 30, pp. 64–75, Dec. 1992.

[2] F. Najm, "Transition density, a stochastic measure of activities in digital circuits," *Proc. of DAC*, pp. 644–649, 1991.

[3] A. Ghosh, S. Devedas, K. Keutzer, and J. White, "Estimation of average switching activity," *Proc. DAC*, pp. 253–259, 1992.

[4] A. P. Chandrakasan, S. Sheng, and R. W. Brodersen, "Low-power digital CMOS design," *IEEE J. Solid-State Circ.*, vol. 27, pp. 473–484, Apr. 1992.

[5] A. P. Chandrakasan and R. W. Brodersen, "Minimizing power consumption in digital CMOS circuits," *Proc. IEEE*, Dec. 1994, to be published.

[6] A. P. Chandrakasan, M. Potkonjak, J. Rabaey, and R. Brodersen, "HYPER-LP: A design system for optimizing power using architectural transformations," in *IEEE/ACM Int. Conf. Computer-Aided Design*, Nov. 1992, pp. 300–303.

[7] D. A. Patterson and J. L. Hennessy, *Computer Architecture: A Quantitative Approach.* San Mateo, CA: Morgan Kaufman, 1989.

[8] T. Burd, "Low-power CMOS library design methodology," Masters thesis, ERL Univ. Calif., Berkeley, Aug. 1994.

[9] J. Yuan and C. Svensson, "High-speed CMOS circuit technique," *IEEE Solid-State Circ.*, vol. 24, pp. 62–70, Feb. 1989.

[10] K. R. Rao and P. Yip, *Discrete Cosine Transform.* New York: Academic, 1990.

[11] A. P. Chandrakasan, R. Allmon, A. Stratakos, and R. W. Brodersen, "Design of portable systems," in *Custom Integrat. Circuits Conf.*, May 1994, pp. 259–266.

[12] T. Miki, Y. Nakamura, M. Nakaya, S. Asai, Y. Akasaka, and Y. Horiba, "An 80-MHz 8-bit CMOS D/A converter," *IEEE J. Solid-State Circ.*, vol. SC-21, pp. 983–988, Dec. 1986.

[13] T. Miki, Y. Nakamura, Y. Nishikawa, K. Okada, and Y. Horiba, "A 10 bit 50 MS/s CMOS D/A converter with 2.7 V power supply," in *1992 Symp. VLSI Circuits*, 1992, pp. 92–93.

[14] J. M. Fournier and P. Senn, "A 130 MHz 8-b CMOS video DAC for HDTV applications," *IEEE J. Solid-State Circ.*, vol. 26, pp. 1073–1077, July 1991.

[15] P. Gray and R. Meyer, *Analysis and Design of Analog Integrated Circuits*, second ed. New York: Wiley, 1984.

[16] A. Stratakos, S. Sanders, and R. Brodersen, "A low-voltage CMOS DC-DC converter for a portable battery-operated system," in *Proc. IEEE Power Electron. Specialists Conf.*, 1994, pp. 619–626.

A Portable Real-Time Video Decoder
for Wireless Communication

TERESA H. MENG, BENJAMIN M. GORDON, AND ELY K. TSERN

DEPARTMENT OF ELECTRICAL ENGINEERING
STANFORD UNIVERSITY

Abstract—Our present ability to work with video has been confined to a wired environment, requiring both the video encoder and decoder to be physically connected to a power supply and a wired communication link. This chapter presents a portable video-on-demand system capable of delivering high-quality image and video data in a wireless communication environment. The discussion will focus on both the architectural and circuit design techniques for implementing a high-performance, error-resistant video compression/decompression system at power levels that are two orders of magnitude below existing solutions. This low-power video compression system not only provides a compression efficiency better than industry standards, but it also embeds a high degree of error tolerance in the compression algorithm itself to guard against transmission errors often encountered in wireless communication.

INTRODUCTION

Today, compression standards, such as JPEG (Joint Photographic Experts Group) for still images and MPEG (Moving Pictures Experts Group) for video, dominate commercial systems. While offering good compression performance, these standards may not, however, be optimal for portable applications, because of their large hardware requirements, typically power-expensive implementations, and sensitivity to bit corruption.

Our compression algorithm is based on subband decomposition and pyramid vector quantization (PVQ) [1,2], which performs as well as the standard JPEG compression in terms of image quality and compression efficiency, while incurring much less hardware complexity and exhibiting a high degree of error resiliency [3].

Under most error conditions, our subband/PVQ algorithm performs better than JPEG with or without protection, as expected from the facts that our algorithm outperforms JPEG even without channel error, and that the subband/PVQ algorithm does not suffer from the bandwidth overhead incurred by the use of error-correcting codes. We have demonstrated an entirely fixed-rate compression scheme based on subband/PVQ that both achieves the compression performance of the variable rate JPEG standard and exhibits resiliency to channel error [4].

Compared with the C-Cube JPEG decoder [5], implemented in 1.2μm COMS technology and dissipating approximately 1W while decoding 30 frames of video per second, our subband/PVQ decoder is more than 100 times more power efficient, not accounting the power dissipated in accessing off-chip memory necessary in the JPEG decoding operation. Within this factor of 100, a factor of ten can be easily obtained by voltage scaling of the power supply. Reduced supply voltage, however, increases circuit delay. To maintain the same throughput, or real-time performance, the hardware must be duplicated seven times, increasing the total chip area by at least the same amount. The fact that our video-rate decoder can be implemented with less than 700,000 transistors (including 300,000 transistors for on-chip memory) operating at a power-supply voltage less than 1.5V, without requiring any off-chip memory support, indicates the efficiency of our decoding procedure. How we designed the decoder module to achieve minimal power consumption will be the focus of this paper.

1. LOW-POWER DESIGN GUIDELINES

The dynamic power consumption of a CMOS gate with a capacitive load C_{load} is given by

$$P_{dynamic} = p_f C_{load} V^2 f, \qquad [1]$$

where f is the frequency of operation, p_f the activity factor, and V the voltage swing [6]. Reducing power consumption amounts to the reduction of one or more of these factors. Analysis in [7] indicated that for energy-efficient design, we should seek to minimize the power-delay product of the circuit, or the energy consumed per operation.

A more complete discussion on low-power CMOS circuit design is given in [8]. This section only briefly describes the guidelines used in designing our low-power circuit library. CMOS circuits operated at extremely low supply voltages allow a very small noise margin. A safe design is therefore of ultimate importance. The static CMOS logic style was chosen for its reliability and relatively good noise immunity. The transistors were sized in the ratio of 2:1 for PMOS and NMOS, multiplied by the number of transistors in series. A library of standard cells has been designed with these specifications in mind.

1.1. Scaling Factor in a Buffer Chain

Well-known analysis of a multistage buffer chain indicates that the optimal scaling factor of the inverters in a buffer chain should be e (the exponential constant). This number is usually too small for practical use, however, as many stages of inverter buffers would be needed to drive a large capacitive load, such as clock lines, while consuming more power than necessary. Through simulations we have found that a scaling factor of 5 to 6 is optimal for low-power buffer chains operated at 1V to 1.5V, where the total delay would be comparable to that of a buffer chain with a scaling factor of e while consuming only two-thirds of its power.

Table 1 shows the energy consumed by a buffer chain at a supply voltage of 1.5V with different scaling factors. The energy consumption of a buffer chain with a scaling factor of 6 remains a constant fraction of the energy consumed by buffer chains with a scaling factor of 4 or less over a wide range of loads, and its delay relative to the buffer chains of smaller scaling factors only increases slightly for large loads (over 5pF). This is attributed to the fact that if we can save one or more stages in a buffer chain by increasing the scaling factor by a reasonable margin, the net result is a lower-power design with a comparable delay.

With a larger scaling factor, the slope of the voltage transfer function of each buffer is smaller. This may potentially increase power consumption due to short-circuit currents when both the PMOS and NMOS branches may be simultaneously on for a longer period of time. However, hSPICE simulations indicated that the factor of short-circuits is negligible, consuming less than 5% of the total power.

1.2 Minimizing Occurrence of Spurious Switching

Spurious switching, which consumes power but does not generate correct output signals, was quoted to account for from 20–70% of total circuit switching power. Spurious switching is caused by unmatched delays such as glitches in combinational circuits and input signals not arriving at the same time. Power-optimized logic synthesis and careful circuit design can limit spurious switching to a minimum.

In designing our library cells, extra care was taken to ensure that unwanted switching is never allowed to drive a buffer chain, as the power consumed is further amplified by the large capacitive load in successive buffer stages. To prevent spurious switching from driving a buffer chain, the input signal to the first buffer is either latched or gated. The former limits the switching to occur only when the latch is on, holding the signal constant when it is off. The latter is used if a completion signal can be identified from the logic module that drives a buffer chain to gate its outputs.

1.3 Equalizing Delays between Logic Paths

In our design, we tried to equalize the delays of all logic paths so that there is no single critical path. The reason is that if the delays of different logic paths are not matched, some logic paths will operate faster than others. As a result, energy will be wasted in those paths with shorter delays by delivering a current larger than necessary. Since switching speed is a function of transistor sizing, which determines switching currents and capacitive loads, we can effectively "slow down" the faster paths by using smaller transistors and fewer buffer stages to reduce both driving currents and transistor-capacitive loads.

1.4 Design Example: The Adder

We use the design of an adder to illustrate the performance trade-off between low-power and high-speed designs. The choice of an adder design is determined by the area, speed, and power budget available. Carry-propagate adders are compact and useful for adding numbers of short wordlengths (fewer than 8 bits), but too slow for adding numbers of wider wordlengths. On the other hand, tree adders perform both carry and summation computations in parallel, each carry taking only $\log_2 n$ gate delays, where n is the wordlength. The area needed for a full tree adder, however, is relatively large, requiring long interconnects and large capacitive loads. Hence, tree adders only yield significant advantage for adders with more than 32 bits. As our system requires adders of a wordlength between 10 and 20 bits, the adders to be considered are carry-select and carry-lookahead adders. By paying close attention to layout and the loading on signal lines that an adder introduces, we can identify the most energy-efficient adder class meeting a given speed requirement.

The adder used in our decoder design, shown in Fig. 1, is a variant of the standard carry-select adder. Each adder block has two speed-optimized ripple-carry chains, one with a zero

Table 1. Comparison of Buffer Chains with Different Scaling Factors (in 0.8μm CMOS Technology).

| C_{load} (pF) | Scaling Factor = e | | | Scaling Factor = 4 | | | Scaling Factor = 6 | | | Normalized to Factor e | | Normalized to Factor 4 | |
	Number of Buffer Stages	Energy/ Switch (pJ)	Avg. Delay (ns)	Number of Buffer Stages	Energy/ Switch (pJ)	Avg. Delay (ns)	Number of Buffer Stages	Energy/ Switch (pJ)	Avg. Delay (ns)	Energy	Delay	Energy	Delay
0.5	4	1.15	4.44	3	0.90	4.07	2	0.75	3.83	0.66×	0.86×	0.84×	0.94×
5	6	11.69	7.27	4	8.99	6.24	3	7.66	6.24	0.66×	0.86×	0.85×	1.00×
20	7	47.09	8.62	5	36.22	7.96	4	30.67	8.63	0.66×	1.00×	0.85×	1.08×
50	8	116.87	9.88	6	90.57	9.69	5	76.64	10.11	0.66×	1.02×	0.85×	1.04×

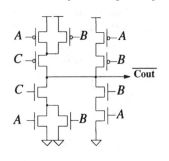

(a) A 1-bit carry circuit (complementary function is obtained by inverting the inputs).

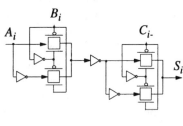

(b) A 1-bit sum circuit using 2 transmission-gate XORs.

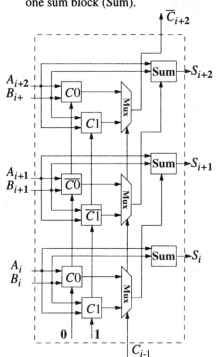

(c) A 3-bit carry select adder using 2 carry-chain blocks (C0 & C1) and one sum block (Sum).

Figure 1 Schematics of the carry-select adder.

carry-in, the other with a one carry-in. The output carry bit is chosen by the actual carry-in to the adder block, then used in the final XOR to produce the sum bit. The transistor sizing in each adder block was designed to hide the delay of ripple-carry chains, guaranteeing that all output carry bits will be ready by the time the carry-in reaches the adder block, eliminating the probability of spurious switching mentioned earlier. By simulation, the different adder block sizes are chosen to be 2, 4, 5, 6, and 8 bits respectively. HSPICE simulations of a 17-bit adder indicated a worst case delay of 35ns at 1.5V, with an energy consumption of only 19pJ per add operation.

A comparative study of various adder classes is given in Table 2, illustrating the types of adders compared and the sizes of their layouts in MAGIC. The performance comparisons were obtained by simulations of the extracted layouts using hSPICE in 0.8μm CMOS technology (the BSIM model). Fig. 2 (a) and (b) compare the delays of the sum and carry circuits respectively, while Fig. 2 (c) graphs the energy consumed per add operation. Fig. 2 (d), which shows the energy-delay product, is a measure of the efficiency of the various adders.

Our carry-select adder (Cs17) differs from the traditional carry-select adder (Csf17) by selecting carry bits instead of sum bits. By eliminating one sum gate, which is equivalent to 2 transmission-gate XORs, our design consumes less power and is faster. By not computing both sums from carry 0 and carry 1, the only additional delay for the whole adder is a final XOR gate, negligible compared to the critical path of the carry-select circuit. Furthermore, removing one sum gate reduces area over-

head and lowers capacitive loads on both the inputs and output carry bits, resulting in faster switching.

From the performance comparisons shown in Fig. 2, it can be seen that our carry-select adder is the fastest among the various adders to compute a 16- or 17-bit sum. The carry-lookahead adder has a slightly faster carry-out delay because it is optimized for generating fast carries. Its sum delay is larger, however, and the layout is asymmetric because of the non-uniform carry functions across the bits. The tree adder delivers slower performance, as its design is more suited for adding numbers of wider wordlengths.

Table 2. Various Adder Classes in Comparison.

Name	Number of Bits	Description of Adders	Layout Area (in 0.8μm CMOS)
Cp16	16	16-bit carry-propagate adder	$890\lambda \times 190\lambda$
Man 16	16	16-bit static Manchester carry adder, in 4-bit blocks with 4-bit carry-bypass	$1284\lambda \times 265\lambda$
Tree16	16	16-bit tree adder (Brent-Kung)	$1292\lambda \times 545\lambda$
Cla16	16	16-bit carry-lookahead adder, in 4-bit blocks	$1260\lambda \times 484\lambda$
CSf17	17	17-bit carry-select adder (standard), selecting sum, in 2-4-5-6 bit blocks	$1236\lambda \times 440\lambda$
CS17	17	Our implementation of a 17-bit carry-select adder, selecting carry, in 2-4-5-6 bit blocks	$1236\lambda \times 332\lambda$

420

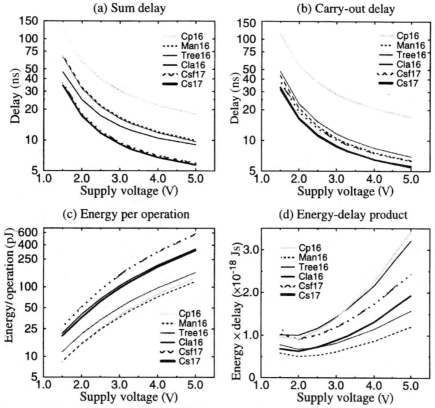

Figure 2 Performance comparisons among different adder implementations.

Our design is not the most energy-efficient one. The reason for not choosing the more efficient Manchester carry-chain adder is to meet the timing constraint of a 50ns clock cycle at 1.5V. This choice is a trade-off between maintaining the throughput required by the system and the energy needed to achieve that throughput. A slower adder implies not only that more adders need to be implemented to meet a throughput requirement, but also that other faster logic blocks will be idle waiting for some addition to complete, resulting in unmatched delays in logic paths.

1.5 Low-Power Memory Design

Within a large system, the memory can potentially be the largest power-consuming element, thus requiring special attention to its design. In addition to the power dissipation, the memory speed is also important, as the memory access time may be in the critical path, determining overall system performance. The first design criterion is of course to guarantee correct circuit operation, especially with regard to the sense amps, at supply voltages down to 1V. As with most memory designs, this requires extensive modeling and simulation.

For low-power memory design, only those memory cells actually being accessed should be activated. This can be accomplished by using a local word line that activates only the bank of memory being used. The width of a memory bank should be selected to match the bandwidth requirement of the overall system, delivering multiple words per access, thus saving on access power overhead dissipated by global word line activation. Next, the power consumed in the path from memory cells to their input/output

ports needs to be minimized. A self-timed circuit can be used to control the amount of voltage swing on the bit lines from memory cells, ensuring that the energy consumed in the bit lines will remain constant independent of operation frequency. Further savings can be achieved by using a small swing differential bus that connects all memory banks together. This bus, which runs the entire width of the memory and connects to every memory bank, can have significant capacitance—reducing the swing of this differential bus is vital to keeping the power consumption low. The required power consumption, however, is data dependent, as the activity of a full-swing single-ended bus will depend upon how often the data value changes.

From our decoder design, measurements on the actual memory design were taken. For eight banks of memory each of 128×20 bits operated at a supply voltage of 1.5V, our memory design delivers an access time of approximately 36ns and consumes 17pJ per bit for a write followed by a read operation.

1.6 Performance and Power Consumption of Library Cells

Table 3 summarizes the hSPICE simulations of the energy consumed per operation in our cell library. The energy per operation numbers shown in Table 3 give us a guideline for eliminating some operations in favor of others in designing our decoder architecture. For example, a SRAM write access consumes 9 times,

Table 3. Energy per Operation at 1.5V Supply in 0.8μm CMOS Technology.

Module/Operation	Word Size	Energy/Operation (pJ)	Normalized to Adder
Carry-select adder	16 bits	18	1
Multiplier	16 bits	64	3.6
Latch	16 bits	4	0.22
8×128×16 SRAM (read)	16 bits	80	4.4
8×128×16 SRAM (write)	16 bits	160	9
External I/O access[a]	16 bits	180	10

[a]Estimate of the off-chip access energy is based on a total capacitance of 20pF per pin. 180pJ was obtained by multiplying the 720pJ (CV^2) by an activity factor of 1/4. This activity factor is based on a uniform distribution of 1s and 0s. The capacitance values were estimated from [33].

a chip I/O access 10 times, and a multiplication approximately 4 times more energy than a simple add operation. Given that it is possible to implement an algorithm in many different ways, the motivation is to replace memory accesses with computation, as computation dissipates less energy per operation. This trade-off will be crucial on architectural design, as will be discussed next.

The relatively large energy dissipated by off-chip I/O accesses is the motivation for minimizing external data communication. In our design, we eliminated external frame buffers and placed memory on-chip whenever possible. As high-quality video implies high-pixel bandwidth, with at least millions of pixel operations per second, external frame buffers lead to huge energy waste and overwhelm all other low-power improvements made. Requiring no off-chip memory support is one of the main factors enabling our portable decoder to deliver high-quality video at a minimal power level.

In the next two sections of this paper, we will discuss the low-power design strategies used in implementing a portable video decoder based on our subband/PVQ algorithm to be operated at a power level two orders of magnitude below existing JPEG decoders in comparable technology.

2. A 1.2mW VIDEO-RATE 2D COLOR SUBBAND DECODER

Subband decomposition divides each video frame into several subbands by passing the frame through a series of 2-D low-pass and high-pass filters. Each level of subband decomposition di-

vides the image into four subbands. We can hierarchically decompose the image by further subdividing each subband. We refer to the pixel values in each subband as subband coefficients. The choice of low-pass and high-pass filter coefficients is determined by their ability to accurately reproduce the original image and cancel frequency aliasing between subbands.

The single-chip subband decoder [9] described here implements subband decoding used in many video-compression algorithms [2] with very low power consumption. Previously designed subband decoders were not optimized for low power operation and additionally required large external memory support which consumes even more power [10–12]. The single-chip decoder described here does not require any external hardware support such as off-chip memory or video control for the delivery of real-time video signals to a color display. Also, this chip, unlike other designs, supports both two-dimensional data and multiple levels of subband decomposition.

In designing the subband decoder chip, we emphasized a low-power implementation without introducing noticeable degradation in decompressed video quality. As memory accessing is by far the most power-consuming operation, the main design strategy has been to eliminate memory accesses in favor of on-chip computation. Table 4 lists the major architectural decisions made for the implementation of a low-power subband decoder.

First, the energy per operation is reduced by using a lower supply voltage which provides just enough performance to compensate for the increased gate delay. Next, the computation operation throughput is minimized by careful selection of the subband filter. Finally, data ordering and representation limit the I/O requirement as well as the amount of memory accesses. In addition, with a system-wide view of power consumption, low-power implementations of numerous external support functions were incorporated into the decoder chip, including a video controller, color converter, and output frame buffer.

Figure 3 illustrates the overall chip architecture. A small input buffer stores the subband data and passes it to the horizontal filter unit. The line-delay memory stores the horizontal filter outputs and sends them to the vertical filter followed by the scale unit. Lower-level results are temporarily stored in the intermediate-store memory where they are passed back to the input buffer for reconstruction of the next level. Top-level subband results are stored in the final-result buffer before conversion from the YUV

Table 4. Architecture Trade-Offs in Designing a Low-Power Subband Decoder.

Technique	Description	Benefit
Filter choice	Use a 4-tap filter with coefficients (3,6,2,−1).	3× power savings versus using a 9-tap filter
External accesses	Reconstructed LL subband coefficients are kept on-chip, requiring no frame buffer. High-frequency subband coefficients are zero run-length encoded.	3× reduction in the number of external memory accesses
Zero processing	Skip processing of zeros in high-frequency subbands.	1.5× reduction in processing cycles
Word size	Rounding to 10 bits gives the same quantization error as truncation to 12 bits (from simulation).	1.2× reduction in memory and datapath wide
Internal memory	Interleave subband levels so that only one line of horizontal filter results and two lines of vertical filter results are stored.	2× reduction in the internal memory size
Color conversion	Simplify coefficients to allow a multiplier-free implementation.	Over 5× reduction in converter area and power

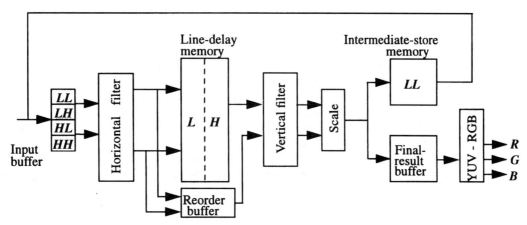

Figure 3 Architecture of the subband decoder chip.

to the color space. The RGB results are sent off chip to a digital analog convertor (DAC) and then to the display.

2.1 Supply Scaling

The design exploits the natural parallelism of the subband algorithm to achieve high peak performance, providing excess throughput that can be traded off for lower power by reducing the supply voltage. Lowering the supply from 5V to 1.5V reduces the energy per operation by a factor of 12 for a constant processing throughput. The gate delay increases by 7 times, for which the excess performance compensates. The design uses concurrent high- and low-pass horizontal and vertical filters which are efficiently pipelined to allow for a faster clock rate. All of the circuit designs, except for the memory, use static CMOS because of their ability to function correctly at low voltages with a minimal leakage current.

2.2 Filter Implementation

A major consideration in subband coding is the choice of filters. This determines the amount of computation required and the number of line delays needed, as well as affecting the algorithm's compression performance. Extensive simulations were performed to select the filter, resulting in a short 4-tap asymmetric wavelet filter (3,6,2,–1) as the low-pass kernel filter [13]. This filter performs comparably to a 9-tap quadrature mirror filter (QMF) but uses only one third of the power and greatly reduces the line-delay memory requirements.

The filter implementation, as shown in Fig. 4, uses shifts and adds to implement multiplications of the simple filter coefficients, requiring only a single 3-2 adder. Only two inputs are needed when applied to the 4-tap filter since the unsampling fills in two zeros between the data values. A small input buffer holds two horizontally consecutive inputs from each of the four subband types.

The same hardware also implements the high-pass filter by reversing the coefficients and negating the $(6, -1)$ coefficient pair. The reversal is accomplished by switching the order of the two input values. The negation occurs when the low-pass and high-pass values, still in their carry-save forms, are combined with a 4-2 adder and then a carry-select adder. The rounded results are stored in the line-delay memory and passed to the vertical filter, which operates nearly identically to the horizontal filter but with one input from the line-delay memory and the

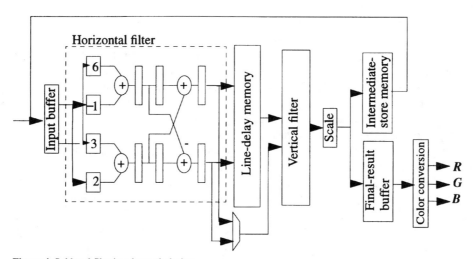

Figure 4 Subband filtering datapath design.

423

other from the output of the horizontal filter, forming two vertically oriented values.

In digital signal-processing design, the word size of internal data representations is an important parameter determining the accuracy of the final output. As video signals at a SNR higher than 46dB usually appear nearly perfect, a word size of 12 bit is usually used. Out simulations indicated that, when using rounding instead of truncation in quantization, a word size 2 bits smaller will achieve a similar output SNR. This allows a word size of only 10 bits in our internal data representation, reducing the power and area of all datapath circuits, including the internal memory which uses the most silicon area. This 10-bit rounding strategy provides a power reduction of nearly 20% over the 12-bit one and a power reduction of 40% over the standard 16-bit approach with no perceptive loss in video quality.

2.3 Memory

The size of the line-delay, intermediate-store, and final-result memories is critical for achieving a low-power implementation. First, the power consumed in the memory increases with its size. Second, the size of these memory units determines if they can be kept on chip, greatly reducing the I/O power. For instance, holding the intermediate reconstructed results on-chip eliminates the need to write these values out only to be read in later. Also, a power-expensive frame buffer is eliminated by generating the output data in the raster scan order and storing it in the final-result buffer. The data can therefore be passed directly to the display device through a DAC. As previously mentioned, the short filter length reduces the number of lines of subband data required for the vertical filtering and, by using the values just created by the horizontal filter, only a single delay line is required.

The ordering of subband level processing also affects the amount of memory needed. Processing one line at the lowest level of the subband hierarchy generates two lines of 1-D results and two lines of intermediate results. If both lines of the intermediate results are processed, 4 lines of data at the next level are generated, followed by 8, and finally 16 lines at the top level. In our implementation, the order of levels is interleaved such that processing of the next line from a given level only occurs when both lines at the higher level have already been processed. This allows memory to be re-used when the next results are generated. Thus, only two lines of line-delay memory and two lines of intermediate-store memory are required per level. Additionally, the chip decodes two chrominance bands along with the luminance data. The chrominance bands are also interleaved and inserted into the luminance band stream so that the data sets maintain synchronization.

The total amount of memory for the line delays is 20 kbits, organized as 8 banks of 128 bits high by two words of 10 bits. The intermediate-store memory, totalling 10 kbits ($128 \times 4 \times 20$), holds two lines of subband coefficients for each level and then feeds them back to the input buffer for processing of the next level. The final-result buffer was increased to 4 lines to prevent buffer underflow and holds 20 kbits ($128 \times 8 \times 20$) of output data.

A local word line activates only the desired memory bank, saving the unnecessary power of activating other banks.

The intermediate and final-result memories also perform a transpose of the data from the vertical filter since the two values generated in a cycle are oriented vertically, but need to be accessed horizontally. Two vertically filtered outputs are combined before each horizontal pair is written to its own bank. The memory also has separate read and write buses, eliminating the problem of disabling the read output before the write data is activated. Furthermore, with dedicated buses, as opposed to shared buses, only the necessary capacitance is switched, reducing the power consumption.

2.4 External Accesses

External data and memory accesses represent another major energy-consuming operation that need to be minimized. Intermediate reconstructed LL subband data are therefore stored internally and not written off chip. This reduces the number of external reads per pixel from 1.98 to 1.5 and the number of external writes per pixel from 3.48 to 3. Also, as the design does not require a frame buffer, an additional 3 external accesses per pixel are saved.

To reduce the communication bandwidth between the PVQ decoder and the subband decoder chips, zero run-length encoding is used to take advantage of the large number of zeros in the three highest-frequency subbands. In these high-frequency subbands, the PVQ uses a 32-dimensional horizontal vector which, when decoded, will contain only 3 non-zero coefficients. The PVQ decoder transmits to the subband decoder the values of non-zero coefficients and a zero run-length to the next non-zero coefficient. Consequently power is saved by reducing the size of the PVQ output buffer and the number of external accesses. By accepting zero run-lengths between non-zero coefficients, the number of external reads per pixel for the subband decoder chip is further reduced from 1.5 to 0.57. Without considering external access reductions, the total power of the subband chip would have been 40% more, not counting the power dissipated in accessing large external memory.

2.5 Color Conversion

The chip includes the YUV-to-RGB color conversion operation because of its potentially high power dissipation. The conversion algorithm is greatly simplified but with no visual degradation in image quality [15]. This provides an efficient implementation requiring only 5 carry-select adds and 1 carry-save add per pixel per RGB component. Our color conversion circuit consumes only 90μW at a 1V supply for three components at 1.27Mpixels/s.

A video timing controller included in the chip regulates reading the YUV data from the final-result buffer and the generation of the RGB outputs. Programmable timing parameters for vertical and horizontal synch and blanking intervals specify the video synchronization signal needed by the display device. When the final-result buffer is full, the chip executes power-down stall cycles, using gated clocks, until buffer space is available. Stalling occurs during vertical and horizontal blanking periods since no data is read from the buffer. Since the data is buffered and formed in the raster scan order, the RGB output can be sent directly through a DAC to the display device without the need for a high power frame buffer.

2.6 Zero Processing

When processing the highest frequency bands, the design takes advantage of the large percentage of zeros in these bands. The input data for the top-level luminance LH and HH bands are zero run-length encoded. The top-level luminance HH as well as the top-level chrominance bands are all set to zero. This reduces the number of external inputs by almost a factor of 4, lowering the overall system power by reducing the power dissipated in the external input source. The controller uses the zero run-lengths to determine when a value is zero and stores this information in a shift register. Processing is skipped when two consecutive values of the HL band are zero. When this occurs, the LL and LH data are processed as usual, but the HL and HH data are skipped, moving to the next LL and LH data.

With this process skipping, the average number of clock cycles required per output pixel, including generation of all intermediate and final image data for all three components, is reduced from 1.98 to 1.23, resulting in a 15% reduction of total chip power by increasing the number of stall cycles. Because the processing of multiple levels does not generate an even flow of output results, the chip, operating at twice the pixel rate, may not meet the original 1.98cycles/pixel requirement. Therefore, the reduction in processing cycles is also important to ensure that output buffer underflow cannot occur.

2.7 Control Flow

Application of the above power-saving techniques, however, tremendously increases the level of complexity in the decoder's controller design. The flow-control circuit consists of four finite-state machines controlling zero run-length decoding, idle-cycle stalls, interleaving levels of subband decomposition, and generating video synchronization signals to the display device.

The run-length section deals with keeping track of the zero data positions and determines when processing is required. The zero-process skipping further complicates the pipeline control since now two different operation sequences are possible. One sequence consists of alternating between all four band types, while the other only alternates between the LL and LH types. The current state must be passed down the pipeline to ensure the correct operation of both the computational and memory units.

The main state machine keeps track of the band type, coefficient position within a line, the color component, and the current level in the interleaving scheme. Additionally, it generates addresses for the memories. The video controller determines the display timing and controls reading from the final-result buffer. This read address is compared to the state controller's final-result buffer write address to determine whether the buffer is full and the chip must be stalled.

2.8 Performance

The subband decoder chip implemented in 0.8μm CMOS technology, occupies a chip area of 9.5×8.7 mm^2 (with an active area of 44 mm^2) and contains 415K transistors. At a 1V supply voltage, the chip runs at a maximum frequency of 4MHz and delivers one output pixel of three color components at every two clock cycles. To satisfy the throughput requirement of 1.27Mpixels/s (176×240pixels/frame at 30frames/sec), the subband decoder chip only needs to run at 3.2MHz. This fairly low operating frequency is an indication of the efficiency of our compression algorithm. Even for video of the standard SIF format (352×240pixels/frame at 30frames/s), the subband decoder chip only needs to run at 6.4MHz, while most JPEG decoding chips are required to run at a frequency between 30MHz and 50MHz.

The peak performance at 5V generates 60Mpixels/s of RGB components with a 120MHz operating frequency while dissipating 1.2W. For a target rate of 3.2MHz for use with a small portable display, the chip provides significant excess throughput, allowing it to meet the requirements with a 1V supply while dissipating under 1.2mW. Figure 5 illustrates the power dissipation at the maximum operating frequency for various supply voltages.

Table 5 displays the breakdown of total power consumption among the different sections of the chip at the 3.2MHz rate. As expected the computation, I/O, and memory dominate the power, each consuming approximately one third of the total power. Power usage in the control section remains a small percentage despite the increased complexity required to implement all of the power-saving strategies. The estimated power numbers were derived by determining the operation throughputs multiplied by simulated energy per operation numbers. These estimates were used in making high-level algorithmic design decisions and are very close to the measured numbers, including the validity of this design approach.

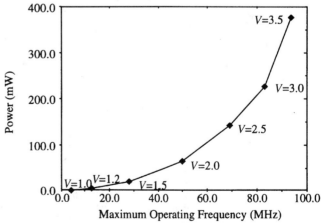

Figure 5 Power consumption of the subband decoder chip as a function of clock speed.

Table 5. Power Breakdown of the Subband Decoder Chip at 1V and 3.2MHz.

Operations	Mops/Sec	Energy/op (pJ)	Estimated Power (mW)	Measured Power (mW)	% of Total
Total datapath			0.35	0.34	29
Add (16 bits)	17.7	7	0.12		
3-2 add (16 bits)	25	2	0.05		
Latching (16 bits)	100	1.8	0.18		
Internal memory			0.26	0.39	33
Internal read (16 bits)	2.4	36	0.09		
Internal write (16 bits)	2.4	71	0.17		
External access (16 bits)	2.7	80	0.22	0.38	31
Control			0.13	0.09	7
Total			0.96	1.2	100

For higher-resolution images, multiple chips would be cascaded, each operating on a maximum of 256-pixel wide slice, producing a final image without boundary artifacts. Table 6 illustrates the power dissipation of multiple-subband decoder chips at the required clock frequency when used for decompressing high-resolution images. The operating voltages are determined by the real-time computation requirements. This additional chip level parallelism keeps the operating frequency and thus the supply voltage low, resulting in extremely low-power dissipation even for HDTV applications.

Figure 6 is a chip micrograph of the subband decoder chip in a 0.8μ, three-level metal, CMOS technology with a threshold voltage of 0.7V. The three memory blocks dominate the area, emphasizing the importance of minimizing their size.

Figure 7 illustrates the power dissipation of the subband decoder chip at 3.2MHz over various supply voltages. At our target supply voltage of 1V, the subband decoder chip consumes only 1.2mW to perform real-time video-rate decoding. At a slightly higher supply voltage (e.g., at 1.5V) the subband decoder chip would consume 2.64mW.

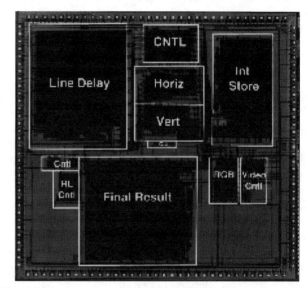

Figure 6 Die photo of the subband decoder chip.

3. A LOW-POWER VIDEO-RATE PYRAMID VQ DECODER

Pyramid vector quantization (PVQ) is a vector-quantization technique that groups multiple data values and scales them onto a multi-dimensional pyramid surface. PVQ is a compression technique that provides both compression efficiency and error resiliency and is well-suited for portable video applications. This section decribes a PVQ decoder chip used for real-time video decompression with low-power operation [17]. The chip performs decompression by converting PVQ codewords into data values and integrates all functionalities requiring no external hardware support or memory.

Unlike standard VQ schemes, which require codebook storage, product PVQ relies on intensive arithmetic computation to perform encoding and decoding. The product PVQ encoding process is as follows: a data vector, formed with L values, is encoded by scaling the vector onto an L-dimensional pyramid surface and finding the nearest lattice point on the pyramid (Fig. 8). Both the scaling factor and an index that corresponds to that lattice point are transmitted. The decoding process converts the in-

Table 6. Subband Decoders for High-Resolution Configurations.

Format	Size	Number of Chips	Clock Frequency (MHz)	Power (mW)
Sharp LCD	$176 \times 240 \times 30$	1	3.2 (1V)	1.2
CD-I (TV)	$352 \times 240 \times 30$	2[a]	3.2 (1V)	2.4
CCIR video	$704 \times 480 \times 30$	3	8.4 (1.2V)	12.6
HDTV	$1920 \times 1035 \times 75$	8	19 (1.5V)	122
Hires monitor	$1024 \times 768 \times 75$	4	37 (2.0V)	192

[a]Multiple-chip power dissipation is extrapolated from single-chip measurements at the specified voltage and frequency.

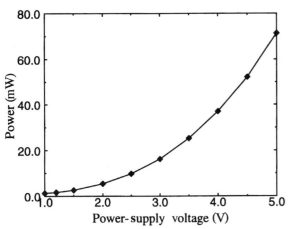

Figure 7 Total power consumption of the subband decoder chip at 3.2MHz over various supply voltages.

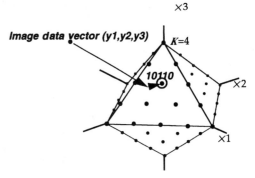

Figure 8 PVQ encoding on a 3-D pyramid surface.

dex back into an L-dimensional vector and scales that vector using the scaling factor. Since the lattice points on the pyramid are regularly spaced and are described by recursive equations of combinatorial values, encoding and decoding PVQ indices are performed with arithmetic computation, using shifts, subtracts, look-ups, and compares.

3.1 Chip Architecture

The overall block diagram in Fig. 9 shows the general dataflow of the chip. The PVQ decoder is divided into four processing blocks. The *stream parser* parses incoming 16-bit words into PVQ indices and scale factors using a series of two 32-bit wide barrel shifters. The various word lengths of indices and scales are stored in a ROM. The *index pre-decoder* (16-bit datapath) decodes each index into four intermediate indices which describe the vector. It decodes the first index by comparing the PVQ index to the offset values stored in a ROM, and the remaining three indices are parsed using a barrel shifter. The *vector decoder* (16-bit datapath) inputs these intermediate indices and generates a data vector by iteratively comparing and subtracting these indices from pre-computed combinatorial offsets stored in a ROM. Each of these comparisons increments a counter which sets the decoded vector value. The resulting non-zero values are then negated, as needed.

Finally a 6×8-bit pipelined multiplier performs the final scaling of each decoded vector element using the scaling factor sent with each PVQ index. The multiplier is composed of a 4-2 adder tree with a carry-propagate adder. FIFO buffers separate the four processing blocks and regulate dataflow between them.

While the 16-bit internal datapaths allow for any vector dimensions up to 256, the chip only stores the required offset values in its ROM tables to decode vectors of dimensions 4 to 32, as required by our particular optimized bit allocation algorithm.[1] Simple modifications to ROM tables of this chip would allow for other combinations of vector dimensions and pyramid radii.

In addition to operating at a low supply voltage, the PVQ decoder chip employs several key architectural strategies to minimize its power consumption as listed in Table 7.

3.2 Key Architectural Strategies for Power Minimization

First, we achieve maximum throughput using parallelism and pipelining, so that excess performance can be traded-off for lower power consumption through voltage scaling. The chip incorporates four independent processing blocks, each indi-

---

[1]For vector dimension of 4, the chip decodes pyramids with "height" (L1 norm) up to 25, which provides bit rates up to 5 bits per vector element. For vector dimension of 32, the chip decodes pyramids with height up to 3, which provides bit rates up to 0.344 bits per vector element.

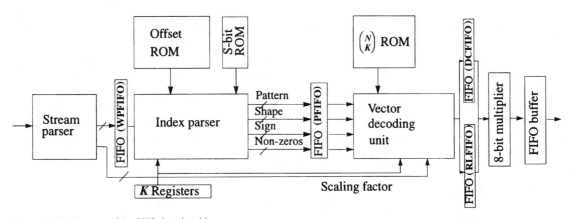

Figure 9 Architecture of the PVQ decoder chip.

Table 7. Architecture Strategies in Designing a Low-Power PVQ Decoder.

Technique	Description	Benefit
Independent processing elements (PEs)	Each PE operates independently, distributing processing cycles to match the non-deterministic nature of the decoding process.	2× reduction in total power by using gated clocks
Fast block search in vector decoding	Fast block search (four at a time) reduces processing cycles in the critical-path vector decoding unit.	2× reduction in processing cycles and the size of the output buffer
Minimal external accesses	All combinatorial and indexing data, required for the decoding process, are stored on-chip.	2× reduction in the number of external accesses
Run-length encoding	Decoder outputs of the high-frequency subbands are zero run-length encoded to compress the output bit stream.	10× reduction in the number of internal buffer accesses 3× reduction in the size of output buffer storage
Data reordering	Interleaved data between subband levels trades off additional control complexity for reduced overall system memory requirements.	Eliminates the off-chip frame buffer

vidually pipelined. Because the PVQ decoding algorithm is inherently non-deterministic, i.e., the number of steps to decode an index depends on vector data, the latency in the *index pre-decoder* and *vector decoder* units is also non-deterministic. Dividing the chip into four processors separates the dependencies between the various blocks and maximizes the chip throughput. Each processor is separated by FIFOs and only continues when its input FIFO is not empty and its output FIFO is not full. When idle, each processor enters a stand-by mode, and the clock to that entire unit is gated. More than half of the chip's total clock capacitance lies in gated clocks. With the exception of the *vector decoder,* whose idle time is typically 10%, the other processing blocks are typically idle 50% to 60%. This high idle-time probability leads to savings in total clock power by a factor of 2.

Second, because the most energy-expensive operations are off-chip I/O accesses and memory accesses, significant effort was made to minimize these operations and keep all data storage on-chip. All stream parsing data and pre-computed factorial and combinatorial offset values, required for index parsing and decoding, are stored in on-chip ROMs. All intermediate results are stored on-chip in register-based FIFO buffers.

The FIFO buffering regulates clock gating of each processor, helps smooth out otherwise erratic dataflow between processors, and helps guarantee constant dataflow at the chip output, which is required by the subband decoder chip that follows. In order to guarantee constant dataflow at the chip output, behavioral simulation using Verilog was performed using worst-case input data to carefully determine the FIFO sizes. In addition, output data vectors that contain a large number of zeros, commonly found in high-frequency subband data, are zero run-length encoded to losslessly compress the representation of consecutive zeros. This encoding reduces the amount of external chip accesses by a factor of 3, output buffering by a factor of 3, and the number of internal buffer accesses by up to a factor of 10. Here, we traded off additional control complexity to perform zero run-length encoding for lower I/O power and on-chip data buffering.

Third, the chip's critical path, found in the *vector decoder,* is optimized to increase throughput. This allows for lower-voltage operation and reduces the output buffering required to meet

real-time constraints. Unlike the direct algorithm implementation, which uses a linear search to locate the correct index offset, improved throughput is achieved by searching and processing a block of four combinatorial offsets at a time. For typical image data, this reduces the average number of search iterations from 15 to 3 and halves the number of processing cycles and the amount of output buffering.

Fourth, the chip incorporates an energy-efficient register-based FIFO design, which uses a pointer-based scheme (Figure 10). When the FIFO is accessed, only the pointer, stored in a single-bit shift register, moves, not the data. Compared to other register-based FIFO designs, where significant power can be consumed in shifting data between registers, this scheme minimizes power by minimizing data switching. Additionally, internal clocking within the FIFO is also turned off when the FIFO is idle. This design performs simultaneous read and write, which allows each processor to operate entirely independent from other processors. To interface with a subband reconstruction chip described in the previous section that operates at half the clock frequency, a special output FIFO buffer was designed for the chip's output section that accommodates two clocks (full and half frequencies) simultaneously. For these output FIFOs, the read-pointer shift-register control operates with the half-frequency clock, while the write-pointer shift-register control operates with the regular clock.

Finally, when used in tandem with the subband decoder chip, the PVQ decoder chip incorporates additional control to allow for data interleaving between subband levels, a feature which eliminates the need for an off-chip frame buffer. An individual controller for each processing unit keeps track of its location in the interleaved bitstream and in which subband it is operating. While this feature requires additional control logic, it was considered a good trade-off for the significant power savings achieved in the overall system by eliminating the frame buffer.

3.3 Clocking Methodology

Because clocking comprises a significant portion of the total chip power, careful design of the clock methodology, distribution, and

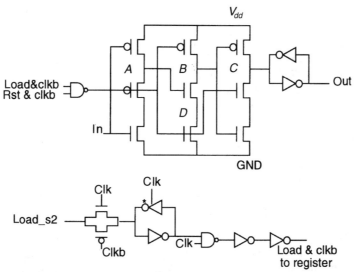

Figure 10 Edge-triggered register circuit and datapath clock driver with gating logic.

gating was performed. The PVQ chip utilizes a single-phase clocking scheme. The basic register cell used throughout the design is a variation of the one proposed by Larson and Svensson [16], as shown in Fig. 10. When the clock is low, the first stage of the register (master) is on, the input value is stored on node A, and node B is pre-charged high. When the clock switches high, the first stage is shut off, and node B discharges or stays high, depending on the input data. This value is then inverted through the last stage (slave) to node C and stored in the feedback inverter configuration. In the standby mode, the clock remains low.

Each register cell has its own logic for load and reset. Since both the load and reset signals are guaranteed to be stable when the clock is high, the clock is gated using the logic shown in Fig. 11. This gating, when load_s2 is low, is used to put the entire datapath in the standby power-saving mode.

The clock distribution is a standard tree configuration with a central clock-buffer chain located at the center of the chip driving the global chip clock to each processor datapath and con-

trol. Because of the large chip size, the total capacitance of this global clock node is quite large (41pF). The clocks to each datapath are gated, but the clocks to the control sections are not, since they must continuously stay on to monitor the FIFO buffer or chip input status.

3.4 External Accesses and Zero Run-Length Encoding

Combinatorial calculation and index offsets, required for the vector decoding unit, were stored on-chip in low-power ROMs. In addition, parsing information for subband bit allocation was also stored in this ROM. This scheme essentially trades off greater programmability for less-external accesses to off-chip memory.

As described earlier, zero run-length encoding of high-frequency subband coefficients reduces the amount of external accesses by a factor of three. More significantly for the PVQ

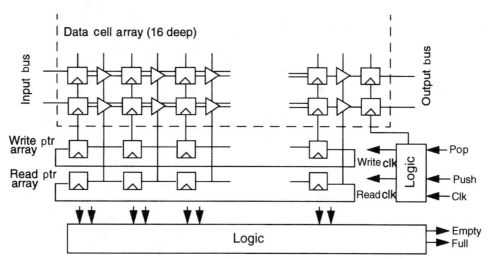

Figure 11 16-cell FIFO block diagram.

decoder chip, because zero run-length encoding compresses the representation of consecutive zero coefficients, it also reduces the amount of output buffering on the PVQ chip by a factor of three and the number of internal buffer accesses by up to a factor of ten. Here again, we traded off additional control to perform zero run-length encoding for lower I/O power and smaller on-chip data buffering.

3.5 Throughput

The PVQ decoder chip meets the throughput requirement of real-time video decoding at 176×240 pixels/frame at 30 frames/s. From simulations of buffer occupancy of different image sequences, the worst-case sizes of various FIFO buffers between processing elements were found, each of a different size between 4 and 192. The different FIFO sizes indicate the relative speeds of production and consumption of data between different processing elements. By ensuring that all buffers are never full nor empty while there is incoming data to the PVQ decoder, the system is guaranteed to operate at maximum performance. From this performance figure, we calculated the maximum delay allowed to meet the video-rate throughput requirement at the lowest power consumption. A clock frequency of 6.4MHz was found to be adequate, although a PVQ decoder chip achieving a maximum clock frequency of 18MHz at 1.5V has been designed in 0.8μm CMOS technology.

3.6 Measured Performance

The PVQ decoder chip occupies a chip area of 9.7×13 mm^2 (with an active area of 74.9mm^2). Because the PVQ decoder chip needs to operate at a frequency twice that of the subband decoder chip, to meet the video-rate throughput at 1.27 Mpixels/s, the PVQ decoder chip needs to run at 6.4MHz, powered by a 1.35V supply. At this rate, the total power consumption of the PVQ decoder chip was measured to be 6.63mW. The breakdown of the total chip power is shown in Table 8, where the operations included are arithmetic computation, control, FIFOs, clocking, internal ROM accesses, and external accesses.

Figure 12 shows the measured power at maximum frequencies for various supply voltages. The chip operates at 1.35V at 6.4MHz to perform real-time video decoding for a display of 176×240 YUV pixels^2 at a peak computation rate of 21 MOPS and dissipates 6.7mW (with an output load of approximately 4pF per pin).

The chip, shown in Fig. 13, contains 272K transistors and is implemented in 0.8μm triple-metal CMOS with typical threshold voltages ($V_{tn} = 0.75$, $V_{tp} = 0.90$).

---

^{2}176×240 represents the resolution of a 4″ color LCD display, which was used in a portable video prototype incorporating this chip. This resolution also represents a single *field;* two interlaced fields form a full frame, which is 352×240. The YUV-RGB conversion is performed on the subband reconstruction chip as described in the previous section.

Table 8. Power Breakdown of the PVQ Decoder Chip at 6.4MHz.

Operations	% of Total Power
Total datapath	17.9%
Control	20.0%
FIFOS	
Local clock	19.9%
Registers/control	15.6%
Global clock	19.0%
ROM accesses	4.1%
External accesses	3.5%
Total	100%

The PVQ decoder consumes roughly five times the power of the subband decoder chip. The doubled clock frequency of the PVQ decoder explains much of this difference. A third of the chip power is consumed by the register FIFO storage with approximately 55% of this power dissipated on local clock lines. Using a memory-based design instead of the register-based FIFO storage would have helped reduce this power as well as area. It is also interesting to note that the datapath power is roughly equal to the control power. There are several reasons for this: (1) the datapaths use local gated clocks, while the control sections do not—a significant factor in that each processing unit may be idle up to 40% of the time; and (2) the control sections require greater complexity due to the non-deterministic nature of the PVQ decoding algorithm and the data reordering scheme. Because of the relatively large design, extensive global clock routing results in significant power consumption. Global and local clocking in the control, datapaths, and FIFOS makes up about 50% of the chip power. Finally, the power consumed in external accesses is a relatively small fraction, because the design limits off-chip accesses to only compressed data, at both the input and output.

In this section, a pyramid vector quantization decoder chip used for real-time video decompression was described. Maximizing throughput using pipelining and parallelism makes it possible to operate at low supply voltages and still meet the real-time processing requirement. In addition, architectural and circuit optimizations that minimize energy-expensive operations, such as I/O, memory accesses, and idle clocking, significantly reduce power consumption. These techniques, combined with a computation-efficient algorithm, result in very low power implementation.

4. CONCLUSIONS

The design of low-power electronics systems, especially portable systems, requires a vertical integration of the design process at all levels, from algorithm development, system architecture, to circuit layout. System performance need not be sacrificed for lower power consumption, if the design of algorithms and hardware can be considered concurrently.

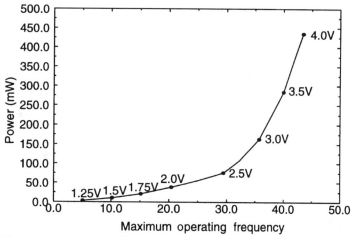

Figure 12 Power dissipation vs. maximum frequency.

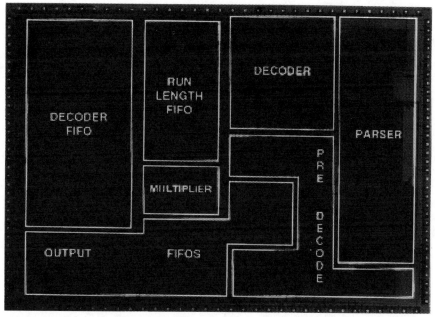

Figure 13 Die photo of the PVQ decoder chip.

From designing this portable video-on-demand system, we learned that power reduction can be best attained through algorithm and architecture innovations, guided by the knowledge of underlying hardware and circuit properties. This hardware-driven algorithm design approach is key to the design of future portable systems under stringent power budgets. Without appropriate high-level synthesis tools, this vertical-integration task is by no means trivial. The designer is required to be well-informed at all levels of the design process, willing to compromise, and make sensible trade-offs in order to reach a globally-optimal solution.

References

[1] E. K. Tsern and T. H. Meng, "Image coding using pyramid vector quantization of subband coefficients," *Proc. of IEEE ICASSP 1994,* Apr. 1994, pp. V-601–604.

[2] J. Woods, editor, *Subband Image Coding,* Boston MA: Kluwer Academic Publishers, 1991.

[3] A. C. Hung and T. H. Meng, "Error-resilient pyramid vector quantization for image compression," *Proc. International Conference on Image Compression,* Nov. 1994, pp. I.583–587.

[4] T. H. Meng, B. M. Gordon, E. K. Tsern, and A. C. Hung, "Portable video-on-demand in wireless communication," invited, *Proceedings of IEEE,* vol. 83, no. 4, Apr. 1995, pp. 659–680.

[5] S. Purcell, "C-Cube CL550 image processor," *Proc. HOT Chip Symposium,* Aug. 1990.

[6] N. Weste and K. Eshragian, *Principles of CMOS VLSI Design: A Systems Perspective,* 2nd Ed., Boston, MA: Addison-Wesley, 1992.

[7] A. Chandrakasan, S. Sheng, and R. W. Brodersen, "Low power CMOS digital design," *IEEE Journal of Solid-State Circuits,* Apr. 1992, pp. 685–691.

[8] A. P. Chandrakasan and R. W. Brodersen, "Minimizing power consumption in digital CMOS circuits," *IEEE Proceedings,* vol. 83, no. 4, Apr. 1995.

[9] B. M. Gordon and T. H. Meng, "A 1.2mW video-rate 2D color sub-band decoder," *IEEE Journal of Solid-States Circuits,* vol. 30, no. 12, Dec. 1995, pp. 1510–1516.

[10] M. Winzker, et al., "VLSI chip set for 2D HDTV subband filtering with on-chip line memories," *IEEE Journal of Solid-State Circuits,* vol. 28, Dec. 1993, pp. 1354–1361.

[11] M. Vishwanath and C. Chakrabarti, "A VLSI architecture for real-time hierarchical encoding/decoding of video using the wavelet transform," *Proceedings ICASSP 1994,* vol. 2, April 1994, pp. 401–404.

[12] A. Lewis and G. Knowles, "VLSI architecture for 2-D Daubechies wavelet transform without multipliers," *Electronic Letters,* Jan. 1991, pp. 171–173.

[13] A. Akansu, "Multiplierless suboptimal PR-QMF design," *SPIE,* vol. 1818, Nov. 1992, pp. 723–734.

[14] T. Senoo and B. Girod, "Vector quantization for entropy coding of image subbands," *IEEE Transactions on Image Processing,* Oct. 1992, pp. 526–533.

[15] B. Gordon, N. Chaddha, and T. Meng, "A low power multiplier-less YUV to RGB converter based on human vision perception," *1994 IEEE Workshop on VLSI Signal Processing,* Oct. 1994, pp. 408–417.

[16] P. Larsson and C. Svensson, "Clock slope constraints in true single phase clocked (TSPC) CMOS circuits," *Proceedings of the Third Eurochip Workshop on VLSI Design Training,* Sept. 1992, pp. 197–202.

[17] E. K. Tsern and T. H. Meng, "A low power video-rate pyramid VQ decoder," *IEEE Journal of Solid-States Circuits,* vol. 31, no. 11, Nov. 1996, pp. 1789–1794.

Low-Power Design of Memory Intensive Functions

David B. Lidsky, Jan M. Rabaey

University of California, Berkeley, California 94720 USA
lidsky@eecs.berkeley.edu, jan@eecs.berkeley.edu

Abstract

Much of the recent efforts into low power design techniques neglects the impact memory accesses have on power consumption. This paper provides a summary of the techniques used in the design of a low-power vector quantizer (VQ), many or all of which can be applied to other memory intensive applications.

Design Approach

High level algorithmic decisions tend to have the most profound effect on power, but architectural and circuit level optimizations are also critical. For memory dominant applications, memory power can be reduced at the expense of slight increases in computation power.

A video image is vector quantized by breaking it into blocks (vectors) of pixels that are mapped to a codebook of probable vectors using mean squared error as the distortion measure [1]. The VQ was designed for the portable battery-operated Info-Pad terminal [2], which employs a codebook of 256 vectors. To process 30 frames/sec, a 4x4 pixel-vector must be compressed every 17.3 μsec.

Algorithm: Reducing the number of operations saves power by reducing switched capacitance and reducing the critical path, enabling operation at lower voltages and frequency. Table 1 shows the computation complexity for three methods of quantization. A full search through the codebook requires intensive computation which can be significantly reduced through the use of a binary tree search (TSVQ), with only a minor cost in accuracy. The number of computations can be further reduced by combining terms of the differential search a priori, as indicated in Figure 1 [3]. The differential tree search was used in all of the implementations.

Architecture: In one approach, a centralized memory, processing element, and controller are time multiplexed as shown in Figure 2. This method requires 18 cycles to process one node of the tree, consuming a total of 146 clock cycles per vector at a frequency of 8.3MHz.

A technique often used in low-power designs is to read bytes of words in parallel to reduce throughput constraints enabling the reduction of clock speed and voltage [2]. This architecture cannot be used to reduce speed requirements in TSVQ since the location of the vector to be chosen is dependent on the results at the previous node of the tree. This same architecture (Figure 3), however, reduces the switched capacitance by reducing the number of memory accesses. Table 3 shows power savings by using parallel access for different memory and byte sizes.

In TSVQ each level of a tree has specific codevectors that are found only at that level. Therefore the memory can be partitioned into separate memories for each level of the tree [4].

Associated with each memory are identical processing elements and controllers. Since this architecture can be pipelined the critical path can be reduced drastically (Figure 4), allowing clock frequency to be reduced by eight, and the supply voltage to be dropped from 3.0V to 1.1V.

The distributive memory architecture switches significantly less capacitance reading its codevectors than the centralized case since there is less overhead in reading from smaller memories (Figure 5). Though there are now eight controllers and eight processors on chip, they are clocked at one eighth the frequency, so the energy dissipated per vector does not change. Interblock power is negligible since it is clocked at an eighteenth of the clocking frequency.

Circuit: All circuits were designed using UC Berkeley's low-power cell library which employs minimal transistor sizes to reduce parasitic capacitance. To minimize transition activity, registers with gated clocks hold the previous inputs to adders during idle clock cycles. To reduce clock power, a single clock is used and registers are implemented using minimum sized True Single Phase Clock Registers (TSPCR).Memory is implemented with a low-power, on-chip SRAM whose bit-lines are pre-charged to an NMOS threshold voltage below the supply voltage to reduce swing [2].

To illustrate the multi-level approach to power reduction, three architectures were implemented: single memory with serial access, and single and distributive memory TSVQ encoders with parallel memory access. At this time, complete testing has not been accomplished, so all numbers are from extensive IRSIM and SPICE simulations. Power and area numbers are summarized in Table 2. Modifying the centralized design to include parallel memory accesses reduced the switching capacitance by 60%. The capacitance, as well as the critical path, is further reduced by memory partitioning.

To reduce power in memory intensive functions, proper choice of algorithm and architecture is essential. Algorithmic level decisions have been shown to reduce computational complexity and to reduce the number of memory accesses by a factor of 30. Architectural optimizations further reduced power by a factor of 17.

REFERENCES

[1] A. Gersho, *et al.*, "Vector quantization and signal compression," Kluwer Academic Publishers, Boston, MA, 1992.

[2] A. Chandrakasan, *et al.*, "A low power chipset for portable multimedia applications," ISSCC, March 1994

[3] W. Fang, *et al.*, "A systolic tree-searched vector quantizer for real-time image compression," *Proc.VLSI Signal Processing IV*, IEEE Press, NY, 1990.

[4] R. Kolagotla, *et al.*, "VLSI implementation of a tree searched vector quantizer," *IEEE Trans. on Signal Proc.*, vol. 41, no 2, February 1993.

Reprinted from *1994 IEEE Symposium on Low Power Electronics*, pp. 16-17, 1994.

$$MSE = \sum_{i=0}^{15} \left(C_i - X_i\right)^2 \qquad \begin{array}{l} C_i \text{ -- Code-vector} \\ X_i \text{ -- Input-vector} \end{array}$$

$$MSE_{ab} = \sum_{i=0}^{15} \left(C_{ai} - X_i\right)^2 - \sum_{i=0}^{15} \left(C_{bi} - X_i\right)^2$$

$$MSE_{ab} = \sum_{i=0}^{15} C_{ai}^2 - 2X_i C_{ai} + X_i^2 - \left(C_{bi}^2 - 2X_i C_{bi} + X_i^2\right)$$

$$MSE_{ab} = \underbrace{\sum_{i=0}^{15} \left(C_{ai}^2 - C_{bi}^2\right)}_{\text{Calculated a priori}} + \sum_{i=0}^{15} X_i 2\underbrace{\left(C_{bi} - C_{ai}\right)}_{\text{Calculated a priori}}$$

Figure 1: Mathematical Optimization

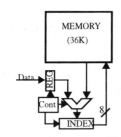

Figure 2: CentralizedMemory

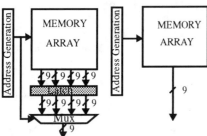

Figure 3: Parallel vs. Serial Memory Access

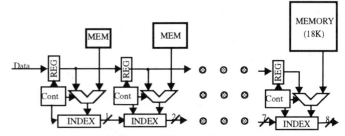

Figure 4: Distributive Memory

TABLE 3.   Energy per vector access vs. Memory Organization

	4 access + 0 multiplex	2 access + 2 multiplex	1 access + 4 multiplex
16 Kbits	360.4 pJ	320.0 pJ	290 pJ
8 Kbits	243.2 pJ	200.2 pJ	186.3 pJ
4 Kbits	186.0 pJ	150.6 pJ	135.5 pJ

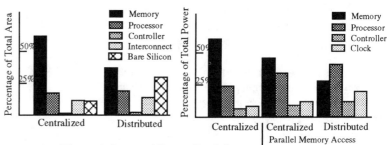

Figure 5: Area and Power Breakdown

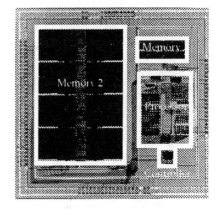

Figure 6: Centralized Memory Chip

TABLE 1.   Computational Complexity

	Memory Access	Multiplic-ation	Add/Subtract
Full Search	4096	4096	8448
Tree Search	256	256	520
Differential Tree Search	**136**	**128**	**136**

TABLE 2.   Simulation Results

	Memory Power			Core Area	Operating Voltage	Clock Frequency
	Power	% of Total	Absolute			
Centralized Serial Access	6.96 mW	**60%**	**4.15mW**	53.3 mm^2	3.0 Volts	8.30 MHz
Centralized	4.28 mW	**46%**	**1.95mW**	46.6 mm^2	3.0 Volts	8.30 MHz
Distributed	412 μW	**28%**	**114 μW**	92.9 mm^2	1.1 Volts	1.04 MHz

A 16b Low-Power-Consumption Digital Signal Processor

KATSUHIKO UEDA, TOSHIO SUGIMURA, MINORU OKAMOTO, SHINICHI MARUI,
TOSHIHIRO ISHIKAWA[1], AND MIKIO SAKAKIHARA[2]

MATSUSHITA ELECTRIC IND. CO., LTD., OSAKA[1]

MATSUSHITA COMMUNICATION IND. CO., LTD., YOKOHAMA[2]

MATSUSHITA ELECTRONICS CORP., KYOTO, JAPAN

This 16b digital signal processor (DSP), realizes a 11.2kb/s vector sum-excited linear predictive (VSELP) speech CODEC chosen as the digital cellular standard in Japan. 70mW power consumption at 3.5V VDD is achieved using a double-speed multiply-accumulate system, by improving logical and transistor circuits, and by using a 0.8μm double-metal-layer CMOS process and low VDD supply.

A VSELP CODEC needs over 20 MIPS DSP because of the complexity of the algorithm[1]. High-speed DSPs, however, consume much power and are not suitable for battery-energized equipment such as portable phones. To resolve this dilemma, the architecture has high performance with relatively low operating frequency.

Figure 1 shows the block diagram of this DSP, based on a previously-reported 16b DSP, targeted for 16kb/s adaptive predictive coding with adaptive bit allocation (APC-AB) CODEC [2]. Two units permit high performance with low operation frequency. One of these is a double-speed multiply-accumulate (MAC) and the other is an add-compare-select/block floating (ACS/BF) accelerator. These are used for three reasons: 1) MAC operation speed determines DSP operation frequency. Furthermore this frequency is very high in VSELP, because MAC operations represent more than 50% of all arithmetic and logical processing operations in the 11.2kb/s VSELP CODEC; 2) Previous DSPs cannot process Viterbi decoding efficiently; 3) Previous DSPs with equal MAC and ALU cycle times and without dedicated Viterbi decoding, need high operation frequency to realize a VSELP CODEC. A previous 24b DSP (24b version of the 16b DSP mentioned above), for example, required 14MHz to realize a 11.2kb/s VSELP CODEC. Furthermore, since the previous DSPs also execute MAC operations with a multiplier, a general ALU separated from the multiplier, general registers, and data buses connecting these components, they use extra power in MAC operation.

Figure 2 shows the mechanism of double-speed MAC operation. A multiplier in the MAC unit multiplies 16b data and output 32b product every half instruction cycle, then a 40b adder in the MAC unit, with input connected to the multiplier output through a pipeline register, accumulates 32b products with 40b accuracy every half instruction cycle. As a result, the MAC unit performs MAC operations every half instruction cycle. The 16b multiplicand and 16b multiplier are stored in the memory X (MX) and the memory Y (MY). These memories, are accessed every instruction cycle. The data stored in even address cells and odd address cells are accessed simultaneously and are stored in temporary registers. The temporary registers feed data to the MAC unit every half machine cycle. Figure 3 shows the block diagram of the ACS/BF accelerator. In ACS operation, the 16b ALU simultaneously adds two sets of 8b path metrics stored in MY and two sets of 8b branch metrics stored in MX, and outputs two sets of new 8b path metrics to the REG and ACS/BF accelerator. The comparator in the ACS/BF accelerator compares these new 8b path metrics and outputs a decision bit that indicates which metric is smaller. The shift register gets in a series of decision bits, and the program control unit outputs control signals depending on the decision bits that transfer a smaller metric from REG to MY as a new path metric. This DSP adopts not floating point but block floating point number system for die size and power consumption. However, facilities are provided to prevent increasing of dynamic program step size in order to maintain data accuracy. Those facilities are the priority encoder (PE) and the ACS/BF accelerator. The PE counts the number of leading 0s (for plus data)/1s (for minus data) of series of input data. The comparator in the ACS/BF accelerator compares the new count outputted from the PE with the previous one stored in a temporary register, and stores the smaller one to the temporary register. By iterating this process, the exponent of block data is finally recorded in the temporary register.

The REGisters in the ALU unit have 2 write ports and 3 read ports. Two read and one write port are for ALU operation and the rest are for data transfer between REG and MX/MY through the M-BUS in parallel with ALU operation. This structure reduces operation frequency.

As a result of these architectural strategies, the instruction cycle time needed to realize a 11.2kb/s VSELP CODEC can be increased to 93ns (operating frequency, 10.7MHz), 25% slower than previous 24b DSP. Lower frequency reduces power. One of the strategies is discarding memory sense amplifiers, required in high-speed DSPs. The sizes of transistors also can be reduced with reduction of operating frequency. The supply voltage also can be reduced under 5V. The shmoo plot in Figure 4 shows these results.

Figure 5 shows the power consumption is about 70mW at instruction cycle 93ns and 3.5V VDD. This is about 14% and about 20% on current base compared to a previous 24b DSP. These values are reasonable for portable phones. Figure 6 is a micrograph and Table 1 summarizes features of this DSP.

Acknowledgments

The authors thank M. Kanno, H. Mori, K. Honma, Y. Fujimoto, H. Tokimori, T. Uno, M. Sanya, T. Suzuki, Y. Nakahara and M. Yasutome.

References

[1] Sunwoo, M. H., S. Park, K. Terry, "A Real-Time Implementation of Key VSELP Routines on a 16-bit Chip", IEEE Dig. Tech. Pap. Int., Conf. Consumer Electronics, pp. 332-333, 1991.

[2] Ueda, K., T. Sakao, T. Suzuki, O. Nishijima, "A High Performance Digital Signal Processor VLSI : MN1900 Series", IEEE Proc. Int. Conf., Acoustics, Speech, and Signal Processing, pp. 2175-2178, 1986.

Reprinted from *IEEE International Solid State Circuits Conference*, pp. 28-29, February 1993.

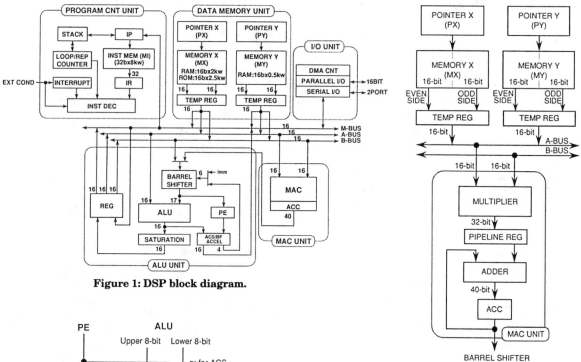

Figure 1: DSP block diagram.

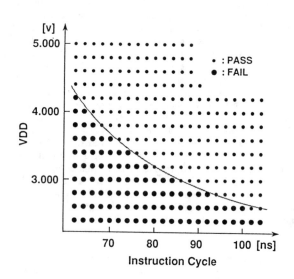

Figure 2: Block diagram of MAC and data memories.

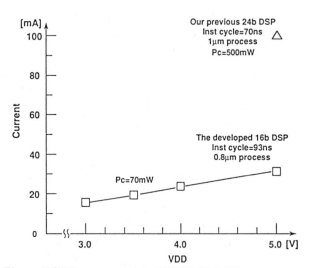

Figure 3: ACS/BF accelerator.

Figure 4: Shmoo plot of VDD vs. instruction cycle.

Figure 5. VDD vs. current for VSELP CODEC program.

Figure 6: See page 256.

Operating frequency	10.7MHz (21.5MHz clock input)
MAC performance	21.5MOPS peak
Process technology	0.8μm CMOS 2-aluminum layer
Die size	9.30x9.09mm²
Number of transistors	644k (Logic 94k, RAM 250k, ROM 300k)
Package	128-pin QFP
Power consumption	70mW at 3.5V VDD

Table 1: Features.

436

A 1.8V 36mW DSP for the Half-Rate Speech CODEC

Taketora Shiraishi, Koji Kawamoto, Kazuyuki Ishikawa, Hisakazu Sato, Fumiyasu Asai, Eiichi Teraoka,
Toru Kengaku, Hidehiro Takata, Takeshi Tokuda, Kouichi Nishida, and Kazunori Saitoh
System LSI Laboratory, Mitsubishi Electric Corporation
4-1 Mizuhara, Itami, Hyogo, 664 Japan

Abstract

A low-power 16-bit DSP has been developed to realize a low bit-rate speech CODEC. A dual datapath architecture and low-power circuit design techniques are employed to reduce power consumption. The PDC half-rate speech CODEC is implemented in the DSP with 36 mW at 1.8V.

Introduction

The PDC half-rate system with both the PSI-CELP (Pitch Synchronous Innovation Code Excited Linear Prediction) algorithm and the VSELP (Vector Sum Excited Linear Prediction) algorithm has gone into operation commercially in Japan. To reduce the bit-rate to half of the full-rate VSELP (6.7kbps to 3.45kbps), the PSI-CELP requires higher performance (approximately 23.4MOPS) than the VSELP (approximately 7.8MOPS) [1]. DSPs for the PSI-CELP CODEC have already been developed [2][3]; however, power consumption of these DSPs (100~200mW) is several times higher than the VSELP CODEC. Reduction of power consumption is achieved by combining the throughput enhancement and supply voltage reduction [4]. Because over 50% of instruction cycles are multiply-accumulate (MAC) operations in the PSI-CELP, parallelism in the datapath, such as parallel datapath [2] or pipelined datapath [3][5], is effective for throughput enhancement. We have developed a new low-power 16-bit DSP employing a dual datapath architecture and low-power circuit design techniques. This paper describes the low-power design and evaluation results of the DSP.

Architecture Overview

Fig. 1 shows the block diagram of the DSP. The DSP core is constructed of a program control unit (PCU), an address arithmetic unit (AAU), and a data arithmetic & logical operation unit (DALU).

The DALU has a dual datapath which includes four 16-bit input registers, two 17-bit × 17-bit multipliers, a 32-bit ALU, a 32-bit add-subtract unit, and four 32-bit accumulators.

The program memories comprise 16K words of ROM and 128 words of RAM. The data memories are organized into 6 banks: four 1K word RAMs, one 6K word ROM, and one 12K word ROM.

The DSP has a 32-bit instruction width. The instruction set includes sequence, move, load, bit-manipulation, and operation instructions. With orthogonally encoded fields of the operation instructions, it is possible to specify memory access, data operation, data transfer, and sequence control.

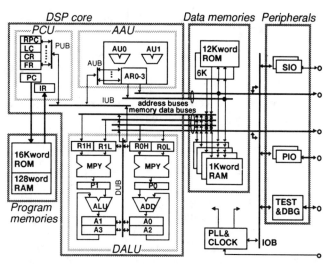

Fig.1 Block Diagram of DSP

Dual Datapath Architecture

The dual datapath architecture allows the DSP to reduce the power supply voltage while maintaining the high throughput.

In this architecture an arbitrary bank of data memory can be connected to any of four 16-bit memory data buses through the memory access field of the operation instructions. Four input registers (R1H, R1L, R0H, R0L), which are directly connected to these four memory data buses, can supply operands to any input of either multipliers, ALU, and add-subtract unit according to the data operation field. Using the inside data bus (DUB) of the DALU, data transfer among four input registers and four accumulators is controlled by the data transfer field. With this architecture, high throughput dual MAC operations can be achieved for single precision, double precision, and mixed precision operations which are necessary to obtain the required precision of the PSI-CELP algorithm in a 16-bit DSP implementation.

Typical examples for MAC operations for FIR filtering using the dual datapath are shown in Fig. 2. Fig. 2(a) shows the data flow for a single precision operation of FIR filtering with coefficients h(i), input x(n), and output y(n). In the operation, the instruction fields of memory access, data operation, and register transfer specify data flow as follows: (i) data load from two memory banks to R1H and R0H; (ii) two multiply and accumulate operations (R1H × R1L and R1H × R0H); and (iii) data transfer from R0H to R1L. With this organization of the instruction fields, one datapath calculates y(n) (see Eq. 1), and the other

Reprinted from *Proceedings of the IEEE Custom Integrated Circuits Conference*, pp. 371-374, May 1996.

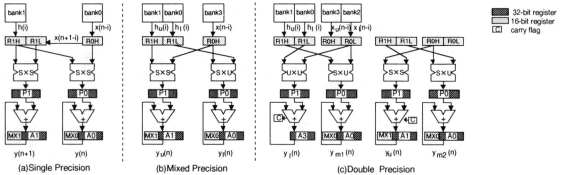

(a)Single Precision   (b)Mixed Precision   (c)Double Precision

Fig. 2 Multiply-Accumulate Data Flow

calculates y(n+1) (see Eq. 2). Thus two single precision FIR filter operations are realized in N clock cycles.

$$y(n) = \sum_{i=0}^{N-1} h(i) \cdot x(n-i) \qquad (1)$$

$$y(n+1) = \sum_{i=0}^{N-1} h(i) \cdot x(n+1-i) \qquad (2)$$

where N is the number of taps.

Fig. 2(b) shows a mixed precision operation. In this operation, the DSP executes: (i) data load from three memory banks to R1H, R1L, and R0H; and (ii) two multiply and accumulate operations (R1H $\times$ R0H and R0H $\times$ R1L). Thus, one datapath calculates $y_u(n)$ (see Eq. 3), and the other calculates $y_l(n)$ (see Eq. 4). Then an add operation is executed to maintain alignment of the binary point. Thus a mixed precision FIR filter operation is realized in N clock cycles.

$$y_u(n) = \sum_{i=0}^{N-1} h_u(i) \cdot x(n-i) \qquad (3)$$

$$y_l(n) = \sum_{i=0}^{N-1} h_l(i) \cdot x(n-i) \qquad (4)$$

where h_u holds the upper 16 bits of the coefficient, and h_l the lower 16 bits of the coefficient.

Fig. 2(c) shows a double precision operation. In this operation, two sets of instructions are executed. First, the DSP performs: (i) data load from four memory banks to R1H, R1L, R0H, and R0L; and (ii) two multiply and accumulate operations (R1L $\times$ R0L and R1H $\times$ R0L). Then, the DSP executes two multiply and accumulate operations (R0H $\times$ R1H and R0H $\times$ R1L). In this way, $y_l(n)$ (see Eq. 5), $y_{m1}(n)$ and $y_{m2}(n)$ (see Eq. 6), and $y_u(n)$ (see Eq. 7) are calculated. After that, an add operation is executed to maintain alignment of the binary point. Thus a double precision FIR filter operation is realized in 2N clock cycles.

$$y_l(n) = \sum_{i=0}^{N-1} h_l(i) \cdot x_l(n-i) \qquad (5)$$

$$y_m(n) = y_{m1}(n) + y_{m2}(n)$$
$$= \sum_{i=0}^{N-1} h_u(i) \cdot x_l(n-i) + \sum_{i=0}^{N-1} h_l(i) \cdot x_u(n-i) \qquad (6)$$

$$y_u(n) = \sum_{i=0}^{N-1} h_u(i) \cdot x_u(n-i) \qquad (7)$$

where h_u holds the upper 16 bits of the coefficient, h_l the lower 16 bits of the coefficient, x_u the upper 16 bits of the data, and x_l the lower 16 bits of the data.

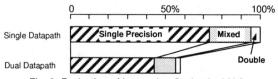

Fig. 3   Reduction of Instruction Cycles for MAC in PSI-CELP Algorithm

The reduction of instruction cycles with dual MAC operation is shown in Fig. 3. 86.3% of the total MAC operations of the PSI-CELP algorithm are executed in parallel in this dual datapath architecture, thus reducing the total instruction cycles of MAC operations to 56.9% of that for a single datapath.

Furthermore, there are two other operations which are executed efficiently in a dual datapath. One is square distance operation for the codebook search: a subtract operation and a square-accumulate operation are executed in parallel. So only one instruction cycle is required for execution of square distance operation. The other operation is the add compare select operation for the Viterbi decoding. Owing to parallel execution of an addition and a subtraction in the dual datapath, six instruction cycles are required for add compare select operation. These cycles are almost the same as those of dedicated hardware for add compare select operation [6].

Owing to the dual datapath with a powerful instruction set, the PSI-CELP CODEC is performed at 22.5 MHz. The area overhead for the dual datapath is only 2.4 mm², 2.4% of the total area.

Low-Power Circuit Design

Analysis of power consumption of a 24-bit floating point DSP for full-rate speech CODEC [7][8], shows that 50% of the total power is used for memory and clock circuits. So, low-power circuit design has been focused on memory design and clock distribution.

A. Low-Power RAM

The DSP has four 1K word RAMs. Each RAM is divided into 4 blocks of 256 words to reduce power consumption by decreasing the activated portion. The circuit diagram of each RAM is shown in Fig. 4. Because sequential memory accesses comprise more than 90% of all data memory access in the PSI-CELP CODEC, we have

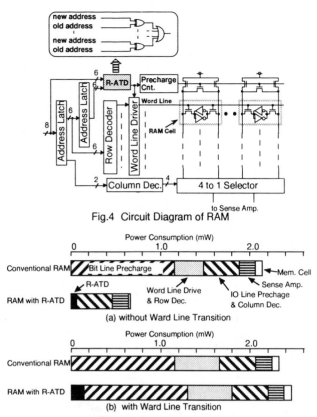

Fig.4 Circuit Diagram of RAM

Conventional RAM — Bit Line Precharge — Mem. Cell

R-ATD — Word Line Drive & Row Dec. — IO Line Precharge & Column Dec. — Sense Amp.

RAM with R-ATD

(a) without Ward Line Transition

Conventional RAM

RAM with R-ATD

(b) with Ward Line Transition

Fig. 5 Simulated Power Consumption for 256×16-bit RAM

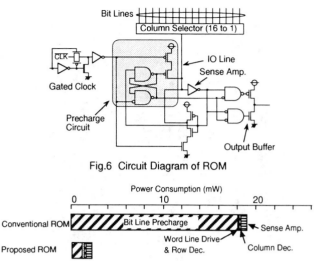

Fig.6 Circuit Diagram of ROM

Conventional ROM — Bit Line Precharge — Sense Amp.

Proposed ROM — Word Line Drive & Row Dec. — Column Dec.

Fig. 7 Simulated Power Consumption for 2K×16-bit ROM

focused on sequential address accessing. With the proposed row-address data transition detector (R-ATD), bit lines are precharged and the active word line is changed only when the row decoder address data changes. Higher address data (6 bits) drives the row decoder, and lower address data (2 bits) the column decoder. Thus, the number of times that bit lines are precharged and the active word line is changed are reduced to 1/4 in the sequential address case.

The result of the circuit simulation is shown in Fig. 5. By using R-ATD, power consumption is reduced to 29.8% when memory is accessed without word line transition, while power consumption increases by 6.2% when memory is accessed with word line transition. Consequently, a power reduction of 49.8% is achieved in the sequential address case.

B. Low-Power ROM

Similarly to the RAM, every ROM is divided into 2K word blocks. Thus, the program ROM is divided into 8 blocks and the two data ROMs are divided into 3 and 6 blocks, respectively. The circuit diagram of each ROM is shown in Fig. 6. As only bit lines selected by the column selector are precharged through the IO lines, the number of bit lines precharged is reduced from 256 lines to 16 lines. And with feedback to the precharge circuit from the output of the sense amplifier, the voltage of a bit line is limited to the threshold voltage of the inverter which is used as a

sense amplifier, so that current for extra precharging is reduced. The result of the circuit simulation is shown in Fig. 7. A power reduction of 89.3% is achieved.

C. Gated Clock

The circuit diagram of a gated clock is shown in Fig. 8. A gated clock is generally used in order to reduce clock lines capacitance. But the effect of low-power consumption by the gated clock depends on activation rate and the number of latches to which the gated clock are applied. To estimate power consumption, we have used a circuit simulation model of a gated clock which considers clock line lengths and the layout. The result of the simulation, the relation between power reduction rate and activation rate, is shown in Fig. 9. When Pgate/P0 (Pgate = power consumption with gated clock; P0 = power consumption without gated clock) is less than 1, a gated clock is useful for power reduction. By estimating activation rate of latches in the PSI-CELP CODEC, a gated clock is applied when Pgate/P0 is less than 1.

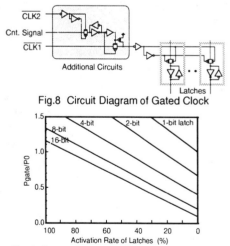

Fig.8 Circuit Diagram of Gated Clock

Fig. 9. Power Reduction Effect by Gated clock
Pgate: Power Consumption with gated clock
P0 : Power Consumption without gated clock

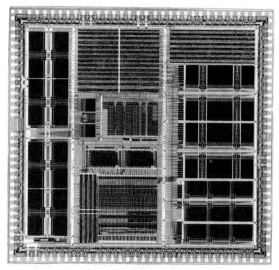

Fig. 10. Photomicrograph of the DSP

Table1 Features

Fabrication technology	0.5 μm double-metal CMOS Vtn=0.55V Vtp=-0.55V
Power Dissipation	36mW at 1.8V VDD
Die Size	9.8mm×10.3mm
Number of Transistors	1.46M (Logic 134k, RAM 455k, ROM 868k)
Package	100-pin QFP

Evaluation Results

The DSP is fabricated using 0.5 μm single-poly, double-metal CMOS technology. The photomicrograph of the DSP is shown in Fig. 10. The LSI characteristics of the DSP are summarized in Table 1. It integrates 1.46 million transistors (134k transistors for logic, 455k transistors for RAM and 868k transistors for ROM) in a 9.8mm× 10.3mm die size, and is packaged in a 100-pin quad-flat-package (QFP). PSI-CELP CODEC and VSELP CODEC are implemented in the program and data ROMs. A shmoo plot, which shows the instruction cycle time vs. supply voltage VDD at room temperature is shown in Fig. 11. Fig. 12 shows power consumption when PSI-CELP CODEC (including noise canceling and error correct coding/decoding) is executed at 22.5MHz. At 1.8V VDD, power consumption is 36mW.

Conclusion

A low-power consumption 16-bit DSP has been developed to implement the PDC half-rate CODEC. By employing a dual datapath architecture, throughput of a MAC operation is enhanced and power consumption is reduced by decreasing power supply voltage. Based on an analysis of the speech CODEC algorithm, power reducing techniques focused on memory design and clock distribution are applied extensively in the circuit design. The DSP, fabricated in 0.5 μm single-poly, double-metal CMOS technology, integrates 1.46million transistors in a 9.8mm × 10.3mm die size. The PSI-CELP algorithm for the PDC half-rate CODEC can operate perfectly at 22.5MHz, resulting in a low-power consumption of 36mW with a 1.8V power supply.

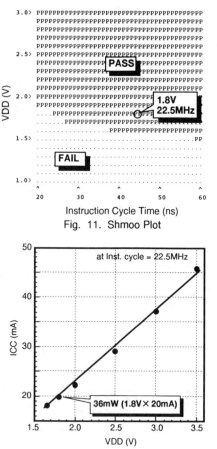

Fig. 11. Shmoo Plot

Fig.12 ICC vs. VDD for PSI-CELP CODEC

Acknowledgments

The authors would like to acknowledge the technical contributions of Y. Seguchi, H. Wakada, Y. Arima, T. Asai, K. Aota, S. Yamazaki, T. Higuchi, K. Yamazaki, S. Furubeppu, M. Nakajima, M. Yamaguchi, H. Momono, T. Inada, M. Nozaki, and Y. Shimazu. We would also like to thank Dr. Y. Horiba for their encouragement.

References

[1]F. Masui and M.Oguchi, "Activity of the Half Rate Speech Codec algorithm Selection for the Personal Digital Cellular System", Technical Report of IEICE, RCS93-77, pp. 55-62, 1993.
[2]Y. Okumura, T. Ohya, Y. Miki, and T, Miki, "A Study on DSP Circuits Applied to Speech CODEC for Digital Mobile Communications", Proceedings of the IEICE fall conference, p. 2-294, 1993.
[3]H. Kabuo, et al., "A 16bit Low-Power-Consumption Digital Signal Processor Using an 80 MOPS Redundant Binary MAC", IEEE Digest of Technical Papers Symposium on VLSI Circuits, pp.63-64, 1995.
[4]A. P. Chandrakasan, S. Sheng, and R. W. Brodersen, "Low-Power CMOS Digital Design", IEEE Journal of Solid-State Circuits. Vol. 27, NO.4, April 1992.
[5]K. Ueda, et al., "A 16b Low-Power-Consumption Digital Signal Processor", IEEE Digest of Technical Papers ISSCC, pp. 28-29, 1993.
[6]Texas Instruments, "TMS320C540 User's Guide"
[7]T. Kengaku, et al., "A Single-Chip DSP for Mobile Telecommunication Applications", Proceedings of IEEE CICC, 1991.
[8]T. Tokuda, et al., "A Mixed-Signal Digital Signal Processor for Single-Chip Speech Codec", IEICE Trans. Electron., Vol.E75-C, No.10 October 1992.

Design of a 1-V Programmable DSP
for Wireless Communications

PAUL LANDMAN, WAI LEE, BROCK BARTON, SHIGESHI ABIKO[1], HIROSHI TAKAHASHI[1], HIROYUKI MIZUNO[1], SHIGETOSHI MURAMATSU[1], KENICHI TASHIRO[1], MASAHIRO FUSUMADA[1], LUAT PHAM[2], FREDERICK BOUTAUD[3], EMMANUEL EGO[3], GIROLAMO GALLO[4], HIEP TRAN, CARL LEMONDS, ALBERT SHIH, MAHALINGAM NANDAKUMAR, BOB EKLUND, IH-CHIN CHEN

CORPORATE R&D, TEXAS INSTRUMENTS INCORPORATED, DALLAS, TEXAS 75243 USA
[1]TEXAS INSTRUMENTS JAPAN LTD., TOKYO, JAPAN
[2]TEXAS INSTRUMENTS INCORPORATED, STAFFORD, TEXAS USA
[3]TEXAS INSTRUMENTS FRANCE, NICE, FRANCE
[4]TEXAS INSTRUMENTS ITALY, AVEZZANO, ITALY

Abstract—In an effort to extend battery life, the manufacturers of portable consumer electronics are continually driving down the supply voltages of their systems. For example, next-generation cellular phones are expected to utilize a 1-V power supply for their digital components. To address this market, an energy-efficient, programmable DSP chip that operates from a 1-V supply has been designed, fabricated, and tested. The DSP features an instruction set and micro-architecture that are specifically targeted at wireless-communication applications and that have been carefully optimized to minimize power consumption without sacrificing performance. The design utilizes a 0.35-μm dual-V_t technology with 0.25-μm gate lengths that enables good performance at 1V. Specifically, the chip dissipates 17mW at 1V, achieving 63MHz operation with a power-performance metric of 0.21mW/MHz.

1. INTRODUCTION

The portable consumer electronics market is currently undergoing a period of rapid growth. Wireless communication devices, such as cellular phones, represent an important segment of this market. One of the keys to success in this highly-competitive arena is to keep products lightweight and small. This places tight constraints on the size of the batteries that can be used and makes it extremely important to employ energy-efficient circuits in the design of these products.

As shown in Fig. 1, wireless systems rely heavily on embedded DSP processors to perform critical communication functions such as speech coding, channel coding, demodulation, and equalization. These functions require a DSP that offers a high computational throughput. This requirement is often at odds with the goal of minimizing power consumption. This paper describes a fixed-point programmable DSP designed for wireless-communication applications that addresses these dual constraints of lower power and higher throughput. This is achieved by operating at low voltage (1V) while using a dual-V_t process to maintain high performance.

The use of multiple threshold voltages to maintain a balance between speed and standby current is not new. Several processors have been designed which incorporate both high and low threshold-voltage transistors on a single chip [1,2]. Relative to the design presented here, however, these earlier low-power DSPs have either incorporated much less functionality (<70K transistors) [1] or have had higher power per unit throughput, in the range of 2mW/MHz [2]. The chip described in this paper has a power-performance metric of 0.21mW/MHz and is a full-function, 1.6-million transistor DSP with a large amount of on-chip memory and a rich set of peripherals for interfacing to external devices.

The design is described in several sections. First, Section 2 presents the architecture of the DSP. Next, Section 3 delves deeper into the implementation of the design, describing the power management strategy and detailing the design of key circuits such as the clock generation and distribution network, the multiply-accumulate unit, and the memory subsystem. The multi-threshold process that enables low voltage operation is then described in Section 4. Finally, Sections 5 and 6 present measured results and concluding remarks.

2. ARCHITECTURE

The CPU uses a modified Harvard architecture as shown in Fig. 2. In order to maximize throughput and energy efficiency, the architecture relies heavily on both pipelining and parallelism. For example, the CPU employs a six-stage instruction pipeline divided as follows: (1) program pre-fetch, (2) program fetch, (3) instruction decode, (4) operand address generation, (5) operand read, and (6) execute/write. Three separate data buses and one

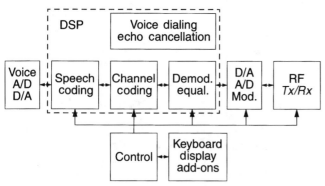

Figure 1 Diagram illustrating the use of DSP in a cellular phone application.

program bus coupled with two data-address generators and one program-address generator facilitate a high degree of parallelism. For instance, two reads and one write operation can be performed in a single cycle. This exploitation of concurrent processing techniques results in a highly-efficient instruction set, which in turn increases the energy efficiency of the processor [3].

The CPU contains a 40-bit ALU, which can selectively feed one of two 40-bit accumulators. The accumulators are divided into three parts: a 16-bit low-order word, a 16-bit high-order word, and eight guard bits. By setting a status register bit, the ALU can function as a single unit or as two 16-bit ALUs operating in parallel. One of the ALU inputs can be taken from a 40-bit barrel shifter, allowing the processor to perform numerical scaling, bit extraction, extended arithmetic, and overflow prevention. Together with the exponent detector, the shifter enables single-cycle normalization of values in an accumulator.

The CPU also features a multiply-accumulate block capable of a 17×17-bit 2's-complement multiplication and a 40-bit addition in a single instruction cycle. The multiplier and ALU can operate in parallel, allowing simultaneous execution of multiply-accumulate (MAC) and arithmetic operations.

The DSP has a rich instruction set including single-instruction repeat and block repeat operations, block memory move in-

structions, instructions with two- or three-operand reads, conditional store instructions, and arithmetic instructions with a parallel load and store. To improve performance and power on typical DSP algorithms, several dedicated instructions are available, including euclidean distance calculation, as well as FIR- and LMS-filtering operations.

In addition to these built-in filtering instructions, the DSP also contains a compare, select, and store unit (CSSU) which accelerates the Viterbi computation required by many communication algorithms [4]. The hardware components that support the Viterbi operator are shown in Fig. 3. This diagram is best understood by examining the code for a Viterbi butterfly operation, which is given in Fig. 4 along with the corresponding state transition diagram. In this example, two old states transition to two new states. For each new state, the objective is to find which of two possible routes results in the maximum path metric. By configuring the ALU in dual 16-bit mode, the four add/subtracts required to calculate the candidate path metrics can be performed in only two cycles. Next, for both new states, the path resulting in the largest metric must be selected and its metric stored. The CSSU allows the two compare, select, and store operations to be carried out in only two cycles. Finally, the decision is recorded in the test/control (TC) flag bit of the status register and in the 16-bit transition shift register (TRN). The information contained in TRN can later be used by a back-tracking routine to find the optimal path through the trellis. The dual-ALU mode and the CSSU allow the Viterbi butterfly to be completed in 4 cycles as opposed to the 20 required by a previous architecture. This represents a substantial savings of both time and energy.

In addition to the CPU, the DSP also contains a substantial amount of on-chip memory, including a 6K×16b SRAM and 48K×16b ROM divided into separate program and data spaces. Several on-chip peripherals are available as well, including a 16-bit timer, a synchronous full-duplex serial port, a buffered full-duplex serial port, and an 8-bit parallel host interface. The DSP also contains emulation and test circuitry which implements the JTAG (IEEE 1149.1) standard and allows access to all on-chip resources.

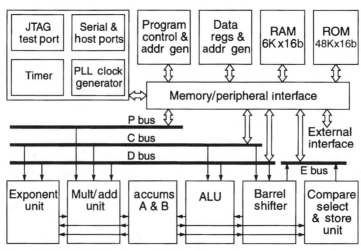

Figure 2 High-level block diagram of DSP architecture.

442

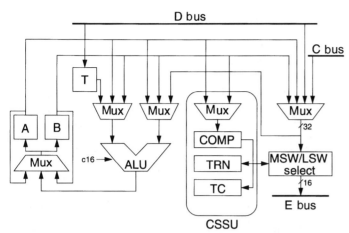

Figure 3 Architectural components supporting compare, select, and store operation used in Viterbi algorithm.

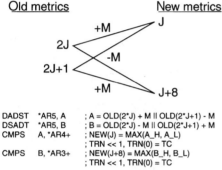

```
DADST   *AR5, A    ; A = OLD(2*J) + M || OLD(2*J+1) - M
DSADT   *AR5, B    ; B = OLD(2*J) - M || OLD(2*J+1) + M
CMPS    A, *AR4+   ; NEW(J) = MAX(A_H, A_L)
                   ; TRN << 1, TRN(0) = TC
CMPS    B, *AR3+   ; NEW(J+8) = MAX(B_H, B_L)
                   ; TRN << 1, TRN(0) = TC
```

Figure 4 State transition diagram and corresponding DSP code for a Viterbi butterfly operation.

3. CIRCUIT IMPLEMENTATION

In this section, we provide more details of the DSP design, beginning with the power management strategy in Section 3-A. This leads into a description of the clock generation and distribution network in Section 3-B. Finally, Sections 3-C and 3-D cover the MAC unit and the memory subsystem, respectively.

A. Power Management Strategy

During active operation, power savings are realized by extensive use of gated clocks (Fig. 5). In particular, latches are only clocked when useful data is available at their inputs. This is achieved by locally gating the master clock with a data-ready signal. This signal is generated automatically by local control logic using information from the instruction decoder. The gated slave logic registers when data has been clocked into the master latch, and it outputs an enable signal that gates the slave clock. Thus, functional blocks are only activated when they have valid data to process. Global clock gating is also available and is controlled by the user through three power-down instructions: IDLE1, IDLE2, and IDLE3. The IDLE1 instruction shuts down the CPU, the IDLE2 instruction shuts down the CPU and the on-chip peripherals (however, the PLL remains active), and the IDLE3 instruction shuts down the entire processor. Thus, the

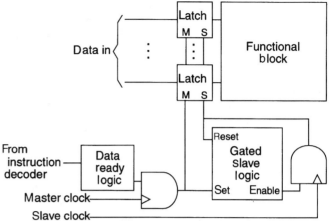

Figure 5 Example illustrating the use of gated slave logic to reduce unnecessary switching activity.

chip utilizes a hybrid power-management strategy that is partially automated and partially under user control.

B. Clock Generation and Distribution

The DSP uses an on-chip PLL for clock generation and multiplication. Traditionally, the PLLs used for this purpose have been largely analog [5]. There are several reasons, however, to consider using a more digital solution. Digital logic, for example, exhibits much better noise immunity and is less sensitive to process variations than analog circuitry. Furthermore, unlike digital logic, the energy efficiency of analog circuitry is actually worsened by aggressive voltage scaling. Moreover, the reduction in available headroom makes analog design much more difficult at low voltage levels. All of these factors led us to opt for a more digital approach.

PLL Architecture. The basic architecture of the PLL is shown in Fig. 6. The reference input can be driven by either an external clock source or a crystal resonator. In addition, it can be used directly or can be divided down in frequency by a factor of two or four. The output clock can be taken either from the PLL or from one of these divided-down versions of the reference.

The core loop of the PLL is structurally similar to that of an analog PLL, with the exception that a digital loop filter replaces the traditional analog filter, and a digitally-controlled oscillator (DCO) replaces the voltage-controlled oscillator (VCO). The programmable-frequency divider in the feedback path enables the PLL to generate an output frequency that is a factor of 1 to 15 higher than the reference. The combination of feed-forward and feedback frequency dividers allows the PLL to generate 15 different integer-frequency multiples (1, . . . ,15), eight different half-frequency multiples (0.5, 1.5, . . . , 7.5), and eight different quarter-frequency multiples (0.25, 0.75, . . . , 3.75).

The operation of the PLL is as follows. The phase detector compares the position of the reference-clock edge to that of the divided-down DCO edge. Depending on whether the DCO edge lags or leads the reference, the phase detector outputs a signal telling the DCO to increase or decrease its frequency. If the two

443

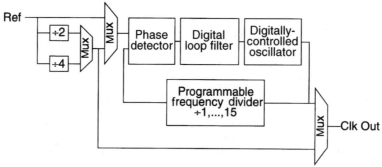

Figure 6 High-level architecture of digital PLL.

edges are aligned within the tolerance of the phase detector, then no correction pulses are generated.

Next, the loop filter examines the phase detector output and determines how to alter the DCO frequency. Since it is a synchronous digital circuit, a local clock signal is required. Therefore, the first stage of the circuit is a clock-generation unit that creates a loop-filter clock pulse whenever it sees a corresponding pulse on the phase detector output. This clock drives the main loop filter logic, which realizes a first-order low-pass filter designed to guarantee loop stability. The output of the loop filter is a 6-bit digital control word that specifies at what frequency the DCO should operate. The final stage of the loop filter is a synchronizer. The purpose of this circuitry is to prevent the DCO frequency from changing mid-cycle, which could give rise to glitches in the output clock.

The synchronized output of the loop filter then passes to the DCO. The DCO produces a square wave whose period is linearly related to the DCO control word. This signal is then fed back to the phase detector through a programmable-frequency divider. The frequency locking action of the feedback loop guarantees that $f_{dco}/N = f_{ref}$, resulting in a DCO frequency equal to N times the reference.

Digitally-Controlled Oscillator. Since the DCO period is controlled by a quantized digital input, the PLL cannot generate a continuous range of frequencies, but rather produces a finite number of discrete frequencies. Since the quantization granular-

ity of the DCO period sets some fundamental limits on the PLL's achievable jitter, it is desirable to have a fairly small step size. Traditional DCOs generate their output by dividing down a fixed, high-frequency clock source [6]. This type of implementation makes it extremely difficult to achieve a small step size since it is limited by how fast the clock source and the associated divider can operate. For example, to achieve a 200ps step between periods would require a 5-GHz oscillator and counter!

The DCO used in this design overcomes this difficulty by directly synthesizing a signal with the desired frequency rather than deriving it from a high-frequency source. The schematic for the DCO is shown in Fig. 7. The circuit uses a binarily-weighted switched capacitor array to control the frequency of oscillation. More specifically, inverter U2 drives a variable load, the size of which is controlled by turning on or off the capacitors labeled $2^0 C$ through $2^{n-1} C$, where n is the number of bits in the loop-filter output (six in our design). The capacitors are realized as NMOS transistors with their source and drain tied together. The gate areas of the transistors are binarily weighted such that if the on-capacitance of the smallest transistor is C, the capacitance of transistor i is roughly $2^i C$. A capacitor is turned on by pulling the corresponding digital control bit, d_i, to V_{dd}. This grounds the source/drain and creates an inversion layer in the NMOS channel contributing $2^i C$ to the capacitive load on node X. The capacitor is turned off by pulling d_i to ground, placing the transistor in the depletion region and significantly reducing

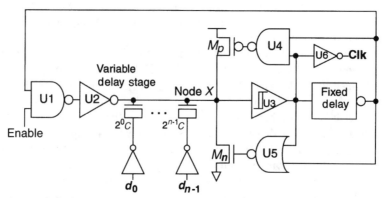

Figure 7 Schematic of digitally-controlled oscillator (DCO) with switched-capacitor frequency control.

its gate capacitance to $C_{off} = C/k$ where $k \gg 1$. Therefore, the load capacitance on node X is roughly given by:

$$C_X = C_p + dC + (2^n - d)C_{off} = C_p' + \left(\frac{k-1}{k}\right)dC \qquad [1]$$

where $d = (d_0 2^0 + d_1 2^1 + \ldots + d_{n-1} 2^{n-1})$ is the value of the digital control word, C_p is the node's parasitic capacitance, and $C_p' = C_p + (2^n/k)C$.

This causes the delay of inverter U2 to be linearly related to the digital control word. The step size between adjacent oscillation periods is controlled by sizing C and U2; the range is controlled by C, U2, and the number of bits, n, in the control word. Schmitt trigger U3 is used to sense when node X has reached its high or low thresholds, at which point a sharp edge is generated. The signal then travels through a fixed-delay stage (which contributes to the offset of the frequency range) and is fed back to NAND U1, completing the loop. Finally, reset devices U4, U5, M_p, and M_n are included, to ensure that Node X is at the rails at the start of each charging cycle.

The result is a DCO capable of excellent frequency resolution whose period is given by $T \simeq T_{offset} + dT_{step}$. Measurements from silicon indicate a typical rms phase jitter of about 45.5ps as shown in Fig. 8.

Clock Distribution Network. The clock distribution network is carefully tuned to minimize power dissipation. To reduce the significant short-circuit currents typical of large clock drivers, a hierarchical clock-buffering scheme is used. Local clock buffers are placed as close as possible to the clock signal destination points to minimize skew. After the required rise and fall times of the clock signal at the inputs to SRLs (scannable register latches) are specified, the driver and interconnect widths are tuned to satisfy the propagation-delay requirements with minimal power dissipation. Power is further reduced by careful design of the SRLs to minimize clock loading as shown in Fig. 9. In the new

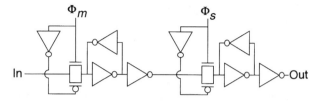

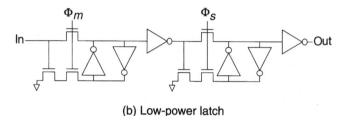

Figure 9 Comparison of (a) conventional CMOS latch to (b) low-power latch with reduced clock loading.

design, the clock signals drive two NMOS devices instead of feeding two NMOS *and* two PMOS devices, as in the conventional version. A differential driving strategy is used to ensure good performance. A 15% reduction in power was obtained, while achieving a 15% speed improvement and a 50% reduction in area. The large area reduction is due to the use of primarily NMOS devices, which reduces isolation-spacing requirements.

C. Multiply-Accumulate Unit

The multiply-accumulate (MAC) unit on this chip is composed of a 17×17-bit Wallace tree multiplier coupled with a 40-bit carry-lookahead adder. This MAC unit is capable of performing a non-pipelined multiply-accumulate operation in a single clock cycle. The multiplier can operate on signed, unsigned, and signed/unsigned 16-bit operands. Both integer and fractional modes, as well as 2's-complement rounding and overflow detection, are supported in this MAC unit.

The Wallace tree architecture was chosen not only because of its inherent speed advantage over the array multiplier, but also because of its lower dynamic power consumption [7]. The speed advantage of the Wallace tree comes from the reduction in the height of the carry-save adder tree—from nine to five stages in our case. The improvement in dynamic power consumption comes from a large reduction in switching events at the internal nodes of the multiplier—also resulting from the reduced tree height. The architecture of the Wallace multiplier used on this chip consists of a 3:2 counter stage, followed by two consecutive 4:2 counter stages. The 4:2 counter is constructed by cascading two carry-save adders. A 23% power reduction and 25% speed improvement are achieved by the Wallace tree architecture over the array architecture using the same carry-save adder.

The carry-save adder makes extensive use of transmission gate logic to minimize transistor count and power consumption. The transistor sizes were carefully tuned so that the variations in delay for different input signal combinations are minimized.

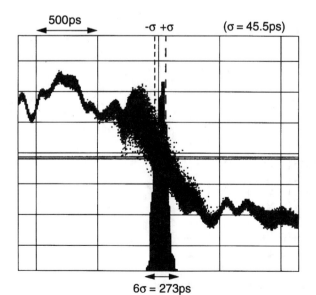

Figure 8 Histogram of typical phase jitter for digital PLL.

The tight output delay distribution helps to minimize the overall dynamic power consumption in the multiplier [7]. SPICE simulations at 1V reveal a worst-case delay of under 5ns.

D. Memory Subsystem

The on-chip memory subsystem includes a 6K×16b SRAM and 48K×16b ROM divided into separate program and data spaces.

Random-Access Memory. The memory subsystem allows two data operand reads or a long (32-bit) read from any one of the three 2K×16b on-chip SRAM blocks in a single machine cycle. If the program bus is used for coefficient access, three operands can be read in a single cycle. A divided word line architecture is used in the SRAM to reduce power [8]. Each 2K×16b SRAM block is divided into two banks as shown in Fig. 10. The banks in turn are sub-divided into four 256×16b arrays, each with its own local word-line driver. Consequently, each memory access only activates 1/8 of the entire array, thereby reducing the power consumption. The sense amplifier employs a latch-type design to reduce the static power dissipation and to ensure proper operation even at very low supply voltages.

High-V_t transistors are used in the six-transistor memory cell to keep the standby current to a minimum, while low-V_t transistors are used in the sense amplifier and all peripheral circuitry to allow for high-speed operation at 1V and below. SPICE simulations reveal that the speed of the SRAM is not significantly degraded by using threshold voltages as high as 0.3V or 0.4V in the memory cell. The speed of the SRAM is strongly dependent, however, on the threshold voltage used in the peripheral circuitry, which is therefore kept low. For a low and high V_t of 0.2V and 0.4V, respectively, SPICE simulations indicate an access time of about 6ns.

Read-Only Memory. There are two large ROMs in the DSP: a 32K×16b program-store ROM and a 16K×16b coefficient-data ROM. Both ROMs use the same basic architecture and cell design. They are constructed using multiple occurrences of a basic building block—an 8K×16b ROM, illustrated in Fig. 11.

These are diffusion-programmed ROMs incorporating a number of design features to reduce power dissipation. They utilize selective precharge of bit lines, for essentially zero-standby power dissipation and low-operating power consumption. A 16:1 column mux ratio minimizes bit-line capacitance and, therefore, access time and power consumption. The diffusion-programming feature results in a 40% memory array area savings with respect to via- or contact-programmed cores of the same capacity, further lowering bit-line and word-line capacitance and, thus, power. The ROMs feature dynamic page selection, which allows either 16-bit or 32-bit access, to meet system architecture requirements. The cell bit-line contact is shared by two adjacent cells to optimize both bit-line precharge/discharge time and power consumption.

To improve ROM performance at low voltage, additional steps were taken. For example, all peripheral circuitry employs low-V_t devices; however, as in the SRAM, high-V_t devices are used in the array to minimize standby current. In addition, horizontal metal-2 straps were added every 32 cells along the word lines. Vertical metal-1 ground lines were connected to each other through metal-2 jumper lines to provide a more solid ground to the array. And finally, the row-deselection circuitry was optimized to reduce word-line recovery time, permitting a shorter cycle time.

4. PROCESS

The chip was fabricated using a 0.35-μm, dual-V_t, twin-well CMOS process with three layers of metal. The minimum length of patterned gate poly is 0.25μm. The gate oxide thickness is 5nm. To fabricate both low- and high-V_t transistors on the same chip with minimal process overhead, a surface counter-doping implantation was performed on the channels of the low-V_t tran-

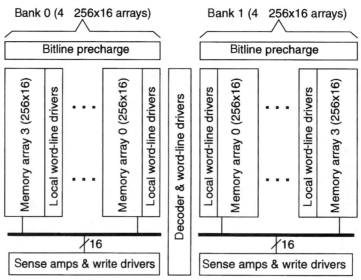

Figure 10 Block diagram of on-chip SRAM architecture illustrating divided word line strategy for power reduction.

446

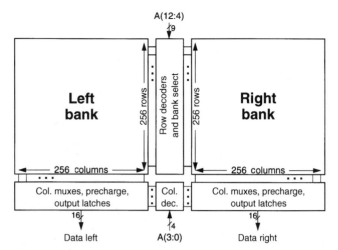

A(12:4)
↓9

| Left bank | 256 rows | Row decoders and bank select | 256 rows | Right bank |

← 256 columns → ← 256 columns →

| Col. muxes, precharge, output latches | Col. dec. | Col. muxes, precharge, output latches |

16↓   ↓4   16↓

Data left   A(3:0)   Data right

Figure 11 Architecture of 8K × 16b building block used to construct data and program ROMs.

sistors in addition to the regular channel implant [9]. The nominal threshold voltages (defined as the gate voltage at which the drain current reaches $0.1\mu A \times W/L$) are 0.4V and 0.2V. Note that an earlier publication [10] using a different V_t definition ($I_d = 0.01\mu A \times W/L$) quoted V_t's of 0.3V and 0.1V. The leakage current of the high- and low-V_t devices are below $1nA/\mu m$ and $1\mu A/\mu m$, respectively. The drive current of the low-V_t transistors is typically twice that of the high-V_t devices. Key process characteristics are summarized in Table 1.

Table 1. Key Process Technology Parameters.

Technology	0.35-μm, twin-well, dual-V_t, triple-metal, CMOS
Gate length	0.25-μm
Gate oxide	5nm
High V_t	0.4V
Low V_t	0.2V

5. RESULTS

The chips measures 5.65×5.50mm² and contains 1.6 million transistors. A photomicrograph of the die is shown in Fig. 12. The DSP was functional on first silicon and was packaged in a

Figure 12 Die photo of a 1-V DSP implemented in 0.35-μm, dual-V_t CMOS technology.

128-pin TQFP. Since the power consumption of the DSP depends on the instructions being executed, it is necessary to specify the benchmark used during testing. The results presented here were obtained while repeatedly executing a 16-tap FIR filtering application. This algorithm heavily exercises the MAC unit, resulting in a fair—if not slightly pessimistic—measurement of power consumption. Furthermore, the program was configured to execute from on-chip memory, ensuring that the critical (SRAM access) path was exercised.

The performance of the DSP as a function of supply voltage is shown in Fig. 13. For comparison purposes, curves are given for both the 1-V (0.35μm) part described in this paper and for a similar DSP implemented in a 3.3-V (0.6μm) technology with a single standard threshold voltage. The 0.35-μm DSP is functional down to 0.6V, reaching 63MHz at 1V and 100MHz at 1.35V. This amounts to roughly a performance improvement by a factor of ten relative to the 0.6-μm part for the same 1-V supply.

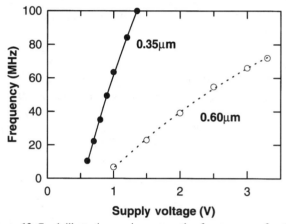

Figure 13 Graph illustrating maximum operating frequency as a function of supply voltage.

Figure 14 gives the power versus supply voltage curves for the 0.35-μm DSP. All data points were taken with the DSP operating at the maximum frequency for that supply voltage (see Fig. 13). At 1V, the DSP operates at 63MHz and dissipates 17mW. The approximate distribution of this power among the various chip components is shown in Table 2. Note that the

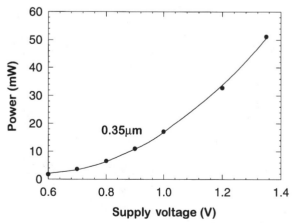

Figure 14 Graph of power consumption as a function of supply voltage when the DSP is operated at the maximum performance level.

Table 2. Approximate Power Breakdown of DSP.

Component	Approximate power consumption (mW)	Fraction of total (%)
Clock	4.4	26%
Datapath	6.5	38%
Control	3.1	18%
Memory	3.1	18%
Total	17	100%

power dissipated in the I/O pads has been excluded since it is heavily dependent on the board-level environment of the chip. By comparison, when the 3.3-V part is operated at 63MHz, the power consumption is higher by a factor of about 15.

A common power-performance metric used to compare different processors is "mW/MHz." This is just the slope of the power-versus-frequency graph for a fixed, nominal supply voltage. From Fig. 15, we see that the 1-V implementation has a power-performance metric of 0.21mW/MHz, which is a improvement by a factor of 19 over the 4.0mW/MHz of the 3.3-V part.

This metric, however, accounts only for dynamic power. Due to extensive use of low-V_t transistors, the standby power is quite high at 4mW. To reduce standby power in future releases, we could take advantage of the power-down modes to control high-V_t devices in series with the supplies. The sizing of these devices would be dictated by the peak-current requirements of the various modules in the chip. A second chip fabricated with all high-V_t devices has a standby power of less than 50μW, which gives some indication of the improvement that this approach would offer.

Another useful metric for evaluating the trade-off between energy efficiency and performance is the so-called energy-delay product. Figure 16 shows the energy-delay product for this chip as a function of supply voltage. The minimum is achieved at 1V, indicating that this is, indeed, the optimal operating voltage for this DSP. The energy-delay product for the 3.3-V DSP is more than an order of magnitude higher.

Some of the important chip statistics presented in this section are summarized in Table 3.

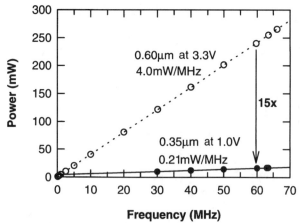

Figure 15 Graph showing power consumption as a function of frequency for a fixed supply voltage.

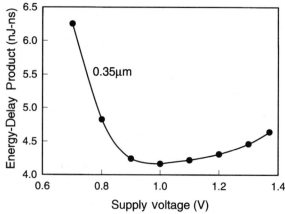

Figure 16 Energy-delay product of DSP showing optimum energy efficiency at 1V.

Table 3. Chip Statistics and Performance Data.

Chip size	$5.65 \times 5.50\text{mm}^2 = 31.1\text{mm}^2$
Transistor count	1.6 million
Package	128-pin TQFP
Speed at room temperature	63MHz at 1V, 100MHz at 1.35 V
Power	17mW at 1V, 63MHz
Power-performance metric	0.21mW/MHz at 1V

6. CONCLUSIONS

This paper has described a low-power, fixed-point programmable DSP optimized for wireless-communication applications. From the outset, the DSP was architected for low power consumption, employing a bus and memory structure that facilitate a high degree of parallelism and an instruction set with special instructions that enable efficient execution of common communication algorithms such as the Viterbi butterfly. Furthermore, the DSP includes a rich set of peripherals and a large amount of on-chip memory (54K×16b) that help to minimize chip count and off-chip I/O, thus saving power at the system level.

The chip utilizes a hybrid power-management strategy with automatic local clock gating and user-controlled gating of global clocks. A flexible PLL-based clock generator/multiplier gives the user fine-grain control over the clock frequency as well. To facilitate efficient low-voltage operation, the traditional analog PLL has been replaced with an almost entirely digital implementation.

The design makes use of a 0.35-μm dual-V_t technology with 0.25-μm gate lengths to enable good performance at 1V. At this voltage, the chip dissipates 17mW and operates at 63MHz with a power-performance metric of 0.21mW/MHz. Moreover, the device is functional down to 0.6V and reaches 100MHz at 1.35V. These results are indicative of the energy efficiency that can be achieved using low-voltage digital design techniques. As cellular phone manufacturers continue to drive down the system supply voltage, the trend toward migrating more functionality to the digital domain is likely to continue. This will fuel the demand for high performance, low-power DSPs. The data presented here demonstrate that high performance and low power consumption can be achieved simultaneously through innovative technology and digital-design techniques.

ACKNOWLEDGMENTS

The authors would like to thank D. Nowlen for mask generation; E. Born, G. Stacey, and B. Garcia for assistance in testing; B. Strong and B. Fleck for SRAM design; A. Amerasekera and C. Duvvury for advice on ESD design; and B. Hewes, A. Shah, P. Yang, S. Shichijo, and G. Frantz for their encouragement and support.

References

[1] M. Izumikawa et al., "A 0.9V 100MHz 4mW 2mm² 16b DSP core," *ISSCC Digest of Technical Papers,* Feb. 1995, pp. 84–85.

[2] S. Mutoh et al., "A 1V multi-threshold voltage CMOS DSP with an efficient power management technique for mobile phone application," *ISCC Digest of Technical Papers,* Feb. 1996, pp. 168–169.

[3] A. Chandrakasan, S. Sheng, and R. Brodersen, "Low-power CMOS digital design," *IEEE J. Solid-State Circuits,* vol. 27, no. 4, Apr. 1992, pp. 473–483.

[4] E. Lee and D. Messerschmitt, *Digital Communication,* Kluwer Academic Publisher, 1988.

[5] I. Young, J. Greason, and K. Wong, "A PLL clock generator with 5 to 100MHz of lock range for microprocessors," *IEEE J. Solid-State Circuits,* vol. 27, no. 11, Nov. 1992, pp. 1599–1606.

[6] R. Best, *Phase-Locked Loops: Design, Simulation, and Applications,* McGraw-Hill, 1996.

[7] T. Sakuta, W. Lee, and P. Balsara, "Delay balanced multipliers for low power/low voltage DSP core," *Symposium on Low Power Electronics Digest of Technical Papers,* Oct. 1995, pp. 36–37.

[8] M. Yoshimoto et al., "A divided word-line structure in the static RAM and its application to 64K full CMOS RAM," *IEEE J. Solid-State Circuits,* vol. 18. no. 5, Oct. 1983, pp. 479–485.

[9] M. Nandakumar, A. Chatterjee, G. Stacey, and I.-C. Chen, "A 0.25μm CMOS technology for 1V low power applications—device design and power/performance considerations," *Symposium on VLSI Technology Digest of Technical Papers,* June 1996, pp. 68–69.

[10] W. Lee et al., "A 1V DSP for wireless communications," *ISSCC Digest of Technical Papers,* Feb. 1997, pp. 92–93.

Stage-Skip Pipeline: A Low Power Processor Architecture Using a Decoded Instruction Buffer

Mitsuru Hiraki, Raminder S. Bajwa*, Hirotsugu Kojima*, Douglas J. Gorny*,
Ken-ichi Nitta**, Avadhani Shridhar*, Katsuro Sasaki*, and Koichi Seki

Central Research Laboratory, Hitachi, Ltd., Kokubunji, Tokyo 185, Japan
*Semiconductor Research Laboratory, Hitachi America, Ltd., San Jose, CA 95134, USA
**Semiconductor and Integrated Circuits Division, Hitachi, Ltd., Kodaira, Tokyo 187, Japan
E-mail: hiraki@crl.hitachi.co.jp, rbajwa@hmsi.com

Abstract

This paper presents a new pipeline structure that dramatically reduces the power consumption of multimedia processors by using the commonly observed characteristic that most of the execution cycles of signal processing programs are used for loop executions. In our pipeline, the signals obtained by decoding the instructions included in a loop are temporarily stored in a small-capacity RAM that we call decoded instruction buffer (DIB), and are reused at every cycle of the loop iterations. The power saving is achieved by stopping the instruction fetch and decode stages of the processor during the loop execution except its first iteration. The result of our power analysis shows that about 40% power saving can be achieved when our pipeline structure is incorporated into a digital signal processor or RISC processor. The area of the DIB is estimated to be about 0.7mm^2 assuming triple-metal $0.5\mu\text{m}$ CMOS technology.

I. Introduction

Multimedia capabilities are being incorporated into many kinds of computer systems, such as personal computers, personal digital assistants, digital cellular phones, car navigation systems, and video games. To handle these computationally intensive applications, the required LSI performance is increasing greatly, and this is a major factor in the increased power consumption of LSIs. This is especially true for processors because more and more multimedia applications are being handled by software as the performance of various kinds of processors improves drastically [1], [2].

Thus, it is evident that low power design methodologies directed toward multimedia processors are desired. So far, low power design research has been made extensively in various areas, including digital architecture, circuit design, process/device technology, and CAD. Although many of the research results for the latter three can be directly applied to multimedia processors, those for digital architecture are of little use in this area. This is because most of the recent research into low power architectures [3]-[5] assumed a dedicated LSI designed for a specific purpose rather than a programmable processor. We have thus focused on a low power architecture for multimedia processors whose functions can be changed by programming.

In this paper, we present a new pipeline structure that dramatically reduces the power consumption of multimedia processors by using characteristics commonly observed in signal processing programs. In Section II, the power wasted in conventional pipeline structures is briefly discussed to explain our motive for this research. The concept of our low power pipeline structure is presented in Section III. Section IV presents application examples of our pipeline structure to demonstrate its power reduction capability. Some variants of our pipeline structure are discussed in Section V. Section VI summarizes the key points of this paper.

II. Problem of Conventional Pipelines

The majority of multimedia applications use digital signal processing. In processors handling signal processing, most of the execution cycles are used for executing loops because the typical core routines of these programs perform multiply-accumulate operations (given by (1) for example),

$$c = \sum_{n=1}^{N} x(n)y(n) \qquad (1)$$

and these operations are accomplished by iterating multiplication and addition many times (N times in the case of (1)).

Processors are generally pipelined to increase throughput. Fig. 1 (a) illustrates the concept of a conventional pipeline. The pipeline is conceptually divided into the following three parts.

1) Instruction memory (or instruction fetch part): provides instruction code stored at the instruction address to which program controller points.
2) Instruction decoder: decodes the instruction code and provides the decoded signal.
3) Execution units: execute the instruction according to the decoded signal.

All three parts of a conventional pipeline are sequentially activated to execute any instruction. From the viewpoint of low power design, however, this is not desirable for loop executions, which cover most of the execution cycles of signal processing programs. This is because instruction fetching and instruction decoding waste precious power budget each time the loop is iterated after the first loop because the codes and decoded signals are the same for each iteration of the loop no matter how many times the loop is iterated.

Reprinted from *International Symposium on Low Power Electronics and Design*, pp. 353-358, August 1996.

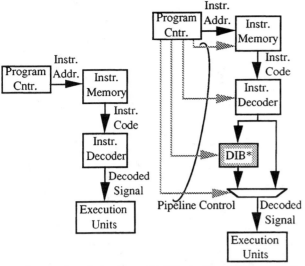

(a) Conventional pipeline   (b) SS pipeline** (proposed)

*DIB: Decoded Instruction Buffer
**SS: Stage-Skip

Fig. 1. Concept of conventional and proposed pipelines.

III. Concept of the Stage-Skip Pipeline

Our proposed pipeline structure dramatically reduces the power consumption of multimedia processors by using the characteristic that most execution cycles in signal processing programs are used for loop executions. The concept of our pipeline is illustrated in Fig. 1 (b). The following two elements are inserted between the instruction decoder and execution units:

- Decoded Instruction Buffer (DIB): stores decoded signals for a whole loop.
- Selector: selects the output from either the instruction decoder or DIB.

The decoded instruction signals for a loop are temporarily stored in the DIB and are reused in each iteration of the loop. The power wasted in the conventional pipeline is saved in our pipeline by stopping the instruction fetching and decoding for each loop execution. Since power consumption is reduced by virtually skipping the instruction fetching and decoding stages, we call it a stage-skip pipeline (SS pipeline).

The SS pipeline has three types of operations as shown in Fig. 2. One of these operations is dynamically selected depending on the current execution state of the processor, as follows.

i) *Outside a Loop*

When the processor is executing instructions outside a loop, the pipeline operates exactly like a conventional pipeline. The instructions are fetched, decoded, and sent to the execution units. The DIB is inactive.

ii) *1st Iteration of a Loop*

When the processor executes the first iteration of a loop, the decoded signals for each instruction of the loop are directly provided to the execution units, and at the same time, also written into the DIB. Thus, the decoded signals for the entire loop are stored in the buffer in sequence.

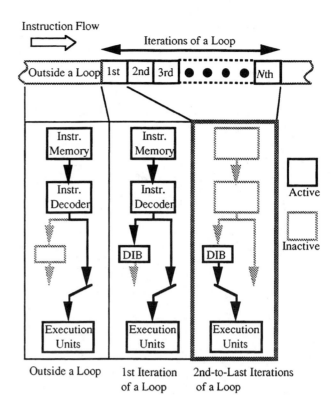

Fig. 2. Operation of SS pipeline.

iii) *2nd-to-Last Iterations of a Loop*

When the processor is executing the 2nd-to-last iterations of a loop, the DIB provides the decoded signals to the execution units in the same sequence as for the first iteration. Since fetching and decoding are not required, power is saved in the SS pipeline.

The pipeline is controlled by a program controller (denoted as Program Cntr. in Fig. 1 (b)), which knows the current execution state as follows. Suppose that the current execution state is *Outside a Loop*, where the program controller increments the instruction addresses one by one. When the processor executes a specific instruction evoking loop execution, the program controller begins to point repeatedly at the addresses for the group of instructions specified in the loop-evoking instruction. When this occurs, the program controller detects that the execution state has changed to *1st Iteration of a Loop*. The number of times for the loop to be iterated is specified in the loop-evoking instruction, and is copied to an iteration count register. Every time the loop is iterated, this register is decremented. The program controller detects that the execution state has changed to *2nd-to-Last Iterations* when the register is decremented for the first time. When the value of the register reaches zero, the program controller detects the end of loop execution and once again begins to increment the instruction addresses one by one. The program controller detects that the execution state has returned to *Outside a Loop* when this occurs.

IV. Application Examples

To demonstrate the power saving capability of the SS pipeline, the power that would be saved by using an SS

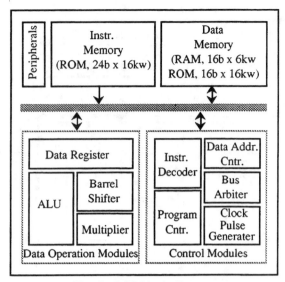

Fig. 3. DSP block diagram.

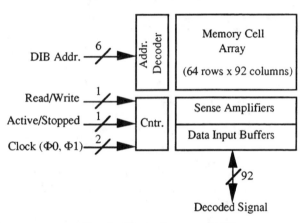

Fig. 4. Block diagram of DIB.

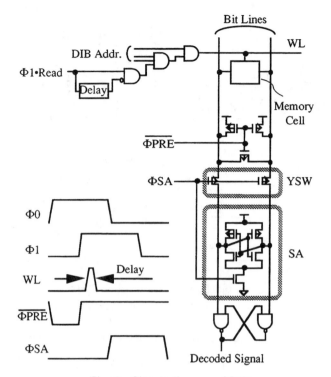

Fig. 5. Circuit diagram of DIB.

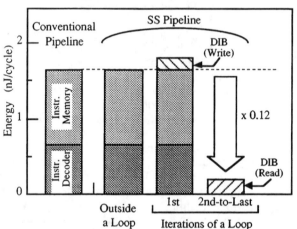

Fig. 6. Energy needed to provide decoded signals.

pipeline was analyzed for two processors, one of which is a digital signal processor (DSP) and the other is a RISC processor. First, we will briefly describe these two processors. Then, we will consider the design issues of a DIB for the processors and analyze the behavior of various signal processing programs. Finally, we will present the results of our power analysis for the processors into which an SS pipeline is incorporated.

A. Outline of the Processors

One of the processors that we analyzed is a 24-bit-instruction DSP that we experimentally designed previously. It has an instruction set applicable to a wide range of signal processing applications [6], [7], although its data length was shortened to 16 bits for use as a speech CODEC LSI. A block diagram of the DSP chip is shown in Fig. 3.

The other is a MIPS R3000-like RISC core that was experimentally implemented to perform a power analysis on it [8]. This chip includes integer datapath (register file, ALU, and shifter), next-PC module, controller, I-cache (2kB), and D-cache (2kB).

B. DIB implementation

We studied the design of a DIB, and estimated the area needed for the buffer and the power consumed by the buffer itself. The estimated results are described below.

A block diagram of the DIB for use in the DSP is shown in Fig. 4. To reduce the area needed for the DIB, we use an SRAM structure. The number of columns in the memory cell array (92) corresponds to the number of pins for the decoded signals in the DSP. The number of rows in the memory cell array (64) corresponds to the maximum number of instructions whose decoded signals can be stored in the buffer. The memory cell is composed of six MOS transistors. The area of the DIB is estimated to be about 0.7 mm², assuming 0.5μm CMOS technology with a triple-metal process. Note that this corresponds to only 0.8% of the whole chip.

452

Table 1. Loop size in application programs.

Program	Loop Size (words)
FIR Filter	
single-precision	1
double-precision	3
complex	4
Biquad IIR Filter	
single-precision	4
double-precision	13
LMS Adaptive Filter	
single-precision	2
double-precision	7
complex	8
All Zero Latice Filter	2
All Pole Latice Filter	3
VSELP	1 to 38
LPC coefficient computation	2 to 7
2D 8 x 8 DCT	53

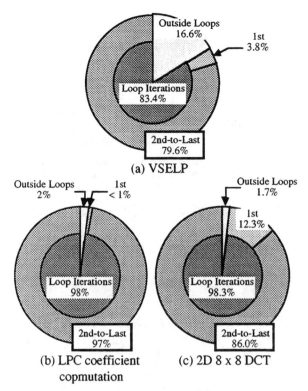

(a) VSELP

(b) LPC coefficient copmutation

(c) 2D 8 x 8 DCT

Fig. 8. Execution cycles within and outside loops.

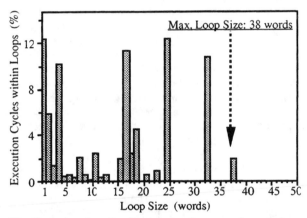

Fig. 7. Loop-size distribution for the case of executing VSELP.

To reduce the power consumed, we used the following power reduction techniques for its read circuits (Fig. 5) [9].

   i) To reduce cell current, the word line (WL) is driven by a short pulse.

   ii) To reduce power dissipation in the sense amplifiers, latch-type sense amplifiers (SA) are used.

   iii) To prevent power dissipation which the latch-type sense amplifiers would cause by driving the bit lines, the bit lines are isolated from the sense amplifiers by the switches (YSW) when the sense amplifiers are activated.

We estimated the power needed for both a conventional pipeline and the SS pipeline to provide decoded signals in the case of the DSP (Fig. 6). Here, SPICE simulation was used to estimate the power consumed in the instruction memory (fetching) and DIB, and the power consumed by the instruction decoder was estimated using power analysis data reported in [10]. Outside a loop, the required power is the same for both. For the 1st iteration of a loop, there is

an overhead of about 10% for the SS pipeline to write to the DIB. However, in the 2nd-to-last iterations of a loop, where instruction fetching and decoding are inactive and only the DIB is activated, the required power is reduced to as little as 12% of that of the conventional pipeline. This is because the DIB uses an extremely small-capacity RAM (only 64words x 92bits), so it uses much less power than is used for fetching and decoding.

In the case of the R3000-like RISC core, the DIB takes up only about 2% of the die area. The power of the DIB is estimated to be about 11% of the power consumed in the controller and I-cache together. This estimation was done based on the information regarding the schematic, chip size, and power analysis of the RISC core, which are reported in [8].

C. Program Analysis

Since the SS pipeline saves power by using a small-capacity DIB to replace the instruction fetching and decoding during loop execution, the following two points are essential from the viewpoint of program characteristics.

   i) For most (preferably all) cases, the number of instructions in a loop (here, it is called loop size) should be small enough for the entire loop to be stored in the DIB.

   ii) Most execution cycles should be loop executions.

We verified these points by extensively analyzing signal processing programs coded for the DSP. The analysis results are described below.

Table 1 shows the loop sizes in various signal processing programs. Fig. 7 shows the loop-size distribution when executing vector sum excited linear

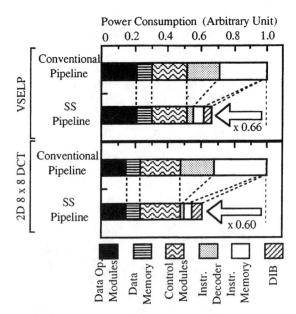

Power Consumption (Arbitrary Unit)

Fig. 9. Improvement in the power consumption
of the DSP.

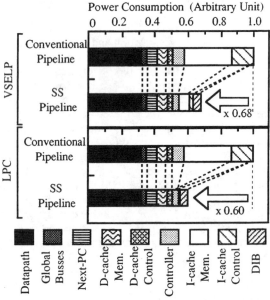

Power Consumption (Arbitrary Unit)

Fig. 10. Improvement in the power consumption
of the R3000-like RISC core.

prediction (VSELP) speech CODEC. The largest loop size
among the programs that we analyzed was 53 words, which
appeared in the 2D 8x8 discrete cosine transform (DCT).
The loops in any of these signal processing programs are
thus small enough to fit in a 64-word DIB.

Fig. 8 shows the execution cycles within and outside
loops when executing VSELP, linear prediction coding
(LPC) coefficient computation, and 2D 8x8 DCT. In any
of these cases, more than 80% of the execution cycles were
used for loop iterations, especially for the 2nd-to-last
iterations that are the key to power saving in the SS
pipeline.

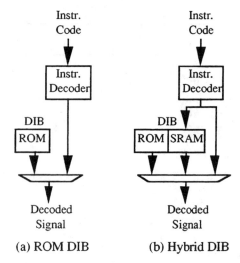

(a) ROM DIB   (b) Hybrid DIB

Fig. 11. Block diagram of the ROM and hybrid DIB.

D. Power Analysis

Based on these studies, we estimated the power that
would be saved by incorporating the SS pipeline into the
DSP or RISC processor.

Fig. 9 compares the power consumption for the DSP
with the SS pipeline and that for a conventional pipeline
for executing VSELP and 2D 8x8 DCT (Our analysis
process is described in the Appendix). The results are
dramatic. The power for the SS pipeline is reduced to
between 60% and 66% of that for a conventional pipeline.
The power of the instruction memory and instruction
decoder in the SS pipeline is greatly reduced since they are
stopped during the many cycles used for loop execution.
The power overhead for the DIB is so small that it is
almost negligible.

Fig. 10 shows the power saving of the R3000-like
RISC core with the SS pipeline for executing VSELP and
LPC coefficient computation. The results are quite similar
to those of the DSP. The power for the SS pipeline is
reduced to between 60% and 68% of that for a conventional
pipeline.

V. Variants of the DIB

In the case of embedded applications where the DSP
processor has to run a small set of applications, cost is a
very big constraints and so it is important to keep code size
small so that memory (RAM/ROM) requirements are kept
down. A variant of the DIB approach can be used to reduce
code size as well as power. The numbers in Fig. 8 and
inspection of the code indicate that the same code structure
is repeated very often. In such cases, it would be possible
to isolate these code structures and install decoded versions
in an on-chip ROM which can be treated as a library to be
used at run time by the application program. ROMs are
approximately four times denser than SRAMs and so for
the same area overhead larger storage can be used. This
approach also presents the opportunity to further reduce the
power consumption by reducing traffic to off-chip
memories. Finally, a hybrid structure which includes a
ROM and SRAM in the DIB can also be used to benefit
from both code size reduction and power reduction. Fig. 11

Table 2. Consumed energy and active cycle rate of each module of the DSP.

Module, i		Energy, E_i (pJ/cycle)	Active Cycle Rate, N_i			
			VSELP		2D 8 x 8 DCT	
			conv.	SS	conv.	SS
Data Op. Modules						
ALU		300	50.25%	←	48.32%	←
Multiplier		349	39.88%	←	25.96%	←
Barrel Shifter		668	8.82%	←	0.00%	←
Data Register						
Data Op.	Load/Store					
double	double	664	27.05%	←	0.00%	←
double	single	538	5.17%	←	1.85%	←
double	none	412	3.42%	←	5.56%	←
single	double	434	3.11%	←	0.00%	←
single	single	308	3.67%	←	25.96%	←
single	none	182	22.51%	←	33.37%	←
none	double	252	10.52%	←	3.71%	←
none	single	126	8.85%	←	25.96%	←
Control Modules						
Data Addr. Cntr.		445	56.71%	←	58.63%	←
Program Cntr.		192	100.00%	←	100.00%	←
Instr. Decoder		634	100.00%	20.36%	100.00%	14.02%
Bus Arbiter		70	100.00%	←	100.00%	←
Clock Pulse Generater		208	100.00%	←	100.00%	←
Data Memory						
RAM (Read)		320	76.89%	←	39.63%	←
RAM (Write)		540	12.85%	←	22.25%	←
ROM		716	0.00%	←	0.00%	←
Instr. Memory		990	100.00%	20.36%	100.00%	14.02%
DIB						
Read		196	0.00%	79.64%	0.00%	85.98%
Write		163	0.00%	3.81%	0.00%	12.28%
Peripherals		669	0.00%	←	0.00%	←

where E_{Total} is the overall consumed energy, E_i is the energy consumed per cycle in module i, and N_i is the number of active cycles for module i. The E_i for each module is listed in Table 2, where the energy values for the on-chip memories and DIB were extracted using SPICE and those for the rest of the modules were extracted using a gate-level power analysis tool [10]. The operating voltage is 3.3 V. The N_i for each module is also listed in Table 2 for the execution of VSELP and 2D 8x8 DCT using the DSP with a conventional pipeline and one with an SS pipeline. The N_i for the conventional pipeline can be obtained by simply counting the number of times each instruction is used because the activated modules for each cycle can be specified from the instruction being executed in that cycle. The N_i for the SS pipeline can be obtained by converting the results based on the loop execution statistics shown in Fig. 8.

Acknowledgments

We are indebted to E. Takeda for his support and encouragement, and to S. Bhattacharyya, Y. Hatano, T. Nakagawa, and A. Kiuchi for their comments, suggestions, and overall assistance in preparation of this paper.

References

[1] H. Sasaki, "Multimedia Complex on a Chip," in *ISSCC Dig. Tech. Papers*, pp. 16-19, Feb. 1996.
[2] D. Epstein, "Chromatic Raises the Multimedia Bar," *Microprocessor Report*, pp. 23-27, Oct. 23, 1995.
[3] A. Chandrakasan, A. Burstein, and R. W. Broderson, "A Low Power Chipset for Portable Multimedia Applications," in *ISSCC Dig. Tech. Papers*, pp. 82-83, Feb. 1994.
[4] D. B. Lidsky and J. M. Rabaey, "Low-Power Design of Memory Intensive Functions," in *Symp. Low Power Electronics, Dig. Tech. Papers*, pp. 16-17, Oct. 1994.
[5] B. M. Gordon, T. H. Meng, and N. Chaddha, "A 1.2mW Video-Rate 2D Color Subband Decoder," in *ISSCC Dig. Tech. Papers*, pp. 290-291, Feb. 1995.
[6] T. Baji, et al., "HX24 24-bit Fixed Point Digital Signal Processor," in *Proc. ICSPAT*, pp. 622-629, Oct. 1993.
[7] A. Kiuchi, T. Nakagawa, T. Baji, and K. Kaneko, "Digital Signal Processor Supporting Two Types of Instruction Sets," *Electronics and Communications in Japan*, Part 3, Vol. 78, No. 6, pp. 20-29, 1995.
[8] T. Burd and B. Peters, "A Power Analysis of a Microprocessor: A Study of an Implementation of the MIPS R3000 Architecture," Technical Report ERL, University of California, Berkeley, May 1994. http://infopad.eecs.berkeley.edu:80/˜burd/gpp/r3000/total.html.
[9] Y. Shimazaki, et al., "An Automatic-Power-Save Cache Memory for Low-Power RISC Processors," in *Symp. Low Power Electronics Dig. Tech. Papers*, pp. 58-59, Oct. 1995.
[10] H. Kojima, D. J. Gorny, K. Nitta, and K. Sasaki, "Power Analysis of a Programmable DSP for Architecture /Program Optimization," in *Symp. Low Power Electronics Dig. Tech. Papers*, pp. 26-27, Oct. 1995.

shows the block diagrams for these cases.

These variants would require careful assembly code generation and can benefit from compiler optimization techniques that can generate reusable segments of code for such a library-based approach. Several small loops may be consolidated whenever possible to reduce overheads as long as this can be done without register spilling.

VI. Conclusion

We have proposed a stage-skip pipeline (SS pipeline) to reduce the power consumption of multimedia processors. The SS pipeline saves power by stopping instruction fetching and decoding during loop execution and using a small-capacity RAM that we call decoded instruction buffer (DIB) to take over for them. This greatly reduces the power used by processors handling multimedia applications because the overwhelming majority of execution cycles in signal processing programs are used for loop execution. Our case study showed that power consumption is reduced about 40% when an SS pipeline is incorporated into a digital signal processor or RISC processor. The increase in silicon area due to the DIB is estimated to be about 0.7 mm^2, assuming that triple-metal 0.5μm CMOS technology is used.

Appendix

The overall consumed energy was assumed to be given by (2).

$$E_{Total} = \sum_i E_i \times N_i \qquad (2)$$

Part VIII

Computer-Aided Design Tools

Transition Density: A New Measure of Activity in Digital Circuits

Farid N. Najm, *Member, IEEE*

Abstract—Reliability assessment is an important part of the design process of digital integrated circuits. We observe that a common thread that runs through most causes of runtime failure is the extent of circuit *activity*, i.e., the rate at which its nodes are switching. We propose a new measure of activity, called the *transition density*, which may be defined as the "average switching rate" at a circuit node. Based on a stochastic model of logic signals, we also present an algorithm to propagate density values from the primary inputs to internal and output nodes. To illustrate the practical significance of this work, we demonstrate how the density values at internal nodes can be used to study circuit reliability by estimating 1) the average power and ground currents; 2) the average power dissipation; 3) the susceptibility to electromigration failures; and 4) the extent of hot-electron degradation. The density propagation algorithm has been implemented in a prototype *density simulator*. Using this, we present experimental results to assess the validity and feasibility of the approach. In order to obtain the same circuit activity information by traditional means, the circuit would need to be simulated for thousands of input transitions. Thus this approach is very efficient, and makes possible the analysis of VLSI circuits, which are traditionally too big to simulate for long input sequences.

I. INTRODUCTION

A MAJOR portion of the design time of digital integrated circuits is dedicated to functional verification and reliability assessment. Of these two, reliability assessment is a more recent problem the severity of which has steadily increased in proportion to chip density. As a result, CAD tools that evaluate the susceptibility of a design to runtime failures are becoming increasingly important.

Chip runtime failures can occur due to a variety of reasons, such as excessive power dissipation, electromigration, hot-electron degradation, voltage drop, aging, and others. In CMOS logic circuits, the rate at which node transitions occur is a good indicator of the circuit's susceptibility to runtime failures. For example, both power dissipation and electromigration in the power lines are directly related to the power supply current, which, in CMOS is nonzero only during transitions. Hot-electron degradation is related to the MOSFET's substrate current, which for CMOS is also significant only during transitions. Thus, the rate at which node transitions occur, i.e.,

the extent of circuit *activity*, may be thought of as a measure of a failure-causing *stress*. However, there has traditionally been no way of *quantifying* this activity because logic signals are, in general, nonperiodic and thus have no fixed switching *frequency*.

This paper proposes a novel measure of activity that we call the *transition density*, along with a simulation technique to compute the density at every circuit node. The transition density may be defined as the "average switching rate"; a more rigorous definition will be given in Section II. Preliminary results of this work have appeared in [1].

To further motivate the notion of transition density, consider the problem of estimating the average power drawn by a CMOS gate. If the gate has output capacitance C and generates a simple clock signal with frequency f, then the average power dissipated is $CV_{dd}^2 f$, where V_{dd} is the power supply voltage. In general, since logic signals may not be periodic, the notion of frequency cannot be used. Instead, one may compute the power as follows. If $x(t)$ is the logic signal at the gate output and $n_x(T)$ is the number of transitions of $x(t)$ in the time interval $(-T/2, +T/2]$, then the average power is

$$P_{av} = \lim_{T \to \infty} V_{dd} \frac{CV_{dd} n_x(T)/2}{T} = \frac{1}{2} CV_{dd}^2 \left\{ \lim_{T \to \infty} \frac{n_x(T)}{T} \right\}.$$

$$(1.1)$$

In the next section we define the transition density to be the last (limit) term in (1.1).

Naturally, one can approximate $\lim_{T \to \infty} n_x(T)/T$ by simulating the circuit for a "large enough" number of input transitions while monitoring $n_x(T)$ at every node. The ambiguity in the phrase "large enough" is precisely the problem with this traditional approach. It is impossible to determine *a priori* how long the simulation should be. Furthermore, long simulations of large circuits are very expensive. However, we will show that if the transition densities at the circuit primary inputs are given, they can be efficiently propagated into the circuit to give the transition density at every internal and output node. In other words, we use the *limits* $\lim_{T \to \infty} n_x(T)/T$ at the circuit inputs to *directly* compute the corresponding *limits* inside the circuit.

The propagation algorithm involves a single pass over the circuit and computes the transition densities at all the nodes. It may be thought of as a *simulation* of the circuit

Manuscript received August 2, 1991; revised March 20, 1992. This paper was recommended by Associate Editor J. White.

The author was with the Texas Instruments Semiconductor Process & Design Center, Dallas, TX. He is now with the Electrical Engineering Department, University of Illinois at Urbana-Champaign, Urbana, IL 61801.

IEEE Log Number 9201146.

Reprinted from *IEEE Transactions on Computer Aided Design of Integrated Circuits and Systems*, pp. 310-323, February 1993.

in which one studies the density of its internal signals that correspond to input signals with specified densities; it has been implemented in a prototype *density simulator*, called DENSIM. In order to obtain the same circuit activity information by traditional means, the circuit must be simulated for thousands of input transitions. Thus this approach is very efficient and makes possible the analysis of VLSI circuits, which are traditionally too big to simulate for long input sequences.

It turns out to be highly beneficial, in terms of the theoretical results to be presented, to cast the problem in a stochastic (probability theory) setting. Thus, in the following two sections we will start with definitions of "idealized logic signals" and then present a stochastic model of logic signals that is essential to the density propagation theorem. Based on these concepts, we then show in Section IV how the transition density can be efficiently propagated from inputs to outputs. In Section V, we study a number of practical applications of the density concept, namely, we demonstrate how the density values at internal nodes can be used to estimate 1) the average power and ground currents; 2) the average power dissipation; 3) the susceptibility to electromigration failures; and 4) the extent of hot electron degradation. Experimental results are presented in Section VI and Section VII contains a summary and conclusions.

Appendix A presents the existence proofs of the equilibrium probability and transition density. Appendix B presents a new application for binary decision diagrams (BDD's) in computing the *probability* of a Boolean function.

II. IDEAL LOGIC SIGNALS

Let $x(t)$, $t \in (-\infty, +\infty)$, be a function of time that takes the values 0 or 1. We use such time functions to model *logic signals* in digital circuits. This ideal model neglects waveform details such as the rise/fall times, glitches, over/under-shoots, etc.

Definition 1: The equilibrium probability of $x(t)$, to be denoted by $P(x)$, is defined as

$$P(x) \triangleq \lim_{T \to \infty} \frac{1}{T} \int_{-T/2}^{-T/2} x(t) \, dt \qquad (2.1)$$

The reason for the name "equilibrium probability" will become clear later on. It is easy to observe, however, that $P(x)$ is the fraction of time that $x(t)$ is in the 1 state. It is also the average value of $x(t)$, over all time. Thus, for instance, a 25% duty cycle clock signal, i.e., one that is high for 1/4th of its period, has $P(x) = 0.25$. The following proposition guarantees that the equilibrium probability is always well defined.

Proposition 1: For a logic signal $x(t)$, the limit in (2.1) always exists.

Proof: See Appendix A.

The discontinuity points of $x(t)$ represent *transitions* in the logic signal. Let $n_x(T)$ be the number of transitions of $x(t)$ in the time interval $(-T/2, +T/2]$.

Definition 2: The transition density of a logic signal $x(t)$, $t \in (-\infty, +\infty)$, is defined as

$$D(x) \triangleq \lim_{T \to \infty} \frac{n_x(T)}{T}. \qquad (2.2)$$

The reason for the name "transition density" will become clear later on. It should be clear, however, that $D(x)$ is the average number of transitions per unit time. Thus, a 10-MHz clock signal has $D(x) = 20 \times 10^6$. The power of the $P(x)$ and $D(x)$ concepts is that they apply equally well to both periodic (clock) and nonperiodic signals. In the remainder of this section, we study the existence of the limit in (2.2).

The time between two consecutive transitions of $x(t)$ will be referred to as an *intertransition time*. Let μ be the average value of all the intertransition times of $x(t)$. Likewise, let $\mu_1(\mu_0)$ be the average of the high (low), i.e., corresponding to $x(t) = 1(0)$, intertransition times of $x(t)$. It should be clear that $\mu = (1/2)(\mu_0 + \mu_1)$. In general, there is no guarantee of the existence of μ, μ_0, and μ_1. If the number of transitions in positive time is *finite*, then we say that there is an *infinite* intertransition time following the last transition, and $\mu = \infty$. A similar convention is made for negative time. We define μ_f to be the average of all the *finite* intertransition times of $x(t)$. In general, there is also no guarantee of the existence of μ_f.

Proposition 2: Two parts:

i) If μ_f exists and is nonzero, then $D(x)$ exists.

ii) If μ_0 and μ_1 exist, and $\mu \neq 0$, then $D(x)$ exists and we have

$$P(x) = \frac{\mu_1}{\mu_0 + \mu_1} \qquad (2.3a)$$

and

$$D(x) = \frac{2}{\mu_0 + \mu_1} \qquad (2.3b)$$

Proof: See Appendix A.

In order to guarantee that the density is always well defined, we make the following basic assumption about every logic signal $x(t)$:

Basic Assumption: The average finite intertransition time μ_f exists and is nonzero.

Logic signals for which this assumption does not hold are considered pathological, and are excluded from the analysis. It can be shown (see Appendix A) that another more stringent sufficient condition for the existence of (2.2) is that there be a nonzero lower bound (however small) on the intertransition times of $x(t)$. This condition is easily satisfied in all practical cases, so that our basic assumption is very mild indeed.

III. THE COMPANION PROCESS OF LOGIC SIGNALS

We will use **bold font** to represent random quantities. We denote the probability of an event A by $\mathcal{P}\{A\}$ and, if x is a random variable, we denote its mean (or expected value) by $E[x]$ and its distribution function by $F_x(a) \triangleq \mathcal{P}\{x \leq a\}$.

Let $x(t)$, $t \in (-\infty, +\infty)$, be a *stochastic process* [2] that takes the values 0 or 1, transitioning between them at *random* transition times. Such a process is called a *0-1 process* ([3, pp. 38–39]). A logic signal $x(t)$ can be thought of as a *sample* of a 0-1 stochastic process $x(t)$, i.e., $x(t)$ is one of an infinity of possible signals that comprise the family $x(t)$.

A stochastic process is said to be *strict-sense stationary* (SSS) if its statistical properties are invariant to a shift of the time origin [2]. Among other things, the mean $E[x(t)]$ of such a process is a constant, independent of time, and will be denoted by $E[x]$. It will be shown below that a logic signal is always a sample of a SSS 0-1 process.

Let $n_x(T)$ denote the (random) number of transitions of $x(t)$ in $(-T/2, +T/2)$. If $x(t)$ is SSS, then $E[n_x(T)]$ depends only on T, and is independent of the location of the time origin.

Proposition 3: If $x(t)$ is SSS, then the mean $E[n_x(T)/T]$ is a constant, independent of T.

Proof: Let $t_1 < t_2 < t_3$ be three arbitrary points along the time axis. Let n_1 be the number of transitions in $(t_1, t_2]$, n_2 be the number of transitions in $(t_2, t_3]$, and n_3 be the number of transitions in $(t_1, t_3]$. Then $n_3 = n_1 + n_2$, and $E[n_3] = E[n_1] + E[n_2]$. Let $T_1 = t_2 - t_1$ and $T_2 = t_3 - t_2$. Since $x(t)$ is SSS, then $E[n_1] = E[n_x(T_1)]$, $E[n_2] = E[n_x(T_2)]$, and $E[n_3] = E[n_x(T_1 + T_2)]$. Hence $E[n_x(T_1 + T_2)] = E[n_x(T_1)] + E[n_x(T_2)]$. Since this is true arbitrary T_1 and T_2, it means that, in general, $E[n_x(T)] = kT$, where k is a positive constant, which completes the proof.

A constant-mean stochastic process $x(t)$ is said to be *mean-ergodic* [2] if

$$\lim_{T \to \infty} \frac{1}{T} \int_{-T/2}^{+T/2} x(t)\, dt \overset{1}{=} E[x] \qquad (3.1)$$

where we have used the symbol " $\overset{1}{=}$ " to denote *convergence with probability 1*. The reader is referred to [2, pp. 188–191], for a discussion of the different stochastic convergence modes. We reserve the symbol " $=$ " to indicate *convergence everywhere* for random quantities. It will be shown below that a logic signal is always a sample of a SSS mean-ergodic 0-1 process.

Let $\tau \in (-\infty, +\infty)$ be a random variable with the probability distribution function $F_\tau(t) = 1/2$ for any finite t, and with $F_\tau(-\infty) = 0$ & $F_\tau(+\infty) = 1$. If $F_{\tau T}(t)$ is the uniform distribution over $[-T/2, +T/2]$, then $\lim_{T \to \infty} F_{\tau T} = F_\tau$. Thus, one might say that τ is uniformly distributed over the whole real line $\mathcal{R}$.

We now use τ to build from $x(t)$ an important 0-1 process $x(t)$, defined as follows.

Definition 3: Given a logic signal $x(t)$ and a random variable τ, uniformly distributed over $\mathcal{R}$, define a 0-1 stochastic process $x(t)$, called the *companion process of $x(t)$*, given by:

$$x(t) \triangleq x(t + \tau). \qquad (3.2)$$

For any given $t = t_1$, $x(t_1)$ is the random variable $x(t_1 + \tau)$ — a function of τ. Thus the stochastic process $x(t)$

is well defined. Intuitively, $x(t)$ is a family of shifted copies of $x(t)$, each shifted by a value of the random variable τ. Thus not only is $x(t)$ a sample of $x(t)$, but we also have the following.

Proposition 4: If $x(t)$ is the companion process of a logic signal $x(t)$, then the following "convergence everywhere" results are true:

$$\lim_{T \to \infty} \frac{1}{T} \int_{-T/2}^{+T/2} x(t)\, dt = P(x) \qquad (3.3)$$

$$\lim_{T \to \infty} \frac{n_x(T)}{T} = D(x). \qquad (3.4)$$

Proof: To prove (3.3), we need to show that for any given finite $\tau_1 \in \mathcal{R}$, the difference

$$\Delta_P \triangleq \frac{1}{T} \int_{-T/2}^{+T/2} x(t + \tau_1)\, dt - \frac{1}{T} \int_{-T/2}^{+T/2} x(t)\, dt$$

tends to zero as $T \to \infty$. This can be written as

$$\Delta_P = \frac{1}{T} \int_{-T/2 + \tau_1}^{+T/2 + \tau_1} x(t)\, dt - \frac{1}{T} \int_{-T/2}^{+T/2} x(t)\, dt$$

$$= \frac{1}{T} \int_{+T/2}^{+T/2 + \tau_1} x(t)\, dt - \frac{1}{T} \int_{-T/2}^{-T/2 + \tau_1} x(t)\, dt. \qquad (3.5)$$

Since $x(t) \in \{0, 1\}$, then $|\Delta_P| \leq |\tau_1|/T$ must go to 0 as $T \to \infty$.

Likewise, to prove (3.4) we must show that the difference $\Delta_D \triangleq 1/T$ {"the number of transitions in $((-T/2) + \tau_1, (+T/2) + \tau_1]$ − "that in $(-T/2, +T/2]$"}, goes to 0 as $T \to \infty$. Note that

$$\frac{n_x(T - 2|\tau_1|)}{T} - \frac{n_x(T)}{T}$$

$$\leq \Delta_D \leq \frac{n_x(T + 2|\tau_1|)}{T} - \frac{n_x(T)}{T}. \qquad (3.6)$$

If $x(t)$ has a finite number of transitions, then $\lim_{T \to \infty} n_x(T \pm 2|\tau_1|) = \lim_{T \to \infty} n_x(T) < \infty$, and $\lim_{T \to \infty} \Delta_D = 0$. Otherwise, notice that $1/D(x) = \lim_{T \to \infty}(T \pm 2|\tau_1|)/n_x(T \pm 2|\tau_1|) = \lim_{T \to \infty} T/n_x(T \pm 2|\tau_1|)$. Rewriting this as $\lim_{T \to \infty} n_x(T \pm 2|\tau_1|)/T = D(x)$ shows that Δ_D must go to 0 as $T \to \infty$.

Since this is true for any $\tau_1 \in \mathcal{R}$, then the convergence is *everywhere*, in the sense that *every* value of τ will lead to convergence.

Theorem 1: The companion process $x(t)$ of a logic signal $x(t)$ is SSS and mean-ergodic with $E[x] = \mathcal{P}\{x(t) = 1\} = P(x)$ and

$$E\left[\frac{n_x(T)}{T}\right] = D(x) \qquad (3.7)$$

Proof: At $t = 0$, we have $E[x(0)] = E[x(\tau)]$. An interesting property of τ is that if a is a constant then $a + \tau$ has the same distribution as τ. Indeed, if $F_{a + \tau}(t)$ is the distribution function of $a + \tau$, then $F_{a + \tau}(t) = \mathcal{P}\{a + \tau \leq t\} = \mathcal{P}\{\tau \leq t - a\} = 1/2 = F_\tau(t)$. Therefore, since $t + \tau$ and τ are identically distributed, we have $E[x(t +$

$\tau)] = E[x(\tau)]$, which means that $x(t)$ is a *constant-mean* process with

$$E[x(t)] = E[x(0)] = E[x(\tau)] \text{ for any time } t. \quad (3.8)$$

Let $\mathcal{R}_a$ be a subset of the real line $\mathcal{R}$ defined by $\mathcal{R}_a \triangleq \{t \in \mathcal{R}: x(t) = 1, x(t + a) = 1\}$. It is clear that $\mathcal{P}\{x(\tau) = 1, x(\tau + a) = 1\} = \mathcal{P}\{\tau \in \mathcal{R}_a\}$. Likewise, $\mathcal{P}\{x(t + \tau) = 1, x(t + \tau + a) = 1\} = \mathcal{P}\{t + \tau \in \mathcal{R}_a\}$. However, since τ and $t + \tau$ are identically distributed, the two probabilities $\mathcal{P}\{\tau \in \mathcal{R}_a\}$ and $\mathcal{P}\{t + \tau \in \mathcal{R}_a\}$ must be equal, which leads to:

$$\mathcal{P}\{x(t) = 1, x(t + a) = 1\}$$
$$= \mathcal{P}\{x(0) = 1, x(a) = 1\}$$
$$= \mathcal{P}\{\tau \in \mathcal{R}_a\} \text{ for any time } t. \quad (3.9)$$

Consequently, the joint distribution of $x(t)$ and $x(t + a)$, i.e., $F_{x(t), x(t+a)}(x_1, x_2)$, is independent of t, and depends only on a, which makes $x(t)$ wide-sense stationary (WSS) [2]. By extending this argument to $a_1, \cdots, a_n$, it follows that $F_{x(t), x(t-a_1), \cdots, x(t+a_n)}(x_1, \cdots, x_n)$ is independent of t, and $x(t)$ is strict-sense stationary (SSS).

To prove mean-ergodicity, and in view of (3.3), it suffices to show that $E[x] = P(x)$. Consider the random variable

$$\eta_T \triangleq \frac{1}{T} \int_{-T/2}^{+T/2} x(t) \, dt.$$

From (3.3) we have $\lim_{T \to \infty} \eta_T = P(x)$, where this is *convergence everywhere*. Therefore $\lim_{T \to \infty} E[\eta_T] = P(x)$. By linearity of the expected value operator, this can be rewritten

$$\lim_{T \to \infty} \frac{1}{T} \int_{-T/2}^{+T/2} E[x(t)] \, dt = P(x). \quad (3.10)$$

But $E[x(t)]$ is a constant. Therefore, the left-hand side of (3.10) is simply $E[x]$, and mean-ergodicity follows, with $E[x] = \mathcal{P}\{x(t) = 1\} = P(x)$.

To complete the proof, we will prove (3.7) by repeating the argument used for η_T. By (3.4), the random variable $n_x(T)/T$ converges *everywhere* to $D(x)$. Therefore, its mean must also converge to $D(x)$. Since, by Proposition 3 its mean is a constant, independent of T, then (3.7) follows. $\qquad \square$

We are now in a position to comment on the names "equilibrium probability" for $P(x)$ and "transition density" for $D(x)$. For a 0-1 process, $\mathcal{P}\{x(t) = 1\} = E[x(t)]$. Thus by (3.3) and since $x(t)$ is mean-ergodic, $P(x)$ is the constant probability that $x(t) = 1$. The name "equilibrium probability" is inspired from the special case when the intertransition times of a 0-1 process $x(t)$ are independent exponentially distributed random variables. In that case, the process is the well-known two-state continuous-time Markov process (see [2, pp. 392–393]) whose state probability *tends to an equilibrium value* for $t \to \infty$, at which time it becomes SSS (see [4, pp. 272–273]).

By (3.7), $D(x)$ is the expected "average number of transitions per unit time," which we compactly refer to as "transition density." This name is inspired by the *density* of random Poisson points (see [2, page 58]). If a large number of points are chosen on the time axis at random, then the "number of points in a given interval" is a random variable with a Poisson distribution whose *density* parameter λ is the "expected number of points per unit time." The *points* that we are concerned with in this paper are the time points at which transitions occur, but we make no assumption about their distribution. The remark about Poisson points is meant only to motivate the terminology.

IV. Density Simulation

A digital circuit provides a mapping from the logic signals at its primary input nodes to those at its internal and output nodes. In the following, we use the term "internal nodes" to refer to the primary output nodes as well as other proper internal circuit nodes.

If we consider the companion process of each such logic signal, the circuit may be seen as mapping stochastic processes at its inputs to similar processes at its internal nodes. The statistics (such as density and probability) of the internal processes are completely determined by those at the primary inputs. In fact, we will demonstrate in this section that the density and probability of internal processes can be efficiently *computed* from those at the primary inputs.

We assume that the primary input processes are *mutually independent*. Therefore, since these inputs are individually SSS, they are also *jointly* SSS. In terms of the underlying logic signals $x(t)$, this assumption means that the signal values are not *correlated*, so that if one of them is 1, then the average fraction of time that another is 1 (or 0) is unaltered.

Given the density and probability values of the companion processes at the primary inputs, we will present an algorithm to propagate them into a circuit to derive the corresponding values at internal nodes. We consider the circuit to be an interconnection of *logic modules*, each representing a certain (combinational) Boolean function and possessing certain delay characteristics. The propagation of density and probability will then proceed on a per-module basis from primary inputs to primary outputs, a process that we refer to as *density simulation*.

A. Propagation Through a Single Module

Consider a multi-input multi-output logic module M, whose outputs are Boolean functions of its inputs, as shown in Fig. 1. M may be a single logic gate or a higher level circuit block. We assume that the inputs to M are mutually independent companion processes. The validity of this assumption will be discussed in Section IV-B.

We use a *simplified timing model* of circuit behavior, as follows. We assume that an input transition that *does* get transmitted to an output node is delayed by a propa-

462

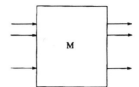

Fig. 1. Logic module M.

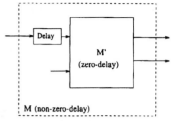

Fig. 2. Decoupling of delays.

gation delay time of τ_p. Different propagation delays may be associated with different input–output node pairs. Implicit in this model is the simplifying assumption that the propagation delay is independent of the values at other inputs of M.

In effect, we decouple the delays inside M from its Boolean function description by introducing a special-purpose delay block to model the delays between every pair of input and output nodes, as shown in Fig. 2. The block M' is a zero-delay logic module that implements the same Boolean function as M.

Since the input signals are SSS, then the output of the delay block has the same statistics as its input, and therefore has the same probability and density. As for the zero-delay module M', we now consider the problem of propagating equilibrium probabilities and transition densities from its inputs to its outputs.

Since $P(x) = \mathcal{P}\{x(t) = 1\}$ (by Theorem 1) and M' has zero delay, then the problem of propagating equilibrium probabilities through it is identical to that of propagating *signal probabilities* through logic circuits, which has been well studied [5]-[9]. Since the internal structure of M' is now known, the problem is actually even more generic than that, and can be expressed as "given a Boolean function $f(x_1, \cdots, x_n)$ and that each x_i can be high with probability $P(x_i)$, what is the probability that f is high?" Any number of published techniques can be used to solve this problem. However, we have chosen (for reasons that will become clear below) to investigate a new approach based on Binary Decision Diagrams (BDD's) [10], [11] which have recently become popular in the verification and synthesis areas. Appendix B describes how we use BDD's to compute the probability of a Boolean function.

We consider next the density propagation problem. Recall the concept of *Boolean Difference*: if y is a Boolean function that depends on x, then the Boolean difference of y with respect to x is defined as

$$\frac{\partial y}{\partial x} \triangleq y|_{x=1} \oplus y|_{x=0} = y(x) \oplus y(\bar{x}) \quad (4.1)$$

where $\oplus$ denotes the EXCLUSIVE-OR operation. Note that, if x is an input and y is an output of M', then $\partial y/\partial x$ is a Boolean function that does *not* depend on x, but may depend on all other inputs of M'. Therefore, $\partial y/\partial x$ and x are independent. A crucial observation is that if $\partial y/\partial x$ is 1, then a transition at x will cause a (simultaneous) transition at y, otherwise *not*. Since the input processes are SSS, then $\partial y/\partial x$ is also SSS; in fact it is a companion process with equilibrium probability $P(\partial y/\partial x)$. We are now ready to prove the following:

Theorem 2: If the inputs $x_i(t)$, $i = 1, \cdots, n$, of a zero-delay logic module are independent companion processes with transition densities $D(x_i)$, then the densities at its outputs $y_j(t)$, $j = 1, \cdots, m$ are given by

$$D(y_j) = \sum_{i=1}^{n} P\left(\frac{\partial y_j}{\partial x_i}\right) D(x_i). \quad (4.2)$$

Proof: Let t_{ik}, $k = 1, 2, \cdots, n_{xi}(T)$, be the sequence of transition time points of $x_i(t)$ in

$$\left(\frac{-T}{2}, \frac{+T}{2}\right].$$

Consider the sequence of random variables $\partial y_j/\partial x_i(t_{ik})$, $k = 1, 2, \cdots, n_{xi}(T)$, defined for every input–output pair (x_i, y_j) of M'.

Since $(\partial y_j/\partial x_i)(t)$ is SSS and independent of $x_i(t)$, then

$$\mathcal{P}\left\{\frac{\partial y_j}{\partial x_i}(t_{ik}) = 1\right\} = P\left(\frac{\partial y_j}{\partial x_i}\right)$$

is the same for any k. Therefore, $(\partial y_j/\partial x_i)(t_{ik})$, $k = 1, 2, \cdots, n_{xi}(T)$, is a sequence of identically distributed (not necessarily independent) random variables, with mean $P(\partial y_j/\partial x_i)$.

Since $(\partial y_j/\partial x_i)(t_{ik}) = 1$ if and only if the kth transition of $x_i(t)$ is transmitted to $y_j(t)$, then the number of transitions of $y_j(t)$ in $(-T/2, +T/2)$ is given by

$$n_{yj}(T) = \sum_{i=1}^{n} \sum_{k=1}^{n_{xi}(T)} \frac{\partial y_j}{\partial x_i}(t_{ik}). \quad (4.3)$$

Taking the expected value of both sides gives

$$E[n_{yj}(T)] = \sum_{i=1}^{n} E\left[\sum_{k=1}^{n_{xi}(T)} \frac{\partial y_j}{\partial x_i}(t_{ik})\right] \quad (4.4)$$

Since $(\partial y_j/\partial x_i)(t)$ is independent of $x_i(t)$, and if n is some positive integer, then,

$$E\left[\frac{\partial y_j}{\partial x_i}(t_{ik}) \bigg| n_{xi}(T) = n\right] = E\left[\frac{\partial y_j}{\partial x_i}(t_{ik})\right] = P\left(\frac{\partial y_j}{\partial x_i}\right). \quad (4.5)$$

Using [2, p. 183], these facts lead to

$$E[n_{yj}(T)] = \sum_{i=1}^{n} P\left(\frac{\partial y_j}{\partial x_i}\right) E[n_{xi}(T)] \quad (4.6)$$

463

which, dividing by T and using (3.7), leads to the required result (4.2). ∎

If the Boolean difference is available, then evaluating $P(\partial y_i/\partial x_i)$ is no more difficult than evaluating the probability of a Boolean function knowing those of its inputs. Note that if M is a 2-input AND gate with inputs x_1 and x_2, and output y, then $P(\partial y/\partial x_1) = P(x_2)$. In more complex situations, the "COMPOSE" and "XOR" functions of the BDD package [11] can be used to evaluate the Boolean difference using (4.1). The BDD-based algorithm given in the appendix (for computing the probability of a Boolean function) can then be used to compute $P(\partial y_j/\partial x_i)$.

B. Global Propagation Strategy

The assumption was made at the beginning of Section IV-A that the inputs to a module are *independent*. Even though this is true at the primary inputs (as we have assumed), it may not be true for internal nodes. Circuit topologies that include reconvergent fanout and feedback will cause internal nodes to be correlated, and destroy the independence property. This problem is central to any circuit analysis based on a statistical representation of signals, and can usually be taken care of by using heuristics that tradeoff accuracy for speed [5]–[9].

Based on our previous experience with the propagation of probability waveforms [12], we have found that if the modules are large enough so that tightly coupled nodes (such as in latches or small cells) are kept inside the same module, then the coupling outside the modules is sufficiently low to justify an independence assumption. While this does take into account the correlations inside a module, it may create inaccuracies because internal delays are lumped together. Furthermore, performance may be sacrificed because the BDD's can become too large. Section VI will investigate this speed–accuracy tradeoff.

V. PRACTICAL APPLICATIONS

Once the density at every internal node has been computed, these values can be used in a postprocessing step to investigate various circuit properties. We present here four different applications of the density concept in CMOS circuits.

A. Average Power/Ground Bus Currents

Consider the problem of computing the average current in the power or ground bus branches. We will consider only the case of the power bus, since that of the ground bus is identical.

A convenient approximation is to view the bus as an interconnection of lumped resistors, with lumped capacitors to ground, i.e., a linear RC network. Some nodes of this network are connected to the external V_{dd} power supply, while others (referred to as *contacts*) are connected to the various circuit components, e.g., CMOS gates, drawing power supply current. Let $i_k(t)$, $k = 1, 2, \cdots,$ n, be the various current waveforms that the circuit draws at these contact nodes. Let $i_j(t)$, $u = 1, 2, \cdots, m$, be

the various current waveforms in the bus branches. The bus can now be viewed as a *linear time-invariant* (LTI) system whose outputs $i_j(t)$ are related to its inputs $i_k(t)$ by the convolutions:

$$i_j(t) = \sum_{k=1}^{n} h_{jk}(t) * i_k(t) = \sum_{k=1}^{n} \int_{-\infty}^{+\infty} h_{jk}(r) i_k(t-r) \, dr,$$
$$j = 1, \cdots, m \qquad (5.1)$$

where $h_{jk}(t)$ are the impulse response functions, and "$*$" denotes the convolution operation.

Let

$$I_j \triangleq \lim_{T \to \infty} \frac{1}{T} \int_{-T/2}^{+T/2} i_j(t) \, dt$$

be the average current in the jth bus branch. Combining this with (5.1) and exchanging the order of the integrals, we get

$$I_j = \sum_{k=1}^{n} h_{jk}(t) * I_k, \qquad j = 1, \cdots, m \qquad (5.2)$$

where we have made use of the fact that

$$\lim_{T \to \infty} \frac{1}{T} \int_{-T/2}^{+T/2} i_k(t - r_1) \, dt$$

is equal to

$$I_k \triangleq \lim_{T \to \infty} \frac{1}{T} \int_{-T/2}^{+T/2} i_k(t) \, dt$$

for any given r_1. The proof of this is identical to that of (3.3) and assumes the existence of an arbitrary, but finite, upper bound on $i_k(t)$.

In other words, if the time-averages of the contact currents are themselves applied at the contacts, and the bus is solved (i.e., simulated) as a resistive network (dc solution), the resulting branch currents *are* the required time-averages of the bus currents. To complete the solution, we will now express the time-average contact currents I_k in terms of the transition densities inside the circuit.

Let $D(x)$ be the transition density at the output node x of a CMOS gate that draws power supply current $i(t)$ whose time-average is I. Furthermore, let $C_n (C_p)$ be the total capacitance from x to the ground (power) bus connection. These capacitances are the sum of i) any lumped capacitance tied to the gate output; ii) MOSFET drain and source capacitances in the gate output stage; and iii) MOSFET gate capacitances in any logic gates driven by x. As such, they are related to both load capacitance and transistor strength. It has been established [13] that a good estimate of the supply current $i(t)$ can be obtained by looking only at its capacitive charging/discharging component. Since the charge drawn from the supply whenever the gate switches low-to-high (high-to-low) is

$V_{dd} C_n (V_{dd} C_p)$, it follows that:

$$I \triangleq \lim_{T \to \infty} \frac{1}{T} \int_{-T/2}^{+T/2} i(t) \, dt$$

$$= \lim_{T \to \infty} \frac{\frac{1}{2} n_x(T) V_{dd} C_n + \frac{1}{2} n_x(T) V_{dd} C_p}{T}$$

$$= \frac{V_{dd} C}{2} D(x) \qquad (5.3)$$

where $C \triangleq C_n + C_p$ is the total capacitance at the output node.

Equations (5.3) and (5.2) provide an efficient technique for computing the average current in every branch of the bus, given the transition densities at all circuit nodes. It is significant that this requires only a single DC simulation of the resistive network representing the power bus; no transient simulation is required, and the bus capacitance is irrelevant.

B. Average Power Dissipation

As a direct consequence of the above results, it should be clear that the overall average power dissipation is given by $P_{av} = 1/2 V_{dd}^2 \Sigma C_i D(x_i)$, summing over all circuit nodes x_i.

C. Electromigration Failures

Electromigration [14], [15] is a major reliability problem caused by the transport of atoms in a metal line due to the electron flow. Under persistent current stress this can cause deformations of the metal, leading to either short or open circuits. The time-to-failure is a lognormally distributed random variable. It is usually characterized by the *median (or mean) time-to-failure* (MTF) [15], which depends on the current density in the metal line.

The models for MTF prediction under pulsed-dc or ac current stress are still controversial. Some recent models [16] predict that, at least under pulsed-dc conditions, the average current is sufficient to predict the MTF, as follows:

$$MTF = \frac{A}{I^2} \qquad (5.4)$$

where a is a parameter that does not depend on the current and I is the average current. However, other recent studies [17] show that the situation is much more complicated.

In any case, even if I is not sufficient by itself to estimate the MTF, it represents a *first-order* approximation of the current stress in the wire. Thus (5.2) and (5.3), based on the transition density, provide the required average current values I, and help identify potential electromigration problems in the power/ground bus branches.

D. Hot-Electron Degradation

As MOSFET devices are scaled down to very small dimensions, certain physical mechanisms start to cause degradation in the device parameters, causing major reliability problems. One such mechanism is the injection of "hot electrons" (or, in general, hot carriers) into the MOS gate oxide layer [14]. Trapping of these carriers in the gate insulator layer causes degradation in the transistor transconductance and/or threshold voltage.

It is widely accepted that the MOSFET substrate current is a good indicator of the severity of the degradation. In fact one can write an expression for the "age" of a transistor (i.e., how far it is down the degradation path) that has been operating for time T as follows [18]

$$\text{Age}(T) = \int_{-T/2}^{+T/2} \frac{I_{ds}}{WH} \left[\frac{I_{sub}}{I_{ds}} \right]^m dt \qquad (5.5)$$

where $I_{ds}(t)$ and $I_{sub}(t)$ are the MOSFET drain-to-source and substrate currents, W is the channel width, and H and m are parameters that do not depend on the transistor currents.

In order to see how this can be used in a CMOS circuit, consider a MOSFET in a CMOS inverter whose output node is x. It can be shown that the both $I_{sub}(t)$ and $I_{ds}(t)$ are nonzero only when the inverter is switching (this also holds for any CMOS gate). Whenever the inverter switches, it generates two current pulses $I_{sub}(t)$ and $I_{ds}(t)$. The pulses resulting from different switching events are identical except for a dependence on the rise–fall at the inverter input. If one assumes a certain nominal rise/fall time at the input, then using (5.5) one can compute the incremental aging due to $0 \to 1$ and $1 \to 0$ transitions at the inverter output, call these A_{lh} and A_{hl}. Then (5.5) may be written:

$$\text{Age}(T) = (A_{lh} + A_{hl}) \frac{n_x(T)}{2}. \qquad (5.6)$$

Degradation due to hot-carriers takes years to manifest itself. In other words, T and $n_x(T)$ are very large, which (using (2.2)) permits the approximation $n_x(T) \approx TD(x)$, and leads to:

$$\text{Age}(T) = \left[\frac{A_{lh} + A_{hl}}{2} \right] TD(x). \qquad (5.7)$$

Thus, if CMOS gates are precharacterized to estimate the incremental damage to their transistors due to a single output transition, then the transition density provides the means to predict transistor aging over extended periods using (5.7).

VI. EXPERIMENTAL RESULTS

We have implemented this approach in a prototype *density simulator*, called DENSIM, that takes a description of a circuit in terms of its Boolean modules and gives the transition density at every node. It also accepts values for transition density and equilibrium probability at the primary inputs. Our current implementation is restricted to combinational (nonfeedback) circuits. Every Boolean module should be an instance of a *model* from a simulation library built by a separate *model compiler* called MODCOM. MODCOM uses an input specification in the form of Boolean equations to build a BDD representation of the module outputs and the relevant Boolean differences, and stores this in a model file that DENSIM can use.

We present below the results of a number of test cases that were used to investigate the accuracy and efficiency of this technique. In order to assess the accuracy of the results, we have devised a test by which randomly generated logic waveforms are fed to the circuit primary inputs and propagated into the circuit (by logic simulation based on the BDD's). The logic simulator uses assignable nonzero delays, scaling them based on the fan-out load at every module output. The input waveforms must have the same probability and density values given to DENSIM, and are generated as follows. Starting with $P(x)$ and $D(x)$ values, we solve for μ_0 and μ_1 from (2.3a) and (2.3b). We then use (arbitrarily) an exponentially distributed random number generator to produce sequences of inter-transitional times that have the means μ_0 and μ_1 (the theory presented above holds for any distribution of intertransition times). Starting from arbitrary initial values, the waveforms are built using these sequences. From the logic simulation results, we estimate the average number of transitions per unit time for every circuit node. For a large number of input transitions, this number should converge to the transition density, according to (2.2). We also estimate the fraction of time that the signal spends in the high state and check if that converges to the equilibrium probability, in accordance with (2.1).

In the first few test cases to be presented, the modules were chosen to contain all reconvergent fan-out. Thus all signals are independent and the results from DENSIM should agree exactly with those from logic simulation. We will then move on to other test cases where signal correlation does become an issue and will study the speed–accuracy tradeoff involved.

As a first test case, consider a single logic module with eight inputs and one output that implements the Boolean function $Z = ABFD + CFD + ABHD + CHD + ABFG + CFG + ABHG + CHG + AFE + ADE + CFE + CDE$. Using input values of $P = 0.5$ and $D = 2.0$, DENSIM gives $P(Z) = 0.476562$ and $D(Z) = 3.71875$. The results of the logic simulation run, showing the correct convergent behavior at the output Z, are shown in Fig. 3.

The horizontal axis in this figure is the CPU time elapsed during the logic simulation run, and the vertical axis is the cumulative values of density and probability at the output node. The two horizontal dashed lines are the values of density and probability computed by DENSIM and the vertical dashed line indicated by the arrow shows the total CPU time required by the DENSIM run. The other vertical line indicates the CPU time required to observe 1000 logic transitions at node Z.

The second test case is the 4-bit ALU/function generator SN54181 from the TI TTL data book. This circuit has 75 logic gates and is shown in Fig. 4.

If we consider the whole circuit as a single Boolean module, then the effects of all internal node correlations

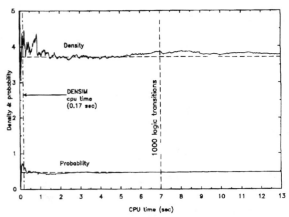

Fig. 3. Density and probability convergence plot at node Z.

are taken care of, and the DENSIM results should, again, be exact. It takes MODCOM 6.53 CPU seconds (SUN SparcStation 1) to build and store the 6092-node BDD model in this case, and DENSIM requires 0.84 CPU seconds (SUN) to run on it. The DENSIM results for the two output nodes $F3$ and X are shown in Figs. 5 and 6, respectively.

The preceding test cases show that even for single-module circuits, computing the density values using DENSIM instead of traditional logic simulation is accurate, much faster, and avoids lengthy simulations involving thousands of logic transitions. This observation will be further enforced by the results presented below.

Moving on to multimodule circuits, consider a 32-bit binary ripple adder. In this case, we chose the full adders to be our Boolean modules. This again leads to a situation where all reconvergent fan-out and signal correlation is inside the modules, and where DENSIM results should be exact. DENSIM takes only 0.46 CPU seconds (SUN), as opposed to the five minutes required for the logic simulation results to converge, as shown in Figs. 7 and 8, respectively.

An interesting feature of the result in Fig. 7 is the prolonged "flat" part of the curve around 1000 transitions. This illustrates the point made in the introduction that it is impossible to tell beforehand exactly when a logic simulation run should be terminated. In this case, if one were monitoring the density values from logic simulation with the intention of terminating the run when the density "converged to something," one might terminate the run at 1000 transitions, getting the wrong result.

We now move on to a consideration of the effects of signal correlation caused by reconvergent fanout. As pointed out in Section IV-B, one can accurately handle these effects by keeping all reconvergent fan-out within the Boolean modules. However, since large BDD's are expensive to build and maintain, this can become impractical and leads to a speed–accuracy tradeoff. To illustrate this point, we again consider the ALU circuit in Fig. 4. We partition the circuit into the 19 smaller modules

functional block diagram

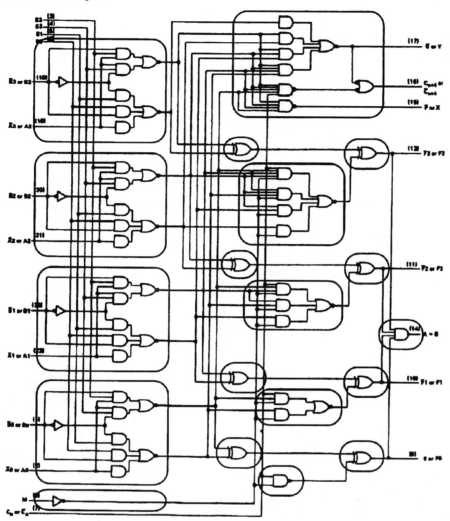

Fig. 4. ALU/function generator circuit.

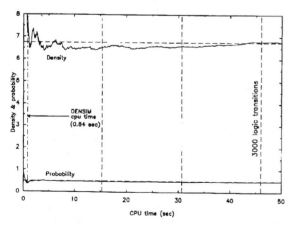

Fig. 5. Results for node $F3$ of the ALU.

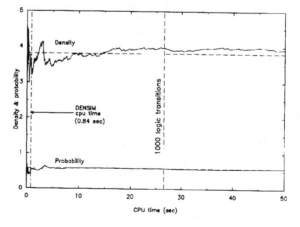

Fig. 6. Results for node X of the ALU.

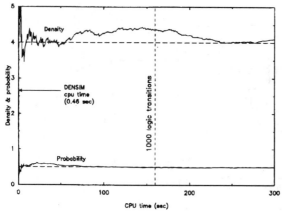

Fig. 7. Results for node *n*2 of the adder.

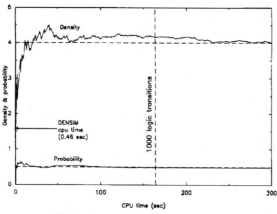

Fig. 8. Results for node *n*129 of the adder.

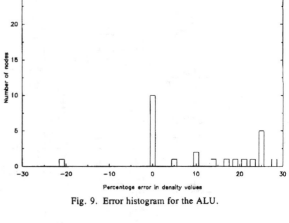

Fig. 9. Error histogram for the ALU.

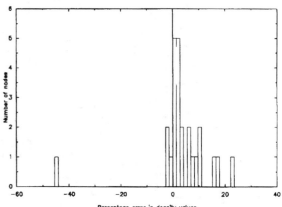

Fig. 10. Second error histogram for the ALU.

shown in the figure and examine the resultant density values at all nodes that are module outputs. By comparing these to the values obtained from the single-module run on this circuit, we get the error histogram shown in Fig. 9. In this case there was a less than 29% loss in accuracy for a 15X gain in speed.

For a further comparison, we ran a logic simulation on the ALU using its gate-level representation, and compared the resulting densities to those observed in the above 19-module run. The error histogram in this case is shown in Fig. 10. All but one of the densities are within 23%. The single point of poor agreement is at node AB which is a reconvergent node for all four ALU outputs F0–F3.

Finally, we present some results obtained for the ISCAS-85 benchmark circuits [19]. In this case we used a "lowest level partitioning" in which every logic gate was represented as a separate Boolean module. This provides the fastest, but potentially the least accurate, DEN-SIM run. The 10 ISCAS circuits, their sizes, and the total DENSIM CPU time (on a CONVEX c240) are shown in Table I.

The execution times are excellent, taking under 10 s even for the largest circuit. It becomes exceedingly diffi-

TABLE I
EXECUTION TIME RESULTS FOR THE ISCAS-85 BENCHMARK CIRCUITS

Circuit Name	Size (# gates)	Total Time (CPU sec.)
c432	160	0.52
c499	202	0.58
c880	383	1.06
c1355	546	1.39
c1908	880	2.00
c2670	1193	3.45
c3540	1669	3.77
c5315	2307	6.41
c6288	2406	5.67
c7552	3512	9.85

cult to assess the accuracy for large circuits because the BDD's become unacceptably large. Even though BDD's for these circuits *have* been built by other researchers, the BDD's that we require are much larger because they must include the Boolean function at every internal node as well as the output nodes, along with all the associated Boolean difference terms. Thus we are reduced to having to assess the accuracy by obtaining a best possible estimate of the densities from long logic simulation runs. Even then, it is practically impossible to examine the density plot for

Circuit Name	Avg. Density (DENSIM)	Avg. Density (time) (logic simulation)		% Error
c432	3.46	3.39	(62.8)	+2.1%
c499	11.36	8.57	(241.1)	+29.8%
c880	2.78	3.25	(131.7)	−14.5%
c1355	4.19	6.18	(407.9)	−32.2%
c1908	2.97	5.01	(463.9)	−40.7%
c2670	3.50	4.00	(618.5)	−12.5%
c3540	4.47	4.49	(1082.0)	−0.4%
c5315	3.52	4.79	(1616.0)	−26.5%
c6288	25.10	34.17	(31 057.0)	−26.5%
c7552	3.85	5.08	(2713.0)	−24.2%

every internal node to determine whether the run was long enough for it to converge. Based on several test cases, however, we found that an average of 1000 transitions per input node seems to be enough to approximate most node densities. Such logic simulation runs were performed on all ten circuits. In order to tabulate the results, we show the *average density* values (averaged over all circuit nodes) in Table II.

The third column in the table also lists the total CPU time required (on the CONVEX) to finish the logic simulation run in each case. Even for the smallest circuits, such long simulation runs meant that hundreds of thousands of internal events had to be simulated. Comparing the execution times between Tables I and II clearly demonstrates the speed advantage of this approach (e.g., 5.67 s versus 8 h 38 min for c6288). As for the average density values, the agreement is very good for c432 and c3540, acceptable for c880 and c2670, and poor for the other circuits. These results highlight the need to better account for signal correlation if one is to obtain consistently good results in the general case.

In general, the problem of estimating equilibrium probabilities, let alone transition densities, is $\mathcal{NP}$-hard. As a result, no single *efficient* solution will work well in *all* cases. The partitioning strategy in general cases, and the speed–accuracy tradeoff, are the focus of our continuing research efforts in this area.

VII. SUMMARY AND CONCLUSIONS

To summarize, we have observed that a common thread that runs through most causes of runtime failure is the extent of circuit *activity*, i.e., the rate at which its nodes are switching. We have defined a new measure of circuit activity, called the *transition density*. Based on a stochastic model of logic signals, we have also presented an algorithm to propagate the density from the primary inputs to internal nodes.

To illustrate the practical significance of these results, we have considered four ways in which the density values can be used to study circuit reliability by estimating 1) the average power and ground currents; 2) the average power dissipation; 3) the susceptibility to electromigration failures; and 4) the extent of hot electron degradation. We

have also presented experimental results that demonstrate the practical significance and power of this approach. We envision that the computation of density values inside the circuit can be used as a preprocessing step and the resulting information applied to these and possibly other reliability problems.

APPENDIX A
EXISTENCE OF $P(x)$ AND $D(x)$

A. Existence of $P(x)$

Recall the definition (2.1) of the equilibrium probability:

$$P(x) \triangleq \lim_{T \to \infty} \frac{1}{T} \int_{-T/2}^{+T/2} x(t) \, dt.$$

For convenience, we also repeat the statement of Proposition 1:

Proposition 1: For a logic signal $x(t)$, the limit in (2.1) always exists.

Proof: Let

$$\bar{x}_T \triangleq \frac{1}{T} \int_0^T x(t) \, dt$$

be the time average of $x(t)$ over $[0, T]$; it suffices to show that $\lim_{T \to \infty} \bar{x}_T$ always exists. Notice that $\bar{x}_T \in [0, 1]$, and

$$\frac{d\bar{x}_T}{dT}(T) = \frac{x(T)}{T} - \frac{\bar{x}_T}{T}. \tag{A.1}$$

Since both $x(T)$ and $\bar{x}_T$ are bounded, then

$$\lim_{T \to \infty} \frac{d\bar{x}_T}{dT}(T) = 0.$$

By the mean value theorem, for any $\Delta > 0$, there exists a $\gamma \in [T, T + \Delta]$ such that

$$\bar{x}_{T+\Delta} - \bar{x}_T = \frac{d\bar{x}_T}{dT}(\gamma)\Delta.$$

Therefore,

$$\lim_{T \to \infty} \{\bar{x}_{T+\Delta} - \bar{x}_T\} = \Delta \lim_{\gamma \to \infty} \frac{d\bar{x}_T}{dT}(\gamma) = 0 \tag{A.2}$$

which means that $\lim_{T \to \infty} \bar{x}_T$ exists.

B. Existence of $D(x)$

Recall the definition (2.2) of the transition density:

$$D(x) \triangleq \lim_{T \to \infty} \frac{n_x(T)}{T}.$$

We also recall a few other definitions: The time between two consecutive transitions of $x(t)$ is called an *intertransition time*; μ is the average value of all the intertransition times of $x(t)$; and μ_1 (μ_0) is the average of the high (low), i.e., corresponding to $x(t) = 1(0)$, intertransition times of $x(t)$. It should be clear that $\mu = (1/2)(\mu_0 + \mu_1)$. In general, there is no guarantee of the existence of μ, μ_0, and μ_1. If the number of transitions in positive time is *finite*, then we say that there is an *infinite* intertransition time following the last transition, and $\mu = \infty$. A similar convention is made for negative time. μ_f is the average of all the *finite* intertransition times of $x(t)$. In general, there is also no guarantee of the existence of μ_f. It should be clear, however, that if μ exists, then μ_f also exists and $\mu_f = \mu$. We are now ready to prove Proposition 2, which we restate for convenience.

Proposition 2: Two parts:

i) If μ_f exists and is nonzero, then $D(x)$ exists.

ii) If μ_0 and μ_1 exist, and $\mu \neq 0$, then $D(x)$ exists and we have:

$$P(x) = \frac{\mu_1}{\mu_0 + \mu_1} \tag{A.3a}$$

and

$$D(x) = \frac{2}{\mu_0 + \mu_1}. \tag{A.3b}$$

Proof: i) Suppose that $\mu_f \neq 0$ exists. We first dispose of the special case when $x(t)$ has a finite number of transitions. In that case, $\lim_{T \to \infty} n_x(T)$ is a finite integer value, and $D(x) = 0$.

Another special case is when $x(t)$ has an infinite number of transitions in only one time direction. Without loss of generality, consider that $x(t) = 0$ for all $t < t_0$. If we build another signal $x'(t)$ so that $x'(t) = x(t)$, for $t > t_0$, and $x'(t) = x(t_0 + (t_0 - t))$, for $t < t_0$, then $x'(t)$ has an infinity of transitions in both time directions and it can be shown that $D(x) = (1/2)D(x')$. Thus the existence of $D(x)$ is covered by the general case of a signal with an infinity of transitions in both time directions, to be considered next.

In the general case of an infinity of transitions in both time directions, $x(t)$ cannot have an infinite intertransition time, so that $\mu_f = \mu$. It will simplify the discussion below to refer to μ rather than μ_f. Consider Fig. 11 where, for every T, t_1 is the time of the last transition of $x(t)$ before $-T/2$, t_2 is that of the first transition after $-T/2$, t_3 is that of the last transition before $+T/2$, and t_4 is that of the first transition after $+T/2$.

There are $n_x(T)$ transitions between $-T/2$ and $+T/2$, including t_2 and t_3. Thus there are $(n_x(T) - 1)$ intertransition time intervals between t_2 and t_3. Since $\lim_{T \to \infty} n_x(T) = \infty$, we have

$$\mu = \lim_{T \to \infty} \frac{t_3 - t_2}{n_x(T) - 1} = \lim_{T \to \infty} \left\{ \frac{n_x(T)}{n_x(T) - 1} \right\} \frac{t_3 - t_2}{n_x(T)}$$

$$= \lim_{T \to \infty} \frac{t_3 - t_2}{n_x(T)}. \tag{A.4}$$

Likewise,

$$\mu = \lim_{T \to \infty} \frac{t_4 - t_1}{n_x(T) + 1} = \lim_{T \to \infty} \frac{t_4 - t_1}{n_x(T)}. \tag{A.5}$$

We now observe that $t_3 - t_2 \leq T \leq t_4 - t_1$, which

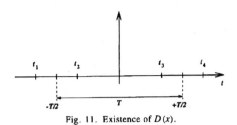

Fig. 11. Existence of $D(x)$.

gives

$$\lim_{T \to \infty} \frac{t_3 - t_2}{n_x(T)} \leq \lim_{T \to \infty} \frac{T}{n_x(T)} \leq \lim_{T \to \infty} \frac{t_4 - t_1}{n_x(T)}. \tag{A.6}$$

Using (A.4) and (A.5), we seen that

$$\lim_{T \to \infty} \frac{T}{n_x(T)} = \mu$$

exists. Since $\mu = \mu_f \neq 0$, then $D(x) = 1/\mu$ exists.

ii) If μ_0 and μ_1 exist, and $\mu = (\mu_0 + \mu_1)/2$ is nonzero, then μ_f exists and is nonzero and (2.2) exists. Existence of μ_0 and μ_1 also means that $x(t)$ has no infinite intertransition times, so that $D(x) = 1/\mu$, and we directly get (A.3b)

$$D(x) = \frac{2}{\mu_0 + \mu_1}. \tag{A.7}$$

To obtain (A.3a), let $n_1(T)$ be the number of (whole) 1-pulses of $x(t)$ in $(-T/2, +T/2]$. It is easy to verify that $|n_1(T) - (n_x(T)/2)| \leq 1$, which gives $\lim_{T \to \infty} (n_1(T)/T) = (1/2)D(x)$. Consider Fig. 12 where, for every T, t_1 is the time of the last $0 \to 1$ transition of $x(t)$ before $-T/2$, t_2 is that of the first $0 \to 1$ transition after $-T/2$, t_3 is that of the last $1 \to 0$ transition before $+T/2$, and t_4 is that of the first $1 \to 0$ transition after $+T/2$.

By definition of μ_1, we have

$$\lim_{T \to \infty} \frac{1}{n_1(T)} \int_{t_2}^{t_3} x(t) \, dt = \mu_1 \tag{A.8}$$

and

$$\mu_1 = \lim_{T \to \infty} \frac{1}{n_1(T) + 2} \int_{t_1}^{t_4} x(t) \, dt$$

$$= \lim_{T \to \infty} \left\{ \frac{n_1(T)}{n_1(T) + 2} \right\} \frac{1}{n_1(T)} \int_{t_1}^{t_4} x(t) \, dt \tag{A.9}$$

which gives

$$\lim_{T \to \infty} \frac{1}{n_1(T)} \int_{t_1}^{t_4} x(t) \, dt = \mu_1. \tag{A.10}$$

We now observe that

$$\int_{t_2}^{t_3} x(t) \, dt \leq \int_{-T/2}^{+T/2} x(t) \, dt < \int_{t_1}^{t_4} x(t) \, dt$$

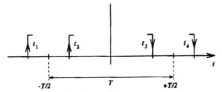

Fig. 12. Deriving the expression for $P(x)$.

which gives

$$\lim_{T \to \infty} \frac{1}{n_1(T)} \int_{t2}^{t3} x(t)\, dt \leq \lim_{T \to \infty} \frac{1}{n_1(T)} \int_{-T/2}^{+T/2} x(t)\, dt$$

$$\leq \lim_{T \to \infty} \frac{1}{n_1(T)} \int_{t1}^{t4} x(t)\, dt. \tag{A.11}$$

Using (A.8) and (A.10), we see that

$$\lim_{T \to \infty} \frac{1}{n_1(T)} \int_{-T/2}^{+T/2} x(t)\, dt = \mu_1.$$

Since

$$\lim_{T \to \infty} \frac{n_1(T)}{T} = \frac{1}{2} D(x)$$

we find that

$$\mu_1 = \lim_{T \to \infty} \frac{T}{n_1(T)} \frac{1}{T} \int_{-T/2}^{+T/2} x(t)\, dt = \frac{2}{D(x)} P(x) \tag{A.12}$$

which leads to (A.3a)

$$P(x) = \frac{\mu_1}{\mu_0 + \mu_1} \tag{A.13}$$

and the proof is complete ∎

In order to illustrate how mild the condition of Proposition 2 is, one can prove another (more stringent) sufficient condition for the existence of $D(x)$, namely, that there exists a non-zero *lower bound* $\delta_x > 0$ on the intertransition times. The proof is as follows: Consider the logic signal $x_\delta(t)$ built as follows: $x_\delta(t)$ is 0 everywhere, except on intervals of width δ_x centered at every transition time point of $x(t)$, where it is 1. It is clear that

$$\left| n_x(T) - \frac{1}{\delta_x} \int_{-T/2}^{+T/2} x_\delta(t)\, dt \right| < 1.$$

Therefore,

$$\lim_{T \to \infty} \frac{n_x(T)}{T} = \frac{1}{\delta_x} \lim_{T \to \infty} \frac{1}{T} \int_{-T/2}^{+T/2} x_\delta(t)\, dt. \tag{A.14}$$

By Proposition 1, and since $\delta_x > 0$, the density exists. This condition can be easily satisfied in all practical cases.

Appendix B
Using BDD's for probability Propagation

We will briefly review the concept of a BDD [10], [11] and then present a new application for BDD's as tools for computing the probability of a Boolean function.

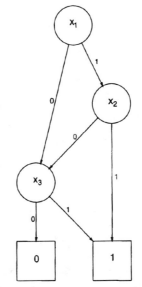

Fig. 13. Example BDD representation.

Consider the Boolean function $y = x_1 \cdot x_2 + x_3$, which can be represented by the BDD shown in Fig. 13. The Boolean variables x_1 are *ordered*, and each *level* in the BDD corresponds to a single variable. Each level may contain one or more BDD nodes at which one can branch in one of two directions, depending on the value of the relevant variable. For example, suppose that $x_1 = 1$, $x_2 = 0$, and $x_3 = 1$. To evaluate y, we start at the top node, branch to the right since $x_1 = 1$, then branch to the left since $x_2 = 0$, and finally branch to the right since $x_3 = 1$ to reach the terminal node "1." Thus the corresponding value of y is 1.

The importance of the BDD representation is that it is *canonical*, i.e., that it does not depend on the Boolean expression used to express the function. In our case, if the function was expressed as $y = x_3 + x_1 \cdot (x_2 + x_3)$ (an equivalent representation), it would have the same BDD. BDD's have been found to be an efficient representation for manipulating Boolean functions, both in terms of memory and execution time. For example, checking if a Boolean function is satisfiable can be done in time that is linear in the number of variables.

Let $y = f(x_1, \cdots, x_n)$ be a Boolean function. We will show that, given signal probabilities for the variables x_i and that these variables are independent (random variables), then the probability of the function f can be obtained in *linear time* (in the size of its BDD representation). By Shannon's expansion:

$$y = x_1 f_{x_1} + \overline{x_1} f_{\overline{x_1}} \tag{B.1}$$

where $f_{x_1} = f(1, x_2, \cdots, x_n)$ and $f_{\overline{x_1}} = f(0, x_2, \cdots, x_n)$ are the *cofactors* of f with respect to x_1. Since $x_1 \overline{x_1} = 0$, then,

$$P(y) = P(x_1 f_{x_1}) + P(\overline{x_1} f_{\overline{x_1}}). \tag{B.2}$$

Since the cofactors of x_i do not depend on x_i, and since all variables are independent, then

$$P(y) = P(x_1)P(f_{x_1}) + P(\overline{x_1})P(f_{\overline{x_1}}). \qquad (B.3)$$

This equation shows how the BDD is to be used to evaluate $P(y)$. The two nodes that are descendants of y in the BDD correspond to the cofactors of f. The probability of the cofactors can then be expressed in the same way in terms of their descendants. Thus a depth-first transversal of the BDD, with a postorder evaluation of $P(\cdot)$ at every node is all that is required. We have implemented this using the "scan" function of the BDD package [11].

ACKNOWLEDGMENT

The author would like to acknowledge the support of Dr. J-H. Chern and Dr. P. Yang at Texas Instruments. He is also thankful to Dr. P. K. Mozumder, also at Texas Instruments, for many fruitful technical discussions. Finally, thanks are due to Karl Brace of Carnegie-Mellon University for providing the BDD package, and to the Microelectronics Center of North Carolina (MCNC) for supplying the ISCAS-85 benchmark circuits.

REFERENCES

[1] F. Najm, "Transition density, a stochastic measure of activity in digital circuits," in *Proc. 28th ACM/IEEE Design Automation Conf.*, June 1991, pp. 644–649.
[2] A. Papoulis, *Probability, Random Variables, and Stochastic Processes*, 2nd Edition. New York: McGraw-Hill, 1984.
[3] E. Parzen, *Stochastic Processes*. San Francisco, CA: Holden–Day, 1962.
[4] D. R. Cox and H. D. Miller, *The Theory of Stochastic Processes*. New York: Wiley, 1968.
[5] K. P. Parker and E. J. McCluskey, "Probabilistic treatment of general combinational networks," *IEEE Trans. Computers*, vol. 24, pp. 668–670, June 1975.
[6] S. C. Seth, L. Pan, and V. D. Agrawal, "PREDICT—probabilistic estimation of digital circuit testability," in *Proc. IEEE 15th Ann. Int. Symp. Fault-Tolerant Computing*, June 1985, pp. 220–225.
[7] J. Savir, G. S. Ditlow, and P. H. Bardell, "Random pattern testability," *IEEE Trans. Computers*, vol. 33, pp. 79–90, Jan. 1984.
[8] G. Markowsky, "Bounding signal probabilities in combinational circuits," *IEEE Trans. Computers*, vol. 35, pp. 1247–1251, Oct. 1987.
[9] S. Ercolani, M. Favalli, M. Damiani, P. Olivo, and B. Riccó, "Estimate of signal probability in combinational logic networks," *1989 IEEE European Test Conf.*, 1989, pp. 132–138.
[10] R. E. Bryant, "Graph-based algorithms for Boolean function manipulation," *IEEE Trans. Computer-Aided Design*, vol. 6, pp. 677–691, Aug. 1986.
[11] K. S. Brace, R. L. Rudell, and R. E. Bryant, "Efficient implementation of a BDD package," in *Proc. 27th ACM/IEEE Design Automation Conf.*, June 1990, pp. 40–45.
[12] F. N. Najm, R. Burch, P. Yang, and I. N. Hajj, "Probabilistic simulation for reliability analysis of CMOS VLSI circuits," *IEEE Trans. Computer-Aided Design*, vol. 9, pp. 439–450, Apr. 1990. (Errata in July 1990.)
[13] H. J. M. Veendrick, "Short-circuit dissipation of static CMOS circuitry and its impact on the design of buffer circuits," *IEEE J. Solid-State Circuits*, vol. SC-19, pp. 468–473, Aug. 1984.
[14] M. H. Woods, "MOS VLSI reliability and yield trends," *Proc. IEEE*, vol. 74, pp. 1715–1729, Dec. 1986.
[15] J. R. Black, "Electromigration failure modes in aluminum metallization for semiconductor devices," *Proc. IEEE*, vol. 57, pp. 1587–1594, Sep. 1969.
[16] B. K. Liew, N. W. Cheung, and C. Hu, "Electromigration interconnect lifetime under AC and pulsed DC stress," in *Proc. IEEE Int. Reliability Physics Symp.*, Apr. 1989, pp. 215–219.
[17] J. S. Suehle and H. A. Schafft, "Current density dependence of electromigration t_{50} enhancement due to pulsed operation," in *Proc. IEEE Int. Reliability Physics Symp.*, Apr. 1990, pp. 106–110.
[18] C. Hu, "Reliability issues of MOS and bipolar ICs," in *Proc. IEEE Int. Conf. Computer Design*, 1989, pp. 438–442.
[19] F. Brglez and H. Fujiwara, "A neutral netlist of 10 combinational benchmark circuits and a target translator in Fortran," in *Proc. IEEE Int. Symp. Circuits and Systems*, June 1985, pp. 695–698.

Estimation of Average Switching activity in Combinational and Sequential Circuits

ABHIJIT GHOSH, SRINIVAS DEVADAS[1], KURT KEUTZER[2] AND JACOB WHITE[1]

MITSUBISHI ELECTRONICS AMERICA, MIT CAMBRIDGE[1], SYNOPSYS[2]

Abstract

We address the problem of estimating the average power dissipated in VLSI combinational and sequential circuits, under random input sequences. Switching activity is strongly affected by gate delays and for this reason we use a general delay model in estimating switching activity. Our method takes into account correlation caused at internal gates in the circuit due to reconvergence of input signals. In sequential circuits, the input sequence applied to the combinational portion of the circuit is highly correlated because some of the inputs to the combinational logic are flip-flop outputs representing the state of the circuit. We present methods to probabilistically estimate switching activity in sequential circuits. These methods automatically compute the switching rates and correlations between flip-flop outputs.

1 Introduction

We address the problem of estimating the average power dissipated in combinational and sequential VLSI circuits given random input sequences. This measure can be used to make architectural or design-style decisions during the VLSI synthesis process.

Probabilistic methods for power or current estimation are attractive because statistical estimates can be obtained without recourse to time-consuming exhaustive simulation. In the past, probabilistic peak current estimation methods (e.g., [5]) that compute probabilistic current waveforms for combinational circuits have been developed. Estimation of worst-case power dissipation (e.g., [10], [7]) is a difficult problem requiring a branch-and-bound search and these methods have been applied to small to moderate sized circuits.

The problem of estimating average power in combinational circuits can be reduced to one of computing signal probabilities [1] of a multilevel circuit derived from the original circuit by a process of symbolic simulation. The work closest to our own is the work on transition density calculation by Najm [11]. Transition densities correspond to average switching rates for gates in the circuit. In [11], an interconnection of combinational logic modules, each with a certain delay, makes up a circuit. Transition densities are propagated through combinational logic modules without regard to their structure. Correlations between internal lines due to reconvergence are ignored during propagation. It is possible to take into account correlations by lumping all the modules into one large module, but in this case the information regarding the delay of the individual modules is lost. Cirit in [8] gives methods to calculate dynamic power dissipation based on approximate signal probability evaluation procedures.

Our work improves upon the state-of-the-art in several ways. We use a general delay model for combinational logic in our symbolic simulation method, which correctly computes the Boolean conditions that cause *glitching* (multiple transitions at a gate) in the circuit. In some cases, glitching may account for a significant percentage of the dissipated power or switching activity. Symbolic simulation produces a set of Boolean functions that represent the conditions for switching at different time points, for each gate in the circuit. Given input switching rates, we can use various exact or approximate methods to compute the probability of each gate switching at any particular time point. We then sum these probabilities over all the gates to obtain the expected switching activity in the entire circuit over all the time points corresponding to a clock cycle. Our method takes into account correlation caused at internal gates in the circuit due to reconvergence of input signals (reconvergent fanout).

Most VLSI circuits are sequential. We make an important extension to our estimation algorithms to handle sequential circuits. The combinational part of the circuit receives primary inputs as well as inputs from flip-flop outputs. Given the pin restrictions on VLSI chips, the number of flip-flops in large sequential circuits can be many times the number of primary inputs. Previous methods for power estimation assume uniform switching rates for all the inputs to the combinational logic or expect that switching rates are pre-computed using methods like random simulation. Further, the correlation [1] between the flip-flop outputs between one time frame and the next is ignored. Using inaccurate switching rates and ignoring this correlation can cause substantial errors in power estimation, especially for datapath circuits. We develop a method that *automatically computes the switching rates* for each of the flip-flop inputs, *taking into account this correlation between the flip-flops.*

---

[1]For example, flip-flop A may only switch from $0 \rightarrow 1$ when flip-flop B switches from $1 \rightarrow 0$.

Reprinted from *IEEE/ACM Design Automation Conference*, pp. 253-259, June 1992.

2 Preliminaries

2.1 A Power Dissipation Model

The amount of energy dissipated by a CMOS logic gate each time its output changes is roughly equal to the change in energy stored in the gate's output capacitance. If the gate is part of a synchronous digital system controlled by a global clock, it follows that the average power dissipated by the gate is given by:

$$P_{avg} = 0.5 \times C_{load} \times (V_{dd}^2 / T_{cyc}) \times E(transitions) \quad (1)$$

where P_{avg} denotes the average power, C_{load} is the load capacitance, V_{dd} is the supply voltage, T_{cyc} is the global clock period, and $E(transitions)$ is the *expected value* of the number of gate output transitions per global clock cycle[11], or equivalently the average number of gate output transitions per clock cycle. All of the parameters in (1) can be determined from technology or circuit layout information except $E(transitions)$, which depends on both the logic function being performed and the statistical properties of the primary inputs.

2.2 Static Probabilities

Consider the case of dynamic CMOS logic (e.g. Domino). At the beginning of each clock cycle, all the gates are precharged, and gates make transitions only if their associated Boolean functions are satisfied. For example, a three-input AND-OR gate's Boolean function might be

$$(i_1 \cdot i_2) \vee (i_2 \cdot i_3), \quad (2)$$

where $i_1, i_2,$ and i_3 are primary inputs. In this case, the expected value of the number of transitions at this gate's output is

$$E(transitions) = 2 \times P((i_1 \cdot i_2) \vee (i_2 \cdot i_3) = 1) \quad (3)$$

where $P(x)$ is defined as probability that x is true, and the factor of two in the equation accounts for the reset transition during precharge.

To evaluate (3), it is necessary to determine the primary input probabilities. We assume that *primary inputs are uncorrelated*, and that each is a waveform in time whose value is either zero or one, changing instantaneously at global clock edges. Assuming ergodicity without further comment, the probability of a particular input i_j being one at a given point in time, denoted $p_j{}^{one}$, is given by

$$p_j{}^{one} = \lim_{N \to \infty} \frac{\sum_{k=1}^{N} i_j(k)}{N} \quad (4)$$

where N is the total number of global clock cycles and $i_j(k)$ is the value of input i_j during clock cycle k. Clearly, the probability that i_j is zero at a given point in time, denoted $p_j{}^{zero}$ is

$$p_j{}^{zero} = 1 - p_j{}^{one}$$

We refer to $p_j{}^{zero}$ and $p_j{}^{one}$ as static probabilities. Note that

$$P((i_1 \cdot i_2) \vee (i_2 \cdot i_3) = 1) \neq p_1^{one} p_2^{one} + p_2^{one} p_3^{one}$$

because the first and second product terms are not independent. Rather,

$$P((i_1 \cdot i_2) \vee (i_2 \cdot i_3) = 1) =$$

$$P((i_1 \cdot i_2) \vee (\overline{i_1} \cdot i_2 \cdot i_3) = 1) = p_1^{one} p_2^{one} + p_1^{zero} p_2^{one} p_3^{one}$$

where the second equality holds because $i_1 \cdot i_2$ is disjoint from $\overline{i_1} \cdot i_2 \cdot i_3$. In this example, $(i_1 \cdot i_2) \vee (\overline{i_1} \cdot i_2 \cdot i_3)$ represents a disjoint cover for the logic function, and the terms $i_1 \cdot i_2$ and $\overline{i_1} \cdot i_2 \cdot i_3$ are referred to as *cubes* in the cover[2]. The equivalent logical expression, $(i_1 \cdot i_2) \vee (i_2 \cdot i_3)$, does not represent a disjoint cover because $i_1 \cdot i_2 \cdot i_3$ is contained in both cubes $i_1 \cdot i_2$ and $i_2 \cdot i_3$.

In general, given a disjoint cover for a Boolean function of uncorrelated inputs described by static probabilities, it is easy to determine the probability of the function evaluating to a 1. The procedure is given in the following two theorems whose proof follows directly from elementary probability[12].

Theorem 2.1 : *Given any disjoint cover for a Boolean function, the probability of the function evaluating to a 1 is equal to the sum of the probabilities of each of the cubes in the cover evaluating to a 1.*

Theorem 2.2 : *Given a logical function of uncorrelated inputs in the form*

$$f = i_{\alpha_1} \cdot i_{\alpha_2} \cdots \cdots i_{\alpha_M} \cdot \overline{i_{\beta_1}} \cdot \overline{i_{\beta_2}} \cdots \cdots \overline{i_{\beta_N}}$$

where the i_{α_j}'s are nonnegated inputs and the $\overline{i_{\beta_k}}$'s are negated inputs, then

$$P(f = 1) = p_{\alpha_1}^{one} \cdot p_{\alpha_2}^{one} \cdots \cdots p_{\alpha_M}^{one} \cdot p_{\beta_1}^{zero} \cdot p_{\beta_2}^{zero} \cdots \cdots p_{\beta_N}^{zero}$$

2.3 Transition Probabilities

For static combinational CMOS logic, a gate output can only change when its inputs change, and then only if the Boolean function describing the gate evaluates differently. For example, a 2-input AND gate's output will change between clock cycle t and $t + 1$ if

$$(i_1(t) \cdot i_2(t)) \oplus (i_1(t + 1) \cdot i_2(t + 1)) \quad (5)$$

evaluates to 1, where $i_1(t)$, $i_2(t)$ and $i_1(t+1)$, $i_2(t+1)$ are the inputs to the gate at clock cycle t and $t + 1$ respectively. The disjoint cover for (5) is

$$(i_1(t) \cdot i_2(t)) \cdot \overline{(i_1(t+1))} \vee (i_1(t) \cdot i_2(t)) \cdot (i_1(t+1) \cdot \overline{i_2(t+1)}) \vee$$

$$\overline{i_1(t)} \cdot (i_1(t+1) \cdot i_2(t+1)) \vee (i_1(t) \cdot \overline{i_2(t)})(i_1(t+1) \cdot i_2(t+1)).$$
$$(6)$$

It is not possible to use Theorem (2.2) to evaluate the probability of (6), because an input at time $t+1$ is correlated to its behavior at time t (as in a sequential circuit). Instead, transition probabilities for the transitions $0 \to 0$, $0 \to 1$, $1 \to 0$ and $1 \to 1$ must be used. We denote these transition probabilities by $p_i{}^{00}$, $p_i{}^{01}$, $p_i{}^{10}$, and $p_i{}^{11}$, respectively, where for example $p_i{}^{10}$ is defined by:

$$ p_j{}^{10} = \lim_{N \to \infty} \frac{\sum_{k=1}^{N} i_j(k)\overline{i_j(k+1)}}{N} \qquad (7) $$

The other transition probabilities follow similarly.

Static probabilities can be computed from transition probabilities, but *not* vice-versa, because of correlation between one time frame and the next. So in general it is necessary to specify transition probabilities. The relations between static probabilities and transition probabilities follow directly from the definitions in (4) and (7), specifically,

$$ p_j{}^{zero} = p_j{}^{00} + p_j{}^{01} \qquad (8) $$

$$ p_j{}^{one} = p_j{}^{11} + p_j{}^{10} $$

Both static and transition probabilities are used to compute $E(transitions)$ for static logic circuits, as can be seen from the expression for the probability that (6) evaluates to a 1,

$$ p_1^{10} \cdot p_2^{one} + p_1^{11} \cdot p_2^{10} + p_1^{01} \cdot p_2^{one} + p_1^{11} \cdot p_2^{01} \qquad (9) $$

which for this example is also $E(transitions)$. For all primary inputs, it may be assumed that successive input vectors are uncorrelated and a 1 or a 0 are equally likely. Therefore, all transition probabilities may be be assumed to be 0.25, and all static probabilities to be 0.5.

2.4 General Combinational Networks

The algorithm for computing the average power dissipated in a dynamic CMOS combinational network follows directly from the approach in Section (2.2). That is, for each gate g_i in a logical network, we first determine the Boolean function of the gate, f_i, in terms of the network's primary inputs. Then we find a disjoint cover for f_i, and use Theorem (2.1) and (2.2) to evaluate $P(f_i = 1)$.

The average power dissipation for the network is then

$$ P_{avg} = (V_{dd}^2/T_{cyc}) \times \sum C_i \times E(transitions\ of\ g_i) $$

$$ = (V_{dd}^2/T_{cyc}) \times \sum C_i \times P(f_i = 1) \qquad (10) $$

where C_i is the load capacitance of the ith gate and the summation is over all gates in the circuit. Note that (10) follows from the fact that the average value of a sum is equal to the sum of the average values, regardless of correlation[12].

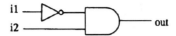

Figure 1: Glitching in a Static CMOS Circuit

A similar overall approach can be used to compute the average power dissipated in a static CMOS combinational network. However, as described in Section (2.3), a two-vector input sequence is required to stimulate activity in static gates. Therefore, the Boolean function for static gate output activity is different than that for a dynamic gate, and the probability of the function being satisfied requires transition probabilities for evaluation. In particular, assuming negligible gate delays, a static CMOS combinational logic gate's output will switch with a change in the primary input vector from $V0$ to Vt if:

$$ f_i = ((h_i(V0) = 0) \wedge (h_i(Vt) = 1)) \bigvee $$
$$ ((h_i(V0) = 1) \wedge (h_i(Vt) = 0)) \qquad (11) $$

is satisfied, where h_i is the logic function corresponding to gate g_i's output.

The average power dissipated in the static network is then computed by using (10), with f_i being given by (11), and $P(f_i = 1)$ is evaluated using both transition and static probabilities, as described in Section (2.3).

Instead of enumerating disjoint covers, Binary Decision Diagrams (BDDs) [3] can be used for the calculation of signal probability. It has been shown in [6] and [11] that exact signal probability calculation for a given function can be performed by a linear traversal of a BDD representation of a logic function. We have implemented methods for signal probability calculation using BDDs.

3 Gate Delay Effects

As mentioned above, for static CMOS circuits, switching activity must be analyzed based on considering an input vector pair, denoted $< V0, Vt >$. If the gates have appreciable delays, there may be output glitches that can contribute significantly to dissipated power.

For instance, consider the circuit of Figure 1. Assuming that the delays of the inverter and the AND gate are both 1 time unit, if we apply the vector $i_1 = 0$, $i_2 = 0$, followed by $i_1 = 1$, $i_2 = 1$, we will obtain a glitch at the output *out*, which could cause power dissipation. Below, we present a symbolic simulation method that can be used to generate a multiple-output function that represents total switching activity over any possible input vector pair, assuming a general delay model for the gates in the circuit.

3.1 Symbolic Simulation

3.1.1 An Example

Consider the multilevel combinational circuit shown in Figure 2.

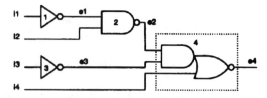

Figure 2: A multilevel combinational circuit

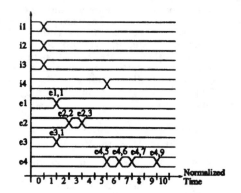

Figure 3: Signal waveforms for the primary inputs and the outputs of the logic gates

It has four primary inputs and one output and consists of four CMOS gates. Figure 3 shows the signal transitions at the primary input nodes as well as the possible transitions at the internal nodes of the network. The time points are in normalized units. The first three inputs, i_1, i_2, and i_3, switch simultaneously between time periods 0 and 1. The fourth input, i_4, is a late arriving signal that switches between the time points 5 and 6. In this example, the delays of both Gate 1 and Gate 3 are one time unit. Gate 2 has a delay of two units while Gate 4 has a delay of four units.

Also shown in Figure 3 are the waveforms, e_i's, representing the signals at the outputs of ith logic gates. Each of the possible transitions, $e_{i,j}$, represents either a low-to-high or high-to-low signal transition between $[j]^{th}$ and $[j+1]^{th}$ time points. The number of all possible transitions at a gate output may equal the sum of all possible transitions at the gate inputs. These transitions are delayed by the gate's propagation delay.

3.1.2 Unit-Delay Model

Even under the idealization of a unit-delay model, the gate output nodes of a multilevel network can have multiple transitions in response to a two-vector input sequence. In fact, it is possible for a gate output to have as many transitions as levels in the network.

We construct the Boolean functions describing the gate outputs at the discrete points in time implied by the unit-delay model. That is, we consider only discrete times t, $t+1$, ..., $t+l$, where t is the time when the inputs change from $V0$ to Vt, and l is the number of levels in the network. For each gate output i, we use symbolic simulation to construct the $l+1$ Boolean functions $f_i(t+j)$, $j \in 0$, ..., l which evaluate to 1 if the gate's output is 1 at time $t+j$. Note that as we assume no gate has zero delay and that the network has settled before the inputs are changed from $V0$ to Vt, $f_i(t)$ is the logic function performed for $V0$ at the ith gate output. Finally, we can determine whether a transition occurs at a boundary between discrete time intervals $t+j$ and $t+j+1$ by exclusive-OR'ing $f_i(t+j)$ with $f_i(t+j+1)$.

3.1.3 General Delay Model

A gate with a large fanin may have several times the delay of an inverter. If one uses normalized time units, one can always introduce unit-delay buffers at the output of gates in a circuit, which have a delay greater than unity, in order to model differing delays among logic gates.

Further, all gates have *inertial delays*, i.e., all gates require energy to switch their output nodes. For a transition at an input terminal to propagate through the gate, the state of the input signal before and after the transition must persist for a minimum duration equal to the delay of the gate. Any high frequency pulse at the inputs whose pulse duration is shorter than the gate delay (or some fraction of the gate delay) will be filtered out.

Our general symbolic simulator is able to simulate circuits with arbitrary gate transport and inertial delays (without introducing any unit delay buffers in the circuit). The simulator processes one gate at a time, moving from the primary inputs to the primary outputs of the circuit. For each gate, the possible transition times of its inputs are first obtained. Then, possible transitions at the output of the gate is derived, taking into account transport and inertial delays. If the inertial delay of a gate is non-zero, then all inputs to the gate must remain the same for a period equal to the inertial delay (or a fraction of it) for the output of the gate to make a transition.

4 Sequential Circuits

4.1 Introduction and Motivation

VLSI circuits are sequential, i.e., they contain memory elements or flip-flops. This sequential nature of VLSI circuits makes the estimation of power dissipation more complicated than for combinational circuits.

A common model for a sequential circuit is shown in Figure 4. We will denote the vector pair applied to the combinational logic as $< V0, Vt >$. $V0$ and Vt have a primary input part and a present-state part. $V0$ is denoted $I0@P0$ and Vt is denoted $It@Pt$, where $I0$ and It correspond to the primary input parts, and $P0$ and Pt correspond to the present-state parts.

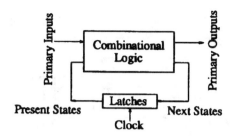

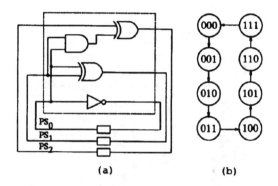

Figure 4: A General Synchronous Sequential Circuit

Figure 5: A 8-State Counter

One can ignore the feedback corresponding to the next-state lines and present-state lines, and estimate the power dissipated by the combinational logic using the methods presented in the previous sections. This strategy is a relatively crude approximation for two reasons as we will illustrate with a simple example.

Consider the STG of Figure 5(b). It represents the behavior of an 8-state autonomous counter, whose logic implementation is shown in Figure 5(a). Assuming that the counter can begin from any state with uniform probability, the probability of line PS_2 making a $1 \rightarrow 0$ transition is 0.125, and probabilities for $0 \rightarrow 1$, $0 \rightarrow 0$ and $1 \rightarrow 0$ transitions are 0.125, 0.375 and 0.375, respectively. On the other hand, the probability of the line PS_0 making a $1 \rightarrow 0$ transition is 0.5, and the probabilities for a $0 \rightarrow 1$, $0 \rightarrow 0$ and $1 \rightarrow 1$ are 0.5, 0, and 0, respectively. The transition probabilities for the different present-state lines are different.

To make matters worse, the transitions on one present-state line are correlated to the transitions on the other present-state lines. For instance, in Figure 5, it is easy to see that PS_2 makes a $0 \rightarrow 1$ transition only when PS_0 makes a $1 \rightarrow 0$ transition. Simply computing the correct switching rates does not necessarily result in exact power dissipation estimation if we ignore this correlation of the transitions. Note however, that in this example, the present-state lines are uncorrelated, i.e., given the static probability of each present-state line, the probability of a particular state can be computed correctly.

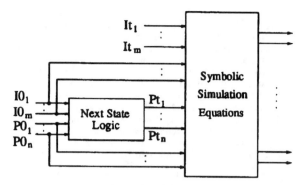

Figure 6: Taking Correlation Into Account

4.2 Power Estimation Method

We will assume that the circuit, upon power-up, can begin in any of the 2^n states, where n is the number of flip-flops. Further, we assume that the STG of the machine is strongly connected, and that an arbitrary number of cycles after power-up, the probability that the machine is in any of its 2^n states is uniform.

Consider the circuit of Figure 6. It has two blocks: the first is the combinational logic implementing the next-state function of the given machine, and is used to derive the next state from the present state. The first block feeds a second block, which represents the Boolean function obtained by symbolic simulation of the combinational logic of the machine.

The decomposition of Figure 6 implies that the gate output switching activity can be determined given the vector pair $< I0, It >$ for the primary inputs, but only $P0$ for the states. Therefore, to compute gate output transition probabilities, we require the transition probabilities for the primary inputs, but only require the static probabilities for the present state. This use of the next-state logic generates Boolean equations which model the correlation between the present and the next states, whereby the transition probabilities are automatically computed, taking into account the correlation between transitions. However, to use the average power estimation techniques described in Section 2, we must assume that the present-state lines are uncorrelated. We discuss the validity of this assumption in the next section.

4.3 STG Characteristics

The state bits in a synchronous sequential circuit are uncorrelated if the machine is equally likely to be in any of its 2^n states an arbitrary number of cycles after power-up (in which case the static probability of each present-state line is 0.5). There are many reasons why this might not be true. In particular if the entire state space is not strongly connected, the machine might operate only in a strongly connected portion of the state space. In general, after an arbitrarily long period of time, the machine will be in a strongly connected portion of its STG, which might contain just a single state. If the assumption above is not valid, then static probability of 0.5 cannot be assumed for the present-state lines. Moreover, because of the correlation between the present-state lines, the power estimate obtained is approximate.

Given a set of states in a strongly connected portion of the STG in which the machine can be, the probability of being in any particular state within the set is not necessarily uniform. In particular, it might depend on the *in-degree* [2] of a state. States with a large in-degree may have greater probabilities of being entered. The exact probabilities of being in particular states can be derived from the Chapman-Kolmogorov equations for discrete-time Markov Chains [12]. However, the number of equations will be exponential in the number of latches and inputs in the circuit, and therefore cannot be solved for most practical circuits.

In practice, large datapath circuits satisfy the assumptions because of their highly connected STGs that include almost all of 2^n states, and states with similar in-degrees. Thus, the assumption made regarding the structure of the STG is quite realistic for such circuits. However, controllers might violate these assumptions. In that case, because of the correlation between present-state lines, we will only be able to obtain an approximate power estimate. We can increase the accuracy of the power estimate by calculating the static probabilities of the present-state lines more accurately. Techniques to compute the reachable set of states, or the strongly connected portion of the STG of a sequential machine, can be developed, based on the strategies of [9, 4]. These techniques are viable for machines with up to approximately 50 flip-flops. Thereafter, the static probability of each present-state line can be calculated. For example, if a machine has states 00, 01, and 10, and it is in these states with probabilities 0.1, 0.5, 0.4, respectively, then the probability of the first present-state line being 1 is 0.4 and that of the second is 0.5. Finally, during signal probability calculation, we have to exclude the set of unreachable states. This is performed by computing the logical AND of the BDD representing the set of reachable states with the BDD representing the switching function.

5 Experimental Results

Throughout this section, we will be measuring the average power dissipation of the circuit by using (1) summed over all the gates in the circuit. The N_G values are computed for the gates in the circuit under different delay models. The capacitance values of the gates are assumed to be given, or can be computed using the fanin and fanout numbers and the size (in terms of the number of transistors) for each gate, or can be obtained after mapping.

The statistics of the examples used are shown in Table 1. All of the examples except the last two belong to the ISCAS-89 Sequential Benchmark set. Example add16 is a 16-bit adder and accumulator, where the

[2] The number of edges in the STG that enter this state.

CKT	Inputs	Outputs	Latches	Gates
s27	4	1	3	10
s298	3	6	14	119
s349	9	11	15	150
s386	7	7	6	159
s420	19	2	16	196
s510	19	7	6	211
s641	35	24	19	379
s713	35	23	19	393
s838	35	2	32	390
s1238	14	14	18	508
s1494	8	19	6	647
s5378	35	49	179	2779
add16	33	17	16	288
max16	33	16	16	154

Table 1: Statistics of examples

CKT	Combinational		Sequential	
	Power	Time	Power	Time
s27	14	0.1	15	0.1
s298	212	0.9	259	1.8
s349	280	1.6	258	2.9
s386	264	1.5	428	2.1
s420	287	1.8	336	2.8
s510	425	2.2	425	3.7
s641	582	3.5	580	5.7
s713	650	4.3	649	6.3
s838	568	4.4	672	4.6
s1238	914	6.2	904	6.4
s1494	1002	8.3	1185	13.8
s5378	4752	11.0	4823	45.1
add16	58	0.4	58	0.5
max16	56	0.4	73	0.7

Table 2: Power estimation for dynamic circuits

output of the adder is fed back to one set of inputs. Example max16 is similar, with the adder replaced by maximum finder. Some of these circuits have over 200 inputs and outputs (*e.g.*, s5378). All the circuits considered are technology mapped circuits.

If the circuits are treated as purely combinational dynamic circuits, then the average-case power dissipation is shown in the second column of Table 2, while the time required to obtain the power dissipation is shown in the third column. These times have been obtained on a DEC-station 5900 with 440Mb of memory, and are in seconds. The power estimates, in microWatts, have been obtained assuming a clock frequency of 20MHz. Treating the circuit as a dynamic sequential circuit, the average power dissipation and the time required to derive the power dissipation are shown in columns four and five respectively of Table 2. Note that taking correlations of state transitions into account, as described in Section 4.2, produces significant differences in the power estimate.

For the remainder of this section, we consider the circuits to be static CMOS circuits. In Table 3, we treat the circuits as purely combinational and show the effects of various delay models on the power estimate. In the zero delay model, all gates have zero delay and therefore they switch instantaneously. In the unit delay model, all gates have one unit delay. Using the zero delay model ignores glitches in the circuit, and therefore power dissipation due to glitches are not

CKT	Zero Delay Power	Unit Delay Power	Variable Delay	
			Power	Time
s27	15	19	17	0.1
s298	190	215	198	1.6
s349	238	358	308	4.7
s386	252	288	271	3.8
s420	108	137	128	8.2
s510	271	346	304	3.1
s641	481	665	596	41.0
s713	513	669	637	45.1
s838	184	253	214	49.3
s1238	622	790	716	31.8
s1494	1032	1280	1194	20.0
s5378	3677	4513	4099	2022.2
add16	58	102	102	8.1
max16	56	85	85	6.4

Table 3: Power estimation for combinational circuits

CKT	Zero Delay Power	Unit Delay Power	Variable Delay	
			Power	Time
s27	11	15	14	0.2
s298	168	193	173	5.2
s349	225	275	262	6.5
s386	300	352	333	5.4
s420	126	152	150	10.5
s510	211	265	237	7.8
s641	399	539	493	40.4
s713	429	537	528	50.1
s838	230	278	275	62.4
s1238	613	781	706	45.0
s1494	1072	1498	1303	45.1
s5378	3644	4620	3978	226.8
add16	71	129	129	11.3
max16	46	75	75	8.5

Table 4: Power estimation for sequential circuits

taken into account. The unit delay model takes into account glitches, but does not take the inertial delay of gates into account and therefore may overestimate the power dissipation. The variable delay model with inertial gate delays is the most realistic one, and its estimates, as expected, lie between the zero delay and the unit delay model. Only the times required to obtain the power estimate for the variable delay model is shown in the last column.

The results obtained by treating the circuits as sequential, using the method of Section 4.2, are shown in Table 4. While considering these circuits as sequential, we have taken into account the correlation between two successive values on a present-state line using the method of Section 4.2. This again results in significant differences in the power estimate, as can be seen by comparing Tables 3 and 4.

For each circuit, we compared the power estimates obtained by our method to that obtained by simulating the circuit. The power estimate was obtained by logic simulation, until convergence, with randomly generated input sequence applied to the sequential circuit. The same power dissipation model was used. We obtained excellent agreement (to within 2%) between the power estimate using our method and the random simulation estimate. However, power estimation by simulation took anywhere between 2 to 100 times longer.

References

[1] P. H. Bardell, W. H. McAnney, and J. Savir. *Built-In Test for VLSI: Pseudorandom Techniques.* John Wiley and Sons, New York, New York, 1987.

[2] R. Brayton, G. Hachtel, C. McMullen, and A. Sangiovanni-Vincentelli. *Logic Minimization Algorithms for VLSI Synthesis.* Kluwer Academic Publishers, 1984.

[3] R. Bryant. Graph-Based Algorithms for Boolean Function Manipulation. In *IEEE Transactions on Computers*, volume C-35, pages 677–691, August 1986.

[4] J. Burch, E. Clarke, K. McMillan, and D. Dill. Sequential Circuit Verification Using Symbolic Model Checking. In *Proceedings of the 27th Design Automation Conference*, pages 46–51, June 1990.

[5] R. Burch, F. Najm, P. Yang, and D. Hocevar. Pattern-independent Current Estimation for Reliability Analysis of CMOS Circuits. In *Proceedings of the 25th Design Automation Conference*, pages 294–299, June 1988.

[6] S. Chakravarty. On the Complexity of Using BDDs for the Synthesis and Analysis of Boolean Circuits. In *Proceedings of the 27th Annual Allerton Conference on Communication, Control, and Computing*, pages 730–739, September 1989.

[7] S. Chowdhury and J. S. Barkatullah. Estimation of Maximum Currents in MOS IC Logic Circuits. In *IEEE Transactions on Computer-Aided Design*, volume 9, pages 642–654, June 1990.

[8] M. A. Cirit. Estimating Dynamic Power Consumption of CMOS Circuits. In *Proceedings of the Int'l Conference on Computer-Aided Design*, pages 534–537, November 1987.

[9] O. Coudert, C. Berthet, and J. C. Madre. Verification of Sequential Machines Using Boolean Functional Vectors. In *IMEC-IFIP Int'l Workshop on Applied Formal Methods for Correct VLSI Design*, pages 111–128, November 1989.

[10] S. Devadas, K. Keutzer, and J. White. Estimation of Power Dissipation in CMOS Combinational Circuits. In *Proceedings of the Custom Integrated Circuits Conference*, pages 19.7.1–19.7.6, May 1990.

[11] F. Najm. Transition Density, A Stochastic Measure of Activity in Digital Circuits. In *Proceedings of the 28th Design Automation Conference*, pages 644–649, June 1991.

[12] A. Papoulis. *Probability, Random Variables and Stochastic Processes.* McGraw Hill, 1984.

Power Estimation Methods
for Sequential Logic Circuits

Chi-Ying Tsui, José Monteiro, Massoud Pedram, *Member, IEEE,* Srinivas Devadas, *Member, IEEE,*
Alvin M. Despain, *Member, IEEE,* and Bill Lin

Abstract— Recently developed methods for power estimation have primarily focused on combinational logic. We present a framework for the efficient and accurate estimation of average power dissipation in sequential circuits.

Switching activity is the primary cause of power dissipation in CMOS circuits. Accurate switching activity estimation for sequential circuits is considerably more difficult than that for combinational circuits, because the probability of the circuit being in each of its possible states has to be calculated. The Chapman–Kolmogorov equations can be used to compute the exact state probabilities in steady state. However, this method requires the solution of a linear system of equations of size 2^N where N is the number of flip-flops in the machine.

We describe a comprehensive framework for exact and approximate switching activity estimation in a sequential circuit. The basic computation step is the solution of a nonlinear system of equations which is derived directly from a logic realization of the sequential machine. Increasing the number of variables or the number of equations in the system results in increased accuracy. For a wide variety of examples, we show that the approximation scheme is within 1–3% of the exact method, but is orders of magnitude faster for large circuits. Previous sequential switching activity estimation methods can have significantly greater inaccuracies.

I. INTRODUCTION

FOR MANY consumer electronic applications low average power dissipation is desirable and for certain special applications low power dissipation is of critical importance. For applications such as personal communication systems and hand-held mobile telephones, low-power dissipation may be the tightest constraint in the design. More generally, with

Manuscript received June 15, 1994; revised February 20, 1995 and March 31, 1995. The work of C.-Y. Tsui and A. M. Despain was supported in part by the Advanced Research Projects Agency under Contract J-FBI-91-194. The work of M. Pedram was supported in part by the Advanced Research Projects Agency under Contract F33615-95-C-1627, and by the SRC under Contract 94-DJ-559. The work of J. Monteiro's and S. Devadas was supported in part by the Advanced Research Projects Agency under Contract DABT63-94-C-0053, and in part by a NSF Young Investigator Award with matching funds from Mitsubishi Corporation.

C.-Y. Tsui is with the Department of Electrical Engineering, Hong Kong University of Science and Technology, Hong Kong.

M. Pedram and A. Despain are with the Department of Electrical Engineering, University of Southern California, Los Angeles, CA 90089 USA.

J. Monteiro and S. Devadas are with the Department of Electrical Engineering and Computer Science, Massachusetts Institute of Technology, Cambridge, MA 02139 USA.

B. Lin is with IMEC, Belgium, France.

IEEE Log Number 9413459.

the increasing scale of integration, we believe that power dissipation will assume greater importance, especially in multi-chip modules where heat dissipation is one of the biggest problems.

Power dissipation of a circuit, like its area or speed, may be significantly improved by changing the circuit architecture or the base technology [3]. However, once these architectural or technological improvements have been made, it is the switching of the logic that will ultimately determine the power dissipation.

Methods for the power estimation of logic-level combinational circuits based on switching activity estimation have been presented previously (e.g., [2], [4], [7], [9], [10], [13]). Power and switching activity estimation for sequential circuits is significantly more difficult, because the probability of the circuit being in any of its possible states has to be computed. Given a circuit with N flip-flops, there are 2^N possible states. These state probabilities are, in general, not uniform. As an example, consider the sequential circuit of Fig. 1 and the example State Transition Graph of Fig. 2. Assuming that the circuit was in state **R** at time 0, and that at each clock cycle random inputs are applied, at time ∞ (i.e., steady state) the probabilities of the circuit being in state **R**, **A**, **B**, **C** are $\frac{1}{6}$, $\frac{1}{3}$, $\frac{1}{4}$, and $\frac{1}{4}$, respectively. These *state probabilities* have to be taken into account during switching activity estimation of the combinational logic part of the machine. Power dissipation and switching activity of CMOS combinational logic are modeled by randomly applied vector pairs. In the case of sequential circuits, the vector pair $\langle v_1, v_2 \rangle$ applied to the combinational logic is composed of a primary input part and a present state part (see Fig. 1), namely $\langle i_1 @ s_1, i_2 @ s_2 \rangle$. Given $i_1 @ s_1$, the next state s_2 is uniquely determined given the functionality of the combinational logic. For example, if i_1 happens to be 0 and the machine of Fig. 2 is in state **R**, the machine will move to state **B**. This *correlation* between the applied vector pairs has to be taken into account in order to obtain accurate estimates of the switching activity in sequential circuits.

A first attempt at estimating switching activity in logic-level sequential circuits was presented in [4]. This method can accurately model the correlation between the applied vector pairs, but assumes that the state probabilities are all uniform. Extensions of this method can produce accurate estimates for acyclic sequential circuits such as pipelines, but not for more general cyclic circuits [8].

Reprinted from *IEEE Transactions on VLSI Systems*, pp. 404-416, September 1995.

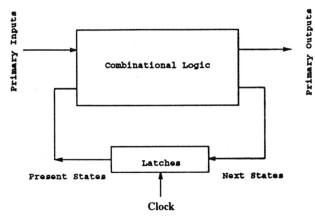

Fig. 1. A synchronous sequential circuit.

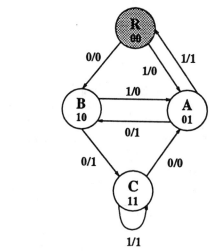

Fig. 2. Example state transition graph.

In this paper, we present results obtained by using the Chapman–Kolmogorov equations for discrete-time Markov Chains [12] to compute the exact state probabilities of the machine. The Chapman–Kolmogorov method requires the solution of a linear system of equations of size 2^N, where N is the number of flip-flops in the machine. Thus, this method is limited to circuits with relatively small number of flip-flops, since it requires the explicit consideration of each state in the circuit.

We next describe an approximate method for switching activity estimation in sequential circuits. The basic computation step is the solution of a nonlinear system of equations which is derived directly from the logic realization of the next state logic of the machine under consideration. Increasing the number of variables or the number of equations in the system results in increased accuracy. For a wide variety of examples, we show that the approximation scheme is within 1–3% of the exact method, but is orders of magnitude faster for large circuits. Previous sequential switching activity estimation methods can have significantly greater inaccuracies.

The rest of this paper is organized as follows. In Section II we briefly review the physical model for power estimation and summarize the combinational estimation method of [4]. In Section III, we describe an exact switching activity estimation method for sequential circuits. In Section IV, we first provide the basis for the approximation schemes we have developed and formulate the problem of estimating switching activity as that of solving a nonlinear system of equations. We describe a scheme based on the notion of a k-unrolled network that can be used to improve the accuracy of estimation in Section V. We describe a different method to improve the accuracy based on the notion of a m-expanded network in Section VI. In Section VII we describe methods to solve the nonlinear system of equations, namely, the Picard–Peano and the Newton–Raphson methods. In Section VIII, we show that purely combinational logic estimation methods can provide inaccurate estimates, whereas the developed approximation methods produce accurate estimates while being applicable to large circuits.

II. Preliminaries

A. A Power Dissipation Model

Under a simplified model of the energy dissipation in CMOS circuits, the energy dissipation of a CMOS circuit is directly related to the switching activity.

In particular the three simplifying assumptions are:

- The only capacitance is at the output node of a CMOS gate (this capacitance includes the source–drain capacitance of the gate itself and the input capacitances of the fanout gates).
- Current is flowing either from V_{DD} to the output capacitor or from the output capacitor to ground (that is, there is no short-circuit current).
- Any change in a logic-gate output voltage is a change from V_{DD} to ground or vice-versa (that is, there are no stable intermediate voltage levels).

These assumptions are reasonably justified for well-designed CMOS gates [5] and when combined, imply that the energy dissipated by a CMOS logic gate each time its output changes is roughly equal to the change in energy stored in the output capacitance seen by the gate. If the gate is part of a synchronous digital system controlled by a global clock, it follows that the average power dissipated by the gate is given by:

$$P_{avg} = 0.5 \times C_{load} \times \left(\frac{V_{dd}^2}{T_{cyc}} \right) \times E(transitions) \quad (1)$$

where P_{avg} denotes the average power, C_{load} is the load capacitance, V_{dd} is the supply voltage, T_{cyc} is the global clock period, and $E(transitions)$ is the *expected value* of the number of gate output transitions per global clock cycle [9], or equivalently the average number of gate output transitions per clock cycle. All of the parameters in (1) can be determined from technology or circuit layout information except $E(transitions)$, which depends on the logic function being performed and the statistical properties of the primary inputs.

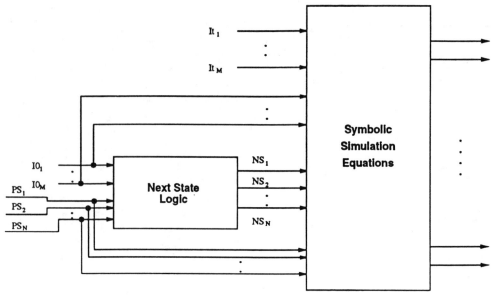

Fig. 3. Taking correlation into account.

Equation (1) is used by the power estimation techniques such as [4], [9] to relate switching activity to power dissipation.

B. Combinational Circuits

Average power can be estimated for combinational circuits by computing the average switching activity at every gate in the circuit.

It is assumed that we are given *transition probabilities* at each of the primary inputs to the circuit. That is, for every primary input the probability of the primary input staying at $0 \, (0 \rightarrow 0)$, staying at $1 \, (1 \rightarrow 1)$, making a $0 \rightarrow 1$ transition and making a $1 \rightarrow 0$ transition are given. Given these probabilities, the average switching activity at each gate in the circuit can be calculated.

A symbolic simulation method that performs this computation was given in [4]. Under the chosen gate delay model, the method first constructs a Boolean function representing the logical value at any gate output at each time point $\geq t$ based on the primary input variables $I0$ applied at time 0 and It applied at time t. For instance, one may compute the functions $f_i(t+1)$ and $f_i(t+2)$ for a particular gate g_i. The Boolean conditions at the inputs that correspond to a $0 \rightarrow 1$ transition on g_i between times $t+1$ and $t+2$ are represented by the function $\overline{f_i(t+1)} \cdot f_i(t+2)$. The probability of a $0 \rightarrow 1$ transition occurring between time $t+1$ and $t+2$ given the transition probabilities at the primary inputs is the probability of the Boolean function $\overline{f_i(t+1)} \cdot f_i(t+2)$ evaluating to a 1. (This probability can be evaluated exactly using Binary Decision Diagrams [1] or approximately using Monte Carlo simulation.) For each gate, probabilities of transitions occurring at any time point can be evaluated efficiently, and these probabilities are summed over all the time points to obtain the average switching activity (at each gate).

Under the zero delay, unit delay, or a general delay model (where delays are obtained from library cells), the symbolic

simulation method takes into account the correlation due to reconvergence of input signals and accurately measures switching activity.

The same computation can be performed more efficiently, although not exactly, using probabilistic simulation techniques such as [10] and [13] or Monte-Carlo simulation [2]. In the remainder of this paper, whenever we need to perform the above computation, we will refer to the symbolic simulation equations (which provide the exact solution). It should however be made clear that any other solution technique (probabilistic simulation, Monte-Carlo simulation, etc.) can be used instead.

III. THE EXACT METHOD

A. Modeling Correlation

To model the correlation between the two vectors in a randomly applied vector pair, we have to augment the combinational estimation method described in Section II-B. This augmentation is summarized in Fig. 3.

In Fig. 3, we have a block corresponding to the symbolic simulation equations for the combinational logic of the general sequential circuit shown in Fig. 1. The symbolic simulation equations have two sets of inputs, namely $\langle I0, It \rangle$ for the primary inputs and $\langle PS, NS \rangle$ for the present state lines. However, given $I0$ and PS, NS is uniquely determined by the functionality of the combinational logic. This is modeled by prepending the next state logic to the symbolic simulation equations.

The configuration of Fig. 3 implies that the gate output switching activity can be determined given the vector pair $\langle I0, It \rangle$ for the primary inputs, but only PS for the state lines. Therefore, to compute gate output transition probabilities, we require the transition probabilities for the primary input lines, and the static probabilities for the present state lines. This configuration was originally proposed in [4].

482

B. State Probability Computation

The static probabilities for the present state lines marked PS in Fig. 3 are spatially correlated. We therefore require knowledge of *present state probabilities* as opposed to present state line (PS) probabilities in order to exactly calculate the switching activity in the sequential machine. The state probabilities are dependent on the connectivity of the State Transition Graph (STG) of the circuit.

For each state s_i, $1 \leq i \leq K$ in the STG, we associate a variable $prob(s_i)$ corresponding to the steady-state probability of the machine being in state s_i at $t = \infty$. For each edge e in the STG, we have $e.Current$ signifying the state that the edge fans out from, $e.Next$ signifying the state that the edge fans out to, and $e.Input$ signifying the input combination corresponding to the edge. Given static probabilities for the primary inputs to the machine, we can compute $prob(Input)$, the probability of the combination $Input$ occurring.[1] We can compute $prob(e.Input)$ using:

$$prob(e.Input) = prob(e.Current) \times prob(Input)$$

For each state s_i we can write an equation:

$$prob(s_i) = \sum_{\forall\, e\, such\, that\, e.Next = s_i} prob(e.Input)$$

Given K states, we obtain K equations out of which any one equation can be derived from the remaining $K - 1$ equations. We have a final equation:

$$\sum_{i=1}^{K} prob(s_i) = 1.$$

This linear set of K equations can be solved to obtain the different $prob(s_i)$'s.

This system of equations is known as the Chapman–Kolmogorov equations for a discrete-time discrete-transition Markov process. Indeed, if the Markov process satisfies the conditions that it has a finite number of states, its essential states form a single-chain and it contains no periodic-states, then the above system of equations will have a unique solution [12].

For example, for the State Transition Graph of Fig. 2 we will obtain the following equations assuming a probability of 0.5 for the primary input being a 1.

$$prob(\mathbf{R}) = 0.5 \times prob(\mathbf{A})$$
$$prob(\mathbf{A}) = 0.5 \times prob(\mathbf{R})$$
$$\qquad\qquad + 0.5 \times prob(\mathbf{B})$$
$$\qquad\qquad + 0.5 \times prob(\mathbf{C})$$
$$prob(\mathbf{B}) = 0.5 \times prob(\mathbf{R})$$
$$\qquad\qquad + 0.5 \times prob(\mathbf{A}).$$

The final equation is:

$$prob(\mathbf{R}) + prob(\mathbf{A}) + prob(\mathbf{B}) + prob(\mathbf{C}) = 1.$$

Solving this linear system of equations results in the state probabilities, $prob(\mathbf{R}) = \frac{1}{6}$, $prob(\mathbf{A}) = \frac{1}{3}$, $prob(\mathbf{B}) = \frac{1}{4}$, and $prob(\mathbf{C}) = \frac{1}{4}$.

[1] Static probabilities can be computed from specified transition probabilities.

C. Power Estimation Given Exact State Probabilities

We now describe a power estimation method that utilizes the exact state probabilities obtained using the Chapman–Kolmogorov method. As described in Section II-B, the symbolic equations express the exact switching conditions for each gate in the circuit under the unit or general delay models. Prepending the next state logic block as illustrated in Fig. 3 accounts for the correlation between the present and next states. Finally, computing the exact state probabilities models the steady-state behavior of the circuit.

As described in Section II-B, power estimation of a given combinational logic circuit can be carried out by creating a set of symbolic functions such that summing the signal probabilities of the functions corresponds to the average switching activity in the original combinational circuit. Some of the inputs to the created symbolic functions are the present state lines of the circuit and the others are primary input lines. Each binary combination of the present state lines is a state in the circuit and we have a number corresponding to the state probability for each state after solving the Chapman–Kolmogorov equations.

The signal probability calculation procedure has to appropriately weight these combinations according to the given probabilities. Suppose n is a disjoint cover of the function f, i.e.,

$$f = \bigvee_{m\, \in\, Disjoint\_Cover(n)} C_m \qquad (2)$$

where the C_m's are cubes of the disjoint cover. Each C_m is a function of the present state lines and primary inputs. We partition the inputs to C_m into two groups: the symbolic state support SS_m which includes all states s_i that have set the appropriate state bits, and the primary input support I_m which includes the PI inputs of C_m. Hence $C_m = SS_m I_m$. The signal probability of n is thus given by:

$$prob(n) = \sum_{m\, \in\, Disjoint\_Cover(n)} prob(C_m). \qquad (3)$$

Since the primary inputs are independent of the state that the machine is currently in and states of the FSM are distinct, we can write

$$\begin{aligned} prob(C_m) &= prob(I_m)prob(SS_m) \\ &= prob(I_m) \sum_{s_i \in SS_m} prob(s_i). \end{aligned} \qquad (4)$$

From (3) and (4), we have:

$$prob(n) = \sum_{m\, \in\, Disjoint\_Cover(n)} prob(I_m) \sum_{s_i \in SS_m} prob(s_i). \qquad (5)$$

As an example, consider the following disjoint cover of a function whose signal probability is to be computed.

$$f = i_1 \wedge ps_1 \vee i_1 \wedge \overline{ps_1} \wedge ps_2.$$

Assume that the probability of i_1 being a 1 is 0.5, and state probabilities are $prob(00) = \frac{1}{6}$, $prob(01) = \frac{1}{3}$, $prob(10) = \frac{1}{4}$ and $prob(11) = \frac{1}{4}$. (The first bit corresponds to ps_1 and the second to ps_2.) The probability of the first cube is

$$
\begin{aligned}
prob(i_1 \wedge ps_1) &= prob(i_1) \times [prob(10) + prob(11)] \\
&= 0.5 \times (\tfrac{1}{4} + \tfrac{1}{4}) \\
&= \tfrac{1}{4}.
\end{aligned}
$$

Similarly the probability of the second cube is:

$$
\begin{aligned}
prob(i_1 \wedge \overline{ps_1} \wedge ps_2) &= prob(i_1) \times prob(01) \\
&= 0.5 \times \tfrac{1}{3} \\
&= \tfrac{1}{6}.
\end{aligned}
$$

Finally we have:

$$
prob(n) = \tfrac{1}{4} + \tfrac{1}{6} = \tfrac{5}{12}.
$$

Note that (5) requires explicit enumeration of the states and is very costly. In [14], a method which employs a partially implicit enumeration of states using OBDDs is described. The estimation method still has *average-case* exponential complexity—the probability of each state (respectively, groups of states) is computed, and the number of states (respectively, such groups) can be exponential in the number of flip-flops in the circuit. However, for the circuits that this method is applicable to, the estimates provided by the method can serve as a basis for comparison among different approximation schemes.

IV. BASIS OF APPROXIMATION STRATEGIES

Consider a machine with two flip-flops whose states are 00, 01, 10, and 11 have state probabilities $prob(00) = \frac{1}{6}$, $prob(01) = \frac{1}{3}$, $prob(10) = \frac{1}{4}$ and $prob(11) = \frac{1}{4}$. We can calculate the present state line probabilities as shown below, where ps_1 and ps_2 are the first and second present state lines.

$$
\begin{aligned}
prob(ps_1 = 0) &= prob(00) + prob(01) \\
&= \tfrac{1}{6} + \tfrac{1}{3} = \tfrac{1}{2} \\
prob(ps_1 = 1) &= prob(10) + prob(11) \\
&= \tfrac{1}{4} + \tfrac{1}{4} = \tfrac{1}{2} \\
prob(ps_2 = 0) &= prob(00) + prob(10) \\
&= \tfrac{1}{6} + \tfrac{1}{4} = \tfrac{5}{12} \\
prob(ps_2 = 1) &= prob(01) + prob(11) \\
&= \tfrac{1}{3} + \tfrac{1}{4} = \tfrac{7}{12}.
\end{aligned}
$$

Note that because ps_1 and ps_2 are correlated, $prob(ps_1 = 0) \times prob(ps_2 = 0) = \frac{5}{24}$ is not equal to $prob(00) = \frac{1}{6}$.

We carried out the following experiment on 52 sequential circuit benchmark examples for which the exact state probabilities could be calculated. These benchmarks included finite state machine controllers, datapaths[2] as well as pipelines. First, the power dissipation of the circuit was calculated using the exact state probabilities as described in Section III-C. Next, given the exact state probabilities, the line probabilities were determined as described in the previous paragraph. Using the topology of Fig. 3 and the computed present state line probabilities for the PS lines, approximate power dissipations were calculated for each circuit. The average error[3] in the power dissipation measures obtained using the line probability approximation over all the circuits was only 2.8%. The maximum error for any one example was 7.3%. Assuming uniform line probabilities of 0.5 as in [4] results in significant errors of over 40% for some examples.

The above experiment leads us to conclude that if accurate line probabilities can be determined then using line probabilities rather than state probabilities is a viable alternative. We only have to determine N numbers for a N flip-flop machine, one for each present state line, rather than 2^N numbers, one for each possible state.

A. Computing Present State Line Probabilities

In our approximation framework we *directly determine line probabilities without recourse to State Transition Graph extraction*. The approximation framework is based on solving a nonlinear system of equations to compute the state line probabilities. This system of equations is given by the combinational logic implementing the next state function of the sequential circuit.

Consider the set of functions below corresponding to the next state lines.

$$
\begin{aligned}
ns_1 &= f_1(i_1, i_2, \cdots, i_M, ps_1, ps_2, \cdots, ps_N) \\
ns_2 &= f_2(i_1, i_2, \cdots, i_M, ps_1, ps_2, \cdots, ps_N) \\
&\cdots \\
ns_N &= f_N(i_1, i_2, \cdots, i_M, ps_1, ps_2, \cdots, ps_N)
\end{aligned}
$$

We can write:

$$
\begin{aligned}
prob(ns_1) &= prob[f_1(i_1, i_2, \cdots, i_M, ps_1, ps_2, \cdots, ps_N)] \\
prob(ns_2) &= prob[f_2(i_1, i_2, \cdots, i_M, ps_1, ps_2, \cdots, ps_N)] \\
&\cdots \\
prob(ns_N) &= prob[f_N(i_1, i_2, \cdots, i_M, ps_1, ps_2, \cdots, ps_N)]
\end{aligned}
$$

where $prob(ns_i)$ corresponds to the probability that ns_i is a 1, and $prob[f_i(i_1, i_2, \cdots, i_M, ps_1, ps_2, \cdots, ps_N)]$ corresponds to the probability that $f_i(i_1, i_2, \cdots, i_M, ps_1, ps_2, \cdots, ps_N)$ is a 1, which is of course dependent on the $prob(ps_j)$ and the $prob(i_k)$.

We are interested in the steady state probabilities of the present and next state lines implying that:

$$
prob(ps_i) = prob(ns_i) = p_i \quad 1 \le i \le N.
$$

A similar relationship was used in the Chapman–Kolmogorov (cf. Section III).

[2] We were restricted to 8-bit datapaths since the state probability computation requires explicitly enumerating the states of the machine.

[3] This error is caused by ignoring the correlation between the present state lines.

The set of equations given the values of $prob(i_k)$ becomes:

$$y_1 = p_1 - g_1(p_1, p_2, \cdots, p_N) = 0$$
$$y_2 = p_2 - g_2(p_1, p_2, \cdots, p_N) = 0$$
$$\cdots$$
$$y_N = p_N - g_N(p_1, p_2, \cdots, p_N) = 0 \qquad (6)$$

where the g_i's are nonlinear functions of the p_i's. We will denote the above equations as $Y(P) = 0$ or as $P = G(P)$. In general the Boolean function f_i can be written as a list of minterms over the i_k and ps_j and the corresponding g_i function can be easily derived. For example, given

$$f_1 = i_1 \wedge ps_1 \wedge \overline{ps_2} \vee i_1 \wedge \overline{ps_1} \wedge ps_2$$

and $prob(i_1) = 0.5$, we have

$$g_1 = 0.5 \cdot [p_1 \cdot (1 - p_2) + (1 - p_1) \cdot p_2]. \qquad (7)$$

We can solve the equation set $Y(P) = 0$ or find a fixed point of $P = G(P)$ to obtain the present state line probabilities. We describe the use of the Picard–Peano method to obtain a fixed point of $P = G(P)$, and the use of the Newton–Raphson method to solve $Y(P) = 0$ in Section VII. The uniqueness or the existence of the solution is not guaranteed for an arbitrary system of nonlinear equations. However, since in our application we have a correspondence between the nonlinear system of equations and the State Transition Graph of the sequential circuit, there will exist at least one solution to the nonlinear system. Further, convergence is guaranteed under mild assumptions for our application.

B. Inaccuracy in Formulation

The above formulation does not capture the correlation between the state line probabilities. Let us consider the example State Transition Graph of Fig. 2. The equations for the next state logic are:

$$ns_1 = i \cdot ps_1 \cdot ps_2 + \overline{i} \cdot \overline{ps_1} + \overline{i} \cdot ps_1\overline{ps_2}$$
$$ns_2 = ps_1 + i \cdot \overline{ps_1} \cdot \overline{ps_2}.$$

Assuming the probability of input i being a 1 is 0.5 we obtain the nonlinear equations (after simplification):

$$n_1 = 0.5 - 0.5p_1 - 0.5p_2$$
$$n_2 = p_1 + 0.5(1 - p_1)(1 - p_2).$$

Setting $n_1 = p_1$ and $n_2 = p_2$ and solving the above equations gives us $p_1 = 0.191$ and $p_2 = 0.424$. However, if we obtain the exact line probabilities using the exact state probabilities as shown in the first paragraph of Section IV, we find that these approximate line probabilities are in error.

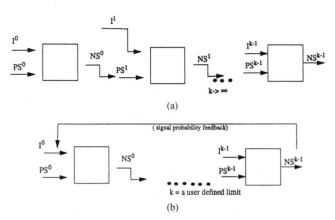

Fig. 4. k-unrolling of the next state logic.

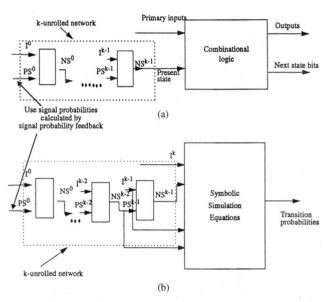

Fig. 5. Calculation of signal and transition probabilities by network unrolling.

The above example is small (4 states) and contrived, and significant errors may be obtained for such examples. The state line probabilities obtained using the approximation method of this section are on average close to the exact line probabilities, and they typically result in switching activity estimates that are close to the exact method for most real-life examples (cf. Section VIII). Nevertheless, it is worthwhile to explore ways to increasing the accuracy. We describe two such mechanisms in Section V and Section VI.

V. IMPROVING ACCURACY USING k UNROLLED NETWORKS

A. State Line Probability Computation

In the formulation of Section IV, the nonlinear equations correspond to a single stage of next state logic. Consider the *unrolled* network of Fig. 4(a). The next state logic has been unrolled k times. As illustrated in Fig. 4(b), we can construct a set of nonlinear equations corresponding to this k-unrolled network, which will partially take into account the

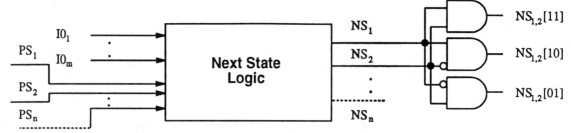

Fig. 6.   An m-Expanded network with $m = 2$.

correlation between the state lines, when computing the state line probabilities.

The exact present state line probabilities can be obtained by unrolling the next state logic ∞ times (Fig. 4(a)). This is however impractical. We thus approximate the signal probabilities by unrolling the next state logic k times where k is a user defined parameter.

The equations corresponding to $k = 2$ will be:

$$ns_1^1 = f_1(i_1^1, \cdots, i_M^1, ps_1^1, \cdots, ps_N^1)$$
$$= f_1(i_1^1, \cdots, i_M^1, ns_1^0, \cdots, ns_N^0)$$
$$= f_1[i_1^1, \cdots, i_M^1, f_1(i_1^0, \cdots, i_M^0, ps_1^0, \cdots, ps_N^0),$$
$$\cdots, f_N(i_1^0, \cdots, i_M^0, ps_1^0, \cdots, ps_N^0)]$$
$$\cdots$$
$$ns_N^1 = f_N[i_1^1, \cdots, i_M^1, f_1(i_1^0, \cdots, i_M^0, ps_1^0, \cdots, ps_N^0),$$
$$\cdots, f_N(i_1^0, \cdots, i_M^0, ps_1^0, \cdots, ps_N^0)].$$

The number of equations is the same. The number of primary input variables has increased, but the probabilities for these variables are known.

Fig. 5(a) shows the method used to calculate signal probability of the internal nodes of the FSM using the k-unrolled network with signal probability feedback.

B. Switching Activity Computation

The topology of Fig. 3 was proposed as a means of taking into account the correlation between the applied input vector pair when computing the transition probabilities. This method takes one cycle of correlation into account.

It is possible to take multiple cycles of correlation into account by prepending the symbolic simulation equations with the k-unrolled network. This is illustrated in Fig. 5(b). Instead of connecting the next state logic network to the symbolic simulation equations, we unroll the next state logic network k times and connect the next state lines of the kth stage of the unrolled network, the next state lines of the $(k-1)$th stage, and the primary input of the $(k-1)$th stage to the symbolic simulation equations.

VI. IMPROVING ACCURACY USING m-EXPANDED NETWORKS

A. State Line Probability Computation

We describe a different method to improve the accuracy of the basic approximation strategy outlined in Section IV. This method models the correlation between m-tuples of present state lines. The method is pictorially illustrated in Fig. 6 for $m = 2$.

The number of equations in the case of $m = 2$ is $3N/2$. We have:

$$ns_{i,\,i+1}[11] = ns_i \wedge ns_{i+1} = f_i \wedge f_{i+1}$$
$$ns_{i,\,i+1}[10] = ns_i \wedge \overline{ns_{i+1}} = f_i \wedge \overline{f_{i+1}}$$
$$ns_{i,\,i+1}[01] = \overline{ns_i} \wedge ns_{i+1} = \overline{f_i} \wedge f_{i+1}.$$

We have to solve for $prob(ns_{i,\,i+1}[11])$, $prob(ns_{i,\,i+1}[10])$, and $prob(ns_{i,\,i+1}[01])$ [rather than $prob(ns_i)$ and $prob(ns_{i+1})$ as in the case of $m = 1$]. We use:

$$prob(ps_i \wedge ps_{i+1}) = prob(ns_{i,\,i+1}[11])$$
$$prob(ps_i \wedge \overline{ps_{i+1}}) = prob(ns_{i,\,i+1}[10])$$
$$prob(\overline{ps_i} \wedge ps_{i+1}) = prob(ns_{i,\,i+1}[01])$$

in the evaluation of the $prob(f_i)$'s.

The signal probability evaluation methods of Section VII-C can be easily augmented to use the above probabilities. In the case of the OBDD-based method placing each ps_i and ps_{i+1} pair adjacent in the chosen ordering allows signal probability computation by a linear-time traversal.

The number of equations for $m = 3$ is $7N/3$. When $m = N$, the number of equations will become 2^N and the method will degenerate to the Chapman–Kolmogorov method.

The choice of the m-tuples of present and next state lines is made by grouping next state lines that have the maximal amount of shared logic into each m-tuple. Note that the accuracy of line probability estimation will depend on the choice of the m-tuples.

B. Switching Activity Computation

To estimate switching activity given m-tuple present state line probabilities, the topology of Fig. 3 is used as before. The difference is that for $m = 2$ the $prob(ps_i \wedge ps_{i+1})$, $prob(ps_i \wedge \overline{ps_{i+1}})$ and $prob(\overline{ps_i} \wedge ps_{i+1})$ values are used to calculate the switching activities.

VII. SOLVING THE NONLINEAR SYSTEM OF EQUATIONS

We describe two methods to solve the nonlinear system of equations obtained using k-unrolled or m-expanded networks. We will assume that the nonlinear system can be represented as $P = G(P)$ or as $Y(P) = 0$ as described in Section IV.

A. Picard–Peano Method

The Picard–Peano method is used to find a fixed point of the $P = G(P)$ system. This system is reproduced below.

$$p_1 = g_1(p_1, p_2, \cdots, p_N)$$
$$p_2 = g_2(p_1, p_2, \cdots, p_N)$$
$$\cdots$$
$$p_N = g_N(p_1, p_2, \cdots, p_N).$$

We can start with an initial guess P^0, and iteratively compute $P^{k+1} = G(P^k)$ until convergence is reached. Convergence is deemed to be achieved if $P^{k+1} - P^k$ is sufficiently small. The above iteration is known as the Picard–Peano iteration for finding a fixed-point of a system of nonlinear equations.

We are only given the Boolean functions $f_i(i_1, i_2, \cdots, i_M, ps_1, ps_2, \cdots, ps_N)$. There exist several methods to compute $g_i(p_1, p_2, \cdots, p_N) = prob[f_i(i_1, i_2, \cdots, i_M, ps_1, ps_2, \cdots, ps_N)]$ for given $p_j = prob(ps_j)$'s and $prob(i_k)$'s. We describe these methods in Section VII-C.

Theorem 7.1: [6] If G is contractive, i.e., $|\partial g_i / \partial p_j| < 1$, for all i, j, then the Picard–Peano iteration method converges at least linearly to a unique solution P^*.

Theorem 7.2: If each next state line is a nontrivial logic function of at least two present state lines, then g_i is contractive on the domain (0, 1).

Proof: Choose any p_j. In order to perform the evaluation of $\partial g_i / \partial p_j$ we cofactor f_i with respect to ps_j.

$$f_i = ps_j \wedge f_{i\,ps_j} \vee \overline{ps_j} \wedge f_{i\,\overline{ps_j}}$$

$f_{i\,ps_j}$ and $f_{i\,\overline{ps_j}}$ are the cofactors of f with respect to ps_j, and are Boolean functions independent of ps_j. We can write:

$$g_i = p_j \cdot prob(f_{i\,ps_j}) + (1 - p_j) \cdot prob(f_{i\,\overline{ps_j}}).$$

Differentiating with respect to p_j gives:

$$\frac{\partial g_i}{\partial p_j} = prob(f_{i\,ps_j}) - prob(f_{i\,\overline{ps_j}}).$$

Since we are considering the domain (0, 1), which is not inclusive of 0 and 1, and the ns_i's are nontrivial Boolean functions of at least two present state lines for every i, this partial differential is strictly less than one, because we are guaranteed that $prob(f_{i\,ps_j}) > 0$ and $prob(f_{i\,\overline{ps_j}}) > 0$. ∎

From Theorems 7.1 and 7.2, we can see that the iterated signal probability calculation is guaranteed to converge to a solution, provided some mild assumptions are made with respect to the functionality of the next state logic.

B. Newton–Raphson Method

The Newton–Raphson method can be used to solve a nonlinear system of equations given an initial guess at the solution. The advantage of the Newton–Raphson method is the quadratic rate of convergence. However, each iteration is more computationally expensive than the Picard–Peano method.

Given $Y(P) = 0$ and a column matrix corresponding to an initial guess P^0, we can write the kth Newton iteration as the linear system solve shown below.

$$J(P^k) \times P^{k+1} = J(P^k) \times P^k - Y(P^k) \qquad (8)$$

where J is the $N \times N$ Jacobian matrix of the system of equations. Each entry in J corresponds to a $\partial y_i / \partial p_j$ evaluated at P^k. The P^{k+1} correspond to the variables in the linearized system and after solving the system P^{k+1} is used as the next guess. Convergence is deemed to be achieved if each entry in $Y(P^k)$ is sufficiently small.

We use the methods of Section VII-C to evaluate:

$$g_i(p_1, p_2, \cdots, p_N)$$
$$= prob[f_i(i_1, i_2, \cdots, i_M, ps_1, ps_2, \cdots, ps_N)]$$

for given $p_j = prob(ps_j)$'s and $prob(i_k)$'s. The $Y(P^k)$ of (8) can easily be evaluated using the p_j^k values and using (6).

We need to also evaluate $J(P^k)$. As mentioned earlier, each entry of J corresponds to $\partial y_i / \partial p_j$ evaluated at P^k. If $i \neq j$, then $\partial y_i / \partial p_j$ equals $-\partial g_i / \partial p_j$, and $\partial y_i / \partial p_i$ equals $1 - \partial g_i / \partial p_i$.

In order to perform the evaluation of $\partial g_i / \partial p_j$ we use the method in the proof of Theorem 7.2.

$$\frac{\partial g_i}{\partial p_j} = prob(f_{i\,ps_j}) - prob(f_{i\,\overline{ps_j}}).$$

We can evaluate $prob(f_{i\,ps_j})$ and $prob(f_{i\,\overline{ps_j}})$ for a given P^k using the methods of Section VII-C.

As an example consider:

$$f_1 = i_1 \wedge ps_1 \wedge \overline{ps_2} \vee i_1 \wedge \overline{ps_1} \wedge ps_2$$
$$\frac{\partial g_1}{\partial p_1} = prob(i_1 \wedge \overline{ps_2}) - prob(i_1 \wedge ps_2)$$
$$\frac{\partial g_1}{\partial p_1} = 0.5 \cdot (1 - p_2) - 0.5 \cdot p_2 = 0.5 - p_2$$

which is exactly what we would have obtained had we differentiated (7) with respect to p_1.

Theorem 7.3: [11] The Newton iterates:

$$P^{k+1} = P^k - J(P^k)^{-1} Y(P^k), \quad k = 0, 1, \cdots,$$

are well-defined and converge to a solution P^* of $Y(P) = 0$ if the following conditions are satisfied:

1) Y is F-differentiable.
2) $\|J(A) - J(B)\| \leq \gamma \|A - B\|, \forall A, B \in D_0$ where D_0 is the domain $0 \leq p_i \leq 1, \forall i$.
3) There exists $P^0 \in D_0$ such that $\|J(P^0)^{-1}\| \leq \beta, \eta \geq \|J(P^0)^{-1} Y(P^0)\|$ and $\alpha = \beta \gamma \eta \leq \frac{1}{2}$.

Condition 1 of the theorem is satisfied in our application because the y_i functions are continuous and differentiable. We need to prove that the parameter γ is finite to show that Condition 2 is satisfied.

Theorem 7.4: If Y is given by (6), then $\gamma \leq 2$.

Proof: In order to show that:

$$\|J(A) - J(B)\| \leq \gamma\|A - B\|, \forall A, B \in D_0$$

is satisfied for $\gamma = 2$, we will show that the derivative of each entry of J is less than or equal to 2.

Recall that J is a matrix with each entry corresponding to $\partial y_i / \partial p_j$. Using the equations provided in the proof of Theorem 7.2 we can write:

$$\frac{\partial y_i}{\partial p_j} = prob(f_{i\,\overline{ps_j}}) - prob(f_{i\,ps_j}) \quad i \neq j.$$

Differentiating with respect to p_k we have:

$$\frac{\partial^2 y_i}{\partial p_j \partial p_k} = prob(f_{i\,\overline{ps_j}\,ps_k}) - prob(f_{i\,\overline{ps_j}\,\overline{ps_k}})$$
$$- prob(f_{i\,ps_j ps_k}) + prob(f_{i\,ps_j \overline{ps_k}}).$$

Given that the probabilities are between 0 and 1, we have:

$$\left|\frac{\partial^2 y_i}{\partial p_j \partial p_k}\right| \leq 2.$$

∎

Condition 3 in Theorem 7.3 is a constraint on the initial guess for the Newton iteration, and this initial guess can be picked appropriately, provided γ is finite. Essentially, we have to choose P^0 such that $\|Y(P^0)\|$ is small.

B. Signal Probability Evaluation

In the nonlinear equation solver, regardless of whether we are using the Picard–Peano method or the Newton–Raphson method, we have to repeatedly evaluate the signal probability of a Boolean function given input probabilities, i.e., compute $prob[f_i(i_1, i_2, \cdots, i_M, ps_1, ps_2, \cdots, ps_N)]$ given the $prob(i_k)$'s and the $prob(ps_j)$'s.

There exist several methods to evaluate signal probability. An exact method corresponds to using Ordered Binary Decision Diagrams (OBDD's) [1]. If an OBDD can be created for f_i, then $prob(f_i)$ can be evaluated in linear time in the size of the OBDD for f_i. OBDD's can be cofactored in linear time, allowing for the efficient evaluation of the Jacobian entries.

An alternative is to use Monte Carlo simulation. Approximate signal probabilities can be computed using random logic simulation on the multilevel network corresponding to f_i. Our experience has been that the signal probabilities quickly converge to the exact results obtained using OBDD's. In order to evaluate a particular Jacobian entry, the appropriate input to f_i has to be set to 0 (1) and random simulation is performed on the remaining inputs.

VIII. EXPERIMENTAL RESULTS

In this section we present experimental results that illustrate the following points:

- Exact and explicit computation of state probabilities is possible for controller type circuits. However, it is not viable for data path circuits. Purely combinational logic estimates result in significant inaccuracies.

- Assuming uniform probabilities for the present state line probabilities and state probabilities as in [4] can result in significant inaccuracies in power estimates.

- Computing the present state *line* probabilities using the technique presented in the previous sections results in 1) accurate switching activity estimates for all internal nodes in the network implementing the sequential machine; 2) accurate, robust and computationally efficient power estimate for the sequential machine.

In Table I, results are presented for several circuits. In the table, *combinational* corresponds to the purely combinational estimation method of [4] and *uniform-prob* corresponds to the sequential estimation method of [4] that assumes uniform state probabilities. The column *line-prob* corresponds to the technique of Section IV and using the Newton–Raphson method with a convergence criterion of 0.0001% to solve the equations. These equations correspond to $k = 0$ or $m = 1$. Finally, *state-prob* corresponds to the exact state probability calculation method of Section III. The zero delay model was assumed, however, any other delay model could have been used instead.

The first set of circuits corresponds to finite state machine controllers. These circuits typically have the characteristic that the state probabilities are highly nonuniform. Restricting oneself to combinational power dissipation (*combinational*) or assuming uniform state probabilities (*uniform-prob*) results in significant errors. However, the line probability method of Section IV produces highly accurate estimates when compared to exact state probability calculation.

TABLE I
COMPARISON OF SEQUENTIAL POWER ESTIMATION METHODS

Circuit Name	#lit	#ff	Combinational			Uniform Prob.			Line Prob.			State Prob.	
			power	err	cpu	power	err	cpu	power	err	cpu	power	cpu
cse	132	4	610.0	58.7	1s	578.1	50.3	7s	380.3	1.0	9s	384.4	11s
dk16	180	5	1077.5	3.1	1s	1097.2	5.0	10s	1045.0	0.0	13s	1044.8	15s
dfile	119	5	923.2	32.5	1s	701.5	0.6	7s	701.4	0.6	8s	696.8	10s
keyb	169	5	749.8	43.3	1s	724.9	38.6	12s	517.6	1.0	14s	523.0	15s
mod12	25	4	245.2	21.7	0s	195.9	2.7	1s	199.1	1.1	1s	201.4	1s
planet	327	6	1640.6	2.5	2s	1709.4	1.5	17s	1685.9	0.1	24s	1683.9	28s
sand	336	5	1446.0	33.1	2s	1165.5	7.2	24s	1078.2	0.7	27s	1086.4	34s
sreg	9	3	127.5	1.4	0s	129.4	0.0	0s	129.4	0.0	0s	129.4	0s
styr	313	5	1394.8	45.3	2s	1208.2	25.8	22s	996.9	3.8	28s	959.9	30s
tbk	478	5	1958.1	24.1	4s	1903.6	20.7	48s	1538.2	2.4	52s	1577.0	71s
accum4	45	4	360.9	3.5	0s	374.3	0.0	2s	374.3	0.0	2s	374.3	5s
accum8	89	8	720.6	4.2	1s	752.6	0.0	7s	752.6	0.0	8s	752.6	875s
accum16	245	16	1521.2	-	2s	1596.3	-	234s	1596.3	-	239s	unable	
count4	19	4	256.2	20.1	0s	213.3	0.0	1s	213.3	0.0	1s	213.3	2s
count7	35	7	474.2	12.2	0s	422.6	0.0	2s	422.6	0.0	3s	422.6	5s
count8	40	8	560.1	10.2	0s	507.9	0.0	3s	507.9	0.0	4s	507.9	8s
cbp32.4	489	223	8719.1	12.2	15s	8731.9	12.3	45s	7745.4	0.3	119s	7769.1	84s
add16	214	98	3772.3	5.1	3s	3780.5	5.4	13s	3568.0	0.5	22s	3586.5	23s
mult8	176	87	5985.6	22.8	12s	5962.6	22.4	82s	4866.9	0.1	110s	4871.1	344s
s953	418	29	762.4	76.8	1s	672.7	56.0	10s	438.7	1.7	12s	431.1	15s
s1196	529	18	2557.6	-	4s	2538.4	-	484s	2293.8	-	488s	unable	
s1238	508	18	2709.4	-	4s	2688.3	-	156s	2439.2	-	151s	unable	
s1423	657	74	6017.1	-	251s	4734.2	-	271s	7087.1	-	289s	unable	
s5378	4212	164	12457.4	-	74s	12415.1	-	455s	6496.0	-	478s	unable	
s13207	11241	669	37842.1	-	5m	27186.4	-	11m	10572.7	-	338m	unable	
s15850	13659	597	40016.2	-	8m	23850.7	-	14m	10534.1	-	167m	unable	
s35932	28269	1728	122131.2	-	20m	118475.3	-	36m	62292.0	-	152m	unable	
s38584	32910	1452	112705.6	-	24m	85842.1	-	44m	63995.1	-	922m	unable	

The second set of circuits corresponds to datapath circuits, such as counters and accumulators. The exact state probability evaluation method requires huge amounts of CPU time for even the medium-sized circuits, and cannot be applied to the large circuits. For all the circuits that the exact method is viable for, our *line-prob* method produces identical estimates. The *uniform-prob* method does better for the datapath circuits—in the case of counters for instance, it can be shown that the state probabilities are all uniform, and therefore the *uniform-prob* method will produce the right estimates. Of course, this assumption is not always valid.

The third set of circuits corresponds to pipelined adders and a pipelined multiplier. For pipelined circuits, exact power estimation is possible without resort to Chapman–Kolmogorov equation solving . The fourth set corresponds to mixed datapath/control circuits from the ISCAS-89 benchmark set. Exact state probability evaluation is not possible for these circuits.

The CPU times in the table corresponds to seconds (s) or (m) on a SUN SPARC-2. The CPU times correspond to times required for symbolic simulation to estimate combinational activity plus the time required for the calculation of state/line probabilities. For all the circuits BDD's were used to obtain the line probabilities. However, Monte-Carlo simulation was used for combinatorial activity estimation for the large ISCAS-89 circuits.

In Table II, present state line probability estimates for the benchmark circuits are presented. The error value provided in each column shows the absolute error (i.e., absolute value of the difference between exact and approximate values) of the signal probabilities averaged over all present state lines in the circuit. The exact values were calculated by the method described in Section III. (We could not generate the exact values for circuits in Groups 3 and 4, as the size of Chapman–Kolmogorov system of equations becomes too large.) It is evident from these results that the error averaged over all benchmark circuits is well below 0.05 (see the *line-prob* column entries which correspond to the method described in Section IV). Note that this error is due to ignoring correlation as exemplified in Section IV-B, and not due to convergence error of the Newton–Raphson method. The convergence criterion for line probabilities was set to 0.0001% to generate these results.

We present the switching activity errors for the benchmark circuits in Table III. Again, the error value provided in each column represents the absolute error averaged over all internal nodes in the circuit. It can be seen that this error is quite small. These two tables demonstrate that the approximate procedure provided in Section IV leads to very accurate estimates for both the present state line probabilities and for the switching activity values for all circuit lines.

Next, we present results comparing the Picard–Peano and Newton–Raphson methods to solve the nonlinear equations of Section IV. These results are summarized in Table IV. The number of iterations required for the Picard–Peano and Newton–Raphson methods are given in Table IV under the appropriate columns, as are the CPU times per iteration and the total CPU time. Newton–Raphson typically takes fewer iterations, but each iteration requires the evaluation of the

TABLE II
ABSOLUTE ERRORS IN PRESENT STATE LINE PROBABILITIES
AVERAGED OVER ALL PRESENT STATE LINES

Circuit Name	Combinational err	Uniform Prob. err	Line Prob. err
cse	0.427	0.427	0.00788
dk16	0.0782	0.0782	0.0125
dfile	0.075	0.075	0.047
keyb	0.414	0.414	0.0133
mod12	0	0	0.03
planet	0.031	0.031	0.09
sand	0.12	0.12	0.044
sreg	0	0	0
styr	0.3138	0.3138	0.0357
tbk	0.2614	0.2614	0.026
accum4	0	0	0
accum8	0	0	0
accum16	0	0	0
count4	0	0	0
count7	0	0	0
count8	0	0	0
cbp32.4	-	-	-
add16	-	-	-
mult8	-	-	-
s953	-	-	-
s1196	-	-	-
s1238	-	-	-

Jacobian and is more expensive than the Picard iteration. The results obtained by the two methods are identical, since the convergence criterion used was the same.

To generate the results in Table IV, the convergence criterion allowed a maximum error of 1% in the line probabilities. In this case, the Picard–Peano method out performs the Newton–Raphson method for virtually all the examples. If the convergence criterion is tightened, e.g., to allow for a maximum error of .01%, the Picard–Peano method requires substantially more iterations than the Newton–Raphson and in several examples, the Newton–Raphson method outperforms the Picard–Peano method. However, since the error due to ignoring correlation (cf. Section IV-B) can be more than 1%, in practice it does not make sense to tighten the convergence criterion beyond a 1% allowed error.

In some pathological examples, where the conditions of Theorem 7.1 are not satisfied, the Picard–Peano method may exhibit oscillatory behavior, and will not converge. In these cases, the strategy we adopt is to use Picard–Peano for several

TABLE III
ABSOLUTE ERRORS IN SWITCHING ACTIVITY
AVERAGED OVER ALL CIRCUIT LINES

Circuit Name	Combinational err	Uniform Prob. err	Line Prob. err
cse	0.402	0.053	0.003
dk16	0.354	0.020	0.010
dfile	0.268	0.019	0.015
keyb	0.363	0.067	0.009
mod12	0.387	0.149	0.156
planet	0.375	0.034	0.034
sand	0.400	0.015	0.010
sreg	0	0	0
styr	0.415	0.058	0.022
tbk	0.423	0.020	0.008
accum4	0.084	0	0
accum8	0.086	0	0
accum16	0.096	0	0
count4	0.169	0	0
count7	0.189	0	0
count8	0.192	0	0
cbp32.4	-	-	-
add16	-	-	-
mult8	-	-	-
s953	-	-	-
s1196	-	-	-
s1238	-	-	-

TABLE IV
COMPARISON OF PICARD–PEANO AND NEWTON–RAPHSON

Circuit Name	Picard-Peano			Newton-Raphson		
	#iter	cpu/iter	total cpu	#iter	cpu/iter	total cpu
cse	5	0.1	0.5	3	1	3
dk16	4	0.18	0.7	3	1	3
dfile	5	0.12	0.6	2	1.5	3
keyb	10	0.07	0.7	6	0.33	2
mod12	3	0.03	0.1	2	0.1	0.2
planet	11	0.13	1.4	3	2.33	7
sand	6	0.22	1.3	3	1	3
sreg	1	0.1	0.1	1	0.1	0.1
styr	7	0.2	1.4	3	2	6
tbk	4	0.5	2.0	3	1.33	4
accum4	1	0.1	0.1	1	0.1	0.1
accum8	1	0.3	0.3	1	1	1
accum16	1	1.0	1.0	1	6	6
count4	1	0.1	0.1	1	0.1	0.1
count7	1	0.2	0.2	1	1	1
count8	1	0.2	0.2	1	1	1
cbp32.4	3	0.8	2.4	4	18.5	74
add16	3	0.3	0.9	3	3	9
mult8	2	3.25	6.5	4	9.25	37
s953	30	0.04	1.1	4	0.5	2
s1196	2	1.1	2.2	2	2	4
s1238	2	1.15	2.3	2	2.5	5

iterations, and if oscillation is detected, the Newton–Raphson method is applied. The Newton–Raphson method does not require the domain to be contractive, however, the initial guess has to be "close" to the solution P^* in a manner quantified by Theorem 7.3.

In Table V, we present results that indicate the improvement in accuracy in power estimation when k-unrolled or m-expanded networks are used. Results are presented for the finite state machine circuits of Table I for $0 \leq k \leq 2$ and $1 \leq m \leq 4$.[4] The percentage differences in power from the exact power estimate are given. In general, if $k \rightarrow \infty$, the error will reduce to 0%, however, increasing k when k is small is not guaranteed to reduce the error in total power estimates (e.g., consider styr). This phenomenon can be explained as follows. The total power estimate is obtained by summing power consumptions of all nodes in the circuit. The individual power estimates may be under- or over-estimated, yet when

they are added together, the overall error may become small due to error cancelation. Increasing k improves the accuracy of power estimates for individual nodes (see Table VI), but does not necessarily improve the accuracy of power estimate for the circuit due to the unpredictability of the error cancelation during the summing step. The m-expansion-based method behaves more predictably for this set of examples, however, again no guarantees can be made regarding the improvement in accuracy (of total power estimates) on increasing m, except that when m is set to the number of flip-flops in the machine, the method produces the Chapman–Kolmogorov equations, and therefore the exact state probabilities are obtained. The Newton–Raphson method with a convergence criterion of 0.0001% was used to obtain the line probabilities in Tables V and VI.

The CPU times for power estimation are in seconds on a SUN SPARC-2. These times can be compared with those listed in Table I under the "Line Prob." column as those times correspond to $k = 0$ and $m = 1$. Based on these results, we conclude that $k = 1$ and $m = 2$ provide a good compromise between accuracy and run-time.

During the synthesis process, we often want to know the switching activity of individual nodes instead of a single power consumption figure. Table VI presents the percentage error in

---

[4] The initial error for dk16 and sreg benchmarks is 0, thus, there is no need to improve the accuracy by using larger values of k and m.

TABLE V
RESULTS OF POWER ESTIMATION BASED ON
k-UNROLLED AND m-EXPANDED NETWORKS

Circuit Name	Initial Error	k-Unrolled Error				m-Expanded Error			
		$k=1$		$k=2$		$m=2$		$m=4$	
		err	cpu	err	cpu	err	cpu	err	cpu
cse	1.06	0.33	18	0.02	51	0.42	10	0.00	10
dfile	0.67	0.20	16	0.20	29	0.23	9	0.17	10
keyb	1.02	0.02	44	0.04	53	1.01	14	0.32	14
mod12	1.13	0.85	2	0.30	3	1.13	1	0.00	2
planet	0.11	0.15	40	1.72	45	0.10	25	0.08	25
sand	0.76	0.61	64	0.29	109	0.64	28	0.43	30
styr	3.85	0.16	67	0.41	113	0.58	29	0.52	29
tbk	2.46	1.52	207	0.12	597	2.17	58	0.12	59

TABLE VI
PERCENTAGE ERROR IN SWITCHING ACTIVITY ESTIMATES
AVERAGED OVER ALL NODES IN THE CIRCUIT

Circuit Name	average % error					
	$k=0$	$k=1$	$k=2$	$m=1$	$m=2$	$m=4$
cse	6.79	2.26	0.57	6.79	3.40	0.00
dfile	14.05	5.37	3.10	14.05	4.82	3.56
keyb	7.18	1.68	0.70	7.18	7.09	2.25
mod12	10.24	6.36	5.00	10.24	10.05	0.00
planet	43.08	30.22	28.97	43.08	41.26	35.22
sand	16.65	12.20	11.78	16.65	14.02	9.42
styr	43.51	12.99	6.31	43.51	6.55	5.97
tbk	18.04	4.48	2.95	18.04	15.91	1.88

individual node's switching activity from the exact values as a function of k and m, averaged over all the nodes in the circuit. It is seen that the accuracy of switching activity estimates consistently increases with the value of k and m. For example, the error in switching activity estimates for styr decreases from 13% to 6.3% when k increases from 1 to 2 and from 6.6% to 6.0% when m increases from 2–4. A similar trend exists with respect to the maximum error and the root-mean-squared error criteria.

IX. CONCLUSIONS AND ONGOING WORK

We presented a framework for sequential power estimation in this paper. In this framework, state probabilities can be computed using the Chapman–Kolmogorov equations, and present state line probabilities are computed by solving a system of nonlinear equations. We have shown that the latter is significantly more efficient for medium to large circuits, and does not sacrifice accuracy.

Given the present state line probabilities, the switching activity and power dissipation of the circuit can be accurately computed. Any combinational logic estimation method that can accurately model the correlation between the applied input vector pairs can be used.

REFERENCES

[1] R. Bryant, "Graph-based algorithms for Boolean function manipulation," *IEEE Trans. Comput.,* vol. C-35, pp. 677–691, Aug. 1986.
[2] R. Burch, F. Najm, P. Yang, and T. Trick." McPOWER: A Monte Carlo approach to power estimation," in *Proc. Int. Conf. Computer-Aided Design,* Nov. 1992, pp. 90–97.
[3] A. Chandrakasan, T. Sheng, and R. W. Brodersen, "Low power CMOS digital design," in *J. Solid State Circuits,* pp. 473–484, Apr. 1992.
[4] A. Ghosh, S. Devadas, K. Keutzer, and J. White, "Estimation of average switching activity in combinational and sequential circuits," in *Proc. 29th Des. Automat. Conf.,* June 1992, pp. 253–259.
[5] L. Glasser and D. Dobberpuhl, *The Design and Analysis of VLSI Circuits.* Reading, MA: Addision-Wesley, 1985.
[6] H. M. Lieberstein, *A Course in Numerical Analysis.* New York: Harper & Row, 1968.
[7] R. Marculescu, D. Marculescu, and M. Pedram, "Logic level power estimation considering spatiotemporal correlations," in *Proc. Int. Conf. Computer-Aided Design,* Nov. 1994, pp. 294–299.
[8] J. Monteiro, S. Devadas, and A. Ghosh, "Retiming sequential circuits for low power," in *Proc. Int. Conf. Computer-Aided Design,* Nov. 1993, pp. 398–402.
[9] F. Najm, "Transition density, A stochastic measure of activity in digital circuits," in *Proc. 28th Des. Automat. Conf.,* June 1991, pp. 644–649.
[10] F. N. Najm, R. Burch, P. Yang, and I. Hajj, "Probabilistic simulation for reliability analysis of CMOS VLSI circuits," *IEEE Trans. Computer-Aided Design,* pp. 439–450, Apr. 1990, vol. 9.
[11] J. M. Ortega and W. C. Rheinboldt, *Interative Solution of Nonlinear Equations in Several Variables.* Boston, MA: Academic, 1970.
[12] A. Papoulis, *Probability, Random Variables and Stochastic Processes.* New York: McGraw-Hill, 1991, 3rd ed.
[13] C. Y. Tsui, M. Pedram, and A. Despain, "Efficient estimation of dynamic power dissipation under a real delay model," in *Proc. Int. Conf. Computer-Aided Design,* June 1993, pp. 224–228.
[14] _____, "Exact and approximate methods for calculating signal and transition probabilities in FSMs," Elec. Eng.-Syst. Dept., Univ. So. CA, Tech. Rep. CNEG 93-42, Oct. 1993.

A Monte Carlo Approach for Power Estimation

Richard Burch, Farid N. Najm, *Member, IEEE*, Ping Yang, *Fellow, IEEE*, and Timothy N. Trick, *Fellow, IEEE*

Abstract— Excessive power dissipation in integrated circuits causes overheating and can lead to soft errors and/or permanent damage. The severity of the problem increases in proportion to the level of integration, so that power estimation tools are badly needed for present-day technology. Traditional simulation-based approaches simulate the circuit using test/functional input pattern sets. This is expensive and does not guarantee a meaningful power value. Other recent approaches have used probabilistic techniques in order to cover a large set of input patterns. However, they trade off accuracy for speed in ways that are not always acceptable. In this paper, we investigate an alternative technique that combines the accuracy of simulation-based techniques with the speed of the probabilistic techniques. The resulting method is *statistical* in nature; it consists of applying randomly-generated input patterns to the circuit and monitoring, with a simulator, the resulting power value. This is continued until a value of power is obtained with a desired *accuracy*, at a specified *confidence* level. We present the algorithm and experimental results, and discuss the superiority of this new approach.

I. INTRODUCTION

EXCESSIVE power dissipation in integrated circuits causes overheating and can lead to soft errors and permanent damage. The severity of the problem increases in proportion to the level of integration. The advent of VLSI has led to much recent work on the estimation of power dissipation *during* the design phase, so that designs can be modified before manufacturing.

Perhaps the most significant obstacle in trying to estimate power dissipation is that the power is *pattern dependent*. In other words, it strongly depends on the input patterns being applied to the circuit. Thus the question "what is the power dissipation of this circuit?" is only meaningful when accompanied with some information on the circuit inputs.

A direct and simple approach of estimating power is to simulate the circuit. Indeed, several *circuit simulation* based techniques have appeared in the literature [1], [2]. Given the speed of circuit simulation, these techniques can not afford to simulate large circuits for long-enough input vector sequences to get meaningful power estimates. In order to simplify the problem and improve the speed, the power supply voltage is often assumed to be the same throughout the chip. Thus the power estimation problem is reduced to that of estimating the power supply *currents* that are drawn by the different circuit components. Fast timing or logic simulation can then be used to estimate these currents [3].

Manuscript received May 5, 1992; revised August 27, 1992 and November 9, 1992.

R. Burch and P. Yang are with the Semiconductor Process & Design Center, Texas Instruments Inc., Dallas, TX 75265.

F. N. Najm and T. N. Trick are with the Electrical Engineering Department, University of Illinois at Urbana-Champaign Urbana, IL 61801.

IEEE Log Number 9206235.

We call these approaches *strongly* pattern dependent because they require the user to specify *complete* information about the input patterns. Recently, other approaches have been proposed [4], [5] that only require the user to specify *typical* behavior at the circuit inputs using *probabilities*. These may be called *weakly* pattern dependent. With little computational effort, these techniques allow the user to cover a huge set of possible input patterns. However, in order to achieve good accuracy, one must model the correlations between internal node values, which can be very expensive. As a result, these techniques usually trade off accuracy for speed. The resulting loss of accuracy is a significant issue that may not always be acceptable to the user.

In this paper, we investigate an alternative approach that combines the accuracy of simulation-based approaches with the weak pattern dependence of probabilistic approaches. The resulting approach is *statistical* in nature; it consists of applying randomly generated input patterns to the circuit and monitoring, with a simulator, the resulting power value. This is continued until a value of power is obtained with a desired *accuracy*, at a specified *confidence* level. Since it uses a *finite* number of patterns to estimate the power, which really depends on the *infinite* set of possible input patterns, this method belongs to the general class of so-called *Monte Carlo* methods. A most attractive property of Monte Carlo techniques is that they are *dimension independent*, meaning that the *number* of samples required to make a good estimate is independent of the problem size. We will show that this property indeed holds for our approach (see Table IV in Section V).

Both [4] and [5] use probabilities to compute the power consumed by *individual gates*, which are then summed up to give the total power. In this context, it was observed in [5] that it would be too expensive to estimate the individual gate powers using a simulation with randomly generated inputs. The key to the efficiency of our new approach is that, if one monitors the *total* power directly during the random simulation, sufficient accuracy is obtained in much less time than is required to compute the individual gate powers. The excellent speed performance and the simplicity of the implementation make this a very attractive approach for power estimation.

An approach similar to this was independently proposed in [6], but the treatment there is not very rigorous and overlooks some important issues. Furthermore, no comparisons were performed with other approaches to show the superiority of the approach. In this paper, we present a rigorous treatment that provides the theoretical justification of this method. We also present experimental results of our implementation and compare it to probabilistic approaches.

Reprinted from *IEEE Transactions on VLSI Systems*, pp. 63-71, March 1993.

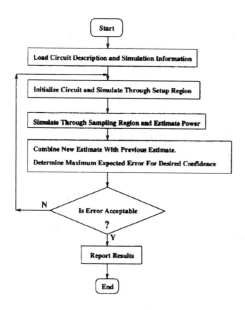

Fig. 1. Block diagram overview.

II. OVERVIEW

In this section, we provide an overall view of our technique, and discuss its superiority to the probabilistic approaches previously proposed [4], [5].

A. Overview of Monte Carlo Power Estimation

The block diagram in Fig. 1 gives an overall view of the technique. The *setup* and *sample* blocks are parts of the same logic simulation run, in which the input patterns are *randomly-generated*. The power value at the end of a sampling phase is noted and used to decide whether to *stop* the process or to do another setup-sample run. The decision is made based on the mean and standard deviation of the power values observed at the end of a number of successive iterations.

The power is found as the average value of the instantaneous power drawn throughout the sample phase, and *not* during the setup phase. However, the setup phase is a critical component of our approach, and serves two purposes.

1) In the beginning of the simulation run, the circuit does not switch as often as it typically would at a later time, when switching activity has had time to spread to all the gates. Thus the circuit is simulated until all nodes are switching at a stable rate during setup. This argument will be made more precise in Section III, where we also derive an exact value for the setup time.

2) The values of power observed at the end of successive sample intervals should be samples of independent random variables. This is required in order for the stopping criterion to be correct, and is guaranteed by restarting the random input waveforms at the beginning of the setup phase. The details are given in Section III.

Thus the setup phase guarantees that we are indeed measuring *typical* power, and ensures the *correctness* of the statistical stopping criterion.

B. Comparison with Probabilistic Techniques

There are two distinct advantages of the Monte Carlo approach that make it an excellent choice for power estimation over probabilistic techniques. These are: 1) it achieves *desired* accuracy in reasonable time, avoiding the speed/accuracy tradeoff of probabilistic techniques and 2) the simplicity of the algorithm makes it very easy to implement in existing logic or timing simulation environments.

Probabilistic methods [4], [5] suffer from a speed/accuracy tradeoff because they must resolve the correlations between internal circuit nodes. If these correlations *are* taken into account, these methods can be very accurate. This, however, is computationally very expensive and impractical. As a result, fast implementations of these techniques are necessarily inaccurate. It is the aim of this paper to show that the proposed Monte Carlo method is very fast (solving circuits with thousands of gates in a matter of seconds) and also highly accurate (easily within 5% of the total power). Tables III and V compare the accuracy of Monte Carlo and probabilistic methods for power estimation.

We also should make the point that the accuracy level in our approach is predictable up-front : the program will work to achieve any level of accuracy desired by the user. Naturally, as higher accuracy is desired, the computational cost starts to increase. However, we will show in Section V that accuracy levels of 5% are easily and efficiently attainable.

III. DETAILED APPROACH

This section describes the details of the approach. We start out with a rigorous formulation of the problem and show how it reduces to the well-known problem of *mean estimation* in statistics. We then discuss the stopping criterion, and the normality assumption required for it to work. We conclude with a discussion of the setup and sample phases and the applicability to sequential circuits.

A. Problem Formulation

Consider a digital circuit with m internal nodes (gate outputs). Let $x_i(t)$, $t \in (-\infty, +\infty)$, be the logic signal at node i and $n_{x_i}(T)$ be the number of transitions of x_i in the time interval $(-\frac{T}{2}, +\frac{T}{2}]$. If, in accordance with [7], we consider only the contribution of the charging/discharging current components, the average power dissipated at node i during that interval is $\frac{1}{2} V_{dd}^2 C_i \frac{n_{x_i}(T)}{T}$, where C_i is the total capacitance at i. The total average power dissipated in the circuit during the same interval is

$$P_T = \frac{V_{dd}^2}{2} \sum_{i=1}^{m} C_i \frac{n_{x_i}(T)}{T}. \tag{1}$$

The power rating of a circuit usually refers to its average power dissipation over extended periods of time. We therefore define the *average power dissipation* P of the circuit as

$$P = \lim_{T \to \infty} P_T = \frac{V_{dd}^2}{2} \sum_{i=1}^{m} C_i \lim_{T \to \infty} \frac{n_{x_i}(T)}{T}. \tag{2}$$

The essence of our approach is to estimate P, corresponding to *infinite* T, as the *mean* of several P_T values, each measured over a *finite* time interval of length T. In order to see how this *mean estimation* problem comes about, we must consider a random representation of logic signals as follows.

Corresponding to every logic signal $x_i(t)$, $t \in (-\infty, +\infty)$, we construct a *stochastic process* $\boldsymbol{x}_i(t)$ as a family of the logic signals $x_i(t + \tau)$, where τ is a random variable. This process has been called the *companion process* of $x_i(t)$ in [5], [8], where the reader may find more details on its construction. For each τ, $x_i(t + \tau)$ is a *shifted* copy of $x_i(t)$. Therefore, observing P_T for $x_i(t+\tau)$ corresponds to measuring the power of $x_i(t)$ over an interval of length T centered at τ, rather than at 0. We can then talk of the *random power* of $\boldsymbol{x}_i(t)$ over the interval $(-\frac{T}{2}, +\frac{T}{2}]$, to be denoted by

$$P_T = \frac{V_{dd}^2}{2} \sum_{i=1}^{m} C_i \frac{\boldsymbol{n}_{x_i}(T)}{T} \qquad (3)$$

where $\boldsymbol{n}_{x_i}(T)$ is now a random variable. It was shown in [5], [8] that $\boldsymbol{x}_i(t)$ is *stationary* [12] so that, for *any* T, the expected average number of transitions per second is a constant:

$$D(x_i) = E\left[\frac{\boldsymbol{n}_{x_i}(T)}{T}\right] = \lim_{T \to \infty} \frac{n_{x_i}(T)}{T} \qquad (4)$$

where $E[\cdot]$ denotes the *expected value* (mean) operator. In [5] and [8], $D(x_i)$ was called the *transition density* of $x_i(t)$; it is the average number of transitions per second, equal to twice the average frequency. As a result of (4), $E[P_T]$ is the same for any T, and the average power can be expressed as a mean:

$$P = E[P_T]. \qquad (5)$$

Thus the power estimation problem has been reduced to that of *mean estimation*, which is a frequently encountered problem in statistics.

In order to apply the above theory, we must ensure that the signals $x_i(t)$ observed throughout the $(\frac{-T}{2}, \frac{+T}{2}]$ interval are *samples* of the *stationary* processes $\boldsymbol{x}_i(t)$. This requirement will be addressed in Section III-D.

B. Stopping Criterion

To determine stopping criteria, some assumptions about the distribution of P_T must be made. Therefore, let us <u>assume</u> that P_T is *normally distributed* for any T. The theoretical justification and experimental evidence for this assumption will be discussed in Section III-C, and Section IV will discuss the applicability of Monte Carlo methods when this assumption is invalid. Suppose also that we perform N different simulations of the circuit, each of length T, and form the *sample average* η_T and *sample standard deviation* s_T of the N different P_T values found. Therefore, we have $(1 - \alpha) \times 100\%$ *confidence* that $|\eta_T - E[P_T]| < t_{\alpha/2} s_T/\sqrt{N}$, where $t_{\alpha/2}$ is obtained from the t distribution [9] with $(N - 1)$ degrees of freedom. This result can be rewritten as

$$\frac{|P - \eta_T|}{\eta_T} < \frac{t_{\alpha/2} s_T}{\eta_T \sqrt{N}}. \qquad (6)$$

Therefore, for a desired *percentage error* ϵ in the power estimate, and for a given confidence level $(1 - \alpha)$, we must simulate the circuit until

$$\frac{t_{\alpha/2} s_T}{\eta_T \sqrt{N}} < \epsilon. \qquad (7)$$

We can use this relation to illustrate the important *dimension independence* property of this approach, common to most Monte Carlo methods, as follows. If N^* is the (smallest) number of iterations that satisfies (7), then,

$$N^* \approx \left(\frac{t_{\alpha/2} s_T}{\epsilon \, \eta_T}\right)^2. \qquad (8)$$

By dimension independence, we mean that N^* should be roughly independent of the circuit size (number of nodes). In (7), $t_{\alpha/2}$ is a small number, typically between 2.0 and 5.0, and ϵ is a constant. We, therefore, look to the ratio s_T^2/η_T^2 to learn of the general behavior of N^*. There is very little one can say in general about this ratio. Nevertheless, and in view of (3), it is instructive to consider the following. If $y = \sum_{i=1}^{m} x_i$ is the sum of m independent identically distributed (IID) random variables, then $\sigma_y^2/\mu_y^2 = (\sigma_{x_i}^2/\mu_{x_i}^2) \times (1/m)$, where μ and σ^2 denote the mean and variance. Thus if the $C_i \boldsymbol{n}_{x_i}(T)/T$ terms of (3) are iid, then N^* should *decrease* with circuit size. Even when the x_i's are *not* independent, we have $\sigma_y^2/\mu_y^2 \leq (\sigma_x^2/\mu_{x_i}^2)$, a constant, which suggests that N^* should typically decrease or remain constant with increasing circuit size. This is indeed the observed behavior in Table IV.

An important consequence of this result is that, since each iteration of the Monte Carlo approach takes roughly linear time (in the size of the circuit), then the overall process should also take linear time. Probabilistic methods that do not take correlation into account also depend linearly on circuit size. However, if correlation is taken into account in order to improve the accuracy, their dependence is frequently super-linear.

In order to use the stopping criterion in practice, we must ensure that the observed P_T values are samples from *independent* P_T random variables. This requirement will be addressed in Section III-D.

C. Normality

It cannot be proven that P_T is normally distributed for finite T; however, P_T can frequently be approximated as normal. This section will discuss sufficient conditions for the normality of P_T that make the approximation sensible, and show experimental evidence that the approximation is good on a variety combinational benchmark circuits.

A sufficient condition for the normality of P_T is that (i) m is large and (ii) $\boldsymbol{n}_{x_i}(T)/T$ are independent. This is true under fairly general conditions irrespective of the individual $\boldsymbol{n}_{x_i}(T)/T$ distributions (see [10, pp. 188–189]), and for any value of T.

Another sufficient condition that holds even for small m, but for large T, is as follows. If (i) the consecutive times between identical transitions of $x_i(t)$ are independent (which, using renewal theory (see [11, pp. 62–63]), means that $\boldsymbol{n}_{x_i}(T)/T$

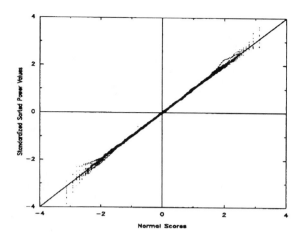

Fig. 2. Normal scores plot for the ISCAS-85 circuits.

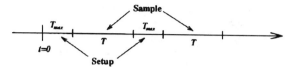

Fig. 3. Successive setup and sample phases.

is normally distributed for large T) and (ii) the $n_{x_i}(T)/T$ are independent (so that they are also jointly normal (see [12], p. 126) for large T) then P_T is normal for large T (see [12, p. 144]).

To the extent that these conditions are approximately met in practice, the power should be approximately normal. In both cases, the condition that the $n_{x_i}(T)/T$ are independent is not met; however, if a significant subset of the $n_{x_i}(T)/T$ are approximately independent, then it is reasonable to expect that the power should be approximately normal. We have found that for a number of benchmark digital circuits [13], the normality assumption is very good, as shown in the *normal scores plots* [9] in Fig. 2. The plot for each circuit corresponds to 1000 evaluations of the average power over a 2.5-μs interval. Each evaluation covered an average of 50 transitions per primary input. It cannot be proven that the power of all circuits is approximately normal. The consequences of deviations from normality are discussed in Section IV.

D. Setup and Sample

This section deals with the mechanics of how the input patterns are to be generated, when to start and stop measuring a P_T value, and how different P_T values should be obtained. We start by observing that, by stationarity of $x_i(t)$, the (finite) intervals of width T, over which the P_T values will be measured, need not be centered at the origin. A P_T value may be obtained from any interval of width T, henceforth called a *sampling interval*. However, the following two requirements must be met:

1) (i) Throughout a sampling interval, the signals $x_i(t)$ must be samples of the *stationary* processes $x_i(t)$.
2) (ii) The different P_T samples must be samples from *independent* P_T random variables.

We will now describe a simulation process that guarantees both of these requirements. Suppose that the circuit primary inputs are at 0 from $-\infty$ to time 0, and then become samples of the stationary processes $x_i(t)$ in positive time. Consider a primary input driving an inverter with delay t_d. Since its input is a stationary process for $t \geq 0$, its output must be stationary for $t \geq t_d$. By using a simplified timing model for every gate

as in [5], we can repeat this argument enough times to obtain the following conclusion: If the maximum delay (along any path) from the primary inputs to node i is $T_{max,i}$, then the process $x_i(t)$ becomes stationary for $t \geq T_{max,i}$.

If the maximum delay (along any path) in the circuit is $T_{max} = \max_i (T_{max,i})$, then the sampling interval may start only after $t \geq T_{max}$. This guarantees that requirement (i) is met. From that time onwards, all internal processes are stationary, and the circuit is in (probabilistic) steady state. We will call the time interval from 0 to T_{max} the *setup* phase. Intuitively, the circuit needs to be simulated until all nodes are switching at stable rates before a reliable sample of power may be taken and, as we have shown, the minimum time required to achieve that is T_{max}. Finding T_{max} in combinational circuits is straightforward; the case of sequential (feedback) circuits is discussed in Section III-E.

In order to guarantee requirement (ii), we simply *restart* the simulation (with an empty event queue) at the beginning of every setup phase. As a result, the time axis is divided into successive regions of setup and sampling, as shown in Fig. 3.

We have discussed the length of the setup region. Now, consider the length of the sampling region, T. Two factors must be considered: the normality approximation and the overall simulation time. The length of the sampling region can affect the error in the normality approximation. The minimum value of T that yields approximately normal power is heavily dependent upon the circuit and can range from nearly zero to nearly infinity. To minimize the simulation time, minimize $N \times (T_{max} + T)$. N depends upon s_T, and s_T decreases as T increases. The speed of this decrease is heavily dependent upon the circuit. These two factors indicate that no general solution for optimum T exists. We chose a value for T (see results) that worked well experimentally.

The only remaining task is to describe how the inputs are to be generated. This has to be done in such a way that the input processes, after the start of every setup phase, are independent of the past. This can be done as follows, for every input signal x_i. At the beginning of a setup phase, we use a random number generator to select a logic value for x_i, with appropriate signal probability $P(x_i)$. We then use another random number generator to decide how long x_i stays in that state before switching. This must assume some distribution for the duration of stay in that state. Once x_i has switched, we use another random number generator to decide how long it will stay in the other state, again using some distribution. Let $F_{x_i}^1(t)$ be the distribution of times spent in the 1 state, and $F_{x_i}^0(t)$ be that of the 0 state. Since computer implementations of random number generators produce sequences of independent random variables, independence between the successive sampling phases is guaranteed.

The starting signal probability $P(x_i)$ and distributions $F_{x_i}^1(t)$ and $F_{x_i}^0(t)$ should be supplied by the user. In fact, these parameters represent the way in which the approach is weakly pattern dependent. They also provide the mechanism by which the user can specify any information about typical behavior at the circuit inputs. In order to simplify the user interface, our current implementation does not require the user to actually specify distributions. Rather, we require only two parameters: the average time that an input is high, denoted by $\mu_{x_i}^1$, and the average time that it is low, denoted by $\mu_{x_i}^0$. Based on this, it can be shown [5, 8] that $P(x_i) = \mu_{x_i}^1/(\mu_{x_i}^1 + \mu_{x_i}^0)$ and $D(x_i) = 2/(\mu_{x_i}^1 + \mu_{x_i}^0)$. As for the distributions, our implementation uses *exponential* distributions [12] so as to allow the comparisons with probabilistic methods to be given in Section V. We emphasize, however, that the stopping criterion and the overall Monte Carlo algorithm are valid for *any* distribution. In fact, in our implementation, the choice of distribution can be easily modified by the user. Furthermore, experiments with different input distributions have shown that the average power of many circuits is not sensitive to the form of the distributions.

Previously, we have assumed that the primary inputs are independent. Because of the expense of handling correlation, this is necessary for probabilistic methods. If the correlation can be expressed as conditional probabilities, then logical waveforms can be generated that take the correlation into account and Monte Carlo methods can be used to estimate the power effectively.

E. Sequential Circuits

The Monte Carlo method presented in this paper is valid for both combinational and sequential circuits. The only aspect of the problem that is specific to sequential circuits is the computation of the setup time T_{max}. Strictly speaking, since T_{max} is the longest delay along *any* path, then $T_{max} = \infty$ for sequential circuits. Recall that it is *sufficient* to wait for T_{max} before starting a sampling interval in order to guarantee stationarity. It is not clear, however, whether that condition is also necessary. Heuristics must be developed to estimate the setup lengths for sequential circuits. The quality of a heuristic can be tested by examining the expected power for sampling regions of different lengths. If the expected power is constant, then the heuristic does a good job of predicting the length of the setup region. Table I shows that this is true for combinational circuits, which agrees with our assertion in Section III that the power would be stationary after T_{max}. Future implementations of this approach will include heuristics to allow it to handle sequential circuits.

IV. DEVIATIONS FROM NORMALITY

Monte Carlo method can be applied to circuits that have nonnormal power distributions without adversely affecting the accuracy of the results. In cases of severe deviations from normality, some modifications of the basic approach may be required. It is important to note at the outset that the normality assumption was required only to formulate the stopping criterion. Given enough samples, one ultimately

TABLE I
If The Setup Region is Chosen Correctly, Then The Process is Stationary and The Expected Power is The Same for Any Sampling Region. This is Illustrated for Combinational Circuits with Sampling Regions of 625 ns, 1.25 μs and 2.5 μs

Circuit Name	Power		
	$T = 625$ ns	$T = 1.25$ μs	$T = 2.5$ μs
c432	1.13 mW	1.13 mW	1.12 mW
c499	2.04 mW	2.04 mW	2.05 mW
c880	2.75 mW	2.74 mW	2.75 mW
c1355	5.45 mW	5.45 mW	5.45 mW
c1908	9.23 mW	9.25 mW	9.22 mW
c2670	10.78 mW	10.79 mW	10.80 mW
c3540	14.57 mW	14.62 mW	14.64 mW
c5315	23.14 mW	23.10 mW	23.10 mW
c6288	70.38 mW	70.36 mW	70.32 mW
c7552	37.57 mW	37.50 mW	37.50 mW

converges to the desired power value, *whatever* the power distribution. This is true because of (5) and the *strong law of large numbers* (see [11, p. 26]). Furthermore, we are only concerned with deviations from normality for small values of T. For large T, and since P_T tends to a constant as $T \to \infty$, the variance of P_T goes to 0, and its distribution must become bell-shaped, approaching a normal distribution.

Let P_T have any distribution. The *central limit theorem* [9] implies that if N samples of P_T are averaged, then the distribution of the averages approaches normal as $N \to \infty$. Convergence criteria exist that are similar to those derived for normally distributed power. For a desired percentage error ϵ in the power estimate and for a given confidence level $(1 - \alpha)$, we must simulate the circuit until

$$\frac{z_{\alpha/2}\sigma_T}{\eta_T\sqrt{N}} < \epsilon$$

σ_T, the standard deviation of P_T, is not known. When σ_T is replaced with the sample standard deviation s_T, error is introduced and some assumption about the distribution of P_T is necessary. We will assume that when $\frac{t_{\alpha/2}s_T}{\eta_T\sqrt{N}} < \epsilon$, then $\frac{z_{\alpha/2}\sigma_T}{\eta_T\sqrt{N}} < \epsilon$. This is equivalent to assuming that P_T is normal; however, it allows us to consider the consequences of nonnormal distributions. If s_T is large, then many samples must be taken before convergence, and with a large enough number of samples, $s_T \approx \sigma_T$. If s_T is small enough to allow convergence for small N, then it is likely that σ_T is also small. If the distribution has a large concentration of values around one power and a small number of values at a much higher (or lower) power, then errors may result because s_T is likely to be much smaller than σ_T. The rest of this section considers various nonnormal distributions, including a *bimodal* distribution that illustrates the error in the normality assumption. The final part of this section suggests an approach to identify the types of circuits that would cause an error.

Circuits do exist that have nonnormal power for small T. An example would be a circuit with an enable signal whose value strongly affects the power drawn by the whole circuit. When the enable signal is low, the circuit would have one power distribution, and when it is high it would have a different

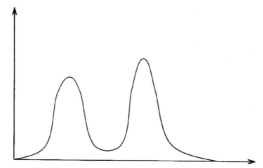

Fig. 4. A bimodal distribution.

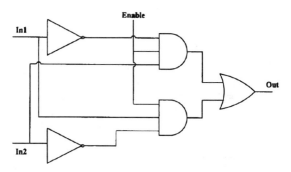

Fig. 5. A circuit with enable.

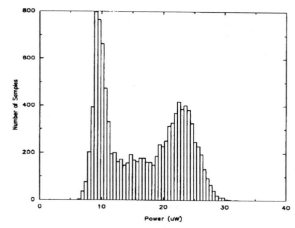

Fig. 6. Power distribution for the XOR circuit.

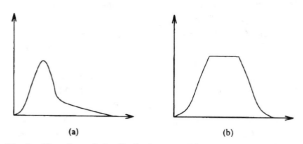

Fig. 7. Distortions of the distributions. (a) Tailed normal. (b) Chopped normal.

distribution. If these two distributions had different means, then over a small T interval the overall distribution would not be normal, but would be a so-called *bimodal* distribution, as shown in Fig. 4. When each of the two distributions is normal, we will refer to the overall distribution as a *double normal*.

As a concrete example, consider the simple XOR circuit in Fig. 5. The enable signal allows the output stage to switch, drawing much more power than otherwise. If the transition density (see equation (4)) at the enable line is low compared to the other two inputs, then, over a short time interval, only one of the two modes of operation would be observed. Using transition densities of 2e7 on the inputs and 2e5 on the enable line, a sampling interval of $T = 2.5\mu sec$, and with 11 000 samples, we get the histogram shown in Fig. 6.

This is one of many ways in which the distribution can deviate from normality. We have also considered two other ways in which the distribution can be distorted. We will refer to distributions with elongated tails, as shown in Fig. 7(a), as *tailed normal* distributions. Those with chopped tops, as in Fig. 7(b), will be called *chopped normal* distributions.

We have examined the performance of our stopping criteria for each of the above varieties of distorted normals. The nonnormal power values were artificially obtained from customized random number generators. The parameters for the stopping criterion were set to 5% accuracy ($\epsilon = 0.05$) with 99% confidence ($\alpha = 0.01$).

Since the distributions were not normal, one would expect the resulting accuracy to be somewhat worse than 5%. In all but a few cases, better than 10% accuracy was achieved with 99% confidence. The only examples that showed worse

than 10% accuracy were distributions with very long tails, and double normal distributions with widely separated means. Even the distributions with very long tails had better than 15% accuracy with 99% confidence. The double normal distributions, however, had very large errors if the first few samples were all centered around one hump of the distribution. When this occurred, the stopping criteria erroneously terminated the simulation. This is exactly as anticipated.

In the cases of the drastic double normal distributions, we feel that they can be treated as follows. When a node has a high fanout, making it a potential cause of the double normal, then the length of each sampling region should be changed so that that node transitions several times in a sampling region. This will prevent the problems associated with an enable signal and should prevent problems with any double normal distributions.

Having said all this, and before leaving this section, it is important to reiterate that the normality assumption holds very well for all the benchmark circuits that we have considered, as discussed in Section III and shown in Fig. 2.

V. EXPERIMENTAL RESULTS

The Monte Carlo methods presented in this paper were implemented based on a simple variable delay logic simulator. This program will be referred to as McPOWER. The test circuits to be used in this section are the benchmarks presented at ISCAS in 1985 [13]. These circuits are combinational logic

TABLE II
THE ISCAS-85 BENCHMARK CIRCUITS

Circuit	#inputs	#outputs	#gates
c432	36	7	160
c499	41	32	202
c880	60	26	383
c1355	41	32	546
c1908	33	25	880
c2670	233	140	1193
c3540	50	22	1669
c5315	178	123	2307
c6288	32	32	2406
c7552	207	108	3512

circuits and Table II presents the number of inputs, outputs, and gates in each.

We will compare the performance of McPOWER to that of probabilistic methods and substantiate the claims of Section II-B that McPOWER has better accuracy and competitive simulation times. DENSIM [5], [8] is an efficient probabilistic simulation program that gives the *average switching frequency* (called *transition density* in [5], [8]) at every circuit node. These density values can be used to give an estimate of the total power dissipation. DENSIM does not take into account the correlation between internal circuit nodes. While it is known that this causes inaccuracy in the density values, it is prohibitively expensive to take all correlation into account for large circuits.

Table III compares the performance of DENSIM, when used to estimate total power, to that of McPOWER. In both programs, every primary input had a signal probability of 0.5 and a transition density of 2e7) transitions per second (corresponding to an average frequency of 10 MHz). For McPOWER, a maximum error of 5% with 99% confidence was specified. As mentioned in Section III, McPOWER performs one long simulation that is broken into setup and sampling regions. The delays of the circuit determine the length of each setup region; however, the length of a sampling region is specified by the user. For Table III, the sampling region was set to 2.5 μs, which allows an average of 50 transitions per sampling interval on each input. The column labeled LOGSIM gives our best estimates of the power dissipation of these circuits, obtained from one very long logic simulation for each circuit. As seen from the table, McPOWER is consistently and highly accurate, while DENSIM has significant errors for some circuits. Although DENSIM is frequently faster, McPOWER's reliable accuracy makes it a more attractive approach for power estimation.

The typical, rapid convergence of McPOWER is illustrated in Fig. 8. The figure shows the power from three different iterations converging to the average power for c6288, one of the most complex ISCAS circuits. A similar plot is shown for c5315 in Fig. 9.

Care must be taken in drawing conclusions from a single run of McPOWER. Since it uses random input vectors, the speed of convergence and the error in the power estimate depend on the initialization of the random number generator. This is

TABLE III
POWER AND TIME RESULTS FOR THE ISCAS-85 CIRCUITS. TIME IS IN CPU SECONDS ON A SUN SPARCSTATION1. MCPOWER IS BASED ON 5% ERROR, 99% CONFIDENCE AND, 2.5μs SAMPLING REGION

Circuit Name	Power			CPU Time	
	DENSIM	LOGSIM	McPOWER	DENSIM	McPower
c432	0.974 mW	1.165 mW	1.17 mW	0.7 s	2.5 s (3.6X)
c499	1.977 mW	2.048 mW	2.13 mW	0.8 s	2.2 s (2.8X)
c880	2.086 mW	2.829 mW	2.81 mW	1.4 s	3.8 s (2.7X)
c1355	3.695 mW	5.735 mW	5.65 mW	1.9 s	3.6 s (1.9X)
c1908	5.154 mW	9.734 mW	9.77 mW	3.1 s	5.6 s (1.8X)
c2670	7.319 mW	11.438 mW	11.24 mW	4.5 s	7.1 s (1.6X)
c3540	9.235 mW	15.328 mW	15.25 mW	5.8 s	12.2 s (2.1X)
c5315	15.471 mW	24.102 mW	23.66 mW	8.5 s	21.9 s (2.6X)
c6288	31.941 mW	78.883 mW	75.53 mW	7.5 s	40.4 s (5.4X)
c7552	23.156 mW	40.006 mW	38.78 mW	12.4 s	24.7 s (2.0X)

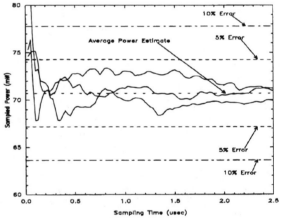

Fig. 8.   McPOWER convergence results for c6288.

illustrated in Table IV, which shows the statistics obtained from 1000 McPOWER runs. The minimum, maximum, and average number of iterations required per run for 5% accuracy with 99% confidence are given. Notice that the average number of iterations required to converge does *not* increase with the circuit size. This confirms the *dimension independence* property of this approach which, as pointed out in the introduction, is a common feature of Monte Carlo methods. Also shown in the table are the percentage number of runs for which the error was greater than 5%, which, as expected, is less than 1% in all cases.

Even smaller execution times are possible if the desired accuracy and confidence levels are relaxed. Table V compares DENSIM to a single run of McPOWER with 95% confidence, 20% accuracy, and a sampling region of 625 ns. As in

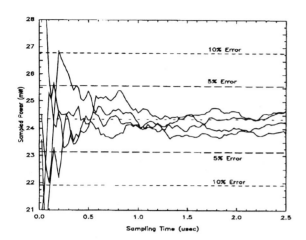

Fig. 9. McPOWER convergence results for c5315.

TABLE IV
STATISTICS FROM 1000 McPOWER RUNS

Name	Min	Max	Avg	% > 5% Err	Avg CPU Time
c432	3	15	8.0	0.9%	3.2 s
c499	3	7	3.9	0.0%	2.9 s
c880	3	14	6.7	0.4%	3.9 s
c1355	3	7	4.0	0.0%	4.8 s
c1908	3	11	5.6	0.3%	10.3 s
c2670	3	10	5.2	0.1%	12.4 s
c3540	3	13	6.0	0.0%	18.3 s
c5315	3	7	4.1	0.0%	22.4 s
c6288	3	6	3.8	0.0%	50.4 s
c7552	3	10	5.5	0.0%	44.4 s

Table III, each input has a signal probability of 0.5 and a transition density of 2e7 transitions per second. With these parameters, McPOWER shows competitive speed and still exhibits superior accuracy.

VI. CONCLUSIONS

In this paper, we have considered the problem of power estimation of a VLSI circuit. We assumed that standby power was negligible; although, standby power can be treated in a very similar manner. Furthermore, we only considered the charging/discharging current since this is the most significant component [7]. Other sources of transient current can occur and are handled identically to the charging/discharging current. They were not considered to avoid complication without additional insight.

We have presented a Monte Carlo based power estimation method. Randomly generated input waveforms are applied to the circuit using a logic/timing simulator and the cumulative value of total power is monitored. The simulation is stopped when sufficient accuracy is obtained with specified confidence. The statistical stopping criterion was discussed, along with experimental results from our prototype implementation McPOWER.

TABLE V
POWER AND TIME RESULTS FOR THE ISCAS-85 CIRCUITS. TIME IS IN
CPU SECONDS ON A SUN SPARCSTATION1. MCPOWER IS BASED
ON 20% ERROR, 95% CONFIDENCE. AND, 625 NS SAMPLING REGION

Circuit Name	Power			CPU Time	
	DENSIM	LOGSIM	McPOWER	DENSIM	McPower
c432	0.974 mW	1.165 mW	1.10 mW	0.7 s	0.7 s (1.0X)
c499	1.977 mW	2.048 mW	2.04 mW	0.8 s	0.9 s (1.1X)
c880	2.086 mW	2.829 mW	2.91 mW	1.4 s	1.0 s (0.7X)
c1355	3.695 mW	5.735 mW	5.41 mW	1.9 s	1.5 s (0.8X)
c1908	5.154 mW	9.734 mW	9.27 mW	3.1 s	2.2 s (0.7X)
c2670	7.319 mW	11.438 mW	10.83 mW	4.5 s	2.7 s (0.6X)
c3540	9.235 mW	15.328 mW	14.28 mW	5.8 s	4.8 s (0.8X)
c5315	15.471 mW	24.102 mW	22.65 mW	8.5 s	6.4 s (0.8X)
c6288	31.941 mW	78.883 mW	71.48 mW	7.5 s	13.2 s (1.8X)
c7552	23.156 mW	40.006 mW	37.88 mW	12.4 s	9.2 s (0.7X)

We have shown that Monte Carlo methods are, in general, better than probabilistic methods for the estimation of power since they achieve superior accuracy with comparable speeds. They are also easier to implement and can be added to existing timing or logic simulation tools. Furthermore, the accuracy can be specified up-front with any desired confidence. Although the type of circuit may affect the amount of power drawn or the number of samples needed to converge, it will not affect the accuracy or reliability of these methods.

Feedback circuits present a severe problem for probabilistic methods. Monte Carlo methods are based on simple timing or logic simulation techniques and, therefore, experience very few difficulties with feedback circuits. The only unresolv ed problem is to determine the length of the setup region, but we feel that good heuristics can be developed for this. Future research will focus on developing such heuristics, thus generalizing the Monte Carlo technique to handle any logic circuit.

Although we have clearly demonstrated the superiority of Monte Carlo methods for power estimation, it is not clear that they will be better than probabilistic methods for other applications, such as estimating the power supply current waveforms. Future research will be aimed at exploring this and other applications of the Monte Carlo approach.

REFERENCES

[1] S. M. Kang, "Accurate simulation of power dissipation in VLSI circuits," *IEEE J. Solid-State Circuits*, vol. SC-21, pp. 889–891, Oct. 1986.
[2] G. Y. Yacoub and W. H. Ku, "An accurate simulation technique for short-circuit power dissipation based on current component isolation," *IEEE Int. Symp. on Circuits and Systems*, pp. 1157–1161, 1989.
[3] A.-C. Deng, Y-C. Shiau, and K-H. Loh, "Time domain current waveform simulation of CMOS circuits," *IEEE Int. Conf. on Computer-Aided Design*, Santa Clara, CA, pp. 208–211, Nov. 7–10, 1988.

[4] M. A. Cirit, "Estimating dynamic power consumption of CMOS circuits," *IEEE Int. Conf. on Computer-Aided Design*, pp. 534–537, Nov. 9–12, 1987.

[5] F. Najm, "Transition density, a stochastic measure of activity in digital circuits," *28th ACM/IEEE Design Automation Conf.*, San Francisco, CA, pp. 644–649, June 17–21, 1991.

[6] C. M. Huizer, "Power dissipation analysis of CMOS VLSI circuits by means of switch-level simulation," *IEEE European Solid State Circuits Conf.*, pp. 61–64, Grenoble, France, 1990.

[7] H. J. M. Veendrick, "Short-circuit dissipation of static CMOS circuitry and its impact on the design of buffer circuits," *IEEE J. Solid-State Circuits*, vol. SC-19, pp. 468–473, Aug. 1984.

[8] F. Najm, "Transition density, a new measure of activity in digital circuits," *IEEE Trans. Computer-Aided Design*, vol. 12, pp. 310–323, Feb. 1993.

[9] I. R. Miller, J. E. Freund, and R. Johnson, *Probability and Statistics for Engineers*. Englewood Cliffs, NJ: Prentice Hall, 1990.

[10] A. Hald, *Statistical Theory with Engineering Applications*. New York: Wiley, 1952.

[11] S. M. Ross, *Stochastic Processes*. New York: Wiley, 1983.

[12] A. Papoulis, *Probability, Random Variables, and Stochastic Processes*, 2nd edition. New York: McGraw-Hill, 1984.

[13] F. Brglez and H. Fujiwara, "A neutral netlist of 10 combinational benchmark circuits and a target translator in Fortran," *IEEE Int. Symp. on Circuits and Systems*, pp. 695–698, June 1985.

500

Stratified Random Sampling for Power Estimation*

Chih-Shun Ding   Cheng-Ta Hsieh   Qing Wu   Massoud Pedram

Department of Electrical Engineering - Systems
University of Southern California
Los Angeles, CA 90089

Abstract

In this paper, we present new statistical sampling techniques for performing power estimation at the circuit level. These techniques first transform the power estimation problem to a survey sampling problem, then apply stratified random sampling to improve the efficiency of sampling. The stratification is based on a low-cost predictor, such as zero delay power estimates. We also propose a two-stage stratified sampling technique to handle very long initial sequences. Experimental results show that the efficiency of stratified random sampling and two-stage stratified sampling techniques are 3-10X higher than that of simple random sampling and the Markov-based Monte Carlo simulation techniques.

1 Introduction

With the continuing reduction in the minimum feature size, chip density and operating frequency of today's ICs are increasing. As a result, power dissipation has become an important concern in IC design. Higher power dissipation in ICs increases the packaging cost and degrades the circuit reliability. Another driver for low power is the class of portable electronic devices ranging from laptop computers to personal communicators. Here, the objective is to minimize the power consumption so as to extend the battery life. To minimize power, one needs to estimate it first. As a result, there is an increasing need for accurate and efficient power estimation tools.

Existing power estimation techniques at the gate level and the circuit level can be classified into two classes: *static* and *dynamic*. Static techniques [1, 2, 3, 4] rely on statistical information (such as the mean activity of the input signals and their correlations) about the input stream to estimate the internal switching activity of the circuit. While these are very efficient, their main limitation is that they cannot accurately capture factors such as slew rates, glitch generation and propagation, dc fighting etc. Dynamic techniques [5, 6, 7, 8] explicitly simulate the circuit under a "typical" input vector stream. They can be applied at both the gate and the circuit levels. Their main shortcoming is however that they are very slow. Moreover, their results are highly dependent on the simulated sequence. To produce a meaningful power estimate, the required number of simulated vectors is usually high, which further exacerbates the run time problem.

To address this problem, a Monte Carlo simulation technique was proposed in [9]. This technique uses an input model based on a Markov process to generate the input

stream for simulation. The simulation is performed in an iterative fashion. In each iteration, a vector sequence of fixed length (called sample) is simulated. The simulation results are monitored to calculate the mean value and variance of the samples. The iteration terminates when some stopping criterion is met. This approach suffers from four major shortcomings. First, since the simulation vectors are generated internally based on statistics of the input stream, a large number of vectors needs to be examined to extract reliable statistics. Second, when the vectors are regenerated for simulation, the spatial correlations among various inputs cannot be adequately captured, which may lead to inaccuracy in the power estimates. Third, the required number of samples, which directly impacts the simulation run time, is approximately proportional to the ratio between the sample variance and square of sample mean value. For certain input sequences, this ratio becomes large, thus significantly increasing the simulation run time. Finally there is a concern about the normality assumption on the sample distribution. Since the stopping criterion is derived based on the normality assumption, if the sample distribution significantly deviate from normal distribution, the simulation may terminate prematurely. Difficult distributions that cause premature termination include bi-modal, multi-modal and distribution with long or asymmetric tails.

In this paper we address the power estimation problem from a survey sampling perspective. We assume a sequence of vectors are provided to estimate the power consumption of a given combinational circuit with certain statistical constraints, such as error and confidence levels. We transform the power estimation problem to a survey sampling problem by dividing the vector sequence into small units, e. g. consecutive vectors, to constitute the population for the survey. Power consumption is the characteristic under study. The average power consumption is estimated by simulating the circuit by a number of samples drawn from the population, a procedure referred to as sampling, using a simulator such as PowerMill [8] or VERILOG-XL. Our objective is to design a sampling procedure that will significantly reduce the number of simulated vectors while satisfying the given error and confidence levels.

Stratified sampling techniques have been widely used for surveys because of their efficiency. The purpose of stratification is to partition the population into disjoint subpopulations so that the power consumption characteristic within each subpopulation is more homogeneous than in the original population. The partitioning is based on a low-cost predictor that needs to be efficiently calculated for each member in the population. In this paper, we use the zero delay power estimate as the predictor. Compared

---

*This research was supported in part by DARPA under contract no. F336125-95-C1627, SRC under contract no. 94-DJ-559, and NSF under contract number no. MIP-9457392.

Reprinted from *IEEE/ACM International Conference on Computer-Aided Design*, pp. 576-582, 1996.

to the technique proposed in [9], the proposed technique offers the following advantages: 1) It performs sampling directly on the population and the estimation results are unbiased, 2) It is more efficient, and 3) The sample distributions are more likely to be a normal distribution. When the population size is large, we propose a two-stage stratified sampling procedure to reduce the overhead of predictor calculation and stratification.

The organization of the paper is as follows. In Section 2, we describe the basic principles of survey sampling and its connection with power estimation. In Section 3 and 4, we present a stratified sampling technique for power estimation and discuss its design issues. A two-stage stratified sampling technique is presented in Section 5. Experimental results are presented in Section 6 followed by concluding remarks which are given in Section 7.

2 Background

We first give some useful notation and definitions:

U the population
N number of units in the population
u_i the ith unit in the population
y_i value of the characteristic under study for u_i
$\overline{Y}$ mean value of y_i in the population
n sample size

We are given a collection (called *population*), $U = \{u_1, u_2, \ldots, u_N\}$ of objects (called *elements* or *units*), of which some property (called *characteristic*) y_i is defined for each u_i. The survey sampling problem deals with ways of selecting samples, i. e. , sequences (or collections) of units from the population, to estimate the mean value of the characteristic under study in the population, denoted by $\overline{Y}$, where

$$\overline{Y} = \frac{1}{N} \sum_{i=1}^{N} y_i.$$

For the sake of the simplicity, we will refer to $\overline{Y}$ as *population mean*. The variance of the characteristic under study in the population is simply referred to as *population variance* and is denoted by $V(y)$. The *relative variance* is defined as the ratio between the variance and square of the mean value of a statistic. Number of units included in a sample is referred to as the *sample size* and is denoted by n. An *estimator* (of population mean) is defined as a function of sample characteristic values that estimates the population mean. An estimator is a random variate and may take different values from sample to sample. The difference between the estimator t and $\overline{Y}$, is called *error*. t is said to be an *unbiased* estimator for $\overline{Y}$ if $E(t) = \overline{Y}$. otherwise biased. If an estimator is biased, the bias is given by $B(t) = E(t - \overline{Y})$. The *estimator variance* of t is defined as $V(t) = E[(t - E(t))^2]$. The *mean-square error* (MSE) is defined as $MSE(t) = E[(t - \overline{Y})^2]$. The relation among $MSE(t)$, $V(t)$, and $B(t)$ is: $MSE(t) = V(t) + B^2(t)$.

Given two estimators t_1 and t_2, t_1 is said to be more *efficient* than t_2 if the mean-square error of t_1 is less than that of t_2. The *relative efficiency* of t_1 as compared to t_2 is defined as the reciprocal of the ratio of their estimator variance when the same number of samples are taken in both estimators. As it will be explained in more detail later in this section, the more efficient an estimator is, the fewer the number of samples required to achieve the same level of error. Furthermore, the ratio between the number of required samples is roughly equal to the reciprocal of their relative efficiency.

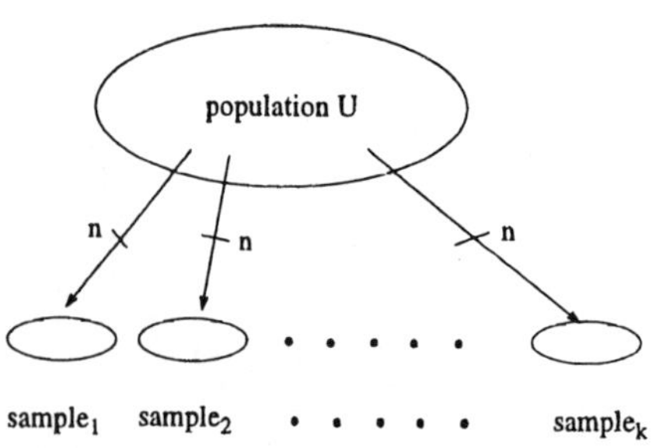

Figure 1: Simple random sampling.

2.1 Power Estimation Problem

We are given a vector trace $(v_1, v_2, \ldots, v_M)$ to estimate the average power consumption of a combinational circuit using an accurate simulator. This problem can be easily transformed into a survey sampling problem by grouping a fixed number l of consecutive vectors with overlap occurring only at the group boundaries, and each group becomes an element (unit) of the population in the survey. For example, if $l = 2$ and $M = N + 1$, the grouping will be:

$$(v_1, v_2), (v_2, v_3), \ldots, (v_N, v_{N+1})$$

The power consumption estimated for the vector sequence in each group becomes the characteristic under study. The mean value of the characteristic in the population gives the average power consumption. By doing so, we assume the state of the circuit (when simulating each group (v_i, v_{i+1})) can be completely described by v_i and v_{i+1}. In practice, state of the circuit (on internal capacitances, for instance) may depend on the v_{i-1}, v_{i-2}, etc. In those cases, we can increase l and/or use a warm-up sequence [9] to reduce the error induced in the transformation procedure.

2.2 Simple Random Sampling

In this subsection we will use *simple random sampling* to illustrate some basic concepts in designing a survey sampling procedure. Simple random sampling is a method of selecting n units out of a population by giving equal probability to all units. If a unit is selected and noted, and then returned to the original population before the next drawing is made, and this procedure is repeated n times, then the selected units $u_1 \ldots u_n$ constitute a simple random sample of n units. This sampling procedure is called simple random sampling with replacement(wr).

Given a sample of n units, $u_1, u_2, \ldots, u_n$ with characteristic values $y_1, y_2, \ldots, y_n$, $\overline{Y}$ is estimated by

$$\overline{y}_{sr} = \sum_{i}^{n} y_i/n,$$

where the subscript sr denotes simple random sampling and $\overline{y}_{sr}$ is referred to as the *sample value*.

Theorem 2.1 *In simple random sampling (wr), the sample value $\overline{y}_{sr}$ is an unbiased estimator of $\overline{Y}$ and its sampling variance is given by:*

$$V(\overline{y}_{sr}) = \frac{V(Y)}{n}. \tag{1}$$

Equation (1) shows that the sampling variance is inversely proportional to sample size. In addition, if we select k samples and use the mean of the sample values as the estimator t, the variance is further reduced by k as each sample selection is independent of the other. Thus, the sampling variance of this estimator is inversely proportional to uk, the total number of units drawn from the population. In the following discussion, for the sake of generality, we assume that our estimator t is the mean value of k samples, each with sample size n.

A few noteworthy remarks should to be made here:

1. For large n (usually, $n \geq 30$ is adequate) [10], the distribution of sample values approaches normal distribution. Therefore, if the population variance is known, so is the variance of the estimator, and one can easily derive the error and confidence level of the estimator using normal distribution curves. When the population variance is not known, one may calculate the variance among the k selected samples use t-distribution with degree $(k-1)$ to derive the confidence interval [10]. Thus, given a confidence level, $(1 - \alpha)$, and a relative error level, ϵ, one can simulate the circuit iteratively until the total number of simulated samples, k, satisfies the follow inequality:

$$k > (\frac{t_{\alpha/2} s}{\epsilon \, \eta})^2 \qquad (2)$$

where $t_{\alpha/2}$ is defined so that the area to its right under a t-distribution with degree $(k-1)$ is equal to $\alpha/2$, and η and s are the mean and standard deviation of the simulated samples, respectively. When the population size is infinite, this procedure is also commonly known as the Monte Carlo simulation approach [9].

2. $V(\overline{y}_{sr})$ is proportional to the population variance. Given two populations with the same population mean, in order to achieve the same error level, the number of samples for these populations are inversely proportional to the population variances. In general, we can use the relative variance of the characteristic under study as an indicator for the degree of difficulty in the corresponding survey sampling problem.

In general, the efficiency of simple random sampling is not very high. This can be explained in the context of power estimation as follows. Consider the case where the distribution of the population characteristic (i.e., power consumption) is bi-modal – that is, the characteristic of half of the population is distributed around a right peak, and the characteristic of the other half is distributed around a left peak. Since the selected unit could either come from the right peak or the left peak, the sampling variance is very high.

If, on the other hand, we divide the population into two halves, those units whose characteristic values are around the right peak are put into one subpopulation while the remaining units are put into the other subpopulation, and select the samples in such a way that half of the units in a sample are selected from each of the subpopulations, then the sampling variance will be significantly reduced. In order to divide the population into subpopulations, a predictor is often used. This predictor need not have a linear relationship with the characteristic under study, as it is only used to divide the population into subpopulations and is not directly used to calculate the power estimates. In the following sections, we will describe a more efficient sampling procedure based on this scheme called *stratified sampling*.

3 Stratified Random Sampling

In the stratified random sampling, the population U is partitioned into k disjoint subpopulations, called *strata*.

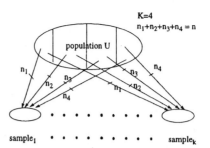

Figure 2: An example of stratified random sampling.

The main objective of stratification is to give a better cross-section of the population so as to gain a higher degree of relative efficiency. The stratification should be done in such a way that strata are homogeneous within themselves w. r. t. the characteristic under study.

There are a number of ways to construct strata. For instance, in the agricultural surveys, the strata is in general constructed based on geographical information. For power estimation, we use the following procedure to construct the strata. The zero delay power estimate is used as the predictor. Let the predictor value of unit u_i be x_i. Population U is first sorted according to the x_i value of each unit. Let the new order be $u_1, u_2, \ldots, u_n$. Then $K - 1$ separators, $p_1, p_2, \ldots, p_{K-1}$, are selected such that

$$x_1 < p_1 < p_2 < \ldots < p_{K-1} < x_N$$

All the units whose x_i values are between two consecutive separators are put into the same stratum, and the strata disjointly cover the whole population. Let the size of each stratum be N_i, then

$$N_1 + N_2 + \ldots + N_K = N$$

Units in a sample are drawn from each stratum independently so that the sample size within the ith stratum become n_i $(i = 1, 2, \ldots, K)$ and

$$n_1 + n_2 + \ldots + n_K = n$$

If the sample is taken randomly from each stratum, the procedure is known as *stratified random sampling*. One way to tell how well the x_i behaves in relation to y_i is from the scatter plot of y_i and x_i. The zero delay power estimation produces poor accuracy on a vector-by-vector basis. However, in the stratified sampling, x_i is only used for forming the strata so that units of close y_i values can be put in the same stratum, the inaccuracy of the zero delay power estimates is acceptable.

3.1 Population Parameter Estimation

In this subsection, we mainly focus on the derivation of an estimators for population mean and study of its properties. We first give some useful notations and definitions:

K number of strata
N_i size of the ith stratum
W_i N_i/N, stratum weight of ith stratum
n_i number of units in a sample falling in ith stratum
y_{ij} characteristic y of jth unit in ith strata in a sample
$\overline{Y_i}$ stratum mean of the ith stratum, $\overline{Y_i} = \sum_j^{N_i} y_{ij}/N_i$
S_i^2 stratum variance of the ith stratum, $S_i^2 = \sum_j^{N_i} (y_{ij} - \overline{Y_i})^2/N_i$
$\overline{y_i}$ $\sum_j^{n_i} y_{ij}/n_i$, sample mean within strata i

The stratification estimator $\overline{y}_{st}$ is formulated as:

$$\overline{y}_{st} = \sum_i^K W_i \overline{y}_i$$

where the subscript st denotes the stratified sampling method[1].

Theorem 3.1 *If units in a sample are drawn independently in different strata, then $\overline{y}_{st}$ is an unbiased estimator of the population mean and its sampling variance is given by*

$$V(\overline{y}_{st}) = \sum_i^K W_i^2 \frac{S_i^2}{n_i} \qquad (3)$$

4 Stratification Scheme Design

From (3), $V(\overline{y}_{st})$ depends on n_i, W_i and S_i. Our objective is to minimize $V(\overline{y}_{st})$ under a given n value. The problem of selecting n_i for ith stratum is referred to as the *sample size allocation* problem. W_i and $V(\overline{y}_i)$ are related and they are determined by strata separators p_i, $i = 1, 2 \ldots, K - 1$, when the strata are formed. The problem of finding the optimal values of the separators p_i is referred to as the *stratum selection* problem. These two problems collectively will be referred to as the *stratification scheme design* problem.

4.1 Sample Size Allocation in Strata

The sampling variance of stratified sampling is not necessarily smaller than that of simple random sampling. However, as long as some guidelines are followed, it is guaranteed that the variance of stratified sampling is no larger than that of simple random sampling. In the following, we describe two important sample size allocation methods that guarantee this:

1. Proportional allocation [12]: In proportional allocation, $n_i = W_i n$, and the sampling variance is given by:

$$V(\overline{y}_{prop}) = \sum_i^K W_i S_i^2 / n$$

This is a very popular scheme since the only information required is W_i's.

2. Minimum variance allocation [13]: It has been shown that allocating $n_i = n \frac{W_i S_i}{\sum_i^K W_i S_i}$ will give the minimum variance among all possible allocation methods; its sampling variance is given by

$$V(\overline{y}_{min}) = (\sum_i^K W_i S_i)^2 / n - \sum_i^K W_i S_i^2 / N$$

$$\approx (\sum_i^K W_i S_i)^2 / n \quad \text{if } n \ll N$$

The difficulty in using the above criterion is that S_i is not known before sampling. One solution is to use the variance of x_i as an approximation of S_i^2. However, this places a condition on the choice x_i since it implies that x_i must be selected judiciously so as to exhibit a variance proportional to that of y_i.

It can be shown that $V(\overline{y}_{min}) \leq V(\overline{y}_{prop}) \leq V(\overline{y}_{sr})$.

---

[1]Proofs are found in [11].

4.2 Stratum Selection Problem

For determining the value K, considering the case where y_i has a uniform distribution and the population is partitioned perfectly into K equal size strata using equally spaced separators p_i's, then $V(\overline{y}_{sr})/V(\overline{y}_{st}) = K^2$. That is, K^2 times improvement over simple random sampling is achieved. In general when we increase K, $V(\overline{y}_{st})$ decreases. However, in practice, after K reaches a certain value, the reduction on $V(\overline{y}_{st})$ becomes less significant and sometimes could even increase slightly. Some statisticians [14] have suggested that an increase in k beyond 6 would seldom be profitable.

After determining the value of K, we need to find the strata separator values p_i's. There are several selection criteria being proposed in the literature. Among them, Danlenius and Gurney [15] suggested that the construction of strata on the basis of equalization of $W_i S_i$ and equal sample size allocation to the strata would lead to optimum stratification. Again, this method is not convenient as it requires the knowledge of S_i. Our method is based a variation of this scheme, which will be described in more detail in the experimental result section.

4.3 Normality of Sample Distribution

As we have mentioned earlier, the objective of stratification is to make the strata homogeneous within themselves. Thus, it is easier to make the sample distribution follow a normal distribution than in a simple random sampling. For example, in *Bi-modal or multi-modal population distribution*, if the population modes are widely separated and if we assume that the predictor behave relatively well, stratification can break up the modes by using a number of strata to cover each mode. Even when one of the strata happens to remain bi-modal, the modes are less likely to be as widely separated as in the original population. In any case, the bimodality behavior only effects the sample units drawn from this stratum and not the whole. Similar arguments hold for population distributions with long or asymmetric tails.

4.4 Cost Comparison

The cost of simple random sampling is due to the calculation of y_i for all units in the samples, that is, nk times the average simulation time of PowerMill for one vector pair. If we assume the latter to be C_{pwm}, then the cost of simple random sampling can be written as:

$$C_{sr} = a_{sr} + nkC_{pwm}$$

where a_{sr} is the constant overhead in simple random sampling.

For stratified random sampling, the cost comprises of two parts: 1) the calculation of x_i (zero delay power estimate) for all units in the population and stratification (such as sorting and strata selection), and 2) the simulation time of PowerMill for all units in the samples. If we assume the amortized cost of x_i calculation is C_1 and the number of samples is selected so as to achieve the same variance as that of simple random sampling, then the cost of stratified random sampling can be written as:

$$C_{st} = a_{st} + NC_1 + \frac{1}{\eta} nkC_{pwm} \qquad (4)$$

where η is the relative efficiency of stratified random sampling vs. simple random sampling.

The constant overheads of both sampling methods are small. Therefore the stratified sampling is more cost-effective if

$$\eta > \frac{1}{1 - \frac{NC_1}{nkC_{pwm}}}$$

In practice, we find that $\frac{C_1}{C_{pwm}} \cong \frac{1}{4000}$. If we assume $N = 4000$ and $nk = 200$, we conclude that when $\eta > 1.005$, stratified random sampling is more cost-effective than simple random sampling.

When the population size is very large, the overhead of calculating the predictor may become significant. To reduce this overhead, a two-stage stratified sampling is proposed in the next section.

5 Two-Stage Stratified Sampling

In the first stage, a sub-population of size $M \ll N$ is first randomly sampled from the original population U. In the second stage, stratified sampling is applied to this sub-population to select a sample of size n. Since stratified sampling is applied to the second stage only, the overhead of calculating x is restricted to the sub-population size. If k samples need to be selected, k subpopulations are randomly selected, and a sample of size n is drawn from each of the selected subpopulation.

The selection of a sample in two-stage sampling consist of two steps. The first step is to select a subpopulation. Once a subpopulation is selected, the selected sample from this subpopulation is an unbiased estimator of the subpopulation mean. Therefore, to calculate the variance of a two-stage sampling, one needs to consider the variance introduced in both steps. The following theorem can be applied to derive the variance of a two-stage estimator.

Theorem 5.1 *[16, p.109] The variance of a random variate X is the sum of the variance of the conditional expected value and the expected value of the conditional variance. Symbolically,*

$$V(X) = V_1(E_2(X|Z)) + E_1(V_2(X|Z)) \qquad (5)$$

where E_1 stands for expectation of X over the space Z, E_2 stands for the conditional expectation of X for a given Z, V_1 stands for the variance of X over the space Z, and V_2 stands for the conditional variance of X for a given Z.

Using the above theorem and conditioning on the subpopulations, the sampling variance of a two-stage sample can be written as the sum of 1) the variance of the subpopulation means between all possible sub-populations (the first term on the right hand side of (5)), and 2) the mean value of the sampling variance within each subpopulation. The former is referred to as the first stage variance, denoted as S_b^2 and the latter as the second stage variance, denoted as S_w^2, where 'b' stands for "between" subpopulations and 'w' stands for "within" subpopulations.

The difficulty in deriving the sampling variance in a two-stage stratified sampling is that there is a dependency between the stratification and the subpopulation. As a result, there is no closed form for it. However, the dependency decreases when M increases. When M is large, the probability distribution within a subpopulation will be very close to the original population. Therefore we can assume the relative efficiency η of stratified sampling over simple random sampling is nearly the same in all subpopulations, and can thus approximate the sampling variance using η and the sampling variance in two-stage simple random sampling, in which simple random sampling is employed at both stages.

The S_b^2 in two-stage simple random sampling is exactly $\frac{1}{M}V(Y)$. If the sampling size is 1, the two-stage simple random sampling is equivalent to a single-stage simple random sampling with sample size 1, in which the sampling variance will be $V(Y)$. Therefore

$$S_w^2 = (1 - \frac{1}{M})V(Y)$$

If the sample size is n and k samples are drawn, the estimator variance can be written as:

$$V(t) = \frac{1}{Mk}V(Y) + \frac{1 - \frac{1}{M}}{nk}V(Y)$$

Therefore the estimator variance of a two-stage stratified sampling can be approximated by:

$$V(t_{2st}) \cong \frac{1}{Mk}V(Y) + \frac{1 - \frac{1}{M}}{nk\eta}V(Y)$$

$$\cong (\frac{1}{Mk} + \frac{1}{nk\eta})V(Y) \quad \text{for large M} \quad (6)$$

where $2st$ denotes the two-stage stratified sampling.

In the next subsection, we describe how to select M values to maximize the efficiency of the two-stage stratified sampling techniques.

5.1 Selection of Subpopulation Size

A simple formulation of the sampling cost of the proposed two-stage stratified sampling technique can be written as :

$$C = a_{2st} + kMC_1 + knC_{pwm}$$

where a_{2st} is the constant overhead cost, C_1 and C_{pwm} is the same as in (4). Compared with (6), one may notice that $V(y_{2st})$ decreases with an increase in k, M, and n, while the cost increases. To maximize the sampling efficiency under a given cost constraint, we need to find the optimal values for M, and n. If we apply the Lagrange multipliers method, the optimal ratio of M and n is:

$$\frac{M}{n} = \sqrt{\frac{\eta C_{pwm}}{C_1}}$$

The ratio of the costs in the first stage (kMC_1) and the second stage (knC_{pwm}) is given by:

$$\frac{kMC_1}{knC_{pwm}} = \sqrt{\frac{\eta C_1}{C_{pwm}}}$$

If we assume $\eta = 2$, then $\frac{M}{n} \cong 90$, $\frac{kMC_1}{knC_{pwm}} \cong \frac{1}{45}$. Therefore the impact of M on the total cost is insignificant when the ratio of $\frac{C_{pwm}}{C_1}$ is large. So is its impact on the estimator variance.

5.2 Normality of the Sample Distribution

When using t-distribution to calculate the error and confidence levels, it is very important that the sample distribution is normal or near-normal. When k samples are taken from k different subpopulations, it is natural to ask if the sample distribution follows a normal distribution. If we look at the limiting case where $M \to \infty$, performing stratification on the subpopulations is very close to performing stratification on the original population. With n large enough, the samples will follow a normal distribution. In practice, we find $M \geq 4000$ is adequate for general distributions. For multi-modal or long-tail distributions, one can perform stratification in the first stage using predictors such as the number of bit changes at circuit inputs, or by increasing M.

Table 1: Run time and power estimation results of PowerMill

ckts	#gate	PowerMill run time (sec.)		avg. power @13.3Mhz (mW)	
		biased seq.	random seq.	biased seq.	random seq.
C432	274	1005	1941	2.07	3.94
C880	480	1930	3834	3.95	7.44
C1355	603	2418	4885	5.68	10.19
C1908	565	3075	4907	6.49	9.76
C2670	867	5103	8783	9.52	16.90
C3540	1464	8076	13536	17.24	29.01
C5315	1950	11522	21737	22.50	42.13
C6288	3385	46544	129156	73.88	240.69
C7552	2550	23386	40428	46.57	79.94

Table 2: Results of 100,000 simulation runs on the random sequence.

ckts	exp. rel. eff.	simulated vec. pairs				improv. of (2) vs. (1)	
		(1) simple		(2) stratified		vect. based	act. based
		max	avg	max	avg		
C432	1.76	960	408	690	271	1.50	1.41
C880	2.06	720	305	450	201	1.51	1.47
C1355	1.57	360	155	270	135	1.15	1.15
C1908	1.32	540	226	450	196	1.15	1.13
C2670	1.67	420	164	300	139	1.18	1.18
C3540	2.14	600	244	450	171	1.43	1.42
C5315	1.57	360	150	270	131	1.15	1.12
C6288	1.49	360	160	300	141	1.13	1.14
C7552	3.53	900	367	450	180	2.04	2.02
avg						1.36	1.34

6 Experimental Results

The proposed techniques have been implemented in C and tested on ISCAS85 benchmarks. We perform two experiments. In the first experiment, we compare the efficiency of the stratified and simple random sampling on a random and a biased sequences. In the second experiment, we compare the efficiency of two-stage stratified sampling to the technique proposed in [9] for infinite-size population (i.e., no initial population was given). The results are presented as follows.

6.1 Experiment I

We performed this experiment on two type of sequences, each of length 4000. The first sequence (random sequence) is generated randomly by assuming 0.5 signal and transition probabilities at circuit inputs. The second sequence (biased sequence) is a non-random one obtained from real applications. The circuits were mapped to a library with NAND, NOR, inverter and XOR gates. Since performing simulation on PowerMill is very time-consuming, we simulate whole sequence once to extract the power consumption on each consecutive vectors. The run times (on a SUN SS-20) and power estimation results are listed in Table 1. The zero delay power estimates are calculated using a bit-parallel algorithm. The average run time of this algorithm on a Sun SS-20 is 5M gate-vector/sec.

The stratification scheme we used is as follows. Our objective is to equalize $W_i S_i$. The sample size allocation is equal-size allocation. The units of the population are first sorted according to their zero delay power estimates and put in 400 buckets. Adjacent buckets are merged iteratively until K strata are formed and $W_i S_i$ are within 125% and 75% of each other. The sample size for both

sampling methods is 30.

The results are summarized in Tables 2, 3 and 4. We first evaluate the 'theoretical' (expected) efficiency improvement based on (1) and (3) when $K = 10$, as listed in the "exp. rel. eff." columns. After performing more experiments we found that the improvement on relative efficiency is not very significant when K is increased beyond 10. Therefore, we used $K = 10$ in the simulation runs. We performed 100,000 simulation runs with 0.99 confidence level and 5% error levels on each circuit. The maximum and average numbers of required vector pairs that satisfy (2) are reported.

We also calculate the projected run time improvement on PowerMill based on the following two criteria: 1) vector-based: use the ratio of the average number of simulated vectors, and 2) activity-based: use the ratio of the average of sum of the activity in each simulated vector during simulation runs. The rationale for the second criterion is that the run time of PowerMill follows the circuit activity very closely, as shown in Table 1, and in stratified sampling the activity of each unit in a sample may be given different weights when calculating the sample value. Therefore the vector-based comparison may not be appropriate. The results are listed in last two columns of Table 2 and 3. The activity-based run time improvement is slightly less than the vector-based one and both of them are smaller than the numbers in the "exp. rel. eff." column. The first result is due to unequal weights in (3). The second result is because of premature simulation termination in simple random sampling.

We observed two different error violation characteristic. One for violating the specified error level. The other one is intentionally set to a higher level to detect the long tails in the sample distribution. We set this level to be 20%. The percentages of error violation are summarized in Table 4. It shows that the high error violation does not exist for this set of circuits when using stratified sampling. This demonstrates that stratified sampling can handle difficult population distributions. As expected, there is no high error violation observed in the random sequence.

The efficiency improvement of stratified sampling in the random sequence is much smaller than that in biased sequence which is explained as follows. Since the transition probability of each circuit input was assumed to be 0.5, from the law of large numbers, the distribution of the number of input bit changes has a high peak around a value equal to half of the circuit input count. Therefore the power distribution is already very homogeneous.

6.2 Experiment II

In this experiment, we compare the efficiency of the proposed two-stage stratified sampling technique to the technique proposed in [9], which is based on a Markov process model. Since this experiment cannot be performed in a reasonable time on PowerMill, we use a real-delay gate-level power estimator instead. The signal and transition probabilities at the circuit inputs was assumed to be 0.5 and 0.25, respectively. The subpopulation size and the sample size for the two-stage sampling approach was set to 4,000 and 30, respectively. In the first stage of the two-stage sampling approach, the random number generator is run twice for each circuit input to generate the vector sequence for each unit in the selected subpopulation. The first run determines the initial input value while the second run determines if the input changes. In the Markov-based approach, only the initial value of the first unit in a sample

506

Table 3: Results of 100,000 simulation runs on the biased sequence.

ckts	exp. rel. eff.	simulated vec. pairs				improv. of (2) vs. (1)	
		(1) simple		(2) stratified		vect. bas.	act. bas.
		max	avg	max	avg		
C432	4.9	4530	2765	1770	1169	2.3	2.1
C880	22.2	4650	2955	540	221	13.2	9.6
C1355	16.2	2580	1293	450	177	7.3	6.5
C1908	7.6	2370	1241	570	252	4.9	4.6
C2670	16.0	3900	2328	570	234	10.0	7.8
C3540	23.1	3450	1989	450	182	10.9	9.2
C5315	39.9	3690	2094	330	152	13.8	12.0
C6288	23.8	8730	6141	780	333	18.4	9.1
C7552	7.7	4140	2418	1230	755	3.2	3.0
avg						9.3	7.1

Table 4: Error violation percentages.

ckts	error violation (> x%) percent. (%)					
	biased seq.				random seq.	
	simple random		stratified random		simple random	stratified random
	> 5	> 20	> 5	> 20	> 5	> 5
C432	1.46	0.07	0.95	0	1.55	0.75
C880	1.35	0.09	0.39	0	1.05	0.22
C1355	1.96	0.02	0.12	0	0.02	0.00
C1908	2.02	0.01	0.64	0	0.44	0.23
C2670	1.48	0.06	0.50	0	0.03	0.01
C3540	1.70	0.05	0.14	0	0.58	0.07
C5315	2.70	0.07	0.06	0	0.06	0.01
C6288	1.03	0.10	0.86	0	0.03	0.01
C7552	2.16	0.06	0.85	0	1.89	0.18

Table 5: Results of Monte Carlo simulation vs. two-stage stratified sampling.

ctks	simulated vec. pairs				error violation		improv.
	(1) Markov based		(2) two-stage stratified		(1)	(2)	
	max	avg	max	avg	5%	5%	vec. bas.
C432	1890	1083	1020	467	2.3	0.9	2.3
C880	1230	726	540	193	1.8	0.3	3.8
C1355	600	246	330	171	0.3	0.3	1.4
C1908	1510	729	660	307	2.6	0.7	2.4
C2670	510	254	210	126	0.7	0.0	2.0
C3540	1230	694	480	257	2.1	0.4	2.7
C5315	540	256	270	152	1.1	0.0	1.7
C6288	1230	442	420	195	2.0	0.4	2.3
C7552	1320	558	270	144	2.3	0.0	3.9
avg							2.5

is randomly generated, the initial values of the remaining units are assumed to follow the previous unit, as described by the Markov process.

We performed 1,000 simulation runs with 0.99 confidence and 5% error levels. Results are summarized in Table 5. There is no high error violation observed in either technique. The proposed two-stage stratified sampling technique is more than twice as efficient as the Markov-based technique. The improvement under the activity-based criterion is very close to that under the vector-based criterion and is omitted in the table. Since the error violation percentage is below 1% for the two-stage stratified sampling, the normality assumption made in Subsection 5.2 is acceptable.

7 Conclusion

We have presented new statistical sampling techniques for circuit-level power estimation. Compared with existing statistical power estimation techniques, not only the efficiency of sampling is improved, but also difficult population distributions can be handled more effectively. Although we have used zero delay power estimates for stratification, other predictors can be used, depending on the trade-off between computation overhead and efficiency improvement. The proposed two-stage stratified sampling technique can be easily extended to multi-stage sampling. However, the predictors used at different stages should be as uncorrelated as possible.

References

[1] F. N. Najm, R. Burch, P. Yang, and I. N. Hajj. Probabilistic simulation for reliability analysis of cmos circuits. IEEE Transactions on Computer-Aided Design of Integrated Circuits and Systems, 9(4):439–450, April 1990.

[2] A. A. Ghosh, S. Devadas, K. Keutzer, and J. White. Estimation of average switching activity in combinational and sequential circuits. In Proceedings of the 29th Design Automation Conference, pages 253–259, June 1992.

[3] C-Y. Tsui, M. Pedram, and A. M. Despain. Efficient estimation of dynamic power dissipation under a real delay model. In Proceedings of the IEEE International Conference on Computer Aided Design, pages 224–228, November 1993.

[4] R. Marculescu, D. Marculescu, and M. Pedram. Logic level power estimation considering spatiotemporal correlations. In Proceedings of the IEEE International Conference on Computer Aided Design, pages 294–299, November 1994.

[5] S. M. Kang. Accurate simulation of power dissipation in VLSI circuits. IEEE Journal of Solid State Circuits, 21(5):889–891, Oct. 1986.

[6] G. Y. Yacoub and W. H. Ku. An accurate simulation technique for short-circuit power dissipation. In Proceedings of the International Symposium on Circuits and Systems, pages 1157–1161, 1989.

[7] C. M. Huizer. Power dissipation analysis of CMOS VLSI circuits by means of switch-level simulation. In IEEE European Solid State Circuits Conf., pages 61–64, 1990.

[8] C. Deng. Power analysis for CMOS/BiCMOS circuits. In Proceedings of 1994 International Workshop on Low Power Design, pages 3–8, April 1994.

[9] R. Burch, F. N. Najm, P. Yang, and T. Trick. A Monte Carlo approach for power estimation. IEEE Transactions on VLSI Systems, 1(1):63–71, March 1993.

[10] I. R. Miller, J. E.Freund, and R. Johnson. Probability and statistics for engineers. Prentice Hall, 1990.

[11] C-S. Ding, C-T. Hsieh, Q. Wu, and M. Pedram. Statistical techniques for power evaluation. In CENG Technical Report No. 96-11, University of Southern California, 1996.

[12] A. L. Bowley. Measurement of precision attained in sampling. Bull. Inter. Statist. Inst., 22:1–26, 1926.

[13] J. Neyman. On the two different aspects of the representative method. J. R. Statist. Soc., 97:558–606, 1934.

[14] V. K. Sethi. A note on optimum stratification for estimating the population means. Aust. J. Statist., 5:20–33, 1963.

[15] T. Dalenius and M. Gurney. The problem of optimum stratification ii. Skand. Akt., 34:133–148, 1951.

[16] B. W. Lindgren. Statistical Theory. Chapman & Hall, 1993.

A Survey of High-Level Power Estimation Techniques

PAUL LANDMAN

CORPORATE R&D

TEXAS INSTRUMENTS INCORPORATED

DALLAS, TEXAS 75243 USA

Abstract—The growing demand for portable electronic devices has led to an increased emphasis on power consumption within the semiconductor industry. As a result, designers are now encouraged to consider the impact of their decisions not only on speed and area, but also on power throughout the entire design process. In order to evaluate how well a particular design variant meets power constraints, engineers often rely on CAD tools for power estimation. While tools have long existed for analyzing power consumption at the lower levels of abstraction (e.g., SPICE and PowerMill) only recently have efforts been directed toward developing a high-level power estimation capability. This paper surveys the state of the art in high-level power estimation, addressing techniques that operate at the architecture, behavior, instruction, and system levels of abstraction.

1. INTRODUCTION

In the early part of this decade, it became clear that power consumption was becoming a problem. In the high performance arena, microprocessors began to appear that consumed tens of watts, and the trend was toward even higher power consumption. This placed stringent demands on packaging and cooling systems, as well as being a major cost and reliability issue. At the same time, the portable consumer-electronics market entered a period of rapid growth. For these battery-operated products, there was similar pressure to reduce power consumption and extend battery life. These factors rapidly led to the emergence of low-power design as a key technology for the 1990s.

As designers began to place increasing emphasis on power as a figure of merit, it became clear that while there were tools to assist in estimating performance and area, relatively few addressed power. In the last several years, the picture has improved dramatically. Circuit- and gate-level power analysis and estimation tools are now offered by nearly every major EDA vendor.

While the situation has clearly improved, the problem is still far from solved. Both academic and industrial experts have noted that after exploiting the obvious technology and circuit level optimizations, we are left orders of magnitude from where we need to be. These additional reductions must come from op-

timizations made at the higher abstraction levels—namely, the architecture, algorithm, and system levels.

Unfortunately, EDA vendors offer few solutions to aid in the exploration of the power domain at these levels. Clearly it is not feasible (nor desirable) to specify/synthesize every design alternative down to the gate level. Tools are needed that operate inherently at the architecture level and above. This has been an active area of university research for some time now, and while there may be little to choose from commercially, academic approaches have begun to appear.

This paper will describe the current state of the art in high-level power estimation. We will begin with a proposal for an analysis-based low-power design methodology. The remainder of the paper will cover emerging architecture-, behavior-, instruction-, and system-level power-estimation techniques that support this methodology. In comparing the work of different researchers, quantitative comparisons of accuracy and speed can be misleading since different assumptions are often made in deriving those results. Therefore, although some numerical results are presented in this paper, the primary focus will be to give a qualitative feel for how the various approaches differ—and from this some conclusions about relative performance can be inferred.

2. AN ANALYSIS-BASED DESIGN METHODOLOGY

While logic synthesis has gained widespread acceptance among the industrial design community, high-level synthesis has found a foothold in a relatively narrow range of applications (most notably, DSP). One of the reasons for this is that in order to make the problem tractable, most high-level synthesis systems are forced to assume some fixed architectural template that is unlikely to be optimum for all applications. This suggests that a better approach is to rely on the designer to specify system partitionings and architectural configurations, with the primary function of the tools being to provide feedback on the quality of a particular solution.

The result is an analysis-based low-power design methodology as illustrated in Fig. 1. Working top-down, the designer be-

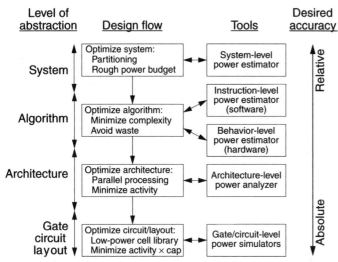

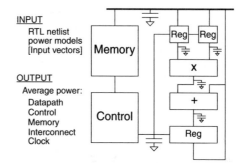

Figure 1 Design flow and supporting tools for an analysis-based design methodology.

Figure 2 Illustration of the architectural power-estimation problem.

gins at the system level, partitioning the design into off-the-shelf and custom components. Here, the function of the tools is to aid in producing a power budget indicating which parts of the system will likely be the major power consumers. Rough power estimates at this stage of design can save a lot of time wasted later on optimizing the wrong part of the system. With an initial partitioning and power budget in hand, the designer can focus on the individual components of the system, which may be realized as software or hardware. Instruction-level power models of programmable processors will be useful here in helping the designer optimize the software portions of the system. Similarly, behavior- and RT-level power estimators will provide a much needed classification of the power efficiency of different algorithms and architectures that might be used for the dedicated hardware. At all these levels, relative accuracy in the power estimates is much more important than absolute accuracy, since what we really want to know is whether one alternative is better than another.

In the final design stages, a more traditional flow applies, utilizing software and hardware compilers along with schematic entry, logic synthesis, layout, and place and route tools. The primary function of gate- and circuit-level power analysis and simulation tools would then be to provide back-end verification of power consumption with sign-off accuracy.

As mentioned above, much of the technology to support this low-power design methodology now exists, either commercially or as university research. In the next few sections of this paper we describe the strategies that have been proposed to realize architecture-, behavior-, instruction-, and system-level power estimation.

3. ARCHITECTURE-LEVEL POWER ESTIMATION

The lowest level we will consider is the architecture, or register-transfer, level. At this level of abstraction, the primitives are functional blocks such as adders, multipliers, controllers, register files, and SRAMs (see Fig. 2). The difficulty in estimating

power at this level stems from the fact that the gate-, circuit-, and layout-level details of the design may not have been specified. Moreover, a floorplan may not be available, making analysis of interconnect and clock distribution networks difficult.

The strategies proposed, thus far, for RT-level power estimation can be divided into two classes: analytical methods and empirical methods.

A. Analytical Methods

Analytical methods attempt to relate the power consumption of a particular RTL description to fundamental quantities that describe the physical capacitance and activity of a design. Since design complexity is a good first-order measure of physical capacitance we can roughly divide the techniques presented in this section into complexity-based and activity-based models.

Complexity-Based Models. One strategy relies on the fact that the complexity of a chip architecture can be described roughly in terms of "gate equivalents." Basically, the gate-equivalent count of a design specifies the approximate number of reference gates (e.g., 2-input NANDs) that would be required to implement a particular function (e.g., a 16×16 multiplier). This number can be specified in a library database or provided by the user. The power required for each functional block can then be estimated by multiplying the approximate number of gate equivalents by the average power consumed per gate. An example of this technique is given by the Chip Estimation System (CES) [1], which uses the following expression for average power:

$$P = \sum_{i \in \left\{ \substack{\text{functional} \\ \text{blocks}} \right\}} GE_i (E_{typ} + C_L^i V_{dd}^2) f A_{int}^i \qquad [1]$$

where GE_i is the gate equivalent count for functional block i, E_{typ} is the average energy consumed by an equivalent gate when active, C_L^i is the average capacitive load per gate including fan-out and wiring, f is the clock frequency, and A_{int}^i is the average percentage of gates switching each clock cycle within functional block i.

One disadvantage of this technique is that all power estimates are based on the energy consumption of a single reference gate. This does not take into account different circuit styles, clocking strategies, or layout techniques. The approximation is particularly inaccurate for specialized blocks, such as memories.

Liu and Svensson improved the situation by applying customized estimation techniques to the different design entities: logic, memory, interconnect, and clock [2]. For example, the power consumed by a memory-cell array (such as the one shown in Fig. 3) is modeled as:

$$P_{memcell} = \frac{2^k}{2} \left(c_{int} l_{column} + 2^{n-k} C_{tr} \right) V_{dd} V_{swing} f \qquad [2]$$

where 2^k is the number of cells in a row, c_{int} is the wire capacitance per unit length, l_{column} is the memory column length, 2^{n-k} is the number of cells in a column, C_{tr} is the minimum-size transistor drain capacitance, and V_{swing} is the bitline voltage swing.

The logic component of power is estimated in a manner conceptually similar to CES. The basic switching energy is based on a three-input AND gate and is calculated from fundamental technology parameters (e.g., minimum gate width, gate length, and oxide thickness). The total chip logic power is estimated (as before) by multiplying the estimated gate equivalent count by the basic gate energy and the activity factor. The activity factor is provided by the user and assumed fixed across the entire chip.

Finally, interconnect length and capacitance are modeled by a derivative of Rent's Rule. The clock capacitance is based on the assumption of an H-tree distribution network.

The advantage of these complexity-based estimation techniques is that they require very little information. Basically, just a few technology parameters, memory sizes, and a count of gate equivalents are needed.

One disadvantage of the complexity-based methods is that they do not model circuit activity accurately. An overall (fixed) activity factor is typically assumed and, in fact, must often be provided by the user. In reality, activity factors will vary with block functionality and with the data being processed. So even if the user provides an activity factor that results in a good estimate of the total chip power, the predicted breakdown of power between modules is likely to be incorrect, making it difficult to perform meaningful architectural trade-offs.

Activity-Based Models. Activity-based models address this issue to some extent. So far all efforts in this area have focused on using the concept of entropy from information theory as a measure of the average activity in a circuit [3,4]. The basic idea is to try to relate the power that a functional block consumes to the amount of computational work it performs. Entropy is a useful metric from information theory for measuring computational work.

Najm [3], for example, observes that power is proportional to the product of physical capacitance and activity. He then uses area as a measure of physical capacitance and entropy as a measure of activity:

$$P \propto \text{Capacitance} \times \text{Activity} \propto \text{Area} \times \text{Entropy} \qquad [3]$$

Leveraging off previous work [5,6], the area of a block's average minimized implementation is related to the number of boolean inputs, n, and to the total entropy of its m outputs, H_o:

$$\text{Area} \propto \begin{cases} \dfrac{2^n}{n} H_o & \text{as } n \to \infty \\ 2^n H_o & \text{for } n \le 10 \end{cases} \qquad [4]$$

In this expression, the output entropy, H_o, is given by:

$$H_o = \sum_{i=1}^{2^m} p_i \log_2 \frac{1}{p_i} \qquad [5]$$

where p_i is the probability that the output takes on the ith of its 2^m possible values.

Using the approximation that entropy decreases quadratically with logic depth, Najm is able to estimate the average entropy of all the nodes in a functional block from its input and output entropies:

$$\text{Entropy} \approx \frac{2/3}{n + m} (H_i + 2H_o) \qquad [6]$$

where H_i is defined in a manner analogous to (5).

Najm's power-estimation methodology then consists of running an RTL simulation of the design to measure the input and output entropies of the functional blocks, using Eq. (3) through (6) to translate these measurements into a prediction of average power. Najm notes that the approach has some significant hurdles to overcome. First, no timing information enters into the above calculations and, therefore, glitching power is not accounted for in any way. Also, there is the implicit assumption in (3) that capacitance is uniformly distributed over all nodes.

Clearly, the accuracy of these techniques is limited; however, they may prove useful for making *relative* architectural comparisons, which as mentioned before is the main function of high-level power-estimation tools. Still, these information-theoretical approaches are in their infancy and much work needs to be done to demonstrate their value in practice.

All of the analytical power-estimation methods described in this section (both complexity- and activity-based) have the advantage of requiring very little information as input. In some sense, this is also a disadvantage in that it is difficult to capture the power attributes of different functional blocks using only parameters such as gate-equivalent count or entropy. Therefore, power predictions based on these techniques may not have a strong connection to the actual power consumed by real designs.

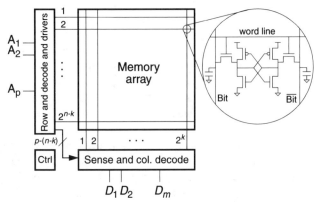

Figure 3 The structure of a typical on-chip SRAM (adapted from [2]).

B. Empirical Methods

The empirical models discussed in this section offer one possible solution to this problem. Rather than trying to relate the power consumption of RTL components to fundamental parameters, the strategy here is to "measure" the power consumption of existing implementations and produce a model based on those measurements. In other words, these techniques employ a *macromodeling* approach to architectural power estimation.

Clearly, this approach is best suited for designs that will be built using a library-based approach, but this is not a necessity. For example, even if a designer intends to build the functional blocks for his architecture from scratch, it is still likely that models based on previous implementations will give good estimates of power figures. If no previous data is available for a particular block, then analytical models may be more appropriate.

The techniques that fall into the category of empirical methods can further be subdivided into those that assume fixed signal activities and those that account for variations in data and instruction statistics.

Fixed-Activity Models. The first proposal for a fixed-activity macromodeling strategy was the Power Factor Approximation (or PFA) method [7]. The power consumed by a given architecture is approximated by the expression:

$$P = \sum_{i \in \{\text{all blocks}\}} \varkappa_i G_i f_i \qquad [7]$$

where each functional block i is characterized by a PFA proportionality constant $\varkappa_i$, a measure of hardware complexity G_i, and an activation frequency f_i.

For example, the hardware complexity of a multiplier is related to the square of the input word length, so $G_{mult} = N^2$. The activation frequency is simply the frequency with which multiplies are performed by the algorithm, f_{mult}. Finally, the PFA constant $\varkappa_{mult}$ is extracted empirically from past multiplier designs (taken from ISSCC proceedings) and shown to be about 15 fW/bit²–Hz for a 1.2-μm technology at 5V. The resulting power model is:

$$P_{mult} = \varkappa_{mult} N^2 f_{mult} \qquad [8]$$

Although the authors only explicitly discuss models for multipliers, memories, and I/O drivers, the PFA method can be viewed as a general technique for individually characterizing an entire library of RT-level functional blocks. The power models can be parameterized in terms of whatever complexity parameters are appropriate for that block. For instance, for the memory the storage capacity in bits is used, and for the drivers the number of I/O bits alone is sufficient.

The weakness of fixed-activity models, of course, is that they do not account for the influence that data activity can have on power consumption. For example, the PFA constant $\varkappa_{mult}$ is intended to capture the intrinsic internal activity associated with a multiplier unit; however, since it is taken to be a constant, there is the implicit assumption that the inputs do not affect the multiplier activity, which is not the case. As an example of this phenomenon, Fig. 4 (which will be discussed fully in the next

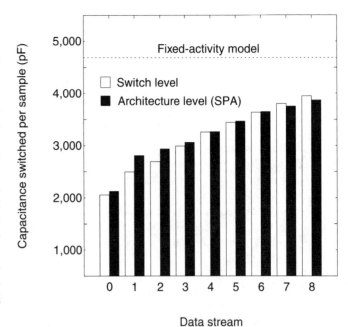

Figure 4 Architectural power estimates (from SPA) vs. switch-level simulations of an LMS noise canceller for different data streams.

subsection) shows how the power consumption of an LMS noise-cancellation filter varies for different data streams.

Activity-Sensitive Models. Activity-sensitive empirical power models have been developed in an attempt to address this situation. These models endeavor to account in some way for the influence that data activity statistics can have on power.

A simple example is the RT-level power estimation tool called ESP [8,9]. Although ESP relies for the most part on a fixed-activity model, it does provide some capability of measuring vector-dependent power. ESP is fundamentally a cycle-based simulator targeted at a RISC processor. As the object code is executed, ESP monitors which block in the architecture are activated, adding a fixed-power contribution for each (Fig. 5). The

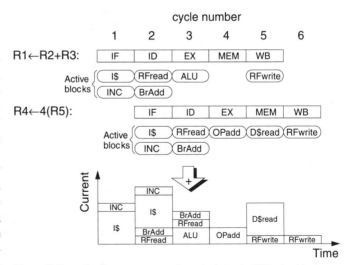

Figure 5 Example of current consumption calculation by ESP using block-activation data from a cycle-based simulator (adapted from [9]).

implicit assumption is that the power consumed by each component has been empirically measured prior to simulation. The datapath power model accounts to some extent for input vector activity by using a power model that has a constant portion and a portion that is proportional to the number of bit transitions n in the input vector:

$$P = P_{const} + n \cdot P_{change} \qquad [9]$$

Another activity-sensitive architectural power-analysis tool called SPA has also been developed [10–12]. The approach is based on the concept of activity profiling. Specifically, prior to power analysis, an RT-level simulation (e.g., VHDL) of the design in question is carried out for typical instruction and data inputs (Fig. 6). During this simulation, the activity of the design entities and signals in the data and control paths are monitored and recorded. These activity statistics are then fed into power models that explicitly account for activity as well as complexity. Two different activity models have been described—one for the datapath and one for the control path.

The datapath activity model is referred to as the dual-bit type, or DBT, model. It is based on the observation that fixed-point, 2's-complement data streams are characterized by two distinct activity regions (Fig. 7). The data bits (LSBs) exhibit activity similar to uniformly-distributed white noise. The activity of the sign bits (MSBs) depends on the sign transition probability, which is related to the temporal data stream correlation, $\rho = \mathrm{cov}(X_{t-1}, X_t)/\sigma^2$, where σ^2 is the data stream variance. Different empirically derived coefficients are used to characterize the capacitance switched in the data (C_U) and sign (C_S) regions of various functional blocks:

$$P = (N_U C_U + N_S C_S)V_{dd}^2 f \qquad [10]$$

The expression can be extended to more complex multi-parameter models using vector notation with arrays of capacitance (C) and complexity (N) parameters.

In the control path the activity-based control, or ABC, model is used. Unlike datapath words which have very definite structure, words in the control path often are formed by concatenation of a number of independent fields or boolean flags. Thus, we cannot rely *a priori* on any particular structure when deriv-

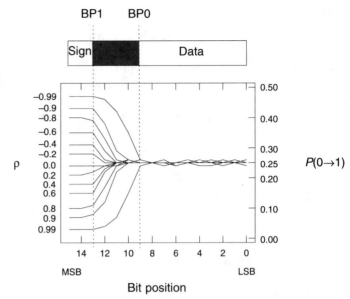

Figure 7 Bit-transition activities for data streams with varying temporal correlation, ρ.

ing an activity model for control streams. As a result, the ABC model falls back on more traditional measures of activity: transition probability, α, and signal probability, P. Combining these activities with complexity parameters such as the number of inputs (N_I), outputs (N_O), and min-terms (N_M) for a finite state machine (FSM) block, we can derive power models for various controller-implementation styles. The model for an FSM implemented in standard cells, for example, might be:

$$P = (C_I \alpha_I N_I N_M + C_O \alpha_O N_O N_M)V_{dd}^2 f \qquad [11]$$

where C_I and C_O are capacitance coefficients empirically extracted from previous standard-cell controller implementations. In addition to datapath and control, SPA also includes custom models for interconnect and memory.

A commercial tool called WattWatcher/Architect (offered by Sente, Inc.) relies on techniques similar to those used by SPA, particularly in the area of datapath power modeling [13]. In addition to simulation-based activity profiling, however, it also offers probabilistic activity propagation. It is capable of analyzing a 120,000 gate architecture in 34 minutes.

One of the advantages of empirical models is that they have a strong link to real implementations. Figure 4 compares RT-level predictions from SPA to switch-level simulation of a fully laid-out LMS noise canceller. The figure also shows the advantage activity-sensitive modeling can have over fixed-activity techniques.

4. BEHAVIOR-LEVEL POWER ESTIMATION

As we move up in abstraction level, power estimation becomes even more difficult. Much less is known about a design at the behavior or algorithm level than was known at the architecture level. The typical approach is to assume some architectural style or template and produce power estimates based on it. Some

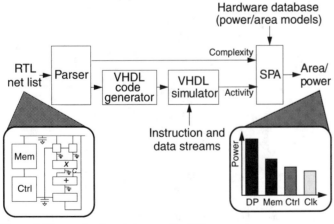

Figure 6 Power estimation strategy used by SPA.

of the numerous unknowns that must be predicted include the foreground/background memory configuration, the number of memory accesses, the bus architecture and average wire length, the number of bus transactions, the control path complexity, and the control line activity.

In studying this list it becomes apparent that some of these parameters relate to the physical capacitance of the resources being accessed, while others describe the activity of those resources. Activity prediction is perhaps the more interesting of the two problems. Behavioral power-estimation techniques can be roughly divided into two camps—those that use static-activity prediction and those that use dynamic-activity prediction.

A. Static Activity Prediction

The access frequency of a resource is important since the more often a resource is activated, the more power it will consume. The object of static-activity prediction is to produce an estimate of the access frequency for different hardware resources by analysis of the behavioral description of the function to be implemented. This description could be in the form of a C, Verilog, or VHDL program, or it could be represented as a control-data flow graph (CDFG)—as is common in high-level synthesis systems. Since only one pass through the program is required, a key advantage of the static-profiling approach is its speed.

For programs with no data dependencies, the analysis is quite straightforward and yields the desired access counts for the different operations required by the algorithm. In the more typical case where data-dependent conditionals, branches, and loops are present, the situation is more complicated and we must resort either to statistical approximations or dynamic profiling techniques.

Mehra and Chandrakasan et al. have developed a behavioral power-estimation strategy using static profiling in the context of the HYPER-LP high-level synthesis system [14,15]. The power required to execute a behavior is expressed as:

$$P = \sum_{r \in \left\{ \begin{smallmatrix} \text{all datapath, control path,} \\ \text{memory, and bus resources} \end{smallmatrix} \right\}} f_r C_r V_{dd}^2 \qquad [12]$$

where f_r is the access frequency of resource r as determined by static analysis of the CDFG. The capacitance C_r switched when resource r is activated is determined using empirical fixed-activity models. For example, the control path model was built by benchmarking the switching capacitance of controllers for 46 different design examples (Fig. 8). From the figure it's clear that while power models at this of abstraction don't offer a high degree of absolute accuracy, they do capture general trends; as stated before, this is the primary goal of high-level power estimation.

Power-Profiler uses a similar strategy for behavioral power estimation [16]. One key difference is that rather than producing a single average estimate of power consumption, Power-Profiler produces a profile of power versus time (for both individual modules as well as the complete design). As shown in Fig. 9, this gives the designer some feel for peak, as well as average, power

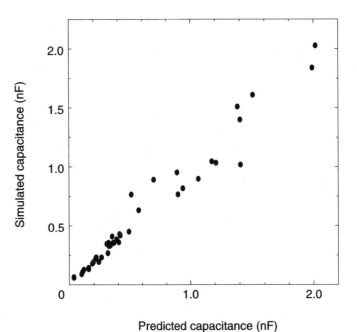

Figure 8 Predicted controller capacitance vs. switch-level simulation (adapted from [14]).

consumption. It should be noted, however, that an averaging window of one clock cycle is used, which tends to smooth out the power peaks substantially. Therefore, while the tool gives the designer a rough feel for where power peaks might occur, the exact instantaneous amplitude of those peaks is not reported.

B. Dynamic Activity Prediction

Dynamic profiling is another technique for determining the activation frequencies of various resources. In this approach, a simulation of the desired behavior is performed for a user-supplied set of inputs. During this simulation, activity statistics are gathered regarding the frequency of various types of operations

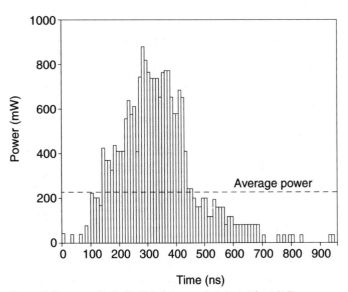

Figure 9 Power profile for DCT-design example (adapted from [16]).

and memory accesses. These frequencies are then plugged into a model similar to (12) to obtain a power estimate. The advantage, of course, is that data dependencies are easily handled. The disadvantages are that it's much slower than the static approach and that it requires the user to supply a set of typical input vectors.

One example of the dynamic approach is the Profile-Driven Synthesis System (PDSS) [17]. The input to the system is a behavioral subset of VHDL. Prior to simulation, the system automatically inserts activity probes. The activation statistics for each datapath operation type are then plugged into what amounts to a library of fixed-activity empirical-power models. The controller FSM is assumed to be of the PLA type and empirical-power models based on simulations of a randomly-generated FSM benchmark set are used to predict its power. Table 1 gives average power results from PDSS for several design examples.

Table 1. PDSS Results for Several Design Examples [17].

Design	Estimated Power (mW)	Actual Power (mW)	Deviation (%)
Decompress	4.14	3.68	12.5%
Compress	4.08	3.56	14.4%
Find	3.80	3.40	11.7%
FIFO	2.45	2.73	10.3%
TLC	2.80	3.11	9.8%
Shuffle	45.09	51.01	11.6%

5. INSTRUCTION-LEVEL POWER ESTIMATION

Most behavior-level power estimators assume architectural models corresponding to dedicated hardware implementations. It is also possible to realize a given behavior in software on a programmable instruction-set processor. In this case, an instruction-level power model is appropriate.

Tiwari and coworkers proposed just such a model for embedded general-purpose and DSP processors [18,19]. The strategy they describe is most similar to the empirical macromodeling approach described earlier. Each available instruction is placed in a loop and executed on the target processor. During this process, current measurements are taken, and the average current drawn by each instruction is stored in a table of *base costs*.

The model also handles what are referred to as inter-instruction effects. In a real program, the change of circuit state between two instructions leads to a current consumption that is higher than predicted by the base cost. Therefore, an additional fixed circuit-state overhead current must be added to the base cost for each instruction. The magnitude of this correction factor can be determined by executing pairs of instructions while measuring current. Additional effects such as pipeline stalls and cache misses are also considered in the model. The complete estimation methodology is illustrated in Fig. 10.

To date, the model has been used to characterize the power consumption of the Intel 486DX2, the Fujitsu SPARClite 934, and a Fujitsu embedded DSP processor. The authors note that while accurate for most instructions, the estimates produced

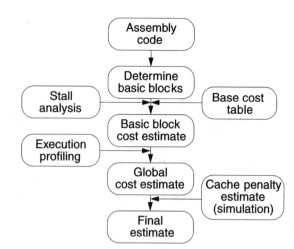

Figure 10 Instruction-level power estimation-methodology (adapted from [18]).

for the DSP MAC instruction can err significantly. In fact, while the base cost for a packed MAC:LAB instruction is 36.9mA, the overhead cost can vary from as little as 1.4mA to as much as 33.0mA [19]. The variation is caused by the impact of operand activity on the multiplier. In order to account for this, an activity-sensitive model similar to those described earlier would be needed.

6. SYSTEM-LEVEL POWER ESTIMATION

At the earliest stages of design specification we can consider performing system-level power estimation. Here the goal is to come up with a rough power budget accounting for all the components in a system. This should include the analog, digital, mixed signal, and even electromechanical portions of a system. A power exploration tool at this level of abstraction would be quite useful for identifying bottlenecks before any time is wasted optimizing the wrong part of the system. It would also be helpful in determining the best way to partition the desired functionality into individual components. The partitioning and level of integration of a system can have a profound effect on the overall power consumption.

A tool called PowerPlay has been recently developed that encompasses these capabilities [20]. PowerPlay is available as an Internet application (Fig. 11) and employs a spreadsheet-like interface to facilitate hierarchical design entry and rapid exploration of design partitionings and parameter variations (e.g., supply voltage and clock frequency).

The power models used by PowerPlay leverage off the existing high-level power estimation literature and currently are primarily empirical in nature. The models are placed in a hardware library and are shared among users. New models can be created easily using user-defined parameterized expressions or values taken straight from product data sheets. The PowerPlay framework has been used successfully to model the power of the Info-Pad system, a portable multimedia terminal developed at U.C. Berkeley [21]. This illustrates PowerPlay's ability to model het-

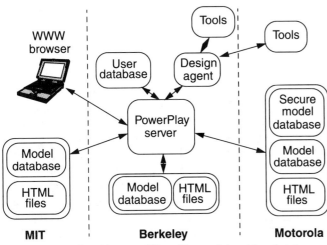

Figure 11 Network architecture of PowerPlay, an Internet-based system-level power estimation tool (adapted from [20]).

erogeneous components such as programmable processors, ASICs, memories, FPGAs, radio modems, and LCD displays.

7. CONCLUSIONS

The development of tools to support a low-power design methodology has been an area of active research for the last several years. While much of the literature deals with circuit- and gate-level techniques, a significant amount has also been published on high-level power estimation. This paper has made an attempt to gather the applicable research into one place and describe how the individual pieces relate to one another.

Power-estimation strategies are now available that operate at the architecture, behavior, instruction, and even system levels of abstraction. While the majority are from academic circles, some tools have also begun to appear commercially, and many more are sure to be offered in the near future.

Even so, the field of high-level power estimation is in its infancy. Much remains to be done in order to demonstrate the robustness and applicability of the techniques in a realistic industrial setting. It is clear, however, that high-level analysis tools fill a significant gap in current low-power design methodologies, allowing designers to make more informed decisions from the earliest stages of system implementation.

ACKNOWLEDGMENTS

The author would like to thank Jan Rabaey for his help during the preparation of this manuscript. Some of the research described in this paper was funded by ARPA grant J-FBI 93-153 and by a fellowship from the National Science Foundation.

References

[1] K. Müller-Glaser, K. Kirsch, and K. Neusinger, "Estimating essential design characteristics to support project planning for ASIC design management," *1991 IEEE International Conference on Computer-Aided Design,* Los Alamitos, CA, November 1991, pp. 148–151.

[2] D. Liu and C. Svensson, "Power consumption estimation in CMOS VLSI chips," *IEEE Journal of Solid-State Circuits,* Jun. 1994, pp. 663–670.

[3] F. Najm, "Towards a high-level power estimation capability," *1995 International Symposium on Low-Power Design,* April 1995, pp. 87–92.

[4] D. Marculescu, R. Marculescu, and M. Pedram, "Information theoretic measures of energy consumption at register transfer level," *1995 International Symposium on Low-Power Design,* April 1995, pp. 81–86.

[5] N. Pippenger, "Information theory and the complexity of boolean functions," *Mathematical Systems Theory,* vol. 10, 1977, pp. 129–167.

[6] K-T Cheng and V. Agrawal, "An entropy measure for the complexity of multi-output boolean functions," *27th ACM/IEEE Design Automation Conference,* June 1990, pp. 302–305.

[7] S. Powell and P. Chau, "Estimating power dissipation of VLSI signal processing chips: The PFA technique," *VLSI Signal Processing IV,* 1990, pp. 250–259.

[8] T. Sato, Y. Ootaguro, M. Nagamatsu, and H. Tago, "Evaluation of architecture-level power estimation for CMOS RISC processors," *1995 IEEE Symposium on Low-Power Electronics,* Oct. 1995, pp. 44–45.

[9] T. Sato, M. Nagamatsu, and H. Tago, "Power and performance simulator: ESP and its application for 100MIPS/W class RISC design," *1994 IEEE Symposium on Low-Power Electronics,* Oct. 1994, pp. 46–47.

[10] P. Landman, *Low-Power Architectural Design Methodologies,* Ph.D. Dissertation, U.C. Berkeley, Aug. 1994.

[11] P. Landman and J. Rabaey, "Architectural power analysis: The dual bit type method," *IEEE Transactions on VLSI Systems,* June 1995, pp. 173–187.

[12] P. Landman and J. Rabaey, "Activity-sensitive architectural power analysis," *IEEE Transactions on CAD,* Jun. 1996, pp. 571–587.

[13] WattWatcher Product Sheet, Sente Corp., Chelmsford, MA, 1996.

[14] R. Mehra and J. Rabaey, "High-level power estimation and exploration," *1994 International Workshop on Low Power Design,* Apr. 1994, pp. 197–202.

[15] A. Chandrakasan, M. Potkonjak, J. Rabaey, and R. Brodersen, "Optimizing power using transformations," *IEEE Transactions on Computer-Aided Design,* Jan. 1995, pp. 12–31.

[16] R. San Martin and J. Knight, "Optimizing power in ASIC behavioral synthesis," *IEEE Design & Test of Computers,* Summer 1996, pp. 58–70.

[17] N. Kumar, S. Katkoori, L. Rader, and R. Vemuri, "Profile-driven behavioral synthesis for low-power VLSI systems," *IEEE Design & Test of Computers,* Fall 1995, pp. 70–84.

[18] V. Tiwari, S. Malik, and A. Wolfe, "Power analysis of embedded software: A first step towards software power minimization," *IEEE Transactions on VLSI Systems,* December 1994, pp. 437–445.

[19] M. T.-C. Lee, V. Tiwari, S. Malik, and M. Fujita, "Power analysis and minimization techniques for embedded DSP software," to appear in *IEEE Transactions on VLSI Systems,* March 1997, pp. 123–135.

[20] D. Lidsky and J. Rabaey, "Early power exploration: A world wide web application," *33rd Design Automation Conference,* Jun. 1996, pp. 27–32.

[21] S. Sheng, A. Chandrakasan, and R. Brodersen, "A portable multimedia terminal," *IEEE Communications Magazine,* December 1992, pp. 64–75.

Activity-Sensitive Architectural Power Analysis

Paul E. Landman, *Member, IEEE*, and Jan M. Rabaey, *Fellow, IEEE*

Abstract—Prompted by demands for portability and low-cost packaging, the electronics industry has begun to view power consumption as a critical design criteria. As such there is a growing need for tools that can accurately predict power consumption early in the design process. Many high-level power analysis models don't adequately model activity, however, leading to inaccurate results. This paper describes an activity-sensitive power analysis strategy for datapath, memory, control path, and interconnect elements. Since datapath and memory modeling has been described in a previous publication, this paper focuses mainly on a new Activity-Based Control (ABC) model and on a hierarchical interconnect analysis strategy that enables estimates of chip area as well as power consumption. Architecture-level estimates are compared to switch-level measurements based on net lists extracted from the layouts of three chips: a digital filter, a global controller, and a microprocessor. The average power estimation error is about 9% with a standard deviation of 10%, and the area estimates err on average by 14% with a standard deviation of 6%.

I. INTRODUCTION

CURRENTLY, the portable consumer electronics market is undergoing a period of rapid growth. With portability comes a new set of design requirements. In particular, the constraints of battery operation have forced designers to focus on power considerations as well as speed and area. Furthermore, the high-cost of packaging and cooling power-hungry devices has led to increasing efforts aimed at minimizing power consumption even in high performance, nonportable systems.

This trend further complicates the design process as engineers must now consider joint optimization not only of area and speed, but also of power. CAD tools can help manage this complexity by providing feedback about the impact of various design decisions on these three important parameters. Analysis tools such as SPICE [1] and PowerMill [2] can be useful in this capacity, but they both require a transistor-level netlist as input and, therefore, they can only be applied toward the end of the design flow when major changes are difficult and expensive to implement.

This paper describes techniques for estimating area and power consumption given an architecture-level description of a system. The paper builds on research presented in [3], which described how to estimate the power of individual architectural blocks in a datapath. Here we extend that work by introducing

Manuscript received December 1, 1995. This work was supported by ARPA Grant J-FBI 93-153 and by a Fellowship from the National Science Foundation. This paper was recommended by Guest Editors M. Pedram and M. Fujita.
P. E. Landman is with the DSP R&D Center, Texas Instruments, Dallas, TX 75265 USA.
J. M. Rabaey is with the EECS Department, University of California, Berkeley, CA 94720 USA.
Publisher Item Identifier S 0278-0070(96)04856-7.

techniques for control path and interconnect analysis, and by presenting results obtained from several design examples.

Fig. 1 gives an overview of the power analysis strategy that we propose in this paper. The inputs from the user are a description of a candidate architecture at the register-transfer level and a set of data and instruction inputs for which a power analysis is desired. Rather than attempting to find a single model for the entire chip, we take the approach of identifying four basic classes of components: datapath, memory, control, and interconnect. The modeling of power consumption for each class is addressed separately. For datapath and memory analysis, Section III reviews a word-level data model known as the Dual Bit Type (or DBT) model. An architectural model for control path power consumption, the Activity-Based Control (ABC) model, is presented in Section IV. Section V describes techniques for estimating the physical capacitance of interconnect (while producing chip area estimates as a side effect). It also describes how to combine these physical capacitances with the appropriate activity measures to obtain estimates of control and data bus power consumption. Section VI brings together the four classes of models, describing how they can be integrated to enable architectural power analysis for entire chips. This section also describes how the complexity and activity parameters required by the DBT and ABC models can be derived. In Section VII, the models are verified using several realistic examples including a programmable microprocessor.

II. PREVIOUS WORK

The majority of the available literature on power estimation deals with transistor- or gate-level modeling [2], [4]–[9]. As stated above, however, we are interested in tools that operate at the architecture level. In the terminology of this paper, architecture refers to the register-transfer level of abstraction, where the primitives are blocks such as multipliers, adders, memories, and finite state machines. Two principal strategies have been proposed for estimating power at this level.

The first technique is based on the concept of gate equivalents. In this method, gate equivalent counts are used to roughly describe the complexity of modules within a chip. The gate equivalent count specifies the approximate number of reference gates (e.g., two-input NAND's) that are required to implement a particular function (e.g., a 16-b counter). The power required for that function can then be estimated by multiplying the approximate number of gate equivalents required by the average power consumed per gate. This is the essence of the strategy used in the Chip Estimation System [10]. Svensson and Liu also model logic power consumption using gate equivalents, but attempt to improve overall

Reprinted from *IEEE Transactions on Computer-Aided Design*, pp. 571–587, June 1996.

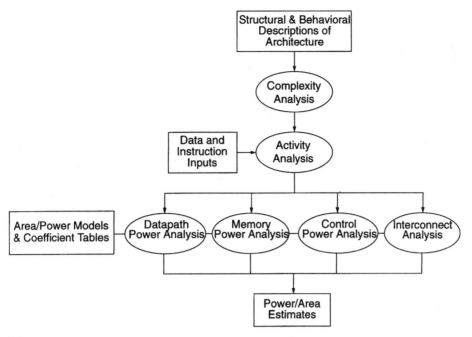

Fig. 1. Overview of architectural power analysis strategy.

accuracy by using custom models to handle memories [11]. Both approaches use derivatives of Rent's Rule to estimate interconnect length and capacitance [12].

The second principal modeling strategy can be classified as a precharacterized cell library approach. Under this scheme, instead of using a single gate-equivalent model for all "logic" blocks, a separate model is supplied for each block in the library: multipliers, adders, buffers, etc. These custom models better reflect how the complexity of a specific block influences its power. This technique was first proposed by Powell and Chau who termed it the Power Factor Approximation method [13].

The advantage of both these approaches is that they require minimal design information as input and that they are straightforward to apply. This simplicity, however, comes at the expense of accuracy. For, while both techniques do a satisfactory job of relating design complexity to power, they do not adequately account for the impact of activity on power. In both cases, power models are characterized assuming fixed activity levels—typically, corresponding to an assumption of random, uniformly distributed white noise (UWN) data. Often this assumption is not justified as demonstrated by Fig. 2. In this example, the divider power based on switch-level simulation of an extracted layout varies by more than a factor of two as input data statistics change. In the following sections, we propose architectural power analysis techniques that are sensitive to activity.

III. DATAPATH AND MEMORY MODELING

Datapath and *data* memory components perform and store the results of the numerical computations required by an algorithm. In previous papers, we have described architectural power analysis techniques for these elements in detail [3], [14],

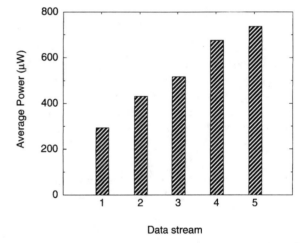

Fig. 2. Divider power from switch-level simulation of various data streams.

[15]. This section will briefly review the principal results of those papers.

The idea is to produce a black-box model of the capacitance switched in each module for various types of input activity. If desired, these capacitance estimates can be converted to an equivalent energy, $E = CV^2$, or power, $P = CV^2f$. The models must take into account not only the complexity of the module being characterized, but also the switching activity within the module.

Complexity is handled by allowing the library designer to specify appropriate complexity parameters for each module. These parameters can then be used in a capacitance model to describe precisely how the physical capacitance of each module should scale with its "size" or complexity. For instance, the capacitance model for a ripple-carry adder is given by

$$C_T = C_{eff}N \tag{1}$$

517

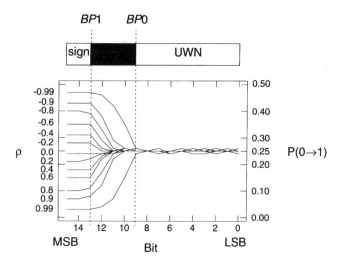

Fig. 3. Bit transition activity for data streams with varying temporal correlation.

where N is the word length and C_{eff} is a capacitive coefficient describing the effective (activity-dependent) capacitance switched per bit. If necessary the library designer can specify more complex expressions. For a typical SRAM

$$C_T = C_0 + C_1 W + C_2 N + C_3 W N \qquad (2)$$

where W is the number of rows in the currently active block and N is the number of columns. The dominant $C_3 W N$ term stems from charging and discharging bit lines in the cell array. The remaining terms pertain to the row/column decoders, the word line drivers, and the sense amps. The equation can be expressed equally well in vector notation

$$C_T = \mathbf{C}_{eff} \cdot \mathbf{N}. \qquad (3)$$

Here, the scalar capacitive coefficient and complexity parameters of (2) have simply been replaced by vectors

$$\mathbf{C}_{eff} = [C_0 \quad C_1 \quad C_2 \quad C_3]^T$$

and

$$\mathbf{N} = [1 \quad W \quad N \quad WN]^T.$$

The model also accounts for activity as well as complexity. Instead of having a single C_{eff} coefficient per module based on a white noise activity assumption, the model employs several capacitive coefficients for each module, corresponding to different input activity types. For fixed-point two's-complement data, we use a dual bit type (DBT) model that accounts for two classes of bits: sign and data. Fig. 3 confirms that two's-complement data is, indeed, characterized by two distinct activity regions. The data bits (LSB's) exhibit the activity of uniform white noise (UWN), while sign bit activity, in contrast, depends on the exact sequence of sign transitions. The likelihood of sign toggling is characterized by the temporal data stream correlation $\rho = \text{cov}\,(X_{t-1}, X_t)/\sigma^2$.

The DBT method accounts for these differing activities by using a separate capacitive coefficient for each transition type. The data region uses a single UWN capacitive coefficient, C_{UU}. In contrast, characterizing all possible sign transitions requires several coefficients of the form, C_{SS}, where S can

be positive or negative: C_{++}, C_{+-}, C_{-+}, and C_{--}. A module that operates on two or more input data streams requires additional capacitive coefficients. Each coefficient can be either a scalar or a vector depending on the number of terms in the capacitance model: for the above SRAM $\mathbf{C}_{\text{UU}} = [C_{0,\,\text{UU}} \quad C_{1,\,\text{UU}} \quad C_{2,\,\text{UU}} \quad C_{3,\,\text{UU}}]^T$.

Capacitive coefficient values are derived for each module during a *library characterization* step. This is a one-time process, not required during power analysis, but instead performed whenever a new cell is added to the library. *Pattern generation* is the first step in the three-stage characterization process. During this phase, white noise and sign input patterns are generated. Next, *simulation* is used to measure the capacitance switched for these patterns. In order to characterize the influence of complexity, as well as activity, the module may be characterized for several complexity parameter values (e.g., word length, storage capacity, etc.). Finally, during *coefficient extraction*, the capacitance models are fit to the simulated capacitance data to produce a set of "best fit" capacitive coefficients.

The power analysis process itself consists of decomposing the modules into white noise and sign regions and then estimating the effective capacitance switched within each region. The size of the regions depends on the positions of the model breakpoints ($BP0$ and $BP1$) in Fig. 3. Analytical expressions were derived in [15] that express the breakpoints as a function of data stream statistics such as mean (μ), variance (σ^2), and correlation (ρ)

$$BP1 = \log_2 \left(|\mu| + 3\sigma\right) \qquad (4)$$

$$BP0 = \log_2 \sigma + \Delta BP0 \qquad (5)$$

$$\Delta BP0 = \log_2 \left[\sqrt{1 - \rho^2} + \frac{|\rho|}{8}\right]. \qquad (6)$$

This information can be used to calculate the effective capacitance switched in each region during a typical module access

$$C_T = \frac{N_{\text{U}}}{N} [\mathbf{C}_{\text{UU}} \cdot \mathbf{N}]$$
$$+ \frac{N_{\text{S}}}{N} \left[\sum_{\text{SS} \in \left(\begin{smallmatrix} ++, & +-, \\ -+, & -- \end{smallmatrix}\right)} \text{P}\,(\text{SS}) \mathbf{C}_{\text{SS}} \cdot \mathbf{N} \right] \qquad (7)$$

where N_{U}/N and N_{S}/N are the fraction of UWN and sign bits, respectively, and $\text{P}(\text{SS})$ represents the probability of the various sign transitions. These probabilities along with the mean, variance, and correlation statistics can be measured during a functional simulation of the candidate architecture. Since this can be a register-transfer level (RTL) simulation, the required activity statistics for the entire chip can be gathered rapidly and efficiently.

Results have shown that this method can be used to produce RT-level power estimates within 15% of switch-level simulations based on circuits extracted from layout. Thus, it is possible to accurately model datapath and memory power consumption at the architecture level by accounting for the impact of both activity and complexity on power. The DBT

518

Input Table:		Output Table:	
Present State	IN	Next State	OUT
0	0	0	0
0	1	1	1
1	0	1	1
1	1	0	0

(a)

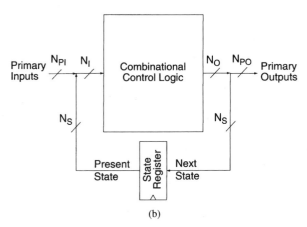

(b)

Fig. 4. Behavioral (a) and structural (b) descriptions of T flip-flop FSM.

model is appropriate for fixed-point two's-complement data; other representations might require new activity models.

It should also be clarified that in this section "memory" has referred to storage elements containing numerical *data*. Of course, memory can also contain control path information, such as instructions. In that case, it is necessary to characterize the memories under a different activity model. In particular, the activity model described in the next section is more appropriate for memories containing instructions or control information.

IV. CONTROL PATH MODELING

Controllers direct the sequence of operations to be executed by the datapath, initiate memory accesses, and coordinate data transfers over interconnect. The behavior of a typical controller can be described by a state transition graph (STG) or, equivalently, by a *control table* as shown in Fig. 4(a). Controllers are often implemented using the finite state machine (FSM) structure of Fig. 4(b). This structure uses combinational logic to generate the next state and the primary outputs given the present state and the primary inputs. The *implementation style* of the combinational logic can take many forms: e.g., ROM, PLA, or random logic (i.e., standard cells).

The task of architectural controller power analysis is to produce an estimate of the final implementation power given only the target implementation style and a description of the state machine to be realized, say, in the form of a control table. It is also possible to envision power estimation based solely on the FSM behavior; however, behavioral power estimation [16]–[19], though useful for rough preliminary predictions, necessarily offers fairly limited accuracy.

Even at the architecture level, accurately estimating controller power is a nontrivial task. In the datapath case, the circuitry of library elements (such as adders and multipliers) is often known *a priori*, in effect, fixing the physical capacitance

of the modules. In contrast, the physical capacitance of a controller depends on the contents of the control table.

Since it is impractical to characterize each class of controller for all possible control table contents we, instead, characterize for random control tables. This results in a fixed, average physical capacitance somewhere between the extremes of an "empty" (all zero) table and a "full" (all one) table. Then, for each implementation style, prototype controllers of different complexities can be synthesized *a priori* and characterized for various activity levels. Since the method accounts explicitly for the effect of activity on power consumption we refer to it as the Activity-Based Control, or ABC, model.

The discussion of controller power modeling will be divided into two parts. Section IV-A will describe model parameters that influence power regardless of implementation style, while Section IV-B will present target-specific power models. This will be followed by Section IV-C which will discuss techniques for library characterization and Section IV-D which will review the controller power analysis method being proposed here.

A. Target-Independent Parameters

Two classes of parameters influence controller power regardless of the target implementation style: complexity parameters and activity parameters. First, the complexity (or size) of a controller directly influences its physical capacitance and, therefore, its power consumption. Fig. 4 suggests that combinational logic block complexity can be measured to some extent by the number of inputs, N_I, and outputs, N_O. An increase in $N_O = N_S + N_{PO}$ requires additional logic to generate the larger number of next state bits, N_S, and/or primary outputs, N_{PO}. Similarly, a larger $N_I = N_S + N_{PI}$ means more input decoding due to an increase in the number of present state bits, N_S, and/or primary inputs, N_{PI}. Actually, N_I is only a good measure of input-plane complexity when exhaustive "address" decoding is used (as in a ROM). In other cases, the number of min-terms, N_M, in the logic-minimized control table is a better measure of input-plane complexity. Section VI-A describes techniques for estimating N_I, N_M, and N_O.

Complexity gives us some indication of the physical capacitance contained in a controller, but if the capacitance is not switched, no power is consumed. Since at the architecture level we treat combinational logic blocks as black boxes, activity can best be described by external measures. The *input activity* tells us something about how much switching occurs in the input plane, or "address" decoding, portion of the combinational logic. For static logic, the transition activity, α_I, is the proper measure of circuit activity in the input plane. It is equal to the fraction of input bits (including state) that switch each cycle. For dynamic logic, the signal probabilities of the inputs—that is, the probability that an input bit is one (P_I) or zero ($1 - P_I$)—determine circuit activity, since precharging to a known state each clock cycle negates the influence of the previous signal value on current power consumption. The input activity parameters (α_I, P_I) tell only half of the story, however—we also require a measure of *output*

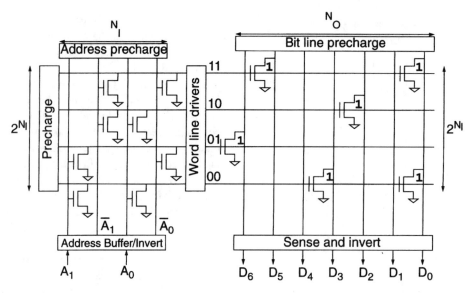

Fig. 5. Basic structure of prototype ROM (4 × 7 in this example).

activity: (α_O, P_O). Section VI-B will describe techniques for acquiring the necessary activity parameters through functional simulation.

B. Target-Specific Capacitance Models

The ABC method allows the user to specify a capacitance model that reflects how the effective capacitance of a particular controller class scales with changes in complexity and activity. This section illustrates how to construct a capacitance model using three case studies: a ROM-based controller, a PLA-based controller, and a random logic controller. The same concepts are readily applicable to other implementation styles.

1) ROM-Based Controllers: The exact form of a ROM capacitance model can vary from one implementation to another. To address this issue, we use a library-based approach, allowing the user to define a unique capacitance model for each implementation in the library. If the ROM structure is not known, the model can be based on measurements taken from data books or previous implementations. In this example, we will use the ROM structure of Fig. 5. The average capacitance switched during a single access is given by

$$C_T = C_0 + C_1 N_I 2^{N_I} + C_2 P_O N_O 2^{N_I} \\ + C_3 P_O N_O + C_4 N_O. \qquad (8)$$

The terms in this expression relate to power consumed in the input plane (address decoding) and the output plane (bit lines). We first analyze the complexity and activity of the input plane. The complexity is proportional to the product of the number of columns and rows in the input plane: $N_I 2^{N_I}$. Since both true and complement address lines are present, the input-plane activity is independent of external address activity—i.e., during evaluation, half of the precharged lines will remain high and the other half will discharge, regardless of external input activity. Thus, no explicit activity factor is present in the $C_1 N_I 2^{N_I}$ term. It might seem that a separate N_I term would be needed to model the address buffers, but remember that the load being driven by these buffers is proportional to

2^{N_I}. Thus, the single $C_1 N_I 2^{N_I}$ term properly models both the address buffers and decoders.

In the output plane, the power consumption is dominated by charging and discharging the bit lines whose lengths (and capacitances) are proportional to the number of rows in the array 2^{N_I}. Initially high, the bit lines corresponding to 1 output bits (on average, $P_O N_O$ of them) discharge during an access and then precharge prior to the next read cycle. This explains the $C_2 P_O N_O 2^{N_I}$ term. The buffering circuitry for the N_O outputs gives rise to the $C_3 P_O N_O$ term, but there is also an activity-independent term $C_4 N_O$ since the sense circuitry produces both true and complemented signals.

Combining terms yields the ROM capacitance model of (8), where C_0, C_1, C_2, C_3, and C_4 are capacitive coefficients dependent on the exact circuitry and technology used by the ROM. These coefficients are extracted through a library characterization process that will be described in Section IV-C.

As a means of validation, a comparison was made between a switch-level simulation of the ROM (using IRSIM-CAP) and the ABC model after characterization for a 1.2 μm technology using random control table contents. IRSIM-CAP [14] is a modified version of the switch-level simulator IRSIM [20] with improved capacitance measurement capabilities. The simulations were performed for random input streams with distributions chosen to exercise a variety of input and output activity levels. As with all IRSIM-CAP results described in this paper, the simulations were allowed to continue until the average capacitance switched satisfied a convergence criteria of 5%. This strategy allows us to handle sequential circuits with reasonable accuracy.

For this example, the results of the validation process are shown in Fig. 6. The ABC model exhibits an rms estimation error (relative to IRSIM-CAP) of about 2.5% and a maximum error of 4.5%. The arrows in the figure denote results for controllers of fixed complexity for which the output signal probability, P_O, varies from zero to one. The fact that power consumption varies significantly with P_O is a strong argument in favor of a modeling strategy (such as the ABC technique)

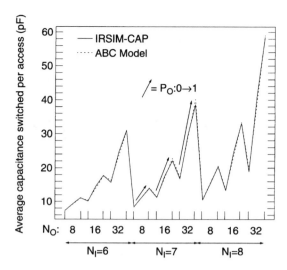

Fig. 6.   ROM-based controller: IRSIM-CAP versus ABC model.

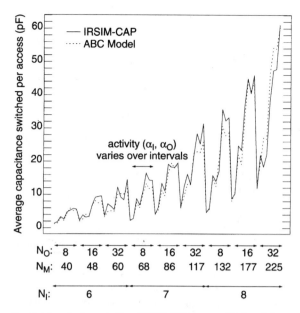

Fig. 7.   Random logic controller: IRSIM-CAP versus ABC model.

which accounts for activity. It is also of interest to observe the magnitude of errors that can be expected for nonrandom control table contents. Section VII-B will show a controller with highly redundant (nonrandom) control table entries for which the power for three implementations differs on average by 12.6% from the ABC prediction and at most by 29%.

2) PLA-Based Controllers: Conceptually, the structure of a PLA is quite similar to that of a ROM. The principal difference between a PLA and a ROM is that the input plane of the ROM performs a full decoding of all possible addresses, while a PLA uses logic minimization to reduce the amount of decoding required. As a result, the height of the input and output planes in a PLA is given by the number of unique min-terms, N_M, which will usually be less than 2^{N_I}. For a sample PLA with a static input plane and a dynamic output plane, the capacitance model was found to follow

$$C_T = C_0 \alpha_I N_I N_M + C_1 P_O N_O N_M \\ + C_2 P_O N_O + C_3 N_O N_M + C_4 N_M. \qquad (9)$$

We can compare this model to the ROM expression of (8). As expected, 2^{N_I} has given way to N_M. Also, since this PLA happens to use static input decoding, an α_I term has been added to model the effect of input transitions on power. In the output plane, since no differential signaling is used, the $C_4 N_O$ term disappears. Furthermore, this PLA uses a clocked virtual ground node in the output plane that charges to V_{dd} each cycle, regardless of the output signal values. This gives rise to the activity-independent terms: $C_3 N_O N_M$ and $C_4 N_M$. The model resulting from characterization of the PLA (in a 1.2 μm technology) has an rms and maximum error of 2.5% and 6.3%, respectively.

3) Random Logic Controllers: Since a random logic implementation is much less regular than a PLA or ROM, it is more difficult to come up with an accurate capacitance model, but an approximate model for a standard cell controller implemented in static logic might be given by

$$C_T = C_0 \alpha_I N_I N_M + C_1 \alpha_O N_O N_M. \qquad (10)$$

This expression contains two components—one relating to the input plane capacitance and the other relating to the output plane. Since this example is based on static logic, the appropriate activity measures are α_I and α_O, respectively. For the input plane, the complexity is given by the product of the number of inputs to that plane N_I and the number of outputs that plane produces N_M. The same is true for the output plane, except in this case there are N_M inputs to the plane and N_O outputs. Since the exact equation depends on the synthesis tools being used, the library maintainer is free to tune the model as needed.

The two capacitive coefficients are derived during a characterization phase and will be a function of the standard cell library being used and, to some extent, the logic minimization, placement, and routing tools being applied. A comparison of the model to switch-level simulations for a 1.2 μm cell library is shown in Fig. 7. The results are for control tables with random contents that have been minimized using espresso [21] and synthesized using MIS [22] and the Lager IV silicon assembly system [23]. While the agreement is not as good as the ROM and PLA models, the rms error is still an acceptable 15.1%.

C. Characterization Method

Before the controller capacitance models can be used, circuit- and technology-dependent values of the capacitive coefficients must be measured through a *characterization* process. For a given class of controller, the idea is to actually measure the capacitance switched within implementations of varying complexities for different input and output activities. The observations are then used to find capacitive coefficients that give the best fit to the measured data. The characterization process occurs in three phases: pattern generation, simulation, and coefficient extraction.

1) Pattern Generation: In the first phase, two distinct sets of data patterns are generated: the patterns stored in the control

TABLE I
"ADDRESS/DATA" PATTERNS FOR GENERATING DESIRED CONTROLLER ACTIVITIES

Label	Address A[7:2]	Map A[1:0]	Data
$A(0,0)$	000000	00	00000000
$A(0,\frac{1}{2})$		01	01010101
$A(0,1)$		11	11111111
$A(\frac{1}{2},0)$	010101	00	00000000
$A(\frac{1}{2},\frac{1}{2})$		01	01010101
$A(\frac{1}{2},1)$		11	11111111
$A(1,0)$	111111	00	00000000
$A(1,\frac{1}{2})$		01	01010101
$A(1,1)$		11	11111111

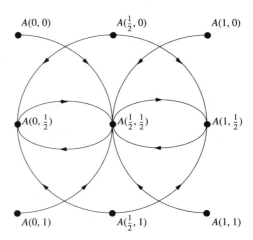

Fig. 8. Graphical representation of address sequences for $(\alpha_I, P_O) = (1/2, 1/2)$.

table that the FSM implements and the patterns applied as input to the controller during simulation. Since the control table patterns can affect the physical capacitance of the controller and since it would be impossible to characterize for all possible control tables, we instead generate patterns that result in some sort of average physical capacitance. A reasonable approximation is to use a random output table with a uniform distributions of zeros and ones. In real controllers, however, not all outputs affect system behavior in all states. Setting a fraction (say, half) of the output table entries to don't-cares models this effect. PLA and random logic synthesizers can then exploit the don't-cares by using logic minimization algorithms (such as espresso [21]).

The second phase of pattern generation entails producing input streams that exercise the controller over a full range of activities from 0–100%. Three evenly spaced activity levels (0, 1/2, and 1) can be realized by correct sequencing of three basic patterns: $00 \cdots 00$, $01 \cdots 01$, and $11 \cdots 11$. The input activity can be controlled by using the patterns as input "addresses" to the combinational logic block of the controller. The output activity can be independently controlled by storing the same three data patterns at each of these three "addresses." While three different values cannot be referenced by a single address, they can, however, be referenced by "similar" addresses. As exemplified in Table I, the two least significant address bits can be used to map the three different data patterns to similar, but unique, addresses. The left-most column provides a convenient label for each (input, output) activity pair. In summary, while the output table is still filled primarily with random data, it also contains nine deterministic values that allow precise control of input and output activities during characterization.

Using this strategy, input and output activities can be controlled fairly independently. For instance, desired input and output *signal probabilities* can be chosen from nine possibilities, $(P_I, P_O) \in \{0, 1/2, 1\} \times \{0, 1/2, 1\}$, simply by accessing address $A(P_I, P_O)$. It is also possible to generate any of nine different *transition* activities of the form $(\alpha_I, \alpha_O) \in \{0, 1/2, 1\} \times \{0, 1/2, 1\}$ by accessing an address sequence $A(\alpha_I^0, \alpha_O^0) \to A(\alpha_I^1, \alpha_O^1)$ that satisfies $\alpha_I = |\alpha_I^0 - \alpha_I^1|$ and $\alpha_O = |\alpha_O^0 - \alpha_O^1|$.

Desired address sequences can be represented graphically by associating the ordered activity pairs with a coordinate system as shown in Fig. 8, which shows address transitions corresponding to the activity pair $(\alpha_I, P_O) = (1/2, 1/2)$. As the figure demonstrates, the input transition activity determines how many columns each edge traverses, and the output signal probability determines at which row each edge terminates. Recalling (9) from Section IV-B-2, input transition activity α_I, and output signal probability P_O are the activity parameters appropriate for the PLA-based controller. A full characterization of the PLA would include nine activity pairs $(\alpha_I, P_O) \in \{0, 1/2, 1\} \times \{0, 1/2, 1\}$. The standard cell controller requires nine as well, $(\alpha_I, \alpha_O) \in \{0, 1/2, 1\} \times \{0, 1/2, 1\}$, but the ROM capacitance model requires only three activity values, $P_O \in \{0, 1/2, 1\}$. In each case, the simulated capacitance observations for individual input transitions are averaged together to produce an aggregate switching capacitance for each I/O activity level and each implementation style.

Clearly, this strategy for selecting representative control table and input/output patterns is an approximation. It is difficult to precisely gauge the sensitivity of the model accuracy to pattern selection, since pathological cases can always be constructed. These cases would not say much, however, about what range of accuracies could be expected in practice. For this information, it is perhaps best to turn to the real controller designs that have been characterized with this method. These are discussed in Section VII where we will see, for example, that the maximum ABC prediction error for any of the implementations shown was 29%.

2) Simulation and Coefficient Extraction: During the next phase of characterization, a module corresponding to a given set of complexity parameters (e.g., number of inputs, min-terms, and outputs) is synthesized and simulated for the data patterns generated by the aforementioned procedure. The simulation can be performed with either a circuit- or gate-level tool (e.g., SPICE [1], PowerMill [2], IRSIM [20]) depending on the time the designer wishes to allow for characterization and the accuracy desired. As mentioned above, the simulations described in this paper were all performed with IRSIM-CAP using a convergence criteria of 5%. Simulating to convergence, rather than for a fixed number of steps, allows the method to handle sequential circuits with reasonable accuracy. The output of the simulation process is a number of capacitance observations—one for each controller style, complexity, and activity level.

Coefficient extraction refers to the process of deriving best-fit model coefficients from the raw simulation data. This can be achieved using techniques such as least-squares regression, which minimizes the mean-squared error of the model. In vector notation, this amounts to solving the following matrix equation for the capacitive coefficient vector $\mathbf{C}_{eff}$

$$\mathbf{C}_{sim} = \mathbf{P}\mathbf{C}_{eff} + \mathbf{e} \tag{11}$$

where $\mathbf{C}_{sim}$ is a vector of simulated capacitance observations, $\mathbf{P}$ is a matrix of complexity and activity parameter values corresponding to the observations, and e is the modeling error.

D. ABC Power Analysis Method

Analyzing controller power requires a capacitance model, a set of capacitive coefficients, and the appropriate complexity and activity parameters. Given these inputs, the combinational logic capacitance model can be evaluated (in vector form) as

$$C_T^{CL} = \mathbf{C}_{eff}^{CL} \cdot \mathbf{N} \tag{12}$$

where $\mathbf{C}_{eff}^{CL}$ is a vector of capacitive coefficients and $\mathbf{N}$ is a vector of complexity and activity parameters for a given module. Aside from the combinational logic block, the state register also contributes to the overall power consumption of the controller. Since N_S state bits must be stored, the capacitance model for the state register will have the following basic form

$$C_T^{reg} = \alpha_S C_0 N_S \tag{13}$$

where α_S is the activity of the state bits. The total controller capacitance per access, then, is $C_T = C_T^{CL} + C_T^{reg}$. If many accesses are involved, the total capacitance switched over a number of input control transitions N_{CT} can be computed using the following expression: $C_T|_{multi-cycle} = N_{CT} \cdot C_T|_{single-cycle}$.

E. Control Path Summary

The ABC model presented here explicitly accounts for activity and provides a general framework for modeling different controller structures. Three specific examples—based on ROM's, PLA's, and random logic—were presented in this section, but the same general techniques could be used to model other implementation styles. Using the ABC method, analyzing controller power amounts to plugging the appropriate activity and complexity parameters into an equation that weights these parameters by technology- and implementation-dependent capacitive coefficients. The result is an accurate architecture-level estimate that reflects both the physical capacitance and the circuit activity of the controller being analyzed.

V. INTERCONNECT MODELING

The final class of component in a typical chip is interconnect. This section addresses the problem of estimating the power consumed in charging and discharging interconnect capacitance.

A. Interconnect Activity

The activity of a wire depends on the type of signal that wire carries: data or control. The activity of a data signal can be described by the DBT model. Using this model, the total capacitance switched during a series of N_{DT} data transitions on a bus driven by static logic is given by

$$\text{static: } C_{DBT} = N_{DT}[\tfrac{1}{4}C_w N_\text{U} + P(+-)C_w N_\text{S}] \tag{14}$$

where 1/4 is the probability that a UWN bit transitions from zero to one, C_w is the physical capacitance of the wires, N_U is the number of UWN bits in the data model, N_S is the number of sign bits, and $P(+-)$ is the probability of a positive to negative transition between two successive samples in the data stream. For a control wire driven by static logic, the appropriate activity is given by the ABC parameter α, the fraction of bits in the control signal that transition. The total bus capacitance switched during N_{CT} control word transitions is

$$\text{static: } C_{ABC} = N_{CT}[\tfrac{1}{2}\alpha C_w N] \tag{15}$$

where α is the probability that a bit in the control word makes a transition, $1/2\alpha$ is the probability of a zero-to-one transition, and N is the number of bits required to represent all values that the control word can assume. The potentially strong correlation between the control signals making up the control word is handled by the fact that activities are estimated using a simulation-based approach as described in Section VI-B.

Equations (14) and (15) apply to static buses. Some buses, however, use precharged logic. In other words, each clock cycle they are precharged to V_{dd} and then the 0 bits discharge during an evaluation phase. For these precharged buses, the appropriate DBT effective capacitance equation is

$$\text{dynamic: } C_{DBT} = N_{clk}[\tfrac{1}{2}C_w N_\text{U} + P(+)C_w N_\text{S}] \tag{16}$$

where N_{clk} is the number of clock cycles being evaluated and $P(+)$ is the probability that a data sample is positive. Similarly, the ABC capacitance model becomes

$$\text{dynamic: } C_{ABC} = N_{clk}[(1 - P)C_w N] \tag{17}$$

where P is the ABC signal probability parameter.

These expressions rely on the availability of DBT/ABC activity parameters and the physical wire capacitance. Section VI-B will describe how to derive activity parameters. The remainder of this section will cover physical capacitance estimation.

B. Physical Capacitance

In the ideal case, the physical capacitance of the interconnect network is known and can be used directly in architectural

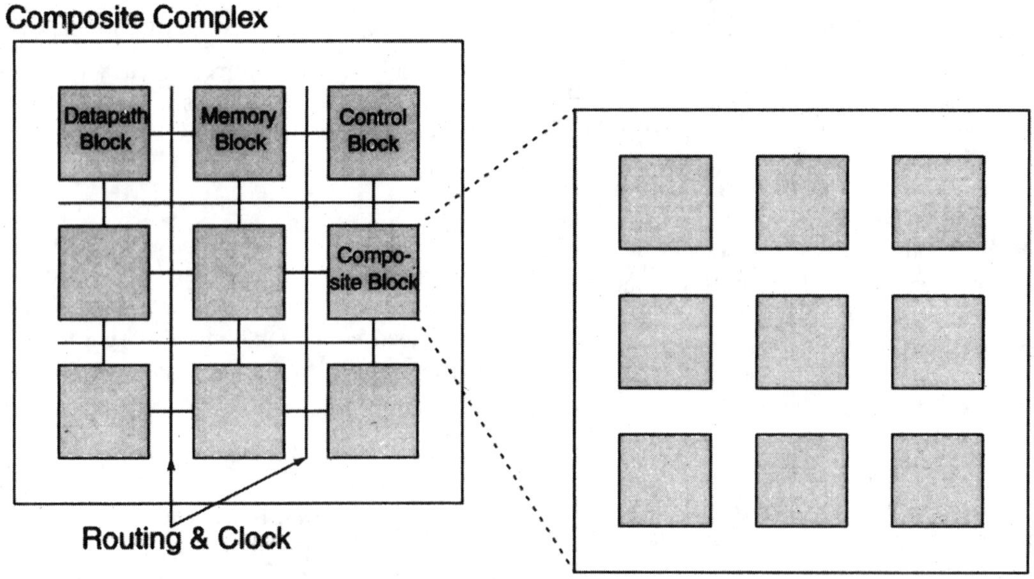

Fig. 9. Hierarchical structure of chip.

power estimation. More typically, however, analysis occurs prior to layout and wire capacitances must be estimated. Given a process technology and a set of design rules, it is possible to make fairly accurate estimates of average interconnect capacitance per unit length. Estimating interconnect length is difficult, however, and depends intimately on design-specific layout considerations.

The topic has received a good deal of attention in the past. Feuer [24] and Donath [25] have presented derivations based on Rent's Rule [12], which give rise to expressions for average lengths proportional to a power $(p < 1)$ of design area. Sorkin [26] confirms this general rule noting that empirical data support average lengths proportional to the square root of the chip area (i.e., $p = 1/2$). Donath also makes use of hierarchical partitioning and placement. Our techniques rely heavily on this notion, as well as the empirical observations of Sorkin resulting in an intuitive, yet reasonably accurate, analysis strategy.

C. Wire Length Estimation Strategy

The input to this process is a hierarchical RTL description of the chip being analyzed, which may contain four classes of blocks (see Fig. 9). A *composite* block is used to introduce further hierarchy into the design. The other three block types are named for the type of structures within the block: datapath, memory, and control.

Estimating wire length at a given level of hierarchy requires area estimates for each component block. This suggests a recursive analysis scheme using a depth-first traversal strategy. Datapath, memory, and control blocks are treated as leaf cells and are handled by dedicated analysis routines. This hierarchical approach preserves partitioning clues supplied by the designer. This enables block-by-block estimates of wire length as opposed to a single chip-wide average. Sections V-D–G describe the analysis strategies for each of the four block types: composite, datapath, memory, and control.

D. Composite Blocks

Since we are interested in early estimates of wire length, precise placement and routing information is not available. Instead, we must predict what the average length is likely to be for a "good" placement. One option is to perform an early floorplanning step, but if this process is deemed too expensive there are other reasonable alternatives.

The approach advocated here relies on the empirical observation that the quality of a "good" placement often differs from a random placement by a constant factor, k [25], [26]. Since the average wire length for a random placement on a square array is 1/3 the side-length of the complex [27], the average length for a "good" placement is approximately

$$L = k \frac{\sqrt{A}}{3}. \tag{18}$$

The exact value of the k factor will depend on the characteristics of the placement and routing tools being used; however, an oft-quoted conversion factor is approximately 3/5. Since this value results in the formula $L = \sqrt{A}/5$, it has been called the "1/5 rule" in the literature [26].

The area of a composite complex is the sum the areas of the component blocks A_B and the area occupied by wires A_w

$$\begin{aligned} A &= A_w + A_B \\ &= A_w + \sum_{i \in \{\text{Blocks}\}} A_i. \end{aligned} \tag{19}$$

The component of area related to routing depends on the total number of wires in the complex (N_w), the average pitch at which they are routed (W_p), and their average length (L)

$$A_w = N_w W_p L. \tag{20}$$

Taken simultaneously, (18)–(20) result in a quadratic expression that can be solved to yield

$$L = \frac{k^2 N_w W_p + \sqrt{(k^2 N_w W_p)^2 + 36 k^2 A_B}}{18}. \tag{21}$$

524

The above expressions can be used to recursively estimate the area and average interconnect length of composite complexes. For a parallel discussion of clock length estimation the reader is referred to [15].

E. Datapath Blocks

Since signals often flow through a datapath in a fairly linear fashion, datapath blocks often employ a one- rather than two-dimensional placement of modules, where the modules are formed by tiling up N individual bit-slices. Thus, interconnect length is proportional to the length, and not the area, of the datapath

$$L_x = k \frac{L_{DP}}{3}. \tag{22}$$

As before, we can use a k factor (specific to the datapath tiler) to convert from a random placement to a "good" one. The length of the datapath L_{DP} in turn is determined by module lengths and by the routing channels separating the modules

$$L_{DP} = L_R + \sum_{i \in \{\text{Blocks}\}} L_i. \tag{23}$$

Module lengths can be read from a library database, but routing channel lengths must be estimated from wiring pitch W_p and the number of vertical wiring tracks between modules, which is close to the total number of I/O terminals on all the modules

$$L_R = W_p \sum_{i \in \{\text{Modules}\}} N_{IO_i}. \tag{24}$$

In order to connect the terminals of two modules in the datapath, the wire must first be routed vertically from the terminals to the selected feedthrough. The length of these vertical wiring components is related to the width W_{BS} of a datapath bit slice

$$L_y = 2k \frac{W_{BS}}{3}. \tag{25}$$

Combining L_x and L_y gives us our average interconnect length $L = L_x + L_y$.

Finally, the area of the datapath, required for analyzing the next level up in the hierarchy, can be approximated as

$$\begin{aligned} A &= W_{DP} L_{DP} \\ &= N W_{BS} L_{DP}. \end{aligned} \tag{26}$$

All these calculations are based on the availability of raw area data for primitive modules (e.g., adders, shifters, etc.). Unlike power, however, the area of primitive modules is a deterministic function of their complexity parameters [15]. Thus, the area data can be entered into the hardware database from direct measurement of layout dimensions.

F. Memory Blocks

Since the power models for memories already account for internal wiring, interconnect analysis is not required, per se, for these modules. Area estimates are required, however, to analyze parent blocks in the hierarchy. As with datapath modules, deterministic formulas can be derived relating complexity parameters (i.e., storage capacity) to expected layout area [15].

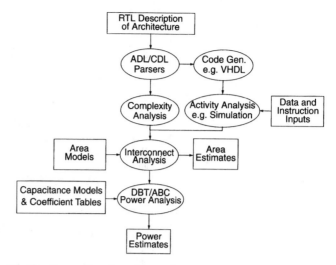

Fig. 10. Process flow for architectural power/area analysis.

G. Control Blocks

As for memory blocks, internal wiring is already included in the power models for control blocks when these blocks are characterized. So, once again, all that is required are area models for the three basic classes of controllers. The same model fitting techniques that were used for power can be abstracted to area modeling; however, since activity does not come into play, the area models will, in general, be more straightforward than the power models. Details are available in [15].

H. Interconnect Summary

To review, we propose a hierarchical approach to interconnect analysis that takes advantage of partitioning clues provided by the designer. This allows a distinction to be made between local and global interconnects. At each level of hierarchy, the length of the wires depended largely on the area of the block being analyzed. Once ascertained, the wire length estimates can be combined with DBT/ABC activity models to predict the power consumed in driving interconnect and clock networks.

The accuracy of these area and interconnect estimation techniques will be discussed more fully in Section VII. As a brief preview, the area estimate for the microprocessor example presented there is within 8% of the actual area. Moreover, the average interconnect and clock length estimates are off by 7% and 22%, respectively. Finally, the datapath-specific interconnect estimate errs by 17%.

VI. SPA: AN ARCHITECTURAL POWER/AREA ANALYZER

The previous sections have presented models and methods for analyzing the power consumption of the four basic classes of components: datapath, memory, control, and interconnect. These techniques have been integrated into an architectural power/area analysis environment called SPA. The SPA analysis flow consists of several phases as shown in Fig. 10. The primary input to SPA is a hierarchical RT-level description of the architecture under consideration. Currently, SPA uses

```
(inputs (State_Type     state))
(outputs (Counter_Fn_Type a_reg)
          ...
        (State_Type next_state))

(output-table
; state     | a_reg b_reg mux0 xi_reg yi_reg mux1 next_state
; -----     | ----- ----- ---- ------ ------ ---- ----------
  (LDB   )  (NOP   LD    SEL0 NOP    NOP    SEL1 INV)
  (INV   )  (LD    NOP   SEL0 LD     NOP    SEL1 DIV0a)
  (DIV0a )  (NOP   NOP   SEL1 NOP    LD     SEL0 DIV0b)
  (DIV0b )  (NOP   NOP   SEL1 LD     NOP    SEL2 DIV1a)
  (DIV1a )  (NOP   NOP   SEL1 NOP    LD     SEL0 DIV1b)
  (DIV1b )  (NOP   NOP   SEL1 LD     NOP    SEL2 DIV2a)
  (DIV2a )  (NOP   NOP   SEL1 NOP    LD     SEL0 DIV2b)
  (DIV2b )  (NOP   NOP   SEL1 LD     NOP    SEL2 DIV3a)
  (DIV3a )  (NOP   NOP   SEL1 NOP    LD     SEL0 DIV3b)
  (DIV3b )  (NOP   NOP   SEL1 LD     NOP    SEL2 MULT)
  (MULT  )  (NOP   NOP   SEL1 LD     NOP    SEL1 LDB)
)
```

Fig. 11. Excerpt from CDL description of iterative hardware divider architecture.

a textual architectural description language (ADL) for this purpose; however, the same information could be provided by a graphical schematic capture interface. The control path is described using a control description language (CDL) that employs control tables, which specify how the next state and outputs of each control module relate to the present state and inputs. In order to maintain a relatively high level of abstraction, CDL allows the user to specify control signals and states as enumerated (symbolic) types rather than bit vectors (see Fig. 11). With the structure and behavior of the architecture fully defined, the next step in the process is to derive the complexity and activity parameters required by the power and area analysis models.

A. Complexity Analysis

The exact complexity parameters required and the method for calculating them differs depending on whether the entity falls under the DBT or ABC models. DBT complexity parameters, such as word length N, specify the "size" of datapath and memory elements. DBT complexity analysis consists of stepping through the various entities in the structural description and reading off the values specified by the user in the ADL "parameter" list of each module.

ABC complexity analysis reduces to ascertaining the "size" of control buses and modules. In order to maintain a high level of abstraction, all control signals are specified as enumerated types that take on symbolic, rather than binary, values. For example, if a control bus carries the function input to an ALU, the type of the bus might be defined as

$$ \text{ALU\_TYPE} \equiv \{\text{ADD, SUB, SHIFT, CMP}\}. \quad (27) $$

The word length N (in bits) required to represent an enumerated type T which can take on $|T|$ different values is

$$ N = \lceil \log_2 |T| \rceil. \quad (28) $$

For instance, two bits are required for a binary encoding of an ALU_TYPE control bus; however, for a "one-hot" encoding

strategy, the appropriate complexity formula would simply be $N = |T|$. One way to handle multiple encoding schemes would be to have the user associate an *encoding style attribute* with each ABC data type or entity.

Two important complexity parameters for a control module are the number of input bits N_I and the number of output bits N_O for the combinational logic block of each controller. This block may take many control buses as input and may feed multiple control buses at its output. The total input and output widths are calculated by summing individual control field widths

$$ N_{\text{I/O}} = \sum_{i \in \{\text{I/Os}\}} N_{\text{I/O}_i} $$
$$ = \sum_{i \in \{\text{I/Os}\}} \lceil \log_2 |\text{I/O\_TYPE}_i| \rceil. \quad (29) $$

The number of min-terms N_M in the minimized control table is another important complexity parameter. This is more difficult to estimate since the amount of minimization that can be performed depends on the binary encoding of the control table, which is probably not available at this stage of design. It is more likely that the control table is specified in terms of symbolic values as shown in Fig. 11. We can approximate N_M by assuming some binary mapping (e.g., random), performing logic minimization (e.g., using espresso [21]), and using the number of min-terms in the minimized table as an estimate of N_M.

In summary, the complexity parameters can be determined from the architectural description of the design in question. The DBT parameters can be read directly from user specified parameter lists and the ABC parameters can be computed using a few simple expressions. Finally, the user is free to add custom complexity parameters to individual modules when appropriate as long as their values are specified when instantiating the module.

526

B. Activity Analysis

The next step in the process is to derive the ABC activity statistics for the design. This is accomplished by a functional simulation of the architecture. Many RT-level simulators would be suitable for this task, but currently SPA uses a VHDL simulator provided by Synopsys. A code generator is used to produce structural VHDL for each ADL block and behavioral VHDL for each CDL block. A collection of activity monitors attached to the buses and modules in the VHDL description accumulate activity statistics during simulation. If the chip requires data or instruction inputs, these must be supplied by the user who is, therefore, able to characterize the power consumption for various input patterns and operating conditions. Since functional simulation is quite fast, run time is usually not an important concern and is often on the order of seconds or minutes.

The activity parameters for the DBT model are the word-level statistics discussed in Section III. These statistics include: mean (μ), variance (σ^2), correlation (ρ), sign transition probabilities, and access count (N_{DT}). For a synchronous system, this access count may be less than or equal to the number of clock cycles simulated; or, in the presence of block-level glitching, it may actually be higher.

ABC activity parameters can pertain either to a control bus or to a control module. For a control bus we must monitor the control word transition count, N_{CT}. We must also determine two other ABC activity parameters: signal probability P and transition activity α. If the binary encoding of symbols is known, exact values for P and α can be calculated during simulation by monitoring control word transitions. Otherwise, we can assume a random encoding (i.e., $P = 1/2$ and $\alpha = 1/2$).

Control modules require two sets of (α, P) activity statistics: one for the combinational logic block input and one for the output. The input and output, in turn, may consist of several control buses bundled together. A transition of the controller input word leads to an increment of N_{CT} and is defined to occur when *any* of the component input signals transition, since this will initiate a reevaluation of the combinational logic block. If we assume a random encoding, we again expect half of the input and output bits to be one, so $P_I = P_O = 1/2$. As for the transition activities, the following expression applies

$$\alpha_{\text{I/O}} = \frac{\sum_{i \in \{\text{I/Os}\}} \frac{1}{2} N_{\text{I/O}_i} N_{CT_i}}{N_{\text{I/O}} N_{CT}} \quad (30)$$

where i refers to the individual control buses that make up the input/output words. For the case of input activity, this equation can be explained as follows: $1/2 N_{I_i}$ is the number of bits that switch on average when input i makes a transition. This is then weighted by the total number of transitions that input i makes, N_{CT}, to yield the number of bits within input i that toggle over the entire simulation. This is then summed across all of the individual inputs to yield the total number of bits that switch in the full input word during the simulation. Dividing this by the number of controller input transitions, N_{CT}, and the total number of input bits, N_I, results in the fraction of input bits that toggle during a typical input transition—that is, α_I.

To review, the ABC activity parameters depend on the encoding of the control words as binary values. If the user specifies the actual mapping, then simulated activities can be quite accurate; however, more likely the user will specify abstract symbolic types for the control signals and state machine control tables. In this case, a random assignment can be assumed in order to allow estimates of the control signal activities to be formed.

C. Power/Area Analysis

The complexity and activity parameters are then fed to the core area and power analysis routines along with the parsed architectural description. SPA then steps through each datapath, memory, control, and interconnect entity in the design, applying the appropriate power and area analysis models. Finally, SPA dumps its results, categorizing area and power consumption in terms of the four classes of components. This points the designer to the most power-intensive portions of the implementation and provides useful guidance for further optimizations.

VII. RESULTS

In this section, we present results gathered using the SPA power/area analysis tool. The first example is a Quadrature Mirror Filter which shows SPA's applicability to datapath-intensive applications. The second example is a control-intensive design—specifically, a finite-state machine that implements the global control function for a speech recognition front-end. The final case study is a programmable instruction set processor that demonstrates SPA's ability to handle a real-world example containing significant datapath and control components. As the microprocessor employs on-chip instruction and data memories, this example will also serve as a verification of SPA's efficacy for memory-intensive designs.

A. Datapath Intensive: A Quadrature Mirror Filter

SPA allows the designer to efficiently explore the design space, searching for low-power solutions. This example will demonstrate a design flow that employs SPA to minimize the power consumed by a Quadrature Mirror Filter (QMF) such as might be used in a subband coding algorithm [28]. A quadrature mirror filter takes an input signal and splits it into two bands: a low-pass band, $H_{LP}(\omega)$, and a high-pass band, $H_{HP}(\omega)$. The sample rate chosen for the filter would, in general, depend on the application. For the purposes of this example, we select a sample period of 0.3 μs or about 3.33 MHz. Four candidate architectures were explored using SPA.

The first version is a direct, naive implementation of the algorithm. The power and area predictions provided by SPA are shown in the "Initial" column of Table II. Four costly array multipliers are required to meet the throughput requirements of the algorithm at 5 V and this leads to a large die size of 95.9 mm^2.

In the second version of the design, the expensive array multipliers are replaced by shift-add operations, reducing the chip area to a more reasonable 17.2 mm^2. SPA reveals that this

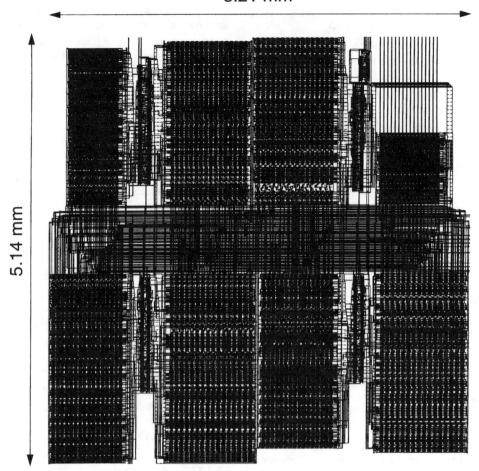

Fig. 12. Layout of retimed QMF filter design.

TABLE II
SPA POWER/AREA PREDICTIONS FOR QMF EXAMPLE

	Initial	Shift-Add	Retimed	Pipelined
V_{dd} (V)	5	5	1.5	1.25
Power (mW)	348.8	149.5	21.0	7.0
Area (mm^2)	95.9	17.2	24.4	115.4

version of the chip consumes 57% less power than the initial version, while at the same time occupying 82% less area.

A third implementation of the QMF example was generated by applying retiming, which reduces the critical path to 60 ns at 5 V. This allows us to lower the supply voltage of the implementation to about 1.5 V while still meeting the sampling rate constraint. Analysis using SPA shows a 7.1× reduction in power. Additional hardware requirements, however, increase the implementation area from 17.2 mm^2 to 24.4 mm^2.

A fourth version of the filter can be generated by pipelining the algorithm enabling a fully parallel implementation. This further reduces the critical path and allows the voltage supply to be reduced to 1.25 V. SPA confirms an additional power reduction of 3× for an overall reduction (from version one to version four) of 50×. Interestingly enough voltage reduction accounts for only 46% of the power saved by going from the retimed to the pipelined design. Fully 54% of the power saved by pipelining can be attributed to a distributed architec-ture which preserves signal correlations and, thus, minimizes switching activity. SPA is able to model these effects, but traditional estimators based on white-noise activity models are not.

While the pipelined example at 7 mW consumes less power than the 21 mW retimed design, it is at the cost of a 4.7× area increase. As a result, the retimed example, which is still 17× lower power than the initial solution and requires only 24.4 mm^2 is probably a more desirable solution.

To verify that SPA provided accurate area and power estimates, this version of the filter has been synthesized (down to layout), extracted, and simulated with actual speech samples as inputs. The chip plot is shown in Fig. 12. The predicted area of 24.4 mm^2 is within 9% of the actual 26.8 mm^2 area. A comparison of the SPA power predictions to switch-level simulation using IRSIM-CAP is given in Fig. 13. The figure shows the average power consumed by the chip for data streams corresponding to increasing input signal powers. SPA's estimates are within 5% to 14% of IRSIM-CAP for all data streams. Estimates based on the white-noise model are also included. The white-noise estimates do not track signal statistics and, therefore, err by as much as 71% for some of the data streams.

By using SPA to compare four candidate architectures we were able to significantly reduce design time. The four filter

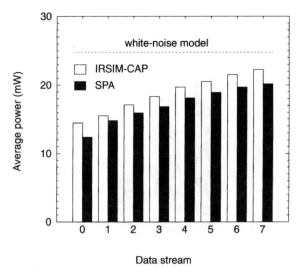

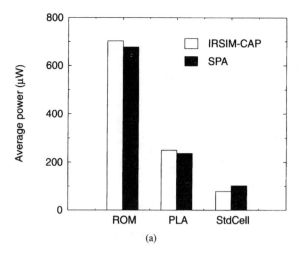

(a)

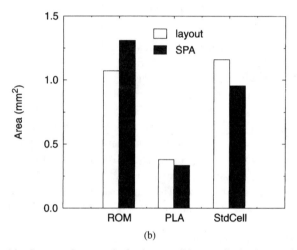

(b)

Fig. 13. Comparison of SPA to IRSIM-CAP for retimed QMF.

versions described here were synthesized using the HYPER high-level synthesis system [29] and analyzed with SPA in 5 min on a Sun SPARCstation 10. In contrast, laying out and simulating the retimed version took 3.2 h. Laying out and analyzing all four designs using low-level power analysis tools would have required 13 h or more.

B. Control Intensive: A Speech Recognition FSM

This example will demonstrate the use of SPA to aid in the design of a global finite state machine taken from the front-end of a speech recognition system [30]. Since this particular speech recognition chip-set is targeted at mobile applications, minimizing power consumption is a significant consideration. The state machine contains over 100 states, 10 inputs, and 25 outputs. The majority of the control table entries are redundant, making the FSM a good test of how well the estimation models can handle nonrandom control table contents.

Fig. 14 shows the estimates provided by SPA for three possible implementation styles: ROM, PLA, and standard cell. The estimates are for a system clock of 3.3 MHz and a supply voltage of 1.5 V. The results show the area and average power for each candidate implementation. In order to verify that these predictions are reliable, all three implementations have been laid out, extracted, and simulated. The results from these physical designs have been included in Fig. 14 for comparison. The average error in the area estimates is 17.3% and the maximum error is 22%. The average error in the power estimates is 12.6%, while the maximum error is 29% (standard cell case). The standard cell case is more difficult since random logic is less regular and is more influenced by the binary contents of the particular control table being implemented. SPA still provides reasonable results, however, and more importantly it correctly tracks the influence of implementation style on area and power.

Based on the SPA results the designer could immediately eliminate the ROM-based solution. It consumes by far the most power and is also very large since it does not take advantage of redundancies in the control table. Both the PLA and standard cell implementations use logic minimization to reduce the

Fig. 14. Power and area results for three possible controller implementations. (a) Predicted versus actual power. (b) Predicted versus actual area.

amount of input decoding required. The higher regularity of the PLA allows it to be significantly smaller than the standard cell design, but the precharged nature of the unit leads to a higher activity (and power consumption) than the static logic standard cell implementation.

The time that can be saved by exploring these kinds of trade-offs at the architecture level is significant. For this example, on a Sun SPARCstation 10, SPA analyzed the area and power of the three controller implementations in about 2 min. In contrast, generating and extracting the layout for the three controllers took about 5 h. Simulation required another 30 min. Without high-level analysis tools it would not be practical to thoroughly explore all implementation possibilities.

C. Memory Intensive: A Programmable Microprocessor

The previous two examples have demonstrated SPA's accuracy and flexibility by showing how it can be used to analyze both datapath- and control-intensive applications. In this example, we tackle a programmable instruction set processor with on-chip instruction and data memories. Therefore, this case study demonstrates SPA's ability to handle memory-intensive, as well as datapath- and control-intensive, designs.

Fig. 15 depicts the architecture of the simple microcoded processor under consideration. The processor can execute the

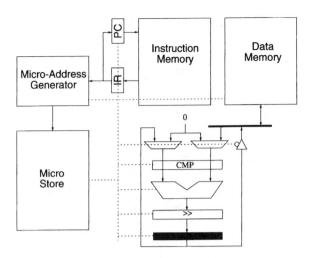

Fig. 15. Architecture of microcoded instruction set processor.

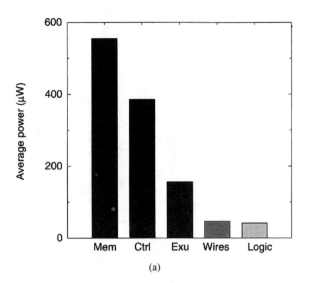

(a)

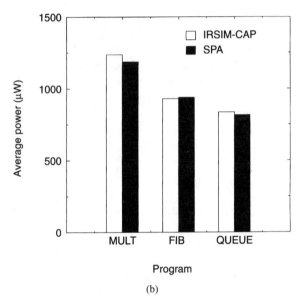

(b)

Fig. 16. SPA power results for microprocessor. (a) Power breakdown by hardware class. (b) Power for different instruction/data streams.

following nine basic instructions: NOP, LDA, STA, ADD, SHR, CMP, JMP, BLT, and BGE. Given the ADL and CDL descriptions of the architecture, SPA is able to make predictions regarding the area and power consumed by the proposed design while executing a given program on a given data stream. Fig. 16(a) contains the power breakdowns as predicted by SPA for a multiplication program running on the processor with a 1.5 V supply and a 10 MHz clock rate. The instruction and data memory accesses account for 47% of the total power consumption. This information could be used by the designer to select an appropriate focus for optimization efforts.

SPA can also be used to analyze the influence of instruction stream on average power consumption. This is made possible by the activity profiling approach used by SPA. As described in Section VI-B, SPA actually runs RT-level simulations profiling hardware activity for all instruction streams provided by the user. For example, Fig. 16(b) shows the average power while running three different programs: a multiplication program (MULT), a fibonacci sequence generator (FIB), and a circular queue (QUEUE). The power differs between these programs by as much as 38% making a prediction tool which can account for the influence of instruction statistics on power quite valuable.

In order to confirm the accuracy of these predictions, the processor has been implemented down to the layout level. The chip plot is shown in Fig. 17. The extracted layout was simulated at the switch level for the three programs and the results are included in Fig. 16(b). The error in the SPA estimates are: −4.1%, 0.88%, and −2.4%, respectively.

We can also use the layout data to verify the interconnect/area analysis strategy. Table III gives the area analysis for the design, showing that the estimated area is within 8% of the actual area. The estimated areas of the various blocks typically err by less than 20%. The micro sequencer is a notable exception. In the architectural description, this unit was defined as three distinct standard cell control blocks, while in the final implementation they were merged into a single block. This explains the overestimate and demonstrates the important point that architecture-level estimates are always subject to

TABLE III
PREDICTED VERSUS ACTUAL AREA FOR SIMPLE MICROPROCESSOR CHIP

Block	Actual (mm^2)	Predicted (mm^2)	Error (%)
Instruction memory	0.16	0.16	0%
Instruction regs	0.05	0.04	-20%
Micro sequencer	0.08	0.14	+75%
Micro store	0.13	0.11	-15%
Data memory	0.25	0.25	0%
Datapath	0.57	0.52	-9%
Wiring/overhead	0.90	0.74	-18%
Total	2.14	1.96	-8%

limitations imposed by an imperfect knowledge of the final implementation details.

Based on the area estimates, the average interconnect and clock length are estimated at 554 μm and 2,174 μm. The actual lengths are 595 μm and 2,796 μm, implying estimation errors of −7% and −22%. To demonstrate that the interconnect

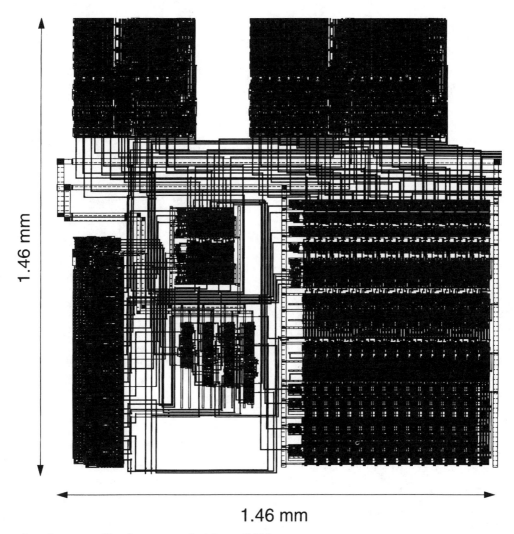

Fig. 17. Implementation of programmable microprocessor in 1.2 μm CMOS.

models are valid for other design examples, a hardware divider was analyzed as well. The predicted area was 4.68 mm^2 versus an actual implementation area of 4.40 mm^2—an error of 6%. The measured interconnect and clock wire lengths were 858 μm and 2,124 μm, while the predicted lengths were 856 μm and 2,861 μm for errors of −0.2% and 35%, respectively.

SPA's main advantage is that it requires only a high-level description of an architecture. For this example, the description took only about 2 h to produce, whereas layout required more than 25 h. Furthermore, SPA analyzed the chip power consumption in about 1 min (on a Sun SPARCstation 10), including parsing, simulation, and estimation times. In contrast, extraction and simulation of the layout consumed about 45 min.

VIII. CONCLUSION

This paper introduced a collection of techniques for analyzing the power consumption of chips at the architecture or RT level of abstraction. Power analysis for datapath and memory components was achieved using the Dual Bit Type (DBT) model. A new technique called the Activity-Based Control (ABC) model was introduced to handle power estimation for

the control path. Both the DBT and ABC models improve upon contemporary architectural power analysis techniques such as gate-equivalent estimation by accurately reflecting the effect of activity, as well as complexity, on power consumption. Finally, a strategy for analyzing interconnect power consumption was presented. A useful side-effect of this analysis was a hierarchical area breakdown for the targeted design.

The above power and area analysis models have been implemented in a tool called SPA, which was used to gather results for several fairly realistic design examples. The three examples were constructed to fully exercise the power models and included datapath, control, and memory intensive designs as represented by a quadrature mirror filter, a speech recognition controller, and a microprocessor, respectively. Relative to switch-level power simulations of extracted layouts, these case studies revealed an average error of 9% with a standard deviation of 10%. Area estimation errors were of similar magnitude averaging 14% with a standard deviation of 6%.

Operating at the architecture level allows designers to obtain results much earlier in the design process and orders of magnitude faster than can be obtained using lower level analysis tools. This should allow designers to explore the area-speed-power trade-offs of several alternative architectures, while maintaining acceptable design times. Ideally, architecture-level analysis tools could be combined with existing gate- and circuit-level tools to form a seamless, integrated design environment that allows the user to consider optimizations at all levels ranging from system to circuit.

ACKNOWLEDGMENT

The authors wish to thank A. Abnous for his significant contributions to the SPA project—in particular, his implementation of the ADL and CDL parsers and of the VHDL code generator.

REFERENCES

[1] L. W. Nagel, "SPICE2: A computer program to simulate semiconductor circuits," Univ. California, Berkeley, Tech. Rep. ERL-M520, 1975.
[2] C. Huang, B. Zhang, A. Deng, and B. Swirski, "The design and implementation of PowerMill," in *Proc. Int. Symp. Low Power Design,* Dana Point, CA, Apr. 1995, pp. 105–110.
[3] P. Landman and J. Rabaey, "Architectural power analysis: The dual bit type method," *IEEE Trans. VLSI Syst.,* pp. 173–187, June 1995.
[4] A. Deng, Y. Shiau, and K. Loh, "Time domain current waveform simulation of CMOS circuits," in *Proc. Int. Conf. Computer-Aided Design,* 1988, pp. 208–211.
[5] M. A. Cirit, "Estimating dynamic power consumption of CMOS circuits," in *Proc. IEEE Int. Conf. Computer Aided Design,* Nov. 1987, pp. 534–537.
[6] F. Najm, I. Hajj, and P. Yang, "Probabilistic simulation for reliability analysis of CMOS VLSI circuits," *IEEE Trans. Computer-Aided Design,* vol. 9, pp. 439–450, Apr. 1990.
[7] F. Najm, "Transition density, a stochastic measure of activity in digital circuits," in *Proc. 28th Design Automation Conf.,* June 1991, pp. 644–649.
[8] A. Ghosh, S. Devadas, K. Keutzer, and J. White, "Estimation of average switching activity in combinational and sequential circuits," in *Proc. 29th Design Automation Conf.,* June 1992, pp. 253–259.
[9] C.-Y. Tsui, M. Pedram, and A. Despain, "Efficient estimation of dynamic power consumption under a real delay model," in *Proc. Int. Conf. Computer-Aided Design,* 1993, pp. 224–228.
[10] K. Muller-Glaser, K. Kirsch, and K. Neusinger, "Estimating essential design characteristics to support project planning for ASIC design management," in *Proc. IEEE Int. Conf. Computer-Aided Design,* Los Alamitos, CA, Nov. 1991, pp. 148–151.
[11] C. Svensson and D. Liu, "A power estimation tool and prospects of power savings in CMOS VLSI chips," in *Proc. Int. Workshop Low-Power Design,* Napa Valley, CA, Apr. 1994, pp. 171–176.
[12] B. Landman and R. Russo, "On a pin versus block relationship for partitions of logic graphs," *IEEE Trans. Computing,* vol. C-20, pp. 1469–1479, Dec. 1971.
[13] S. R. Powell and P. M. Chau, "Estimating power dissipation of VLSI signal processing chips: The PFA technique," *VLSI Signal Processing IV,* pp. 250–259, 1990.
[14] P. Landman and J. Rabaey, "Black-box capacitance models for architectural power analysis," in *Proc. Int. Workshop Low Power Design,* Napa Valley, CA, Apr. 1994, pp. 165–170.
[15] P. Landman, "Low-power architectural design methodologies," Ph.D. dissertation, Univ. of California, Berkeley, Aug. 1994.
[16] J. Ward *et al.,* "Figures of merit for VLSI implementations of digital signal processing algorithms," *Proc. IEE,* vol. 131, Part F, pp. 64–70, Feb. 1984.
[17] S. Powell and P. Chau, "A model for estimating power dissipation in a class of DSP VLSI chips," *IEEE Trans. Circuits Syst.,* vol. 38, pp. 646–650, June 1991.
[18] R. Mehra, "High-level power estimation and exploration," in *Proc. Int. Workshop Low Power Design,* Apr. 1994, pp. 197–202.
[19] A. Chandrakasan, M. Potkonjak, R. Mehra, J. Rabaey, and R. W. Brodersen, "Optimizing power using transformations," *IEEE Trans. Computer-Aided Design,* vol. 14, pp. 12–31, Jan. 1995.
[20] A. Salz and M. Horowitz, "IRSIM: An incremental MOS switch-level simulator," in *Proc. 26th Design Automation Conf.,* 1989, pp. 173–178.
[21] R. Brayton, G. Hachtel, C. McMullen, and A. Sangiovanni-Vincentelli, *Logic Minimization Algorithms for VLSI Synthesis.* Boston, MA: Kluwer Academic, 1984.
[22] L. Lavagno, S. Malik, R. Brayton, and A. Sangiovanni-Vincentelli, "MIS-MV: Optimization of multi-level logic with multiple-valued inputs," in *Proc. IEEE Int. Conf. Computer-Aided Design,* Santa Clara, CA, Nov. 1990, pp. 560–563.
[23] R. W. Brodersen, Ed., *Anatomy of a Silicon Compiler.* Kluwer Academic, 1992.
[24] M. Feuer, "Connectivity of random logic," *IEEE Trans. Comput.,* vol. C-31, pp. 29–33, Jan. 1982.
[25] W. Donath, "Placement and average interconnection lengths of computer logic," *IEEE Trans. Circuits Syst.* vol. CAS-26, pp. 272–277, Apr. 1979.
[26] G. Sorkin, "Asymptotically perfect trivial global routing: A stochastic analysis," *IEEE Trans. Computer-Aided Design,* vol. CAD-6, p. 820, 1987.
[27] B. Preas and M. Lorenzetti, Eds., *Physical Design Automation of VLSI Systems.* Menlo Park, CA: Benjamin/Cummings, 1988.
[28] N. Jayant and P. Noll, *Digital Coding of Waveforms.* Englewood Cliffs, NJ: Prentice-Hall (Signal Processing Series), 1984.
[29] J. M. Rabaey, C. Chu, P. Hoang, and M. Potkonjak, "Fast prototyping of datapath-intensive architectures," *IEEE Design Test of Comput.,* pp. 40–51, June 1991.
[30] S. Stoiber, "Low-power digital signal processing for speech recognition," Master's degree thesis, Univ. of California, Berkeley, Dec. 1994.

Power Analysis of Embedded Software: A First Step Towards Software Power Minimization

Vivek Tiwari, Sharad Malik, and Andrew Wolfe

Abstract—Embedded computer systems are characterized by the presence of a dedicated processor and the software that runs on it. Power constraints are increasingly becoming the critical component of the design specification of these systems. At present, however, power analysis tools can only be applied at the lower levels of the design—the circuit or gate level. It is either impractical or impossible to use the lower level tools to estimate the power cost of the software component of the system. This paper describes the first systematic attempt to model this power cost. A power analysis technique is developed that has been applied to two commercial microprocessors—Intel 486DX2 and Fujitsu SPARClite 934. This technique can be employed to evaluate the power cost of embedded software. This can help in verifying if a design meets its specified power constraints. Further, it can also be used to search the design space in software power optimization. Examples with power reduction of up to 40%, obtained by rewriting code using the information provided by the instruction level power model, illustrate the potential of this idea.

I. INTRODUCTION

EMBEDDED COMPUTER systems are characterized by the presence of a dedicated processor which executes application specific software. Recent years have seen a large growth of such systems. This growth is driven by several factors. The first is an increase in the number of applications as illustrated by the numerous examples of "smart electronics" around us. A notable example is automobile electronics where embedded processors control each aspect of the efficiency, comfort and safety of the new generation of cars. The second factor leading to their growth is the increasing migration from application specific logic to application specific code running on existing processors. This in turn is driven by two distinct forces. The first is the increasing cost of setting up and maintaining a fabrication line. At over a billion dollars for a new line, the only components that make this affordable are high volume parts such as processors, memories and possibly FPGA's. Application specific logic is getting increasingly expensive to manufacture and is the solution only when speed constraints rule out programmable alternatives. The second force comes from the application houses, which are facing increased pressures to reduce the time to market as well as

to have predictable schedules. Both of these can be better met with software programmable solutions made possible by embedded systems. Thus, we are seeing a movement from the logic gate being the basic unit of computation on silicon, to an instruction running on an embedded processor.

A large number of embedded computing applications are power critical, i.e., power constraints form an important part of the design specification. This has led to a significant research effort in power estimation and low power design. However, there is very little available in the form of design tools to help embedded system designers evaluate their designs in terms of the power metric. At present, power measurement tools are available for only the lower levels of the design-at the circuit level and to a limited extent at the logic level. At the least these are very slow and impractical to use to evaluate the power consumption of embedded software, and often cannot even be applied due to lack of availability of circuit and gate level information of the embedded processors. The embedded processors currently used in designs take two possible shapes. The first is "off the shelf" microprocessors or digital signal processors (DSP's). The second is in the form of embedded cores which can be incorporated in a larger silicon chip along with program/data memory and other dedicated logic. In the first case, the processor information available to the designer is whatever the manufacturer cares to make available through data books. In the second case, the designer has logic/timing simulation models to help verify the designs. In neither case is there lower level information available for power analysis.

This paper describes a power analysis technique for embedded software. The goal of this research is to present a methodology for developing and validating an instruction level power model for any given processor. Such a model can then be provided by the processor vendors for both off the shelf processors as well as embedded cores. This can then be used to evaluate embedded software, much as a gate level power model has been used to evaluate logic designs. The technique has so far been applied to two commercial microprocessors—the Intel 486DX2 and the Fujitsu SPARClite 934. This paper uses the former as a basis for illustrating the technique. The application of this technique for the latter is described in [9]. The ability to evaluate software in terms of the power metric helps in verifying if a design meets its specified power constraints. In addition, it can also be used to search the design space in software power optimization. Examples with power reduction of up to 40% on the 486DX2, obtained by rewriting code using the information provided by

Manuscript received June 15, 1994; revised August 23, 1994. The work of V. Tiwari was supported by an IBM Graduate Fellowship. The work of S. Malik was supported by an IBM Faculty Development Award.

The authors are with the Department of Electrical Engineering, Princeton University, Princeton, NJ 08540 USA.

IEEE Log Number 9406371.

Reprinted from *IEEE Transactions on VLSI Systems*, pp. 437–445, December 1994.

533

the instruction level power model, illustrate the potential of this idea.

II. EXPERIMENTAL METHOD

The power consumption in microprocessors has been a subject of intense study lately. Attempts to model the power consumption in processors often adopt a "bottom-up" approach. Using detailed physical layouts and sophisticated power analysis tools, isolated power models are built for each of the internal modules of the processor. The total power consumption of the processor is then estimated using these individual models. No systematic attempt, however, has been made to relate the power consumption of the processor to the software that executes on it. Thus, while it is generally recognized that the power consumption of a processor varies from program to program, there is a complete lack of models and tools to analyze this variation. This is also the reason why the potential for power reduction through modification of software is so far unknown and unexploited. The goal of our work is to overcome these deficiencies by developing a methodology that would provide a means for analyzing the power consumption of a processor as it executes a given program. We want to provide a method that makes it possible to talk about the "power/energy cost of a given program on a given processor." This would make it possible to very accurately evaluate the power cost of the programmable part of an embedded system.

We propose the following hypothesis that forms the basis for meeting the above goal: *By measuring the current drawn by the processor as it repeatedly executes certain instructions or certain short instruction sequences, it is possible to obtain most of the information that is needed to evaluate the power cost of a program for that processor.*

The intuition that guides this hypothesis is as follows: Modern microprocessors are extremely complex systems consisting of several interacting functional blocks. However, this internal complexity is hidden behind a simple interface—the instruction set. Thus to model the energy consumption of this complex system, it seems intuitive to consider individual instructions. Further, each instruction involves specific processing across various units of the processor. This can result in circuit activity that is characteristic of each instruction and can vary with instructions. If a given instruction is executed repeatedly, then the power consumed by the processor can be thought of as the power cost of that instruction. In a given program, certain inter-instruction effects also occur, such as the effect of circuit state, pipeline stalls and cache misses. Repeatedly executing certain instruction sequences during which these effects occur may provide a way to isolate the power cost of these effects. Thus the sum of the power costs of the each instruction that is executed in a program enhanced by the power cost of the inter-instruction effects can be an estimate for the power cost of the program.

The above hypothesis, however, is of no use until it is validated. We have empirically validated the hypothesis for two commercial microprocessors using actual physical measurements of the current drawn by them. The validation of the

hypothesis, and based on it, the derivation of the parameters of an instruction level power model for the Intel 486DX2, is the subject of the next few sections.

Given that the above hypothesis has been validated for two processors using physical measurements, there is an alternative way for deriving the parameters of the instruction level power model. Instead of physically measuring the current drawn by the CPU, it can be estimated using accurate, simulation based power analysis tools. The execution of the given instruction/instruction sequence is simulated on lower level (circuit or layout) models of the CPU, and the power analysis tool provides an estimate of the current drawn. The advantage of this method is that since detailed internal information of the CPU is available, it may be possible to relate the power cost of the instructions to the micro-architecture of the CPU. This could provide cues to the CPU designer for optimizing the designs for low power.

However, in the case of embedded system design, detailed layout information of the CPU is often not available to the designer of the system. Even if it is available, the simulation based tools and techniques are expensive and difficult to apply. A methodology based on laboratory measurements, like the one described below, is inexpensive and practical, and often may be the only option available. Given a setup to measure the current being drawn by the microprocessor, the only other information required can be obtained from the widely available manuals and handbooks specific to that microprocessor. The specifics of the measurement methodology are described next.

A. Power and Energy

The average power consumed by a microprocessor while running a certain program is given by: $P = I \times V_{CC}$, where P is the average power, I is the average current and V_{CC} is the supply voltage. Since power is the rate at which energy is consumed, the energy consumed by a program is given by: $E = P \times T$, where T is the execution time of the program. This in turn is given by: $T = N \times \tau$, where N is the number of clock cycles taken by the program and τ is the clock period.

In common usage the terms power consumption and energy consumption are often interchanged, as has been done in the above discussion. However it is important to distinguish between the two in the context of programs running on mobile applications. Mobile systems run on the limited energy available in a battery. Therefore the energy consumed by the system or by the software running on it determines the length of the battery life. Energy consumption is thus the focus of attention. We will attempt to maintain a distinction between the two terms in the rest of the paper. However, in certain cases the term power may be used to refer to energy, in adherence to common usage.

B. Current Measurement

For this study, the processor used was a 40 MHz Intel 486DX2-S Series CPU. The CPU was part of a mobile personal computer evaluation board with 4 MB of DRAM memory. The reason for the choice of this processor was that its board setup allowed the measurement of the CPU and

DRAM subsystem current in isolation from the rest of the system. *We would like to emphasize that while the numbers we report here are specific to this processor and board, the methodology used by us in developing the model is widely applicable.* The current was measured through a standard off the shelf, dual-slope integrating digital ammeter. Execution time of programs was measured through detection of specific bus states using a logic analyzer.

If a program completes execution in a short time, a current reading cannot be obtained visually. To overcome this, the programs being considered were put in infinite loops and current readings were taken. The current consumption in the CPU will vary in time depending on what instructions are being executed. But since the chosen ammeter averages current over a window of time (100 ms), if the execution time of the program is much less than the width of this window, a stable reading will be obtained.

The main limitation of this approach is that it will not work for programs with larger execution times since the ammeter may not show a stable reading. However, in this study, the main use of this approach was in determining the current drawn while a particular instruction (instruction sequence) was being executed. A program written with several instances of the targeted instruction (instruction sequence) executing in a loop, has a periodic current waveform which yields a steady reading on the ammeter. This inexpensive approach works very well for this. However, the main concepts described in this paper are independent of the actual method used to measure average current. If sophisticated data acquisition based measurement instruments are available, the measurement method can be based on them.

For our setup, V_{CC} was 3.3 V and τ was 25 ns, corresponding to the 40 MHz internal frequency of the CPU. Thus, if the average current for an instruction sequence is I A, and the number of cycles it takes to execute is N, the energy cost of the sequence is given by: $E = I \times V_{CC} \times N \times \tau$, which equals: $(8.25 \times 10^{-8} \times I \times N)$ J. Throughout the rest of the paper, in order to specify the energy cost of an instruction (instruction sequence), the average current will be specified. The number of cycles will either be explicitly specified, or will be clear from the context.

III. INSTRUCTION LEVEL MODELING

Based on the hypothesis described in Section II, an instruction level energy model has been developed and validated for the 486DX2. Under this model each instruction in the instruction set is assigned a fixed energy cost called the *base energy cost*. The variation in base costs of a given instruction due to different operand and address values is then quantified. The base energy cost of a program is based on the sum of the base energy costs of each executed instruction. However, during the execution of a program, certain inter-instruction effects occur whose energy contribution is not accounted for if only base costs are considered. The first type of inter-instruction effect is the effect of circuit state. The second type is related to resource constraints that can lead to stalls and cache misses. The energy cost of these effects is also modeled and used to obtain the total energy cost of a program.

Fig. 1.   Internal pipelining in the 486DX2

The instruction-level energy model described here is based on actual measurements and evolved as a result of extensive experimentation. The various components of this model are described in the subsections below.

A. Base Energy Cost

The base cost for an instruction is determined by constructing a loop with several instances of the same instruction. The average current being drawn is then measured. This current multiplied by the number of cycles taken by each instance of the instruction is proportional to the total energy as described in Section II.

While this method seems intuitive if the CPU executes only one instruction at a given time, most modern CPU's, including the 486DX2, process more than one instruction at a given time due to pipelining. However, the following discussion shows that the concept of a base energy cost per instruction and its derivation remains unchanged.

The 486DX2 CPU has a five-stage pipeline as shown in Fig. 1 [6]. Let Ej_{I_k} be the average energy consumed by pipeline stage j, when instruction I_k executes in that stage. Pipeline stages are separated from each other by latches. Thus, if we ignore the effect of circuit state and resource constraints for now, the energy consumption of different stages is independent of each other. Let us assume that in a given cycle, instruction I_1 is being processed by stage 1, I_2 by stage 2, and so on. The total energy consumed by the CPU in that cycle would be: $E_{\text{cycle}} = E1_{I_1} + E2_{I_2} + E3_{I_3} + E4_{I_4} + E5_{I_5}$. On the other hand, the total energy consumed by a given instruction I_1, as it moves through the various stages is: $E_{\text{ins}} = \sum_j Ej_{I_1}$. This quantity actually refers to the base cost in the sense described above. Our method of forming a loop of instances of instruction I_1, results in $E_{\text{cycle}} = E_{\text{ins}}$, since in that case, $I_1 = I_2 = I_3 = I_4 = I_5$. The average current in this case is $\sum_j Ej_{I_1}/(V_{CC} \times \tau)$, which is the same as the ammeter reading obtained.

Some instructions take multiple cycles in a given pipeline stage. All stages are then stalled. The reasoning applied above, however, remains unchanged. The base energy cost of the instruction is just the observed average current value multiplied by the number of cycles taken by the instruction in that stage. For instance, consider a loop of instruction I_1, where I_1 takes m cycles in the 4th stage. Therefore, $E4_{I_1}$ is spread over m cycles. Energy consumption in any of the stalled stages can be considered as a part of $E4_{I_1}$. Then the current value observed on the ammeter will be $\sum_j Ej_{I_1}/(V_{CC} \times \tau \times m)$. This quantity multiplied by m yields $\sum_j Ej_{I_1}/(V_{CC} \times \tau)$, the base energy cost of the instruction. m represents the "number of cycles" parameter specified in instruction timing tables in microprocessor manuals.

TABLE I
SUBSET OF THE BASE COST TABLE FOR THE 486DX2[1]

Number	Instruction	Current (mA)	Cycles
1	NOP	275.7	1
2	MOV DX,BX	302.4	1
3	MOV DX,[BX]	428.3	1
4	MOV DX,[BX][DI]	409.0	2
5	MOV [BX],DX	521.7	1
6	MOV [BX][DI],DX	451.7	2
7	ADD DX,BX	313.6	1
8	ADD DX,[BX]	400.1	2
9	ADD [BX],DX	415.7	3
10	SAL BX,1	300.8	3
11	SAL BX,CL	306.5	3
12	LEA DX,[BX]	364.4	1
13	LEA DX,[BX][DI]	345.2	2
14	JMP label	373.0	3
15	JZ label	375.7	3
16	JZ label	355.9	1
17	CMP BX,DX	298.2	1
18	CMP [BX],DX	388.0	2

TABLE II
BASE COSTS OF MOV BX, DATA

data	0	0F	0FF	0FFF	0FFFF
No. of 1's	0	4	8	12	16
Current(mA)	309.5	305.2	300.1	294.2	288.5

Table I is a sample table of CPU base costs for some 486DX2 instructions. The numbers in Column 3 are the observed average current values. The overall base energy cost of an instruction is the product of the numbers in Columns 3 and 4 and the constants V_{CC} and τ. A rigorous confidence interval was not determined for the current measurement apparatus. However, it was observed that repeated runs of an experiment at different times resulted in only a very small variation in the observed average current values. The variation was in the range of ± 1 mA.

Care should be taken in designing the experiments used to determine the base costs. The loops that are used to determine the base costs of instructions have to satisfy certain size constraints. As more of the target instructions are put in the loop, the impact of the branch statement at the bottom of the loop is minimized. The observed current value thus converges with increasing loop size. Thus, the loop size should be large enough in order to obtain the converged value. Very large loops, on the other hand, may cause cache misses, which are undesirable during determination of base costs. A loop size of 120, which satisfies both the above constraints, was chosen. Only the target instructions should execute on the CPU during experiments, and thus system effects like multiple time-sharing applications and interrupts cannot be allowed during the experiments.

Variations in Base Cost: As Table I shows, instructions with differing functionality and different addressing modes can have very different energy costs. This is to be expected since different functional blocks are being affected in different ways by these instructions. Within the same family of instructions, there is variability in base costs depending on the value of operands used. For example, consider the MOV

---

[1] All instructions are executed in "Real Mode". All registers contain 0, except in entry 11, where CL contains 1. Entry 15 is a "taken" jump while entry 16 is "fall through". Entries 5, 6, and 9 show *normalized* costs [11].

register,immediate family. Use of different *registers* results in insignificant variation since the register file is probably a symmetric structure. Variation in the *immediate* value, however, leads to measurable variation. For example, for MOV BX, *immediate*, the costs seem to be almost a linear function of the number of 1's in the binary representation of the immediate data—the more the 1's, the lesser the cost. Table II illustrates this through some sample values. Similarly, for the ADD instruction, the base costs are a function of the two numbers being added. The range of variation in all cases, however, is small. It is observed to be about 14 mA, which corresponds to less than a 5% variation.

For instructions involving memory operands, there is a variation in the base cost depending upon the address of the operand. The variation is of two kinds. The first is due to operands that are misaligned [6]. Mis-aligned accesses lead to cycle penalties and thus energy penalties that are added to the base cost. Within aligned accesses there is variation in the base cost depending upon the value of the address. For example, for MOV DX, [BX], the base cost can be greater than the cost shown in Table I by about 3.5%. This variation is a function of the number of, and position of, 1's in the binary representation of the address.

Given the operand value and address, exact base costs can be obtained through direct measurements. However, these exact values will be of little use since typically a data or address value can be known only at run-time. Thus, from the point of view of program energy cost estimation, the only alternative is to use average base cost values. This is reasonable given that the variation in base costs is small and thus the discrepancy between the average and actual values will be limited.

B. Inter-Instruction Effects

When sequences of instructions are considered, certain inter-instruction effects come into play, which are not reflected in the cost computed solely from base costs. These effects are discussed below.

Effect of Circuit State: The switching activity in a circuit is a function of the present inputs and the previous state of the circuit. Thus, it can be expected that the actual energy cost of executing an instruction in a program may be different from the instruction's base cost. This is because the previous instruction in the given program and in the program used for base cost determination may be different. For example, consider a loop of the following pair of instructions:

```
XOR BX,1
ADD AX, DX
```

The base costs of the XOR and ADD instructions are 319.2 mA and 313.6 mA. The expected base cost of the pair, using the individual base costs would be their average, i.e., 316.4

536

TABLE III
AN EXAMPLE INSTRUCTION SEQUENCE

Number	Instruction	Current(mA)	Cycles
1	MOV CX,1	309.6	1
2	ADD AX,BX	313.6	1
3	ADD DX,8[BX]	400.2	2
4	SAL AX,1	308.3	3
5	SAL BX,CL	306.5	3

mA, while the actual cost is 323.2 mA. It is greater by 6.8 mA. The reason is that the base costs are determined while executing the same instruction again and again. Thus each instruction executes in what we expect is a context of least change. At least, that is what the observations consistently seem to indicate. When a pair of two different instructions is considered, the context is one of greater change. The cost of a pair of instructions is always greater than the base cost of the pair and the difference is termed as the *circuit state overhead*.

As another example, consider the sequence of instructions shown in Table III. The current cost and the number of cycles of each instruction is listed alongside. The measured cost for this sequence is 332.8 mA (avg. current over 10 cycles). Using base costs we get:

$$(309.6 + 313.6 + 400.2 \times 2 + 308.3 \times 3 + 306.5 \times 3)/10$$
$$= 326.8 \tag{1}$$

The circuit state overhead is thus 6.0 mA.

It is possible to get a closer estimate if we consider the circuit state overhead between each pair of consecutive instructions. This is done as follows. Consider a loop of the targeted pair, e.g., instructions 2 and 3. The estimated cost for the pair is $(2 \times 400.2 + 313.6 \times 1)/3 = 371.3$ mA, while the measured cost is 374.8 mA. Thus, the circuit state overhead is 3.5 mA. Now the overhead occurs twice in every 3 cycles, once between instructions 2 & 3, and once between 3 & 2. Since these two different cases cannot be resolved, let us assume that they are the same. Thus, the overhead each time it occurs would be $3.5 \times \frac{3}{2} = 5.25$ mA. Similarly, the overhead between the pairs 1 and 2, 3 and 4, 4 and 5, and 5 and 1 is found to be 17.9, 12.25, 3.3, and 17.2 mA, respectively. When these overheads are added to the numerator in (1), we get an estimated cost of 332.38 mA, which is within 0.12% of the measured value.

This example illustrates that by determining costs of pairs of instructions, it is possible to improve upon the results of the estimation obtained with base costs alone. However, extensive experiments with pairs of instructions revealed that the circuit state overhead has a limited range—between 5.0 mA and 30.0 mA and most frequently occurred in the vicinity of 15.0 mA. This motivates an efficient yet fairly accurate way to account for the circuit state overhead. Calculate the average current for the program using the base costs. Then, add 15.0 mA to it, to account for circuit state overhead.

A specific manifestation of the effect of circuit state is the effect of switching that occurs on address and data lines. Our experiments revealed that the overall impact of this effect was small. For back-to-back data reads from the cache, greater switching of the address values led to at most a 3% increase in the energy cost. For back-to-back data writes (which go to both the cache and the memory bus since the cache is write-through), the impact of greater switching of the address values was less than 5%.

The limited variation in the circuit state overhead is contrary to popular belief. In fact, a recent work [8], talks about scheduling instructions to reduce this overhead. But as our experiments reveal, the methods described in this work will not have much impact for the 486DX2. The probable explanation for the limited variation in circuit state overhead is that a major part of the circuit activity in a complex processor like the 486DX2, is common to all instructions, e.g., instruction prefetch, pipeline control, clocks etc. While the circuit state may cause significant variation within certain modules, its impact on the overall energy cost is swamped by the much greater common cost. However, we would not like to rule out the impact of circuit state overhead for all processors. It may well be the case that it is a significant part of the energy consumption in processors like RISC's (Reduced Instruction Set Computers) DSP's, and processors with complex power management features. An investigation of this issue is the subject of our future study.

Effect of Resource Constraints: Resource constraints in the CPU can lead to stalls e.g. pipeline stalls and write buffer stalls [6], [7]. These can be considered as another kind of inter-instruction effect. They cause an increase in the number of cycles needed to execute a sequence of instructions. For example, a sequence of 120 MOV DX, [BX] instructions takes about 164 cycles to execute, instead of 120 due to prefetch buffer stalls. While determining the base cost of instructions, it is important to avoid stalls, since they represent a condition that ought not to be reflected in the base cost. Thus, for MOV DX, [BX] a sequence consisting of 3 MOV instructions followed by a NOP is used since there are no stalls during its execution [7]. Knowing the cost of the NOP and the measured value for the sequence, the base cost of the MOV is determined.

The energy cost of each kind of stall is experimentally determined through experiments that isolate the particular kind of stall. For example, an average cost of 250 mA for stall cycles was determined for the prefetch buffer stall.

It has been observed that the cost of stalls can show some variation depending upon the instructions involved in the stall. Through extensive experimentation it may be possible to subdivide each stall type into specific cases and to assign a cost to each case. However, in general, the use of a single average cost value for each stall type suffices.

To account for the energy cost of the above stalls during program cost estimation, the number of stall cycles has to be multiplied by the experimentally determined stall energy cost. This product is then added to the base cost of the program. The number of stall cycles is estimated through a traversal of the program code.

Effect of Cache Misses: Another inter-instruction effect is the effect of cache misses. The instruction timings listed in manuals give the cycle count assuming a cache hit. For a cache miss, a certain cycle penalty has to be added to the instruction execution time. Along the same lines, the base costs for instructions with memory operands are determined in the context of cache hits. A cache miss will lead to extra cycles being consumed, which leads to an energy penalty. For experimentation purposes, a cache miss scenario is created by accessing memory addresses in an appropriate order. An average penalty of 216 mA for cache miss cycles has been experimentally obtained. This has to be multiplied by the average number of miss penalty cycles to get the average energy penalty for one miss. The average penalty multiplied by the cache miss rate is added to the base cost estimate to account for the cache misses during execution of a program.

IV. ESTIMATION FRAMEWORK

In this section we describe a framework for energy estimation of programs using the instruction level power model outlined in the previous section. We start by illustrating this estimation process for the program shown in Table IV. The program has three basic blocks as shown in the figure[2]. The average current and the number of cycles for each instruction are provided in two separate columns. For each basic block, the two columns are multiplied and the products are summed up over all instructions in the basic block. This yields a value that is proportional to the base energy cost of one instance of the basic block. The values are 1713.4, 4709.8, and 2017.9, for B1, B2, and B3, respectively. B1 is executed once, B2 4 times and B3 once. The jmp main statement has been inserted to put the program in an infinite loop. Cost of the jl L2 statement is not included in the cost of B2 since its cost is different depending on whether the jump is taken or not. It is taken 3 times and not taken once. Multiplying the base cost of each basic block by the number of times it is executed and adding the cost of the unconditional jump jl L2, we get a number proportional to the total energy cost of the program. Dividing it by the estimated number of cycles (72) gives us an average current of 369.1 mA. Adding the circuit state overhead offset value of 15.0 mA we get 384.0 mA. The actual measured average current is 385.0 mA. This program does not have any stalls and thus no further additions to the estimated cost are required. If in the real execution of this program, some cold-start cache misses are expected, their energy overhead will have to be added.

To validate the estimation model described in the previous section, experiments were conducted with several programs. A close correspondence between the estimated and measured cost was obtained. It was observed that the main reasons for the discrepancy in the estimated and actual cost are as follows: First, for instructions that require operands, the operand values and addresses are often not known until runtime. Thus, average base costs may have to be used instead of exact costs. Second, the circuit state overhead for pairs of consecutive instructions in the program may differ from the default value used. Third,

[2]A basic block is defined as a contiguous section of code with exactly one entry and exit point.

TABLE IV
ILLUSTRATION OF THE ESTIMATION PROCESS

Program	Current(mA)	Cycles
; Block B1		
main:		
mov bp,sp	285.0	1
sub sp,4	309.0	1
mov dx,0	309.8	1
mov word ptr -4[bp],0	404.8	2
;Block B2		
L2:		
mov si,word ptr -4[bp]	433.4	1
add si,si	309.0	1
add si,si	309.0	1
mov bx,dx	285.0	1
mov cx,word ptr _a[si]	433.4	1
add bx,cx	309.0	1
mov si,word ptr _b[si]	433.4	1
add bx,si	309.0	1
mov dx,bx	285.0	1
mov di,word ptr -4[bp]	433.4	1
inc di, 1	297.0	1
mov word ptr -4[bp],di	560.1	1
cmp di,4	313.1	1
jl L2	405.7(356.9)	3(1)
;Block B3		
L1:		
mov word ptr _sum,dx	521.7	1
mov sp,bp	285.0	1
jmp main	403.8	3

the penalty due to stalls and cache misses is difficult to predict statically. As discussed in Section III, the first two effects are limited in their impact on the overall cost. The inability to predict the penalty due to stalls and cache misses, on the other hand, can potentially have a greater impact on the accuracy of the estimate. However, for programs with no stalls and cache misses, the maximum difference between the estimated and the measured cost was less than 3% of the measured cost.

A. Overall Flow

The overall flow of the estimation procedure is shown in Fig. 2. Given an assembly or machine level program, it is first split up into basic blocks. The base cost of each instance of the basic block is determined by adding up the base costs of the instructions in the block. These costs are provided in a base cost table. The energy overhead due to pipeline, write buffer and other stalls is estimated for each basic block and added to the basic block cost. Next, the number of times each basic block is executed has to be determined. This depends on the path that the program follows and is dynamic, run-time information that is obtained from a program profiler. Given this information, each basic block is multiplied by the number of times it will be executed. The circuit state overhead is added to the overall sum at this stage, or alternatively, it could have been determined for each basic block using a table of energy costs for pairs of instructions. An estimated cache penalty is added to get the final estimate. The cache penalty overhead

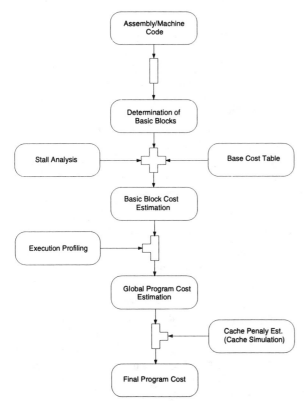

Fig. 2. Software energy consumption estimation methodology.

computation needs an estimate of the miss ratio, which is obtained through a cache simulator.

V. MEMORY SYSTEM MODELING

The energy consumption in the memory system is also a function of the software being executed. The salient observations regarding the DRAM system current on our experimental setup are briefly described here. Details are provided in [11].

The DRAM system draws constant current when no memory access is taking place. This current value was determined to be 77.0 mA or 5.3 mA, depending on whether page mode was active or not.[3] Greater current is drawn during a memory access. The exact value of this current depends on the address of the present and previous memory access. For example, for writes, when both the previous and the present access map to the same page, i.e., for a page hit, the cost is 122.8 mA (for 3 cycles including 1 wait state). For a page miss, the cost is 247.8 mA (for 6 cycles including 4 wait states). For page hits, a smaller variation was observed depending on the number of bits that change from the previous address to the present.

Let X be the sum of the energy costs of each individual memory access. Let n and m be the number of memory idle cycles during which the page mode is active and inactive, respectively. The total memory system energy cost is given by, $(X + (77.0 \times n + 5.3 \times m) \times 8.25 \times 10^{-8}) J$. As discussed above, the quantity X depends on the location and sequence of memory accesses made by the program. Along with n and m, this is dynamic, run-time information, which can only

[3] "Page mode active" refers to the condition when the row address has been latched and the Row Address Strobe (RAS) signal is active.

be loosely estimated by static analysis. Thus, modeling of memory system energy consumption is difficult if only static analysis is used. However, as the above discussion shows, analysis of this consumption is feasible. This is significant, given that for systems with tight energy budgets, it is important to understand all sources of energy consumption.

VI. SOFTWARE POWER OPTIMIZATION

In recent years there has been a spurt of research activity targeted at reducing the energy consumption in systems. This research, however, has by and large not recognized the potential energy savings achievable through optimization of software. This was mainly due to the lack of practical techniques for analyzing the energy consumption of programs. This deficiency has been alleviated by the measurement and estimation methodology described in the previous sections. The growing trend towards tight energy budgets necessitates identification and exploration of every possible source of energy reduction, forcing us to examine the design of energy efficient software.

The energy formula described in Section II-A shows that the energy cost of a program is proportional to the product of the average current and the running time of the program. Thus, the value of this product has to be reduced in order to reduce the energy cost. This section examines some alternatives for this, in the context of the 486DX2, using the results of the instruction level analysis that was described earlier.

A. Instruction Reordering

A recent work [8] presents a technique for scheduling instructions on an experimental RISC processor in such a way that the switching on the control path is minimized. In terms of the energy formula, this technique is trying to reduce the average current for the program through instruction reordering.

Our experiments based on actual energy measurements on the 486DX2, however, reveal that this technique does not translate into very significant overall energy reduction. This technique is trying to reduce what we termed as the circuit state overhead. As we saw earlier, this quantity is bounded by a small range and does not show a great amount of variation. In fact, it was observed that different reorderings of several sequences of instructions showed a variation of only up to 2% in their current cost. It can be concluded that this technique is not very effective for the 486DX2. However, its effectiveness on other architectures and processors should be investigated further.

B. Generation of Energy Efficient Code

While reordering of a given set of instructions in a piece of code may have only a limited impact on the energy cost, the actual choice of instructions in the generated code can significantly affect the cost. As a specific example, an inspection of the energy costs of 486DX2 instructions reveals that instructions with memory operands have very high average current compared to instructions with register operands. Instructions using only register operands cost in the vicinity of 300 mA. Memory reads that hit the cache cost upwards of 430

TABLE V
RESULTS OF ENERGY OPTIMIZATION OF SORT AND CIRCLE

Program	hlcc.asm	hht1.asm	hht2.asm	hht3.asm
Avg. Current (mA)	525.7	534.2	507.6	486.6
Execution Time (μsec)	11.02	9.37	8.73	7.07
Energy ($10^{-6}J$)	19.12	16.52	14.62	11.35
Program	clcc.asm	cht1.asm	cht2.asm	cht3.asm
Avg. Current (mA)	530.2	527.9	516.3	514.8
Execution Time (μsec)	7.18	5.88	5.08	4.93
Energy ($10^{-6}J$)	12.56	10.24	8.65	8.37

mA. Memory writes cost upwards of 530 mA and also incur a memory system current cost since the cache is write-through. Thus, reduction in the number of memory operands can lead to a reduction in average current.

The reduction in energy cost, i.e., in the product of average current and running time would be greater still, since use of memory operands incurs more cycles. For example, ADD DX, [BX] takes two cycles, even in the case of a cache hit, while ADD DX, BX takes just one cycle. Potential pipeline stalls, misaligned accesses, and cache misses, further add to the running time. Reduction in number of memory operands can be achieved by adopting suitable code generation policies, e.g. saving the least amount of context during function calls. However, the most effective way of reducing memory operands is through better utilization of registers. This entails techniques akin to optimal global register allocation of temporaries and frequently used variables [1] [2].

The impact of the above ideas on the energy cost of programs is illustrated here using examples. The first program considered is a *heapsort* program in C, called "sort" [3]. hlcc.asm is the assembly code for this program generated by lcc, an ANSI C compiler [4]. The sum of the observed average CPU and memory currents is given in the table above. The program execution times and overall energy costs are also reported. lcc is a general purpose compiler and while it produces good code, it leaves room for further improvement of running time. Hand tuning of the code for shorter running time (hht1) leads to a 15% reduction in running time. The average current goes up a little since of all the instructions that were eliminated, a greater proportion had lower average currents. However, due to the reduction in running time, the overall energy cost goes down by 13.5%. So far only temporary variables had been allocated to registers. In hht2, 3 local variables are allocated to registers and the appropriate memory operands are replaced by register operands. Even though redundant instructions are not removed, there is a 5% reduction in current and a 7% reduction in running time. In hht3, 2 more local variables are allocated to registers and all redundant instructions are removed. Compared to hlcc, hht3 has 40.6% lower energy consumption. Results for another program derived from the circle program [5] are also shown in Table V. Significant energy reduction, about 33%, are observed for this program too.

The specific optimizations used in the above examples were prompted by the results of the instruction level analysis of the 486DX2. They are discussed in greater detail in [10]. In general, the ideas used for energy efficient code for one processor may not hold for another. An instruction level analysis, using the methodology described earlier, should therefore be performed for each processor under consideration. That methodology provides a way for assigning energy costs to instructions. The idea behind energy driven code generation is to select instructions using these costs, such that the overall energy cost of a program is minimized. An investigation of this issue for different architectural styles will be pursued further as part of research in the area of software power optimization.

VII. ANALYSIS OF SPARClite 934

The previous sections describe the application of the power analysis methodology for the 486DX2, a CISC processor. To verify the general applicability of this methodology, it was decided to apply the methodology to a processor with a different architectural style. The Fujitsu SPARClite 934, a RISC processor targeted for embedded applications was chosen for this purpose. A power analysis of this processor has been performed using the measurement and experimentation techniques described in the previous sections. The basic model of a base energy cost per instruction, enhanced by the inter-instruction effects remains valid for this processor, though the actual costs differ in value. The details of this analysis are described in [9].

VIII. SUMMARY AND FUTURE WORK

This paper presents a methodology for analyzing the energy consumption of embedded software. It is based on an instruction level model that quantifies the energy cost of individual instructions and of the various inter-instruction effects. The motivation for the analysis methodology is three-fold. It provides insights into the energy consumption in processors. It can be used to help verify if an embedded design meets its energy constraints and it can also be used to guide the design of embedded software such that it meets these constraints. Initial attempts at code re-writing demonstrate significant power reductions—justifying the motivation for such a power analysis technique.

The methodology has so far been applied to two commercial processors, a CISC and a RISC. Future work will extend this to other architecture styles to characterize and contrast their energy consumption models. DSPs, superscalar processors, and processors with internal power management will be considered. Finally, we hope to use this analysis in automatic techniques for the reduction of power consumption in embedded software.

ACKNOWLEDGMENT

We would like to thank D. Singh, S. Rajgopal, and T. Rossi of Intel for providing us with the 486DX2 evaluation board; M. Tien-Chien Lee, M. Fujita, and D. Maheshwari of Fujitsu for helping make the SPARClite analysis possible; C. Fraser of AT&T Bell Labs and D. Hanson of Princeton University for the 486 code generator.

REFERENCES

[1] A. V. Aho, R. Sethi, and J. D. Ullman, *Compilers, Principles, Techniques and Tools.* Reading, MA: Addison Wesley, 1988.

[2] M. Benitez and J. Davidson, "A retargetable integrated code improver," *Tech. Rep. CS-93-64*, Univ. of Virginia, Dept of Computer Sci., Nov. 1993.

[3] Press *et al.*, *Numerical Recipes in C.* Cambridge, MA: Cambridge Univ., 1988.

[4] C. W. Fraser and D. R. Hanson, "A retargetable compiler for ANSI C," *SIGPLAN Notices*, pp. 29–43, Oct., 1991.

[5] R. Gupta, "Co-synthesis of hardware and software for digital embedded systems," Ph.D. dissertation, Dept. of Electrical Eng., Stanford University, CA, 1993.

[6] Intel Corp., *i486 Microprocessor, Hardware Reference Manual*, 1990.

[7] Intel Corp., *Intel486 Microprocessor Family, Programmer's Reference Manual*, 1992.

[8] C. L. Su, C. Y. Tsui, and A. M. Despain, "Low power architecture design and compilation techniques for high-performance processors," in *IEEE COMPCON*, Feb. 1994.

[9] V. Tiwari, T. C. Lee, M. Fujita, and D. Maheshwari, "Power analysis of the SPARClite MB86934," *Tech. Rep. FLA-CAD-94-01*, Fujitsu Labs of America, Aug. 1994.

[10] V. Tiwari, S. Malik, and A. Wolfe, "Compilation techniques for low energy: An overview," in *Proc. 1994 Symp. Low Power Electron.*, Oct. 1994.

[11] V. Tiwari, S. Malik, A. Wolfe, "Power analysis of the Intel 486DX2," *Tech. Rep. CE-M94-5*, Princeton Univ., Dept. of Elect. Eng., June, 1994.

541

Technology Mapping for Low Power

Vivek Tiwari
Dept. of EE, Princeton Univ

Pranav Ashar
C&CRL, NEC USA

Sharad Malik
Dept. of EE, Princeton Univ. *

Abstract

The last couple of years have seen the addition of a new dimension in the evaluation of circuit quality – its power requirements. Low power circuits are emerging as an important application domain, and synthesis for low power is demanding attention. The research presented in this paper addresses one aspect of low power synthesis. It focuses on the problem of mapping a technology independent circuit to a technology specific one, using gates from a given library, with power as the optimization metric. Several issues in modeling and measuring circuit power, as well as algorithms for technology mapping for low power are presented here. Empirically, it is observed that a significant variation in the power consumption is possible just by varying the choice of gates. Technology mapping for low power provides circuits with up to 24% lower power requirements than those obtained by technology mapping for area.

1 Introduction

There is a growing demand for low power circuits from two distinct sources: (1) circuits that have significantly high power requirements which necessitate expensive packaging and cooling (*e.g.* the 21064 (Alpha) processor from DEC consuming 25 Watts), (2) portable applications requiring long battery life.

The need for low power circuits is being actively addressed on several fronts. At the micro-architecture level, several techniques have been presented that reduce the power requirements (e.g. [3, 2]). At the circuit level, a popular technique is to turn off the system clock for parts of the circuit that are not active. At the device level, work is being done to reduce the peak voltage needed for switching, thus directly resulting in reduced power consumption. Some work has recently been done in the area of low power logic synthesis (automatic design of logic) [6]. Several interesting ideas are presented in this piece of research. However, one shortcoming of that work is that it is applied to technology independent circuits, where gate models for power are very inaccurate. As a result it is difficult to predict if any of the reduction seen at the technology independent level will hold up after a final technology mapping step. Similar problems of meaningful modeling have been faced in logic synthesis in the past in the context of area and delay optimization. Partly as a consequence of this experience, the focus of our research in low power logic synthesis is the technology mapping stage where, we believe, it is possible to have reasonable power models for gates in the library.

A related topic, which has also been the focus of recent research [5, 9, 8, 4], is one of accurately and efficiently measuring the power dissipated in a circuit. In our work, we use a power measuring technique akin to that reported by Ghosh *et al.* [5]. The cost function for technology mapping for low power is derived from this power measurement technique. The gate power models and power measurement techniques used in our work are reviewed in Section 2.

The most successful techniques in technology mapping for the traditional metrics of area and delay have followed a similar paradigm. Exact and efficient algorithms, using dynamic programming, were first developed for tree circuits, these were then extended

to take care of circuits with internal fanout. We follow a similar paradigm. We first present an exact algorithm for tree mapping. This algorithm is novel in as much as it has the flavor of both area and delay mapping. Power consumption, like area, is additive; and like delay, it depends on the load. For non-tree circuits, the mapping problem is NP-hard. Thus, recourse has to be taken to efficient heuristics that will work well for most practical instances. One of the techniques used in area and delay mapping has been the use of tree mapping for parts of the circuits that are trees and considering some matches across the fanout points. We adopt a similar strategy, with some variations in the heuristics. Tree mapping for low power, and its extensions to handle non-tree circuits is the subject of Section 3. We believe that the effectiveness of the mapping algorithm can be improved by an appropriate choice of the starting point for low power mapping. We present some ideas in that direction.

Experimental results shown in Section 4 on a large set of benchmark circuits indicate a potential for reduction in the power requirements by optimal technology mapping. Typical reductions are more than 10% with cases where the improvement is even 24%.

2 Power Dissipation Model

In a CMOS gate, the most significant part of the power consumption occurs only during transitions at the output when the output is charging/discharging. In effect, the average power consumed by a CMOS gate is given by $\frac{1}{2} \times C_{output} \times V_{dd}^2 \times N$, where C_{output} is the output capacitance of the gate, V_{dd} is the supply voltage, and N is the expected number of transitions at the output per clock cycle. Besides V_{dd}, which we assume is fixed, there are two different parameters in the above expression, C_{output} and N. Let us consider models for each.

2.1 Transition Probabilities

We can think of N as the probability of a transition occurring at the output of a gate during one clock cycle. This probability depends upon the Boolean function being computed, the probabilities of transitions at the primary inputs, and the delay model being used.

The *signal probability* p_g at the output of a gate g is the probability of a logical 1 at the output of this gate. As an example of computing p_g, suppose the minterms for which the function at gate g evaluates to a 1 are $a_1 a_2' a_3$, $a_1' a_2 a_3$ and $a_1 a_2 a_3$. Further let p_{a_1}, p_{a_2} and p_{a_3} be the mutually independent signal probabilities of the primary inputs a_1, a_2 and a_3, respectively. Then the probability of the output of g being 1 in the steady state is given by: $p_g = p_{a_1}(1-p_{a_2})p_{a_3} + (1-p_{a_1})p_{a_2}p_{a_3} + p_{a_1}p_{a_2}p_{a_3}$

The probability of 0 being the stable value is $(1 - p_g)$. Thus the probability of a change in the stable state of the output as a result of change in the primary inputs is $p_t = 2 \times p_g \times (1 - p_g)$ corresponding to the sum of the probabilities of the changes from 1 to 0 and 0 to 1. This assumes that the consecutive input vectors to the circuit are uncorrelated.

Thus, p_t is the *signal transition probability*, *i.e.* the probability of a transition occurring at the output of the gate, if all the gates had zero delay. In that case we are only dealing with transitions between stable states and filtering out the *glitches*. Note that under the zero-delay approximation, the signal transition probability at a node in the network is dependent only on the Boolean function being computed at the node. Glitches must be considered if we assume non-zero delays at gates. Thus the total power consumed in a circuit

Reprinted with permission from *IEEE/ACM 30th Design Automation Conference*, V. Tiwari and P. Ashar and S. Malik, "Technology Mapping for Low Power," pp. 74-79, June 1993. © ACM.

is the sum of the logical switching power and the glitching power. The logical switching power is the same as the total power in a zero delay model. In this paper, we consider the zero delay power to be a valid approximation of the total power for the following reasons:

- The glitches are caused by unequal path lengths. This effect can be mitigated to some extent by balancing path lengths by buffer insertion as a post-processing step.

- In practice, glitching power forms a relatively small component of the total power dissipation in most circuits [6]. This is especially true when one is interested in average rather than peak power dissipation.

- Glitching is difficult to model accurately. For example, if adjacent transitions occur sufficiently close in time, only a small hump occurs at the output instead of a complete swing to V_{dd} or ground due to the inertial gate delay. In this case, what is logically a glitch consumes little power. Accurate low level information about physical gate delays is required to model this effect correctly.

2.2 The Output Capacitance

The capacitance at the output of a CMOS gate is the sum of three components C_{INT}, C_{WIRE}, and C_{LOAD} [14]. C_{INT} represents the internal capacitance of the gate. This consists largely of the diffusion capacitance of the drain. C_{WIRE} represents the capacitance due to the physical interconnection, and C_{LOAD} represents the sum of gate capacitances of the transistors fed by the output. The values of C_{INT} and C_{LOAD} can be obtained from the data provided for each gate in the technology library. C_{WIRE} is routing dependent and is ignored at the gate level.

Given the capacitances, the power consumed by a gate g is $P_g = k \times (C_{INT_g} + C_{LOAD}) \times p_g$. k is a technology dependent scaling factor. Using this model, it is easy to compute the power consumed by a circuit, once it has been mapped, by traversing the circuit and tallying up the power consumed at each gate.

3 Mapping for Low Power

The overall flow of our algorithm for technology mapping for low power follows along the lines of technology mapping for area and delay as described in [10]. Specifically, a canonical representation (called the subject DAG) is created for the Boolean function to be mapped using a basis function consisting of two-input NAND/NOR gates and inverters. Canonical representations (called patterns) are also obtained for each of the gates in the library using the same basis function. The technology mapping problem is then formulated as one of optimally covering the nodes in the subject DAG with patterns so that the cost function modeling power dissipation is minimized under the constraint that the inputs to a match are available as the outputs of other matches.

3.1 Relationship to Area/Delay Mapping

We now give an intuitive feel for the key differences and similarities between the cost function for power and cost functions for area and delay during technology mapping. As described in Section 2, the average power dissipated at a gate is proportional to $C_g \times p_t$, where C_g is the total capacitance at the gate output and p_t is the probability of a transition at the gate output. *Thus, the cost of a match is dependent on the Boolean function at the output of a node in the subject graph unlike in the case of technology mapping for area and delay.*

Consider the circuit in Figure 1(a). The signal transition probabilities at the outputs of gates G_1, G_2 and G_3 are 0.109, 0.109 and 0.179, respectively.[1] Assume that the gates available in the

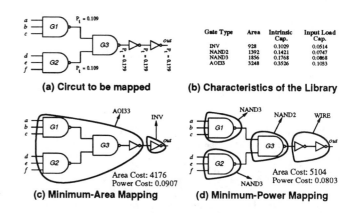

(a) Circut to be mapped (b) Characteristics of the Library

Gate Type	Area	Intrinsic Cap.	Input Load Cap.
INV	928	0.1029	0.0514
NAND2	1392	0.1421	0.0747
NAND3	1856	0.1768	0.0868
AOI33	3248	0.3526	0.1083

(c) Minimum-Area Mapping
Area Cost: 4176
Power Cost: 0.0907

(d) Minimum-Power Mapping
Area Cost: 5104
Power Cost: 0.0803

Figure 1: Example Illustrating the Difference Between Technology Mapping for Power and Area

library for technology mapping are inverters, two-input and three-input NAND gates and AOI33 gates. The areas and capacitances in standard units associated with these gates are shown in Figure 1(b).[2] (Since the point being made by this example is independent of the load driven by the *out* signal, assume that that load is zero.) The minimum-area mapping for this circuit is shown in Figure 1(c), and minimum-power mapping in Figure 1(d). The area and power cost of each mapping is shown in the corresponding figure. The power cost of a mapping is the sum of the power costs of each match in the mapping. Using the power models, the total power cost of the minimum-area mapping is 0.0907 while that for the minimum power mapping is 0.0803. Note that since the signal transition probability at the output of gate G_3 is much higher than the signal transition probabilities at the outputs of gates G_1 and G_2, it is less expensive in terms of power dissipation to have large gates feeding the outputs of gates G_1 and G_2 and small gates feeding the output of gate G_3, than have large gates feed the output of gate G_3. It turns out that the minimum-area mapping has a large gate and an inverter feeding the wires with high signal transition probability, and is therefore expensive in terms of power dissipation. As we can see from the figure, the power cost of the minimum-power mapping is found to be more than 10% lower than the power cost of the minimum-area mapping.

The similarity between technology mapping for power and delay is that the cost of a match is, in part, dependent on the load driven by the match. In the most general case of the use of a non-zero-delay model in the computation of signal transition probabilities,[3] the best match at a node cannot be determined until the matches at the fanout nodes and the matches at their transitive fanins have been chosen. In the specialized case of the zero-delay model which we use in this paper, the signal transition probability at the output of a node is independent of the match chosen for that node (*cf* Section 2). Therefore, the power dissipated in the load capacitance is the same for all matches at the node. Consequently, the best match at a node can be determined before the matches at the fanout nodes have been chosen as in the case of technology mapping for area. Even so, the actual cost of the best match at a node can only be determined once the matches at the fanout nodes are known. This point will be elaborated in subsequent sections.

Finally, as in the case of technology mapping for area, and unlike technology mapping for delay, fanout optimization is a non-issue in technology mapping for low power under the zero-delay model. Essentially, there is no notion of slack which can be distributed between the fanout branches of a node. An interesting observation is that if a non-zero delay model was used to calculate the signal probabilities, the insertion of the fanout tree might be beneficial since it can be used to balance the path lengths, and thereby reduce glitching. Therefore, fanout optimization could play a role in the technology mapping for power once glitching power is incorporated

---

[1]The signal transition probabilities are computed by assuming that each primary input takes on a value of 1 or 0 with a probability of 0.5, independent of the values on the other primary inputs.

[2]Derived from the `lib2.genlib` library in the SIS [11] distribution.

[3]Equivalent to considering power dissipation due to glitching (*cf* Section 2).

into the model.

3.2 Computing Transition Probabilities

In the zero delay model used in this paper, the signal transition probability at the output of a node is dependent only on the Boolean function at the node output in terms of the primary inputs. The nature of implementation of the logic feeding the node does not affect the signal transition probability. Therefore, the signal transition probabilities at the node outputs in the subject graph only need to be computed once in the beginning. We use the same approach for computing signal transition probabilities as proposed in [5]. The probabilities are computed under the assumption that each primary input takes on a value of 1 or 0 with a probability of 0.5, independent of the values on the other primary inputs. This is typically true for datapath circuits, but may need refinement for control circuits. The probability that a signal makes the transition from 1 to 0 ($p_{1 \to 0}$) or from 0 to 1 ($p_{0 \to 1}$), is given by $p_1 \times (1 - p_1)$, or equivalently by $p_0 \times (1 - p_0)$. p_0 and p_1 are the probabilities that the signal takes the value 0 and 1, respectively. p_0 (p_1) can be computed efficiently by counting the number of minterms in the off-set (on-set) of the Boolean function at the wire in terms of the primary inputs using Binary Decision Diagram [1] based algorithms. The same algorithm can also be applied to compute the signal transition probabilities when the signal probabilities of the primary inputs are different from 0.5.

3.3 Tree Covering for Low Power

As in the case of area and delay, the DAG covering problem is approximated by posing it as a composition of a number of tree covering problems by first partitioning the subject DAG into a forest of trees. As in the case of technology mapping for area and delay, this approach is motivated by the existence of efficient tree covering algorithms for low power. The algorithm for optimum tree covering targeting low power is described below. We assume familiarity with tree covering algorithms for area and delay mapping.

The key property that makes efficient tree covering algorithms for area/delay possible holds in the case of tree covering for power also. Given a match, m, at a node in the subject graph, the power costs of the best matches at the inputs to m are independent of each other. It follows that the match, m, with the minimum power cost at the root of a tree is the match that minimizes $power(m) + \sum_{v_i \in inputs(m)} min\_power(v_i)$, where the v_i are nodes in the subject graph input to the match m and $min\_power(v_i)$ is the cost of the minimum-cost match at the node v_i. $power(m)$ is proportional to $p_t \times (C_{INT} + C_{LOAD})$, where p_t is the signal transition probability at the node, C_{INT} is the intrinsic capacitance of the match, and C_{LOAD} is the load capacitance seen by the match. Since p_t is the same for all matches, the $p_t \times C_{load}$ component is the same for all matches at the node. Therefore, the best match at the node can be obtained without knowing the value of C_{load}. The actual cost (which includes the $p_t \times C_{load}$ term) of the best match at a node is needed in order to determine the best match at the fanout of the node. This can be computed once the match at the fanout node is known.

An algorithm for tree covering for low power traverses the tree once from the leaves to the root, visiting each node exactly once. When a node is visited, all the matches at that node are enumerated and from among those, the match with the minimum power cost is stored. As in the case of tree covering for area and delay, the complexity of this approach is the cost of generating all the pattern matches at the nodes of the subject graph since the cost function evaluation at each node is simple.

The pseudo-code for the tree-covering algorithm is given below. The argument to it is the root of the tree to be covered.

optimal-power-cover(*node*)
{
    if (*node* has been visited before) {
        /* Optimal cover for *node* is known. */
        return optimal cover for *node* ;
    }

    /* Find the optimal cover at *node* */
    $min\_cost = \infty$;
    foreach *match* at *node* {
        $cost$ = cost of *match* ;
        foreach *input* to *match* {
            optimal-power-cover(*input*) ;
            $cost = cost +$ cost of optimal cover at *input* ;
        }
        if ($cost < min\_cost$) {
            $min\_cost = cost$; $node\_match = match$
        }
    }
    return optimal cover for *node* ;
}

3.4 Tree Covering for Low Power Under a Delay Constraint

The algorithm for tree covering for low power (for the zero-delay model) under a delay constraint proceeds in much the same manner as tree covering for area under a delay constraint. The delay constraint is specified as a required time at the root of the tree. Since the required time cannot be propagated towards the leaves unless the load being driven by each node in the subject graph is known, the solutions with minimum power cost for each possible required time are stored at every node during the first pass through the tree from the leaves to the root. In the second pass from the root to the leaves, the best match is picked for each node visited. As soon as a match is picked, the required time can be propagated to the inputs of the match. Discretization of required times is used to reduce the number of solutions that need to be stored as in the case of tree covering for area under a delay constraint. (This part is currently being implemented.)

3.5 DAG Covering for Low Power

Most real life circuits have nodes with multiple fanout, and are therefore not trees. For non-tree circuits, the technology mapping problem becomes NP-hard. Intuitively, the complexity increases because the matches at the inputs to a match are, in general, no longer independent of each other. For instance, if a match requires that a multiple-fanout node be internal to it, then the logic for the multiple-fanout node must be duplicated if all its fanout points are not contained in the match.

A good approximation to DAG covering is to pose it as a composition of tree covering problems. The DAG is broken into trees at each multiple-fanout point and each tree is optimally mapped, separately. This heuristic has proven effective for area and delay optimization in the past and can also be used for power. However this approach has some problems. First, no matches across fanout points are allowed and thus tree overlap cannot occur (Figures 2 (a) and (b)). Usually, overlapping of trees can give better results for the three metrics, area, delay and power. Second, if tree boundaries are enforced while covering the subject DAG, the phase of the root and the leaves of each tree is fixed, reducing the flexibilty available to the matching algorithm. One way to solve this problem is to perform global phase optimization after an initial technology mapping pass. As an alternative, a cross-tree-phase heuristic is described in [10] to solve this problem and it has been implemented in the technology mapping package in SIS [11].

An alternative approach which can partly alleviate both the previous problems is to not restrict the algorithm to trees. The library is allowed to have non-tree patterns, and the subject graph can be a general DAG. Starting from the primary inputs, subject DAG nodes are traversed in a depth first[4] manner. All patterns (including ones which span multiple-fanout nodes) which match at a node are enumerated. Like in tree mapping, the minimum-cost match is stored for a node.

The issues to be addressed now are the evaluation of the cost of matches that are rooted at, or go across, multiple-fanout points and

---

[4]By depth first we mean that a node is seen only after all the nodes in its fanin have been seen.

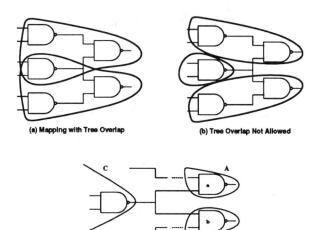

(a) Mapping with Tree Overlap   (b) Tree Overlap Not Allowed

(C) What is the Cost of a Match?

Figure 2: Ways of Handling Fanout

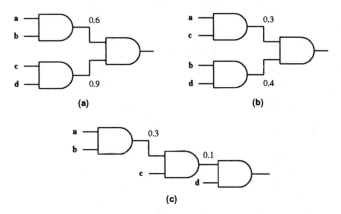

(a)   (b)

(c)

Figure 3: Effect of Decomposition on Transition Probabilities

the propagation of costs beyond such points. The latter problem occurs for the case of area and is illustrated by Figure 2(c). Here the cost of the match at node C has to be propagated to both nodes A and B, without it being considered twice. A reasonable heuristic is to propagate half the cost to both A and B. For the case of delay, evaluation of the cost of matches that are fed by multiple-fanout points, is a problem, as illustrated by Figure 2(c). The cost (arrival time) of matches at neither node A nor node B can be determined before the other. Thus neither can be mapped before the other. Touati [13] discusses this problem and proposes several heuristics to handle it.

Another problem is that if tree overlaps are allowed, the number and location of multiple-fanout points can change arbitrarily during the mapping process. Touati [13] recognizes this effect but ignores it.

General DAG covering for low power faces all of the above problems because of the nature of the cost measure used. Like area, it is additive and like delay it depends upon the load value of the fanouts. The experience with generalized area and delay mapping, however, suggests that looking across tree boundaries can be beneficial. DAG mapping was implemented for power in spite of the potential problems. To handle multiple-fanouts, several heuristics, some of them similar to the ones used in area and delay were tried out. The effectiveness of the DAG covering approach will be evaluated in Section 4.2.

3.6 Subject DAG Generation

The subject DAG corresponding to a technology independent circuit is not unique. Several different "technology decompositions" exist that result in internal signals having differing transition probabilities. Consider Figure 3 which shows three different decompositions into two input AND gates of the same function. The numbers on the internal signals are the transition probabilities. We see that differences in pin assignment and topology can lead to different transition probabilites at the internal nodes. Thus, it is possible that the technology decomposition step can lead to a significant difference in the power cost of the final mapped circuit. We do not have a good solution to this problem yet. We are exploring this issue further.

4 Experimental Results

The algorithms and heuristics described in the previous sections have been implemented and integrated with the technology mapping package in SIS [11]. Several experiments were conducted on circuits from the MCNC and ISCAS benchmark suites. These experiments can be classified into two types. Those dealing strictly with tree covering, and those dealing with the generalized DAG covering. The experiments were carried out on a SUN SPARC 2 platform. Except for the additional time required to form BDDs and compute signal transition probabilities, the CPU time requirements were of the same order as for technology mapping for area.

The library used was lib2.genlib, a large, relatively accurately calibrated library with a wide variety of gates distributed along with SIS. The libraries that are available do not specify the internal capacitance of the gates explicitly. We estimated the capacitances associated with a gate from the α, β and γ provided for each gate in lib2.genlib.[5] If explicit capacitance values were supplied with the libraries, they could be used directly.

4.1 Tree Covering Results

The first experiment we carried out was designed to estimate the optimization potential of technology mapping for low power. The benchmark circuits were first mapped for minimum power, and then for maximum power. The variation in the power consumption in these two cases provides a measure of the flexibility in the circuit structure towards the power metric.

For the experiment to be useful, it had to be conducted under a rigid and well understood framework. Since DAG covering heuristics have varying behavior, strict tree covering was used to map the circuits, i.e., the circuit was partitioned into a forest of trees and optimal tree-mapping used for each tree.

The results for maximum and minimum power dissipation show an average difference of around 46%, with individual differences going as high as 67%. The results indicate that power variations during technology mapping are significant and thus this provides a motivation for our work.

Traditionally area has been used as a rough, first order estimate of the power consumption. To determine the validity of this assumption during technology mapping, the next phase of the experiment involved comparing the minimum power results with the power dissipation observed in circuits mapped for minimum area. For a fair comparison, strict tree mapping was again used for area minimization. The results show that the minimum power dissipation is lower than the power dissipation of a circuit mapped for area by an average of about 12%.

For detailed results of these experiments, please refer to [12].

---

[5]Given a gate g, α is the intrinsic (load independent) delay of g, β is the delay per unit load factor, and γ is load being driven by g. The intrinsic capacitance of g was estimated as being proportional to α/β, and the input capacitance was estimated as being proportional to γ. This is justified since α has the dimensions RC and β has the dimensions R. Thus, their ratio has the dimensions of capacitance. Since α is the intrinsic delay of the gate and β is also internal to the gate, their ratio can be looked upon as the intrinsic capacitance of the gate. The power values given in the tables are meant for relative comparison rather than as absolute measures of power.

Circuit	Power Comparison			Area Comparison		
	Min Power	Area Power	% Lower	Power Area	Area Area	% Penalty
5xp1	8.559	9.697	11.74	145696	135488	7.53
5xp1-hdl	4.771	5.317	10.27	73776	71456	3.25
9sym	14.068	15.416	8.74	234320	199520	17.44
9symml	11.848	13.278	10.77	210656	187456	12.38
9sym-hdl	5.937	6.180	3.93	123424	115072	7.26
alu2	17.348	19.891	12.78	452864	376768	20.20
alu4	31.326	38.134	17.85	824528	722912	14.06
alupla	7.118	7.841	9.22	134096	122496	9.47
apex6	35.969	45.553	21.08	908512	717344	26.65
apex7	13.060	15.284	14.55	299744	253808	18.10
b9	7.371	8.379	12.03	139200	131312	6.01
bw	15.037	16.145	6.86	264944	226896	16.77
c8	10.675	11.817	9.66	206480	176784	16.80
con1	1.135	1.500	24.33	22272	20416	9.09
duke2	30.764	35.380	13.05	703888	600880	17.14
f2	1.527	1.767	13.58	24128	24128	0.00
f51m	8.429	8.956	5.88	141984	129920	9.29
f51m-hdl	4.663	5.081	8.23	71920	67744	6.16
frg2	63.402	75.914	16.48	1623536	1361376	19.26
misex1	4.708	5.321	11.52	83520	72848	14.65
misex2	5.587	6.784	17.64	162400	138736	17.06
misex3	41.510	46.368	10.48	972544	846800	14.85
misex3c	33.645	37.662	10.67	697856	589280	18.43
pair	86.235	96.600	10.73	1932096	1629568	18.56
sao2	8.334	9.387	11.22	161936	153120	5.76
sao2-hdl	12.052	12.886	6.47	257520	238960	7.77
rd53	3.130	3.346	6.46	50112	44544	12.50
rd53-hdl	2.446	2.450	0.16	44080	40368	9.20
rd73	9.486	10.417	8.94	178176	154976	14.97
rd73-hdl	4.098	4.119	0.51	78416	68208	14.97
rd84	19.696	21.218	7.17	381872	313664	21.75
rd84-hdl	4.917	4.886	-0.63	103472	90480	14.36
rot	37.667	42.479	11.33	762816	665840	14.56
vda	28.614	31.862	10.19	1325184	1147008	15.53
vg2	10.483	12.218	14.20	261696	193952	34.93
x1	19.790	23.744	16.65	411568	362848	13.43
x3	55.053	61.449	10.41	1071376	922432	16.15
x4	27.412	32.711	16.20	548448	541952	1.20
z4ml	4.108	4.570	10.11	64496	67280	-4.14
z4ml-hdl	2.901	3.095	6.27	50112	45936	9.09
C1355	27.498	28.260	2.70	735904	713632	3.12
C17	0.466	0.466	0.00	8352	8352	0.00
C1908	29.806	31.435	5.18	840304	610160	37.72
C432	11.210	12.687	11.64	239888	234784	2.17
C499	23.336	26.365	11.49	403680	458432	-11.94
Average			10.19			12.30

Table 1: Results of Technology Mapping Using DAG Covering Heuristics for Non-Tree Circuits

4.2 DAG Covering Results

The results of technology mapping using DAG covering are shown in Table 1. The second column shows the minimum power obtained using a variety of heuristics, some of which are mentioned in Section 3.5. The third column shows the power for the circuits that have been mapped for the minimum area using the SIS technology mapper. The power cost of the circuits optimized for power is, on the average, lower than the power of the area optimized circuits by more than 10%, with individual cases showing a difference of up to 24%. Note that for one case (rd84-hdl), the power of the area optimized circuit is slightly lower than the power of the circuit optimized for power. This shows the non-optimality of the DAG-covering heuristics, and suggests that there is scope for further improvement.

The comparison between the areas of the power optimized circuits and the area optimized circuits is also useful. It is shown in the last three columns of Table 1. The last column shows the area penalty paid to achieve the reduction in power. The results show an average area increase of about 12%. Again it is interesting to see that in some cases the area of the power optimized circuits is

lower than the area of the area optimized circuits, pointing to the non-optimality of the generalized DAG covering for area as well.

We judged the effectiveness of the DAG covering heuristics for power minimization by comparing the results obtained with the results for strict tree covering. On the average, DAG covering results in a 12% improvement over tree-covering.

5 Summary and Future Directions

There are several contributions made by this paper. The first is the development of power models for library gates. The second is the use of these models in a technology mapping algorithm that optimizes for power. Using tree circuits, for which we know that the technology mapping algorithms are optimal, we have demonstrated that there is significant difference in the minimum and maximum possible values of the power consumptions. In addition, there is a difference between the power consumption of the power optimized circuit and the area optimized circuit; showing that the least area circuit is not necessarily the least power circuit. These two results

546

show that there is potential to reduce the power consumption during the technology mapping stage.

While the potential for power reduction has been demonstrated; not all of it has been tapped yet. There are several issues that are currently under exploration that we believe hold the key to even greater reductions in power requirements. The first of these deals with the generation of the subject DAG, which is the starting point for the mapping process. More work needs to be done in improving this starting point since the final result can be very sensitive to this. Trying to take care of internal fanout points in non-tree circuits has been a problem for both area and delay mapping, it continues to be a bother in power mapping. Two different directions are being pursued here. The first is doing DAG covering using binate covering. This is likely to be slow. We do not have much hope for this for general use, but would like to use it to see what the exact results are for at least some small circuits. The second direction is to explore more heuristics for handling the fanout points in an attempt to find one that works well. Minimization of glitching power needs to be considered explicitly, possibly as a post-mapping step by selectively balancing path lengths. Finally, it would be interesting to examine the impact of library sizes on the variation in power requirements. Prior experience with varying libraries in the case of area mapping has shown that increasing the library size does result in a significant improvement in the quality of the results until a certain point after which the additional gates provide little advantage [10, 7]. We would like to examine this aspect in the context of mapping for low power. Based on preliminary experiments, we have reason to believe that larger libraries will result in a more marked difference between the power consumption of a circuit mapped for area and that of a circuit mapped for low power.

References

[1] R. Bryant. Graph-Based Algorithms for Boolean Function Manipulation. *IEEE Transactions on Computers*, C-35:677–691, Aug. 1986.

[2] A. Chandrakasan, M. Potkonjak, J. Rabaey, and R. Broderson. HYPER-LP: A System for Power Minimization Using Architectural Transformations. In *Proceedings of the International Conference on Computer-Aided Design*, pages 300–303, November 1992.

[3] A. Chandrakasan, S. Sheng, and R. Broderson. Low-Power CMOS Digital Design. *IEEE Journal of Solid-State Circuits*, 27(4):473–484, April 1992.

[4] M. Cirit. Estimating Dynamic Power Consumption of CMOS Circuits. In *Proceedings of the International Conference on Computer-Aided Design*, pages 534–537, November 1987.

[5] A. Ghosh, S. Devadas, K. Keutzer, and J. White. Estimation of average switching activity in combinational and sequential circuits. In *The Proceedings of the Design Automation Conference*, pages 253–259, June 1992.

[6] A. Ghosh, A. Shen, S. Devadas, and K. Keutzer. On Average Power Dissipation and Random Pattern Testability. In *Proceedings of the International Conference on Computer-Aided Design*, November 1992.

[7] K. Keutzer, K. Kolwicz, and M. Lega. Impact of Library Size on the Quality of Automated Synthesis. In *Proceedings of the International Conference on Computer-Aided Design*, pages 120–123, November 1987.

[8] H. Kriplani, F. Najm, and I. Hajj. Maximum Current Estimation in CMOS Circuits. In *The Proceedings of the Design Automation Conference*, pages 2–7, June 1992.

[9] F. Najm. Transition Density, A Stochastic Measure of Activity in Digital Circuits. In *The Proceedings of the Design Automation Conference*, pages 644–649, June 1991.

[10] R. Rudell. *Logic Synthesis for VLSI Design*. PhD thesis, University of California, Berkeley, 1989.

[11] E. M. Sentovich, K. J. Singh, C. Moon, H. Savoj, R. K. Brayton, and A. Sangiovanni-Vincentelli. Sequential circuit design using synthesis and optimization. In *Proceedings of the International Conference on Computer Design*, pages 328–333, October 1992.

[12] V. Tiwari, P. Ashar, and S. Malik. Technology Mapping for Low Power. *NEC CCRL Technical Report*, (92-C019-4-5509-2), October 1992.

[13] H. Touati. *Performance Oriented Technology Mapping*. PhD thesis, University of California, Berkeley, 1990.

[14] N. Weste and K. Eshraghian. *Principles of CMOS VLSI Design, A Systems Perspective*. Addison-Wesley Publishing Company, 1985.

POSE: Power Optimization and Synthesis Environment

Sasan Iman, Massoud Pedram

Department of Electrical Engineering - Systems
University of Southern California
Los Angeles, CA 90089

Abstract

Recent trends in the semiconductor industry have resulted in an increasing demand for low power circuits. POSE is a step in providing the EDA community and academia with an environment and tool suite for automatic synthesis and optimization of low power circuits. POSE provides a unified framework for specifying and maintaining power relevant circuit information and means of estimating power consumption of a circuit using different load models. POSE also gives a set of options for making are-power trade-offs during logic optimization.

1. Introduction

In the past, the main objective of designers has been to design faster and denser circuits. In response to this demand, design tools have been developed to help automate the design process for achieving maximum speed and minimum area. These design automation tools have been used extensively in the industry and are an integral part of any design cycle.

With the increased popularity of portable devices, battery size and lifetime are becoming important factors in the design process. At the same time, the amount of data to be processed is increasing at a rapid pace. This also calls for faster digital devices which in turn increases power consumption. The circuit power is also becoming one of the limiting factors in the amount of logic that can be placed in a VLSI chip and the determining factor in the packaging cost. These considerations have resulted in a growing need for minimizing power consumption in today's digital systems.

The demand for low power digital systems has motivated significant research in the area of power estimation and power optimization. Power estimation and optimization techniques have been proposed at all stages of the design process. Power estimation techniques have been proposed by researchers at the gate level [3], [8], [15], [17], [21]. Power optimization techniques have also been proposed at all levels of the design abstraction. Many optimization approaches have been proposed at the behavioral, RT and logic level [2], [5], [9], [10], [16], [19], [20]. Reference [18] contains a detailed survey.

Even though considerable effort has been made in creating new techniques for power estimation and optimization, a unified framework for designing low power digital systems has not yet been developed. The void created by the absence of such a framework has presented designers with serious problems. Optimization algorithms that target low power circuits use the frameworks designed for synthesizing minimum

area and delay circuits. This means that the critical information needed for power estimation and optimization is not available when low power techniques are applied. In most cases, minimal information is available to the power conscious design procedures which in turn results in reducing the potentials of these procedures. A more significant problem caused by the lack of a unified framework for low power design is that designers are forced to use low power techniques as isolated procedures. This creates a major obstacle in developing a methodology for effective and efficient power specification, estimation and optimization. The lack of a methodology in turn results in a limited understanding of the applicability of existing techniques which contributes to holding back the state of the art low power technology. At the same time, many optimization techniques are only applicable and relevant when applied in conjunction with other optimization approaches. Therefore without a unified framework, many new techniques will not be discovered.

In this paper we address this challenge by presenting a methodology for designing low power digital circuits at the RT and logic levels. In doing so, we present POSE, the Power Optimization and Synthesis Environment. POSE is the first step in creating a complete and unified framework for design and analysis of low power digital circuits. POSE provides an easy to use interactive environment which is an extension of the familiar environment provided by the SIS package [1].

This paper is organized as follows. In section 2 we present the power model used during optimization. In section 3 we describe the low power design methodology presented in this paper. In section 4 and 5 we discuss issues behind design specification for power and power estimation. Results and conclusions are presented in sections 6 and 7.

2. Power Model

Power dissipation in CMOS circuits is caused by four sources: 1) the leakage current which consists of reverse bias current in the parasitic diodes formed between source and drain diffusions and the bulk region in a MOS transistor as well as the subthreshold current that arises from the inversion charge that exists at the gate voltages below the threshold voltage, 2) the stand-by current which is the current drawn continuously from V_{dd} to ground which happens, for example, when the tri-stated input of a CMOS gate leaks away to a value between V_{dd} and ground, 3)the short-circuit (rush-through) current which is due to the DC path between the supply rails during output transitions and 4) the capacitive current due to charging and discharging of capacitive loads during logic changes.

In well-designed CMOS circuits, the dominant source of power dissipation is due to capacitive currents due to charging and discharging of the gate capacitances (also referred to as the dynamic power dissipation) and is given by:

$$P_i = \frac{1}{2} \cdot V^2 \cdot C_i \cdot f \cdot E_i \qquad (1)$$

where V is the supply voltage, f is the clock frequency, C_i is the capacitance seen by gate n_i and E_i (referred to as the

* This research was supported in part by the Semiconductor Research Corp. under contract no. 94-DJ-559 and by ARPA under contract number F336125-95-C1627.

switching activity) is the expected number of transitions at the output of n_i per clock cycle. The product of C_i and E_i is often referred as the *switched capacitance*.

At the time of logic synthesis, all architectural and technology related issues for the network being optimized have been decided. This means that the values for V and f have already been selected. Therefore the main issues in computing a power estimate are in computing the load and switching activity values for nodes in the network. At the logic level, our optimization goal is also to minimize the total switched capacitance of the circuit.

3. Low Power Design Methodology

A low power design methodology can only be developed when a number of key components are made available to the designer. These components fall into the following categories: *Design Specification, Design verification* and *Synthesis Procedures*. Complete design specification is necessary in order to provide the synthesis environment with maximum information necessary for the optimization process. For example, when designing a circuit for minimum area, the design specification for this synthesis process should include a measure of the area for the gates in the target library. Similarly, a synthesis environment for low power should include all power modeling and estimation information that is necessary during the synthesis and validation procedure. An important issue to also consider is that design specification for power may not be the same at different levels of abstraction and therefore a set of consistent specification standards need to be available at different levels. For example, input data activity at RT level may be given in terms of word-level probability density functions while the same information may be provided at logic level in terms of bit-level switching activity. Design specification should describe the possible forms of data and provide means of translation into the other forms.

Design validation is also an important part of any design methodology. Final design has to comply with the design specifications. For conventional logic synthesis, this validation is in the form of checking functional correctness and checking that design meets the area/delay requirements. For power consideration, a design has to meet the target power budget. Power estimation procedures are therefore necessary to check design compliance with the given specifications. At the same time, power estimation is necessary to check the quality of the synthesis steps. Power estimation is also a crucial part of interactive optimization techniques.

Synthesis procedures are the basic steps used to incrementally change the circuit structure while optimizing the target cost function. The combination of these steps guided by the power estimation procedures are used to develop a power optimization methodology.

Figure 1 presents the power optimization methodology used in POSE. The highlighted boxes specify the information that has to be provided by the designer. Input to POSE is a state transition graph describing the finite state machine for the circuit. For FSMs and combinational circuits, input can also be provided as a Boolean network and the statistical information for the primary inputs of the circuit. The set of power information required by design specifications (see section 4) is also provided at this time. State assignment is then used to assign an state encoding to states of the finite state machine. The conventional cost function for state assignment algorithms has been minimum area.

State assignment of a finite state machine has a signifi-

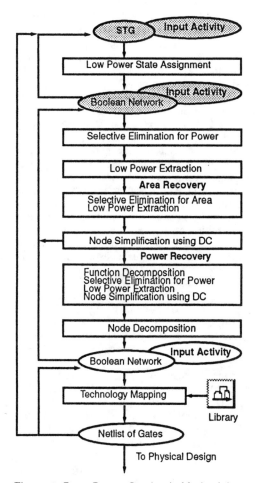

Figure 1: Pose Power Synthesis Methodology

cant impact on the area of its final logic implementation. In the past, many researchers have addressed the encoding problem for minimum area of two-level or multi-level logic implementations. These techniques can be modified to minimize the power dissipation. An effective approach [19] is to consider the complexity of the combinational logic resulting from the state assignment and to directly modify the objective functions used in conventional encoding schemes such as NOVA [22] and JEDI [12] to achieve lower power dissipation. This approach is used in POSE to perform state assignment. The output of the state assignment program is a Boolean network. A Boolean network is a directed graph where each node represents either a Boolean function or a latch element. The Boolean network is then optimized using a set of logic synthesis operations. The design entry into POSE can also be in form of a Boolean network.

Low power optimization algorithms provided in POSE consist of three categories of optimization techniques: 1)Low power algebraic restructuring techniques, 2) Low power simplification and 3)Low power technology mapping. Algebraic restructuring techniques and function simplification are used during the logic synthesis process while the technology mapping is used to map the optimized networks to gates in the target technology.

Algebraic logic restructuring techniques have been used in the past to minimize the area of a Boolean network by taking advantage of the common sub-functions between different nodes and within the same node. These operations include common sub-function extraction, function decomposition and factorization. A selective collapse procedure is used to remove nodes from the network which don't contribute to minimizing the cost of the network.

POSE includes operations to perform low power logic restructuring. The low power algebraic operations implemented in POSE are discussed in detail in [10]. The basic approach here is to use extraction, factorization, decomposition and selective collapse operations to reduce load on high activity nodes by introducing new nodes into the network that have low output activity and also removing nodes that are not good candidates for reducing the network power cost.

It should be noted that power optimization techniques are developed to make area-power trade-offs. In the methodology given in Figure 1, the initial logic restructuring have been performed for minimum power. This means that network area may not have maximally decreased during this operation. Therefore the methodology includes a number of operations for recovering some area. This is called "area recovery". A "power recovery" stage is also included before the technology mapping process. The reason for this is that decomposition for minimum power does not maximally decompose all node functions. In other words it leaves some area redundancy in the network that may still be taken advantage of for power optimization. The "power recovery" stage is included to use this flexibility.

After the Boolean network is optimized for power, it is then mapped to the gates in the target technology. This operation will generate a netlist of gates in the library. The goal of low power technology mapping is to generate a mapped network where high activity nodes are hidden inside complex gates. The procedure for low power technology mapping has been discussed in detail in [20]. After technology mapping is performed, it is possible to swap the symmetric inputs to a gate. The power after technology mapping can be further reduced by swapping pins where inputs with high input load are driven by lower activity inputs. By taking advantage of this flexibility, power can be further reduced.

Note that at some points during the optimization, the quality of the results are checked using the power estimation utilities. This is to check how much reduction in power has been obtained after each iteration. The synthesis process is terminated if no further improvement can be obtained or the resulting power estimate is within the design specifications. The next section presents a detailed discussion on the design specification requirement and power estimation.

4. Design Specification for Power

In the past, logic synthesis has concentrated on minimizing the area of a circuit while meeting the timing constraints. The design specification for logic synthesis therefore consisted mainly of providing the functional description of the circuit, the timing constraints and the area/delay characteristics of the target library. As logic synthesis environments are extended to take into account power consumption, the conventional design specification techniques prove to be inadequate. In this section we discuss information that is necessary for effective power estimation and optimization at the logic synthesis stage.

4.1. Input Switching Activity

The power consumption of a CMOS circuit is a function of the expected number of times the logic signals in the circuit change values. Unlike conventional logic synthesis where the

circuit performance (area and delay) can be calculated deterministically, the statistical behavior of the input data has a significant impact on the power consumption of the circuit. In order to calculate the circuit power, it is necessary to provide a sequence of bit vectors applied at the primary inputs of the network. This sequence of bit-vectors may in turn be used by *statistical or probabilistic power estimation techniques* to calculate the expected switching rate of gates in a network. POSE relies on the expected behavior of the circuit at the primary inputs in terms of probability values to calculate the expected switching rates of the internal gates in a circuit.

Given a logic value v in a Boolean network, SP^v_1 (SP^v_0), the *signal 1(0) probability* for v, gives the probability that v evaluates to value 1 (0); $TP^v_{x \rightarrow y}$, the *transition probability* for v, gives the probability that v evaluates to x at time t and evaluates to y at time $(t-dt)$. E_v, the switching activity for v is then defined as ($E_v = TP^v_{0 \rightarrow 1} + TP^v_{1 \rightarrow 0}$). Assuming temporally independent[1] input vectors, this equation simplifies to:

$$E_v = 2 \cdot SP^v_1 \cdot SP^v_0 = 2 \cdot SP^v_1 \cdot \left(1 - SP^v_1\right) \qquad (2)$$

4.2. Library Load Values

Libraries used during logic synthesis only provide information on area and delay of each gate in the library. More information is however required to accurately measure the power consumption of a gate in a technology mapped network. The power consumption of a gate consists of the power consumed at the output of the gate and the power internal to the gate.

The power at the output of a gate g is a function of the load seen at the output of this gate. This load is a combination of the input loading of the output gates and also the *self-loading capacitance* for the gate itself. The self loading capacitance for a gate is defined as the load driven by the gate when the gate output is left open and is due to the source/drain diffusion capacitances of the gate. Experimental results show that self-loading capacitances contribute up to 20% of the total power consumption in CMOS circuits. Ignoring these capacitances will no doubt affect the accuracy of power estimation and the optimality of power optimization. The *internal power consumption* of a gate refers to the power required to charge and discharge the internal capacitances of this gate. Therefore a power optimization procedure requires knowledge of the internal capacitances (diffusion capacitances) of all gates in the library to be able to compute the power dissipation due to the self-loading capacitance and the internal parasitic capacitances (see Figure 2).

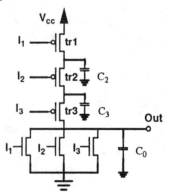

Figure 2: Parasitic Capacitances for a 3 input NOR Gate

1. Temporal independence of variable x implies that value of x at time t is independent of its value at time t+dt

5. Power Estimation

Technology independent power optimization requires a good estimate of the contribution of each node in the technology independent network to the power consumption of the mapped network. Previous methods for technology independent power estimation have used the number of fanouts for a node as an estimate of the load at the output of a node. Experimental results show the number of fanouts in the factored form (see section 5.1) to be a better load estimate for nodes in a technology independent network. At the same time some power optimization algorithms (e.g. algebraic restructuring techniques) operate on the sum-of-products form of nodes. This suggests that technology independent power analysis will be most effective when a number of different load models are considered during technology independent power estimation. This is similar to the area optimization process where a combination of number of literals in the sum-of-products form and number of literals in the factored form are used to minimize the area of the network.

5.1. Load Models

We define the power consumption and power contribution of a node in terms of its input power, internal power and output power. Note that all power values are reported as switched capacitance values where supply voltage and clock frequency are assumed to be fixed.

$$Power(n_i) = InternalPower(n_i) + OutputPower(n_i) \quad (3)$$

$$PowerContribution(n_i) = InputPower(n_i) \\ + InternalPower(n_i) + OutputPower(n_i) \quad (4)$$

The input and output power estimates for a node in a Boolean network are estimated as given below.

$$InputPower(n_i) = \sum_{n_j \in fanins(n_i)} E(n_j) \cdot L(n_j, n_i) \quad (5)$$

$$OutputPower(n_i) = E(n_i) \cdot \left(\sum_{n_k \in fanouts(n_i)} L(n_i, n_k) \right) \quad (6)$$

In this equation $L(n_j, n_i)$ gives load on gate n_j due to its fanout node n_i. We define four load models:1)simple load model, 2)factored form load model and 3)sum-of-products form load model and 4)library load. For simple load model the value $L(n_j, n_i)$ if always equal to 1. For factored form load, $L(n_j, n_i)$ gives the number of times variables n_j is used in the factored form representation of node n_i. For example, in Figure 3, $L(n_1, n_3) = 2$ where the total load in the factored form on node n_1 is equal to 5. For sum-of-products form $L(n_j, n_i)$

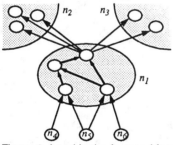

Figure 3: Load in the factored form

gives the number of times variables n_j is used in the sum-of-products form representation of node n_i. For library loads, $L(n_j, n_i)$ is computed using the library parameters.The internal power consumption in the simple load model is assumed to be

zero, for factored form load, the internal power is computed by first finding the factored form implementation of the node and then finding the power on the outputs of this factored form implementation. The same technique is applied for sum-of-products form load model. For library load, the internal load is computed using the diffusion capacitance information as discussed in section 4.2.

5.2. Switching Activities Under a Zero-Delay Model

Under a zero-delay model and assuming temporal independence at the circuit primary inputs, the switching activity of each node in the network can be computed by finding the signal probability of the node and then using equation 2 to compute its switching activity. A number of issues need to be considered when computing the signal probability of an internal node. In the following we discuss accuracy-speed trade-offs for computing node signal probability values. We also discuss how different logic synthesis algorithms may impact the signal probability and therefore the switching activity of nodes in a network.

5.2.1 Speed/Accuracy Trade-offs

The immediate fanins of an internal node are in general spatially correlated[1]. This correlation may be present even if primary inputs are spatially uncorrelated. Under these conditions, the spatial correlation at the immediate fanins of a node n is due to the reconvergant fanout regions in the transitive fanin cone of n. This means that an exact calculation of the signal probability for an internal node n requires that this signal probability be computed using the global function of n. BDDs have provided a more feasible approach for representing the global function of nodes in a Boolean network. Reference [4] presents an efficient procedure for computing the signal probability of a function from its BDD representation. Therefore BDD based techniques are a good candidate for computing the signal probability of nodes in a network.

Representing the global function of nodes in some circuits may however become too expensive even when BDDs are used to represent the global functions. Therefore it is necessary to provide a mechanism for making speed-accuracy trade-offs when computing signal probabilities. In the following, we describe and justify our technique for speeding up the procedure for computing these signal probability values.

5.2.2 Using Semi-local BDDs

Local BDD for a node n is defined as a BDD where immediate fanins of node n are used in building the BDD. The semi-local BDD for a node n is defined as the BDD where the nodes used as the BDD variables create a cut in the transitive fanin cone of n. Note that the global and local BDD for a node are special cases of the semi-local BDDs for that node.

The main reason for using global BDDs when computing the signal probability for a node n is to take into account the spatial correlation at the immediate fanins of the node. Figure 4 shows a shallow and a deep Reconvergant fanout region in the fanin cone of node n. A shallow Reconvergant region spans over less number of levels than a deep reconvergant region. It can be stated that a shallow reconvergant region, in general, results in more spatially correlated fanins to a node. At the same time, a deep reconvergant region will in general result in lower spatial correlation at the immediate fanins of node n due to randomizing effect of the side inputs to the reconvergant path. Therefore it is possible to capture most of the spatial correlations of inputs of a node n by using the semi-

---

1. Spatial correlation between nodes x and y in a network means that the values of x and y in the same clock cycle are dependent.

local BDD for that node. The semi-local BDDs will also take into account the spatial correlation due to shallow reconvergant region. POSE uses semi-local BDDs as a compromise between efficiency and accuracy [7].

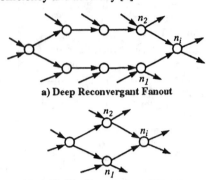

a) Deep Reconvergant Fanout

a) Shallow Reconvergant Fanout

Figure 4: Reconvergent Fanout Regions

The definition for deep and shallow regions is implementation dependent. Experimental results however show that in a network that is decomposed into 2-input gates, taking into account Reconvergant regions that span over 5 levels capture a significant part of the spatial correlation at the input of nodes in the network.

Note that as the network is being optimized, the network structure is modified and therefore the definition of semi-local bdds for each node will change. Therefore, if an optimization step changes the network structure drastically, the node switching activity values should be recomputed so that the new activity values reflect the new structure of the network.

5.3. Effect of Optimization on Switching Activities

A correct methodology for low power synthesis requires that the signal probability and switching activity values computed and stored on each node remain correct as the network is being restructured during the logic synthesis process. Most operations on a Boolean network are algebraic in nature. This means that global function of nodes and therefore the zero-delay signal probability of nodes do not change for these operations. Some logic synthesis operations, however, do modify the global function of the node being optimized and other nodes in the network. These operations will result in changing the signal probability of nodes in the network without having modified these nodes. Figure 5 shows an example where simplifying the function of a node using its observability don't care has resulted in changing the signal probability of the node and nodes in its transitive fanout.

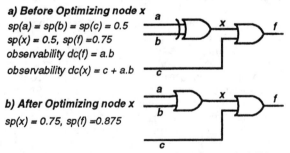

a) Before Optimizing node x

$sp(a) = sp(b) = sp(c) = 0.5$
$sp(x) = 0.5$, $sp(f) = 0.75$
observability dc(f) = a.b
observability dc(x) = c + a.b

b) After Optimizing node x

$sp(x) = 0.75$, $sp(f) = 0.875$

Figure 5: Effect of optimizations on signal probabilities

If switching activity values are estimated using a real delay model, then after changing a node n, the delay of the network changes and therefore the switching activity of all nodes in transitive fanout of n, and the switching activity of all nodes in the transitive fanouts of the immediate fanins of node n change. This is because changing the function of a node, changes the load on its immediate fanins which changes the delay for fanout nodes of the immediate fanin nodes.

6. Results

POSE has been implemented and results have been generated for minimum power circuits. Table 1 gives the power consumption of multi-level circuits in the MCNC benchmark set when they are optimized and mapped for minimum area and when they are optimized and mapped for minimum power. Columns 2, 3 and 4 give the area, delay and power (switched capacitance) of the technology mapped circuits when they are optimized for minimum area. Columns 5, 6 and 7 give the area, delay and power (normalized with respect to values in the first three columns) for the same circuits when they are optimized and technology mapped for minimum power consumption using the methodology presented in section 3. As the results show, POSE has been able to reduce the circuit power consumption on average by 26% at the expense of increasing the area of the circuits by 23%. This clearly shows a trade-off between area and power. The delay of the circuits have only been increased by 4%. Table 2 gives the same statistics for two-level benchmarks in the MCNC benchmark set. The power consumption of the technology mapped circuits has been reduced on average by 29% at the expense of increasing the area on average by 30%. Again, this clearly shows that POSE has been able to successfully find a minimal power solution by trading off area for power.

7. Conclusion

In this paper we have presented a methodology for designing low power digital circuits. POSE has been developed as a unified framework for specifying and maintaining power relevant circuit information. POSE also provides power estimation and optimization procedures which facilitate the low power design methodology.

The emphasis in this paper has been in identifying an effective low power design methodology. In addition POSE has been developed and tested with the proposed methodology where it is shown that the power estimation techniques provided a good method for verifying the quality of the incrementally power optimized circuits. An optimization script has also been developed based on the methodology presented in this paper.

POSE is a first step in developing a complete environment for low power circuit synthesis. Many optimization algorithms at different levels of abstraction are yet to be incorporated into the POSE environment. At the same time, power estimation techniques should also be included at different levels of the design abstraction. In accordance with this philosophy, POSE has been developed to allow researchers to easily incorporate new power estimation and optimization techniques into POSE. This will then allow POSE to be used as a platform for future research on low power CAD tools. The first release of POSE has been made available (http://atrak.usc.edu/~pose) and is being used both as an optimization/estimation tool and as a platform for implementing and experimenting with new low power design techniques.

Ex	Minimum Area			Minimum Power		
	area	delay	power	area	delay	power
C1355	389760	24.4	16.1	1.12	1.16	0.93
C1908	434768	39.0	14.8	1.12	1.13	0.84
C432	170288	43.5	6.65	1.20	0.97	0.84
C5315	1365088	41.5	65.9	1.25	0.99	0.85
alu2	305312	40.4	9.67	1.19	1.20	0.64
alu4	592528	48.7	9.83	1.43	1.17	0.70
b9	110896	9.34	4.17	1.13	1.09	0.78
dalu	760960	50.3	23.8	1.26	1.06	0.68
des	2865200	163	112	1.34	0.78	0.81
frg1	115536	18.9	4.98	1.25	0.86	0.71
frg2	692752	36.4	19.8	1.18	1.13	0.68
i8	797616	39.8	20.5	1.35	0.95	0.68
k2	1011056	32.8	12.3	1.44	1.13	0.65
rot	594848	26.1	22.8	1.20	1.35	0.78
sct	75168	31.4	2.51	1.00	0.44	0.60
t481	612944	30.8	7.89	1.15	0.90	0.68
term1	149408	13.5	5.17	1.03	1.10	0.71
ttt2	193952	17.7	6.32	1.01	0.99	0.61
9symml	159152	22.4	6.99	1.68	1.26	0.77
Average				1.23	1.04	0.74

Table 1. Minimum power solutions for multi-level examples

Ex	Minimum Area			Minimum Power		
	area	delay	power	area	delay	power
apex1	1308944	30.4	31.7	1.33	1.34	0.68
apex5	676512	34.2	12.9	1.40	0.70	0.72
b12	75168	11.0	2.19	1.35	1.10	0.85
bw	130848	33.1	4.41	1.49	0.49	0.67
clip	118320	20.3	5.60	1.17	1.17	0.71
cps	1050960	36.7	22.7	1.34	1.07	0.58
duke2	392544	33.1	10.2	1.13	0.72	0.61
e64	293248	110	4.22	1.00	1.01	0.63
ex4	402288	12.4	16.0	1.39	1.26	0.83
misex1	46864	13.5	1.70	1.43	0.80	0.81
misex2	91872	10.9	2.52	1.34	1.07	0.85
misex3	575360	32.1	15.6	1.40	1.20	0.69
pdc	341040	21.2	12.1	1.40	1.06	0.83
rd73	58928	17.8	1.35	1.25	0.98	0.68
rd84	128992	18.2	4.70	1.24	1.33	0.65
spla	545200	24.2	15.8	1.35	1.15	0.62
vg2	85376	11.0	3.42	1.24	1.47	0.71
5xp1	102544	30.9	4.05	0.99	0.53	0.66
9sym	178640	19.1	9.06	1.48	0.96	0.78
Average				1.30	1.02	0.71

Table 2. Minimum power solutions for two-level examples

8. References

[1] "SIS: A system for sequential circuit synthesis," Report M92/41, UC Berkeley, 1992.

[2] J. B. Burr. Stanford ultra low power CMOS. In *Proceedings of Hot Chips Symposium V*, pages 583–588, June 1993.

[3] R. Burch, F. N. Najm, P. Yang, and T. Trick. " A Monte Carlo approach for power estimation. " IEEE *Transactions on VLSI Systems*, 1(1):63–71, March 1993.

[4] S. Chakravarty. " On the complexity of using BDDs for the synthesis and analysis of boolean circuits. " In *Proceedings of the 27th Annual Allerton Conf. on Communication, Control and Computing*, pages 730–739, 1989.

[5] A. Chandrakasan, S. Sheng, and R. W. Brodersen, "Low-power CMOS design. " *IEEE Journal of Solid-State Circuits*, pages 472-484, April 1992.

[6] A. Chandrakasan, M. Potkonjak, J. Rabaey and R. W. Brodersen, " HYPER-LP: A System for Power Minimization Using Architectural Transformation. "In *Proceedings of the I EEE International Conference on Computer Aided Design*, pages 300-303, November 1992.

[7] C-S. Ding and M. Pedram. "Tagged probabilistic simulation provides accurate and efficient power estimates at the gate level . " In *Proceedings of the Symposium on Low Power Electronics*, September 1995.

[8] A. Ghosh, S. Devadas, K. Keutzer, and J. White. " Estimation of average switching activity in combinational and sequential circuits. " In *Proceedings of the 29th Design Automation Conference*, pages 253–259, June 1992.

[9] S. Iman and M. Pedram. " Multi-level network optimization for low power. " In *Proceedings of the I EEE International Conference on Computer Aided Design*, pages 372–377, November 1994.

[10] S. Iman and M. Pedram. " Logic extraction and decomposition for low power. " In *Proceedings of the 32nd Design Automation Conference*, June 1995.

[11] B. Lin and A. R. Newton. " Synthesis of multiple-level logic from symbolic high-level description languages. " In *Proceedings of IFIP International Conf. on VLSI*, pages 187-196, August 1989.

[12] R. Marculescu, D. Marculescu, and M. Pedram. " Logic level power estimation considering spatiotemporal correlations. " In *Proceedings of the IEEE International Conference on Computer Aided Design*, pages 294–299, November 1994.

[13] R. Marculescu, D. Marculescu, and M. Pedram. Efficient power estimation for highly correlated input streams. In Proceedings of the Design Automation Conference, pages 628-634, June 1995.

[14] J. Monteiro, S. Devadas and A. Ghosh. Retiming sequential circuits for low power. In Proceedings of the IEEE International Conference on CAD, pages 398-402, November 1993.

[15] F. Najm Transition density: A new measure of activity in digital circuits. IEEE transactions on CAD of Integrated Circuits and Systems, 13(9):1123-1131, September 1994.

[16] M. Pedram. "Power minimization in IC design: principles and applications," Invited Paper. ACM Transactions on Design Automation of Electronic Systems, Vol. 1, No. 1 (1996), pages 1-54.

[17] C-Y. Tsui, M. Pedram, C-H. Chen, and A. M. Despain. " Low power state assignment targeting two- and multi-level logic implementations. " In *Proceedings of the* IEEE *International Conference on Computer Aided Design*, pages 82–87, November 1994.

[18] C-Y. Tsui, M. Pedram, and A. M. Despain. " Power efficient technology decomposition and mapping under an extended power consumption model. " *IEEE Transactions on Computer-Aided Design of Integrated Circuits and Systems*, 13(9), September 1994.

[19] C-Y. Tsui, M. Pedram, and A. M. Despain. Efficient estimation of dynamic power dissipation under a real delay model. In Proceedings of the International Conference on CAD, pages 224-228, November 1993.

[20] T. Villa and A. Sangiovanni-Vincentelli. " NOVA: State assignment of finite state machines for optimal two-level logic implementations. " IEEE *Transactions on Computer-Aided Design of Integrated Circuits and Systems*, 9: 905-924, September 1990.

Transformation and synthesis of FSMs for low-power gated-clock implementation

Luca Benini and Giovanni De Micheli
Center for Integrated Systems
Stanford University
Stanford, CA, 94305

Abstract

We present a technique that automatically synthesizes finite state machines with gated clocks to reduce the power dissipation of the final implementation.

We describe a new transformation for general incompletely specified Mealy-type machines that makes them suitable for gated clock implementation. The transformation is probabilistic-driven, and leads to the synthesis of an optimized combinational logic block that stops the clock with high probability.

A prototype tool has been implemented and its performance, although strongly influenced by the initial structure of the finite state machine, shows that sizable power reductions can be obtained with our technique.

1 Introduction

The majority of the currently published work in the area of automatic synthesis for low power focuses on the reduction of the level of activity in some portion of the circuit [3, 4, 5, 6], since in the dominant CMOS technology the most important fraction of the power is dissipated during switching events.

In synchronous circuits, a very promising technique is based on selectively stopping the clock in portions of the circuit where active computation is not being performed. Local clocks that are conditionally enabled are called *gated clocks*, because a signal from the environment is used to qualify (gate) the global clock signal. Gated clocks are commonly used by designers of complex power-constrained systems [10, 7]. It should be noticed, however, that it is usually responsibility of the designer to find the conditions that disable the clock.

Some attempts have been made to automate the generation of signals that can be used to gate the global clock. In [1] a *Precomputation-based approach* has been described that focuses mainly on data-path circuits, while in [2] the

authors have described a method to generate gated clocks for systems described as finite state machines.

Our previous work [2] exploits the concept of *self-loop*, an idle condition for a Moore machine. If the machine is in a self-loop, the next state and the output do not change, therefore clocking the FSM only wastes power. Obviously, detecting self-loop conditions requires some computation to be performed by additional circuitry. This computation dissipates power, and sometimes it will be too expensive to detect all self-loop conditions. It is therefore very important to select a subset of all self-loops that are taken with high probability during the operation of the FSM.

We have improved in several directions and extended the validity of the techniques discussed in our previous work [2]. First, applicability of our techniques has been increased, and the limitation to Moore finite state machines has been removed. Our new method deals with a very general model of sequential circuit, the *incompletely specified Mealy machine*. Second, we adopt a novel probabilistic approach, that can selectively individuate and exploit the idle conditions that occur with high probability. Moreover, our new algorithms for the synthesis of the clock-stopping logic are more accurate an powerful, and are able to find exact and heuristic solutions as well.

A tool has been implemented and applied to a number of benchmark circuits. We have embedded our tool in a complete path from high-level specification to transistor-level implementation, and we have verified our results using accurate switch-level simulation.

For some circuits, more than 100% improvement in average power dissipation have been obtained. Notice that the quality of the results is strongly dependent on the type of finite state machine we start with. In particular, our method is well suited for FSMs that behave as *reactive systems*: they wait for some input event to occur and they produce a response, but for a large fraction of the total time they are idle. These FSMs are common in portable devices where low power consumption is important.

2 Background

In this work we will assume a single clock scheme with edge triggered flip-flops, as shown in Figure 1 (a). This is not a limiting assumption. We have discussed the applicability of our methods to different clocking schemes in [2]

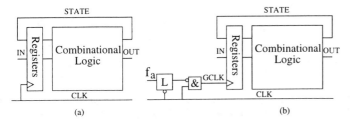

Figure 1: (a) Single clock, flip-flop based finite state machine. (b) Gated clock version.

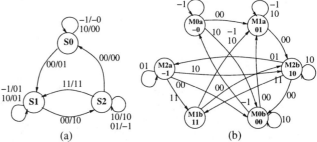

Figure 2: (a) STG of a Mealy machine. (b) STG of the equivalent Moore machine.

where we used transparent latches and multiphase clocks.

With a gated clock implementation, we need to slightly modify the structure in Figure 1 (a). We define a new signal called *activation function* (f_a) whose purpose is to selectively stop the local clock of the FSM, when the machine is idle and do not perform state transitions. When $f_a = 1$ the clock will be stopped. The modified structure is shown in Figure 1 (b). The block labeled "L" represents level-sensitive register, transparent when the global clock signal CLK is low. Notice that the presence of L is needed for a correct behavior, because possible glitches on f_a must be filtered out when the global clock signal is high.

We assume that the activation function f_a becomes valid before the raising edge of the global clock. At this time the clock signal is low and L is transparent. If the f_a signal becomes high, the upcoming edge of the global clock will not filter trough the AND gate, therefore the FSM will not be clocked and GCLK will remain low. Note that when the global clock is high, L is not transparent, therefore the GCLK signal is forced low at least up to the next falling edge of the global clock.

When the local clock is stopped, no power is consumed in the FSM combinational logic, on the clock line and in the sequential elements (differently from the scheme proposed in [1] where enabling signals are used). Notice however that the delay of the logic for the computation of f_a is on the critical path of the circuit, and its effect must be taken into account during timing verification.

Our technique automatically generates the activation function in form of a combinational logic block that uses as its inputs the primary input IN and the state lines STATE of the FSM. The input data for our algorithm is the behavioral description of the FSM and the probability distribution of the input signals.

In the following subsections we will describe some basic concepts from automata and probability theory that will be useful for the understanding of our algorithms. Refer to [14, 8] for a more detailed treatment.

2.1 Models of finite state systems

A Mealy-type FSM can be described by a six-tuple $(X, Y, S, s_0, \delta, \lambda)$ where X is the set of inputs, Y is the set of outputs, S is the set of states, and s_0 is the initial (reset) state. The next state function δ is given by:

$$s_{t+1} = \delta(X, s_t) \qquad (1)$$

The output function λ is defined as:

$$y_t = \lambda(X, s_t) \qquad (2)$$

If the machine is incompletely specified δ and λ are partial functions. For a Moore FSM the output *does not* depend on the input, therefore λ_M is defined as:

$$y_t = \lambda_M(s_t) \qquad (3)$$

Conceptually, Mealy and Moore machines are equivalent, in the sense that it is always possible to specify a Moore machine whose input-output behavior is equal to a given Mealy machine behavior, and viceversa [11]. Practically, however, there is an important difference. The Mealy model is usually more compact than the Moore model, in fact the transformation from Mealy to Moore involves a state splitting procedure that may significantly increase the number of states and state transitions [11].

Example 1 *In Figure 2 (a), a Mealy machine is represented in form of state transition graph (STG). If it is transformed in the equivalent Moore machine (using the procedure outlined in [11]), the new STG is shown in Figure 2 (b). The higher complexity in terms of states and edges of the Moore representation is evident. Notice that both FSMs are incompletely specified.*

2.2 Probabilistic models of FSMs

We model the probabilistic behavior of a general FSM using a Markov chain [8], as done in [9, 6, 16]. This model can be described by a weighted directed graph with a structure isomorphic to the STG of the machine. For a transition from state s_i to state s_j, the weight $p_{i,j}$ on the corresponding edge represents the *conditional probability* of the transition (i.e., the probability of a transition to state s_j *given* that the machine was in state s_i). Symbolically this can be expressed as:

$$p_{i,j} = Prob(Next = s_j | Present = s_i) \qquad (4)$$

The $p_{i,j}$ are collected in a matrix **P** and depend on the probability distribution of the inputs, that is initially known. However, using the conditional probability as an estimate of the total transition probability can lead to large errors, because the probability of a transition strongly depends on the probability for the machine to be in the state tail of the transition.

In order to find the probability of a transition without any condition, we need to know the *state probabilities* q_i, that represent the probability for the machine to be in a given state i. Namely, the *total transition probabilities* we are looking for are

$$r_{i,j} = p_{i,j} q_i \qquad (5)$$

Many methods have been proposed to calculate the state probabilities [8, 9]. In this work we have used the *Power Method*. Using this method, the state probability vector $\mathbf{q} = [q_1, q_2, ..., q_{|S|}]^T$ can be computed using the iteration:

$$\mathbf{q}_{n+1}^T = \mathbf{q}_n^T \mathbf{P} \qquad (6)$$

with the normalization condition $\sum_{i=1}^{|S|} q_i = 1$ until convergence is reached. The convergence properties of this method are discussed in [9]. The power method has been chosen because of its simplicity and its applicability (if sparse matrix manipulation or symbolic formulation are used [9]) to FSMs with a very large number of states. In the following sections we assume that the state probability vector and the total transition probabilities have already been computed.

If we now consider a Boolean function f with inputs the state and input variables of the machine, we can compute its probability (the probability for the function to be 1) in an exact fashion. If f is specified in cover form (a list of cubes that cover the ON-set of the function), the probability of f can be calculated using the following steps.

- Make the cover disjoint.
- Compute the probability of the disjoint cubes.
- Sum the disjoint cube probabilities.

Notice that this calculation is of vital importance in our algorithm, that performs a search based on the probability of the activation function.

3 Problem formulation

Given the knowledge of the FSM structure and its probabilistic model, we first want to identify the idle conditions where the clock may be stopped. If the machine is a Moore one, this is a simple task. For each state s we identify all the input conditions such that $\delta(x, s) = s$. We therefore define a set of *self-loop state function* $Self_s : X \to \{0, 1\}$ such that $Self_s = 1 \ \forall x \in X \ where \ \delta(x, s) = s$.

We then encode the machine. After the encoding step every state will have a unique code: $s_i \leftrightarrow e_i$ and $e_i = (e_{i,1}, e_{i,2}, ..., e_{i,|V|})$, where V is the set of the state variables used in the encoding.

Finally, the *activation function* is defined as $f_a : X \times V \to \{0, 1\}$:

$$f_a = \sum_{i=1,2,...,|S|} Self_{s_i} \cdot e_i \qquad (7)$$

These definitions can be clarified using an example.

Example 2 *For the Moore machine in Example 1, the self-loop state function for state M5 is $Self_{M5} = in_0' in_1$ Similarly, all the other self-loop state functions can be obtained. We encode*

the states using three state variables, v_1, v_2, v_3. The encodings are: $M0 \to v_1' v_2' v_3'$, $M1 \to v_1' v_2 v_3'$, $M2 \to v_1 v_2 v_3'$, $M3 \to v_1 v_2 v_3$, $M4 \to v_1' v_2 v_3$, $M5 \to v_1' v_2' v_3$. The activation function is therefore: $f_a = in_2 v_1' v_2' v_3' + in_2 v_1' v_2 v_3' + in_1 in_2' v_1' v_2 v_3' + in_1 in_2' v_1 v_2 v_3' + in_1 in_2' v_1 v_2 v_3 + in_1' in_2 v_1' v_2' v_3$.

If the machine is Mealy-type, the problem is substantially more complex. The knowledge of the state and the input is not sufficient to individuate the conditions when the clock can be stopped. If only the next state lines and the inputs are available for the computation of the activation function, we do not have a way to determine what was the output. This is a direct consequence of the Mealy model: since the outputs are on the edges of the STG, we may have the same next state for many different outputs. The important consequence is that, even if we know that the state in not going to change, we cannot guarantee that the output too will remain constant, therefore we cannot safely stop the clock.

There are two ways to solve this problem. The simpler way is to use the outputs of the FSM as additional inputs for the activation function. The other approach is to transform the STG in such a way that the FSM will be functionally compatible with the original one, but only the input and state lines will be sufficient to compute the activation function.

We decided to investigate the second method for two main reasons. First, since for many FSMs the number of output signals is large, it is likely that adding all the output signals to the inputs of the activation function will produce poor results because of the high complexity of the activation function itself. Second, in the present implementation our tool operates using state transition tables as input, therefore we still have the freedom to modify the number of states and the STG structure (this is not the case if we start from a synchronous network that is an implementation of the STG).

The simplest transformation that enables us to use only input and state signals as inputs of the activation function f_a, is a Mealy to Moore transformation. The algorithm that performs this conversion is well known [11] and its implementation is simple, but it may sensibly increase the number of states and edges (correlated with the complexity of the FSM implementation).

3.1 Locally-Moore machines

We now define and study a new kind of FSM transformation that enables us to use a Moore-like activation function without a large penalty in increased complexity of the FSM. A *Moore state* is a state such that all incoming transitions have the same output field. Formally: $s \in S \mid \forall x \in X, r \in S, \delta(x, r) = s \Rightarrow \lambda(x, r) = const$.

Proposition 1 *A Mealy-state s with k different values of the output fields on the edges that have s as a destination can be transformed in k Moore-states. No other state splitting is required.*

We could transform the FSM simply applying the Mealy to Moore transformations locally to states that have self loops. The local Moore transformation has the advantage that it allows us to concentrate only on states with self-loops, avoiding the useless state splitting on the states without self-lops. Still, this is not enough, because for many examples all the states will have self-loops and the local transformation will produce the complete Moore equivalent machine.

Our next step is to further localize the transformation. Consider an incompletely specified Mealy-machine. In general we may have many different outputs for different inputs, even if the next state is always the same. Intuitively we want split the Mealy state with self-loops simply in a couple of states. One of the two states will be Moore-type with a self-loop that has maximum probability.

We define the *maximum probability state self-loop function* $MPself_s : X \rightarrow \{0,1\}$. Its ON-set represents the set of input conditions for a state that are on self-loops and produce compatible outputs (two outputs fields are compatible if they differ only in entries where at least one of the two is don't care) and are taken with maximum probability.

To find $MPself_s$ we group the self-loops from state s in (possibly overlapping) compatibility classes, we then compute the probability of each compatible class and we choose the class with maximum probability as the ON set of $MPself_s$.

Example 3 *In the Mealy machine of Example 1, if we consider state S2, we have two self loops: $in_1 in_2'$ with output 10 and $in_1' in_2$ with output -1. The two output fields are not compatible, therefore we have two compatible classes (the same two functions). We will choose the class that is more probable. In this particular example, we assumed equiprobable and independent inputs and both functions have the same probability, therefore one of the two is randomly chosen.*

Once the $MPself_s$ functions have been found for all the states with self-loops, the second step of our transformation algorithms is performed. If a state s is not a Moore-state, it is split in two states s_o and s_l. The first state (s_o) is the original one, but its edges corresponding to the self-loops included in $MPself_s$ become transitions from s_o to s_l. The second state (s_l) is reached only from s_o and has a self-loop corresponding to $MPself_s(x)$. All the outgoing edges that leave s_o are replicated for s_l. The s_l state is now Moore-type, because all the edges that have s_l as tail have the same output field.

This procedure is advantageous for many reasons. First, the increase in the number of states is tightly controlled (in the worst case, if all the states are Mealy-type and have self-loops, we can have a twofold increase in the number of states). Second, the self-loops with maximum probability are selected. Third, if we really want to limit the increase in the number of states, we may define a threshold: only the first k states in a list ordered for decreasing *total probability* of $MPself_s$ are duplicated.

We call the FSM obtained after the application of this procedure *locally-Moore* FSM, because in general only a subset of the states is Moore-type.

Example 4 *The application of our procedure on the Mealy machine of Example 1 produces the locally-Moore FSM shown in Figure 3. The shaded areas enclose states that have been split. The Moore-states with self-loops are drawn with bold lines. The*

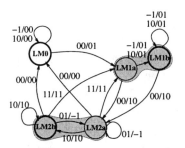

Figure 3: STG of the locally-Moore FSM

number of states and edges of the locally Moore machine is smaller than those that we obtained with the complete Mealy to Moore transformation.

If we restrict our consideration to Moore-states, we can generate an activation function that uses as inputs only the state and primary input of the FSM. The next step is to generate an activation function that produces a final implementation with minimum power dissipation.

3.2 Optimal activation function

If we call $S' \subseteq S$ the set of Moore-states in a FSM, the complete activation function is given by

$$f_a = \sum_{i=1,2,\ldots,|S'|} Self_{s_i} \cdot e_i \tag{8}$$

where e_i is the encoding of the states in S'. The simplest approach is to try to use the complete f_a as activation function. This is seldom the best solution, because the complexity if f_a can be too high, and the power dissipated by its implementation may reduce or nullify the power reduction that we obtain stopping the clock. It is therefore necessary to be able to choose a simpler function contained in f_a whose implementation dissipates minimum power, but whose efficiency in stopping the clock is maximum.

In [2] we proposed a simple greedy algorithm that will be shortly outlined. First, the f_a is two-level minimized, and a minimum cover is obtained. Then, the larger cubes in the cover are greedily selected until the number of literals in the partial cover exceed a user-specified literal threshold. The rationale of this approach is that generally large cubes have high probability and the primes that compose a minimum cover are as large as possible.

The solution proposed in [2] is highly heuristic and can be improved. We have devised a new strategy for the synthesis of reduced activation functions that exploits the knowledge of the probability of the self-loops. Moreover, we have studied and solved the new combinational synthesis problem that arises when we want to find a minimum complexity function that is active with a pre-fixed probability.

We do not describe in detail our new algorithms because of space limitations. Intuitively, we find the optimal activation function using a branch-and-bound algorithm that select a minimum-literal-count cover of a function $F_a \subseteq f_a$ that is guaranteed to be active (one) with

557

probability $P(F_a) \geq \alpha P(f_a)$. The parameter $\alpha \leq 1$ is user-defined.

We can now briefly summarize the full procedure used for the synthesis of our low-power gated clock FSMs.

- The Mealy machine is transformed in an equivalent locally-Moore machine.

- The complete activation function f_a is extracted from the Moore-states of the locally-Moore machine.

- The probability of the complete f_a is computed.

- The branch-and-bound algorithm finds the minimum literal count solution F_a whose probability is a pre-specified fraction α of the probability of f_a

- F_a is used as additional DC set for the combinational logic of the FSM.

The last step (more thoroughly described in [2]) can sensibly improve the quality of the results, in particular if F_a is large. Unfortunately its effect is to greatly increase the theoretic complexity of the problem, because it is very hard to foresee the effect of F_a used as DC set. Sometimes it may be convenient to choose a F_a that is not minimal in the sense discussed above, if it allows a large simplification in the combinational part of the FSM. Our heuristic approach is to try different F_a of decreasing size (and complexity), in an attempt to explore the trade-off curve. We generate a set of solutions using different values of α, in such a way that the possible range of solutions is uniformly sampled. The details of this approach will not be discussed here for space reasons.

4 Experimental results

We implemented the described algorithms as a part of a tool-set for low-power synthesis that we are developing. The tool reads the state transition table of the FSM. The first step is the transformation of the Mealy machine to a Locally-Moore machine and the extraction of the self-loops from the Moore-states. We then apply the power method to compute the exact state probabilities given an input probability distribution

Once the state codes have been assigned (using JEDI [17]), our probabilistic-driven procedure for the selection of the activation function can start. First, all the primes of the activation function are generated using symbolic methods [15], then the probability of the complete activation function f_a is computed starting from a minimized cover (obtained with ESPRESSO [12])

The user specifies the number of activation functions that the procedure should generate, and the branch-and-bound algorithms finds reduced activation functions as many times as it is requested. Surprisingly, for all the MCNC benchmarks this step has never been the bottle-neck. This is certainly due to the fact that the majority of the FSM MCNC benchmarks do not have a large number of self-loops (in particular the larger ones).

The combinational logic of the Locally-Moore FSM is then optimized in SIS [12] using the additional DC set

Circuit	Original		Locally-M.		Gated		
	Size	P	Size	P	Size	P	%
bbara	330	67	422	72	408	34	97
bbsse	640	121	742	137	736	119	2
bbtas	142	56	138	57	164	44	27
keyb	721	128	754	132	820	114	12
lion9	188	60	226	60	248	52	15
s298	7492	899	7496	900	7502	810	11
s420	544	132	544	132	602	108	22
scf	3222	437	3222	437	3169	400	9
styr	1474	159	2468	230	2534	208	0
test	348	73	442	76	374	32	128

Table 1: Results of our procedure applied to MCNC benchmarks. Power is in μW.

given by the activation function. The DC-based minimization of the combinational logic using the activation functions is the main bottleneck of our procedure. In our tool the user has the possibility to specify a CPU-time limit for each minimization attempt. This of course limits the possible improvements obtainable on large FSMs.

The activation functions are also optimized using SIS, then the alternative solutions are mapped with CERES [13], and the gated clocking circuitry is generated. Finally the alternative gated clock implementations and the implementation of the original Mealy FSM are simulated with a large number of test patterns using a switch level simulator (IRSIM [18]) modified for power estimation.

The quality of the results strongly depends on two factors. First, how much state splitting has been needed to transform the machine in a locally-Moore one. Second, for what percentage of the total operation time the FSM is in a self-loop condition (this depends on the FSM structure and on the input probability distribution). For machines with a very small number of self-loops or a very low-probability complete activation function, the area of improvement is almost null. This is the case for many MCNC benchmarks for which the final improvement is negligible. As for the first problem, it may be worthy to investigate if, in case the state duplication is too high, using an activation function with the outputs of the FSM as additional inputs may lead to better results.

Table 1 reports the performance of our tools on a sub-set of the MCNC benchmarks. The first six columns show the area (number of transistors) and the power dissipation of the normal Mealy FSM, the locally-Moore FSM without gated clock and the locally-Moore machine with gated clock. The last column shows the power improvement, computed as $(100(P_{mealy}/P_{gated} - 1))$. Notice that, if there is no power improvement this number is set to 0.

The tool is able to process all the benchmarks, but in the table we list examples representative of various classes of possible results. The benchmarks bbara and test are reactive FSMs. The high number and probability of the self-loops allow an impressive reduction of the total power dissipation, even if the area penalty can be not negligible.

In contrast, for bbsse and styr there is no power reduction or even a power increase. The bbsse benchmark is representative of a class of machines where the number and probability of the self-loops is too small for our procedure to obtain substantial power savings. The `styr` benchmark has many self-loops, but they all have very low probability. Moreover, the transformation to locally-Moore machine is paid in this case with a too large area overhead.

For the other examples in the table the power savings vary between 10% and 30%. For some of these machines (s420 and `scf`), the self-loops are only on Moore states and there is no area overhead for the locally-Moore transformation. We included some of the large examples in the benchmark suite (s298 and `scf`) to show the applicability of our method to large FSMs.

From the observation of the results, it is quite clear that several complex trade-offs are involved. First, the transformation to locally-Moore machine can sometimes be very expensive in terms of area overhead. Second, the choice of the best possible activation function is paramount for good results. In fact, for many examples, the complete activation function was too large, and reduced activation functions gave better results.

5   Conclusions and future work

We have described a technique for the automatic synthesis of gated clocks for FSMs of a very general class. We want to emphasize that our method is a complete procedure, from the FSMs high-level specification to the fully mapped network, and it has been tested with accurate power estimation tools. The quality of our results depends on the initial structure of the FSM, but we obtain substantial power savings for a large class of finite state machines.

We have presented a new transformation for Mealy FSMs that makes them suitable for gated-clock implementation, and outlined a complete procedure for the synthesis of an optimized combinational logic function whose purpose is to stop the clock with high probability during the operation of the machine.

Future research will concentrate on the implementation of fully symbolic algorithm for the synthesis of the activation function and on the application of our techniques to large synchronous networks.

6   Acknowledgements

This research is supported by NSF and ARPA under contract number 9115432.

References

[1] M. Alidina, et al., "Precomputation-based sequential logic optimization for low power," in *ICCAD, Proceedings of the International Conference on Computer-Aided Design*, pp. 74–80, Nov. 1994.

[2] L. Benini, P. Siegel and G. De Micheli, "Automatic synthesis of gated clocks for power reduction in sequential circuits" *IEEE Design and Test of Computers*, pp. 32–40, Dic. 1994.

[3] A. Shen, A. Ghosh, S. Devadas, and K. Keutzer, "On average power dissipation and random pattern testability of CMOS combinational logic networks," in *ICCAD, Proceedings of the International Conference on Computer-Aided Design*, pp. 402–407, Nov. 1992.

[4] C. Tsui, M. Pedram, and A. Despain, "Technology decomposition and mapping targeting low power dissipation," in *DAC, Proceedings of the Design Automation Conference*, pp. 68–73, June 1993.

[5] K. Roy and S. Prasad, "Circuit activity based logic synthesis for low power reliable operations," *IEEE Transactions on Very Large Scale Integration (VLSI) Systems*, vol. 1, no. 4, pp. 503–513, Dec. 1993.

[6] L. Benini and G. De Micheli, "State assignment for low power dissipation," in *CICC, Proceedings of the IEEE Custom Integrated Circuits Conference*, pp. 136–139, May 1994.

[7] B. Suessmith and G. Paap III, "PowerPC 603 microprocessor power management," *Communications of the ACM*, no. 6, pp. 43–46, June 1994.

[8] K. Trivedi. *Probability and statistics with reliability, queuing and computer science applications.* Prentice-Hall, 1982.

[9] G. Hachtel, E. Macii, A. Pardo and F. Somenzi "Symbolic algorithms to calculate Steady-State probabilities of a finite state machine," in *Proc. of IEEE European Design and Test Conf.*, pp. 214–218, Feb. 1994.

[10] J. Schutz, "A 3.3V 0.6µm BiCMOS superscalar microprocessor," in *IEEE International Solid-State Circuits Conference*, pp. 202–203, Feb. 1994.

[11] J. Hartmanis and H. Stearns, *Algebraic Structure Theory of Sequential Machines.* Prentice-Hall, 1966.

[12] E. Sentovich, et al., "Sequential circuit design using synthesis and optimization," in *ICCD, Proceedings of the International Conference on Computer Design*, pp. 328–333, Oct. 1992.

[13] F. Mailhot and G. De Micheli, "Algorithms for technology mapping based on binary decision diagrams and on Boolean operations," *IEEE Transactions on CAD/ICAS*, pp. 599–620, May 1993.

[14] G. De Micheli. *Synthesis and optimization of digital circuits.* McGraw-Hill, 1994.

[15] O. Coudert and C. Madre, "Implicit and incremental computation of primes and essential primes of Boolean functions," in *DAC, Proceedings of the Design Automation Conference*, pp. 36–39, June 1992.

[16] R. Marculescu, D. Marculescu and M. Pedram, "Switching activity analysis considering spatiotemporal correlations," in *ICCAD, Proceedings of the International Conference on Computer-Aided Design*, pp. 294–299, Nov. 1994

[17] B. Lin and A. R. Newton, "Synthesis of multiple-level logic from symbolic high-level description languages," in *Proc. of IEEE Int. Conf. On Computer Design*, pp. 187–196, Aug. 1989.

[18] A. Salz and M. Horowitz, "IRSIM: an incremental MOS switch-level simulator," in *DAC, Proceedings of the Design Automation Conference*, pp. 173–178, June 1989.

Precomputation-Based Sequential Logic Optimization for Low Power

Mazhar Alidina,* José Monteiro, Srinivas Devadas
Department of EECS
MIT, Cambridge, MA

Abhijit Ghosh
MERL
Sunnyvale, CA

Marios Papaefthymiou
Department of EE
Yale University, CT

Abstract

We address the problem of optimizing logic-level sequential circuits for low power. We present a powerful sequential logic optimization method that is based on selectively *precomputing* the output logic values of the circuit one clock cycle before they are required, and using the precomputed values to reduce internal switching activity in the succeeding clock cycle. We present two different precomputation architectures which exploit this observation.

We present an automatic method of synthesizing precomputation logic so as to achieve maximal reductions in power dissipation. We present experimental results on various sequential circuits. Upto 75% reductions in average switching activity and power dissipation are possible with marginal increases in circuit area and delay.

1 Introduction

Average power dissipation has recently emerged as an important parameter in the design of general-purpose and application-specific integrated circuits. Optimization for low power can be applied at many different levels of the design hierarchy. For instance, algorithmic and architectural transformations can trade off throughput, circuit area, and power dissipation [5], and logic optimization methods have been shown to have a significant impact on the power dissipation of combinational logic circuits [12].

In CMOS circuits, the probabilistic average switching activity of a circuit is a good measure of the average power dissipation of the circuit. Average power dissipation can thus be computed by estimating the average switching activity. Several methods to estimate power dissipation for CMOS combinational circuits have been developed (e.g., [7, 10]). More recently, efficient and accurate methods of power dissipation estimation for sequential circuits have been developed [9, 13].

In this work, we are concerned with the problem of optimizing logic-level sequential circuits for low power. Previous work in the area of sequential logic synthesis for low power has focused on state encoding (e.g.,

[11]) and retiming [8] algorithms. We present a powerful sequential logic optimization method that is based on selectively *precomputing* the output logic values of the circuit one clock cycle before they are required, and using the precomputed values to reduce internal switching activity in the succeeding clock cycle.

The primary optimization step is the synthesis of the precomputation logic, which computes the output values for a *subset* of input conditions. If the output values can be precomputed, the original logic circuit can be "turned off" in the next clock cycle and will not have any switching activity. Since the savings in the power dissipation of the original circuit is offset by the power dissipated in the precomputation phase, the selection of the subset of input conditions for which the output is precomputed is critical. The precomputation logic adds to the circuit area and can also result in an increased clock period.

Given a logic-level sequential circuit, we present an automatic method of synthesizing the precomputation logic so as to achieve a maximal reduction in switching activity. We present experimental results on various sequential circuits. For some circuits, 75% reductions in average power dissipation are possible with marginal increases in circuit area and delay.

The model we use to relate switching activity to power dissipation can be found in [7]. In Section 2 we describe two different precomputation architectures. An algorithm that synthesizes precomputation logic so as to achieve power dissipation reduction is presented in Section 3. In Section 4 we describe a method for multiple-cycle precomputation. In Section 5 we describe additional precomputation architectures which are the subject of ongoing research. Experimental results are presented in Section 6.

2 Precomputation Architectures

We describe two different precomputation architectures and discuss their characteristics in terms of their impact on power dissipation, circuit area and circuit delay.

2.1 First Precomputation Architecture

Consider the circuit of Figure 1. We have a combinational logic block **A** that is separated by registers R_1

---

*Currently at AT&T Bell Laboratories, Allentown, PA

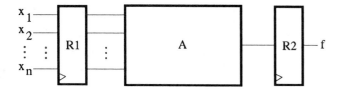

Figure 1: Original Circuit

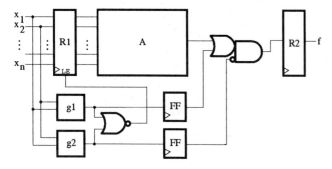

Figure 2: First Precomputation Architecture

and R_2. While R_1 and R_2 are shown as distinct registers in Figure 1 they could, in fact, be the same register. We will first assume that block **A** has a single output and that it implements the Boolean function f.

The first precomputation architecture is shown in Figure 2. Two Boolean functions g_1 and g_2 are the *predictor* functions. We require:

$$g_1 = 1 \;\Rightarrow\; f = 1 \qquad (1)$$

$$g_2 = 1 \;\Rightarrow\; f = 0 \qquad (2)$$

Therefore, during clock cycle t if either g_1 or g_2 evaluates to a 1, we set the load enable signal of the register R_1 to be 0. This means that in clock cycle $t+1$ the inputs to the combinational logic block **A** do not change. If g_1 evaluates to a 1 in clock cycle t, the input to register R_2 is a 1 in clock cycle $t+1$, and if g_2 evaluates to a 1, then the input to register R_2 is a 0. Note that g_1 and g_2 cannot both be 1 during the same clock cycle due to the conditions imposed by Equations 1 and 2.

A power reduction in block **A** is obtained because for a subset of input conditions corresponding to $g_1 + g_2$ the inputs to **A** do not change implying zero switching activity. However, the area of the circuit has increased due to additional logic corresponding to g_1, g_2, the two additional gates shown in the figure, and the two flip-flops marked **FF**. The delay between R_1 and R_2 has increased due to the addition of the AND-OR gate. Note also that g_1 and g_2 add to the delay of paths that originally ended at R_1 but now pass through g_1 or g_2 and the NOR gate before ending at the load enable signal of the register R_1. Therefore, we would like to apply this transformation on non-critical logic blocks.

The choice of g_1 and g_2 is critical. We wish to include as many input conditions as we can in g_1 and g_2. In other words, we wish to maximize the probability of g_1 or g_2 evaluating to a 1. In the extreme case this probability can be made unity if $g_1 = f$ and $g_2 = \overline{f}$. However, this would imply a duplication of the logic block **A** and

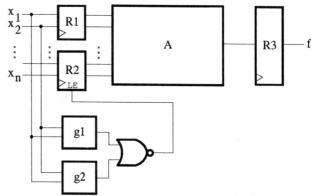

Figure 3: Second Precomputation Architecture

no reduction in power with a twofold increase in area! To obtain reduction in power with marginal increases in circuit area and delay, g_1 and g_2 have to be significantly less complex than f. One way of ensuring this is to make g_1 and g_2 depend on significantly fewer inputs than f. This leads us to the second precomputation architecture of Figure 3.

2.2 Second Precomputation Architecture

In the architecture of Figure 3, the inputs to the block **A** have been partitioned into two sets, corresponding to the registers R_1 and R_2. The output of the logic block **A** feeds the register R_3. The functions g_1 and g_2 satisfy the conditions of Equations 1 and 2 as before, but g_1 and g_2 only depend on a subset of the inputs to f. If g_1 or g_2 evaluates to a 1 during clock cycle t, the load enable signal to the register R_2 is turned off. This implies that the outputs of R_2 during clock cycle $t+1$ do not change. However, since the outputs of register R_1 are updated, the function f will evaluate to the correct logical value. A power reduction is achieved because only a subset of the inputs to block **A** change which should produce reduced switching activity in most cases.

As before, g_1 and g_2 have to be significantly less complex than f and the probability of $g_1 + g_2$ being a 1 should be high in order to achieve substantial power gains. The delay of the circuit between R_1/R_2 and R_3 is unchanged, allowing precomputation of logic that is on the critical path. However, the delay of paths that originally ended at R_1/R_2 has increased.

The choice of inputs to g_1 and g_2 has to be made first, and then the particular functions that satisfy Equations 1 and 2 have to be selected. A method to perform this selection is described in Section 3.

2.3 An Example

We give an example that illustrates the fact that substantial power gains can be achieved with marginal increases in circuit area and delay. The circuit we are considering is a n-bit comparator that compares two n-bit numbers C and D and computes the function $C > D$. The optimized circuit with precomputation logic is shown in Figure 4. The precomputation logic is as follows.

$$g_1 \;=\; C\langle n-1 \rangle \;\cdot\; \overline{D\langle n-1 \rangle}$$

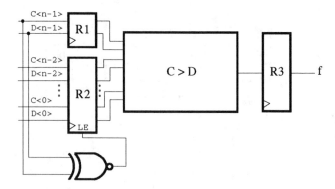

Figure 4: A Comparator Example

$$g_2 = \overline{C\langle n-1 \rangle} \cdot D\langle n-1 \rangle$$

Clearly, when $g_1 = 1$, C is greater than D, and when $g_2 = 1$, C is less than D. We have to implement

$$\overline{g_1 + g_2} = C\langle n-1 \rangle \otimes D\langle n-1 \rangle$$

where $\otimes$ stands for the exclusive-nor operator.

Assuming a uniform probability for the inputs [1], the probability that the XNOR gate evaluates to a 1 is 0.5, regardless of n. For large n, we can neglect the power dissipation in the XNOR gate, and therefore, we can expect to achieve a power reduction of close to 50%. The reduction will depend upon the relative power dissipated by the vector pairs with $C\langle n-1 \rangle \otimes D\langle n-1 \rangle = 1$ and the vector pairs with $C\langle n-1 \rangle \otimes D\langle n-1 \rangle = 0$. If we add the inputs $C\langle n-2 \rangle$ and $D\langle n-2 \rangle$ to g_1 and g_2 we expect to achieve a power reduction close to 75%.

3 Synthesis of Precomputation Logic

3.1 Introduction

In this section, we will describe methods to determine the functionality of the precomputation logic, and then describe methods to efficiently implement the logic.

We will focus primarily on the second precomputation architecture illustrated in Figure 3. In order to ensure that the precomputation logic is significantly less complex than the combinational logic in the original circuit, we will restrict ourselves to identifying g_1 and g_2 such that they depend on a relatively small subset of the inputs to the logic block **A**.

3.2 Precomputation and Observability Don't-Cares

Assume that we have a logic function $f(X)$, with $X = \{x_1, \cdots, x_n\}$, corresponding to block **A** of Figure 2. Given that the logic function implemented by block **A** is f, then the *observability don't-care set* for input x_i is given by:

$$ODC_i = f_{x_i} \cdot f_{\overline{x_i}} + \overline{f}_{x_i} \cdot \overline{f}_{\overline{x_i}}$$

where f_{x_i} and $f_{\overline{x_i}}$ are the *cofactors* of f with respect to x_i, and similarly for $\overline{f}$.

---
[1] The assumption here is that each $C\langle i \rangle$ and $D\langle i \rangle$ has a 0.5 static probability of being a 0 or a 1.

If we determine that a given input combination is in ODC_i then we can disable the loading of x_i into the register. If we wish to disable the loading of registers x_m, x_{m+1}, $\cdots$, x_N, we will have to implement the function:

$$g = \prod_{i=m}^{N} ODC_i$$

and use $\overline{g}$ as the (active low) load enable signal for the registers corresponding to x_m, x_{m+1}, $\cdots$, x_N.

3.3 Precomputation Logic

Consider the architecture of Figure 3. Assume that the inputs x_1, $\cdots$, x_m, with $m < n$ have been selected as the variables that g_1 and g_2 depend on. We have to find g_1 and g_2 such that they satisfy the constraints of Equations 1 and 2, respectively, and such that $prob(g_1 + g_2 = 1)$ is maximum.

We can determine g_1 and g_2 using universal quantification on f. The *universal quantification* of a function f with respect to a variable x_i is defined as:

$$U_{x_i} f = f_{x_i} \cdot f_{\overline{x_i}}$$

Given a subset of inputs $S = \{x_1, \cdots, x_m\}$, set $D = X - S$. We can define:

$$U_D f = U_{x_{m+1}} \ldots U_{x_n} f$$

Theorem 3.1 $g_1 = U_D f$ *satisfies Equation 1. Further, no function* $h(x_1, \cdots, x_m)$ *exists such that* $prob(h = 1) > prob(g_1 = 1)$ *and such that* $h = 1 \Rightarrow f = 1$.

Proof. By construction, if for some input combination a_1, $\cdots$, a_m causes $g_1(a_1, \cdots, a_m) = 1$, then for that combination of x_1, $\cdots$, x_m and all possible combinations of variables in x_{m+1}, $\cdots$, x_n $f(a_1, \cdots, a_m, x_{m+1}, \cdots, x_n) = 1$. We cannot add any minterm over x_1, $\cdots$, x_m to g_1 because for any minterm that is added, there will be some combination of x_{m+1}, $\cdots$, x_n for which $f(x_1, \cdots, x_n)$ will evaluate to a 0. Therefore, we cannot find any function h that satisfies Equation 1 and such that $prob(h = 1) > prob(g_1 = 1)$. ∎

Similarly, given a subset of inputs S, we can obtain a maximal g_2 by:

$$g_2 = U_D \overline{f} = U_{x_{m+1}} \ldots U_{x_n} \overline{f}$$

We can compute the functionality of the precomputation logic as $g_1 + g_2$.

3.3.1 Selecting a Subset of Inputs

Given a function f we wish to select the "best" subset of inputs S of cardinality k. Given S, we have $D = X - S$ and we compute $g_1 = U_D f$, $g_2 = U_D \overline{f}$. In the sequel, we assume that the best set of inputs corresponds to the inputs which result in $prob(g_1 + g_2 = 1)$ being maximum for a given k. We know that $prob(g_1 + g_2 = 1) = prob(g_1 = 1) + prob(g_2 = 1)$ since g_1 and g_2 cannot

```
SELECT_INPUTS( f, k ):
{
    /* f = function to precompute */
    /* k = # of inputs to precompute with */
    BEST_PROB = 0 ;
    SELECTED_SET = φ ;
    SELECT_RECUR( f, f̄, φ, X, |X| − k ) ;
    return( SELECTED_SET ) ;
}

SELECT_RECUR( fₐ, f_b, D, Q, l ):
{
    if( |D| + |Q| < l )
        return ;
    pr = prob(fₐ = 1) + prob(f_b = 1) ;
    if( pr ≤ BEST_PROB )
        return ;
    else if( |D| == l ) {
        BEST_PROB = pr ;
        SELECTED_SET = X − D ;
        return ;
    }
    choose xᵢ ∈ Q such that i is minimum ;
    SELECT_RECUR( U_{xᵢ}fₐ, U_{xᵢ}f_b,
                  D ∪ xᵢ, Q − xᵢ, l ) ;
    SELECT_RECUR( fₐ, f_b, D, Q − xᵢ, l ) ;

    return ;
}
```

Figure 5: Procedure to Determine the Optimal Set of Inputs

both be 1 on the same input vector. The above cost function ignores the power dissipated in the precomputation logic, but since the number of inputs to the precomputation logic is significantly smaller than the total number of inputs, this is a good approximation.

A branch and bound algorithm is used to determine the optimal set of inputs maximizing the probability of the g_1 and g_2 functions. This algorithm is shown in pseudo-code in Figure 5 and is described in detail in [1].

3.3.2 Implementing the Logic

The Boolean operations of OR and universal quantification required in the input selection procedure can be carried out efficiently using reduced, ordered Binary Decision Diagrams (ROBDDs) [4]. We obtain a ROBDD for the $g_1 + g_2$ function. A ROBDD can be converted into a multiplexor-based network (see [2]) or into a sum-of-products cover. The network or cover can be optimized using standard combinational logic optimization methods that reduce area [3] or those that target low power dissipation [12].

3.4 Multiple-Output Functions

In general, we have a multiple-output function $f_1, \cdots, f_m$ that corresponds to logic block **A** in Fig-

ures 2 and 3. All the procedures described thus far can be generalized to the multiple-output case.

The functions g_{1i} and g_{2i} are obtained using the equations below.

$$g_{1i} = U_D f_i$$
$$g_{2i} = U_D \overline{f_i}$$

where $D = X - S$ as before. The function g whose complement drives the load enable signal is obtained as:

$$g = \prod_{i=1}^{m} (g_{1i} + g_{2i})$$

The function g corresponds to the set of input conditions where the variables in S control the values of *all* the f_i's regardless of the values of variables in $D = X - S$.

3.4.1 Selecting a Subset of Outputs

In general, it is hard to find a set of inputs for which every output of a multiple-output function is precomputable. We have developed an algorithm, which given a multiple-output function, selects a subset of outputs *and* a subset of inputs so as to maximize a given cost function that is dependent on the probability of the precomputation logic and the number of selected outputs. This algorithm is described in pseudo-code in Figure 6 and is described in detail in [1].

Since we are only precomputing a subset of outputs, we may incorrectly evaluate the outputs that we are *not* precomputing as we disable certain inputs during particular clock cycles. If an output that is not being precomputed depends on an input that is being disabled, then the output will be incorrect.

Once a set of outputs $G \subset F$ and a set of precomputation logic inputs $S \subset X$ have been selected, we need to duplicate the registers corresponding to $(support(G) - S) \cap support(F - G)$. The inputs that are being disabled are in $support(G) - S$. Logic in the $F - G$ outputs that depends on the set of duplicated inputs has to be duplicated as well. It is precisely for this reason that we maximize $prG \times gates(G)/total\_gates$ rather than prG in the output-selection algorithm as we want to reduce the amount of duplication as much as possible.

4 Multiple Cycle Precomputation

4.1 Basic Strategy

It is possible to precompute output values that are not required in the succeeding clock cycle, but required 2 or more clock cycles later. We give an example illustrating multiple-cycle precomputation.

Consider the circuit of Figure 7. The function f computes $(C + D) > (X + Y)$ in two clock cycles. Attempting to precompute $C + D$ or $X + Y$ using the methods of the previous section do not result in any savings because there are too many outputs to consider. However, 2-cycle precomputation can reduce switching activity by close to 12.5% if the functions below are used.

$$g_1 = C\langle n-1 \rangle \cdot D\langle n-1 \rangle \cdot \overline{X\langle n-1 \rangle} \cdot \overline{Y\langle n-1 \rangle}$$

SELECT_OUTPUTS($F = \{f_1, \cdots, f_m\}$, k):
{
   /* F = multi-output func. to precompute */
   /* k = # of inputs to precompute with */
   BEST_COST = 0 ;
   SEL_OP_SET = ϕ ;
   SELECT_OREC(ϕ, F, 1, k) ;
   return(SEL_OP_SET) ;
}

SELECT_OREC(G, H, $proldG$, k):
{
   lf = gates$(G \cup H)$/total_gates × $proldG$;
   if($lf \leq$ BEST_COST)
     return ;
   BEST_PROB = total_gates/gates$(G \cup H)$
         × BEST_COST ;
   if($G \neq \phi$)
     if(SELECT_INPUTS(G, k) == ϕ)
       return ;
   prG = BEST_PROB ;
   cost = prG × gates(G)/total_gates ;
   if(cost > BEST_COST) {
     BEST_COST = cost ;
     SEL_OP_SET = G ;
   }
   choose $f_i \in H$ such that i is minimum ;
   SELECT_OREC($G \cup f_i$, $H - f_i$, prG, k) ;
   SELECT_OREC(G, $H - f_i$, prG, k) ;

   return ;
}

Figure 6: Procedure to Determine the Optimal Set of Outputs

$$g_2 = \overline{C\langle n-1\rangle} \cdot \overline{D\langle n-1\rangle} \cdot X\langle n-1\rangle \cdot Y\langle n-1\rangle$$

where g_1 and g_2 satisfy the constraints of Equations 1 and 2, respectively. Since $prob(g_1 + g_2) = \frac{2}{16} = 0.125$, we can disable the loading of registers $C\langle n-2:0\rangle$, $D\langle n-2:0\rangle$, $X\langle n-2:0\rangle$, and $Y\langle n-2:0\rangle$ 12.5% of the time, which results in switching activity reduction. This percentage can be increased to over 45% by using $C\langle n-2\rangle$ through $Y\langle n-2\rangle$. We can additionally use single-cycle precomputation logic (as illustrated in Figure 4) to further reduce switching activity in the > comparator of Figure 7. More examples of this technique can be found in [1].

5 Other Precomputation Architectures

In this section, we describe additional precomputation architectures. We first present an architecture that is applicable to *all* logic circuits and does not require, for instance, that the inputs should be in the observability don't-care set in order to be disabled. This was the case for the architectures shown in Section 2. We also extend precomputation so that it can be used in combinational logic circuits.

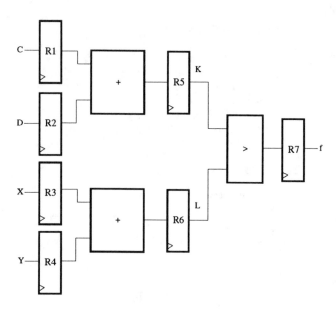

Figure 7: Adder-Comparator Circuit

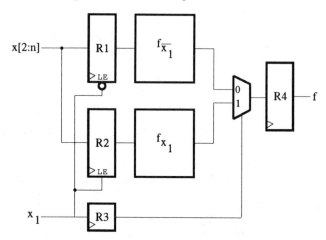

Figure 8: Precomputation Using the Shannon Expansion

5.1 Multiplexor-Based Precomputation

All logic functions can be written in a Shannon expansion. For the function f with inputs $X = \{x_1, \cdots, x_n\}$, we can write:

$$f = x_1 \cdot f_{x_1} + \overline{x_1} \cdot f_{\overline{x_1}} \qquad (3)$$

where f_{x_1} and $f_{\overline{x_1}}$ are the cofactors of f with respect to x_1.

Figure 8 shows an architecture based on Equation 3. We implement the functions f_{x_1} and $f_{\overline{x_1}}$. Depending on the value of x_1, only one of the cofactors is computed while the other is disabled by setting the load-enable signal of its input register. The input x_1 drives the select line of a multiplexor which chooses the correct cofactor.

The main advantage of this architecture is that it applies to *all* logic functions. The input x_1 in the example was chosen for the purpose of illustration. In fact, any

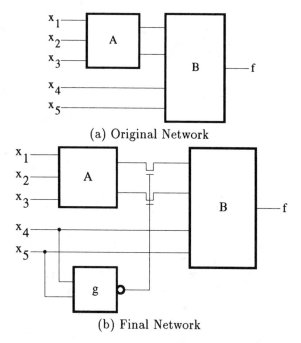

(a) Original Network

(b) Final Network

Figure 9: Combinational Logic Precomputation

input $x_1, \cdots, x_n$ could have been selected. Unlike the architectures described earlier, we do not require that the inputs being disabled should be don't-cares for the input conditions which we are precomputing. In other words, the inputs being disabled do not have to be in the observability don't-care set. A disadvantage of this architecture is that we need to duplicate the registers for the inputs not being used to turn off part of the logic. On the other hand, no precomputation logic functions have been added to the circuit.

The algorithm to select the best input for this architecture is also quite different. We will not discuss this algorithm in detail, except to mention that in this case, we are interested in finding the input that yields the most area efficient f_{x_1} and $f_{\overline{x_1}}$ functions.

5.2 Combinational Logic Precomputation

The architectures described so far apply only to sequential circuits. We now describe precomputation of combinational circuits.

Suppose we have some combinational logic function f composed of two sub-functions **A** and **B** as shown in Figure 9(a). Suppose we also want to precompute this function with the inputs x_4 and x_5. Figure 9(b) shows how this can be accomplished. For simplicity, pass transistors, instead of transmission gates, are shown. The function g with inputs x_4 and x_5 drives the gates of the pass transistors. As in the previous architectures, $g = g_1 + g_2$. Hence, when g is a 0, the pass transistors are turned off and the new values of logic block **A** are prevented from propagating into logic block **B**. The inputs x_4 and x_5 are also inputs to the logic block **B** just as in the original network in order to ensure that the output is set correctly.

For the combinational architecture, there is an implied delay constraint, i.e. the pass transistors should

be off *before* the new values of **A** are computed. In the example shown, the worst-case delay of the g block plus the arrival time of inputs x_4 or x_5 should be less than the best-case delay of logic block **A** plus the arrival time of the inputs x_1, x_2, or x_3. The *arrival time* of an input is defined as the time at which the input settles to its steady state value [6]. If the delay constraint is not met, then it may be necessary to delay the x_1, x_2, and x_3 inputs with respect to the x_4 and x_5 inputs in order to get the switching activity reduction in logic block **B**.

6 Experimental Results

At first we present results on datapath circuits such as carry-select adders, comparators, and interconnections of adders and comparators in Table 1. The precomputation architecture of Figure 3 was used in all examples and the selection of outputs and inputs to use for precomputation was done manually for examples **csa16**, **add_comp16** and **add_max16** and automatically (using the algorithms outlined in Figures 5 and 6) for the rest. For each circuit, the number of literals, levels of logic and power of the original circuit, the number of inputs, literals and levels of the precompute logic, the final power and the percent reduction in power are shown. All power estimates are in micro-Watts and are computed using the techniques described in [7]. A zero delay model and a clock frequency of 20MHz was assumed. The rugged script of **sis** was used to optimize the precompute logic.

Power dissipation decreases for almost all cases. For circuit **comp16**, a 16-bit parallel comparator, the power decreases by as much as 60% when 8 inputs are used for precomputation. Multiple-cycle precomputation results are given for circuits **add_comp16** and **add_max16**. The circuit **add_comp16** is shown in Figure 7, and the circuit **add_max16** is the same circuit with the comparator replaced by a maximum function. For circuit **add_comp16**, for instance, the numbers 4/8 under the fifth column indicates that 4 inputs are used to precompute the adders in the first cycle and 8 inputs are used to precompute the comparator in the next cycle.

Results on random logic circuits are presented in Table 2. The random logic circuits are taken from the MCNC combinational benchmark sets. We have presented results for those examples where significant savings in power was obtained. Once again, the same precomputation architecture and input and output selection algorithms are used as in Table 1 and the columns have the same meaning, except for the second and third columns which show the number of inputs and outputs of each circuit. It is noteworthy that in some cases, as much as 75% reduction in power dissipation is obtained.

The area penalty incurred is indicated by the number of literals in the precomputation logic and is 3% on the average. The extra delay incurred is proportional to the number of levels in the precomputation logic and is quite small in most cases. It should be noted that it may be possible to use the other precomputation architectures for all of the examples presented here. Some of these examples are perhaps better suited to other architectures than the one we used do derive the results, and therefore larger savings in power may be possible. Secondly, the inputs and outputs to be selected and the precomputation logic are determined automatically, making this approach suitable for automatic logic synthesis systems. Finally, the significant power savings obtained for random logic circuits indicate that this approach is not restricted only to certain classes of datapath circuits.

7 Conclusions and Ongoing Work

We have presented a method of precomputing the output response of a sequential circuit one clock cycle before the output is required, and exploited this knowledge to reduce power dissipation in the succeeding clock cycle. Several different architectures that utilize precomputation logic were presented.

In a finite state machine there is typically a single register, whose inputs are combinational functions of the register outputs. The precomputation architectures make no assumptions regarding feedback. For instance, R_1 and R_2 in Figure 2 can be the same register.

Precomputation increases circuit area and can adversely impact circuit performance. In order to keep area and delay increases small, it is best to synthesize precomputation logic which depends on a small set of inputs.

Precomputation works best when there are a small number of complex functions corresponding to the logic block **A** of Figures 2 and 3. If the logic block has a large number of outputs, then it may be worthwhile to selectively apply precomputation-based power optimization to a small number of complex outputs. This selective partitioning will entail a duplication of combinational logic and registers, and the savings in power is offset by this duplication.

Other precomputation architectures are being explored, including the architectures of Section 5, and those that rely on a history of previous input vectors. More work is required in the automation of a logic design methodology that exploits multiplexor-based, combinational and multiple-cycle precomputation.

8 Acknowledgements

Thanks to Anantha Chandrakasan for providing us with information regarding power dissipation in registers. J. Monteiro and S. Devadas were supported in part by the Defense Advanced Research Projects Agency under contract N00014-91-J-1698 and in part by a NSF Young Investigator Award with matching funds from Mitsubishi Corporation.

References

[1] M. Alidina. Precomputation-Based Sequential Logic Optimization for Low Power. Master's thesis, Massachusetts Institute of Technology, May 1994.

[2] P. Ashar, S. Devadas, and K. Keutzer. Path-Delay-Fault Testability Properties of Multiplexor-Based Networks. *INTEGRATION, the VLSI Journal*, 15(1):1–23, July 1993.

[3] R. Brayton, R. Rudell, A. Sangiovanni-Vincentelli, and A. Wang. MIS: A Multiple-Level Logic Optimization System. In *IEEE Transactions on Computer-Aided Design*, volume CAD-6, pages 1062–1081, November 1987.

[4] R. Bryant. Graph-Based Algorithms for Boolean Function Manipulation. *IEEE Transactions on Computers*, C-35(8):677–691, August 1986.

[5] A. Chandrakasan, T. Sheng, and R. W. Brodersen. Low Power CMOS Digital Design. In *Journal of Solid State Circuits*, pages 473–484, April 1992.

[6] S. Devadas, A. Ghosh, and K. Keutzer. *Logic Synthesis*. McGraw Hill, New York, NY, 1994.

[7] A. Ghosh, S. Devadas, K. Keutzer, and J. White. Estimation of Average Switching Activity in Combinational and Sequential Circuits. In *Proceedings of the 29th Design Automation Conference*, pages 253–259, June 1992.

[8] J. Monteiro, S. Devadas, and A. Ghosh. Retiming Sequential Circuits for Low Power. In *Proceedings of the Int'l Conference on Computer-Aided Design*, pages 398–402, November 1993.

[9] J. Monteiro, S. Devadas, and B. Lin. A Methodology for Efficient Estimation of Switching Activity in Sequential Logic Circuits. In *Proceedings of the 31st Design Automation Conference*, pages 12–17, June 1994.

[10] F. Najm. Transition Density, A Stochastic Measure of Activity in Digital Circuits. In *Proceedings of the 28th Design Automation Conference*, pages 644–649, June 1991.

[11] K. Roy and S. Prasad. SYCLOP: Synthesis of CMOS Logic for Low Power Applications. In *Proceedings of the Int'l Conference on Computer Design: VLSI in Computers and Processors*, pages 464–467, October 1992.

[12] A. Shen, S. Devadas, A. Ghosh, and K. Keutzer. On Average Power Dissipation and Random Pattern Testability of Combinational Logic Circuits. In *Proceedings of the Int'l Conference on Computer-Aided Design*, pages 402–407, November 1992.

[13] C-Y. Tsui, M. Pedram, and A. Despain. Exact and Approximate Methods for Switching Activity Estimation in Sequential Logic Circuits. In *Proceedings of the 31st Design Automation Conference*, pages 18–23, June 1994.

Circuit	Original			Precompute Logic			Optimized	
	Lits	Levels	Power	I	Lits	Levels	Power	% Reduction
comp16	286	7	1281	2	4	2	965	25
				4	8	2	683	47
				6	12	2	550	57
				8	16	2	518	60
				10	20	2	538	58
add_comp16	3026	8	6941	4/0	8	2	6346	9
				4/8	24	4	5711	18
				8/0	51	4	4781	31
				8/8	67	6	3933	43
max16	350	9	1744	8	16	2	1281	27
csa16	975	10	2945	2	4	2	2958	0
				4	11	4	2775	6
				6	18	4	2676	9
				8	25	5	2644	10
add_max16	3090	9	7370	4/0	8	2	7174	3
				4/8	24	4	6751	8
				8/0	51	4	6624	10
				8/8	67	6	6116	17

Table 1: Power Reductions for Datapath Circuits

Circuit	Original					Precompute Logic			Optimized	
	I	O	Lits	Levels	Power	I	Lits	Levels	Power	% Reduction
apex2	39	3	395	11	2387	4	4	1	1378	42
cht	47	36	167	3	1835	1	1	1	1537	16
cm138	6	8	35	2	286	3	3	1	153	47
cm150	21	1	61	4	744	1	1	1	574	23
cmb	16	4	62	5	620	5	10	1	353	43
comp	32	3	185	6	1352	6	13	2	627	54
cordic	23	2	194	13	1049	10	18	2	645	39
cps	24	109	1203	9	3726	7	26	3	2191	41
dalu	75	16	3067	24	11048	5	12	2	7344	34
duke2	22	29	424	7	1732	9	24	3	1328	23
e64	65	65	253	32	2039	5	5	1	513	75
i2	201	1	230	3	5606	17	42	5	1943	65
majority	5	1	12	3	173	1	1	1	141	19
misex2	25	18	113	5	976	8	16	3	828	15
misex3	25	18	626	14	2350	2	2	1	1903	19
mux	21	1	54	5	715	1	0	0	557	22
pcle	19	9	71	7	692	3	3	1	486	30
pcler8	27	17	95	8	917	3	3	1	571	38
sao2	10	4	270	17	1191	2	2	1	422	65
seq	42	35	1724	11	6112	2	1	1	2134	65
spla	16	46	634	9	2267	4	6	1	1340	41
term1	34	10	625	9	3605	8	14	3	2133	41
too_large	38	3	491	11	2718	1	1	1	1756	35
unreg	36	16	144	2	1499	2	2	1	1234	18

Table 2: Power Reductions for Random Logic Circuits

Glitch Analysis and Reduction in Register Transfer Level Power Optimization

Anand Raghunathan *
Department of EE
Princeton University
Princeton, NJ 08544

Sujit Dey
C&C Research Labs
NEC USA, Inc.
Princeton, NJ 08540

Niraj K. Jha [†]
Department of EE
Princeton University
Princeton, NJ 08544

ABSTRACT: We present design-for-low-power techniques based on glitch reduction for register-transfer level circuits. We analyze the generation and propagation of glitches in both the control and data path parts of the circuit. Based on the analysis, we develop techniques that attempt to reduce glitching power consumption by minimizing generation and propagation of glitches in the RTL circuit. Our techniques include restructuring multiplexer networks (to enhance data correlations, eliminate glitchy control signals, and reduce glitches on data signals), clocking control signals, and inserting selective rising/falling delays. Our techniques are suited to control-flow intensive designs, where glitches generated at control signals have a significant impact on the circuit's power consumption, and multiplexers and registers often account for a major portion of the total power. Application of the proposed techniques to several examples shows significant power savings, with negligible area and delay overheads.

I. Introduction

Most savings in power consumption can be obtained through a combination of various techniques at different levels of the design hierarchy. We focus on techniques to reduce average power consumption in register-transfer level (RTL) circuits. Power estimation techniques for RTL designs, and high-level synthesis techniques for reducing power consumption have been previously investigated [1, 2]. Several studies have reported the importance of considering glitching power during power estimation and optimization [3, 4]. However, very few automated design and synthesis techniques exist for reducing glitching power consumption. At the architecture and behavior levels, previous work on power estimation and optimization ignores the effects of glitch generation and propagation across the boundaries of blocks in the architecture. While accurate library modeling approaches can be used to account for the effect of glitches within architectural blocks, they typically assume that inputs to these blocks are glitch-free. Most previous work at the architecture and behavior levels has also sought to focus on data-flow intensive designs, where arithmetic units like adders and multipliers account for most of the total power consumption. However, our experiments with control-flow intensive designs reveal that functional units may constitute a much smaller fraction of total power than multiplexer networks and registers.

In this paper, we analyze the generation and propagation of glitches in both the control and data path parts of the circuit, and propose techniques to reduce glitch power consumption. In order to minimize the generation and propagation of glitches from control as well as

*Supported by NEC C&C Research Labs

[†]Supported by NSF under Grant No. MIP-9319269.

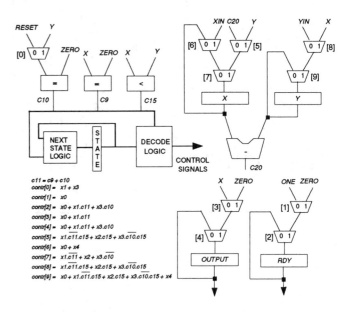

Figure 1: The RTL Architecture of the GCD circuit

data signals, we propose several techniques including restructuring multiplexer networks, clocking control signals, and inserting selective rising/falling delays. These techniques do not rely upon the existence of idle periods for components in a design, *i.e.*, they are also applicable to designs with complete or near-complete resource utilization. In addition, they target power consumption in all parts of the design, including multiplexer networks and registers, not just functional units.

II. Motivation

We motivate our work through the analysis of an example RTL circuit shown in Figure 1, which computes the *greatest common divisor* (GCD) of two numbers. The inputs are applied at XIN and YIN, and the result is written into register $OUTPUT$. Since the number of cycles required for computing the GCD depends on the input values provided, an additional output signal RDY indicates when the result is available. The circuit consists of one subtractor, two *equal-to* ($=$) comparators, one *less-than* ($<$) comparator, registers, multiplexer trees, the controller finite state machine (FSM), and the decode logic. The decode logic generates the control signals that configure the multiplexers in the circuit. We refer to the controller FSM and the decode logic collectively as the control logic of the circuit. The logic expressions for the decode logic are also shown in the figure. Literals $x0$ through $x4$ represent the decoded controller present state lines, while literals $c9$, $c10$, and $c15$ represent results of the three comparators.

The RTL circuit shown in Figure 1 was mapped to the NEC CMOS6 library [5]. An in-house simulation-based power estimation tool, CSIM [6], was used to measure power consumption in the various parts of the design. Table 1 provides the break up of the total power consumption into separate figures for the functional units (subtractor and three comparators), random logic (controller FSM and

Table 1: Power consumption in various parts of the GCD circuit

Block	% of total power
Functional units	9.08%
Random Logic	4.67%
Registers	39.55%
Multiplexers	46.70%

decode logic blocks), registers, and multiplexers. It indicates that most of the power consumption is in the multiplexers and registers. Similar figures were observed for several circuits that implemented other control-flow intensive specifications.

In order to get a feel for the glitching activity in the GCD circuit, we collected data on the transition activities with and without glitches in various parts of the design. Table 2 shows the total bit transitions with and without glitches for all the control signals, and selected data path signals [1]. Control signal *contr[i]* feeds the select input of the multiplexer marked *[i]* in Figure 1. Similarly, data path signal *dpi* corresponds to the output of the multiplexer marked *[i]* in Figure 1. Clearly, a significant portion of the total transition activity at several signals in the circuit is due to glitches. Another interesting observation is that several control signals in the GCD circuit, like *contr[2]* and *contr[4]*, are highly glitchy. We later illustrate that control signal glitches can have a significant effect on the glitching power consumption in the rest of the circuit. We would like to point out here that while CSIM, being a discrete simulator, does not model effects like partial transitions, it does model the attenuation or suppression of glitches due to inertial delays of gates.

Table 2: Activities with/without glitches for various signals of the GCD circuit

Control signal	Activity		Datapath signal	Activity	
	Total	W/O Gl.		Total	W/O Gl.
contr[0]	71	70.5	*dp2[7..0]*	71.5	21.5
contr[1]	22	22	*dp4[7..0]*	92	26
contr[2]	72	20	*dp5[7..0]*	1124.5	247
contr[3]	42	20	*dp7[7..0]*	1044.5	273
contr[4]	72	20	*dp9[7..0]*	321.5	80.5
contr[5]	55.5	54			
contr[6]	22	22			
contr[7]	50	20			
contr[8]	55.5	54			
contr[9]	77	70.5			

The following example illustrates how ignoring glitches can be misleading and result in designs that are sub-optimal in terms of their power consumption. Consider the two RTL circuits shown in Figures 2(a) and 2(b) that implement the simple function: `if(x < y) then z = c + d else z = a + b` in two different ways. ARCHITECTURE 2 uses two adders as opposed to one adder in the case of ARCHITECTURE 1. Power estimation methods that do not take glitches into account would report that ARCHITECTURE 2, in which both adders perform computations in each cycle, consumes more power than ARCHITECTURE 1. However, when accurate power estimation that also considers glitches is performed, it turns out that ARCHITECTURE 2 actually consumes 17.7% less power than ARCHITECTURE 1. The above observation can be explained as follows. The comparator generates glitches at its output though its inputs are glitch-

[1]CSIM counts each $0 \rightarrow 1$ or $1 \rightarrow 0$ transition as half a transition. Hence, the transition numbers that are reported throughout the paper may be fractional

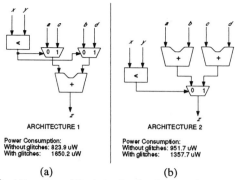

ARCHITECTURE 1

Power Consumption:
Without glitches: 823.9 uW
With glitches: 1650.2 uW

ARCHITECTURE 2

Power Consumption:
Without glitches: 951.7 uW
With glitches: 1357.7 uW

(a)　　　　　　(b)

Figure 2: Alternate architectures that implement the same function: Effect of glitching

free. In the case of ARCHITECTURE 1, these glitches then propagate through the two multiplexers to the inputs of the adder, which causes a significant increase in glitching activity and hence power consumption in the two multiplexers and the adder. In ARCHITECTURE 2, though the comparator generates glitches as before, the effect of these glitches is restricted to the single multiplexer.

III. Glitch Generation in the controller and data path

In this section, we analyze the generation of glitches in RTL circuits. This analysis leads to an understanding that forms the basis for our glitch reduction techniques that we present in Section IV. For clarity, we illustrate glitch generation in the data path blocks (functional units, comparators, and multiplexer trees) and in the control logic separately.

A. Glitch generation in data path blocks

Consider the elements shown in Figure 3 — a subtractor, an equal-to comparator, a less-than comparator, and a 3-to-1 multiplexer tree — as representative data path blocks for studying glitch generation. Each block was mapped to the technology library, and then simulated under long input sequences that consisted of random vectors. Figure 3 is annotated with the total number of bit-transitions including and excluding glitches that were observed at the output of each block. The results clearly indicate significant generation of glitches in various data path blocks. In the equal-to comparator, no glitches were generated due to the fact that all its paths are balanced. However, even in such cases, wiring delays can disturb the balance of delays and thus cause generation of glitches. When data path blocks like those shown in Figure 3 are connected together, the glitches generated by the various blocks propagate through the following blocks, often causing an explosion in glitches and glitching power consumption.

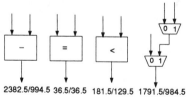

2382.5/994.5　36.5/36.5　181.5/129.5　1791.5/984.5

Figure 3: Glitch generation in various data path blocks

B. Glitch generation in control logic

Though the control logic itself accounts for only a small portion of the total circuit power, it plays an important role in determining the total circuit power because it is responsible for the generation of glitches at the control signals, which in turn have a significant impact on the glitching activity in the rest of the circuit. The inputs to the decode logic (see Figure 1) are fed by the outputs of comparators and the state flip-flops of the controller. The previous subsection has already demonstrated that outputs of comparators can be glitchy. The glitches at comparator outputs can propagate through the decode logic and cause glitches on the control signals. In addition, the decode logic can itself generate a lot of glitches, as shown next. Let us focus on control signal *contr[2]* in the GCD RTL circuit, which is highly glitchy according to the statistics of Table 2. The portion of the decode logic that implements this control signal is shown in Figure 4(a). We observe that though the inputs are nearly glitch-free, significant glitches are generated at AND gates $G1$ and $G2$. After careful analysis,

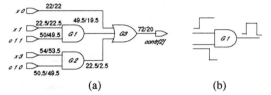

(a)   (b)

Figure 4: (a) Implementation of control signal *contr[2]*, and (b) generation of glitches at gate $G1$

the generation of glitches at $G1$ was attributed to two conditions that are depicted graphically in Figure 4(b):

C1: A rising transition on signal $x1$ was frequently accompanied by a falling transition on $c11$. Thus, the rising transition on $x1$ and the falling transition on $c11$ are highly correlated.

C2: Transitions on signal $x1$ arrive earlier than transitions on $c11$.

Condition C1 arises due to the functionality of the design: most of the times when state s_1 is entered (rising transition on $x1$), comparators feeding $c9$ and $c10$ produce a 0, changing from 1 in the previous state. On the other hand, condition C2 is a result of the delay/temporal characteristics of the design. A similar explanation holds for the output of gate $G2$ being glitchy. Generation of glitches in the control logic has been described in detail in [7].

IV. Glitch Reduction Techniques

In this section, we describe our techniques for reducing glitch power consumption in RTL circuits, by minimizing the generation and propagation of glitches through different blocks of the circuit.

A. Reducing glitch propagation from control signals

As shown before, control signals to the data path can be very glitchy. Our aim is to stop glitches on control signals from propagating as close to their source as possible in order to reap the maximum benefits in terms of power savings. We illustrate each of our techniques separately through examples in this subsection, and later integrate these techniques into a single power optimization framework.

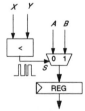

Figure 5: Example circuit used to illustrate the effect of data signal correlations on control signal glitches

Glitchy control signals and data correlations. Consider the circuit shown in Figure 5. A multiplexer selects between two 8-bit data

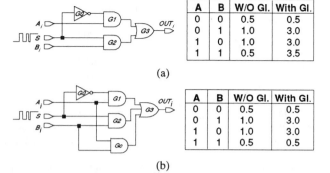

(a)

(b)

Figure 6: (a) Effect of data correlations on select signal glitches, and (b) use of the consensus term to reduce glitch propagation

signals, A and B, depending on whether the expression $X < Y$ evaluates to $True$ or $False$. Its output is written into a register. Suppose that the less-than comparator generates glitches at its output, and that data inputs to the multiplexer are not glitchy and settle to their final value well before the select signal settles. The glitches on the select signal of the multiplexer propagate to its output. In order to study this propagation, consider the gate-level implementation of a bit-slice of the multiplexer that is shown in Figure 6(a). The table shown in Figure 6(a) reports the glitches at the multiplexer output for all possible values of the data signal bits A_i and B_i. In the $< 0, 0 >$ case, glitches on select signal S are killed at AND gates $G1$ and $G2$ due to controlling side inputs that arrive early. When data inputs are $< 0, 1 > (< 1, 0 >)$, glitches on S propagate through gates $G2$ and $G3$ ($G1$ and $G3$). Finally, when data inputs are $< 1, 1 >$, glitches on S propagate through gates $G1$ and $G2$. The output of the multiplexer is glitchy as a result of the interaction of the glitchy signal waveforms at $G1$ and $G2$. The exact manner in which the waveforms interact depends on the propagation and inertial delays of the various wires and gates in the implementation. There are many ways of preventing the propagation of glitches for the $< 1, 1 >$ case. One way is to add an extra gate Gc, as shown in Figure 6(b). Gc realizes $A_i.B_i$ which is the *consensus* of $\bar{S}.A_i$ and $S.B_i$. When data inputs are $< 1, 1 >$, Gc effectively kills any glitches at the other inputs of $G3$ that arrive after the output of Gc settles to a 1, as shown in the table of Figure 6(b). Maximum benefits are derived from the addition of the consensus term when the select signal is very glitchy, the data inputs arrive early compared to the select signal, and the probability of the data inputs being $< 1, 1 >$ is high.

Note that with the addition of the consensus term, glitches do not propagate from the select signal to the multiplexer output if the data values are correlated ($< 0, 0 >$ or $< 1, 1 >$). We next show how to restructure a multiplexer tree so as to maximize data correlations and hence minimize propagation of glitches from its select signals.

Enhancing data correlations by restructuring multiplexer networks. Consider the 3-to-1 multiplexer tree shown in Figure 7(a), that feeds register $OUTPUT$ in the GCD RTL circuit. The select signals are annotated with their cumulative transition counts including and excluding glitches. Functionally, the multiplexer tree can be thought of as an abstract 3-to-1 multiplexer, as shown in Figure 7(b). The conditions under which $OUTPUT$, X and $ZERO$ are selected are represented as C_{OUTPUT}, C_X, and C_{ZERO}, respectively (which must be mutually exclusive). Select signal C_{ZERO} is observed to be glitchy, leading to propagation of glitches to the output of the first 2-to-1 multiplexer in Figure 7(a). Note that data signals $OUTPUT$ and $ZERO$ are highly correlated at the bit level. Hence, in order to minimize the propagation of glitches on C_{ZERO} through the multiplexer

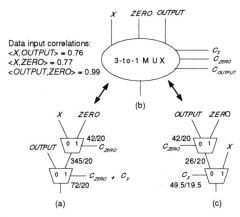

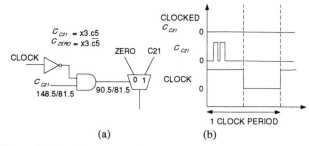

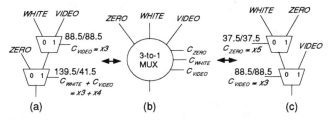

Figure 7: Multiplexer restructuring to enhance data correlations: (a) initial multiplexer network, (b) abstract 3-to-1 multiplexer, and (c) restructured network

Figure 8: Eliminating glitchy control signals: (a) initial multiplexer network, (b) abstract 3-to-1 multiplexer, and (c) restructured network

tree, we transform the multiplexer tree to the implementation shown in Figure 7(c), such that the highly correlated data signals $OUTPUT$ and $ZERO$ become inputs to the first 2-to-1 multiplexer. This significantly lowers the switching activity at the output of the first 2-to-1 multiplexer to $26/20$ from $345/20$ originally.

Restructuring multiplexer networks to eliminate glitchy select signals. In order to implement an abstract n-to-1 multiplexer with n data inputs $(d_1...d_n)$, and n select inputs $(c_1...c_n)$ as a tree of 2-to-1 multiplexers, it can be shown that depending on the exact structure of the implementation, anywhere between $\lceil log_2 n \rceil$ and $n-1$ select expressions of the form $\bigcup_i c_i$ can be used, where $\bigcup$ represents the Boolean OR operation. It is possible that some of $c_1...c_n$ are glitchy, while others are not. Similarly, it is possible that some of the disjunctive expressions in $c_1...c_n$ are glitchy. Our aim is to restructure the multiplexer tree so that as few as possible of the select expressions used are glitchy. This concept is illustrated using the 3-to-1 multiplexer network shown in Figure 8(a) that is part of the RTL circuit implementing a `Barcode` preprocessor [8]. The select signal of the second multiplexer in Figure 8(a) $(x3 + x4)$ is glitchy. An alternative implementation of the 3-to-1 multiplexer network, that does not require the use of any glitchy select signal expressions, is shown in Figure 8(c).

Clocking control signals to kill glitches. When all the methods presented so far to reduce the effect of glitches on control signals do not help, we use the clock to suppress glitches on control signals. For the following example, we assume that the design is implemented using rising-edge-triggered flip-flops and a single phase clock with a duty cycle of 50%. Consider the 2-to-1 multiplexer shown in Figure 9(a), that is part of an Unmanned Auto Vehicle controller (UAV) circuit [9]. Both C_{ZERO} and C_{C21} are glitchy due to the generation of glitches in the less-than comparator that generates signal $c5$. Thus, multiplexer restructuring transformations to eliminate glitchy control

Figure 9: Clocking control signals to kill glitches: (a) multiplexer network with clocked control signal, and (b) sample waveforms

signals cannot be applied here. As shown in Figure 9(a), the original select signal is ANDed with the inverted clock to result in the *clocked select signal*. For the first half of the clock period, when the clock is high, the clocked select signal is forced to 0 in spite of the glitches on the original select signal. Figure 9(b) shows example waveforms for the clock, the original select signal and the clocked select signal. The switching activity numbers shown in Figure 9(a) demonstrate that clocking the control signal significantly reduces its glitching activity.

The technique of clocking control signals needs to be applied judiciously due to the following reasons. By clocking the control signal, we are preventing it from evaluating to its final value until time $\frac{T}{2}$, where T is the clock period. This could lead to an increase in the delay of the circuit, if the control signal needs to settle to its final value before $\frac{T}{2}$ in order to meet the specified timing constraints at the circuit outputs. It should also be noted that clocking control signals may introduce extra transitions on the control signal under certain conditions. Consider a situation where the control signal remains at a steady 1 over a pair of clock cycles. By forcing the control signal to 0 in the first half of both the clock cycles, we are actually introducing extra transitions on the control signal, which can lead to increased power consumption. Thus, the scheme presented in Figure 9(b) leads to most power savings when the probability of the control signal evaluating to a 1 (signal probability) is low. On the other hand, if the signal probability of the control signal is very high, the control signal can be clocked by ORing the original control signal with the clock.

B. Minimizing glitch propagation from data signals

The previous subsection outlined several ways in which the generation and propagation of glitches from control signals can be reduced to save power. The data inputs to a circuit block can also be glitchy, as seen in Section III. In this subsection, we present several techniques to restrict propagation of glitches from data signals.

Glitch reduction using selective rising/falling delays. Consider the 2-to-1 multiplexer shown in Figure 10(a). Both the data inputs to the multiplexer have glitches, which propagate through the multiplexer and then through the adder, causing significant power dissipation. Consider the gate-level implementation of a bit-slice of the multiplexer as shown in Figure 10(b). Consider a pair of consecutive clock cycles c_1 and c_2 such that select signal S makes a $1 \rightarrow 0$ (falling) transition from c_1 to c_2. If this transition is early arriving, there will be an early rising transition at the output of gate $G0$ that implements $\bar{S}$. Consequently, the side input of $G1$ will become non-controlling early, allowing the data input glitches to propagate through $G1$. This propagation can be minimized by delaying the rising transition at the output of $G0$ ($\bar{S}$), *i.e.*, by adding a "rising transition delay" to it. Similarly, to minimize glitch propagation through gate $G2$ when there is an early rising transition at S, it is desirable to delay the rising transition on the fanout branch of S that feeds $G2$. Since we wish to delay selected (either rising or falling, but not both) transitions at

571

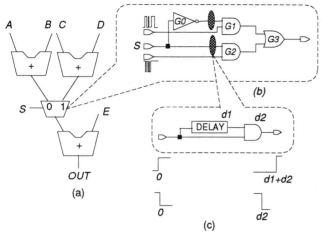

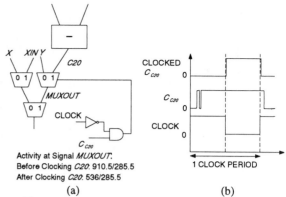

(a)

(b)

Figure 12: Clocking control signals to kill data signal glitches: (a) example circuit, and (b) sample waveforms

Figure 10: (a) Example circuit, (b) multiplexer bit-slice with selective delays inserted, and (c) implementation of a rising delay block

certain signals, we refer to the technique as *selective delay insertion*. The selective rising delay blocks are represented by the shaded ellipses shown in Figure 10(b). One possible implementation of a rising delay block, that uses one AND gate and a delay element, is shown in Figure 10(c). Under a simplified delay model of d_1 ns for the delay block and d_2 ns for the AND gate, it can be seen that a rising transition at the input is delayed by $(d_1 + d_2)$ ns, while a falling transition is delayed by only d_2 ns. A selective falling delay block is similar, except that the AND gate is replaced by an OR gate.

Inserting a rising delay block leads to a reduction in the propagation of glitches through a multiplexer only in the clock cycles in which there is a rising transition at the delay block's input. Thus, the probability of a rising transition at the signal where we desire to insert a selective rising delay block should be high. In addition, to ensure that the circuit delay does not increase, we insert selective delay blocks only at signals that have sufficient slack.

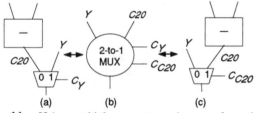

(a)   (b)   (c)

Figure 11: Using multiplexer restructuring transformations for glitchy data signals: (a) initial multiplexer network, (b) abstract 2-to-1 multiplexer, and (c) restructured network

Effect of multiplexer restructuring transformations on glitchy data signals. Multiplexer restructuring transformations can also be used to reduce the propagation of glitches on data signals. We illustrate this concept using a small portion of the GCD RTL circuit, that is shown in Figure 11(a). The subtractor's output, $C20$, has a lot of glitches which propagate through the multiplexer shown in the figure, and also through the logic that it feeds. Let us assume that signal Y is glitch-free. Figure 11(b) shows the equivalent abstract 2-to-1 multiplexer. We utilize the fact that there might be several instances when the value of the select signal is a *don't care* ($C_{C20}+C_Y$ is not a tautology). In the implementation of Figure 11(a), the glitchy operand $C20$ is selected in the don't care cases as well. The transformed implementation of the 2-to-1 multiplexer that is shown in Figure 11(c) ensures that the glitchy data input is selected as infrequently as possible, thus reducing

the propagation of glitches to the multiplexer output.

Clocking control signals to kill data signal glitches. When the techniques presented above to handle glitchy data signals are either not applicable or not adequate, we utilize the technique of clocking control signals to kill data signal glitches as well. Consider the part of the GCD circuit shown in Figure 12(a). The subtractor's output $C20$, which is glitchy, feeds the data input of a 2-to-1 multiplexer. As shown in the figure, this results in significant glitches at the output of the multiplexer. Clocking select signal C_{C20} alleviates this problem. Since the clocked select signal is forced to 0 for the first half of the clock period, the multiplexer selects the value of data input Y for this duration. Thus, the glitches on the subtractor's output are killed at the multiplexer for approximately the first half of the clock period. This leads to a significant decrease in the glitching activity at the multiplexer output, as shown in the figure. Sample waveforms for the clock, original select signal, and the clocked select signal are shown in Figure 12(b).

C. Algorithm

The previous two subsections described the various techniques that we use to minimize the glitching power consumption in RTL circuits. Also, the conditions under which each technique is applicable and most beneficial have been stated. In this section, we give a brief overview of the order of application of the various glitch reduction techniques. The pseudo-code for power optimization procedure is shown in Figure 13. In order to apply each of the techniques, we need information about signal statistics and glitches at various signals in the circuit, including the control signals and the outputs of each RTL unit like functional unit, register, comparator, 2-to-1 multiplexer, *etc*. We first obtain an initial technology mapped gate-level implementation of the RTL circuit, and use a simulator to collect the required information.

A block in the RTL circuit is defined as a functional unit, comparator, or register, together with the multiplexer networks that feed it. We partition the RTL circuit into constituent blocks, and levelize the blocks through a single traversal starting from primary inputs or register outputs to primary outputs or register inputs. We then visit the circuit blocks in increasing order of levels (since applying glitch reduction techniques to a block can affect glitching at other blocks which are at later levels) and use applicable glitch reduction techniques at each step. Within a block, we first attempt to apply multiplexer network restructuring transformations to either eliminate glitchy select signals, or else maximize bit-level correlation between the data inputs of multiplexers whose select signals have a lot of glitches. We also selectively determine which bit-slices, if any, of each multiplexer to add the consensus term to, based on the probability of the data inputs

572

```
Procedure RTL_POWER_REDUCTION(RTL Circuit R)
    EXTRACT_BLOCKS(R);
    LEVELIZE_CIRCUIT(R);
    for each block B in levelized order {
        (*eliminate glitchy control signals,
        enhance data signal correlations,
        select glitchy data signals as infrequently as possible*)
        RESTRUCTURE_MUX_NETWORK(B);
        if (significant glitches remain) {
            ADD_SELECTIVE_DELAYS(B);
            CLOCK_CONTROL_SIGNALS(B);
        }
    }
}
```

Figure 13: Procedure Overview

Table 3: Experimental Results

Circuit	Original			Optimized			Pow Red (%)
	Pow (mw)	Area	Del (ns)	Pow (mw)	Area	Del (ns)	
GCD	8.74	1037	32.3	7.23	1034	31.9	17.27
Bar-code	9.41	1945	49.7	7.77	1968	47.3	17.42
UAV	10.89	1954	83.5	8.02	1967	83.2	26.25
Vend	10.47	1595	70.1	7.72	1617	71.6	26.22

taking on values of $< 1, 1 >$ and the glitchiness of the select signal. If significant glitches are present at the inputs of any RTL unit (functional unit, register, comparator, multiplexer) in the block even after the application of multiplexer restructuring transformations, we attempt to add selective rising/falling delays to multiplexers in order to reduce the propagation of glitches on data signals, or clock control signals as described in Subsections A and B in order to kill glitches on both control signals as well as data signals that feed multiplexers.

The data presented in Section II revealed that register power constituted a significant portion of the total circuit power. Upon further analysis, we found that a major portion of the register power was in turn consumed due to transitions at the clock inputs to registers. The technique of gating clocks has been used by designers to selectively turn off parts of a system. Methods to automatically detect conditions under which the clock inputs to all the registers in a design can be shut off, based on identifying self-loops and unreachable states in the state transition graph (STG), were presented in [10]. However, the techniques in [10] can be applied only to the control and random logic parts of a design for which it is feasible to extract the STG. We have developed a procedure, that is based on a structural analysis of the RTL circuit, to determine the conditions under which transitions on the clock input to a register can be suppressed. Our procedure consists of identifying (structural) self-loops involving a register in the RTL circuit, and analyzing the conditions under which it is logically enabled. Gating the clock input to a register can lead to glitches on the gated clock signal, that not only cause unnecessary power consumption, but may also cause the design to function incorrectly. Our procedure ensures that glitches are not introduced while gating clock signals. Details of the procedure are provided in [7].

V. Experimental Results and Conclusions

We present results of the application of the proposed power reduction techniques to four RTL circuits implementing: GCD, a barcode reader preprocessor (Barcode) [8], the controller for an Unmanned Auto Vehicle (UAV) [9], and a vending machine controller (Vend) [11]. The initial RTL circuits were obtained by synthesizing VHDL behavioral descriptions using the SECONDS high-level synthesis system [9, 12]. Both the original and optimized RTL circuits were mapped to NEC's CMOS6 library [5], and evaluated for area and delay using the logic synthesis system VARCHSYN [13], and for power consumption using the simulation based power estimation tool, CSIM [6]. The vectors used for simulation were obtained for each design by simulating the scheduled behavioral description with a test bench written to generate typical input cases, using a VHDL simulator, to result in a cycle-by-cycle input vector trace. The above

step is important for control-flow intensive designs where the number of clock cycles required to perform the computation varies depending on the input values. Table 3 reports the results of our experiments. The power, area (# of transistor pairs), and delay numbers are obtained after mapping to the technology library used.

The results shown in Table 3 demonstrate that our glitch reduction techniques can significantly reduce power consumption in RTL circuits. Note that these techniques target power reduction solely by reducing the propagation of glitches between various blocks in the RTL circuit. Hence, they can be combined with other power reduction techniques that attempt to suppress transitions that do not correspond to glitches. The area and delay overheads incurred by our power reduction techniques can be seen to be nominal. In some cases, the area and delay are slightly reduced due to the fact that multiplexer restructuring transformations can lead to a simplification in control logic.

References

[1] J. Rabaey and M. Pedram (Editors), *Low Power Design Methodologies.* Kluwer Academic Publishers, Boston, MA, 1996.

[2] M. Pedram, "Power minimization in IC design: principles and applications," *ACM Trans. Design Automation of Electronic Systems*, vol. 1, Jan. 1996.

[3] M. Favalli and L. Benini, "Analysis of glitch power dissipation in CMOS IC's," in *Proc. Int. Symp. Low Power Design*, pp. 123–128, Apr. 1995.

[4] S. Rajagopal and G. Mehta, "Experiences with simulation-based schematic-level power estimation," in *Proc. Int. Wkshp. Low Power Design*, pp. 9–14, Apr. 1994.

[5] *CMOS6 Library Manual.* NEC Electronics, Inc., Dec. 1992.

[6] *CSIM Version 5 Users Manual.* Systems LSI Division, NEC Corp., 1993.

[7] A. Raghunathan, S. Dey, and N. K. Jha, "Register-transfer-level power optimization techniques with emphasis on glitch analysis and optimization," Tech. Rep., NEC C&C Research Labs, Princeton, NJ, Oct. 1995.

[8] High-level synthesis benchmarks, CAD Benchmarking Laboratory, Research Triangle Park, NC. Benchmarks can be downloaded anonymously from http://www.cbl.ncsu.edu.

[9] S. Bhattacharya, S. Dey, and F. Brglez, "Clock period optimization during resource sharing and assignment," in *Proc. Design Automation Conf.*, pp. 195–200, June 1994.

[10] L. Benini, P. Siegel, and G. DeMicheli, "Saving power by synthesizing gated clocks for sequential circuits," *IEEE Design & Test of Computers*, pp. 32–41, Winter 1994.

[11] D. L. Perry, *VHDL.* New York, NY 10020: McGraw-Hill, 1991.

[12] S. Bhattacharya, S. Dey, and F. Brglez, "Performance analysis and optimization of schedules for conditional and loop-intensive specifications," in *Proc. Design Automation Conf.*, pp. 491–496, June 1994.

[13] *VARCHSYN Version 2.0 Users Manual.* Advanced CAD Development Laboratory, NEC Corporation, Nov. 1993.

Exploiting Locality for Low-Power Design

Renu Mehra, Lisa Guerra, and Jan Rabaey

Department of Electrical Engineering and Computer Sciences
University of California at Berkeley
Berkeley, CA 94720

Abstract

We propose a new high-level synthesis technique for the low-power implementation of real-time applications. The technique uses algorithm partitioning to preserve locality in the assignment of operations to hardware units. This results in reduced usage of long high-capacitance buses, fewer accesses to multiplexors and buffers, and more compact layouts. Experimental results show average reductions in bus and multiplexor power of 62.9% and 38.5%, respectively, resulting in an average reduction of 18.5% in total power.

1. Introduction

High-level synthesis is steadily making an inroad into the digital design community. Most work to date has focused on techniques for area and speed optimization [1]. Recently, there has been significant interest in techniques and tools for power optimization. While for area optimization, high resource utilization through hardware sharing is one of the main goals, for power optimization, reduced hardware sharing often gives better results.

Consider Wu's comparison of an automatically-generated maximally time-shared and a manually-generated fully-parallel implementation of a QMF sub-band coder filter [2]. In the manual design, a number of optimizations were used to obtain power savings in the various components. The power consumption of both versions is documented in Table 1. For the same supply voltage, an improvement of a factor of 10.5 was obtained at the expense of a 20% increase in area.

Note that the interconnect elements (buses, multiplexors, and buffers) consume 43% and 28% of the total power in the time-shared and parallel versions, respectively. Further, these elements contribute the most to the power reduction achieved in the parallel version. Power improvement factors of 16.9, 15.1, and 12.5 were obtained for buses, multiplexors, and buffers, respectively, mainly due to dedicated communication and reduced usage of multiplexors and buffers. This points to the large opportunity available for interconnect power reduction and highlights its significance.

While in this example, the fully-parallel implementation resulted in large power gains with low area overhead, this may not always be the case. Parallel implementations may be too

Table 1. Power consumption (mW) in the maximally time-shared and fully-parallel versions of the QMF sub-band coder filter.

Component	Time-shared	Fully-parallel	Improvement factor
Functional units	8.52	1.03	8.3
Registers	9.76	1.08	9.0
Buses	23.69	1.40	16.9
Multiplexors	3.77	0.25	15.1
Buffers	4.36	0.35	12.5
Others	23.99	2.92	8.2
Total	74.09	7.03	10.5

area intensive and may not necessarily result in reduced interconnect power. If the area overhead is too high, the increase in the required bus lengths may offset the power gains due to other factors.

In this work, techniques are presented to achieve low-power designs by reducing the interconnect power while incurring low area overhead. The approach aims to capture some of the optimizations of the above example in an automated way while maintaining a balance between the maximally time-shared and the fully-parallel implementations. The next section illustrates the main idea behind our proposed low-power synthesis technique.

2. The impact of exploiting locality

The main idea behind our approach is to synthesize designs with localized communications. We achieve this by dividing the algorithm into *spatially local clusters* and performing a *spatially local assignment*. A spatially local cluster is a group of algorithm operations that are close to each other in the flowgraph representation. A spatially local assignment is a mapping of the algorithm operations to specific hardware units such that no operations in different clusters share the same hardware. Partitioning the algorithm into spatially local clusters ensures that the majority of the data transfers take place within clusters and relatively few occur between clusters. The spatially local assignment restricts intra-cluster data transfers to buses that are local to a subset of the hardware (local buses); thus only inter-cluster data transfers use buses that are shared by all resources (global buses). The combined result is that local buses which are

Reprinted from *Proceedings of the IEEE Custom Integrated Circuit Conference*, pp. 401-404, May 1996.

shorter are used more frequently than longer highly-capacitive global buses.

Consider the two different assignments for maximum throughput implementations of the fourth-order parallel-form IIR filter shown in Fig. 1. Indicated beside each operation is the hardware resource that it is assigned to. (A_i are adders and M_i are multipliers). In Fig. 1a, the graph is divided into two spatially local clusters and the operations in each cluster are mapped to mutually exclusive sets of hardware resources (A_1, A_2, and M_1 are used for operations in cluster 1 and A_3, A_4, and M_2 are used for those in cluster 2). As a result, a large number of the communications are restricted to only a subset of the hardware. In Fig. 1b, however, the hardware is not partitioned and all communications are global. The number of global data transfers (shown with solid lines in both cases) for the local and the non-local assignments are 2 and 20, respectively.

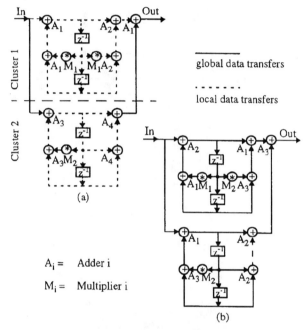

Fig. 1. A fourth-order parallel-form IIR filter: (a) Local assignment, (b) Non-local assignment.

Notice that the local version needs 4 adders and 2 multipliers whereas the non-local assignment requires just 3 adders and 2 multipliers. This increase in the number of functional units does not necessarily translate into a corresponding increase in the overall area since localization of interconnect makes the design more conducive to compact layout. Furthermore, reduced hardware sharing results in additional power savings due to fewer accesses to multiplexors and buffers.

Varying the number of clusters trades off local and global bus power. This is because, as the number of clusters is increased, the number of inter-cluster communications increases while the local bus lengths decrease.

3. Low-power synthesis system

The high-level synthesis process generates an architectural level netlist from a behavioral description and a set of performance constraints. In this section we present a new high-level synthesis strategy based on exploiting the locality of algorithm operations. The core of the approach is a partitioning and assignment scheme. The techniques have been integrated into the *Hyper-LP* system [3].

While the basic synthesis flow of the *Hyper-LP* system is the same as that of the *Hyper* system [4], a new partitioning step is added preceding the assignment phase and the assignment algorithm itself is modified to exploit spatial locality.

3.1 Partitioning methodology

Previous works in partitioning for high-level synthesis have targeted area minimization, with a significant portion of the gains resulting from interconnect reduction [5, 6]. For power minimization, however, it is better to have two global buses each accessed twice rather than one bus accessed six times. The goal for reducing interconnect power, therefore, is to minimize the *number of accesses* to long global buses.

Our partitioning methodology consists of two phases — the first phase generates several candidate partitioning solutions and the second phase evaluates them and selects the best one.

The generation of candidate partitions is based on a spectral partitioning technique used in a number of partitioning algorithms [7, 8, 9]. The technique was introduced by Hall [7] who proved that the second smallest eigenvector of the Laplacian of a graph gives a one-dimensional placement of graph nodes such that the sum of squares of edge lengths is minimized. Large gaps in this ordering are used to delimit the clusters. For example, Fig. 2 shows an eighth order cascade filter and the corresponding eigenvector placement. The spacing between nodes in the placement clearly indicates four distinct clusters which are also evident from the structure. We use $m+3\sigma$ as the threshold for detecting these gaps, where m is the mean of all the distances between the nodes and σ is the standard deviation of the distances. In the cascade example of Fig. 2, this threshold delimits the expected four clusters.

In our scheme, several different candidate partitions are generated by varying the targeted number of clusters. For example, in the cascade filter of Fig. 2, in addition to the 4-cluster partition, a 2-cluster partition may also be proposed. As discussed before, varying the number of clusters trades off between global and local bus power.

In the second phase a rough estimate of the total bus power is used to evaluate and compare the candidate partitions. The metric used as a measure of the global bus power is the number of global data transfers times the estimated global bus

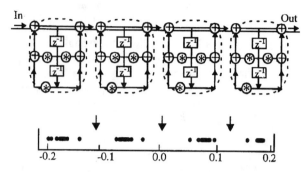

Fig. 2. An eighth-order cascade-form IIR filter and the corresponding eigenvector placement.

length. Similarly, the measure used for the local bus power of each cluster is the number of data transfers local to it times the cluster's estimated bus length.

Since the lengths of the buses have been shown to be proportional to the square root of the area, the cluster and total chip area estimates are used as measures of the local and global bus lengths, respectively. Estimates of these areas are in turn given by the maximum value of the weighted concurrency distribution graph [10]. The distribution graph gives the amount of concurrent hardware needed by the computation in each time slot.

At the culmination of the partitioning phase, the single most promising candidate partition is applied to the algorithm.

3.2 Assignment methodology

Once the partitioning is complete, all operations have an associated cluster number. Also, each data transfer is classified as either an inter- or intra-cluster transfer. The functional unit assignment performs the random assignment with iterative improvement approach of the *Hyper* system [4] and uses the clustering information to ensure that each cluster is assigned to mutually exclusive hardware.

Graph coloring is a commonly used technique to assign data transfers to shared buses such that there are no timing conflicts. In our scheme, we assign data transfers to buses avoiding not only timing but also clustering conflicts. Local buses are shared only among transfers within a cluster. Global buses are used only by inter-cluster data transfers. This ensures that intra-cluster data transfers occur on short local buses and only inter-cluster ones use the long highly-capacitive global buses.

4. Results

In this section we present the results of our partitioning-based synthesis scheme. Implementations generated using the *Hyper-LP* and the *Hyper* systems are compared. The SPA architectural power estimation tool [11] is used for power estimations. Estimates of the total chip area and bus lengths are obtained using models presented in [12]. The bus length

model was enhanced to estimate the local and global bus lengths separately.

4.1 Cascade filter

The first result compares the *Hyper-LP* and *Hyper* implementations of the eighth-order cascade filter (Fig. 2). Given a throughput constraint of 21 clock cycles, the *Hyper* implementation uses 4 adders and 3 shifters while the *Hyper-LP* implementation uses one adder and one shifter for each cluster resulting in a total of eight units. Layouts of the two implementations are shown in Fig. 3. In the *Hyper* implementation, 2 of the 7 functional units are merged by the layout tool. In the *Hyper-LP* implementation the 4 datapaths pictured correspond to each of the 4 clusters.

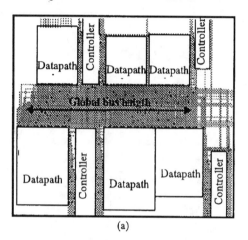

(a)

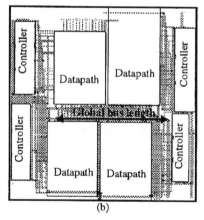

(b)

Fig. 3. Cascade filter layouts: (a) Non-local implementation from *Hyper*, (b) Local implementation from *Hyper-LP*.

Table 2 compares the power dissipated in the two implementations. An overall reduction of 34% in the power consumption was realized by the *Hyper-LP* approach. The average length of the global buses reduced by approximately 55%, from 2100 to 950 microns and the bus power reduced 5-fold, from 2 mW to only 0.4 mW. The multiplexor power reduced by 60% as the reduced time-sharing of units results in lower usage of multiplexors. Notice that the contribution of interconnect (buses, multiplexors, and buffers) to the total power dissipation was reduced from 30% to 17%.

Table 2. Comparison of power consumption (mW) in the *Hyper* and *Hyper-LP* implementations of the cascade filter.

Component	*Hyper*	*Hyper-LP*	Percentage reduction
Buses	2.0	0.4	80.0
Multiplexors	3.7	1.5	59.5
Buffers	1.0	0.9	10.0
Others	8.7	7.4	14.9
Total	15.4	10.2	33.77

4.2 Other examples

This section summarizes our experimental results for the cascade and several other DSP filter and transform examples. Some are in their original form (DCT, FFT, and parallel-form IIR) and others are transformed using either constant multiplication expansion (cascade-form IIR, direct-form IIR, and wavelet) or retiming (wave digital filter).

Table 3. Comparison of power consumption (mW) in the *Hyper* and *Hyper-LP* implementations.

Design	*Hyper*			*Hyper-LP*			
	Bus	Mux	Total	Clusters	Bus	Mux	Total
Cascade	2.0	3.2	21.3	4	0.4	1.5	16.3
Direct form	29.8	38	144.6	3	10.3	21.1	110.4
Wave digital	1.5	3.2	20.4	2	0.5	1.7	18.0
DCT	9.0	3.8	41.5	2	4.5	2.3	37.2
FFT	17.6	4.7	48.6	2	5.2	3.8	36.8
Parallel IIR	15.1	2.8	57.5	2	3.2	2.2	48.8

Table 3 shows the bus, multiplexor, and the overall power dissipation for both implementations of each example. Table 4 summarizes the percentage power improvements. The *Hyper-LP* implementations uniformly dissipate less power than the *Hyper* implementations. Power consumed by buses is reduced drastically in all examples (up to 80%). Up to 60% reduction in multiplexor power is seen due to reduced and more localized hardware sharing. The average reduction in bus, multiplexor, and total power is 62.9%, 38.5%, and 18.5%, respectively.

The power reduction comes at the cost of an increase in the number of units. However, since the communications are localized, the designs are more conducive to compact layout. Further, overhead elements such as multiplexors and buffers are reduced. Table 4 shows the estimated area penalty obtained in the *Hyper-LP* designs.

Table 4. Overall power reduction and area overhead.

Design	Percentage power reduction			Percentage change in area
	Bus	Mux	Total	
Cascade	80.0	59.6	25.9	-15.1
Direct form	65.4	44.5	23.7	+2.1
Wave digital	33.3	46.9	11.8	+7.2
DCT	50.0	39.5	10.4	-23.4
FFT	69.7	19.1	24.3	+0.8
Parallel IIR	78.7	21.4	15.1	+22.1
Average	62.9	38.5	18.5	-1.05

5. Conclusions

We have presented a technique for power reduction based on exploiting the locality in a given application. At the core of the approach is a partitioning and assignment strategy. It was seen that the proposed scheme improves the implementation in a variety of ways. The predominant effect is the reduction of accesses to highly-capacitive global buses. Our results showed average bus, multiplexor, and overall power reductions of 62.9%, 38.5%, and 18.5%, respectively, and low associated area overheads. The partitioning and assignment techniques have been integrated into the *Hyper-LP* system.

6. References

1. D.D. Gajski, *High-Level Synthesis: Introduction to Chip and System Design*, Boston, Kluwer Academic, 1992.

2. S. Wu, "A Hardware Library Representation for the Hyper Synthesis System," *Masters' Thesis*, University of California, Berkeley, Memorandum No. UCB/ERL M94/47, June 1994.

3. A. Chandrakasan, M. Potkonjak, R. Mehra, J. Rabaey, and R. W. Brodersen, "Optimizing Power Using Transformations," *IEEE Trans. on CAD*, Vol. 14, No. 1, Jan. 1995, pp. 12-31.

4. J. M. Rabaey, C. Chu, P. Hoang, and M. Potkonjak, "Fast Prototyping of Datapath-Intensive Architectures," *IEEE Design & Test of Computers*, June 1991, pp. 40-51.

5. M.C. McFarland and T.J. Kowalski, "Incorporating Bottom-up Design into Hardware Synthesis," *IEEE Trans. on CAD*, Vol. 9, No. 9, Sept. 1990, pp. 938-949.

6. E.D. Lagnese and D.E. Thomas, "Architectural Partitioning for System Level Synthesis of Integrated Circuits," *IEEE Trans. on CAD*, Vol. 10, No. 7, July 1991, pp. 847-860.

7. K. M. Hall, "An r-Dimensional Quadratic Placement Algorithm," *Management Science*, Vol. 17, No. 3, Nov. 1970, pp. 219-229.

8. L. Hagen and A. B. Kahng, "New Spectral Methods for Ratio Cut Partitioning and Clustering," *IEEE Trans. on CAD*, Vol. 11, No. 9, Sept. 1992, pp. 1074-1085.

9. B. Hendrickson and R. Leland, "The Chaco User's Guide, V. 1.0," *Tech Report SAND93-2339, Sandia National Lab*, Oct. 1993.

10. P. G. Paulin and J.P. Knight, "Force-Directed Scheduling for Behavioral Synthesis of ASIC's," *IEEE Trans. on CAD*, Vol. 8, No. 6, June 1989, pp. 661-679.

11. P. E. Landman and J. M. Rabaey, "Architectural Power Analysis: The Dual Bit Type Method," *IEEE Trans. on VLSI Systems*, Vol.3, No.2, June 1995, pp. 173-87.

12. R. Mehra and J. M. Rabaey, "Behavioral Level Power Estimation and Exploration," *Proc. of the Int'l. Workshop on Low-Power Design*, April 1994, pp. 197-202.

HYPER-LP: A System for Power Minimization Using Architectural Transformations

Anantha P. Chandrakasan[†]   Miodrag Potkonjak[††]   Jan Rabaey[†]   Robert W. Brodersen[†]

[†]EECS Department, University of California at Berkeley

[††]C & C Research Laboratories, NEC USA, Princeton

ABSTRACT

The increasing demand for "portable" computing and communication, has elevated power consumption to be the most critical design parameter. An automated high-level synthesis system, HYPER-LP, is presented for minimizing power consumption in application specific datapath intensive CMOS circuits using a variety of architectural and computational transformations. The sources of power consumption are reviewed and the effects of architectural transformations on the various power components are presented. The synthesis environment consists of high-level estimation of power consumption, a library of transformation primitives (local and global), and heuristic/probabilistic optimization search mechanisms for fast and efficient scanning of the design space. Examples with varying degree of computational complexity and structures are optimized and synthesized using the HYPER-LP system. The results indicate that an order of magnitude reduction in power can be achieved over current-day design methodologies while maintaining the system throughput; in some cases this can be accomplished while preserving or reducing the implementation area.

1. Introduction

The major VLSI design and research efforts until now have been focused on optimizing speed to realize computationally intensive real-time tasks such as video compression and speech recognition. As a result, many systems have successfully integrated various complex signal processing modules meeting users computation and entertainment demands. While these solutions have provided answers to the real-time problem, they have not addressed the rapidly increasing demand for portable operation. This strict limitation on power dissipation which portability imposes, must be met by the designer while still meeting ever higher computational requirements.

An example of a system requiring portability, moving beyond today's portable computers, is a future personal communications terminal described in [1] that will support speech communication and recognition, data transfer, computing services, and high-quality, full-motion video. The intense computational nature of the terminal functions coupled with the requirement of portability will place severe constraints on the total power being consumed. For example, a basic terminal with speech recognition and video decompression units implemented using current day technology will require about 20lbs of battery for 10hrs of operation [1]. Clearly, more power efficient means for implementing these functions need to be developed. One major degree of freedom available in optimizing design for such applications is that once real-time operation is achieved, there is *no* advantage in making the computation any faster. The goal then becomes one of reducing the power

consumption while maintaining the system throughput.

Fortunately, the increasing density of VLSI systems, due to sub-micron feature size scaling and high-density packaging such as multichip modules, has enabled the development of an architectural strategy which can be used to trade-off area and power for a fixed throughput [2].

In this work, we attack the problem of automatically finding computational structures that results in the lowest power consumption for a specified throughput given a high-level algorithmic specification. The basic approach is to scan the design space by utilizing various flowgraph transformations, high-level power estimation, and efficient heuristic/probabilistic search mechanisms. While transformations have been successfully applied recently in high-level synthesis with the goal of optimizing speed and/or area, the problem of power optimization has not been addressed. It will be shown that optimizing for power using transformations requires a different strategy than those used for speed or area optimization.

2. Sources of Power Consumption

In CMOS technology, there are three sources of power dissipation arising from: switching (dynamic) currents, short-circuit currents, and leakage currents. The switching component, however, is the only one which cannot be made negligible if proper design techniques are followed. The power consumption due to the switching of a CMOS gate with a load capacitor, C_L, is given by the following formula [3]:

$$P_{switching} = p_t (C_L * V_{dd}^2 * f) \qquad \text{(EQ 1)}$$

where f is the clock frequency, V_{dd} is the supply voltage, and p_t is the probability of a power consuming transition (or the activity factor). Probabilistic approaches have been proposed to estimate the internal node activities of a network given the distribution of the input signals (i.e. the probability the value is a "1" or "0") [4,5].

In our analysis and optimizations, we will refer to the energy per computation of a gate or module (e.g. an adder), which is given by:

$$\text{Energy} = P_{total} / f_{clk} = C_{avg} V_{dd}^2 \qquad \text{(EQ 2)}$$

where C_{avg} is the average capacitance being switched per clock cycle (i.e $C_{avg} = p_t * C_{Ltotal}$).

The energy consumed by a logic block per computation is therefore a quadratic function of the operating voltage, as verified experimentally for a number of logic functions and logic styles in [2]. It is clear that operating at the lowest possible voltage is most desirable, however, this comes at the cost of increased delays and thus reduced throughput. This is seen from Figure 1 which shows an experimentally derived plot of normalized delay *vs.* V_{dd} for a typical CMOS gate. Once again, the delay dependence on supply voltage was verified to be relatively independent of various logic

Reprinted from *IEEE/ACM International Conference on Computer-Aided Design*, pp. 300-303, November 1992.

functions and logic styles [2].

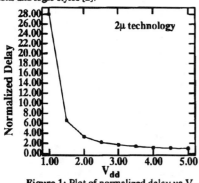

Figure 1: Plot of normalized delay vs. V_{dd}.

By modifying the architecture through a variety of transformations, however, the throughput can be regained, and thus a power savings can be accomplished while retaining the required functionality. It is also possible to reduce the power by choosing an architecture that minimizes the effective capacitance; through reductions in the number of operations, the average transition activity, the interconnect capacitance, and internal bit widths and using operations that require less energy per computation. It is these two strategies which will be pursued to minimize the power dissipation. There is, however, a strong interaction between optimizing capacitance and voltage for a fixed throughput. A lot of transformations will have conflicting effects on these parameters making the optimization a non-trivial task.

3. Transformations for Optimizing Power

Transformations are changes to the computational structure in a manner that the input/output behavior is preserved. The use of transformations makes it possible to explore a number of alternative architectures and to choose those which result in the lowest power. A brief summary of transformations for optimizing power is presented in this section. A detailed discussion of the effects of transformations on power is presented in [6].

3.1 Critical Path Reduction

This is probably the single most important type of transformations for power reduction. It is not only the most common type of transformation, but also often has the strongest impact on power. The basic idea is to reduce the critical path, so that supply voltage can be lowered while keeping the throughput fixed. The reduction of critical path is most often possible due to the exploitation of concurrency. Many transformations profoundly affect the amount of concurrency in the computation including pipelining and loop unrolling.

To illustrate the reduction of power using speedup techniques, consider a module with capacitance C running at a maximum frequency of f @ 5V ($\Rightarrow P_{ref} = C (5)^2 f$). By transforming this structure to a parallel architecture with two identical units (unrolling), the clock frequency can be dropped to half the original rate while maintaining the original throughput. Since the modules have twice the available time as the original case, the voltage can be dropped to 2.9V(where the delays increase by a factor of 2, Figure 1). The power of the transformed solution is $P_{par} = 2C (2.9)^2 f /2$ and a factor of $(5 / 2.9)^2$ reduction in power is achieved without sacrificing performance. The above analysis assumed that there is no overhead for parallelizing. In reality, the overhead due to routing and control must be taken into account. Even so, the quadratic dependence of voltage on power usually more than compensates for the

increase in capacitance resulting in an overall reduction of power. However at very low voltages (< 1.5V), the delays (and hence the overhead circuitry) increase very rapidly, causing the power to increase with further reduction of the supply voltage [2].

3.2 Reducing the Number of Operations

The most obvious approach to capacitance reduction, is to reduce the number of operations (and hence the number of switching events) in the data control flow graph. While this almost always has the effect of reducing the effective capacitance, the effect on critical path is case dependent. Transformations which directly reduce the number of operations in a data control flow graph include: common subexpression elimination, manifest expression calculation, loop merging, and distributivity.

3.3 Reducing the Transition Activity

Designs using static CMOS logic can exhibit spurious transitions due to finite propagation delays from one logic block to the next (called critical races or dynamic hazards). i.e. a node can have multiple transitions in a single clock cycle before settling to the correct logic value. The amount of extra transitions is a complex

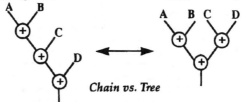

Chain vs. Tree

Figure 2: Reducing the glitching activity.

function of logic depth, input patterns, and skew. To minimize the "extra" transitions and power in a design, it is important to balance all signal paths and reduce the logic depth. For example, consider the two implementations for adding four numbers shown in Figure 2 (assuming a non-pipelined implementation). Assume that all primary inputs arrive at the same time. Since there is a finite propagation delay through the first adder for the chained case, the second adder is computing with the new C input and the previous output of A + B. When the correct value of A + B finally propagates, the second adder recomputes the sum. Similarly, the third adder computes three times per cycle. In the tree implementation, however, the signal paths are more balanced and the amount of extra transitions is reduced. The capacitance switched for a chained implementation is a factor of 1.5 larger than the tree implementation for a four input addition and 2.5 larger for an eight input addition. The above simulations were done on layouts generated by the LagerIV silicon compiler [7] using the IRSIM [8] switch-level simulator over 1000 random input patterns.

3.4 Reducing the Interconnect Capacitance

It is often possible to reduce the required amount of hardware, while preserving the critical path (or number of control steps) [9]. This is possible because after certain transformations, operations are more uniformly distributed over the available time, and thus a denser scheduling (effective utilization of the hardware) can be achieved. These transformations include retiming for resource utilization, associativity, distributivity and commutativity. Smaller capacitance is achieved because there are fewer interconnects and/or fewer functional elements and registers, which are obstacles during floorplanning and routing, which indirectly influence interconnect area and capacitance.

3.5 Operation Substitution

Certain operations inherently require less energy per computation than other operations. A prime example of this is strength reduction, often used in software compilers, in which multiplications are substituted for additions.

Another powerful transformation in this category is converting multiplications with constants into shift-add operations. Since multiplications with fixed coefficients are quite common in many signal processing applications (DCT, FFT, various types of filters, etc.), this transformation can prove to be beneficial.

3.6 Bit-width Optimization

The number of bits used can strongly affects all the key parameters of a design, including speed, area and power. A smaller bit-width typically results in fewer switching events (and hence lower capacitance), faster circuits (and hence lower supply voltage), and smaller area (and hence lower average interconnect length). Certain transformations (e.g. associativity and distributivity) can have a profound impact on the bit-width.

4. Power Estimation

The goal is to develop an objective function that is highly correlated to the final (and unknown) power dissipation of the circuit. The objective function should be very easy to compute since it has to be evaluated many times during the optimization process. Elaborate power estimation, while being much more accurate, will require the hardware mapping and compilation steps to convert a flowgraph to layout, making it impractical during the optimization process. Hence a model correlated to the power must be developed strictly from the flowgraph level.

The goal of power optimization in this work is to keep the throughput constant by allowing the supply voltage to vary. Given that the sample period is fixed, power optimization is equivalent to minimizing the total energy switched, $C_{total} V^2$, where V is appropriate voltage required to meet the initial throughput rate.

4.1 Capacitance estimate

Estimating the total capacitance being switched involves considering four components:

$$C_{total} = C_{exu} + C_{registers} + C_{control} + C_{interconnect} \quad \text{(EQ 3)}$$

The capacitance estimation is built on top of an existing estimation routine in HYPER that determines the bounds and activity of various execution, register and interconnect components as well as the active implementations area [10]. A brief description of the estimation routines is presented below.

4.1.1 Execution Units and Registers

The capacitance contributed due to the execution units is determined by multiplying (over all types of units utilized) the number of times the operation was performed per sample period with the average capacitance of the unit type. The total capacitance is hence given by:

$$C_{exu} = \sum_{i=1}^{numtypes} N_i \cdot C_i \quad \text{(EQ 4)}$$

where *numtypes* is the total number of operation types, N_i the number the times the operation of type *i* is performed per sample period, and C_i is the average capacitance being switched per operation of type *i*. The average capacitance for the various modules has been characterized as a function of bit-width (through SPICE and IRSIM simulations) for a uniformly distributed set of inputs.

In general the probabilities are not uniform, however, this assumption is made to simplify the cost evaluation. The effect of inter-module capacitance (between modules inside a datapath) is taken into account by incorporating an average loading capacitance during the characterization of the leaf-cells. It is important to note that the contribution due to the execution units is relatively independent of resource utilization (or the degree of time-sharing) since the required number of operations must be performed within the sample period. However, the amount of parallelism will affect the interconnect capacitance.

Registers are treated the same way as the execution units. The number of register accesses per sample period (read/write) is multiplied with the capacitance per register access to yield a register contribution given by $N_{registers} * C_{register}$. Once again, while the total number of registers is not important in calculating the register switching capacitance, it will affect floorplanning and therefore the interconnect capacitance.

4.1.2 Interconnect

A relatively accurate model for interconnect capacitance is important when performing power trade-offs since often the interconnects starts to dominate over the logic capacitance and restricts improvement in power that can be achieved. Determining the interconnect capacitance is a difficult task since we have to in essence emulate the partitioning, place, and route.

The goal of interconnect estimation is to estimate the inter-block (between macro blocks, such as between datapaths) routing capacitance. A statistical model based approach is used to predict the inter-block capacitance from high level parameters such as the number of global interconnects, active area and bit-width. This model for estimating the average interconnect capacitance switched requires the number of interconnects, N, the average activity, α, and the average interconnect length, L. The average interconnect length is obtained from statistical estimates of the final routed chip area and typical interconnect length distributions as a function of area [11]. The interconnect capacitance is then estimated as:

$$C_{interconnect} = \alpha * N * L * B * C_L \quad \text{(EQ 5)}$$

where C_L is the capacitance per unit length and B is the bit-width. A more extensive model for the interconnect is currently being developed.

4.1.3 Control Logic

From several circuits, it was observed that there is a strong correlation between the control capacitance switched and the total relevant capacitance (muxes, tri-states, registers, and any other module that requires control signals). Based on this information, a simple model is used to predict the total control capacitance as a function of high-level parameters. Notice that control contribution will be a function of the architecture style used.

4.2 Supply Voltage Estimation

We are interested in computing the power supply voltage at which the transformed flowgraph will meet the timing constraints. The initial flowgraph which meets the timing constraints is typically assumed to be operating at a supply voltage of 5V with a critical path of $T_{initial}$ (the initial voltage will be lower if $T_{initial} < T_{sampling}$). After each move, the critical path is re-estimated, and the new supply voltage at which the transformed flowgraph still meets the time constraint, $T_{sampling}$, is determined. For example, if the initial solution requires 10 control steps running at a supply voltage of 5V, then a transformed solution that requires only 5 control steps can run at a supply voltage of 2.9V while meeting the throughput constraint. This relationship (of delay-V_{dd}) was modeled using Neville's algorithm for rational function interpolation and extrapolation [12].

5. Optimization Algorithm

The transformation mechanism is based on two types of moves, global and local. While global moves optimize the whole DCFG simultaneously, local moves involve applying a transformation only on one or very few nodes in the DCFG. The most important advantage of global moves is, of course, a higher optimization effect; the advantages of local moves is their simplicity and small computational cost. We used the following global transformations (i) retiming and pipelining for critical path reduction (ii) associativity (iii) constant elimination and (iv) loop unrolling. In the library of local moves we have implemented three algebraic transformations: associativity, distributivity and commutativity.

The computational complexity analysis of the power minimization problem showed that even highly simplified versions of the optimization tasks are NP-complete. Two widely used alternatives for the design of high quality suboptimal optimization algorithms are probabilistic and heuristic algorithms. Both heuristic and probabilistic algorithms have several distinctive advantages over each other. While the most important advantage of heuristic algorithms is a shorter run time, probabilistic algorithms are more robust and have stronger mechanisms for escaping local minimas.

The algorithm for power minimization using transformations has both heuristic and probabilistic components. While the heuristic part uses global transformations, the probabilistic component uses local moves. The heuristic part applies global transformations one at the time in order to provide good starting points for the application of the probabilistic algorithm. The probabilistic algorithm conducts a probabilistic search in a broad vicinity of the solution provided by the heuristic part. The underlying search mechanism of the probabilistic part is simulated annealing.

6. Results

A summary of power improvement after applying transformations relative to an initial solution that met the required throughput constraint at 5V for several representative examples is shown in Table 1. The results indicate that a large reduction in power consumption is possible (at the expense of area) compared to present-day methodologies. Also interesting was the fact that the "optimal" final supply voltage for all the examples was much lower than existing and emerging standards and was around 1.5V.

Example	Power Reduction	Area Increase
RGB -> YUV	8	5
FIR Filter	11	1.1
DCT (8 point)	8	5
Speech Filter	8	6.4
Elliptical Filter	9	2.7
Wavelet Filter	10	2

Table 1: Summary of results.

Example	Power Reduction	Area Increase
Volterra Filter	8.6	1

Table 1: Summary of results.

7. Conclusions

The problem of power minimization is becoming a very important problem with the increasing demand for "portable" computing and communication and we have presented a high-level synthesis system, HYPER-LP, for optimizing power consumption in application specific datapath intensive circuits using a variety of architectural and computational transformations. The synthesis approach consisted of applying transformation primitives (from a library of local and global moves) in a well defined manner in conjunction with efficient high-level estimation of power consumption. The results indicate that an order of magnitude reduction in power is possible over current-day design methodologies while maintaining the system throughput, and it was found that the optimal supply voltage for minimizing power was much lower than existing standards (present-day 5V and emerging 3.3V) and was around 1.5V for most of the examples investigated. While this work has addressed some key problems in the automated design of low-power systems, there are still many open research problems like detailed power estimation, module selection, partitioning, and scheduling for power optimization.

Acknowledgments

This project was funded by DARPA. We would like to thank Shan-Hsi Huang for coding and supporting the library of local transformation primitives.

References

[1] A. Chandrakasan, S. Sheng, R.W. Brodersen, "Design Considerations for a Future Multimedia Terminal", in Third Generation Wireless Information Network, edited by D. Goodman and S. Nanda, Kluwer Academic Publishers, 1992.

[2] A. Chandrakasan, S. Sheng, R. Brodersen, "Low-power CMOS Digital Design", IEEE Journal of Solid-state circuit, pp. 473-484, April 1992.

[3] N. Weste and K. Eshragian, Principles of CMOS VLSI Design: A Systems Perspective, Addison-Wesley, MA, 1988.

[4] F. Najm, "Transition Density, A Stochastic Measure of Activities in Digital Circuits", DAC, pp. 644-649, 1991.

[5] A. Ghosh, S. Devadas, K. Keutzer, J. White, "Estimation of Average Switching Activity in Combinational and Sequential Circuits", DAC, pp. 253-259, 1992.

[6] A. Chandrakasan, M. Potkonjak, J. Rabaey, R. Brodersen, "An Approach to Power Minimization Using Transformations", IEEE VLSI Signal Processing Workshop, 1992.

[7] R. W. Brodersen, (ed.), "Anatomy of a Silicon Compiler", Klewer Academic Publishers, 1992.

[8] A. Salz, M. Horowitz, "IRSIM: An Incremental MOS Switch-level Simulator", Proceedings of the 26th ACM/IEEE Design Automation Conference, June 1989, pp. 173-178.

[9] M. Potkonjak and J. Rabaey, "Optimizing the Resource Utilization Using Transformations", Proc. IEEE ICCAD Conference, Santa Clara, pp. 88-91, November 1991.

[10] J. Rabaey, C. Chu, P. Hoang, M. Potkonjak, "Fast Prototyping of Data Path Intensive Architecture", IEEE Design and Test, Vol. 8, No. 2, pp. 40-51, 1991.

[11] D. Schultz, "The Influence of Hardware Mapping on High-Level Synthesis", M.S. report, U.C. Berkeley, 1992.

[12] W.H. Press, B.P. Flannery, S.A. Teukolsky, W.T. Vetterling: "Numerical Recipes in C", Cambridge University Press, 1988.

Scheduling with Multiple Voltages [1]

Salil Raje and M. Sarrafzadeh

Department of Electrical Engineering and Computer Science

Northwestern University

Evanston, IL 60208

{salil, majid}@eecs.nwu.edu

Abstract

This paper presents a low power design technique at the behavioral synthesis stage. A scheduling technique for low power is studied and a theoretical foundation is established. The equation for dynamic power, $P_{dyn} = V_{dd}^2 C_{load} f_{switch}$, is used as a basis. The voltage applied to the functional units is varied, slowing down the functional unit throughput and reducing the power while meeting the throughput constraint for the entire system. The input to our problem is an unscheduled *data flow graph* with a timing constraint. The goal is to establish a voltage value at which each of the operations of the *data flow graph* would be performed, thereby fixing the latency for the operation such that the total timing constraint for the system is met. We give an algorithm to minimize the system's power; the algorithm finds an optimal schedule. The timing constraint for our system could be any value greater than or equal to the critical path.

The experimental results for some High-Level Synthesis benchmarks show considerable reduction in the power consumption. Using 5V and 3V supply voltages we achieve a maximum reduction of approximately 40% given tight timing constraints. Similarly, we obtain a 46% reduction using 5V, 3V and 2.4V supply voltages. For larger timing constraints, the maximum reduction is about 64% using 5V and 3V supply voltages and a maximum reduction of about 74% using 5V, 3V and 2.4V supply voltages.

Keywords: *Data flow graph, Scheduling, Behavioral synthesis, low-power design, multiple voltages.*

---

[1] This work was supported in part by the National Science Foundation under Grant MIP-9207267.

1 Introduction

The advent of portable communication and computing services has stirred a great deal of interest in both commercial and research areas. The minimization of power consumption in these battery operated circuits has become the crucial issue in the design process. We are of the opinion that the power consumption issue should be addressed both at lower levels (logic synthesis and layout) and at the behavioral level. Lower levels provide new insights while decisions made at the higher levels of the design process are likely to have a greater impact on certain aspects of the design quality, power dissipation being one of them. Recent literature has placed a lot of emphasis on power minimization at the logic synthesis level, e.g. see [4], [5], [6], and at the architectural level [8]. A very comprehensive survey of all the low power techniques at the logic and architectural level can be obtained in [15].

There has been significantly less progress at the *behavioral level*. One of the contributions at the behavioral level, [8], relies on algorithmic transformations. The HYPER-LP system minimizes power consumption in application specific datapath intensive CMOS circuits using architectural and computational transformations. [11] minimizes the activity of the functional units by minimizing the changes of their input operands. [12] describes an allocation technique during behavioral synthesis for low power: reducing the power dissipated in the registers. [13] shows how to synthesize energy optimized circuits by multiplexing the signals onto buses such that the switching activity on these buses is minimized. [14] gives a register allocation algorithm based on compatibility of variables to reside in the same register. Switching activity for pairs of variables that could potentially share the same register is obtained. The problem of register sharing to minimize switching is then solved using a max-cost flows algorithm. [16] uses pipelining and module selection to minimize the power consumption subject to timing constraints. Using these and other high-level synthesis techniques they target design to 3.3V libraries and save 56% on power as compared to the original 5V implementation.

The effects of reducing the supply voltage on the speed of the functional module are quite well known. The curve in Figure 1 characterizes this effect for functional modules. The functional module throughput decreases as its supply voltage is reduced. The dynamic power consumed by a functional module depends on the square of the supply voltage applied to it, $P_{dyn} = V_{dd}^2 C_{load} f_{switch}$. Thus, reduction of supply voltage reduces the dynamic power consumed by the functional module but also has the undesirable effect of reducing its throughput. Attempts were made at the logic synthesis level and the architectural level to utilize this quadratic dependence on voltage. The idea is to maintain throughput at reduced supply voltages through hardware duplication or pipelining. By using parallel, identical units, the speed requirements

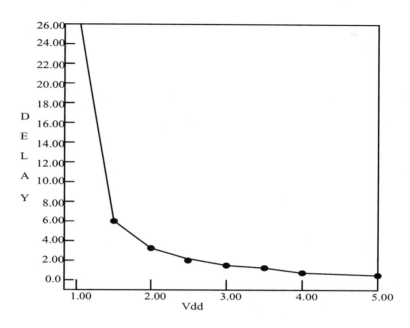

Figure 1: Normalized delay versus Vdd curve, [8]

on each unit are reduced, allowing for a reduction in voltage [8].

Nielsen et. al. in [1], [2] recently showed the practicality of adaptive scaling of supply voltages to reduce power consumption. Their technique combines self-timed circuitry with a mechanism that adaptively adjusts the supply voltage to the smallest possible, taking into account: process variations, operating conditions, and data dependent computation times. They monitor the activity of the circuit as proportionate to the amount of data in the input buffer and scale the supply voltage accordingly to maintain a constant throughput.

Usami and Horowitz in [10] describe a technique to reduce power without changing circuit performance by making use of two supply voltages only. Gates off the critical path are run at the lower supply to reduce power.

We present here a voltage-based technique to minimize power consumption at the *behavioral level*. We exploit the potential parallelism evident among operations in the *data flow graphs* that are used to define systems at this level of abstraction. Given a timing constraint, there are some operations in the *data flow graph* that could be slowed down by applying a smaller supply voltage and the system throughput would still meet the timing constraint. Assignment of supply voltages to each of the operations and thereby defining their latencies to minimize power consumption while still meeting the timing constraint is the problem we address here.

[10] uses two voltages at the logic synthesis level. In their work the power consumed by voltage level converters is significant when compared to the power consumed by gates. They therefore minimize the number of level converters. We, however, use multiple voltages at

the behavioral level to minimize power. Here, the power consumed by level converters is not negligible, as will be shown later, when compared to the power reduction when running the functional units at a lower supply voltage. [16] uses a module selection algorithm to select slower but power optimized modules in a non-critical path. They target designs to 3.3V libraries and multiple voltages are not used in the same design. However, their algorithm could easily be modified to use multiple voltages. The complexity of their algorithm would depend on the size of the module library. The fact that the same implementation could be run at various supply voltages above the threshold voltage could put a severe strain on the complexity of their algorithm.

A simple illustration of how voltage scaling at this level could minimize power is seen in Figure 2.

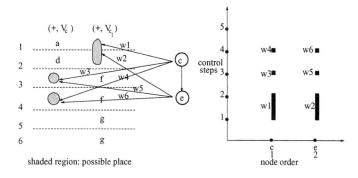

Figure 2: *Data flow graph*, Operation E, F, G, C are slowed down by applying Vdd'.

As can be seen from Figure 2, operation C, E, F and G could be slowed down by applying a smaller voltage Vdd', all the other operations are applied voltage Vdd and hence the total power consumed would reduce from $6Vdd^2 C_{load} f_{switch}$ to $2Vdd^2 C_{load} f_{switch} + 4Vdd'^2 C_{load} f_{switch}$. If $Vdd' = Vdd/2$, the power consumed reduces by $(6/(2 + 4/2^2) = 2)$ half.

The voltage scaling approach could be used at the system level as well. The nodes in the graph could represent different chips and each of these could be running at different supply voltages to minimize power but at the same time maintaining the system throughput. Figure 3(a) shows the voltage scaling approach used at system level and Figure 3(b) shows the voltage scaling approach used at the chip level.

This paper is organized as follows: Section 2 formulates the problems under a particular set of assumptions. Section 3 discusses some of the preliminary notations. Section 4 enumerates the steps in the algorithm without getting into the details of the proofs. Section 5 is a step by step illustration of the algorithm. Section 6 expounds on the proofs for optimality of the algorithm under certain constraints and some details involving each step of the algorithm. Section 7 relaxes some of the assumptions used by us to make scheduling with multiple voltages a feasible option.

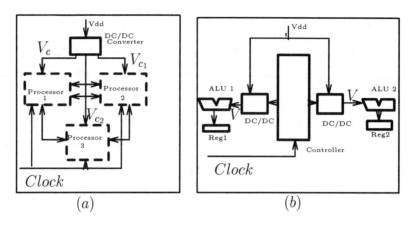

Figure 3: (a) voltage scaling approach at system level, (b) voltage scaling approach used at chip level.

Section 8 describes experimental results and we conclude in Section 8.

2 Problem Formulation

Data flow graph, DFG, represents a partial ordering restrictions of operations. It is represented as a directed acyclic graph $G = (V, E)$. $\mathcal{P}$ is the set of operations and there exists a mapping $Type$ from the nodes to τ, $Type : V \rightarrow \mathcal{P}$. A directed edge from node v_i to node v_j means execution of v_i must precede that of v_j; this is termed as the precedence constraint. A functional unit, F, is a hardware unit in the functional units library that can perform one or more operations, $F \subseteq \mathcal{P}$. The latency of a node, v_i, is the amount of execution time needed for $Type(v_i)$ and is dependent on the supply voltage V_{dd}. The supply voltage versus latency curve is assumed to be the same for all functional units.

A base voltage V_c (e.g., 5V) is fixed and it is the maximum supply voltage applied to any of the functional units. A base latency of all the operations, t_c, is the execution time for the functional units with a supply voltage V_c. A timing constraint for the system is given in terms of an integral multiple of t_c as in kt_c, where k would be an integer.

A set of allowable supply voltages S is obtained from the supply voltage versus latency curve. If $(V_{c_1}, 2t_c)$, $(V_{c_2}, 3t_c)$, $\cdots$, $(V_{c_i}, (i+1)t_c)$ are points on the voltage-latency curve then $S = \{V_c, V_{c_1}, V_{c_2}, \cdots, V_{c_i}\}$. The number of allowable voltages will depend on the technology and the designers preference. The curve could also dictate this choice as some of the voltages at higher latency points could get very close. See Figure 4.

Definition 2.1 *A critical path of a system is defined as the path in the DFG, $\{v_1, v_2, \cdots, v_k\}$, such that the summation of the latencies of the nodes in the path is maximal among all the paths of the DFG. The sum, C_p, is termed as the critical path length.*

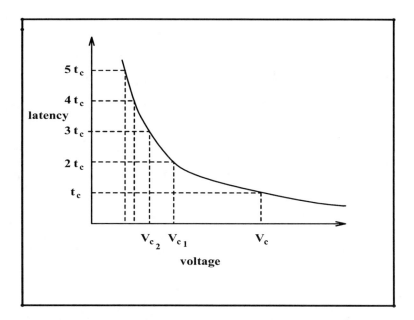

Figure 4: Voltage versus Latency curve and S. The range is 1V-5V in Figure 1.

The problem now could be defined as: Given a $DFG = G(V, E)$, t_c (or V_c), a timing constraint kt_c, where k is an integer, and the number of allowable supply voltages, obtain a mapping $\tau : V \to S$ to minimize

$$\sum_{v_i \in V} \tau(v_i)^2$$

such that the critical path length of the DFG is less than or equal to kt_c.

A voltage curve is said to have a **linear differential (LD)** property if the difference of the squares of consecutive voltages in S is a constant: $V_c^2 - V_{c_1}^2 = V_{c_1}^2 - V_{c_2}^2 = \cdots = V_{c_i}^2 - V_{c_{i+1}}^2 = \Delta V^2$. Our algorithm produces optimal schedules under this property.

3   Preliminaries

For the $DFG = G(V, E)$, we define the set of *fan-in* nodes of a node $v \in V$ as

$$\pi^-(v) = \{v' | (v', v) \in E\}$$

and the set of *fan-out* nodes as

$$\pi^+(v) = \{v' | (v, v') \in E\}.$$

The input node set I is the set of nodes for which $\pi^-(v) = \emptyset$. The output node set O is the set of nodes for which $\pi^+(v) = \emptyset$.

Each node, v, is associated with a delay, $d(v)$, equal in value to its latency. The original delay for all nodes is $d(v) = t_c$. This is because the algorithm will proceed with an initial assignment of V_c to all the nodes of the DFG and then decrement the voltage for some nodes, therefore minimizing the power. At each step of the algorithm, the current voltage assignment to node v is $\tau(v)$.

For the nodes $v \in O$ we can define a required arrival time $t_r(v) = kt_c$ which is the timing constraint for the system. Actual *in-time* for each of the nodes is defined recursively as,

$$t_i(v) = d(v) + max_{z \in \pi^-(v)} t_i(z)$$

Similarly, the actual *out-time* for each of the nodes is defined recursively as,

$$t_o(v) = d(v) + max_{z \in \pi^+(v)} t_o(z).$$

Therefore, the length of the longest path in DFG passing through the node v is given by,

$$l(v) = t_i(v) + t_o(v) - d(v).$$

An efficient way of implementing this procedure is to perform a *depth first search*. For each $v \in V$ we compute *slack*,

$$s(v) = kt_c - l(v).$$

During a slack computation we mark edges as being *critical* or *non-critical*. $(u, v) \in E$ is termed *critical* if either, $t_i(u) - t_i(v) = d(u)$ or $t_o(v) - t_o(u) = d(v)$. A *critical* edge (u, v) essentially means that there is at least one longest delay path in G through some node that goes through this edge.

The DFG is called *safe* if $\forall_{v \in V} s(v) \geq 0$, meaning that the critical path length of the DFG is less than kt_c. An assignment of voltages to nodes such that $\forall_{v \in V} s(v) \geq 0$ is called a *safe assignment*. Also, since for every node v, $\tau(v) \in S$, $l(v)$ is an integral multiple of t_c. $s(v)$ is an integral multiple of t_c as well.

Figure 5 illustrates the computation of slack given a timing constraint of $4t_c$ and an initial assignment of V_c to each of the nodes. The slack values for g, e and f are $s(c) = s(e) = s(g) = s(f) = t_c$ and for all the other nodes they are 0.

Notice that in Figure 5 the nodes such as c, g, e and f have a positive slack value. If the voltage assignment of c, g and e is changed from V_c to V_{c_1} such that $d(c) = d(g) = d(e) = 2t_c$, new slack values for all nodes in the DFG is 0. The DFG is still *safe*. The timing constraint for the system, $4t_c$, is satisfied and power consumption is reduced. Note also that we could have chosen c and f for a voltage assignment of V_{c_1} in which case the reduction in power consumption for the system would have been lesser than in the case described above.

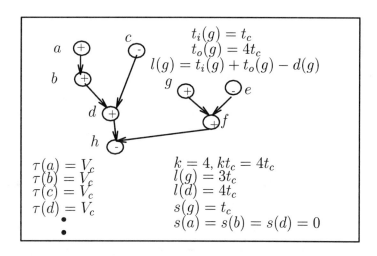

Figure 5: Illustration of slack computation.

4 Algorithm Overview

We present in this section an overview of the algorithm without going into the details of the proofs for optimality. Proofs are given in Section 6. The following are the steps of the algorithm:

- Step 1: *Initialization*

  Each of the nodes, $v \in V$, in $G(V, E)$ is originally assigned a voltage V_c (also denoted by V_{c_0}), $\tau(v) = V_c$. $d(v)$ value is therefore initialized.

- Step 2: *Computation of slack*

  Using *depth first search* calculate $l(v)$ values for each of the nodes $v \in V$ and hence obtain $s(v) = kt_c - l(v)$. Also mark the *critical* edges in this step.

- Step 3: *Constructing TG*

  Construct the *critical transitive closure*, $TG(V, E_c)$, a graph: $E_c = \{(u, v) \in V \times V : there$ *exists a path from u to v in G consisting of only critical edges* $\}$.

- Step 4: *Maximum slack value*

  Identify the maximum slack value s^* in the graph G and all the nodes, v, such that $s(v) = s^*$. If the maximum slack value $s^* = 0$ terminate the algorithm; we have obtained an optimal voltage assignment. The set of nodes with maximal slack value s^*, P, induces a subgraph $G_p(P, E')$ in TG: $E' = \{(u, v) \in P \times P : (u, v) \in E_c\}$.

- Step 5: *Weight assignment*

  If a node $v \in P$ is assigned a voltage $\tau(v) = V_{c_k}$ then assign a weight $W(v) = (V_{c_k}^2 - V_{c_{k+1}}^2)$ in G_p.

- Step 6: *Maximum weighted independent set for G_p*

  Obtain a maximum weighted independent set, $MWIS$, for G_p (this can be solved in polynomial time, since G_p is a transitively oriented graph).

- Step 7: *Reassigning voltages to nodes in the MWIS*

  Reassign voltages to nodes in the maximum weighted independent set obtained in the previous step. If a node in the $MWIS$, v, has a prior voltage assignment $\tau(v) = V_{c_k}$ then change this assignment to $\tau(v) = V_{c_{k+1}}$. The new assignment of voltages to nodes in P changes the delay values ($d(v)$ values) for nodes.

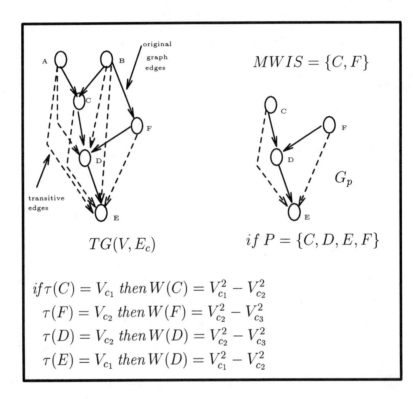

$$MWIS = \{C, F\}$$

$$G_p$$

$$TG(V, E_c)$$

$$if\, P = \{C, D, E, F\}$$

$$if\, \tau(C) = V_{c_1}\ then\, W(C) = V_{c_1}^2 - V_{c_2}^2$$
$$\tau(F) = V_{c_2}\ then\, W(F) = V_{c_2}^2 - V_{c_3}^2$$
$$\tau(D) = V_{c_2}\ then\, W(D) = V_{c_2}^2 - V_{c_3}^2$$
$$\tau(E) = V_{c_1}\ then\, W(D) = V_{c_1}^2 - V_{c_2}^2$$

Figure 6: *Critical transitive closure* of G, weight calculation for G_p and the $MWIS$.

- Step 8: *loop*

  Go to Step 2.

Figure 6 shows the construction of TG from G, weight calculation for the induced graph G_p, $P = \{C, D, E, F\}$ and the $MWIS$.

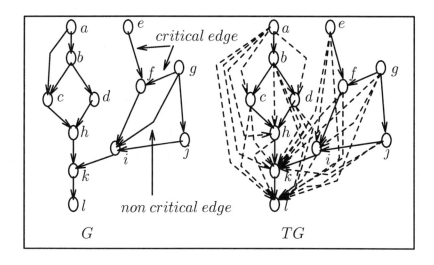

Figure 7: *Critical transitive closure* of G

5 Illustration

In this section we illustrate the flow of the algorithm using a sample DFG.

Figure 7 is the unscheduled DFG. Some of the *critical* and *non-critical edges* are shown. The *critical transitive closure* of G, TG is shown. Let's assume $S = \{5V, 3V, 2.4V\}$. Initialize $\tau(v) = 5V$ $(V_c = 5V)$ $\forall_{v \in V}$. $d(v) = t_c$ $\forall_{v \in V}$. Assume $k = 7$.

Slack computation will yield:

$s(a) = t_c$	$s(b) = t_c$	$s(c) = t_c$
$s(d) = t_c$	$s(h) = t_c$	$s(k) = t_c$
$s(l) = t_c$	$s(e) = 2t_c$	$s(g) = 2t_c$
$s(f) = 2t_c$	$s(i) = 2t_c$	$s(j) = 2t_c$
$s^* = 2t_c.$		

The set of nodes with maximum slack values is $P = \{e, f, g, i, j\}$. G_p is the induced subgraph of TG on P. Figure 8 shows G_p.

The maximum weighted independent set in G_p is $MWIS = \{e, g\}$. The voltage assignment of e and g is changed from $5V$ to $3V$. $d(e) = 2t_c$ and $d(g) = 2t_c$. This concludes one iteration of the loop in the algorithm.

At this point we recompute the slack values. All the slack values are t_c, P is the set V itself, $P = V$. G_p on the set V is the original TG, *critical transitive closure* of G.

The weights on the nodes are computed to be:

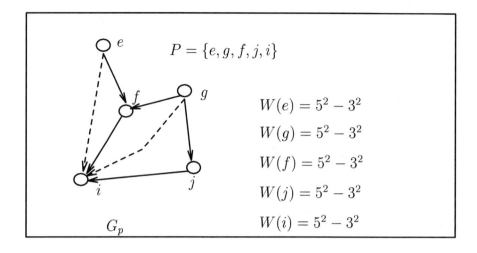

$$P = \{e, g, f, j, i\}$$

$$W(e) = 5^2 - 3^2$$
$$W(g) = 5^2 - 3^2$$
$$W(f) = 5^2 - 3^2$$
$$W(j) = 5^2 - 3^2$$
$$W(i) = 5^2 - 3^2$$

Figure 8: Induced subgraph G_p and weight assignment.

$W(a) = 5^2 - 3^2 = 16$	$W(b) = 16$	$W(c) = 16$
$W(d) = 16$	$W(h) = 16$	$W(k) = 16$
$W(l) = 16$	$W(e) = 3^2 - 2.4^2 = 3.24$	$s(g) = 3.24$
$W(f) = 16$	$W(i) = 16$	$W(j) = 16$

The $MWIS$ can be seen to be $\{c, d, f, j\}$. The voltage assignment for the nodes in the $MWIS$ is then changed. The initial $\tau(c) = 5V$ is changed to $\tau(c) = 3V$, the voltage assignment of all other nodes remains the same. The delay values of all nodes in the $MWIS$ are incremented by t_c. The voltage assignment for all nodes is now shown:

$\tau(a) = 5V$	$\tau(b) = 5V$	$\tau(c) = 3V$
$\tau(d) = 3V$	$\tau(h) = 5V$	$\tau(k) = 5V$
$\tau(l) = 5V$	$\tau(e) = 3V$	$\tau(g) = 3V$
$\tau(f) = 3V$	$\tau(i) = 5V$	$\tau(j) = 3V$

This concludes another iteration of the loop of the algorithm. The slack values of all nodes in G is 0 and the above gives us the final voltage assignment for nodes.

6 Details of the Algorithm and Correctness Proofs

A series of lemmas and certain key observations will be proved. The correctness of the algorithm will be shown in subsection 6.1 and the time complexity of the algorithm along with some discussion of a few well known algorithms used as subroutines in our algorithm will be discussed in subsection 6.2.

6.1 Correctness Proofs

Let's begin by making some key observations:

- Given a voltage assignment $\tau : V \to S$, the path delays are integral multiples of t_c.

- Given a voltage assignment $\tau : V \to S$, the node slacks are integral multiples of t_c.

- Changing the voltage assignment of a node v from $\tau(v) = V_{c_k}$ to $\tau(v) = V_{c_{k+1}}$, the delay associated with v increases by t_c. In other words $d(v) = d(v) + t_c$, $s(v) = s(v) - t_c$.

- The DFG is *safe* if the delay of the longest path, $l(v)$, through each node, v, is less than or equal to kt_c.

For the sake of understanding and notational simplicity we use primes (′) to denote the new values. For example, when we make a change in the voltage assignment of a node v from V_{c_k} to $V_{c_{k+1}}$, we would denote the original assignment as $\tau(v) = V_{c_k}$ and the new assignment as $\tau'(v) = V_{c_{k+1}}$. $d'(v) = d(v) + t_c$, $l'(v) = l(v) + t_c$ and $s'(v) = s(v) - t_c$. A change in the voltage assignment would be *safe* if $\forall_{v \in V} s'(v) \geq 0$.

Lemma 6.1 *If given a voltage assignment, τ, such that $\forall_{v \in V} s(v) \geq 0$, then a change in assignment for a node $w \in V$, from $\tau(w) = V_{c_k}$ to $\tau'(w) = V_{c_{k+1}}$ is safe if and only if $s(w) \geq 0$.*

Proof: The *only if* direction is obvious. We will prove the *if* part: $\forall_{v \in V} s(v) \geq 0$ means that the DFG is originally *safe*. $l(w) \leq kt_c - t_c$ or $l(w) \leq (k-1)t_c$. Changing the assignment from $\tau(w) = V_{c_k}$ to $\tau'(w) = V_{c_{k+1}}$ increments the delay value $d(w)$ by t_c. So, $l'(w) \leq kt_c$, implying $s'(w) \geq 0$. Consider any other node $u \in V$. If there exists a path passing through u that passes through w, then two cases arise:

- Case 1.: The path is a longest path through u; so, $l(u) \leq l(w)$ and therefore $s(u) > 0$. So, $s'(u) = s(u) - t_c \geq 0$.

- Case 2: The path length is shorter than $l(u)$. In which case $l'(u) = l(u)$; so, $s'(u) = s(u) \geq 0$.

If there is no path that goes through both u and w then $s'(u) = s(u) \geq 0$. $\square$

Corollary 6.1 *If $\forall_{v \in V} s(v) \geq 0$, a change in assignment of any node w from $\tau(w) = V_{c_i}$ to $\tau'(w) = V_{c_j}$, $j > i$ is safe if and only if $s(w) \geq (j-i)t_c$.*

Proof: A change from $\tau(w) = V_{c_i}$ to $\tau'(w) = V_{c_j}$, $j > i$, implies $d'(v) = d(v) + (j - i)t_c$. We could use $j - i$ steps to reduce the voltage assignment for w from V_{c_i} to V_{c_j}. In the first step, reduce the voltage assignment to $V_{c_{i+1}}$, in the second step reduce it to $V_{c_{i+2}}$ and so on. Applying Lemma 6.1 at each step proves this corollary. $\square$

At this point we introduce the concept of **critical independence:**

Consider a set $I = \{v_1, v_2, \cdots, v_k\}$ of nodes in G such that there does not exist a single path consisting of all *critical edges* from one node in I to another in I. We will call this set I *critically independent*. Two nodes are termed as *critically independent* if there does not exist a single path in G that goes through both these nodes and consists solely of *critical edges*. A set I is *critically independent* if all nodes in the set are pair-wise *critically independent*.

Lemma 6.2 *If $u, v \in V$ are critically independent, then if there exists a path from u to v in G, then $t_i(u) > t_i(v) + d(u)$ and $t_o(v) > t_o(u) + d(v)$.*

Proof: Since there exists a path from u to v in G we know that $t_i(u) \geq t_i(v) + d(v)$ and $t_o(v) \geq t_o(u) + d(v)$. $t_i(u) = t_i(v) + d(v)$ would imply $(u, v) \in E$ and (u, v) would be a *critical edge*. So, $t_i(u) > t_i(v) + d(u)$. Similarly, $t_o(v) > t_o(u) + d(u)$. $\square$

Corollary 6.2 *Let $I = \{v_1, v_2, \cdots, v_k\}$ be a critically independent set of G. Without loss of generality, if there exists a path in G that goes through the nodes in I in the same order that they appear in I, then $t_i(v_i) > t_i(v_j) + (j - i)t_c$ ($k \leq j > i$) and $t_o(v_j) > t_o(v_i) + (j - i)t_c$.*

Proof: We know that $t_i(v_{l-1}) > t_i(v_l) + d(v_{l-1})$ by Lemma 6.2. $d(v_{l-1}) \geq t_c$ and so $t_i(v_{l-1}) > t_i(v_l) + t_c$ Also, $t_i(v_{l-2}) > t_i(v_{l-1}) + t_c > t_i(v_l) + 2t_c$ and so on. So, $t_i(v_i) > t_i(v_j) + (j - i)t_c$. Similarly, $t_o(v_j) > t_o(v_i) + (j - i)t_c$. $\square$

Lemma 6.3 *If there exists a critically independent set I in G and for all nodes $v \in I$, $s(v) > 0$, then if $\tau(v) = V_{c_k}$, a change $\tau'(v) = V_{c_{k+1}}$ is safe if the DFG is originally safe. Also, if $v \in I$, then $s'(v) = s(v) - t_c$.*

Proof: The change has increased the delay value of each node in I by t_c. Given this, we want to show that the DFG is still safe. We have two parts to this proof: one, showing that for every node in I, the new slack value is at least 0; two, showing that for a node not in I the new slack value is at least 0.

- Consider first any node, $v \in I$. Consider any path $\Phi(v)$ in G that goes through node v. If we can show that the length of this path after the new assignment is less than or equal to kt_c then we have shown that $s(v) \geq 0$. Again, there are two cases:

  - 1. If $\Phi(v)$ does not pass through any other node in I. The path length of $\Phi(v)$ increases by at most t_c and since $s(v) > 0$, $s(v) \geq t_c$ we know that the original path length of $\Phi(v)$ could at most be $kt_c - t_c$; so, the *new path length of* $\Phi(v)$ is at most kt_c.

  - 2. If $\Phi(v) = \{\cdots, v_1, v_2, \cdots, v, \cdots, v_l, \cdots\}$, where $v_1, v_2, \cdots, v_l \in I$. Let v_1 be the first node of $\Phi(v)$ that is also in I, v_2 be the second such node and so on. Without loss of generality, let v be the i th such node, $v = v_i$. The path length of $\Phi(v)$ has been increased by lt_c. We know from Corollary 6.2 that: $t_i(v) > t_i(v_l) + (l - i)t_c$ and $t_o(v) > t_o(v_1) + (i - 1)t_c$. So, $t_i(v) + t_o(v) > t_i(v_l) + t_o(v_1) + (l - 1)t_c$. So, $l(v) > t_i(v_l) + t_o(v_1) + (l - 1)t_c - d(v)$. Notice that $t_o(v_l)$ is greater than the length of the part of the path $\Phi(v)$ above (nodes in $\Phi(v)$ preceding v_l in the order) and including v_l and $t_i(v_1)$ is greater than the part of the path $\Phi(v)$ below and including v_1. So, $t_i(v_1) + t_o(v_l) - d(v)$ is greater than or equal to the length of $\Phi(v)$ itself. So, $l(v) > length$ of $\Phi(v) + (l - 1)t_c$. Since the path length of $\Phi(v)$ has been increased by lt_c after the change; *the new path length of* $\Phi(v)$ is less than $l(v) + t_c$. Implying that the new path length of every such path passing through v and one or more other nodes of I is less than $l(v) + t_c$. Now, the delay value of v has been increased by t_c, $d'(v) = d(v) + t_c$. The longest path through v does not pass through any other node in I. So, $l'(v) = l(v) + t_c$. Therefore, $s'(v) = s(v) - t_c \geq 0$.

- Consider now any node $u \notin I$. Consider any path $\Phi(u)$ in G that goes through node u. Again two cases arise:

  - 1. $\Phi(u)$ does not go through any node in I. So, length of $\Phi(u)$ remains unchanged.

  - 2. Consider that the longest path through u cannot pass through more than one node in I. If it does pass through say $v \in I$ and $w \in I$, then there is a path consisting of solely *critical edges* passing through v and w and so v and w would not be *critically independent*. Given this, we know any longest path through u could increase its length by t_c only. Now consider a path, $\Phi(u)$ that passes through u going through more than one node in I. So, *length of* $\Phi(u) < l(u)$. $\Phi(u) = \{\cdots, v_1, \cdots, v_2, \cdots, v_i, \cdots, u, \cdots v_{i+1}, \cdots, v_l, \cdots\}$, where v_1 is the first node of $\Phi(v)$ that is also in I, v_2 be the second such node and so on. u lies between v_i and v_{i+1}. By the second case in part one of the lemma we know that $t_o(v_i) > t_o(v_1) + (i-1)t_c$, but $t_o(u) \geq t_o(v_i)$. So, $t_o(u) > t_o(v_1) + (i-1)t_c$. Also, $t_i(v_{i+1}) > t_i(v_l) + (l-(i+1))t_c$, but $t_i(u) \geq t_i(v_{i+1})$. So, $t_i(u) > t_i(v_l) + (l-i-1)t_c$. Therefore, $t_o(u) + t_i(u) > t_o(v_1) + t_i(v_l) + (l-2)t_c$. Again, by the second case of the first part of lemma, $l(u) > $ *length of* $\Phi(u) + (l-2)t_c$. Increasing *length of* $\Phi(u)$ by lt_c, would make $l(u) > $ *new length of* $\Phi(u) - 2t_c$. So, $l(u) \geq$ *new length of* $\Phi(u) - t_c$ or *new length of* $\Phi(u) \leq l(u) + t_c$. Since path length of longest path through u also increases by only t_c. $l'(u) \leq l(u) + t_c$. This is an important result and will be used again in Lemma 6.6.

□

Notice that the set I is essentially an independent set in the *critical transitive closure* of G, TG. A *critical transitive closure*, $TG(V, E_c)$ of a graph $G(V, E)$, is constructed as follows: $E' = \{(u, v) \in V \times V :$ *there exists a path from u to v consisting of solely critical edges in G*$\}$. If two nodes u and v are independent in TG (there does not exist edges (u, v) or $(v, u) \in E_c$), it implies that there is no single path in G that passes through both u and v consisting solely of *critical edges*. The set of nodes I would be pairwise independent in TG.

Consider a subgraph, $G_p(P, E_p)$, induced on TG on a set of nodes P; $E_p = \{(u, v) \in P \times P : (u, v) \in E_c\}$.

Lemma 6.4 *If $u, v \in P$, then u, v are independent in G_p ($(u, v) \notin E_p$) if and only if u, v are critically independent in G.*

Proof: If $(u, v) \in E_p$ implies that $(u, v) \in E_c$, meaning that (u, v) are not *critically independent* in G. So, if u, v are *critically independent* in G $(u, v) \in E_p$. Now, if u, v are not *critically independent* in G, implies that $(u, v) \notin E_c$, meaning that $(u, v) \notin E_p$. □

Corollary 6.3 *A set of nodes $I \subseteq P$ are independent in G_p if the set of nodes is critically independent in G.*

Proof: Use Lemma 6.4 for each pair of nodes in I. □

Lemma 6.5 *If I is an independent set of G_p such that for all nodes $v \in I$, $s(v) > 0$, then if $\tau(v) = V_{c_k}$, a change $\tau'(v) = V_{c_{k+1}}$ is safe if the DFG is originally safe.*

Proof: This follows from Corollary 6.3 and Lemma 6.3. □

Lemma 6.6 *If MI is a maximal independent set of G_p such that for all nodes $v \in MI_p$, $s(v) > 0$, then if $\tau(v) = V_{c_k}$, a change $\tau'(v) = V_{c_{k+1}}$ is safe if the TG is originally safe and for all nodes $w \in P$, $s'(w) = s(w) - t_c$.*

Proof: A maximal independent set of a graph is defined as the set of pair wise independent nodes that is maximal in size. By Lemma 6.3 if $v \in MI$, then $s'(v) = s(v) - t_c$. By Lemma 6.3 we know that for every node $w \notin MI$, $s'(w) \leq s(w) - t_c$. Also, we will prove that for every node $w \in P$ and $w \notin MI$, there exists at least one longest delay path in G through w that passes through exactly one node, say $v \in MI$ implying that $s'(w) = s(w) - t_c$.

We will first show that a longest path going through any such node w cannot pass through more than one node in MI. Suppose there exists a path in G going through w that passes through $u, v \in MI$. This implies that there exists a path through u and v in G consisting of only *critical edges*. This is a contradiction since u and v are *critically independent*.

Since MI is maximal there is some node $v \in MI$ such that either $(w, v) \in E_p$ or $(v, w) \in E_p$. Implying that the longest path in G passing through w passes through at least one node $v \in MI$.

Therefore, there exists at least one longest path in G through w that passes through exactly one node $v \in MI$. □

Corollary 6.4 *If P is a set of all nodes, v, such that $s(v) = s^*$, the maximum slack value in G, then if MI is a maximal independent set of G_p such that for all nodes $v \in MI$, $s(v) > 0$, then if $\tau(v) = V_{c_k}$, a change $\tau'(v) = V_{c_{k+1}}$ is safe if the G is originally safe and for all nodes $u \notin P$, $s'(u) = s(u)$.*

Proof: Let's consider a node $w \notin P$ ($s(w) < s^*$). Let $\Phi(u)$ be some path that passes through u in G. Let, as in Lemma 6.3, $\Phi(u) = \{\cdots, v_1, \cdots, v_2, \cdots, v_i, \cdots, u, \cdots, v_{i+1}, \cdots, v_l\}$, where $v_1, v_2, \cdots, v_l \in MI$. As in Lemma 6.3, *length of $\Phi(u) < l(v_i) - (l - 1)t_c$*. Since, $s(u) > s^*$, $l(u) > l(v_i)$ or $l(u) \geq l(v_i) + t_c$. So, *length of $\Phi(u) + (l - 1)t_c + t_c < l(u)$* or *length of $\Phi(u) + lt_c < l(u)$*. Increasing $\Phi(u)$ by lt_c, *new length of $\Phi(u) < l(u)$*. Therefore, $l'(u) = l(u)$, and $s'(u) = s(u)$. □

We have now established that if we use the set of nodes, P, with maximum slack value, $s^* > 0$, and obtain the maximal independent set MI and reduce the voltage assigned to the nodes in MI to a step lower in S then the slack value of all nodes in P decrease exactly by t_c and the slack values of all nodes not in P remain unchanged.

The lemmas and corollaries we have proved upto this point hold for any monotonic voltage versus delay curve, where execution delay for a functional unit increases as supply voltage decreases. We will prove that our algorithm will produce optimal results if the set of allowed voltages, S, obeys the LD property.

LD property : *The difference of the squares of consecutive voltages is a constant. If the constant is ΔV^2, then $V_{c_i}^2 - V_{c_{i+1}}^2 = \Delta V^2$.*

Notice that if S obeys LD property , then the weights that we assign to the nodes in G_p are exactly the same. A maximum unweighted independent set in G_p is, therefore, also a maximum weighted independent set in G_p. The implication is that minimizing $\sum_{v_i \in V} \tau(v_i)^2$ is the same as maximizing the total amount of delay that we assign on the nodes of G. So, the cost function reduces to maximize $\sum_{v_i \in V} d(v_i)$.

Lemma 6.7 *Given, G, such that the maximum slack value is s^*, the maximum slack value in G reduces to $s^* - t_c$, if and only if the delay values of the nodes of at least one of the maximal independent set in G_p are increased by a minimum of t_c.*

Proof: If there is not a single maximal independent set in G_p for which the delay values of all the nodes are increased by t_c, then the longest path through at least one of the nodes, $v \in P$ remains unchanged. Therefore, $s'(v) = s(v) = s^*$. So, maximum slack value in G would not reduce to $s^* - t_c$. On the other hand, by Lemma 6.6, if the delay values of all nodes in one of the maximal independent set is increased by t_c, then the maximum slack value in G would be $s^* - t_c$. $\square$

Corollary 6.5 *Given, G, such that the maximum slack value is s^*, the maximum slack value in G reduces to $s^* - t_c$, if and only if the voltages assigned to the nodes, v, of at least one of the maximal independent set in G_p are reduced to a step lower voltage in S.*

Proof: Follows from Lemma 6.7 and the fact that an assignment of a step lower voltage in S to a node increases the delay value of the node by t_c. $\square$

If we denote $G(s^*)$ as the original graph with maximum slack value of s^*, then to obtain a voltage assignment that will give us zero slack values on all nodes, we could, by Corollary 6.5,

- assign a step lower voltage in S to all nodes in some maximal independent set of $G(s^*)_p$ and

- then obtain an assignment for $G(s^* - t_c)$ to give us zero slack values on all nodes in $G(s^* - t_c)$.

The Corollary 6.5, therefore, suggests the inductive algorithm as proposed by us, wherein, in each step the nodes of P are assigned delay values such that the maximum slack value in the graph G reduces by a unit amount t_c.

Next, we will show that optimizing each step of the inductive algorithm, will indeed give us an optimal algorithm for the case where the voltage versus the delay curve is linear.

Theorem 6.1 *We are given $G(s^*)$ and an assignment τ. Obtain two distinct assignments τ_1' and τ_2' by assigning delay t_c to each of the nodes of two different maximal independent sets, MIS_1 and MIS_2 in $G(s^*)_p$ respectively giving us $G(s^* - t_c)_1$ and $G(s^* - t_c)_2$ respectively. (We do not have two different graphs in the classical sense, only two different assignments to nodes of a graph) The amount of delay we can assign to nodes in $G(s^* - t_c)_1$, such that the assignment is safe, is the same as the amount of delay we can assign to nodes in $G(s^* - t_c)_2$ such that the assignment is safe.*

Proof: We know from Lemma 6.6 that an assignment, such as stated in the Theorem, to nodes in the maximal independent set reduces the slack values of all nodes in P by t_c. Also, by Corollary 6.4, that if $u \notin P$, then $s'(u) = s(u)$. So, in graphs $G(s^* - t_c)_1$ and $G(s^* - t_c)_2$, if $u \notin P$ (P being the set of nodes with slack values s^* in the original graph, $G(s^*)$.) the slack value for u is same. Also, if $v \in P$, then the slack value for v in both graphs is $s^* - t_c$ after the new assignments. Meaning, that the new slack values of each of the nodes is exactly the same in both the graphs. So, any delays that we add on nodes of one of the graphs, if is a *safe* assignment, is also a *safe* assignment for the other graph. So, the total amount of delay we can add on to nodes in one graph such that the assignment is *safe* is equal to the total amount of delay we can add on to nodes in the other graph such that the assignment is safe. □

Theorem 6.2 *Given the LD property for S, the algorithm presented in the previous section gives an optimal solution.*

Proof: By Lemma 6.7, all nodes of at least one maximal independent set in $G(s^*)_p$ are assigned delay t_c. Also, by Theorem 6.1, the choice of any one of the maximal independent set over some other does not affect the total amount of delay we can add on nodes in $G(s^* - t_c)$. Therefore, an algorithm that assigns maximum amount of delay on the nodes should assign delay t_c to nodes in a maximum independent set of $G(s^*)_p$ and inductively find an assignment that maximizes the amount of delay assigned to nodes in $G(s^* - t_c)$. If the set S complies with the LD property, then maximizing the delay assignment to nodes in $G(s^*)$ also minimizes $\sum_{v_i \in V} \tau(v_i)^2$. $\square$

6.2 Time Complexity

The time complexity of our algorithm is $O(knmlog(n^2/m))$, where kt_c is the timing constraint, n the number of nodes and m the number of edges in the DFG. We have shown in the above subsection that the maximum slack value in G is reduced by t_c through each iteration of the loop. Since the maximum slack value of G at the very outset of the algorithm cannot exceed kt_c, there are k iterations of the loop. The initial voltage assignment takes only $O(n)$ time.

The following sub-subsection will be devoted to showing that every single step of each iteration has an asymptotic upper bound of $O(nmlog(n^2/m))$ time:

- Step 2:*Computation of slack*

  We make two calls to DFS to compute t_i and t_o values for all the nodes. *Critical edges* are marked during the DFS. This step, therefore, takes only $O(n + m)$ time.

- Step 3: *Constructing TG*

  The construction of $TG(V, E_c)$ adds *critical transitive edges* to $G(V, E)$. The following psuedocode adds *critical transitive edges* to a graph, G, that has edges marked as *critical* or *non-critical*:

  The **Procedure** $\mathcal{A\_L}$ sets $R(v)$ equal to the set of nodes that can be reached from v by a simple path consisting of solely critical edges. The topological sorting of nodes in G takes $O(n+m)$ time. Let's suppose that the maximum in-degree of a node in G does not exceed d_{in}. Then the number of nodes that would be added to a set $R(v)$ in **Procedure** $\mathcal{A\_L}$ cannot exceed the number of transitive edges to v times d_{in}. So, if m' is the cardinality of E_c, then the loop starting from statement 4 and ending at statement 6 in **Procedure** $\mathcal{A}$ takes $O(m'd_{in})$ time. Since, cardinality of $R(v)$ also does not exceed the number of transitive edges to v in TG, the loop starting from statement 7 and ending at statement 11 in **Procedure** $\mathcal{A}$ takes $O(m'd_{in})$ as well. The time complexity for constructing TG is, therefore, $O(m'd_{in})$. Notice that $m' \leq n^2$, in general, and d_{in} for most $DFGs$ is a small constant.

- Step 4: *Maximum slack value*

  Obtaining the maximum slack value in the current graph and the set P takes linear time, $O(n)$. Construction of G_p from TG takes $O(n + m)$ time.

Procedure $\mathcal{A}$: Add_C_Transitive_Edges()

(1)　*Initialize* the L to be an array of all sinks in G.

(2)　*Initialize* the $R(v) = \{v\}$, for each node v in V

(3)　*Topologically sort* the nodes in G using only *critical edges*

　　　　　　　　　　　　　/∗ Only *critical edges* are followed in the sort.∗/

(4)　**for** (all nodes, v, in G in *reverse topological order*) **do**

(5)　　　C_Reach_L(v);

(6)　**endfor**

(7)　**for** (all nodes v in V) **do**

(8)　　　**for** (all nodes u in $R(v)$ except v) **do**

(9)　　　　　**ADD** (v, u) to E_c

(10)　　　**endfor**

(11)　**endfor**

Procedure $\mathcal{A}\_\mathcal{L}$: C_Reach_L(v)

(1)　　L = Adjacent_Out_C_Nodes(v);

　　　　　　　　　/∗ gives all the nodes that have *critical edges* going out of v ∗/

(2)　　**for** (all nodes u in L) **do**

(3)　　　　$R(v) = R(u) \cup R(v)$;

(4)　　**endfor**

- Step 5: *Weight assignment*

  This also takes $O(n)$ time.

- Step 6: Maximum weighted independent set

  G_p is a transitively oriented graph. Let m' be the number of edges in G_p, number of nodes in G_p is $|P|$ and is less than n. *Maximum weighted independent set* for a transitively oriented graph is computed using network flows, refer to [9]. The following are the steps involved:

  1. Add two new nodes s and t and edges (s, a) and (b, t) for every source a and every sink b of G_p.

  2. These edges and all edges $(u, v) \in E_p$ have lower capacities 0 and upper capacities ∞.

  3. Each node $v_i \in P$ is split into two nodes v_i^1 and v_i^2, such that all incoming edges of v_i have v_i^1 as their endpoint, and all outgoing edges of v_i have v_i^2 as starting point. The nodes v_i^1 and v_i^2 are joined by an edge (v_i^1, v_i^2) with lower capacity $W(v_i)$ (the weight of v_i) and upper capacity ∞.

  An example of this construction is given in Figure 9

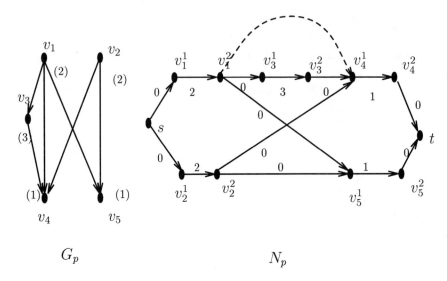

$$G_p \qquad\qquad\qquad N_p$$

Figure 9: A weighted transitively oriented graph, G_p, and its associated flow network, N_p.

Note that all transitive edges (a, b) (i.e. for which there exist edges (a, c) and (b, c)) can without loss of generality be deleted to be obtain the following result:

Theorem 6.3 *The minimum flow value in N_p equals the maximum weighted independent set of G_p.*

For proofs refer to [9]. Once we have the min-flow, (or max-cut, by the min-flow max-cut theorem), we can easily obtain the nodes v_i such that (v_i^1, v_i^2) lies on the max-cut.

After removal of transitive edges, the number of edges in N_p is $O(m)$ only. The algorithm for removal of transitive edges is similar to the one presented for adding transitive edges and will run in $O(m'd_{in})$ time.. The best known min-flow algorithm takes $O(VElogV^2/E)$ time in a graph with V nodes and E edges. Therefore, obtaining $MWIS$ in G_p takes no more than $O(nmlog(n^2/m))$ time.

- Step 7: *Reassigning voltages to nodes in the MWIS*
  This should also take $O(n)$ time.

Thus we have shown that no single step of the loop takes more than $O(nmlog(n^2/m))$ time.

7 Experimental Results

7.1 Gains

Experiments were conducted using some of the High-Level Synthesis benchmarks; results have been tabulated in Table 1. Using the normalized delay versus supply voltage curve in Figure

1 obtained in [8] we select three supply voltages mainly 5V, 3V and 2.4V for purposes of our experiments. These supply voltages do not obey the LD property, but even so, our algorithm has given optimal results for all timing constraints for all benchmarks; this has been proven using an exhaustive search. The increment in the delay in going from 5V to 3V and from 3V to 2.4V are equal; refer to Figure 1 and Figure 4. The functional module delay at a supply voltage of 5V is termed t_c. The timing constraint is kt_c and results were obtained for different values of k for each of the benchmarks. Table 1 shows results obtained for smaller values of k, typically starting from the length of the longest path in the DFG with increments of 1. Table 2 shows results obtained for k equals twice the length of the longest path and thrice the length of the longest path of the DFG. Three sets of results are shown in the tables: power consumed allowing for only 5V to be the supply voltage, power consumed allowing for 5V and 3V to be the supply voltage, and power consumed allowing for 5V, 3V and 2.4V as supply voltages. The power consumed, $P_{dyn} = V_{dd}^2 C_{load} f_{switch}$. Since $C_{load} f_{switch}$ is constant, only $\sum V_{dd}^2$ values are shown. The column x1 shows the percentage reduction in power consumption using 5V and 3V for the benchmark over using a single supply voltage of 5V. The column x2 shows the percentage reduction in power consumption using 5V, 3V and 2.4V for the benchmark over using a single supply voltage of 5V.

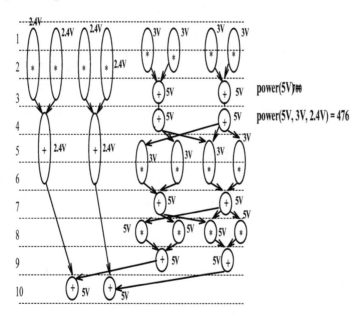

Figure 10: AR-Lattice Filter schedule with $k = 10$, using 5V, 3V, 2.4V

As can be seen from Table 1 there is an average reduction of 33.6% in power for column x1 and an average reduction of 37.26% for column x2. Table 2 shows improved reduction in power consumption: The average reduction in column x1 is 64% and in column x2 it is 72.88%.

Bench-mark	Timing Constraint k	Power using 5V	Power using 5V, 3V	x1 % reduc.	Avg. % reduc.	Power using 5V, 3V, 2.4V	x2 % reduc.	Avg. % reduc.
Diffeq	4 †	275	195	29.1		195	29.1	
	5	275	179	34.91		172.52	37.27	
	6	275	147	46.55	40.73	130.8	52.44	44.56
	7	275	131	52.36		111.56	59.43	
FIR	9 †	525	349	33.53		326.32	37.84	
	10	525	317	39.62		287.84	45.17	
	11	525	301	42.67	40.38	265.36	49.46	46.4
	12	525	285	45.71		246.12	53.12	
AR-Lattice Filter	8 †	700	604	13.71		584.56	16.49	
	9	700	540	22.86		520.56	25.63	
	10	700	476	32.0	27.43	456.56	34.77	30.20
	12	700	412	41.14		392.56	43.92	
EWFilter	15 †	850	690	18.82		677.04	20.35	
	16	850	642	24.47		629.04	25.99	
	17	850	610	28.24	25.88	590.56	30.52	27.88
	18	850	578	32.00		555.32	34.67	

Table 1: Power Consumption Results for smaller Timing Constraints

†: Corresponds to the longest path length for the DFG.

Bench-mark	Timing Constraint k	Power using 5V	Power using 5V, 3V	x1 % reduc.	Avg. % reduc.	Power using 5V, 3V, 2.4V	x2 % reduc.	Avg. % reduc.
Diffeq	8	275	99	64		82.8	69.89	
	12	275	99	64	64	63.36	76.96	73.43
FIR	18	525	189	64		150.12	71.41	
	27	525	189	64	64	120.96	76.96	74.19
AR-Lattice Filter	16	700	252	64		232.56	66.77	
	24	700	252	64	64	161.28	76.96	71.87
EWFilter	30	850	306	64		280.08	67.05	
	45	850	306	64	64	195.84	76.96	72.00

Table 2: Power Consumption Results for larger Timing Constraints

Figure 10 shows the schedule for the AR-Lattice Filter for $k = 10$ and a possible set of voltages - 5V, 3V and 2.4V.

7.2   Overheads

The experimental results were derived based only on the reductions in the total V_{dd}^2. There are some overheads that are incurred in this approach:

Level Converters: As explained in [10] there is a static current flowing at the interface of the circuit running at a lower supply voltage and one that runs at a higher supply voltage. A level converter, shown in Figure 11, is used to block the static current.

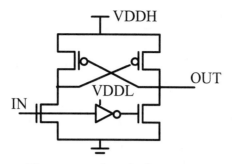

Figure 11: Level Converter

In Figure 11, VDDH is the higher supply voltage and VDDL is the lower supply voltage. A level converter is needed only when the output of a circuit running at a lower supply voltage feeds into one with a higher supply voltage. [10] also estimates the dynamic power consumed by such a level converter for 0.8micron CMOS technology, 100MHz as about 0.5mW. We will give a rough estimate of the power consumed by a 16 bit adder in 0.8micron CMOS technology and show that the overhead caused by adding level converters to outputs of functional units that run at lower supply voltages is compensated by the power reduction due to voltage scaling. A minimum sized inverter in a 0.8micron CMOS technology is about 100fF capacitance with gate delay of 500ps. Say each bit flips 25 gates with average gate 2 times the size of a minimum sized inverter. This gives 50 minimum sized inverters capacitance. At 5V, 100MHz, using $P_{dyn} = V_{dd}^2 C_{load} f_{switch}$:

$P_{dyn} = $ *16 bits* $\times 100ff \times 50$ *gates* $\times 5V \times 5V \times 100Mhz = 200\ mW$.

For a multiplier, use *16 bit* $\times$ *16 bit* because, the area is roughly the square of bit size. This is a very crude estimate, but establishes that the power consumed by functional units is at least an order larger than that consumed by level converters (16 bit level converters would consume $16 \times 0.5\ mW = 8mW$). Therefore, if the overhead due to level converters at the output of a 16 bit adder is about $8/200 \times 100 = 4\%$, the saving on the other hand using a 3V supply instead of a 5V supply is about $(1 - (3/5)^2) \times 100 = 64\%$.

Area: This work has not considered area-power tradeoffs. There are no constraints on the number of resources in our formulation and by allowing non-critical path elements to run slower at lower voltages, we have increased the number of execution units. Functional units using disjoint control steps and running at the same supply voltage could be shared thus reducing the area in some cases. The voltage conversion circuits would also increase the area. Increased area could lead to increased wire lengths, leading to increased interconnect capacitances. The power consumed by the interconnect is neglected in our studies, since it would be hard to estimate this power at the behavioral level.

8 Heuristics for Practicality

Some of the assumptions we use in the paper make it impractical to be used directly for designs. We list here some of these assumptions and try to relax some of them.

Voltage vs. Delay Curve: The voltage versus delay curve is assumed to be the same for all operators. This assumption could be relaxed completely. Each operator would add delay depending on its curve to its original latency in going from a higher supply voltage to a lower supply voltage in S in the algorithm. Given that we have different voltage vs delay curves for different operators, we do not guarantee the optimality of the algorithm even under the LD property for S.

$C_{load} \times f_{switch}$: The product $C_{load} \times f_{switch}$ is assumed to be a constant for all operators. Our experimental results give reduction in the sum of voltage square for all operations. All other factors remaining constant this would reduce the power consumption, but the % power reduction would depend on the $C_{load} \times f_{switch}$ term for each of the operators. We do not have any information on this number at this level of abstraction. This number would depend on where in the data path the operator is placed, the input pattern, interconnect among other things. Given this, we do not make any claims on the final % reduction in power consumption.

Operator Sharing: We have not allowed for operator sharing. This could also be relaxed using one of the two methods:

- 1. Share operators that have the same functionality, supply voltage and that work in disjoint control steps.

- 2. Allow for operator multicycling and switching supply voltages on each of the operators online using the finite state machine.

The second method is the more effective of the two in reducing the number of operators. The number of operators needed in the second method would equal only the maximum number of operators that are being used in any one control step. The controller design in this case is beyond the scope of this paper. Sharing operators would reduce the layout area and therefore reduce the interconnect capacitance. Of course, the number of operators used would still be larger than the number used in traditional High-Level Synthesis.

9   Conclusion

We have presented here a novel technique for reducing power at the behavioral level. Research at this level is particularly important since decisions made at this level have greater impact on the design quality. Ease of change in the design structure can also be cited as one of the advantages of going to a higher level of abstraction.

Our algorithm uses multiple voltages for scheduling the DFG to minimize power consumption. The speed issue is also considered: Timing constraints are satisfied. Given a particular set of voltages shown in Figure 4, we were able to obtain a schedule that minimizes the total V_{dd}^2. The algorithm is time efficient and runs in $O(knmlog(n^2/m))$ time where kt_c is the timing constraint (t_c is typically a clock cycle, so k is the number of control steps), n is the number of nodes and m the number of edges in the DFG. The experimental results for some High-Level Synthesis benchmarks show considerable reduction in the power consumption. For tighter timing constraints the maximum reduction is about 40% by using supply voltages 5V and 3V and a maximum reduction of about 46% using supply voltages 5V, 3V and 2.4V. For larger timing constraints the maximum reduction is about 64% by using supply voltages 5V and 3V and a maximum reduction of about 74% using supply voltages 5V, 3V and 2.4V. Clearly for progressively larger timing constraints and for smaller supply voltages the power consumption reduces giving us a speed versus power consumption tradeoff.

Most classical scheduling algorithms are based on the theoretical foundation of the relaxed scheduling (ignoring resources) algorithm presented in [17]. We have studied the relaxed scheduling problem with voltage scaling and resource constraints are not part of our formulation. Our work lays a strong theoretical foundation for future research in reducing power consumption by voltage scaling at the behavioral level. The notion that level converters would compensate for any power savings due to voltage scaling is shown to be erroneous. It is shown that using only two/three power lines power consumption could be reduced considerably.

Future work in this area should involve scheduling with multiple voltages under resource constraints. It would also be interesting to consider a diverse set of implementations for each of the functional units. This would enable the use of faster yet power intensive modules in the critical path and slower but low-power modules in the non-critical parts of the design. Module selection along with voltage scaling that considers resource constraints would become a viable option in the future for low power design.

References

[1] L. S. Nielsen and J. Sparso, "Low-Power Operation Using Self-Timed Circuits and Adaptive Scaling of the Supply Voltage", *Proc. of the International Workshop on Low Power Design*, April 1994, pp. 99–104.

[2] L. S. Nielsen, C. Niessen, J. Sparso, and K. V. Berkel, "Low-Power Operation Using Self-Timed Circuits and Adaptive Scaling of the Supply Voltage", *IEEE transactions on VLSI Systems*, Vol. 2, No. 4, December 1994, pp. 391–397.

[3] A. Chandrakasan, S. Sheng, and R. Brodersen," Low-power CMOS Digital Design", *IEEE Journal of Solid-state circuit*, April 1992, pp. 473–484.

[4] J. Monteiro, A. Devadas, and A. Ghosh, "Retiming Sequential Circuits for Low Power," *Proc. of the IEEE ICCAD-93*, November 1993, pp. 398–402.

[5] M. Alidina, J. Monteiro, S. Devadas, A. Ghosh, and M. Papaefthymiou, "Precomputation-Based Sequential Logic Optimization for Low Power", *IEEE transactions on VLSI Systems*, Vol. 2, No. 4, December 1994, pp. 426–436.

[6] K. Roy and S. Prasad, "SYCLOP: Synthesis of CMOS Logic for Low Power Applications", *Proc. of the International Conference on Computer Design: VLSI in Computers and Processors*, October 1992, pp. 464–467.

[7] R. A. Walker and R. Camposano, "A Survey of High-Level Synthesis Systems". Kluwer, MA, 1991.

[8] A. P. Chandrakasan, M. Potkonjak, R. Mehra, J. Rabaey, and R. W. Brodersen, "Optimizing Power Using Transformations", *IEEE transactions on CAD*, Vol. 14, No. 1, January 1995, pp. 12–31.

[9] R. H. Mohring, "Graphs and Orders: the Role of Graphs in the Theory of Ordered Sets and Its Applications", *Published by D. Reidel Publishing Company, edited by*, I. Rival, New York and London, May 1984, pp. 41–101.

[10] K. Usami and M. Horowitz, "Clustered Voltage Scaling Technique for Low-Power Design", *Proc. of the International Symposium on Low Power Design*, April 1995, pp. 3–8.

[11] E. Musoll and J. Cortadella, "High-level Synthesis Techniques for Reducing the Activity of Functional Units", *Proc. of the International Symposium on Low Power Design*, April 1995, pp. 99–104.

[12] A. Raghunathan and N. K. Jha, "Behavioral Synthesis for Low Power", *Proc. of the ICCD*, November 1994.

[13] A. Dasgupta and R. Karri, "Simultaneous Scheduling and Binding for Power Minimization During Microarchitecture Synthesis", *Proc. of the International Symposium on Low Power Design*, April 1995, pp. 69–74.

[14] J. M. Chang and M. Pedram, " Register Allocation and Binding for Low Power", *Proc. of the IEEE Design Automation Conference*, June 1995.

[15] S. Devadas and S. Malik, "Tutorial: A Survey of Optimizing Techniques Targeting Low Power VLSI Ciruits", *Proc. of the IEEE Design Automation Conference*, June 1995.

[16] L. Goodby, A. Orailoglu and P.M. Chau, "Microarchitectural Synthesis of Performance-Constrained, Low-Power Designs", *Proc. International Conference on Computer Design*, October 1994, pp. 323–326.

[17] T. C. Hu, "Parallel Sequencing and Assembly Line Problems", *Operations Research*, 9, A5.2, 1961, pp. 841-848.

System-Level Transformations for Low Power Data Transfer and Storage

FRANCKY CATTHOOR,* SVEN WUYTACK, EDDY DE GREEF,
FRANK FRANSSEN, LODE NACHTERGAELE, AND HUGO DE MAN*

IMEC, KAPELDREEF 75, B-3001 LEUVEN, BELGIUM

Abstract—In this paper we present our system level power exploration methodology for data-dominated multi-media applications. This formalized methodology is based on the observation that for this type of applications the power consumption is dominated by the data transfer and storage organization. Hence, the first exploration phase should be to come up with an optimized data transfer and storage organization. In this paper, we concentrate on the upper stage in our proposed script, focusing on system-level transformations.

INTRODUCTION

For most real-time signal processing applications there are many ways to realize them in terms of specific algorithms. As reported by system designers, in practice this choice is mainly based on "cost" measures such as the number of components, performance, pin count, power consumption, and the area of the custom components. Currently, due to design time restrictions, the system designer has to select—on an ad-hoc basis—a single promising path in the huge decision tree from abstract specification to more refined specification. To alleviate this situation, there is a need for fast and early feedback at the algorithm level *without* going all the way to machine code or hardware layout. Only when the design space has been sufficiently explored at a high level and when a limited number of promising candidates have been identified, is a more thorough and accurate evaluation required for the final choice.

TARGET APPLICATION DOMAIN AND STYLE

We cannot achieve this ambitious goal for general applications and target architectural styles. So we have focused ourselves to select a number of reasonable assumptions. Our target domain consists of real-time signal and data processing systems which deal with large amounts of data. This happens in real-time multi-dimensional signal processing (RSMP) applications like video and image processing, which handle indexed array signals (usually in the context of loops). It also happens in sophisticated communication network protocols, which handle large sets of records organized in tables and pointers. Both classes of applications contain many important applications like video coding, medical image archival, multi-media terminals, artificial vision, ATM networks, and LAN/WANs.

Architecture experiments have shown that 50–80% of the area cost in (application-specific) heterogeneous system architectures for real-time multi-dimensional signal processing is due to *memory units,* i.e., single or multi-port RAMs, pointer-addressed memories, and register files [1,2]. Also, the power cost is heavily dominated by storage and transfers [3]. This has been demonstrated both for custom hardware [4] and processors [5].

Hence, we believe that the organization of the global communication and data storage, together with the related algorithmic transformations, form the dominating factors (both for area and power) in the system-level design decisions. Therefore, we have concentrated mainly on the effect of system-level decisions on the access to large (background) memories which requires separate cycles, and on the transfer of data over long "distances" (over long-term main storage). In order to assist the system designer in this, we have developed a formalized system-level exploration methodology, partly supported in our prototype tool environment ATOMIUM [6,7,1]. We have also demonstrated that for our target application domain, it is best to optimize the memory/communication related issues *before* the data-path and control-related issues are tackled [2,8]. Even within the constraints resulting from the memory decisions, it is then still possible to obtain a feasible solution for the data-path organization, and even a near-optimal one if the appropriate transformations are applied at that stage as well [8]. Most of this activity has been aimed at application-specific architecture styles, but recently also predefined (multi-)processors (e.g., DSP cores) are envisioned [9–11]. In this paper, however, we focus on custom realizations. The cost functions which we currently incorporate for the storage and communication resources are both area and power oriented [3,12]. Due to the real-time nature of the targeted applications, the throughput is normally a

*Professor at the Katholieke Universiteit, Leuven, Belgium

constraint. For lack of space, we refer to [10] for a detailed discussion of the on/off-chip memory and data-transfer power models which we use for practical experiments. For embedded SRAMs we use VLSI Technology's single- and dual-port compilers for a 0.6 μm CMOS technology at 5V. If large off-the-shelf SRAMs are needed, we have used the model of a recent Fujitsu low-power memory [13]. It leads to 0.26W for a 1Mbit SRAM operating at 100MHz at 5V.

This paper is organized as follows. First, we describe the related work. Then, we introduce our methodology. In this paper, we concentrate on the upper stage in our proposed script, focusing on system-level transformations. The lower stage is presented in another book [14]. Next, the methodology is illustrated in-depth on a small but realistic test-vehicle. Finally, we discuss other experiments on power exploration for real-life applications.

RELATED WORK

Up until now, little design automation development has been done to help designers with this problem. Commercial EDA tools such as SPW/HDS (Alta/Cadence Design), System Design Station (Mentor Graphics), and the COSSAP environment (CADIS/Synopsys) support system-level specification and simulation, but they are not geared toward design exploration and optimization of memory or communication-oriented designs. Indeed, all of these tools start from a procedural interpretation of the loops where the memory organization is largely fixed. Moreover, the actual memory organization has to be indicated by means of user directives or by a partial netlist. In the CASE area, e.g., Statemate (I-Logix), Matrix-X (ISI), and Workbench (SES), these same issues are not addressed either. In the parallel compiler community, much research has been performed on loop transformations for parallelism improvement (e.g., [15,16]). In the scope of our multi-media target application domain, however, the effect on memory size and bandwidth has been largely neglected or solved with an overly simple model in terms of power or area consequences.

Within the system-level/high-level synthesis research community, the first results on memory management support in a hardware context have been obtained at Philips (Phideo environment [17]) and IMEC (prototype ATOMIUM environment [1,6]), as discussed here. Phideo is mainly oriented to stream-based video applications and focuses on memory allocation and address generation. A few other recent initiatives have been started up as well [18,19], but they focus on point tools.

Most research on power-oriented methodologies has focused on data-path or control logic, clocking, and I/O (see other papers in this book and [12]). As we have shown earlier [3], in principle, for data-dominated applications much more power can be gained by reducing the number of accesses to large frame memories or buffers. Also other groups have made similar observations [24] for video applications. Up until now, however, no systematic approach has been published to target this important

field. Indeed, most effort has been spent either on data-path oriented work (e.g. [20]), control-dominated logic, or programmable processors (see [12] for a good overview). Most designs which have already been published for our test vehicle, namely 2D motion estimation, are related to MPEG video coders [21].[1]

DATA TRANSFER AND STORAGE EXPLORATION

The current starting point of the ATOMIUM methodology is a system specification with accesses on multi-dimensional (M-D) signals which can be statically ordered. The output is a netlist of memories, combined with a transformed specification which is the input for the architecture (high-level) synthesis when custom realizations are envisioned, or for the software compilation stage in the case of predefined processors.

Prior to the system-level data transfer and storage exploration stage, it is crucial to apply a data type refinement stage where the *abstract data types* are mapped into carefully optimized data structures to be used during the implementation of the algorithm. Typically this involves **parameter quantization,** dynamic data type refinement, and data format selection. The **dynamic data type refinement** subtask performs a selection of the detailed data structure realization of dynamic data types in an algorithm, e.g., the decision on potential splitting of keys, potential use of hashing and on the realization as a linked list, an array, or a pointer array for (ordered) sets with access keys [23]. The potentially huge effect on the power budget will be illustrated later for an ATM application.

The **data format selection** subtask performs the selection of the detailed realization of data formats for all signals in the algorithm. This involves both floating-point versus fixed-point and finite word-length decisions. In both cases, significant reductions can be obtained on the realization cost (also in terms of power), due to the reduction in data storage and transfer requirements but also in terms of the reduced arithmetic complexity. Examples include the finite word-length exploration in the DSPDIGEST toolbox [24] for linear DSP applications based on quantization, limit-cycle, overflow and dynamic range requirements, and the decision on data types and quantization characteristics for DSP processor mapping in the Grape-2 toolbox [25].

As discussed above, the next stage for data dominated applications[2] involves data transfer and storage exploration. The re-

---

[1]These are based on a systolic array type approach because of the relatively large frame sizes involved, leading to a large computational requirement on the DCT. However, in the video conferencing case where the computational requirements are lower, this is not needed. An example of this is discussed in [22]. As a result, a power- and area-optimized architecture is not so parallel. Hence, also the multi-dimensional signals should be stored in a more centralized way and not fully distributed over a huge amount of local registers. This storage organization then becomes the bottleneck. Note that the transfer between the required frame memories and the systolic array is also quite power hungry and usually not incorporated in the analysis in previous work.

[2]This happens when no task or data parallelism has to be exploited.

[3]All of these prototype tools operate on models which allow run-time complexities which are dependent in a limited way on system parameters like the size of the loop iterators, as opposed to the scalar-based methods published in conventional high-level synthesis literature.

search results on techniques and prototype tools[3] which have been obtained within the ATOMIUM project are briefly discussed now (see also Fig. 1). As mentioned above, more explanation will be provided only for the system-level transformation related tasks (upper stage). More information on the techniques is available in the cited references and partly also in the discussion of the test-vehicle results.

Memory Oriented Data Flow Analysis and Model Extraction. A novel data/control flow model [6] has been developed, aimed at memory oriented algorithmic reindexing transformations. Originally it was developed to support irregular nested loops with manifest, affine iterator bounds, and index expressions. However, extensions are possible toward WHILE loops and to data-dependent and regular piece-wise linear (modulo) indices [26]. A synthesis backbone with generic kernels and shared software routines is under implementation.

Global Data-Flow Transformations. We have classified and developed a formalized methodology for the set of system-level data-flow transformations that have the most crucial effect on the system exploration decisions [27]. They consist mainly of advanced signal substitution (which especially includes moving conditional scopes), modifying computation order in associative chains, shifting of "delay lines" through the algorithm, and recomputation issues. No design tool support has been studied.

Global Loop and Reindexing Transformations. These are aimed at improving the data access locality for M-D signals and at removing the system-level buffers introduced due to mismatches in production and consumption ordering. The effect of these system-level transformations is exemplified below:

Ex. 1: allocation (assume 2N cycles):

For i:= 1 TO N DO	FOR i:= 1 TO N DO
B[i]:=f(A[i]);	BEGIN
FOR i:= 1 TO N DO	B[i]:=f(A[i]);
C[i]:=g(B[i]);	C[i]:=g(B[i]);
	END;

⇒ 2 background memory ports   ⇒ 1 background port + 1 register

In this first example, the intermediate storage of the B[] signals in the original code is avoided by merging the two loop bodies in a single loop nest at the right. As a result, the background access bandwidth is reduced and only one background memory port is required, next to the register to store intermediate scalars.

Ex. 2: in-place mapping (assume 1 memory):

FOR j:= 1 TO M DO	FOR i:= 1 TO N DO
FOR i:= 1 TO N DO	BEGIN
BEGIN	FOR j:= 1 TO M DO
A[i][j]:=g(A[i][j−1]);	A[i][j]:=g(A[i][j−1]);
END;	OUT[i]:=A[i][M];
FOR i:= 1 TO N DO	END;
OUT[i]:=A[i][M]	

⇒ N locations (background)   ⇒ 1 location (foreground)

In example 2, a function of the A[] values is iteratively computed in the first loop nest, which is expressed by the *j* loop iterating over the columns in any row of A[][]. However, only the end results (*N* values in total) have to be stored and retrieved from background memory for use in the second loop nest. This intermediate storage can be avoided by the somewhat more complex loop reorganization at the right. Now only a single foreground register is sufficient, leading to a large reduction in storage size.

In order to provide design tool support for such manipulations, an interactive loop transformation engine (SYNGUIDE) has been developed that allows both interactive and automated (script-based) steering of source code transformations [28].[4]

In addition, research has been performed on loop transformation steering methodologies. For power, we have developed a

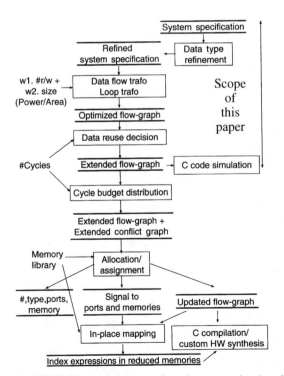

Figure 1 ATOMIUM script for data transfer and storage exploration of the specification, to be used for simulation and hardware/software synthesis. This methodology is partly supported with prototype tools. In front of the ATOMIUM approach, also data type refinement is required for power optimization.

---

[4]It includes a syntax-based check, which captures most simple specification errors, and a user-friendly graphical interface. The transformations are applied by identifying a piece of code and by entering the appropriate parameters for a selected transformation. The main emphasis lies on loop manipulations including both affine (loop interchange, reversal, and skewing) and non-affine (e.g., loop splitting and merging) cases. Also the feasibility of effective system-level validation of the global data-flow and loop transformations has been demonstrated with a formal verification methodology, partly supported by a prototype tool [29].

[5]Also an automatable CAD technique has been developed, partly demonstrated with a prototype tool called MASAI, aiming at total background memory cost reduction with emphasis on transfers and size. An abstract measure for the number of transfers is used as an estimate of the power cost and a measure for the number of locations as estimate for the final area cost [7]. This research is however not mature yet.

script oriented to removing the global buffers which are typically present between subsystems and on creating more data locality [9]. This can be applied manually.[5]

Data Reuse Decision in a Hierarchical Memory Context. In this step, we have to decide on the exploitation of the memory hierarchy [30]. Important considerations here are the distribution of the data (copies) over the hierarchy levels as these determine the access frequency and the size of each resulting memory. Obviously, the most frequently accessed memories should be the smallest ones. This can be fully optimized only if a memory hierarchy is introduced. We have proposed a formalized methodology to steer this, which is driven by estimates on bandwidth and high-level in-place cost [30]. Based on this, the background transfers are partitioned over several hierarchical memory levels to reduce the power and/or area cost. An illustration of this is shown in Fig. 2, which shows two signals, A and B, stored in background memory. Signal A is accessed via two intermediate signal copies (A_temp1 and A_temp2) so its organization involves 3 levels, while signal B is accessed directly from its storage location, i.e., located at level 1. Notice that A and B are stored in the same memory. This illustrates that a certain memory can store signals at a different level in the hierarchy. Therefore we group the memories in partitions (P1, P2, P3, etc.) instead of "levels." The data reuse and memory hierarchy decision step selects how many partitions are needed, in which partition each of the signals and its temporary copies will be stored, and the partition connections.

An important task at this step is to perform transformations which introduce extra transfers between the different memory partitions and which are mainly reducing the power cost. In particular, these involve adding temporary values—to be assigned to a "lower level"—wherever a signal in a "higher level" is read more than once. This clearly involves a trade-off between the power lost by adding these transfers and the power gained by having less frequent access to the larger memories in the higher level. At this stage of the script, the transformed behavioral description can already be used for more efficient simulation, or it can be further optimized in the next steps.

Memory Organization Issues. Given the cycle budget, the goal is to allocate memory units and ports (including their types) from a memory library and to assign the data to the best-suited memory units. This happens in several steps, described in [14] (see also the next section).

In-Place Mapping. Initial methods have been investigated for deciding on in-place storage of multi-dimensional signals [2]. Recently, a solid theoretical foundation has been obtained for the in-place mapping task, and effective heuristics have been developed to solve this very complex problem in a reasonable CPU time in a 2-stage approach [31]: intra-signal windowing, followed by inter-signal placement. These techniques have been implemented in a prototype tool with very promising results [31].

In addition to RMSP applications, we are also investigating extensions of the methodology toward network applications as occurring in Layer 3-4 of ATM [23]. The methodology for the steps related to the bandwidth requirements and balancing the available cycle budget over the different memory accesses in the algorithmic specification (flow-graph balancing for non-hierarchical graphs within a single loop body) during background memory allocation has been worked out in detail [32]. Also the combination with memory allocation and assignment has been explored with very promising results.

SMALL DEMONSTRATOR APPLICATION

Motion Estimation Algorithm

The 2D motion estimation algorithm [33] is used in moving image compression algorithms. It estimates the motion vector of small blocks of successive image frames. We will assume that the images are gray-scaled (for color images, in practice, only the luminance is considered). The version we consider here is the kernel of what is commonly referred to as the "full-search full-pixel" implementation [21].

Each frame is divided into small blocks. For each of these blocks (called *current blocks* or CBs in the sequel), a region in the previous image is defined around the same center coordinates (called the *reference window* or RW in the sequel). This is shown in Fig. 3. Every CB is matched with every possible region (of the same size as the CB) in the RW corresponding to the CB. The matching is done by accumulating the absolute pixel differences between the CB and the considered regions of the RW. The position of the region that results in the smallest difference is assumed to be the previous location of the CB. In this way, a motion vector can be calculated for every CB in a frame.

The algorithm is typically executed in six nested loops, next to the implicit frame loop. The choice of the nesting for these loops is partially open, and there is quite a lot of room for parallelization and (loop) reordering. The frames of $W \times H$ pixels are processed at F frames/s. In our experiments, we used the following typical parameters (QCIF standard): $W = 176$ pixels, $H = 144$ pixels, blocks of $n = 8 \times 8$ pixels with a search range of $2m = 16$ pixels (resulting in a 23×23 search window); $F = 30$ frames/s. This results in an incoming pixel frequency of about 0.76MHz. The pixel are 8-bit grayscale values.

In our target architecture, the number of parallel data-paths needed (depends on the parameters) which together with their local buffers are combined into one large data-path communicating with the distributed frame memory. This is only allowed

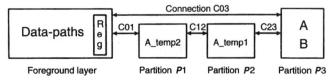

Figure 2 Memory hierarchy illustration: the foreground memory next to the data-paths consists of registers and register files; the intermediate buffers (partition P1 and P2) typically consist of fast embedded synchronous SRAMs with many ports to the lower level where potential "signal copies" from the higher levels are stored; the top layer (partition P3 in this example) consists of a slower mass storage unit with low access bandwidth.

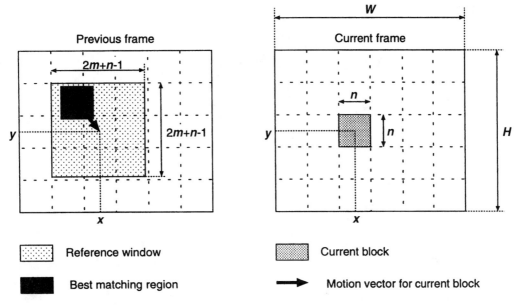

Figure 3 The motion estimation algorithm and its parameters.

if the parallelism is not too large (as is the case for the motion estimator for the QCIF format). Otherwise, more systolic organizations, with memory architectures tuned to that approach, would lead to better results. In practice, we will assume that a maximal F_{cl} of 50MHz[6] is feasible for the on-chip components, which means that we need four parallel data-path processors. In many applications, the memory organization can be assumed to be identical for each of the parallel processors (data-paths) because the parallelism is usually created by "unrolling" one or more of the loops and letting them operate at different parts of the image data [9].

We will now discuss a power-optimized architecture exploration for the motion estimation kernel as a more detailed illustration of the more general data transfer and storage exploration methodology described in Fig. 1 and the previous section.

Application of Low-Power Exploration Methodology

For the background memory organization, experiments have been performed to go from a non-optimized applicative description of the kernel in Fig. 3 to an optimized one for power, tuned to an optimized allocation and internal storage organization. In the latter case, the accesses to the large frame memories are heavily reduced. These accesses take up the majority of the power as we will see later.

Based on our script an optimized memory organization will be derived for the frame memories and for the local memories in the different data-path processors for 2D motion estimation.

STEP 1: Data and Control-Flow Optimization. The first optimization step in our methodology is related to data-flow and loop transformations. For the 2D motion estimation, we will focus on the effect of loop transformations. It is clear that re-

ordering of the loops in the kernel will affect the order of accesses and hence the regularity and locality of the frame accesses. In order to improve this, it is vital to group related accesses in the same loop scope. This means that all important accesses have to be collected in one inner loop in the 2D motion estimation example. The latter is usually done if one starts from a C specification for one mode of the motion estimation, but it is usually not the case if several modes are present. Indeed, most descriptions will then partition the quite distinct functionality over different functions which are not easily combined. Here is a first option to improve the access locality by reorganizing the loop nest order and function hierarchy among the different modes.

Another important class of loop transformations is related to reversal and interchanging the loop iterators in one loop nest. For instance, the four loops corresponding to the reference window and current block traversal in Fig. 3 can be ordered either with the window-based ones as the outer or with the block-based ones as the outer. In this relatively simple case, a straightforward analysis of the required signal storage and the related number of transfers, shows that we have a trade-off. If the traversal over the block is put in the outer loops, the advantage is that for each pixel in the block, we can directly use it to compute all related contributions for all block locations in the window. This avoids a large amount of redundant frame accesses. However, we then need to store the resulting intermediate accumulation for the motion error for the different locations. This buffer will be quite large (16×16 words) and hence, this is not a good option as will be shown below. This situation is depicted in Fig. 4. At this stage of the script, the memory architecture has not yet been defined. Therefore, the signals that have to be stored in memory are represented as ellipses. Once the memory allocation and assignment is done, the signal names will be drawn inside the memories to which they

---

[6]48.66MHz is actually needed as a minimum in this case.

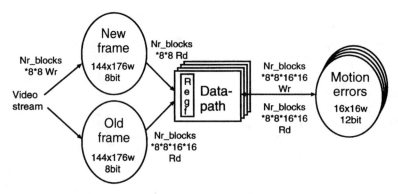

Figure 4 Required signal storage and data transfers when the traversal over the current block is done in the outer loops. The formulas at the arrows indicate the amount of words that are read (Rd) from the memories or written (Wr) to the memories per frame.

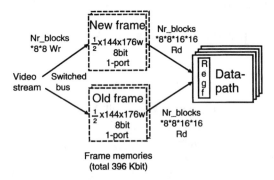

Figure 5 Data flow for the straightforward memory architecture. The formulas at the arrows again indicate the transfers.

have been assigned. These allocated memories will be represented as dashed rectangles in the figures.

The other alternative is to put the block traversal as inner loops, surrounded by the window loops. In that case, the motion error can be directly accumulated in foreground registers, eliminating the costly background (buffer) access. This situation is depicted in Fig. 5. If we compare a direct mapping of this organization on a memory architecture with a direct mapping of the first organization, we find that this one is cheaper in terms of power (1015mW compared to 1200mW, using the power models discussed earlier). In this direct mapping, we assume that each signal is mapped to its own memory module. The frame memories are considered to be off-chip because they are usually too large. The buffer memories are assumed to be on-chip. For the off-chip memories, the power consumption is determined by the access frequency only. The access frequency can be calculated by multiplying the number of accesses per frame, which are indicated on the arrows in the figures, with the number of frames per second. For the on-chip memories, the power depends on the memory size (also indicated in the figures) as well as the access frequency. Using these values the power figures are obtained from the VLSI Technologies library. The power figures of the different memories are summed to arrive at the total power consumption of the memory architecture.

STEP 2: Data Reuse Decision in a Hierarchical Memory Context. In a second step, we have to decide on the exploitation

of the available data reuse possibilities to maximally benefit from a customized memory hierarchy. The distribution of the data (copies) over the hierarchy levels as these determine the access frequency and the size of the resulting memories are important considerations.

The original architecture for the optimized loop order, if we assume that only one layer of background memories is present, is shown in Fig. 5. Note that the required access rate of about 396 blocks $\times$ 8 $\times$ 8 $\times$ 16 $\times$ 16 $\times$ 30 frames/s = 195MHz in this case, is too high for the available frame memories, so two of them should be accessed in parallel (four in total for both frames). Each of these can then be half the size. The memory access power budget related to this is $4 \times \frac{195}{2}$ MHz $\times \frac{260 \text{ mW}}{100 \text{ MHz}}$ = 1015mW.[7] We use this architecture as a reference for our further power exploration experiments.

After the introduction of one extra layer of buffers, both for the current block and the reference window accesses, we arrive at the memory hierarchy shown in Fig. 6. A direct implementation of this organization leads to a memory architecture that consumes about 580mW.

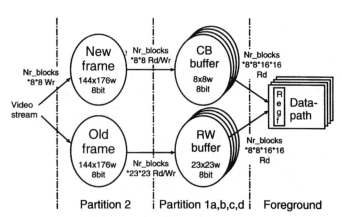

Figure 6 Data transfers between stored signals for the partly optimized memory hierarchy. There are 4 identical partitions on level 1 (1a,b,c,d) because we assume 4 parallel data-paths.

[7]Four memories, each accessed at half of 195MHz (due to doubling of the frame memories) and consuming 260mW per 100MHz.

614

We can do even better, however, if we realize that a large amount of the reference window pixels can be reused from the "previous" current block processing by an appropriate loop transformation to localize them (which has already been done in STEP 1) and a modification of the data transfers. By exploiting this overlap (so-called "inter-copy reuse" [30]), we can reduce the number of write transfers per block for the reference window buffer to $(2m + n - 1) \times n \times 8$ bit with a corresponding reduction also in read accesses to the *oldframe* memory. The result is shown in Fig. 7, with a power budget of about 560mW.

The entire search space for the data reuse decision is much too large to be explored manually or even with the support of tools. Therefore, we have developed a formalized methodology to efficiently explore this search space [30]. Based on this methodology, other interesting alternatives can be identified [30], but we will not go into this further. Note also that due to the parallel data-path architecture target, in the end some extra issues have to be taken into account [9] which are not discussed here either.

STEP 3: Storage Cycle Budget Distribution. At this stage, the data have been partitioned over different "levels" in the memory hierarchy, and all transfers between the different memory partitions are known. We are now ready to optimize the organization of every memory partition. But before doing the actual allocation of memory modules and the assignment of the signals to the memory modules, we have to decide for which signals simultaneous access capability should be provided to meet the real time constraints. This is based on generalized conflict graphs [32]. We call this task storage cycle budget distribution in general and the results for this test-vehicle are provided in [14].

STEP 4: Memory Allocation and Assignment. The next step is to allocate the memory modules for every memory partition and to assign all (intermediate) signals to their memory module. The main input for this task are the conflict graphs obtained during the storage cycle budget distribution step. The memory allocation/assignment step tries to find the cheapest memory organization that satisfies all constraints expressed in the conflict graphs. The final memory organization [14] is illustrated in Fig. 7. It consumes 560mW.

STEP 5: In-Place Mapping Optimization. In a final step, each of the memories—with the corresponding M-D signals assigned to it—should be optimized in terms of storage size ap-

plying so-called in-place mapping for the M-D signals. This will directly reduce the area and indirectly it can also reduce power further if the memory size of frequently accessed memories is reduced. However, in this particular example, no further power reduction is feasible [14].

So, finally we have arrived at a power optimized memory organization that consumes about 560mW. However, because of the relatively small frame size in QCIF, we can now consider putting this combined frame memory of 209Kbit on chip. If low-power embedded RAMs are used,[8] this will reduce the power budget further because the expensive off-chip communication is totally avoided.

Summary of Results

The substantial effect of the different optimizations in our exploration methodology is shown in Table 1. It is assumed that the frame memories are stored in off-chip memories, while all intermediate buffers are considered to be on-chip memories. Note that the initial 1015 and 560mW figures were based on lower bound estimates at that stage, so in reality the data reuse decision stage solution would have required 960mW as indicated between parentheses. A breakdown of the final power figure over the different buffers is shown in Fig. 8.

In addition to the memory architecture optimization, we have also explored the address generation unit and data-path/controller realizations for the original architecture of Fig. 5. The results are shown in Fig. 8. It can be concluded from these experiments that after optimization of the memories, data-paths, and address generation units, the power which goes into the memory accesses (560mW) dominates the other contributions, which are both comparable (less than 90mW for all data-paths and about 140mW for the optimized address generators).[9]

INDUSTRIAL APPLICATION DEMONSTRATORS

In the dynamic data type refinement related stage, we have obtained very good results on power reduction for table-driven network components [23]. An example for an industrial test-vehicle from Alcatel, namely a segment protocol processor

Table 1. Results of Applying Our Exploration Methodology.

Step	Resulting Power (mW)	Off-chip (# words)	On-chip (# words)
Control-flow optimization	1015	50688	0
Data reuse decision	560	50688	2372
Flow graph balancing + Memory allocation/assignment	(960→)560	50688	2372
In-place optimization	560	26752	2372

[8]Due to the low access speed, very low-power circuitry should be feasible.

[9]This is true even when low-power circuits are used in the memories and when power hungry standard cells are used in the data-paths and address units. Moreover, the current figures do not yet include the power of the data transfers themselves which will also consume much power (especially the off-chip ones). These transfers are also within the focus of ATOMIUM and they are equally optimized by reducing the background accesses.

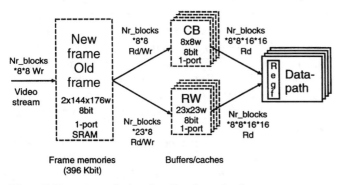

Figure 7 Memory organization after allocation and assignment.

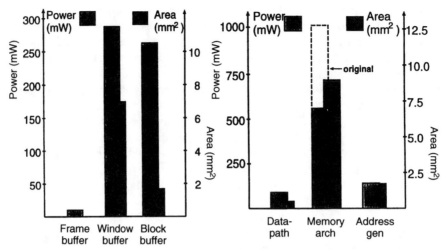

Figure 8 A breakdown of the power consumption for the three memories in the optimized memory architecture (left) and over the different architecture modules (right).

within the common adaptation layer (3-4) of ATM [34] is provided in Fig. 9. The SRAM memory power model has been adopted from U.C. Berkeley for a $1.2\mu m$ CMOS technology. The large search space for possible data structure realizations (horizontal axis) contains many combinations of access key splitting and data structure combinations [e.g., pointer arrays (PA), arrays, or linked lists (LL)]. In the figure, examples are provided of 2 keys for which each a PA or an LL implementation are chosen. The search space can be scanned with our for-

malized methodology which allows us to efficiently select the optimum for a given set based specification without exhaustively searching all candidates. Notice also the very large range of power budgets for different alternatives.

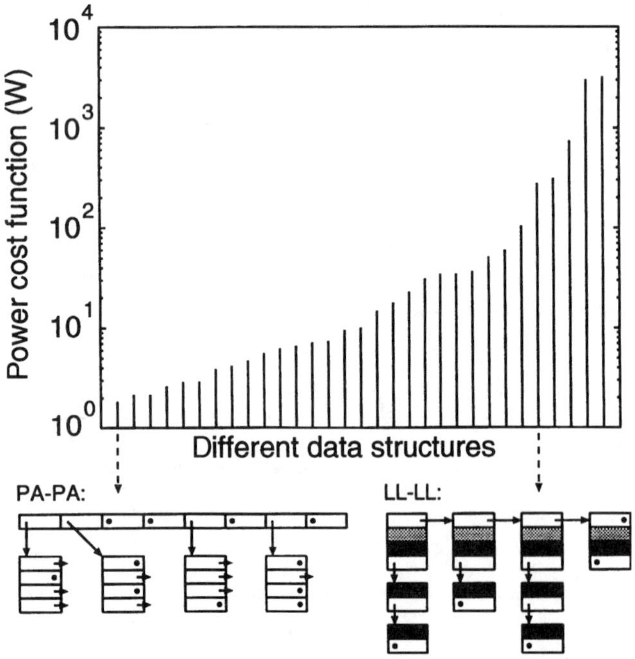

An example of the large effect of the system-level data transfer and storage exploration methodology on power for multimedia applications will now be illustrated first for a realistic medical image processing application (with image size 512 × 512). Both for small (*NrProj* = 12) and high (*NrProj* = 1200) resolution versions, the off-chip memory power results for a customized hardware realization have been obtained [3]. If reuse is made for the high-resolution version of the algorithm code which was optimal for the *NrProj* = 12 case, a quite high power and storage size are required. By again performing system-level loop transformations and memory organization decisions from the initial code, we can reduce the size from 77 Mbit to 2 times 2.25 Mbit and the system power budget from 20W to 1.56W [3]. This clearly shows that large gains can be achieved with our methodology and that no simple reuse from a "code library" should be attempted for such data-dominated applications.

Also for another demanding test-vehicle, namely the entire video decoder algorithm in the H.263 video conferencing standard, we have obtained large reductions of the power consumption associated with the background memory accesses [35]. The starting point for our exploration has been a C specification of the video decoder, available in the public domain from Telenor Research. We have transformed the data transfer scheme in the initial system specification and have optimized the distributed memory organization. This results in a memory architecture with significantly reduced power consumption. For the worst-case mode using Predicted (P) frames, memory power consumption is reduced by a factor of seven when compared to a good direct implementation of the reference specification. For the worst-case mode using Predicted and Bi-directional (PB) frames, the maximum memory power consumption is further reduced by a

Figure 9 Estimated power cost for a large range of candidate pointer data structure realizations for a given set based specification of an ATM network component. With our design exploration methodology, the optimal solution can be found without exhaustively scanning the search space.

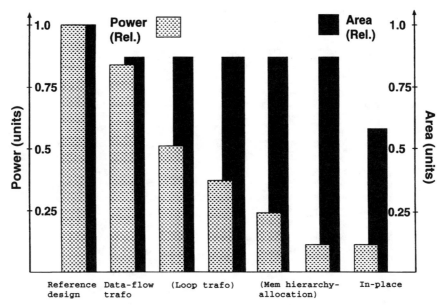

Figure 10 Relative power in continuous PB mode for each optimization stage of H.263 frame access.

factor of nine. To achieve these results, we have used our formalized data transfer and storage exploration methodology, partly supported in the ATOMIUM environment and script.

Figure 10 gives an overview of the relative power consumption for each optimization stage for the PB mode. This is the power consumed by the frame memories when decoding bidirectional B frames with unrestricted motion vectors and overlapped motion compensation. The power consumption is normalized with respect to the power consumption for the reference design. We believe these results to be very convincing for the effect on the global system power of the proposed approach.

Other major driver vehicles during 1993–96 have included a voice coder in a mobile terminal application (auto-correlation, Schur decomposition, and Viterbi coder), a regularity detector kernel for vision [7], a quadrature-structured differential PCM kernel for image processing, a non-linear diffusion algorithm for image enhancement, updating singular value decomposition for beam-forming [36], memory-dominated submodules of the MPEG2 video decoder/encoder [9], and a 2D wavelet coder (forward and backward) for image coding.

ACKNOWLEDGMENT

We wish to thank our colleagues at VSDM Division of IMEC Leuven for the stimulating discussions and the fruitful cooperation. We especially acknowledge the contributions of our former colleagues in the memory management research domain.

Note: more information on the ATOMIUM project can be obtained at our web site:

http://www.imec.be/vsdm/domains/designtechno/
index.html

References

[1] W. Geurts, F. Franssen, M. van Swaaij, F. Catthoor, H. De Man, and M. Moonen, "Memory and data-path mapping for image and video applications," in *Application-driven architecture synthesis,* F. Catthoor and L. Svensson (eds.), Boston, MA: Kluwer, 1993, pp. 143–166.

[2] I. Verbauwhede, F. Catthoor, J. Vandewalle, and H. De Man, "Background memory management for the synthesis of algebraic algorithms on multiprocessor DSP chips," *Proc. VLSI'89, Int. Conf. on VLSI,* Munich, Germany, Aug. 1989, pp. 209–218.

[3] F. Catthoor, F. Franssen, S. Wuytack, L. Nachtergaele, and H. De Man, "Global communication and memory optimizing transformations for low power signal processing systems," in *VLSI Signal Processing VII,* J. Rabaey, P. Chau, and J. Eldon (eds.). New York, NY: IEEE Press, 1994, pp. 178–187.

[4] T. H. Meng, B. Gordon, E. Tsern, and A. Hung, "Portable video-on-demand in wireless communication," special issue on "Low power design" of the *Proceedings of the IEEE,* vol. 83, no. 4, April 1995, pp. 659–680.

[5] V. Tiwari, S. Malik, and A. Wolfe, "Power analysis of embedded software: A first step towards software power minimization," *IEEE Trans. on VLSI Systems,* vol. 2, no. 4, Dec. 1994, pp. 437–445.

[6] M. van Swaaij, F. Franssen, F. Catthoor, and H. De Man, "Automating high-level control flow transformations for DSP memory management," in *VLSI Signal Processing V,* K. Yao, R. Jain, and W. Przytula (eds.), New York, NY: IEEE Press, 1992, pp. 397–406.

[7] F. Franssen, L. Nachtergaele, H. Samsom, F. Catthoor, and H. De Man, "Control flow optimization for fast system simulation and storage minimization," *Proc. 5th ACM/IEEE Europ. Design and Test Conf.,* Paris, France, Feb. 1994, pp. 20–24.

[8] F. Catthoor, W. Geurts, and H. De Man, "Loop transformation methodology for fixed-rate video, image and telecom processing applications," *Proc. Intnl. Conf. on Applic.-Spec. Array Processors,* Aug. 1994, pp. 427–438.

[9] E. De Greef, F. Catthoor, and H. De Man, "Memory organization for video algorithms on programmable signal processors," *Proc. IEEE Int. Conf. on Computer Design,* Austin TX, Oct. 1995, pp. 552–557.

[10] L. Nachtergaele, D. Moolenaar, B. Vanhoof, F. Catthoor, and H. De Man, "System-level power optimization of video codecs on embedded cores: A

systematic approach," *Journal of VLSI Signal Processing,* Boston, MA: Kluwer, 1998.

[11] K. Danckaert, F. Catthoor, and H. De Man, "System-level memory management for weakly parallel image processing, "Lecture notes in computer science" series, *Proc. EuroPar Conference,* Lyon, France, Springer Verlag, August 1996, pp. 217–225.

[12] D. Singh, J. Rabaey, M. Pedram, F. Catthoor, S. Rajgopal, N. Sehgal, and T. Mozdzen, "Power conscious CAD tools and methodologies: A perspective," special issue on "Low power design" of the *Proceedings of the IEEE,* vol. 83, no. 4, April 1995, pp. 570–594.

[13] T. Seki, E. Itoh, C. Furukawa, I. Maeno, T. Ozawa, H. Sano, and N. Suzuki, "A 6-ns 1-Mb CMOS SRAM with latched sense amplifier," *IEEE J. of Solid-state Circuits,* vol. SC-28, no. 4, Apr. 1993, pp. 478–483.

[14] F. Catthoor, S. Wuytack, E. De Greef, F. Balasa, and P. Slock, "System exploration for custom low power data storage and transfer," chapter in "Digital signal processing for multimedia systems," K. Parhi and T. Nishitani (eds.), New York: Marcel Dekker, Inc., 1998.

[15] S. Amarasinghe, J. Anderson, M. Lam, and C. Tseng, "The SUIF compiler for scalable parallel machines," *Proc. of the 7th SIAM Conf. on Parallel Proc. for Scientific Computing,* 1995.

[16] U. Banerjee, R. Eigenmann, A. Nicolau, and D. Padua, "Automatic program parallelisation," *Proc. of the IEEE,* invited paper, vol. 81, no. 2, Feb. 1993.

[17] P. Lippens, J. van Meerbergen, W. Verhaegh, and A. van der Werf, "Allocation of multiport memories for hierarchical data streams," *Proc. IEEE Int. Conf. Comp. Aided Design,* Santa Clara, CA, Nov. 1993.

[18] L. Ramachandran, D. Gajski, and V. Chaiyakul, "An algorithm for array variable clustering," *Proc. 5th ACM/IEEE Europ. Design and Test Conf.,* Paris, France, Feb. 1994, pp. 262–266.

[19] D. Kolson, A. Nicolau, and N. Dutt, "Minimization of memory traffic in high-level synthesis," *Proc. 31st ACM/IEEE Design Automation Conf.,* San Diego, CA, June 1994, pp. 149–154.

[20] A. Chandrakasan, M. Potkonjak, R. Mehra, J. Rabaey, and R. W. Brodersen, "Optimizing power using transformations," *IEEE Trans. on Comp.-aided Design,* vol. CAD-14, no. 1, Jan. 1995, pp. 12–30.

[21] P. Pirsch, N. Demassieux, and W. Gehrke, "VLSI architectures for video compression—a survey," *Proc. of the IEEE,* invited paper, vol. 83, no. 2, Feb. 1995, pp. 220–246.

[22] M. Harrand, M. Henry, P. Chaisemartin, P. Mougeat, Y. Durand, A. Tournier, R. Wilson, J. Herluison, J. Langchambon, J. Bauer, M. Runtz, and J. Bulone, "A single chip videophone encoder/decoder," *Proc. IEEE Int. Solid-State Circ. Conf.,* Feb. 1995, pp. 292–293.

[23] S. Wuytack, F. Catthoor, and H. De Man, "Transforming set data types to power optimal data structures," *IEEE Trans. on Comp.-aided Design,* vol. CAD-15, no. 6, June 1996, pp. 619–629.

[24] L. Claesen, F. Catthoor, H. De Man, J. Vandewalle, S. Note, and K. Mertens, "A CAD environment for the thorough analysis, simulation and characterisation of VLSI implementable DSP systems," *Proc. IEEE Int. Conf. on Computer Design,* Port Chester, NY, Oct. 1986, pp. 72–75.

[25] L. De Coster, M. Engels, R. Lauwereins, and J. Peperstraete, "Global approach for compiled bit-true simulation of DSP systems," "Lecture notes in computer science" series, Lyon, France, Springer Verlag, August 1996, pp. 236–239.

[26] F. Franssen, F. Balasa, M. van Swaaij, F. Catthoor, and H. De Man, "Modeling multi-dimensional data and control flow," *IEEE Trans. on VLSI systems,* vol. 1, no. 3, Sep. 1993, pp. 319–327.

[27] F. Catthoor, M. Janssen, L. Nachtergaele, and H. De Man, "System-level data-flow transformation exploration and power-area trade-offs demonstrated on video codecs," special issue on Systematic trade-off analysis in signal processing systems design" in *J. of VLSI Signal Processing,* 1997.

[28] H. Samsom, L. Claesen, and H. De Man, "SynGuide: an environment for doing interactive correctness preserving transformations," in *VLSI Signal Processing VI,* L. Eggermont, P. Dewilde, E. Deprettere, and J. van Meerbergen (eds.), New York, NY: IEEE Press, 1993, pp. 269–277.

[29] H. Samsom, F. Franssen, F. Catthoor, and H. De Man, "Verification of loop transformations for real time signal processing applications," in *VLSI Signal Processing VII,* J. Rabaey, P. Chau, and J. Eldon (eds.), New York, NY: IEEE Press, 1994, pp. 269–277.

[30] J. P. Diguet, S. Wuytack, F. Catthoor, and H. De Man, "Formalized methodology for data reuse exploration in hierarchical memory mappings," accepted for *Proc. IEEE Intnl. Symp. on Low Power Design,* Monterey, Aug. 1997.

[31] E. De Greef, F. Catthoor, and H. De Man, "Memory size reduction through storage order optimization for embedded parallel multimedia applications," *Intnl. Parallel Proc. Symp. (IPPS)* in Proc. Workshop on "Parallel Processing and Multimedia," Geneva, Switzerland, April 1997.

[32] S. Wuytack, F. Catthoor, G. De Jong, B. Lin, and H. De Man, "Flow graph balancing for minimizing the required memory bandwidth," *Proc. 9th ACM/IEEE Intnl. Symp. on System-Level Synthesis,* 1996, pp. 127–132.

[33] C. Lin and S. Kwatra, "An adaptive algorithm for motion compensated colour image coding," *Proc. IEEE Globecom,* 1984, pp. 47.1.1–4.

[34] Y. Therasse, G. H. Petit, and M. Delvaux, "VLSI architecture of a SMDS/ATM router," *Annales des Télécommunications,* vol. 48, no. 3–4, 1993, pp. 166–180.

[35] L. Nachtergaele, F. Catthoor, B. Kapoor, D. Moolenaar, and S. Janssen, "Low power data transfer and storage exploration for H.263 video decoder system," accepted for special issue on *Very Low-Bit Rate Video Coding of IEEE Journal on Selected Areas in Communications,* vol. 15, 1997.

[36] F. Balasa, F. Catthoor, and H. De Man, "Background memory area estimation for multi-dimensional signal processing systems," *IEEE Trans. on VLSI Systems,* vol. 3, no. 2, June 1995, pp. 157–172.

Author Index

Index